Springer Proceedings in Mathematics & Statistics

Volume 85

For further volumes:
http://www.springer.com/series/10533

Springer Proceedings in Mathematics & Statistics

This book series features volumes composed of selected contributions from workshops and conferences in all areas of current research in mathematics and statistics, including OR and optimization. In addition to an overall evaluation of the interest, scientific quality, and timeliness of each proposal at the hands of the publisher, individual contributions are all refereed to the high quality standards of leading journals in the field. Thus, this series provides the research community with well-edited, authoritative reports on developments in the most exciting areas of mathematical and statistical research today.

Abdenacer Makhlouf · Eugen Paal
Sergei D. Silvestrov · Alexander Stolin
Editors

Algebra, Geometry and Mathematical Physics

AGMP, Mulhouse, France, October 2011

Springer

Editors
Abdenacer Makhlouf
Laboratoire de Mathématiques, Informatique et Applications
Université of Haute Alsace
Mulhouse
France

Eugen Paal
Department of Mathematics
Tallinn University of Technology
Tallinn
Estonia

Sergei D. Silvestrov
Division of Applied Mathematics, The School of Education Culture and Communication (UKK)
Mälardalen University
Västerås
Sweden

Alexander Stolin
Department of Mathematical Sciences
University of Göteborg
Göteborg
Sweden

ISSN 2194-1009 ISSN 2194-1017 (electronic)
ISBN 978-3-642-55360-8 ISBN 978-3-642-55361-5 (eBook)
DOI 10.1007/978-3-642-55361-5
Springer Heidelberg New York Dordrecht London

Library of Congress Control Number: 2014941309

Mathematics Subject Classification (2010): 17-06, 16-06, 81-06, 17A-XX, 17B-XX, 81R-XX

Printed on acid-free paper

Springer is part of Springer Science+Business Media (www.springer.com)

This volume is dedicated to the memory of Jean-Louis Loday

Preface

This work presents the proceedings of the Algebra, Geometry and Mathematical Physics Conference, which was held at the University of Haute Alsace (Mulhouse, France) from 24 to 26 October, 2011. This international conference brought together 126 researchers from 33 different countries, who are working on these topics. In total, there were 12 plenary talks and 10 sessions with 86 contributed talks. We would like to thank all of the conference participants and speakers for making it such a successful and fruitful event.

The specific fields covered by the conference were

- Deformation theory and quantization,
- Hom-algebras and n-ary algebraic structures,
- Hopf algebra and quantum algebra,
- Integrable systems and related math structures,
- Jet theory and Weil bundles,
- Lie theory and applications,
- Noncommutative and Lie algebra,
- Number theoretical methods in string theory,
- Spectral and comp methods in physics, and
- Ternary algebras and applications.

This volume collects contributions which are divided into four main parts: Algebra, Geometry, Dynamical Symmetries and Conservation Laws, and a final part dedicated to Mathematical Physics and Applications. The common denominator of all the contributions is that they are mostly based on algebraic tools.

Part I, which is also the largest, includes contributions on Algebra. It covers topics in ring theory, Lie algebras, ternary algebras, and deformation theory. The "Poincaré Duality for Koszul Algebras" offers a complete study of the consequences of the Poincaré duality versus the AS-Gorenstein property for Koszul algebras (homogeneous and nonhomogeneous). The "Quantized Reduced Fusion Elements and Kostant's Problem" provides a partial solution to Kostant's problem concerning a description of the locally finite endomorphisms of highest weight irreducible modules. Two further chapters deal with commuting elements. The first examines the algebraic dependence in the Weyl algebra and generalizations, while the second focuses on centers in a six-parameter family of quadratically linked quantum plane algebras. A description of the $\mathscr{C}^\infty$-algebra on the cohomology of the free two-nilpotent Lie algebra is also provided, drawing on T. Kadeishvili's homotopy transfer theorem. Moreover, we present a study of subalgebra depths, in generalized triangular matrix algebras, within the path algebra of an acyclic quiver.

The proceedings also include papers dedicated to some classes of Lie algebras, such as the anisotropic, regular, minimal nonabelian, algebras of depth two, and symplectic quadratic Lie algebras related to Poisson algebras. Lie algebras generalize naturally to 3-Lie algebras. We highlight a comparison of the structure and the cohomology spaces of Lie algebras with induced 3-Lie algebras and a description of Peirce decomposition for unitary (1,1)-Freudenthal Kantor triple systems. For Hom-algebras, algebras involving a linear map twisting the usual identities, a universal algebra theory is developed, mainly for Hom-associative algebras. It covers the envelopment problem, operads, and the Diamond Lemma. Furthermore, quadratic n-ary Hom-Nambu algebras are studied and various constructions are presented. Afterwards, there is a long series of chapters concerning deformations, the first of which compares Leibniz and Lie algebra cohomology and deformations of a given Lie algebra. The second chapter studies deformations of finite dimensional current Lie algebras and their rigidity. Then, using a functorial point of view, a deformation theory for diagrams is described and noncommutative varieties are constructed using a polynomial matrix algebra and deformations (noncommutative deformation theory), as well as computations of noncommutative deformation. The last chapter in this series provides a geometric classification of four-dimensional superalgebras, based on the concept of degeneration. The purely algebraic contributions in the first part are rounded out with a survey on distributivity in quasigroup theory and in quandle theory, in connection with knot theory.

Part II is more geometrical, even if it also involves several algebraic structures. The contributions concern differential geometry and projective geometry with an algebraic treatment. The "Torsors and Ternary Moufang Loops Arising in Projective Geometry" deals with torsors and ternary Moufang loops, which arise in projective geometry. Concerning differential geometry, we present a study of connections through a graded q-differential algebra of polynomials, a classification of principal connections on a principal prolongation of a principal bundle, and an interpretation of higher order connections. Utilizing a differential geometrical approach, parallel transport on path spaces is studied using representations of categorical groups. A differential geometry of microlinear Frölicher spaces, which is mainly concerned with jet bundles, is presented. Moreover, this part includes a contribution that collects key material on the generic rank of A-modules for the purposes of differential geometrical applications, and closes with a geometrical approach to ghost fields appearing in quantized gauge theory.

Part III is concerned with dynamical symmetries and conservation laws. The idea of a conservation law can be traced back to the fields of mechanics and physics. Many physical theories and "laws of nature" are usually expressed as systems of nonlinear differential equations. The "Causality from Dynamical

Symmetry: An Example from Local Scale-Invariance" studies causality from dynamical symmetry and provides an example from local scale-invariance. Subsequently, various systems are discussed: reaction-diffusion systems with constant diffusivities, an inverse problem of reconstructing permittivity of an n-sectional diaphragm in a rectangular waveguide, a class of Hamilton–Jacobi–Bellman equations, and a generalized Dullin–Gottwald–Holm equation. Moreover, the heat-mass transfer problem is studied using a group theoretical approach, and Sinykov equations of the geodesic mappings of Riemannian manifolds are analyzed using the curvature operator of the second kind.

The last part concerns various applications of mathematics to physics. It starts with a realization of the affine Lie algebra $A_1^{(1)}$ and the relevant Z-algebra at negative level k in terms of parafermions. Then, invariance and symmetries of cubic and ternary algebras are discussed and a relationship of this construction with the operators defining quark states is demonstrated. We then present a calculation of decay times for simple modules, using the mathematical model of the physical process of decay suggested by Laudal. As an application of number theory to cryptography, an algorithmic study of the detection of permutation polynomials follows. In connection with cosmology, we include a study on scalar-tensor and multiscalar-tensor gravity and cosmological models. The last contribution deals with quantum gravity and the quantum nature of the probes used to unravel spacetime geometry.

One of the plenary speakers was Jean-Louis Loday, who gave a fascinating talk on "Divided power algebras." He has since, to our great sadness, passed away. Jean-Louis Loday was a great mathematician with broad interests in mathematics, such as the study of the interplay between algebraic K-theory and cyclic homology, as well as the applications of the theory of algebraic operads. He was a great mind and had a very generous spirit. We will always remember him, and we would like to dedicate this volume to his memory.

We would like to thank the Region of Alsace, the City of Mulhouse and the University of Haute Alsace for their financial support, as well as the AGMP network for its technical support (http://www.astralgo.com/cweb/agmp). We are grateful to the Faculty of Science and Technology and Dean Christophe Krembel for the use of their facilities, our secretary Liliane Fricker for the great job she did in organizing the conference, and Olivier Elchinger for his valued technical assistance in the preparation of this volume.

Mulhouse, May 2013

Abdenacer Makhlouf
Eugen Paal
Sergei D. Silvestrov
Alexander Stolin

Jean-Louis Loday (1946–2012)

Jean-Louis Loday was a French mathematician born in 1946 in Brittany (France). After attending the high school Clémenceau in Nantes, he went to the famous Lycée Louis-le-Grand to prepare for the competition to join the even more prestigious Ecole Normale Supérieure, which he did in 1965. He passed the *Agrégation* in 1969 before completing his Ph.D. under the guidance of Max Karoubi in 1975. He would later become "Directeur de Recherche" at the CNRS. He spent his entire career at the IRMA, part of the Department of Mathematics of the University of Strasbourg, where he later served as director.

His works deal with Algebra and Topology. He began by studying algebraic K-theory in the 1970s, the golden age for this field. He then worked on algebraic homotopy, most notably with Ronnie Brown. In the early 1980s, his focus shifted to cyclic homology, the additive version of K-theory. Six months after its introduction by Alain Connes, he and Dan Quillen discovered a seminal application in the domain of matrix Lie algebra cohomology (a result found independently by Boris Tsygan). The 1990s brought him into contact with the notion of algebraic operads, which he developed until his death. Thanks to operads, he introduced and studied in detail many types of algebras, including Leibniz algebras, dendriform algebras, and generalized bialgebras. He was very much interested in combinatorial Hopf algebras, like those appearing in renormalization theory, and was fascinated by the Stasheff polytopes, also known as associahedra, which encode associative algebras up to homotopy.

Over the course of his career Jean-Louis published 75 papers and two reference books, one on cyclic homology and the other on algebraic operads. He supervised 15 Ph.D. theses and invited many postdoctoral students to Strasbourg, organizing countless conferences and projects. Having recognized that research in mathematics does not consist in individual researchers working on their own, he was open and generous in sharing his time and ideas. Passionate about his work, not only did he always find time for his students and colleagues, he was also unable to refuse invitations to give a talk or a series of lectures, even if it meant traveling halfway around the globe (e.g. to Montréal, Chile, Kazakhstan, and China over the last few years).

Those of us mathematicians who were fortunate enough to meet him will always remember him as a wonderful person with a great sense of humor, a wealth of humanity, and an enduring love for mathematics.

Bruno Vallette

Contents

Contributors

Mina B. Abd-el-Malek Department of Engineering Mathematics and Physics, Faculty of Engineering, Alexandria University, Alexandria, Egypt; Department of Mathematics and Actuarial Sciences, The American University in Cairo, Cairo, Egypt

Viktor Abramov Institute of Mathematics, University of Tartu, Tartu, Estonia

Pedro Aguilar Institute of Nuclear Sciences, National Autonomous University of Mexico, Mexico City, Mexico

Faouzi Ammar Faculté des Sciences, Université de Sfax, Sfax, Tunisia

Aaron Armour School of Mathematics, Statistics and Computer Science, Victoria University of Wellington, Wellington, New Zealand

Joakim Arnlind Department of Mathematics, Linköping University, Linköping, Sweden

Saïd Benayadi IECL, CNRS UMR 7502, Université de Lorraine, Metz cedex, France

Wolfgang Bertram CNRS, INRIA, Institut Élie Cartan de Lorraine, Université de Lorraine, Vandœuvre-lès-Nancy, France

Yuri Bonder Physics Department, Indiana University, Bloomington, IN, USA

Maria S. Bruzón Department of Mathematics, University of Cádiz, Puerto Real, Cádiz, Spain

Saikat Chatterjee School of Mathematics, Tata Institute of Fundamental Research, Mumbai, Maharashtra, India; Institut des Hautes Études Scientifiques, Bures Sur Yvette, France

Roman Cherniha Institute of Mathematics, Ukrainian National Academy of Sciences, Kyiv, Ukraine; School of Mathematical Sciences, University of Nottingham, Nottingham, UK

Chryssomalis Chryssomalakos Institute of Nuclear Sciences, National Autonomous University of Mexico, Mexico City, Mexico

Vasyl' Davydovych Institute of Mathematics, Ukrainian National Academy of Sciences, Kyiv, Ukraine

Ekaterina D. Derevyanchuk Penza State University, Penza, Russia

Jilan Dong School of Sciences, South China University of Technology, Guangzhou, Guangdong, China

Michel Dubois-Violette Laboratoire de Physique Théorique, UMR 8627, Université Paris XI, Orsay Cedex, France

Fredrik Ekström Centre for Mathematical Sciences, Lund University, Lund, Sweden

Mohamed Elhamdadi University of South Florida, Tampa, FL, USA

Eivind Eriksen Department of Economics, BI Norwegian Business School, Oslo, Norway

Alice Fialowski Eötvös Loránd University Budapest, Budapest, Hungary

Maria Luz Gandarias Department of Mathematics, University of Cádiz, Puerto Real, Cádiz, Spain

Mazen Gharibah Brno University of Technology, Brno, Czech Rep.

Michel Goze Université de Haute Alsace, Mulhouse, France

Medhat M. Helal Department of Engineering Mathematics and Physics, Faculty of Engineering, Zagazig University, Zagazig, Egypt

Lars Hellström Division of Applied Mathematics, The School of Education, Culture and Communication, Mälardalen University, Västerås, Sweden

Malte Henkel Groupe de Physique Statistique, DP2M, Institut Jean Lamour (CNRS UMR 7198), Université de Lorraine Nancy, Vandœuvre-lès-Nancy Cedex, France

Irena Hinterleitner Brno University of Technology, Brno, Czech Republic

Jaroslav Hrdina Faculty of Mechanical Engineering, Institute of Mathematics, Brno University of Technology, Brno, Czech Republic

Olav Gravir Imenes Buskerud University College, Drammen, Norway

Laur Järv Institute of Physics, University of Tartu, Tartu, Estonia

Naihuan Jing Department of Mathematics, North Carolina State University, Raleigh, USA

Marek Jukl Palacky University, Olomouc, Czech Republic

Lars Kadison Departamento de Matematica, Faculdade de Ciencias da Universidade do Porto, Porto, Portugal

Noriaki Kamiya Center for Mathematical Sciences, University of Aizu, Aizu-wakamatsu, Japan

Eugene Karolinsky Department of Mathematics, Kharkov National University, Kharkov, Ukraine

Richard Kerner Laboratoire de Physique Théorique de la Matière Condensée, Université Pierre et Marie Curie (Paris-VI)-CNRS UMR 7600, Paris, France

Michael Kinyon Department of Mathematics, University of Denver, Denver, CO, USA

Abdennour Kitouni Université de Haute-Alsace, Mulhouse, France

Piret Kuusk Institute of Physics, University of Tartu, Tartu, Estonia

Amitabha Lahiri S. N. Bose National Centre for Basic Sciences, Salt Lake, Kolkata, West Bengal, India

Olga Liivapuu Institute of Technology, Estonian University of Life Sciences, Tartu, Estonia

Sami Mabrouk Faculté des Sciences, Université de Sfax, Sfax, Tunisia

Abdenacer Makhlouf Université de Haute-Alsace, Mulhouse, France

Josef Mikeš Palacky University Olomouc, Olomouc, Czech Republic

Daniel Mondoc Centre for Mathematical Sciences, Lund University, Lund, Sweden

Hirokazu Nishimura Institute of Mathematics, University of Tsukuba, Tsukuba, Ibaraki, Japan

Susumu Okubo Department of Physics, University of Rochester, Rochester, NY, USA

Todor Popov Institute for Nuclear Research and Nuclear Energy, Bulgarian Academy of Sciences, Sofia, Bulgaria

Maïdo Rahula Institute of Mathematics, University of Tartu, Tartu, Estonia

Erik Randla Institute of Physics, University of Tartu, Tartu, Estonia

Maria V. Redondo Departamento de Matemáticas, Universidad de Cádiz, Puerto Real, Cádiz, Spain

Elisabeth Remm Université de Haute Alsace, Mulhouse, France

Johan Richter Division of Applied Mathematics, The School of Education Culture and Communication, Mälardalen University, Västerås, Sweden

Ambar N. Sengupta Department of Mathematics, Louisiana State University Baton Rouge, Louisiana, USA

Yuri V. Shestopalov University of Gävle, Gävle, Sweden

Sergei D. Silvestrov Division of Applied Mathematics, The School of Education, Culture and Communication, Mälardalen University, Västerås, Sweden

Arvid Siqveland Faculty of Technology, Buskerud Vestfold University College, Kongsberg, Norway

Yuri G. Smirnov Penza State University, Penza, Russia

Elena Stepanova Palacky University Olomouc, Olomouc, Czech Republic

Alexander Stolin Department of Mathematics, University of Göteborg, Göteborg, Sweden

Daniel Sudarsky Institute of Nuclear Sciences, National Autonomous University of Mexico, Mexico City, Mexico

Vitaly Tarasov St. Petersburg Branch of Steklov Mathematical Institute, St. Petersburg, Russia; Department of Mathematical Sciences, IUPUI, Indianapolis, IN, USA

Jaan Vajakas University of Tartu, Tartu, Estonia

Petr Vašík Faculty of Mechanical Engineering, Institute of Mathematics, Brno University of Technology, Brno, Czech Republic

Jan Vondra Department of Mathematics and Statistics, Masaryk University, Brno, Czech Republic

Christopher J. Young Departamento de Matematica, Faculdade de Ciencias da Universidade do Porto, Porto, Portugal

Yinhuo Zhang Department of Mathematics and Statistics, University of Hasselt, Diepenbeek, Belgium

Pasha Zusmanovich Department of Mathematics, Tallinn University of Technology, Tallinn, Estonia; University of São Paulo, São Paulo, Brazil

Part I
Algebra

Poincaré Duality for Koszul Algebras

Michel Dubois-Violette

Abstract We discuss the consequences of the Poincaré duality, versus AS-Gorenstein property, for Koszul algebras (homogeneous and non homogeneous). For homogeneous Koszul algebras, the Poincaré duality property implies the existence of twisted potentials which characterize the corresponding algebras while in the case of quadratic linear Koszul algebras, the Poincaré duality is needed to get a good generalization of universal enveloping algebras of Lie algebras. In the latter case we describe and discuss the corresponding generalization of Lie algebras. We also give a short review of the notion of Koszulity and of the Koszul duality for N-homogeneous algebras and for the corresponding nonhomogeneous versions.

1 Introduction

Our aim in these notes is to review some important consequences of the Poincaré duality versus AS-Gorenstein property for the Koszul algebras.

We shall first describe the AS-Gorenstein property [1] for graded algebras of finite global dimensions and explain in what sense we consider it as a form of Poincaré duality as well as its connection with the Frobenius property, [10, 28, 34].

We then review the Koszul duality [8] and the notion of Koszulity [3] for homogeneous algebras. We explain that for a homogeneous Koszul algebra the Gorenstein property implies the existence of a homogeneous twisted potential which characterizes algebra completely [18–20].

Lots of examples together with the corresponding twisted potentials (i.e. preregular multilinear forms) are given in [19] and [20]. Here, in these notes, we do not describe them in order to save space and we refer to the above quoted papers.

M. Dubois-Violette (✉)
Laboratoire de Physique Théorique, UMR 8627 Université Paris XI,
Bâtiment 210, 91405 Orsay Cedex, France
e-mail: Michel.Dubois-Violette@u-psud.fr

A. Makhlouf et al. (eds.), *Algebra, Geometry and Mathematical Physics*,
Springer Proceedings in Mathematics & Statistics 85, DOI: 10.1007/978-3-642-55361-5_1,

We pass then to the description of the nonhomogeneous case and to the Poincaré-Birkhoff-Witt (PBW) property and explain why for the quadratic-linear algebras, the Poincaré duality is needed to obtain a good generalization of the universal enveloping algebras of Lie algebras, namely the enveloping algebras of Lie prealgebras [21].

Throughout this paper $\mathbb{K}$ denotes a (commutative) field and all vector spaces, algebras, etc. are over $\mathbb{K}$. By an algebra without other specification we mean a unital associative algebra with unit denoted by **1** whenever no confusion arises. By a graded algebra we mean a $\mathbb{N}$-graded algebra $\mathscr{A} = \oplus_{n\geq 0}\mathscr{A}_n$. We use everywhere the Einstein summation convention over the repeated up-down indices.

2 The AS-Gorenstein Property

In this section we describe our general framework and the AS-Gorenstein property which is our version of the Poincaré duality.

2.1 *Graded Algebras*

We shall be concerned here with graded algebras $\mathscr{A} = \oplus_{n\in\mathbb{N}}\mathscr{A}_n$ of the form $\mathscr{A} = T(E)/I$ where E is a finite-dimensional vector space and where I is a finitely generated graded ideal of the tensor algebra $T(E)$ such that $I = \oplus_{n\geq 2} I_n \subset \oplus_{n\geq 2} E^{\otimes^n}$. This class of graded algebras and the homomorphisms of degree 0 of graded algebras define a category which will be denoted by **GrAlg**.

For such an algebra $\mathscr{A} = T(E)/I \in$ **GrAlg** choosing a basis $(x^\lambda)_{\lambda\in\{1,\dots,d\}}$ of E and a system of homogeneous independent generators $(f_\alpha)_{\alpha\in\{1,\dots,r\}}$ of I with $(f_\alpha) \in E^{\otimes^{N_\alpha}}$ and $N_\alpha \geq 2$ for $\alpha \in \{1,\dots,r\}$, one can also write

$$\mathscr{A} = \mathbb{K}\langle x^1,\dots,x^d\rangle/(f_1,\dots,f_r)$$

where $(f_1,\dots,f_r)$ is the ideal I generated by the f_α. Define $M_{\alpha\lambda} \in E^{\otimes^{N_\alpha-1}}$ by setting $f_\alpha = M_{\alpha\lambda} \otimes x^\lambda \in E^{\otimes^{N_\alpha}}$. Then the presentation of $\mathscr{A}$ by generators and relations is equivalent to the exactness of the sequence of left $\mathscr{A}$-modules

$$\mathscr{A}^r \xrightarrow{M} \mathscr{A}^d \xrightarrow{x} \mathscr{A} \xrightarrow{\varepsilon} \mathbb{K} \to 0 \tag{1}$$

where M means right multiplication by the matrix $(M_{\alpha\lambda})$, x means right multiplication by the column (x^λ) and where ε is the projection onto $\mathscr{A}_0 = \mathbb{K}$, [1]. In more intrinsic notations the exact sequence (1) reads

$$\mathscr{A} \otimes R \to \mathscr{A} \otimes E \xrightarrow{m} \mathscr{A} \xrightarrow{\varepsilon} \mathbb{K} \to 0 \tag{2}$$

where R is the graded subspace of $T(E)$ spanned by the f_α ($\alpha \in \{1, \dots, r\}$), m is the product in $\mathscr{A}$ (remind that $E = \mathscr{A}_1$) and where the first arrow is as in (1).

When R is homogeneous of degree N ($N \geq 2$), i.e. $R \subset E^{\otimes^N}$, then $\mathscr{A}$ is said to be a *N-homogeneous algebra*: for $N = 2$ one speaks of a quadratic algebra, for $N = 3$ one speaks of a cubic algebra, etc. The N-homogeneous algebras form a full subcategory $\mathbf{H_N Alg}$ of $\mathbf{GrAlg}$.

2.2 Global Dimension

The exact sequence (2) of presentation of $\mathscr{A}$ can be extended as a minimal projective resolution of the trivial left module $\mathbb{K}$, i.e. as an exact sequence of left modules

$$\cdots \to M_n \to \cdots \to M_2 \to M_1 \to M_0 \to \mathbb{K} \to 0$$

where the M_n are projective i.e. in this graded case free left-modules [13], which is minimal ; one has $M_0 = \mathscr{A}$, $M_1 = \mathscr{A} \otimes E$, $M_2 = \mathscr{A} \otimes R$ and more generally here $M_n = \mathscr{A} \otimes E_n$ where the E_n are finite-dimensional vector spaces. If such a minimal resolution has finite length $D < \infty$, i.e. reads

$$0 \to \mathscr{A} \otimes E_D \to \cdots \to \mathscr{A} \otimes E \to \mathscr{A} \to \mathbb{K} \to 0 \tag{3}$$

with $E_D \neq 0$, then D is an invariant called the *left projective dimension* of $\mathbb{K}$ and it turns out that D which coincide with the right projective dimension of $\mathbb{K}$ is also the sup of the lengths of the minimal projective resolutions of the left and of the right $\mathscr{A}$-modules [13] which is called the *global dimension* of $\mathscr{A}$. Furthermore it was recently shown [5] that this global dimension D also coincides with the Hochschild dimension in homology as well as in cohomology. Thus for an algebra $\mathscr{A} \in \mathbf{GrAlg}$, there is a unique notion of dimension from a homological point of view which is its global dimension $g\ell \dim(\mathscr{A}) = D$ whenever it is finite.

2.3 Poincaré Duality Versus AS-Gorenstein Property

Let $\mathscr{A} \in \mathbf{GrAlg}$ be of finite global dimension D. Then one has a minimal free resolution

$$0 \to M_D \to \cdots \to M_0 \to \mathbb{K} \to 0$$

with $M_n = \mathscr{A} \otimes E_n$, $\dim(E_n) < \infty$ and $E_2 \simeq R$, $E_1 \simeq E$ and $E_0 \simeq \mathbb{K}$. By applying the functor $\mathrm{Hom}_{\mathscr{A}}(\bullet, \mathscr{A})$ to the chain complex of free left $\mathscr{A}$-module

$$0 \to M_D \to \cdots \to M_0 \to 0 \tag{4}$$

one obtains the cochain complex

$$0 \to M'_0 \to \cdots \to M'_D \to 0 \tag{5}$$

of free right $\mathscr{A}$-modules with $M'_n \simeq E^*_n \otimes \mathscr{A}$ where for any vector space F, one denotes by F^* its dual vector space.

The algebra $\mathscr{A} \in$ **GrAlg** is said to be *AS-Gorenstein* whenever one has

$$\begin{cases} H^n(M') = 0, & \text{for } n \neq D \\ H^D(M') = \mathbb{K} \end{cases}$$

which reads $\text{Ext}^n_{\mathscr{A}}(\mathbb{K}, \mathscr{A}) = \delta^{nD}\mathbb{K}$ by definition ($\delta^{nD} = 0$ for $n \neq D$ and $\delta^{DD} = 1$). This implies that for $0 \leq n \leq D$ one has

$$E^*_{D-n} \simeq E_n \tag{6}$$

which is a version of the Poincaré duality interesting by itself as shown e.g. by Proposition 1.4 of [6]. However as pointed out in [6] (see the counterexample there), this version of the Poincaré duality is not equivalent to the AS-Gorenstein property (which is the version adopted in these notes for the Poincaré duality property).

Notice that one has

$$\text{Ext}^n_{\mathscr{A}}(\mathbb{K}, \mathbb{K}) \simeq E^*_n$$

which follows easily from the definitions. The direct sum $E(\mathscr{A}) = \oplus_n \text{Ext}^n_{\mathscr{A}}(\mathbb{K}, \mathbb{K})$ is a graded algebra, the *Yoneda algebra* of $\mathscr{A}$. One has the following result.

Theorem 2.1 *Assume that $\mathscr{A} \in$ **GrAlg** has finite global dimension. Then $\mathscr{A}$ is AS-Gorenstein if and only if $E(\mathscr{A})$ is a Frobenius algebra.*

This result which is a weak version of a result of [28] is a generalization of a result of [10] which is itself a generalization of a result of [34].

The Yoneda algebra $E(\mathscr{A})$ is the cohomology of a graded differential algebra, so in view of the homotopy transfer theorem [26] (see also in [27]), it has besides its ordinary product $m = m_2$, a sequence of higher order product $m_n : E(\mathscr{A})^{\otimes^n} \to E(\mathscr{A})$ for $n \geq 3$ which satisfy together with m_2 the axioms of A_∞-algebras (introduced in [36]) with $m_1 = 0$.

It is only when it is endowed with its A_∞-structure that one can recover the original algebra $\mathscr{A}$ from $E(\mathscr{A})$. In some cases one has $m_n = 0$ for $n \geq 3$; this is in particular the case when $\mathscr{A}$ is a quadratic Koszul algebra but then the Yoneda algebra $E(\mathscr{A})$ identifies with the Koszul dual $\mathscr{A}^!$ of $\mathscr{A}$ (see below).

3 Homogeneous Algebras

We review here some definitions and basic properties of homogeneous algebras, [3, 8]. Throughout the following N denotes an integer with $N \geq 2$.

3.1 Koszul Duality

Let $\mathcal{A} \in \mathbf{H_N Alg}$ be a N-homogeneous algebra, that is as explained above, an algebra of the form

$$\mathcal{A} = A(E, R) = T(E)/(R)$$

where E is a finite-dimensional vector space, where R is a linear subspace of $E^{\otimes^N}$ and where (R) denotes the two-sided ideal of the tensor algebra $T(E) = \oplus_{n\in\mathbb{N}} E^{\otimes^n}$ of E generated by R. The algebra $\mathcal{A} = A(E, R)$ is a graded connected algebra $\mathcal{A} = \oplus_{n\in\mathbb{N}}\mathcal{A}_n$ generated in degree 1 by $E = \mathcal{A}_1$.

To $\mathcal{A} = A(E, R)$ one associates another N-homogeneous algebra, its *Koszul dual* N-homogeneous algebra $\mathcal{A}^!$ defined by [8]

$$\mathcal{A}^! = A(E^*, R^\perp)$$

where E^* is the dual vector space of E and where $R^\perp \subset (E^*)^{\otimes^N}$ denotes the orthogonal of R

$$R^\perp = \{\omega \in (E^*)^{\otimes^N} | \omega(r) = 0, \;\; \forall r \in R\}$$

with the identification $(E^*)^{\otimes^N} = (E^{\otimes^N})^*$ which is allowed in view of the finite-dimensionality of E.

One has

$$(\mathcal{A}^!)^! = \mathcal{A}$$

so this is a duality in $\mathbf{H_N Alg}$ which is above the usual duality of the finite-dimensional vector spaces. It is straightforward that this duality defines a contravariant involutive endofunctor of $\mathbf{H_N Alg}$. This is the direct generalization of the usual Koszul duality of quadratic algebras (case $N = 2$) [29, 30].

3.2 The Koszul N-complex $K(\mathcal{A})$

Let $\mathcal{A} = A(E, R)$ be a N-homogeneous algebra with Koszul dual $\mathcal{A}^! = \oplus_n \mathcal{A}^!_n$. Then the dual vector space $\mathcal{A}^{!*}_n$ of $\mathcal{A}^!_n$ is given by

$$\mathcal{A}^{!*}_n = E^{\otimes^n}$$

for $n < N$ and by

$$\mathcal{A}^{!*}_n = \cap_{r+s=n-N} E^{\otimes^r} \otimes R \otimes E^{\otimes^s} \tag{7}$$

for $n \geq N$. Consider the sequence of homomorphisms of (free) left $\mathcal{A}$-modules

$$\cdots \xrightarrow{d} \mathcal{A} \otimes \mathcal{A}^{!*}_{n+1} \xrightarrow{d} \mathcal{A} \otimes \mathcal{A}^{!*}_n \xrightarrow{d} \cdots \xrightarrow{d} \mathcal{A} \to 0 \tag{8}$$

where the homomorphism $d : \mathscr{A} \otimes \mathscr{A}^{!*}_{n+1} \to \mathscr{A} \otimes \mathscr{A}^{!*}_n$ is induced by the homomorphism $d : \mathscr{A} \otimes E^{\otimes^{n+1}} \to \mathscr{A} \otimes E^{\otimes^n}$ defined by

$$d(a \otimes (e_0 \otimes e_1 \otimes \cdots \otimes e_n)) = ae_0 \otimes (e_1 \otimes \cdots \otimes e_n) \tag{9}$$

for $a \in \mathscr{A}$, $e_0 \otimes \cdots \otimes e_n \in E^{\otimes^{n+1}}$ and where ae_0 is the product in $\mathscr{A}$ of a and e_0. In view of (7), one has $\mathscr{A}^{!*}_n \subset R \otimes E^{\otimes^{n-N}}$ for $n \geq N$ which implies

$$d^N = 0 \tag{10}$$

so the sequence (8) is a chain N-complex of left $\mathscr{A}$-modules which is referred to as the *Koszul N-complex of $\mathscr{A}$* and is denoted by $K(\mathscr{A})$.

By applying the functor $\mathrm{Hom}_{\mathscr{A}}(\bullet, \mathscr{A})$ to (8) one obtains a cochain N-complex of right $\mathscr{A}$-module

$$0 \to \mathscr{A} \stackrel{d^*}{\to} E^* \otimes \mathscr{A} \stackrel{d^*}{\to} \cdots \stackrel{d^*}{\to} \mathscr{A}^!_n \otimes \mathscr{A} \stackrel{d^*}{\to} \mathscr{A}^!_{n+1} \otimes \mathscr{A} \stackrel{d^*}{\to} \cdots \tag{11}$$

where d^* is the right multiplication by $\theta_\lambda \otimes x^\lambda$ where (x^λ) is a basis of E with dual basis (θ_λ). This N-complex of right $\mathscr{A}$-module is denoted by $L(\mathscr{A})$.

3.3 The Koszul Complexes $\mathscr{K}(\mathscr{A},\mathbb{K})$ and $\mathscr{K}(\mathscr{A},\mathscr{A})$

From a N-complex like $K(\mathscr{A})$ one obtains ordinary complexes called *contractions* by starting at some place and applying alternatively arrows d^k and d^{N-k} $(1 \leq k < N)$. Remembering that, see (2), the presentation of the N-homogeneous algebra $\mathscr{A} = A(E, R)$ by generators and relation is equivalent to the exactness of

$$\mathscr{A} \otimes R \stackrel{d^{N-1}}{\to} \mathscr{A} \otimes E \stackrel{d}{\to} \mathscr{A} \stackrel{\varepsilon}{\to} \mathbb{K} \to 0 \tag{12}$$

it is natural to consider the particular contraction extending

$$\cdots \stackrel{d^{N-1}}{\to} \mathscr{A} \otimes \mathscr{A}^{!*}_{N+1} \stackrel{d}{\to} \mathscr{A} \otimes \mathscr{A}^{!*}_N \stackrel{d^{N-1}}{\to} \mathscr{A} \otimes E \stackrel{d}{\to} \mathscr{A} \to 0$$

this is a chain complex of free left $\mathscr{A}$-modules which will be denoted by $\mathscr{K}(\mathscr{A},\mathbb{K})$ and called the *left $\mathscr{A}$-module Koszul complex of $\mathscr{A}$* or simply the *Koszul complex of $\mathscr{A}$*. One has

$$\begin{cases} \mathscr{K}_{2m}(\mathscr{A},\mathbb{K}) = \mathscr{A} \otimes \mathscr{A}^{!*}_{Nm} \\ \mathscr{K}_{2m+1}(\mathscr{A},\mathbb{K}) = \mathscr{A} \otimes \mathscr{A}^{!*}_{Nm+1} \end{cases} \tag{13}$$

the differential δ of $\mathscr{K}(\mathscr{A},\mathbb{K})$ is given by

$$\begin{cases} \delta = d : \mathscr{K}_{2m+1}(\mathscr{A},\mathbb{K}) \to \mathscr{K}_{2m}(\mathscr{A},\mathbb{K}) \\ \delta = d^{N-1} : \mathscr{K}_{2(m+1)}(\mathscr{A},\mathbb{K}) \to \mathscr{K}_{2m+1}(\mathscr{A},\mathbb{K}) \end{cases} \tag{14}$$

and it turns out that this complex identifies canonically with the Koszul complex introduced originally in [3]. Moreover this complex is the only contraction of the Koszul N-complex $K(\mathscr{A})$ whose acyclicity in positive degrees does not lead to a trivial algebra $\mathscr{A}$ as shown in [8].

In the case $N = 2$, i.e. when $\mathscr{A}$ is quadratic, one has of course $\mathscr{K}(\mathscr{A},\mathbb{K}) = K(\mathscr{A})$.

By reversing the order of tensor product in Sequence (8), one obtains similarly the N-complex $K'(\mathscr{A})$ of free right $\mathscr{A}$-modules

$$\cdots \stackrel{d'}{\to} \mathscr{A}^{!*}_{n+1} \otimes \mathscr{A} \stackrel{d'}{\to} \mathscr{A}^{!*}_{n} \otimes \mathscr{A} \stackrel{d'}{\to} \cdots \stackrel{d'}{\to} \mathscr{A} \to 0 \tag{15}$$

with an obvious definition of d'.

On the sequence of bimodules $(\mathscr{A} \otimes \mathscr{A}^{!*}_n \otimes \mathscr{A})_{n\in\mathbb{N}}$, one has the two commuting N-differentials $d \otimes I_{\mathscr{A}}$ and $I_{\mathscr{A}} \otimes d'$ which will be simply denoted again by d and d'. Following [4] one defines the *bimodule Koszul complex of* $\mathscr{A}$ denoted by $\mathscr{K}(\mathscr{A},\mathscr{A})$ to be the chain complex of free $(\mathscr{A},\mathscr{A})$-bimodules (i.e. of free $\mathscr{A}\otimes\mathscr{A}^{opp}$-modules) defined by

$$\begin{cases} \mathscr{K}_{2m}(\mathscr{A},\mathscr{A}) = \mathscr{A} \otimes \mathscr{A}^{!*}_{Nm} \otimes \mathscr{A} \\ \mathscr{K}_{2m+1}(\mathscr{A},\mathscr{A}) = \mathscr{A} \otimes \mathscr{A}^{!*}_{Nm+1} \otimes \mathscr{A} \end{cases} \tag{16}$$

with differential δ' of $\mathscr{K}(\mathscr{A},\mathscr{A})$ defined by

$$\begin{cases} \delta' = d - d' : \mathscr{K}_{2m+1}(\mathscr{A},\mathscr{A}) \to \mathscr{K}_{2m}(\mathscr{A},\mathscr{A}) \\ \delta' = \sum_{r=0}^{N-1} d^r (d')^{N-r-1} : \mathscr{K}_{2(m+1)}(\mathscr{A},\mathscr{A}) \to \mathscr{K}_{2m+1}(\mathscr{A},\mathscr{A}) \end{cases} \tag{17}$$

the identity $\delta'^2 = 0$ following from $0 = d^N - d'^N = (d-d')\sum_{r=0}^{N-1} d^r (d')^{N-r-1}$.

Notice that the presentation of $\mathscr{A}$ by generators and relations is also equivalent to the exactness of

$$\mathscr{A} \otimes R \otimes \mathscr{A} \stackrel{\delta'}{\to} \mathscr{A} \otimes E \otimes \mathscr{A} \stackrel{\delta'}{\to} \mathscr{A} \otimes \mathscr{A} \stackrel{m}{\to} \mathscr{A} \to 0 \tag{18}$$

where the last arrow m is the multiplication in $\mathscr{A}$.

Finally, by applying the functor $\mathrm{Hom}_{\mathscr{A}}(\bullet,\mathscr{A})$ to the chain complex of free left $\mathscr{A}$-modules $\mathscr{K}(\mathscr{A},\mathbb{K})$, one obtains the cochain complex of free right $\mathscr{A}$-modules $\mathscr{K}^*(\mathscr{A},\mathbb{K}) = \mathscr{L}(\mathscr{A},\mathbb{K})$

$$\cdots \stackrel{\delta^*}{\to} \mathscr{L}^n(\mathscr{A},\mathbb{K}) \stackrel{\delta^*}{\to} \mathscr{L}^{n+1}(\mathscr{A},\mathbb{K}) \stackrel{\delta^*}{\to} \cdots \tag{19}$$

which is of course a contraction of the N-complex $L(\mathscr{A})$.

3.4 N-Koszul Algebras

One has the following result [3].

Theorem 3.1 *Let $\mathscr{A}$ be a N-homogeneous algebra. Then the following properties (i) and (ii) are equivalent:*

(i) *The complex $\mathscr{K}(\mathscr{A},\mathbb{K})$ is acyclic in degrees ≥ 1,*
(ii) *The complex $\mathscr{K}(\mathscr{A},\mathscr{A})$ is acyclic in degrees ≥ 1.*

When $\mathscr{A}$ is such that the above equivalent properties are satisfied, $\mathscr{A}$ is said to be a *N-Koszul algebra* or simply a *Koszul algebra.*

In view of the exact sequences (12) and (18), if $\mathscr{A}$ is Koszul then

$$\mathscr{K}(\mathscr{A},\mathbb{K}) \stackrel{\varepsilon}{\to} \mathbb{K} \to 0 \tag{20}$$

is a free resolution of the trivial left $\mathscr{A}$-module $\mathbb{K}$ while

$$\mathscr{K}(\mathscr{A},\mathscr{A}) \stackrel{m}{\to} \mathscr{A} \to 0 \tag{21}$$

is a free resolution of the $(\mathscr{A},\mathscr{A})$-bimodule $\mathscr{A}$. These resolutions are minimal projective in the graded category.

This last point is important since if $\mathscr{M}$ is a bimodule on the Koszul algebra $\mathscr{A}$ then the chain complex $\mathscr{M}\otimes_{\mathscr{A}\otimes\mathscr{A}^{opp}}\mathscr{K}(\mathscr{A},\mathscr{A})$ computes the Hochschild homology $H_\bullet(\mathscr{A},\mathscr{M})$, (i.e. its homology is $H_\bullet(\mathscr{A},\mathscr{M})$), while the cochain complex $\mathrm{Hom}_{\mathscr{A}\otimes\mathscr{A}^{opp}}(\mathscr{K}(\mathscr{A},\mathscr{A}),\mathscr{M})$ computes the Hochschild cohomology $H^\bullet(\mathscr{A},\mathscr{M})$, (i.e. its cohomology is $H^\bullet(\mathscr{A},\mathscr{M})$), in view of the interpretations of $H_\bullet(\mathscr{A},\mathscr{M})$ as $\mathrm{Tor}^{\mathscr{A}\otimes\mathscr{A}^{opp}}(\mathscr{M},\mathscr{A})$ and of $H^\bullet(\mathscr{A},\mathscr{M})$ as $\mathrm{Ext}^\bullet_{\mathscr{A}\otimes\mathscr{A}^{opp}}(\mathscr{A},\mathscr{M})$. In particular when $\mathscr{A}$ has finite global dimension D, these complexes are "small" of lenght D.

Warning. For $N=2$, that is for $\mathscr{A}$ quadratic, it is easy to show that $\mathscr{A}$ is Koszul (i.e. 2-Koszul) if and only if its Koszul dual $\mathscr{A}^!$ is Koszul. However for $N\geq 3$, the Koszul dual $\mathscr{A}^!$ of a N-Koszul algebra $\mathscr{A}$ is generally not N-Koszul.

3.5 The A_∞-structure of $E(\mathscr{A})$

Let $\mathscr{A}$ be a N-Koszul algebra.

If $N=2$, that is if $\mathscr{A}$ is quadratic, then $E(\mathscr{A})=\mathscr{A}^!$ and there are no non trivial higher order products in the A_∞-structure of $E(\mathscr{A})$.

Let us assume now that $N\geq 3$. In this case, the Yoneda algebra $E(\mathscr{A})$ can be extracted from the Koszul dual $\mathscr{A}^!$ of $\mathscr{A}$ in the following manner as show in [10].

One sets $E(\mathscr{A}) = \oplus_{n\in\mathbb{N}} E_n(\mathscr{A})$ with

$$\begin{cases} E_{2m}(\mathscr{A}) &= \mathscr{A}^!_{Nm} \\ E_{2m+1}(\mathscr{A}) &= \mathscr{A}^!_{Nm+1} \end{cases} \tag{22}$$

and the product m_2 of $E(\mathscr{A})$ is defined in terms of the product $(x, y) \mapsto xy$ of $\mathscr{A}^!$ by

$$m_2(x, y) = xy$$

if x or y is of even degree in $E(\mathscr{A})$ which means of degree multiple of N in $\mathscr{A}^!$, and by

$$m_2(x, y) = 0$$

otherwise. Concerning the A_∞-structure of $E(\mathscr{A})$, the only nontrivial higher order product is the product m_N which is given by

$$m_N(x_1, \dots, x_N) = x_1 \dots x_N$$

if all the x_k are of odd degrees in $E(\mathscr{A})$ and

$$m_N(x_1, \dots, x_N) = 0$$

otherwise [25]. As an A_∞-algebra, $E(\mathscr{A})$ is generated in degree 1.

4 Twisted Potentials and Algebras

In this section we recall the construction of algebras associated to preregular multilinear forms or which is the same to twisted potentials. We consider only the homogeneous case here.

4.1 Multilinear Forms and Twisted Potentials

Let V be a vector space and let $n \geq 1$ be a positive integer, then a $(n + 1)$-linear form w on V is said to be *preregular* [18, 19] iff it satisfies the following conditions (i) and (ii).

(i) If $X \in V$ is such that $w(X, X_1, \dots, X_n) = 0$ for any $X_1, \dots, X_n \in V$, then $X = 0$.

(ii) There is an element $Q_w \in GL(V)$ such that one has

$$w(X_0, \dots, X_{n-1}, X_n) = w(Q_w X_n, X_0, \dots, X_{n-1})$$

for any $X_0, \dots, X_n \in V$.

It follows from (i) that Q_w as in (ii) is unique. Property (i) when combined with (ii) implies the stronger property (i').

(i') For any $0 \leq k \leq n$, if $X \in V$ is such that

$$w(X_1, \dots, X_k, X, X_{k+1}, \dots, X_n) = 0$$

for any $X_1, \dots, X_n \in V$, then $X = 0$.

Property (i') will be referred to as 1-*site nondegeneracy* while (ii) will be referred to as *twisted cyclicity* with *twisting element* Q_w. Thus a preregular multilinear form is a multilinear form which is 1-site nondegenerate and twisted cyclic.

Note that then, by applying n times the relations of (ii) one obtains the invariance of w by Q_w that is

$$w(X_0, \dots, X_n) = w(Q_w X_0, \dots, Q_w X_n)$$

for any $X_0, \dots, X_n \in V$.

Let w be an arbitrary Q-invariant m-linear form on V (with $Q \in GL(V)$) then, assuming $m \neq 0$ in $\mathbb{K}$, the m-linear form $\pi_Q(w)$ defined by

$$\pi_Q(w)(X_1, \dots, X_m) = \frac{1}{m} \sum_{k=1}^{m} w(QX_k, \dots, QX_m, X_1, \dots, X_{k-1})$$

for any $X_1, \dots, X_m \in V$ is twisted cyclic with twisting element Q, (in short is Q-*cyclic*).

Let E be a finite-dimensional vector space, then an element w of $E^{\otimes^m}$ is the same thing as a m-linear form on the dual E^* of E. To make contact with the terminology of [23] we will say that w is a *twisted potential* of degree m on E if the corresponding m-linear form on E^* is preregular.

4.2 Algebras Associated with Twisted Potentials

Let $w \in E^{\otimes^m}$ be a twisted potential and let $w_{\lambda_1 \dots, \lambda_m}$ be its components in the basis $(x^\lambda)_{\lambda \in \{1, \dots, \dim(E)\}}$ of E, i.e. one has $w = w_{\lambda_1 \dots \lambda_m} x^{\lambda_1} \otimes \cdots \otimes x^{\lambda_m}$. Let (θ_λ) be the dual basis of (x^λ), the corresponding preregular multilinear form on E^* is given by $w(\theta_{\lambda_1}, \dots, \theta_{\lambda_m}) = w_{\lambda_1 \dots \lambda_m}$ and we denote by Q_w the twisting element. One has $Q_w \in GL(E^*)$ and $Q_w^t \in GL(E)$ where $Q \mapsto Q^t$ denotes the transposition.

Assume that m is such that $m \geq 2$ and let N be an integer such that $m \geq N \geq 2$. One defines the N-homogeneous algebra $\mathscr{A} = \mathscr{A}(w, N)$ to be the graded algebra generated in degree 1 by the elements x^λ with relations

$$w_{\lambda_1 \dots \lambda_{m-N} \mu_1 \dots \mu_N} x^{\mu_1} \dots x^{\mu_N} = 0 \tag{23}$$

for $\lambda_k \in \{1, \ldots, \dim(E)\}$, $1 \leq k \leq m - N$. In other words

$$\mathscr{A} = A(E, R_{wN}) = T(E)/(R_{wN})$$

where R_{wN} is the subspace of $E^{\otimes^N}$ generated by the elements

$$w_{\lambda_1 \ldots \lambda_{m-N} \mu_1 \ldots \mu_N} x^{\mu_1} \otimes \cdots \otimes x^{\mu_N}$$

with $\lambda_k \in \{1, \ldots, \dim(E)\}$, $1 \leq k \leq m - N$. The algebra $\mathscr{A} = \mathscr{A}(w, N)$ will be refered to as the *N-homogeneous algebra associated with w*. The relations of $\mathscr{A}$ are given by "the $(m - N)$-th derivatives" of w. Notice that the twisted cyclicity of w, or more precisely its preregularity, implies that the relations (23) of $\mathscr{A} = \mathscr{A}(w, N)$ read equivalently for any $1 \leq p \leq m - N$ as

$$w_{\lambda_p \cdots \lambda_{m-N} \mu_1 \ldots \mu_N \lambda_1 \ldots \lambda_{p-1}} x^{\mu_1} \ldots x^{\mu_N} = 0$$

for $\lambda_k \in \{1, \ldots, \dim(E)\}$, $1 \leq k \leq m - N$.

Let us define the subspaces $\mathscr{W}_n \subset E^{\otimes^n}$ for $m \geq n \geq 0$ by

$$\begin{cases} \mathscr{W}_n = E^{\otimes^n} \text{ for } N - 1 \geq n \geq 0 \\ \mathscr{W}_n = \sum_{(\lambda)} \mathbb{K} w_{\lambda_1 \ldots \lambda_{m-n} \mu_1 \ldots \mu_n} x^{\mu_1} \otimes \cdots \otimes x^{\mu_n} \text{ for } m \geq n \geq N \end{cases} \tag{24}$$

so one has in particular $\mathscr{W}_m = \mathbb{K}w$, $\mathscr{W}_N = R_{wN}$, $\mathscr{W}_1 = E$ and $\mathscr{W}_0 = \mathbb{K}$. The twisted cyclicity of w and (7) imply the following result.

Theorem 4.1 *The sequence*

$$0 \to \mathscr{A} \otimes \mathscr{W}_m \stackrel{d}{\to} \mathscr{A} \otimes \mathscr{W}_{m-1} \stackrel{d}{\to} \cdots \stackrel{d}{\to} \mathscr{A} \to 0 \tag{25}$$

is a sub-N-complex $W(\mathscr{A})$ of the Koszul N-complex $K(\mathscr{A})$ of $\mathscr{A}$.

4.3 The Complexes $\mathscr{W}(\mathscr{A},\mathbb{K})$ and $\mathscr{W}(\mathscr{A},\mathscr{A})$

In the case $N = 2$, the sequence (25) is a complex which is a subcomplex of the Koszul complex and, from the very definition (24), one has the isomorphisms of vector spaces $\mathscr{W}^*_{m-n} \stackrel{\simeq}{\to} \mathscr{W}_n$ defined by

$$\dot{\zeta} \mapsto \zeta^{\lambda_1 \ldots \lambda_{m-n}} w_{\lambda_1 \ldots \lambda_{m-n} \mu_1 \ldots \mu_n} x^{\mu_1} \otimes \ldots x^{\mu_n}$$

where $\zeta = \zeta^{\lambda_1 \ldots \lambda_{m-n}} \theta_{\lambda_1} \otimes \cdots \otimes \theta_{\lambda_{m-n}}$ is any element of $E^{*\otimes^{m-n}}$ which projects onto $\dot{\zeta} \in \mathscr{W}^*_{m-n}$.

In the case $N \geq 3$, to obtain a similar situation, one has to "jump" over the appropriate degrees as for the definition of the Koszul complex $\mathscr{K}(\mathscr{A}, \mathbb{K})$ and to assume that $m = Np + 1$ for some integer $p \geq 1$. One then define the complex $\mathscr{W}(\mathscr{A}, \mathbb{K})$ by setting

$$\begin{cases} \mathscr{W}_{2k}(\mathscr{A}, \mathbb{K}) \quad = \mathscr{A} \otimes \mathscr{W}_{Nk} \\ \mathscr{W}_{2k+1}(\mathscr{A}, \mathbb{K}) = \mathscr{A} \otimes \mathscr{W}_{Nk+1} \end{cases} \tag{26}$$

so that one has $\mathscr{W}_n(\mathscr{A}, \mathbb{K}) \subset \mathscr{K}_n(\mathscr{A}, \mathbb{K})$.
One verifies that $\delta \mathscr{W}_{n+1}(\mathscr{A}, \mathbb{K}) \subset \mathscr{W}_n(\mathscr{A}, \mathbb{K})$ and therefore $\mathscr{W}(\mathscr{A}, \mathbb{K})$ is a subcomplex of the Koszul complex with

$$\mathscr{W}_n(\mathscr{A}, \mathbb{K}) = \mathscr{A} \otimes \mathscr{W}_{\nu_N(n)}$$

where $\nu_N(2k) = Nk$ and $\nu_N(2k+1) = Nk + 1$.

One observes then that the complex $\mathscr{W}(\mathscr{A}, \mathbb{K})$ is of length $2p + 1$ and that one has the canonical isomorphisms of vector spaces

$$\mathscr{W}^*_{\nu_N(2p+1-n)} \overset{\simeq}{\to} \mathscr{W}_{\nu_N(n)} \tag{27}$$

similar to the ones of the case $N = 2$.

Similarily one defines in the same conditions a subcomplex $\mathscr{W}(\mathscr{A}, \mathscr{A})$ of the bimodule Koszul complex $\mathscr{K}(\mathscr{A}, \mathscr{A})$ by setting

$$\mathscr{W}_n(\mathscr{A}, \mathscr{A}) = \mathscr{A} \otimes \mathscr{W}_{\nu_N(n)} \otimes \mathscr{A}$$

and verifying that $\delta' \mathscr{W}_{n+1}(\mathscr{A}, \mathscr{A}) \subset \mathscr{W}_n(\mathscr{A}, \mathscr{A})$.

Notice that in view of (27) these complexes satisfy a Poincaré duality condition similar to the one corresponding to (6) for AS-Gorenstein algebras. Furthermore the complex of free bimodules $\mathscr{W}(\mathscr{A}, \mathscr{A})$ is self dual in an obvious sense, see [11] .

Finally by applying $\mathrm{Hom}_{\mathscr{A}}(\bullet, \mathscr{A})$ to $\mathscr{W}(\mathscr{A}, \mathbb{K})$, one obtains the cochain complex of free right $\mathscr{A}$-modules $\mathscr{W}^*(\mathscr{A}, \mathbb{K})$

$$\dots \overset{\delta^*}{\to} \mathscr{W}^*_{\nu_N(n)} \otimes \mathscr{A} \overset{\delta^*}{\to} \mathscr{W}^*_{\nu_N(n+1)} \otimes \mathscr{A} \overset{\delta^*}{\to} \dots$$

which is a subcomplex of $\mathscr{L}(\mathscr{A}, \mathbb{K})$.

The self duality of $\mathscr{W}(\mathscr{A}, \mathscr{A})$ corresponds precisely to the duality between $\mathscr{W}(\mathscr{A}, \mathbb{K})$ and $\mathscr{W}^*(\mathscr{A}, \mathbb{K})$.

4.4 Automorphisms σ_w of $\mathscr{A}^!$ and σ^w of $\mathscr{A}$, Modular Property of σ_w and Pre-Frobenius Structure of $\mathscr{A}^!$

Let $\mathscr{A} = \mathscr{A}(w, N)$ be as in Sect. 4.2 and let $Q_w \in GL(E^*)$ be the corresponding twisting element of w, ($E = \mathscr{A}_1$). Then Q_w induces an automorphism of degree 0 of $T(E^*)$ which preserves $R^\perp_{wN} \subset E^{*\otimes^N}$ while $Q^t_w \in GL(E)$ induces an automorphism of degree 0 of $T(E)$ which preserves $R_{wN} \subset E^{\otimes^N}$. It follows that Q_w induces an automorphism σ_w of the graded algebra $\mathscr{A}^!$ while Q_w^t induces an automorphism σ^w of the graded algebra $\mathscr{A}$.

One has $w \in \mathscr{A}_m^{!*}$ since $W(\mathscr{A})$ is a sub-N-complex of $K(\mathscr{A})$ and one defines a linear form ω_w on the algebra $\mathscr{A}^!$ by setting

$$\omega_w = w \circ p_m \tag{28}$$

where $p_m : \mathscr{A}^! \to \mathscr{A}_m^!$ is the canonical projection onto the degree m component. One has the following theorem [19].

Theorem 4.2 *The linear form ω_w and the automorphism σ_w are connected by*

$$\omega_w(xy) = \omega_w(\sigma_w(y)x) \tag{29}$$

for any $x, y \in \mathscr{A}^!$. The subset of $\mathscr{A}^!$

$$\mathscr{I} = \{y \in \mathscr{A}^! | \omega_w(xy) = 0, \quad \forall x \in \mathscr{A}^!\}$$

is a two-sided ideal of $\mathscr{A}^!$ and the quotient algebra $\mathscr{F}(w, N) = \mathscr{A}^!/\mathscr{I}$ endowed with the linear form induced by ω_w is a graded Frobenius algebra.

The relation (29) is just a reformulation of the preregularity of w, it reflects the modular property of σ_w with respect to ω_w. One clearly has $\mathscr{F}(w, N) = \oplus_{n=0}^m \mathscr{F}_n(w, N)$ so $\mathscr{F}(w, N)$ is finite-dimensional and the pairing induced by $(x, y) \mapsto \omega_w(xy)$ is nondegenerate by construction and is a Frobenius pairing on $\mathscr{F}(w, N)$.

Let ${}^w\mathscr{A}$ be the $(\mathscr{A}, \mathscr{A})$-bimodule which coincides with $\mathscr{A}$ as right $\mathscr{A}$-module but whose left $\mathscr{A}$-module structure is given by the left multiplication by $(-1)^{(m-1)n}$ $(\sigma^w)^{-1}(a)$ for $a \in \mathscr{A}_n$. Thus ${}^w\mathscr{A}$ is a twisted version of the bimodule $\mathscr{A}$. For $N = 2$, one has the following result [19].

Proposition 4.1 *For $N = 2$, that is for $\mathscr{A} = \mathscr{A}(w, 2)$, the element $\mathbf{1} \otimes w$ of $\mathscr{A}^{\otimes^{m+1}}$ is canonically a nontrivial ${}^w\mathscr{A}$-valued Hochschild m-cycle on $\mathscr{A}$.*

In this proposition $\mathbf{1} \in \mathscr{A}$ is interpreted as an element of ${}^w\mathscr{A}$. This proposition for $N = 2$ gives the interpretation of $\mathbf{1} \otimes w$ as a twisted volume element since for $Q_w = (-1)^{m-1}$ it would represent an element of $HH_m(\mathscr{A})$.

4.5 *N*-Koszul AS-Gorenstein Algebras

For N-Koszul algebras of finite global dimension which are AS-Gorenstein one has the following result [18, 19], see also in [12] for the case $N = 2$.

Theorem 4.3 *Let $\mathscr{A}$ be a N-Koszul algebra of finite global dimension D which is AS-Gorenstein. Then $\mathscr{A} = \mathscr{A}(w, N)$ for some twisted potential of degree m on*

$E = \mathscr{A}_1$. For $N = 2$ one has $m = D$ while for $N \geq 3$ one has $m = Np + 1$ and $D = 2p + 1$ for some integer $p \geq 1$.

Under the assumptions of this theorem, the N-complex $W(\mathscr{A})$ of Sect. 4.2 coincides with the Koszul N-complex $K(\mathscr{A})$ which implies that the Koszul resolution of the trivial left $\mathscr{A}$-module $\mathbb{K}$ reads

$$0 \to \mathscr{A} \otimes \mathscr{W}_{\nu_N(D)} \stackrel{\delta}{\to} \cdots \stackrel{\delta}{\to} \mathscr{A} \otimes \mathscr{W}_{\nu_N(k)} \stackrel{\delta}{\to} \cdots \stackrel{\delta}{\to} \mathscr{A} \stackrel{\varepsilon}{\to} \mathbb{K} \to 0 \qquad (30)$$

with $\nu_N(2n) = Nn$ and $\nu_N(2n+1) = Nn + 1$ for $n \in \mathbb{N}$ and where δ is as in (14), that is

$$\mathscr{W}(\mathscr{A}, \mathbb{K}) \stackrel{\varepsilon}{\to} \mathbb{K} \to 0$$

with the notations of Sect. 4.2. One has

$$\dim(\mathscr{W}_{\nu_N(k)}) = \dim(\mathscr{W}_{\nu_N(D-k)}) \text{ for } 0 \leq k \leq D$$

since as observed in Sect. 4.2, one has the isomorphisms $\mathscr{W}^*_{\nu_N(D-k)} \simeq \mathscr{W}_{\nu_N(k)}$ for $0 \leq k \leq D$. In particular $\mathscr{W}_{\nu_N(D)} = \mathbb{K}w$ so $\mathbf{1} \otimes w$ is the generator of the top free module of the Koszul resolution of $\mathbb{K}$.

Remark Under the assumptions of Theorem 4.3 the Yoneda algebra $E(\mathscr{A})$ of $\mathscr{A}$ is a Frobenius algebra in view of Theorem 2.1, (endowed with its ordinary product m_2). If $N = 2$, one has $E(\mathscr{A}) = \mathscr{A}^!$ and therefore $E(\mathscr{A}) = \mathscr{F}(w, 2)$, however for $N \geq 3$ the Frobenius algebras $E(\mathscr{A})$ and $\mathscr{F}(w, N)$ are completely different, (we use here the notations of Theorem 4.2). Indeed in the case $N \geq 3$, $E(\mathscr{A})$ is obtained from $\mathscr{A}^!$ by dropping terms of degrees ν with $Np + 1 < \nu < N(p + 1)$ with $p \geq 1$, while $\mathscr{F}(w, N)$ is a quotient of $\mathscr{A}^!$ by a graded ideal (which vanishes in some cases such as for the Yang-Mills algebra of [15] and some generalizations [16]).

As observed in [11], there is a sort of converse in the sense that the acyclicity in degrees ≥ 1 of $\mathscr{W}(\mathscr{A}, \mathbb{K})$ or of $\mathscr{W}(\mathscr{A}, \mathscr{A})$ implies that $\mathscr{A}$ is Koszul of global dimension D and is AS-Gorenstein. Thus Theorem 4.3 admits the following refinement.

Theorem 4.4 *Let N, m and D be as in Theorem 4.3 that is either $N = 2$ with $D = m$ or $N \geq 3$ with $m = Np + 1$ and $D = 2p + 1$ for some integer $p \geq 1$. Then the following conditions* (i), (ii) *and* (iii) *are equivalent for a N-homogeneous algebra $\mathscr{A}$:*

(i) *$\mathscr{A}$ is N-Koszul of finite global dimension D and is AS-Gorenstein (or twisted Calabi-Yau),*
(ii) *$\mathscr{A} = \mathscr{A}(w, N)$ for some twisted potential w of degree m and $\mathscr{W}(\mathscr{A}, \mathbb{K})$ is acyclic in degrees ≥ 1,*
(iii) *$\mathscr{A} = \mathscr{A}(w, N)$ for some twisted potential w of degree m and $\mathscr{W}(\mathscr{A}, \mathscr{A})$ is acyclic in degrees ≥ 1.*

Under these equivalent conditions one has $\mathscr{W}(\mathscr{A}, \mathbb{K}) = \mathscr{K}(\mathscr{A}, \mathbb{K})$ and $\mathscr{W}(\mathscr{A}, \mathscr{A}) = \mathscr{K}(\mathscr{A}, \mathscr{A})$.

In practice, the acyclicity condition for $\mathscr{W}(\mathscr{A}, \mathbb{K})$ or $\mathscr{W}(\mathscr{A}, \mathscr{A})$ is hard to verify and implies very nontrivial nondegeneracy conditions for w. For instance, in the case $m = N+1$ the condition $\mathscr{W}(\mathscr{A}, \mathbb{K}) = \mathscr{K}(\mathscr{A}, \mathbb{K})$ is equivalent to the condition of 3-regularity as shown in [19] (Proposition 16) which is a subtle 2-sites nondegeneracy condition.

It is worth noticing here that, as pointed out in [10], for $\mathscr{A}$ of global dimension $D = 2$ or $D = 3$ the AS-Gorenstein condition implies already that $\mathscr{A}$ is N-homogeneous and N-Koszul with $N = 2$ for $D = 2$.

Notice also that "N-Koszul of finite global dimension and AS-Gorenstein" is equivalent to "N-Koszul of finite global dimension and twisted Calabi-Yau" [11]. This is connected with the equivalence (ii) $\Leftrightarrow$ (iii) of Theorem 4.4 together with the self duality of $\mathscr{W}(\mathscr{A}, \mathscr{A})$.

5 Nonhomogeneous Algebras

All the nonhomogeneous algebras considered in this article will be obtained by starting with homogeneous relations, say N-homogeneous, and by adding second members of lower degrees to the homogeneous relations. We always assume that these algebras are finitely generated with a finite presentation. This means that such an algebra $\mathfrak{A}$ is of the form

$$\mathfrak{A} = T(E)/(\{r - \varphi(r) | r \in R\}) \tag{31}$$

where E is a finite-dimensional vector space, R is a linear subspace of $E^{\otimes^N}$ ($N \geq 2$) and $\varphi : R \to \oplus_{n=0}^{N-1} E^{\otimes^n}$ is a linear mapping of R into the space of tensors of degrees strictly smaller than N.

5.1 The Poincaré-Birkhoff-Witt Property

Let $\mathfrak{A}$ be the nonhomogeneous algebra given by (31). Then $\mathfrak{A}$ is not naturally graded since its relations are not homogeneous but it inherits a filtration $F^n(\mathfrak{A})$ induced by the natural filtration $F^n(T(E)) = \oplus_{k\leq n} E^{\otimes^k}$ of the tensor algebra associated to its graduation.

There are two natural graded algebras associated to $\mathfrak{A}$:

1. the graded algebra
$$gr(\mathfrak{A}) = \oplus_n F^n(\mathfrak{A})/F^{n-1}(\mathfrak{A}) \tag{32}$$
refered to as *the associated graded algebra to the filtered algebra* $\mathfrak{A}$,
2. the N-homogeneous algebra
$$\mathscr{A} = A(E, R) \tag{33}$$

obtained by switching in the relations of $\mathfrak{A}$ the terms of degrees strictly smaller than N ; $\mathscr{A}$ is refered to as the N-homogeneous part of $\mathfrak{A}$ or simply as *the homogeneous part* of $\mathfrak{A}$.

We use the convention that $F^p(\mathfrak{A}) = 0$ whenever $p < 0$. One has a canonical surjective homomorphism of graded algebra

$$can : \mathscr{A} \to \mathrm{gr}(\mathfrak{A}) \tag{34}$$

which maps linearly $\mathscr{A}_1 = E$ onto $F^1(\mathfrak{A})/F^0(\mathfrak{A}) = E$.

The nonhomogeneous algebra $\mathfrak{A}$ is said to have *the Poincaré-Birkhoff-Witt property (PBW property)* whenever the canonical homomorphism *can* is an isomorphism. If $\mathfrak{A}$ has the PBW property and if its homogeneous part is N-Koszul, then $\mathfrak{A}$ is said to be a nonhomogeneous *Koszul algebra*, [9]. One has the following result [22], see also in [9] for a more general context.

Theorem 5.1 *Let us decompose* φ *as* $\varphi = \sum_{n=0}^{N-1} \varphi_n$ *with* $\varphi_n : R \to E^{\otimes^n}$ *and set* $\mathscr{V}_{N+1} = (R \otimes E) \cap (E \otimes R)$. *Assume that* $\mathfrak{A}$ *has the* PBW *property then one has the following relations*

(a) $(\varphi_{N-1} \otimes I - I \otimes \varphi_{N-1})(\mathscr{V}_{N+1}) \subset R$,
(b) $(\varphi_n(\varphi_{N-1} \otimes I - I \otimes \varphi_{N-1}) + \varphi_{n-1} \otimes I - I \otimes \varphi_{n-1})(\mathscr{V}_{N+1}) = 0$
for $1 \leq n \leq N-1$, *and*
(c) $\varphi_0(\varphi_{N-1} \otimes I - I \otimes \varphi_{N-1})(\mathscr{V}_{N+1}) = 0$

where I *is the identity mapping of* E *onto itself.*

Conversely, if the homogeneous part $\mathscr{A}$ *of* $\mathfrak{A}$ *is* N*-Koszul and if the above relations are satisfied then* $\mathfrak{A}$ *has the* PBW *property.*

The assumption that $\mathscr{A}$ is N-Koszul is natural but not completely optimal for the converse in the above theorem. In any case, this theorem implies that $\mathfrak{A}$ is a nonhomogeneous Koszul algebra if and only if its homogeneous part $\mathscr{A}$ is N-Koszul and the relations (i), (ii), (iii) of the theorem are satisfied.

Notice that one has $\mathscr{V}_{N+1} = \mathscr{A}^{!*}_{N+1}$, (see in Sect. 3.2).

Instructive examples (with $N > 2$) of nonhomogeneous Koszul algebras obtained by application of Theorem 5.1 are given in [9] and in [7].

5.2 Nonhomogeneous Koszul Duality for $N = 2$

In the following, we shall be concerned only with the case $N = 2$ and we call *nonhomogeneous quadratic algebra* an algebra of the form (31) with $R \subset E \otimes E$ and $\varphi : R \to E \oplus \mathbb{K}$ (here, $E^{\otimes^0}$ is identified with $\mathbb{K}$).

Let $\mathfrak{A}$ be a nonhomogeneous quadratic algebra with quadratic part $\mathscr{A} = A(E, R)$, and let $\varphi_1 : R \to E$ and $\varphi_0 : R \to \mathbb{K}$ be as in 5.1 the decomposition $\varphi = \varphi_1 + \varphi_0$.

Consider the transposed $\varphi_1^t : E^* \to R^*$ and $\varphi_0^t : \mathbb{K} \to R^*$ of φ_1 and φ_0 and notice that one has by definition of $\mathscr{A}^!$ that $\mathscr{A}^!_1 = E^*, \mathscr{A}^!_2 = R^*$ and $\mathscr{A}^!_3 = (R\otimes E \cap E \otimes R)^*$, so one can write (the minus sign is put here to match the usual conventions)

$$-\varphi_1^t : \mathscr{A}^!_1 \to \mathscr{A}^!_2, \ -\varphi_0^t(1) = F \in \mathscr{A}^!_2 \tag{35}$$

and one has the following result [32].

Theorem 5.2 *Conditions* (a), (b) *and* (c) *of Theorem 5.1 are equivalent for* $N = 2$ *to the following conditions* (a′), (b′) *and* (c′)*:*

(a′) $-\varphi_1^t$ *extends as an antiderivation* δ *of* $\mathscr{A}^!$
(b′) $\delta^2(x) = [F, x], \ \ \forall x \in \mathscr{A}^!$
(c′) $\delta(F) = 0$.

A graded algebra equipped with an antiderivation δ of degree 1 and an element F of degree 2 satisfying the conditions (b′) and (c′) above is refered to as a *curved graded differential algebra* [32].

Thus the correspondence $\mathfrak{A} \mapsto (\mathscr{A}^!, \delta, F)$ define a contravariant functor from the category of nonhomogeneous quadratic algebras satisfying the conditions (a), (b) and (c) of Theorem 5.1 (for $N = 2$) to the category of curved differential quadratic algebras (with the obvious appropriate notions of morphism). One can summarize the Koszul duality of [32] for non homogeneous quadratic algebras by the following.

Theorem 5.3 *The above correspondence defines an anti-isomorphism between the category of nonhomogeneous quadratic algebras satisfying Conditions* (a), (b) *and* (c) *of Theorem 5.1 (for* $N = 2$*) and the category of curved differential quadratic algebras which induces an anti-isomorphism between the category of nonhomogeneous quadratic Koszul algebras and the category of curved differential quadratic Koszul algebras.*

There are two important classes of nonhomogeneous quadratic algebras $\mathfrak{A}$ satisfying the conditions (a), (b) and (c) of Theorem 5.1. The first one corresponds to the case $\varphi_0 = 0$ which is equivalent to $F = 0$ while the second one corresponds to $\varphi_1 = 0$ which is equivalent to $\delta = 0$. An algebra $\mathfrak{A}$ of the first class is called a *quadratic-linear algebra* [31] and corresponds to a differential quadratic algebra $(\mathscr{A}^!, \delta)$ while an algebra $\mathfrak{A}$ of the second class corresponds to a quadratic algebra $\mathscr{A}^!$ equipped with a central element F of degree 2.

5.3 Examples

1. *Universal enveloping algebras of Lie algebras.* Let $\mathfrak{g}$ be a finite-dimensional Lie algebras then its universal enveloping algebra $\mathfrak{A} = U(\mathfrak{g})$ is Koszul quadratic-linear. Indeed one has $\mathscr{A} = S\mathfrak{g}$ which is a Koszul quadratic algebra of finite global dimension $D = \dim(\mathfrak{g})$ while the PBW property is here the classical

PBW property of $U(\mathfrak{g})$. The corresponding differential quadratic algebra $(\mathscr{A}^!, \delta)$ is $(\wedge\mathfrak{g}^*, \delta)$, i.e. the exterior algebra of the dual vector space $\mathfrak{g}^*$ of $\mathfrak{g}$ endowed with the Koszul differential δ. Notice that this latter differential algebra is the basic building block to construct the Chevalley-Eilenberg cochain complexes. Notice also that $\mathscr{A} = S\mathfrak{g}$ is not only Koszul of finite global dimension but is also AS-Gorenstein (Poincaré duality property).

2. *Adjoining a unit element to an associative algebra.* Let A be a finite-dimensional associative algebra and let

$$\mathfrak{A} = \tilde{A} = T(A)/(\{x \otimes y - xy, y \in A\})$$

be the algebra obtained by adjoining a unit $\mathbf{1}$ to A ($\tilde{A} = \mathbb{K}\mathbf{1} \oplus A$, etc.). This is again a Koszul quadratic-linear algebra. Indeed the PBW property is here equivalent to the associativity of A while the quadratic part is $\mathscr{A} = T(A^*)^!$ which is again $\mathbb{K}\mathbf{1} \oplus A$ as vector space but with a vanishing product between the elements of A and is a Koszul quadratic algebra. The corresponding differential quadratic algebra $(\mathscr{A}^!, \delta)$ is $(T(A^*), \delta)$ where δ is the antiderivation extension of minus the transposed $m^t : A^* \to A^* \otimes A^*$ of the product m of A. Again $(T_+(A^*), \delta)$ is the basic building block to construct the Hochschild cochain complexes. Notice however that $\mathscr{A} = T(A^*)^!$ is not AS-Gorenstein (no Poincaré duality).

3. *A deformed universal enveloping algebra.* Let $\mathfrak{A}$ be the algebra generated by the 3 elements $\nabla_0, \nabla_1, \nabla_2$ with relations

$$\begin{cases} \mu^2\nabla_2\nabla_0 - \nabla_0\nabla_2 = \mu\nabla_1 \\ \mu^4\nabla_1\nabla_0 - \nabla_0\nabla_1 = \mu^2(1+\mu^2)\nabla_0 \\ \mu^4\nabla_2\nabla_1 - \nabla_1\nabla_2 = \mu^2(1+\mu^2)\nabla_2. \end{cases} \tag{36}$$

This is again a Koszul quadratic-linear algebra with homogeneous part $\mathscr{A}$ which is Koszul of global dimension $D = 3$ [24, 37] and is AS-Gorenstein. The corresponding differential quadratic algebra $(\mathscr{A}^!, \delta)$ is the algebra $\mathscr{A}^!$ generated by $\omega_0, \omega_1, \omega_2$ with quadratic relations

$$\begin{cases} \omega_0^2 = 0, \omega_1^2 = 0, \omega_2^2 = 0 \\ \omega_2\omega_0 + \mu^2\omega_0\omega_2 = 0 \\ \omega_1\omega_0 + \mu^4\omega_0\omega_1 = 0 \\ \omega_2\omega_1 + \mu^4\omega_1\omega_2 = 0 \end{cases} \tag{37}$$

endowed with the differential δ given by

$$\begin{cases} \delta\omega_0 + \mu^2(1+\mu^2)\omega_0\omega_1 = 0 \\ \delta\omega_1 + \mu\omega_0\omega_2 = 0 \\ \delta\omega_2 + \mu^2(1+\mu^2)\omega_1\omega_2 = 0 \end{cases} \tag{38}$$

which corresponds to the left covariant differential calculus on the twisted $SU(2)$ group of [38].

4. *Canonical commutation relations algebra.* Let $E = \mathbb{K}^{2n}$ with basis (q^λ, p_μ), $\lambda, \mu \in \{1, \dots, n\}$ and let $i\hbar \in \mathbb{K}$ with $i\bar{h} \neq 0$. Consider the nonhomogeneous quadratic algebra $\mathfrak{A}$ generated by the q^λ, p_μ with relations

$$q^\lambda q^\mu - q^\mu q^\lambda = 0,\ \ p_\lambda p_\mu - p_\mu p_\lambda = 0,\ \ q^\lambda p_\mu - p_\mu q^\lambda = i\hbar\delta^\lambda_\mu \mathbf{1}$$

for $\lambda, \mu \in \{1, \dots, n\}$. The quadratic part of $\mathfrak{A}$ is the symmetric algebra $\mathscr{A} = SE$ which is Koszul of global dimension $D = 2n$. One has $\varphi_1 = 0$ and φ_0 is such that its transposed φ_0^t is given by

$$-\varphi_0^t(1) = F = -(i\hbar)^{-1} q^*_\lambda \wedge p^{\lambda^*}$$

which is central in $\mathscr{A}^! = \wedge(E^*)$ where $(q^*_\lambda, p^{\mu *})$ is the dual basis of (q^λ, p_μ). This implies that $\mathfrak{A}$ has the PBW property and therefore is Koszul.

5. *Clifford algebra (C.A.R. algebra).* Let $E = \mathbb{K}^n$ with canonical basis (γ_λ), $\lambda \in \{1, \dots, n\}$ and consider the nonhomogeneous quadratic algebra $\mathfrak{A} = C(n)$ generated by the elements γ_λ, $\lambda \in \{1, \dots, n\}$ with relations

$$\gamma_\mu \gamma_\nu + \gamma_\nu \gamma_\mu = 2\delta_{\mu\nu}\mathbf{1}$$

for $\mu, \nu \in \{1, \dots, n\}$. The quadratic part of $\mathfrak{A}$ is then the exterior algebra $\mathscr{A} = \wedge E$ which is Koszul. One has again $\varphi_1 = 0$ and φ_0^t is given by

$$-\varphi_0^t(1) = F = -\frac{1}{2}\sum \gamma^{\lambda *} \vee \gamma^{\lambda *}$$

which is a central element of $\mathscr{A}^! = SE^*$ (which is commutative). It again follows that $\mathfrak{A}$ is Koszul (i.e. PBW + $\mathscr{A}$ Koszul).

6. *Remarks on the generic case.* Let $\mathscr{A}$ be a (homogeneous) quadratic algebra which is Koszul. In general (for generic $\mathscr{A}$) any nonhomogeneous quadratic algebra $\mathfrak{A}$ which has $\mathscr{A}$ as quadratic part and has the PBW property is such that one has both $\varphi_1 \neq 0$ and $\varphi_0 \neq 0$ or is trivial in the sense that it coincides with $\mathscr{A}$, i.e. $\varphi_1 = 0$ and $\varphi_0 = 0$. This is the case for instance when $\mathscr{A}$ is the 4-dimensional Sklyanin algebra [14, 17, 33, 35] for generic values of its parameters [2].
Thus, Examples 1, 2, 3, 4, 5 above are rather particular from this point of view. However the next section will be devoted to a generalization of Lie algebra which has been introduced in [21] and which involves quadratic-linear algebras, i.e. for which $\varphi_0 = 0$.

6 A Generalization of Lie Algebras

6.1 Prealgebras

By a (finite-dimensional) *prealgebra* we here mean a triple (E, R, φ) where E is a finite-dimensional vector space, $R \subset E \otimes E$ is a linear subspace of $E \otimes E$ and $\varphi : R \to E$ is a linear mapping of R into E. Given a supplementary R' to R in $E \otimes E$, $R \oplus R' = E \otimes E$, the corresponding projector P of $E \otimes E$ onto R allows to define a bilinear product $\varphi \circ P : E \otimes E \to E$, i.e. a structure of algebra on E. The point is that there is generally no natural supplementary of R. Exception are $R = E \otimes E$ of course and $R = \wedge^2 E \subset E \otimes E$ for which there is the canonical $GL(E)$-invariant supplementary $R' = S^2 E \subset E \otimes E$ which leads to an antisymmetric product on E, (e.g. case of the Lie algebras).

Given a prealgebra (E, R, φ), there are two natural associated algebras :

1. The nonhomogeneous quadratic algebra

$$\mathfrak{A}_E = T(E)/(\{r - \varphi(r) \mid r \in R\})$$

which will be called its *enveloping algebra*.

2. The quadratic part $\mathscr{A}_E$ of $\mathfrak{A}_E$

$$\mathscr{A}_E = T(E)/(R),$$

where the prealgebra (E, R, φ) is also simply denoted by E when no confusion arises.

The enveloping algebra $\mathfrak{A}_E$ is a filtered algebras as explained before but it is also an augmented algebra with augmentation

$$\varepsilon : \mathfrak{A}_E \to \mathbb{K}$$

induced by the canonical projection of $T(E)$ onto $T^0(E) = \mathbb{K}$. One has the surjective homomorphism

$$can : \mathscr{A}_E \to \mathrm{gr}(\mathfrak{A}_E)$$

of graded algebras.

In the following we shall be mainly interested on prealgebras such that their enveloping algebras are quadratic-linear. If (E, R, φ) is such a prealgebra, to $\mathfrak{A}_E$ corresponds the differential quadratic algebra $(\mathscr{A}^!_E, \delta)$ (as in Sect. 5) where δ is the antiderivation extension of minus the transposed φ^t of φ.

Notice that if $\mathfrak{A}_E$ has the PBW property one has

$$E = F^1(\mathfrak{A}_E) \cap \mathrm{Ker}(\varepsilon)$$

so that the canonical mapping of the prealgebra E into its enveloping algebra $\mathfrak{A}_E$ is then an injection.

6.2 Lie Prealgebras

A prealgebra (E, R, φ) will be called a *Lie prealgebra* [21] if the following conditions (1) and (2) are satisfied :

(1) The quadratic algebra $\mathscr{A}_E = A(E, R)$ is Koszul of finite global dimension and is AS-Gorenstein (Poincaré duality).
(2) The enveloping algebra $\mathfrak{A}_E$ has the PBW property.

If $E = (E, R, \varphi)$ is a Lie prealgebra then $\mathfrak{A}_E$ is a Koszul quadratic linear algebra, so to (E, R, φ) one can associate the differential quadratic algebra $(\mathscr{A}^!_E, \delta)$ and one has the following theorem [21]:

Theorem 6.1 *The correspondence* $(E, R, \varphi) \mapsto (\mathscr{A}^!_E, \delta)$ *defines an anti-isomorphism between the category of Lie prealgebra and the category of differential quadratic Koszul Frobenius algebras.*

This is a direct consequence of Theorem 5.3 and of the Koszul Gorenstein property of $\mathscr{A}_E$ by using [34].

Let us remind that a *Frobenius algebra* is a finite-dimensional algebra $\mathscr{A}$ such that as left $\mathscr{A}$-modules $\mathscr{A}$ and its vector space dual $\mathscr{A}^*$ are isomorphic (the left $\mathscr{A}$-module structure of $\mathscr{A}^*$ being induced by the right $\mathscr{A}$-module structure of $\mathscr{A}$). Concerning the graded connected case one has the following classical useful result.

Proposition 6.1 *Let* $\mathscr{A} = \oplus_{m\geq 0}\mathscr{A}_m$ *be a finite-dimensional graded connected algebra with* $\mathscr{A}_D \neq 0$ *and* $\mathscr{A}_n = 0$ *for* $n > D$. *Then the following conditions* (i) *and* (ii) *are equivalent:*

(i) $\mathscr{A}$ *is Frobenius,*
(ii) $\dim(\mathscr{A}_D) = 1$ *and* $(x, y) \mapsto (xy)_D$ *is nondegenerate, where* $(z)_D$ *denotes the component on* $\mathscr{A}_D$ *of* $z \in \mathscr{A}$.

6.3 Some Representative Cases

1. *Lie algebras*. It is clear that a Lie algebra $\mathfrak{g}$ is canonically a Lie prealgebra $(\mathfrak{g}, R, \varphi)$ with $R = \wedge^2\mathfrak{g} \subset \mathfrak{g}\otimes\mathfrak{g}$, $\varphi = [\bullet, \bullet]$, $\mathfrak{A}_\mathfrak{g} = U(\mathfrak{g})$ and $\mathscr{A}_\mathfrak{g} = S\mathfrak{g}$, (see Example 1 in Sect. 5.3).
2. *Associative algebras are not Lie prealgebras*. An associative algebra A is clearly a prealgebra $(A, A\otimes A, m)$ with enveloping algebra $\mathfrak{A}_A = \tilde{A}$ as in Example 2 of Sect. 5.3 but $\mathscr{A}_A = T(A^*)^! = \mathbb{K}\mathbb{1}\oplus A$ is not AS-Gorenstein although it is Koszul

as well as $\mathfrak{A}_A = \tilde{A}$, (see the discussion of Example 2 in Sect. 5.3). The missing item is here the Poincaré duality.

3. *A deformed version of Lie algebras.* The algebra $\mathfrak{A}$ of Example 3 of Sect. 5.3 is the enveloping algebra of a Lie prealgebra (E, R, φ) with $E = \mathbb{K}^3$, $R \subset E \otimes E$ generated by
$r_1 = \mu^2\nabla_2 \otimes \nabla_0 - \nabla_0 \otimes \nabla_2$
$r_0 = \mu^4\nabla_1 \otimes \nabla_0 - \nabla_0 \otimes \nabla_1$
$r_2 = \mu^4\nabla_2 \otimes \nabla_1 - \nabla_1 \otimes \nabla_2$
and φ given by
$$\varphi(r_1) = \mu\nabla_1, \quad \varphi(r_0) = \mu^2(1+\mu^2)\nabla_0, \quad \varphi(r_2) = \mu^2(1+\mu^2)\nabla_2$$
where $(\nabla_0, \nabla_1, \nabla_2)$ is the canonical basis of E.
4. *Differential calculi on quantum groups.* More generally most differential calculi on the quantum groups can be obtained via the duality of Theorem 6 from Lie prealgebras. In fact the Frobenius property is generally straightforward to verify, what is less obvious to prove is the Koszul property.

6.4 Generalized Chevalley-Eilenberg Complexes

Throughout this section, $E = (E, R, \varphi)$ is a fixed Lie prealgebra, its enveloping algebra is simply denoted by $\mathfrak{A}$ with quadratic part denoted by $\mathscr{A}$ and the associated differential quadratic Koszul Frobenius algebra is $(\mathscr{A}^!, \delta)$.

A *left representation* of the Lie prealgebra $E = (E, R, \varphi)$ is a left $\mathfrak{A}$-module. Let V be a left representation of $E = (E, R, \varphi)$, let (x^λ) be a basis of E with dual basis (θ_λ) of $E^* = \mathscr{A}^!_1$. One has
$$x^\mu x^\nu \Phi \otimes \theta_\mu\theta_\nu + x^\lambda \Phi \otimes \delta\theta_\lambda = 0$$
for any $\Phi \in V$. This implies that one defines a differential of degree 1 on $V \otimes \mathscr{A}^!$ by setting
$$\delta_V(\Phi \otimes \alpha) = x^\lambda \Phi \otimes \theta_\lambda \alpha + \Phi \otimes \delta\alpha$$
so $(V \otimes \mathscr{A}^!, \delta_V)$ is a cochain complex. These cochain complexes generalize the Chevalley-Eilenberg cochain complexes. Given a right representation of E, that is a right $\mathfrak{A}$-module W, one defines similarily the chain complex $(W \otimes \mathscr{A}^{!*}, \delta_W)$, remembering that $\mathscr{A}^{!*}$ is a graded coalgebra.

One has the isomorphisms
$$\begin{cases} H^\bullet(V \otimes \mathscr{A}^!) \simeq \mathrm{Ext}^\bullet_{\mathfrak{A}}(\mathbb{K}, V) \\ H_\bullet(W \otimes \mathscr{A}^{!*}) \simeq \mathrm{Tor}^{\mathfrak{A}}_\bullet(W, \mathbb{K}) \end{cases}$$

which implies that one has the same relation with the Hochschild cohomology and the Hochschild homology of $\mathfrak{A}$ as the relation of the (co-)homology of a Lie algebra with the Hochschild (co-)homology of its universal enveloping algebra.

7 Conclusion

In these notes, we have only considered algebras which are quotient of tensor algebras of finite-dimensional vector spaces. One can extend the results described here in much more general frameworks. For instance in [11] the results of [19] concerning the homogeneous case have been extended to the quiver case. An even more general framework has been adopted in [9] for the nonhomogeneous Koszul algebras. Namely the algebras considered in [9] are quotient of tensor algebras of bimodules over von Neumann regular rings. This latter context seems quite optimal.

Acknowledgments It is a pleasure to thank Roland Berger for his kind advices.

References

1. Artin, M., Schelter, W.F.: Graded algebras of global dimension 3. Adv. Math. **66**, 171–216 (1987)
2. Bellon, M., Connes, A., Dubois-Violette, M.: Noncommutative finite-dimensional manifolds. III. Suspension functor and higher dimensional spherical manifolds. To appear
3. Berger, R.: Koszulity for nonquadratic algebras. J. Algebra **239**, 705–734 (2001)
4. Berger, R.: Koszulity for nonquadratic algebras II. math. QA/0301172 (2002)
5. Berger, R.: Dimension de Hochschild des algèbres graduées. C.R. Acad. Sci. Paris Ser. I **341**, 597–600 (2005)
6. Berger, R. (2008) Algèbres Artin-Schelter régulières (nouvelle version du chapitre 7 d'un cours de Master 2 à Lyon 1). http://webperso.univ-st-etienne.fr/rberger/mes-textes.html
7. Berger, R., Dubois-Violette, M.: Inhomogeneous Yang-Mills algebras. Lett. Math. Phys. **76**, 65–75 (2006)
8. Berger, R., Dubois-Violette, M., Wambst, M.: Homogeneous algebras. J. Algebra **261**, 172–185 (2003)
9. Berger, R., Ginzburg, V.: Higher symplectic reflection algebras and non-homogeneous N-Koszul property. J. Algebra **305**, 577–601 (2006)
10. Berger, R., Marconnet, N.: Koszul and Gorenstein properties for homogeneous algebras. Algebras Represent. Theory **9**, 67–97 (2006)
11. Bocklandt, R., Schedler, T., Wemyss, M.: Superpotentials and higher order derivations. J. Pure Appl. Algebra **214**, 1501–1522 (2010)
12. Bondal, A.I., Polishchuk, A.E.: Homological properties of associative algebras: the method of helices. Russian Acad. Sci. Izv. Math. **42**, 219–260 (1994)
13. Cartan, H.: Homologie et cohomologie d'une algèbre graduée. Séminaire Henri Cartan **11**(2), 1–20 (1958)
14. Connes, A., Dubois-Violette, M.: Noncommutative finite-dimensional manifolds. I. Spherical manifolds and related examples. Commun. Math. Phys. **230**, 539–579 (2002)
15. Connes, A., Dubois-Violette, M.: Yang-Mills algebra. Lett. Math. Phys. **61**, 149–158 (2002)

16. Connes, A., Dubois-Violette, M.: Yang-Mills and some related algebras. In: Rigorous Quantum Field Theory, Progress in Mathematics, vol. 251, pp. 65–78. Birkhaüser (2007)
17. Connes, A., Dubois-Violette, M.: Noncommutative finite-dimensional manifolds. II. Moduli space and structure of noncommutative 3-spheres. Commun. Math. Phys. **281**(1), 23–127 (2008)
18. Dubois-Violette, M.: Graded algebras and multilinear forms. C. R. Acad. Sci. Paris Ser. I **341**, 719–724 (2005)
19. Dubois-Violette, M.: Multilinear forms and graded algebras. J. Algebra **317**, 198–225 (2007)
20. Dubois-Violette, M.: Noncommutative coordinate algebras. In: Blanchard, E. (ed.) Quanta of Maths, dédié à A. Connes, Clay Mathematics Proceedings, vol. 14, pp. 171–199. Clay Mathematics Institute (2010)
21. Dubois-Violette, M., Landi, G.: Lie prealgebras. In: Connes, A. et al. (eds.), Noncommutative Geometry and Global Analysis, Contemporary Mathematics, vol. 546, pp. 115–135. American Mathematical Society (2011)
22. Fløystad, G., Vatne, J.E.: PWB-deformations of N-Koszul algebras. J. Algebra **302**, 116–155 (2006)
23. Ginzburg, V.: Calabi-Yau algebras. math.AG/0612139
24. Gurevich, D.I.: Algebraic aspects of the quantum Yang-Baxter equation. Algebra i Analiz (Transl. in Leningrad Math. J. 2 801–828 (1991)), **2**, 119–148 (1990)
25. He, J.-W., Lu, D.M.: Higher Koszul algebras and A-infinity algebras. J. Algebra **293**, 335–362 (2005)
26. Kadeishvili, T.V.: On the theory of homology of fiber spaces. Russ. Math. Surv. **35**, 231–238 (1980)
27. Loday, J.L., Vallette, B.: Algebraic operads, Grundlehren der mathematischen Wissenschaften. vol. 346, Springer (2012)
28. Lu, D.M., Palmieri, J.H., Wu, Q.S., Zhang, J.J.: A_∞-algebras for ring theorists. Algebra Colloquium **11**(1), 91–128 (2004)
29. Manin, Yu.I.: Some remarks on Koszul algebras and quantum groups. Ann. Inst. Fourier Grenoble **37**, 191–205 (1987)
30. Manin, Yu.I.: Quantum Groups and Non-Commutative Geometry. CRM Université de, Montréal (1988)
31. Polishchuk, A., Positselski, L.: Quadratic algebras, University Lecture Series, American Mathematical Society, vol. 37, Providence, RI (2005)
32. Positselski, L.: Nonhomogeneous quadratic duality and curvature. Func. Anal. Appl. **27**, 197–204 (1993)
33. Sklyanin, E.K.: Some algebraic structures connected with the Yang-Baxter equation. Func. Anal. Appl. **16**, 263–270 (1982)
34. Smith, S.P.: Some finite dimensional algebras related to elliptic curves. CMS Conf. Proc. Proc. **19**, 315–348 (1996)
35. Smith, S.P., Stafford, J.T.: Regularity of the four dimensional Sklyanin algebra. Compos. Math. **83**, 259–289 (1992)
36. Stasheff, J.D.: Homotopy associativity of H-spaces. I. II. Trans. Amer. Math. Soc. **108**, 275–312 (1963)
37. Wambst, M.: Complexes de Koszul quantiques. Ann. Inst. Fourier Grenoble **43**, 1083–1156 (1993)
38. Woronowicz, S.L.: Twisted SU(2) group. an example of noncommutative differential calculus. Publ. RIMS Kyoto Univ. **23**, 117–181 (1987)

Quantized Reduced Fusion Elements and Kostant's Problem

Eugene Karolinsky, Alexander Stolin and Vitaly Tarasov

Abstract We find a partial solution to the problem of Kostant concerning description of the locally finite endomorphisms of highest weight irreducible modules. The solution is obtained by means of its reduction to an extension of the quantization problem. While the classical quantization problem consists in finding $\star$-product deformations of the commutative algebras of functions, we consider the q-case when the initial object is already a noncommutative algebra.

1 Introduction

Let $\check{U}_q\mathfrak{g}$ be the quantized universal enveloping algebra "of simply connected type" [8] that corresponds to a finite dimensional split semisimple Lie algebra $\mathfrak{g}$. Let $L(\lambda)$ be the irreducible highest weight $\check{U}_q\mathfrak{g}$-module of highest weight λ. The aim of this paper is to show that for certain values of λ, the action map $\check{U}_q\mathfrak{g} \to \big(\operatorname{End} L(\lambda)\big)_{\text{fin}}$ is surjective. Here $(\operatorname{End} L(\lambda))_{\text{fin}}$ stands for the locally finite part of $\operatorname{End} L(\lambda)$ with respect to the adjoint action of $\check{U}_q\mathfrak{g}$. For the Lie-algebraic case ($q = 1$), this problem

E. Karolinsky (✉)
Department of Mathematics, Kharkov National University, 4 Svobody Sq., Kharkov 61077, Ukraine
e-mail: eugene.a.karolinsky@univer.kharkov.ua

A. Stolin
Department of Mathematics, University of Göteborg, 412 96 Göteborg, Sweden
e-mail: astolin@math.chalmers.se

V. Tarasov
St. Petersburg Branch of Steklov Mathematical Institute, Fontanka 27, St. Petersburg 191023, Russia
e-mail: vt@pdmi.ras.ru; vt@math.iupui.edu

V. Tarasov
Department of Mathematical Sciences, IUPUI, Indianapolis, IN 46202, USA

A. Makhlouf et al. (eds.), *Algebra, Geometry and Mathematical Physics*, Springer Proceedings in Mathematics & Statistics 85, DOI: 10.1007/978-3-642-55361-5_2,

is known as the classical Kostant's problem, see [6, 7, 12, 15, 16]. The complete answer to it is still unknown even in the $q = 1$ case. However, there are examples of λ for which the action map $U(\mathfrak{g}) \to (\mathrm{End}\, L(\lambda))_{\mathrm{fin}}$ is not surjective. Such examples exist even in the case $\mathfrak{g}$ is of type A [17].

The main idea of our approach to Kostant's problem, both in the Lie-algebraic and quantum group cases, is that $(\mathrm{End}\, L(\lambda))_{\mathrm{fin}}$ has two other presentations. First, it follows from the results of [11] that $(\mathrm{End}\, L(\lambda))_{\mathrm{fin}}$ is canonically isomorphic to $\mathrm{Hom}_U(L(\lambda), L(\lambda) \otimes F)$, where U is $U(\mathfrak{g})$ (resp. $\check{U}_q\mathfrak{g}$), and F is the algebra of (quantized) regular functions on the connected simply connected algebraic group G corresponding to the Lie algebra $\mathfrak{g}$. In other words, F is spanned by matrix elements of finite dimensional representations of U with an appropriate multiplication.

One more presentation of the algebra $(\mathrm{End}\, L(\lambda))_{\mathrm{fin}}$ comes from the fact that $\mathrm{Hom}_U(L(\lambda), L(\lambda) \otimes F)$ is isomorphic as a vector space to a certain subspace F' of F. The subspace F' can be equipped with a $\star$-multiplication obtained from the multiplication on F by applying the so-called reduced fusion element. Then $(\mathrm{End}\, L(\lambda))_{\mathrm{fin}}$ is isomorphic as an algebra to F' with this new multiplication. For certain values of λ, the same $\star$-multiplication on F' can be defined by applying the universal fusion element, that yields the affirmative answer to Kostant's problem in such cases.

More exactly, consider the triangular decomposition $U = U^-U^0U^+$. We have $L(\lambda) = M(\lambda)/K_\lambda \mathbf{1}_\lambda$, where $M(\lambda)$ is the corresponding Verma module, $\mathbf{1}_\lambda$ is the generator of $M(\lambda)$, and $K_\lambda \subset U^-$. Consider also the opposite Verma module $\widetilde{M}(-\lambda)$ with the lowest weight $-\lambda$ and the lowest weight vector $\widetilde{\mathbf{1}}_{-\lambda}$. Then its maximal U-submodule is of the form $\widetilde{K}_\lambda \cdot \widetilde{\mathbf{1}}_{-\lambda}$, where $\widetilde{K}_\lambda \subset U^+$. We have $F' = F[0]^{K_\lambda + \widetilde{K}_\lambda}$— the subspace of U^0-invariant elements of F annihilated by both K_λ and $\widetilde{K}_\lambda$. The $\star$-product on $F[0]^{K_\lambda + \widetilde{K}_\lambda}$ has the form

$$f_1 \star_\lambda f_2 = \mu\left(J^{\mathrm{red}}(\lambda)(f_1 \otimes f_2)\right),$$

where μ is the multiplication on F, and the reduced fusion element $J^{\mathrm{red}}(\lambda) \in U^- \widehat{\otimes}\, U^+$ is computed in terms of the Shapovalov form on $L(\lambda)$. Notice that for generic λ the element $J^{\mathrm{red}}(\lambda)$ is equal up to an U^0-part to the fusion element $J(\lambda)$ related to the Verma module $M(\lambda)$, see for example [4].

We also investigate limiting properties of $J(\lambda)$. In particular, for some values of λ_0 we can guarantee that $f_1 \star_\lambda f_2 \to f_1 \star_{\lambda_0} f_2$ as $\lambda \to \lambda_0$. Also, for any λ_0 having a "regularity property" of this kind, the action map $U \to (\mathrm{End}\, L(\lambda_0))_{\mathrm{fin}}$ is surjective. This gives the affirmative answer to the (quantum version of) Kostant's problem.

For some values of λ, the subspace $F[0]^{K_\lambda + \widetilde{K}_\lambda}$ is a subalgebra of $F[0]$, and can be considered as (a flat deformation of) the algebra of regular functions on some Poisson homogeneous space G/G_1. In those cases, the algebra $\left(F[0]^{K_\lambda + \widetilde{K}_\lambda}, \star_\lambda\right)$ is an equivariant quantization of the Poisson algebra of regular functions on G/G_1.

This paper is organized as follows. In Sect. 2 we recall the definition of the version of quantized universal enveloping algebra used in this paper, and some related constructions that will be useful in the sequel. In Sect. 3 we construct an isomorphism

$\mathrm{Hom}_U\big(L(\lambda), L(\lambda)\otimes F\big) \simeq F[0]^{K_\lambda+\widetilde{K}_\lambda}$ and, as a corollary, provide a construction of a star-product on $F[0]^{K_\lambda+\widetilde{K}_\lambda}$ in terms of the Shapovalov form on $L(\lambda)$. In Sect. 4 we study limiting properties of fusion elements and the corresponding star-products. Namely, in Sect. 4.1 we introduce the notion of a J-regular weight. In a neighborhood of a J-regular weight the fusion element behaves nicely, which allows one to give a solution to the Kostant's problem for such weights (see Theorem 4.1). We also provide non-trivial examples of J-regular weights. Finally, in Sect. 4.2 we apply limiting properties of fusion elements to quantize explicitly certain Poisson homogeneous spaces (see Theorem 4.4).

In this short version of the paper the proofs are omitted. A complete version with proofs will be published elsewhere.

2 Algebra $\check{U}_q\mathfrak{g}$

Let $\Bbbk$ be the field extension of $\mathbb{C}(q)$ by all fractional powers $q^{1/n}$, $n \in \mathbb{N} = \{1, 2, 3, \ldots\}$. We use $\Bbbk$ as the ground field.

Let (a_{ij}) a finite type $r \times r$ Cartan matrix. Let d_i be relatively prime positive integers such that $d_i a_{ij} = d_j a_{ji}$. For any positive integer k, define

$$[k]_i = \frac{q^{kd_i} - q^{-kd_i}}{q^{d_i} - q^{-d_i}}, \quad [k]_i! = [1]_i\,[2]_i\,\ldots\,[k]_i.$$

The algebra $U = \check{U}_q\mathfrak{g}$ is generated by the elements $t_i, t_i^{-1}, e_i, f_i, i = 1, \ldots, r$, subject to the relations

$$t_i t_i^{-1} = t_i^{-1} t_i = 1$$
$$t_i e_j t_i^{-1} = q^{d_i\delta_{ij}} e_j,$$
$$t_i f_j t_i^{-1} = q^{-d_i\delta_{ij}} f_j,$$

$$e_i f_j - f_j e_i = \delta_{ij}\frac{k_i - k_i^{-1}}{q^{d_i} - q^{-d_i}}, \quad \text{where } k_i = \prod_{j=1}^{r} t_j^{a_{ij}},$$

$$\sum_{m=0}^{1-a_{ij}} \frac{(-1)^m}{[m]_i!\,[1-a_{ij}-m]_i!} e_i^m e_j e_i^{1-a_{ij}-m} = 0 \text{ for } i \neq j,$$

$$\sum_{m=0}^{1-a_{ij}} \frac{(-1)^m}{[m]_i!\,[1-a_{ij}-m]_i!} f_i^m f_j f_i^{1-a_{ij}-m} = 0 \text{ for } i \neq j$$

Notice that $k_i e_j k_i^{-1} = q^{d_i a_{ij}} e_j$, $k_i f_j k_i^{-1} = q^{-d_i a_{ij}} f_j$.

The algebra U is a Hopf algebra with the comultiplication Δ, the counit ε, and the antipode σ given by

$$\begin{aligned}
&\Delta(t_i) = t_i \otimes t_i, && \varepsilon(t_i) = 1,\ \sigma(t_i) = t_i^{-1}\\
&\Delta(e_i) = e_i \otimes 1 + k_i \otimes e_i, && \varepsilon(e_i) = 0,\ \sigma(e_i) = -k_i^{-1} e_i\\
&\Delta(f_i) = f_i \otimes k_i^{-1} + 1 \otimes f_i, && \varepsilon(f_i) = 0,\ \sigma(f_i) = -f_i k_i.
\end{aligned}$$

In what follows we will sometimes use the Sweedler notation for comultiplication.

Let U^0 be the subalgebra of U generated by the elements $t_1, \dots, t_r, t_1^{-1}, \dots, t_r^{-1}$. Let U^+ and U^- be the subalgebras generated respectively by the elements $e_1, \dots, e_r$ and $f_1, \dots, f_r$. We have a triangular decomposition $U = U^- U^0 U^+$. Denote by θ the involutive automorphism of U given by $\theta(e_i) = -f_i, \theta(f_i) = -e_i, \theta(t_i) = t_i^{-1}$. Notice that θ gives an algebra isomorphism $U^- \to U^+$. Set $\omega = \sigma\theta$, i.e., ω is the involutive antiautomorphism of U given by $\omega(e_i) = f_i k_i, \omega(f_i) = k_i^{-1} e_i, \omega(t_i) = t_i$.

Let $(\mathfrak{h}, \Pi, \Pi^\vee)$ be a realization of (a_{ij}) over $\mathbb{Q}$, that is, $\mathfrak{h}$ is (a rational form of) a Cartan subalgebra of the corresponding semisimple Lie algebra, $\Pi = \{\alpha_1, \dots, \alpha_r\} \subset \mathfrak{h}^*$ the set of simple roots, $\Pi^\vee = \{\alpha_1^\vee, \dots, \alpha_r^\vee\} \subset \mathfrak{h}$ the set of simple coroots. Let $\mathbf{R}$ be the root system, $\mathbf{R}_+$ the set of positive roots, and W the Weyl group.

Let $u_1, \dots, u_r \in \mathfrak{h}$ be the simple coweights, i.e., $\langle \alpha_i, u_j \rangle = \delta_{ij}$. We denote by ρ the half sum of the positive roots.

Let

$$Q_+ = \sum_{\alpha \in \Pi} \mathbb{Z}_+ \alpha.$$

For $\lambda, \mu \in \mathfrak{h}^*$ we set $\lambda \geq \mu$ iff $\lambda - \mu \in Q_+$.

Denote by T the multiplicative subgroup generated by $t_1, \dots, t_r$. Any $\lambda \in \mathfrak{h}^*$ defines a character $\Lambda : T \to \Bbbk$ given by $t_i \mapsto q^{d_i \langle \lambda, u_i \rangle}$. We will write $\Lambda = q^\lambda$. Notice that $q^\lambda(k_i) = q^{d_i \langle \lambda, \alpha_i^\vee \rangle}$. We extend q^λ to the subalgebra U^0 by linearity. We say that an element $x \in U$ is of weight λ if $txt^{-1} = q^\lambda(t)x$ for all $t \in T$.

For a U-module V, we denote by

$$V[\lambda] = \{v \in V \mid tv = q^\lambda(t)v \quad \text{for all } t \in T\}$$

the weight subspace of weight λ. We call the module V admissible if V is a direct sum of finite-dimensional weight subspaces $V[\lambda]$.

The Verma module $M(\lambda)$ over U with highest weight λ and highest weight vector $\mathbf{1}_\lambda$ is defined in the standard way:

$$M(\lambda) = U^- \mathbf{1}_\lambda, \qquad U^+ \mathbf{1}_\lambda = 0, \qquad t\mathbf{1}_\lambda = q^\lambda(t)\mathbf{1}_\lambda, \quad t \in T.$$

The map $U^- \to M(\lambda)$, $y \mapsto y\mathbf{1}_\lambda$ is an isomorphism of U^--modules.

Set $U_+^\pm = \operatorname{Ker} \varepsilon|_{U^\pm}$ and denote by $x \mapsto (x)_0$ the projection $U \to U^0$ along $U_+^- \cdot U + U \cdot U_+^+$. For any $\lambda \in \mathfrak{h}^*$ consider $\pi_\lambda : U^+ \otimes U^- \to \Bbbk$, $\pi_\lambda(x \otimes y) =$

$q^{\lambda}((\sigma(x)y)_0)$, and $\mathbb{S}_\lambda : U^-\otimes U^- \to \Bbbk$, $\mathbb{S}_\lambda(x\otimes y) = \pi_\lambda(\theta(x)\otimes y) = q^{\lambda}((\omega(x)y)_0)$. We call $\mathbb{S}_\lambda$ the Shapovalov form on U^- corresponding to λ. We can regard $\mathbb{S}_\lambda$ as a bilinear form on $M(\lambda)$.

Set

$$K_\lambda = \{y \in U^- \mid \pi_\lambda(x \otimes y) = 0 \quad \text{for all } x \in U^+\},$$

$$\widetilde{K}_\lambda = \{x \in U^+ \mid \pi_\lambda(x \otimes y) = 0 \quad \text{for all } y \in U^-\}.$$

Clearly, K_λ is the kernel of $\mathbb{S}_\lambda$, $\widetilde{K}_\lambda = \theta(K_\lambda)$. Notice that $K(\lambda) = K_\lambda \cdot \mathbf{1}_\lambda$ is the largest proper submodule of $M(\lambda)$, and $L(\lambda) = M(\lambda)/K(\lambda)$ is the irreducible U-module with highest weight λ. Denote by $\overline{\mathbf{1}}_\lambda$ the image of $\mathbf{1}_\lambda$ in $L(\lambda)$.

Let $F = \Bbbk[G]_q$ be the quantized algebra of regular functions on a connected simply connected algebraic group G that corresponds to the Cartan matrix (a_{ij}) (see [8, 14]). We can consider F as a Hopf subalgebra in the dual Hopf algebra $U^\star$. We will use the left and right regular actions of U on F defined respectively by the formulae $(\overrightarrow{a} f)(x) = f(xa)$ and $(f \overleftarrow{a})(x) = f(ax)$. Notice that F is a sum of finite-dimensional admissible U-modules with respect to both regular actions of U (see [14]).

3 Star Products and Fusion Elements

3.1 Algebra of Intertwining Operators

Let us denote by $U_{\mathrm{fin}} \subset U$ the subalgebra of locally finite elements with respect to the right adjoint action of U on itself. We will use similar notation for any (right) U-module.

For any (left) U-module M we equip F with the left regular U-action and consider the space $\mathrm{Hom}_U(M, M \otimes F)$. For any $\varphi, \psi \in \mathrm{Hom}_U(M, M \otimes F)$ define

$$\varphi * \psi = (\mathrm{id} \otimes \mu) \circ (\varphi \otimes \mathrm{id}) \circ \psi, \tag{1}$$

where μ is the multiplication in F. We have $\varphi * \psi \in \mathrm{Hom}_U(M, M \otimes F)$, and this definition equips $\mathrm{Hom}_U(M, M \otimes F)$ with a unital associative algebra structure.

Consider the map $\Phi : \mathrm{Hom}_U(M, M \otimes F) \to \mathrm{End}\, M$, $\varphi \mapsto u_\varphi$, defined by $u_\varphi(m) = (\mathrm{id} \otimes \varepsilon)(\varphi(m))$; here $\varepsilon(f) = f(1)$ is the counit in F. Consider U_{fin}, $\mathrm{Hom}_U(M, M \otimes F)$ and $\mathrm{End}\, M$ as right U-module algebras: U_{fin} via right adjoint action, $\mathrm{Hom}_U(M, M \otimes F)$ via right regular action on F (i.e., $(\varphi \cdot a)(m) = (\mathrm{id} \otimes \overleftarrow{a})(\varphi(m)))$, and $\mathrm{End}\, M$ in a standard way (i.e., $u \cdot a = \sum_{(a)} \sigma(a_{(1)})_M u a_{(2)}{}_M$). Then $\mathrm{Hom}_U(M, M \otimes F)_{\mathrm{fin}} = \mathrm{Hom}_U(M, M \otimes F)$, and $\Phi : \mathrm{Hom}_U(M, M \otimes F) \longrightarrow (\mathrm{End}\, M)_{\mathrm{fin}}$ is an isomorphism of right U-module algebras (see [11, Proposition 6]).

Now we apply this to $M = M(\lambda)$ and $M = L(\lambda)$. Since $U_{\mathrm{fin}} \to (\mathrm{End}\, M(\lambda))_{\mathrm{fin}}$ is surjective (see [8, 9]), we have the following commutative diagram

$$\begin{array}{ccc} \mathrm{Hom}_U(M(\lambda), M(\lambda)\otimes F) & \longrightarrow & \mathrm{Hom}_U(L(\lambda), L(\lambda)\otimes F) \\ \Big\downarrow \Phi_{M(\lambda)} & & \Big\downarrow \Phi_{L(\lambda)} \\ (\mathrm{End}\, M(\lambda))_{\mathrm{fin}} & \longrightarrow & (\mathrm{End}\, L(\lambda))_{\mathrm{fin}} \end{array}$$

(see [11, Proposition 9]).

For any $\varphi \in \mathrm{Hom}_U(L(\lambda), L(\lambda)\otimes F)$ the formula

$$\varphi(\overline{\mathbf{1}}_\lambda) = \overline{\mathbf{1}}_\lambda \otimes f_\varphi + \sum_{\mu<\lambda} v_\mu \otimes f_\mu,$$

where v_μ is of weight μ, defines a map

$$\Theta : \mathrm{Hom}_U(L(\lambda), L(\lambda)\otimes F) \to F[0], \qquad \varphi \mapsto f_\varphi.$$

Theorem 3.1 *Θ is an embedding, and its image equals $F[0]^{K_\lambda+\widetilde{K}_\lambda}$.*

3.2 Reduced Fusion Elements

In this subsection we describe $\Theta^{-1} : F[0]^{K_\lambda+\widetilde{K}_\lambda} \to \mathrm{Hom}_U(L(\lambda), L(\lambda)\otimes F)$ explicitly in terms of the Shapovalov form. Recall that we can regard $\mathbb{S}_\lambda$ as a bilinear form on $M(\lambda)$. Denote by $\overline{\mathbb{S}}_\lambda$ the corresponding bilinear form on $L(\lambda)$. For any $\beta \in Q_+$ denote by $\overline{\mathbb{S}}_\lambda^\beta$ the restriction of $\overline{\mathbb{S}}_\lambda$ to $L(\lambda)[\lambda-\beta]$. Let $y_\beta^i \cdot \overline{\mathbf{1}}_\lambda$ be an arbitrary basis in $L(\lambda)[\lambda-\beta]$, where $y_\beta^i \in U^-[-\beta]$.

Take $f \in F[0]^{K_\lambda+\widetilde{K}_\lambda}$ and set $\varphi = \Theta^{-1}(f)$,

$$\varphi(\overline{\mathbf{1}}_\lambda) = \sum_{\beta\in Q_+}\sum_i y_\beta^i \overline{\mathbf{1}}_\lambda \otimes f^{\beta,i}.$$

For $\beta = 0$ we have $y_\beta^i = 1$ and $f^{\beta,i} = f$.

Proposition 3.1 $f^{\beta,i} = \sum_j \left(\overline{\mathbb{S}}_\lambda^\beta\right)^{-1}_{ij} \overrightarrow{\theta\left(y_\beta^j\right)} f.$

For any $\lambda \in \mathfrak{h}^*$ consider

$$J^{\mathrm{red}}(\lambda) = \sum_{\beta\in Q_+}\sum_{i,j} \left(\overline{\mathbb{S}}_\lambda^\beta\right)^{-1}_{ij} y_\beta^i \otimes \theta\left(y_\beta^j\right). \qquad (2)$$

One can regard $J^{\mathrm{red}}(\lambda)$ as an element in a certain completion of $U^-\otimes U^+$.

Remark 3.1 This element $J^{\text{red}}(\lambda)$ is not uniquely defined (e.g., because $U^- \to L(\lambda)$ has a kernel), but this does not affect our further considerations.

Remark 3.2 For $f \in F[0]^{K_\lambda+\widetilde{K}_\lambda}$ and $\varphi = \Theta^{-1}(f)$ one has $\varphi(\bar{\mathbf{1}}_\lambda) = J^{\text{red}}(\lambda)(\bar{\mathbf{1}}_\lambda \otimes f)$.

Let us define an associative product $\star_\lambda$ on $F[0]^{K_\lambda+\widetilde{K}_\lambda}$ by means of Θ, i.e., for any $f_1, f_2 \in F[0]^{K_\lambda+\widetilde{K}_\lambda}$ we define $f_1 \star_\lambda f_2 = \Theta(\varphi_1 * \varphi_2)$, where $\varphi_1 = \Theta^{-1}(f_1)$, $\varphi_2 = \Theta^{-1}(f_2)$, and $*$ is the product on $\text{Hom}_U(L(\lambda), L(\lambda) \otimes F)$ given by (1). By this definition, we get a right U-module algebra $(F[0]^{K_\lambda+\widetilde{K}_\lambda}, \star_\lambda)$.

Theorem 3.2 *We have*

$$f_1 \star_\lambda f_2 = \mu\left(\overrightarrow{J^{\text{red}}(\lambda)}(f_1 \otimes f_2)\right). \tag{3}$$

Remark 3.3 Theorem 23 together with results of [11] implies that $\text{Hom}_U(L(\lambda), L(\lambda) \otimes F)$, $(\text{End}\, L(\lambda))_{\text{fin}}$, and $\left(F[0]^{K_\lambda+\widetilde{K}_\lambda}, \star_\lambda\right)$ are isomorphic as right Hopf module algebras over U.

4 Limiting Properties of the Fusion Element

We say that $\lambda \in \mathfrak{h}^*$ is *generic* if $\langle \lambda + \rho, \beta^\vee \rangle \notin \mathbb{N}$ for all $\beta \in \mathbf{R}_+$. In this case $L(\lambda) = M(\lambda)$, and we set $J(\lambda) = J^{\text{red}}(\lambda)$. Notice that $J(\lambda)$ up to a U^0-part equals the fusion element related to the Verma module $M(\lambda)$ (see, e.g., [4]).

4.1 Regularity

Let $\lambda_0 \in \mathfrak{h}^*$. Since $J(\lambda)$ is invariant w. r. to $\tau(\theta \otimes \theta)$ (where τ is the tensor permutation), one can easily see that the following conditions on λ_0 are equivalent: 1) for any U^--module M the family of operators $J(\lambda)^M : M \otimes F[0]^{\widetilde{K}_{\lambda_0}} \to M \otimes F$ naturally defined by $J(\lambda)$ is regular at $\lambda = \lambda_0$, 2) for any U^+-module N the family of operators $J(\lambda)_N : F[0]^{K_{\lambda_0}} \otimes N \to F \otimes N$ naturally defined by $J(\lambda)$ is regular at $\lambda = \lambda_0$. We will say that λ_0 is *J-regular* if these conditions are satisfied. Clearly, any generic λ_0 is J-regular.

The following theorem collects some general properties of J-regular weights. In particular, for J-regular weights the answer to Kostant's question is affirmative.

Theorem 4.1 *Assume that $\lambda_0 \in \mathfrak{h}^*$ is J-regular. Then*

(1) $F[0]^{K_{\lambda_0}} = F[0]^{\widetilde{K}_{\lambda_0}} = F[0]^{K_{\lambda_0}+\widetilde{K}_{\lambda_0}}$,
(2) the natural map $\text{Hom}_U(M(\lambda_0), M(\lambda_0) \otimes F) \to \text{Hom}_U(L(\lambda_0), L(\lambda_0) \otimes F)$ *is surjective,*

(3) (Kostant's problem) *the action map* $U_{\mathrm{fin}} \to (\operatorname{End} L(\lambda_0))_{\mathrm{fin}}$ *is surjective,*
(4) for any $f, g \in F[0]^{K_{\lambda_0}}$ *we have* $\overrightarrow{J(\lambda)}(f \otimes g) \to \overrightarrow{J^{\mathrm{red}}(\lambda_0)}(f \otimes g)$ *as* $\lambda \to \lambda_0$,
(5) for any $f, g \in F[0]^{K_{\lambda_0}}$ *we have* $f_1 \star_\lambda f_2 \to f_1 \star_{\lambda_0} f_2$ *as* $\lambda \to \lambda_0$.

The following two theorems provide examples of J-regular weights.

Theorem 4.2 *Let* $\alpha \in \mathbf{R}_+$. *Consider* $\lambda_0 \in \mathfrak{h}^*$ *that satisfies* $\langle \lambda_0 + \rho, \alpha^\vee \rangle \in \mathbb{N}$, $\langle \lambda_0 + \rho, \beta^\vee \rangle \notin \mathbb{N}$ *for all* $\beta \in \mathbf{R}_+ \setminus \{\alpha\}$. *Then* λ_0 *is* J*-regular.*

Theorem 4.3 *Let* $\Gamma \subset \Pi$. *Consider* $\lambda_0 \in \mathfrak{h}^*$ *that satisfies* $\langle \lambda_0 + \rho, \alpha_i^\vee \rangle \in \mathbb{N}$ *for all* $\alpha_i \in \Gamma$, $\langle \lambda_0 + \rho, \beta^\vee \rangle \notin \mathbb{N}$ *for all* $\beta \in \mathbf{R}_+ \setminus \operatorname{Span} \Gamma$. *Then* λ_0 *is* J*-regular.*

4.2 Application to Poisson Homogeneous Spaces

Let $\Gamma \subset \Pi$. Assume that $\lambda \in \mathfrak{h}^*$ is such that $\langle \lambda, \alpha^\vee \rangle = 0$ for all $\alpha \in \Gamma$, and $\langle \lambda + \rho, \beta^\vee \rangle \notin \mathbb{N}$ for all $\beta \in \mathbf{R}_+ \setminus \operatorname{Span} \Gamma$. By Theorem 4.3, λ is J-regular. In particular, $F[0]^{K_\lambda + \widetilde{K}_\lambda} = F[0]^{K_\lambda}$.

In what follows it will be more convenient to write F_q, J_q, and $K_{q,\lambda}$ instead of F, J, and K_λ. We will also need the classical limits $F_1 = \lim_{q\to 1} F_q$ and $K_{1,\lambda} = \lim_{q \to 1} K_{q,\lambda}$. They can be defined in the same way as in ([8], Sects. 3.4.5 and 3.4.6).

Clearly, F_1 is the algebra of regular functions on the connected simply connected group G, whose Lie algebra is $\mathfrak{g}$. Let $\mathfrak{k}$ be a reductive subalgebra of $\mathfrak{g}$ which contains $\mathfrak{h}$ and is defined by Γ, K the corresponding subgroup of G, and $F(G/K)$ the algebra of regular functions on the homogeneous space G/K. According to [11, Theorem 33], we have $F(G/K) = F_1[0]^{K_{1,\lambda}}$. Therefore we get

Proposition 4.1 $\lim_{q\to 1} F_q[0]^{K_{q,\lambda}} = F(G/K)$.

Furthermore, since $F_q[0]^{K_{q,\lambda}}$ is a Hopf module algebra over U, G/K is a Poisson homogeneous space over G equipped with the Poisson-Lie structure defined by the Drinfeld-Jimbo classical r-matrix $r_0 = \sum_{\alpha \in \mathbf{R}_+} e_\alpha \wedge e_{-\alpha}$.

All such structures on G/K were described in [10]. It follows from [10] that any such Poisson structure on G/K is uniquely determined by an an intermediate Levi subalgebra $\mathfrak{n}$ satisfying $\mathfrak{k} \subset \mathfrak{n} \subset \mathfrak{g}$ and some $\lambda \in \mathfrak{h}^*$ which satisfies certain conditions, in particular, $\langle \lambda, \alpha^\vee \rangle = 0$ for $\alpha \in \Gamma$ and $\langle \lambda, \beta^\vee \rangle \notin \mathbb{Z}$ for $\beta \in \operatorname{Span} \Gamma_{\mathfrak{n}} \setminus \operatorname{Span} \Gamma$. Here $\Gamma_{\mathfrak{n}}$ is the set of simple roots defining $\mathfrak{n}$.

Now we can describe the Poisson bracket on G/K defined by $\star_\lambda$-multiplication on $F_q[0]^{K_{q,\lambda}}$.

Theorem 4.4 *Assume that* $\langle \lambda_0, \alpha^\vee \rangle = 0$ *for* $\alpha \in \Gamma$ *and* $\langle \lambda_0, \beta^\vee \rangle \notin \mathbb{Z}$ *for* $\beta \in \mathbf{R}_+ \setminus \operatorname{Span} \Gamma$. *Then the classical limit of* $(F_q[0]^{K_{q,\lambda_0}}, \star_{\lambda_0})$ *is the algebra* $F(G/K)$ *of regular functions on* G/K *equipped with the Poisson homogeneous structure defined by* $\mathfrak{n} = \mathfrak{g}$ *and* λ_0.

Notice that an analogous result for simple Lie algebras of classical type was obtained in [18] using reflection equation algebras.

Proposition 4.1 and Theorem 4.4 suggest a conjecture which we formulate below.

Let G be a connected Poisson affine algebraic group, $\mathfrak{g}$ the corresponding Lie bialgebra with the co-bracket δ, X a Poisson homogeneous G-variety, Y an affine Zariski open dense subset of X. Consider the Poisson algebra $F(Y)$ of regular functions on Y. Let $U_q\mathfrak{g}$ be a quantized universal enveloping algebra corresponding to $\mathfrak{g}$.

Conjecture 4.1 *There exists a Hopf module algebra over $U_q\mathfrak{g}$ whose classical limit is $F(Y)$.*

Let us show another example which confirms this conjecture. Consider the case $X = G$. Let $D(\mathfrak{g})$ be the classical double of $\mathfrak{g}$. According to [2], Poisson G-homogeneous structures on G are in one-to-one correspondence with Largangian subalgebras of $D(\mathfrak{g})$ transversal to $\mathfrak{g} \subset D(\mathfrak{g})$. Consider such a Lagrangian subalgebra $\mathfrak{l} \subset D(\mathfrak{g})$, which corresponds to a certain Poisson G-homogeneous structure on G. It is well known [1] that $\mathfrak{l}$ also induces a new Poisson-Lie structure on G, which differs from the original one by a so-called classical twist. Hence we obtain a new Lie bialgebra structure δ_1 on the Lie algebra $\mathfrak{g}$.

The following conjecture was made in [12] and later published in [13].

Conjecture 4.2 *There exists an element T in a certain completion of $(U_q\mathfrak{g})^{\otimes 2}$ which satisfies*

$$T^{12}(\Delta \otimes \mathrm{id})(T) = T^{23}(\mathrm{id} \otimes \Delta)(T) \tag{4}$$

and $(\varepsilon \otimes \mathrm{id})(T) = (\mathrm{id} \otimes \varepsilon)(T) = 1$ such that the Hopf algebra $U_{q,T}\mathfrak{g}$ quantizes $(\mathfrak{g}, \delta_1)$. Here $U_{q,T}\mathfrak{g}$ and $U_q\mathfrak{g}$ are isomorphic as algebras, and the co-multiplication on $U_{q,T}\mathfrak{g}$ is given by $\Delta_T(a) = T\Delta(a)T^{-1}$.

This conjecture was proved in [3, 5].

Now let $F_q(G)$ be the restricted dual of $U_q\mathfrak{g}$. It is well known that $F_q(G)$ quantizes $F(G)$. Let us equip $F_q(G)$ with a new product defined by $f_1 \star_T f_2 = \mu\left(\overrightarrow{T}(f_1 \otimes f_2)\right)$. According to (4), $\star_T$ is associative. Hence we get

Corollary 4.1 *The algebra $(F_q(G), \star_T)$ is a Hopf module algebra over $U_q\mathfrak{g}$ which quantizes the Poisson homogeneous structure on G defined by $\mathfrak{l}$.*

Acknowledgments The authors are grateful to Maria Gorelik, Jiang-Hua Lu, and Catharina Stroppel for useful discussions on the topic of the paper. E. K. was supported in part by the Royal Swedish Academy of Sciences, A. S.—by the Crafoord Prize research grant, V. T.–by NSF grant DMS-0901616.

References

1. Drinfeld, V.G.: Hamiltonian structures on Lie groups, Lie bialgebras and the geometric meaning of the classical Yang-Baxter equations. Sov. Math. Dokl. **27**, 68–71 (1983)
2. Drinfeld, V.G.: On Poisson homogeneous spaces of Poisson-Lie groups. Theor. Math. Phys. **95**, 524–525 (1993)
3. Enriquez, B., Halbout, G.: Quantization of coboundary Lie bialgebras. Ann. Math. **2**(171), 1267–1345 (2010)
4. Etingof, P., Schiffmann, O.: Lectures on the dynamical Yang-Baxter equations. In: Quantum groups and Lie theory (Durham, 1999), pp. 89–129. London Mathematical Society Lecture Note Series, 290. Cambridge University Press, Cambridge (2001)
5. Halbout, G.: Formality theorem for Lie bialgebras and quantization of twists and coboundary r-matrices. Adv. Math. **207**, 617–633 (2006)
6. Jantzen, J.C.: Einhüllende Algebren halbeinfacher Lie-Algebren. Springer, Berlin (1983)
7. Joseph, A.: Kostant's problem, Goldie rank and the Gelfand-Kirillov conjecture. Invent. Math. **56**, 191–213 (1980)
8. Joseph, A.: Quantum groups and their primitive ideals. Springer, Berlin (1995)
9. Joseph, A., Letzter, G.: Verma modules annihilators for quantized enveloping algebras. Ann. Sci. École Norm. Sup. **28**, 493–526 (1995)
10. Karolinsky, E., Muzykin, K., Stolin, A., Tarasov, V.: Dynamical Yang-Baxter equations, quasi-Poisson homogeneous spaces, and quantization. Lett. Math. Phys. **71**, 179–197 (2005)
11. Karolinsky, E., Stolin, A., Tarasov, V.: Irreducible highest weight modules and equivariant quantization. Adv. Math. **211**, 266–283 (2007)
12. Khoroshkin, S. M., Pop., I., Stolin, A., Tolstoy, V. N.: On some Lie bialgebra structures on polynomial algebras and their quantization. Preprint No. 21, 2003/2004, Mittag-Leffler Institute, Sweden (2004)
13. Khoroshkin, S.M., Pop, I., Samsonov, M.E., Stolin, A., Tolstoy, V.N.: On some Lie bialgebra structures on polynomial algebras and their quantization. Comm. Math. Phys. **282**, 625–662 (2008)
14. Korogodski, L., Soibelman, Y.: Algebras of functions on quantum groups. American Mathematical Society, Providence (1998)
15. Mazorchuk, V.: A twisted approach to Kostant's problem. Glasgow Math. J. **47**, 549–561 (2005)
16. Mazorchuk, V., Stroppel, C.: Categorification of (induced) cell modules and the rough structure of generalised Verma modules. Adv. Math. **219**, 1363–1426 (2008)
17. Mazorchuk, V., Stroppel, C.: Categorification of Wedderburn's basis for $\mathbb{C}[S_n]$. Arch. Math. **91**, 1–11 (2008)
18. Mudrov, A.: Quantum conjugacy classes of simple matrix groups. Comm. Math. Phys. **272**, 635–660 (2007)

Commutants and Centers in a 6-Parameter Family of Quadratically Linked Quantum Plane Algebras

Fredrik Ekström and Sergei D. Silvestrov

Abstract We consider a family of associative algebras, defined as the quotient of a free algebra with the ideal generated by a set of multi-parameter deformed commutation relations between four generators consisting of five quantum plane relations between pairs of generators and one sub-quadratic relation inter-linking all four generators. For generic parameter vectors, the center and the commutants of the two of the generators are described and conditions on the parameters for these commutants to be itself commutative or non-commutative are obtained.

1 Introduction

Commuting elements in non-commutative algebra are important for representation theory, classifications, interplay with harmonic analysis and spectral theory, topology and algebraic geometry, operator algebras and applications in Physics and Engineering. For example, commutants or centralisers, maximal commutative subalgebras of crossed product C^*-algebras and von Neumann algebras play a central role in investigation of representations, classifications and in structure of state space [1–6]. In particular, maximal commutative subalgebras are essential objects for the famous Kadison-Singer conjecture stated in a pioneering 1959—paper by Kadison and Singer [7], equivalent to the paving conjecture [8, 9] and several conjectures important for wavelets and frames analysis and applications in signal and image processing, one of them the well-known Feichtinger conjecture [10] in frame theory. For the

F. Ekström (✉)
Centre for Mathematical Sciences, Lund University, Box 118, 22100 Lund, Sweden
e-mail: Fredrik.Ekstrom@math.lth.se

S. D. Silvestrov
Division of Applied Mathematics, The School of Education, Culture and Communication, Mälardalen University, Box 883, 72123 Västerås, Sweden
e-mail: sergei.silvestrov@mdh.se

A. Makhlouf et al. (eds.), *Algebra, Geometry and Mathematical Physics*, Springer Proceedings in Mathematics & Statistics 85, DOI: 10.1007/978-3-642-55361-5_3,

key role of maximal commutative subalgebras for establishing interplay between Kadison-Singer conjecture, properties of projections, topological dynamical systems and compactifications of topological spaces see for example [11]. Commutants and maximal commutative subalgebras in generalized crossed product algebras arising from non-invertible dynamics and actions are used in the important ways in the general operator and spectral theory approach to wavelets analysis and investigation of wavelets on fractals [12–16]. The description of commuting elements and of corresponding commuting operators in the representing operator algebra, or in other words the problem of explicit description of commutative subalgebras is important in description and classifications of operator representations and applications of non-commutative algebras [17–27]. The commuting operators and commuting elements in rings and algebras also are important in study of integrable systems and non-linear equations. Further discussions in connection to this topic and numerous references can be found for instance in the book [28] devoted to commuting elements in the algebra defined by the q-deformed Heisenberg relations (see also [29, 30]).

The centers and commutants of elements or subsets in non-commutative algebras are fundamentally important subsets of an algebra or a ring in this context (see for example [31–39] and references therein). The center consists of elements commuting with all elements in the algebra, is the intersection of the commutants of all elements in the algebra and so is always a commutative subalgebra. The commutants of elements or subsets of elements in an associative algebra are subalgebras which contain the center of the whole algebra as its subalgebra, but may be commutative or may be not depending on the structure of the algebra and the subset for which the commutant is considered.

In this article we consider the centers and commutants for an interesting multi-parameter family of associative algebras generated by four generators and six sub-quadratic relations involving six deformation parameters. The five of these relations are the famous quantum plane relations playing important role in quantum groups, q-calculus and quantum mechanics, operator algebras and non-commutative geometry (rotation algebras, non-commutative tori, etc.). The sixth relation is interconnecting the four generators by a special q-deformed quadratic relation expressing the sum of two generators as q-commutator of the other two of the generators:

$$\begin{cases} AB - q_0BA = S + T & (a) \\ AT - q_1TA = 0 & (b) \\ BS - q_2SB = 0 & (c) \\ AS - q_3SA = 0 & (d) \\ BT - q_4TB = 0 & (e) \\ ST - q_5TS = 0 & (f) \end{cases} \tag{1}$$

where $\mathbf{q} = (q_0, q_1, q_2, q_3, q_4, q_5) \in \mathbb{C}^6$.

All of (1b)–(1f) are of the type $XY - qYX = 0$, which is the so called quantum plane relation studied in non-commutative geometry. Equation (1a) resembles the Sylvester equation $AX - XB = C$ and the Lyapunov equation $AX + XA^* =$

$-Q$, both of which are encountered in control theory. Specializing $S = \lambda I$ and $T = (1-\lambda)I$ (where I denotes the multiplicative identity) and $q_1, \ldots, q_5 = 1$, the relations (1b)–(1f) become trivial and (1a) becomes $AB - q_0BA = I$. This is a generalization of the Heisenberg canonical commutation ($q_0 = 1$) and anti-commutation ($q_0 = -1$) relations, which are used in quantum mechanics to describe systems with one degree of freedom. More on the algebras defined by q-deformed Heisenberg relations (called also q-Weil relations) and commuting elements in such algebras can be found in the monograph [28] and references there.

It is a well known interesting issue whether it is possible to realize a given family of commutation relations in one or another way using matrices or differential operators or other types of linear operators, or any objects for which (1) makes sense, for example elements of some associative algebra. When the realization by the operators of a specific type is possible, further description and classifications of the representations of the relations by the operators of such type arise and often becomes a problem of great interest. It often requires insights both in the algebraic structure of the commutation relations and in the properties of the involved classes of operators. In algebraic contexts it often leads to interesting combinatorial identities and problems, while in the context of $*$-representations (involutive representations) and operator algebras it involves also spectral theory of possibly unbounded operators in the finite-dimensional or infinite-dimensional spaces.

The relations (1) provide an interesting example in this respect. For a first taste of what can happen in (1a)–(1c) when A, B, S, T are complex $(n \times n)$-matrices, consider the case when A and B are hermitian and q_0 lies on the unit circle. Note that $\|X\|_F^2 = \operatorname{tr}(X^*X)$ defines a norm $\|X\|_F$ on $\mathbb{C}^{n\times n}$ (this is the so called Frobenius norm). Since A, B are hermitian and $q_0q_0^* = 1$, $(AB-q_0BA)^* = -q_0^*(AB-q_0BA)$, and thus

$$\begin{aligned}\|AB - q_0BA\|_F^2 &= -q_0^* \operatorname{tr}((AB - q_0BA)^2) = -q_0^* \operatorname{tr}((AB - q_0BA)(S+T)) \\ &= -q_0^* \left(\operatorname{tr}(ABS) + \operatorname{tr}(ABT) - q_0 \operatorname{tr}(BAS) - q_0 \operatorname{tr}(BAT)\right).\end{aligned}$$

Using (1b,c) and the fact that $\operatorname{tr}(XYZ) = \operatorname{tr}(ZXY)$ for all $(n \times n)$-matrices X, Y, Z, this implies that

$$\|AB - q_0BA\|_F^2 = -q_0^*((q_2 - q_0)\operatorname{tr}(ASB) + (1 - q_0q_1)\operatorname{tr}(BTA)).$$

Thus, if $q_2 = q_0$ and $q_0q_1 = 1$, A, B must satisfy $AB - q_0BA = 0$, and S, T must satisfy $S = -T$. In particular, if $q_0 = q_1 = q_2 = 1$, A and B must commute. It is not difficult to see that for many other conditions on the parameters this argument breaks down. This could be interpreted as an indication that the algebraic structures defined by these relations and their representations might have rich dependence on the interplay between the values of the six deformation parameters.

In this article, we provide further indication of richness of the structure of this family of algebras depending on the values of the deformation parameters, by considering some important properties of the algebra with generators and relations (1), especially focussing on centers and commutants. As these algebras are defined as the quotient algebra $\mathscr{F}/\mathscr{J}(\mathbf{q})$ of a free algebra on six generators by the ideal associated to the commutation relations (1), in order to be able to compute in this algebra it is particularly important to be able to decide the equality of the elements in the algebra, since using the relations (1), the same element can be expressed in many ways, and it is not obvious whether or not two given expressions are equal. To handle this, one needs a normal form for elements in $\mathscr{F}/\mathscr{J}(\mathbf{q})$, which for the relations of the type (1) amounts to finding a basis for $\mathscr{F}/\mathscr{J}(\mathbf{q})$ for various choices of parameters. In Sect. 2, we indicate that relations are more subtle than it may seem on the first sight as there are more relations than generators, and for many values of parameters these relations imply some further much more special relations between generators bringing significant restrictions on the size or the structure of the bases and thus on various further properties and computations in the algebra. Finding in a systematic way bases for the algebras for various choices of parameters becomes an elaborate task requiring non-trivial use of the Bergman's diamond lemma and relations (1) as well as some symmetries of the relations and their consequences for case reductions of various subtle parameter subsets. It appears in the course of this analysis, that the basis takes a somewhat simpler form for a large subset of parameters given by a system of certain inequalities. This set of "generic" parameters, as we call it, and the bases yield useful grading structures, used in Sect. 3 to describe the commutants of the main generators A and B by describing the spanning sets. In Sect. 4, the results from the preceding sections combined with further computations are used to describe explicitly the center by providing its basis depending on the deformation parameters. While the center of an algebra is always a commutative subalgebra, as an intersection of commutants of all elements of the algebra, the commutants of elements or non-trivial subsets of a non-commutative algebra are subalgebras which are not necessarily themselves commutative. For some classes of algebras it is possible to prove that commutants of the elements are commutative. Investigating whether this is a case and finding examples and counterexamples for such commutativity property for a particular family of algebras defined by generators and specific relations is an important problem which is often highly non-trivial, especially so when the defining relations are dependent on parameters. In Sect. 5, we provide necessary and sufficient conditions on $\mathbf{q}$ within the set of "generic" parameters, for $\mathscr{C}(A)$ and $\mathscr{C}(B)$ to be commutative, thus also providing necessary and sufficient conditions on "generic" parameters for when these commutants are not-commutative. The results of these paper suggest that the description and further in-depth analysis of the structure of the commutants of these and other elements and subsets for the family of algebras considered in this paper both for "generic" as well as for non-generic parameters is an interesting and rich open problem.

2 First Steps: Reordering and Basis

Let $\mathscr{F}$ be the free unital associative algebra over $\mathbb{C}$ generated by the set $\{A, B, S, T\}$. For $\mathbf{q} = (q_0, q_1, q_2, q_3, q_4, q_5) \in \mathbb{C}^6$ let $\mathscr{J}(\mathbf{q})$ be the ideal generated by the set

$$G(\mathbf{q}) = \{AB - q_0 BA - S - T,\ AT - q_1 TA,\ BS - q_2 SB,\\ AS - q_3 SA,\ BT - q_4 TB,\ ST - q_5 TS\}, \tag{2}$$

or equivalently the ideal in $\mathscr{F}$ generated by the relations

$$\begin{aligned} AB - q_0 BA &\equiv S + T \\ AT - q_1 TA &\equiv 0 \\ BS - q_2 SB &\equiv 0 \\ AS - q_3 SA &\equiv 0 \\ BT - q_4 TB &\equiv 0 \\ ST - q_5 TS &\equiv 0, \end{aligned} \tag{3}$$

where $\equiv$ denotes equivalence modulo $\mathscr{J}(\mathbf{q})$.

From (3) it is not too hard to derive the additional relations

$$(1 - q_2 q_3) S^2 \equiv (q_2 q_3 q_5 - 1) TS \tag{4}$$

$$(1 - q_1 q_4) T^2 \equiv (q_1 q_4 - q_5) TS. \tag{5}$$

The implications of these relations depend on which of the involved scalar expressions are zero and which are non-zero. There are also a few more expressions in the parameters that change the situation if they are zero. Only the generic case will be considered here.

Definition 2.1 A parameter vector $\mathbf{q} = (q_0, q_1, q_2, q_3, q_4, q_5) \in \mathbb{C}^6$ is *generic* if

$$\begin{cases} q_0, q_1, q_2, q_3, q_4, q_5 \neq 0, \\ 1 - q_5, 1 - q_1 q_4, 1 - q_2 q_3, q_1 - q_3, q_2 - q_4 \neq 0, \text{ and} \\ q_1 q_4 - q_5 \neq 0 \text{ or } q_2 q_3 q_5 - 1 \neq 0. \end{cases}$$

For generic $\mathbf{q}$ it follows from (3), (4) and (5) that

$$XY \equiv 0 \text{ for all } X \in \{S^2,\ ST,\ TS,\ T^2\}, Y \in \{A,\ B,\ S,\ T\}. \tag{6}$$

Since (3) can be used to put the symbols in the monomials in the order T, S, B, A, this means that any monomial that has two symbols from $\{S, T\}$ and additionally one symbol from $\{A, B, S, T\}$ is $\equiv 0$. Thus it seems plausible that the set

$$\mathscr{B} = \{B^b A^a,\ S B^b A^a,\ T B^b A^a,\ TS;\ a, b \in \mathbb{N}\}$$

is a basis for $\mathscr{F}/\mathscr{J}(\mathbf{q})$. This can be shown using the Diamond Lemma for ring theory [40].

Sums of the form $\sum_{i=0}^{n-1} q^i$ will often appear in what follows, so it is convenient to have a more compact notation for them.

Definition 2.2 For $q \in \mathbb{C}$ and $n \in \mathbb{N}$, let

$$\{n\}_q = \sum_{i=0}^{n-1} q^i.$$

$\{n\}_q$ is called the *n:th q-natural number.*

To express the product of two general elements in the basis $\mathscr{B}$, it is necessary to be able to rewrite monomials of the form $A^m B^n$ so that the B:s are moved to the left of the A:s.

Lemma 2.1 *Let* $\mathbf{q} = (q_0, \ldots, q_5) \in \mathbb{C}^6$ *be generic. Then the following formula holds in* $\mathscr{F} / \mathscr{J}(\mathbf{q})$ *for* $m, n \geq 1$, $(m, n) \neq (2, 2)$.

$$\begin{aligned} A^m B^n &\equiv q_0^{mn} B^n A^m \\ &+ q_0^{(m-1)n} \{m\}_{\frac{q_3}{q_0}} \{n\}_{q_0 q_2} S B^{n-1} A^{m-1} \\ &+ q_0^{(m-1)n} \{m\}_{\frac{q_1}{q_0}} \{n\}_{q_0 q_4} T B^{n-1} A^{m-1}. \end{aligned} \tag{7}$$

The formula can be proved for most (m, n) by induction first on m and then on n, or by induction first on n and then on m. The exceptional point $(m, n) = (2, 2)$ makes it necessary to use both orders of induction to cover all $(m, n) \neq (2, 2)$. We omit the elaborate details of the proof.

Equation (7) does not hold for $(m, n) = (2, 2)$. The reordering formula for $(m, n) = (2, 2)$ is instead

$$\begin{aligned} A^2 B^2 &\equiv q_0^4 B^2 A^2 + q_0(q_0 + q_3)(1 + q_0 q_2) S B A + q_0(q_0 + q_1)(1 + q_0 q_4) T B A \\ &+ (1 - q_5) \frac{q_1 - q_3 + q_0 q_1 q_4 - q_0 q_2 q_3 + q_1 q_3 q_4 - q_1 q_2 q_3}{(1 - q_1 q_4)(1 - q_2 q_3)} T S. \end{aligned} \tag{8}$$

This formula agrees with (7) except for the extra TS-term on the right side.

Let $\mathscr{M}$ be the set of monomials in $\mathscr{F}$ and define for $Y \in \mathscr{M}$

$$\begin{aligned} \deg_{A,S,T}(Y) &= \#A : s + \#S : s + \#T : s \text{ that occur in } Y \\ \deg_{B,S,T}(Y) &= \#B : s + \#S : s + \#T : s \text{ that occur in } Y. \end{aligned}$$

Then $\mathscr{F}$ has an $\mathbb{N}^2$-gradation $\{A_{(m,n)}\}_{(m,n) \in \mathbb{N}^2}$ given by

$$A_{(m,n)} = \text{Span}\{Y \in \mathscr{M};\ \deg_{A,S,T}(Y) = m,\ \deg_{B,S,T}(Y) = n\}.$$

All elements in the generating set of $\mathscr{J}(\mathbf{q})$ are homogeneous in this gradation and thus the induced gradation of $\mathscr{F} / \mathscr{J}(\mathbf{q})$ is well defined. If $m, n \geq 1$ and $(m, n) \neq (2, 2)$ then a basis for the homogeneous component of degree (m, n) is given by

$$\{B^n A^m,\ S B^{n-1} A^{m-1},\ T B^{n-1} A^{m-1}\}.$$

3 Commutants of A and B

The *commutant* of an element $X \in \mathscr{F}/\mathscr{J}(\mathbf{q})$ is the set of all elements that commute with X. It will be denoted by $\mathscr{C}(X)$. In this section, spanning sets for $\mathscr{C}(A)$ and $\mathscr{C}(B)$ are described for generic $\mathbf{q}$.

In general, a commutant of a homogeneous element is spanned by the homogeneous elements of the commutant. This means that it is enough to find all *homogeneous* elements that commute with A or B. Let $X_{m,n}$ denote a general homogeneous element of degree (m, n). For $m, n \geq 1$, $(m, n) \neq (2, 2)$, such an element can be uniquely written as

$$X_{m,n} \equiv c_1 B^n A^m + c_2 S B^{n-1} A^{m-1} + c_3 T B^{n-1} A^{m-1}, \tag{9}$$

with $c_1, c_2, c_3 \in \mathbb{C}$. Since TS commutes with every $Y \in \{A, B, S, T\}$ (in fact $YTS \equiv TSY \equiv 0$ by (6)), it is general enough to consider $X_{m,n}$ of the form (9) when $(m, n) = (2, 2)$ as well. When $m = 0$ or $n = 0$, the homogeneous elements of degree (m, n) are of the form $X_{0,n} = c_1 B^n$ and $X_{m,0} = c_1 A^m$ respectively.

Using the defining relations (3) and the reordering formula (7), the commutators of $X_{m,n}$ with A and B can be computed. For $m, n \geq 1$, $(m, n) \neq (1, 2)$, the commutator of $X_{m,n}$ with A is

$$\begin{aligned}[X_{m,n}, A] \equiv\ & c_1(1 - q_0^n) B^n A^{m+1} \\ & + \left(-c_1 \{n\}_{q_0 q_2} + c_2(1 - q_0^{n-1} q_3)\right) S B^{n-1} A^m \\ & + \left(-c_1 \{n\}_{q_0 q_4} + c_3(1 - q_0^{n-1} q_1)\right) T B^{n-1} A^m .\end{aligned}$$

If $(m, n) = (1, 2)$ then there is an additional term

$$\frac{(1 - q_5)}{(1 - q_2 q_3)(1 - q_1 q_4)} (c_2 q_3 (1 - q_1 q_4) - c_3 q_1 (1 - q_2 q_3)) T S$$

on the right side. For $m = 0$, the commutator is

$$[X_{0,n}, A] \equiv c_1(1 - q_0^n) B^n A - c_1 \{n\}_{q_0 q_2} S B^{n-1} - c_1 \{n\}_{q_0 q_4} T B^{n-1}$$

and $[X_{m,0}, A] \equiv 0$ for all m.

For $m, n \geq 1$, $(m, n) \neq (2, 1)$, the commutator of $X_{m,n}$ with B is

$$\begin{aligned}[X_{m,n}, B] \equiv\ & c_1(q_0^m - 1) B^{n+1} A^m \\ & + \left(c_1 q_0^{m-1} q_2^n \{m\}_{q_3/q_0} + c_2(q_0^{m-1} - q_2)\right) S B^n A^{m-1} \\ & + \left(c_1 q_0^{m-1} q_4^n \{m\}_{q_1/q_0} + c_3(q_0^{m-1} - q_4)\right) T B^n A^{m-1} .\end{aligned}$$

If $(m,n)=(2,1)$ then there is an additional term

$$-\frac{(1-q_5)}{(1-q_2q_3)(1-q_1q_4)}(c_2(1-q_1q_4)-c_3(1-q_2q_3))TS$$

on the right side. For $n=0$, the commutator is

$$[X_{m,0},B]\equiv c_1(q_0^m-1)BA^m+c_1q_0^{m-1}\{m\}_{\frac{q_3}{q_0}}SA^{m-1}+c_1q_0^{m-1}\{m\}_{\frac{q_1}{q_0}}TA^{m-1}.$$

and $[X_{0,n},B]\equiv 0$ for all n.

The computations are summarised in the following lemma.

Lemma 3.1 *Let* $\mathbf{q}=(q_0,\ldots,q_5)\in\mathbb{C}^6$ *be generic.*

$\mathscr{C}(A)$ is the linear subspace of $\mathscr{F}/\mathscr{J}(\mathbf{q})$ spanned by the elements listed in the following table.

Element (m,n range over $\mathbf{N}^+$)	Conditions
I	—
TS	—
A^m	—
B^n	$1-q_0^n=\{n\}_{q_0q_2}=\{n\}_{q_0q_4}=0$
$c_1B^nA^m+(c_2S+c_3T)B^{n-1}A^{m-1}$	$K_{(m,n)}\begin{bmatrix}c_1 & c_2 & c_3\end{bmatrix}^T=0$

Here,

$$K_{(m,n)}=\begin{bmatrix}1-q_0^n & 0 & 0\\ -\{n\}_{q_0q_2} & 1-q_0^{n-1}q_3 & 0\\ -\{n\}_{q_0q_4} & 0 & 1-q_0^{n-1}q_1\end{bmatrix}$$

for $(m,n)\neq(1,2)$ and

$$K_{(1,2)}=\begin{bmatrix}1-q_0^2 & 0 & 0\\ -(1+q_0q_2) & 1-q_0q_3 & 0\\ -(1+q_0q_4) & 0 & 1-q_0q_1\\ 0 & q_3(1-q_1q_4) & -q_1(1-q_2q_3)\end{bmatrix}.$$

$\mathscr{C}(B)$ is the linear subspace of $\mathscr{F}/\mathscr{J}(\mathbf{q})$ spanned by the elements listed in the following table.

Element (m,n range over $\mathbf{N}^+$)	Conditions
I	—
TS	—
A^m	$q_0^m-1=\{m\}_{q_1/q_0}=\{m\}_{q_3/q_0}=0$
B^n	—
$c_1B^nA^m+(c_2S+c_3T)B^{n-1}A^{m-1}$	$L_{(m,n)}\begin{bmatrix}c_1 & c_2 & c_3\end{bmatrix}^T=0$

Here,

$$L_{(m,n)} = \begin{bmatrix} q_0^m - 1 & 0 & 0 \\ q_0^{m-1} q_2^n \{m\}_{q_3/q_0} & q_0^{m-1} - q_2 & 0 \\ q_0^{m-1} q_4^n \{m\}_{q_1/q_0} & 0 & q_0^{m-1} - q_4 \end{bmatrix}$$

for $(m, n) \neq (2, 1)$ *and*

$$L_{(2,1)} = \begin{bmatrix} q_0^2 - 1 & 0 & 0 \\ q_2(q_0 + q_3) & q_0 - q_2 & 0 \\ q_4(q_0 + q_1) & 0 & q_0 - q_4 \\ 0 & 1 - q_1 q_4 & -(1 - q_2 q_3) \end{bmatrix}.$$

4 The Center of $\mathscr{F} / \mathscr{J}(\mathbf{q})$

The *center* of $\mathscr{F} / \mathscr{J}(\mathbf{q})$ is the set of elements that commute with every element of $\mathscr{F} / \mathscr{J}(\mathbf{q})$. It will be denoted by $\mathscr{Z}$. In this section, $\mathscr{Z}$ is described for generic $\mathbf{q}$.

Lemma 4.1 *Let* $\mathbf{q} = (q_0, \ldots, q_5) \in \mathbb{C}^6$ *be generic. Then* $\mathscr{Z} = \mathscr{C}(A) \cap \mathscr{C}(B)$.

Proof Let X be a general homogeneous element that commutes with both A and B. Then X commutes with $S + T$ by (3a). It will be shown that X commutes with S and T as well. There are four cases depending on the degree of X.

If X has degree $(0, n)$, then $X \equiv c_1 B^n$ for some $c_1 \in \mathbb{C}$. Then

$$[X, S + T] \equiv 0 \implies c_1(q_2^n - 1)SB^n + c_1(q_4^n - 1)TB^n \equiv 0 \implies$$
$$c_1(q_2^n - 1) = c_1(q_4^n - 1) = 0 \implies [X, S] \equiv [X, T] \equiv 0.$$

If X has degree $(m, 0)$, then $X \equiv c_1 A^m$ for some $c_1 \in \mathbb{C}$. Then

$$[X, S + T] \equiv 0 \implies c_1(q_3^m - 1)SA^m + c_1(q_1^m - 1)TA^m \equiv 0 \implies$$
$$c_1(q_3^m - 1) = c_1(q_1^m - 1) = 0 \implies [X, S] \equiv [X, T] \equiv 0.$$

If X has degree $(1, 1)$, then $X \equiv c_1 BA + c_2 S + c_3 T$ for some $c_1, c_2, c_3 \in \mathbb{C}$. Then

$$[X, S + T] \equiv c_1(q_2 q_3 - 1)SBA + c_1(q_1 q_4 - 1)TBA + (c_3 - c_2)(1 - q_5)TS.$$

The right side can be zero only if $c_1 = c_3 - c_2 = 0$ since $\mathbf{q}$ is generic. But then $X \equiv c_3(S + T)$, so

$$[X, BA] \equiv c_3(1 - q_2 q_3)SBA + c_3(1 - q_1 q_4)TBA.$$

Since X is assumed to commute with A and B, the right side must be zero, which implies that $c_3 = 0$ since $\mathbf{q}$ is generic. Thus $X \equiv 0$, and so X commutes with S and T.

Finally, if X has degree (m, n) with $m, n \neq 0$ and $(m, n) \neq (1, 1)$ then

$$X \equiv c_1 B^n A^m + c_2 S B^{n-1} A^{m-1} + c_3 T B^{n-1} A^{m-1} + c_4 T S,$$

for some $c_1, c_2, c_3, c_4 \in \mathbb{C}$ (with $c_4 = 0$ unless $(m, n) = (2, 2)$). Then

$$[X, S+T] \equiv 0 \implies c_1(q_2^n q_3^m - 1) S B^n A^m + c_1(q_1^m q_4^n - 1) T B^n A^m \equiv 0 \implies$$
$$c_1(q_2^n q_3^m - 1) = c_1(q_1^m q_4^n - 1) = 0 \implies [X, S] \equiv [X, T] \equiv 0.$$

Theorem 4.1 *Let* $\mathbf{q} = (q_0, \ldots, q_5) \in \mathbb{C}^6$ *be generic and suppose that* q_0 *is not a root of unity. Then a basis for* $\mathscr{Z}$ *is given by the elements listed in the following table.*

Element (m, n range over $\mathbb{N}^+$)	Conditions
I	—
TS	—
$SB^{n-1}A^{m-1}$	$1 - q_0^{n-1} q_3 = q_0^{m-1} - q_2 = 0$
$TB^{n-1}A^{m-1}$	$1 - q_0^{n-1} q_1 = q_0^{m-1} - q_4 = 0$

Moreover, this basis contains at most one element of the form $SB^{n-1}A^{m-1}$ *and at most one element of the form* $TB^{n-1}A^{m-1}$*. Thus,* $\mathscr{Z}$ *has dimension at most four.*

Proof By Lemma 4.1, it is enough to show that the listed elements form a basis for $\mathscr{C}(A) \cap \mathscr{C}(B)$. Spanning sets for $\mathscr{C}(A)$ and $\mathscr{C}(B)$ are given by Lemma 3.1; denote them by $\mathscr{B}(A)$ and $\mathscr{B}(B)$ respectively. Then $\mathscr{C}(A) \cap \mathscr{C}(B)$ is the linear space spanned by $\mathscr{B}(A) \cap \mathscr{B}(A)$. Now, $I, TS \in \mathscr{B}(A) \cap \mathscr{B}(B)$ always. For $m, n \geq 1$, $A^m \notin \mathscr{B}(B)$ and $B^n \notin \mathscr{B}(A)$ since q_0 is not a root of unity, and thus $A^m, B^n \notin \mathscr{B}(A) \cap \mathscr{B}(B)$. An element in $\mathscr{B}(A) \cap \mathscr{B}(B)$ of the form $c_1 B^n A^m + c_2 S B^{n-1} A^{m-1} + c_3 T B^{n-1} A^{m-1}$ with $m, n \geq 1$ must have $c_1 = 0$ since q_0 is not a root of unity. The coefficient c_2 may be non-zero if and only if $1 - q_0^{n-1} q_3 = q_0^{m-1} - q_2 = 0$. Since q_0 is not a root of unity, this can happen for at most one value of (m, n). Similarly, c_3 may be non-zero if and only if $1 - q_0^{n-1} q_1 = q_0^{m-1} - q_4 = 0$, and this can happen for at most one value of (m, n). Thus the listed elements span $\mathscr{C}(A) \cap \mathscr{C}(B)$, and since they are linearly independent they form a basis.

Theorem 4.2 *Let* $\mathbf{q} = (q_0, \ldots, q_5) \in \mathbb{C}^6$ *be generic and suppose that* q_0 *is a root of unity. Let* d *be the smallest positive integer such that* $q_0^d = 1$*. Then a basis for* $\mathscr{Z}$ *is given by the elements listed in the following table.*

Element (m, n range over $\mathbb{N}^+$)	Conditions
I	—
TS	—
A^m	$d\|m, q_1^m = q_3^m = 1, q_1/q_0, q_3/q_0 \neq 1$
B^n	$d\|n, q_2^n = q_4^n = 1, q_0q_2, q_0q_4 \neq 1$
$SB^{n-1}A^{m-1}$	$d\|m-1-r_2, d\|n-1+r_3, q_2 = q_0^{r_2}, q_3 = q_0^{r_3}$
$TB^{n-1}A^{m-1}$	$d\|n-1+r_1, d\|m-1-r_4, q_1 = q_0^{r_1}, q_4 = q_0^{r_4}$
$B^nA^m + (c_SS + c_TT)B^{n-1}A^{m-1}$	$d\|m, d\|n, q_1^mq_4^n = q_2^nq_3^m = 1$ and in addition $q_2 + 1/q_2 = q_4 + 1/q_4$ if $(m, n) = (1, 2)$; $q_1 + 1/q_1 = q_3 + 1/q_3$ if $(m, n) = (2, 1)$

Here,

$$c_S = \begin{cases} \frac{1-q_2^n}{(1-q_0q_2)(1-q_3/q_0)} = \frac{1-1/q_3^m}{(1-q_0q_2)(1-q_3/q_0)} & \textit{if } 1-q_0q_2, 1-q_3/q_0 \neq 0 \\ \frac{-m}{1-q_0q_2} & \textit{if } 1-q_0q_2 \neq 0, 1-q_3/q_0 = 0 \\ \frac{n}{1-q_3/q_0} & \textit{if } 1-q_0q_2 = 0, 1-q_3/q_0 \neq 0 \end{cases}$$

and

$$c_T = \begin{cases} \frac{1-q_4^n}{(1-q_0q_4)(1-q_1/q_0)} = \frac{1-1/q_1^m}{(1-q_0q_4)(1-q_1/q_0)} & \textit{if } 1-q_0q_4, 1-q_1/q_0 \neq 0 \\ \frac{-m}{1-q_0q_4} & \textit{if } 1-q_0q_4 \neq 0, 1-q_1/q_0 = 0 \\ \frac{n}{1-q_1/q_0} & \textit{if } 1-q_0q_4 = 0, 1-q_1/q_0 \neq 0. \end{cases}$$

Proof By Lemma 4.1, it is enough to show that the listed elements form a basis for $\mathscr{C}(A) \cap \mathscr{C}(B)$. Spanning sets for $\mathscr{C}(A)$ and $\mathscr{C}(B)$ are given by Lemma 3.1; denote them by $\mathscr{B}(A)$ and $\mathscr{B}(B)$ respectively. Then $\mathscr{C}(A) \cap \mathscr{C}(B)$ is the linear space spanned by $\mathscr{B}(A) \cap \mathscr{B}(B)$. Now, $I, TS \in \mathscr{B}(A) \cap \mathscr{B}(B)$ always, and

$$\begin{aligned} A^m \in \mathscr{B}(A) \cap \mathscr{B}(B) &\iff q_0^m - 1 = \{m\}_{q_1/q_0} = \{m\}_{q_3/q_0} = 0 \\ &\iff q_0^m = q_1^m = q_3^m = 1, q_1/q_0, q_3/q_0 \neq 1 \end{aligned}$$

and

$$\begin{aligned} B^n \in \mathscr{B}(A) \cap \mathscr{B}(B) &\iff 1 - q_0^n = \{n\}_{q_0q_2} = \{n\}_{q_0q_4} = 0 \\ &\iff q_0^n = q_2^n = q_4^n = 1, q_0q_2, q_0q_4 \neq 1. \end{aligned}$$

Assume now that $m, n \geq 1$, $(m, n) \neq (1, 2)$ and $(m, n) \neq (2, 1)$. An element of the form $c_1B^nA^m + c_2SB^{n-1}A^{m-1} + c_3TB^{n-1}A^{m-1}$ lies in $\mathscr{B}(A) \cap \mathscr{B}(B)$ if and only if $K_{(m,n)}[c_1\, c_2\, c_3]^T = 0$ and $L_{(m,n)}[c_1\, c_2\, c_3]^T = 0$, where $K_{(m,n)}$ and $L_{(m,n)}$ are defined as in Lemma 3.1. Regrouping these equations gives the equivalent equation system

$$\begin{cases} (1-q_0^n)c_1 = (q_0^m - 1)c_1 = 0 & \text{(first rows of } K_{(m,n)} \text{ and } L_{(m,n)}) \\ M_S \begin{bmatrix} c_1 & c_2 \end{bmatrix}^T = 0 & \text{(second rows of } K_{(m,n)} \text{ and } L_{(m,n)}) \\ M_T \begin{bmatrix} c_1 & c_3 \end{bmatrix}^T = 0 & \text{(third rows of } K_{(m,n)} \text{ and } L_{(m,n)}), \end{cases} \tag{10}$$

where

$$M_S = \begin{bmatrix} -\{n\}_{q_0 q_2} & 1 - q_0^{n-1} q_3 \\ q_0^{m-1} q_2^n \{m\}_{q_3/q_0} & q_0^{m-1} - q_2 \end{bmatrix} \quad M_T = \begin{bmatrix} -\{n\}_{q_0 q_4} & 1 - q_0^{n-1} q_1 \\ q_0^{m-1} q_4^n \{m\}_{q_1/q_0} & q_0^{m-1} - q_4 \end{bmatrix}.$$

There are two types of possible solutions to (10): Those with $c_1 = 0$ and those with $c_1 \neq 0$. Note that $[0\, c_2\, c_3]^T$ satisfies (10) if and only if $[0\, c_2\, 0]^T$ and $[0\, 0\, c_3]^T$ do. Thus for the case $c_1 = 0$, it is enough to consider elements of the forms $SB^{n-1}A^{m-1}$ and $TB^{n-1}A^{m-1}$ separately, rather that a general linear combination $c_2 SB^{n-1}A^{m-1} + c_3 TB^{n-1}A^{m-1}$.

Now, $SB^{n-1}A^{m-1} \in \mathscr{B}(A) \cap \mathscr{B}(B)$ if and only if $M_S[0\, 1]^T = 0$, that is, iff

$$q_0^{n-1} q_3 = 1 \text{ and } q_0^{m-1} = q_2. \tag{11}$$

This implies (by raising both sides of the equations to the power of d) that $q_2^d = q_3^d = 1$, so q_2 and q_3 are d:th roots of unity and thus $q_2 = q_0^{r_2}$ and $q_3 = q_0^{r_3}$ for some $r_2, r_3 \in \{0, \dots, d-1\}$ (since q_0 generates the group of d:th roots of unity). Then (11) holds if and only if $q_0^{n-1+r_3} = q_0^{m-1-r_2} = 1$, that is, d divides both $n - 1 + r_3$ and $m - 1 - r_2$.

Similarly, $TB^{n-1}A^{m-1} \in \mathscr{B}(A) \cap \mathscr{B}(B)$ if and only if $M_T[0\, 1]^T = 0$, that is, iff

$$q_0^{n-1} q_1 = 1 \text{ and } q_0^{m-1} = q_4. \tag{12}$$

This implies that $q_1^d = q_4^d = 1$ and thus that $q_1 = q_0^{r_1}$ and $q_4 = q_0^{r_4}$ for some $r_1, r_4 \in \{0, \dots, d-1\}$. Then (12) holds if and only if $q_0^{n-1+r_1} = q_0^{m-1-r_4} = 1$, that is, d divides both $n - 1 + r_1$ and $m - 1 - r_4$.

If there is a solution of (10) with $c_1 \neq 0$ then $q_0^m = q_0^n = 1$ so $d|m$ and $d|n$. In addition, $\det(M_S) = \det(M_T) = 0$, which is equivalent (using $q_0^m = q_0^n = 1$ and $\{k\}_q (1-q) = 1 - q^k$) to $q_2^n q_3^m = q_1^m q_4^n = 1$. When M_S is singular, either of the equations of the system $M_S[c_1\, c_2]^T = 0$ can be used to solve for c_2. One gets

$$c_2 = \begin{cases} \frac{1-q_2^n}{(1-q_0q_2)(1-q_3/q_0)} c_1 = \frac{1-1/q_3^m}{(1-q_0q_2)(1-q_3/q_0)} c_1 & \text{if } 1 - q_0q_2,\, 1 - q_3/q_0 \neq 0 \\ \frac{-m}{1-q_0q_2} c_1 & \text{if } 1 - q_0q_2 \neq 0,\, 1 - q_3/q_0 = 0 \\ \frac{n}{1-q_3/q_0} c_1 & \text{if } 1 - q_0q_2 = 0,\, 1 - q_3/q_0 \neq 0 \end{cases} \tag{13}$$

(the case $1 - q_0q_2 = 1 - q_3/q_0 = 0$ is excluded since $q_2q_3 \neq 1$ when $\mathbf{q}$ is generic). Similarly, when M_T is singular, one can use either of the equations of the system $M_T[c_1\, c_3]^T = 0$ to solve for c_3. One gets

$$c_3 = \begin{cases} \frac{1-q_4^n}{(1-q_0q_4)(1-q_1/q_0)} c_1 = \frac{1-1/q_1^m}{(1-q_0q_4)(1-q_1/q_0)} c_1 & \text{if } 1 - q_0q_4, 1 - q_1/q_0 \neq 0 \\ \frac{-m}{1-q_0q_4} c_1 & \text{if } 1 - q_0q_4 \neq 0, 1 - q_1/q_0 = 0 \\ \frac{n}{1-q_1/q_0} c_1 & \text{if } 1 - q_0q_4 = 0, 1 - q_1/q_0 \neq 0 \end{cases} \tag{14}$$

(the case $1 - q_0q_4 = 1 - q_1/q_0 = 0$ is excluded since $q_1q_4 \neq 1$ when $\mathbf{q}$ is generic).

When $(m, n) = (1, 2)$ or $(m, n) = (2, 1)$ then (10) is still a necessary condition for $c_1 B^n A^m + c_2 S B^{n-1} A^{m-1} + c_3 T B^{n-1} A^{m-1}$ to lie in $\mathscr{C}(A, B)$, but c_2, c_3 must also satisfy

$$\begin{cases} c_2 q_3 (1 - q_1 q_4) = c_3 q_1 (1 - q_2 q_3) & \text{if } (m, n) = (1, 2) \\ c_2 (1 - q_1 q_4) = c_3 (1 - q_2 q_3) & \text{if } (m, n) = (2, 1). \end{cases} \tag{15}$$

If $c_1 = 0$ in any of these cases then either $c_2 = 0$ or $c_3 = 0$, since if $c_2, c_3 \neq 0$ then (10) would imply $q_1 = q_3 \,(= 1/q_0^{n-1})$ and $q_2 = q_4 \,(= q_0^{m-1})$. But if one of c_2, c_3 is $= 0$, then so is the other by (15). Thus, there are no non-trivial solutions with $c_0 = 0$ when $(m, n) = (1, 2)$ or $(m, n) = (2, 1)$. The element $B^n A^m + c_2 S B^{n-1} A^{m-1} + c_3 T B^{n-1} A^{m-1}$ lies in $\mathscr{C}(A, B)$ if and only if $[1\, c_2\, c_3]^T$ satisfies both (10) and (15). Using the expressions (13) and (14) for c_2 and c_3 together with $q_0 = q_1^m q_4^n = q_2^n q_3^m = 1$ (that is, using the conditions that have just been shown to be equivalent to (10); note that when $m = 1$ or $n = 1$, $d|m$ and $d|n$ iff $q_0 = 1$. Also note that $q_1^m q_4^n = q_2^n q_3^m = 1$ implies $q_2, q_4 \neq 1$ when $(m, n) = (1, 2)$ and $q_1, q_3 \neq 1$ when $(m, n) = (2, 1)$ since $\mathbf{q}$ is generic), (15) can be simplified to

$$\begin{cases} q_2 + 1/q_2 = q_4 + 1/q_4 & \text{if } (m, n) = (1, 2) \\ q_1 + 1/q_1 = q_3 + 1/q_3 & \text{if } (m, n) = (2, 1). \end{cases}$$

Thus, the elements listed in the theorem form a spanning set for $\mathscr{C}(A) \cap \mathscr{C}(B)$. To see that they are linearly independent, note that for a fixed (m, n) it is impossible that both $B^n A^m + c_2 S B^{n-1} A^{m-1} + c_3 T B^{n-1} A^{m-1}$ and $S B^{n-1} A^{m-1}$ lie in $\mathscr{C}(A) \cap \mathscr{C}(B)$, for that would imply

$$\left.\begin{array}{l} m \equiv n \pmod d \\ m - 1 - r_2 \equiv n - 1 + r_3 \pmod d \end{array}\right\} \implies -r_2 \equiv r_3 \pmod d \implies q_2 q_3 = 1.$$

Similarly, it is impossible that both $B^n A^m + c_2 S B^{n-1} A^{m-1} + c_3 T B^{n-1} A^{m-1}$ and $T B^{n-1} A^{m-1}$ lie in $\mathscr{C}(A) \cap \mathscr{C}(B)$, for that would imply $q_1 q_4 = 1$. Thus, the listed elements are linearly independent, and so they form a basis for $\mathscr{C}(A) \cap \mathscr{C}(B)$.

5 Commutativity of $\mathscr{C}(A)$ and $\mathscr{C}(B)$

This section gives, for generic $\mathbf{q}$, necessary and sufficient conditions on $\mathbf{q}$ for $\mathscr{C}(A)$ and $\mathscr{C}(B)$ to be commutative. As before, it is enough to consider homogeneous elements of $\mathscr{C}(A)$ and $\mathscr{C}(B)$, and the TS-components of the homogeneous elements of degree (2, 2) can be disregarded since $TS \in \mathscr{Z}$. Thus, $\mathscr{C}(A)$ is commutative if and only if the set

$$\mathscr{H}_A = \{\, H \in \mathscr{C}(A);\ H \text{ is homogeneous and has no } TS\text{-component}\}$$

is, and $\mathscr{C}(B)$ is commutative if and only if the set

$$\mathscr{H}_B = \{\, H \in \mathscr{C}(B);\ H \text{ is homogeneous and has no } TS\text{-component}\}$$

is. Define further the sets

$$\begin{aligned}
\mathscr{X}_A &= \left\{X \in \mathscr{H}_A;\ X \text{ is homogeneous of degree } (m,n) \text{ and } q_0^n = 1\right\}\\
\mathscr{Y}_A &= \left\{Y \in \mathscr{H}_A;\ Y \text{ is homogeneous of degree } (m,n) \text{ and } q_0^n \neq 1\right\} \cup \{0\}\\
\mathscr{X}_B &= \left\{X \in \mathscr{H}_B;\ X \text{ is homogeneous of degree } (m,n) \text{ and } q_0^m = 1\right\}\\
\mathscr{Y}_B &= \left\{Y \in \mathscr{H}_B;\ Y \text{ is homogeneous of degree } (m,n) \text{ and } q_0^m \neq 1\right\} \cup \{0\}.
\end{aligned}$$

Then $\mathscr{H}_A = \mathscr{X}_A \cup \mathscr{Y}_A$ and $\mathscr{H}_B = \mathscr{X}_B \cup \mathscr{Y}_B$, and $\mathscr{X}_A \cap \mathscr{Y}_A = \mathscr{X}_B \cap \mathscr{Y}_B = \{0\}$, since 0 is homogeneous of all degrees. (The reason for explicitly including 0 in $\mathscr{Y}_A$ and $\mathscr{Y}_B$ is to make sure that they always contain 0: If 0 were not explicitly included then the case $q_0 = 1$ would be exceptional.) These sets do, of course, depend on $\mathbf{q}$; when this dependence needs to be emphasised the notation will be $\mathscr{X}_A(\mathbf{q})$, $\mathscr{Y}_A(\mathbf{q})$ and so on.

Consider two general homogeneous elements

$$\begin{aligned}
X_{k,l} &\equiv b_1 B^l A^k + b_2 S B^{l-1} A^{k-1} + b_3 T B^{l-1} A^{k-1}\\
X_{m,n} &\equiv c_1 B^n A^m + c_2 S B^{n-1} A^{m-1} + c_3 T B^{n-1} A^{m-1}
\end{aligned}$$

of degrees (k, l) and (m, n) respectively. A somewhat lengthy calculation using the reordering formula (7) and Eq. (6) shows that

$$\begin{aligned}
&X_{k,l} X_{m,n} \equiv \dots\\
&\equiv b_1 c_1 q_0^{kn} B^{l+n} A^{k+m}\\
&\quad + q_0^{(k-1)(n-1)} \left(b_1 c_1 q_0^{k-1} q_2^l \{k\}_{\frac{q_3}{q_0}} \{n\}_{q_0 q_2} + b_1 c_2 q_0^{n-1} q_2^l q_3^k + b_2 c_1 q_0^{k-1} \right)\\
&\quad S B^{l+n-1} A^{k+m-1}\\
&\quad + q_0^{(k-1)(n-1)} \left(b_1 c_1 q_0^{k-1} q_4^l \{k\}_{\frac{q_1}{q_0}} \{n\}_{q_0 q_4} + b_1 c_3 q_0^{n-1} q_1^k q_4^l + b_3 c_1 q_0^{k-1} \right)\\
&\quad T B^{l+n-1} A^{k+m-1}.
\end{aligned}$$

This holds for all $k, l, m, n \geq 1$ except for $k = l = m = n = 1$ (if $(k, l) = (2, 2)$ or $(m, n) = (2, 2)$ then there is an additional TS-term in the expression for $X_{k,l}$ or

$X_{m,n}$, but that makes no difference for the product because of (6)). The commutator of $X_{k,l}$ and $X_{m,n}$ then is

$$\begin{aligned}
&[X_{k,l}, X_{m,n}] \equiv b_1c_1(q_0^{kn} - q_0^{ml})B^{l+n}A^{k+m} \\
&+ \left(q_0^{(k-1)(n-1)}\left(b_1c_1q_0^{k-1}q_2^l\,\{k\}_{\frac{q_3}{q_0}}\,\{n\}_{q_0q_2} + b_1c_2q_0^{n-1}q_2^lq_3^k + b_2c_1q_0^{k-1}\right)\right. \\
&\left.- q_0^{(m-1)(l-1)}\left(b_1c_1q_0^{m-1}q_2^n\,\{m\}_{\frac{q_3}{q_0}}\,\{l\}_{q_0q_2} + b_2c_1q_0^{l-1}q_2^nq_3^m + b_1c_2q_0^{m-1}\right)\right) \\
&\quad SB^{l+n-1}A^{k+m-1} \\
&+ \left(q_0^{(k-1)(n-1)}\left(b_1c_1q_0^{k-1}q_4^l\,\{k\}_{\frac{q_1}{q_0}}\,\{n\}_{q_0q_4} + b_1c_3q_0^{n-1}q_1^kq_4^l + b_3c_1q_0^{k-1}\right)\right. \\
&\left.- q_0^{(m-1)(l-1)}\left(b_1c_1q_0^{m-1}q_4^n\,\{m\}_{\frac{q_1}{q_0}}\,\{l\}_{q_0q_4} + b_3c_1q_0^{l-1}q_1^mq_4^n + b_1c_3q_0^{m-1}\right)\right) \\
&\quad TB^{l+n-1}A^{k+m-1}. \qquad (16)
\end{aligned}$$

The switch $A \leftrightarrow B$ will be used in the proofs below. Assume that $q_0 \neq 0$ (as is the case when **q** is generic) and let $f_{AB} : \mathscr{F} \to \mathscr{F}$ be the isomorphism defined by

$$f_{AB}(A) = B, \quad f_{AB}(B) = A, \quad f_{AB}(S) = -q_0S, \quad f_{AB}(T) = -q_0T.$$

The image under f_{AB} of $G(\mathbf{q})$, defined in (2), is

$$\begin{aligned}
&\{-q_0(AB - \frac{1}{q_0}BA - S - T),\ -q_0(BT - q_1TB),\ -q_0(AS - q_2SA), \\
&-q_0(BS - q_3SB),\ -q_0(AT - q_4TA),\ q_0^2(ST - q_5TS)\}.
\end{aligned}$$

Thus $f_{AB}(G(\mathbf{q}))$ generates the ideal $\mathscr{J}(\hat{\mathbf{q}})$ where $\hat{\mathbf{q}} = (\frac{1}{q_0}, q_4, q_3, q_2, q_1, q_5)$, and consequently $f_{AB}(\mathscr{J}(\mathbf{q})) = \mathscr{J}(\hat{\mathbf{q}})$. This makes it possible to define an isomorphism $h_{AB} : \mathscr{F}/\mathscr{J}(\mathbf{q}) \to \mathscr{F}/\mathscr{J}(\hat{\mathbf{q}})$ by

$$h_{AB}(X + \mathscr{J}(\mathbf{q})) = f_{AB}(X) + \mathscr{J}(\hat{\mathbf{q}}). \qquad (17)$$

It is easily checked that $\hat{\mathbf{q}}$ is generic whenever **q** is.

Also the switch $S \leftrightarrow T$ will be used below. Let $f_{ST} : \mathscr{F} \to \mathscr{F}$ be the isomorphism defined by

$$f_{ST}(A) = A, \quad f_{ST}(B) = B, \quad f_{ST}(S) = T, \quad f_{ST}(T) = S.$$

The image under f_{ST} of $G(\mathbf{q})$ is (assuming that $q_5 \neq 0$)

$$\{AB - q_0BA - S - T,\ AS - q_1SA,\ BT - q_2TB,$$

$$AT - q_3TA,\ BS - q_4SB,\ -q_5(ST - \frac{1}{q_5}TS)\}.$$

Thus $f_{ST}(G(\mathbf{q}))$ generates the ideal $\mathscr{J}(\tilde{\mathbf{q}})$, where $\tilde{\mathbf{q}} = (q_0, q_3, q_4, q_1, q_2, \frac{1}{q_5})$, so $f_{ST}(\mathscr{J}(\mathbf{q})) = \mathscr{J}(\tilde{\mathbf{q}})$ and an isomorphism $h_{ST} : \mathscr{F}/\mathscr{J}(\mathbf{q}) \to \mathscr{F}/\mathscr{J}(\tilde{\mathbf{q}})$ can be defined by

$$h_{ST}(X + \mathscr{J}(\mathbf{q})) = f_{ST}(X) + \mathscr{J}(\tilde{\mathbf{q}}). \tag{18}$$

Again, it is easily checked that $\tilde{\mathbf{q}}$ is generic whenever $\mathbf{q}$ is.

Lemma 5.1 *Let* $\mathbf{q} = (q_0, \ldots, q_5) \in \mathbb{C}^6$ *be generic. Then the sets* $\mathscr{X}_A$ *and* $\mathscr{X}_B$ *are commutative.*

Proof Pick any $X_1, X_2 \in \mathscr{X}_A$. Then $X_1, X_2 \in \mathscr{C}(A)$ and they are homogeneous, say of degrees (k, l) and (m, n) respectively with $q_0^l = q_0^n = 1$. If $k = 0$ then $X_1 \equiv b_1B^l$ and Lemma 3.1 implies that $X_1 \equiv 0$ or $q_2^l = q_4^l = 1$. In either case, $X_1 \in \mathscr{Z}$, so in particular X_1 commutes with X_2. Similarly, if $m = 0$ then $X_2 \in \mathscr{Z}$ and thus commutes with X_1. If $l = 0$ then $X_1 = b_1A^k$ and if $n = 0$ then $X_2 = c_1A^m$; in both cases X_1 and X_2 commute. Thus it may be assumed that $k, l, m, n \geq 1$, so that X_1 and X_2 can be written as

$$\begin{aligned} X_1 &\equiv b_1B^lA^k + b_2SB^{l-1}A^{k-1} + b_3TB^{l-1}A^{k-1} \qquad k, l \geq 1 \\ X_2 &\equiv c_1B^nA^m + c_2SB^{n-1}A^{m-1} + c_3TB^{n-1}A^{m-1} \qquad m, n \geq 1 \end{aligned}$$

with coefficients that satisfy

$$K_{(k,l)}[b_1\, b_2\, b_3]^T = 0, \quad K_{(m,n)}[c_1\, c_2\, c_3]^T = 0, \tag{19}$$

where $K_{(k,l)}$, $K_{(m,n)}$ are defined as in Lemma 3.1. If $k = l = m = n = 1$ then X_1 and X_2 are parallel, because $K_{(1,1)}$ has rank at least two (using that $q_1 \neq q_3$ since $\mathbf{q}$ is generic). Hence it may be assumed that at least one of k, l, m, n is ≥ 2, so that the commutator $[X_1, X_2]$ is given by (16). Since $q_0^l = q_0^n = 1$, the coefficient of $B^{l+n}A^{k+m}$ in (16) is 0—it has to be shown that the coefficients of $SB^{l+n-1}A^{k+m-1}$ and $TB^{l+n-1}A^{k+m-1}$ are 0 as well.

If $q_3 \neq q_0$ then (19) implies that

$$b_2 = \frac{\{l\}_{q_0q_2}}{1 - q_3/q_0}b_1, \qquad c_2 = \frac{\{n\}_{q_0q_2}}{1 - q_3/q_0}c_1$$

and the $SB^{l+n-1}A^{k+m-1}$-coefficient in (16) can be simplified to

$$\begin{aligned} &b_1c_1q_2^l\frac{1 - (q_3/q_0)^k}{1 - q_3/q_0}\{n\}_{q_0q_2} + b_1c_1q_0^{-k}q_2^lq_3^k\frac{\{n\}_{q_0q_2}}{1 - q_3/q_0} + b_1c_1\frac{\{l\}_{q_0q_2}}{1 - q_3/q_0} \\ &- b_1c_1q_2^n\frac{1 - (q_3/q_0)^m}{1 - q_3/q_0}\{l\}_{q_0q_2} - b_1c_1q_0^{-m}q_2^nq_3^m\frac{\{l\}_{q_0q_2}}{1 - q_3/q_0} - b_1c_1\frac{\{n\}_{q_0q_2}}{1 - q_3/q_0} \end{aligned}$$

$$= \frac{b_1 c_1}{1 - q_3/q_0} \left(q_2^l \{n\}_{q_0 q_2} + \{l\}_{q_0 q_2} - q_2^n \{l\}_{q_0 q_2} - \{n\}_{q_0 q_2} \right) = 0. \tag{20}$$

The last equality holds by the identity $\{a\}_q + q^a \{b\}_q = \{a+b\}_q$ for q-natural numbers, since $q_2^l = (q_0 q_2)^l$, $q_2^n = (q_0 q_2)^n$. Similarly if $q_1 \neq q_0$ then

$$b_3 = \frac{\{l\}_{q_0 q_4}}{1 - q_1/q_0} b_1, \qquad c_3 = \frac{\{n\}_{q_0 q_4}}{1 - q_1/q_0} c_1$$

and the $TB^{l+n-1}A^{k+m-1}$-coefficient in (16) can be simplified to

$$\begin{aligned} b_1 c_1 q_4^l \frac{1 - (q_1/q_0)^k}{1 - q_1/q_0} \{n\}_{q_0 q_4} + b_1 c_1 q_0^{-k} q_1^k q_4^l \frac{\{n\}_{q_0 q_4}}{1 - q_1/q_0} + b_1 c_1 \frac{\{l\}_{q_0 q_4}}{1 - q_1/q_0} \\ - b_1 c_1 q_4^n \frac{1 - (q_1/q_0)^m}{1 - q_1/q_0} \{l\}_{q_0 q_4} - b_1 c_1 q_0^{-m} q_1^m q_4^n \frac{\{l\}_{q_0 q_4}}{1 - q_1/q_0} - b_1 c_1 \frac{\{n\}_{q_0 q_4}}{1 - q_1/q_0} \\ = \frac{b_1 c_1}{1 - q_1/q_0} \left(q_4^l \{n\}_{q_0 q_4} + \{l\}_{q_0 q_4} - q_4^n \{l\}_{q_0 q_4} - \{n\}_{q_0 q_4} \right) = 0. \end{aligned} \tag{21}$$

Now there are three cases to consider ($q_1 = q_3 = q_0$ is impossible since $\mathbf{q}$ is generic).

1. If $q_1, q_3 \neq q_0$ then the coefficients of $SB^{l+n-1}A^{k+m-1}$ and $TB^{l+n-1}A^{k+m-1}$ in (16) are 0 by (20) and (21).

2. If $q_1 = q_0$, $q_3 \neq q_0$, then the computation (20) is still valid, that is, the coefficient of $SB^{l+n-1}A^{k+m-1}$ in (16) is 0. Further, (19) implies that

$$\{l\}_{q_0 q_4} b_1 = \{n\}_{q_0 q_4} c_1 = 0$$

and thus also

$$(1 - q_4^l) b_1 = (1 - q_4^n) c_1 = 0.$$

Then the $TB^{l+n-1}A^{k+m-1}$-coefficient in (16) can be simplified to

$$b_1 c_3 q_4^l + b_3 c_1 - b_3 c_1 q_4^n - b_1 c_3 = -b_1 c_3 (1 - q_4^l) + b_3 c_1 (1 - q_4^n) = 0.$$

3. If $q_1 \neq q_0, q_3 = q_0$ then the computation (21) is still valid, that is, the coefficient of $TB^{l+n-1}A^{k+m-1}$ in (16) is 0. The condition (19) implies that

$$\{l\}_{q_0 q_2} b_1 = \{n\}_{q_0 q_2} c_1 = (1 - q_2^l) b_1 = (1 - q_2^n) c_1 = 0,$$

and the $SB^{l+n-1}A^{k+m-1}$-coefficient becomes

$$b_1 c_2 q_2^l + b_2 c_1 - b_2 c_1 q_2^n - b_1 c_2 = -b_1 c_2 (1 - q_2^l) + b_2 c_1 (1 - q_2^n) = 0.$$

Thus it has been shown that $\mathscr{X}_A$ is commutative.

In order to see that $\mathscr{X}_B = \mathscr{X}_B(\mathbf{q})$ is commutative, let $h_{AB} : \mathscr{F}/\mathscr{J}(\mathbf{q}) \to \mathscr{F}/\mathscr{J}(\hat{\mathbf{q}})$ be the isomorphism defined as in (17). Then $\mathscr{X}_A(\hat{\mathbf{q}})$ is commutative by the above proof, so $\mathscr{X}_B(\mathbf{q}) = h_{AB}^{-1}(\mathscr{X}_A(\hat{\mathbf{q}}))$ is commutative as well.

Lemma 5.2 *Let* $\mathbf{q} = (q_0, \ldots, q_5) \in \mathbb{C}^6$ *be generic. Then the sets* $\mathscr{Y}_A$ *and* $\mathscr{Y}_B$ *are commutative.*

Proof Pick any $Y_1, Y_2 \in \mathscr{Y}_A$. Then $Y_1, Y_2 \in \mathscr{C}(A)$ and they are homogeneous, say of degrees (k, l) and (m, n) respectively with $q_0^l, q_0^n \neq 1$. It is then impossible that $l = 0$ or $n = 0$, and by Lemma 3.1 it cannot be that $Y_1 = b_1 B^l$ or $Y_2 = c_1 B^n$ with $b_1, c_1 \neq 0$. Thus it may be assumed that $k, l, m, n \geq 1$, and Y_1, Y_2 can be written as

$$Y_1 \equiv b_1 B^l A^k + b_2 S B^{l-1} A^{k-1} + b_3 T B^{l-1} A^{k-1} \qquad k, l \geq 1$$
$$Y_2 \equiv c_1 B^n A^m + c_2 S B^{n-1} A^{m-1} + c_3 T B^{n-1} A^{m-1} \qquad m, n \geq 1$$

with coefficients that satisfy

$$K_{(k,l)}[b_1\, b_2\, b_3]^T = 0, \quad K_{(m,n)}[c_1\, c_2\, c_3]^T = 0, \tag{22}$$

where $K_{(k,l)}$, $K_{(m,n)}$ are defined as in Lemma 3.1. If $k = l = m = n = 1$ then Y_1 and Y_2 are parallel, because $K_{(1,1)}$ has rank at least two (using that $q_1 \neq q_3$ since $\mathbf{q}$ is generic). Hence it may be assumed that at least one of k, l, m, n is ≥ 2. Since $q_0^l, q_0^n \neq 1$, (22) implies that $b_1 = c_1 = 0$. But then $Y_1 Y_2 \equiv Y_2 Y_1 \equiv 0$ by (6), so Y_1, Y_2 commute.

To see that $\mathscr{Y}_B = \mathscr{Y}_B(\mathbf{q})$ is commutative, let $h_{AB} : \mathscr{F}/\mathscr{J}(\mathbf{q}) \to \mathscr{F}/\mathscr{J}(\hat{\mathbf{q}})$ be the isomorphism defined as in (17). Then $\mathscr{Y}_A(\hat{\mathbf{q}})$ is commutative by the above proof, so $\mathscr{Y}_B(\mathbf{q}) = h_{AB}^{-1}(\mathscr{Y}_A(\hat{\mathbf{q}}))$ is commutative as well.

Because of Lemma 5.1 and Lemma 5.2, it is enough to check whether the elements of $\mathscr{X}_A$ commute with the elements of $\mathscr{Y}_A$ to see if $\mathscr{C}(A)$ is commutative, and to check whether the elements of $\mathscr{X}_B$ commute with the elements of $\mathscr{Y}_B$ to see if $\mathscr{C}(B)$ is commutative.

Theorem 5.1 *Let* $\mathbf{q} = (q_0, \ldots, q_5) \in \mathbb{C}^6$ *be generic and suppose that* q_0 *is not a root of unity. Then* $\mathscr{C}(A)$ *and* $\mathscr{C}(B)$ *are commutative.*

Proof Since q_0 is not a root of unity,

$$\mathscr{X}_A = \{cA^m;\ c \in \mathbb{C},\ m \in \mathbf{N}\}.$$

Thus, every element of $\mathscr{X}_A$ commutes with every element of $\mathscr{Y}_A$, and it follows from Lemma 5.1 and Lemma 5.2 that $\mathscr{C}(A)$ is commutative. Similarly,

$$\mathscr{X}_B = \{cB^n;\ c \in \mathbb{C},\ n \in \mathbf{N}\};$$

thus every element of $\mathscr{X}_B$ commutes with every element of $\mathscr{Y}_B$ and thus $\mathscr{C}(B)$ is commutative.

Theorem 5.2 *Let $\mathbf{q} = (q_0, \ldots, q_5) \in \mathbb{C}^6$ be generic, and suppose that q_0 is a root of unity with d being the smallest positive integer such that $q_0^d = 1$. Then $\mathscr{C}(A)$ is commutative if and only if*

$$\left(\{d\}_{q_3/q_0} \neq 0 \text{ or } q_2^d = 1 \text{ or } (q_1 = q_0 \text{ and } q_4^d \neq 1)\right) \tag{23}$$

and

$$\left(\{d\}_{q_1/q_0} \neq 0 \text{ or } q_4^d = 1 \text{ or } (q_3 = q_0 \text{ and } q_2^d \neq 1)\right), \tag{24}$$

and $\mathscr{C}(B)$ is commutative if and only if

$$\left(\{d\}_{q_0 q_2} \neq 0 \text{ or } q_3^d = 1 \text{ or } (q_4 = q_0^{-1} \text{ and } q_1^d \neq 1)\right) \tag{25}$$

and

$$\left(\{d\}_{q_0 q_4} \neq 0 \text{ or } q_1^d = 1 \text{ or } (q_2 = q_0^{-1} \text{ and } q_3^d \neq 1)\right). \tag{26}$$

Proof It follows from Lemma 5.1 and Lemma 5.2 that $\mathscr{C}(A)$ is commutative if and only if every element of $\mathscr{X}_A$ commutes with every element of $\mathscr{Y}_A$. A non-zero element of $\mathscr{Y}_A$ cannot have degree $(m, 0)$ since $q_0^0 = 1$, and it cannot have degree $(0, n)$ by Lemma 3.1. Thus any non-zero $Y \in \mathscr{Y}_A$ is of the form

$$Y \equiv c_1 B^n A^m + c_2 S B^{n-1} A^{m-1} + c_3 B^{n-1} A^{m-1}$$

for some n with $q_0^n \neq 1$ and with coefficients that satisfy

$$\begin{bmatrix} 1 - q_0^n & 0 & 0 \\ -\{n\}_{q_0 q_2} & 1 - q_0^{n-1} q_3 & 0 \\ -\{n\}_{q_0 q_4} & 0 & 1 - q_0^{n-1} q_1 \end{bmatrix} \begin{bmatrix} c_1 \\ c_2 \\ c_3 \end{bmatrix} = 0. \tag{27}$$

Since $q_1 \neq q_3$ (because $\mathbf{q}$ is generic), the matrix in (27) has one of the forms

$$\begin{bmatrix} * & 0 & 0 \\ ? & * & 0 \\ ? & 0 & * \end{bmatrix}, \quad \begin{bmatrix} * & 0 & 0 \\ ? & 0 & 0 \\ ? & 0 & * \end{bmatrix}, \quad \begin{bmatrix} * & 0 & 0 \\ ? & * & 0 \\ ? & 0 & 0 \end{bmatrix},$$

where $*$ indicates a non-zero element and ? indicates an element that may or may not be zero. Thus the solutions to (27) are

$$\begin{array}{ll} c_1 = c_2 = c_3 = 0 & \text{if } q_0^{n-1}q_3, q_0^{n-1}q_1 \neq 1 \\ c_1 = c_3 = 0 & \text{if } q_0^{n-1}q_3 = 1, q_0^{n-1}q_1 \neq 1 \\ c_1 = c_2 = 0 & \text{if } q_0^{n-1}q_3 \neq 1, q_0^{n-1}q_1 = 1. \end{array}$$

Since it is impossible that both c_2 and c_3 are non-zero, Y must actually have the form $c_2SB^{n-1}A^{m-1}$ or $c_3TB^{n-1}A^{m-1}$. Thus $\mathscr{Y}_A$ can be decomposed as $\mathscr{Y}_A^S \cup \mathscr{Y}_A^T \cup \{0\}$, where

$$\mathscr{Y}_A^S = \left\{cSB^{n-1}A^{m-1} \in \mathscr{Y}_A;\ q_0^n \neq 1,\ q_0^{n-1}q_3 = 1\right\}$$
$$\mathscr{Y}_A^T = \left\{cTB^{n-1}A^{m-1} \in \mathscr{Y}_A;\ q_0^n \neq 1,\ q_0^{n-1}q_1 = 1\right\}$$

(here, c ranges over $\mathbb{C}$ and m, n range over $\mathbf{N}^+$), and $\mathscr{C}(A)$ is commutative if and only if every element of $\mathscr{X}_A$ commutes with every element of $\mathscr{Y}_A^S$ and $\mathscr{Y}_A^T$.

Consider first $\mathscr{Y}_A^S$ and the conditions (23). If $\{d\}_{q_3/q_0} \neq 0$ then $q_3^d \neq 1$ or $q_3 = q_0$ and it cannot be that $q_0^n \neq 1$ and $q_0^{n-1}q_3 = 1$; thus $\mathscr{Y}_A^S = \emptyset$. Otherwise, there is an $r_3 \in \{2, \ldots, d\}$ such that $q_3 = q_0^{r_3}$, and

$$\mathscr{Y}_A^S = \left\{cSB^{n-1}A^{m-1};\ q_0^{n-1+r_3} = 1\right\}$$

is non-empty. Now pick any $X \in \mathscr{X}_A$. If $X = b_1A^k$ then X obviously commutes with every element in $\mathscr{Y}_A^S$. Otherwise, X has one of the forms

$$b_1B^l, \quad b_1B^lA^k + b_2SB^{l-1}A^{k-1} + b_3TB^{l-1}A^{k-1}$$

with $d|l$. Then the commutator of X with an element of $\mathscr{Y}_A^S$ is (note that $l = n = 1$ is impossible and use (6))

$$[X, cSB^{n-1}A^{m-1}] \equiv b_1c(q_2^l - 1)SB^{l+n-1}A^{k+m-1}. \tag{28}$$

Thus if $\{d\}_{q_3/q_0} \neq 0$, $q_2^d = 1$ then X commutes with everything in $\mathscr{Y}_A^S$. Finally, if $q_1 = q_0$ and $q_4^d \neq 1$ then Lemma 3.1 implies that $b_1 = 0$, so that the commutator (28) is 0, and again X commutes with everything in $\mathscr{Y}_A^S$.

On the other hand, if none of the conditions

$$q_3^d \neq 1,\ q_3 = q_0,\ q_2^d = 1,\ (q_1 = q_0 \text{ and } q_4^d \neq 1)$$

is satisfied, then

$$B^dA^2 + \frac{\{d\}_{q_0q_2}}{1 - q_0^{d-1+r_3}}SB^{d-1}A + b_TTB^{d-1}A \in \mathscr{X}_A,$$

where $r_3 \in \{2, \ldots, d\}$ is such that $q_3 = q_0^{r_3}$, and

$$b_T = \begin{cases} \{d\}_{q_0 q_4}/(1 - q_0^{d-1} q_1) & \text{if } q_1 \neq q_0 \\ \text{arbitrary} & \text{if } q_1 = q_0 \end{cases}$$

(b_T arbitrary if $q_1 = q_0$ works because then $q_4^d = 1$ and thus $\{d\}_{q_0 q_4} = 0$ since $\mathbf{q}$ generic implies $q_4 \neq q_0^{-1}$). This element of $\mathscr{X}_A$ does not commute with SB^{2d-r_3} in $\mathscr{Y}_A^S$; their commutator is $(q_2^d - 1)SB^{3d-r_3}A^2$. Thus it has been shown that every element of $\mathscr{X}_A$ commutes with every element of $\mathscr{Y}_A^S$ if and only if (23) is satisfied.

To see that every element of $\mathscr{X}_A$ commutes with every element of $\mathscr{Y}_A^T$ consider the isomorphism $h_{ST} : \mathscr{F}/\mathscr{J}(\mathbf{q}) \to \mathscr{F}/\mathscr{J}(\tilde{\mathbf{q}})$, where $\tilde{\mathbf{q}} = (q_0, q_3, q_4, q_1, q_2, \frac{1}{q_5})$, as defined in (18). Note that $\tilde{\mathbf{q}}$ is generic, $\tilde{q}_0$ is a root of unity with d being the smallest positive integer such that $(\tilde{q}_0)^d = 1$ and $\tilde{\mathbf{q}}$ satisfies (23) if and only if $\mathbf{q}$ satisfies (24). Furthermore, $h_{ST}(\mathscr{X}_A(\mathbf{q})) = \mathscr{X}_A(\tilde{\mathbf{q}})$ and $h_{ST}(\mathscr{Y}_A^T(\mathbf{q})) = \mathscr{Y}_A^S(\tilde{\mathbf{q}})$. Thus, using what has already been proved,

$$\begin{aligned} &\mathbf{q} \text{ satisfies (24)} \iff \tilde{\mathbf{q}} \text{ satisfies (23)} \iff \\ &\text{every } \tilde{X} \in \mathscr{X}_A(\tilde{\mathbf{q}}) \text{ commutes with every } \tilde{Y} \in \mathscr{Y}_A^S(\tilde{\mathbf{q}}) \iff \\ &\text{every } X \in \mathscr{X}_A(\mathbf{q}) \text{ commutes with every } Y \in \mathscr{Y}_A^T(\mathbf{q}). \end{aligned}$$

This concludes the proof of the first part of the theorem, namely that $\mathscr{C}(A)$ is commutative if and only if (23) and (24) are satisfied.

For the second part of the theorem, consider the isomorphism $h_{AB} : \mathscr{F}/\mathscr{J}(\mathbf{q}) \to \mathscr{F}/\mathscr{J}(\hat{\mathbf{q}})$, where $\hat{\mathbf{q}} = (\frac{1}{q_0}, q_4, q_3, q_2, q_1, q_5)$, as defined in (17). Note that $\hat{\mathbf{q}}$ is generic, $\hat{q}_0$ is a root of unity with d being the smallest positive integer such that $(\hat{q}_0)^d = 1$ and $\hat{\mathbf{q}}$ satisfies (23) and (24) if and only if $\mathbf{q}$ satisfies (25) and (26). Moreover, $h_{AB}(\mathscr{C}(B)(\mathbf{q})) = \mathscr{C}(A)(\hat{\mathbf{q}})$ and thus using what has already been proved,

$$\begin{aligned} &\mathbf{q} \text{ satisfies (25) and (26)} \iff \hat{\mathbf{q}} \text{ satisfies (23) and (24)} \iff \\ &\mathscr{C}(A)(\hat{\mathbf{q}}) \text{ is commutative} \iff \mathscr{C}(B)(\mathbf{q}) \text{ is commutative.} \end{aligned}$$

Acknowledgments This research was supported in part by the Swedish Research Council (621-2007-6338), Swedish Foundation for International Cooperation in Research and Higher Education (STINT), Swedish Royal Academy of Sciences, Crafoord Foundation and the Nordforsk network "Operator algebra and dynamics" (grant # 11580).

References

1. Bratteli, O., Robinson, D.W.: Operator Algebras and Statistical Mechanics. Springer, New York, I 1981, (1979)
2. Dixmier, J.: Les C^*-algèbres et Leurs Représentations, (French), Cahiers Scientifiques, Fasc. XXIX, Gauthier-Villars, Paris (1964). Russian translation: Nauka, Moscow (1974). English, translation: North-Holland, Amsterdam (1982)

3. Sakai, S.: Operator Algebras in Dynamical Systems. Cambridge University Press, Cambridge (1991)
4. Silvestrov, S.D., Tomiyama, Y.: Topological dynamical systems of type i. Expositiones Mathe. **20**, 117–142 (2002)
5. Tomiyama, J.: Invitation to the C^*-algebras and topological dynamics. World Scientific, Singapore (1987)
6. Tomiyama, J.: C^*-algebras and topological dynamical systems. Rev. Math. Phys. **8**, 741–760 (1996)
7. Kadison, R.V., Singer, I.M.: Extensions of pure states. Amer. J. Math. **81**, 383–400 (1959)
8. Anderson, J.: Extensions, restrictions and representations of states on c*-algebras. Trans. Am. Math. Soc. **249**, 303–329 (1979)
9. Anderson, J.: Extreme points in sets of linear maps on b(h). J. Func. Anal. **31**, 195–217 (1979)
10. Casazza, P.G., Fickus, M., Tremain, J.C., Weber, E.: The Kadison-Singer Problem in Mathematics and Engineering:A Detailed Account, Operator Theory, Operator Algebras, and Applications, vol. 414, pp. 299–355. Contemporary Mathematics—American Mathematical Society, Providence, RI (2006)
11. Paulsen, V.I.: A dynamical systems approach to the kadison-singer problem. J. Funct Anal. **255**(1), 120–132 (2008)
12. Bratteli, O., Jorgensen, P.E.T.: Wavelets Through a Looking Glass. Birkhäuser Verlag, Basel (2002)
13. Dutkay, D.E., Silvestrov, S.: Reducibility of the wavelet representation associated to the Cantor set. Proc. Am. Math. Soc. **139**(10), 3657–3664 (2011) MR2813395
14. Dutkay, D.E., Larson, D.R., Silvestrov, S.: Irreducible wavelet representations and ergodic automorphisms on solenoids. Oper. Matrices **5**(2), 201–219 (2011) MR2830593
15. Dutkay, D.E., Jorgensen, P.E.T., Silvestrov, S.: Decomposition of wavelet representations and martin boundaries. J. Func. Anal. **262**(3), 1043–1061 (2012)
16. Jorgensen, P.E.T.: Ruelle Operators: Functions which are Harmonic with Respect to a Transfer Operator. The University of Iowa, Iowa City (1998). (Preprint)
17. Davidson, K.R.: C*-algebras by example. Fields Institute Monographs, vol. 6. American Mathematical Society, Providence, RI (1996)
18. Jorgensen, P.E.T.: Operators and representation theory. Canonical models for algebras of operators arising in quantum mechanics. North-Holland Mathematical Studies, 147, Notas de Matematica 120, p. 337. North-Holland Publishing Company, Amsterdam (1988)
19. Li, B.R.: Introduction To Operator Algebras. World Scientific, Singapore (1992)
20. Mackey, G.W.: Induced Representations of Groups and Quantum Mechanics. W.A. Benjamin New York. Boringhieri Editore, Turin (1968)
21. Mackey, G.W.: The Theory of Unitary Group Representations. The University of Chicago Press, Chicago (1976)
22. Mackey, G.W.: Unitary Group Representations in Physics, Probability, and Number Theory. Addison-Wesley, Boston (1989)
23. Nazaikinskii, V.E., Shatalov, V.E., Yu, B.: Sternin, Methods of Noncommutative Analysis. Theory and Applications, De Gruyter Studies in Mathematics, vol .22. De Gruyter, Berlin (1976)
24. Ostrovskyi, V., Yu, S.: Introduction to the theory of representations of finitely presented *-algebras. I. Representations by bounded operators. Reviews in Mathematics and Mathematical Physics, 11, pt.1. Harwood Academic Publishers, Amsterdam (1999)
25. Pedersen, G.K.: C^*-algebras and their automorphism group. Academic Press, Salt Lake City (1979)
26. Samoĭlenko, Y.S.: Spectral Theory of Families of Selfadjoint Operators. Naukova Dumka, Kiev (1984) (Russian). Extended English transl.: Kluwer Academic Publication, Boston (1991)
27. Schmüdgen, K.: Unbounded Operator Algebras and Representation Theory. Birkhäuser Verlag, Basel (1990)
28. Hellström, L., Silvestrov, S.D.: Commuting Elements in q-deformed Heisenberg Algebras, p. 256. World Scientific, Singapore (2000). ISBN:981-02-4403-7

29. de Jeu, M., Svensson, C., Silvestrov, S.: Algebraic Curves for Commuting Elements in the q-deformed Heisenberg Algebra. J. Algebra **321**(4), 1239–1255 (2009). (Preprints in Mathematical Sciences 2007:23, LUTFMA-5089-2007, Centre for Mathematical Sciences, p. 17. Lund Institute of Technology, Lund University, Lund (2007). arXiv:0710.2748
30. Hellström, L., Silvestrov, S.D.: Ergodipotent maps and commutativity of elements in noncommutative rings and algebras with twisted intertwining. J. Algebra **314**(1), 17–41 (2007)
31. Carlsen, T.M., Silvestrov, S.: On the exel crossed product of topological covering maps. Acta Applicandae Mathe. **108**(3), 573–583 (2009)
32. Nystedt, P., Öinert, J.: Simple skew category algebras associated with minimal partially defined dynamical systems. Discrete Continuous Dyn. Syst. Ser. A (DCDS-A), **33**(9), 4157–4171 (2013)
33. Öinert, J., Silvestrov, S.D.: Commutativity and ideals in algebraic crossed products. J. Gen. Lie Theory. Appl. **2**(4), 287–302 (2008)
34. Öinert, J., Silvestrov, S., Theohari-Apostolidi, Th, Vavatsoulas, H.: Commutativity and ideals in strongly graded rings. Acta Applicandae Math. **108**(3), 585–602 (2009)
35. Öinert, J., Silvestrov, S.: Commutativity and ideals in pre-crystalline graded rings. Acta Applicandae Math. **108**(3), 603–615 (2009)
36. Öinert, J., Richter, J., Silvestrov, S.: Maximal ommutative subrings and simplicity of Ore extensions J. Algebra Appl. **12**(4), 1250192-1-16 (2013)
37. Persson, T., Silvestrov, S.: Commuting operators for representations of commutation relations defined by dynamical systems. Numer. Funct. Anal. Optim. **33**(7–9) (2012)
38. Svensson, C., Silvestrov, S., de Jeu, M.: Dynamical systems and commutants in crossed products. Int. J. Math. **18**(4), 455–471 (2007)
39. Svensson, C., Silvestrov, S., de Jeu, M.: Dynamical systems associated with crossed products. Acta Applicandae Mathe. **108**(3), 547–559 (2009)
40. Bergman, G.M.: The diamond lemma for ring theory. Avd. Math. **29**, 178–218 (1978)

Burchnall-Chaundy Theory for Ore Extensions

Johan Richter

Abstract We begin by reviewing a classical result on the algebraic dependence of commuting elements in the Weyl algebra. We proceed by describing generalizations of this result to various classes of Ore extensions, including both results that are already known and one new result.

1 Introduction

Let R be a commutative ring and S an R-algebra. Let a, b be two commuting elements of S. We are interested in the question whether they are algebraically dependent over R. i.e., does there exist a non-zero polynomial $f(s,t) \in R[s,t]$ such that $f(a,b) = 0$? Furthermore, can we find a proper subring F of R such that a, b are algebraically dependent over F?

In this article S will typically be an Ore extension of R. We start by introducing the notations and conventions we will use in this article and define what an Ore extension is. After that we review without giving proofs results obtained by other authors for the case that S is a differential operator ring (a special case of Ore extensions). We then proceed to describe results obtained by the present author and his collaborators and we finish by describing a strengthening of these results we recently obtained.

1.1 Notation and Conventions

$\mathbb{R}$ will denote the field of real numbers, $\mathbb{C}$ the field of complex numbers. $\mathbb{Z}$ will denote the integers.

J. Richter (✉)
Centre for Mathematical Sciences, Lund University, Box 118, 22100 Lund, Sweden
e-mail: johanr@maths.lth.se

A. Makhlouf et al. (eds.), *Algebra, Geometry and Mathematical Physics*, Springer Proceedings in Mathematics & Statistics 85, DOI: 10.1007/978-3-642-55361-5_4,

If R is a ring then $R[x_1, x_2, \dots x_n]$ denotes the ring of polynomials over R in central indeterminates $x_1, x_2, \dots, x_n$.

By a ring we will always mean an associative and unital ring. All morphisms between rings are assumed to map the multiplicative identity element to the multiplicative identity element.

By an ideal we shall mean a two-sided ideal.

If R is a ring we can regard it as a module (indeed algebra) over $\mathbb{Z}$ by defining $0r = 0$, $nr = \sum_{i=1}^{n} r$ if $n > 0$ and $nr = -(-n)r$ if n is a negative integer. If there is a positive integer n such that $n1_R = 0$, we call the smallest such positive integer the *characteristic* of R. If no such integer exists we set say that the characteristic is zero.

Let R be a commutative ring and S an R-algebra. Two commuting elements, $p, q \in S$, are said to be *algebraically dependent* (over R) if there is a non-zero polynomial, $f(s,t) \in R[s,t]$, such that $f(p,q) = 0$, in which case f is called an annihilating polynomial.

If S is a ring and a is an element in S, the *centralizer* of a, denoted $C_S(a)$, is the set of all elements in S that commute with a.

This article studies a class of rings called Ore extensions. For general references on Ore extensions, see e.g. [9, 14]. We shall briefly recall the definition. If R is a ring and σ is an endomorphism of R, then an additive map $\delta : R \to R$ is said to be a *σ-derivation* if

$$\delta(ab) = \sigma(a)\delta(b) + \delta(a)b$$

holds for all $a, b \in R$.

Definition 1.1 Let R be a ring, σ an endomorphism of R and δ a σ-derivation. The *Ore extension* $R[x; \sigma, \delta]$ is defined as the ring generated by R and an element $x \notin R$ such that $1, x, x^2, \dots$ form a basis for $R[x; \sigma, \delta]$ as a left R-module and all $r \in R$ satisfy

$$xr = \sigma(r)x + \delta(r). \quad (1)$$

Such a ring always exists and is unique up to isomorphism (see [9]). From $\delta(1 \cdot 1) = \sigma(1) \cdot 1 + \delta(1) \cdot 1$ we get that $\delta(1) = 0$, and since $\sigma(1) = 1$ we see that 1_R will be a multiplicative identity for $R[x; \sigma, \delta]$ as well.

Any element r of R such that $\sigma(r) = r$ and $\delta(r) = 0$ will be called a *constant*. In any ring with an endomorphism σ and a σ-derivation δ the constants form a subring.

If $\sigma = \mathrm{id}_R$, then a σ-derivation is simply called a derivation and $R[x; \mathrm{id}_R, \delta]$ is called a *differential operator ring*.

An arbitrary non-zero element $P \in R[x; \sigma, \delta]$ can be written uniquely as $P = \sum_{i=0}^{n} a_i x^i$ for some $n \in \mathbb{Z}_{\geq 0}$, with $a_i \in R$ for $i \in \{0, 1, \dots, n\}$ and $a_n \neq 0$. The *degree* of P will be defined as $\deg(P) := n$. We set $\deg(0) := -\infty$.

2 Burchnall-Chaundy Theory for Differential Operator Rings

We shall begin by describing some results on the algebraic dependence of commuting elements in differential operator rings. As the title of this subsection suggests, this sort of question has its origin in a series of papers by the British mathematicians Joseph Burchnall and Theodore Chaundy [2–4].

Proposition 2.1 *Let R be a ring and $\delta : R \to R$ a derivation. Let C be the set of constants of δ. Then*

(i) $1 \in C$;
(ii) *C is a subring of R, called the ring of constants;*
(iii) *for any $c \in C$ and $r \in R$ we have*

$$\delta(cr) = c\delta(r),$$
$$\delta(rc) = \delta(r)c.$$

Proof We skip the simple calculational proof.

As expected any derivation satisfies a version of the quotient rule.

Proposition 2.2 *Let R be a ring with a derivation, δ, and let a be any invertible element of R. Then*

$$\delta(a^{-1}) = -a^{-1}\delta(a)a^{-1}.$$

Proof

$$0 = \delta(1) = \delta(a^{-1}a) = a^{-1}\delta(a) + \delta(a^{-1})a \Rightarrow \delta(a^{-1}) = -a^{-1}\delta(a)a^{-1}.$$

Corollary 2.1 *Let R be a ring with a derivation δ and C its ring of constants. If a is an invertible element that lies in C, then so does a^{-1}. If R is a field, then C is a subfield of R.*

Example 2.1 As the ring R we can take $C^\infty(\mathbb{R}, \mathbb{C})$, the ring of all infinitely many times differentiable complex-valued functions on the real line. For δ we can take the usual derivative. The ring of constants in this case will consist of the constant functions.

With R and δ as in Example 2.1 we can form the differential operator ring $R[x; \mathrm{id}_R, \delta]$. We will show that the name "differential operator ring" is apt by constructing a ring of concrete differential operators that is isomorphic to $R[x; \mathrm{id}_R, \delta]$.

The ring $R = C^\infty(\mathbb{R}, \mathbb{C})$ can be seen as a vector space over $\mathbb{C}$, with operations defined pointwise. So we can consider the ring $\mathrm{End}_\mathbb{C}(R)$ of all linear endomorphisms of R. (Note that the endomorphisms are not required to be multiplicative.) $\mathrm{End}_\mathbb{C}(R)$ is in turn an algebra over R. One of the operators in $\mathrm{End}_\mathbb{C}(R)$ is the derivation operator, which we denote by D. Furthermore, for any $f \in R$ there is the multiplication operator

M_f that maps any function $g \in R$ to fg. The operator D and all the M_f together generate a subalgebra of $\mathrm{End}_{\mathbb{C}}(R)$, which we denote by T.

It is clear that the set of all M_f, for $f \in R$, is a subalgebra of T, isomorphic to R. Thus we abuse notation and identify M_f with f. By doing this we can write any element of T as a finite sum, $\sum_{i=0}^{n} a_i D^i$, where each a_i is a function in $C^\infty(\mathbb{R}, \mathbb{C})$. Furthermore such a decomposition is unique, or in other words: the powers of D form a basis for T as a free module over R.

We now compute the commutator of D and f for any $f \in R$. We temporarily revert to writing M_f for the element in T to make our calculations easier to understand. Let g be an arbitrary function in R. We find that

$$\begin{aligned}(DM_f - M_fD)(g) &= DM_f(g) - M_fD(g) = D(fg) - M_f(g') \\ &= f'g + fg' - fg' = f'g = M_{\delta(f)}(g).\end{aligned}$$

Hence

$$DM_f - M_fD = M_{\delta(f)}.$$

Relapsing into our abuse of notation we write this as $Df - fD = \delta(f)$ or equivalently as $Df = fD + \delta(f)$.

Denote the identity function on the real line by y. Then $Dy - yD = 1$, a relation known as the Heisenberg relation. The elements y and D together generate a subalgebra of T known as the Weyl algebra or the Heisenberg algebra, which is of interest in quantum mechanics, among other areas.

Any element, P, of T can be written as $P = \sum_{i=0}^{n} p_i D^i$, for some non-negative integer n and some $p_i \in C^\infty(\mathbb{R}, \mathbb{C})$. Conversely every such sum is an element of T. Thus T is isomorphic to $R[x; \mathrm{id}_R, \delta]$ with R and δ defined as in Example 2.1.

In a series of papers in the 1920s and 1930s [2–4], Burchnall and Chaundy studied the properties of commuting pairs of ordinary differential operators. In our terminology they may be said to study the properties of pairs of commuting elements of T. (They do not specify what function space their differential operators are supposed to act on.) The following theorem is essentially found in their papers.

Theorem 2.1 *Let $P = \sum_{i=0}^{n} p_i D^i$ and $Q = \sum_{j=0}^{m} q_j D^j$ be two commuting elements of T with constant leading coefficients. Then there is a non-zero polynomial $f(s, t)$ in two commuting variables over $\mathbb{C}$ such that $f(P, Q) = 0$. Note that the fact that P and Q commute guarantees that $f(P, Q)$ is well-defined.*

The result of Burchnall and Chaundy was rediscovered independently during the 1970s by researchers in the area of PDEs. It turns out that several important PDEs are equivalent to the condition that a pair of differential operators commute. These differential equations are completely integrable as a result, which roughly means that they possess an infinite number of conservation laws.

Burchnall's and Chaundy's work rely on analytical facts, such as the existence theorem for solutions of linear ordinary differential equations. However, it is possible to give algebraic proofs for the existence of the annihilating polynomial. This was done later by authors such as Amitsur [1] and Goodearl [5, 8]. Once one casts Burchnall's and Chaundy's results in an algebraic form one can also generalize them to a broader class of rings.

More specifically, one can prove Burchnall's and Chaundy's result for certain differential operator rings.

Amitsur [1, Theorem 1] (following work of Flanders [7]) studied the case when R is a field of characteristic zero and δ is an arbitrary derivation on R. He obtained the following theorem.

Theorem 2.2 *Let k be a field of characteristic zero with a derivation δ. Let F denote the subfield of constants. Form the differential operator ring $S = k[x; \mathrm{id}, \delta]$, and let P be an element of S of degree n. Denote by by $F[P]$ the ring of polynomials in P with constant coefficients, $F[P] = \{\sum_{j=0}^{m} b_j P^j \mid b_j \in F\}$. Then $C_S(P)$ is a commutative subring of S and a free $F[P]$-module of rank at most n.*

The next corollary can be found in [1, Corollary 2].

Corollary 2.2 *Let P and Q be two commuting elements of $k[x; \mathrm{id}, \delta]$, where k is a field of characteristic zero. Then there is a nonzero polynomial $f(s,t)$, with coefficients in F, such that $f(P,Q) = 0$.*

Proof Let P have degree n. Since Q belongs to $C_S(P)$ we know that $1, Q, \ldots, Q^n$ are linearly dependent over $F[P]$ by Theorem 2.2. But this tells us that there are elements $\phi_0(P), \phi_1(P), \ldots \phi_n(P)$, in $F[P]$, of which not all are zero, such that

$$\phi_0(P) + \phi_1(P)Q + \cdots + \phi_n(P)Q^n = 0.$$

Setting $f(s,t) = \sum_{i=0}^{n} \phi_i(s)t^i$ the corollary is proved.

Remark 2.1 Note that F, the field of constants, equals the center of $R[x; \mathrm{id}_R, \delta]$.

In [8] Goodearl has extended the results of Amitsur to a more general setting. The following theorem is contained in [8, Theorem 1.2].

Theorem 2.3 *Let R be a semiprime commutative ring with derivation δ and assume that its ring of constants is a field, F. If P is an operator in $R[x; \mathrm{id}_R, \delta]$ of positive degree n, where n is invertible in F, and has an invertible leading coefficient, then $C_S(P)$ is a free $F[P]$-module of rank at most n.*

We recall that a commutative ring is semiprime if and only if it has no nonzero nilpotent elements.

Goodearl notes that if R is a semiprime ring of positive characteristic such that the ring of constants is a field, then R must be a field. In this case he proves the following theorem [8, Theorem 1.11].

Theorem 2.4 *Let R be a field, with a derivation δ, and let F be its subfield of constants. If P is an element of $S = R[x; \mathrm{id}_R, \delta]$ of positive degree n and with invertible leading coefficient, then $C_S(P)$ is a free $F[P]$-module of rank at most n^2.*

As before we get the following corollary (of both Theorem 2.3 and Theorem 2.4), which is found in [8, Theorem 1.13].

Corollary 2.3 *Let P and Q be commuting elements of $R[x; \mathrm{id}_R, \delta]$, where R is a semiprime commutative ring, with a derivation δ such that the subring of constants is a field. Suppose that the leading coefficient of P is invertible. Then there exists a non-zero polynomial $f(s, t) \in F[s, t]$ such that $f(P, Q) = 0$.*

Note that Amitsur's work does not quite generalize Burchnall's and Chaundy's results since $C^\infty(\mathbb{R}, \mathbb{C})$ is not a field. Theorem 2.3 does however imply their results since $C^\infty(\mathbb{R}, \mathbb{C})$ is certainly commutative, does not have any nonzero nilpotent elements and its ring of constants is a field (isomorphic to $\mathbb{C}$). The only point to notice is that Theorem 2.3 requires P to have positive degree. If P is an element of degree zero and with constant leading coefficient however, it is itself a constant. Then $f(s, t) = s - P$ will be an annihilating polynomial for P and any Q.

An earlier paper by Carlson and Goodearl, [5], contains results similar to Theorems 2.3 and 2.4, in a different setting. Part of the theorem labelled Theorem 1 in [5] can be formulated as follows.

Theorem 2.5 *Let R be a commutative ring, with a derivation δ such that the ring of constants is a field, F, of characteristic zero. Assume that, for all $a \in R$, if the set $\{r \in R \mid \delta(r) = ar\}$ contains a nonzero element, then it contains an invertible element. Let P be an element of $R[x; \mathrm{id}_R, \delta]$ of positive degree n with invertible leading coefficient. Then $C_S(P)$ is a free $F[P]$-module of rank at most n. As before, this implies that if Q commutes with P, there exists a nonzero polynomial $f(s, t) \in F[s, t]$ such that $f(P, Q) = 0$.*

Note that the ring R in Example 2.1 satisfies the conditions of the theorem.

3 Burchnall-Chaundy Theory for Ore Extensions

Let k be a field and q a nonzero element of that field, not a root of unity. Set $R = k[y]$, a polynomial ring in one variable over k. There is an endomorphism σ of R such that $\sigma(y) = qy$ and the restriction of σ to k is the identity. For this σ there exists a unique σ-derivation δ such that $\delta(y) = 1$ and $\delta(\alpha) = 0$ for all $\alpha \in k$. The Ore extension $R[x; \sigma, \delta]$ for this choice of R, σ and δ is known as the (first) q-Weyl algebra. (An alternative name is the q-Heisenberg algebra.)

Silvestrov and collaborators [6, 10, 12] have extended the result of Burchnall and Chaundy to the q-Weyl algebra. The cited references contain two different proofs of the fact that any pair of commuting elements of $R[x; \sigma, \delta]$ are algebraically dependent over k. In [6] an algorithm is given to compute an annihilating polynomial explicitly.

The algorithm is a variation of one presented by Burchnall and Chaundy in their original papers and consists of forming a certain determinant that when evaluated gives the annihilating polynomial.

Mazorchuk [13] has presented an alternative approach to showing the algebraic dependence of commuting elements in q-Weyl algebras. He proves the following theorem.

Theorem 3.1 *Let k be a field and q an element of k. Set $R = k[y]$ and suppose that $\sum_{i=0}^{N} q^i \neq 0$ for any natural number N. Let P be an element of $S = R[x; \sigma, \delta]$ of degree at least 1. Then $C_S(P)$ is a free $k[P]$-module of finite rank.*

If P is as in the theorem and Q is any element of $R[x; \sigma, \delta]$ that commutes with P, then there is an annihilating polynomial $f(s, t)$ with coefficients in k. This is proven in the same way as Corollary 2.2. The methods used to obtain Theorem 3.1 have been generalized by Hellström and Silvestrov in [11].

In [15, Theorem 3] Silvestrov and the present author extend the algorithmic method of [6] to more general Ore extensions.

Theorem 3.2 *Let R be an integral domain with an injective endomorphism σ and a σ-derivation δ. Let a, b be two commuting elements of $R[x; \sigma, \delta]$. Then there exists a nonzero polynomial $f(s, t) \in R[s, t]$ such that $f(a, b) = 0$.*

Note that if we apply this theorem to the q-Weyl algebra with $R = k[y]$ we get a weaker result than the one stated above. We would like to be able to conclude that if a, b are commuting elements of $k[y][x; \sigma, \delta]$ then there is a polynomial $f(s, t)$ in $k[s, t]$ such that $f(a, b) = 0$.

Under certain assumptions on σ we have been able to prove this and we now proceed to describe how. We begin with a general theorem that we use as a lemma.

Theorem 3.3 *Let R be an integral domain, σ an injective endomorphism of R and δ a σ-derivation on R. Suppose that the ring of constants, F, is a field. Let a be an element of $S = R[x; \sigma, \delta]$ of degree n and assume that if b and c are two elements in $C_S(a)$ such that* $\deg(b) = \deg(c) = m$, *then $b_m = \alpha c_m$, where b_m and c_m are the leading coefficients of b and c respectively, and α is some constant.*

Then $C_S(a)$ is a free $F[a]$-module of rank at most n.

The proof we give is the same as used in [8] to prove Theorem 2.3.

Proof Denote by M the subset of elements of $\{0, 1, \dots, n-1\}$ such that an integer $0 \leq i < n$ is in M if and only if $C_S(a)$ contains an element of degree equivalent to i modulo n. For $i \in M$ let p_i be an element in $C_S(a)$ such that $\deg(p_i) \equiv i (\mathrm{mod}\, n)$ and p_i has minimal degree for this property. Take $p_0 = 1$.

We will show that $\{p_i | i \in M\}$ is a basis for $C_S(a)$ as a $F[a]$-module.

Since R is an integral domain and σ is injective, the degree of a product of two elements in $R[x; \sigma, \delta]$ is the sum of the degrees of the two elements.

We start by showing that the p_i are linearly independent over $F[a]$. Suppose $\sum_{i \in M} f_i p_i = 0$ for some $f_i \in F[a]$. If $f_i \neq 0$ then $\deg(f_i)$ is divisible by n, in which case

$$\deg(f_i p_i) = \deg(f_i) + \deg(p_i) \equiv \deg(p_i) \equiv i (\mathrm{mod}\, n). \tag{2}$$

If $\sum_{i \in M} f_i p_i = 0$ but not all f_i are zero, we must have two nonzero terms, $f_i p_i$ and $f_j p_j$, that have the same degree despite $i, j \in M$ being distinct. But this is impossible since $i \not\equiv j (\text{mod } n)$.

We now proceed to show that the p_i span $C_S(a)$. Let W denote the submodule they do span. We use induction on the degree to show that all elements of $C_S(a)$ belong to W. If e is an element of degree 0 in $C_S(a)$ we find by the hypothesis on a applied to e and $p_0 = 1$ that $e = \alpha$ for some $\alpha \in F$. Thus $e \in W$.

Now assume that W contains all elements in $C_S(a)$ of degree less than j. Let e be an element in $C_S(a)$ of degree j. There is some i in M such that $j \equiv i (\text{mod } n)$. Let m be the degree of p_i. By the choice of p_i we now that $m \equiv j (\text{mod } n)$ and $m \leq j$. Thus $j = m + qn$ for some non-negative integer q. The element $a^q p_i$ lies in W and has degree j. By hypothesis, the leading coefficient of e equals the leading coefficient of $a^q p_i$ times some constant α. The element $e - \alpha a^q p_i$ then lies in $C_S(a)$ and has degree less than j. By the induction hypothesis it also lies in W, and hence so does e.

We aim to use Theorem 3.3 when $R = k[y]$. To that end we have obtained the following proposition.

Proposition 3.1 *Let k be a field and set $R = k[y]$. Let σ be an endomorphism of R such that $\sigma(\alpha) = \alpha$ for all $\alpha \in k$ and $\sigma(y) = p(y)$, where $p(y)$ is a polynomial of degree (in y) greater than* 1. *Let δ be a σ-derivation such that $\delta(\alpha) = 0$ for all $\alpha \in k$. Form the Ore extension $S = R[x; \sigma, \delta]$. We note that its ring of constants is k. Let $a \notin k$ be an element of $R[x; \sigma, \delta]$. Assume that b, c are elements of S such that* $\deg(b) = \deg(c) = m$ *(here the degree is taken with respect to x) and b, c both belong to $C_S(a)$. Then $b_m = \alpha c_m$, where b_m, c_m are the leading coefficients of b and c respectively, and α is some constant.*

The author wishes to thank Fredrik Ekström for contributing a crucial idea to the following proof.

Proof Let a_n be the leading coefficient of a. By comparing the leading coefficient of ab and ba we see that

$$a_n \sigma^n(b_m) = b_m \sigma^m(a_n). \tag{3}$$

Similarly

$$a_n \sigma^n(c_m) = c_m \sigma^m(a_n). \tag{4}$$

By dividing Eq. 3 by Eq. 4 we see that

$$\frac{\sigma^n(b_m)}{\sigma^n(c_m)} = \frac{b_m}{c_m}. \tag{5}$$

We can perform such a division by passing to the quotient field of $k[y]$.

It thus suffices to prove that if f, g, p are polynomials in $k[y]$, with $\deg(p) > 1$, and

$$f(y)g(p(y)) = f(p(y))g(y), \tag{6}$$

then $f(y) = \alpha g(y)$ for some $\alpha \in k$.

So suppose that such f, g and p are given. We will also assume that k is algebraically closed, which can be done without loss of generality. If f and g have a common factor h we write $f(y) = h(y)\hat{f}(y)$ and similarly for g. We find that

$$\hat{f}(y)h(y)h(p(y))\hat{g}(p(y)) = \hat{f}(p(y))h(p(y))h(y)\hat{g}(y) \tag{7}$$

$$\Rightarrow \hat{f}(y)\hat{g}(p(y)) = \hat{f}(p(y))\hat{g}(y). \tag{8}$$

So we can assume without loss of generality that f and g are co-prime. It follows that the composite polynomials $f \circ p$ and $g \circ p$ are also co-prime. For if $f \circ p$ and $g \circ p$ had the common factor $l(y)$ it would follow that $f \circ p$ and $g \circ p$ had a common zero since k is algebraically closed. This would imply that f and g had a common zero, contradicting their co-primeness.

From Eq. 6 we see that f must divide $f \circ p$ and g must divide $g \circ p$. So write $f(p(y)) = e(y)f(y)$ and $g(p(y)) = \hat{e}(y)g(y)$. From (6) we see that $e = \hat{e}$. But this implies that e is a constant polynomial, since otherwise $f \circ p$ and $g \circ p$ would be co-prime. On the other hand $\deg(f \circ p) = \deg(p) \cdot \deg(f)$, which is a contradiction unless $\deg(f) = 0$. The proposition follows.

Proposition 3.2 *Let k, σ, δ, a be as in Proposition 3.1. Then $C_S(a)$ is a free $k[a]$-module of finite rank.*

Proof This follows directly from Theorem 3.3.

The following theorem, which as far as the author knows is a new result, follows from what we proved above.

Theorem 3.4 *Let k be a field. Let σ be an endomorphism of $k[y]$ such that $\sigma(y) = p(y)$, where $\deg(p) > 1$, and let δ be a σ-derivation. Suppose that $\sigma(\alpha) = \alpha$ and $\delta(\alpha) = 0$ for all $\alpha \in k$. Let a, b be two commuting elements of $k[y][x; \sigma, \delta]$. Then there is a nonzero polynomial $f(s,t) \in k[s,t]$ such that $f(a,b) = 0$.*

Proof Using the reasoning in the proof of Corollary 2.2 this follows from Theorem 3.3 and Proposition 3.1.

Note that the center of $k[y][x; \sigma, \delta]$ coincides with k and thus we have a parallel with, for example, Corollary 2.2. We would like to generalize Theorem 3.4 to obtain general conditions under which two commuting elements of $S = R[x; \sigma, \delta]$ are algebraically dependent over the center of S. An example of a result in that direction can be found in [10] where Hellström and Silvestrov prove the following theorem.

Theorem 3.5 ([10], Theorem 7.5). *Let $R = k[y]$, $\sigma(y) = qy$ and $\delta(y) = 1$, where $q \in k$ and q is a root of unity. Form $S = R[x; \sigma, \delta]$ and let C be the center of S. If a, b are commuting elements of S then there is a nonzero polynomial $f(s,t) \in C[s,t]$ such that $f(a,b) = 0$.*

This theorem can not be strengthened to give algebraic dependence over k. Indeed, suppose that $q^n = 1$. One can check that x^n and y^n commute (in fact they both belong to the center) but they are not algebraically dependent over k.

Acknowledgments The author wishes to thank Fredrik Ekström, Johan Öinert and Sergei Silvestrov for helpful comments. The author is also grateful to the Dalhgren and Lanner foundations for support. This research was also supported in part by the Swedish Foundation for International Cooperation in Research and Higher Education (STINT), The Swedish Research Council (grant no. 2007-6338), The Royal Swedish Academy of Sciences, The Crafoord Foundation and the NordForsk Research Network "Operator Algebras and Dynamics" (grant no. 11580).

References

1. Amitsur, S.A.: Commutative linear differential operators. Pacific J. Math. **8**, 1–10 (1958)
2. Burchnall, J.L., Chaundy, T.W.: Commutative ordinary differential operators. Proc. London Math. Soc. **3**(21), 420–440 (1923)
3. Burchnall, J.L., Chaundy, T.W.: Commutative ordinary differential operators. Proc. R. Soc. Lond. Ser. A Math. Phys. Eng. Sci. **118**, 557–583 (1928)
4. Burchnall, J.L., Chaundy, T.W.: Commutative ordinary differential operators ii, the identity $p^n = q^m$. Proc. R. Soc. Lond. Ser. A Math. Phys. Eng. Sci. **134**, 471–485 (1931)
5. Carlson, R.C., Goodearl, K.R.: Commutants of ordinary differential operators. J. Diff. Equat. **35**, 339–365 (1980)
6. de Jeu, M., Svensson, C., Silvestrov, S.: Algebraic curves for commuting elements in the q-deformed heisenberg algebra. J. Algebra **321**, 1239–1255 (2009)
7. Flanders, H.: Commutative linear differential operators. Technical report 1, University of California, Berkely (1955)
8. Goodearl, K.R.: Centralizers in differential, pseudodifferential, and fractional differential operator rings. Rock. Mt. J. Math. **13**, 573–618 (1983)
9. Goodearl, K.R., Warfield, R.B.: An introduction to noncommutative Noetherian rings, 2nd edn. London Mathematical Society Student Texts, vol. 61. Cambridge University Press, Cambridge (2004)
10. Hellström, L., Silvestrov, S.: Commuting Elements in q-deformed Heisenberg Algebras. World Scientific, Singapore (2000)
11. Hellström, L., Silvestrov, S.: Ergodipotent maps and commutativity of elements in noncommutative rings and algebras with twisted intertwining. J. Algebra **314**, 17–41 (2007)
12. Larsson, D., Silvestrov, S.: Burchnall-Chaundy theory for q-difference operators and q-deformed Heisenberg algebras. J. Nonlin. Math. Phys. **10**, 95–106 (2003)
13. Mazorchuk, V.: A note on centralizers in q-deformed Heisenberg algebras. AMA Algebra Montpellier announcements paper 2, 6 pp. (2001)
14. Rowen, L.H.: Ring Theory, vol. I. Pure and Applied Mathematics, vol. 127. Academic Press Inc., Boston (1988)
15. Richter, J., Silvestrov, S.: Burchnall-Chaundy annihilating polynomials for commuting elements in Ore extension rings. In: Abramov, V. et al. (eds.) Algebra, Geometry, and Mathematical Physics 2010, vol. 346. IOP Publishing, Bristol. J. Phys. Conf. Ser. (2012)

Homotopy Commutative Algebra and 2-Nilpotent Lie Algebra

Michel Dubois-Violette and Todor Popov

Abstract The homotopy transfer theorem due to Tornike Kadeishvili induces the structure of a homotopy commutative algebra, or C_∞-algebra, on the cohomology of the free 2-nilpotent Lie algebra. The latter C_∞-algebra is shown to be generated in degree one by the binary and the ternary operations.

1 Introduction

Every Universal Enveloping Algebra (UEA) $U\mathfrak{g}$ of a finite dimensional positively graded Lie algebra $\mathfrak{g}$ belongs to the class of Artin-Schelter regular algebras(see e.g. [4]). As every finitely generated graded connected algebra, $U\mathfrak{g}$ has a free minimal resolution which is canonically built from the data of its Yoneda algebra $\mathscr{E} := \mathrm{Ext}_{U\mathfrak{g}}(\mathbb{K}, \mathbb{K})$. By construction the Yoneda algebra $\mathscr{E}$ is isomorphic (as algebra) to the cohomology of the Lie algebra (with coefficients in the trivial representation provided by the ground field $\mathbb{K}$)

$$\mathscr{E} = \mathrm{Ext}^\bullet_{U\mathfrak{g}}(\mathbb{K}, \mathbb{K}) \cong H^\bullet(\mathfrak{g}, \mathbb{K}) \qquad (1)$$

equipped with wedge product between cohomological classes in $H^\bullet(\mathfrak{g}, \mathbb{K})$.

The homotopy transfer theorem of Kadeishvili [7] implies that the Yoneda algebra $\mathscr{E} = \mathrm{Ext}^\bullet_{U\mathfrak{g}}(\mathbb{K}, \mathbb{K})$ has the structure of homotopy associative algebra, or A_∞-algebra.

M. Dubois-Violette (✉)
Laboratoire de Physique Théorique, UMR 8627, Université Paris XI, Bâtiment 210, 91405 Orsay Cedex, France
e-mail: Michel.Dubois-Violette@th.u-psud.fr

T. Popov
Institute for Nuclear Research and Nuclear Energy, Bulgarian Academy of Sciences, 72 Tsarigradsko chaussée, 1784 Sofia, Bulgaria
e-mail: tpopov@inrne.bas.bg

A. Makhlouf et al. (eds.), *Algebra, Geometry and Mathematical Physics*, Springer Proceedings in Mathematics & Statistics 85, DOI: 10.1007/978-3-642-55361-5_5,

Since $\mathscr{E} \cong H^\bullet(\mathfrak{g}, \mathbb{K})$ is the cohomology of the exteriour algebra $\bigwedge \mathfrak{g}^*$ which is graded-commutative, it has the structure of homotopy commutative and associative algebra, or C_∞-algebra.

Throughout the text $\mathfrak{g}$ will be the free 2-nilpotent graded Lie algebra, with degree one generators in the finite dimensional vector space V over a field $\mathbb{K}$ of characteristic 0,

$$\mathfrak{g} = V \oplus \bigwedge^2 V .$$

The UEA $U(V \oplus \wedge^2 V)$ arises naturally in physics in the universal Fock-like space of the parastatistics algebra introduced by Green [5] (see also [3]). Here we will concentrate on the case when V is an ordinary (even) vector space V, when the algebra $U\mathfrak{g}$ is the parafermionic algebra.

The aim of this note is to describe the Yoneda algebra $\mathscr{E}$ of the UEA $U\mathfrak{g}$, i.e., the cohomology $H^\bullet(\mathfrak{g}, \mathbb{K})$ with its C_∞-structure induced by the isomorphism (1) through the homotopy transfer.

The cohomology space $H^\bullet(\mathfrak{g}, \mathbb{K})$ has a natural $GL(V)$-action. The decomposition of the $GL(V)$-module $H^\bullet(\mathfrak{g}, \mathbb{K})$ into irreducible Schur modules V_λ is known since the work of Józefiak and Weyman [6]; it contains all $GL(V)$-modules with self-conjugated Young diagrams $\lambda = \lambda'$ once and exactly once. The decomposition of $\mathscr{E} = H^\bullet(\mathfrak{g}, \mathbb{K})$ into Schur modules provides a powerful tool to handle its C_∞-algebra structure.

2 Artin-Schelter Regularity

Let $\mathfrak{g}$ be the 2-nilpotent graded Lie algebra $\mathfrak{g} = V \oplus \bigwedge^2 V$ generated by the finite dimensional vector space V having Lie bracket

$$[x, y] := \begin{cases} x \wedge y & x, y \in V \\ 0 & \text{otherwise} \end{cases} . \tag{2}$$

We denote the Universal Enveloping Algebra $U\mathfrak{g}$ by PS and will refer to it as *parastatistics algebra* (by some abuse[1]). The parastatistics algebra $PS(V)$ generated in V is graded

$$PS(V) := U\mathfrak{g} = U(V \oplus \bigwedge^2 V) = T(V)/([[V, V], V]) .$$

We shall write simply PS when the space of generators V is clear from the context.

Artin and Schelter [1] introduced a class of regular algebras sharing some "good" homological properties with the polynomial algebra $\mathbb{K}[V]$. These algebras were dubbed Artin-Schelter regular algebras (AS-regular algebra for short).

[1] Strictly speaking $PS(V)$ is the creation parastatistics algebra, closed by creation operators alone.

Definition 2.1 (AS-regular algebras) A connected graded algebra $\mathscr{A} = \mathbb{K} \oplus \mathscr{A}_1 \oplus \mathscr{A}_2 \oplus \ldots$ is called Artin-Schelter regular of dimension d if

(i) $\mathscr{A}$ has finite global dimension d,
(ii) $\mathscr{A}$ has finite Gelfand-Kirillov dimension,
(iii) $\mathscr{A}$ is Gorenstein, i.e., $\mathrm{Ext}^i_{\mathscr{A}}(\mathbb{K}, \mathscr{A}) = \delta^{i,d}\mathbb{K}$.

A general theorem claims that the UEA of a finite dimensional positively graded Lie algebra is an AS-regular algebra of global dimension equal to the dimension of the Lie algebra [4]. Hence the parastatistics algebra PS is AS-regular of global dimension $d = \frac{\dim V(\dim V+1)}{2}$. In particular the finite global dimension of PS implies that the ground field $\mathbb{K}$ has a minimal resolution $P_\bullet$ by projective left PS-modules P_n

$$P_\bullet: \quad 0 \to P_d \to \cdots \to P_n \to \cdots \to P_2 \to P_1 \to P_0 \xrightarrow{\epsilon} \mathbb{K} \to 0\,. \tag{3}$$

Here $\mathbb{K}$ is a trivial left PS-module, the action being defined by the projection ϵ onto $PS_0 = \mathbb{K}$. Since PS is positively graded and, in the category of positively graded modules over connected locally finite graded algebras, projective module is the same as free module [2], we have $P_n \cong PS \otimes E_n$ where E_n are finite dimensional vector spaces.

The minimal projective resolution is unique (up to an isomorphism). Minimality implies that the complex $\mathbb{K} \otimes_{PS} P_\bullet$ has "zero differentials" hence

$$H_\bullet(\mathbb{K} \otimes_{PS} P_\bullet) = \mathbb{K} \otimes_{PS} P_\bullet = E_n\,.$$

One can calculate the derived functor $\mathrm{Tor}_n^{PS}(\mathbb{K}, \mathbb{K})$ using the resolution $P_\bullet$, it yields

$$\mathrm{Tor}_n^{PS}(\mathbb{K}, \mathbb{K}) = E_n\,. \tag{4}$$

The data of a minimal resolution of $\mathbb{K}$ by free PS-modules provides an easy way to find $\mathrm{Tor}_n^{PS}(\mathbb{K}, \mathbb{K})$. Conversely if the spaces $\mathrm{Tor}_n^{PS}(\mathbb{K}, \mathbb{K})$ are known, then one can constuct a minimal free resolution of $\mathbb{K}$.

The Gorenstein property guarantees that when applying the functor $\mathrm{Hom}_{PS}(-, PS)$ to the minimal free resolution $P_\bullet$ we get another minimal free resolution $P^\bullet := \mathrm{Hom}_{PS}(P_\bullet, PS)$ of $\mathbb{K}$ by right PS-modules

$$P^\bullet: \quad 0 \leftarrow \mathbb{K} \leftarrow P_d^{'} \leftarrow \cdots \leftarrow P_n^{'} \leftarrow \cdots \leftarrow P_2^{'} \leftarrow P_1^{'} \leftarrow P_0^{'} \leftarrow 0 \tag{5}$$

with $P_n^{'} \cong E_n^* \otimes PS$. Note that by construction $E_n^* = \mathrm{Ext}^n_{PS}(\mathbb{K}, \mathbb{K})$, thus one has vector space isomorphisms [2]

$$E_n \cong E_n^* \cong \mathrm{Tor}_n^{PS}(\mathbb{K}, \mathbb{K}) \cong \mathrm{Ext}^n_{PS}(\mathbb{K}, \mathbb{K})\,. \tag{6}$$

The Gorenstein property is an analog of the Poincaré duality, it implies $E^*_{d-n} \cong E_n$. The finite global dimension d of PS and the Gorenstein condition imply that its Yoneda algebra

$$\mathscr{E}^\bullet := \mathrm{Ext}^\bullet_{PS}(\mathbb{K}, \mathbb{K}) \cong \bigoplus_{n=0}^{d} E^*_n$$

is Frobenius [10]. More on Gorenstein property you can find in the first autor's lecture "Poincaré duality for Koszul algebras" in the present volume.

3 Homology and Cohomology of $\mathfrak{g}$

A non-minimal projective (in fact free) resolution of $\mathbb{K}$, $C(\mathfrak{g}) \xrightarrow{\epsilon} \mathbb{K}$ is given by the standard Chevalley-Eilenberg chain complex $C_\bullet(\mathfrak{g}) = (U\mathfrak{g} \otimes_{\mathbb{K}} \wedge^p \mathfrak{g}, d_p)$ with differential maps

$$\begin{aligned} d_p(u \otimes x_1 \wedge \ldots \wedge x_p) = &\sum_i (-1)^{i+1} u x_i \otimes x_1 \wedge \ldots \wedge \hat{x}_i \wedge \ldots \wedge x_p \\ &+ \sum_{i<j} (-1)^{i+j} u \otimes [x_i, x_j] \wedge \\ &x_1 \wedge \ldots \wedge \hat{x}_i \wedge \ldots \wedge \hat{x}_j \wedge \ldots \wedge x_p. \end{aligned} \tag{7}$$

The resolution $C_\bullet(\mathfrak{g})$ calculates the homologies of the derived complex $\mathbb{K} \otimes_{PS} C_\bullet(\mathfrak{g})$

$$E_n = \mathrm{Tor}^{PS}_n(\mathbb{K}, \mathbb{K}) \cong H_n(\mathbb{K} \otimes_{PS} C_\bullet(\mathfrak{g})) = H_n(\mathfrak{g}, \mathbb{K}) \ ,$$

coinciding with the homologies $H_n(\mathfrak{g}, \mathbb{K})$ of the Lie algebra $\mathfrak{g}$ with trivial coefficients. The derived complex $\mathbb{K} \otimes_{PS} C_\bullet(\mathfrak{g})$ is the chain complex with degrees $\bigwedge^\bullet \mathfrak{g} = \mathbb{K} \otimes_{PS} PS \otimes \bigwedge^\bullet \mathfrak{g}$ and differentials $\partial_p := id \otimes_{PS} d_p : \bigwedge^p \mathfrak{g} \to \bigwedge^{p-1} \mathfrak{g}$. One has

$$\bigwedge^p \mathfrak{g} = \bigwedge^p (V \oplus \bigwedge^2 V) = \bigoplus_{s+r=p} \bigwedge^s (\bigwedge^2 V) \otimes \bigwedge^r (V) \tag{8}$$

and differentials $\partial_{p=r+s} : \bigwedge^s(\bigwedge^2 V) \otimes \bigwedge^r(V) \to \bigwedge^{s+1}(\bigwedge^2 V) \otimes \bigwedge^{r-2}(V)$ are given by

$$\begin{aligned} \partial_p : \ & e_{i_1 j_1} \wedge \ldots \wedge e_{i_s j_s} \ \otimes \ e_1 \wedge \ldots \wedge e_r \mapsto \\ & \sum_{i<j} (-1)^{i+j} e_{ij} \wedge e_{i_1 j_1} \wedge \ldots \wedge e_{i_s j_s} \ \otimes \ e_1 \wedge \ldots \wedge \hat{e}_i \wedge \ldots \wedge \hat{e}_j \wedge \ldots \wedge e_r \ . \end{aligned}$$

The differential ∂ is induced by the Lie bracket $[\,\cdot\,,\,\cdot\,] : \bigwedge^2 \mathfrak{g} \to \mathfrak{g}$, it identifies a pair of degree 1 generators $e_i, e_j \in$ with one degree 2 generator $e_{ij} := (e_i \wedge e_j) = [e_i, e_j]$. The differential ∂_p is the extension of $\partial_2 := -[\,\cdot\,,\,\cdot\,]$ as coderivation on $\bigwedge^p \mathfrak{g}$.

The dual cochain complex $\mathrm{Hom}_{PS}(C(\mathfrak{g}), \mathbb{K}) = (\bigwedge^\bullet \mathfrak{g}^*, \delta)$ calculates cohomology[2]

$$E_n^* = \mathrm{Ext}^n_{PS}(\mathbb{K}, \mathbb{K}) \cong H^n(\mathrm{Hom}_{PS}(C(\mathfrak{g}), \mathbb{K})) = H^n(\mathfrak{g}, \mathbb{K}) \,. \tag{9}$$

The coboundary map $\delta^p : \bigwedge^p \mathfrak{g}^* \to \bigwedge^{p+1} \mathfrak{g}^*$ is transposed to the differential ∂_{p+1}

$$\delta^p : \; e^*_{i_1 j_1} \wedge \ldots \wedge e^*_{i_s j_s} \; \otimes \; e^*_{l_1} \wedge \ldots \wedge e^*_{l_r} \mapsto \tag{10}$$

$$\sum_{k=1}^{s} \sum_{i_k < j_k} (-1)^{i+j} e^*_{i_1 j_1} \wedge \ldots \wedge \hat{e}^*_{i_k j_k} \wedge \ldots \wedge e^*_{i_s j_s} \; \otimes \; e^*_{i_k} \wedge e^*_{j_k} \wedge e^*_{l_1} \wedge \ldots \wedge \ldots \wedge e^*_{l_r} \,,$$

it is (up to a conventional sign) the extension as derivation of the dualization of the Lie bracket $\delta^1 := [\,\cdot\,,\,\cdot\,]^* : \mathfrak{g}^* \to \bigwedge^2 \mathfrak{g}^*$. Thus the algebra $(\bigwedge^\bullet \mathfrak{g}^*, \delta)$ equipped with δ is a *(graded-)commutative* DGA.

4 Homology of $\mathfrak{g}$ as a $GL(V)$-Module

An irreducible polynomial $GL(V)$-module V_λ is called Schur module, it has a basis labelled by semistandard Young tableaux which are fillings of the Young diagram λ with the numbers of the set $\{1, \ldots, \dim V\}$. The action of the linear group $GL(V)$ on the space V of the generators of the Lie algebra $\mathfrak{g}$ induces a $GL(V)$-action on the UEA $PS = U\mathfrak{g} \cong S(V \oplus \Lambda^2 V)$ and on the space $\bigwedge^\bullet \mathfrak{g} \cong \bigwedge^\bullet (V \oplus \bigwedge^2 V)$.

In the presence of metric g one has an identification $V \overset{g}{\cong} V^*$, and $\bigwedge^\bullet \mathfrak{g} \overset{g}{\cong} \bigwedge^\bullet \mathfrak{g}^*$. The adjoint operator $\partial_p^* : \bigwedge^p \mathfrak{g} \to \bigwedge^{p+1} \mathfrak{g}$ is defined by $g(\partial_p^* v, w) = g(v, \partial_{p+1} w)$. It turns out that the action of ∂_p^* always takes the form (similar to the action of δ^p)

$$\partial_p^* : \; e_{i_1 j_1} \wedge \ldots \wedge e_{i_s j_s} \; \otimes \; e_{l_1} \wedge \ldots \wedge e_{l_r} \mapsto \tag{11}$$

$$\sum_{k=1}^{s} \sum_{i_k < j_k} (-1)^{i+j} e_{i_1 j_1} \wedge \ldots \wedge \hat{e}_{i_k j_k} \wedge \ldots \wedge e_{i_s j_s} \; \otimes \; e_{i_k} \wedge e_{j_k} \wedge e_{l_1} \wedge \ldots \wedge \ldots \wedge e_{l_r} \,,$$

It is obvious that the maps ∂ and ∂^* both commute with the $GL(V)$-action. The *Laplacian* $\Delta = \oplus_{p \geq 0} \Delta_p$ of the pair $(\mathfrak{g}, g)$ is defined to be the self-adjoint operator

$$\Delta_p = \partial_{p+1} \partial^*_{p+1} + \partial^*_p \partial_p \in \mathrm{End}(\bigwedge^{\mathrm{p}} \mathfrak{g}) \,.$$

[2] In the presence of metric one has $\delta := \partial^*$(see below).

Its kernel is a complete set of representatives for the homology classes in $H_p(\mathfrak{g}, \mathbb{K})$

$$\ker \Delta_p \cong H_p(\mathfrak{g}, \mathbb{K}) .$$

The decomposition of the $GL(V)$-module $H_n(\mathfrak{g}, \mathbb{K})$ into irreducible polynomial representations V_λ is given by the following theorem;

Theorem 4.1 (Józefiak and Weyman [6], Sigg [11]) *The homology $H_\bullet(\mathfrak{g}, \mathbb{K})$ of the 2-nilpotent Lie algebra $\mathfrak{g} = V \oplus \bigwedge^2 V$ decomposes into irreducible $GL(V)$-modules*

$$H_n(\mathfrak{g}, \mathbb{K}) = H_n(\bigwedge^\bullet \mathfrak{g}, \partial) \cong \mathrm{Tor}_n^{PS}(\mathbb{K}, \mathbb{K})(V) \cong \bigoplus_{\lambda:\lambda=\lambda'} V_\lambda \tag{12}$$

where the sum is over self-conjugate Young diagrams λ such that $n = \frac{1}{2}(|\lambda| + r(\lambda))$.

The data $H_n(\mathfrak{g}, \mathbb{K}) = \mathrm{Tor}_n^{PS}(\mathbb{K}, \mathbb{K})$ encodes the minimal free resolution $P_\bullet$ (cf. (3)).

The Euler characteristics of $P_\bullet$ implies an identity about the $GL(V)$-characters

$$ch\, PS(V)\,.\,ch\left(\bigoplus_{\lambda:\lambda=\lambda'} (-1)^{\frac{1}{2}(|\lambda|+r(\lambda))} V_\lambda\right) = 1\,.$$

The character of a Schur module V_λ is the Schur function, $ch\, V_\lambda = s_\lambda(x)$. Due to the Poincaré-Birkhoff-Witt theorem $ch\, PS(V) = ch\, S(V \oplus \bigwedge^2 V)$ thus the identity reads

$$\prod_i \frac{1}{(1-x_i)} \prod_{i<j} \frac{1}{(1-x_i x_j)} \sum_{\lambda:\lambda=\lambda'} (-1)^{\frac{1}{2}(|\lambda|+r(\lambda))} s_\lambda(x) = 1\,. \tag{13}$$

But the latter identity is nothing but rewriting of the Littlewood identity [6]. The moral is that the Littlewood identity reflects a homological property of the algebra PS, namely the above particular structure of the minimal projective (free) resolution of $\mathbb{K}$ by PS-modules.

5 Homotopy Algebras A_∞ and C_∞

Definition 5.1 (A_∞-algebra) A homotopy associative algebra, or A_∞-algebra, over $\mathbb{K}$ is a $\mathbb{Z}$-graded vector space $A = \bigoplus_{i\in\mathbb{Z}} A^i$ endowed with a family of graded mappings (operations)

$$m_n : A^{\otimes n} \to A, \qquad \deg(m_n) = 2 - n \qquad n \geq 1$$

satisfying the Stasheff identities **SI(n)** for $n \geq 1$

$$\sum_{r+s+t=n} (-1)^{r+st} m_{r+1+t}(Id^{\otimes r} \otimes m_s \otimes Id^{\otimes t}) = 0 \qquad \mathbf{SI(n)}$$

where the sum runs over all decompositions $n = r + s + t$.

Here we assume the Koszul sign convention $(f \otimes g)(x \otimes y) = (-1)^{|g||x|} f(x) \otimes g(y)$. We define the shuffle product $Sh_{p,q} : A^{\otimes p} \otimes A^{\otimes q} \to A^{\otimes p+q}$ throughout the expression

$$(a_1 \otimes \ldots \otimes a_p) \, \Delta \, (a_{p+1} \otimes \ldots \otimes a_{p+q}) = \sum_{\sigma \in Sh_{p,q}} sgn(\sigma) \, a_{\sigma^{-1}(1)} \otimes \ldots \otimes a_{\sigma^{-1}(p+q)}$$

where the sum runs over all (p, q)-shuffles $Sh_{p,q}$, i.e., over all permutations $\sigma \in S_{p+q}$ such that $\sigma(1) < \sigma(2) < \ldots < \sigma(p)$ and $\sigma(p+1) < \sigma(p+2) < \ldots < \sigma(p+q)$.

Definition 5.2 (C_∞-algebra [7]) A homotopy commutative algebra, or C_∞-algebra, is an A_∞-algebra $\{A, m_n\}$ such that each operation m_n vanishes on non-trivial shuffles

$$m_n \left((a_1 \otimes \ldots \otimes a_p) \, \Delta \, (a_{p+1} \otimes \ldots \otimes a_n) \right) = 0 \, , \qquad 1 \le p \le n-1 \, . \qquad (14)$$

In particular for m_2 we have $m_2(a \otimes b \pm b \otimes a) = 0$, so a C_∞-algebra such that $m_n = 0$ for $n \geq 3$ is a (super-)commutative DGA.

A morphism of two A_∞-algebras A and B is a family of graded maps $f_n : A^{\otimes n} \to B$ for $n \geq 1$ with $\deg f_n = 1 - n$ such that the following conditions hold

$$\sum_{r+s+t=n} (-1)^{r+st} f_{r+1+t}(Id^{\otimes r} \otimes m_s \otimes Id^{\otimes r}) = \sum_{1 \le r \le n} (-1)^S m_r(f_{i_1} \otimes f_{i_2} \otimes \ldots \otimes f_{i_r})$$

where the sum is on all decompositions $i_1 + \ldots + i_r = n$ and the sign on RHS is determined by $S = \sum_{k=1}^{r-1} (r-k)(i_k - 1)$. The morphism f is a *quasi-isomorphism of A_∞-algebras* if f_1 is a quasi-isomorphism. It is strict if $f_i = 0$ for all $i \neq 1$. The identity morphism of A is the strict morphism f such that f_1 is the identity of A.

A morphism of C_∞-algebras is a morphism of A_∞-algebras vanishing on non-trivial shuffles $f_n \left((a_1 \otimes \ldots \otimes a_p) \, \Delta \, (a_{p+1} \otimes \ldots \otimes a_n) \right) = 0 \, , \quad 1 \le p \le n-1 \, .$

6 Homotopy Transfer Theorem

Lemma 6.1 *Every cochain complex (A, d) of vector spaces over a field $\mathbb{K}$ has its cohomology $H^\bullet(A)$ as a deformation retract.*

One can always choose a vector space decomposition of the cochain complex (A, d) such that $A^n \cong B^n \oplus H^n \oplus B^{n+1}$ where H^n is the cohomology and B^n is the space of coboundaries, $B^n = dA^{n-1}$. We choose a homotopy $h : A^n \to A^{n-1}$ which identifies B^n with its copy in A^{n-1} and is 0 on $H^n \oplus B^{n+1}$. The projection p to the cohomology and the cocycle-choosing inclusion i given by $A^n \underset{i}{\overset{p}{\rightleftarrows}} H^n$ are chain homomorphisms (satisfying the additional conditions $hh = 0$, $hi = 0$ and $ph = 0$). With these choices done the complex $(H^\bullet(A), 0)$ is a deformation retract of (A, d)

$$h \circlearrowright (A, d) \underset{i}{\overset{p}{\rightleftarrows}} (H^\bullet(A), 0) \ , \qquad pi = Id_{H^\bullet(A)} \ , \qquad ip - Id_A = dh + hd \ .$$

Let now (A, d, μ) be a DGA, i.e., A is endowed with an associative product μ compatible with d. The cochain complexes (A, d) and its contraction $H^\bullet(A)$ are homotopy equivalent, but the associative structure is not stable under homotopy equivalence. However the associative structure on A can be transferred to an A_∞-structure on a homotopy equivalent complex, a particular interesting complex being the deformation retract $H^\bullet(A)$. For a friendly introduction to homotopy transfer theorems in much boarder context we send the reader to the textbook [9], see Chap. 9.

Theorem 6.1 (Kadeishvili [7]) *Let (A, d, μ) be a (commutative) DGA over a field $\mathbb{K}$. There exists a A_∞-algebra (C_∞-algebra) structure on the cohomology $H^\bullet(A)$ and a $A_\infty(C_\infty)$-quasi-isomorphism $f_i : (\otimes^i H^\bullet(A), \{m_i\}) \to (A, \{d, \mu, 0, 0, \ldots\})$ such that the inclusion $f_1 = i : H^\bullet(A) \to A$ is a cocycle-choosing homomorphism of cochain complexes. The differential m_1 on $H^\bullet(A)$ is zero ($m_1 = 0$) and m_2 is strictly associative operation induced by the multiplication on A. The resulting structure is unique up to quasi-isomorphism.*

Kontsevich and Soibelman [8] gave an explicit expressions for the higher operations of the induced A_∞-structure as sums over decorated planar binary trees with one root where all leaves are decorated by the inclusion i, the root by the projection p the vertices by the product μ of the (commutative) DGA (A, d, μ) and the internal edges by the homotopy h. The C_∞-structure implies additional symmetries on trees. We will make use of the graphic representation for the binary operation on $H^\bullet(A)$

$m_2(x, y) := p\mu(i(x), i(y))$ or $m_2 =$

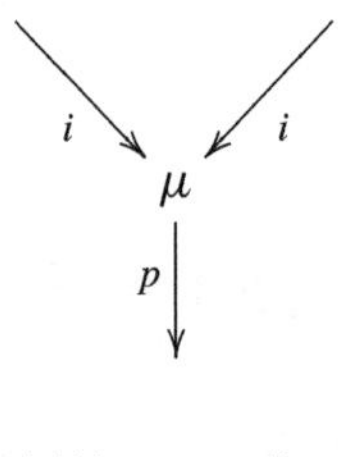

and the ternary one $m_3(x, y, z) = p\mu(i(x), h\mu(i(y), i(z))) - p\mu(h\mu(i(x), i(y)), i(z))$ being the sum of two planar binary trees with three leaves

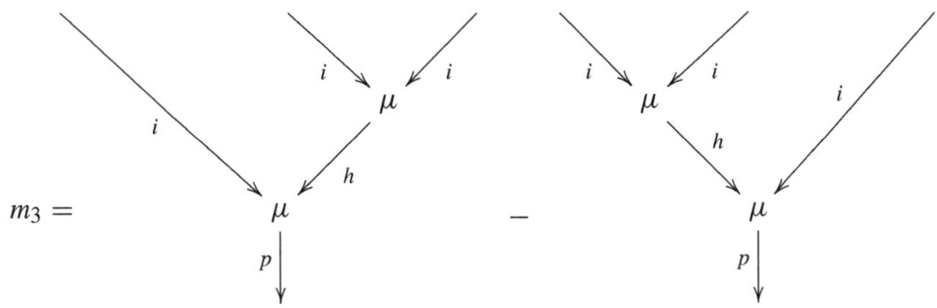

Theorem 6.2 *The cohomology $H^\bullet(\mathfrak{g}, \mathbb{K}) \cong \mathrm{Ext}^\bullet_{PS}(\mathbb{K}, \mathbb{K})$ of the 2-nilpotent graded Lie algebra $\mathfrak{g} = V \otimes \bigwedge^2 V$ is a homotopy commutative algebra which is generated in degree 1 (i.e., in $H^1(\mathfrak{g}, \mathbb{K})$) by the operations m_2 and m_3.*

Sketch of the proof Let us choose a metric $g(\cdot, \cdot) = \langle \cdot, \cdot \rangle$ on the vector space V and an orthonormal basis $\langle e_i, e_j \rangle = \delta_{ij}$. The choice induces a metric on $\bigwedge^\bullet \mathfrak{g} \overset{g}{\cong} \bigwedge^\bullet \mathfrak{g}^*$.

Due to the isomorphisms $\mathrm{Tor}_n^{PS}(\mathbb{K}, \mathbb{K}) \cong \mathrm{Ext}^n_{PS}(\mathbb{K}, \mathbb{K})$ (see Eq. 6) and $V \cong V^*$ the theorem 4.1 implies the decomposition of $H^\bullet(\mathfrak{g}, \mathbb{K})$ into irreducible $GL(V)$-modules

$$H^n(\mathfrak{g}, \mathbb{K}) \cong H^n(\bigwedge \mathfrak{g}^*, \delta) \cong \mathrm{Ext}^n_{PS}(\mathbb{K}, \mathbb{K})(V^*) \cong \bigoplus_{\lambda:\lambda=\lambda'} V_\lambda$$

where the sum is over self-conjugate diagrams λ such that $n = \frac{1}{2}(|\lambda| + r(\lambda))$.

In the presence of metric g the differential δ is identified with the adjoint of ∂, $\delta \overset{g}{:=} \partial^*$ while ∂ plays the role of a homotopy. In view of lemma 1 we have the cohomology $H^\bullet(\bigwedge^\bullet \mathfrak{g}^*, \delta^\bullet)$ as deformation retract of the complex $(\bigwedge^\bullet \mathfrak{g}^*, \delta^\bullet)$,

$$pi = Id_{H^\bullet(\bigwedge^\bullet \mathfrak{g}^*)}\,, \qquad ip - Id_{\bigwedge^\bullet \mathfrak{g}^*} = \delta\delta^* + \delta^*\delta\,, \qquad \delta^* \overset{g}{=} \partial\,.$$

Here the projection p identifies the subspace $\ker \delta \cap \ker \delta^*$ with $H^\bullet(\bigwedge^\bullet \mathfrak{g}^*)$, which is the orthogonal complement of the space of the coboundaries $\mathrm{im}\delta$. The cocycle-choosing homomorphism i is Id on $H^\bullet(\bigwedge^\bullet \mathfrak{g}^*)$ and zero on coboundaries.

We apply the Kadeishvili homotopy transfer Theorem 6.1 for the commutative DGA $(\bigwedge^\bullet \mathfrak{g}^*, \mu, \delta^\bullet)$ and its deformation retract $H^\bullet(\bigwedge^\bullet \mathfrak{g}^*) \cong H^\bullet(\mathfrak{g}, \mathbb{K})$ and conclude that the cohomology $H^\bullet(\mathfrak{g}, \mathbb{K})$ is a C_∞-algebra.

The Kontsevich and Soibelman tree representations of the operations m_n provide explicit expressions. Let us take μ to be the super-commutative product $\wedge$ on the DGA $(\bigwedge^\bullet \mathfrak{g}^*, \delta^\bullet)$. The projection p maps onto the Schur modules V_λ with $\lambda = \lambda'$.

The binary operation on the degree 1 generators $e_i \in H^1(\mathfrak{g}, \mathbb{K})$ is trivial, one gets

$$m_2(e_i, e_j) = p(e_i \wedge e_j) = 0 \qquad p(V_{(1^2)}) = 0.$$

Hence $H^\bullet(\mathfrak{g}, \mathbb{K})$ could not be generated in $H^1(\mathfrak{g}, \mathbb{K})$ as algebra with product m_2.

The ternary operation m_3 restricted to $H^1(\mathfrak{g}, \mathbb{K})$ is nontrivial, indeed one has

$$\begin{aligned} m_3(e_i, e_j, e_k) &= p\left\{e_i \wedge \partial(e_j \wedge e_k) - \partial(e_i \wedge e_j) \wedge e_k\right\} = p\left\{e_{ij} \wedge e_k - e_i \wedge e_{jk}\right\} \\ &= p\left\{(e_{ij} \wedge e_k + e_{jk} \wedge e_i + e_{ki} \wedge e_j) - e_{ki} \wedge e_j\right\} \\ &= e_{ik} \wedge e_j \in H^2(\mathfrak{g}, \mathbb{K}) \end{aligned}$$

The completely antisymmetric combination in the brackets $(\ldots)$ spans the Schur module $V_{(1^3)}$, $p(e_{ij} \wedge e_k + e_{jk} \wedge e_i + e_{ki} \wedge e_j) = 0$ yields a Jacobi-type identity. The monomials $e_{ij} \wedge e_k$ modulo $V_{(1^3)}$ span a Schur module $V_{(2,1)} \in H^2(\mathfrak{g}, \mathbb{K})$ with basis in bijection with the semistandard Young tableaux $e_{ik} \wedge e_j \leftrightarrow \begin{array}{|c|c|}\hline i & j \\ \hline k & \\ \cline{1-1}\end{array}$ and $e_{ij} \wedge e_k \leftrightarrow \begin{array}{|c|c|}\hline i & k \\ \hline j & \\ \cline{1-1}\end{array}$. We check the symmetry condition on ternary operation m_3 in C_∞-algebra; indeed m_3 vanishes on the (signed) shuffles $Sh_{1,2}$ and $Sh_{2,1}$

$$\begin{aligned} m_3(e_i \,\Delta\, e_j \otimes e_k) = m_3(e_i, e_j, e_k) &- m_3(e_j, e_i, e_k) \\ &+ m_3(e_j, e_k, e_i) = 0 = m_3(e_i \otimes e_j \,\Delta\, e_k). \end{aligned}$$

It is important that in the complexes $(\bigwedge^p \mathfrak{g}, \partial_p)$ and $(\bigwedge^p \mathfrak{g}^*, \delta^p)$ two different degees are involved; one is the homological degree $p := r + s$ counting the number of $\mathfrak{g}$-generators, while the second is the tensor degree $t := 2s + r$(also called weight). The differentials ∂ and δ preserve the tensor degree t but the spaces $H_n(\mathfrak{g}, \mathbb{K})$ and $H^n(\mathfrak{g}, \mathbb{K})$ are not homogeneous in t. The operation m_n is bigraded by homological and tensor gradings of bidegree $(p, t) = (2 - n, 0)$. The bi-grading impose the vanishing of many higher products.

On the level of Schur modules the ternary operation glues three fundamental $GL(V)$-representations $V_\square$ into a Schur module $V_{(2,1)}$. By iteration of the process of gluing boxes we generate all elementary hooks $V_k := V_{(k+1,1^k)}$,

$$m_3(V_\square, V_\square, V_\square) = V_{(2,1)}, \quad m_3\left(V_{(2,1)}, V_\square, V_\square\right) = V_{(3,1,1)}, \ldots, m_3(V_k, V_0, V_0) = V_{k+1}\,.$$

In our context the more convenient notation for Young diagrams is due to Frobenius: $\lambda := (a_1, \ldots, a_r | b_1, \ldots b_r)$ stands for a diagram λ with a_i boxes in the i-th row on the right of the diagonal, and with b_i boxes in the i-th column below the diagonal and the rank $r = r(\lambda)$ is the number of boxes on the diagonal.

For self-dual diagrams $\lambda = \lambda'$, i.e., $a_i = b_i$ we set $V_{a_1,\ldots,a_r} := V_{(a_1,\ldots,a_r|a_1,\ldots a_r)}$ when $a_1 > a_2 > \ldots > a_r \geq 0$ (and set the convention $V_{a_1,\ldots,a_r} := 0$ otherwise). Any two elementary hooks V_{a_1} and V_{a_2} can be glued together by the binary operation m_2, the decomposition of $m_2(V_{a_1}, V_{a_2}) \cong m_2(V_{a_2}, V_{a_1})$ is given by

$$m_2(V_{a_1}, V_{a_2}) = V_{a_1,a_2} \oplus (\bigoplus_{i=1}^{a_2} V_{a_1+i,a_2-i}) \qquad a_1 \geq a_2$$

where the "leading" term V_{a_1,a_2} has the diagram with minimal height. Hence any m_2-bracketing of the hooks $V_{a_1}, V_{a_2} \ldots, V_{a_r}$ yields[3] a sum of $GL(V)$-modules

$$m_2(\ldots m_2(m_2(V_{a_1}, V_{a_2}), V_{a_3}), \ldots, V_{a_r}) = V_{a_1,\ldots,a_r} \oplus \ldots$$

whose module with minimal height is precisely $V_{a_1,\ldots,a_r}$. We conclude that all elements in the C_∞-algebra $H^\bullet(\mathfrak{g}, \mathbb{K})$ can be generated in $H^1(\mathfrak{g}, \mathbb{K})$ by m_2 and m_3. □

Acknowledgments We are grateful to Jean-Louis Loday for many enlightening discussions and his encouraging interest. The work was supported by the French-Bulgarian Project Rila under the contract Egide-Rila N112.

References

1. Artin, M., Schelter, W.F.: Graded algebras of global dimension 3. Adv. Math. **66**, 171–216 (1987)
2. Cartan, H.: Homologie et cohomologie d' une algèbre graduée. Séminaire Henri Cartan **11**, 1–20 (1958)
3. Dubois-Violette, M., Popov, T.: Homogeneous algebras. Statistics and Combinatorics, LMP **61**, 159–170 (2002)
4. Floystad, G., Vayne, J.E.: Artin-schelter regular algebras of dimension five. Algebra Geom. Math. Phys. Banach Cent. Publ. **93**, 19–39 (2011)
5. Green, H.S.: A generalized method of field quantization. Phys. Rev. **90**, 270–273 (1953)
6. Józefiak, T., Weyman, J.: Representation-theoretic interpretation of a formula of D. E. Littlewood. Math. Proc. Cambridge Phil. Soc. **103**, 193–196 (1988)
7. Kadeishvili, T.: The a_∞-algebra structure and cohomology of hochschild and harrison. Proc. Tbil. Math. Inst. **91**, 19–27 (1988)
8. Kontsevich, M., Soibelman, Y.: Deformations of Algebras Over Operads and the Deligne Conjecture, Mathematical PhysicsStudies, vol. 21, pp. 255–307. kluwer Academic Publishers, Dordrecht (2000)
9. Loday, J.-L., Vallette, B.: Algebraic Operads. Grundlehren der Mathematischen Wis- senschafte, vol. 346. Springer, Heidelberg (2012)
10. Lu, D.-M., Palmieri, J.H., Wu, Q.-S., Zhang, J.J.: Koszul Equivalences in A_∞-Algebras. New York J. Math (2008). arXiv:0710.5492 [math.RA]
11. Sigg, S.: Laplacian and homology of free 2-step nilpotent lie algebras. J. Algebra **185**, 144–161 (1996)

[3] The operation m_2 is associative thus the result does not depend on the choice of the bracketing.

Subalgebra Depths Within the Path Algebra of an Acyclic Quiver

Lars Kadison and Christopher J. Young

Abstract Constraints are given on the depth of diagonal subalgebras in generalized triangular matrix algebras. The depth of the top subalgebra $B \cong A/\mathrm{rad}\, A$ in a finite, connected, acyclic quiver algebra A over an algebraically closed field $\mathbb{K}$ is then computed. Also the depth of the primary arrow subalgebra $1\mathbb{K} + \mathrm{rad}\, A = B$ in A is obtained. The two types of subalgebras have depths 3 and 4 respectively, independent of the number of vertices. An upper bound on depth is obtained for the quotient of a subalgebra pair.

1 Introduction

Given a subalgebra pair, one extracts a (minimum) depth from a comparison of n-fold tensor products of the subalgebra pair with one another in a meaningful way. The interesting case is when an $(n+1)$-fold tensor product divides a multiple of the n-fold tensor product in the sense of Krull-Schmidt unique factorization into indecomposable bimodules, or more generally as a bimodule isomorphism with a direct summand. The bimodule structures on the n-fold tensor products are naturally any one of four possibilities as left and right modules over the subalgebra or overalgebra. The least restrictive of these conditions is two-sided over the subalgebra and we fix the depth in the situation mentioned above to be $2n + 1$; for mixed bimodules, we have the left and right depth $2n$ conditions [4]. The most stringent condition, as

L. Kadison (✉) · C. J. Young
Departamento de Matematica, Faculdade de Ciencias da Universidade do Porto,
687 Rua Campo Alegre, 4169-007 Porto, Portugal
e-mail: lkadison@fc.up.pt

C. J. Young
e-mail: booloon_b457@hotmail.com

A. Makhlouf et al. (eds.), *Algebra, Geometry and Mathematical Physics*,
Springer Proceedings in Mathematics & Statistics 85, DOI: 10.1007/978-3-642-55361-5_6,

bimodules of the overalgebra, is H-depth $2n - 1$ [17], and is useful to ordinary depth gauging as well when the overalgebra has nice bimodules such as a separable algebra (see Proposition 2.1 below).

Comparing the tensor-square of an algebra extension with the overalgebra as mixed bimodules leads to a characterization of the Galois extension [7, 15, 16]. Thus not unexpectedly the depth two condition placed on Hopf subalgebras is equivalent to the normality condition with respect to the adjoint actions [3]. The depth three condition is satisfied by a subalgebra $B \subseteq A$ when, in a suitably nice category of bimodules, A contains all B^e-indecomposables that can possibly appear up to isomorphism in decompositions of tensor products $A \otimes_B \cdots \otimes_B A$ [3, 6]. Semisimple complex subalgebra pairs of each depth $n \in \mathbb{N}$ are noted in [5] via bipartite graphs and inclusion matrices for $K_0(B) \to K_0(A)$.

In the paper [4] it was shown that the depth of a finite group algebra extension is bounded by twice the index of the normalizer of the subgroup in the group. In the papers [4, 5, 11–13] the depth of certain group algebra extensions are computed; for example, [13] computes the depth of all the subgroups of $PSL(2, q)$ viewed as complex group algebras. In [5] the complex group algebras associated to the permutation groups are shown to have depth $d(S_n, S_{n+1}) = 2n - 1$; in [4], this same result is shown to not depend on the ground ring.

It was noted in the paper [6] that a subalgebra B in a finite-dimensional algebra A has finite depth $d(B, A)$ if B^e has finite representation type; below we note that this holds if A^e has finite representation type. In addition it is possible in algebras without involution that a subalgebra having left depth $2n$ may not have right depth $2n$. Moreover, the matrix power inequality characterizing depth n subalgebra pairs of semisimple complex algebras in [5, 11] breaks down in the presence of indecomposables of length greater than one. For these reasons, it becomes interesting to begin a study of depth of subalgebras in path algebras of quivers. A reasonable place to start is with acyclic quivers for whose path algebras there is a classic theorem about which have finite representation type in terms of Dynkin diagrams and the underlying graphs [1]. This paper computes the depth of the top and arrow subalgebras of the path algebra of a finite, connected, acyclic quiver. In Sect. 3 we note constraints on the depth of a diagonal subalgebra of a generalized matrix ring. We also note an inequality of depth in case the subalgebra contains ideals of the overalgebra, perhaps useful in computing depth of certain subalgebras of bounded quiver algebras. In the last Sect. 6 of concluding remarks we discuss other subalgebras of certain quiver algebras and their depth.

2 Preliminaries on Depth

Given a unital associative ring R and unital R-modules M and N, we say that M divides N and write $M \mid N$ if $N \cong M \oplus *$ as R-module for some (unnamed) complementary module. If there are natural numbers r and s such that $N \mid rM = M \oplus \cdots \oplus M$ and $M \mid sN$, then M and N are H-equivalent (or similar), as R-modules;

denoted by $M \sim N$. Note that this is indeed an equivalence relation. In this case their endomorphism rings End M_R and End N_R are Morita equivalent with Morita context bimodules Hom (M_R, N_R) and Hom (N_R, M_R) (with module actions and Morita pairings given by composition).

If M and N are in a category of finitely generated R-modules having unique factorization into indecomposables, then M and N have the same indecomposable constituents if and only if M and N are H-equivalent modules. If F is an additive endofunctor of the category of R-modules, then $M \sim N$ implies $F(M) \sim F(N)$; which in practice means that H-equivalent bimodules may replace one another in certain H-equivalences of tensor products. In addition, $M \sim N$ and $U \sim V$ implies $M \oplus U \sim N \oplus V$.

Throughout this paper, let A be a unital associative ring and $B \subseteq A$ a subring where $1_B = 1_A$. Note the natural bimodules ${}_BA_B$ obtained by restriction of the natural A-A-bimodule (briefly A-bimodule) A, also to the natural bimodules ${}_BA_A$, ${}_AA_B$ or ${}_BA_B$, which are referred to with no further notation. Equivalently we denote the proper ring extension $A \supseteq B$ occasionally by $A \mid B$. (Often results are valid as well for a ring homomorphism $B \to A$ and its induced bimodules on A.)

Let $C_0(A, B) = B$, and for $n \geq 1$,

$$C_n(A, B) = A \otimes_B \cdots \otimes_B A \quad (n \text{ times } A)$$

For $n \geq 1$, the $C_n(A, B)$ has a natural A-bimodule structure given by $a(a_1 \otimes \cdots \otimes a_n)a' = aa_1 \otimes \cdots \otimes a_na'$. Of course, this bimodule structure restricts to B-A-, A-B- and B-bimodule structures as we may need them. Let $C_0(A, B)$ denote the natural B-bimodule B itself. Recall from [4, 6] that a subring $B \subseteq A$ has right depth $2n$ if

$$C_{n+1}(A, B) \sim C_n(A, B) \tag{1}$$

as natural A-B-bimodules; left depth $2n$ if the same condition holds as B-A-bimodules; if both left and right conditions hold, it has depth $2n$; and depth $2n+1$ if the same condition holds as B-bimodules. If condition (1) holds in its strongest form as A-A-modules for $n \geq 1$ the subring $B \subseteq A$ is said to have H-depth $2n-1$; H-depth is investigated in [17].

Note that if the subring has left or right depth $2n$, it automatically has depth $2n+1$ by restriction to B-bimodules. Also note that if the subring has depth $2n+1$, it has depth $2n+2$ by tensoring the H-equivalence by $- \otimes_B A$ or $A \otimes_B -$. The *minimum depth* (or just depth when the context makes it clear) is denoted by $d(B, A)$; if $B \subseteq A$ has no finite depth, write $d(B, A) = \infty$. There is hidden in this a subtlety: if there is a subring $B \subseteq A$ of left depth $2n$ but not of right depth $2n$, then it has depth $2n+1$, left and right depth $2n+2$, and nevertheless its minimum depth is $2n$. There is not a published example of such a subring at present (but a search for this must occur outside the class of QF extensions [6, Theorem 2.4]). Note too that if $B \subseteq A$ has H-depth $2n-1$, it has depth $2n$ by restriction.

In practice one only need check half of the condition in (1) to establish depth $2n$ or $2n+1$ of a ring extension $A \supseteq B$. This is due to the fact that it is always the case that $C_n(A, B) \mid C_{n+1}(A, B)$ for $n \geq 1$ via appropriate face and degeneracy maps in the relative homological bar complex; e.g. the A-A-epimorphism $a_1 \otimes a_2 \mapsto a_1 a_2$ is split by the B-A-monomorphism $a \mapsto 1 \otimes_B a$, whence $C_1(A, B) \mid C_2(A, B)$ as B-A-bimodules.

For a k-algebra B let B^e denote $B \otimes_k B^{\mathrm{op}}$. For a finite dimensional algebra A let n_A denote the cardinal number of isomorphism classes of indecomposable finitely generated A-modules. Of course each of the B^e-modules $C_n(A, B)$ are finitely generated when A is a finite dimensional algebra.

Proposition 2.1 *Let $B \subseteq A$ be a subring pair of finite dimensional algebras. If B^e has finite representation type, then $d(B, A) \leq 1 + 2n_{B^e}$. If A^e has finite representation type, then $d(B, A) \leq 2n_{A^e}$. If $A \otimes B^{\mathrm{op}}$ has finite representation type, then $d(B, A) \leq 2n_{A \otimes B^{\mathrm{op}}}$.*

Proof If B^e has finite representation type, it is shown in [6] that subring depth $d(B, A)$ is finite based on two basic facts. First, a finitely generated module M over a finite dimensional algebra divides a multiple of another module N if and only if their Krull-Schmidt unique factorization into indecomposable modules possess the indecomposable constituents satisfying Indec $(M) \subseteq$ Indec (N); then M and N are H-equivalent iff Indec $(M) =$ Indec (N). Secondly, from $C_n(A, B) \mid C_{n+1}(A, B)$ we obtain Indec $C_n(A, B)$ as sequence of subsets of a finite number of indecomposables that grows with n.

If A^e has finite representation type, then one applies the same argument with growing Indec $C_n(A, B)$, this time as A-A-bimodules, which shows that $C_{N+1}(A, B)$ and $C_N(A, B)$ are H-equivalent after at most $N = n_{A^e}$ steps. Then the minimum H-depth $d_H(B, A) \leq 2N - 1$, and one notes by restricting modules that $d(B, A) \leq 2N$. The last statement is proven similarly using the definition of even depth.

Corollary 2.1 *Suppose $B \subseteq A$ is a subalgebra pair where either A or B is a separable algebra. Then depth $d(B, A)$ is finite.*

3 Constraints on Subring Depth in Triangular Matrix Rings

Let R and S be unital associative rings. Suppose ${}_S M_R$ is a unital S-R-bimodule as suggested by the notation. There is a triangular matrix ring, denoted by A, associated with this data,

$$A := \begin{pmatrix} R & 0 \\ M & S \end{pmatrix} \tag{2}$$

with the obvious matrix addition and multiplication, which defines a well-known class of examples in the demonstration of independence of axioms in ring theory such as left and right noetherian property of rings.

Note the subring of diagonal matrices in A is isomorphic (and identified) with $R \times S$. The obvious split epimorphism of rings $A \to R \times S$ is denoted by

$\pi : \begin{pmatrix} r & 0 \\ m & s \end{pmatrix} \mapsto (r, s)$. The mapping π is of course an isomorphism if $M = 0$. Also note the orthogonal idempotents $e_1 = (1_R, 0)$ and $e_2 = (0, 1_S)$, where $A = e_1 A \oplus e_2 A e_1 \oplus A e_2$.

Let R' be a unital subring of R, and S' a unital subring of S. Then $B := R' \times S'$ is a subalgebra of diagonal matrices in A. We will be interested in the depth $d(B, A)$. At first we will dispose of the case $M = 0$ and note that $d(R' \times S', R \times S) = \max\{d(R', R), d(S', S)\}$. (This proposition should be compared with [5, Prop. 3.15].)

Proposition 3.1 *The depth of a subalgebra of a direct product of rings is given by*

$$d(R' \times S', R \times S) = \max\{d(R', R), d(S', S)\}.$$

Proof Let $A = R \times S$ and $B = R' \times S'$. Note that the central orthogonal idempotents $e_1, e_2 \in B \subseteq A$. It follows that there is the following isomorphism of n-fold tensor products (any $n \in \mathbb{N}$),

$$C_n(A, B) \cong C_n(R, R') \oplus C_n(S, S') \tag{3}$$

as B-B-, A-B- and B-A-bimodules up to a trivial extension of for example R-module to A-module by $S \cdot x = 0$, all elements x in the module. Such a decomposition holds as well for bimodule homomorphisms between n- and $n+1$-fold tensor products.

Let $2m + 1 \geq \max\{d(R', R), d(S', S)\}$. Then the righthand-side of (3) where $n = m + 1$ divides a multiple of the m-fold tensor product of the same form, then so does the lefthand-side. Hence $d(B, A) \leq 2m + 1$. If both depths $d(R', R)$ and $d(S', S)$ are even, the same argument replacing $2m + 1$ with $2m$ suffices to establish $d(B, A) \leq \max\{d(R', R), d(S', S)\}$. Note that the argument works for 0-fold tensor product and depth one case too. The reverse inequality follows from applying the central idempotents to $C_n(A, B) \sim C_{n+1}(A, B)$.

Next we continue the notation $B = R' \times S'$ and A as the triangular matrix ring formed from the rings R, S and the bimodule ${}_S M_R \neq 0$. Let $\mathcal{M}$ denote a category of modules or bimodules, where left and right subscripts denote the rings in action.

Lemma 3.1 *As abelian categories,*

$${}_B\mathcal{M}_B \cong {}_{R'}\mathcal{M}_{R'} \oplus {}_{R'}\mathcal{M}_{S'} \oplus {}_{S'}\mathcal{M}_{R'} \oplus {}_{S'}\mathcal{M}_{S'}$$

Proof This isomorphism is induced on objects by ${}_B V_B \mapsto e_1 V e_1 \oplus e_1 V e_2 \oplus e_2 V e_1 \oplus e_2 V e_2$. Conversely, an object (W_1, W_2, W_3, W_4) on the right side is sent to a matrix $\begin{pmatrix} W_1 & W_2 \\ W_3 & W_4 \end{pmatrix}$ with left action by row vectors (r, s) and right action by column vectors $\begin{pmatrix} r' \\ s' \end{pmatrix}$. A B-bimodule homomorphism $f : V \to W$ commutes with e_1, e_2 from left and right, so that f sends $e_i V e_j$ into $e_i W e_j$ for all $i, j = 1, 2$. Conversely, a morphism of 2×2 matrices as before commutes with row and column vectors, and so is a B-bimodule homomorphism.

We now apply the lemma to the B-bimodules, the n-fold tensor products of the triangular matrix ring A over the diagonal subalgebra B.

Lemma 3.2 *For integer $n \geq 1$, $e_1 C_n(A, B)e_1 = C_n(R, R')$, $e_1 C_n(A, B)e_2 = 0$, and $e_2 C_n(A, B)e_2 = C_n(S, S')$; also*

$$e_2 C_n(A, B)e_1 = \sum_{r=0}^{n-1} \oplus\, C_r(S, S') \otimes_{S'} M \otimes_{R'} C_{n-1-r}(R, R') \qquad (4)$$

Proof For $a_1, \ldots, a_n \in A$, the computations follow from $e_1 a_1 \otimes_B \cdots \otimes_B a_n = e_1 a_1 e_1 \otimes \cdots \otimes_B a_n = \cdots = e_1 a_1 \otimes_B \cdots \otimes_B e_1 a_n$; moreover, $a_1 \otimes_B \cdots \otimes_B a_n e_2 = a_1 \otimes_B \cdots \otimes_B e_2 a_n e_2 = \cdots = a_1 e_2 \otimes_B \cdots \otimes_B a_n e_2$; furthermore, $e_1 a_1 \otimes_B \cdots \otimes_B a_n e_2 = 0$ by referring to the last computation and noting $e_1 A e_2 = 0$. Naturally, $C_n(e_1 A, B) = C_n(R, R')$ since $B = R' \times S'$ and S' acts as zero, so the relative tensor product is given by factoring out by only the nonzero relations; the same is true of $C_n(Ae_2, B) = C_n(S, S')$.

Finally, the last equation follows from $e_2 a_1 \otimes_B \cdots \otimes_B a_n e_1 = (e_2 a_1 e_2 + e_2 a_1 e_1) \otimes_B \cdots \otimes_B (e_2 a_n e_1 + e_1 a_n e_1) = \cdots = \sum_{i=1}^n a_1 e_2 \otimes_B \cdots \otimes e_2 a_i e_1 \otimes_B \cdots \otimes_B e_1 a_n$. This follows from cancellations of the type $\cdots \otimes a_i e_1 \otimes_B \cdots \otimes_B e_2 a_j \otimes_B \cdots = 0$ since $e_1 a_k = e_1 a_k e_1$, $a_k e_2 = e_2 a_k e_2$ for all $a_k \in A$ and of course $e_1 e_2 = 0$.

Let $d_{\text{odd}}(B, A)$ be the smallest odd number greater than or equal to $d(B, A)$, which we call the odd depth of the subring $B \subseteq A$. If the depth is finite and already odd, then $d_{\text{odd}}(B, A) = d(B, A)$, and otherwise $d_{\text{odd}}(B, A) = d(B, A) + 1$. In other words, a ring extension $A \mid B$ has $d_{\text{odd}}(B, A) = 2n+1$ if the natural B-B-bimodules $C_{n+1}(A, B) \sim C_n(A, B)$ and n is the smallest such natural number.

Theorem 3.1 *The odd depth $d_{\text{odd}}(B, A)$ satisfies the inequalities,*

$$d(B, R \oplus S) \leq d_{\text{odd}}(B, A) \leq d_{\text{odd}}(R', R) + d_{\text{odd}}(S', S) + 1 \qquad (5)$$

Proof If $B \subseteq A$ has depth $2n + 1$, then there is $q \in \mathbb{N}$ such that $C_{n+1}(A, B) \oplus V \cong qC_n(A, B)$ for some B-B-bimodule V. It follows that $e_i C_{n+1}(A, B)e_i \oplus e_i V e_i \cong q e_i C_n(A, B)e_i$ for $i = 1, 2$, so that $C_{n+1}(R, R') \mid qC_n(R, R')$ and $C_{n+1}(S, S') \mid qC_n(S, S')$. It follows that $R' \subseteq R$ and $S' \subseteq S$ both have depth $2n + 1$. Then $\max\{d(R', R), d(S', S)\} \leq d_{\text{odd}}(B, A)$. This completes the proof of the first of the two inequalities.

Next let $R' \subseteq R$ and $S' \subseteq S$ have depths $2n + 1$ and $2m + 1$ respectively. This means that for each integer $s \geq 1$ and $r \geq 0$ there is $q \in \mathbb{N}$ such that $C_{n+s}(R, R') \mid qC_{n+r}(R, R')$ as B-B-bimodules (and similarly for $S' \subseteq S$). Consider $C_{n+m+2}(A, B)$ as a natural B-B-bimodule. By the lemma, $C_{n+m+2}(A, B) \cong$

$$C_{n+m+2}(R, R') \oplus C_{n+m+2}(S, S') \oplus \sum_{i=0}^{n+m+1} \oplus C_i(S, S') \otimes_{S'} M \otimes_{R'} C_{n+m+1-i}(R, R')$$

which divides as B-B-bimodules (due to the depth hypotheses) a multiple of

$$C_{n+m+1}(R, R') \oplus C_{n+m+1}(S, S') \oplus \sum_{j=0}^{n+m} C_j(S, S') \otimes_{S'} M \otimes_{R'} C_{n+m-j}(R, R'),$$

which is isomorphic to a multiple of $C_{n+m+1}(A, B)$. Hence $B \subseteq A$ has depth $2(n+m+1)+1 = 2n+2m+3$. This establishes that $d(B, A) \leq d_{\text{odd}}(B, A) \leq d_{\text{odd}}(R', R) + d_{\text{odd}}(S', S) + 1$.

Note that the proof shows that if $R' \subseteq R$ and $S' \subseteq S$ are subrings of finite depth, then so is $B \subseteq A$, and conversely.

3.1 Quotient Algebras and Depth Bounds

Let $B \subseteq A$ be an arbitrary algebra extension and let $I \subseteq B$ be an A-ideal. For purposes of expedient notation we write $B_I := B/I$ and similarly for A_I. The main purpose of this section is to give some depth bounds for $B_I \subseteq A_I$ as another algebra extension. It turns out that if $d(B, A)$ is finite, then so is $d(B_I, A_I)$.

Recall that if the extension $B \subseteq A$ has odd depth $2n+1$ (even depth $2n$) then

$$C_{n+1}(A, B) \sim C_n(A, B)$$

as B-bimodules (A-B- and B-A-bimodules), which is in general equivalent to saying that there're two B-B-homomorphisms (A-B- and B-A-homomorphisms) $f: C_{n+1}(A, B) \to mC_n(A, B)$ and $g: mC_n(A, B) \to C_{n+1}(A, B)$ such that $g \circ f = id$.

Lemma 3.3 (**π and σ properties**) *Suppose that $B \subseteq A$ and $I \subseteq B$ are as above. We define the following maps:*

$$\begin{aligned} \pi &: C_n(A, B) \to C_n(A_I, B_I) \\ &: a_1 \otimes \ldots \otimes a_n \mapsto \overline{a_1} \otimes \ldots \otimes \overline{a_n}. \end{aligned}$$

$$\begin{aligned} \sigma &: C_{n+1}(A, B) \to C_{n+1}(A_I, B_I) \\ &: a_1 \otimes \ldots \otimes a_{n+1} \mapsto \overline{a_1} \otimes \ldots \otimes \overline{a_{n+1}} \end{aligned}$$

These two maps are well-defined and will be k-linear as well as satisfying

$$\pi(x \heartsuit y) = \overline{x}\pi(\heartsuit)\overline{y} \text{ and } \sigma(x \diamondsuit y) = \overline{x}\sigma(\diamondsuit)\overline{y},$$

$\forall x, y \in A$, $\forall \heartsuit \in C_n(A, B)$ *and* $\forall \diamondsuit \in C_{n+1}(A, B)$.

As will be necessary in our next result we "raise π to the mth power" in that we define $\pi' : mC_n(A, B) \to mC_n(A_I, B_I)$ in the obvious way:

$$(\heartsuit_i) \mapsto (\pi(\heartsuit_i)).$$

The important thing to note however is that $\pi'(x\heartsuit_i y) = \overline{x}\pi'(\heartsuit_i)\overline{y}$, where $x, y \in A$ and $\heartsuit_i \in mC_n(A, B)$, furthermore π' is k-linear over elements of $mC_n(A, B)$.

Theorem 3.2 *Suppose that $B \subseteq A$ is an algebra extension with depth $2n + 1$ ($2n$), suppose also that $I \subseteq B \subseteq A$ is an A-ideal. Then $B_I \subseteq A_I$ also has depth $2n + 1$ ($2n$). Indeed we can say $d(B_I, A_I) \leq d(B, A)$.*

Proof We prove the odd case because it involves B-bimodules and the proof can be extended to the even case with A-B-bimodules. First, because $B \subseteq A$ has depth $2n+1$ we have B-bimodule maps $f : C_{n+1}(A, B) \to mC_n(A, B)$ and $g : mC_n(A, B) \to C_{n+1}(A, B)$ such that $g \circ f = id$, where $m \geq 1$. We'd like first to find a B_I-bimodule map

$$\widetilde{f} : C_{n+1}(A_I, B_I) \to mC_n(A_I, B_I)$$

and secondly another B_I-bimodule map

$$\widetilde{g} : mC_n(A_I, B_I) \to C_{n+1}(A_I, B_I)$$

such that $\widetilde{g} \circ \widetilde{f} = id$. This enforcing the depth $2n + 1$ condition on $B_I \subseteq A_I$.

We define $\widetilde{f}$ as follows:

$$\widetilde{f}(\overline{a_1} \otimes \ldots \otimes \overline{a_n}) := \pi' \circ f(a_1 \otimes \ldots \otimes a_n) \tag{6}$$

We must show that $\widetilde{f}$ is well-defined, and to that end with some $1 \leq p \leq n$ let $\overline{a_p} = \overline{y}$, that is $a_p = y + t$, for $t \in I$. Thus

$$\begin{aligned}\widetilde{f}(\overline{a_1} \otimes \ldots \otimes \overline{a_p} \otimes \ldots \otimes \overline{a_n}) &= \pi' f(a_1 \otimes \ldots \otimes y + t \otimes \ldots \otimes a_n) \\ &= \pi' f(a_1 \otimes \ldots \otimes y \otimes \ldots \otimes a_n) \\ &\quad + \pi' f(a_1 \otimes \ldots \otimes t \otimes \ldots \otimes a_n) \\ &= \pi' f((a_1 \otimes \ldots \otimes y \otimes \ldots \otimes a_n)) \\ &= \widetilde{f}(\overline{a_1} \otimes \ldots \otimes \overline{y} \otimes \ldots \otimes \overline{a_n})\end{aligned}$$

since $\pi' f(a_1 \otimes \ldots \otimes a_{p-1} \otimes t \otimes a_{p+1} \otimes \ldots \otimes a_n) = \pi' f(a_1 \otimes \ldots \otimes t_1 \otimes 1 \otimes a_{p+1} \otimes \ldots \otimes a_n)$ etc until we have $\pi'(t_p f(1 \otimes \ldots \otimes 1 \otimes a_{p+1} \otimes \ldots \otimes x_n)) = \overline{t_p}(\pi' f(1 \otimes \ldots \otimes a_n)) = 0$ (where each $t_i \in I$). This all follows because $I \subseteq B$ is an A-ideal with the properties of lemma (3.3) in effect. Repeating such a process over all $1 \leq p \leq n$ the map will be well-defined.

Now we describe $\widetilde{g}$:

$$\widetilde{g}((\overline{a_1} \otimes \ldots \otimes \overline{a_{n+1}})_i) := \sigma \circ g((a_1 \otimes \ldots \otimes a_{n+1})_i) \tag{7}$$

Proving that $\widetilde{g}$ is well-defined is so similar to the (6) case it can be considered a minor exercise. Furthermore we should notice that $\widetilde{g} \circ \pi' = \sigma \circ g$ straight off. Using (6) and (7) we demonstrate that $\widetilde{g} \circ \widetilde{f} = id$:

$$\begin{aligned}\widetilde{g} \circ \widetilde{f}(\overline{a_1} \otimes \ldots \otimes \overline{a_n}) &= \widetilde{g} \circ \pi' \circ f(a_1 \otimes \ldots \otimes a_n)\\ &= \sigma \circ g \circ f(a_1 \otimes \ldots \otimes a_n)\\ &= \sigma \circ id(a_1 \otimes \ldots \otimes a_n)\\ &= \overline{a_1} \otimes \ldots \otimes \overline{a_n}\end{aligned}$$

Corollary 3.1 *Given a chain of A-ideals $J_0 \subseteq J_1 \subseteq \ldots \subseteq B$ we have*

$$1 \leq \cdots \leq d(B_{J_1}, A_{J_1}) \leq d(B_{J_0}, A_{J_0}) \leq d(B, A)$$

Proof The second isomorphism theorem tells us that $(B/J_0)/(J_1/J_0) \cong B/J_1$. Apply our last theorem to see that the depth of $(B/J_0)/(J_1/J_0) \subseteq (A/J_0)/(J_1/J_0)$ is less than or equal to the depth of $(B/J_0) \subseteq (A/J_0)$, but then we're done.

4 Depth of Top Subalgebra in Path Algebra of Acyclic Quiver

Let $Q = (V, E, s, t)$ denote a finite connected acyclic quiver with vertices V of cardinality $|V| = n$ and oriented edges E such that $|E| < \infty$, where an oriented edge or arrow is denoted by $\alpha : a \to b$, or $(a|\alpha|b) \in E$, where $a = s(\alpha)$ and $b = t(\alpha)$ define the source and target mappings $E \to V$, respectively. Since Q is acyclic, there is no loop in E, i.e., no arrow $\beta \in E$ such that $s(\beta) = t(\beta)$; moreover, there are no other cycles, i.e., paths $(a|\alpha_1, \ldots, \alpha_r|a)$ of length $r > 1$ beginning at a vertex a and ending there (where all $\alpha_i \in E$ and $s(\alpha_{i+1}) = t(\alpha_i)$, $i = 1, \ldots, r-1$).

Let $\mathbb{K}$ be an algebraically closed field and let $A = \mathbb{K}Q$ be the path algebra on the quiver A [1, 8] with basis the set of all paths, including stationary paths denoted by $\varepsilon_a = (a||a)$ for each $a \in V$, such that the product of two basis elements is given by the following concatenation formula:

$$(a|\alpha_1, \ldots, \alpha_r|b)(c|\beta_1, \ldots, \beta_s|d) = \delta_{bc}(a|\alpha_1, \ldots, \alpha_r, \beta_1, \ldots, \beta_s|d). \tag{8}$$

The product on A is given by this formula and linearization, which clearly makes A into a graded algebra where A_s denotes the $\mathbb{K}$-vector subspace spanned by paths of length s, a complete set of primitive orthogonal idempotents are $\{\varepsilon_a | a \in V\} \in A_0$ and the radical ideal is rad $A = A_1 \oplus A_2 \oplus \cdots$, also known as the arrow ideal.

There is always a numbering of the vertices from $1, \ldots, n$ such that $(i|\alpha|j) \in E$ implies $i > j$ [8, Corollary 8.6]. The vertex n is then a source and 1 a sink. With such a numbering the algebra $A = \mathbb{K}Q$ is embeddable in a lower triangular matrix algebra [1, Lemma 1.12] of the form,

$$A = \begin{pmatrix} \varepsilon_1(\mathbb{K}Q)\varepsilon_1 & 0 & \cdots & 0 \\ \varepsilon_2(\mathbb{K}Q)\varepsilon_1 & \varepsilon_2(\mathbb{K}Q)\varepsilon_2 & \cdots & 0 \\ \vdots & \vdots & & \vdots \\ \varepsilon_n(\mathbb{K}Q)\varepsilon_1 & \varepsilon_n(\mathbb{K}Q)\varepsilon_2 & \cdots & \varepsilon_n(\mathbb{K}Q)\varepsilon_n \end{pmatrix} \tag{9}$$

Note that $\varepsilon_i(\mathbb{K}Q)\varepsilon_i \cong K$ for each $i = 1, \ldots, n$ since there are no cycles. For example, if the quiver Q has no multiple arrows between vertices and its underlying graph is a tree, then there is at most one path between two points $i > j$, so that $\dim \varepsilon_i(\mathbb{K}Q)\varepsilon_j \leq 1$, and $A = \mathbb{K}Q$ is isomorphic to a subalgebra of the full triangular matrix algebra $T_n(\mathbb{K}) = \sum_{n\geq i\geq j\geq 1} \mathbb{K}\, e_{ij}$ (in terms of matrix units e_{ij}).

Another example: if $Q = (V, E)$ where $V = \{1, 2\}$ and $E = \{\alpha, \beta : 2 \to 1\}$, then

$$A = \mathbb{K}Q = \begin{pmatrix} \mathbb{K} & 0 \\ \mathbb{K}^2 & \mathbb{K} \end{pmatrix} \tag{10}$$

From the result of the previous section, we note that with $M = \mathbb{K}^2$, and $B = \mathbb{K}\varepsilon_1 + \mathbb{K}\varepsilon_2$, the depth of B in A is bounded by

$$1 \leq d(B, A) \leq 3. \tag{11}$$

For this algebra, one constructs from nilpotent Jordan blocks of order m an infinite sequence of indecomposable A-modules [1, pp. 75–76], a tame Kronecker algebra [2, V111.7]. The algebra $A = \mathbb{K}Q$ has finite representation type if and only if the underlying (multi-) graph of Q is one of the Dynkin diagrams $A_n (n \geq 1)$, D_n $(n \geq 4)$, E_6, E_7, E_8: see for example [1, Gabriel's Theorem, 5.10] or [2, VIII.5.2].

Coming back to the algebra A in (9), note that A has n augmentations $\rho_i : A \to \mathbb{K}$ given by $\rho_i(\lambda_1, \ldots, \lambda_n) = \lambda_i$. Let A_i^+ denote $\ker \rho_i$, and for a subalgebra $B \subseteq A$, let B_i^+ denote $\ker \rho_i \cap B$. Denote the n A-simples of dimension one by ${}_{\rho_i}\mathbb{K}$, and the n^2 A^e-simples by $\mathbb{K}_{ij}$ where $a \cdot 1 \cdot b = \rho_i(a)\rho_j(b)1$ for all $a, b \in A$ and $i, j = 1, \ldots, n$. We have the following

Lemma 4.1 *Suppose $B \subseteq A$ is a subalgebra of an algebra with augmentations $\rho_1, \ldots, \rho_n$. If $B \subseteq A$ has right depth 2, then $AB_i^+ \subseteq B_i^+ A$ for each $i = 1, \ldots, n$. If $B \subseteq A$ has left depth 2, then $B_i^+ A \subseteq AB_i^+$ for each $i = 1, \ldots, n$.*

Proof We prove the statement about a subalgebra having left depth two, namely, $A \otimes_B A \mid qA$ as B-A-bimodules. To this apply the additive functor $- \otimes_A {}_{\rho_i}\mathbb{K}$, which results in $A/AB_i^+ \mid q\mathbb{K}$ as left B-modules. The annihilator of $q\mathbb{K}$ restricted to B is of course B_i^+, which then also annihilates A/AB_i^+, so $B_i^+ A \subseteq AB_i^+$. This holds for each $i = 1, \dots, n$. The opposite inclusion is similarly shown to be satisfied by a right depth 2 extension of augmented algebras.

The next theorem computes the depth $d(B, A)$ of the top subalgebra $A/\mathrm{rad}\, A \cong \mathbb{K}^n$, or subalgebra of diagonal matrices, in the path algebra A of an acyclic quiver as given in (9).

Theorem 4.1 *Suppose the number of vertices $n > 1$ in the quiver Q, $A = \mathbb{K}Q$ and $B = \mathbb{K}^n$. Then depth $d(B, A) = 3$.*

Proof If the subalgebra in question has depth 1, it has depth 2. But if it has left depth 2, the lemma above applies, so that $B_i^+ A \subseteq AB_i^+$ for each $i = 1, \dots, n$. Note that AB_i^+ are all the lower triangular matrices of the form in (9) having only 0's on column i; similarly, $B_i^+ A$ are the triangular matrices having only zeroes on row i. It follows that $\varepsilon_j A \varepsilon_i = 0$ for each $j = i+1, \dots, n$. But $\varepsilon_j(\mathbb{K}Q)\varepsilon_i$ consists of all the paths from j to i. Since this holds for each i, Q consists of n points with no edges; thus we have contradicted the assumption that Q is connected. The same contradiction is reached assuming $B \subset A$ has right depth 2.

Next it is shown that ${}_BA \otimes_B A_B$ divides a multiple of ${}_BA_B$. Let $\dim \varepsilon_i A \varepsilon_j = n_{ij}$. Then it is clear from (9) and simple matrix arithmetic that ${}_BA_B \cong \oplus_{n \geq i \geq j \geq 1} n_{ij} \mathbb{K}_{ij}$.

Now

$$A \otimes_B A = \oplus_{i,j=1}^n \oplus_{i \geq k \geq j} \varepsilon_i A \varepsilon_k \otimes_B \varepsilon_k A \varepsilon_j$$

since each $\varepsilon_j \in B$ and for each $r \neq k$, $\varepsilon_k \varepsilon_r = 0$. It follows that ${}_BA \otimes_B A_B \cong \oplus_{n \geq i \geq j \geq 1} m_{ij} \mathbb{K}_{ij}$ where $m_{ij} = \sum_{i \geq k \geq j} n_{ik} n_{kj}$. Since $n_{ii} = 1$ for each i, it follows that $m_{ij} \geq n_{ij}$; moreover, $n_{ij} = 0$ implies $m_{ij} = 0$, since otherwise there is a path from i to j via some k such that $i \geq k \geq j$.

From the last remark it follows that there is $q \in \mathbb{N}$ such that $A \otimes_B A \,|\, qA$ as B-B-bimodules. Thus the minimum depth $d(B, A) = 3$.

5 Depth of Arrow Subalgebra in Acyclic Quiver Algebra

In this section we compute the depth of the primary arrow subalgebra $B = \mathbb{K}1_A \oplus A_1 \oplus A_2 \oplus \cdots = \mathbb{K}1_A + \mathrm{rad}\, A$ in the path algebra A of an acyclic quiver Q, which is of the form

$$A = \begin{pmatrix} \mathbb{K} & 0 & \cdots 0 \\ \varepsilon_2(\mathbb{K}Q)\varepsilon_1 & \mathbb{K} & \cdots 0 \\ \vdots & \vdots & \vdots \\ \varepsilon_n(\mathbb{K}Q)\varepsilon_1 & \varepsilon_n(\mathbb{K}Q)\varepsilon_2 & \cdots \mathbb{K} \end{pmatrix} \tag{12}$$

Note that B is a local algebra and augmented algebra with one augmentation $\varepsilon : B \to \mathbb{K}$ equal to the canonical quotient map $B \to B/\mathrm{rad}\, B \cong \mathbb{K}$. We denote the B-simple by $\mathbb{K}_\varepsilon$ as a pullback module. Again there are n augmentations of A denoted by ρ_i defining n simple A-B-bimodules denoted by ${}_i\mathbb{K}_\varepsilon$, $i = 1, \dots, n$.

Lemma 5.1 *The natural B-B-bimodule A is indecomposable.*

Proof It suffices to show that $\mathrm{End}\,{}_BA_B$ is a local ring [1, 8]. Let $F \in \mathrm{End}\,{}_BA_B$ and choose an ordered basis of A given by $I = \langle \varepsilon_1, \dots, \varepsilon_n, \alpha_1, \dots, \alpha_m \rangle$ where the length of the path α_i is less than or equal to the length of α_{i+1}, all $i = 1, \dots, m-1$. Consider the matrix with $\mathbb{K}$-coefficients, $M = (M^\alpha_\beta)_{\alpha,\beta \in I}$ of F relative to I; then $F(\alpha) = \sum_{\beta \in I} M^\alpha_\beta \beta$.

Given a path of length $r \geq 1$, $(i|\alpha|j) \in A_r$, note that $F(\alpha) = \alpha F(\varepsilon_j) = F(\varepsilon_i)\alpha$, so that

$$\sum_{\beta \in I} M^\alpha_\beta \beta = \sum_{\gamma \in I} M^{\varepsilon_j}_\gamma \alpha\gamma = \sum_{\delta \in I} M^{\varepsilon_i}_\delta \delta\alpha.$$

It follows that $M^{\varepsilon_j}_\gamma = 0$ for paths $(j|\gamma|k)$ and $M^{\varepsilon_i}_\delta = 0$ for all paths $(\ell|\delta|i)$. Also $M^\alpha_\beta = 0$ for all path $\beta \notin \varepsilon_i A \varepsilon_j$, i.e. not a path from i to j. Finally deduce that $M^\alpha_\beta = 0$ if $\beta \in \varepsilon_i A \varepsilon_j$ but $\beta \neq \alpha$ and $M^\alpha_\alpha = M^{\varepsilon_i}_{\varepsilon_i} = M^{\varepsilon_j}_{\varepsilon_j}$.

For $i \neq j$ and $\alpha \in \varepsilon_k A \varepsilon_i$, note that $\alpha F(\varepsilon_j) = F(\alpha\varepsilon_j) = 0$, so that $\sum_{\beta \in I} M^{\varepsilon_j}_\beta \alpha\beta = 0$ implies $M^{\varepsilon_j}_\beta = 0$ whenever $s(\beta) = i$. In particular, $M^{\varepsilon_j}_{\varepsilon_i} = 0$. It follows that the set of $F \in \mathrm{End}\,{}_BA_B$ has the form of a triangular matrix algebra with constant diagonal, like B, and is a local algebra.

Theorem 5.1 *The depth of the primary arrow subalgebra B in the path algebra A defined above is $d(B, A) = 4$.*

Proof We first compute $A \otimes_B A$ and show $d(B, A) > 3$. Note that two paths of nonzero length, α, β where $s(\alpha) = i$ satisfy $\alpha \otimes_B \beta = \varepsilon_i \otimes_B \alpha\beta$, which is zero unless $t(\alpha) = s(\beta)$. It follows that

$$A \otimes_B A = \oplus_{i=1}^n \mathbb{K}\varepsilon_i \otimes_B \varepsilon_i \oplus_{i=2}^n \oplus_{j=1}^{i-1} \varepsilon_i \otimes_B \varepsilon_i A \varepsilon_j \oplus_{i \neq j} \mathbb{K}\varepsilon_i \otimes_B \varepsilon_j.$$

It is obvious that the first two summations above are isomorphic as B-B-bimodules to ${}_BA_B$. Note that when $i \neq j$, for all paths α, β,

$$\alpha\varepsilon_i \otimes_B \varepsilon_j = 0 = \varepsilon_i \otimes_B \varepsilon_j \beta$$

since $\alpha\varepsilon_i \in B$ is either zero or a path ending at i, whence $\alpha\varepsilon_i\varepsilon_j = 0$. It follows that $A \otimes_B A \cong A \oplus n(n-1)_\varepsilon \mathbb{K}_\varepsilon$ as B-B-bimodules; moreover, as A-B-bimodules, we note for later reference

$$ {}_AA \otimes_B A_B \cong {}_AA_B \oplus \oplus_{i=1}^n (n-1)_i \mathbb{K}_\varepsilon \tag{13}$$

By lemma, ${}_BA_B$ is an indecomposable, but the B-B-bimodule $A \otimes_B A$ contains another nonisomorphic indecomposable, in fact ${}_\varepsilon\mathbb{K}_\varepsilon$, so that as B-bimodules, $A \otimes_B A \oplus * \ncong qA$ for any multiple q by Krull-Schmidt.

Now we establish that the subalgebra $B \subseteq A$ has right depth 4 by comparing (13) with the computation below:

$$A \otimes_B A \otimes_B A = \oplus_{i=1}^n \mathbb{K}\varepsilon_i \otimes \varepsilon_i \otimes \varepsilon_i \oplus_{i=2}^n \oplus_{j=1}^{i-1} \varepsilon_i \otimes \varepsilon_i \otimes \varepsilon_i A \varepsilon_j \oplus_{i \neq j \neq k} \mathbb{K}\varepsilon_i \otimes \varepsilon_j \otimes \varepsilon_k$$

$$\cong A \oplus (n^2 - 1)\, {}_1\mathbb{K}_\varepsilon \oplus \cdots \oplus (n^2 - 1)\, {}_n\mathbb{K}_\varepsilon$$

as A-B-bimodules, where $i \neq j \neq k$ symbolizes $i \neq j$, $j \neq k$ or $i \neq k$. It is clear that since no new bimodules appear in a decomposition of ${}_AA \otimes_B A \otimes_B A_B$ as compared with ${}_AA \otimes_B A_B$, that there is $q \in \mathbb{N}$ (in fact $q = n + 1$ will do) such that $A \otimes_B A \otimes_B A \mid qA \otimes_B A$ as A-B-bimodules. It follows that the minimum depth $d(B, A) = 4$.

It is easy to see from the proof that as natural B-A bimodules $A \otimes_B A \otimes_B A \mid (n + 1)A \otimes_B A$ for very similar reasons. Note the general fact that ${}_AA_B$ or ${}_BA_A$ are indecomposable modules if End ${}_AA_B \cong A^B$, the centralizer subalgebra of B in A, is a local algebra.

6 Concluding Remarks

It is well-known and easily computed from (12) that the path algebra $\mathbb{K}Q$ of the quiver

$$Q: \; n \longrightarrow n-1 \longrightarrow \cdots \longrightarrow 2 \longrightarrow 1$$

is the lower triangular matrix algebra $T_n(\mathbb{K})$. Then we have shown above that for the subalgebras $B_1 = D_n(\mathbb{K})$ equal to the set of diagonal matrices, and $B_2 = U_n(\mathbb{K})$ defined by

$$U_n(\mathbb{K}) = \{ \begin{pmatrix} a & 0 & 0 & \cdots 0 \\ a_{21} & a & 0 & \cdots 0 \\ a_{31} & a_{32} & a & \cdots 0 \\ \vdots & & & \vdots \\ a_{n1} & a_{n2} & a_{n3} & \cdots a \end{pmatrix} | a, a_{ij} \in \mathbb{K} \} \quad (14)$$

the depths are given by $d(D_n(\mathbb{K}), T_n(\mathbb{K})) = 3$ and $d(U_n(\mathbb{K}), T_n(\mathbb{K})) = 4$. Both are not dependent on the order n of matrices.

This situation is different for another interesting series of subalgebras within $T_n(\mathbb{K})$ given by

$$J_n(\mathbb{K}) = \{ \begin{pmatrix} a_1 & 0 & 0 & \cdots 0 \\ a_2 & a_1 & 0 & \cdots 0 \\ a_3 & a_2 & a_1 & \cdots 0 \\ \vdots & & & \vdots \\ a_n & a_{n-1} & a_{n-2} & \cdots a_1 \end{pmatrix} | a_1, \ldots, a_n \in \mathbb{K} \} \quad (15)$$

also known as the Jordan algebra. This is isomorphic as algebras to $\mathbb{K}[x]/(x^n)$, a Gorenstein dimension zero local ring. Notice that $U_2(\mathbb{K}) = J_2(\mathbb{K})$, so

$$d(J_2(\mathbb{K}), T_2(\mathbb{K})) = 4.$$

The interesting fact worth mentioning here is that $d(J_3(\mathbb{K}), T_3(\mathbb{K})) \geq 6$. This is based on computations comparing $A \otimes_B A$ and $A \otimes_B A \otimes_B A$ as B-B-bimodules, since a new 2-dimensional indecomposable turns up in the tensor-cube of the ring extension.

The following seems to be an interesting problem not accessible by the techniques of the previous sections:

$$d(J_n(\mathbb{K}), T_n(\mathbb{K})) = ? \tag{16}$$

Acknowledgments The authors would like to thank Sebastian Burciu for visiting Porto in May 2012 and discussing topics related to this paper. Research on this paper was partially funded by the European Regional Development Fund through the programme COMPETE and by the Portuguese Government through the FCT under the project PE-C/MAT/UI0144/2011.

References

1. Assem, I., Simson, D., Skowroński, A.: Elements of the representation theory of associative algebras. Lond. Math. Soc. Stud. Texts **1**, 65 (2006)
2. Auslander, M., Reiten, I., Smalö, S.: Representation theory of artin algebras. Cambridge Stud. Adv. Math. **36** (1995)
3. Boltje, R., Külshammer, B.: On the depth 2 condition for group algebra and Hopf algebra extensions. J. Algebra **323**, 1783–1796 (2010)
4. Boltje, R., Danz, S., Külshammer, B.: On the depth of subgroups and group algebra extensions. J. Algebra **335**, 258–281 (2011)
5. Burciu, S., Kadison, L., Külshammer, B.: On subgroup depth, I.E.J.A. **9**, 133–166 (2011)
6. Kadison, L.: Subring depth, Frobenius extensions and towers. Int. J. Math. Sci. (2012). article 254791
7. Kadison, L., Szlachanyi, K.: Bialgebroid actions on depth two extensions and duality. Adv. in Math. **179**, 75–121 (2003)
8. Pierce, R.S.: Associative algebras, Grad. Texts math, vol. 88. Springer, Berlin (1982)

Further Reading

9. Boltje, R., Külshammer, B.: Group algebra extensions of depth one. Algebra Number Theory **5**, 63–73 (2011)
10. Burciu, S.: Depth one extensions of semisimple algebras and Hopf subalgebras, Alg. Rep. Theory (to appear)
11. Burciu, S., Kadison, L.: Subgroups of depth three. Surv. Diff. Geom. **15**, 17–36 (2011)
12. Danz, S.: The depth of some twisted group algebra extensions. Comm. Algebra **39**, 1–15 (2011)
13. Fritzsche, T.: The depth of subgroups of Psl(2, q). J. Algebra **349**, 217–233 (2011)

14. Kadison, L.: New examples of Frobenius extensions, University Lecture Series 14, Am. Math. Soc. (Providence) **1999**, 63–73 (2011)
15. Kadison, L.: Infinite index subalgebras of depth two. Proc. A.M.S. **136**, 1523–1532 (2008)
16. Kadison, L.: Finite depth and Jacobson-Bourbaki correspondence. J. Pure Appl. Alg. **212**, 1822–1839 (2008)
17. Kadison, L.: Odd H-depth and H-separable extensions. Cent. Eur. J. Math. **10**, 958–968 (2012)

Lie Algebras with Given Properties of Subalgebras and Elements

Pasha Zusmanovich

Abstract We study finite-dimensional Lie algebras with given properties of subalgebras (like all proper subalgebras being abelian) and elements (like all elements being semisimple). We get results on both the structure of the whole class of algebras with the given property, and the structure of individual algebras in the class.

We study the following classes of Lie algebras: anisotropic (i.e., Lie algebras for which each adjoint operator $\mathsf{ad}\, x$ is semisimple), regular (i.e., Lie algebras in which each nonzero element is regular), minimal nonabelian (i.e., nonabelian Lie algebras all whose proper subalgebras are abelian), and algebras of depth 2 (i.e., Lie algebras all whose proper subalgebras are abelian or minimal nonabelian).

All algebras, Lie and associative, are assumed to be finite-dimensional and defined over a fixed field of characteristic zero (though some of the results, in a weaker form or under additional restrictions, will hold also in positive characteristic). We stress that the base field is not assumed to be algebraically closed (all the things considered here are collapsing to vacuous trivialities in the case of an algebraically closed base field).

Our notations are standard and largely follow Bourbaki [2]. The symbols $\dot{+}$, $\oplus$, and $\in\!\!\!+$ denote direct sum of vector spaces, direct sum of Lie algebras, and semidirect sum of Lie algebras (the first summand acting on the second), respectively.

P. Zusmanovich (✉)
Department of Mathematics, Tallinn University of Technology, Ehitajate tee 5, 19086 Tallinn, Estonia
e-mail: pasha.zusmanovich@ttu.ee

P. Zusmanovich
University of São Paulo,São Paulo, Brazil
e-mail: pasha@ime.usp.br

A. Makhlouf et al. (eds.), *Algebra, Geometry and Mathematical Physics*, Springer Proceedings in Mathematics & Statistics 85, DOI: 10.1007/978-3-642-55361-5_7,

1 Anisotropic Algebras

It is shown in [3, Propostion 1.2] that any anisotropic solvable Lie algebra is abelian. From this and the Levi–Malcev decomposition follows that any anisotropic Lie algebra is reductive.

Theorem 1.1 *For a reductive Lie algebra L the following are equivalent:*

(i) *L is anisotropic;*
(ii) *all proper subalgebras of L are anisotropic;*
(iii) *all proper subalgebras of L are reductive;*
(iv) *all 2-dimensional subalgebras of L are abelian;*
(v) *L does not contain subalgebras isomorphic to* $\mathsf{sl}(2)$.

Proof (i) $\Rightarrow$ (ii). If S is a subalgebra of L, then for any $x \in S$, $\mathsf{ad}_S\, x$ is a restriction of $\mathsf{ad}_L\, x$, hence the semisimplicity of the latter implies the semisimplicity of the former.

(ii) $\Rightarrow$ (iii) follows from the observation above that any anisotropic Lie algebra is reductive.

(iii) $\Rightarrow$ (iv) follows from the obvious fact that a 2-dimensional reductive Lie algebra is abelian.

(iv) $\Rightarrow$ (v) follows from the obvious fact that $\mathsf{sl}(2)$ contains a 2-dimensional nonabelian subalgebra.

(v) $\Rightarrow$ (i). Write L as a direct sum $L = \mathfrak{g} \oplus A$, where $\mathfrak{g}$ is semisimple and A is abelian. Suppose $\mathfrak{g}$ is not anisotropic. As $\mathfrak{g}$ contains semisimple and nilpotent components of each of its elements ([2, Chap. I, Sect. 6, Theorem 3]), $\mathfrak{g}$ contains a nonzero nilpotent element, and by the Jacobson–Morozov theorem ([2, Chap. VIII, Sect. 11, Proposition 2]) $\mathfrak{g}$ contains $\mathsf{sl}(2)$ as a subalgebra, a contradiction. Hence $\mathfrak{g}$ is anisotropic and L is anisotropic.

Though the proof is elementary, and all the necessary ingredients are contained in [3] anyway (in particular, the implication (i) $\Rightarrow$ (iv) is noted in [3, Sect. 1], and the equivalence (i) $\Leftrightarrow$ (v) in the case of semisimple L is proved, with a slightly different argument, in [3, Theorem 2.1]), we find this explicit formulation of Theorem 1.1 interesting enough. There are many works in the literature devoted to study of *minimal non-*$\mathscr{P}$ Lie algebras, i.e. Lie algebras not satisfying $\mathscr{P}$ and such that all their proper subalgebras satisfy $\mathscr{P}$, where $\mathscr{P}$ is a certain "natural" property of Lie algebras (abelianity, nilpotency, solvability, simplicity, modularity of the lattice of subalgebras, ...). In all the cases studied so far, the class of minimal non-$\mathscr{P}$ algebras turns out to be highly nontrivial (without further assumptions about the base field, such as algebraic or quadratic closedness, triviality of the Brauer group, etc.), with lot of simple objects. To the contrary, from the Levi–Malcev decomposition and Theorem 1.1 it follows that the class of minimal nonanisotropic Lie algebras is relatively trivial: those are exactly solvable minimal nonabelian Lie algebras. One may ask a "philosophical" question: what makes the condition of being anisotropic different in that regard from other conditions? Where is a borderline for a property $\mathscr{P}$ which makes the class of minimal non-$\mathscr{P}$ Lie algebras small and "simple" (or even empty)?

Corollary 1.1 *A simple Lie algebra all whose proper subalgebras are not simple, is either minimal nonabelian, or isomorphic to* $\mathsf{sl}(2)$.

Proof Let L be a reductive Lie algebra all whose proper subalgebras are not simple. By implication (v) $\Rightarrow$ (iii) of Theorem 1.1, either L is isomorphic to $\mathsf{sl}(2)$, or all proper subalgebras of L are reductive. As any nonabelian reductive Lie algebra contains a simple subalgebra, in the latter case all proper subalgebras of L are abelian.

In [16, Theorem 2.2] a statement similar to the corollary is proved about simple Lie algebras, all whose proper subalgebras are supersolvable.

Theorem 1.2 *Let* $\mathscr{L}$ *be a nonempty class of Lie algebras satisfying the following properties:*

(i) $\mathscr{L}$ *is closed with respect to subalgebras;*
(ii) *if each proper subalgebra of a reductive Lie algebra L belongs to* $\mathscr{L}$*, then L belongs to* $\mathscr{L}$*;*
(iii) *solvable Lie algebras belonging to* $\mathscr{L}$ *are abelian.*

Then $\mathscr{L}$ *is the class of all anisotropic Lie algebras.*

Proof Any class of Lie algebras satisfying conditions (i) and (iii) consists of anisotropic algebras. Indeed, from the Levi–Malcev decomposition and condition (iii) it follows that any algebra in $\mathscr{L}$ is reductive. Then from implication (iii) $\Rightarrow$ (i) of Theorem 1.1 and condition (i), it follows that any algebra in $\mathscr{L}$ is anisotropic.

Now, suppose that there is an anisotropic Lie algebra not belonging to $\mathscr{L}$, and consider such algebra L of the minimal possible dimension. Then all proper subalgebras of L belong to $\mathscr{L}$, and by condition (ii) L itself belongs to $\mathscr{L}$, a contradiction.

2 Regular Algebras

If N is a nilpotent subalgebra of a Lie algebra L, by $L^0(N)$ is denoted the Fitting 0-component with respect to the N-action on L (i.e., the set of all elements of L on which N acts nilpotently).

Recall ([2, Chap. VII, Sect. 2.2]) that *rank* $\operatorname{rk} L$ of a Lie algebra L is the minimal possible non-vanishing power of the characteristic polynomial of $\mathsf{ad}\, x$, $x \in L$, and elements of L for which this minimal number is attained are called *regular*. Another characterization of $x \in L$ to be a regular element is the equality $\dim L^0(x) = \operatorname{rk} L$.

If each nonzero element of L is regular, then L itself is called *regular*.

It is clear that any nilpotent Lie algebra is regular, with rank equal the dimension of the algebra. If a regular Lie algebra L is not semisimple, i.e., contains a nonzero abelian ideal I, then for any $x \in I$, $(\mathsf{ad}\, x)^2 = 0$, hence each element in L is nilpotent, and by the Engel theorem L is nilpotent. It is clear also that a regular semisimple Lie algebra is simple (see [2, Chap. VII, Sect. 2.2, Proposition 7]), and that a regular simple Lie algebra is anisotropic (see [2, Chap. VII, Sect. 2.4, Corollary 2]).

Theorem 2.1 *For a simple Lie algebra L the following are equivalent:*

(i) *L is regular;*
(ii) *all proper subalgebras of L are regular;*
(iii) *all proper subalgebras of L are either simple, or abelian.*

Proof (i) $\Rightarrow$ (ii) follows from the fact that if S is a subalgebra of L, and $x \in S$ is a regular element in L, then x is a regular element in S ([2, Chap. VII, Sect. 2.2, Proposition 9]).

(ii) $\Rightarrow$ (iii). By the observation above, any proper subalgebra of L is either simple, or nilpotent. Hence L does not contain a 2-dimensional nonabelian Lie algebra, and by implication (iv) $\Rightarrow$ (iii) of Theorem 1.1, all proper subalgebras of L are reductive, and all its nilpotent subalgebras are abelian.

(iii) $\Rightarrow$ (i). By implication (iii) $\Rightarrow$ (i) of Theorem 1.1, L is anisotropic. In any Lie algebra, Cartan subalgebras are exactly nilpotent subalgebras N such that $L^0(N) = N$ ([2, Chap. VII, Sect 2.1, Proposition 4]). But nilpotent subalgebras of L are abelian, and $L^0(N)$ coincides with the centralizer of N, so Cartan subalgebras of L are exactly abelian subalgebras coinciding with their own centralizer. For an arbitrary nonzero element $x \in L$, its centralizer $Z_L(x)$ cannot be simple, hence it is abelian. But, obviously, $Z_L(x)$ coincides with its own centralizer, hence $Z_L(x)$ is a Cartan subalgebra of L, $\dim Z_L(x) = \dim L^0(x) = \mathrm{rk}\, L$, and x is regular.

Note that similar to the anisotropic case, minimal nonregular Lie algebras are exactly solvable minimal nonnilpotent Lie algebras.

Theorem 2.2 *Let $\mathscr{L}$ be a nonempty class of Lie algebras satisfying the following properties:*

(i) *$\mathscr{L}$ is closed with respect to subalgebras;*
(ii) *if each proper subalgebra of a Lie algebra L belongs to $\mathscr{L}$, then L belongs to $\mathscr{L}$;*
(iii) *non-semisimple Lie algebras belonging to $\mathscr{L}$ are nilpotent.*

Then $\mathscr{L}$ is the class of all regular Lie algebras.

Proof Any class of Lie algebras satisfying conditions (i) and (iii) consists of regular algebras. Indeed, from the Levi–Malcev decomposition and condition (iii) it follows that any algebra L in $\mathscr{L}$ is either semisimple, or nilpotent. In the former case, write $L = \mathfrak{g}_1 \oplus \dots \oplus \mathfrak{g}_n$ as the direct sum of simple components. If $n > 1$, by condition (i) the subalgebra of L of the form $\mathfrak{g}_1 \oplus Kx$, where x is an arbitrary nonzero element of $\mathfrak{g}_2$, belongs to $\mathscr{L}$, and by condition (iii) it is nilpotent, a contradiction. Hence $n = 1$, that is, L is simple. By conditions (i) and (iii) L does not contain 2-dimensional nonabelian subalgebra, and by implication (iv) $\Rightarrow$ (iii) of Theorem 1.1, all subalgebras of L are reductive. This, together with conditions (i) and (iii) again, implies that all subalgebras of L are either simple, or abelian, and by implication (iii) $\Rightarrow$ (i) of Theorem 2.1, L is regular.

Now, the same elementary reasoning utilizing condition (ii) as at the end of the proof of Theorem 1.2, shows that any regular Lie algebra belongs to $\mathscr{L}$.

3 Minimal Nonabelian Algebras

It follows from the Levi–Malcev decomposition that any minimal nonabelian Lie algebra is either simple, or solvable. Solvable minimal nonabelian Lie algebras (even in a slightly more general minimal nonnilpotent setting) were described in [7], [14], and [15]. A simple minimal nonabelian Lie algebra is regular. Simple minimal nonabelian Lie algebras were studied in [4] and [6], but their full description remains an open problem.

Recall that an algebra is called *central* if its centroid coincides with the base field. For simple algebras this is equivalent to the condition that the algebra remains simple under extension of the base field.

Theorem 3.1 *There are no central simple minimal nonabelian Lie algebras of types* B_l *($l \geq 2$),* C_l *($l \geq 3$, $l \neq 2^k$),* D_l *($l \geq 5$, $l \neq 2^k$),* G_2*, and* F_4.

Proof The proof follows from the known classification of central simple Lie algebras of these types (see, for example, [13, Chap. IV]).

Types B–D. Each central simple Lie algebra of this type (with the exception of D_4) is isomorphic to a Lie algebra of J-skew-symmetric elements $\mathrm{S}^-(A, J) = \{x \in A \mid J(x) = -x\}$, where A is a central simple associative algebra of dimension $n^2 > 16$ with involution J of the first kind (smaller dimensions of A are covered by "occasional" isomorphisms between "small" algebras of different types, including type A). By a known description of such algebras (see, for example, [10, Theorem 5.1.23]), A is isomorphic to $M_m(D)$, a matrix algebra of size $m \times m$ over a central division algebra D with involution j, and J has the form

$$(d_{k\ell})_{k,\ell=1}^m \mapsto \operatorname{diag}(g_1, \dots, g_m)(j(d_{k\ell}))^\top \operatorname{diag}(g_1^{-1}, \dots, g_m^{-1})$$

for some $g_1, \dots, g_m \in D$ such that $j(g_k) = g_k$, $k = 1, \dots, m$.

If D coincides with the base field, i.e. A is a full matrix algebra, than the Lie algebra $\mathrm{S}^-(A, J)$ is split and, obviously, contains a lot of proper nonabelian subalgebras. Hence we may assume $\dim D \geq 4$. From the description above it is clear that, provided $m > 1$, the subalgebra B of A of all matrices with vanishing last row and column, is isomorphic to $M_{m-1}(D)$ and is stable under J, hence $\mathrm{S}^-(B, J)$ is a Lie subalgebra of $\mathrm{S}^-(A, J)$. Since $\dim A = m^2 \dim D \geq 25$, we have $\dim B = (m-1)^2 \dim D = s^2 \geq 9$, and this subalgebra is a central simple Lie algebra of dimension $\frac{s(s-1)}{2}$ or $\frac{s(s+1)}{2}$. Therefore, if $m > 1$, $\mathrm{S}^-(A, J)$ contains proper nonabelian subalgebras, and it remains to consider the case where $A = D$ is a division algebra.

Since D has an involution, its exponent is equal to 2, and its dimension n^2 is equal to some power of 4. This excludes all the types mentioned in the statement of the theorem.

Type G_2. Each central simple Lie algebra of this type is a derivation algebra of a 8-dimensional Cayley algebra $\mathbb{O}$. The latter is obtained by the doubling (Cayley–Dickson) process from the 4-dimensional associative quaternion algebra $\mathbb{H}$, and it is known that each derivation of $\mathbb{H}$ can be extended to a derivation of $\mathbb{O}$ (see, for example [12, Theorem 2]). Thus, $\mathrm{Der}(\mathbb{O})$ always contains a 3-dimensional central simple Lie algebra $\mathrm{Der}(\mathbb{H})$ as a subalgebra, and hence cannot be minimal nonabelian.

Type F_4. Each central simple Lie algebra of this type is a derivation algebra of a 27-dimensional exceptional simple Jordan algebra $\mathbb{J}$. It is known that derivations of $\mathbb{J}$ mapping a cubic subfield of $\mathbb{J}$ to zero form a central simple Lie algebra of type D_4 (see, for example, [9, Chap. IX, Sect. 11, Exercise 5]).

We conjecture that the remaining types not covered by Theorem 3.1—C_{2^k} and D_{2^k}—cannot occur as well.

Conjecture 3.1 *There are no central simple minimal nonabelian Lie algebras of types* B–D *(except of* D_4*).*

Let us provide some evidence in support of this conjecture.

Lemma 3.1 *Let D be a central division algebra of dimension n^2 over a field K with involution J of the first kind, such that $\mathrm{S}^-(D,J)$ is a minimal nonabelian Lie algebra. Then for any J-symmetric or J-skew-symmetric element x in D, not lying in K, one of the following holds:*

(i) x is J-symmetric and of degree 2;
(ii) $K[x]$ is of degree $\frac{n}{2}$, and $\dim_{K[x]} \mathrm{C}_D(x) = 4$;
(iii) $K[x]$ is a maximal subfield of D.

Proof The associative centralizer of x in D, $\mathrm{C}_D(x)$, is a proper simple associative subalgebra of D. By the Double Centralizer Theorem (see, for example, [11, Sect. 12.7]),

$$\dim K[x] \cdot \dim \mathrm{C}_D(x) = n^2, \tag{1}$$

and the associative center of $\mathrm{C}_D(x)$ coincides with $K[x]$.

As $\mathrm{C}_D(x)$ is stable under J, $\mathrm{S}^-(\mathrm{C}_D(x),J)$ is a Lie subalgebra of $\mathrm{S}^-(D,J)$. If it coincides with the whole $\mathrm{S}^-(D,J)$, then $\mathrm{S}^-(D,J) \subseteq \mathrm{C}_D(x)$, and by (1), $\dim K[x] \leq \frac{n^2}{\frac{n(n-1)}{2}} < 3$, hence $\dim K[x] = 2$, i.e. $K[x]$ is a quadratic extension of K, the case (i). Note that in this case x cannot be J-skew-symmetric, as otherwise it lies in the Lie center of $\mathrm{S}^-(D,J)$, a contradiction.

If $\mathrm{S}^-(\mathrm{C}_D(x),J)$ is a proper subalgebra of $\mathrm{S}^-(D,J)$, then it is abelian, and by [8, Theorem 2.2], $\mathrm{C}_D(x)$ is either commutative (i.e., a subfield of D), or is 4-dimensional over its center $K[x]$. In the former case, since the degree (= dimension) over K of each intermediate field between K and D is $\leq n$ (actually, a divisor of n), and since $K[x] \subseteq \mathrm{C}_D(x)$, we have $\dim K[x] = \dim \mathrm{C}_D(x) = n$, and $\mathrm{C}_D(x) = K[x]$, the case (iii). In the latter case, from (1) we have $\dim K[x] = \frac{n}{2}$ and $\dim \mathrm{C}_D(x) = 2n$, the case (ii).

For example, if the division algebra D is cyclic (what always happens over number fields), then, considering the conditions of the lemma simultaneously for a J-skew-symmetric element x generating a cyclic extension of the base field, and even powers of x (which are J-symmetric), one quickly arrives to a contradiction.

For the remaining exceptional types, the question seems to be much more difficult, and it is treated in [5] using the language and technique of algebraic groups and Galois cohomology. There are central simple minimal nonabelian Lie algebras of types D_4 and E_8. For types E_6 and E_7 partial answers are available.

Central simple minimal nonabelian Lie algebras of type A of the form $D^{(-)}/K1$ (i.e., quotient of D, considered as a Lie algebra subject to commutator $[a, b] = ab - ba$, by the 1-dimensional center spanned by the unit 1 of D), where D is a central division associative algebra, were studied in [6]. A necessary, but not sufficient condition for such Lie algebra to be minimal nonabelian is D to be minimal noncommutative (i.e., all proper subalgebras of D are commutative). In this connection the following observation is of interest:

Theorem 3.2 *Let D be a central division associative algebra. Then the Lie algebra $D^{(-)}/K1$ is regular if and only if D is a minimal noncommutative algebra.*

Proof Let the dimension of D over the base field K is equal to n^2, so $\dim D^{(-)}/K1 = n^2 - 1$. The Lie algebra $D^{(-)}/K1$ is regular if and only if the Lie centralizer of any nonzero element $\overline{x} \in D^{(-)}/K1$ is a Cartan subalgebra of dimension $n - 1$, what, in associative terms, is equivalent to the condition that the associative centralizer $\mathrm{C}_D(x)$ of any element $x \in D\backslash K$, is a maximal subfield of D of dimension n over K. Taking this into account, the proof is an easy application of the Double Centralizer Theorem, with reasonings similar to those used in the proof of the lemma above.

The "only if" part. Suppose that for any $x \in D\backslash K$, $\mathrm{C}_D(x)$ is a maximal subfield of D. Consider a subfield $K[x] \subseteq \mathrm{C}_D(x)$ of D. We have $\mathrm{C}_D(x) = \mathrm{C}_D(K[x])$, and by the Double Centralizer Theorem, $\dim K[x] \cdot \dim \mathrm{C}_D(x) = n^2$. But the degree (= dimension) over K of each intermediate field between K and D is $\leq n$ (actually, a divisor of n), hence $\dim K[x] = \dim \mathrm{C}_D(x) = n$, and $\mathrm{C}_D(x) = K[x]$. That means that there are no intermediate fields between K and the maximal subfields of D.

If A is a noncommutative proper subalgebra of D, then, obviously, A is a division algebra. Its center $Z(A)$, being a field extension of K, either coincides with K, or is a maximal subfield of D. In the former case A is central of dimension m^2, where $1 < m < n$, and its maximal subfield has degree m over K, a contradiction. In the latter case, we have $\dim A > \dim Z(A) = n$. Applying again the Double Centralizer Theorem, we have $\dim A \cdot \dim \mathrm{C}_D(A) = n^2$. Since $Z(A) \subseteq \mathrm{C}_D(A)$, we have $\dim \mathrm{C}_D(A) \geq \dim Z(A) = n$, a contradiction.

The "if" part. Suppose D is minimal noncommutative. For an arbitrary $x \in D$ not lying in the base field K, its centralizer $\mathrm{C}_D(x)$ is a subfield of D. By the Double Centralizer Theorem, $\mathrm{C}_D(\mathrm{C}_D(x))$ is a simple subalgebra of D (and, hence, is also a subfield), and $\dim \mathrm{C}_D(x) \cdot \dim \mathrm{C}_D(\mathrm{C}_D(x)) = n^2$. By the same argument as above about degrees of intermediate fields between K and D, $\dim \mathrm{C}_D(x) = \dim \mathrm{C}_D(\mathrm{C}_D(x)) = n$. Since $\mathrm{C}_D(x) \subseteq \mathrm{C}_D(\mathrm{C}_D(x))$, this implies $\mathrm{C}_D(x) = \mathrm{C}_D(\mathrm{C}_D(x))$, and $\mathrm{C}_D(x)$ is a maximal subfield of D.

4 Algebras of Depth 2

Define the *depth* of a Lie algebra in the following inductive way: a Lie algebra has depth 0 if and only if it is abelian, and has depth $n > 0$ if and only if it does not have depth $< n$ and all its proper subalgebras have depth $< n$. Thus, minimal nonabelian Lie algebras are exactly algebras of depth 1.

Many of the algebras considered below arise as semidirect sums $L \inplus V$ of a Lie algebra L and an L-module module V (in such a situation, we will always assume that V is an abelian ideal: $[V, V] = 0$). It is clear that the depth of such semidirect sums is related to depth of L and the maximal length of chains of subspaces of V invariant under action of subalgebras of L, though the exact formulation in the general case seems to be out of reach. In the particular case where L is 1-dimensional, the depth of such semidirect sum is equal to the maximal length of chains in V of invariant subspaces with nontrivial L-action.

The following can be considered as an extension of the corresponding results from [7], [14], and [15].

Theorem 4.1 *A non-simple Lie algebra of depth* 2 *over a field K is isomorphic to one of the following algebras:*

(i) *A* 4*-dimensional solvable Lie algebra having the basis $\{x, y, z, t\}$ and the following multiplication table:*

$$[x, y] = z, \quad [x, z] = 0, \quad [y, z] = 0, \quad [z, t] = 0,$$

with $\mathrm{ad}\, t$ *acting on the space $Kx \dotplus Ky$ invariantly, without nonzero eigenvectors, and with trace zero.*

(ii) *A* 4*-dimensional solvable Lie algebra having the basis $\{x, y, z, t\}$ and the following multiplication table:*

$$[x, y] = z, \quad [x, z] = 0, \quad [y, z] = 0, \quad [z, t] = z,$$

with $\mathrm{ad}\, t$ *acting on the space $Kx \dotplus Ky$ invariantly, without nonzero eigenvectors, and with trace* 1.

(iii) *A direct sum of a simple minimal nonabelian Lie algebra and* 1*-dimensional algebra.*

(iv) *A semidirect sum $S \inplus V$, where S is either the* 2*-dimensional nonabelian Lie algebra, or a* 3*-dimensional simple minimal nonabelian Lie algebra, and V is an S-module such that each nonzero element of S acts on V irreducibly.*

(v) *A semidirect sum $S \inplus V$, where S is an abelian* 1*- or* 2*-dimensional Lie algebra, and V is an S-module such that for each nonzero element $x \in S$, the maximal length of chains of x-invariant subspaces of V is equal to* 2 *(what is equivalent to saying that any proper x-invariant subspace does not contain proper x-invariant subspaces).*

Proof It is a straightforward verification that in each of these cases the corresponding Lie algebras have depth 2, so let us prove that each non-simple Lie algebra L of depth 2 has one of the indicated forms.

Note that L cannot be semisimple. For, in this case it is decomposed into the direct sum of simple components: $L = \mathfrak{g}_1 \oplus \cdots \oplus \mathfrak{g}_n$, $n > 1$, and any subalgebra of the form $\mathfrak{g}_1 \oplus Kx$, $x \in \mathfrak{g}_2$, $x \neq 0$, is not minimal nonabelian.

Suppose that L is non-semisimple and non-solvable, and let $L = \mathfrak{g} \inplus \mathrm{Rad}(L)$ be its Levi–Malcev decomposition. Then $\mathfrak{g}$ is minimal nonabelian and hence is simple. Further, $\mathrm{Rad}(L)$ abelian, as otherwise $\mathfrak{g} \inplus [\mathrm{Rad}(L), \mathrm{Rad}(L)]$ is a proper subalgebra of L which is not minimal nonabelian. Suppose now that $\mathrm{rk}\,\mathfrak{g} > 1$, and $\mathfrak{g}$ acts on $\mathrm{Rad}(L)$ nontrivially. Then taking $x \in \mathfrak{g}$ with a nontrivial action on $\mathrm{Rad}(L)$, and the Cartan subalgebra H of $\mathfrak{g}$ of dimension > 1 containing x, we get a subalgebra $H \inplus \mathrm{Rad}(L)$ of L which is not minimal nonabelian. Hence in the case $\mathrm{rk}\,\mathfrak{g} > 1$, $\mathrm{Rad}(L)$ is a trivial (and then, obviously, 1-dimensional) $\mathfrak{g}$-module, and we arrive at case (iii). If $\mathrm{rk}\,\mathfrak{g} = 1$, then $\mathfrak{g}$ is 3-dimensional. If some nonzero $x \in \mathfrak{g}$ acts on $\mathrm{Rad}(L)$ trivially, then so is $[x, \mathfrak{g}]$, and, since $\mathfrak{g}$ is generated by the latter subspace, the whole $\mathfrak{g}$ acts on $\mathrm{Rad}(L)$ trivially, a case covered by (iii). Assume that any nonzero $x \in \mathfrak{g}$ acts on $\mathrm{Rad}(L)$ nontrivially. The Lie subalgebra $Kx \inplus \mathrm{Rad}(L)$ contains, in its turn, a subalgebra $Kx \inplus V$ for any proper $\mathrm{ad}\,x$-invariant subspace V of $\mathrm{Rad}(L)$, what shows that x acts trivially on V. Letting here V to be the Fitting 1-component with respect to the x-action on $\mathrm{Rad}(L)$, we see that $\mathrm{Rad}(L) = V$, what means that x acts on $\mathrm{Rad}(L)$ nondegenerately, and hence, irreducibly. We arrive at case (iv).

It remains to consider the case of L solvable. Take any subspace A of L of codimension 1 containing $[L, L]$, and a complimentary 1-dimensional subspace:

$$L = Kt \dotplus A, \tag{2}$$

$\mathrm{ad}\,t$ acts on A. Since A is a proper ideal of L, it is either abelian or minimal nonabelian. In the former case, we arrive at the semidirect sum $Kt \inplus A$, and it is easy to see that any proper nonabelian subalgebra of L is isomorphic to the semidirect sum $Kt \inplus V$, where V is a proper $\mathrm{ad}\,t$-invariant subspace of A. Thus, for L to be of depth 2 is equivalent to the condition described in case (v) (with S 1-dimensional).

Suppose now that A is minimal nonabelian. According to [7, Theorem 4] (also implicit in [14] and [15]), each solvable minimal nonabelian Lie algebra is either isomorphic to the 3-dimensional nilpotent Lie algebra, or to the semidirect sum $Kx \inplus V$ such that $\mathrm{ad}\,x$ acts on V irreducibly (in particular, $\mathrm{ad}\,x|_V$ is nondegenerate). Further, $\mathrm{ad}\,t$ is a derivation of A, and subtracting from t an appropriate element of A, we may assume that either t is central, i.e. (2) is the direct sum of A and 1-dimensional algebra, or $\mathrm{ad}\,t$ is an outer derivation of A.

Suppose first that A is 3-dimensional nilpotent, i.e., has a basis $\{x, y, z\}$ with multiplication table $[x, y] = z$, $[x, z] = [y, z] = 0$. If t is central, we arrive at a particular case of (i). Straightforward computation shows that each outer derivation of A is equivalent to a derivation d which acts invariantly on the space $Kx \dotplus Ky$, and either $d|_{Kx \dotplus Ky}$ has trace zero, and $d(z) = 0$, or $d|_{Kx \dotplus Ky}$ has trace 1, and $d(z) = z$. These two cases correspond to the cases (i) and (ii) respectively, with the condition of absence of nonzero eigenvectors to ensure the absence of subalgebras which are not minimal nonabelian.

Suppose now that $A = Kx \inplus V$, $\mathsf{ad}\, x$ acts on V irreducibly. If t is central, $L \simeq Kx \inplus (V \dotplus Kt)$ (with $\mathsf{ad}\, x$ acting on t trivially), a case covered by (v) (with S 1-dimensional). Straightforward computation shows that each outer derivation of A is equivalent to a derivation d which acts on V invariantly, and either $[\mathsf{ad}\, x, d] = 0$ in the Lie algebra $\mathsf{gl}(V)$, and $d(x) = 0$, or $[\mathsf{ad}\, x, d] = \mathsf{ad}\, x$ and $d(x) = x$. These two cases correspond to the cases (v) and (iv) respectively (with S 2-dimensional), with the respective conditions to ensure the absence of subalgebras which are not minimal nonabelian.

Corollary 4.1 (to Theorems 1.1 and 2.1) *A simple Lie algebra of depth* 2 *is either isomorphic to* $\mathsf{sl}(2)$, *or regular.*

Proof It is clear that $\mathsf{sl}(2)$ has depth 2. Hence a simple Lie algebra L of depth 2 is either isomorphic to $\mathsf{sl}(2)$, or does not contain $\mathsf{sl}(2)$ as a proper subalgebra. In the latter case, by implication (v) $\Rightarrow$ (iii) of Theorem 1.1, all subalgebras of L are reductive. But as each minimal nonabelian Lie algebra is either simple, or solvable, all subalgebras of L are either simple, or abelian, and by implication (iii) $\Rightarrow$ (i) of Theorem 2.1, L is regular.

In group theory, a notion analogous to depth in the class of finite p-groups is called $\mathscr{A}_n$*-groups*, see [1, Sect. 65] for their discussion and for a partial description of $\mathscr{A}_2$-groups.

Acknowledgments This work was essentially done more than 20 years ago, while being a student at the Kazakh State University under the guidance of Askar Dzhumadil'daev, and has been reported at a few conferences in the Soviet Union at the end of 1980s. However, I find the results interesting and original enough even today to be put in writing, with some minor additions and modifications. During the final write-up I was supported by grants ERMOS7 (Estonian Science Foundation and Marie Curie Actions) and ETF9038 (Estonian Science Foundation).

References

1. Berkovich, Y., Janko, Z.: Groups of Prime Power Order, Vol. 2, Walter de Gruyter, New York (2008)
2. Bourbaki, N.: Lie Groups and Lie Algebras. Chapters 1–3, Chapters 7–9, Springer, 1989, 2005 (translated from the original French editions: Hermann, 1972, 1975 and Masson, 1982)
3. Farnsteiner, R.: On ad-semisimple Lie algebras. J. Algebra **83**, 510–519 (1983)
4. Farnsteiner, R.: On the structure of simple-semiabelian Lie algebras. Pacific J. Math. **111**, 287–299 (1984)
5. Garibaldi, S., Gille, P.: Algebraic groups with few subgroups. J. London Math. Soc. **80**, 405–430 (2009)
6. Gein, A.G.: Minimal noncommutative and minimal nonabelian algebras. Comm. Algebra **13**, 305–328 (1985)
7. Gein, A.G., Kuznetsov, S.V., Mukhin, Yu.N.: On minimal nonnilpotent Lie algebras, Mat. Zapiski Uralsk. Gos. Univ. 8, $\mathscr{N}$3, 18–27 (1972) (in Russian)

8. Herstein, I.N.: Topics in Ring Theory. The University of Chicago Press, Chicago (1969)
9. Jacobson, N.: Structure and Representations of Jordan Algebras. AMS, Providence (1968)
10. Jacobson, N.: Finite-Dimensional Division Algebras over Fields. Springer, Berlin (1996)
11. Pierce, R.S.: Associative Algebras. Springer, Newyork (1982)
12. Schafer, R.D.: On the algebras formed by the Cayley-Dickson process. Amer. J. Math. **76**, 435–446 (1954)
13. Seligman, G.B.: Modular Lie Algebras. Springer, Newyork (1967)
14. Stitzinger, E.L.: Minimal nonnilpotent solvable Lie algebras. Proc. Amer. Math. Soc. **28**, 47–49 (1971)
15. Towers, D.A.: Lie algebras all whose proper subalgebras are nilpotent. Lin. Algebra Appl. **32**, 61–73 (1980)
16. Towers, D.A.: Minimal non-supersolvable Lie algebras. Algebras Groups Geom. **2**, 1–9 (1985)

Construction of Symplectic Quadratic Lie Algebras from Poisson Algebras

Saïd Benayadi

Abstract We introduce the notion of quadratic (resp. symplectic quadratic) Poisson algebras and we show how one can construct new interesting quadratic (resp. symplectic quadratic) Lie algebras from quadratic (resp. symplectic quadratic) Poisson algebras. Finally, we give inductive descriptions of symplectic quadratic Poisson algebras.

1 Introduction

In this paper, we consider finite dimensional algebras over a commutative field K of characteristic zero.

Recall that the Lie algebra $\mathscr{G}$ of a Lie group $\mathfrak{G}$ which admits a bi-invariant pseudo-Riemannian structure is quadratic (i.e. $\mathscr{G}$ is endowed with a symmetric non degenerate invariant (or associative) bilinear form B). Conversely, any connected Lie group whose Lie algebra $\mathscr{G}$ is quadratic, is endowed with bi-invariant pseudo-Riemannian structure [14]. The semisimple Lie algebras are quadratic. Many solvable Lie algebras are also quadratic. Quadratic Lie algebras appear, in particular, in connection with Lie bialgebras and physical models based on Lie algebras. Recall that quadratic Lie algebras are precisely the Lie algebras for which a Sugawara construction exists [9]. Several papers provided interesting results on the structure of quadratic Lie algebras [4, 5, 8–10, 12, 13].

In [13], Medina and Revoy have introduced the concept of double extension in order to give an inductive description of quadratic Lie algebras. This concept is also a tool to construct a new quadratic Lie algebra from a quadratic Lie algebra $(\mathfrak{g}_1, B_1)$ and a Lie algebra $\mathfrak{g}_2$ (not necessarily quadratic) which acts on $\mathfrak{g}_1$ by skew-symmetric derivations with respect to B_1. Let us remark that the non-trivial new quadratic

S. Benayadi (✉)
IECL, CNRS UMR 7502, Université de Lorraine, Ile du Saulcy, 57045 Metz cedex, France
e-mail: said.benayadi@univ-lorraine.fr

A. Makhlouf et al. (eds.), *Algebra, Geometry and Mathematical Physics*,
Springer Proceedings in Mathematics & Statistics 85, DOI: 10.1007/978-3-642-55361-5_8,

Lie algebra will be obtained if $\mathfrak{g}_2$ acts by non-inner skew-symmetric derivations on $(\mathfrak{g}_1, B_1)$. In general, it is difficult to find a Lie algebra $\mathfrak{g}_2$ of dimension upper or equal to 2. In the first part of this paper, we will show how from quadratic Poisson-admissible algebra $(\mathscr{A}, B)$ we can find a Lie algebra $\mathfrak{g}_2$ of dimension upper or equal to 2 acting on a quadratic Lie algebra $(\mathfrak{g}_1, B_1)$ by non-inner skew-symmetric derivations.

In addition, we introduce the concept of symplectic quadratic Poisson algebra and we show how one constructs interesting symplectic quadratic Lie algebras from symplectic quadratic Poisson algebras. Let us recall that the Lie algebra of a Lie group which admits a bi-invariant pseudo-Riemannian metric and also a left-invariant symplectic form is a symplectic quadratic Lie algebra. These Lie groups are nilpotent and their geometry (and, consequently, that of their associated homogeneous spaces) is very rich. In particular, they carry two left-invariant affine structures: one defined by the symplectic form and another which is compatible with a left-invariant pseudo-Riemannian metric. Moreover, if the symplectic form is viewed as a solution r of the classical Yang Baxter equation of Lie algebras (i.e. r is an r-matrix), then the Poisson-Lie tensor $\pi = r^+ - r^-$ and the geometry of double Lie groups $D(r)$ can be nicely described in [7]. In addition, symplectic quadratic Lie algebras were described by methods of double extensions in [1, 2]. Further, in [2], it is proved that every symplectic quadratic Lie algebra $(\mathscr{G}, B, \omega)$, over an algebraically closed field $\mathbb{K}$, may be constructed by T^*−extension of nilpotent Lie algebra which admits an invertible derivation.

In the last section, we study structures of symplectic quadratic Poisson algebras and we give inductive descriptions of symplectic quadratic Poisson algebras over an algebraically closed field with characteristic zero by using some results of [2, 3].

2 Definitions and Preliminary Results

Definition 2.1 Let $\mathscr{A}$ be a vector space endowed with two bilinear operations $[\ ,\ ]$ and $\circ$. $(\mathscr{A}, [\ ,\ ], \circ)$ is called a Poisson algebra if $(\mathscr{A}, [\ ,\ ])$ is a Lie algebra and $(\mathscr{A}, \circ)$ is a commutative associative algbra (not necessarily unital) such that

$$[a, b \circ c] = [a, b] \circ c + b \circ [a, c], \quad \forall a, b, c, \in \mathscr{A} \quad \text{(Leibniz rule)}.$$

Definition 2.2 Let $\mathscr{A}$ be an algebra, we denote by . its multiplication. On the underlying vector space of $\mathscr{A}$ one can defined the two following new products:

$$[x, y] := x.y - y.x; \quad x \circ y := \frac{1}{2}(x.y + y.x), \quad \forall x, y \in \mathscr{A}.$$

$\mathscr{A}$ (or .) is called Poisson-admissible if $(\mathscr{A}, [\ ,\ ], \circ)$ is a Poisson algebra.

We denote by $\mathscr{A}^-$ (resp. $\mathscr{A}^+$) the algebra $(\mathscr{A}, [\ ,\ ])$ (resp.$(\mathscr{A}, \circ)$).

Definition 2.3 1. Let $(\mathscr{A}, .)$ be an algebra and $B : \mathscr{A} \times \mathscr{A} \to \mathbb{K}$ be a bilinear form. We say that B is associative (or invariant) if

$$B(a.b, c) = B(a, b.c), \ \forall a, b, c \in \mathscr{A}.$$

2. Let $(\mathfrak{g}, [\ ,\ ])$ be a Lie algebra and $B : \mathfrak{g} \times \mathfrak{g} \to \mathbb{K}$ be a bilinear form. $(\mathfrak{g}, B)$ is called a quadratic Lie algebra if B is symmetric, non-degenerate and invariant. In this case, B is called an invariant scalar product on $\mathfrak{g}$.
3. Let $(\mathscr{A}, \circ)$ be an associative algebra and $B : \mathscr{A} \times \mathscr{A} \to \mathbb{K}$ be a bilinear form. $(\mathscr{A}, B)$ is called symmetric algebra if B is symmetric, non-degenerate and associative. In this case, B is called an invariant scalar product on $\mathscr{A}$.
4. Let $(\mathscr{A}, .)$ be a Poisson-admissible algebra and $B : \mathscr{A} \times \mathscr{A} \to \mathbb{K}$ be a bilinear form. $(\mathscr{A}, B)$ will be called quadratic if B is symmetric, non-degenerate and associative. In this case, B is called an invariant scalar product on $\mathscr{A}$.
5. Let $(\mathscr{A}, [\ ,\ ], \circ)$ be a Poisson algebra and $B : \mathscr{A} \times \mathscr{A} \to \mathbb{K}$ be a bilinear form. $(\mathscr{A}, B)$ will be called quadratic if B is symmetric, non-degenerate such that:-

$$B([a, b], c) = B(a, [b, c]) \text{ and } B(a \circ b, c) = B(a, b \circ c), \ \forall a, b, c \in \mathscr{A}.$$

Remark 2.1 1. Let $(\mathscr{A}, .)$ be a Poisson-admissible algebra and $B : \mathscr{A} \times \mathscr{A} \to \mathbb{K}$ be a bilinear form. It is clear that $(\mathscr{A}, B)$ is quadratic if and only if $(\mathscr{A}^-, B)$ is a quadratic Lie algebra and $(\mathscr{A}^+, B)$ is a symmetric algebra.
2. $(\mathscr{A}, ., B)$ is a quadratic Poisson-admissible algebra if and only if $(\mathscr{A}, [\ ,\ ], \circ, B)$ is a quadratic Poisson algebra (where $[x, y] := x.y - y.x$ and $x \circ y := \frac{1}{2}(x.y + y.x), \forall x, y \in \mathscr{A}$).

Now, we are going to give some examples of quadratic Poisson-admissible (or Poisson) algebras

1. Let $(\mathscr{A}, [\ ,\ ], \circ)$ be a Poisson algebra and $\mathscr{A}^*$ is the dual vector space of underlying vector space of $\mathscr{A}$. An easy computation prove that the following bracket $[\ ,\ ]_\sim$ and multiplication $\star$ define a Poisson algebra structure on the vector space $\mathscr{A} \oplus \mathscr{A}^*$:

$$[x + f, y + h]_\sim : = [x, y] - h \circ \mathrm{ad}x + f \circ \mathrm{ad}y;$$

$$(x + f) \star (y + h) : = x \circ y + h \circ L_x + f \circ L_y, \ \forall (x, f), (y, h) \in \mathscr{A} \times \mathscr{A}^*,$$

where L_x is the left multiplication by x in the algebra $(\mathscr{A}, \circ)$.

Moreover, if we consider the bilinear form $B : (\mathscr{A} \oplus \mathscr{A}^*) \times (\mathscr{A} \oplus \mathscr{A}^*) \to \mathbb{K}$ defined by:

$$B(x + f, y + h) : = f(y) + h(x), \ \forall (x, f), (y, h) \in \mathscr{A} \times \mathscr{A}^*,$$

then $(\mathscr{A} \oplus \mathscr{A}^*, B)$ is a quadratic Poisson algebra.

Let us remak that if $(\mathscr{A}, .)$ is a Poisson-admissible algebra, then the following multiplication $\bowtie$ on the vector space $\mathscr{A} \oplus \mathscr{A}^*$ define a Poisson-admissible structure on $\mathscr{A} \oplus \mathscr{A}^*$:

$$(x + f) \bowtie (y + h) := x.y + h \circ R_x + f \circ L_y, \ \forall (x, f), (y, h) \in \mathscr{A} \times \mathscr{A}^*,$$

where L_y (resp. R_x) is the left (resp. right) multiplication by y (resp. y) in the algebra $(\mathscr{A}, .)$.

In addition, the bilinear form $B : (\mathscr{A} \oplus \mathscr{A}^*) \times (\mathscr{A} \oplus \mathscr{A}^*) \to \mathbb{K}$ defined by:

$$B(x + f, y + h) := f(y) + h(x), \ \forall (x, f), (y, h) \in \mathscr{A} \times \mathscr{A}^*,$$

is an invariant scalar product on $\mathscr{A} \oplus \mathscr{A}^*$ (ie. $(\mathscr{A}, B)$ is a quadratic Poisson-admissible algebra).

2. Let $(\mathscr{A}, ., B)$ be a quadratic Poisson-admissible algebra and $(\mathfrak{H}, \star, \varphi)$ be a symmetric commutative algebra.

The commutativity and the associativity of $\star$ imply that the vector space $\mathscr{A} \otimes \mathfrak{H}$ with the multiplication:

$$(a \otimes x) \bullet (b \otimes y) := a.b \otimes x \star y, \ \forall (a, x), (b, y) \in \mathscr{A} \times \mathfrak{H},$$

is the Poisson-admissible algebra.

Moreover, the bilinear form $B \otimes \varphi : (\mathscr{A} \otimes \mathfrak{H}) \times (\mathscr{A} \otimes \mathfrak{H}) \to \mathbb{K}$ defined by:

$$B \otimes \varphi(a \otimes x, b \otimes y) := B(a, b)\varphi(x, y), , \ \forall (a, x), (b, y) \in \mathscr{A} \times \mathfrak{H},$$

define a quadratic structure on the Poisson-admissible algebra $(\mathscr{A} \otimes \mathfrak{H}, \bullet)$.

Definition 2.4 Let $(\mathscr{A}, .)$ be an algebra and $\omega : \mathscr{A} \times \mathscr{A} \to \mathbb{K}$ be a bilinear form. We say that $(\mathscr{A}, \omega)$ is a symplectic algebra (or ω is a symplectic structure on $(\mathscr{A}, .)$) if:

1. $\omega(x, y) = -\omega(y, x) \ \forall x, y \in \mathscr{A}$, (ie. ω is skew-symmetric);
2. ω is non-degenerate;
3. $\omega(x.y, z) + \omega(y.z, x) + \omega(z.x, y) = 0, \forall x, y, z \in \mathscr{A}$.

Definition 2.5 If $(\mathscr{A}, .)$ is an algebra, B an associative scalar product on $\mathscr{A}$ and ω is a symplectic structure on $\mathscr{A}$, we say that $(\mathscr{A}, B, \omega)$ is a symplectic quadratic algebra.

If $(\mathscr{A}, .)$ is an associative algebra, we can also say that $(\mathscr{A}, B, \omega)$ is a symplectic symmetric algebra.

Proposition 2.1 *If $(\mathscr{A}, B)$ is a quadratic algebra, ω is a symplectic structure on $\mathscr{A}$ if and only if there exists a unique skew-symmetric (with respect to B) invertible derivation of $(\mathscr{A}, ., B)$ such that:*

$$\omega(x, y) = B(D(x), y), \ \forall x, y \in \mathscr{A}.$$

Proof It is straightforward calculation considering $\omega(x, y) = B(D(x), y)$, for all $x, y \in \mathscr{A}$.

We finish this section by showing how to construct symplectic quadratic Poisson-admissible algebras from an arbitrary Poisson-admissible algebras.

Let $(\mathscr{P}, .)$ be a Poisson-admissible algebra. Let $\mathscr{O} := X\mathbb{K}[X]$ be the ideal of $\mathbb{K}[X]$ generated by X and $\mathscr{R} := \mathscr{O}/X^n\mathscr{O}$, where $n \in \mathbb{N}^*$. $\mathscr{R}$ is a commutative and associative algebra and $\{\bar{X}, \bar{X}^2, \ldots, \bar{X}^n\}$ is a basis of the underlying vector space of $\mathscr{R}$. The vector space $\tilde{\mathscr{P}} := \mathscr{P} \otimes \mathscr{R}$ endowed with the multiplication defined by:

$$(x \otimes \bar{P}) \bullet (y \otimes \bar{Q}) := x.y \otimes \bar{PQ}, \ \forall x, y \in \mathscr{P}, \forall P, Q \in \mathscr{O},$$

is a nilpotent Poisson-admissible algebra. Next, $(\mathscr{A} := \tilde{\mathscr{P}} \oplus \tilde{\mathscr{P}}^*, \bowtie, B)$ is a quadratic Poisson-admissible algebra, where:

$$(x + f) \bowtie (y + h) := x \bullet y + h \circ R_x + f \circ L_y,$$

and

$$B(x + f, y + h) := f(y) + h(x), \ \forall (x, f), (y, h) \in \tilde{\mathscr{P}} \times \tilde{\mathscr{P}}^*.$$

Now, let us consider the endomorphism D of $\tilde{\mathscr{P}}$ defined by:

$$D(x \otimes \bar{X}^i) := ix \otimes \bar{X}^i, \ \forall x \in \mathscr{P}, \forall i \in \{1, \ldots, n\},$$

is an invertible derivation of $\tilde{\mathscr{P}}$. It easy to verify that the endomorphism $\tilde{D}$ of $\mathscr{A}$ defined by:

$$\tilde{D}(x + f) := D(x) - f \circ D, \ \forall (x, f) \in \tilde{\mathscr{P}} \times \tilde{\mathscr{P}}^*,$$

is an invertible derivation of $\mathscr{A}$ which is skew-symmetric with respect to B. Consequently, the bilinear form ω on $\mathscr{A}$ defined by:

$$\omega(x + f, y + h) := B(\tilde{D}(x + f), y + h), \ \forall (x, f), (y, h) \in \tilde{\mathscr{P}} \times \tilde{\mathscr{P}}^*,$$

is a symplectic structure on $\mathscr{A}$. Then, $(\mathscr{A}, B, \omega)$ is symplectic quadratic Poisson-admissible algebra.

3 Construction of Quadratic (Resp. Symplectic Quadratic) Lie Algebras from Quadratic (Resp. Symplectic Quadratic) Poisson-Admissible Algebras

First, let us recall the concept of the double extension in the case of quadratic Lie algebras.

Let $(\mathfrak{g}_1, [\ ,\ ]_1, B_1)$ be a quadratic Lie algebra and $(\mathfrak{g}_2, [\ ,\ ]_2)$ be a Lie algebra which is not necessarily quadratic such that there exists a morphism of Lie algebras $\varphi : \mathfrak{g}_2 \to \mathrm{Der}_a(\mathfrak{g}_1, B_1)$ where $\mathrm{Der}_a(\mathfrak{g}_1, B_1)$ is the set of the skew-symmetric derivations with respect to B_1, this set is a Lie subalgebra of $\mathfrak{gl}(\mathfrak{g}_1)$. Since $\varphi(\mathfrak{g}_2) \subseteq \mathrm{Der}_a(\mathfrak{g}_1, B_1)$, then the bilinear map $\psi : \mathfrak{g}_1 \times \mathfrak{g}_1 \to (\mathfrak{g}_2)^*$ is a $2-$cocycle where $(\mathfrak{g}_2)^*$ is considered as a trivial $\mathfrak{g}_1-$module. Consequently, the vector space $\mathfrak{g}_1 \oplus (\mathfrak{g}_2)^*$ endowed with the multiplication:

$$[X_1 + f, Y_1 + h]_c := [X_1, Y_1]_1 + \psi(X_1, Y_1),\ \forall X_1, Y_1 \in \mathfrak{g}_1, f, h \in (\mathfrak{g}_2)^*,$$

is a Lie algebra. This Lie algebra is the central extension of $\mathfrak{g}_1$ by means of ψ.

Let π be the co-adjoint representation of $\mathfrak{g}_2$. If $X_2 \in \mathfrak{g}_2$, an easy computation prove that the endomorphism $\bar{\varphi}(X_2)$ of $\mathfrak{g}_1 \oplus (\mathfrak{g}_2)^*$ defined by: $\bar{\varphi}(X_2)(X_1 + f) := \varphi(X_2)(X_1) + \pi(X_2)(f), \forall X_1 \in \mathfrak{g}_1, f \in (\mathfrak{g}_2)^*$, is a derivation of Lie algebra $(\mathfrak{g}_1 \oplus (\mathfrak{g}_2)^*, [\ ,\ ]_c)$. Next, it is easy to see that the linear map $\bar{\varphi} : \mathfrak{g}_2 \to \mathrm{Der}(\mathfrak{g}_1 \oplus (\mathfrak{g}_2)^*)$ is a morphism of Lie algebras. Therefore, one can consider $\mathfrak{g} := \mathfrak{g}_2 \ltimes_{\bar{\varphi}} (\mathfrak{g}_1 \oplus (\mathfrak{g}_2)^*)$ the semi-direct product of $\mathfrak{g}_1 \oplus (\mathfrak{g}_2)^*$ by $\mathfrak{g}_2$ by means of $\bar{\varphi}$. As vector space $\mathfrak{g} = \mathfrak{g}_2 \oplus \mathfrak{g}_1 \oplus (\mathfrak{g}_2)^*$ and the bracket of the Lie algebra $\mathfrak{g}$ is gien by:

$$\begin{aligned}[X_2 + X_1 + f, Y_2 + Y_1 + h] = {} & [X_2, Y_2]_2 + \Big([X_1, Y_1]_1 + \varphi(X_2)(Y_1) - \varphi(Y_2)(X_1)\Big) \\ & + \Big(\pi(X_2)(h) - \pi(Y_2)(f) + \psi(X_1, Y_1)\Big),\end{aligned}$$

$\forall (X_2, X_1, f), (Y_2, Y_1, h) \in \mathfrak{g}_2 \times \mathfrak{g}_1 \times (\mathfrak{g}_2)^*$. Moreover, if $\gamma : \mathfrak{g}_2 \times \mathfrak{g}_2 \to \mathbb{K}$ is an invariant, symmetric bilinear form on $\mathfrak{g}_2$, it is easy to see that the bilinear form $B_\gamma : \mathfrak{g} \times \mathfrak{g} \to \mathbb{K}$ defined by:

$$B_\gamma(X_2 + X_1 + f, Y_2 + Y_1 + h) := \gamma(X_2, Y_2) + B(X_1, Y_1) + f(Y_2) + h(X_2),$$

$\forall (X_2, X_1, f), (Y_2, Y_1, h) \in \mathfrak{g}_2 \times \mathfrak{g}_1 \times (\mathfrak{g}_2)^*$, is an invariant scalar product on $\mathfrak{g}$. $\mathfrak{g}$ (or $(\mathfrak{g}, B_0)$) is called the double extension of $(\mathfrak{g}_1, [\ ,\ ]_1, B_1)$ by $\mathfrak{g}_2$ by means of φ.

Now, we are going to construct quadratic Lie algebras from quadratic Poisson-admissible algebras by using this concept of double extension.

Let $(\mathscr{A}, ., B)$ be a quadratic Poisson-admissible algebra. Then, $(\mathscr{A}^-, [., .], B)$ is a quadratic Lie algebra and $(\mathscr{A}^+, \circ, B)$ is a symmetric commutative algebra. Let us consider the three-dimensional Lie algebra $\mathfrak{sl}(2)$. Recall that there exists a basis

$\{H, E, F\}$ of $\mathfrak{sl}(2)$ such that $[H, E] = E, [H, F] = -F, [E, F] = 2H$. The vector space $\mathfrak{sl}(2) \otimes \mathscr{A}^+$ with the bracket $[.,.]_1$ defined by:

$$[x \otimes a, y \otimes b]_1 := [x, y] \otimes a \circ b, \ \forall (x, a), (y, b) \in \mathfrak{sl}(2) \times \mathscr{A},$$

is a Lie algebra. Moreover, the bilinear form $B_1 : (\mathfrak{sl}(2)\otimes\mathscr{A}^+)\times(\mathfrak{sl}(2)\otimes\mathscr{A}^+) \to \mathbb{K}$ defined by:

$$B_1(x \otimes a, y \otimes b) := \mathscr{K}(x, y)B(a, b), \ \forall (x, a), (y, b) \in \mathfrak{sl}(2) \times \mathscr{A},$$

is an invariant scalar product on the Lie algebra $(\mathfrak{sl}(2) \otimes \mathscr{A}^+, [.,.]_1)$ (ie. $(\mathfrak{sl}(2) \oplus \mathscr{A}^+, [.,.]_1, B_1)$ is a quadratic Lie algebra) where $\mathscr{K}$ is the Killing form of $\mathfrak{sl}(2)$.

It is cleat that if D is a derivation of $(\mathscr{A}^+, \circ)$, then the endomorphism $\bar{D} := \mathrm{id}_{\mathfrak{sl}(2)} \otimes D$ of $\mathfrak{sl}(2) \otimes \mathscr{A}^+$ defined by:

$$\bar{D}(x \otimes a) := x \otimes D(a), \ \forall (x, a) \in \mathfrak{sl}(2) \times \mathscr{A},$$

is a derivation of the Lie algebra $(\mathfrak{sl}(2) \otimes \mathscr{A}^+, [.,.]_1)$. In addition, if D is skew-symmetric with respect to B, then $\bar{D}$ is skew-symmetric with respect to B_1. In fact, let $(x, a), (y, b)$ be two elements of $\mathfrak{sl}(2) \times \mathscr{A}$,

$$\begin{aligned} B_1(\bar{D}(x \otimes a), y \otimes b) &= \mathscr{K}(x, y)B(D(a), b) \\ &= -\mathscr{K}(x, y)B(a, D(b)) = -B_1(x \otimes a, \bar{D}(y \otimes b)). \end{aligned}$$

Claim $\bar{D}$ is an inner derivation of the Lie algebra $(\mathfrak{sl}(2) \otimes \mathscr{A}^+, [.,.]_1)$ if and only if $D = 0$.

Proof of claim Let us suppose that the derivation $\bar{D}$ is inner, then

$$\bar{D} = \mathrm{ad}(H \otimes a_1) + \mathrm{ad}(E \otimes a_2) + \mathrm{ad}(F \otimes a_3),$$

where $a_1, a_2, a_3 \in \mathscr{A}$. Let $a \in \mathscr{A}$, then $H \otimes D(a) = -E \otimes a \circ a_2 + F \otimes a \circ a_3$, so $D(a) = 0$. We conclude that $D = 0$.

Since $(\mathscr{A}, .)$ is a Poisson-amissible algebra, then for all $X \in \mathscr{A}$ we have $\delta_X := \mathrm{ad}_{\mathscr{A}^-} X$ is a derivation of $(\mathscr{A}^+, \circ)$ and in addition this derivation is skew-symmetric with respect to B because B is associative. Therefore for all $X \in \mathscr{A}$, $\bar{\delta}_X$ is a skew-symmetric derivation of $(\mathfrak{sl}(2) \otimes \mathscr{A}^+, [.,.]_1, B_1)$ and $\bar{\delta}_X$ is not inner if $\mathrm{ad}_{\mathscr{A}^-} X \neq 0$ (ie. X is not in the center of $\mathscr{A}^-$). Then we can consider $\mathfrak{g}(\mathscr{A}) := \mathscr{A}^- \oplus (\mathfrak{sl}(2) \otimes \mathscr{A}^+) \oplus (\mathscr{A}^-)^*$ the double extension of $(\mathfrak{sl}(2) \otimes \mathscr{A}^+, [.,.]_1, B_1)$ by the Lie algebra $\mathscr{A}^-$ by means the morphism of Lie algebra $\varphi : \mathscr{A}^- \to \mathrm{Der}_a(\mathfrak{sl}(2) \otimes \mathscr{A}^+, B_1)$ defined by: $\varphi(X) := \bar{\delta}_X, \ \forall X \in \mathscr{A}$. Let us remark that the dimension of this quadratic Lie algebra obtained by double extension is $5n$ where n is the dimension of $\mathscr{A}$. Recall that the bilinear form $T_0 : \mathfrak{g}(\mathscr{A}) \times \mathfrak{g}(\mathscr{A}) \to \mathbb{K}$ defined by:

$$T_0(X + s \otimes a + f, Y + s' \otimes b + h) := \mathscr{K}(s, s')B(a, b) + f(Y) + h(X),$$

for all $X, Y, a, b \in \mathscr{A}$, $f, h \in \mathscr{A}^*$, is an invariant scalar product on $\mathfrak{g}(\mathscr{A})$.

Remark 3.1 In the construction above, one can replace $\mathfrak{sl}(2)$ by an arbitrary simple Lie algebra.

In [2], symplectic quadratic Lie algebras are studied. Now, we are going to show how we can construct symplectic quadratic Lie algebras from symplectic quadratic Poisson-admissible algebras.

By easy computation, we prove the following lemma.

Lemma 3.1 *If D is a derivation of a quadratic Poisson-admissible algebra $(\mathscr{A}, ., B)$, then the endomorphism $\tilde{D}$ of $\mathfrak{g}(\mathscr{A})$ defined by:*

$$\tilde{D}(x) := D(x),\ \tilde{D}(f) := -f \circ D;\ \tilde{D}(s \otimes a) := s \otimes D(a),$$
$$\forall a, x \in \mathscr{A}, f \in \mathscr{A}^*, s \in \mathfrak{sl}(2),$$

is a derivation of Lie algebra $\mathfrak{g}(\mathscr{A})$. Moreover, if D is invertible (resp. skew-symmetric with respect to B), then $\tilde{D}$ is invertible (resp. skew-symmetric with respect to T_0).

Consequently, if $(\mathscr{A}, B, \omega)$ is a symplectic quadratic Poisson-admissible algebras and D the skew-symmetric (with respect to B) invertible derivation of $\mathscr{A}$ such that $\omega(x, y) = B(D(x), y)$, $\forall x, y \in \mathscr{A}$, then $(\mathfrak{g}(\mathscr{A}), T, \Omega)$ is a symplectic quadratic Lie algebra where Ω is the symplectic structure on Lie algebra $\mathfrak{g}(\mathscr{A})$ defined by:

$$\Omega(X, Y) := T(\tilde{D}(X), Y),\ \forall X, Y \in \mathfrak{g}(\mathscr{A}).$$

3.1 Structures of Symplectic Quadratic Poisson-Admissible Algebras

Recall that if $(\mathscr{A}, \circ, B)$ is a commutative associative algebra, we denote by $\mathrm{End}_s(\mathscr{A}, B)$ the set of symmetric endomorphisms of the vector space $\mathscr{A}$ with respect to B. It is clear that $\mathrm{End}_s(\mathscr{A}, B)$ is a subalgebra of $\mathrm{End}(\mathscr{A})$ (the associative algebra of the endomorphisms of $\mathscr{A}$).

Let $(\delta, a_0) \in \mathrm{End}_s(\mathscr{A}, B) \times \mathscr{A}$. In [3], (δ, a_0) is called a pre-admissible element of $\mathrm{End}_s(\mathscr{A}, B) \times \mathscr{A}$ if

$$\delta \circ L_x = L_x \circ \delta \text{ and } \delta^2 = L_{a_0} \text{ (ie. } \delta(x \circ y) = x \circ \delta(y),\ \delta^2(x) = a_0 \circ x,\ \forall x, y \in \mathscr{A}).$$

Now, Let $(\mathscr{W}, ., B)$ be a quadratic Poisson-admissible algebra. Let $\Delta \in \mathrm{Der}_a(\mathscr{W}^-, B)$ and (δ, a_0) be a pre-admissible element of $\mathrm{End}_s(\mathscr{W}^+, B) \times \mathscr{W}^+$. Then,

1. The vector space $\mathscr{A} := \mathbb{K}e \oplus \mathscr{W} \oplus \mathbb{K}e^*$ with the skew-symmetric bilinear map $[\,,\,] : \mathscr{A} \times \mathscr{A} \to \mathscr{A}$ defined by:

$$[x, y] := [x, y]_{\mathscr{W}^-} + B(\Delta(x), y)e^*; \quad [e, x] := \Delta(x); \quad [e^*, \mathscr{A}] = \{0\},$$

$\forall x, y \in \mathscr{W}$, is a Lie algebra.
2. The vector space $\mathscr{A} := \mathbb{K}e \oplus \mathscr{W} \oplus \mathbb{K}e^*$ with the symmetric bilinear map $\star :$ $\mathscr{A} \times \mathscr{A} \to \mathscr{A}$ defined by:

$$x \star y := x \circ y + B(\delta(x), y)e^*; \quad e \star x := \delta(x) + B(a_0, x)e^*;$$
$$e \star e := a_0 + ke^*; \quad e^* \star \mathscr{A} = \{0\},$$

$\forall x, y \in \mathscr{W}$, is an associative commutative algebra.

Moreover, if we consider the symmetric bilinear form $T : \mathscr{A} \times \mathscr{A} \to \mathbb{K}$ defined by:

$$T_{|\mathscr{W}\times\mathscr{W}} := B; \; T(e, e^*) = 1; \; T(e, \mathbb{K}e \oplus \mathscr{W}) = T(e^*, \mathbb{K}e^* \oplus \mathscr{W}) = \{0\},$$

then $(\mathscr{A}, [\,,\,], T)$ is a quadratic Lie algebra (called the double extension of $(\mathscr{W}^-,$ $[\,,\,]_{\mathscr{W}^-}, B)$ by the one-dimensional Lie algebra by means of D (see [13])) and $(\mathscr{A}, \star, T)$ is a symmetric commutative associative algebra (called generalized double extension of $(\mathscr{W}^+, \circ, B)$ by the one dimensional algebra with null product by means of (δ, a_0) (see [3])).

In addition, if we suppose that

$$\Delta \circ \delta = \delta \circ \Delta = \frac{1}{2}\mathrm{ad}_{\mathscr{W}^-}(a_0) \text{ (ie. } = \frac{1}{2}[a_0, .]_{\mathscr{W}^-}); \; \Delta(a_0) = 0;$$

$$\Delta \in \mathrm{Der}(\mathscr{W}^+); \quad \delta([x, y]_{\mathscr{W}^-}) = \Delta(x) \circ y + [x, \delta(y)]_{\mathscr{W}^-}, \; \forall x, y \in \mathscr{W},$$

then $(\mathscr{A}, [\,,\,], \star)$ is a Poisson algebra, so $(\mathscr{A}, [\,,\,], \star, T)$ is a quadratic Poisson algebra. We call this quadratic Poisson algebra the double extension of the quadratic Poisson algebra $(\mathscr{W}, [\,,\,]_{\mathscr{W}^-}, \circ, B)$ by means of (Δ, δ, a_0, k).

Now, the vector space $\mathscr{A}$ with the product:

$$X \boxtimes Y := \frac{1}{2}[X, Y] + X \star Y, \; \forall \; X, Y \in \mathscr{A},$$

is Poisson-admissible algebra. Then $(\mathscr{A}, \boxtimes, T)$ is a quadratic Poisson-admissible algebra called the double extension of the quadratic Poisson-admissible algebra $(\mathscr{W}, ., B)$ by means of (Δ, δ, a_0, k).

More precisely, the product $\boxtimes$ is given by:

$$x \boxtimes y = x.y + B(\Omega(x), y)e^*,$$

$$e \boxtimes x = \Omega(x) + B(a_0, x)e^*, \;\; x \boxtimes e = \Omega^*(x) + B(a_0, x)e^*,$$

$$e \boxtimes e = e \star e := a_0 + ke^*;\; e^* \star \mathscr{A} = \mathscr{A} \star e^* = \{0\}, \forall x, y \in \mathscr{W},$$

where $\Omega := \frac{1}{2}\Delta + \delta$ and $\Omega^* := -\frac{1}{2}\Delta + \delta$.

Let us consider a symplectic quadratic Poisson-admissible algebra $(\mathscr{W}, ., B, \omega)$. Then there exists a unique invertible skew-symmetric derivation of $\mathscr{W}$ such that $\omega(x, y) = B(D(x), y)$, $\forall x, y \in \mathscr{W}$. Next, we consider a double extension $(\mathscr{A}, \boxtimes, T)$ of $(\mathscr{W}, ., B)$ by means (Δ, δ, a_0, k).

By [2, 3], if there exist $ł \in \mathbb{K}$ and $c_0 \in \mathscr{W}$ such that:

$$[D, \Delta] + ł\Delta = \mathrm{ad}_{\mathscr{W}^-}(c_0),$$

$$[D, \delta] + ł\delta = L_{c_0} \;\; and \;\; \delta(c_0) = ła_0 + \frac{1}{2}D(a_0),$$

Then the endomorphism Γ of $\mathscr{A}$ defined by:

$$\Gamma(x) := D(x) - B(c_0, x)e^*;\quad \Gamma(e^*) := łe^*;\quad \Gamma(e) := c_0 - łe^*,$$

is an invertible derivation of $(\mathscr{A}, \boxtimes)$ and Γ is skew-symmetric with respect to T. It follows that the skew-symmetric bilinear form $\diamondsuit : \mathscr{A} \times \mathscr{A} \to \mathbb{K}$ defined by:

$$\diamondsuit(X, Y) := T(\Gamma(X), Y),\;\; \forall X, Y \in \mathscr{A},$$

is a symplectic structure on $(\mathscr{A}, \boxtimes)$. Therefore, $(\mathscr{A}, \boxtimes, T, \diamondsuit)$ is a symplectic quadratic Poisson-admissible algebra called the double extension of $(\mathscr{W}, ., B, \omega)$ (by means of $(\Delta, \delta, a_0, c_0, k, ł)$).

Proposition 3.1 *Let $(\mathscr{A}, \boxtimes, T)$ be a quadratic Poisson-admissible algebra. Suppose that there exists $e^* \in Ann(\mathscr{A})\backslash\{0\}$ such that $B(e^*, e^*) = 0$. Then, $(\mathscr{A}, \boxtimes, T)$ is a double extension of a quadratic Poisson-admissible algebra $(\mathscr{W}, ., B)$. More precisely, $\mathscr{W} := (\mathbb{K}e^*)^\perp / \mathbb{K}e^*$ and*

$$(x + \mathbb{K}e^*).(y + \mathbb{K}e^*) := (x \boxtimes y) + \mathbb{K}e^*,$$

$$B(x + \mathbb{K}e^*, (y + \mathbb{K}e^*) := T(x, y),\;\; \forall x, y \in (\mathbb{K}e^*)^\perp.$$

Proof Since B is non-degenerate, then there exists $e \in \mathscr{A}$ such that $B(e^*, e) = 1$. Consequently, if $\mathscr{W} := (\mathbb{K}e^* \oplus \mathbb{K}e)^\perp$ denotes the orthogonal of $\mathbb{K}e^* \oplus \mathbb{K}e$ with respect to T, with the Poisson-admissible structure . induced by the one of $(\mathbb{K}e^*)^\perp / \mathbb{K}e^*$, one easily verifies that $B := T_{|\mathscr{W} \times \mathscr{W}}$ is an invariant scalar product on Poisson-admissible algebra $(\mathscr{W}, .)$

Let us remark that there exit $a_0 \in \mathscr{W}$ and $k \in \mathbb{K}$ such that $e \boxtimes e = a_0 + ke^*$ because $B(e \boxtimes e, e^*) = B(e, e \boxtimes e^*) = 0$,

Let us consider Ω the endomorphism of $\mathscr{W}$ defined by:

$$\Omega(x) := P_{\mathscr{W}}(e \boxtimes x), \ \forall x \in \mathscr{W},$$

where $P_{\mathscr{W}}: \mathscr{A} \to \mathscr{W}$ is the natural projection.

Next, we consider $\Delta := \Omega - \Omega^*$, $\delta := \frac{1}{2}(\Omega + \Omega^*)$, where Ω^* the endomorphism of $\mathscr{W}$ defined by $B(\Omega(x), y) = B(x, \Omega^*(y))$, for all $x, y \in \mathscr{W}$ (ie. Ω^* is the adjoint of Ω with respect to B). It easy to verify that $(\mathscr{A}, \boxtimes, T)$ is the double extension of the quadratic Poisson-admissible algebra $(\mathscr{W}, ., B)$ by means of (Δ, δ, a_0, k).

Lemma 3.2 *Let $(\mathscr{A}, .)$ be a Poisson-admissible algebra. If $\mathscr{A}$ admits an invertible derivation, then $Ann(\mathscr{A}) \neq \{0\}$.*

Proof If $(\mathscr{A}, .)$ admits an invertible derivation Γ then Γ is either an invertible derivation of $(\mathscr{A}^-, [\,,\,])$ and an invertible derivation of $(\mathscr{A}^+, \circ)$. Consequently, by [11] (rep. by [3]), $(\mathscr{A}^-, [\,,\,])$ is a nilpotent Lie algebra (resp. $(\mathscr{A}^+, \circ)$ is a nilpotent associative algebra). It follows that $\mathfrak{z}(\mathscr{A}^-) \neq \{0\}$ and $\mathrm{Ann}(\mathscr{A}^+) \neq \{0\}$. Since $\mathrm{ad}_{\mathscr{A}^-} X$ is a derivation of $(\mathscr{A}^+, \circ)$, for all $X \in \mathscr{A}$, then $\mathrm{Ann}(\mathscr{A}^+)$ is an ideal of $(\mathscr{A}^-, [\,,\,])$. Therefore $\mathrm{Ann}(\mathscr{A}^+) \cap \mathfrak{z}(\mathscr{A}^-) \neq \{0\}$ because $(\mathscr{A}^-, [\,,\,])$ is a nilpotent Lie algebra. The fact that $\mathrm{Ann}(\mathscr{A}^+) \cap \mathfrak{z}(\mathscr{A}^-) \subseteq \mathrm{Ann}(\mathscr{A})$ implies that $\mathrm{Ann}(\mathscr{A}) \neq \{0\}$.

Theorem 3.1 *If $\mathbb{K}$ is algebraically closed, then every non-zero symplectic quadratic Poisson-admissible algebra $(\mathscr{A}, \boxtimes, T, \diamondsuit)$ is a double extension of a symplectic quadratic Poisson-admissible algebra $(\mathscr{W}, ., B, \omega)$*

Proof Let $(\mathscr{A}, \boxtimes, T, \diamondsuit)$ is a non-zero symplectic quadratic Poisson-admissible. There exists a unique skew-symmetric (with respect to T) invertible derivation Γ of $(\mathscr{A}, \boxtimes)$ such that $\diamondsuit(X, Y) = T(\Gamma(X), Y)$, for all $X, Y \in \mathscr{A}$. Then, By Lemma 3.2, $\mathrm{Ann}(\mathscr{A}) \neq \{0\}$. Since Γ is a derivation of $(\mathscr{A}, \boxtimes)$, then $\Gamma(\mathrm{Ann}(\mathscr{A})) \subseteq \mathrm{Ann}(\mathscr{A})$. It follows that there exist $e^* \in \mathrm{Ann}(\mathscr{A}) \backslash \{0\}$ and $\text{ł} \in \mathbb{K} \backslash \{0\}$ such that $\Gamma(e^*) = \text{ł} e^*$. The fact that Γ is skew-symmetric with respect to T implies that $T(e^*, e^*) = 0$. By, Proposition 3.1, $(\mathscr{A}, \boxtimes, T)$ is a double extension of a quadratic Poisson-admissible algebra $(\mathscr{W} = \mathbb{K}e^*)^{\perp}/\mathbb{K}e^*, ., B)$ by means of (Δ, δ, a_0, k) (see the proof of Proposition 3.1 for definitions of Δ, δ, a_0 and k). Therefore, $\mathscr{A} = \mathbb{K}e^* \oplus \mathscr{W} \oplus \mathbb{K}e$ with $T(e, e^*) = 1$ and $\mathscr{W} = (\mathbb{K}e^* \oplus \mathbb{K}e)^{\perp}$.

Since the ideal $\mathbb{K}e^*$ of $(\mathscr{A}, \boxtimes)$ is invariant by the skew-symmetric derivation Γ, so is its orthogonal (with respect to T) $\mathbb{K}e^* \oplus \mathscr{W}$. Now, if $p : \mathbb{K}e^* \oplus \mathscr{W} \to \mathscr{W}$ denotes the projection $p(te^* + x) := x$, for $t \in \mathbb{K}, x \in \mathscr{W}$, then one can easily verify that $D := p \circ \Gamma_{|\mathscr{W}}$ is an invertible skew-symmetric derivation of $(\mathscr{W}, ., B)$. Since Γ is skew-symmetric with respect to T and $T(e^*, e) = 1$, one immediately obtains that there exists $c_0 \in \mathscr{W}$ such that $\Gamma(e) := c_0 - \text{ł} e^*$ and $\Gamma_{|\mathscr{W}} = D - B(c_0, .)e^*$. Since Γ is a derivation of $(\mathscr{A}, \boxtimes)$, then Γ is either a derivation of $\mathscr{A}^-$ and a derivation of $\mathscr{A}^+$, one easily deduces that

$$[D, \Delta] + \text{ł}\Delta = \mathrm{ad}_{\mathscr{W}^-}(c_0),$$

$$[D, \delta] + \text{ł}\delta = L_{c_0} \ \ and \ \ \delta(c_0) = \text{ł} a_0 + \frac{1}{2} D(a_0).$$

Therefore $(\mathscr{A}, \boxtimes, T, \diamond)$ is the double extension of $(\mathscr{W}, ., B, \omega := B(D(.), .))$ (by means of $(\Delta, \delta, a_0, c_0, k, \text{ł}))$.

Now, we denote by $\mathscr{M}$ the 2−dimensional Poisson-admissible algebra with zero product. If $\{e, f\}$ is a basis of the vector space $\mathscr{M}$, the symmetric (resp. skew-symmetric) bilinear form B_0 (resp. ω_0) of $\mathscr{M}$ defined by $B_0(e, e) = B_0(f, f) = 1$, $B_0(e, f) = 0$ (resp. $\omega_0(e, f) = 1$), is quadratic (resp. symplectic) structure on $\mathscr{M}$. By Theorem 3.1, The following result follows easily:

Corollary 3.1 *If $\mathbb{K}$ is algebraically closed, then every non-zero symplectic quadratic Poisson-admissible algebra $(\mathscr{A}, \boxtimes, T, \diamond)$ is obtained from the 2−dimensional symplectic quadratic Poisson-admissible algebra $(\mathscr{M}, B_0, \omega_0)$ by a sequence of double extensions.*

References

1. Aubert, A.: Structures affines et pseudo-métriques invariantes à gauche sur des groupes de Lie. Thesis, Université Montpellier II (1996)
2. Bajo, I., Benayadi, S., Medina, A.: Symplectic structures on quadratic Lie algebras. J. Algebra **316**(1), 174–188 (2007)
3. Baklouti, A., Benayadi, S.: Symmetric symplectic associative commutative algebras and related Lie algebras. Algebra colloq. **18**(1), 973–986 (2011)
4. Benayadi, S.: Structures de certaines algèbres de Lie quadratiques. Commun. Algebra **23**(10), 3867–3887 (1995)
5. Bordemann, M.: Nondegenerate invariant bilinear forms on nonassociative algebras. Acta Math. Univ. Com. **LXVIP**(2), 151–201 (1997)
6. Dardié, J.-M., Medina, A.: Double extension symplectique d'un group de Lie symplectique. Adv. Math. **117**(2), 208–227 (1996)
7. Diatta, A., Medina, A.: Classical Yang-Baxter equation and left invariant affine geometry on Lie groups. Manuscr. Math. **114**(4), 477–486 (2004)
8. Favre, G., Santaroubane, L.: Symmetric invariant non-degenerate bilinear form on a Lie algebra. J. Algebra **105**, 451–464 (1987)
9. Figueroa-O'Farrill, J.M., Stanciu, S.: On the structure of symmetric self-dual Lie algebras. J. Math. Phys. **37**(8), 4121–4134 (1996)
10. Hofmann, K.H., Keith, V.S.: Invariant quadratic forms on finite dimensional Lie algebras. Bull. Austral. Math. Soc. **33**, 21–36 (1986)
11. Jacobson, N.: A note on automorphisms and derivations of Lie algebras. Proc. Amer. Math. Soc. **6**, 33–39 (1955)
12. Lu, C.: Finite-dimensional solvable Lie algebras with nondegenerate invariant bilinear forms. J. Algebra **311**, 178–201 (2007)
13. Medina, A., Revoy, Ph.: Algèbres de Lie et produit scalaire invariant. Ann. Scient. Ec. Norm. Sup. 4ème série **18**, 556–561 (1985)
14. O'Neill B.: Semi-riemannian geometry with applications to relativity. Academic Press, New York (1983)

Structure and Cohomology of 3-Lie Algebras Induced by Lie Algebras

Joakim Arnlind, Abdennour Kitouni, Abdenacer Makhlouf and Sergei Silvestrov

Abstract The aim of this paper is to compare the structure and the cohomology spaces of Lie algebras and induced 3-Lie algebras.

1 Introduction

Lie algebras have held a very important place in mathematics and physics for a long time. Ternary Lie algebras appeared first in Nambu's generalization of Hamiltonian mechanics [11] which uses a generalization of Poisson algebras with a ternary bracket. The algebraic formulation is due to Takhtajan. The structure of n-Lie algebra was studied by Filippov [8] and Kasymov [10].

The Lie algebra cohomology complex is well known under the name of Chevalley-Eilenberg cohomology complex. The cohomology of n-Lie algebras was first introduced by Takhtajan [13] in its simplest form, later a complex adapted to the study of formal deformations was introduced by Gautheron [9], then reformulated by Daletskii and Takhtajan [5] using the notion of base Leibniz algebra of a n-Lie algebra.

J. Arnlind
Department of Mathematics, Linköping University, 581 83 Linköping, Sweden
e-mail: joakim.arnlind@liu.se

A. Kitouni (✉) · A. Makhlouf
Université de Haute-Alsace, 4 rue des Frères Lumière, 68093 Mulhouse, France
e-mail: abdennour.kitouni@uha.fr

A. Makhlouf
e-mail: abdenacer.makhlouf@uha.fr

S. Silvestrov
Mälardalens Högskola, Box 883, 721 23 Västerås, Sweden
e-mail: sergei.silvestrov@mdh.se

A. Makhlouf et al. (eds.), *Algebra, Geometry and Mathematical Physics*,
Springer Proceedings in Mathematics & Statistics 85, DOI: 10.1007/978-3-642-55361-5_9,

In [3], the authors introduce a realization of the quantum Nambu bracket in terms of matrices (using the commutator and the trace of matrices). This construction is generalized in [1] to the case of any Lie algebra where the commutator is replaced by the Lie bracket, and the matrix trace is replaced by linear forms having similar properties, which we call 3-Lie algebras induced by Lie algebras. This construction is generalized to the case of n-Lie algebras in [2]. We investigate in this paper the connections between the structure properties (solvability, nilpotency,...) and the cohomology of a Lie algebra and its induced 3-Lie algebra.

The paper is organized as follows: in Sect. 2 we recall main definitions and results concerning n-Lie algebras, and construction of $(n+1)$-Lie algebras induced by n-Lie algebras. In Sect. 3 we study some structural properties of 3-Lie algebras induced by Lie algebras, in particular: common subalgebras and ideals, solvability and nilpotency. In Sect. 4, we recall the cohomology complexes for Lie algebras and 3-Lie algebras, then we study relations between 1 and 2 cocycles of a Lie algebra and the induced 3-Lie algebra. In Sect. 5, we give definitions of central extensions of Lie algebras and n-Lie algebras, then we study the relation between central extension of a Lie algebra and those of a 3-Lie algebra it induces. In Sect. 6 we give a method to recognize 3-Lie algebras that are induced by some Lie algebra, and applying it, we determine all 3-Lie algebras induced by Lie algebras up to dimension 5, based on classifications given in [4, 8], then we give a list of Lie algebras up to dimension 4 and all the possible induced 3-Lie algebras. In Sect. 7 we present computations on 4 chosen Lie algebras together with a trace map. We compare the set of 1-cocycles and 1 coboundaries of the Lie algebras and the induced 3-Lie algebras using the computer algebra software Mathematica; the algorithm is briefly explained there too.

2 n-Lie Algebras

In this paper, all considered vector spaces are over a field $\mathbb{K}$ of characteristic 0. n-Lie algebras were introduced in [8], then deeper investigated in [10]. Let us recall some basic definitions.

Definition 2.1 A n-Lie algebra $(A, [\cdot, \ldots, \cdot])$ is a vector space together with a skew-symmetric n-linear map $[\cdot, \ldots, \cdot] : A^n \to A$ such that:

$$\big[x_1, \ldots, x_{n-1}, [y_1, \ldots, y_n]\big] = \sum_{i=1}^{n} \big[y_1, \ldots, \big[x_1, \ldots, x_{n-1}, y_i\big], \ldots, y_n\big]. \quad (1)$$

for all $x_1, \ldots, x_{n-1}, y_1, \ldots, y_n \in A$. This condition is called the fundamental identity or Filippov identity. For $n = 2$, the identity (1) becomes the Jacobi identity and we get the definition of a Lie algebra.

Definition 2.2 Let $(A, [\cdot, \ldots, \cdot])$ be a n-Lie algebra, and I a subspace of A. We say that I is an ideal of A if, for all $i \in I, x_1, \ldots, x_{n-1} \in A$, it holds that $\left[i, x_1, \ldots, x_{n-1}\right] \in I$.

Lemma 2.1 *Let $(A, [\cdot, \ldots, \cdot])$ be a n-Lie algebra, and $I_1, \ldots, I_n$ be ideals of A, then $I = [I_1, \ldots, I_n]$ is an ideal of A.*

Definition 2.3 Let $(A, [\cdot, \ldots, \cdot])$ be a n-Lie algebra, and I an ideal of A. Define the derived series of I by:

$$D^0(I) = I \text{ and } D^{p+1}(I) = \left[D^p(I), \ldots, D^p(I)\right].$$

and the central descending series of I by:

$$C^0(I) = I \text{ and } C^{p+1}(I) = \left[C^p(I), I, \ldots, I\right].$$

Definition 2.4 Let $(A, [\cdot, \ldots, \cdot])$ be a n-Lie algebra, and I an ideal of A. The ideal I is said to be solvable if there exists $p \in \mathbb{N}$ such that $D^p(I) = \{0\}$. It is said to be nilpotent if there exists $p \in \mathbb{N}$ such that $C^p(I) = \{0\}$.

Definition 2.5 A n-Lie algebra $(A, [\cdot, \ldots, \cdot])$ is said to be simple if $D^1(A) \neq \{0\}$ and if it has no ideals other than $\{0\}$ and A. A direct sum of simple n-Lie algebras is said to be semi-simple.

In [1, 2] a construction of a 3-Lie algebra from a Lie algebra, and more generally a $(n+1)$-Lie algebra from a n-Lie algebra was introduced. We recall the main definitions and results.

Definition 2.6 Let $\phi : A^n \to A$ be a n-linear map and τ be a linear map from A to $\mathbb{K}$. Define $\phi_\tau : A^{n+1} \to A$ by:

$$\phi_\tau(x_1, \ldots, x_{n+1}) = \sum_{k=1}^{n+1} (-1)^k \tau(x_k) \phi(x_1, \ldots, \hat{x}_k, \ldots, x_{n+1}), \tag{2}$$

where the hat over $\hat{x}_k$ on the right hand side means that x_k is excluded, that is ϕ is calculated on $(x_1, \ldots, x_{k-1}, x_{k+1}, \ldots, x_{n+1})$.

We will not be concerned with just any linear map τ, but rather maps that have a generalized trace property. Namely:

Definition 2.7 For $\phi : A^n \to A$ we call a linear map $\tau : A \to \mathbb{K}$ a *ϕ-trace (or trace)* if $\tau(\phi(x_1, \ldots, x_n)) = 0$ for all $x_1, \ldots, x_n \in A$.

Lemma 2.2 *Let $\phi : A^n \to A$ be a skew-symmetric n-linear map and τ a linear map $A \to \mathbb{K}$. Then ϕ_τ is a $(n+1)$-linear totally skew-symmetric map. Furthermore, if τ is a ϕ-trace then τ is a ϕ_τ-trace.*

Theorem 2.1 *Let (A, ϕ) be a n-Lie algebra and τ a ϕ-trace, then (A, ϕ_τ) is a $(n+1)$-Lie algebra. We shall say that (A, ϕ_τ) is induced by (A, ϕ). In particular, let $(A, [.,.])$ be a Lie algebra and $\tau : A \to \mathbb{K}$ be a trace map, the ternary bracket $[.,.,.]$ given by: $[x, y, z] = \circlearrowleft_{x,y,z} \tau(x)[y, z]$ defines a 3-Lie algebra; we refer to A when considering the Lie algebra and A_τ when considering induced 3-Lie algebra.*

3 Structure of 3-Lie Algebras Induced by Lie Algebras

Let $(A, [.,.])$ be a Lie algebra, τ a $[.,.]$-trace and $(A, [.,.,.]_\tau)$ the induced 3-Lie algebra.

Proposition 3.1 *If B is a subalgebra of $(A, [.,.])$ then B is also a subalgebra of $(A, [.,.,.]_\tau)$.*

Proof Let B be a subalgebra of $(A, [.,.])$ and $x, y, z \in B$:

$$[x, y, z]_\tau = \tau(x)[y, z] + \tau(y)[z, x] + \tau(z)[x, y],$$

which is a linear combination of elements of B and then belongs to B. □

Proposition 3.2 *Let J be an ideal of $(A, [.,.])$. Then J is an ideal of $(A, [.,.,.]_\tau)$ if and only if:*

$$[A, A] \subseteq J \text{ or } J \subseteq \ker \tau.$$

Proof Let J be an ideal of $(A, [.,.])$, and let $j \in J$ and $x, y \in A$, then we have:

$$[x, y, j]_\tau = \tau(x)[y, j] + \tau(y)[j, x] + \tau(j)[x, y].$$

We have that $\tau(x)[y, j] + \tau(y)[j, x] \in J$, then, to have $[x, y, j]_\tau \in J$ it is necessary and sufficient to have $\tau(j)[x, y] \in J$, which is equivalent to $\tau(j) = 0$ or $[x, y] \in J$. □

Theorem 3.1 *Let $(A, [.,.])$ be a Lie algebra, τ a $[.,.]$-trace and $(A, [.,.,.]_\tau)$ the induced 3-Lie algebra. The 3-Lie algebra $(A, [.,.,.]_\tau)$ is solvable, more precisely $D^2(A_\tau) = 0$ i.e. $\left(D^1(A_\tau) = [A, A, A]_\tau, [.,.,.]_\tau\right)$ is abelian.*

Proof Let $x, y, z \in [A, A, A]_\tau$, $x = [x_1, x_2, x_3]_\tau$, $y = [y_1, y_2, y_3]_\tau$ and $z = [z_1, z_2, z_3]_\tau$, then:

$$\begin{aligned}[x, y, z]_\tau &= \tau([x_1, x_2, x_3]_\tau)\left[[y_1, y_2, y_3]_\tau, [z_1, z_2, z_3]_\tau\right] \\ &\quad + \tau\left([y_1, y_2, y_3]_\tau\right)\left[[z_1, z_2, z_3]_\tau, [x_1, x_2, x_3]_\tau\right] \\ &\quad + \tau([z_1, z_2, z_3]_\tau)\left[[x_1, x_2, x_3]_\tau, [y_1, y_2, y_3]_\tau\right] \\ &= 0.\end{aligned}$$

Because $\tau\left([.,.,.]\right)=0$. □

Remark 3.1 ([8]) Let $(A,[.,.,.])$ be a 3-Lie algebra. If we fix $a\in A$, the bracket

$$[.,.]_a=[a,.,.]$$

is skew-symmetric and satisfies Jacobi identity. Indeed, we have, for $x,y,z\in A$:

$$\begin{aligned}\left[x,[y,z]_a\right]_a&=[a,x,[a,y,z]]\\&=[[a,x,a],y,z]+[a,[a,x,y],z]+[a,y,[a,x,z]]\\&=[a,[a,x,y],z]+[a,y,[a,x,z]]\\&=\left[[x,y]_a,z\right]_a+\left[x,[y,z]_a\right]_a.\end{aligned}$$

Proposition 3.3 *Let $(A,[.,.])$ be a Lie algebra, τ be a trace and $(A,[.,.,.]_\tau)$ the induced algebra, let $(C^p(A))$ be the central descending series of $(A,[.,.])$, and $(C^p(A_\tau))$ be the central descending series of $(A,[.,.,.]_\tau)$. Then we have:*

$$C^p(A_\tau)\subset C^p(A),\forall p\in\mathbb{N}.$$

If there exists $i\in A$ such that $[i,x,y]_\tau=[x,y],\forall x,y\in A$ then:

$$C^p(A_\tau)=C^p(A),\forall p\in\mathbb{N}.$$

Proof We proceed by induction over $p\in\mathbb{N}$. The case of $p=0$ is trivial, for $p=1$ we have:

$$\forall x=[a,b,c]_\tau\in C^1(A_\tau),x=\tau(a)\,[b,c]+\tau(b)\,[c,a]+\tau(c)\,[a,b],$$

which is a linear combination of element of $C^1(A)$ and then is an element of $C^1(A)$. Suppose now that there exists $i\in A$ such that $[i,x,y]_\tau=[x,y],\forall x,y\in A$, then for $x=[a,b]\in C^1(A)$, $x=[i,a,b]_\tau$ and then is an element of $C^1(A_\tau)$.

Now, we suppose this proposition is true for some $p\in\mathbb{N}$, and let $x\in C^{p+1}(A_\tau)$, then $x=[a,u,v]_\tau$ with $u,v\in A$ and $a\in C^p(A_\tau)$

$$x=[a,u,v]_\tau=\tau(u)\,[v,a]+\tau(v)\,[a,u]\qquad(\tau(a)=0)$$

which is an element of $C^{p+1}(A)$ because $a\in C^p(A_\tau)\subset C^p(A)$. If there exists $i\in A$ such that $[i,x,y]_\tau=[x,y],\forall x,y\in A$ then, if $x\in C^{p+1}(A)$ then $x=[a,u]$ with $a\in C^p(A)$ and $u\in A$ and we have:

$$x=[a,u]=[i,a,u]_\tau=[a,u,i]_\tau\in C^{p+1}(A_\tau).$$ □

Remark 3.2 It also results from the preceding proposition that $D^1(A_\tau)=[A,A,A]_\tau\subset D^1(A)=[A,A]$, and that if there exists $i\in A$ such that $[i,x,y]_\tau=$

$[x, y], \forall x, y \in A$, then $D^1(A_\tau) = D^1(A)$. For the rest of the derived series, we have obviously the first inclusion by Theorem 3.1.

Theorem 3.2 *Let* $(A, [.,.])$ *be a Lie algebra,* τ *be a trace and* $(A, [.,.,.]_\tau)$ *the induced 3-Lie algebra, then we have:*

$$(A, [.,.]) \text{ is nilpotent of class } p \implies (A, [.,.,.]_\tau) \text{ is nilpotent of class at most } p.$$

Moreover, if there exists $i \in A$ *such that* $[i, x, y]_\tau = [x, y], \forall x, y \in A$ *then:*

$$(A, [.,.]) \text{ is nilpotent of class } p \iff (A, [.,.,.]_\tau) \text{ is nilpotent of class } p.$$

Proof
1. Suppose that $(A, [.,.])$ is nilpotent of class $p \in \mathbb{N}$, then $C^p(A) = \{0\}$. By the preceding proposition, $C^p(A_\tau) \subseteq C^p(A) = \{0\}$, therefore $(A, [.,.,.]_\tau)$ is nilpotent of class at most p.
2. We suppose now that $(A, [.,.,.]_\tau)$ is nilpotent of class $p \in \mathbb{N}$, and that there exists $i \in A$ such that $[i, x, y]_\tau = [x, y], \forall x, y \in A$, then $C^p(A_\tau) = \{0\}$. By the preceding proposition, $C^p(A) = C^p(A_\tau) = \{0\}$. Therefore $(A, [.,.])$ is nilpotent, since $C^{p-1}(A) = C^{p-1}(A_\tau) \neq \{0\}$, $(A, [.,.,.]_\tau)$ and $(A, [.,.])$ have the same nilpotency class.

□

4 Lie and 3-Lie Algebras Cohomology

In this section, we study the connections between the Chevalley-Eilenberg cohomology for Lie algebras and the cohomology of 3-Lie algebras induced by Lie algebras.

Now, let us recall the main definitions of Lie algebra and n-Lie algebra cohomology, for reference and further details, see [5, 6, 9, 13].

Definition 4.1 Let $(A, [.,.])$ be a Lie algebra, ρ a representation of A in a vector space M. A M-valued p-cochains on A is a skew-symmetric p-linear map $\varphi : A^p \to M$, the set of M-valued p-cochains is denoted by $C^p(A, M)$.

The coboundary operator is the linear map $\delta^p : C^p(A, M) \to C^{p+1}(A, M)$ given by:

$$\delta^p \varphi(x_1, \dots, x_{p+1}) = \sum_{j=1}^{p+1} (-1)^{j+1} \rho(x_k) \varphi(x_1, \dots, \hat{x}_j, \dots, x_{p+1}) + \sum_{j=1}^{p+1} \sum_{k=j+1}^{p+1} (-1)^{j+k} \varphi([x_j, x_k], x_1, \dots, \hat{x}_j, \dots, \hat{x}_k, \dots, x_{p+1}).$$

We will study two particular cases, the adjoint cohomology $M = A$, $\rho = ad$ and the scalar cohomology $M = \mathbb{K}$, $\rho = 0$.

Definition 4.2 Let $(A, [., ., .])$ be a 3-Lie algebra. An A-valued p-cochain is a linear map $\psi : (\wedge^2 A)^{\otimes p-1} \wedge A \to A$.

Definition 4.3 The coboundary operator for the adjoint action is given by:

$$\begin{aligned} d^p \psi(x_1, \ldots, x_{2p+1}) = & \sum_{j=1}^{p} \sum_{k=2j+1}^{2p+1} (-1)^j \psi(x_1, \ldots, \hat{x}_{j-1}, \hat{x}_j, \ldots ad_{a_j} x_k, \ldots, x_{2p+1}) \\ & + \sum_{k=1}^{p} (-1)^{k-1} ad_{a_k} \psi(x_1, \ldots, \hat{x}_{2k-1}, \hat{x}_{2k}, \ldots, x_{2p+1}) \\ & + (-1)^{p+1} \left[x_{2p-1}, \psi(x_1, \ldots, x_{2p-2}, x_{2p}), x_{2p+1}\right] \\ & + (-1)^{p+1} \left[\psi(x_1, \ldots, x_{2p-1}, x_{2p}, x_{2p+1}\right], \end{aligned}$$

where $a_k = (x_{2k-1}, x_{2k})$.

Definition 4.4 Let $(A, [., ., .])$ be a 3-Lie algebra. A $\mathbb{K}$-valued p-cochain is a linear map $\psi : (\wedge^2 A)^{\otimes p-1} \wedge A \to \mathbb{K}$.

Definition 4.5 The coboundary operator for the trivial action is given by:

$$d^p \psi(x_1, \ldots, x_{2p+1}) = \sum_{j=1}^{p} \sum_{k=2j+1}^{2p+1} (-1)^j \psi(x_1, \ldots, \hat{x}_{j-1}, \hat{x}_j, \ldots ad_{a_j} x_k, \ldots, x_{2p+1}),$$

where $a_k = (x_{2k-1}, x_{2k})$.

The elements of $Z^p(A, M) = \ker \delta^p$ are called p-cocycles, those of $B^n(A, M) = \operatorname{Im} \delta^{p-1}$ are called coboundaries. $H^p(A, M) = \frac{Z^p(A,M)}{B^n(A,M)}$ is the p-th cohomology group. We sometimes add in subscript the representation used in the cohomology complex, for example $Z^p_{ad}(A, A)$ denotes the set of p-cocycle for the adjoint cohomology and $Z^p_0(A, \mathbb{K})$ denotes the set of p-cocycle for the scalar cohomology.

In particular, the elements of $Z^1(A, A)$ are the derivations. Recall that a derivation of a n-Lie algebra is a linear map $f : A \to A$ satisfying:

$$f([x_1, \ldots, x_n]) = \sum_{i=1}^{n} [x_1, \ldots, f(x_i), \ldots, x_n], \forall x_1, \ldots, x_n \in A.$$

4.1 Derivations and 2-Cocycles Correspondence

Let $(A, [., .])$ be a Lie algebra, τ a $[., .]$-trace and $(A, [., ., .]_\tau)$ the induced 3-Lie algebra. Then we have the following correspondence between 1 and 2-cocycles of $(A, [., .])$ and those of $(A, [., ., .]_\tau)$.

Lemma 4.1 *Let $f : A \to A$ be a Lie algebra derivation, then $\tau \circ f$ is a $[.,.]$-trace.*

Proof for all $x, y \in A$, we have:

$$\tau\left(f\left([x, y]\right)\right) = \tau\left([f(x), y] + [x, f(y)]\right) = \tau\left([f(x), y]\right) + \tau\left([x, f(y)]\right) = 0.$$

□

Theorem 4.1 *Let $f : A \to A$ be a derivation of the Lie algebra A, then f is a derivation of the induced 3-Lie algebra if and only if:*

$$[x, y, z]_{\tau \circ f} = 0, \forall x, y, z \in A.$$

Proof Let f be a derivation of A and $x, y, z \in A$:

$$\begin{aligned} f\left([x, y, z]_\tau\right) &= \tau(x) f([y, z]) + \tau(y) f([z, x]) + \tau(z) f([x, y]) \\ &= \tau(x)[f(y), z] + \tau(y)[f(z), x] + \tau(z)[f(x), y] \\ &\quad + \tau(x)[y, f(z)] + \tau(y)[z, f(x)] + \tau(z)[x, f(y)] \\ &\quad + \tau(f(x))[y, z] + \tau(f(y))[z, x] + \tau(f(z))[x, y] \\ &\quad - \tau(f(x))[y, z] + \tau(f(y))[z, x] + \tau(f(z))[x, y] \\ &= [f(x), y, z]_\tau + [x, f(y), z]_\tau + [x, y, f(z)]_\tau - [x, y, z]_{\tau \circ f}. \end{aligned}$$

□

Theorem 4.2 *Let $\varphi \in Z^2_{ad}(A, A)$ and $\omega : A \to \mathbb{K}$ be a linear map satisfying:*

1. $\tau(x)\omega(y) = \tau(y)\omega(x)$,
2. $\omega([x, y]) = 0$,
3. $\circlearrowleft_{x,y,z} \omega(x)\, \tau\left(\varphi(y, z)\right) = 0.$

Then $\psi(x, y, z) = \circlearrowleft_{x,y,z} \omega(x)\, \varphi(y, z)$ is a 2-cocycle of the induced 3-Lie algebra.

Proof Let $\varphi \in Z^2_{ad}(A, A)$ and $\omega : A \to \mathbb{K}$ a linear map satisfying conditions 1, 2 and 3 above, and let $\psi(x, y, z) = \circlearrowleft_{x,y,z} \omega(x)\, \varphi(y, z)$, then we have:

$$\begin{aligned} d^2\psi(x_1, x_2, y_1, y_2, z) &= \psi\left(x_1, x_2, [y_1, y_2, z]_\tau\right) - \psi\left([x_1, x_2, y_1]_\tau, y_2, z\right) \\ &\quad - \psi\left(y_1, [x_1, x_2, y_2]_\tau, z\right) - \psi\left(y_1, y_2, [x_1, x_2, z]_\tau\right) + [x_1, x_2, \psi(y_1, y_2, z)]_\tau \\ &\quad - [\psi(x_1, x_2, y_1), y_2, z]_\tau - [y_1, \psi(x_1, x_2, y_2), z]_\tau - [y_1, y_2, \psi(x_1, x_2, z)]_\tau \\ &= \tau(y_1)\, \psi(x_1, x_2, [y_2, z]) + \tau(y_2)\, \psi(x_1, x_2, [z, y_1]) + \tau(z)\, \psi(x_1, x_2, [y_1, y_2]) \\ &\quad - \tau(x_1)\, \psi(y_1, y_2, [x_2, z]) - \tau(x_2)\, \psi(y_1, y_2, [z, x_1]) - \tau(z)\, \psi(y_1, y_2, [x_1, x_2]) \\ &\quad - \tau(x_1)\, \psi([x_2, y_1]\, y_2, z,) - \tau(x_2)\, \psi([y_1, x_1]\, y_2, z,) - \tau(y_1)\, \psi([x_1, x_2]\, y_2, z,) \\ &\quad - \tau(x_1)\, \psi(y_1, [x_2, y_2], z) - \tau(x_2)\, \psi(y_1, [y_2, x_1], z) - \tau(y_2)\, \psi(y_1, [x_1, x_2], z) \\ &\quad + \tau(x_1)\, [x_2, \psi(y_1, y_2, z)] + \tau(x_2)\, [\psi(y_1, y_2, z), x_1] + \tau(\psi(y_1, y_2, z))\, [x_1, x_2] \\ &\quad - \tau(y_1)\, [y_2, \psi(x_1, x_2, z)] - \tau(y_2)\, [\psi(x_1, x_2, z), y_1] - \tau(\psi(x_1, x_2, z))\, [y_1, y_2] \end{aligned}$$

$$
\begin{aligned}
&-\tau(\psi(x_1,x_2,y_1))[y_2,z]-\tau(y_2)[z,\psi(x_1,x_2,y_1)]-\tau(z)[\psi(x_1,x_2,y_1),y_2]\\
&-\tau(y_1)[\psi(x_1,x_2,y_2),z]-\tau(\psi(x_1,x_2,y_2))[z,y_1]-\tau(z)[y_1,\psi(x_1,x_2,y_2)]\\
=&\ \tau(y_1)\Big(\omega(x_1)\varphi(x_2,[y_2,z])+\omega(x_2)\varphi([y_2,z],a)-\omega(y_2)\varphi(z,[x_1,x_2])\\
&-\omega(z)\varphi([x_1,x_2],y_2)-\omega(x_1)[y_2,\varphi(x_2,z)]-\omega(x_2)[y_2,\varphi(z,x_1)]\\
&-\omega(z)[y_2,\varphi(x_1,x_2)]-\omega(x_1)[\varphi(x_2,y_2),z]-\omega(x_2)[\varphi(y_2,x_1),z]\\
&-\omega(y)[\varphi(x_1,x_2),z]\Big)\\
&+\tau(y_2)\Big(\omega(x_1)\varphi(x_2,[z,y_1])+\omega(x_2)\varphi([z,y_1],x_1)-\omega(x_1)[\varphi(x_2,z),y_1]\\
&-\omega(x_2)[\varphi(z,x_1),y_1]-\omega(z)[\varphi(x_1,x_2),y_1]-\omega(x_1)[z,\varphi(x_2,y_1)]\\
&-\omega(x_2)[z,\varphi(y_1,x_1)]-\omega(y_1)[z,\varphi(x_1,x_2)]-\omega(y_1)\varphi([x_1,x_2],z)\\
&-\omega(z)\varphi(y_1,[x_1,x_2])\Big)\\
&+\tau(z)\Big(\omega(x_1)\varphi(x_2,[y_1,y_2])+\omega(x_2)\varphi([y_1,y_2],x_1)-\omega(y_1)\varphi(y_2,[x_1,x_2])\\
&-\omega(y_2)\varphi([x_1,x_2],y_1)-\omega(x_1)[\varphi(x_2,y_1),y_2]-\omega(x_2)[\varphi(y_1,x_1),y_2]\\
&-\omega(y_1)[\varphi(x_1,x_2),y_2]-\omega(x_1)[y_1,\varphi(x_2,y_2)]-\omega(x_2)[y_1,\varphi(y_2,x_1)]\\
&-\omega(y_2)[y_1,\varphi(x_1,x_2)]\Big)\\
&+\tau(x_1)\Big(\omega(y_1)[x_2,\varphi(y_2,z)]+\omega(y_2)[x_2,\varphi(z,y_1)]+\omega(z)[x_2,\varphi(y_1,y_2)]\\
&-\omega(y_1)\varphi(y_2,[x_2,z])-\omega(y_2)\varphi([x_2,z],y_1)-\omega(y_2)\varphi(z,[x_2,y_1])\\
&-\omega(z)\varphi([x_2,y_1],y_2)-\omega(y_1)\varphi([x_2,y_2],z)-\omega(z)\varphi(y_1,[x_2,y_2])\Big)\\
&+\tau(x_2)\Big(\omega(y_1)[\varphi(y_2,z),x_1]+\omega(y_2)[\varphi(z,y_1),x_1]+\omega(z)[\varphi(y_1,y_2),x_1]\\
&-\omega(y_1)\varphi(y_2,[z,x_1])-\omega(y_2)\varphi([z,x_1],y_1)-\omega(y_2)\varphi(z,[y_1,x_1])\\
&-\omega(z)\varphi([y_1,x_1],y_2)-\omega(y_1)\varphi([y_2,x_1],z)-\omega(z)\varphi(y_1,[y_2,x_1])\Big)\\
&+\Big(\omega(y_1)\tau(\varphi(y_2,z))+\omega(y_2)\tau(\varphi(z,y_1))+\omega(z)\tau(\varphi(y_1,y_2))\Big)[x_1,x_2]\\
&-\Big(\omega(x_1)\tau(\varphi(x_2,y_1))+\omega(x_2)\tau(\varphi(y_1,x_1))+\omega(y_1)\tau(\varphi(x_1,x_2))\Big)[y_2,z]\\
&-\Big(\omega(x_1)\tau(\varphi(x_2,z))+\omega(x_2)\tau(\varphi(z,x_1))+\omega(z)\tau(\varphi(x_1,x_2))\Big)[y_1,y_2]\\
&-\Big(\omega(x_1)\tau(\varphi(x_2,y_2))+\omega(x_2)\tau(\varphi(y_2,x_1))+\omega(y_2)\tau(\varphi(x_1,x_2))\Big)[z,y_1]\\
=&-\tau(y_1)\omega(x_1)\delta^2\varphi(z,y_2,x_2)-\tau(y_1)\omega(x_2)\delta^2\varphi(y_2,z,x_1)\\
&-\tau(y_2)\omega(x_1)\delta^2\varphi(y_1,z,x_2)-\tau(y_2)\omega(x_2)\delta^2\varphi(z,y_1,x_1)\\
&-\tau(z)\omega(x_1)\delta^2\varphi(y_2,y_1,x_2)-\tau(z)\omega(x_2)\delta^2\varphi(y_1,y_2,x_1)\\
&+\Big(\omega(y_1)\tau(\varphi(y_2,z))+\omega(y_2)\tau(\varphi(z,y_1))+\omega(z)\tau(\varphi(y_1,y_2))\Big)[x_1,x_2]\\
&-\Big(\omega(x_1)\tau(\varphi(x_2,y_1))+\omega(x_2)\tau(\varphi(y_1,x_1))+\omega(y_1)\tau(\varphi(x_1,x_2))\Big)[y_2,z]\\
&-\Big(\omega(x_1)\tau(\varphi(x_2,z))+\omega(x_2)\tau(\varphi(z,x_1))+\omega(z)\tau(\varphi(x_1,x_2))\Big)[y_1,y_2]\\
&-\Big(\omega(x_1)\tau(\varphi(x_2,y_2))+\omega(x_2)\tau(\varphi(y_2,x_1))+\omega(y_2)\tau(\varphi(x_1,x_2))\Big)[z,y_1].
\end{aligned}
$$

Since

$$\circlearrowleft_{x,y,z}\omega(x)\tau(\varphi(y,z))=0,\ \forall x,y,z\in A,$$

we get

$$d^2\psi = 0.$$

□

Theorem 4.3 *Every* 1*-cocycle for the scalar cohomology of* $(A, [.,.])$ *is a* 1*-cocycle for the scalar cohomology of the induced* 3*-Lie algebra.*

Proof Let ω be a 1-cocycle for the scalar cohomology of $(A, [.,.])$, then

$$\forall x, y \in A, \delta^1\omega(x, y) = \omega([x, y]) = 0,$$

which is equivalent to $[A, A] \subset \ker\omega$. By Remark 3.2 $[A, A, A]_\tau \subset [A, A]$ and then $[A, A, A]_\tau \subset \ker\omega$, that is

$$\forall x, y, z \in A, \omega\left([x, y, z]_\tau\right) = d^1\omega(x, y, z) = 0,$$

which means that ω is a 1-cocycle for the scalar cohomology of $(A, [.,.,.]_\tau)$. □

Theorem 4.4 *Let* $\varphi \in Z_0^2(A, \mathbb{K})$ *and* $\omega : A \to \mathbb{K}$ *a linear map satisfying:*

1. $\tau(x)\omega(y) = \tau(y)\omega(x)$,
2. $\omega([x, y]) = 0$,
3. $\omega(y_2)\left(\tau(x_1)\varphi([x_1, z]x_2) + \tau(x_2)\varphi([z, y_1]x_1)\right) = 0.$

Then $\psi(x, y, z) = \circlearrowleft_{x,y,z} \omega(x)\varphi(y, z)$ *is a* 2*-cocycle of the induced* 3*-Lie algebra.*

Proof Let $\varphi \in Z_0^2(A, \mathbb{K})$ and $\omega : A \to \mathbb{K}$ a linear map satisfying conditions 1, 2 and 3 above, and let $\psi(x, y, z) = \circlearrowleft_{x,y,z} \omega(x)\varphi(y, z)$, then we have:

$$\begin{aligned}
d^2\psi(x_1, x_2, y_1, y_2, z) &= \psi\left(x_1, x_2, [y_1, y_2, z]_\tau\right) - \psi\left([x_1, x_2, y_1]_\tau, y_2, z\right) \\
&\quad - \psi\left(y_1, [x_1, x_2, y_2]_\tau, z\right) - \psi\left(y_1, y_2, [x_1, x_2, z]_\tau\right) \\
&= \tau(y_1)\psi(x_1, x_2, [y_2, z]) + \tau(y_2)\psi(x_1, x_2, [z, y_1]) + \tau(z)\psi(x_1, x_2, [y_1, y_2]) \\
&\quad - \tau(x_1)\psi([x_2, y_1], y_2, z) - \tau(x_2)\psi([y_1, x_1], y_2, z) - \tau(y_1)\psi([x_1, x_2], y_2, z) \\
&\quad - \tau(x_1)\psi(y_1, [x_2, y_2], z) - \tau(x_2)\psi(y_1, [y_2, x_1], z) - \tau(y_2)\psi(y_1, [x_1, x_2], z) \\
&\quad - \tau(x_1)\psi(y_1, y_2, [x_2, z]) - \tau(x_2)\psi(y_1, y_2, [z, x_1]) - \tau(z)\psi(y_1, y_2, [x_1, x_2]) \\
&= \tau(y_1)\left(\omega(x_1)\varphi(x_2, [y_2, z]) + \omega(x_2)\varphi([y_2, z], x_1)\right) \\
&\quad + \tau(y_2)\left(\omega(x_1)\varphi(x_2, [y_1, z]) + \omega(x_2)\varphi([y_1, z], x_1)\right) \\
&\quad + \tau(z)\left(\omega(x_1)\varphi(x_2, [y_1, y_2]) + \omega(x_2)\varphi([y_1, y_2], x_1)\right) \\
&\quad - \tau(x_1)\left(\omega(y_2)\varphi(z, [x_2, y_1]) + \omega(z)\varphi([x_2, y_1], y_2)\right) \\
&\quad - \tau(x_2)\left(\omega(y_2)\varphi(z, [y_1, x_1]) + \omega(z)\varphi([y_1, x_1]y_2)\right) \\
&\quad - \tau(y_1)\left(\omega(y_2)\varphi(z, [x_1, x_2]) + \omega z\varphi([x_1, x_2], y_2)\right) \\
&\quad - \tau(x_1)\left(\omega(y_1)\varphi([x_2, y_2], z) + \omega(z)\varphi(y_1, [x_2, y_2])\right) \\
&\quad - \tau(x_2)\left(\omega(y_1)\varphi([y_2, x_1], z) + \omega(z)\varphi(y_1, [y_2, x_1])\right) \\
&\quad - \tau(y_2)\left(\omega(y_1)\varphi([x_1, x_2], z) + \omega(z)\varphi(y_1, [x_1, x_2])\right)
\end{aligned}$$

$$
\begin{aligned}
&-\tau(x_1)(\omega(y_1)\varphi(y_2,[x_2,z])+\omega(y_2)\varphi([x_2,z],y_1))\\
&-\tau(x_2)(\omega(y_1)\varphi(y_2,[z,x_1])+\omega(y_2)\varphi([z,x_1],y_1))\\
&-\tau(z)(\omega(y_1)\varphi(y_2,[x_1,x_2])+\omega(y_2)\varphi([x_1,x_2],y_1))\\
&=\tau(x_1)\omega(y_1)\delta^2\varphi(y_2,z,x_2)+\tau(x_1)\omega(y_2)\delta^2\varphi(z,y_1,x_2)\\
&+\tau(x_2)\omega(y_1)\delta^2\varphi(z,y_2,x_1)+\tau(x_2)\omega(y_2)\delta^2\varphi(x_1,y_1,z)\\
&+\tau(x_1)\omega(z)\delta^2\varphi(y_1,y_2,x_2)+\tau(x_2)\omega(z)\delta^2\varphi(y_2,y_1,x_1)\\
&-2\omega(y_2)(\tau(x_1)\varphi([y_1,z],x_2)+\tau(x_2)\varphi([z,y_1],x_1)).
\end{aligned}
$$

Since

$$\omega(y_2)(\tau(x_1)\varphi([y_1,z],x_2)+\tau(x_2)\varphi([z,y_1],x_1))=0,$$

it follows that $d^2\psi=0$. □

Remark 4.1 Condition 1 in Theorems 4.2 and 4.4 are equivalent to $\omega=\lambda\tau$, $\lambda\in\mathbb{K}$, and therefore one may remove condition 2, which is redundant.

Lemma 4.2 *Let $\alpha\in C^1(A,\mathbb{K})$. Then:*

$$d^1\alpha(x,y,z)=\circlearrowleft_{x,y,z}\tau(x)\delta^1\alpha(y,z),\ \forall x,y,z\in A.$$

Proof Let $\alpha\in C^1(A,\mathbb{K})$, $x,y,z\in A$, then we have:

$$d^1\alpha(x,y,z)=\alpha([x,y,z])=\circlearrowleft_{x,y,z}\tau(x)\alpha([y,z])=\circlearrowleft_{x,y,z}\tau(x)\delta^1\alpha(y,z)$$

□

Proposition 4.1 *Let $\varphi_1,\varphi_2\in Z_0^2(A,\mathbb{K})$ satisfying conditions of Theorem 4.4. If φ_1,φ_2 are in the same cohomology class then ψ_1,ψ_2 defined by:*

$$\psi_i(x,y,z)=\circlearrowleft_{x,y,z}\tau(x)\varphi_i(y,z),\ i=1,2$$

are in the same cohomology class.

Proof Let $\varphi_1,\varphi_2\in Z_0^2(A,\mathbb{K})$ be two cocycles in the same cohomology class, that is

$$\varphi_2-\varphi_1=\delta^1\alpha,\ \alpha\in C^1(A,\mathbb{K})$$

satisfying conditions of Theorem 4.4, and

$$\psi_i(x,y,z)=\circlearrowleft_{x,y,z}\tau(x)\varphi_i(y,z):i=1,2,$$

then we have:

$$\begin{aligned}\psi_2(x,y,z)-\psi_1(x,y,z) &= \circlearrowleft_{x,y,z} \tau(x)\varphi_2(y,z) - \circlearrowleft_{x,y,z} \tau(x)\varphi_1(y,z)\\ &= \circlearrowleft_{x,y,z} \tau(x)(\varphi_2-\varphi_1)(y,z)\\ &= \circlearrowleft_{x,y,z} \tau(x)\delta^1\alpha(y,z)\\ &= d^1\alpha(x,y,z),\end{aligned}$$

which means that ψ_1 and ψ_2 are in the same cohomology class. □

5 Central Extensions of 3-Lie Algebras Induced by Lie Algebras

Definition 5.1 Let A, B, C be n-Lie algebras ($n \geq 2$). An extension of B by A is a short sequence:

$$A \xrightarrow{\lambda} C \xrightarrow{\mu} B,$$

such that λ is an injective homomorphism, μ is a surjective homomorphism, and $\operatorname{Im}\lambda \subset \ker\mu$. We say also that C is an extension of B by A.

Definition 5.2 Let A, B be n-Lie algebras, and $A \xrightarrow{\lambda} C \xrightarrow{\mu} B$ be an extension of B by A.

- The extension is said to be trivial if there exists an ideal I of C such that $C = \ker\mu \oplus I$.
- It is said to be central if $\ker\mu \subset Z(C)$.

We may equivalently define central extensions by a 1-dimensional algebra (we will simply call it central extension) this way:

Definition 5.3 Let A be a n-Lie algebra, we call central extension of A the space $\bar{A} = A \oplus \mathbb{K}c$ equipped with the bracket:

$$\forall x_1,\ldots,x_n \in A, [x_1,...,x_n]_c = [x_1,\ldots,x_n] + \omega(x_1,\ldots,x_n)\,c \text{ and } \left[x_1,\ldots,x_{n-1},c\right]_c = 0,$$

where ω is a skew-symmetric n-linear form such that $[\cdot,\ldots,\cdot]$ satisfies the fundamental identity (or Jacobi identity for $n = 2$).

Proposition 5.1 ([6])

1. *The bracket of a central extension satisfies the fundamental identity (resp. Jacobi identity) if and only if ω is a 2-cocycle for the scalar cohomology of n-Lie algebras (resp. Lie algebras).*
2. *Two central extensions of a n-Lie algebra (resp. Lie algebra) A given by two maps ω_1 and ω_2 are isomorphic if and only if $\omega_2 - \omega_1$ is a 2-coboundary for the scalar cohomology of n-Lie algebras (resp. Lie algebras).*

Now, we look at the question of whether a central extension of a Lie algebra may give a central extension of the induced 3-Lie algebra (by some trace τ), the answer is given by the following theorem:

Theorem 5.1 *Let* $(A, [.,.])$ *be a Lie algebra,* τ *be a trace and* $(A, [.,.,.]_\tau)$ *be the induced* 3-*Lie algebra. If* $\left(\bar{A}, [.,.]_c\right)$ *is a central extension of* $(A, [.,.])$ *where*

$$\bar{A} = A \oplus \mathbb{K}c \text{ and } [x, y]_c = [x, y] + \omega(x, y)\, c,$$

and we extend τ *to* $\bar{A}$ *by assuming* $\tau(c) = 0$*. Then* $\left(\bar{A}, [.,.,.]_{c,\tau}\right)$ *the* 3-*Lie algebra induced by* $\left(\bar{A}, [.,.]_c\right)$*, is a central extension of* $(A, [.,.,.]_\tau)$.

Proof Let $x, y, z \in A$:

$$\begin{aligned}
[x, y, z]_{c,\tau} &= \tau(x)[y, z]_c + \tau(y)[z, x]_c + \tau(z)[x, y]_c \\
&= \tau(x)([y, z] + \omega(y, z)c) + \tau(y)([z, x] + \omega(z, x)c) + \tau(z)([x, y] + \omega(x, y)c) \\
&= (\tau(x)[y, z] + \tau(y)[z, x] + \tau(z)[x, y]) \\
&\quad + (\tau(x)\omega(y, z) + \tau(y)\omega(z, x) + \tau(z)\omega(x, y))\, c. \\
&= [x, y, z]_\tau + \omega_\tau(x, y, z)\, c.
\end{aligned}$$

The map $\omega_\tau(x, y, z) = \tau(x)\omega(y, z) + \tau(y)\omega(z, x) + \tau(z)\omega(x, y)$ is a skew-symmetric 3-linear form, and $[.,.,.]_{c,\tau}$ satisfies the fundamental identity, we have also:

$$\begin{aligned}
[x, y, c]_{c,\tau} &= \tau(x)[y, c]_c + \tau(y)[c, x]_c + \tau(c)[x, y]_c \\
&= 0. \quad \left([y, c]_c = [c, x]_c = 0 \text{ and } \tau(c) = 0.\right)
\end{aligned}$$

Therefore $\left(\bar{A}, [.,.,.]_{c,\tau}\right)$ is a central extension of $(A, [.,.,.]_\tau)$. □

Example 5.1 Consider the 4-dimensional Lie algebra $(A, [.,.])$ with basis $\{e_1, e_2, e_3, e_4\}$ defined by:

$$[e_2, e_4] = e_3\ ;\ [e_3, e_4] = e_3,$$

(remaining brackets are either obtained by skew-symmetry or zero), and let ω be a skew-symmetric bilinear form on A. ω is fully defined by the scalars

$$\omega_{ij} = \omega\left(e_i, e_j\right), 1 \leq i < j \leq 4.$$

By solving the equations for ω to be a 2-cocycle:

$$\delta^2\omega\left(e_i, e_j, e_k\right) = 0, 1 \leq i < j < k \leq 4,$$

we get the conditions:

$$\omega_{13} = 0 \text{ and } \omega_{23} = 0.$$

Now, let α be a linear form on A, defined by $\alpha(e_i) = \alpha_i$, $1 \leq i \leq 4$, we find that $\delta^1\alpha(e_2, e_4) = \delta^1\alpha(e_3, e_4) = \alpha_3$ and $\delta^1\alpha(e_i, e_j) = 0$ for other values of i and j $(i < j)$. Now consider the trace map τ such that $\tau(e_1) = 1$ and $\tau(e_i) = 0, i \neq 1$, and the 2-cocycles λ and μ defined by:

$$\lambda(e_1, e_2) = 1$$

and

$$\mu(e_2, e_4) = 1\ ;\ \mu(e_3, e_4) = -1.$$

Central extensions of $(A, [.,.])$ by λ and μ are respectively given by ($\bar{A} = A \oplus \mathbb{K}c$):

$$[e_1, e_2]_\lambda = c\ ;\ [e_2, e_4]_\lambda = e_3\ ;\ [e_3, e_4]_\lambda = e_3$$

and

$$[e_2, e_4]_\mu = e_3 + c\ ;\ [e_3, e_4]_\mu = e_3 - c.$$

3-Lie algebras induced by $(A, [.,.])$ and by these central extensions are given by:

$$[e_1, e_2, e_4]_\tau = e_3\ ;\ [e_1, e_3, e_4]_\tau = e_3,$$

$$[e_1, e_2, e_4]_{\tau,\lambda} = e_3\ ;\ [e_1, e_3, e_4]_{\tau,\lambda} = e_3$$

and

$$[e_1, e_2, e_4]_{\tau,\mu} = e_3 + c\ ;\ [e_1, e_3, e_4]_{\tau,\mu} = e_3 - c.$$

We can see that, here, the central extension given by λ induces a trivial one, while the one given by μ induces a non-trivial one. This example shows also that the converse of Lemma 4.1 is, in general, not true.

6 3-Lie Algebras Induced by Lie Algebras in Low Dimensions

In this section, we give a list of all 3-Lie algebras induced by Lie algebras in dimension $d \leq 5$, based on the classifications given in [4, 8]. For this, we shall use the following result:

Proposition 6.1 *Let* $(A, [., ., .])$ *be a* 3*-Lie algebra,* $(e_i)_{1\le i\le d}$ *a basis of* A*. If there exists* e_{i_0} *in this base, such that the multiplication table of* $(A, [., ., .])$ *is given by:*

$$\left[e_{i_0}, e_j, e_k\right] = x_{jk};\ j \neq i_0, k \neq i_0, k \neq j,$$

with e_{i_0} *and* x_{jk} *linearly independent, then* $(A, [., ., .])$ *is induced by a Lie algebra*

Proof We define a bilinear skew-symmetric map $[., .]$ on A and a form $\tau : A \to \mathbb{K}$ by:

$$\left[e_j, e_k\right] = x_{jk}, j \neq i_0, k \neq i_0, k \neq j \text{ and } \left[e_{i_0}, e_j\right] = 0$$

and

$$\tau(x) = \tau\left(\sum_{k=0}^{d} x_k e_k\right) = x_{i_0}.$$

The bracket $[., .]$ satisfies the Jacobi identity:

$$\begin{aligned}\left[e_j, \left[e_k, e_l\right]\right] &= \left[e_{i_0}, e_j, \left[e_{i_0}, e_k, e_l\right]\right]\\ &= \left[\left[e_{i_0}, e_j, e_{i_0}\right], e_k, e_l\right] + \left[e_{i_0}, \left[e_{i_0}, e_j, e_k\right], e_l\right] + \left[e_{i_0}, e_k, \left[e_{i_0}, e_j, e_l\right]\right]\\ &= \left[\left[e_j, e_k\right], e_l\right] + \left[e_k, \left[e_j, e_l\right]\right]\end{aligned}$$

The obtained Lie bracket $[., .]$ and the trace τ given above indeed induce the ternary bracket considered above:

$$\begin{aligned}\left[e_{i_0}, e_j, e_k\right]_\tau &= \tau(e_{i_0})\left[e_j, e_k\right] + \tau(e_j)\left[e_k, e_{i_0}\right] + \tau(e_k)\left[e_{i_0}, e_j\right]\\ &= \tau(e_{i_0})\left[e_j, e_k\right]\\ &= x_{jk}\\ &= \left[e_{i_0}, e_j, e_k\right]\end{aligned}$$

for $i \neq i_0$:

$$\left[e_i, e_j, e_k\right]_\tau = \tau(e_i)\left[e_j, e_k\right] + \tau(e_j)\left[e_k, e_i\right] + \tau(e_k)\left[e_i, e_j\right] = 0 = \left[e_i, e_j, e_k\right].$$

□

Theorem 6.1 ([8] 3 -Lie algebras of dimension less than or equal to 4) *Any* 3*-Lie algebra* A *of dimension less than or equal to* 4 *is isomorphic to one of the following algebras: (omitted brackets are obtained by skew-symmetry,* $(e_i)_{1\le i\le dim A}$ *is a basis of* A*)*

1. *If* $dim A < 3$ *then* A *is abelian.*
2. *If* $dim A = 3$*, then we have 2 cases:*

a. *A is abelian.*
b. $[e_1, e_2, e_3] = e_1$.

3. *if* $dim A = 4$ *then we have the following cases:*

   a. *A is abelian.*
   b. $[e_2, e_3, e_4] = e_1$.
   c. $[e_1, e_2, e_3] = e_1$.
   d. $[e_1, e_2, e_4] = ae_3 + be_4; [e_1, e_2, e_3] = ce_3 + de_4$, *with* $C = \begin{pmatrix} a & b \\ c & d \end{pmatrix}$ *an invertible matrix. Two such algebras, defined by matrices* C_1 *and* C_2*, are isomorphic if and only if there exists a scalar* α *and an invertible matrix* B *such that* $C_2 = \alpha B C_1 B^{-1}$.
   e. $[e_2, e_3, e_4] = e_1; [e_1, e_3, e_4] = ae_2; [e_1, e_2, e_4] = be_3$ $(a, b \neq 0)$.
   f. $[e_2, e_3, e_4] = e_1; [e_1, e_3, e_4] = ae_2; [e_1, e_2, e_4] = be_3; [e_1, e_2, e_3] = ce_4$ $(a, b, c \neq 0)$.

Theorem 6.2 ([4] 5-dimensional 3-Lie algebras) *Let* $\mathbb{K}$ *be an algebraically closed field. Any 5-dimensional 3-Lie algebra A defined with respect to a basis* $\{e_1, e_2, e_3, e_4, e_5\}$ *is isomorphic to one of the algebras listed below, where* A^1 *denotes* $[A, A, A]$*:*

1. *If* $dim A^1 = 0$ *then A is abelian.*
2. *If* $dim A^1 = 1$*, let* $A^1 = \langle e_1 \rangle$*, then we have:*

   a. $A^1 \subseteq Z(A)$*:* $[e_2, e_3, e_4] = e_1$.
   b. $A^1 \nsubseteq Z(A)$*:* $[e_1, e_2, e_3] = e_1$.

3. *If* $dim A^1 = 2$*, let* $A^1 = \langle e_1, e_2 \rangle$*, then we have:*

   a. $[e_2, e_3, e_4] = e_1; [e_3, e_4, e_5] = e_2$.
   b. $[e_2, e_3, e_4] = e_1; [e_2, e_4, e_5] = e_2; [e_1, e_4, e_5] = e_1$.
   c. $[e_2, e_3, e_4] = e_1; [e_1, e_3, e_4] = e_2$.
   d. $[e_2, e_3, e_4] = e_1; [e_1, e_3, e_4] = e_2; [e_2, e_4, e_5] = e_2; [e_1, e_4, e_5] = e_1$.
   e. $[e_2, e_3, e_4] = \alpha e_1 + e_2; [e_1, e_3, e_4] = e_2$.
   f. $[e_2, e_3, e_4] = \alpha e_1 + e_2; [e_1, e_3, e_4] = e_2; [e_2, e_4, e_5] = e_2; [e_1, e_4, e_5] = e_1$.
   g. $[e_1, e_3, e_4] = e_1; [e_2, e_3, e_4] = e_2$.

   where $\alpha \in \mathbb{K} \setminus \{0\}$
4. *If* $dim A^1 = 3$*, let* $A^1 = \langle e_1, e_2, e_3 \rangle$*, then we have:*

   a. $[e_2, e_3, e_4] = e_1; [e_2, e_4, e_5] = -e_2; [e_3, e_4, e_5] = e_3$.
   b. $[e_2, e_3, e_4] = e_1; [e_3, e_4, e_5] = e_3 + \alpha e_2; [e_2, e_4, e_5] = e_3; [e_1, e_4, e_5] = e_1$.
   c. $[e_2, e_3, e_4] = e_1; [e_3, e_4, e_5] = e_3; [e_2, e_4, e_5] = e_2; [e_1, e_4, e_5] = 2e_1$.
   d. $[e_2, e_3, e_4] = e_1; [e_1, e_3, e_4] = e_2; [e_1, e_2, e_4] = e_3$.
   e. $[e_1, e_4, e_5] = e_1; [e_2, e_4, e_5] = e_3; [e_3, e_4, e_5] = \beta e_2 + (1 + \beta)e_3$, $\beta \in \mathbb{K} \setminus \{0, 1\}$.
   f. $[e_1, e_4, e_5] = e_1; [e_2, e_4, e_5] = e_2; [e_3, e_4, e_5] = e_3$.

g. $[e_1, e_4, e_5] = e_2$; $[e_2, e_4, e_5] = e_3$; $[e_3, e_4, e_5] = se_1 + te_2 + ue_3$. *And* 3-*Lie algebras corresponding to this case with coefficients* s, t, u *and* s', t', u' *are isomorphic if and only if there exists a non-zero element* $r \in K$ *such that:*

$$s = r^3 s'; t = r^2 t'; u = ru'.$$

5. *If* $dim A^1 = 4$, *let* $A^1 = \langle e_1, e_2, e_3, e_4 \rangle$, *then we have:*

   a. $[e_2, e_3, e_4] = e_1$; $[e_3, e_4, e_5] = e_2$; $[e_2, e_4, e_5] = e_3$; $[e_2, e_3, e_5] = e_4$.
   b. $[e_2, e_3, e_4] = e_1$; $[e_1, e_3, e_4] = e_2$; $[e_1, e_2, e_4] = e_3$; $[e_1, e_2, e_3] = e_4$.

The 3-Lie algebras which are induced by Lie algebras are given by the following proposition:

Proposition 6.2 *Let* $\mathbb{K}$ *be an algebraically closed field of characteristic* 0. *According to Theorems 6.1 and 6.2, the* 3*-Lie algebras induced by Lie algebras of dimension* $d \leq 5$ *are:*

- $d = 3$ *Theorem 6.1: 2.*
- $d = 4$ *Theorem 6.1: 3.: a,b,c,d,e.*
- $d = 5$ *Theorem 6.2: 1. 2. 3. 4.*

Proof By applying Proposition 6.1, the algebras given in Theorem 6.1, 6.2, and 6.3: a,b,c,d,e and Theorem 6.2: 1, 2, 3 and 4. are all induced by Lie algebras, the remaining algebras have derived algebras which are not abelian and then they cannot be induced by Lie algebras (Theorem 3.1). □

6.1 From Lie Algebras to 3-Lie Algebras

We list, below, all 3 and 4-dimensional Lie algebras and all 3-Lie algebras they may induce; 3-dimensional algebras are classified in [12] and 4-dimensional ones, partially, in [7]. For every Lie algebra, we compute all the trace maps and the induced 3-Lie algebras using these trace maps.

Theorem 6.3 (3-dimensional Lie algebras [12]) *Let* $\mathfrak{g}$ *be a Lie algebra and* $\{e_1, e_2, e_3\}$ *a basis of* $\mathfrak{g}$, *then* $\mathfrak{g}$ *is isomorphic to one of the following algebras: (Remaining brackets are either obtained by skew-symmetry or zero)*

1. *The abelian Lie algebra* $[x, y] = 0, \forall x, y \in \mathfrak{g}$.
2. $L(3, -1) : [e_1, e_2] = e_2$.
3. $L(3, 1) : [e_1, e_2] = e_3$.
4. $L(3, 2, a) : [e_1, e_3] = e_1; [e_2, e_3] = ae_2; 0 < |a| \leq 1$.
5. $L(3, 3) : [e_1, e_3] = e_1; [e_2, e_3] = e_1 + e_2$.
6. $L(3, 4, a) : [e_1, e_3] = ae_1 - e_2; [e_2, e_3] = e_1 + ae_2; a \geq 0$.
7. $L(3, 5) : [e_1, e_2] = e_1; [e_1, e_3] = -2e_2; [e_2, e_3] = e_3$.
8. $L(3, 6) : [e_1, e_2] = e_3; [e_1, e_3] = -e_2; [e_2, e_3] = e_1$.

Remark 6.1 The classification given above is for the ground field $\mathbb{K} = \mathbb{R}$, if $\mathbb{K} = \mathbb{C}$ then $L\left(3, 2, \frac{x-i}{x+i}\right)$ is isomorphic to $L(3, 4, x)$ and $L(3, 5)$ is isomorphic to $L(3, 6)$.

Theorem 6.4 (Solvable 4-dimensional Lie algebras [7]) *Let $\mathfrak{g}$ be a solvable Lie algebra, and $\{e_1, e_2, e_3, e_4\}$ a basis of $\mathfrak{g}$, then $\mathfrak{g}$ is isomorphic to one of the following algebras: (Remaining brackets are either obtained by skew-symmetry or zero)*

1. *The abelian Lie algebra* $[x, y] = 0, \forall x, y \in \mathfrak{g}$.
2. M^2: $[e_1, e_4] = e_1$; $[e_2, e_4] = e_2$; $[e_3, e_4] = e_3$.
3. M^3_a: $[e_1, e_4] = e_1$; $[e_2, e_4] = e_3$; $[e_3, e_4] = -ae_2 + (a+1)e_3$.
4. M^4: $[e_2, e_4] = e_3$; $[e_3, e_4] = e_3$.
5. M^5: $[e_2, e_4] = e_3$.
6. $M^6_{a,b}$: $[e_1, e_4] = e_2$; $[e_2, e_4] = e_3$; $[e_3, e_4] = ae_1 + be_2 + e_3$.
7. $M^7_{a,b}$: $[e_1, e_4] = e_2$; $[e_2, e_4] = e_3$; $[e_3, e_4] = ae_1 + be_2$ *(*$a = b \neq 0$ *or* $a = 0$ *or* $b = 0$*).*
8. M^8: $[e_1, e_2] = e_2$; $[e_3, e_4] = e_4$.
9. M^9_a: $[e_1, e_4] = e_1 + ae_2$; $[e_2, e_4] = e_1$; $[e_1, e_3] = e_1$; $[e_2, e_3] = e_2$ *(*$X^2 - X - a$ *has no root in* $\mathbb{K}$*).*
10. M^{11}: $[e_1, e_4] = e_1$; $[e_3, e_4] = e_3$; $[e_1, e_3] = e_2$.
11. M^{12}: $[e_1, e_4] = e_1$; $[e_2, e_4] = e_2$; $[e_3, e_4] = e_3$; $[e_1, e_3] = e_2$.
12. M^{13}_a: $[e_1, e_4] = e_1 + ae_3$; $[e_2, e_4] = e_2$; $[e_3, e_4] = e_1$; $[e_1, e_3] = e_2$.
13. M^{14}_a: $[e_1, e_4] = ae_3$; $[e_3, e_4] = e_1$; $[e_1, e_3] = e_2$. *(*M^{14}_a *is isomorphic to* M^{14}_b *if and only if* $a = \alpha^2 b$ *for some* $\alpha \neq 0$*).*

Lemma 6.1 *Let $\mathfrak{g}$ be a non solvable 4-dimensional Lie algebra. Then $[\mathfrak{g}, \mathfrak{g}]$ is simple.*

Proof $\mathfrak{g}$ is not semi-simple ($dim\,\mathfrak{g} = 4$, $\mathfrak{g}$ cannot be simple, and cannot be a direct sum of simple Lie algebras, $4 = 3+1$ or $2+2$), if $dim\,[\mathfrak{g}, \mathfrak{g}] \leq 2$ then $\mathfrak{g}$ is solvable, if $dim\,\mathfrak{g} = 3$ then from the classification above, if it is not simple (isomorphic to $L(3, 5)$ or $L(3, 6)$) then it is solvable, and then $\mathfrak{g}$ is solvable too. □

Proposition 6.3 *Let $\mathfrak{g}$ be a non solvable 4-dimensional Lie algebra, then $\mathfrak{g} \cong S \times \mathbb{K}$ where S is a 3-dimensional simple Lie algebra.*

Proof From the preceding lemma, $[\mathfrak{g}, \mathfrak{g}]$ is simple, which implies that $\mathfrak{g}$ is reductive, that is, a product of a semi-simple and an abelian Lie algebra. □

In the following, we will give all the traces τ on the Lie algebras listed above, and the induced 3-Lie algebras: (for a Lie algebra $\mathfrak{g}$, $(e_i)_{1 \leq i \leq dim\,\mathfrak{g}}$ is a basis of $\mathfrak{g}$, and for $x \in \mathfrak{g}$, $(x_i)_{1 \leq i \leq dim\,\mathfrak{g}}$ are its coordinates in this basis).

Lie algebra	**Trace**	**Induced 3-Lie algebra**
Abelian Lie algebra	All linear forms	Abelian 3-Lie algebra
$L(3,-1)$	$\tau(x)=t_1x_1+t_3x_3$	$[e_1,e_2,e_3]=t_3e_2$
$L(3,1)$	$\tau(x)=t_1x_1+t_2x_2$	Abelian 3-Lie algebra
$L(3,2,a)$, $L(3,3)$		
$L(3,4,a)$	$\tau(x)=t_3x_3$	Abelian 3-Lie algebra
$L(3,5)$, $L(3,6)$	$\tau(x)=0$	Abelian 3-Lie algebra
M^2, M^3_a		
$M^6_{a,b}$; $M^7_{a,b}$ $(a\neq 0)$	$\tau(x)=t_4x_4$	Abelian 3-Lie algebra
M^3_0	$\tau(x)=t_2x_2+t_4x_4$	$[e_1,e_2,e_4]=-t_2e_1$
M^4	$\tau(x)=t_1x_1+t_2x_2+t_4x_4$	$[e_1,e_2,e_4]=t_1e_3$
		$[e_1,e_3,e_4]=t_1e_3$
		$[e_2,e_3,e_4]=t_2e_3$
M^5	$\tau(x)=t_1x_1+t_2x_2+t_4x_4$	$[e_1,e_2,e_4]=t_1e_3$
M^6_{0b}	$\tau(x)=t_1x_1+t_4x_4$	$[e_1,e_2,e_4]=t_1e_3$
		$[e_1,e_3,e_4]=t_1(be_2+e_3)$
M^7_{0b}	$\tau(x)=t_1x_1+t_4x_4$	$[e_1,e_2,e_4]=t_1e_3$
		$[e_1,e_3,e_4]=t_1be_2$
M^8	$\tau(x)=t_1x_1+t_3x_3$	$[e_1,e_2,e_3]=t_3e_2$
		$[e_1,e_3,e_4]=t_1e_4$
M^9_a	$\tau(x)=t_3x_3+t_4x_4$	$[e_1,e_3,e_4]=-t_3(e_1+ae_2)+t_4e_1$
		$[e_2,e_3,e_4]=t_3e_1+t_4e_2$
M^{11}	$\tau(x)=t_4x_4$	$[e_1,e_3,e_4]=t_4e_2$
		$[e_2,e_3,e_4]=t_4e_1$
M^{12}, M^{13}_a		
M^{14}_a, $a\neq 0$	$\tau(x)=t_4x_4$	$[e_1,e_3,e_4]=t_4e_2$
M^{13}_0	$\tau(x)=t_3x_3+t_4x_4$	$[e_1,e_3,e_4]=-t_3e_1+t_4e_2$
		$[e_2,e_3,e_4]=-t_3e_2$
M^{14}_0	$\tau(x)=t_3x_3+t_4x_4$	$[e_1,e_3,e_4]=t_4e_2$
$\mathfrak{gl}_2(\mathbb{K})$	$\tau(x)=t_4x_4$	$[e_1,e_2,e_4]=2t_4e_2$
		$[e_1,e_3,e_4]=-2t_4e_3$
		$[e_2,e_3,e_4]=t_4e_1$
$E_3\times\mathbb{K}$ $(\mathbb{K}=\mathbb{R})$	$\tau(x)=t_4x_4$	$[e_1,e_2,e_4]=t_4e_3$
		$[e_1,e_3,e_4]=-t_4e_2$
		$[e_2,e_3,e_4]=t_4e_1$

where E_3 denotes the 3-dimensional Euclidean space equipped with the cross product.

7 Examples

7.1 Adjoint Representation 1*-Cocycles and Coboundaries*

Here, we give the set of 1-cocycles/coboundaries of 4 chosen Lie algebras ($\mathfrak{gl}_2(\mathbb{K}$ and M^4, M^5 and M^8) in the classification above and 1-cocycles/coboundaries of the induced algebras using a chosen trace map for each one; the computations were done using the computer algebra software Mathematica.

Shortly explained, the computation goes this way:

Let $(A, [.,.])$ be a Lie algebra of dimension n with a basis $B = \{e_1, \dots, e_n\}$, τ a trace and $(A, [.,.,.]_\tau)$ the induced algebra. Denote the structure constants of $(A, [.,.])$ with respect to this basis B by $\left(c_{ij}^k\right)_{1\le i,j,k\le n}$ and by $\left(ct_{ijk}^q\right)_{1\le i,j,k,q\le n}$ those of $(A, [.,.,.]_\tau)$. The linear form τ is represented by the one-line matrix $T = (t_i)_{1\le i\le n}$. A given linear map $f : A \to A$ (1-cochain) may be represented by a $n \times n$ matrix, $Z = \left(z_{ij}\right)_{1\le i,j\le n}$. In terms of structure constants, the condition for f (represented by the matrix Z) to be a cocycle writes for the Lie algebra:

$$\sum_{k=1}^{n} \left(c_{ij}^k z_{qk} - c_{kj}^q z_{ki} - c_{ik}^q z_{k,j}\right) = 0, \forall i, j, q,$$

and for the induced ternary algebra

$$\sum_{p=1}^{n} \left(ct_{ijk}^p z_{qp} - ct_{pjk}^q z_{pi} - ct_{ipk}^q z_{pj} - ct_{ijp}^q z_{pk}\right) = 0, \forall i, j, k, q.$$

By solving these equations, we get a set of conditions, that we apply to Z to get the matrices listed in the tables below, under "Cocycle" and "Ternary cocycle" respectively.

Matrices listed under "Coboundary" and "Ternary coboundary" are obtained by putting in column j respectively $\left[y, e_j\right]$ or $\left[x, y, e_j\right]_\tau$, where $x = (x_1, \dots, x_n)$ and $y = (y_1, \dots, y_n)$, and $x_{ij} = x_i y_j - x_j y_i$.

- $\mathfrak{gl}_2(\mathbb{K})$:

Cocycle	Coboundary	$dimH^1$
$\begin{pmatrix} 0 & z_{12} & z_{13} & 0 \\ -2z_{13} & z_{22} & 0 & 0 \\ -2z_{12} & 0 & z_{22} & 0 \\ 0 & 0 & 0 & z_{44} \end{pmatrix}$	$\begin{pmatrix} 0 & -y_3 & y_2 & 0 \\ -2y_2 & 2y_1 & 0 & 0 \\ 2y_3 & 0 & -2y_1 & 0 \\ 0 & 0 & 0 & 0 \end{pmatrix}$	1
Trace:	$\tau(x) = x_4$	
Ternary cocycle	Ternary cobounary	$dimH_\tau^1$
$\begin{pmatrix} z_{11} & z_{12} & z_{13} & z_{14} \\ -2z_{13} & z_{22} & 0 & z_{24} \\ -2z_{12} & 0 & 2z_{11} - z_{22} & z_{34} \\ 0 & 0 & 0 & -z_{11} \end{pmatrix}$	$\begin{pmatrix} 0 & x_{34} & x_{42} & x_{23} \\ -2x_{42} & -2x_{14} & 0 & 2x_{12} \\ -2x_{34} & 0 & 2x_{14} & -2x_{13} \\ 0 & 0 & 0 & 0 \end{pmatrix}$	1

- M^4:

Cocycle	Coboundary	$dimH^1$
$\begin{pmatrix} z_{11} & z_{12} & 0 & z_{14} \\ z_{21} & z_{22} & 0 & z_{24} \\ -z_{21} & z_{32} & z_{22}+z_{32} & z_{34} \\ 0 & 0 & 0 & 0 \end{pmatrix}$	$\begin{pmatrix} 0 & 0 & 0 & 0 \\ 0 & 0 & 0 & 0 \\ 0 & -y_4 & -y_4 & y_2+y_3 \\ 0 & 0 & 0 & 0 \end{pmatrix}$	6
Trace:	$\tau(x) = x_1 + x_2 + x_4$	
Ternary cocycle		
$\begin{pmatrix} z_{11} & z_{12} & 0 & z_{14} \\ z_{21} & z_{11}-z_{12}+z_{21} & 0 & z_{24} \\ z_{31} & z_{32} & z_{11}-z_{12}-z_{31}+z_{32} & z_{34} \\ z_{41} & z_{41} & 0 & -z_{11}-z_{21} \end{pmatrix}$		
Ternary cobounary	$dimH^1_\tau$	
$\begin{pmatrix} 0 & 0 & 0 & 0 \\ 0 & 0 & 0 & 0 \\ x_{24}+x_{34} & x_{34}-x_{14} & -x_{14}-x_{24} & x_{12}+x_{13}+x_{23} \\ 0 & 0 & 0 & 0 \end{pmatrix}$	6	

- M^5:

Cocycle	Coboundary	$dimH^1$
$\begin{pmatrix} z_{11} & z_{12} & 0 & z_{14} \\ 0 & z_{22} & 0 & z_{24} \\ z_{31} & z_{32} & z_{33} & z_{34} \\ 0 & z_{42} & 0 & z_{33}-z_{22} \end{pmatrix}$	$\begin{pmatrix} 0 & 0 & 0 & 0 \\ 0 & 0 & 0 & 0 \\ 0 & -y_4 & 0 & y_2 \\ 0 & 0 & 0 & 0 \end{pmatrix}$	8
Trace:	$\tau(x) = x_1$	
Ternary cocycle	Ternary cobounary	$dimH^1_\tau$
$\begin{pmatrix} -z_{22}+z_{33}-z_{44} & z_{12} & 0 & z_{14} \\ z_{21} & z_{22} & 0 & z_{24} \\ z_{31} & z_{32} & z_{33} & z_{34} \\ z_{41} & z_{42} & 0 & z_{44} \end{pmatrix}$	$\begin{pmatrix} 0 & 0 & 0 & 0 \\ 0 & 0 & 0 & 0 \\ x_{24} & -x_{14} & 0 & x_{12} \\ 0 & 0 & 0 & 0 \end{pmatrix}$	9

- M^8:

Cocycle	Coboundary	$dimH^1$
$\begin{pmatrix} 0 & 0 & 0 & 0 \\ z_{21} & z_{22} & 0 & 0 \\ 0 & 0 & 0 & 0 \\ 0 & 0 & z_{43} & z_{44} \end{pmatrix}$	$\begin{pmatrix} 0 & 0 & 0 & 0 \\ -y_2 & y_1 & 0 & 0 \\ 0 & 0 & 0 & 0 \\ 0 & 0 & -y_4 & y_3 \end{pmatrix}$	0
Trace:	$\tau(x) = x_1 + x_3$	
Ternary cocycle	Ternary cobounary	$dimH^1_\tau$
$\begin{pmatrix} -z_{33} & 0 & z_{13} & 0 \\ z_{21} & z_{22} & z_{23} & 0 \\ z_{31} & 0 & z_{33} & 0 \\ z_{41} & 0 & z_{43} & z_{44} \end{pmatrix}$	$\begin{pmatrix} 0 & 0 & 0 & 0 \\ x_{23} & -x_{13} & x_{12} & 0 \\ 0 & 0 & 0 & 0 \\ x_{34} & 0 & -x_{14} & x_{13} \end{pmatrix}$	4

References

1. Arnlind J., Makhlouf A., Silvestrov S.: Ternary Hom-Nambu-Lie algebras induced by Hom-Lie algebras. J. Math. Phys. **51**, 043515, 11 (2010)
2. Arnlind J., Makhlouf A., Silvestrov S.: Construction of n-Lie algebras and n-ary Hom-Nambu-Lie algebras, J. Math. Phys. **52**, 123502, 13 (2011)
3. Awata H., Li M., Minic D., Yoneya T.: On the quantization of Nambu brackets, J. High Energy Phys. **2**, Paper 13, 17 (2001)
4. Bai, R., Song, G., Zhang, Y.: On classification of n-Lie algebras. Front. Math. China **6**, 581–606 (2011)
5. Daletskii, Y.L., Takhtajan, L.A.: Leibniz and Lie algebra structures for Nambu algebra. Lett. Math. Phys. **39**, 127–141 (1997)
6. de Azcárraga, J.A., Izquierdo, J.M.: n-ary algebras: a review with applications. J. Phys. **A43**, 293001-1–117 (2010)
7. de Graaf, W.A.: Classification of solvable Lie algebras. Experiment. Math. **14**, 15–25 (2005)
8. Filippov, V.T.: n-Lie algebras. Siberian Math. J. **26**, 879–891 (1985)
9. Gautheron, Ph: Some remarks concerning Nambu mechanics. Lett. Math. Phys. **37**(1), 103–116 (1996)
10. Kasymov, Sh. M.: Theory of n-Lie algebras. Algebra Logic **26**, 155–166 (1987)
11. Nambu, Y.: Generalized Hamiltonian dynamics. Phys. Rev. D **3**(7), 2405–2412 (1973)
12. Patera, J., Sharp, R.T., Winternitz, P., Zassenhaus, H.: Invariants of real low dimension Lie algebras. J. Math. Phys. **17**, 986–994 (1976)
13. Takhtajan, L.A.: Higher order analog of Chevalley-Eilenberg complex and deformation theory of n-gebras. St. Petersburg Math. J. **6**(2), 429–438 (1995)

A Review of Peirce Decomposition for Unitary $(-1, -1)$-Freudenthal Kantor Triple Systems

Noriaki Kamiya, Daniel Mondoc and Susumu Okubo

Abstract In this paper we discuss a Peirce decomposition for unitary $(-1, -1)$-Freudenthal Kantor triple systems.

1 Introduction

Lie algebra is rich in algebraic structures and provides an important common ground for various branches of mathematics, not only for geometry and analysis, but also for physics. The historical background of our study goes back to Freudenthal [9], Tits [54], Kantor [36–38] and Koecher [41] who studied constructions of Lie algebras from nonassociative algebras and triple systems, in particular Jordan algebras. Allison [1, 2] defined the concept of structurable algebras, containing Jordan algebras. We have studied constructions of Lie algebras and superalgebras from triple systems [24, 26, 29–31].

As a continuation of [29, 30, 32] we are interested to characterize the structure properties [7, 19, 27, 34, 40] of the subspace L_{-1} of the five graded Lie (super)algebra $L(\varepsilon, \delta) := L_{-2} \oplus L_{-1} \oplus L_0 \oplus L_1 \oplus L_2$, $[L_i, L_j] \subseteq L_{i+j}$, associated with an (ε, δ)-Freudenthal Kantor triple system. Thus we discuss here a Peirce decomposition for unitary $(-1, -1)$-Freudenthal Kantor triple systems. A Peirce decomposition for

N. Kamiya (✉)
Center for Mathematical Sciences, University of Aizu, Aizuwakamatsu 965-8580, Japan
e-mail: kamiya@u-aizu.ac.jp

D. Mondoc
Centre for Mathematical Sciences, Lund University, 22100 Lund, Sweden
e-mail: daniel.mondoc@math.lu.se

S. Okubo
Department of Physics, University of Rochester, Rochester, NY 14627, USA
e-mail: okubo@pas.rochester.edu

A. Makhlouf et al. (eds.), *Algebra, Geometry and Mathematical Physics*,
Springer Proceedings in Mathematics & Statistics 85, DOI: 10.1007/978-3-642-55361-5_10,

$(-1, 1)$-Freudenthal Kantor triple systems was discussed in [40] and examples were given in [34].

Jordan and Lie (super) algebras [12, 53] play an important role in many mathematical and physical subjects [5, 10–13, 15, 25, 28, 39, 48, 49, 56, 57]. We also note that the construction and characterization of these algebras can be expressed in terms of triple systems [19, 22, 23, 27, 40, 50] by using the standard embedding method [21, 42, 43, 51, 55] and motivating the study of triple systems.

Summarizing the content we give the introduction in Sect. 1, definitions and preamble in Sect. 2, the main theorem in Sect. 3, examples in Sect. 4 and concluding remark in Sect. 5.

2 Definitions and Preamble

2.1 (ε, δ)-Freudenthal Kantor Triple Systems

We are concerned in this paper with triple systems which have finite dimension over a field Φ of characteristic $\neq 2$ or 3.

In order to render this paper as self-contained as possible, we recall first the definition of a generalized Jordan triple system of second order (for short GJTS of 2nd order).

A vector space V over a field Φ endowed with a trilinear operation $V \times V \times V \to V$, $(x, y, z) \longmapsto (xyz)$ is said to be a *GJTS of 2nd order* if the following conditions are fulfilled:

$$(ab(xyz)) = ((abx)yz) - (x(bay)z) + (xy(abz)), \tag{1}$$

$$K(K(a, b)x, y) - L(y, x)K(a, b) - K(a, b)L(x, y) = 0, \tag{2}$$

where $L(a, b)c := (abc)$ and $K(a, b)c := (acb) - (bca)$.

A *Jordan triple system* (for short JTS) satisfies (1) and the identity $(abc) = (cba)$.

We can generalize the concept of GJTS of 2nd order as follows (see [13, 14, 17–21, 55] and the earlier references therein). For $\varepsilon = \pm 1$, $\delta = \pm 1$, a triple product that satisfies

$$(ab(xyz)) = ((abx)yz) + \varepsilon(x(bay)z) + (xy(abz)), \tag{3}$$

$$K(K(a, b)x, y) - L(y, x)K(a, b) + \varepsilon K(a, b)L(x, y) = 0, \tag{4}$$

where

$$L(a, b)c := (abc), \quad K(a, b)c := (acb) - \delta(bca), \tag{5}$$

is called an (ε, δ)-*Freudenthal Kantor triple system* (for short (ε, δ)-FKTS).

Remark We note that the concept of GJTS of 2nd order coincides with that of $(-1,1)$-FKTS. Thus we can construct the simple Lie algebras by means of the standard embedding method [6, 13–17, 21, 24, 26, 38, 55].

For an (ε,δ)-FKTS U we denote

$$A(a,b) := L(a,b) - \varepsilon L(b,a), \tag{6}$$

where $L(a,b)$ is defined by (5). Then $A(a,b)$ is an anti-derivation of U [30], that is

$$[A(a,b), L(c,d)] = L(A(a,b)c, d) - L(c, A(a,b)d). \tag{7}$$

An (ε,δ)-FKTS U is called *unitary* if the identity map Id is contained in $\kappa := K(U,U)$ i.e., if there exist $a_i, b_i \in U$, such that $\Sigma_i K(a_i,b_i) = Id$.

Remark We note that a balanced triple system (i.e. it fulfills $K(x,y) =< x|y >' Id$, where $< \mid >'$ is a symmetric bilinear form) is unitary, since $Id \in \kappa = K(U,U)$.

We show in the following remark the equivalence between the balanced notion defined above and the one of [7].

Remark We note that for a triple system U with product (xyz), $x,y,z \in U$, the notion of balanced $(-1,-1)$-FKTS is equivalent to saying that the triple system satisfies (3) and

$$(xxy) = (xyx) =< x|x > y, \quad x,y \in U, \tag{8}$$

where $< \mid >$ is a symmetric bilinear form.

Indeed, if U is a $(-1,-1)$-FKTS then (3) is fulfilled and, by (4),

$$K(K(a,b)x,y) - L(y,x)K(a,b) - K(a,b)L(x,y) = 0, \quad a,b,x,y \in U,$$

that is $K(x,y) = L(y,x) + L(x,y) =< x|y >' Id$, since $K(x,y) =< x|y >' Id$, hence

$$(xwy) + (ywx) = (xyw) + (yxw) =< x|y >' w, \quad x,y,w \in U.$$

If we put now $x = y$ in the last line follows $2(xwx) = 2(xxw) =< x|x >' w$, that is (8) is valid for the symmetric form $< \mid >= \frac{1}{2} < \mid >'$.

Opposite, by linearizing (8), we have

$$(xzy) + (yzx) = (xyz) + (zyx) =< x|z > y+ < z|x > y, \quad x,y,z \in U,$$

hence, by (5), $K(x,z) = 2 < x|z >$, $x,z \in U$, so $K(x,y) =< x|y >' Id$, $x,y \in U$.

Remark We note that for an unitary $(-1,-1)$-FKTS U we have, by [30] (Proposition 2.17 for $\varepsilon = -1$), that $K(a,b) = A(a,b)$, for all $a,b \in U$, thus by (6), for $\delta = -1$, follows

$$K(a, b) = L(a, b) + L(b, a), \tag{9}$$

where $L(a, b)$ is defined by (5). Moreover, by [30] (Proposition 2.17),

$$K(u, v)K(x, y) + K(x, y)K(u, v) = K(K(u, v)x, y) + K(x, K(u, v)y), \quad u, v, x, y \in U, \tag{10}$$

thus, by [30] (Proposition 2.18), the commutative product in κ defined by

$$K(u, v) * K(x, y) = K(u, v)K(x, y) + K(x, y)K(u, v)$$

defines a Jordan algebra.

Let U be a triple system with product $(xyz), x, y, z \in U$. An element $e \in U$ is called a *left unit element* if

$$(eex) = x. \tag{11}$$

An element $e \in U$ is called a *tripotent* if

$$(eee) = e. \tag{12}$$

We denote

$$L(x) := (eex), \quad Q(x) := (exe), \quad R(x) := (xee), \quad x \in U. \tag{13}$$

2.2 δ-Structurable Algebras

Within the framework of (ε, δ)-FKTSs $(\varepsilon, \delta = \pm 1)$ and the standard embedding Lie (super) algebra construction [6, 7, 13–15, 26] we defined δ-structurable algebras [29] as a class of nonassociative algebras with involution which coincides with the class of structurable algebras for $\delta = 1$ as introduced in [1, 2]. Structurable algebras are a class of nonassociative algebras with involution related to $(-1, 1)$-FKTSs as introduced in [36, 37] (and further studied in [3, 4, 35, 44–47, 52]). Their importance lies with constructions of five graded Lie algebras $L(-1, 1)$. For $\delta = -1$ the anti-structurable algebras [29] are a class of nonassociative algebras that may similarly shed light on $(-1, -1)$-FKTSs hence (by [6, 7]) on the construction of Lie superalgebras and Jordan algebras.

Let $(\mathscr{A}, ^-)$ be a finite dimensional nonassociative unital algebra with involution (involutive anti-automorphism, i.e. $\overline{\overline{x}} = x, \overline{xy} = \overline{y}\,\overline{x}, x, y \in \mathscr{A}$) over Φ. Since char$\Phi \neq 2$, by [1] we have $\mathscr{A} = \mathscr{H} \oplus \mathscr{S}$, where $\mathscr{H} = \{a \in \mathscr{A} | \overline{a} = a\}$ and $\mathscr{S} = \{a \in \mathscr{A} | \overline{a} = -a\}$.

Suppose $x, y, z \in \mathscr{A}$. Put $[x, y] := xy - yx$ and $[x, y, z] := (xy)z - x(yz)$ so $\overline{[x, y, z]} = -[\overline{z}, \overline{y}, \overline{x}]$. The operators L_x and R_x are defined by $L_x(y) := xy$, $R_x(y) := yx$.

For $\delta = \pm 1$ and $x, y \in \mathscr{A}$ define

$$^{\delta}V_{x,y} := L_{L_x(\bar{y})} + \delta(R_x R_{\bar{y}} - R_y R_{\bar{x}}), \tag{14}$$

$$^{\delta}B_{\mathscr{A}}(x, y, z) :=^{\delta} V_{x,y}(z) = (x\bar{y})z + \delta[(z\bar{y})x - (z\bar{x})y], x, y, z \in \mathscr{A}. \tag{15}$$

$^{+}B_{\mathscr{A}}(x, y, z)$ is called the *triple system obtained from the algebra* $(\mathscr{A}, ^{-})$. We will call $^{-}B_{\mathscr{A}}(x, y, z)$ the *anti-triple system obtained from the algebra* $(\mathscr{A}, ^{-})$. We write for short

$$V_{x,y} :=^{\delta} V_{x,y}, \quad B_{\mathscr{A}} := (^{\delta}B_{\mathscr{A}}, \mathscr{A}). \tag{16}$$

A unital non-associative algebra with involution $(\mathscr{A}, ^{-})$ is called a *structurable algebra* if

$$[V_{u,v}, V_{x,y}] = V_{V_{u,v}(x),y} - V_{x,V_{v,u}(y)}, \tag{17}$$

for $V_{u,v} = {}^{+}V_{u,v}, V_{x,y} = {}^{+}V_{x,y}, u, v, x, y \in \mathscr{A}$, and we call $(\mathscr{A}, ^{-})$ an *anti-structurable algebra* if the identity (17) is fulfilled for $V_{u,v} = ^{-}V_{u,v}, V_{x,y} = ^{-}V_{x,y}$.

If $(\mathscr{A}, ^{-})$ is structurable then, by [37], the triple system $B_{\mathscr{A}}$ is called a *generalized Jordan triple system* (abbreviated GJTS) and by [8], $B_{\mathscr{A}}$ is a GJTS of 2nd order.

3 Main Theorem

We give first two lemmas.

Lemma 3.1 *Let U be an unitary $(-1, -1)$-FKTS with a tripotent $e \in U$. Then*

$$Q(x) = L(x) = L^2(x), \ RL(x) = LR(x), R^2(x) - L^2(x) + LR(x) = R(x), x \in U,$$

where $L(x), Q(x), R(x)$ are defined by (13). Moreover the following decomposition is valid

$$U = U_0 \oplus U_1, \quad U_i := \{x \in U | L(x) = ix\}, i = 0, 1. \tag{18}$$

Proof We remark first that by (5), for $\delta = -1$, it follows $K(e, e)x = 2Q(x)$, for all $x \in U$. Further, by (9), it follows $K(e, e)x = 2L(x)$ hence $Q(x) = L(x)$, for all $x \in U$. Moreover, by (5), for $\delta = -1$, and (12) it follows $K(e, e)e = 2e$. Then, by (10), it follows

$$2K(e, e)(K(e, e)x) = K(K(e, e)e, e)x + K(e, K(e, e)e)x,$$

hence $Q^2(x) = Q(x)$ for all $x \in U$. Then $L^2(x) = L(x)$ since $Q(x) = L(x)$, for all $x \in U$.

Further, by (3),

$$(ee(xee)) = ((eex)ee) - (x(eee)e) + (xe(eee)),$$

or equivalently, by (12),

$$(ee(xee)) = ((eex)ee) - (xee) + (xee),$$

that is $LR(x) = RL(x)$, for all $x \in U$.

Now, by (3), it follows $(xe(eee)) = ((xee)ee) - (e(exe)e) + (ee(xee))$, or equivalently, by (12), $R(x) = R^2(x) - Q^2(x) + LR(x)$, that is $R(x) = R^2(x) - L^2(x) + LR(x)$ since $Q(x) = L(x)$, for all $x \in U$.

Finally, since $L^2 = L$ then the decomposition (18) is straightforward.

Remark Note that, by (18), $e \in U_1$ since $L(e) = (eee) = e$, by (12).

Lemma 3.2 *Let U be an unitary $(-1,-1)$-FKTS with a tripotent $e \in U$ and decomposition (18). If $x_i \in U_i$, $i = 0, 1$, then $R^2(x_0) = R(x_0)$, $R^2(x_1) = x_1$, where $R(x)$ is defined by (13).*

Proof Since, by Lemma 3.1, $R^2(x_i) - L^2(x_i) + LR(x_i) = R(x_i), x_i \in U_i$ then the assertions follow straightforward from the definition of the decomposition (18).

Theorem 3.1 *Let U be an unitary $(-1,-1)$-FKTS with a tripotent $e \in U$. Then we have a Peirce decomposition $U = U_0 \oplus U_1$, $U_i := \{x \in U | L(x) = ix\}, i = 0, 1$, such that*

$$\begin{array}{l} (U_1U_1U_1) \subseteq U_1,\ (U_0U_0U_0) \subseteq U_0,\ (U_0U_1U_1) \subseteq U_0, \\ (U_1U_1U_0) \subseteq U_0,\ (U_0U_0U_1) \subseteq U_1,\ (U_1U_0U_0) \subseteq U_1, \\ (U_0U_1U_0) = 0, \quad (U_1U_0U_1) = 0. \end{array} \tag{19}$$

Proof The existence of the Peirce decomposition has been proved in Lemma 3.1.

Let now $x_i \in U_i, y_j \in U_j, z_k \in U_k, i, j, k \in \{0, 1\}$. Then, by (3), it follows

$$(ee(x_iy_jz_k)) = ((eex_i)y_jz_k) - (x_i(eey_j)z_k) + (x_iy_j(eez_k)).$$

Since $L(x) = ix$, $i = 0, 1$, then from the last identity it follows

$$L(x_iy_jz_k) = (i - j + k)(x_iy_jz_k), \quad i, j, k \in \{0, 1\}.$$

It is then a straightforward calculation to show (19), since e.g. $L(x_0y_0z_0) = 0$, but also remarking that $L(x_0y_1z_0) = -(x_0y_1z_0) = 0$, $L(x_1y_0z_1) = 2(x_0y_1z_0) = 0$.

Corollary 3.1 *Let U be an unitary $(-1,-1)$-FKTS with a tripotent $e \in U$. Then the subspace U_1 defined by (18) is an unitary $(-1,-1)$-FKTS such that $e \in U_1$ and moreover $L(x_1) = Q(x_1) = R^2(x_1) = x_1, x_1 \in U_1$, where $L(x), Q(x), R(x)$ are defined by (13).*

Remark By [33], Theorem 3.2, a $(-1,-1)$-FKTS U with product $(xyz), x, y, z \in U$, and left unit element e can be determined in terms of the bilinear product of U defined by

$$x \cdot y = (exy), \quad x, y \in U, \tag{20}$$

while by [40], a $(-1, 1)$-FKTS U with product $(xyz), x, y, z \in U$, and left unit element e can be determined in terms of the bilinear product of U defined by $x \circ y = (xey), x, y \in U$, that is, by Theorem 3.3 [40], $U = U_{11}^{+} \oplus U_{11}^{-} \oplus U_{13}^{+} \oplus U_{13}^{-}$ with the product

$$(xyz) = (Q^{-1}(y) \circ x) \circ z + x \circ (Q^{-1}(y) \circ z) - Q^{-1}(y) \circ (x \circ z),$$

where $Q(x) = \begin{cases} \pm x, \text{ if } x \in U_{11}^{\pm} \\ \pm 3x, \text{ if } x \in U_{13}^{\pm} \end{cases}$ and $U_{1i} := \{x \in U | R(x) = ix\}$.

4 Examples

4.1 *Balanced* $(-1, -1)$-*FKTSs*

Let U be a balanced $(-1, -1)$-FKTS with product

$$(xyz) =< z|x > y- < z|y > x+ < x|y > z, x, y, z \in U,$$

where $< .|. >: U \otimes U \to \Phi$ is a symmetric bilinear form. Then, by Sect. 2.1, U is an unitary $(-1, -1)$-FKTS. Further, if $U = \{e_1, \dots, e_n\}_{span}$ such that $< e_i|e_j >= \delta_{ij}, i, j = 1 \dots, n$ and $e := e_i$ then the following decomposition is valid

$$U = U_0 \oplus U_1, \text{ where } U_0 = \{0\} \text{ and } U_1 = U = \{x \in U | L(x) = (eex) = x\}.$$

In fact, $R(x) = (xee) = 2 < e|x > e - x$. Then $R(x) = x, x \in U$, if and only if $x \in \{e_i\}_{span}$ and $R(x) = -x, x \in U$, if and only if $x \in \{e_1, \dots, e_{i-1}, e_{i+1}, \dots, e_n\}_{span}$. Thus we have a decomposition of U_1 with respect to $R(x)$ as follows

$$\begin{aligned} U_1 = U_{1,1} \oplus U_{1,-1}, \text{ where } U_{1,1} = \{e_i\}_{span} \text{ and } U_{1,-1} \\ = \{e_1, \dots, e_{i-1}, e_{i+1}, \dots, e_n\}_{span}. \end{aligned}$$

4.2 *Anti-Structurable Algebras*

Let $\mathcal{M}_{m,n}(\Phi)$ denote the vector space of $m \times n$ matrices over Φ and for $x \in \mathcal{M}_{m,n}(\Phi)$ denote by $x^\top$ the transposed matrix. Then $U := \mathcal{M}_{n,n}(\Phi)$ with the product

$$(xyz) = xy^\top z - zy^\top x + zx^\top y, \quad x, y, z \in \mathcal{M}_{n,n}(\Phi) \tag{21}$$

is an anti-structurable algebra and a $(-1, -1)$-FKTS [33]. Further, by (5), straightforward calculations give $K(x, y)z = (xy^\top + (xy^\top)^\top)z, x, y, z \in U$, hence U is unitary.

If $e = E_n$ is the identity matrix of order n then the following decomposition is valid

$$U = U_0 \oplus U_1, \text{ where } U_0 = \{0_n\} \text{ and } U_1 = U = \{x \in U | L(x) = (eex) = x\}.$$

Remark If $e = \left\{ \left(\begin{array}{c|c} E_l & 0 \\ \hline 0 & 0 \end{array} \right) | E_l \in \mathcal{M}_{l,l}(\Phi), l < n \right\}$ and

$$U = \left\{ \left(\begin{array}{c|c} A & B \\ \hline C & D \end{array} \right) | A \in \mathcal{M}_{l,l}(\Phi), B \in \mathcal{M}_{l,n-l}(\Phi), C \in \mathcal{M}_{n-l,l}(\Phi), D \in \mathcal{M}_{n-l,n-l}(\Phi) \right\}$$

then the following decomposition is valid

$$U = U_0 \oplus U_1, \text{ where } U_0 = \left\{ \left(\begin{array}{c|c} 0 & 0 \\ \hline C & D \end{array} \right) \right\} \text{ and } U_1 = \left\{ \left(\begin{array}{c|c} A & B \\ \hline 0 & 0 \end{array} \right) \right\}.$$

4.3 Non Unitary Example

Let U be a vector space with product

$$(xyz) =< y|z > x, x, y, z \in U,$$

where $< .|. >: U \otimes U \to \Phi$ is a symmetric bilinear form. Then U is a $(-1, -1)$-FKTS but not balanced and not unitary.

5 Concluding Remark

By [26], the following construction of Lie superalgebras is obtained by the standard embedding method. If $U(-1, -1) := \mathcal{M}_{2n,m}(\Phi)$ with the product (21) then the corresponding standard embedding Lie superalgebra is $\mathbf{osp}(2n|2m) = D(n, m)$ (as defined by [12]), hence the standard embedding Lie superalgebra of the anti-structurable algebra $\mathcal{M}_{2n,2n}(\Phi)$ is $\mathbf{osp}(2n|4n)$.

Similarly, if $U(-1, -1) := \mathcal{M}_{2n+1,m}(\Phi)$ with the product (21) then the corresponding standard embedding Lie superalgebra is $\mathbf{osp}(2n + 1|2m) = B(n, m)$ [12], hence the standard embedding Lie superalgebra of the anti-structurable algebra $\mathcal{M}_{2n+1,2n+1}(\Phi)$ is $\mathbf{osp}(2n+1|4n+2)$. Hence the construction of Lie (super)algebras from triple systems motivates the study of the structure properties of triple systems. Finally and briefly describing, it seems that the concept of Peirce decomposition is closely related to the inner structure within the quarks theory in physics as well as the structure of Lie algebras or superalgebras.

Acknowledgments The authors N. Kamiya acknowledge partial supported for this research by Grant-in-Aid for Scientific Research (No. 19540042 (C),(2)), Japan Society for the Promotion of Science and S. Okubo acknowledge support by U.S. Department of Energy Grant No. DE-FG02-91ER40685.

References

1. Allison, B.N.: A class of nonassociative algebras with involution containing the class of jordan algebras. Math. Ann. **237**(2), 133–156 (1978)
2. Allison, B.N.: Models of isotropic simple lie algebras. Comm. Algebra **7**(17), 1835–1875 (1979)
3. Asano, H., Yamaguti, K.: A construction of lie algebras by generalized jordan triple systems of second order. Nederl. Akad. Wetensch. Indag. Math. **42**(3), 249–253 (1980)
4. Asano, H.: Classification of non-compact real simple generalized jordan triple systems of the second kind. Hiroshima Math. J. **21**, 463–489 (1991)
5. Bertram W.: Complex and quaternionic structures on symmetric spaces—correspondence with Freudenthal-Kantor triple systems. In: Miyaoka, R., Tamaru, H. (eds.) Theory of lie groups and manifolds, Sophia Kokyuroku in mathematics, vol. 45, pp. 61–80 (2002)
6. Elduque, A., Kamiya, N., Okubo, S.: Simple $(-1,-1)$ balanced freudenthal kantor triple systems. Glasg. Math. J. **11**(2), 353–372 (2003)
7. Elduque, A., Kamiya, N., Okubo, S.: $(-1,-1)$ balanced freudenthal kantor triple systems and noncommutative jordan algebras. J. Algebra **294**(1), 19–40 (2005)
8. Faulkner, J.R.: Structurable triples, lie triples, and symmetric spaces. Forum Math. **6**, 637–650 (1994)
9. Freudenthal, H.: Beziehungen der e7 und e8 zur oktavenebene I. Indag. Math. **16**, 218–230 (1954)
10. Jacobson, N.: Lie and jordan triple systems. Am. J. Math. **71**, 149–170 (1949)
11. Jacobson N.: Structure and representations of Jordan algebras, vol. 39, pp. 452–453. American Mathematical Society Colloquium Publications, Providence (1968)
12. Kac, V.G.: Lie superalgebras. Adv. Math. **26**(1), 8–96 (1977)
13. Kamiya, N.: A structure theory of freudenthal-kantor triple systems. J. Algebra **110**(1), 108–123 (1987)
14. Kamiya, N.: A construction of anti-lie triple systems from a class of triple systems. Mem. Fac. Sci. Shimane Univ. **22**, 51–62 (1988)
15. Kamiya, N.: A structure theory of freudenthal-kantor triple systems II. Comment. Math. Univ. St. Paul. **38**(1), 41–60 (1989)
16. Kamiya, N.: On (ε, δ)-Freudenthal-Kantor triple systems. Nonassociative algebras and related topics (Hiroshima, 1990) pp. 65–75. World Scientific Publishing, River Edge (1991)
17. Kamiya, N.: The construction of all simple Lie algebras over **C** from balanced Freudenthal-Kantor triple systems. Contributions to general algebra (Vienna, 1990) vol. 7, pp. 205–213. Hölder-ichler Tempsky, Vienna (1991)
18. Kamiya N. On Freudenthal-Kantor triple systems and generalized structurable algebras. Non-associative algebra and its applications (Oviedo, 1993) pp. 198–203. Mathematics and its Applications, vol. 303. Kluwer Academic Publishers, Dordrecht (1994)
19. Kamiya, N.: On the peirce decompositions for freudenthal-kantor triple systems. Commun. Algebra **25**(6), 1833–1844 (1997)
20. Kamiya, N.: On a realization of the exceptional simple graded lie algebras of the second kind and freudenthal-kantor triple systems. Bull. Polish Acad. Sci. Math. **46**(1), 55–65 (1998)
21. Kamiya, N., Okubo, S.: On δ-lie supertriple systems associated with (ε, δ)-freudenthal-kantor supertriple systems. Proc. Edinb. Math. Soc. **43**(2), 243–260 (2000)

22. Kamiya N., Okubo S.: A construction of Jordan superalgebras from Jordan-Lie triple systems. Lecture Notes in Pure and Applied Mathematics, vol. 211. Non-Associative Algebra and its Applications (ed. Costa, Peresi, etc.) pp. 171–176. Marcel Dekker Inc (2002)
23. Kamiya N., Okubo S.: A construction of simple Jordan superalgebra of F type from a Jordan-Lie triple system. Ann. Mat. Pura Appl. **181**(3), 339–348 (2002)
24. Kamiya, N., Okubo, S.: Construction of lie superalgebras $d(2, 1; \alpha)$, $g(3)$ and $f(4)$ from some triple systems. Proc. Edinb. Math. Soc. **46**(1), 87–98 (2003)
25. Kamiya N., Okubo S.: On generalized Freudenthal-Kantor triple systems and Yang-Baxter equations. In: Proceedings of 22nd international colloquium on group theoretical methods in physics, IPCS, vol. 173, pp. 815–818 (2003)
26. Kamiya, N., Okubo, S.: A construction of simple lie superalgebras of certain types from triple systems. Bull. Aust. Math. Soc. **69**(1), 113–123 (2004)
27. Kamiya N.: Examples of Peirce decomposition of generalized Jordan triple system of second order-Balanced cases. Noncommutative geometry and representation theory in mathematical physics, vol. 391, pp. 157–165. Contemporary Mathematics/American Mathematical Society, Providence (2005)
28. Kamiya N., Okubo S.: Composition, quadratic, and some triple systems. Non-associative algebra and its applications, vol. 246, pp. 205–231. Lecture Notes in Pure and Applied Mathematics, Chapman & Hall/CRC, Boca Raton, FL (2006)
29. Kamiya, N., Mondoc, D.: A new class of nonassociative algebras with involution. Proc. Jpn. Acad. Ser. A **84**(5), 68–72 (2008)
30. Kamiya, N., Mondoc, D., Okubo, S.: A structure theory of $(-1, -1)$-freudenthal kantor triple systems. Bull. Aust. Math. Soc. **81**, 132–155 (2010)
31. Kamiya, N., Mondoc, D.: On anti-structurable algebras and extended dynkin diagrams. J. Gen. Lie Theor. Appl. **3**(3), 185–192 (2009)
32. Kamiya, N., Mondoc, D., Okubo, S.: A review on δ-structurable algebras. In: Abramov, V., Fuchs, J., Paal, E., Stolin, A., Tralle, A., Urbanski, P. (eds.) Proceedings of algebra, geometry and mathematical physics 5, 2009. Banach Center Publication, vol. 93, pp. 59–67 (2011)
33. Kamiya, N., Mondoc, D., Okubo, S.: A characterization of $(-1, -1)$-freudenthal-kantor triple systems. Glasgow Math. J. **53**, 727–738 (2011)
34. Kamiya, N., Mondoc, D.: Examples of peirce decomposition for compact simple kantor triple systems. Algebras Groups Geom. **24**(3), 325–347 (2007)
35. Kaneyuki, S., Asano, H.: Graded lie algebras and generalized jordan triple systems. Nagoya Math. J. **112**, 81–115 (1988)
36. Kantor, I.L.: Graded lie algebras. Trudy Sem. Vect. Tens. Anal. **15**, 227–266 (1970)
37. Kantor, I.L.: Some generalizations of jordan algebras. Trudy Sem. Vect. Tens. Anal. **16**, 407–499 (1972)
38. Kantor, I.L.: Models of exceptional lie algebras. Sov. Math. Dokl. **14**(1), 254–258 (1973)
39. Kantor I.L.: A generalization of the Jordan approach to symmetric Riemannian spaces. The Monster and Lie algebras (Columbus, OH, 1996) vol. 7, pp. 221–234. Ohio State University Mathematical Research Institute Publications, de Gruyter, Berlin (1998)
40. Kantor, I.L., Kamiya, N.: A peirce decomposition for generalized jordan triple systems of second order. Commun. Algebra **31**(12), 5875–5913 (2003)
41. Koecher M.: Embedding of Jordan algebras into Lie algebras I, II. Am. J. Math. **89** (1967), 476–510, 787–816 and 90 (1968)
42. Lister, W.G.: A structure theory of lie triple systems. Trans. Am. Math. Soc. **72**, 217–242 (1952)
43. Meyberg K.: Lectures on algebras and triple systems. Lecture Notes, pp. v+226. The University of Virginia, Charlottesville, Va (1972)
44. Mondoc, D.: Models of compact simple kantor triple systems defined on a class of structurable algebras of skew-dimension one. Commun. Algebra **34**(10), 3801–3815 (2006)
45. Mondoc, D.: On compact realifications of exceptional simple kantor triple systems. J. Gen. Lie Theor. Appl. **1**(1), 29–40 (2007)
46. Mondoc, D.: Compact realifications of exceptional simple kantor triple systems defined on tensor products of composition algebras. J. Algebra **307**(2), 917–929 (2007)

47. Mondoc, D.: Compact exceptional simple kantor triple systems defined on tensor products of composition algebras. Commun. Algebra **35**(11), 3699–3712 (2007)
48. Neher E.: Jordan triple systems by the grid approach. Lecture Notes in Mathematics, vol. 1280, pp. 133–193. Springer, Berlin (1987)
49. Okubo S.: Introduction to octonion and other non-associative algebras in physics. Montroll Memorial Lecture Series in Mathematical Physics, vol. 2, pp. 133–136. Cambridge University Press, Cambridge (1995)
50. Okubo, S., Kamiya, N.: Jordan-lie superalgebra and jordan-lie triple system. J. Algebra **198**(2), 388–411 (1997)
51. Okubo, S., Kamiya, N.: Quasi-classical lie superalgebras and lie supertriple systems. Commun. Algebra **30**(8), 3825–3850 (2002)
52. Okubo, S.: Symmetric triality relations and structurable algebras. Linear Algebra Appl. **396**, 189–222 (2005)
53. Scheunert M.: The theory of Lie superalgebras. An introduction. Lecture Notes in Mathematics, vol. 716, pp. x+271. Springer, Berlin (1979)
54. Tits, J.: Une classe d'algèbres de lie en relation avec les algèbres de jordan. Nederl. Acad. Wet. Proc. Ser. A **65**, 530–535 (1962)
55. Yamaguti, K., Ono, A.: On representations of Freudenthal-Kantor triple systems $U(\varepsilon, \delta)$. Bull. Fac. Sch. Edu. Hiroshima Univ. **7**(2), 43–51 (1984)
56. Zhevlakov, K.A., Slinko, A.M., Shestakov, I.P., Shirshov, A.I.: Rings that are nearly associative, pp. xi+371. Academic Press Inc, New York (1982)
57. Zelmanov E.: Primary Jordan triple systems. Sibirsk. Math. Zh. **24**(4), 23–37 (1983)

Universal Algebra Applied to Hom-Associative Algebras, and More

Lars Hellström, Abdenacer Makhlouf and Sergei D. Silvestrov

Abstract The purpose of this paper is to discuss the universal algebra theory of hom-algebras. This kind of algebra involves a linear map which twists the usual identities. We focus on hom-associative algebras and hom-Lie algebras for which we review the main results. We discuss the envelopment problem, operads, and the Diamond Lemma; the usual tools have to be adapted to this new situation. Moreover we study Hilbert series for the hom-associative operad and free algebra, and describe them up to total degree equal 8 and 9 respectively.

1 Introduction

Abstract algebra is a subject that may be investigated on many different levels of maturity. At the most elementary level that still meets the standards of mathematical rigor, the investigator simply postulates some set of axioms (usually in the form of a definition) and then goes on to derive random consequences of these axioms, hopefully topping it off with examples to illustrate the range of possible outcomes for the results that are stated (as there have been some spectacular instances of mathematical theories that died due to having no nontrivial examples where they were applicable). This level of investigation may produce a nicely whole theory of something, but in the hands of an immature investigator it runs a significant risk of

L. Hellström (✉) · S. D. Silvestrov
Division of Applied Mathematics, The School of Education, Culture and Communication, Mälardalen University, Box 883, 721 23 Västerås, Sweden
e-mail: Lars.Hellstrom@residenset.net

S. D. Silvestrov
e-mail: sergei.silvestrov@mdh.se

A. Makhlouf
LMIA, University of Haute Alsace, 4 rue des Frères Lumière,68093 Mulhouse, France
e-mail: Abdenacer.Makhlouf@uha.fr

A. Makhlouf et al. (eds.), *Algebra, Geometry and Mathematical Physics*,
Springer Proceedings in Mathematics & Statistics 85, DOI: 10.1007/978-3-642-55361-5_11,

ending up as a random collection of facts that don't combine to anything greater than themselves; the whole of a good theory should be greater than the sum of its parts.

One way of reaching a higher level can be to investigate matters using the techniques of universal algebra, since these combine looking at concrete examples with the generality of investigating the generic case. Another way is to employ the language of category theory to investigate matters on a level that is even more abstract. Indeed, category theory has become so fashionable that modern presentations of universal algebra may treat it as a mere application of the categorical formalism. This has the advantage of allowing definitions of for example free algebras to be given that do not presuppose a specific construction machinery, but on the other hand it runs the risk of losing itself in the heavens of abstraction, because the difficulties have been postponed rather than taken care of; doing any nontrivial example may bring them all back with a vengeance. Therefore we were glad to see how Yau in [44] would proceed from an abstract categorical definition to concrete constructions of many free algebras of relevance to hom-associative and hom-Lie algebras—glad, but also a bit curious as to why the constructions were not more systematic.

For better or worse, there is probably a simple reason for someone doing *ad hoc* constructions rather than the standard systematic ones here: even though the systematic constructions are well known within the Formal languages, Logic, and Discrete mathematics communities, they are *not* so within the Algebra community. Therefore one aim for us in writing this paper has been to bring to the attention of the Algebra community this veritable treasure-trove of methods and techniques that universal algebra and formal languages have to offer. Another aim was of course to find out more about hom-algebras, as what as come so far is only the beginning of the exploration of these.

The first motivation to study nonassociative hom-algebras comes from quasi-deformations of Lie algebras of vector fields, in particular q-deformations of Witt and Virasoro algebras [1, 6, 8–10, 13, 15, 25, 27, 32]. The deformed algebras arising in connection with σ-derivation are no longer Lie algebras. It was observed in the pioneering works that in these examples a twisted Jacobi identity holds. Motivated by these examples and their generalisation on the one hand, and the desire to be able to treat within the same framework such well-known generalisations of Lie algebras as the color and Lie superalgebras on the other hand, quasi-Lie algebras and subclasses of quasi-hom-Lie algebras and hom-Lie algebras were introduced by Hartwig, Larsson and Silvestrov in [19, 29–31].

The hom-associative algebras play the role of associative algebras in the hom-Lie setting. They were introduced by Makhlouf and Silvestrov in [35]. Usual functors between the categories of Lie algebras and associative algebras were extended to hom-setting, see [44] for the construction of the enveloping algebra of a hom-Lie algebra. Likewise, many classical structures as alternative, Jordan, Malcev, graded algebras and n-ary algebras of Lie and associative type, were considered in this framework, see [2–5, 7, 34, 36–39, 41, 46–50]. Notice that Hom-algebras over a PROP were defined and studied in [51] and deformations of hom-type of the Associative operad from the point of view of the confluence property discussed in [26].

The main feature of all these algebras is that classical identities are twisted by a homomorphism. Pictorially, drawing the multiplication m as a circle and the linear map α as a square, hom-associativity may be written as

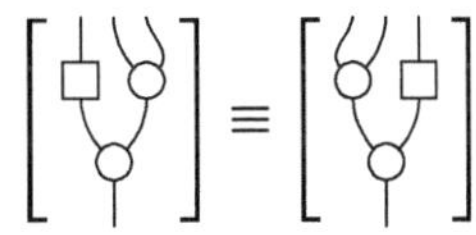

(1)

In this paper, we summarize the basics of hom-algebras in the first section. We emphasize on hom-associative and hom-Lie algebras. We show first the paradigmatic example of q-deformation of $\mathfrak{sl}_2$ using σ-derivations, leading to an interesting example of hom-Lie algebra. We provide the general method and some other procedures to construct examples of hom-associative or hom-Lie algebras. We describe the free hom-nonassociative algebra constructed by Yau. It leads to free hom-associative algebra and to the enveloping algebra of a hom-Lie algebra. In Sect. 3 we recall the basic concepts in universal algebra as signature Ω, Ω-algebra, formal terms, normal form, rewriting system, and quotient algebra. We emphasize on hom-associative algebras and discuss the envelopment problem. Section 4 is devoted to operadic approach. We discuss this concept and universal algebra for operads. We provide a diamond lemma for operads and discuss ambiguities for symmetric operads. Then we focus on hom-associative algebras operad for which attempt to resolve the ambiguities. Likewise we study congruence modulo hom-associativity and Hilbert series in this case. Moreover we study Hilbert series for the hom-associative operad and compute several dozen terms of it exactly using techniques from formal languages (notably regular tree languages).

2 Hom-Algebras: Definitions, Constructions and Examples

We summarize in this section the basics about hom-associative algebras and hom-Lie algebras.

The hom-associative identity $\alpha(x) \cdot (y \cdot z) = (x \cdot y) \cdot \alpha(z)$ is a generalisation of the ordinary associative identity $x \cdot (y \cdot z) = (x \cdot y) \cdot z$. Study of it could be motivated simply by the creed that "one should always generalise", and in Sect. 4.5 we will briefly consider the view that hom-associativity (in a rather abstract setting) can be considered as homogenisation of ordinary associativity, but historically the hom-associative identity was first suggested by an application; the line of thought went from σ-derivations, then to hom-*Lie* algebras, before finally touching upon hom-associative algebras. We sketch the σ-derivation development in the first section below, but the rest of the text does not depend on the material presented there, so the reader who prefers to skip to Sect. 2.2 now should have no problem doing so.

2.1 q-Deformations and σ-Derivations

Let A be an associative $\mathbb{K}$-algebra with unity 1. Let σ be an endomorphism on A. By a *twisted derivation* or *σ-derivation* on A, we mean a $\mathbb{K}$-linear map $\Delta\colon A \longrightarrow A$ such that a σ-twisted product rule (Leibniz rule) holds:

$$\Delta(ab) = \Delta(a)b + \sigma(a)\Delta(b). \tag{2}$$

The ordinary derivative $(\partial\, a)(t) = a'(t)$ on the polynomial ring $A = \mathbb{K}[t]$ is a σ-derivation for $\sigma = \mathrm{id}$. If on a superalgebra $A = A_0 \oplus A_1$ one defines $\sigma(a) = a$ for $a \in A_0$ but $\sigma(a) = -a$ for $a \in A_1$, then (2) precisely captures the parity adjustments of the product rule that derivations in such settings tend to exhibit, and it does so in a manner that unifies the even and odd cases. Returning to the polynomial ring $A = \mathbb{K}[t]$, the σ-derivation concept offers a unified framework for various derivation-*like* operators, perhaps most famously the Jackson q-derivation operator $(D_q a)(t) = \frac{1}{(q-1)t}\bigl(a(qt) - a(t)\bigr)$ for some $q \in \mathbb{K}$, that has the ordinary derivative as the $q \to 1$ limit and the product rule $D_q(ab)(t) = D_q(a)(t)\, b(t) + a(qt)\, D_q(b)(t)$; this is thus a σ-derivation for $\sigma(a)(t) = a(qt)$, which acts on the standard basis for $\mathbb{K}[t]$ as $\sigma(t^n) = q^n t^n$. (See [24] and references therein.)

The big algebraic insight about derivations is that they form Lie algebras, from which one can go on to universal enveloping algebras and exploit the connections to formal groups and Lie groups. What about twisted derivations, then? A quick calculation will reveal that they do not form a Lie algebra in the usual way, but there can still be a Lie-algebra-like structure on them.

We let $\mathfrak{D}_\sigma(A)$ denote the set of σ-derivations on A. As with vector fields in differential geometry, one may define the product of some $a \in A$ and $\Delta \in \mathfrak{D}_\sigma(A)$ to be the $a \cdot \Delta \in \mathfrak{D}_\sigma(A)$ defined by $(a \cdot \Delta)(b) = a\,\Delta(b)$ for all $b \in A$; hence $\mathfrak{D}_\sigma(A)$ can be regarded as a left A-module. The *annihilator* $\mathrm{Ann}(\Delta)$ of some $\Delta \in \mathfrak{D}(A)$ is the set of all $a \in A$ such that $a \cdot \Delta = 0$. By [19, Theorem 4], if A is a commutative unique factorisation domain then $\mathfrak{D}_\sigma(A)$ is as a left A-module free and of rank one, which lets us use the following construction to exhibit a Lie-algebra-like structure on $\mathfrak{D}_\sigma(A)$.

Theorem 2.1 ([19, Theorem 5]) *Let A be a commutative associative $\mathbb{K}$-algebra with unit 1 and let $\sigma : A \longrightarrow A$ be an algebra homomorphism other than the identity map. Fix some $\Delta \in \mathfrak{D}_\sigma(A)$ such that $\sigma\bigl(\mathrm{Ann}(\Delta)\bigr) \subseteq \mathrm{Ann}(\Delta)$. Define a binary operation $[\cdot,\cdot]_\sigma$ on the left A-module $A \cdot \Delta$ by*

$$[a \cdot \Delta, b \cdot \Delta]_\sigma = (\sigma(a) \cdot \Delta) \circ (b \cdot \Delta) - (\sigma(b) \cdot \Delta) \circ (a \cdot \Delta) \quad \text{for all } a, b \in A, \tag{3}$$

where $\circ$ denotes composition of functions. This operation is well-defined and satisfies the two identities

$$[a\cdot\Delta, b\cdot\Delta]_\sigma = (\sigma(a)\Delta(b)-\sigma(b)\Delta(a))\cdot\Delta, \tag{4}$$
$$[b\cdot\Delta, a\cdot\Delta]_\sigma = -[a\cdot\Delta, b\cdot\Delta]_\sigma \tag{5}$$

for all $a, b \in A$. If there in addition is some $\delta \in A$ such that

$$\Delta(\sigma(a)) = \delta\sigma(\Delta(a)) \qquad \text{for all } a \in A, \tag{6}$$

then $[\cdot,\cdot]_\sigma$ satisfies the deformed six-term Jacobi identity

$$\circlearrowleft_{a,b,c}\big([\sigma(a)\cdot\Delta,[b\cdot\Delta,c\cdot\Delta]_\sigma]_\sigma + \delta\cdot[a\cdot\Delta,[b\cdot\Delta,c\cdot\Delta]_\sigma]_\sigma\big) = 0 \tag{7}$$

for all $a, b, c \in A$.

The algebra $A\cdot\Delta$ in the theorem is then a quasi-hom-Lie algebra with, in the notation of [29], $\alpha(a\cdot\Delta) = \sigma(a)\cdot\Delta$, $\beta(a\cdot\Delta) = (\delta a)\cdot\Delta$, and $\omega = -\mathrm{id}_{A\cdot\Delta}$. For $\delta \in \mathbb{K}$, as is the case with $\Delta = D_q$, (7) further simplifies to the deformed three-term Jacobi identity (12) of a hom-Lie algebra.

As example of how the method in Theorem 1.1 ties in with more basic deformation approaches, we review the results in [30, 31] concerned with this quasi-deformation scheme when applied to the simple Lie algebra $\mathfrak{sl}_2(\mathbb{K})$. Recall that the Lie algebra $\mathfrak{sl}_2(\mathbb{K})$ can be realized as a vector space generated by elements H, E and F with the bilinear bracket product defined by the relations

$$[H,E] = 2E, \qquad [H,F] = -2F, \qquad [E,F] = H. \tag{8}$$

A basic starting point is the following representation of $\mathfrak{sl}_2(\mathbb{K})$ in terms of first order differential operators acting on some vector space of functions in a variable t:

$$E \mapsto \partial, \qquad H \mapsto -2t\partial, \qquad F \mapsto -t^2\partial.$$

To quasi-deform $\mathfrak{sl}_2(\mathbb{K})$ means that we firstly replace ∂ by some twisted derivation Δ in this representation. At our disposal as deformation parameters are now A (the "algebra of functions") and the endomorphism σ. After computing the bracket on $A\cdot\Delta$ by Theorem 1.1 the relations in the quasi-Lie deformation are obtained by pullback.

Let A be a commutative, associative $\mathbb{K}$-algebra with unity 1, let t be an element of A, and let σ denote a $\mathbb{K}$-algebra endomorphism on A. As above, $\mathfrak{D}_\sigma(A)$ denotes the linear space of σ-derivations on A. Choose an element Δ of $\mathfrak{D}_\sigma(A)$ and consider the $\mathbb{K}$-subspace $A\cdot\Delta$ of elements on the form $a\cdot\Delta$ for $a \in A$. The elements $e := \Delta$, $h := -2t\cdot\Delta$, and $f := -t^2\cdot\Delta$ span a $\mathbb{K}$-linear subspace

$$\mathcal{S} := \mathrm{LinSpan}_{\mathbb{K}}\{\Delta, -2t\cdot\Delta, -t^2\cdot\Delta\} = \mathrm{LinSpan}_{\mathbb{K}}\{e,h,f\}$$

of $A\cdot\Delta$. We restrict the multiplication (4) to $\mathcal{S}$ without, at this point, assuming closure. Now, $\Delta(t^2) = \Delta(t\cdot t) = \sigma(t)\Delta(t) + \Delta(t)t = \big(\sigma(t)+t\big)\Delta(t)$. Under the natural (see [30]) assumptions $\sigma(1) = 1$ and $\Delta(1) = 0$, (4) leads to

$$[h,f] = 2\sigma(t)t\Delta(t) \cdot \Delta, \tag{9a}$$
$$[h,e] = 2\Delta(t) \cdot \Delta, \tag{9b}$$
$$[e,f] = -\big(\sigma(t) + t\big)\Delta(t) \cdot \Delta, \tag{9c}$$

hence as long as σ and Δ, similarly to their untwisted counterparts, yield that the degrees of t in the expressions on the right hand side remain among those present in the generating set for the $\mathbb{K}$-linear subspace $\mathcal{S}$, it follows that $\mathcal{S}$ indeed is closed under this bracket.

In the particular case that $\sigma(t) = qt$ for some $q \in \mathbb{K}$ and $\Delta = D_q$, we obtain a family of hom-Lie algebras deforming $\mathfrak{sl}_2$, defined with respect to the basis $\{e, f, h\}$ by the brackets and the linear map α as follows:

$$[h,f] = -2qf, \qquad \alpha(f) = q^2 f, \tag{10a}$$
$$[h,e] = 2e, \qquad \alpha(e) = qe, \tag{10b}$$
$$[e,f] = \tfrac{1}{2}(1+q)h, \qquad \alpha(h) = qh. \tag{10c}$$

This is a hom-Lie algebra for all $q \in \mathbb{K}$ but not a Lie algebra unless $q = 1$, in which case we recover the classical $\mathfrak{sl}_2$.

2.2 Hom-Algebras: Lie and Associative

An ordinary Lie or associative algebra may informally be described as an underlying linear space (often assumed to be a vector space, but we will typically allow it to be a more general module) on which is defined some bilinear map m called the *multiplication* (or in the Lie case sometimes the *bracket*). Depending on what identities this multiplication satisfies, the algebra is classified as being associative, commutative, anticommutative, Lie, etc. A *hom-algebra* may similarly be described as an underlying linear space on which is defined two maps m and α. The multiplication m is again required to be bilinear, whereas α is merely a linear map from the underlying set to itself. The 'hom-' prefix is historically because α in many examples turn out to be a *homomorphism* with respect to some operation (not necessarily the m of the hom-algebra, even though that is certainly not uncommon), but the modern understanding is that α may be any linear map.

Practically, the point of incorporating some extra map α in the definition of an algebra is that this can be used to "twist" or "deform" the identities defining a variety of algebras, and thus offer greater opportunities for capturing within an abstract axiomatic framework the many concrete "twisted" or "deformed" algebras that have emerged in recent decades. It was shown in [19] that hom-Lie algebras are closely related to discrete and deformed vector fields and differential calculus and that some

q-deformations of the Witt and the Virasoro algebras have the structure of a hom-Lie algebra. The paradigmatic example (given above) is the $\mathfrak{sl}_2$ Lie algebra which deforms to a new nontrivial hom-Lie algebra by means of σ-derivations. Hom-associative algebras are likewise a generalisation of a usual associative algebras. A common recipe for producing the hom-analogue of a classical identity is to insert α applications wherever some variable is not acted upon by m as many times as the others.

Definition 2.1 Let $\mathcal{R}$ be some associative and commutative unital ring. Formally, an *$\mathcal{R}$-hom-algebra* $\mathcal{A}$ is a triplet (A, m, α), where A is an $\mathcal{R}$-module, $m\colon A \times A \longrightarrow A$ is a bilinear map, and $\alpha\colon A \longrightarrow A$ is a linear map. As usual, the algebra $\mathcal{A}$ and its carrier set A are notationally identified whenever there is no risk of confusion.

The *hom-associative identity* for $\mathcal{A}$ is the formula

$$m\bigl(\alpha(x), m(y,z)\bigr) = m\bigl(m(x,y), \alpha(z)\bigr) \quad \text{for all } x, y, z \in \mathcal{A}. \tag{11}$$

A hom-algebra which satisfies the hom-associative identity is said to be a *hom-associative* algebra. Similarly, the *hom-Jacobi identity* for $\mathcal{A}$ is the formula

$$m\bigl(\alpha(x), m(y,z)\bigr) + m\bigl(\alpha(y), m(z,x)\bigr) + m\bigl(\alpha(z), m(x,y)\bigr) = 0 \quad \text{for all } x, y, z \in \mathcal{A}. \tag{12}$$

For a hom-algebra $\mathcal{A}$ to be a *hom-Lie algebra*, it must satisfy the hom-Jacobi identity and the ordinary anticommutativity (skew-symmetry) identity

$$m(x,x) = 0 \quad \text{for all } x \in \mathcal{A}. \tag{13}$$

A hom-algebra $\mathcal{A}$ is said to be *multiplicative* if α is an endomorphism of the algebra (A, m), i.e., if

$$m\bigl(\alpha(x), \alpha(y)\bigr) = \alpha\bigl(m(x,y)\bigr) \quad \text{for all } x, y \in \mathcal{A}. \tag{14}$$

Now let $\mathcal{A} = (A, m, \alpha)$ and $\mathcal{A}' = (A', m', \alpha')$ be two hom-algebras. A *morphism* $f\colon \mathcal{A} \longrightarrow \mathcal{A}'$ of hom-algebras is a linear map $f\colon A \longrightarrow A'$ such that

$$m\bigl(f(x), f(y)\bigr) = f\bigl(m(x,y)\bigr) \qquad \text{for all } x, y \in A, \tag{15}$$

$$\alpha\bigl(f(x)\bigr) = f\bigl(\alpha(x)\bigr) \qquad \text{for all } x \in A. \tag{16}$$

A linear map $f\colon A \longrightarrow A'$ that merely satisfies the first condition (15) is called a *weak morphism* of hom-algebras.

The concept of weak morphism is somewhat typical of the classical algebra attitude towards hom-algebras: the multiplication m is taken as part of the core structure, whereas the map α is seen more as an add-on. In both universal algebra and the categorical setting, it is instead natural to view m and α as equally important for the hom-algebra concept, even though it is of course also possible to treat weak morphisms (for example with the help of a suitable forgetful functor) within these settings, should weak morphisms turn out to be of interest for the problems at hand. Yau [44] goes one step in the opposite direction and considers hom-algebras as being hom-modules with a multiplication; this makes α part of the core structure whereas m is the add-on.

As usual, the squaring form (13) of the anticommutative identity implies the more traditional

$$m(x, y) = -m(y, x) \quad \text{for all}\, x, y \in \mathcal{A} \tag{17}$$

in any hom-algebra $\mathcal{A}$. The two are equivalent in an algebra over a field of characteristic $\neq 2$, but (17) implies nothing about $m(x, x)$ in an algebra over a field of characteristic equal to 2, and for hom-algebras over other rings more intermediate outcomes are possible.

An example of a hom-Lie algebra was given in the previous section. A similar example of a hom-associative algebra would be:

Example 2.1 Let $\{e_1, e_2, e_3\}$ be a basis of a 3-dimensional linear space A over some field $\mathbb{K}$. Let $a, b \in \mathbb{K}$ be arbitrary parameters. The following equalities

$$\begin{aligned}
&m(e_3, e_2) = m(e_3, e_3) = 0, && m(e_1, e_1) = a\, e_1, && \alpha(e_1) = a\, e_1,\\
&m(e_1, e_2) = m(e_2, e_1) = a\, e_2, && m(e_2, e_3) = b\, e_3, && \alpha(e_2) = a\, e_2,\\
&m(e_1, e_3) = m(e_3, e_1) = b\, e_3, && m(e_2, e_2) = a\, e_2, && \alpha(e_3) = b\, e_3,
\end{aligned}$$

define the multiplication m and linear map α on a hom-associative algebra on $\mathbb{K}^3$. This algebra is not associative when $a \neq b$ and $b \neq 0$, since $m\big(m(e_1, e_1), e_3\big) - m\big(e_1, m(e_1, e_3)\big) = (a - b)be_3$.

Example 2.2 (Polynomial hom-associative algebra [45]) Consider the polynomial algebra $A = \mathbb{K}[x_1, \cdots x_n]$ in n variables. Let α be an algebra endomorphism of A which is uniquely determined by the n polynomials $\alpha(x_i) = \sum \lambda_{i;r_1,\dots,r_n} x_1^{r_1} \cdots x_n^{r_n}$ for $1 \leqslant i \leqslant n$. Define m by

$$m(f, g) = f(\alpha(x_1), \dots, \alpha(x_n))g(\alpha(x_1), \dots, \alpha(x_n)) \tag{18}$$

for f, g in A. Then (A, m, α) is a hom-associative algebra. (This example is a special case of Corollary 1.1.)

Example 2.3 ([47]) Let (A, m, α) be a hom-associative $\mathcal{R}$-algebra. Denote by $\mathrm{M}_n(A)$ the $\mathcal{R}$-module of $n \times n$ matrices with entries in A. Then $(\mathrm{M}_n(A), m', \alpha')$ is also a hom-associative algebra, in which $\alpha' \colon \mathrm{M}_n(A) \longrightarrow \mathrm{M}_n(A)$ is the map that applies α to each matrix element and the multiplication m' is the ordinary matrix multiplication over (A, m).

The following result states that hom-associative algebra yields another hom-associative algebra when its multiplication and twisting map are twisted by a morphism. The following results work as well for hom-Lie algebras and more generally G-hom-associative algebras. These constructions introduced in [45] and generalized in [50] were extended to many other algebraic structures.

Theorem 2.2 *Let $\mathcal{A} = (A, m, \alpha)$ be a hom-algebra and $\beta : \mathcal{A} \longrightarrow \mathcal{A}$ be a weak morphism. Then $\mathcal{A}_\beta = (A, m_\beta, \alpha_\beta)$ where $m_\beta = \beta \circ m$ and $\alpha_\beta = \beta \circ \alpha$ is also a hom-algebra. Furthermore:*

1. *If $\mathcal{A}$ is hom-associative then $\mathcal{A}_\beta$ is hom-associative.*
2. *If $\mathcal{A}$ is hom-Lie then $\mathcal{A}_\beta$ is hom-Lie.*
3. *If $\mathcal{A}$ is multiplicative and β is a morphism then $\mathcal{A}_\beta$ is multiplicative.*

Proof For the hom-associative and hom-Jacobi identities, it suffices to consider what a typical term in these identities looks like. We have

$$\begin{aligned} m_\beta\bigl(\alpha_\beta(x), m_\beta(y,z)\bigr) &= (\beta \circ m)\bigl((\beta \circ \alpha)(x), (\beta \circ m)(y,z)\bigr) \\ &= \beta\Bigl((m \circ \beta \otimes \beta)\bigl(\alpha(x), m(y,z)\bigr)\Bigr) = \beta\Bigl((\beta \circ m)\bigl(\alpha(x), m(y,z)\bigr)\Bigr) \\ &= (\beta \circ \beta)\Bigl(m\bigl(\alpha(x), m(y,z)\bigr)\Bigr) \end{aligned}$$

Hence either side of the hom-associative and hom-Jacobi respectively identities for $\mathcal{A}_\beta$ comes out as $\beta^{\circ 2}$ of the corresponding side of the corresponding identity for $\mathcal{A}$, and thus these identities for $\mathcal{A}_\beta$ follow directly from their $\mathcal{A}$ counterparts. The anticommutativity identity similarly follows from its counterpart, as does the multiplicative identity via

$$\begin{aligned} m_\beta\bigl(\alpha_\beta(x), \alpha_\beta(y)\bigr) &= \beta\Bigl(m\bigl(\beta\bigl(\alpha(x), \beta\bigl(\alpha(y)\bigr)\bigr)\bigr)\Bigr) = \beta^{\circ 2}\Bigl(m\bigl(\alpha(x), \alpha(y)\bigr)\Bigr) \\ &= \beta^{\circ 2}\Bigl(\alpha\bigl(m(x,y)\bigr)\Bigr) = \beta\Bigl(\alpha\bigl((\beta \circ m)(x,y)\bigr)\Bigr) = \alpha_\beta\bigl(m_\beta(x,y)\bigr) \end{aligned}$$

for all $x, y \in A$.

The $\alpha = \mathrm{id}$ special case of Theorem 1.2 yields.

Corollary 2.1 *Let (A, m) be an associative algebra and $\beta : A \longrightarrow A$ be an algebra endomorphism. Then $A_\beta = (A, m_\beta, \beta)$ where $m_\beta = \beta \circ m$ is a multiplicative hom-associative algebra.*

That result also has the following partial converse.

Corollary 2.2 *([18]) Let $\mathcal{A} = (A, m, \alpha)$ be a multiplicative hom-algebra in which α is invertible. Then $\mathcal{A}' = (A, \alpha^{-1} \circ m, \mathrm{id})$ is a hom-algebra. In particular, any multiplicative hom-associative or hom-Lie algebra where α is invertible may be regarded as an ordinary associative or Lie respectively algebra, albeit with an awkwardly defined operation.*

Proof Take $\beta = \alpha^{-1}$ in Theorem 1.2.

An application of that corollary is the identity

$$m\bigl(x_0, m(x_1, x_2)\bigr) = m\Bigl(m\bigl(\alpha^{-1}(x_0), x_1\bigr), \alpha(x_2)\Bigr)$$

which hold in multiplicative hom-associative algebras with invertible α, and generalises to change the "tilt" of longer products. The idea is to rewrite the product in terms of the corresponding associative multiplication $\tilde{m} = \alpha^{-1} \circ m$, with respect to which α and α^{-1} are also algebra homomorphisms, and apply the ordinary associative law to change the "tilt" of the product before converting the result back to the hom-associative product m.

Since many (hom-)Lie algebras of practical interest are finite-dimensional, and injectivity implies invertibility for linear operators on a finite-dimensional space, one might expect hom-Lie algebras to be particularly prone to fall under the domain of that corollary, but the important condition that should not be forgotten is that of the algebra being multiplicative. For example the q-deformed $\mathfrak{sl}_2$ of (10) is easily seen to not be multiplicative for general q.

An identity that may seem conspicuously missing from Definition 1.1 is that of the unit; although they do not make sense in Lie algebras due to contradicting anticommutativity, units are certainly a standard feature of associative algebras, so why has there been no mention of hom-associative unital algebras? The reason is that they, by the following theorem, constitute a subclass of that of hom-associative algebras which is even more restricted than that of the multiplicative hom-associative algebras. Unitality of hom-associative algebras were discussed first in [18].

Theorem 2.3 *Let $\mathcal{A}$ be a hom-associative algebra. If there is some $e \in \mathcal{A}$ such that*

$$m(e, x) = x = m(x, e) \quad \text{for all } x \in \mathcal{A} \tag{19}$$

then

$$m\bigl(\alpha(x), y\bigr) = m\bigl(x, \alpha(y)\bigr) = \alpha\bigl(m(x, y)\bigr) \quad \text{for all } x, y \in \mathcal{A}. \tag{20}$$

Proof For the first equality,

$$m\bigl(\alpha(x), y\bigr) = m\bigl(\alpha(x), m(e, y)\bigr) = m\bigl(m(x, e), \alpha(y)\bigr) = m\bigl(x, \alpha(y)\bigr)$$

by hom-associativity. For the second equality,

$$\begin{aligned} m\bigl(x, \alpha(y)\bigr) &= m\bigl(m(e, x), \alpha(y)\bigr) = m\bigl(\alpha(e), m(x, y)\bigr) \\ &= m\Bigl(e, \alpha\bigl(m(x, y)\bigr)\Bigr) = \alpha\bigl(m(x, y)\bigr) \end{aligned}$$

by hom-associativity and the first equality.

An identity such as (20) has profound effects on the structure of a hom-associative algebra. Basically, it means applications of α are not located in any particular position in a product, but can move around unhindered. At the same time, even a single α somewhere will act as a powerful lubricant that lets the hom-associative identity shuffle around parentheses as easily as the ordinary associative identity. In particular, any product of n algebra elements $x_1, \dots, x_n$ where at least one is in the image of α will effectively be an associative product; probably not the wanted outcome if one's aim is to create new structures through deformations of old ones.

On the other hand, α satisfying (20) obviously have some rather special properties. One may for any algebra $\mathcal{A} = (A, m)$ define the *centroid* $\mathrm{Cent}(\mathcal{A})$ of $\mathcal{A}$ as the set of all linear self-maps $\alpha\colon A \longrightarrow A$ satisfying the condition $\alpha\bigl(m(x, y)\bigr) = m\bigl(\alpha(x), y\bigr) = m\bigl(x, \alpha(y)\bigr)$ for all $x, y \in A$. Notice that if $\alpha \in \mathrm{Cent}(\mathcal{A})$, then we have $m\bigl(\alpha^p(x), \alpha^q(y)\bigr) = (\alpha^{p+q} \circ m)(x, y)$ for all $p, q \geqslant 0$. The construction of hom-algebras using elements of the centroid was initiated in [5] for Lie algebras. We have

Proposition 2.1 *Let $(\mathcal{A}, m)$ be an associative algebra and $\alpha \in \mathrm{Cent}(\mathcal{A})$. Set for $x, y \in \mathcal{A}$*

$$\begin{aligned} m_1(x, y) &= m\bigl(\alpha(x), y\bigr), \\ m_2(x, y) &= m\bigl(\alpha(x), \alpha(y)\bigr). \end{aligned}$$

Then $(\mathcal{A}, m_1, \alpha)$ and $(\mathcal{A}, m_2, \alpha)$ are hom-associative algebras.

Indeed we have

$$\begin{aligned} m_1\bigl(\alpha(x), m_1(y, z)\bigr) &= m\Bigl(\alpha^2(x), m\bigl(\alpha(y), z\bigr)\Bigr) = \alpha\Bigl(m\bigl(\alpha(x), m(\alpha(y), z)\bigr)\Bigr) \\ &= m\Bigl(\alpha(x), \alpha\bigl(m(\alpha(y), z)\bigr)\Bigr) = m\Bigl(\alpha(x), m\bigl(\alpha(y), \alpha(z)\bigr)\Bigr). \end{aligned}$$

Remark 2.1 The definition of unitality which fits with Corollary 1.1 was introduced in [18] and then used for hom-bialgebra and hom-Hopf algebras in [7].

Let $(\mathcal{A}, m, \alpha)$ be a hom-associative algebra. It is said to be unital if there is some $e \in \mathcal{A}$ such that

$$m(e, x) = \alpha(x) = m(x, e) \qquad \text{for all } x \in \mathcal{A}. \tag{21}$$

Therefore, similarly to Corollary 1.1, a unital associative algebra gives rise a unital hom-associative algebra.

2.3 Admissible and Enveloping Hom-Algebras

Two concepts that are of key importance in the theory of ordinary Lie algebras are those of Lie-admissible and enveloping algebras. In the setting of hom-algebras, these concepts are defined as follows, with the classical non-hom concepts arising in the special case $\alpha = \mathrm{id}$.

Definition 2.2 Let a hom-algebra $\mathcal{A} = (A, m, \alpha)$ be given. Define $b(x, y) = m(x, y) - m(y, x)$ to be the commutator (bracket) corresponding to m, and let $\mathcal{A}^-$ be the hom-algebra (A, b, α). The algebra $\mathcal{A}$ is said to be *hom-Lie-admissible* if the hom-algebra $\mathcal{A}^-$ is hom-Lie.

Now let $\mathcal{L}$ be a hom-Lie algebra. $\mathcal{A}$ is said to be an *enveloping algebra* for $\mathcal{L}$ if $\mathcal{L}$ is isomorphic to some hom-subalgebra $\mathcal{B} = \left(B, b_{|B\times B}, \alpha_{|B}\right)$ of $\mathcal{A}^-$ such that B generates $\mathcal{A}$.

It was shown in [35, Proposition 1.6] that any hom-associative algebra is hom-Lie-admissible. On one hand, this becomes another method of constructing new hom-Lie algebras, but it is more interesting when wielded to the opposite end of studying a given hom-Lie algebra through a corresponding enveloping algebra. To explain why this is so, we will briefly review the classical theory of ordinary Lie and associative algebras.

On a Lie group, the exponential map $v \mapsto \exp(v)$ allows transitioning from tangent vectors to non-infinitesimal shifts; $\exp(tv)$ is the point where you end up if travelling from the identity point at velocity v for time t. Under the interpretation that identifies vectors with invariant vector fields, and vector fields with derivations on the ring of scalar-valued functions ("scalar fields", in the physicist terminology), the exponential map may in fact be defined via the elementary power series formula $\exp(v) = \sum_{n=0}^{\infty} \frac{v^n}{n!}$ (where multiplication of vectors is composition of differential operators) and in the Lie group $(\mathbb{R}, +)$ this turns out to be Taylor's formula: $\exp\left(t\frac{d}{dx}\right)$ is the shift operator mapping an analytic function f to the shifted variant $x \mapsto f(x+t)$. When doing the same in a more general Lie group, one must however be careful to note that vector fields need not commute, and that already the degree 2 term of for example $\exp(u + v)$ contains uv and vu terms that need not be equal. The role of the Lie algebra is precisely to keep track of the extent to which vector fields do not commute, so the proper place to do algebra with vector fields to the aim of studying the exponential map must be in an enveloping algebra of the Lie algebra of invariant vector fields on the underlying Lie group.

Conversely, one may start with a Lie algebra $\mathfrak{g}$ and ask oneself what the corresponding Lie group would be like, by studying formal series in the basic vector fields, while keeping in mind that these should satisfy the commutation relations encoded into $\mathfrak{g}$; this leads to the concept of *formal groups*. An important step towards it is the construction of the *(associative) universal enveloping algebra* $\mathrm{U}(\mathfrak{g})$, which starts with the free associative algebra generated by $\mathfrak{g}$ as a module and imposes upon it the relations that

$$xy - yx = [x, y] \qquad \text{for all } x, y \in \mathfrak{g}, \tag{22}$$

where on the left hand side we have multiplication in $\mathrm{U}(\mathfrak{g})$ but on the right hand side the bracket operation of the Lie algebra $\mathfrak{g}$. More technically, the free associative algebra in question can be constructed as the tensor algebra $\mathrm{T}(\mathfrak{g}) = \bigoplus_{n=0}^{\infty} \mathfrak{g}^{\otimes n}$ where the product of $x_1 \otimes \cdots \otimes x_m \in \mathfrak{g}^{\otimes m}$ and $y_1 \otimes \cdots \otimes y_n \in \mathfrak{g}^{\otimes n}$ is $x_1 \otimes \cdots \otimes x_m \otimes y_1 \otimes \cdots \otimes y_n \in \mathfrak{g}^{\otimes (m+n)}$. Imposing the commutation relations can then be done by taking the quotient by the two-sided ideal $J(\mathfrak{g})$ in $\mathrm{T}(\mathfrak{g})$ that is generated by all $xy - yx - [x, y]$ for $x, y \in \mathfrak{g}$, i.e.,

$$\mathrm{U}(\mathfrak{g}) := \mathrm{T}(\mathfrak{g}) / J(\mathfrak{g}) = \mathrm{T}(\mathfrak{g}) \Big/ \Big\langle \big\{ xy - yx - [x, y] \,\big|\, x, y \in \mathfrak{g} \big\} \Big\rangle.$$

With this in mind, it is only natural to generalise this construction to the hom-case, and in [44] Yau does so. Since he in the non-associative case cannot take advantage of familiar concepts such as the tensor algebra, this construction will however involve a few steps more than one might be used to from the non-hom setting. Notably, Yau begins with setting up the free hom-algebra $F_{\mathrm{HNAs}}(\mathfrak{g})$: neither hom-associativity nor ordinary associativity is inherent. Then he goes on to impose hom-associativity by taking a quotient, which results in the free hom-associative algebra $F_{\mathrm{HAs}}(\mathfrak{g})$; this is what corresponds to the tensor algebra $\mathrm{T}(\mathfrak{g})$. Another quotient imposes also the commutation relations, to finally yield the universal enveloping hom-associative algebra $U_{\mathrm{HLie}}(\mathfrak{g})$.

When reading through the technical details of these constructions, which we shall quote below for the reader's convenience, they may seem a daring plunge forward into very general algebra, that harnesses advanced combinatorial objects to achieve a clear picture of the algebra. It may be that they are that, but our main point in the next section is that they are also an entirely straightforward application of the basic methods of universal algebra, so there is in fact very little that was novel in these constructions. The reader who has grasped the material in Sect. 3 will be able to recreate something equivalent to the following (modulo some minor optimisations) from scratch.

For $n \geqslant 1$, let T_n denote the set of isomorphism classes of plane[1] binary trees with n leaves and one root. The first T_n are depicted below.

[1] Yau, like many other algebraists, actually uses the term 'planar' rather than 'plane', but this practice is simply wrong as the two words refer to slightly different graph-theoretical properties: a graph is *planar* if it can be embedded in a genus 0 surface, but *plane* if it is given with such an embedding. To speak of a 'planar tree' is a tautology, because trees by definition contain no cycles, will therefore have no subdivided K_5 or $K_{3,3}$ as subgraph, and thus by Kuratowski's Theorem be planar. What is of utmost importance here is rather that the trees are given with a (combinatorial) *embedding* into the plane, since that specifies a local cyclic order on edges incident with a vertex, which is what the isomorphisms spoken of are required to preserve. As rooted trees, the two elements of T_3 are isomorphic, but as plane rooted trees they are not.

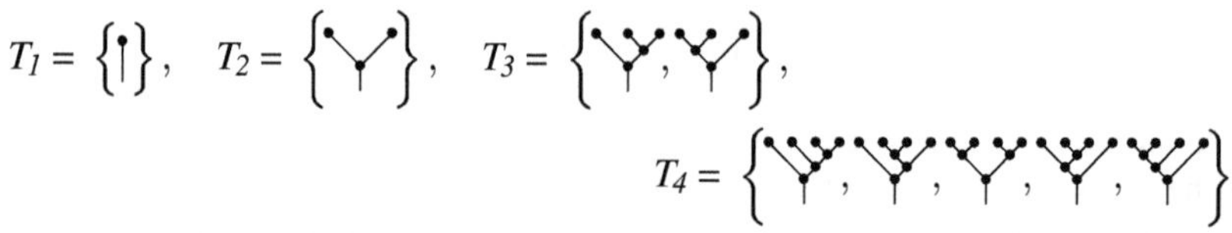

Each dot represents either a leaf, which is always depicted at the top, or an internal vertex. An element in T_n will be called an *n-tree*. The set of nodes (= leaves and internal vertices) in a tree ψ is denoted by $N(\psi)$. The node of an n-tree ψ that is connected to the root (the lowest point in the n-tree) will be denoted by v_{low}. In other words, v_{low} is the lowest internal vertex in ψ if $n \geq 2$ and is the only leaf if $n = 1$.

Given an n-tree ψ and an m-tree φ, their *grafting* $\psi \vee \varphi \in T_{n+m}$ is the tree obtained by placing ψ on the left and φ on the right and joining their roots to form the new lowest internal vertex, which is connected to the new root. Pictorially, we have

$\psi \vee \varphi =$ 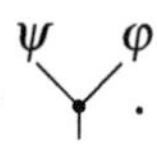 .

Note that grafting is a nonassociative operation. As we will discuss below, the operation of grafting is for generating the multiplication m of a free nonassociative algebra.

To handle hom-algebras, we need to introduce weights on plane trees. A *weighted n-tree* is a pair $\tau = (\psi, w)$, in which $\psi \in T_n$ is an n-tree and w is a function from the set of internal vertices of ψ to the set $\mathbb{N}$ of non-negative integers. If v is an internal vertex of ψ, then we call $w(v)$ the weight of v. The n-tree ψ is called the underlying n-tree of τ, and w is called the weight function of τ. The set of all weighted n-trees is denoted T_n^{wt}. Since the 1-tree has no internal vertex, we have that $T_1 = T_1^{\text{wt}}$. Likewise, the grafting of two weighted trees is defined as above by connecting them to a new root for which the weight is 0. There is also an operation to change the weight; for $\tau = (\psi, w)$, we define $\tau[r] = (\psi, w')$ where $w'(v_{\text{low}}) = w(v_{\text{low}}) + r$ and $w'(v) = w(v)$ for all internal vertices $v \neq v_{\text{low}}$.

Now let an $\mathcal{R}$-module A and a linear map $\alpha\colon A \longrightarrow A$ be given. As a set,

$$F_{\text{HNAs}}(A) = \bigoplus_{n \geq 1} \bigoplus_{\tau \in T_n^{\text{wt}}} A^{\otimes n}.$$

We write $A_\tau^{\otimes n}$ for the component in this direct sum that corresponds to the values n and τ of these summation indices. There is a canonical isomorphism $A_\tau^{\otimes n} \cong A^{\otimes n}$. For any $n \geqslant 1$, $\tau \in T_n^{\text{wt}}$, and $x_1, \ldots, x_n \in A$, we write $(x_1 \otimes \cdots \otimes x_n)_\tau$ for the element of $A_\tau^{\otimes n}$ that corresponds to $x_1 \otimes \cdots \otimes x_n \in A^{\otimes n}$. The linear map α is extended to a linear map $\alpha_F\colon F_{\text{HNAs}}(A) \longrightarrow F_{\text{HNAs}}(A)$ by the rule

$$\alpha_F\big((x_1 \otimes \cdots \otimes x_n)_\tau\big) = (x_1 \otimes \cdots \otimes x_n)_{\tau[1]} \qquad \text{for } \tau \notin T_1$$

and the multiplication m_F on $F_{\text{HNAs}}(A)$ is defined by

$$m_F\big((x_1 \otimes \cdots \otimes x_n)_\tau, (x_{n+1} \otimes \cdots \otimes x_{n+m})_\sigma\big) = (x_1 \otimes \cdots \otimes x_{n+m})_{\tau \vee \sigma}$$

and bilinearity. This $\big(F_{\text{HNAs}}(A), m_F, \alpha_F\big)$ is the free (nonassociative) $\mathcal{R}$-hom-algebra generated by the hom-module (A, α).

From there, the corresponding free hom-associative algebra is constructed as the quotient

$$F_{\text{HAs}}(A) := F_{\text{HNAs}}(A)\big/J^{\infty}$$

where $J^{\infty} = \bigcup_{n\geq 1} J^n$ and $J^1 \subseteq J^2 \subseteq \cdots \subseteq J^{\infty} \subset F_{\text{HNAs}}(A)$ is an ascending chain of two-sided ideals defined by

$$\begin{aligned} J^1 &= \Big\langle \operatorname{Im}\big(m_F \circ (m_F \otimes \alpha_F - \alpha_F \otimes m_F)\big) \Big\rangle, \\ J^{n+1} &= \big\langle J^n \cup \alpha_F(J^n) \big\rangle \qquad \text{for } n \geqslant 1. \end{aligned}$$

The *universal enveloping algebra* of a hom-Lie algebra $(\mathfrak{g}, b, \alpha)$ is similarly obtained as the quotient

$$U_{\text{HLie}}(\mathfrak{g}) := F_{\text{HNAs}}(\mathfrak{g})\big/I^{\infty}$$

where I^{∞} is the two-sided ideal obtained if one starts with

$$I^1 = \Big\langle \operatorname{Im}\big(m_F \circ (m_F \otimes \alpha_F - \alpha_F \otimes m_F)\big) \cup \big\{ m_F(x,y) - m_F(y,x) - b(x,y) \bigm| x, y \in \mathfrak{g} \big\} \Big\rangle$$

and then similarly lets $I^{n+1} = \big\langle I^n \cup \alpha_F(I^n) \big\rangle$ for $n \geqslant 1$ and $I^{\infty} = \bigcup_{n\geqslant 1} I^n$. Since $I^n \supseteq J^n$ for all $n \geqslant 1$, it follows that $U_{\text{HLie}}(\mathfrak{g})$ may alternatively be regarded as a quotient of $F_{\text{HAs}}(\mathfrak{g})$. This further justifies labelling the hom-associative algebra $U_{\text{HLie}}(\mathfrak{g})$ as the hom-analogue for a hom-Lie algebra $\mathfrak{g}$ of the universal enveloping algebra of a Lie algebra.

There is however one important question regarding this U_{HLie} which has not been answered by the above, and in fact seems to be open in the literature: *Is $U_{\text{HLie}}(\mathfrak{g})$ for every hom-Lie algebra $\mathfrak{g}$ an enveloping algebra of $\mathfrak{g}$?* It follows from the form of the construction that there is a linear map $j\colon \mathfrak{g} \longrightarrow U_{\text{HLie}}(\mathfrak{g})$ with the properties that

$$\begin{aligned} j\big([x,y]\big) &= j(x)j(y) - j(y)j(x) && \text{for all } x, y \in \mathfrak{g}, \\ j\big(\alpha(x)\big) &= \alpha\big(j(x)\big) && \text{for all } x \in \mathfrak{g}, \end{aligned}$$

and hence j becomes a morphism of hom-Lie algebras $\mathfrak{g} \longrightarrow U_{\text{HLie}}(\mathfrak{g})^{-}$, but it is entirely unknown whether j is injective. A failure to be injective would obviously render these hom-associative enveloping algebras of hom-Lie algebras less important than the ordinary associative enveloping algebras of ordinary Lie algebras, as they would fail to capture all the information encoded into the hom-Lie algebra.

Another way of phrasing the conjecture that the canonical homomorphism is injective is that the ideal I^{∞} used to construct $U_{\text{HLie}}(\mathfrak{g})$ does not contain any degree 1 elements; such elements would correspond to linear dependencies in $U_{\text{HLie}}(\mathfrak{g})$ between the images of basis elements in $\mathfrak{g}$. A simple argument for this conjecture would be that such dependencies do not occur in the associative case, and since the hom-associative case has "more degrees of freedom" than the associative case, it shouldn't happen here either. An argument *against* it comes from the converse of the Poincaré–Birkhoff–Witt Theorem [43]: *If the canonical homomorphism* $\mathfrak{g} \longrightarrow \mathrm{U}(\mathfrak{g})$ *is injective, then* $\mathfrak{g}$ *is a Lie algebra;* the ordinary universal enveloping algebra construction only manages to envelop the algebra one starts with if that algebra is a Lie algebra. What can be hoped for is of course that the conditions inherent in U_{HLie} have precisely those deformations relative to the conditions of U_{Lie} that makes everything work out for hom-Lie algebras instead, but they could just as well end up going some other way.

To positively resolve the envelopment problem, one would probably have to prove a hom-analogue of the Poincaré–Birkhoff–Witt Theorem. Methods for this—particularly the Diamond Lemma—are available, but the calculations required seem to be rather extensive. To negatively resolve the envelopment problem, it would be sufficient to find one hom-Lie algebra $\mathfrak{g}$ for which the canonical homomorphism $\mathfrak{g} \longrightarrow U_{\mathrm{HLie}}(\mathfrak{g})$ is not injective. Yau does show in [44, Theorem 2] that $U_{\mathrm{HLie}}(\mathfrak{g})$ satisfies an universal property with respect to hom-associative enveloping algebras, so a hom-Lie algebra $\mathfrak{g}$ which constitutes a counterexample cannot arise as a subalgebra of $\mathcal{A}^-$ for any hom-associative algebra $\mathcal{A}$.

3 Classical Universal Algebra: Free Algebras and Their Quotients

3.1 Discrete Free Algebras

A basic concept in universal algebra is that of the *signature*. A signature Ω is a set of formal symbols, together with a function arity: $\Omega \longrightarrow \mathbb{N}$ that gives the arity, or "wanted number of operands", for each symbol. Symbols with arity 0 are called *constants* (or said to be *nullary*), symbols with arity 1 are said to be *unary*, symbols with arity 2 are said to be *binary*, symbols with arity 3 are said to be *ternary*, and so on; one may also speak about a symbol being *n*-ary. A convenient shorthand, used in for example [12], for specifying signatures is as a set of "function prototypes": symbols of positive arity are followed by a parenthesis containing one comma less than the arity, whereas constants are not followed by a parenthesis. Hence $\Omega = \{\mathsf{a}(), \mathsf{m}(,), \mathsf{x}, \mathsf{y}\}$ is the signature of four symbols a, m, x, and y, where a is unary, m is binary, and the remaining two are constants. The signature for a hom-algebra is thus $\{\mathsf{a}(), \mathsf{m}(,)\}$, whereas the signature for a unary hom-algebra would be $\{\mathsf{a}(), \mathsf{m}(,), \mathsf{1}\}$; a unit would be an extra constant symbol.

Given a signature Ω, a set A is said to be an Ω-algebra if it for every symbol $x \in \Omega$ comes with a map $f_x\colon A^{\mathrm{arity}(x)} \longrightarrow A$; these maps are the *operations* of the algebra. Note that no claim is made that the operations fulfill any particular property (beyond matching the respective arities of their symbols), so the Ω-algebra structure is not determined by A unless that set has cardinality 1; therefore one might want to be more formal and say it is $\mathcal{A} = \bigl(A, \{f_x\}_{x\in\Omega}\bigr)$ that is the Ω-algebra, but we shall in what follows generally be concerned with only one Ω-algebra structure at a time on each base set.

What the Ω-algebra concept suffices for, despite imposing virtually no structure upon the object in question, is the definition of an *Ω-algebra homomorphism*: a map $\phi\colon A \longrightarrow B$ is an Ω-algebra homomorphism from $\bigl(A, \{f_x\}_{x\in\Omega}\bigr)$ to $\bigl(B, \{g_x\}_{x\in\Omega}\bigr)$ if

$$\phi\bigl(f_x(a_1, \dots, a_{\mathrm{arity}(x)})\bigr) = g_x\bigl(\phi(a_1), \dots, \phi(a_{\mathrm{arity}(x)})\bigr) \quad \text{for all } a_1, \dots, a_{\mathrm{arity}(x)} \in A \text{ and } x \in \Omega. \tag{23}$$

It is easy to verify that these homomorphisms obey the axioms for being the morphisms in the category of Ω-algebras, so that category Ω−`algebra` is what one gets. One may then define (up to isomorphism) *the free Ω-algebra* as being the free object in this category, or more technically state that $\mathcal{F}_\Omega(X)$ together with $i\colon X \longrightarrow \mathcal{F}_\Omega(X)$ is *the free Ω-algebra generated by* X if there for every Ω-algebra $\mathcal{A}$ and every map $j\colon X \longrightarrow \mathcal{A}$ exists a unique Ω-algebra homomorphism $\phi\colon \mathcal{F}_\Omega(X) \longrightarrow \mathcal{A}$ such that $j = \phi \circ i$. An alternative claim to the same effect is that $\mathcal{F}_\Omega$, interpreted as a functor from `Set` to Ω−`algebra`, is left adjoint of the forgetful functor mapping an Ω-algebra to its underlying set.

Although these definitions may seem frightfully abstract, the objects in question are actually rather easy to construct: $\mathcal{F}_\Omega(X)$ is merely the set $T(\Omega, X)$ of all *formal terms* in $\Omega \mathbin{\dot\cup} X$, where the elements of X are interpreted as symbols of arity 0. Hence the first few elements of $T\bigl(\{\mathsf{a}(), \mathsf{m}(,)\}, \{\mathsf{x}, \mathsf{y}\}\bigr)$ are

$$\mathsf{x}, \mathsf{y}, \mathsf{a}(\mathsf{x}), \mathsf{a}(\mathsf{y}), \mathsf{a}(\mathsf{a}(\mathsf{x})), \mathsf{a}(\mathsf{a}(\mathsf{y})), \mathsf{m}(\mathsf{x}, \mathsf{x}), \mathsf{m}(\mathsf{x}, \mathsf{y}), \mathsf{m}(\mathsf{y}, \mathsf{x}), \mathsf{m}(\mathsf{y}, \mathsf{y}), \dots$$

and the operations $\{f_x\}_{x\in\Omega}$ in the free Ω-algebra $T(\Omega, X)$ merely produce their formal terms counterparts:

$$f_x(t_1, \dots, t_{\mathrm{arity}(x)}) := x(t_1, \dots, t_{\mathrm{arity}(x)}) \quad \text{for all } t_1, \dots, t_{\mathrm{arity}(x)} \in T(\Omega, X) \text{ and } x \in \Omega.$$

Conversely, the unique morphism ϕ of the universal property turns out to evaluate formal terms in the codomain Ω-algebra, so for any given $j\colon X \longrightarrow B$ it can be defined recursively through

$$\phi(t) = \begin{cases} j(x) & \text{if } t = x \in X, \\ g_x\bigl(\phi(t_1), \dots, \phi(t_n)\bigr) & \text{if } t = x(t_1, \dots, t_n) \text{ where } x \in \Omega \end{cases}$$

for all $t \in T(\Omega, X)$.

3.2 Quotient Algebras

Completely free algebras might be cute, but most of the time one is rather interested in something with a bit more structure, in the sense that certain identities are known to hold; in an associative algebra, the associativity identity holds, whereas in a hom-associative algebra the hom-associative identity (11) holds. One approach to imposing such properties on one's algebras is to restrict attention to the subcategory of Ω-algebras which satisfy the wanted identities, and then look at the free object of that subcategory. Another approach is to take a suitable quotient of the free object from the full category.

In general Ω-algebras, the denominator in a quotient is a *congruence relation* on the numerator, and an Ω-algebra congruence relation is an equivalence relation which is preserved by the operations; $\equiv$ is a congruence relation on $\mathcal{A} = \bigl(A, \{f_x\}_{x\in\Omega}\bigr)$ if it is an equivalence relation on A and

$$f_x(a_1,\dots,a_n) \equiv f_x(b_1,\dots,b_n)$$
$$\text{for all } a_1,\dots,a_n,b_1,\dots,b_n \in A, x \in\Omega, \text{ and } n = \operatorname{arity}(x)$$
$$\text{such that } a_1 \equiv b_1,\ a_2 \equiv b_2,\dots, \text{ and } a_n \equiv b_n.$$

The quotient $\bigl(B, \{g_x\}_{x\in\Omega}\bigr) := \mathcal{A}/{\equiv}$ then has B equal to the set of $\equiv$-equivalence classes in A, and operations defined by

$$g_x\bigl([a_1],\dots,[a_{\operatorname{arity}(x)}]\bigr) = \bigl[f_x(a_1,\dots,a_{\operatorname{arity}(x)})\bigr]$$
$$\text{for all } a_1,\dots,a_{\operatorname{arity}(x)} \in A \text{ and } x \in \Omega;$$

congruence relations are precisely those for which this definition makes sense. Conversely, the relation $\equiv$ defined on some Ω-algebra $\mathcal{A}$ by $a \equiv b$ iff $\phi(a) = \phi(b)$ will be a congruence relation whenever ϕ is an Ω-algebra homomorphism.

It should at this point be observed that defining specific congruence relations to that they respect particular identities is not an entirely straightforward matter; it would for example be wrong to expect a simple formula such as '$b \equiv b'$ iff $b = \mathsf{m}\bigl(\mathsf{a}(b_1), \mathsf{m}(b_2,b_3)\bigr)$ and $b' = \mathsf{m}\bigl(\mathsf{m}(b_1,b_2), \mathsf{a}(b_3)\bigr)$ for some $b_1,b_2,b_3 \in \mathfrak{F}_\Omega(X)$' to set up the congruence relation imposing hom-associativity on $\mathfrak{F}_\Omega(X)$, as it actually fails even to define an equivalence relation. Instead one considers the family of *all* congruence relations which fulfill the wanted identities, and picks the smallest of these, which also happens to be the intersection of the entire family; this makes precisely those identifications of elements which would be logical consequences of the given axioms, but nothing more. Thus to construct the free hom-associative $\bigl\{\mathsf{a}(), \mathsf{m}(,)\bigr\}$-algebra generated by X, one would let $\Omega = \bigl\{\mathsf{a}(), \mathsf{m}(,)\bigr\}$ and form $T(\Omega,X)\big/\equiv$, where $\equiv$ is defined by

$$t \equiv t' \iff t \sim t' \text{ for every congruence relation } \sim \text{ on } T(\Omega, X) \text{ satisfying } \mathsf{m}\bigl(\mathsf{a}(t_1), \mathsf{m}(t_2, t_3)\bigr) \sim \mathsf{m}\bigl(\mathsf{m}(t_1, t_2), \mathsf{a}(t_3)\bigr) \text{ for all } t_1, t_2, t_3 \in T(\Omega, X). \tag{24}$$

Another thing that should be observed is that this construction of the free hom-associative algebra is not *effective*, i.e., one cannot use it to implement the algebra on a computer, nor to reliably carry out calculations with pen and paper. The construction does suggest both an encoding of arbitrary algebra elements—since the algebra elements are equivalence classes, just use any element of a class to represent it—and an implementation of operations—just perform the corresponding operation of $\mathcal{F}_{\Omega}(X)$ on the equivalence representatives—but it does then not suggest any algorithm for deciding equality. Providing such an algorithm is of course equivalent to solving the word problem for the algebra/congruence relation in question, so there cannot be a universal method which works for arbitrary algebras, but nothing prevents seeking a solution that works a particular algebra, and indeed one should always consider this an important problem to solve for every class of algebras one considers.

One common form of solutions to the word problem is to device a *normal form map* for the congruence relation $\equiv$: a map $N \colon T(\Omega, X) \longrightarrow T(\Omega, X)$ such that $N(t) \equiv t$ for all $t \in T(\Omega, X)$ and $t \equiv t'$ iff $N(t) = N(t')$; this singles out one element from each equivalence class as being the *normal form* representative of that class, thereby reducing the problem of deciding congruence to that of testing whether the respective normal forms are equal. Normal form maps are often realised as the limit of a system of *rewrite rules* derived directly from the defining relations; we shall return to this matter in Sect. 4.3.

3.3 Algebras with Linear Structure

One thing that has so far been glossed over is that e.g. a hom-associative algebra is not just supposed to have a non-associative multiplication m and a homomorphism a, it is also supposed to have addition and multiplication by a scalar. The general way to ensure this is of course to extend the signature with operations for these, and then impose the corresponding axioms on the congruence relation used, but a more practical approach is usually to switch to a category where the wanted linear structure is in place from the start. As it turns out the free object in the category of algebras with a linear structure can be constructed as the set of formal linear combinations of elements in the free (without linear structure) algebra, our constructions above remain highly useful.

Let $\mathcal{R}$ be an associative and commutative ring with unit. An Ω-algebra $\bigl(A, \{f_x\}_{x\in\Omega}\bigr)$ is *$\mathcal{R}$-linear* if A is an $\mathcal{R}$-module and each operation f_x is $\mathcal{R}$-multilinear, i.e., it is $\mathcal{R}$-linear in each argument. An Ω-algebra homomorphism $\phi\colon \mathcal{A} \longrightarrow \mathcal{B}$ is an *$\mathcal{R}$-linear Ω-algebra homomorphism* if $\mathcal{A}$ and $\mathcal{B}$ are $\mathcal{R}$-linear Ω-algebras and ϕ is an $\mathcal{R}$-module homomorphism. An *$\mathcal{R}$-linear Ω-algebra congruence relation* $\equiv$ is an

Ω-algebra congruence relation on an $\mathcal{R}$-linear Ω-algebra which is preserved also by module operations, i.e., $a_1 \equiv b_1$ and $a_2 \equiv b_2$ implies $ra_1 \equiv rb_1$ (for all $r \in \mathcal{R}$) and $a_1 + a_2 \equiv b_1 + b_2$.

The free $\mathcal{R}$-linear Ω-algebra generated by a set X can be constructed as the set of all formal linear combinations of elements of $T(\Omega, X)$, i.e., as the free $\mathcal{R}$-module with basis $T(\Omega, X)$; we will denote this free algebra by $\mathcal{R}\{\Omega, X\}$ (continuing the notation family $\mathcal{R}[X]$, $\mathcal{R}(X)$, $\mathcal{R}\langle X\rangle$). The universal property it satisfies is that any function $j\colon X \longrightarrow \mathcal{A}$ where $\mathcal{A}$ is an $\mathcal{R}$-linear Ω-algebra gives rise to a unique $\mathcal{R}$-linear Ω-algebra homomorphism $\phi\colon \mathcal{R}\{\Omega, X\} \longrightarrow \mathcal{A}$ such that $j = \phi \circ i$, where i is the function $X \longrightarrow \mathcal{R}\{\Omega, X\}$ such that $i(x)$ is x, or more precisely the linear combination which has coefficient 1 for the formal term x and coefficient 0 for all other terms.

A consequence of the above is that $\mathcal{R}\{\varnothing, X\}$ is the *free* $\mathcal{R}$-module with basis X, which might be seen as restrictive. There is an alternative concept of free $\mathcal{R}$-linear Ω-algebra which is generated by an $\mathcal{R}$-module $\mathcal{M}$ rather than a set X, in which case the above universal property must instead hold for j being an $\mathcal{R}$-module homomorphism $\mathcal{M} \longrightarrow \mathcal{A}$; in more categoric terms, this corresponds to the functor producing the free algebra being left adjoint of not the forgetful functor from $\mathcal{R}$`-linear` Ω`-algebra` to `Set`, but left adjoint of the forgetful functor from $\mathcal{R}$`-linear` Ω`-algebra` to $\mathcal{R}$`-module`. It is however quite possible to get to that also by going via $\mathcal{R}\{\Omega, X\}$, as all one has to do is take $X = \mathcal{M}$ and then consider the quotient by the smallest congruence relation $\equiv$ which has $i(a) + i(b) \equiv i(a + b)$ and $ri(a) \equiv i(ra)$ for all $a, b \in \mathcal{M}$ and $r \in \mathcal{R}$ (it is useful here to make the function $i\colon X \longrightarrow \mathcal{R}\{\Omega, X\}$ figuring in the universal property of $\mathcal{R}\{\Omega, X\}$ explicit, as $\equiv$ would otherwise seem a triviality); the result is the free object in the category of $\mathcal{R}$-linear Ω-algebras that are equipped with an $\mathcal{R}$-module homomorphism i' from $\mathcal{M}$, just like the alternative universal property would require.

No doubt some readers may find this construction wasteful—a separate constant symbol for every element of the module $\mathcal{M}$, with a host of identities just to make them "remember" this module structure, immediately rendering most of the symbols redundant—and would rather prefer to construct the free $\mathcal{R}$-linear Ω-algebra on the $\mathcal{R}$-module $\mathcal{M}$ by direct sums of appropriate tensor products of $\mathcal{M}$ with itself, somehow generalising the tensor algebra construction $\mathrm{T}(\mathcal{M}) = \bigoplus_{n=0}^{\infty} \mathcal{M}^{\otimes n}$. However, from the perspectives of constructive set theory and effectiveness, such constructions are guilty of the exact same wastefulness; they only manage to sweep it under the proverbial rug that is the definition of the tensor product. As is quite often the case, one ends up doing the same thing either way, although the presentation may obscure the correspondencies between the two approaches.

Another stylistic detail is that of whether the denominator in a quotient should be a congruence relation or an ideal. For $\mathcal{R}$-linear Ω-algebras, the equivalence class of 0 turns out to be an ideal, and conversely a congruence relation $\equiv$ is uniquely determined by its equivalence class of 0 since $a \equiv b$ if and only if $a - b \equiv 0$. In our experience, an important advantage of the congruence relation formalism is that it makes the dependency on the signature Ω more explicit, since it is not uncommon to see authors continue to associate "ideal" and/or related concepts with the definition these have in a more traditional setting; particularly continuing to use 'two-sided ideal' and '$\langle S\rangle$' as they would be defined in an $\{\mathsf{m}(,)\}$-algebra even though all objects under consideration are really $\{\mathsf{m}(,), \mathsf{a}()\}$-algebras. To be explicit, an *ideal* $\mathfrak{I}$ in an $\mathcal{R}$-linear Ω-algebra $(A, \{f_x\}_{x\in\Omega})$ is an $\mathcal{R}$-submodule of A with the property that

$$f_x(a_1, \dots, a_{\operatorname{arity}(x)}) \in \mathfrak{I} \quad \text{whenever } \{a_1, \dots, a_{\operatorname{arity}(x)}\} \cap \mathfrak{I} \neq \varnothing, \\ \text{for all } a_1, \dots, a_{\operatorname{arity}(x)} \in A \text{ and } x \in \Omega.$$

Note that for constants x, the left operand of $\cap$ above is always empty, and thus this condition does not require that (the values of) constants would be in every ideal. It does however imply that unary operations map ideals into themselves, and higher arity operations take values within the ideal as soon as any operand is in the ideal.

3.4 Algebra Constructions Revisited

Modulo some minor details, this universal algebra machinery allows us to reproduce quickly the constructions of free hom-nonassociative algebras, free hom-associative algebras, and universal enveloping hom-associative algebras from Sect. 2.3, as well as various others that [44] treat more cursory. The plane binary trees are simply an alternative encoding of formal terms over the signature $\{\mathsf{m}(,)\}$; the correspondence of one to the other is arguably not entirely trivial, but well-known, and it is clearly the binary trees that have the weaker link to the algebra. There is perhaps a slight mismatch in that a formal term would encode an actual constant within each leaf, whereas the binary trees as specified rather take the leaves to mark places where a constant can be inserted, but we shall return to that in the next section.

The weighting added to the trees is a method of encoding also the α operation of a hom-algebra; the unstated idea is that the weight $w(v)$ of a node v specifies how many times α should be applied to the partial result of that node. This is thus why grafting creates new nodes with weight 0—grafting is multiplication, so when the outermost operation was a multiplication, no additional αs are to be applied—and why α raises the weight of the root node v_{low} only. One would like to think of a weighted n-tree as a specification of how n elements in a hom-algebra are being composed—for example the term $\alpha^3(m(\alpha(m(\alpha^2(x_1), x_2)), \alpha^4(x_3)))$ would correspond to the weighted 3-tree

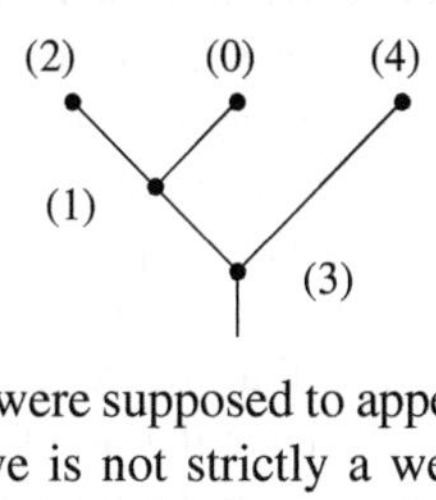

—but there is a catch: weights were supposed to appear only on the internal vertices, not on the leaves, so the above is not strictly a weighted tree as defined in [44]. This choice of disallowing weights on leaves corresponds to the dichotomy in the definition of α_F for F_{HNAs}: as the underlying α on 1-tree terms, but as a shift [1]

on n-tree terms for $n > 1$. This in turn corresponds to the choice of making F_{HNAs} a functor from $\mathcal{R}$-hom-module to $\mathcal{R}$-hom-algebra rather than a functor from $\mathcal{R}$-module to $\mathcal{R}$-hom-algebra; the former produces objects that are less free than those of the latter. It is arguably a strength of the universal algebra method that this distinction appears so clearly, and also a strength that it prefers the more general approach.

What one would do in the universal algebra setting to recover the exact same $F_{\text{HNAs}}(A)$ as Yau defined is to impose $\mathsf{a}(x) \equiv \alpha(x)$ for all $x \in A$ as conditions upon a congruence relation $\equiv$, and then take the quotient by that. Technically, one would start out with the free $\mathcal{R}$-linear Ω-algebra $\mathcal{R}\{\Omega, A\}$ and impose upon it (in addition to $\mathsf{a}(x) \equiv \alpha(x)$) the silly-looking congruences

$$rx \equiv (rx), \qquad x + y \equiv (x + y) \qquad \text{for all } x, y \in A \text{ and } r \in \mathcal{R}; \tag{25}$$

the technical point here is that addition and multiplication in the left hand sides refer to the operations in $\mathcal{R}\{\Omega, A\}$, whereas those on the right hand side refer to operations in A. What happens is effectively the same as in the set-theoretic construction of tensor product of modules. Similarly, to recover the hom-associative $F_{\text{HAs}}(A)$ one would start out with $\mathcal{R}\{\Omega, A\}$ for $\Omega = \{\mathsf{a}(), \mathsf{m}(,)\}$ and quotient that by the smallest $\mathcal{R}$-linear Ω-algebra congruence relation $\equiv$ satisfying the linearity condition (25) and

$$\mathsf{a}(x) \equiv \alpha(x) \qquad \text{for all } x \in A, \tag{26a}$$

$$\mathsf{m}\big(\mathsf{a}(t_1), \mathsf{m}(t_2, t_3)\big) \equiv \mathsf{m}\big(\mathsf{m}(t_1, t_2), \mathsf{a}(t_3)\big) \qquad \text{for all } t_1, t_2, t_3 \in T(\Omega, A). \tag{26b}$$

Finally, in order to recover $U_{\text{HLie}}(\mathfrak{g})$ for the hom-Lie algebra $\mathfrak{g} = (A, b, \alpha)$, one needs only impose also the condition

$$\mathsf{m}(x, y) - \mathsf{m}(y, x) \equiv b(x, y) \qquad \text{for all } x, y \in A \tag{26c}$$

on the congruence relation $\equiv$. What in this step has been noticeably simplified in comparison to the presentation of Sect. 2.3 is that the infinite sequence of alternatingly generating two-sided ideals and applying α_F has been compressed into just one operation, namely that of forming the generated congruence relation. This has not made the whole thing more effective, but it greatly simplifies reasoning about it. For the reader approaching the above as was it a deformation of the associative universal enveloping algebra of a Lie algebra, it might instead be more natural to impose the conditions in the order (26b) first, (26c) second, and (26a) last. Doing so might also raise the question of why one should stop there, as opposed to imposing some additional condition on a, such as $\mathsf{a}\big(\mathsf{m}(t_1, t_2)\big) \equiv \mathsf{m}\big(\mathsf{a}(t_1), \mathsf{a}(t_2)\big)$? The reason not to ask for that particular condition is that it forces the resulting hom-algebra to be multiplicative, and it is easily checked that if $\mathcal{A}$ is a multiplicative hom-algebra, then $\mathcal{A}^-$ is multiplicative as well; doing so would immediately destroy all hope of getting an enveloping algebra, unless the hom-Lie algebra one started with was already multiplicative.

For a hom-Lie algebra presented in terms of a basis, such as the q-deformed $\mathfrak{sl}_2$ of (10), it is usually more natural to seek its U_{HLie} by starting with only the basis elements as constant symbols. In that example one would instead take $X = \{\mathsf{e}, \mathsf{f}, \mathsf{h}\}$ and seek a congruence relation on $\mathbb{K}\{\Omega, X\}$, namely that which satisfies

$$\mathsf{a}(\mathsf{e}) \equiv q\,\mathsf{e}, \qquad \mathsf{a}(\mathsf{f}) \equiv q^2\mathsf{f}, \qquad \mathsf{a}(\mathsf{h}) \equiv q\,\mathsf{h}, \tag{27a}$$

$$\mathsf{m}\bigl(\mathsf{a}(t_1), \mathsf{m}(t_2, t_3)\bigr) \equiv \mathsf{m}\bigl(\mathsf{m}(t_1, t_2), \mathsf{a}(t_3)\bigr) \quad \text{for all } t_1, t_2, t_3 \in T(\Omega, X), \tag{27b}$$

$$\mathsf{m}(\mathsf{e}, \mathsf{f}) - \mathsf{m}(\mathsf{f}, \mathsf{e}) \equiv \tfrac{1}{2}(1+q)\mathsf{h}, \qquad \mathsf{m}(\mathsf{e}, \mathsf{h}) - \mathsf{m}(\mathsf{h}, \mathsf{e}) \equiv -2\mathsf{e}, \qquad \mathsf{m}(\mathsf{h}, \mathsf{f}) - \mathsf{m}(\mathsf{f}, \mathsf{h}) \equiv -2q\,\mathsf{f}. \tag{27c}$$

It suffices to impose hom-associativity for monomial terms (those that can be formed using a, m, and elements of X only) as anything else is a finite linear combination of such terms.

In these equations, it should be observed that (27a) and (27c) are three discrete conditions each, whereas (27b) imposing hom-associativity is an infinite family of conditions. This is mirrored in (26) by the difference in ranges: in (26a), x ranges only over elements of A (i.e., terms that are constants), but in (26b) the variables range over arbitrary terms. Comparing this to presentations of *associative* algebras on the form $\mathcal{R}\langle x, y, z \mid \dots\rangle$, the discrete conditions are like prescribing a relation between the generators x, y, and z, whereas the infinite family used for hom-associativity is like prescribing a Polynomial Identity for the algebra. In rewriting theory, one would rather say (27a) and (27c) are equations of *ground terms* whereas (27b) is an equation involving variables (note that this is a different sense of 'variable' than in 'variable' as generator of $\mathcal{R}\langle x, y, z\rangle$).

The exact same analysis can be carried out for the hom-dialgebras and diweighted trees of [44, Sects. 5–6]; the main point of deviation is merely that one starts out the signature $\bigl\{\mathsf{a}(), \mathsf{l}(,), \mathsf{r}(,)\bigr\}$ (because a dialgebra has separate left multiplication $\dashv$ and right multiplication $\vdash$) rather than the hom-algebra signature $\bigl\{\mathsf{a}(), \mathsf{m}(,)\bigr\}$. The diweighted tree encoding takes another step away from the canonical formal terms by bundling into the weight the left/right nature of each multiplication with the number of αs to apply after it. This is not quite as *ad hoc* as it may seem, because in non-hom dialgebras the associativity-like axioms have the effect that general products of n elements look like $(\cdots(x_1 \vdash x_2) \vdash \cdots) \vdash x_m \dashv (\cdots \dashv (x_{n-1} \dashv x_n)\cdots)$; the left/right nature of a multiplication is pretty much determined by its position in relation to the switchover factor x_m, so there it makes sense to seek a mostly unified encoding of the two. It is however far from clear that the same would be true also for general hom-dialgebras; free hom-associative algebras are certainly far more complicated than free associative algebras.

4 A Newer Setting: Free Operads

One awkward point above is that for example the hom-associativity axiom, despite in some sense being just one identity, required an infinite family of equations to be imposed upon the free hom-associative algebra; shouldn't there be a way of imposing it in just one step? Indeed there is, but it requires broadening one's view, and to think in terms of operads rather than algebras. A programme for this was outlined in [20].

4.1 What Is an Operad?

Nowadays, many introductions to the operad concept are available, for example [33, 40, 42]. What is important for us to stress is the analogy with associative algebras: Operators acting on (say) a vector space can be added together, taken scalar multiples of, and composed; any given set of operators will generate an associative algebra under these operations. When viewed as functions, operators are only univariate however, so one might wonder what happens if we instead consider multivariate functions (still mapping some number of elements from a vector space into that same space)? One way of answering that question is that we get an operad.

Composition in operads work as when one uses dots '$\cdot$' to mark the position of "an argument" in an expression: From the bivariate functions $f(\cdot,\cdot)$ and $g(\cdot,\cdot)$, one may construct the compositions $f(g(\cdot,\cdot),\cdot)$, $f(\cdot,g(\cdot,\cdot))$, $g(f(\cdot,\cdot),\cdot)$, and $g(\cdot,f(\cdot,\cdot))$, which are all trivariate. Note in particular that the "variable-based" style of composition that permits forming e.g. the bivariate function $(x,y)\mapsto f\bigl(g(x,y),y\bigr)$ from f and g is not allowed in an operad, because it destroys multilinearity; $f(x,y)=xy$ is a bilinear map $\mathbb{R}^2\longrightarrow\mathbb{R}$, but $h(x)=f(x,x)=x^2$ is nonlinear.[2] In an expression that composes several operad elements into one, one is however usually allowed to choose where the various arguments are used: $g(f(x_1,x_2),x_3)$, $g(f(x_2,x_1),x_3)$, $g(f(x_3,x_1),x_2)$, etc. are all possible as operad elements. This is formalised by postulating a right action of the group Σ_n of permutations of $\{1,\dots,n\}$ on those operad elements which take n arguments; in function notation one would have $f(x_{\sigma^{-1}(1)},\dots,x_{\sigma^{-1}(n)})=(f\sigma)(x_1,\dots,x_n)$.

More formally, an operad $\mathcal{P}$ is a family $\bigl\{\mathcal{P}(n)\bigr\}_{n\in\mathbb{N}}$ of sets, where $\mathcal{P}(n)$ is "the set of those operad elements which have arity n". Alternatively, an operad $\mathcal{P}$ can be viewed as a set with an arity function, in which case $\mathcal{P}(n)$ is a shorthand for $\bigl\{\,a\in\mathcal{P}\;\;\operatorname{arity}(a)=n\,\bigr\}$. Both approaches are (modulo some formal nonsense) equivalent, and we will employ both since some concepts are easier under one approach and others are easier under the other.

[2] It may then seem serendipitous that Cohn [11, p. 127] citing Hall calls an algebraic structure with the variable-based form of composition a *clone*, since it gets its extra power from being able to "clone" input data, but he explains it as being a contraction of 'closed set of operations'. In the world of Quantum Mechanics, the well-known 'No cloning' theorem forbids that kind of behaviour (essentially because it violates multilinearity), so by sticking to operads we take the narrow road.

Composition can be given the form of composing one element $a \in \mathcal{P}(m)$ with the m elements $b_i \in \mathcal{P}(n_i)$ for $i = 1, \dots, m$ (i.e., one for each "argument" of a) to form $a \circ b_1 \otimes \dots \otimes b_m \in \mathcal{P}\bigl(\sum_{i=1}^{m} n_i\bigr)$; note that the '$\circ$' and the $m-1$ '$\otimes$' are all part of the same operad composition. (There is a more general concept called PROP where $b_1 \otimes \dots \otimes b_m$ would be an actual element, but we won't go into that here.) Operad composition is associative in the sense that the unparenthesized expression

$$a \circ b_1 \otimes \dots \otimes b_\ell \circ c_1 \otimes \dots \otimes c_m$$

is the same whether it is interpreted as

$$(a \circ b_1 \otimes \dots \otimes b_\ell) \circ c_1 \otimes \dots \otimes c_m$$

or as

$$a \circ (b_1 \circ c_1 \otimes \dots \otimes c_{m_1}) \otimes \dots \otimes (b_\ell \circ c_{m_1+\dots+m_{\ell-1}+1} \otimes \dots \otimes c_{m_1+\dots+m_\ell})$$

where $m = \sum_{i=1}^{\ell} m_i$ and $b_i \in \mathcal{P}(m_i)$ for $i = 1, \dots, \ell$.[3]

Since Σ_n acts on the right of each $\mathcal{P}(n)$, this action satisfies $(a\sigma)\tau = a(\sigma\tau)$ for all $a \in \mathcal{P}(n)$ and $\sigma, \tau \in \Sigma_n$. There is also a condition called equivariance that

$$(a\sigma) \circ b_1 \otimes \dots \otimes b_m = (a \circ b_{\sigma^{-1}(1)} \otimes \dots \otimes b_{\sigma^{-1}(m)})\tau$$

where τ is a block version of σ, such that the kth block has size equal to the arity of b_k. Finally, it is usually also required that there is an identity element $\mathrm{id} \in \mathcal{P}(1)$ such that $\mathrm{id} \circ a = a = a \circ \mathrm{id}^{\otimes n}$ for all $a \in \mathcal{P}(n)$ and $n \in \mathbb{N}$.

Example 4.1 For every set A, there is an operad Map_A such that $\mathrm{Map}_A(n)$ is the set of all maps $A^n \longrightarrow A$; in particular, $\mathrm{Map}_A(0)$ may be identified with A. For $a \in \mathrm{Map}_A(m)$ and $b_i \in \mathrm{Map}_A(n_i)$ for $i = 1, \dots, m$, the composition $a \circ b_1 \otimes \dots \otimes b_m$ is defined by

$$\begin{aligned}(a \circ b_1 \otimes \dots \otimes b_m)&(x_{1,1}, \dots, x_{1,n_1}, \dots, x_{m,1}, \dots, x_{m,n_m}) \\ &= a\bigl(b_1(x_{1,1}, \dots, x_{1,n_1}), \dots, b_m(x_{m,1}, \dots, x_{m,n_m})\bigr)\end{aligned}$$

for all $x_{1,1}, \dots, x_{m,n_m} \in A$. $\mathrm{id} \in \mathrm{Map}_A(1)$ is the identity map on A. The permutation action is defined by $(a\sigma)(x_1, \dots, x_n) = a(x_{\sigma^{-1}(1)}, \dots, x_{\sigma^{-1}(n)})$.

An alternative notation for composition is $\gamma(a, b_1, \dots, b_m) = a \circ b_1 \otimes \dots \otimes b_m$; that γ is then called the *structure map*, or *structure maps* if one requires each map to have a signature on the form $\mathcal{P}(m) \times \mathcal{P}(n_1) \times \dots \times \mathcal{P}(n_m) \longrightarrow \mathcal{P}\bigl(\sum_{i=1}^{m} n_i\bigr)$.

---

[3] As the number of ellipses (...) above indicate, the axioms for operads are somewhat awkward to state, even though they only express familiar properties of multivariate functions. The PROP formalism may therefore be preferable even if one is only interested in an operad setting, since the PROP axioms can be stated without constantly going '...'.

An alternative composition *concept* is the *ith composition* $\circ_i$, which satisfies $a \circ_i b = a \circ \mathrm{id}^{\otimes(i-1)} \otimes b \otimes \mathrm{id}^{\otimes(m-i)}$ for $a \in \mathcal{P}(m)$ and $i = 1, \ldots, m$. Note that ith composition, despite being a binary operation, is not at all associative in the usual sense and expressions involving it must therefore be explicitly parenthesized; operad associativity does however imply that subexpressions can be regrouped (informally: "parentheses can be moved around") provided that the position indices are adjusted accordingly.

An *operad homomorphism* $\phi\colon \mathcal{P} \longrightarrow \mathcal{Q}$ is a map that is compatible with the operad structures of $\mathcal{P}$ and $\mathcal{Q}$: $\mathrm{arity}_{\mathcal{Q}}(\phi(a)) = \mathrm{arity}_{\mathcal{P}}(a)$, $\phi(a \circ b_1 \otimes \cdots \otimes b_m) = \phi(a) \circ \phi(b_1) \otimes \cdots \otimes \phi(b_m)$, $\phi(a\sigma) = \phi(a)\sigma$, and $\phi(\mathrm{id}_{\mathcal{P}}) = \mathrm{id}_{\mathcal{Q}}$ for all $a \in \mathcal{P}(m)$, $b_i \in \mathcal{P}$ for $i = 1, \ldots, m$, $\sigma \in \Sigma_m$, and $m \in \mathbb{N}$. A *suboperad* of $\mathcal{P}$ is a subset of $\mathcal{P}$ that is closed under composition, closed under permutation action, and contains the identity element. The operad *generated* by some $\Omega \subseteq \mathcal{P}$ is the smallest suboperad of $\mathcal{P}$ that contains Ω.

Let $\mathcal{R}$ be an associative and commutative unital ring. An operad $\mathcal{P}$ is said to be *$\mathcal{R}$-linear* if (i) each $\mathcal{P}(n)$ is an $\mathcal{R}$-module, (ii) every structure map $(a, b_1, \cdots, b_m) \mapsto a \circ b_1 \otimes \cdots \otimes b_m$ is $\mathcal{R}$-linear in each argument separately, and (iii) each action of a permutation is $\mathcal{R}$-linear.

Example 4.2 The Map_A operad is in general not $\mathcal{R}$-linear, but if A is an $\mathcal{R}$-module, then the suboperad $\mathcal{E}nd_A$ where $\mathcal{E}nd_A(n)$ consists of all $\mathcal{R}$-multilinear maps $A^n \longrightarrow A$ will be $\mathcal{R}$-linear. $\mathcal{E}nd_A(0)$ can also be identified with A.

The operad concept defined above is sometimes called a *symmetric* operad, because of the actions on it of the symmetric groups. Dropping everything involving permutations above, one instead arrives at the concept of a *nonsymmetric* or *non-*Σ operad. Much of what is done below could just as well be done in the non-Σ setting, but we find the symmetric setting to be more akin to classical universal algebra.

4.2 Universal Algebra for Operads

Regarding universal algebra, an interesting thing about operads is that they may serve as generalisations of both the algebra concept and the signature concept. The way that an operad $\mathcal{P}$ may generalise a signature Ω is that a set A is said to be a *$\mathcal{P}$-algebra* if it is given with an operad homomorphism $\phi\colon \mathcal{P} \longrightarrow \mathrm{Map}_A$; the operation f_x of some $x \in \mathcal{P}$ is then simply $\phi(x)$. Being an operad-algebra is however a stronger condition than being a signature-algebra, because the map ϕ will only be a homomorphism if every identity in $\mathcal{P}$ is also satisfied in $\phi(\mathcal{P})$; this can be used to impose "laws" on algebras, and several elementary operads are defined to precisely this purpose: an algebra is an $\mathcal{A}ss$-algebra iff it is associative, a $\mathcal{C}om$-algebra iff it is commutative, a $\mathcal{L}ie$-algebra iff it is a Lie algebra, a $\mathcal{L}eib$-algebra iff it is a Leibniz algebra, and so on. It is therefore only natural that we will shortly construct an operad $\mathcal{HA}ss$ whose algebras are precisely the hom-associative algebras.

Before taking on that problem, we should however give an example of how identities in an operad become laws of its algebras. To that end, consider $\mathbb{N}$ as an operad by making $\operatorname{arity}(n) = n$; this uniquely defines the operad structure, since the arity of any particular composition is given by the axioms, and that in turn determines the value since every $\mathbb{N}(n)$ only has one element. What can now be said about an $\mathbb{N}$-algebra A if $f\colon \mathbb{N} \longrightarrow \mathrm{Map}_A$ is the given operad homomorphism? Clearly $f(2)\colon A^2 \longrightarrow A$ is a binary operation. If $\tau \in \Sigma_2$ is the transposition, one furthermore finds that

$$f(2)(x,y) = f(2\tau)(x,y) = \bigl(f(2)\tau\bigr)(x,y) = f(2)(y,x)$$

for all $x, y \in A$, so $f(2)$ is commutative. Similarly it follows from $2 \circ 1 \otimes 2 = 3 = 2 \circ 2 \otimes 1$ that $f(2)\bigl(x, f(2)(y,z)\bigr) = f(2)\bigl(f(2)(x,y), z\bigr)$ for all $x, y, z \in A$, and thus $f(2)$ is associative. Finally one may deduce from $\mathrm{id} = 1 = 2 \circ 0 \otimes 1$ that $f(0)$ is a unit element with respect to $f(2)$, so in summary any $\mathbb{N}$-operad algebra carries an abelian monoid structure. This is almost the same as $\mathcal{C}om$ is supposed to accomplish, so one might ask whether in fact $\mathcal{C}om = \mathbb{N}$, but traditionally $\mathcal{C}om$, $\mathcal{A}ss$, etc. are taken to be the $\mathcal{R}$-linear (for whatever ring $\mathcal{R}$ of scalars is being considered) operads that impose the indicated laws on their algebras. $\mathcal{C}om$ is thus rather characterised by having $\dim \mathcal{C}om(n) = 1$ for all n, and may if one wishes be constructed as $\mathcal{R} \times \mathbb{N}$.

While specific operads may sometimes be constructed through elementary methods as above, the general approach to constructing an operad that corresponds to a specific set of laws is instead the universal algebraic one, which rather employs the point of view that an operad is a generalisation of an algebra. Obviously any specific Ω-algebra $\bigl(A, \{f_x\}_{x\in\Omega}\bigr)$ gives rise to the operad Map_A, but the operad that more naturally generalises A as an Ω-algebra is the suboperad of Map_A that is generated by $\{f_x\}_{x\in\Omega}$. Conversely, if A is supposed to be some kind of free algebra, one may choose to construct it as the constant component of the corresponding free operad.

An equivalence relation $\equiv$ on an operad $\mathcal{P}$ is an *operad congruence relation* if:

1. $a \equiv a'$ implies $\operatorname{arity}(a) = \operatorname{arity}(a')$,
2. $a \equiv a'$ and $b_i \equiv b_i'$ for $i = 1, \dots, \operatorname{arity}(a)$ implies $a \circ b_1 \otimes \cdots \otimes b_{\operatorname{arity}(a)} \equiv a' \circ b_1' \otimes \cdots \otimes b'_{\operatorname{arity}(a)}$, and
3. $a \equiv a'$ implies $a\sigma \equiv a'\sigma$ for all $\sigma \in \Sigma_{\operatorname{arity}(a)}$.

As for algebras, it follows that the quotient $\mathcal{P}/{\equiv}$ carries an operad structure, and the canonical map $\mathcal{P} \longrightarrow \mathcal{P}/{\equiv}$ is an operad homomorphism. If additionally $\mathcal{P}$ is $\mathcal{R}$-linear and $\equiv$ is an $\mathcal{R}$-module congruence relation on each $\mathcal{P}(n)$, then $\equiv$ is an *$\mathcal{R}$-linear operad congruence relation* and the corresponding *operad ideal* $\mathcal{J}$ is defined by $\mathcal{J}(n) = \bigl\{ a \in \mathcal{P}(n) \ \ a \equiv 0 \bigr\}$ for all $n \in \mathbb{N}$ (note that each $\mathcal{P}(n)$ has a separate 0 element). Equivalently, $\mathcal{J} \subseteq \mathcal{P}$ is an operad ideal if each $\mathcal{J}(n)$ is a submodule of $\mathcal{P}(n)$, each $\mathcal{J}(n)$ is closed under the action of Σ_n, and $a \circ b_1 \otimes \cdots \otimes b_m \in \mathcal{J}$ whenever at least one of $a, b_1, \dots, b_m$ is an element of $\mathcal{J}$.

So far, the operad formalism is very similar to that for algebras, but an important difference occurs when one wishes to impose laws on a congruence. For an algebra, the hom-associativity condition (24) required an infinite family of identities. The corresponding condition in the operad Map_A requires only the single identity $f_{\mathsf{m}} \circ f_{\mathsf{a}} \otimes f_{\mathsf{m}} \equiv f_{\mathsf{m}} \circ f_{\mathsf{m}} \otimes f_{\mathsf{a}}$, as the infinite family is recovered from this using composition on the right: $f_{\mathsf{m}} \circ f_{\mathsf{a}} \otimes f_{\mathsf{m}} \circ t_1 \otimes t_2 \otimes t_3 \equiv f_{\mathsf{m}} \circ f_{\mathsf{m}} \otimes f_{\mathsf{a}} \circ t_1 \otimes t_2 \otimes t_3$. The $\mathcal{A}ss$, $\mathcal{C}om$, $\mathcal{L}eib$, etc. operads can all be seen to be finitely presented, and the same holds for their free algebras if generated as the arity 0 component of an operad, even though they are not finitely presented within the Ω-algebra formalism!

The universal property satisfied by the free operad $\mathcal{F}$ on Ω is that it is given with an arity-preserving map $i\colon \Omega \longrightarrow \mathcal{F}$ such that there for every operad $\mathcal{P}$ and every arity-preserving map $j\colon \Omega \longrightarrow \mathcal{P}$ exists a unique operad homomorphism $\phi\colon \mathcal{F} \longrightarrow \mathcal{P}$ such that $j = \phi \circ i$. A practical construction of that free operad is to let $\mathcal{F}(n)$ be the set of all n-variable *contexts* [12, p. 17], but since we'll anyway need some notation for these, we might as well give an explicit definition based on Polish notation for expressions.

Definition 4.1 A (left-)*Polish term* on the signature Ω is a finite word on $\Omega \cup \{\square_i\}_{i=1}^{\infty}$ (where it is presumed that $\square_i \notin \Omega$ and $\operatorname{arity}(\square_i) = 0$ for all i), which is either $\square_i$ for some $i \geqslant 1$, or $x\mu_1 \cdots \mu_n$ where $x \in \Omega$, $n = \operatorname{arity}(x)$, and $\mu_1, \dots, \mu_m$ are themselves Polish terms on Ω. A Polish term is an *n-context* if each symbol $\square_i$ for $i = 1, \dots, n$ occurs exactly once and no symbol $\square_i$ with $i > n$ occurs at all. For $\square_1, \dots, \square_9$ we will write $1, \dots, 9$ for short. Denote by $\mathcal{Y}_\Omega(n)$ the set of all n-contexts on Ω.

The action of $\sigma \in \Sigma_n$ on $\mathcal{Y}_\Omega(n)$ is that each $\square_i$ is replaced by $\square_{\sigma^{-1}(i)}$. The composition $\mu \circ \nu_1 \otimes \cdots \otimes \nu_n$ is a combined substitution and renumbering: first each $\square_i$ in μ is replaced by the corresponding ν_i, then the $\square_k$'s in the composite term are renumbered so that the term becomes a context—preserving the differences within each ν_i and giving $\square_k$'s from ν_i lower indices than those from ν_j whenever $i < j$.

For any associative and commutative unital ring $\mathcal{R}$, and for every $n \in \mathbb{N}$, denote by $\mathcal{R}\{\Omega\}(n)$ the set of all formal $\mathcal{R}$-linear combinations of elements of $\mathcal{Y}_\Omega(n)$. Extend the action of $\sigma \in \Sigma_n$ on $\mathcal{Y}_\Omega(n)$ to $\mathcal{R}\{\Omega\}(n)$ by linearity. Let $\mathcal{R}\{\Omega\} = \bigcup_{n\in\mathbb{N}} \mathcal{R}\{\Omega\}(n)$. Extend the composition on $\mathcal{Y}_\Omega$ to $\mathcal{R}\{\Omega\}$ by multilinearity. When $\mathcal{Y}_\Omega$ is viewed as a subset of $\mathcal{R}\{\Omega\}$, its elements are called *monomials*.

With $\mathrm{id} = 1 = \square_1$, this makes $\mathcal{Y}_\Omega$ the free operad on Ω and $\mathcal{R}\{\Omega\}$ is the free $\mathcal{R}$-linear operad on Ω.

For $\Omega = \{\mathsf{x}, \mathsf{a}(), \mathsf{m}(,)\}$, one may thus find in $\mathcal{Y}_\Omega(0)$ elements such as x, ax, mxx, amxx, and maxx which in parenthesized notation would rather have been written as x, $\mathsf{a(x)}$, $\mathsf{m(x, x)}$, $\mathsf{a(m(x, x))}$, and $\mathsf{m(a(x), x)}$ respectively. In $\mathcal{Y}_\Omega(1)$ we similarly find 1, $\mathsf{a}1$, $\mathsf{aa}1$, $\mathsf{mx}1$, $\mathsf{m}1\mathsf{x}$, and $\mathsf{maxm}1\mathsf{x}$ which in parenthesized notation could have been written as $\square_1$, $\mathsf{a}(\square_1)$, $\mathsf{a}(\mathsf{a}(\square_1))$, $\mathsf{m}(\mathsf{x}, \square_1)$, $\mathsf{m}(\square_1, \mathsf{x})$, and $\mathsf{m}(\mathsf{a}(\mathsf{x}), \mathsf{m}(\square_1, \mathsf{x}))$. In $\mathcal{R}\{\Omega\}(2)$ there are elements such as $\mathsf{m}12 - \mathsf{m}21$ and $\mathsf{m}12 + \mathsf{m}21$ which would be mapped to 0 by any operad homomorphism f to Map_A for which $f(\mathsf{m})$ is commutative or anticommutative respectively. Finally there is in $\mathcal{R}\{\Omega\}(3)$ the elements $\mathsf{m}1\mathsf{m}23 - \mathsf{mm}123$ and $\mathsf{ma}1\mathsf{m}23 - \mathsf{mm}12\mathsf{a}3$ which have similar roles with respect to associativity and hom-associativity respectively.

A practical problem, which is mostly common to the Polish and the parenthesized notations, is that it can be difficult to grasp the structure of one of these expressions just from a quick glance at the written forms of them; small expressions may be immediately recognised by the trained eye, but larger expressions almost always require a conscious effort to parse. This is unfortunate, as the exact structure is very important when working in a setting this general. The structure can however be made more visible by *drawing* expressions rather than *writing* them; informally one depicts an expression using its abstract syntax tree, but those of a more formalistic persuasion may think of these drawings as graph-theoretical objects underlying the trees (in the sense of [12, pp. 15–16]) of these terms. A few examples can be

m12 = 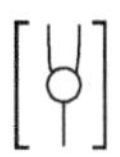 m21 = 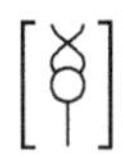 mm312 = 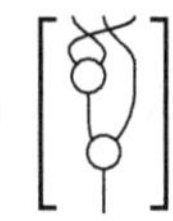

and several more can be found below. A Polish term may even be read as a direct instruction for how to draw these trees: in order to draw $\mu = x\nu_1 \cdots \nu_{\mathrm{arity}(x)}$, first draw a vertex for x as the root, and then draw the subtrees ν_1 through $\nu_{\mathrm{arity}(x)}$ above the x vertex and side by side, letting the order of edges along the top of a vertex show the order of the subexpressions. The "inputs" $\square_k$ of a context are represented by edges to the top side of the drawing, with $\square_1$ being leftmost, $\square_2$ being second to left, and so on.

Definition 4.2 An element of $\mathcal{Y}_\Omega(n)$ is said to be *plane* if the $\square_i$ symbols (if any) occur in ascending order: none to the left of $\square_1$, only $\square_1$ to the left of $\square_2$, and so on. (Equivalently, the drawing procedure described above will not produce any crossing edges.) An element of $\mathcal{R}\{\Omega\}(n)$ is *plane* if it is a linear combination of plane elements. An element of $\mathcal{R}\{\Omega\}(n)$ is *planar* if it is of the form $a\sigma$ for some plane $a \in \mathcal{R}\{\Omega\}(n)$ and $\sigma \in \Sigma_n$. Finally, an ideal in $\mathcal{R}\{\Omega\}$ is said to be *planar* if it is generated by planar elements.

Elements in a planar ideal need not be planar, but every element in a planar ideal can be written as a sum of planar elements that are themselves in the ideal.

4.3 The Diamond Lemma for Operads

This and the following sections rely heavily on results and concepts from [21]. We try to always give a reference, where a concept is first used that will not be explained further here, to the exact definition in [21] of that concept.

Let a signature Ω and an associative and commutative unital ring $\mathcal{R}$ be given. Consider the free $\mathcal{R}$-linear operad $\mathcal{R}\{\Omega\}$ and its suboperad of monomials $\mathcal{Y}_\Omega$. Let $V(i,j)$ be the set of all maps $\mathcal{R}\{\Omega\}(j) \longrightarrow \mathcal{R}\{\Omega\}(i)$ that are on the form

$$a \mapsto \big(\lambda \circ_k (a \circ \nu_1 \otimes \cdots \otimes \nu_j)\big)\sigma \tag{28}$$

where $\nu_r \in \mathcal{Y}_\Omega(n_r)$ for $r = 1,\dots,j$, $\lambda \in \mathcal{Y}_\Omega(\ell)$, $\ell \geqslant k \geqslant 1$, $\sigma \in \Sigma_i$, and $i = \ell - 1 + n_1 + \ldots + n_j$. The family $V = \bigcup_{i,j\in\mathbb{N}} V(i,j)$ is then a category [21,

Definition 6.8], and each $v \in V(i,j)$ is an injection $\mathcal{Y}_\Omega(j) \longrightarrow \mathcal{Y}_\Omega(i)$. Also note that with respect to the tree (drawing) forms of monomials, each $v \in V(i,j)$ defines an embedding of $\mu \in \mathcal{Y}_\Omega(j)$ into $v(\mu)$; this will be important for identifying V-critical ambiguities.

Definition 4.3 A *rewriting system* for $\mathcal{R}\{\Omega\}$ is a set $S = \bigcup_{i\in\mathbb{N}} S(i)$ such that $S(i) \subseteq \mathcal{Y}_\Omega(i) \times \mathcal{R}\{\Omega\}(i)$ for all $i \in \mathbb{N}$. The elements of a rewrite system are called *(rewrite) rules*. The components of a rule s are often denoted μ_s and a_s, meaning $s = (\mu_s, a_s)$ for all rules s.

For a given rewriting system S, define $T_1(S)(i) = \bigcup_{j\in\mathbb{N}} \{t_{v,s}\}_{v\in V(i,j), s\in S(j)}$, where $t_{v,s}$ is the $\mathcal{R}$-linear map $\mathcal{R}\{\Omega\}(i) \longrightarrow \mathcal{R}\{\Omega\}(i)$ which satisfies

$$t_{v,s}(\lambda) = \begin{cases} v(a_s) & \text{if } \lambda = v(\mu_s), \\ \lambda & \text{otherwise,} \end{cases} \qquad \text{for all } \lambda \in \mathcal{Y}_\Omega(i). \tag{29}$$

The elements of $T_1(S)(i)$ are called the *simple reductions* (with respect to S) on $\mathcal{R}\{\Omega\}(i)$. For each $i \in \mathbb{N}$, let $T(S)(i)$ be the set of all finite compositions of maps in $T_1(S)(i)$.

Sometimes, a claim that $t_{v,s}(a) = b$ is more conveniently written as $a \xrightarrow{s} b$ (for example when several such claims are being chained, as in $a \xrightarrow{s_1} b \xrightarrow{s_2} c$). When doing that, we may indicate what v is by inserting parentheses into the Polish term on the tail side of the arrow that is being changed by the simple reduction: the outer parenthesis then surrounds the $\mu_s \circ v_1 \otimes \cdots \otimes v_j$ part, whereas inner parentheses surround the various v_k subterms of it, although these inner parentheses are for brevity omitted where $v_k = \mathrm{id}$. See Example 3.3 for some examples of this.

With respect to $T(S)$, all maps in V are absolutely advanceable [21, Definition 6.1]. The following subsets of $\mathcal{R}\{\Omega\}$ are defined in [21, Definition 3.4], but so important that we include the definitions here:

$$\mathrm{Irr}(S)(i) = \bigl\{ a \in \mathcal{R}\{\Omega\}(i) \bigm| t(a) = a \text{ for all } t \in T(S)(i) \bigr\},$$
$$\mathcal{I}(S)(i) = \sum_{t\in T(S)(i)} \bigl\{ a - t(a) \bigm| a \in \mathcal{R}\{\Omega\}(i) \bigr\}$$

for all $i \in \mathbb{N}$. We write $a \equiv b \pmod{S}$ for $a - b \in \mathcal{I}(S)$. An $a \in \mathrm{Irr}(S)$ is said to be a *normal form* of $b \in \mathcal{R}\{\Omega\}$ if $a \equiv b \pmod{S}$.

$\mathcal{I}(S)$ is the operad ideal in $\mathcal{R}\{\Omega\}$ that is generated by $\{\, \mu_s - a_s \mid s \in S \,\}$ .Irr(S) is what we want to use as model for the quotient $\mathcal{R}\{\Omega\}\big/\mathcal{I}(S)$, and we use Theorem 3.1 below to tell us that it really is. An *ambiguity* [21, Definition 5.9] of $T_1(S)(i)$ is a triplet $(t_{v_1,s_1}, \mu, t_{v_2,s_2})$ such that $v_1(\mu_{s_1}) = \mu = v_2(\mu_{s_2})$. The ambiguity is *plane* if μ is plane.

Theorem 4.1 (Basic Diamond Lemma for Symmetric Operads) *If $P(i)$ is a well-founded partial order on $\mathcal{Y}_\Omega(i)$ such that $a_s \in \mathrm{DSM}\bigl(\mu_s, P(i)\bigr)$ for all $i \in \mathbb{N}$, and moreover for all $i, j \in \mathbb{N}$ every $v \in V(i,j)$ is monotone [21, Definition 6.4] with respect to $P(j)$ and $P(i)$, then the following claims are equivalent:*

(a) *For all $i \in \mathbb{N}$, every ambiguity of $T_1(S)(i)$ is resolvable [21, Definition 5.9].*
(a′) *For all $i \in \mathbb{N}$, every V-critical [21, Definition 6.8] ambiguity of $T_1(S)(i)$ is resolvable.*
(a″) *For all $i \in \mathbb{N}$, every plane V-critical ambiguity of $T_1(S)(i)$ is resolvable.*
(b) *Every element of $\mathcal{R}\{\Omega\}$ is persistently [21, Definition 4.1] and uniquely [21, Definition 4.6] reducible, with normal form map t^S [21, Definition 4.6].*
(c) *Every element of $\mathcal{R}\{\Omega\}$ has a unique normal form, i.e., $\mathcal{R}\{\Omega\}(i) = \mathcal{I}(S)(i) \oplus \operatorname{Irr}(S)(i)$ for all $i \in \mathbb{N}$.*

Proof Taking $\mathcal{M}(i) = \mathcal{R}\{\Omega\}(i)$ and $\mathcal{Y}(i) = \mathcal{Y}_\Omega(i)$, this is mostly a combination of Theorem 5.11, Theorem 6.9, and Construction 7.2 of [21]. Theorem 5.11 provides the basic equivalence of (a), (b), and (c). Theorem 6.9 says (a′) is sufficient, as resolvability implies resolvability relative to P. Construction 7.2 shows the V, P, and $T_1(S)$ defined above fulfill the conditions of these two theorems.

What remains to show is that (a″) implies (a′). Let $(t_{v_1,s_1}, \mu, t_{v_2,s_2})$ be a V-critical ambiguity of some $T_1(S)(i)$, and let $\sigma \in \Sigma_i$ be such that $\mu\sigma$ is plane. Then $w\colon a \mapsto a\sigma$ and $w^{-1}\colon a \mapsto a\sigma^{-1}$ are both elements of $V(i,i)$, and hence $(t_{v_1,s_1}, \mu, t_{v_2,s_2})$ is an absolute shadow of the plane and V-critical ambiguity $(t_{w\circ v_1,s_1}, \mu\sigma, t_{w\circ v_2,s_2})$. The latter is resolvable by (a″), so it follows from [21, Lemma 6.2] that the former is resolvable as well.

Remark 4.1 Theorem 3.1 may also be viewed as a slightly streamlined version of [23, Corollary 10.26], but that approach is probably overkill for readers uninterested in the PROP setting.

It may be observed that $\operatorname{Irr}(S)(i)$ is closed under the action of Σ_i, regardless of S; this is thus a restriction of the applicability of this diamond lemma, as its conditions can never be fulfilled when $\mathcal{R}\{\Omega\}(i)/\mathcal{I}(S)(i)$ is fixed under a non-identity element of Σ_i. All of that is however a consequence of the choice of V, and a different choice of V (e.g. excluding the permutation σ from (28)) will result in a different (but very similar-looking) diamond lemma, with a different set of critical ambiguities and a different domain of applicability.

For an ambiguity $(t_{v_1,s_1}, \mu, t_{v_2,s_2})$ to be V-critical in this basic diamond lemma, it is necessary that the graph-theoretical embeddings into μ of μ_{s_1} and μ_{s_2} have at least one vertex in common (otherwise the ambiguity is a montage) and furthermore these two embeddings must cover μ (otherwise the ambiguity is a proper V-shadow). Enumerating the critical ambiguities formed by two given rules s_1 and s_2 is thus mostly a matter of listing the ways of superimposing the two trees μ_{s_1} and μ_{s_2}.

Example 4.3 (*Ass* operad) Let $\Omega = \{\mathsf{m}(,)\}$. Consider the rewriting system $S = \{s\}$ where $s = (\mathsf{m1m23}, \mathsf{mm123})$. Graphically, this rule takes the form

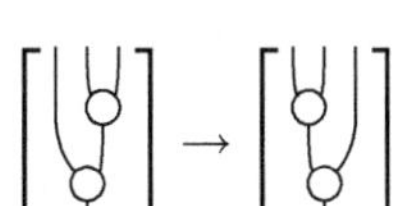

(30)

The (non-unital) *associative operad* $\mathcal{A}ss$ over $\mathcal{R}$ can then be defined as the quotient $\mathcal{R}\{\Omega\}/\mathcal{I}(S)$.

One way of partially ordering trees that will be compatible with this rule is to count, separately for each input, the number of times the path from that input to the root enters an m vertex from the right; denote this number for input i of the tree μ by $h_i(\mu)$. Then define $\mu \geqslant \nu$ in $P'(n)$ if and only if $h_i(\mu) \geqslant h_i(\nu)$ for all $i = 1, \dots, n$, and define a partial order $P(n)$ by $\mu > \nu$ in $P(n)$ if and only if $\mu \geqslant \nu$ in $P'(n)$ and $\mu \not\leqslant \nu$ in $P'(n)$, i.e., let $P(n)$ be the restriction to a partial order of the quasi-order $P'(n)$. For the left hand side of s above one has $h_1 = 0$, $h_2 = 1$, and $h_3 = 2$ whereas the left hand side has $h_1 = 0$, $h_2 = 1$, and $h_3 = 1$, so S is indeed compatible with P. Furthermore $P(n)$ is clearly well-founded; $\sum_{i=1}^{n} h_i(\mu)$ is simply the rank of μ in the poset $(\mathcal{Y}_\Omega(n), P(n))$.

The only plane critical ambiguity of S is $(t_{\nu_1,s}, \mathsf{m1m2m34}, t_{\nu_2,s})$, where $\nu_1(\mu) = \mu \circ_3 \mathsf{m12}$ and $\nu_2(\mu) = \mathsf{m12} \circ_2 \mu$. This is resolved as follows:

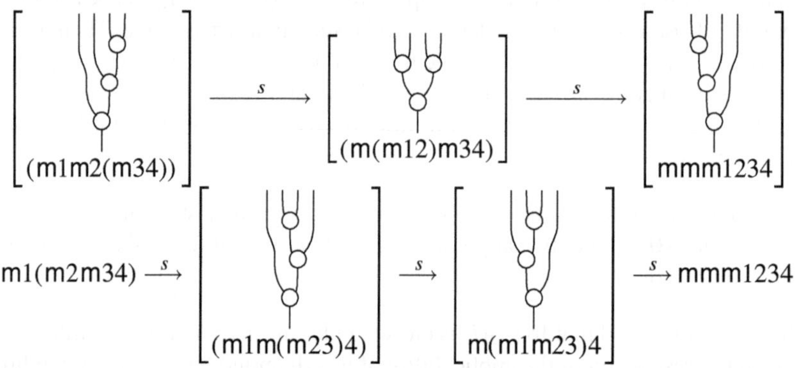

Hence the conditions of Theorem 3.1 are fulfilled, $\mathcal{R}\{\Omega\}(n) = \mathcal{I}(S)(n) \oplus \mathrm{Irr}(S)(n)$ for all $n \in \mathbb{N}$, and $\mathcal{A}ss(n) \cong \mathrm{Irr}(S)(n)$ as $\mathcal{R}$-modules for all $n \in \mathbb{N}$. Since a monomial μ is irreducible iff it does not contain an m as right child of an m, i.e., iff every right child of an m is an input, it follows that the only thing that distinguishes two irreducible elements of $\mathcal{Y}(n)$ is the order of the inputs. On the other hand, every permutation of the inputs gives rise to a distinct irreducible element, so $\dim \mathcal{A}ss(n) = |\Sigma_n| = n!$ for all $n \geqslant 1$, exactly as one would expect.

For $n = 0$ one gets $\dim \mathcal{A}ss(0) = \dim \mathcal{R}\{\Omega\}(0) = \left|\mathcal{Y}(0)\right| = 0$ however, which is perhaps not quite what the textbooks say $\mathcal{A}ss$ should have. The reason it comes out this way is that we took $\mathcal{A}ss$ to be the operad for associative algebras, period; had we instead taken it to be the operad for *unital* associative algebras then $\dim \mathcal{A}ss(n) = n!$ would have held also for $n = 0$. Obviously $\dim \mathcal{R}\{\Omega\}(0) = 0$ because Ω doesn't contain any constants, but requiring a unit introduces such a constant u. Making that constant behave like a unit requires two additional rules (mu1, 1) and (m1u, 1) in the rewriting system however, and we felt the resolution of the resulting ambiguities are perhaps better left as exercises.

Another useful exercise is to similarly construct the $\mathcal{L}eib$ operad, which merely amounts to replacing the rewriting system S with $S' = \{s'\}$, where $s = (\mathsf{m1m23}, \mathsf{mm123} - \mathsf{mm132})$. Using brackets as notation for the operation in a Leibniz algebra, this rule corresponds to the law that $\bigl[x,[y,z]\bigr] = \bigl[[x,y],z\bigr] - \bigl[[x,z],y\bigr]$. Graphically, s' takes the form

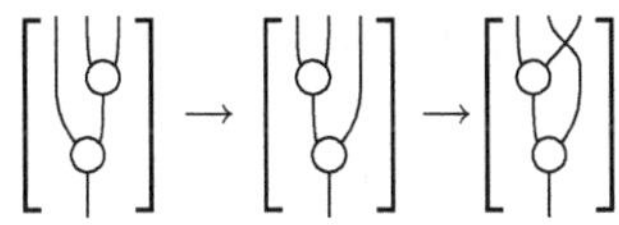

(31)

which unlike associativity is not planar, but that makes no difference for the Diamond Lemma machinery. The left hand side of s' is the same as the left hand side of s, so $\mathrm{Irr}(S') = \mathrm{Irr}(S)$ and both rewriting systems have the same sites of ambiguities. What is different are the resolutions, where the resolution in the Leibniz case is longer since it involves more terms; a compact notation such as the Polish one is highly recommended when reporting the calculations. Still, it is well within the realm of what can be done by hand.

4.4 The Hom-Associative Operad

When pursuing the same approach for the hom-associative identity, one of course needs an extra symbol for the unary operation, so $\Omega = \bigl\{\mathsf{m}(,), \mathsf{a}()\bigr\}$. Drawing m as a circle and a as a square, hom-associativity is then the congruence

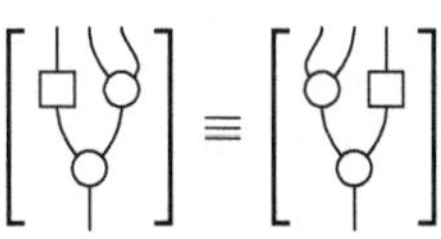

(32)

which can be expressed as a rule $s = (\mathsf{ma1m23}, \mathsf{mm12a3})$. It is thus straightforward to define $\mathcal{HAss} = \mathcal{R}\{\Omega\}/\mathcal{I}\bigl(\{s\}\bigr)$, but not quite so straightforward to decide whether two elements of $\mathcal{R}\{\Omega\}$ are congruent modulo $\mathcal{I}\bigl(\{s\}\bigr)$, because $\{s\}$ is not a complete rewriting system; the ambiguity one has to check fails to resolve:

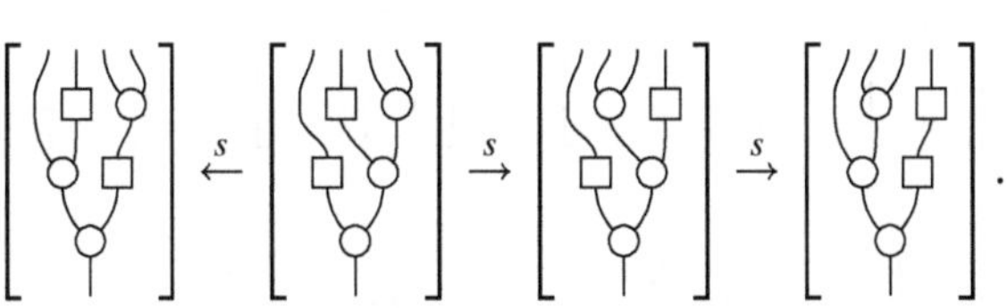

(33)

Failed resolutions should not be taken as disasters however; they are in fact opportunities to learn, since what the above demonstrates is that $\mathsf{mm1a2am34} \equiv \mathsf{mm1m23aa4}$ (mod $\{s\}$) (or as a law: $\bigl(x\alpha(y)\bigr)\alpha(zw) = \bigl(x(yz)\bigr)\alpha\bigl(\alpha(w)\bigr)$ for all x, y, z, w), which was probably not apparent from the definition of hom-associativity. Therefore one's response to this discovery should be to make a new rule out of this new and nontrivial congruence, so that one can use it to better understand congruence modulo hom-associativity.

A problem with this congruence is however that the left and right hand sides are not comparable under the same partial order as worked fine for the associative and Leibniz operads: $(h_0, h_1, h_2, h_3) = (0, 1, 1, 2)$ for the left hand side but $(h_0, h_1, h_2, h_3) = (0, 1, 2, 1)$ for the right hand side; finding a compatible order can be a rather challenging problem for complex congruences. In the case of hom-associativity though, the fact that all inputs are at the same height in the left and right hand sides makes it possible to use something very classical: a lexicographic order. Recursively it may be defined as having $\mu > \nu$ if:

- $\mu = \mathsf{m}\mu'\mu''$ and $\nu = \mathsf{m}\nu'\nu''$, where $\mu' > \nu'$, or
- $\mu = \mathsf{m}\mu'\mu''$ and $\nu = \mathsf{m}\nu'\nu''$, where $\mu' = \nu'$ and $\mu'' > \nu''$, or
- $\mu = \mathsf{a}\mu'$ and $\nu = \mathsf{m}\nu'\nu''$, or
- $\mu = \mathsf{a}\mu'$ and $\nu = \mathsf{a}\nu'$, where $\mu' > \nu'$.

Equivalently, one may define it as the word-lexicographic order on the Polish notation, over the order on letters which has $\mathsf{m} < \mathsf{a}$ and each $\square_i$ unrelated to all other letters. With this order, it is clear that the congruence (33) should be turned into the rule $(\mathsf{mm1a2am34}, \mathsf{mm1m23aa4})$.

In general, the idea to "find all ambiguities, try to resolve them, make new rules out of everything that doesn't resolve, and repeat until everything resolves" is called the *Critical Pairs/Completion* (CPC) procedure; its most famous instance is the Buchberger algorithm for computing Gröbner bases. 'Critical pairs' corresponds to identifying ambiguities, whereas 'completion' is the step of adding new rules; a rewriting system is said to be *complete* when all ambiguities are resolvable.

In the case at hand, the calculations quickly become extensive, so we make use of a program [22] one of us has written that automates the CPC procedure in the operadic setting (actually, in the more general PROP setting). Running it with (32) as input quickly leads to the discovery of (33) and several more identities:

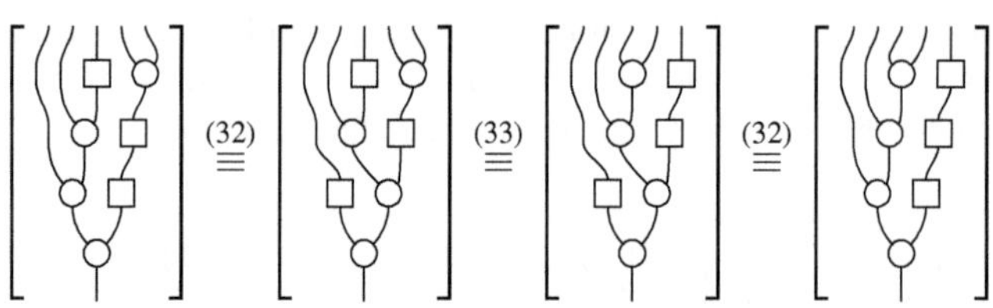

(34)

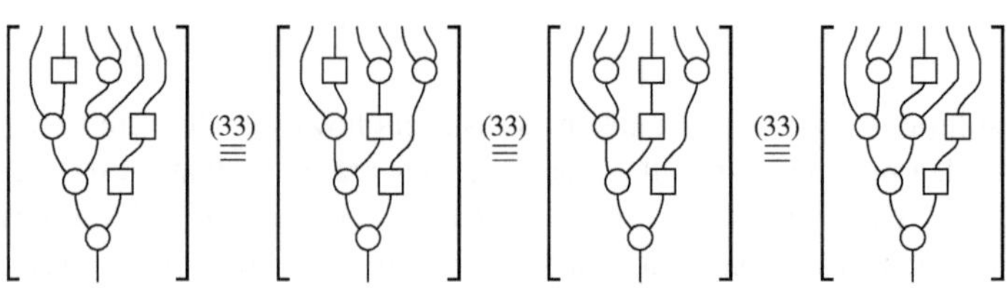

(35)

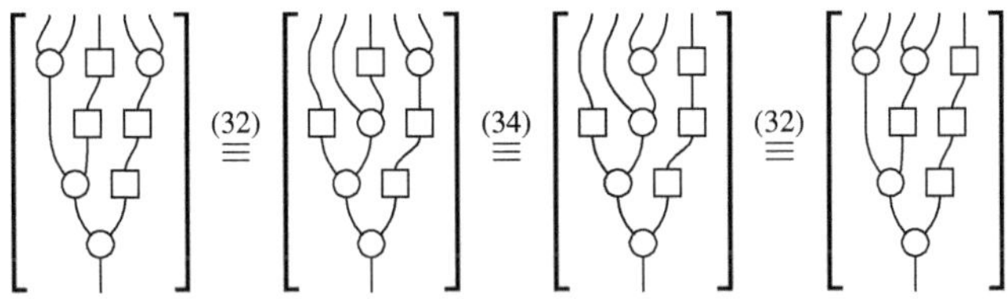

(36)

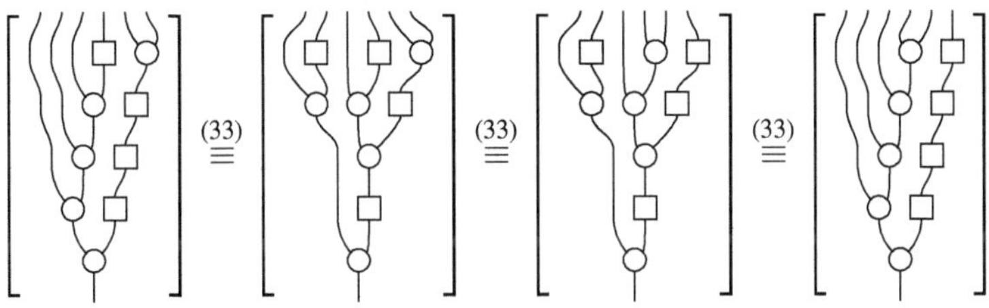

(37)

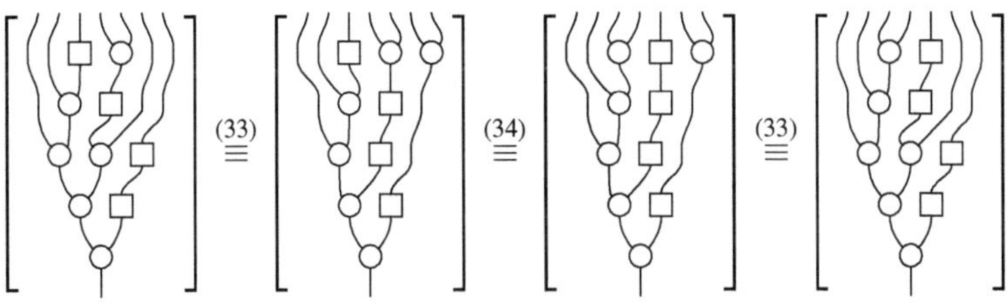

(38)

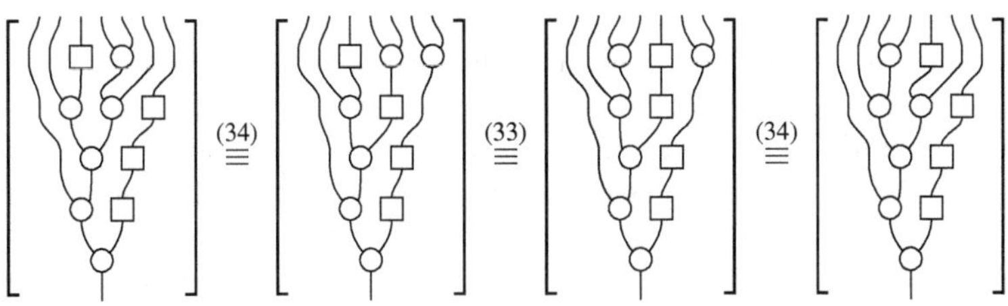

(39)

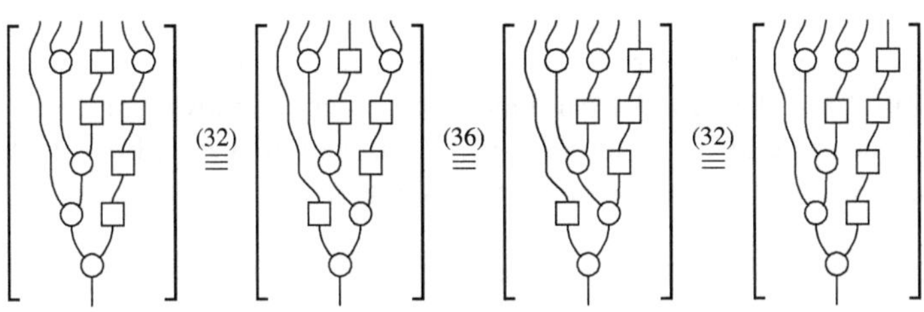

(40)

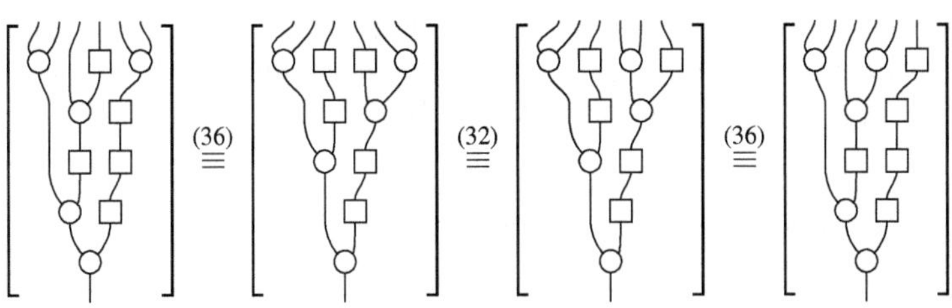

(41)

And so on… When we stopped it, the program had 1 rule (32) of order (number of vertices) 3, 1 rule (33) of order 5, 1 rule (34) of order 7, 2 rules (35, 36) of order 8, 1 rule (37) of order 9, 4 rules (38–41) of order 10, 7 rules of order 11, 12 rules of order 12, 19 rules of order 13, and 38 rules of order 14. Besides those 85 ambiguities that had given rise to new rules, 280 had turned out to be resolvable and 22417 had still not been processed; obviously the program wasn't going to finish anytime soon, and it's a fair guess that the complete rewriting system it sought to compute is in fact infinite. Certainly (32), (33), (34), and (37) look suspiciously like the beginning of an infinite family of rules, and indeed the expected sequence with one tower of m's and another tower of a's continues for as long as we have run the computations.

What is now our next step, when automated deduction has failed to deliver a complete answer? One approach is to try to guess the general pattern for these rules, and from that construct a provably complete rewriting system; we shall return to that problem in a later article. Right here and now, it is however possible to wash out several pieces of hard information even from the incomplete rewriting system presented above.

4.5 Hilbert Series and Formal Languages

A useful observation about the hom-associativity axiom (32) is that it is homogeneous in pretty much every sense imaginable: there are the same number of m's in the left and right hand sides, there are the same number of a's in the left and right hand sides, and the inputs are all at the same height in the left as in the right hand sides. (The last is not even true for the ordinary associativity rule (30), so from a very abstract symbolic point of view, hom-associativity may actually be regarded as a homogenised form of ordinary associativity.) It is a well-known principle in Gröbner basis calculations that CPC procedures working on homogeneous rewriting systems *only generates homogeneous rules* and *never derives smaller rules from larger ones*; once the procedure has processed all ambiguities up to a particular order, one knows for sure that no more rules of that order remain to be discovered. Hence the ten rules shown above are all there are of order 10 or less, and since no advanceable map of those used for Theorem 3.1 can reduce the order, it follows that those rules do effectively describe $\mathcal{HA}ss$ up to order 10. There is of course nothing special about order 10, so we may state these observations more formally as follows.

Lemma 4.1 *Let $Y_{k,\ell}$ be the subset of $\mathcal{Y}_\Omega$ whose elements contain exactly k vertices* a *and exactly ℓ vertices* m*; it follows that $\mathcal{Y}_\Omega(n) = \bigcup_{k=0}^{\infty} Y_{k,n-1}$. Let $S_{k,\ell}$ be the set of rules the CPC procedure has generated from* (ma1m23, mm12a3) *after processing all ambiguities at sites in $\bigcup_{i=0}^{k} \bigcup_{j=0}^{\ell} Y_{i,j}$ but no ambiguities with sites outside this set. Let $S = \bigcup_{k,\ell \in \mathbb{N}} S_{k,\ell}$. Then the following holds:*

1. $S_{1,2} = \{(\mathsf{ma1m23}, \mathsf{mm12a3})\}$.
2. $S_{k,\ell} \subseteq S_{k+1,\ell}$ *and* $S_{k,\ell} \subseteq S_{k,\ell+1}$ *for all* $k, \ell \in \mathbb{N}$.

3. $\operatorname{Irr}(S_{k,\ell}) \supseteq \operatorname{Irr}(S_{k+1,\ell})$ *and* $\operatorname{Irr}(S_{k,\ell}) \supseteq \operatorname{Irr}(S_{k,\ell+1})$ *for all* $k, \ell \in \mathbb{N}$.
4. *All ambiguities of* S *are resolvable.*
5. $\operatorname{Irr}(S) \cap Y_{k,\ell} = \operatorname{Irr}(S_{k,\ell}) \cap Y_{k,\ell}$.
6. *Every element of* $Y_{k,\ell}$ *has a unique normal form modulo* $S_{i,j}$, *for all* $i \geqslant k$ *and* $j \geqslant \ell$.

To finish off, we shall apply a bit of formal language theory to compute the beginning of the *Hilbert series* of $\mathcal{HAss}$. The kind of information encoded in this is, just like the $\dim \mathcal{Ass}(n) = n!$ result mentioned above, basically the numbers of dimensions of the various components of the operad, although in the case of $\mathcal{HAss}$ it is trivial to see that $\dim \mathcal{HAss}(n) = \infty$ for all $n \geqslant 0$ since inserting more a's into an expression does not change its arity. Instead one should partition by both a and m to get finite-dimensional components. Furthermore there is a rather boring factorial factor which is due to the action of Σ_n, so we restrict attention to plane monomials, factor out that factorial, and define the Hilbert series of $\mathcal{HAss}$ to be the formal power series

$$H(a, m) = \sum_{i,j \in \mathbb{N}} \frac{\left|Y_{i,j} \cap \operatorname{Irr}(S)\right|}{(j+1)!} a^i m^j. \tag{42}$$

Note that this is also the Hilbert series of the free hom-associative algebra with one generator on which f_{a} acts freely. Indeed, that algebra is preferably constructed as $\mathcal{R}\{\Omega'\}(0)/\mathcal{I}(S)(0)$ where $\Omega' = \{\mathsf{m}(,), \mathsf{a}(), \mathsf{x}\}$, and since no rule in S changes x in any way, it follows that $\operatorname{Irr}(S) \subseteq \mathcal{R}\{\Omega\}$ is in bijective correspondence to $\operatorname{Irr}(S)(0) \subseteq \mathcal{R}\{\Omega'\}(0)$—just put an x in every input! However, if one prefers have the Hilbert series for the free algebra counting a and x rather than a and m, then it should instead be stated as $xH(a, x)$, since there is always one x more in an element of $\mathcal{Y}_{\Omega'}(0)$ than there are m's. Finally, the Hilbert series for the free hom-associative algebra with k generators $\mathsf{x}_1, \dots, \mathsf{x}_k$ is $kxH(a, kx)$, since there for every constant symbol (which is what x becomes the counting variable for) are k choices of what that symbol should be.

As approximations of $H(a, m)$, we furthermore define

$$H_{k,\ell}(a, m) = \sum_{i,j \in \mathbb{N}} \frac{\left|Y_{i,j} \cap \operatorname{Irr}(S_{k,\ell})\right|}{(j+1)!} a^i m^j \qquad \text{for all } k, \ell \in \mathbb{N}. \tag{43}$$

By claim 3 of Lemma 3.1, $H_{i,j} \geqslant H_{k,\ell}$ coefficient by coefficient whenever $i \leqslant k$ and $j \leqslant \ell$. By claim 5, the coefficient of $a^i m^j$ in $H(a, m)$ is equal to the coefficient in $H_{k,\ell}(a, m)$ whenever $i \leqslant k$ and $j \leqslant \ell$. Therefore, when one wishes to compute the beginning of $H(a, m)$, one may alternatively compute the beginning of $H_{k,\ell}(a, m)$ for sufficiently large k and ℓ.

To get an initial bound, let us first compute $H_{0,0}(a, m)$. From the basic observation that a plane element of $\mathcal{Y}_\Omega$ is either id, $\mathsf{a}1 \circ \nu$ for some plane $\nu \in \mathcal{Y}_\Omega$, or $\mathsf{m}12 \circ \nu_1 \otimes \nu_2$ for some plane $\nu_1, \nu_2 \in \mathcal{Y}_\Omega$, it follows that the language[4] L of all plane elements of

[4] In formal language theory, a 'language' is simply some set of the kind of objects being considered.

$\mathcal{Y}_\Omega$ satisfies the equation $L = \{\mathrm{id}\} \cup (\mathsf{a1} \circ L) \cup (\mathsf{m12} \circ L \otimes L)$, and consequently that $H_{0,0}$ satisfies the functional equation

$$H_{0,0}(a,m) = 1 + aH_{0,0}(a,m) + mH_{0,0}(a,m)^2; \tag{44}$$

the details of this correspondence between combinatorial constructions and functional equations can be found in for example [16, Chap. 1]. Solving that equation symbolically yields

$$H_{0,0}(a,m) = \frac{1 - a - \sqrt{(1-a)^2 - 4m}}{2m} \tag{45}$$

and using Newton's generalised binomial theorem one can even get a closed form formula for the coefficients:

$$\begin{aligned}
H_{0,0}(a,m) &= \frac{1}{2m}\left(1 - a - \sum_{n=0}^{\infty}\binom{\frac{1}{2}}{n}\bigl((1-a)^2\bigr)^{\frac{1}{2}-n}(-4m)^n\right)\\
&= -\frac{1}{2m}\sum_{n=1}^{\infty}\binom{\frac{1}{2}}{n}(1-a)^{1-2n}(-4m)^n \stackrel{(\ell=n-1)}{}\\
&= 2\sum_{\ell=0}^{\infty}\binom{\frac{1}{2}}{\ell+1}(1-a)^{-2\ell-1}(-4m)^{\ell}\\
&= \sum_{\ell=0}^{\infty}\sum_{k=0}^{\infty}2\binom{\frac{1}{2}}{\ell+1}\binom{-2\ell-1}{k}(-a)^k(-4m)^{\ell}\\
&= \sum_{k,\ell\in\mathbb{N}}\frac{1}{\ell+1}\binom{k+2\ell}{k,\ell,\ell}a^k m^{\ell}.
\end{aligned}$$

As expected, the coefficients for $\ell = 0$ are all 1 and the coefficients for $k = 0$ are the Catalan numbers. These remain that way in all $H_{k,\ell}$, but away from the axes the various rules makes a difference. In order to determine how much, it is time to take some rules into account.

For any finite set of rules, it is straightforward to set up a system of equations for the language L_0 of plane monomials that are reducible by at least one of these rules; in the case of $S_{1,2}$, one such equation system is

$$L_0 = (\mathsf{a1} \circ L_0) \cup (\mathsf{m12} \circ L_0 \otimes L_1) \cup (\mathsf{m12} \circ L_1 \otimes L_0) \cup (\mathsf{m12} \circ L_2 \otimes L_3), \tag{46a}$$
$$L_1 = (\mathsf{m12} \circ L_1 \otimes L_1) \cup (\mathsf{a1} \circ L_1) \cup \{\mathrm{id}\}, \tag{46b}$$
$$L_2 = \mathsf{a1} \circ L_1, \tag{46c}$$
$$L_3 = \mathsf{m12} \circ L_1 \otimes L_1 \tag{46d}$$

(where we as usual consider operad composition of sets to denote the sets of operad elements that can be produced by applying the composition to elements of the given sets). A more suggestive presentation might however be as the BNF grammar

$$\begin{aligned}
\langle \textit{reducible} \rangle &::= \mathsf{a}\langle \textit{reducible} \rangle \mid \mathsf{m}\langle \textit{reducible} \rangle\langle \textit{arbitrary} \rangle \mid \mathsf{m}\langle \textit{arbitrary} \rangle\langle \textit{reducible} \rangle \\
&\qquad \mid \mathsf{m}\langle \textit{left} \rangle\langle \textit{right} \rangle \\
\langle \textit{arbitrary} \rangle &::= \mathsf{a}\langle \textit{arbitrary} \rangle \mid \mathsf{m}\langle \textit{arbitrary} \rangle\langle \textit{arbitrary} \rangle \mid \square_i \\
\langle \textit{left} \rangle &::= \mathsf{a}\langle \textit{arbitrary} \rangle \\
\langle \textit{right} \rangle &::= \mathsf{m}\langle \textit{arbitrary} \rangle\langle \textit{arbitrary} \rangle
\end{aligned}$$

whose informal interpretation is that a Polish term is $\langle \textit{reducible} \rangle$ by $S_{1,2}$ if one of the children of the root node is itself $\langle \textit{reducible} \rangle$, or if the root node is an m whose $\langle \textit{left} \rangle$ child is an a and whose $\langle \textit{right} \rangle$ child is an m. This can be trivially extended to larger sets of rules by adding to the formula for L_0 one production for each new rule (describing the root of the m_s of that rule) and one new variable (together with its defining equation) for every internal edge in the m_s of the new rule. Hence if also taking (33) into account, the system grows to

$$\begin{aligned}
L_0 &= (\mathsf{a1} \circ L_0) \cup (\mathsf{m12} \circ L_0 \otimes L_1) \cup (\mathsf{m12} \circ L_1 \otimes L_0) \\
&\qquad \cup (\mathsf{m12} \circ L_2 \otimes L_3) \cup (\mathsf{m12} \circ L_4 \otimes L_5), \\
L_1 &= (\mathsf{m12} \circ L_1 \otimes L_1) \cup (\mathsf{a1} \circ L_1) \cup \{\mathrm{id}\}, \\
L_2 &= \mathsf{a1} \circ L_1, \\
L_3 &= \mathsf{m12} \circ L_1 \otimes L_1, \\
L_4 &= \mathsf{m12} \circ L_1 \otimes L_6, \\
L_5 &= \mathsf{a1} \circ L_7, \\
L_6 &= \mathsf{a1} \circ L_1, \\
L_7 &= \mathsf{m12} \circ L_1 \otimes L_1.
\end{aligned}$$

Smaller systems for the same L_0 are often possible (and can save work in the next step), but here we are content with observing that a finite system exists.

While the system (46) is of the same general type as the equation that was used to derive (44), it would not be correct to simply convert it in the same way to an equation system for $H_{1,2}$, since there is a qualitative difference: the unions in (46) are not in general disjoint, for example because $L_0 \subset L_1$ and thus $\mathsf{m12} \circ L_0 \otimes L_0 \subseteq \mathsf{m12} \circ L_0 \otimes L_1, \mathsf{m12} \circ L_1 \otimes L_0$. This may be possible to overcome through inclusion–exclusion style combinatorics, but we would rather like to attack this issue using tools from formal language theory. In the terminology of [12], an equation system such as (46) defines a nondeterministic finite top-down tree automaton; it is finite because the set of states is {0, 1, 2, 3} (finite) and it is the nondeterminism that can cause the unions to be non-disjoint. By the Subset Construction [12, Thorem 1.1.9] however, there exists an equivalent deterministic finite bottom-up tree automaton whose states are subsets of the set of top-down states; moreover this bottom-up automaton may be regarded as an Ω-algebra $\bigl(A, \{f_x\}_{x\in\Omega}\bigr)$. In the case of (46), this Ω-algebra has

$$A = \bigl\{\{1\}, \{1,2\}, \{1,3\}, \{0,1,3\}, \{0,1,2\}\bigr\}$$

and operations given by the tables

First operand	f_a	f_m when second operad is:				
		$\{1\}$	$\{1,2\}$	$\{1,3\}$	$\{0,1,3\}$	$\{0,1,2\}$
$\{1\}$	$\{1,2\}$	$\{1,3\}$	$\{1,3\}$	$\{1,3\}$	$\{0,1,3\}$	$\{0,1,3\}$
$\{1,2\}$	$\{1,2\}$	$\{1,3\}$	$\{1,3\}$	$\{0,1,3\}$	$\{0,1,3\}$	$\{0,1,3\}$
$\{1,3\}$	$\{1,2\}$	$\{1,3\}$	$\{1,3\}$	$\{1,3\}$	$\{0,1,3\}$	$\{0,1,3\}$
$\{0,1,3\}$	$\{0,1,2\}$	$\{0,1,3\}$	$\{0,1,3\}$	$\{0,1,3\}$	$\{0,1,3\}$	$\{0,1,3\}$
$\{0,1,2\}$	$\{0,1,2\}$	$\{0,1,3\}$	$\{0,1,3\}$	$\{0,1,3\}$	$\{0,1,3\}$	$\{0,1,3\}$

When such an Ω-algebra $\bigl(A, \{f_x\}_{x\in\Omega}\bigr)$ is given, the equation system of generating functions takes the form

$$G_b(a,m) = a \sum_{\substack{c\in A\\ f_\mathrm{a}(c)=b}} G_c(a,m) + m \sum_{\substack{c,d\in A\\ f_\mathrm{m}(c,d)=b}} G_c(a,m)G_d(a,m) + \begin{cases} 1 & \text{if } b=\{1\},\\ 0 & \text{otherwise} \end{cases} \quad \text{for all } b \in A \tag{47}$$

where the extra term for $b = \{1\}$ is because that is the state that inputs are considered to be in. The generating function for reducible plane monomials is the sum of all G_b such that $b \ni 0$, since 0 was the top-down ⟨*reducible*⟩ state, and consequently the generating function for irreducible plane monomials is the sum of all G_b such that $b \not\ni 0$. Thus we have

$$\begin{aligned} H_{1,2}(a,m) &= G_{\{1\}}(a,m) + G_{\{1,2\}}(a,m) + G_{\{1,3\}}(a,m),\\ G_{\{1\}}(a,m) &= 1,\\ G_{\{1,2\}}(a,m) &= aH_{1,2}(a,m),\\ G_{\{1,3\}}(a,m) &= mH_{1,2}(a,m)^2 - mG_{\{1,2\}}(a,m)G_{\{1,3\}}(a,m) \end{aligned}$$

where the definition of $H_{1,2}(a,m)$ was used to shorten the last two right hand sides a bit. Solving as above is still possible, but results in the somewhat messier expression

$$H_{1,2}(a,m) = \frac{1 - a - am^2 - \sqrt{(1 - a - am^2)^2 + 4(1 - am + a^2m)m}}{2(1 - am + a^2m)m}$$
$$= \sum_{k=0}^{\infty} 2\binom{\frac{1}{2}}{k+1}(1 - a - am^2)^{-1-2k}4^k(1 - am + a^2m)^k m^k = \cdots$$

which is probably not so important to put on closed form; the interesting quantity is $H(a, m)$, and the terms in $H_{1,2}$ which coincide with their counterparts in $H(a, m)$ can be determined by an ansatz in the equation system already.

Theorem 4.2 *The Hilbert series $H(a, m)$ for the hom-associative operad $\mathcal{HAss}$ satisfies $H(a, m) = 1 + m + a + 2m^2 + 3am + a^2 + 5m^3 + 9am^2 + 6a^2m + a^3 + 14m^4 + 30am^3 + 26a^2m^2 + 10a^3m + a^4 + 42m^5 + 105am^4 + 110a^2m^3 + 60a^3m^2 + 15a^4m + a^5 + 132m^6 + 378am^5 + 465a^2m^4 + 315a^3m^3 + 120a^4m^2 + 21a^5m + a^6 + 429m^7 + 1386am^6 + 1960a^2m^5 + 1575a^3m^4 + 770a^4m^3 + 217a^5m^2 + 28a^6m + a^7 + 1430m^8 + 5148am^7 + 8232a^2m^6 + 7644a^3m^5 + 4494a^4m^4 + 1680a^5m^3 + 364a^6m^2 + 36a^7m + a^8 + \cdots$. In particular, the difference to the Hilbert series $H_{0,0}(a, m)$ for the free hom-algebra operad is*

$$\begin{aligned}
H_{0,0}(a,m) - H(a,m) = \\
= \quad & am^2 + 4a^2m^2 + 10a^3m^2 + 20a^4m^2 + 35a^5m^2 + 56a^6m^2 + \\
& 5am^3 + 30a^2m^3 + 105a^3m^3 + 280a^4m^3 + 630a^5m^3 + \\
& 21am^4 + 165a^2m^4 + 735a^3m^4 + 2436a^4m^4 + \\
& 84am^5 + 812a^2m^5 + 4368a^3m^5 + \\
& 330am^6 + 3780a^2m^6 + \\
& 1287am^7 + \cdots
\end{aligned}$$

Remark 4.2 The interpretation of for example the term $4368a^3m^5$ above is thus that imposing the hom-associativity identity (32) reduces by 4368 the dimension of the space of plane operad elements that can be formed with 3 operations α and 5 multiplications.

Proof As shown above for $H_{1,2}$, but taking all of (32)–(36) into account, so that one instead considers $S_{5,3} \cup S_{4,4}$ and thus gets all terms of total degree $\leqslant 8$.

Acknowledgments This work was partially supported by The Swedish Foundation for International Cooperation in Research and Higher Education (STINT), The Swedish Research Council, The Royal Swedish Academy of Sciences, The Royal Physiographic Society in Lund, The Crafoord Foundation and The Letterstedtska Föreningen.

References

1. Aizawa, N., Sato, H.: q-deformation of the Virasoro algebra with central extension. Phys. Lett. B **256**(1), 185–190 (1991)
2. Ammar, F., Ejbehi, Z., Makhlouf, A.: Cohomology and deformations of hom-algebras. J. Lie Theory **21**(4), 813–836 (2011)
3. Ammar, F., Makhlouf, A.: Hom-lie algebras and hom-lie admissible superalgebras. J. Algebra **324**(7), 1513–1528 (2010)
4. Ataguema, H., Makhlouf, A., Silvestrov, S.: Generalization of n-ary nambu algebras and beyond. J. Math. Phys. **50**, 1 (2009)
5. Benayadi, S., Makhlouf, A.: Hom-lie algebras with symmetric invariant nondegenerate bilinear form. J. Geom. Phys. **76**, 38–60 (2014)
6. Chaichian, M., Ellinas, D., Popowicz, Z.: Quantum conformal algebra with central extension. Phys. Lett. B **248**(1–2), 95–99 (1990)
7. Caenepeel, S., Goyvaerts, I.: Monoidal hom-hopf algebras. Comm. Alg. **39**, 2216–2240 (2011)
8. Chaichian, M., Isaev, A.P., Lukierski, J., Popowic, Z., Prešnajder, P.: q-deformations of virasoro algebra and conformal dimensions. Phys. Lett. B **262**(1), 32–38 (1991)
9. Chaichian, M., Kulish, P., Lukierski, J.: q-deformed jacobi identity, q-oscillators and q-deformed infinite-dimensional algebras. Phys. Lett. B **237**(3–4), 401–406 (1990)
10. Chaichian, M., Popowicz, Z., Prešnajder, P.: q-virasoro algebra and its relation to the q-deformed kdv system. Phys. Lett. B **249**(1), 63–65 (1990)
11. Cohn, P.M.: Universal Algebra. Harper & Row, New York (1965)
12. Comon, H., Dauchet, M., Gilleron, R., Jacquemard, F., Lugiez, D., Löding, C., Tison, S., Tommasi, M.: Tree Automata Techniques and Applications (web book), version of November 18, 2008; http://tata.gforge.inria.fr/
13. Curtright, T.L., Zachos, C.K.: Deforming maps for quantum algebras. Phys. Lett. B **243**(3), 237–244 (1990)
14. Damaskinsky, E.V., Kulish, P.P.: Deformed oscillators and their applications (Russian), Zap. Nauch. Semin. LOMI **189**, 37–74 (1991); Engl. transl. in J. Soviet Math. **62**(5), 2963–2986 (1992)
15. Daskaloyannis, C.: Generalized deformed virasoro algebras. Mod. Phys. Lett. A **7**(9), 809–816 (1992)
16. Flajolet, P., Sedgewick, R.: Analytic Combinatorics. Cambridge University Press, Cambridge (2009); ISBN 978-0-521-89806-5
17. Fregier, Y., Gohr, A., Silvestrov, S.D.: Unital algebras of hom-associative type and surjective or injective twistings. J. Gen. Lie Theory Appl. **3**(4), 285–295 (2009)
18. Gohr, A.: On hom-algebras with surjective twisting. J. Algebra **324**, 1483–1491 (2010)
19. Hartwig, J.T., Larsson, D., Silvestrov, S.D.: Deformations of lie algebras using σ-derivations. J. Algebra **295**, 314–361 (2006)
20. Hellström, L.: Rewriting in Operads and PROPs. J. Nonlinear Math. Phys. (electronic), **13**, 66–75 (2006); supplement (Proceedings of the Baltic-Nordic Workshop Algebra, Geometry and Mathematical Physics, Tallinn, 2005; edited by E. Paal and S. Silvestrov; ISBN 91-974824-6-3)
21. Hellström, L.: A Generic Framework for Diamond Lemmas (2007); arXiv:0712.1142v1 [math.RA]
22. Hellström, L.: Rewriting library and completion utility for network algebra, web site and computer software (2011); http://www.mdh.se/ukk/personal/maa/lhm03/sw/rewriting
23. Hellström, L.: Network Rewriting I: The Foundation (2012); arXiv:1204.2421v1 [math.RA]
24. Hellström, L., Silvestrov, S.D.: Commuting Elements in q-deformed Heisenberg Algebras. World Scientific, Singapore (2000)
25. Hu, N.: q-witt algebras, q-lie algebras, q-holomorph structure and representations. Algebra Colloq. **6**(1), 51–70 (1999)
26. Iyudu, N., Makhlouf, A.: On Koszulity for operads of Conformal Field Theory (2013) arXiv:1304.0687

27. Kassel, C.: Cyclic homology of differential operators, the virasoro algebra and a q-analogue. Commun. Math. Phys. **146**, 343–351 (1992)
28. Kerdman, F.S.: Analytic moufang loops in the large. Algebra i Logika **18**, 523–555 (1979)
29. Larsson, D., Silvestrov, S.D.: Quasi-hom-lie algebras, central extensions and 2-cocycle-like identities. J. Algebra **288**, 321–344 (2005)
30. Larsson, D., Silvestrov, S.D.: Quasi-Lie algebras. In: Noncommutative Geometry and Representation Theory in Mathematical Physics. Contemporary Mathematics, vol. 391,pp. 241–248. American Mathematical Society, Providence, RI (2005)
31. Larsson, D., Silvestrov, S.D.: Quasi-deformations of $\mathfrak{sl}_2(\mho)$ using twisted derivations. Comm. Algebra **35**, 4303–4318 (2007)
32. Liu, K.Q.: Characterizations of the quantum Witt algebra. Lett. Math. Phys. **24**(4), 257–265 (1992)
33. Loday, J.-L., Vallette, B.: Algebraic Operads. Springer, Berlin (2013)
34. Makhlouf, A.: Paradigm of Nonassociative Hom-algebras and Hom-superalgebras. In: Carmona Tapia, J., Morales Campoy, A., Peralta Pereira, A.M., Ramirez ilvarez, M.I. (eds.) Proceedings of Jordan Structures in Algebra and Analysis Meeting, pp. 145–177. Publishing House, Circulo Rojo (2010)
35. Makhlouf, A., Silvestrov, S.: Hom-algebra structures. J. Generalized Lie Theory Appl. **2**(2), 51–64 (2008)
36. Makhlouf, A., Silvestrov, S.: Hom-Lie admissible Hom-coalgebras and Hom-Hopf algebras, published as Chap. 17. In: Silvestrov, S., Paal, E., Abramov, V., Stolin, A. (eds.) Generalized Lie Theory in Mathematics, Physics and Beyond, pp. 189–206. Springer, Heidelberg (2008)
37. Makhlouf, A., Silvestrov, S.: Notes on formal deformations of hom-associative and hom-lie algebras. Forum Mathematicum **22**(4), 715–759 (2010)
38. Makhlouf, A., Silvestrov, S.: Hom-Algebras and Hom-Coalgebras. J. Algebra Appl. **9**, 553–589 (2010)
39. Makhlouf, A., Yau, D.: Rota-baxter hom-lie-admissible algebras. Commun. Algebra **42**, 1231–1257 (2014)
40. Markl, M., Shnider; S., Stasheff, J.: Operads in algebra, topology and physics. In: Mathematical Surveys and Monographs **96**. American Mathematical Society, Providence, RI (2002)
41. Sheng, Y.: Representations of hom-lie algebras. Algebras Represent. Theory **15**(6), 1081–1098 (2012)
42. Stasheff, J.: What is … an operad? Notices Amer. Math. Soc. **51**(6), 630–631 (2004)
43. Ufnarovskij, V.A.: Combinatorial and asymptotic methods of algebra. In: Kostrikin, A.I., Shafarevich, I.R. (eds.) Algebra VI, Encyclopedia of Mathematical Sciences **57**, pp. 5–196. Springer (1995); ISBN 3-540-54699-5 and 0-387-54699-5. Translated from Russian
44. Yau, D.: Enveloping algebra of hom-lie algebras. J. Gen. Lie Theory Appl. **2**, 95–108 (2008)
45. Yau, D.: Hom-algebras and homology. J. Lie Theory **19**, 409–421 (2009)
46. Yau, D.: The hom-yang-baxter equation, hom-lie algebras, and quasi-triangular bialgebras. J. Phys. A **42**, 165–202 (2009)
47. Yau, D.: Hom-bialgebras and comodule hom-algebras. Int. E. J. Alg. **8**, 45–64 (2010)
48. Yau, D.: The hom-yang-baxter equation and hom-lie algebras. J. Math. Phys. **52**, 053502 (2011)
49. Yau, D.: The classical Hom-Yang-Baxter equation and Hom-Lie bialgebras (2009), e-Print arXiv:0905.1890
50. Yau, D.: Hom-malsev, hom-alternative, and hom-jordan algebras. Int. E. J. Alg. **11**, 177–217 (2012)
51. Yau, D.: Hom-algebras via Prop's (2011), e-Print arXiv:1103.5261

27. Kassel, C.: Cyclic homology of differential operators, the Virasoro algebra and a q-analogue. Commun. Math. Phys. 146, 343–351 (1992)
28. Kerdman, F.S.: Analytic moufang loops in the large. Alg. Log. 18, 523–555 (1980)
29. Larsson, D., Silvestrov, S.D.: Quasi-hom-Lie algebras, central extensions and 2-cocycle-like identities. J. Algebra 288, 321–344 (2005)
30. Larsson, D., Silvestrov, S.D.: Quasi-Lie algebras. In: Noncommutative Geometry and Representation Theory in Mathematical Physics. Contemporary Mathematics, vol. 391, pp. 241–248. American Mathematical Society, Providence, RI (2005)
31. Larsson, D., Silvestrov, S.D.: Quasi-deformations of $sl_2(\mathbb{F})$ using twisted derivations. Comm. Algebra 35, 4303–4318 (2007)
32. Liu, K.Q.: Characterizations of the quantum Witt algebra. Lett. Math. Phys. 24(4), 257–265 (1992)
33. Loday, J.-L., Vallette, B.: Algebraic Operads. Springer, Berlin (2012)
34. Makhlouf, A.: Paradigm of nonassociative Hom-algebras and Hom-superalgebras. In: Proceedings of Jordan Structures in Algebra and Analysis Meeting, pp. 145–177. Publishing House: Circulo Rojo (2010)
35. Makhlouf, A., Silvestrov, S.: Hom-algebra structures. J. Generalized Lie Theory Appl. 2(2), 51–64 (2008)
36. Makhlouf, A., Silvestrov, S.: Hom-Lie admissible Hom-coalgebras and Hom-Hopf algebras. [illegible]
37. [illegible] deformations of Hom-associative and Hom-Lie algebras. Forum Math. 22(4), 715–759 (2010)
38. Makhlouf, A., Silvestrov, S.: Hom-Algebras and Hom-Coalgebras. J. Algebra Appl. 9, 553–589 (2010)
39. Makhlouf, A., Yau, D.: Rota-Baxter Hom-Lie-admissible algebras. Commun. Algebra 42, 1231–1257 (2014)
40. Markl, M., Shnider, S., Stasheff, J.: Operads in Algebra, Topology and Physics. In: Mathematical Surveys and Monographs 96. American Mathematical Society, Providence (2002)
41. Sheng, Y.: Representations of hom-Lie algebras. Algebra Represent. Theory 15(6), 1081–1098 (2012)
42. Stasheff, J.: What is… an operad? Notices Amer. Math. Soc. 51(6), 630–631 (2004)
43. Ufnarovskij, V.A.: Combinatorial and asymptotic methods of algebra. In: Kostrikin, A.I., Shafarevich, I.R. (eds.) Algebra VI. Encyclopedia of Mathematical Sciences 57, pp. 5–196. Springer, Berlin (1995). Translated from Russian
44. Yau, D.: Enveloping algebras of Hom-Lie algebras. J. Gen. Lie Theory Appl. 2, 95–108 (2008)
45. Yau, D.: Hom-algebras and homology. J. Lie Theory 19, 409–421 (2009)
46. Yau, D.: The Hom-Yang-Baxter equation, Hom-Lie algebras, and quasi-triangular bialgebras. J. Phys. A 42, 165–202 (2009)
47. Yau, D.: Hom-bialgebras and comodule Hom-algebras. Int. E. J. Alg. 8, 45–64 (2010)
48. Yau, D.: The Hom-Yang-Baxter equation and Hom-Lie algebras. J. Math. Phys. 52, 053502 (2011)
49. Yau, D.: The classical Hom-Yang-Baxter equation and Hom-Lie bialgebras (2009), e-Print arXiv:0905.1890
50. Yau, D.: Hom-maltsev, Hom-alternative and Hom-Jordan algebras. Int. E. J. Alg. 11, 177–217 (2012)
51. Yau, D.: Hom-algebras via PROPs (2011), e-Print arXiv:1103.5261

Constructions of Quadratic n-ary Hom-Nambu Algebras

Faouzi Ammar, Sami Mabrouk and Abdenacer Makhlouf

Abstract The aim of this paper is provide a survey on n-ary Hom-Nambu algebras and study quadratic n-ary Hom-Nambu algebras, which are n-ary Hom-Nambu algebras with an invariant, nondegenerate and symmetric bilinear forms that are also α-symmetric and β-invariant where α and β are twisting maps. We provide various constructions of quadratic n-ary Hom-Nambu algebras. Also is discussed their connections with representation theory and centroids.

1 Introduction

The main motivations to study n-ary algebras came firstly from Nambu mechanics [34] where a ternary bracket allows to use more than one hamiltonian and recently from string theory and M-branes which involve naturally an algebra with ternary operation called Bagger-Lambert algebra [11]. Also ternary operations appeared in the study of some quarks models see [22–24]. For more general theory and further results see references [5, 6, 15–17, 20, 21, 26, 27, 35, 38].

Algebras endowed with invariant nondegenerate symmetric bilinear form (scalar product) appeared also naturally in several domains in mathematics and physics. Such algebras were intensively studied for binary Lie and associative algebras. The main results are that called double extension given by Medina and Revoy [33] and T*-extension given by Bordemann [14]. These fundamental results were extended to

F. Ammar · S. Mabrouk
Faculté des Sciences, Université de Sfax, Sfax, Tunisia
e-mail: Faouzi.Ammar@fss.rnu.tn

S. Mabrouk
e-mail: Mabrouksami00@yahoo.fr

A. Makhlouf (✉)
Université de Haute-Alsace, 4 rue des Frères Lumière, 68093 Mulhouse, France
e-mail: abdenacer.makhlouf@uha.fr

A. Makhlouf et al. (eds.), *Algebra, Geometry and Mathematical Physics*,
Springer Proceedings in Mathematics & Statistics 85, DOI: 10.1007/978-3-642-55361-5_12,

n-ary algebras in [18]. The extension to Hom-setting for binary case was introduced and studied in [13]. For further results about Hom-type algebras, see refs [2–4, 25, 28, 30–32, 41, 42].

In this paper we summarize in the Sect. 2 definitions of n-ary Hom-Nambu algebras and recall the constructions using twisting principles and tensor product with n-ary algebras of Hom-associative type. In Sect. 3, we introduce the notion of quadratic n-ary Hom-Nambu algebra, generalizing the notion introduced for binary Hom-Lie algebras in [13]. A more general notion called Hom-quadratic n-ary Hom-Nambu algebra is introduced by twisting the invariance identity. In Sect. 4, we show that a quadratic n-ary Hom-Nambu algebra gives rise to a quadratic Hom-Leibniz algebra. A connection with representation theory is discussed in Sect. 5. We deal in particular with adjoint and coadjoint representations, extending the representation theory initiated in [13, 36]. Several procedures to built quadratic n-ary Hom-Nambu algebras are provided in Sect. 6. We use twisting principles, tensor product and T*-extension to construct quadratic n-ary Hom-Nambu algebras. Moreover we show that one may derive from quadratic n-ary Hom-Nambu algebra ones of increasingly higher arities and that under suitable assumptions it reduces to a quadratic $(n-1)$-ary Hom-Nambu algebra. Also real Faulkner construction is used to obtain ternary Hom-Nambu algebras. The last Sect. 7 is dedicated to introduce and study the centroids of n-ary Hom-Nambu algebras and their properties. We supply a construction procedure of quadratic n-ary Hom-Nambu algebras using elements of the centroid.

Most of the results concern n-ary Hom-Nambu algebras. Naturally they are valid and may be stated for n-ary Hom-Nambu-Lie algebras.

2 The n-ary Hom-Nambu Algebras

Throughout this paper, $\mathbb{K}$ is an algebraically closed field of characteristic zero, even though for most of the general definitions and results in the paper this assumption is not essential.

2.1 Definitions

In this section, we summarize definitions of n-ary Hom-Nambu algebras and n-ary Hom-Nambu-Lie algebras, introduced in [7], generalizing n-ary Nambu algebras and n-ary Nambu-Lie algebras (called also Filippov algebras or n-Lie algebras).

Definition 2.1 An *n-ary Hom-Nambu* algebra is a triple $(\mathcal{N}, [\cdot, \dots, \cdot], \widetilde{\alpha})$ consisting of a vector space $\mathcal{N}$, an n-linear map $[\cdot, \dots, \cdot] : \mathcal{N}^n \longrightarrow \mathcal{N}$ and a family $\widetilde{\alpha} = (\alpha_i)_{1 \leq i \leq n-1}$ of linear maps $\alpha_i : \mathcal{N} \longrightarrow \mathcal{N}$, satisfying

$$\begin{aligned}&\big[\alpha_1(x_1), \ldots, \alpha_{n-1}(x_{n-1}), [y_1, \ldots, y_n]\big]\\&=\sum_{i=1}^{n}\big[\alpha_1(y_1), \ldots, \alpha_{i-1}(y_{i-1}), [x_1, \ldots, x_{n-1}, y_i], \alpha_i(y_{i+1}), \ldots, \alpha_{n-1}(y_n)\big], \quad (1)\end{aligned}$$

for all $(x_1, \ldots, x_{n-1}) \in \mathcal{N}^{n-1}$, $(y_1, \ldots, y_n) \in \mathcal{N}^n$.

The identity (1) is called *Hom-Nambu identity*.

Let $x = (x_1, \ldots, x_{n-1}) \in \mathcal{N}^{n-1}$, $\widetilde{\alpha}(x) = (\alpha_1(x_1), \ldots, \alpha_{n-1}(x_{n-1})) \in \mathcal{N}^{n-1}$ and $y \in \mathcal{N}$. We define an adjoint map $ad(x)$ as a linear map on $\mathcal{N}$, such that

$$ad(x)(y) = [x_1, \ldots, x_{n-1}, y]. \quad (2)$$

Then the Hom-Nambu identity (1) may be written in terms of adjoint map as

$$\begin{aligned}&ad(\widetilde{\alpha}(x))([x_n, \ldots, x_{2n-1}])\\&=\sum_{i=n}^{2n-1}[\alpha_1(x_n), \ldots, \alpha_{i-n}(x_{i-1}), ad(x)(x_i), \alpha_{i-n+1}(x_{i+1}), \ldots, \alpha_{n-1}(x_{2n-1})].\end{aligned}$$

Remark 2.1 When the maps $(\alpha_i)_{1\leq i\leq n-1}$ are all identity maps, one recovers the classical n-ary Nambu algebras. The Hom-Nambu Identity (1), for $n = 2$, corresponds to Hom-Jacobi identity (see [29]), which reduces to Jacobi identity when $\alpha_1 = id$.

Let $(\mathcal{N}, [\cdot, \ldots, \cdot], \widetilde{\alpha})$ and $(\mathcal{N}', [\cdot, \ldots, \cdot]', \widetilde{\alpha}')$ be two n-ary Hom-Nambu algebras where $\widetilde{\alpha} = (\alpha_i)_{i=1,\ldots,n-1}$ and $\widetilde{\alpha}' = (\alpha'_i)_{i=1,\ldots,n-1}$. A linear map $f : \mathcal{N} \to \mathcal{N}'$ is an n-ary Hom-Nambu algebras *morphism* if it satisfies

$$\begin{aligned}f([x_1, \cdots, x_n]) &= [f(x_1), \cdots, f(x_n)]'\\ f\circ\alpha_i &= \alpha'_i \circ f \quad \forall i = 1, \ldots, n-1.\end{aligned}$$

Definition 2.2 An n-ary Hom-Nambu algebra $(\mathcal{N}, [\cdot, \ldots, \cdot], \widetilde{\alpha})$ where $\widetilde{\alpha} = (\alpha_i)_{1\leq i\leq n-1}$ is called *n-ary Hom-Nambu-Lie* algebra if the bracket is skew-symmetric that is

$$[x_{\sigma(1)}, \ldots, x_{\sigma(n)}] = Sgn(\sigma)[x_1, \ldots, x_n], \ \ \forall\sigma \in \mathcal{S}_n \text{ and } \forall x_1, \ldots, x_n \in \mathcal{N}. \quad (3)$$

where $\mathcal{S}_n$ stands for the permutation group of n elements.

The condition (1) may be written using skew-symmetry property of the bracket as

$$\begin{aligned}&\big[\alpha_1(x_1), \ldots, \alpha_{n-1}(x_{n-1}), [y_1, \ldots, y_n]\big]\\&=\sum_{i=1}^{n}(-1)^{i+n}\big[\alpha_1(y_1), \ldots, \widehat{y_i}, \ldots, \alpha_{n-1}(y_n), [x_1, \ldots, x_{n-1}, y_i]\big], \quad (4)\end{aligned}$$

In the sequel we deal sometimes with a particular class of n-ary Hom-Nambu algebras which we call n-ary multiplicative Hom-Nambu algebras.

Definition 2.3 An *n-ary multiplicative Hom-Nambu algebra* (resp. *n-ary multiplicative Hom-Nambu-Lie algebra*) is an n-ary Hom-Nambu algebra (resp. n-ary Hom-Nambu-Lie algebra) $(\mathcal{N}, [\cdot, \dots, \cdot], \widetilde{\alpha})$ with $\widetilde{\alpha} = (\alpha_i)_{1 \leq i \leq n-1}$ where $\alpha_1 = \cdots = \alpha_{n-1} = \alpha$ and satisfying

$$\alpha([x_1, \dots, x_n]) = [\alpha(x_1), \dots, \alpha(x_n)], \quad \forall x_1, \dots, x_n \in \mathcal{N}. \tag{5}$$

For simplicity, we will denote the n-ary multiplicative Hom-Nambu algebra as $(\mathcal{N}, [\cdot, \dots, \cdot], \alpha)$ where $\alpha : \mathcal{N} \to \mathcal{N}$ is a linear map. Also by misuse of language an element $x \in \mathcal{N}^n$ refers to $x = (x_1, \dots, x_n)$ and $\alpha(x)$ denotes $(\alpha(x_1), \dots, \alpha(x_n))$.

2.2 Constructions

In this section we recall the construction procedures by twisting principles. The first twisting principle, introduced for binary case in [39], was extend to n-ary case in [7]. The second twisting principle was introduced in [40]. Also we recall a construction by tensor product of symmetric totally n-ary Hom-associative algebra by an n-ary Hom-Nambu algebra given in [7].

The following Theorem gives a way to construct n-ary multiplicative Hom-Nambu algebras starting from a classical n-ary Nambu algebras and algebra endomorphisms.

Theorem 2.1 *([7]). Let $(\mathcal{N}, [\cdot, \dots, \cdot])$ be an n-ary Nambu algebra and $\rho : \mathcal{N} \to \mathcal{N}$ be an n-ary Nambu algebra endomorphism. Then $(\mathcal{N}, \rho \circ [\cdot, \dots, \cdot], \rho)$ is an n-ary multiplicative Hom-Nambu algebra.*

In the following we use the second twisting principal to generate new n-ary Hom-Nambu algebra starting from a given multiplicative n-ary Hom-Nambu algebra.

Theorem 2.2 *Let $(\mathcal{N}, [\cdot, \dots, \cdot], \alpha)$ be a multiplicative n-ary Hom-Nambu algebra. Then $(\mathcal{N}, \alpha^{n-1} \circ [\cdot, \dots, \cdot], \alpha^n)$, for any integer n, is an n-ary multiplicative Hom-Nambu algebra.*

Example 2.1 ([7]). The polynomial algebra $\mathcal{N} = \mathbb{K}[x_1, x_2, x_3]$ of 3 variables x_1, x_2, x_3, with the bracket defined by the functional jacobian:

$$[f_1, f_2, f_3] = \begin{vmatrix} \frac{\delta f_1}{\delta x_1} & \frac{\delta f_1}{\delta x_2} & \frac{\delta f_1}{\delta x_3} \\ \frac{\delta f_2}{\delta x_1} & \frac{\delta f_2}{\delta x_2} & \frac{\delta f_2}{\delta x_3} \\ \frac{\delta f_3}{\delta x_1} & \frac{\delta f_3}{\delta x_2} & \frac{\delta f_3}{\delta x_3} \end{vmatrix},$$

is a ternary Nambu-Lie algebra. By considering a Nambu-Lie algebra endomorphism of such algebra, we construct a Hom-Nambu-Lie algebra on the polynomial algebra of 3 variables x_1, x_2, x_3.

Let $\gamma(x_1, x_2, x_3)$ be a polynomial or more general differentiable transformation of three variables mapping elements of $\mathcal{N}$ to elements of $\mathcal{N}$ and such that the determinant of the functional Jacobian $J(\gamma) = 1$. Any $\rho_\gamma : \mathcal{N} \to \mathcal{N}$, the composition transformation defined by $f \to f \circ \gamma$ for any $f \in \mathcal{N}$, defines an endomorphism of the ternary Nambu-Lie algebra given above. Therefore, for any such transformation γ, the triple $(\mathcal{N}, \rho_\gamma \circ [\cdot, \cdot, \cdot], \rho_\gamma)$ is a ternary Hom-Nambu-Lie algebra.

Now, we define the tensor product of two n-ary Hom-algebras and prove some results involving n-ary Hom-algebras of Lie type and Hom-associative type.

Let A be a $\mathbb{K}$-vector space, μ be an n-linear map on A and $\eta_i, i \in \{1, \dots, n-1\}$, be linear maps on A. A triple $(A, \mu, \widetilde{\eta} = (\eta_1, \dots, \eta_{n-1}))$ is said to be a symmetric n-ary totally Hom-associative algebra over $\mathbb{K}$ if the following identities hold

$$\mu(a_{\sigma(1)}, \dots, a_{\sigma(n)}) = \mu(a_1, \dots, a_n), \ \forall \sigma \in \mathcal{S}_n, \tag{6}$$

$$\begin{aligned} &\mu(\mu(a_1, \dots, a_n), \eta_1(a_{n+1}), \dots, \eta_{n-1}(a_{2n-1})) \\ &= \mu(\eta_1(a_1), \mu(a_2, \dots, a_{n+1}), \eta_2(a_{n+2}), \dots, \eta_{n-1}(a_{2n-1})) \\ &= \dots = \mu(\eta_1(a_1), \dots, \eta_{n-1}(a_{n-1}), \mu(a_n, \dots, a_{2n-1})), \end{aligned} \tag{7}$$

where $a_1, \dots, a_{2n-1} \in A$.

Theorem 2.3 *Let* $(A, \mu, \widetilde{\eta} = (\eta_1, \dots, \eta_{n-1}))$ *be a symmetric n-ary totally Hom-associative algebra and* $(\mathcal{N}, [\cdot, \dots, \cdot]_{\mathcal{N}}, \widetilde{\alpha})$ *be an n-ary Hom-Nambu algebra. Then the tensor product* $A \otimes \mathcal{N}$ *carries a structure of n-ary Hom-Nambu algebra over* $\mathbb{K}$ *with respect to the n-linear operation defined by*

$$[a_1 \otimes x_1, \dots, a_n \otimes x_n] = \mu(a_1, \dots, a_n) \otimes [x_1, \dots, x_n]_{\mathcal{N}}, \quad \text{where} \quad x_l \in \mathcal{N}, a_l \in A, l \in \{1, \dots, n\}, \tag{8}$$

and linear maps $\widetilde{\zeta} = (\zeta_1, \dots, \zeta_{n-1})$ *where* $\zeta_i = \eta_i \otimes \alpha_i$*, for* $i \in \{1, \dots, n-1\}$*, defined by*

$$\zeta_i(a \otimes x) = \eta_i(a) \otimes \alpha_i(x), \ \forall\, a \otimes x \in A \otimes \mathcal{N}. \tag{9}$$

Proof For $a_k \otimes x_k,\ b_l \otimes y_l \in A \otimes \mathcal{N}$, $1 \le k \le n-1$ and $1 \le l \le n$, we have

$$\begin{aligned} &[\zeta_1(a_1 \otimes x_1), \dots, \zeta_{n-1}(a_{n-1} \otimes x_{n-1}), [b_1 \otimes y_1, \dots, b_n \otimes y_n]] \\ &= \mu(\eta_1(a_1), \dots, \eta_{n-1}(a_{n-1}), \mu(b_1, \dots, b_n)) \otimes [\alpha_1(x_1), \dots, \alpha_{n-1}(x_{n-1}), \\ &\quad [y_1, \dots, y_n]_{\mathcal{N}}]_{\mathcal{N}}. \end{aligned}$$

The symmetry and totally associativity of A lead to

$$\begin{aligned}
&[\zeta_1(b_1 \otimes y_1), \dots, [a_1 \otimes x_1, \dots, a_{n-1} \otimes x_{n-1}, b_l \otimes y_l], \dots, \zeta_{n-1}(b_n \otimes y_n)] \\
&= \mu(\eta_1(b_1), \dots, \mu(a_1, \dots, a_{n-1}, b_l), \dots, \eta_{n-1}(b_{n-1})) \\
&\quad \otimes [[\alpha_1(y_1), \dots, [x_1, \dots, x_{n-1}, y_l]_{\mathcal{N}}, \dots, \alpha_{n-1}(y_n)]_{\mathcal{N}} \\
&= \mu(\eta_1(a_1), \dots, \eta_{n-1}(a_{n-1}), \mu(b_1, \dots, b_n)) \\
&\quad \otimes [\alpha_1(y_1), \dots, [x_1, \dots, x_{n-1}, y_l]_{\mathcal{N}}, \dots, \alpha_{n-1}(y_n)]_{\mathcal{N}}.
\end{aligned}$$

Thus

$$\begin{aligned}
&\sum_{l=1}^{n} [\zeta_1(b_1 \otimes y_1), \dots, [a_1 \otimes x_1, \dots, a_{n-1} \otimes x_{n-1}, b_l \otimes y_l], \dots, \zeta_{n-1}(b_n \otimes y_n)] \\
&= \mu(\eta_1(a_1), \dots, \eta_{n-1}(a_{n-1}), \mu(b_1, \dots, b_n)) \\
&\quad \otimes \Big(\sum_{l=1}^{n} [\alpha_1(y_1), \dots, [x_1, \dots, x_{n-1}, y_l]_{\mathcal{N}}, \dots, \alpha_{n-1}(y_n)]_{\mathcal{N}} \Big) \\
&= \mu(\eta_1(a_1), \dots, \eta_{n-1}(a_{n-1}), \mu(b_1, \dots, b_n)) \otimes [\alpha_1(x_1), \dots, \alpha_{n-1}(x_{n-1}), \\
&\quad [y_1, \dots, y_n]_{\mathcal{N}}]_{\mathcal{N}}.
\end{aligned}$$

Corollary 2.1 *Let (A, μ, η) be a multiplicative symmetric n-ary Hom-associative algebra (i.e. $\eta \circ \mu = \mu \circ \eta^{\otimes n}$) and $(\mathcal{N}, [\cdot, \dots, \cdot]_{\mathcal{N}}, \widetilde{\alpha})$ be a multiplicative n-ary Hom-Nambu algebra. Then $A \otimes \mathcal{N}$ is a multiplicative n-ary Hom-Nambu algebra.*

Remark 2.2 Let $(A, \cdot)$ be a binary commutative associative algebra and $(\mathcal{N}, [\cdot, \dots, \cdot]_{\mathcal{N}}, \widetilde{\alpha})$ be an n-ary Hom-Nambu algebra. Then the tensor product $A \otimes \mathcal{N}$ carries a structure of n-ary Hom-Nambu algebra over $\mathbb{K}$ with respect to the n-linear operation defined by

$$[a_1 \otimes x_1, \dots, a_n \otimes x_n] = (a_1 \cdot \dots \cdot a_n) \otimes [x_1, \dots, x_n]_{\mathcal{N}}, \tag{10}$$

and linear maps $\widetilde{\zeta} = (\zeta_1, \dots, \zeta_{n-1})$ where $\zeta_i = id \otimes \alpha_i$, for $i \in \{1, \dots, n-1\}$, defined by

$$\zeta_i(a \otimes x) = a \otimes \alpha_i(x), \ \forall\, a \otimes x \in A \otimes \mathcal{N}. \tag{11}$$

3 Definitions and Examples of Quadratic n-ary Hom-Nambu Algebras

In this section we introduce a class of Hom-Nambu-Lie algebras which possess a scalar product (a nondegenerate symmetric bilinear form which is invariant). This class of algebras is important due to their appearance in a number of physical contexts. They were extensively studied in the case of Lie algebras and Lie superalgebras, see

[9, 10, 14, 33]. The study was extended to Hom-Lie algebras in [13]. See also [18] for 3-Lie algebras.

Definition 3.1 Let $(\mathcal{N}, [., \dots, .], \widetilde{\alpha})$, $\widetilde{\alpha} = (\alpha_i)_{1\leq i\leq n-1}$, be an n-ary Hom-Nambu algebra and $B : \mathcal{N} \times \mathcal{N} \to \mathbb{K}$ be a nondegenerate symmetric bilinear form such that, for all $y, z \in \mathcal{N}$ and $x \in \wedge^{n-1}\mathcal{N}$

$$B([x_1, \cdots, x_{n-1}, y], z) + B(y, [x_1, \cdots, x_{n-1}, z]) = 0, \quad (12)$$

$$B(\alpha_i(y), z) = B(y, \alpha_i(z)), \quad \forall i \in \{1, \dots, n-1\}. \quad (13)$$

The quadruple $(\mathcal{N}, [\cdot, \dots, \cdot], \widetilde{\alpha}, B)$ is called quadratic n-ary Hom-Nambu algebra.

Remark 3.1 If $\alpha_i = Id$ for all $i \in \{1, \dots, n-1\}$, we recover quadratic (metric) n-ary Nambu algebras.

Definition 3.2 An n-ary Hom-Nambu algebra $(\mathcal{N}, [., \dots, .], \widetilde{\alpha})$, $\widetilde{\alpha} = (\alpha_i)_{1\leq i\leq n-1}$, is called *Hom-quadratic* if there exists a pair (B, β) where $B : \mathcal{N} \times \mathcal{N} \to \mathbb{K}$ is a nondegenerate symmetric bilinear form and $\beta \in End(\mathcal{N})$ a linear map satisfying

$$B([x_1, \cdots, x_{n-1}, y], \beta(z)) + B(\beta(y), [x_1, \cdots, x_{n-1}, z]) = 0, \quad (14)$$

We call the identity (14) the β-invariance of B. We recover the quadratic n-ary Hom-Nambu algebras when $\beta = id$ and the identity (12) is called the invariance of B. The tuple $(\mathcal{N}, [\cdot, \dots, \cdot], \widetilde{\alpha}, B, \beta)$ denotes the Hom-quadratic n-ary Hom-Nambu algebra.

Example 3.1 We consider an example of ternary Hom-Nambu algebra given in [40]. Let V be a $\mathbb{K}$-module and $B : V^{\otimes 2} \to \mathbb{K}$ be a nondegenerate symmetric bilinear form. Suppose $\alpha : V \to V$ is an involution, that is $\alpha^2 = id$. Assume that α is B-symmetric, that is $B(\alpha(x), y) = B(x, \alpha(y))$ for all $x, y \in V$. We have also $B(\alpha(x), \alpha(y)) = B(\alpha^2(x), y) = B(x, y)$. Then for any scalar $\lambda \in \mathbb{K}$, the triple product

$$[x, y, z]_\alpha = \lambda(B(y, z)\alpha(x) - B(z, x)\alpha(y)) \quad \text{for all } x,\ y,\ z \in V. \quad (15)$$

gives a Hom-quadratic ternary Hom-Nambu algebra $(V, [., ., .]_\alpha, (\alpha, \alpha))$, α-invariant by the pair (B, α). Indeed, for $x, y, z, t \in V$

$$\begin{aligned} B([x, y, z]_\alpha, \alpha(t)) &= \lambda(B(B(y, z)\alpha(x) - B(z, x)\alpha(y), \alpha(t))) \\ &= \lambda(B(y, z)B(\alpha(x), \alpha(t)) - B(z, x)B(\alpha(y), \alpha(t))) \\ &= \lambda(B(y, z)B(x, t) - B(z, x)B(y, t)) \\ &= \lambda(B(\alpha(y), \alpha(z))B(x, t) - B(\alpha(z), \alpha(x))B(y, t)) \\ &= \lambda(B(\alpha(z), \alpha(y))B(x, t) - B(\alpha(z), \alpha(x))B(y, t)) \end{aligned}$$

$$= \lambda(B(\alpha(z), \alpha(y)B(x,t) - \alpha(x)By,t)))$$
$$= -\lambda(B(\alpha(z), \alpha(x)B(y,t) - \alpha(y)B(t,x)))$$
$$= -B(\alpha(z), [x,y,t]_\alpha).$$

Example 3.2 Let $(\mathcal{N}, [\cdot,\cdot,\cdot], (\alpha_1, \alpha_2))$ be a 3-dimensional ternary Hom-Nambu-Lie algebras, defined with respect to a basis $\{e_1, e_2, e_3\}$ of $\mathcal{N}$ by

$$[e_1, e_2, e_3] = e_1 + 2e_2 + e_3, \tag{16}$$

$$\alpha_1(e_1) = 0,\ \alpha_1(e_2) = \lambda e_1 + \nu e_2,\ \alpha_1(e_3) = \frac{\lambda}{2}e_1 + \frac{\nu}{2}e_2, \tag{17}$$

$$\alpha_2(e_1) = 0,\ \alpha_2(e_2) = 0,\ \alpha_2(e_3) = be_3, \tag{18}$$

where λ, ν, b are parameters with $\lambda\nu \neq 0$. These 3-dimensional ternary algebras admit symmetric bilinear forms B given, with respect to the previous basis, by the following matrix

$$M = \begin{pmatrix} -\frac{\nu}{\lambda} & 1 & (\frac{\nu}{\lambda} - 2) \\ 1 & -\frac{\lambda}{\nu} & (2\frac{\lambda}{\nu} - 1) \\ (\frac{\nu}{\lambda} - 2) & (2\frac{\lambda}{\nu} - 1) & (4 - 4\frac{\lambda}{\nu} - \frac{\nu}{\lambda}) \end{pmatrix}$$

The ternary Hom-Nambu-Lie algebras $(\mathcal{N}, [\cdot,\cdot,\cdot], (\alpha_1, \alpha_2))$, with respect to B, is not quadratic because B is degenerate ($det(M) = 0$).

4 Relationship Between Quadratic *n*-ary Hom-Nambu-Lie Algebra and Quadratic Hom-Leibniz Algebra

In the context of Hom-Lie algebras one gets the class of Hom-Leibniz algebras (see [29]). A Hom-Leibniz algebra is a triple $(V, [\cdot,\cdot], \alpha)$ consisting of a linear space V, a bilinear map $[\cdot,\cdot] : V \times V \to V$ and a homomorphism $\alpha : V \to V$ with respect to $[\cdot,\cdot]$ satisfying

$$[\alpha(x), [y,z]] = [[x,y], \alpha(z)] + [\alpha(y), [x,z]] \tag{19}$$

We fix in the following notations. Let $(\mathcal{N}, [\cdot, \ldots, \cdot], \widetilde{\alpha})$ be an n-ary Hom-Nambu algebra, we define

• a linear map $L : \otimes^{n-1}\mathcal{N} \longrightarrow End(\mathcal{N})$ by

$$L(x) \cdot z = [x_1, \ldots, x_{n-1}, z], \tag{20}$$

for all $x = x_1 \otimes \ldots \otimes x_{n-1} \in \otimes^{n-1}\mathcal{N},\ z \in \mathcal{N}$ and extending it linearly to all $\otimes^{n-1}\mathcal{N}$. Notice that $L(x) \cdot z = ad(x)(z)$.

If the n-ary Hom-Nambu algebra $\mathcal{N}$ is multiplicative, then we define
• a linear map $\hat{\alpha} : \otimes^{n-1}\mathcal{N} \longrightarrow \otimes^{n-1}\mathcal{N}$ by

$$\hat{\alpha}(x) = \alpha(x_1) \otimes \ldots \otimes \alpha(x_{n-1}) \tag{21}$$

for all $x = x_1 \otimes \ldots \otimes x_{n-1} \in \otimes^{n-1}\mathcal{N}$,
• a bilinear map $[\ ,\ ]_\alpha : \otimes^{n-1}\mathcal{N} \times \otimes^{n-1}\mathcal{N} \longrightarrow \otimes^{n-1}\mathcal{N}$ defined by

$$[x, y]_\alpha = L(x) \bullet_\alpha y = \sum_{i=0}^{n-1} \big(\alpha(y_1), \ldots, L(x) \cdot y_i, \ldots, \alpha(y_{n-1})\big), \tag{22}$$

for all $x = x_1 \otimes \ldots \otimes x_{n-1} \in \otimes^{n-1}\mathcal{N},\ y = y_1 \otimes \ldots \otimes y_{n-1} \in \otimes^{n-1}\mathcal{N}$

Lemma 4.1 *Let $(\mathcal{N}, [\cdot, \ldots, \cdot], \widetilde{\alpha})$ be a multiplicative n-ary Hom-Nambu algebra then the map L satisfies*

$$L([x, y]_\alpha) \cdot \alpha(z) = L(\tilde{\alpha}(x)) \cdot \big(L(y) \cdot z\big) - L(\tilde{\alpha}(y)) \cdot \big(L(x) \cdot z\big) \tag{23}$$

for all $x, y \in \mathcal{L}(\mathcal{N}),\ z \in \mathcal{N}$.

If $(\mathcal{N}, [\cdot, \ldots, \cdot], \widetilde{\alpha})$ is a multiplicative n-ary Hom-Nambu-Lie algebra, we denote by $\mathcal{L}(\mathcal{N})$ the space $\wedge^{n-1}\mathcal{N}$ and we call it the fundamental set.

Proposition 4.1 *The triple $\big(\mathcal{L}(\mathcal{N}),\ [\ ,\ ]_\alpha,\ \hat{\alpha}\big)$, where $[\ ,\ ]_\alpha$ and $\hat{\alpha}$ are defined respectively in* (21) *and* (22), *is a Hom-Leibniz algebra.*

Remark 4.1 The invariance identity (12) of an n-ary Nambu algebra with respect to a bilinear form B can be written

$$B(L(x) \cdot y, z) + B(y, L(x) \cdot z) = 0, \tag{24}$$

and β-invariance identity (14) by

$$B(L(x) \cdot y, \beta(z)) + B(\beta(y), L(x) \cdot z) = 0, \tag{25}$$

Proposition 4.2 *Let $(\mathcal{N}, [\cdot, \ldots, \cdot], \alpha, B, \alpha)$ be a Hom-quadratic multiplicative Hom-Nambu-Lie algebra α-invariant and $\big(\mathcal{L}(\mathcal{N}),\ [\ ,\ ]_\alpha,\ \hat{\alpha}\big)$ be its associated Hom-Leibniz algebra, then the natural scalar product on $\mathcal{L}(\mathcal{N})$, $\widehat{B}$ defined by*

$$\widehat{B}(x, y) = B(x_1 \wedge \ldots \wedge x_{n-1}, y_1 \wedge \ldots \wedge y_{n-1}) = \prod_{i=0}^{n-1} B(x_i, y_i) \tag{26}$$

and later extending linearly to all of $\mathcal{L}(\mathcal{N})$, is $\tilde{\alpha}$-invariant. That is, for all $x, y, z \in \mathcal{L}(\mathcal{N})$:

$$\widehat{B}([z, x]_\alpha, \hat{\alpha}(y)) + \widehat{B}(\hat{\alpha}(x), [z, x]_\alpha) = 0. \tag{27}$$

Proof Let $x = (x_1, \dots, x_{n-1}),\ y = (y_1, \dots, y_{n-1})$ and let $z \in \mathscr{L}(\mathscr{N})$. Then using equation (27) we have

$$\begin{aligned}
\widehat{B}([z, x]_\alpha, \hat{\alpha}(y)) &= \widehat{B}(L(z) \bullet_\alpha x, \hat{\alpha}(y)) \\
&= \sum_{i=0}^{n-1} \widehat{B}\big((\alpha(x_1), \dots, L(z) \cdot x_i, \dots, \alpha(x_{n-1})), (\alpha(y_1), \dots, \alpha(y_{n-1}))\big) \\
&= \sum_{i=0}^{n-1} B(L(z) \cdot x_i, \alpha(y_i)) \prod_{j=0\ j\neq i}^{n-1} B(\alpha(x_j), \alpha(x_j)) \\
&= -\sum_{i=0}^{n-1} B(\alpha(x_i), L(z) \cdot y_i) \prod_{j=0\ j\neq i}^{n-1} B(\alpha(x_j), \alpha(y_j)) \\
&= -\sum_{i=0}^{n-1} \widehat{B}\big((\alpha(x_1), \dots, \alpha(x_{n-1})), (\alpha(y_1), \dots, L(z) \cdot y_i, \dots, \alpha(y_{n-1}))\big) \\
&= -\widehat{B}(\hat{\alpha}(x), L(z) \bullet_\alpha y) \\
&= -\widehat{B}(\hat{\alpha}(x), [z, y]_\alpha).
\end{aligned}$$

5 Representations and Quadratic n-ary Hom-Nambu Algebras

In this Section we study in the general case the representation theory of n-ary Hom-Nambu algebras introduced for multiplicative n-ary Hom-Nambu algebras in [1]. We discuss in particular adjoint and coadjoint representations for quadratic n-ary Hom-Nambu algebras. The results obtained in this Section generalize those given for binary case in [13]. The representation theory of Hom-Lie algebras were independently studied in [36].

Definition 5.1 A representation of an n-ary Hom-Nambu algebra $(\mathscr{N}, [\cdot, \dots, \cdot], \tilde{\alpha})$ on a vector space V is a skew-symmetric multilinear map $\rho : \mathscr{N}^{n-1} \longrightarrow End(V)$, satisfying for $x, y \in \mathscr{N}^{n-1}$ the identity

$$\rho(\tilde{\alpha}(x))\circ\rho(y) - \rho(\tilde{\alpha}(y))\circ\rho(x) = \sum_{i=1}^{n-1} \rho(\alpha_1(x_1), \dots, L(y)\cdot x_i, \dots, \alpha_{n-2}(x_{n-1}))\circ\nu \tag{28}$$

where ν is an endomorphism on V. We denote this representation by a triple (V, ρ, μ).

Two representations (V, ρ, μ) and (V', ρ', μ') of $\mathscr{N}$ are *equivalent* if there exists $f : V \to V'$ an isomorphism of vector space such that $f(x \cdot v) = x \cdot' f(v)$ and $f \circ \nu = \nu' \circ f$ where $x \cdot v = \rho(x)(v)$ and $x \cdot' v' = \rho'(x)(v')$ for $x \in \mathscr{N}^{n-1}$, $v \in V$ and $v' \in V'$. Then V and V' are viewed as $\mathscr{N}^{n-1}$-modules.

Example 5.1 Let $(\mathcal{N}, [\cdot, \ldots, \cdot], \tilde{\alpha})$ be an n-ary Hom-Nambu-Lie algebra. The map L defined in (20) is a representation on $\mathcal{N}$, where the endomorphism v is the twist map α_{n-1}. The identity (28) is equivalent to Hom-Nambu identity (1). It is called the adjoint representation.

Proposition 5.1 *Let $(\mathcal{N}, [., \ldots,], \widetilde{\alpha})$ be an n-ary Hom-Nambu algebra and (V, ρ, v) be a representation of $\mathcal{N}$. The triple $(V^*, \rho^*, \widetilde{v})$, where $\rho^* : \mathcal{N}^{n-1} \to End(V^*)$ is given by $\rho^* = -{}^t\rho$ and $\mu^* : V^* \to V^*, \quad f \mapsto v^*(f) = f \circ v$, defines a representation of the n-ary Hom-Nambu-Lie algebra $(\mathcal{N}, [., \ldots,], \widetilde{\alpha})$ if and only if*

$$\rho(x)\circ\rho(\tilde{\alpha}(y))-\rho(y)\circ\rho(\tilde{\alpha}(x)) = \sum_{i=1}^{n-1} v\circ\rho(\alpha_1(x_1), \ldots, L(y)\cdot x_i, \ldots, \alpha_{n-2}(x_{n-1})) \tag{29}$$

Proof Let $f \in \mathcal{N}^*, x, y \in \mathcal{N}^{n-1}$ and $u \in \mathcal{N}$. We compute the right hand side of the identity (28)

$$\begin{aligned}
&\rho^*(\tilde{\alpha}(x)) \circ \rho^*(y)(f)(u) - \rho^*(\tilde{\alpha}(y)) \circ \rho^*(x)(f)(u)\\
&= (\rho^*(\tilde{\alpha}(x))(\rho^*(y)(f)) - \rho^*(\tilde{\alpha}(y))(\rho^*(x)(f)))(u)\\
&= -(\rho^*(y)(f)(\rho(\tilde{\alpha}(x)))(u)) + (\rho^*(x)(f)(\rho(\tilde{\alpha}(y)))(u))\\
&= f(\rho(y)(\rho(\tilde{\alpha}(x))(u))) - f(\rho(x)(\rho(\tilde{\alpha}(y))(u)))\\
&= f(\rho(y)(\rho(\tilde{\alpha}(x))(u)) - \rho(x)(\rho(\tilde{\alpha}(y))(u))).
\end{aligned}$$

In the other hand, the left hand side of (28) writes

$$\begin{aligned}
&(\sum_{i=1}^{n-1} \rho^*(\alpha_1(x_1), \ldots, L(y) \cdot x_i, \ldots, \alpha_{n-2}(x_{n-1})) \circ v^*(f))(u)\\
&= -\sum_{i=1}^{n-1}(v^*(f)(\rho(\alpha_1(x_1), \ldots, L(y) \cdot x_i, \ldots, \alpha_{n-2}(x_{n-1}))(u)))\\
&= -\sum_{i=1}^{n-1} f(v(\rho(\alpha_1(x_1), \ldots, L(y) \cdot x_i, \ldots, \alpha_{n-2}(x_{n-1}))(u)))\\
&= f(-\sum_{i=1}^{n-1} v(\rho(\alpha_1(x_1), \ldots, L(y) \cdot x_i, \ldots, \alpha_{n-2}(x_{n-1}))(u))).
\end{aligned}$$

Therefore we obtain the identity (29).

Corollary 5.1 *Let $(\mathcal{N}, L, \alpha_{n-1})$ be a representation of an n-ary Hom-Nambu algebra $(\mathcal{N}, [\cdot, \ldots, \cdot], \widetilde{\alpha})$. We define the map $\widetilde{L} : \mathcal{N}^{n-1} \to End(\mathcal{N}^*)$, for $x \in \mathcal{N}^{n-1}$, $f \in \mathcal{N}^*$ and $y \in \mathcal{N}$, by $(\widetilde{L}(x) \cdot f)(y) = -f(L(x) \cdot y)$.*
Then $(\mathcal{N}^*, \widetilde{L}, \alpha^*_{n-1})$ is a representation of $\mathcal{N}$ if and only if

$$L(x) \circ L(\tilde{\alpha}(y)) - L(y) \circ L(\tilde{\alpha}(x)) = \sum_{i=1}^{n-1} \alpha_{n-1} \circ L(\alpha_1(x_1), \ldots, L(y) \cdot x_i, \ldots, \alpha_{n-2}(x_{n-1})). \tag{30}$$

We establish now a connection between quadratic n-ary Hom-Nambu algebras and representation theory. We discuss coadjoint representations for quadratic n-ary Hom-Nambu algebras.

Proposition 5.2 *Let $(\mathcal{N}, [\cdot, \ldots, \cdot], \widetilde{\alpha})$ be an n-ary Hom-Nambu algebra. If there exists $B : \mathcal{N} \times \mathcal{N} \to \mathbb{K}$ a bilinear form such that the quadruple $(\mathcal{N}, [\cdot, \ldots, \cdot], \widetilde{\alpha}, B)$ is a quadratic n-ary Hom-Nambu algebra then*

1. *$(\mathcal{N}^*, \widetilde{L}, \alpha^*_{n-1})$ is a representation of $\mathcal{N}$,*
2. *the representations $(\mathcal{N}, L, \alpha_{n-1})$ and $(\mathcal{N}^*, \widetilde{L}, \alpha^*_{n-1})$ are isomorphic.*

Proof To prove the first assertion, we should show that, for any $z \in \mathcal{N}$, we have

$$\begin{aligned} & L(x) \circ L(\tilde{\alpha}(y)) \cdot z - L(y) \circ L(\tilde{\alpha}(x)) \cdot z \\ &= \sum_{i=1}^{n-1} \alpha_{n-1} \circ L(\alpha_1(x_1), \ldots, L(y) \cdot x_i, \ldots, \alpha_{n-2}(x_{n-1}))(z). \end{aligned} \tag{31}$$

Let $u \in \mathcal{N}$

$$\begin{aligned} & B(L(x) \circ L(\tilde{\alpha}(y)) \cdot z - L(y) \circ L(\tilde{\alpha}(x)) \cdot z, u) \\ &= B(L(x) \circ L(\tilde{\alpha}(y)) \cdot z, u) - (L(y) \circ L(\tilde{\alpha}(x)) \cdot z, u) \\ &= B(L(\tilde{\alpha}(y)) \cdot z, L(x) \cdot u) - (L(\tilde{\alpha}(x)) \cdot z, L(y) \cdot u) \\ &= B(z, L(\tilde{\alpha}(y)) \circ L(x) \cdot u) - (z, L(\tilde{\alpha}(x)) \circ L(y) \cdot u) \\ &= B(z, L(\tilde{\alpha}(y)) \circ L(x)(u) - L(\tilde{\alpha}(x)) \circ L(y)(u)). \end{aligned}$$

and

$$\begin{aligned} & B(\alpha_{n-1} \circ L(\alpha_1(x_1), \ldots, L(y) \cdot x_i, \ldots, \alpha_{n-2}(x_{n-1}))(z), u) \\ &= B(L(\alpha_1(x_1), \ldots, L(y) \cdot x_i, \ldots, \alpha_{n-2}(x_{n-1}))(z), \alpha_{n-1}(u)) \\ &= B(z, L(\alpha_1(x_1), \ldots, L(y) \cdot x_i, \ldots, \alpha_{n-2}(x_{n-1})) \circ \alpha_{n-1}(u)) \\ &= B(z, L(\alpha_1(x_1), \ldots, L(y) \cdot x_i, \ldots, \alpha_{n-2}(x_{n-1})) \circ \alpha_{n-1}(u)). \end{aligned}$$

Since B is bilinear, then

$$\begin{aligned} & B(\sum_{i=1}^{n-1} \alpha_{n-1} \circ L(\alpha_1(x_1), \ldots, L(y) \cdot x_i, \ldots, \alpha_{n-2}(x_{n-1}))(z), u) \\ &= B(z, \sum_{i=1}^{n-1} L(\alpha_1(x_1), \ldots, L(y) \cdot x_i, \ldots, \alpha_{n-2}(x_{n-1})) \circ \alpha_{n-1}(u)). \end{aligned}$$

Hence

$$\begin{aligned}
&B(L(x)\circ L(\tilde{\alpha}(y))\cdot z - L(y)\circ L(\tilde{\alpha}(x))\cdot z\\
&\quad -\sum_{i=1}^{n-1}\alpha_{n-1}\circ L(\alpha_1(x_1),\dots,L(y)\cdot x_i,\dots,\alpha_{n-2}(x_{n-1}))(z),u)\\
&= B(z, L(\tilde{\alpha}(y))\circ L(x)\cdot u - L(\tilde{\alpha}(x))\circ L(y)\cdot u\\
&\quad -\sum_{i=1}^{n-1} L(\alpha_1(x_1),\dots,L(y)\cdot x_i,\dots,\alpha_{n-2}(x_{n-1}))\circ\alpha_{n-1}(u))\\
&= 0.
\end{aligned}$$

Since B is nondegenerate then the identity (31) holds.

For the second assertion we consider the map $\psi : \mathcal{N} \to \mathcal{N}^*$ defined by $x \mapsto B(x,\cdot)$ which is bijective since B is nondegenerate and prove that it is also a module morphism.

6 Constructions of Quadratic n-ary Hom-Nambu Algebras

We provide in this section some key constructions of Hom-quadratic n-ary Hom-Nambu-Lie algebras. First we extend twisting principles, then the T^*-extension construction for Hom-quadratic n-ary Hom-Nambu-Lie algebras. Moreover, we show constructions involving tensor product of Hom-quadratic commutative Hom-associa-
tive algebra and Hom-quadratic Hom-Nambu-Lie algebra considered in Theorem 1.3.

6.1 Twisting Principles

Let $(\mathcal{N}, [\cdot,\dots,\cdot], B)$ be a quadratic n-ary Nambu algebras. We denote $Aut_S(\mathcal{N}, B)$ by the set of symmetric automorphisms of $\mathcal{N}$ with respect of B, that is automorphisms $f : \mathcal{N} \to \mathcal{N}$ such that $B(f(x), y) = B(x, f(y)), \forall x, y \in \mathcal{N}$.

Proposition 6.1 *Let* $(\mathcal{N}, [\cdot,\dots,\cdot], B)$ *be a quadratic* n*-ary Nambu algebra and* $\rho \in Aut_S(\mathcal{N}, B)$.

Then $(\mathcal{N}, [\cdot, \ldots, \cdot]_\rho, \tilde{\rho}, B, \rho)$ *where*

$$[\cdot, \ldots, \cdot]_\rho = \rho \circ [\cdot, \ldots, \cdot] \tag{32}$$

is a Hom-quadratic n-ary Hom-Nambu algebra, and $(\mathcal{N}, [\cdot, \ldots, \cdot]_\rho, \tilde{\rho}, B_\rho)$ *where*

$$B_\rho(x, y) = B(\rho(x), y) \tag{33}$$

is a quadratic n-ary Hom-Nambu algebra.

Proof Let $x = (x_1, \ldots, x_{n-1}) \in \mathcal{N}^{\otimes n-1}$ et $y_1, y_2 \in \mathcal{N}$,

$$\begin{aligned} B([x_1, \ldots, x_{n-1}, y_1]_\rho, \rho(y_2)) &= B([\rho(x_1), \ldots, \rho(x_{n-1}), \rho(y_1)], \rho(y_2)) \\ &= -B(\rho(y_1), [\rho(x_1), \ldots, \rho(x_{n-1}), \rho(y_2)]) \\ &= -B(\rho(y_1), \rho \circ [x_1, \ldots, x_{n-1}, y_2]) \\ &= -B(\rho(y_1), [x_1, \ldots, x_{n-1}, y_2]_\rho). \end{aligned}$$

In the other hand we have

$$\begin{aligned} B_\rho([x_1, \ldots, x_{n-1}, y_1]_\rho, y_2) &= B(\rho[\rho(x_1), \ldots, \rho(x_{n-1}), \rho(y_1)], y_2) \\ &= B([\rho(x_1), \ldots, \rho(x_{n-1}), \rho(y_1)], \rho(y_2)) \\ &= -B(\rho(y_1), [\rho(x_1), \ldots, \rho(x_{n-1}), \rho(y_2)]) \\ &= -B(\rho(y_1), \rho \circ [x_1, \ldots, x_{n-1}, y_2]) \\ &= -B_\rho(y_1, [x_1, \ldots, x_{n-1}, y_2]_\rho). \end{aligned}$$

Proposition 6.2 *Let* $(\mathcal{N}, [\cdot, \ldots, \cdot], \alpha, B)$ *be a quadratic multiplicative n-ary Hom-Nambu algebra. Then* $(\mathcal{N}, \alpha^{n-1} \circ [\cdot, \ldots, \cdot], \alpha^n, B, \alpha^{n-1})$ *is a Hom-quadratic n-ary Hom-Nambu algebra and* $(\mathcal{N}, \alpha^{n-1} \circ [\cdot, \ldots, \cdot], \alpha^n, B_\alpha)$*, where*

$$B_\alpha(x, y) = B(\alpha^{n-1}(x), y) = B(x, \alpha^{n-1}(y)), \tag{34}$$

is a quadratic n-ary Hom-Nambu algebra,

Proof Using the second twisting principle construction of Theorem 2.2, $(\mathcal{N}, \alpha^{n-1} \circ [\cdot, \ldots, \cdot], \alpha^n)$ is a n-ary Hom-Nambu algebra. Let now $x_i, y, z \in \mathcal{N}, i \in \{1, \ldots, n-1\}$, we have

$$B_\alpha(\alpha^{n-1}(y), z) = B(\alpha^{2n-2}(y), z) = B(\alpha^{n-1}(y), \alpha^{n-1}(z)) = B_\alpha(y, \alpha^{n-1}(z)).$$

In the other hand, we have

$$\begin{aligned} B_\alpha(\alpha^{n-1} \circ [x_1, \ldots, x_{n-1}, y], z) &= B(\alpha^{n-1} \circ [x_1, \ldots, x_{n-1}, y], \alpha^{n-1}(z)) \\ &= B([\alpha^{n-1}(x_1), \ldots, \alpha^{n-1}(x_{n-1}), \alpha^{n-1}(y)], \alpha^{n-1}(z)) \\ &= -B(\alpha^{n-1}(y), [\alpha^{n-1}(x_1), \ldots, \alpha^{n-1}(x_{n-1}), \alpha^{n-1}(z)]) \end{aligned}$$

$$= -B(\alpha^{n-1}(y), \alpha^{n-1} \circ [x_1, \ldots, x_{n-1}, z)]$$
$$= -B_\alpha(y, \alpha^{n-1} \circ [x_1, \ldots, x_{n-1}, z)].$$

Therefore B_α is invariant.

6.2 T^*-Extension of n-ary Hom-Nambu Algebras

We provide here a construction of n-ary Hom-Nambu algebra $\mathscr{L}$ which is a generalization of the trivial T^*-extension introduced in [14, 33].

Theorem 6.1 *Let $(\mathscr{N}, [\cdot, \ldots, \cdot]_{\mathscr{N}}, B)$ be a quadratic n-ary Nambu-Lie algebra and $\mathscr{N}^*$ be the underlying dual vector space. The vector space $\mathscr{L} = \mathscr{N} \oplus \mathscr{N}^*$ equipped with the following product $[\cdot, \ldots, \cdot]_{\mathscr{L}} : \mathscr{L}^n \to \mathscr{L}$ given, for $u_i = x_i + f_i \in \mathscr{L}$ where $i \in \{1, \ldots, n\}$ by*

$$[u_1, \ldots, u_n]_{\mathscr{L}} = [x_1, \ldots, x_n]_{\mathscr{N}} + \sum_{i=1}^{n} (-1)^{i+n+1} f_i \circ L(x_1, \ldots, \widehat{x_i}, \ldots, x_n), \quad (35)$$

and a bilinear form

$$B_{\mathscr{L}} : \begin{array}{c} \mathscr{L} \times \mathscr{L} \longrightarrow \mathscr{L} \\ B_{\mathscr{L}}(x + f, y + g) = B(x, y) + f(y) + g(x) \end{array} \quad (36)$$

is a quadratic n-ary Nambu algebra.

Proof $\star$) Set $u_i = x_i + f_i \in \mathscr{L}$ and $v_i = y_k + g_k \in \mathscr{L}$. We show the following Nambu identity on $\mathscr{L}$

$$[u_1, \ldots, u_{n-1}, [v_1, \ldots, v_n]_{\mathscr{L}}]_{\mathscr{L}} = \sum_{l=1}^{n} (-1)^{l+n} [v_1, \ldots, \widehat{v_l}, \ldots, v_n, [u_1, \ldots, u_{n-1}, v_l]_{\mathscr{L}}]_{\mathscr{L}}. \quad (37)$$

Let us compute first $[u_1, \ldots, u_{n-1}, [v_1, \ldots, v_n]_{\mathscr{L}}]_{\mathscr{L}}$. This is given by

$$\begin{aligned} &[u_1, \ldots, u_{n-1}, [v_1, \ldots, v_n]_{\mathscr{L}}]_{\mathscr{L}} \\ &= [x_1, \ldots, x_{n-1}, [y_1, \ldots, y_n]_{\mathscr{N}}]_{\mathscr{N}} + \sum_{i=1}^{n-1} (-1)^{i+n+1} f_i \circ L(x_1, \ldots, \widehat{x_i}, \ldots, x_{n-1}, \\ &\quad [y_1, \ldots, y_n]_{\mathscr{N}}) \\ &\quad + \sum_{i=1}^{n} (-1)^{i+n} g_i \circ L(y_1, \ldots, \widehat{y_i}, \ldots, y_n) \circ L(x_1, \ldots, x_{n-1}). \end{aligned}$$

Hence the right hand side of (37) gives, for any $l \in \{1, \ldots, n\}$

$$[v_1,\dots,\widehat{v_l},\dots,v_n,[u_1,\dots,u_{n-1},v_l]_{\mathscr{L}}]_{\mathscr{L}} = [y_1,\dots,\widehat{y_l},\dots,y_n,[x_1,\dots,x_{n-1},y_l]_{\mathscr{N}}]_{\mathscr{N}}$$
$$+\sum_{i=1}^{n-1}(-1)^{i+n} f_i\circ L(x_1,\dots,\widehat{x_i},\dots,x_{n-1},y_l)\circ L(y_1,\dots,\widehat{y_l},\dots,y_n)$$
$$+\sum_{i=1}^{n}(-1)^{i+n+1} g_i\circ L(y_1,\dots,\widehat{y_i},\dots,\widehat{y_l},\dots,y_n,[x_1,\dots,x_{n-1},y_l]_{\mathscr{N}})$$
$$+\,g_l\circ L(x_1,\dots,x_{n-1})\circ L(y_1,\dots,\widehat{y_l},\dots,y_n).$$

- Using the Nambu identity on $\mathscr{N}$, we obtain

$$[x_1,\dots,x_{n-1},[y_1,\dots,\widehat{y_l},\dots,y_n,z]_{\mathscr{N}}]_{\mathscr{N}} =$$
$$\sum_{i=1,i\neq l}^{n}(-1)^{i+n}[y_1,\dots,\widehat{y_i},\dots,\widehat{y_l},\dots,y_{n-1},[x_1,\dots,x_{n-1},y_i]_{\mathscr{N}},z]_{\mathscr{N}}$$
$$+\,[y_1,\dots,\widehat{y_l},\dots,y_{n-1},[x_1,\dots,x_{n-1},z]_{\mathscr{N}}]_{\mathscr{N}}.$$

Equivalently

$$\begin{aligned} &L(y_1,\dots,\widehat{y_l},\dots,y_n)\circ L(x_1,\dots,x_{n-1})\\ &= L(x_1,\dots,x_{n-1})\circ L(y_1,\dots,\widehat{y_l},\dots,y_n)\\ &\quad+\sum_{i=1}^{n-1}(-1)^{i+n+1}L(y_1,\dots,\widehat{y_i},\dots,\widehat{y_l},\dots,y_n,[x_1,\dots,x_{n-1},y_l]). \end{aligned}$$

Thus for any $l\in\{1,\dots n\}$

$$\begin{aligned} &g_l\circ L(y_1,\dots,\widehat{y_l},\dots,y_n)\circ L(x_1,\dots,x_{n-1}) - g_l\circ L(x_1,\dots,x_{n-1})\circ L(y_1,\dots,\widehat{y_l},\dots,y_n)\\ &= \sum_{i=1}^{n-1}(-1)^{i+n} g_l\circ L(y_1,\dots,\widehat{y_i},\dots,\widehat{y_l},\dots,y_n,[x_1,\dots,x_{n-1},y_l]). \end{aligned}$$

- In the other hand we show that, for $k\in\{1,\dots n\}$

$$\begin{aligned} &-\,f_k\circ L(x_1,\dots,\widehat{x_k},\dots,x_{n-1},[y_1,\dots,y_n]_{\mathscr{N}})\\ &= \sum_{i=1}^{n}(-1)^{i+n} f_k\circ L(x_1,\dots,\widehat{x_k},\dots,x_{n-1},y_i)\circ L(y_1,\dots,\widehat{y_i},\dots,y_n). \end{aligned}$$

Using the Namby identity (37) on $\mathscr{N}$ and the invariance of B, we obtain

$$B([x_1,\dots,\widehat{x_k},\dots,x_n,[y_1,\dots,y_n]_{\mathscr{N}}]_{\mathscr{N}},z) = B(x_n,[x_1,\dots,\widehat{x_k},\dots,x_{n-1},\ [y_1,\dots,y_n]_{\mathscr{N}},z]_{\mathscr{N}}).$$

Hence

$$B\Big(\sum_{i=1}^{n}(-1)^{i+n}[y_1,\dots,\widehat{y_i},\dots,y_n,[x_1,\dots,\widehat{x_k},\dots,x_n,y_i]_{\mathscr{N}}]_{\mathscr{N}},z\Big)$$

$$= \sum_{i=1}^{n} (-1)^{i+n} B([y_1, \ldots, \widehat{y_i}, \ldots, y_n, [x_1, \ldots, \widehat{x_k}, \ldots, x_n, y_i]_{\mathcal{N}}]_{\mathcal{N}}, z)$$

$$= -\sum_{i=1}^{n} (-1)^{i+n} B([x_1, \ldots, \widehat{x_k}, \ldots, x_n, y_i]_{\mathcal{N}}, [y_1, \ldots, \widehat{y_i}, \ldots, y_n, z]_{\mathcal{N}})$$

$$= -\sum_{i=1}^{n} (-1)^{i+n} B(x_n, [x_1, \ldots, \widehat{x_k}, \ldots, x_{n-1}, y_i, [y_1, \ldots, \widehat{y_i}, \ldots, y_n, z]_{\mathcal{N}}]_{\mathcal{N}})$$

$$= -B(x_n, \sum_{i=1}^{n} (-1)^{i+n} [x_1, \ldots, \widehat{x_k}, \ldots, x_{n-1}, y_i, [y_1, \ldots, \widehat{y_i}, \ldots, y_n, z]_{\mathcal{N}}]_{\mathcal{N}}).$$

Since B is nondegenerate, then

$$- [x_1, \ldots, \widehat{x_k}, \ldots, x_{n-1}, [y_1, \ldots, y_n]_{\mathcal{N}}, z]_{\mathcal{N}}$$
$$= \sum_{i=1}^{n} (-1)^{i+n} [x_1, \ldots, \widehat{x_k}, \ldots, x_{n-1}, y_i, [y_1, \ldots, \widehat{y_i}, \ldots, y_n, z]_{\mathcal{N}}]_{\mathcal{N}},$$

and equivalently

$$- L(x_1, \ldots, \widehat{x_k}, \ldots, x_{n-1}, [y_1, \ldots, y_n]_{\mathcal{N}})$$
$$= \sum_{i=1}^{n} (-1)^{i+n} L(x_1, \ldots, \widehat{x_k}, \ldots, x_{n-1}, y_i) \circ L(y_1, \ldots, \widehat{y_i}, \ldots, y_n).$$

Finally, the Nambu identity (37) is satisfied. Thus $(\mathcal{L}, [\cdot, \ldots, \cdot]_{\mathcal{L}})$ is an n-ary Nambu algebra.

Theorem 6.2 *Let $(\mathcal{N}, [\cdot, \ldots, \cdot]_{\mathcal{N}}, B)$ be a quadratic n-ary Nambu-Lie algebra where $\alpha \in Aut_S(\mathcal{N}, B)$ is an involution. Then $(\mathcal{L}, [\cdot, \ldots, \cdot]_{\Omega}, \widetilde{\Omega}, B_{\mathcal{L}}, \Omega)$, where $\Omega : \mathcal{L} \to \mathcal{L}, x+f \to \Omega(x+f) = \alpha(x)+f\circ\alpha$ and $[\cdot, \ldots, \cdot]_{\Omega} = \Omega\circ[\cdot, \ldots, \cdot]_{\mathcal{L}}$, is a Hom-quadratic multiplicative n-ary Hom-Nambu algebra.*

Proof Let $x_1, \ldots, x_n \in \mathcal{N}$ and $f_1, \ldots, f_n \in \mathcal{N}^*$,

$$\Omega[x_1 + f_1, \ldots, x_n + f_n]_{\mathcal{L}} = \alpha[x_1, \ldots, x_n]_{\mathcal{N}} + \sum_{i=1}^{n} (-1)^i f_i \circ L(x_1, \ldots, \widehat{x_i}, \ldots, x_n) \circ \alpha,$$

$$[\Omega(x_1 + f_1), \ldots, \Omega(x_n + f_n)]_{\mathcal{L}}$$
$$= [\alpha(x_1), \ldots, \alpha(x_n)]_{\mathcal{N}} + \sum_{i=1}^{n} (-1)^i f_i\circ\alpha \circ L(\alpha(x_1), \ldots, \widehat{x_i}, \ldots, \alpha(x_n)).$$

That is for all $z \in \mathcal{N}$

$$\alpha \circ L(\alpha(x_1), \ldots, \widehat{x_i}, \ldots, \alpha(x_n))(z) = \alpha[\alpha(x_1), \ldots, \widehat{x_i}, \ldots, \alpha(x_n), z]_{\mathcal{N}}$$
$$= [\alpha^2(x_1), \ldots, \widehat{x_i}, \ldots, \alpha^2(x_n), \alpha(z)]_{\mathcal{N}}$$

$$= [x_1, \ldots, \widehat{x_i}, \ldots, x_n, \alpha(z)]_{\mathcal{N}}$$
$$= L(x_1, \ldots, \widehat{x_i}, \ldots, x_n) \cdot \alpha(z).$$

Then $\Omega[x_1 + f_1, \ldots, x_n + f_n]_{\mathcal{L}} = [\Omega(x_1 + f_1), \ldots, \Omega(x_n + f_n)]_{\mathcal{L}}$.
In the following we show that Ω is symmetric with respect to $B_{\mathcal{L}}$.

Indeed, let $x, y \in \mathcal{N}$ and $f, h \in \mathcal{N}$

$$\begin{aligned} B_{\mathcal{L}}(\Omega(x+f), y+h) &= B_{\mathcal{L}}(\alpha(x) + f \circ \alpha, y+h) \\ &= B(\alpha(x), y) + f \circ \alpha(y) + h \circ \alpha(x) \\ &= B(x, \alpha(y)) + f \circ \alpha(y) + h \circ \alpha(x) \\ &= B_{\mathcal{L}}(x+f, \alpha(y) + h \circ \alpha) = B_{\mathcal{L}}(x+f, \Omega(y+h)). \end{aligned}$$

Thus, using Proposition 6.1, $(\mathcal{L}, [\cdot, \ldots, \cdot]_{\Omega}, \widetilde{\Omega}, B_{\mathcal{L}}, \Omega)$ is a Hom-quadratic multiplicative n-ary Hom-Nambu algebra. We have also that $(\mathcal{L}, [\cdot, \ldots, \cdot]_{\Omega}, \Omega, B_{\mathcal{L},\Omega})$, where $B_{\mathcal{L},\Omega}(u, v) = B_{\mathcal{L}}(\Omega(u), v)$, for all $u, v \in \mathcal{L}$, is a quadratic multiplicative n-ary Hom-Nambu algebra.

6.3 Tensor Product

Let $(A, \mu, \widetilde{\eta}, B_A, \beta_A)$ be a Hom-quadratic symmetric n-ary totally Hom-associative algebra, that is a symmetric n-ary totally Hom-associative algebra together with a symmetric nondegenerate form satisfying the following assertions

$$B_A(\eta_i(a), b) = B_A(a, \eta_i(b)), \ \text{for all } i \in \{1, \ldots, n-1\} \tag{38}$$

$$B_A(\mu(a_1, \ldots, a_{n-1}, b), \beta_A(c)) = B_A(\beta_A(a), \mu(a_1, \ldots, a_{n-1}, c)), \\ \text{for all } a_i, b, c \in A, \ i \in \{1, \ldots, n-1\}. \tag{39}$$

We discuss now the tensor product as in Proposition 5.1.

Theorem 6.3 *Let $(\mathcal{N}, [\cdot, \ldots, \cdot]_{\mathcal{N}}, \alpha, B_{\mathcal{N}}, \beta_{\mathcal{N}})$ be a Hom-quadratic n-ary Hom-Nambu algebra, then $(A \otimes \mathcal{N}, [\cdot, \ldots, \cdot], \zeta, \widetilde{B}, \omega)$, where*

$$\widetilde{B}(a \otimes x, b \otimes y) = B_A(a, b) B_{\mathcal{N}}(x, y), \tag{40}$$

$$\omega(a \otimes x) = \beta_A(a) \otimes \beta_{\mathcal{N}}(x), \tag{41}$$

is a Hom-quadratic n-ary Hom-Nambu algebra.

6.4 Hom-quadratic Hom-Nambu-Lie Algebras Induced by Hom-quadratic Hom-Lie Algebras

In [8] the authors provided a construction procedure of ternary Hom-Nambu-Lie algebras starting from a bilinear bracket of a Hom-Lie algebra and a trace function satisfying certain compatibility conditions including the twisting map.

The aim of this section is to prove that this procedure is still true for quadratic ternary Hom-Nambu-Lie algebra. First we recall the result in [8].

Definition 6.1 Let $(V, [\cdot,\cdot])$ be a binary algebra and $\tau : V \to \mathbb{K}$ be a linear form. The trilinear map $[\cdot,\cdot,\cdot]_\tau : V \times V \times V \to V$ is defined as

$$[x, y, z]_\tau = \tau(x)[y, z] + \tau(y)[z, x] + \tau(z)[x, y]. \tag{42}$$

Remark 6.1 If the bilinear multiplication $[\cdot,\cdot]$ is skew-symmetric, then the trilinear map $[\cdot,\cdot,\cdot]_\tau$ is skew-symmetric as well.

Theorem 6.4 *([8]). Let $(V, [\cdot,\cdot], \alpha)$ be a Hom-Lie algebra and $\gamma : V \longrightarrow V$ be a linear map. Furthermore, assume that τ is a trace function on V fulfilling*

$$\tau(\alpha(x))\tau(y) = \tau(x)\tau(\alpha(y)), \tag{43}$$
$$\tau(\gamma(x))\tau(y) = \tau(x)\tau(\gamma(y)), \tag{44}$$
$$\tau(\alpha(x))\gamma(y) = \tau(\gamma(x))\alpha(y), \tag{45}$$

for all $x, y \in V$. Then $(V, [\cdot,\cdot,\cdot]_\tau, (\alpha, \gamma))$ is a ternary Hom-Nambu-Lie algebra, and we say that it is induced by $(V, [\cdot,\cdot], \alpha)$.

Proposition 6.3 *Let $(V, [\cdot,\cdot], \alpha, B, \beta)$ be a Hom-quadratic Hom-Lie algebra satisfying*

$$B(\alpha(x), y) = B(x, \alpha(y)), \tag{46}$$
$$B(\gamma(x), y) = B(x, \gamma(y)), \tag{47}$$
$$\tau(x)B(\beta(y), z) - \tau(y)B(\beta(x), z) = 0 \quad \textit{for all } x, y, z \in V. \tag{48}$$

Then $(V, [\cdot,\cdot,\cdot]_\tau, (\alpha, \gamma), B, \beta)$ is a Hom-quadratic ternary Hom-Nambu-Lie algebra.

Proof Let $x_1, x_2, y_1, y_2 \in V$

$$\begin{aligned} B([x_1, x_2, y_1]_\tau, \beta(y_2)) =& \tau(x_1)B([x_2, y_1], \beta(y_2)) - \tau(x_2)B([x_1, y_1], \beta(y_2)) \\ &+ \tau(y_1)B([x_1, x_2], \beta(y_2)). \end{aligned}$$

$$\begin{aligned} B(\beta(y_1), [x_1, x_2, y_2]_\tau) =& \tau(x_1)B(\beta(y_1), [x_2, y_2]) - \tau(x_2)B(\beta(y_1), [x_1, y_2]) \\ &- \tau(y_2)B(\beta(y_1), [x_1, x_2]). \end{aligned}$$

Since B is symmetric, then

$$B([x_1, x_2, y_1]_\tau, \beta(y_2)) + B(\beta(y_1), [x_1, x_2, y_2]_\tau) = 0.$$

6.5 Quadratic Hom-Nambu Algebras of Higher Arities

The purpose of this section is to observe that every Hom-quadratic multiplicative n-ary Hom-Nambu algebra gives rise to a sequence of quadratic multiplicative Hom-Nambu algebras of increasingly higher arities. The construction of this sequence was given first in [40].

Theorem 6.5 *Let $(\mathcal{N}, [\cdot, \dots, \cdot], \alpha, B, \beta)$ be a Hom-quadratic multiplicative n-ary Hom-Nambu algebra. Define the $(2n-1)$- ary product*

$$[x_1, \dots, x_{2n-1}]^{(1)} = [[x_1, \dots, x_n], \alpha(x_{n+1}), \dots, \alpha(x_{2n-1})] \quad \textit{for } x_i \in V \tag{49}$$

Then $\mathcal{N}^1 = (\mathcal{N}, [\cdot, \dots, \cdot]^{(1)}, \alpha^2, B, \beta')$, where $\beta' = \beta\alpha$, is a Hom-quadratic multiplicative $(2n-1)$-ary Hom-Nambu algebra.

Proof For the proof of the $(2n-1)$-ary Hom-Nambu identity and the multiplicativity for $\mathcal{N}^1$, see [40].

Let $x_1, \dots, x_{2n-2}, y_1, y_2 \in \mathcal{N}$

$$\begin{aligned} B([x_1, \dots, x_{2n-2}, y_1]^{(1)}, \beta'(y_2)) &= B([[x_1, \dots, x_n], \alpha(x_{n+1}), \dots, \alpha(x_{2n-2}), \alpha(y_1)], \\ &\quad \beta(\alpha(y_2))) \\ &= -B(\beta(\alpha(y_1)), [[x_1, \dots, x_n], \alpha(x_{n+1}), \\ &\quad \dots, \alpha(x_{2n-2}), \alpha(y_2)]) \\ &= -B(\beta'(y_1), [x_1, \dots, x_{2n-2}, y_2]^{(1)}). \end{aligned}$$

Hence

$$B([x_1, \dots, x_{2n-2}, y_1]^{(1)}, \beta'(y_2)) + B(\beta'(y_1), [x_1, \dots, x_{2n-2}, y_2]^{(1)}) = 0.$$

Corollary 6.1 *Let $(\mathcal{N}, [\cdot, \dots, \cdot], \alpha, B, \beta)$ be a Hom-quadratic multiplicative n-ary Hom-Nambu algebra. For $k \geq 1$ define the $(2^k(n-1)+1)$- ary product $[\cdot, \dots, \cdot]^{(k)}$ inductively by setting $[\cdot, \dots, \cdot]^{(0)} = [\cdot, \dots, \cdot]$ and*

$$\begin{aligned} &[x_1, \dots, x_{2^k(n-1)+1}]^{(k)} \\ &= [[x_1, \dots, x_{2^{k-1}(n-1)+1}]^{(k-1)}, \alpha^{2^{k-1}}(x_{2^{k-1}(n-1)+2}), \dots, \alpha^{2^{k-1}}(x_{2^k(n-1)+1})]^{(k-1)} \end{aligned} \tag{50}$$

for all $x_i \in \mathcal{N}$.

Then $\mathcal{N}^k = (\mathcal{N}, [\cdot, \dots, \cdot]^{(k)}, \alpha^{2^k}, B, \beta')$, *where* $\beta' = \beta\alpha^{2^{k-1}}$, *is a Hom-quadratic multiplicative* $(2^k(n-1)+1)$*-ary Hom-Nambu algebra.*

6.6 Quadratic Hom-Nambu Algebras of Lower Arities

The purpose of this section is to observe that, under suitable assumptions, a quadratic n-ary Hom-Nambu algebra with $n \geq 3$ reduces to a quadratic $(n-1)$-ary Hom-Nambu algebra. We use the construction given in [40].

Theorem 6.6 *Let* $n \geq 3$ *and* $(\mathcal{N}, [\cdot, \dots, \cdot], \alpha = (\alpha_1, \dots, \alpha_{n-1}), B, \beta)$ *be a Hom-quadratic* n*-ary Hom-Nambu algebra. Suppose* $a \in \mathcal{N}$ *satisfies*

$$\alpha_1(a) = a \ \text{ and } \ [a, x_1, \dots, x_{n-2}, a] = 0 \ \text{ for all } \ x_i \in \mathcal{N}.$$

Then $\mathcal{N}_a = (\mathcal{N}, [\cdot, \dots, \cdot]_a, \alpha_a = (\alpha_2, \dots, \alpha_{n-1}), B, \beta)$, *where*

$$[x_1, \dots, x_{n-1}]_a = [a, x_1, \dots, x_{n-1}] \ \text{ for all } \ x_i \in \mathcal{N},$$

is a Hom-quadratic $(n-1)$*-ary Hom-Nambu algebra.*

Proof Using [40], $\mathcal{N}_a = (\mathcal{N}, [\cdot, \dots, \cdot]_a, \alpha_a = (\alpha_2, \dots, \alpha_{n-1}))$ is an $(n-1)$-ary Hom-Nambu algebra.

Let $x_1, \dots, x_{n-2}, y_1, y_2 \in \mathcal{N}$, then

$$\begin{aligned} B([x_1, \dots, x_{n-2}, y_1]_a, \beta(y_2)) &= B([a, x_1, \dots, x_{n-2}, y_1], \beta(y_2)) \\ &= -B(\beta(y_1), [a, x_1, \dots, x_{n-2}, y_2]) \\ &= -B(\beta(y_1), [x_1, \dots, x_{n-2}, y_2]_a). \end{aligned}$$

Hence

$$B([x_1, \dots, x_{n-2}, y_1]_a, \beta(y_2)) + B(\beta(y_1), [x_1, \dots, x_{n-2}, y_2]_a) = 0.$$

Corollary 6.2 *Let* $(\mathcal{N}, [\cdot, \dots, \cdot], \alpha = (\alpha_1, \dots, \alpha_{n-1}), B, \beta)$ *be a Hom-quadratic* n*-ary Hom-Nambu algebra, with* $n \geq 3$. *Suppose for some* $k \in \{1, \dots, n-2\}$ *there exist* $a_i \in L$ *for* $1 \leq i \leq k$ *satisfying*

$$\alpha_i(a_i) = a_i \ \text{ for } \ 1 \leq i \leq k$$

and

$$[a_1, \dots, a_j, x_{j+1}, \dots, x_{n-1}, a_j] = 0 \ \text{ for } \ 1 \leq j \leq k \ \text{ and all } \ x_l \in \mathcal{N}.$$

Then $\mathcal{N}_k = (\mathcal{N}, [\cdot, \dots, \cdot]_k, \alpha_k = (\alpha_{k+1}, \dots, \alpha_{n-1}), B, \beta)$, *where*

$$[x_1, \ldots, x_{n-1}]_k = [a_1, \ldots, a_k, x_k, \ldots, x_{n-1}] \ \ \textit{for all} \ \ x_i \in \mathcal{N}$$

is a Hom-quadratic $(n-k)$-ary Hom-Nambu algebra.

6.7 Ternary Nambu Algebras Arising from the Real Faulkner Construction

Let $(\mathfrak{g}, [\cdot, \cdot], B)$ be a real finite-dimensional quadratic Lie algebra and $\mathfrak{g}^*$ be the dual of $\mathfrak{g}$. We denote by $\langle -, - \rangle$ the dual pairing between $\mathfrak{g}$ and $\mathfrak{g}^*$.

For all $x \in \mathfrak{g}$ and $f \in \mathfrak{g}^*$ we define an element $\phi(x \otimes f) \in \mathfrak{g}$ by

$$B(y, \phi(x \otimes f)) = \langle [y, x], f \rangle = f([y, x]) \text{ for all } y \in \mathfrak{g}. \tag{51}$$

Extending ϕ linearly, defines a $\mathfrak{g}$-equivariant map $\phi : \mathfrak{g} \otimes \mathfrak{g}^* \longrightarrow \mathfrak{g}$, which is surjective. To lighten the notation we will write $\phi(x, f)$ for $\phi(x \otimes f)$ in the sequel. The $\mathfrak{g}$-equivariance of ϕ is equivalent to

$$[\phi(x, f), \phi(y, g)] = \phi([\phi(x, f), y], g) + \phi(y, \phi(x, f) \cdot g), \tag{52}$$

for all $x, y \in \mathfrak{g}$ and $f, g \in \mathfrak{g}^*$, where $\phi(x, f) \cdot g$ is defined by

$$\langle y, \phi(x, f) \cdot g \rangle = -\langle [y, \phi(x, f)], g \rangle, \text{ for all } y \in \mathfrak{g}. \tag{53}$$

The fundamental identity (52) suggests defining a bracket on $\mathfrak{g} \otimes \mathfrak{g}^*$ by

$$[x \otimes f, y \otimes g] = [\phi(x, f), y] \otimes g + y \otimes \phi(x, f) \cdot g. \tag{54}$$

Proposition 6.4 *([19]). The bracket* (54) *turns $\mathfrak{g} \otimes \mathfrak{g}^*$ into a Leibniz algebra.*

Proposition 6.5 *Let $\alpha \in Aut_S(B, \mathfrak{g})$ be an involution, then $(\mathfrak{g}\otimes\mathfrak{g}^*, [.,.]_\Omega, \Omega, B_\Omega)$, where*

$$\begin{aligned} \Omega(x \otimes f) &= \alpha(x) \otimes f \circ \alpha, && (55)\\ [x \otimes f, y \otimes g]_\Omega &= \Omega \circ [x \otimes f, y \otimes g], && (56)\\ B_\Omega(x \otimes f, y \otimes g) &= \langle \alpha(x), g \rangle \langle \alpha(y), f \rangle, && (57) \end{aligned}$$

is a multiplicative quadratic Hom-Leibniz algebra.

Proof Let $x, y, z \in \mathcal{N}$, $f, g, h \in \mathcal{N}^*$. Using (51) and (53), we have

$$\begin{aligned}B(y,\alpha(\phi(x\otimes f))) &= B(\alpha(y),\phi(x\otimes f))\\ &= \langle[\alpha(y),x],f\rangle\\ &= \langle\alpha([y,\alpha(x)]),f\rangle\\ &= \langle[y,\alpha(x)],f\circ\alpha\rangle\\ &= B(y,\phi(\alpha(x)\otimes f\circ\alpha)),\end{aligned}$$

and

$$\begin{aligned}\langle y,(\phi(x,f)\cdot g)\circ\alpha\rangle &= \langle\alpha(y),\phi(x,f)\cdot g\rangle\\ &= -\langle[\alpha(y),\phi(x,f)],g\rangle\\ &= -\langle\alpha([y,\alpha(\phi(x,f))]),g\rangle\\ &= -\langle[y,\alpha(\phi(x,f))],g\circ\alpha\rangle\\ &= -\langle[y,\phi(\alpha(x),f\circ\alpha)],g\circ\alpha\rangle\\ &= \langle y,\phi(\alpha(x),f\circ\alpha)\cdot(g\circ\alpha)\rangle.\end{aligned}$$

Thus, we obtain the following identity

$$\begin{aligned}&\alpha(\phi(x\otimes f)) = \phi(\alpha(x)\otimes f\circ\alpha),\\ &(\phi(x,f)\cdot g)\circ\alpha = \phi(\alpha(x),f\circ\alpha)\cdot(g\circ\alpha).\end{aligned}$$

Therefore

$$\begin{aligned}&\Omega([x\otimes f,y\otimes g]) = \alpha([\phi(x,f),y])\otimes g\circ\alpha+\alpha(y)\otimes(\phi(x,f)\cdot g)\circ\alpha\\ &= [\alpha(\phi(x,f)),\alpha(y)]\otimes g\circ\alpha+\alpha(y)\otimes(\phi(x,f)\cdot g)\circ\alpha\\ &= [\phi(\alpha(x)\otimes f\circ\alpha),\alpha(y)]\otimes g\circ\alpha+\alpha(y)\otimes\phi(\alpha(x),f\circ\alpha)\cdot(g\circ\alpha)\\ &= [\alpha(x)\otimes f\circ\alpha,\alpha(y)\otimes g\circ\alpha]\\ &= [\Omega(x\otimes f),\Omega(y\otimes g)].\end{aligned}$$

Thus, $\Omega([x\otimes f,y\otimes g]) = [\Omega(x\otimes f),\Omega(y\otimes g)]$. Then using Theorem 2.1, $(\mathfrak{g}\otimes\mathfrak{g}^*,[.,.]_\Omega,\Omega)$ is a multiplicative Hom-Leibniz algebra.

Since α is an involution, then

$$\begin{aligned}&B_\Omega([x\otimes f,y\otimes g]_\Omega,z\otimes h)\\ &= B_\Omega(\alpha([\phi(x,f),y])\otimes g\circ\alpha,z\otimes h)+B_\Omega(\alpha(y)\otimes(\phi(x,f)\cdot g)\circ\alpha,z\otimes h)\\ &= \langle[\phi(x,f),y],h\rangle\langle\alpha(z),g\circ\alpha\rangle+\langle y,h\rangle\langle\alpha(z),(\phi(x,f)\cdot g)\circ\alpha\rangle\\ &= \langle[\phi(x,f),y],h\rangle\langle z,g\rangle+\langle y,h\rangle\langle z,\phi(x,f)\cdot g\rangle\\ &= \langle[\phi(x,f),y],h\rangle\langle z,g\rangle-\langle y,h\rangle\langle[z,\phi(x,f)],g\rangle\\ &= -(\langle[z,\phi(x,f)],g\rangle\langle y,h\rangle-\langle z,g\rangle\langle[\phi(x,f),y],h\rangle)\\ &= -B_\Omega(y\otimes g,[x\otimes f,z\otimes h]_\Omega).\end{aligned}$$

Finally, the bilinear form B_Ω is symmetric nondegenerate and invariant. Then $(\mathfrak{g} \otimes \mathfrak{g}^*, [.,.]_\Omega, \Omega, B_\Omega)$ is a multiplicative quadratic Hom-Leibniz algebra.

The inner product on $\mathfrak{g}$ sets up an isomorphism $\flat : \mathfrak{g} \to \mathfrak{g}^*$ of $\mathfrak{g}$-modules, defined by $x^* = \flat(x) = B(x, \cdot)$.

The map ϕ defined by equation (51) induces a map $T : \mathfrak{g}\otimes\mathfrak{g} \to \mathfrak{g}$, by $T(x\otimes y) = \phi(x \otimes y^*)$. In other words, for all $x, y, z \in \mathfrak{g}$, we have

$$B(T(x \otimes y), z) = B([z, x], y),$$

whence

$$T(x \otimes y) = -T(y \otimes x).$$

This means that T factors through a map also denoted $T : \wedge^2\mathfrak{g} \to \mathfrak{g}$.

Using T we can define a ternary bracket on $\mathfrak{g}$ by

$$[x, y, z] := [T(x \otimes y), z] \tag{58}$$

and $(\mathfrak{g}, [., ., .], B)$ is a quadratic ternary Nambu algebra.

Proposition 6.6 *Let $(\mathfrak{g}, [\cdot,\cdot], B)$ be a real finite-dimensional quadratic Lie algebra and $\alpha \in Aut_S(B, \mathfrak{g})$ be an involution. Then $(\mathfrak{g}, \alpha \circ [\cdot,\cdot,\cdot], (\alpha, \alpha), B_\alpha)$, where the bracket is defined in 58 and $B_\alpha(x, y) = B(\alpha(x), y)$, is a quadratic multiplicative ternary Hom-Nambu algebra.*

7 Centroids, Derivations and Quadratic n-ary Hom-Nambu Algebras

In this section, we first generalize to n-ary Hom-Nambu algebras the notion of centroid and its properties given in [12]. We also generalize to Hom setting the connections between centroid elements and derivations. Finally we construct quadratic n-ary Hom-Nambu algebras involving elements of the centroid of n-ary Nambu algebras.

7.1 Centroids of n-ary Hom-Nambu Algebras

Definition 7.1 Let $(\mathcal{N}, [\cdot, \dots, \cdot], \alpha)$ be a multiplicative n-ary Hom-Nambu algebra and $End(\mathcal{N})$ be the endomorphism algebra of $\mathcal{N}$. Then the following subalgebra of $End(\mathcal{N})$

$$Cent(\mathcal{N}) = \{\theta \in End(\mathcal{N}) : \theta[x_1, \dots, x_n] = [\theta x_1, \dots, x_n],\ \forall x_i \in \mathcal{N}\} \tag{59}$$

is said to be the centroid of the n-ary Hom-Nambu algebra.

The definition is the same for classical case of n-ary Nambu algebra. We may also consider the same definition for any n-ary Hom-Nambu algebra.

Now, let $(\mathcal{N}, [\cdot, \ldots, \cdot], \alpha)$ be a multiplicative n-ary Hom-Nambu algebra. We denote by α^k, where $\alpha \in End(\mathcal{N})$, the k-times composition of α. We set in particular $\alpha^{-1} = 0$ and $\alpha^0 = Id$.

Definition 7.2 An α^k-centroid of a multiplicative n-ary Hom-Nambu algebra $(\mathcal{N}, [\cdot, \ldots, \cdot], \alpha)$ is a subalgebra of $End(\mathcal{N})$ denoted $Cent_{\alpha^k}(\mathcal{N})$, given by

$$Cent_{\alpha^k}(\mathcal{N}) = \{\theta \in End(\mathcal{N}) : \theta[x_1, \ldots, x_n] = [\theta x_1, \alpha^k(x_2) \ldots, \alpha^k(x_n)], \ \forall x_i \in \mathcal{N}. \tag{60}$$

We recover the definition of the centroid when $k = 0$.

If $\mathcal{N}$ is a multiplicative n-ary Hom-Nambu-Lie algebra, then it is a simple fact that

$$\theta[x_1, \ldots, x_n] = [\alpha^k(x_1), \ldots, \theta x_p, \ldots, \alpha^k(x_n)], \ \ \forall\, p \in \{1, \ldots, n\}.$$

Lemma 7.1 *Let $(\mathcal{N}, [\cdot, \ldots, \cdot])$ be an n-ary Nambu-Lie algebra. If $\theta \in Cent(\mathcal{N})$, then for $x_1, \ldots, x_n \in \mathcal{N}$*

1. $[\theta^{p_1} x_1, \ldots, \theta^{p_n} x_n] = \theta^{p_1+\cdots+p_n}[x_1, \ldots, x_n]$, $\forall p_1, \ldots, p_n \in \mathbb{N}$,
2. $[\theta^{p_1} x_1, \ldots, \theta^{p_n} x_n] = Sgn(\sigma)[\theta^{p_1} x_{\sigma(1)}, \ldots, \theta^{p_n} x_{\sigma(n)}]$, $\forall p_1, \ldots, p_n \in \mathbb{N}$ *and* $\forall \sigma \in \mathcal{S}_n$.

Proof Let $\theta \in Cent(\mathcal{N})$, $x_1, \ldots, x_n \in \mathcal{N}$ and $1 \le p \le n$, we have

$$[\theta^p x_1, \ldots, x_n] = \theta[\theta^{p-1} x_1, \ldots, x_n] = \cdots = \theta^p[x_1, \ldots, x_n].$$

Also, observe that for any $k \in \{1, \ldots, n\}$

$$\begin{aligned}&[x_1, \ldots, \theta^p x_k, \ldots, x_n] = -[\theta^p x_k, x_2, \ldots, x_1, \ldots, x_n]\\ &= -\theta^p[x_k, x_2, \ldots, x_1, \ldots, x_n] = \theta^p[x_1, \ldots, x_k, \ldots, x_n].\end{aligned}$$

Then, similarly we have

$$[\theta^{p_1} x_1, \ldots, \theta^{p_n} x_n] = \theta^{p_n}[\theta^{p_1} x_1, \ldots, \theta^{p_{n-1}} x_{n-1}, x_n] = \cdots = \theta^{p_1+\cdots+p_n}[x_1, \ldots, x_n].$$

The second assertion is a consequence of previous calculations and the skew-symmetry of $[\cdot, \ldots, \cdot]$.

Proposition 7.1 *Let $(\mathcal{N}, [\cdot, \ldots, \cdot])$ be an n-ary Nambu-Lie algebra and $\theta \in Cent(\mathcal{N})$.*

Let us fix p and set for any $x_1, \ldots, x_n \in \mathcal{N}$

$$\{x_1, \ldots, x_n\}_p = [\theta x_1, \ldots, \theta x_{p-1}, \theta x_p, x_{p+1}, \ldots, x_n]. \tag{61}$$

Then $(\mathcal{N}, \{\cdot, \dots, \cdot\}_p, \widetilde{\theta} = (\theta, \dots, \theta))$ *is an n-ary Hom-Nambu-Lie algebra.*

Proof For $\theta \in Cent(\mathcal{N})$ and $p \in \{1, \dots, n\}$, we have

$$\begin{aligned}\{\theta x_1, \dots, \theta x_{n-1}, \{y_1, \dots, y_n\}_p\}_p &= [\theta^2 x_1, \dots, \theta^2 x_p, \dots, \theta x_{n-1}, [\theta y_1, \dots, \theta y_p, \dots, y_n]]\\ &= [\theta^2 x_1, \dots, \theta^2 x_p, \dots, \theta x_{n-1}, \theta^p [y_1, \dots, y_n]]\\ &= \theta^{2p+n-1}([x_1, \dots, x_{n-1}, [y_1, \dots, y_n]]).\end{aligned}$$

In the other hand we have

$$\begin{aligned}&\sum_{k=0}^{n} \{\theta y_1, \dots, \{x_1, \dots, x_{n-1}, y_k\}_p, \dots, \theta y_n\}_p\\ &= \sum_{k=0}^{p} \{\theta y_1, \dots, \{x_1, \dots, x_{n-1}, y_k\}_p, \dots, \theta y_n\}_p\\ &\quad + \sum_{k=p}^{n} \{\theta y_1, \dots, \{x_1, \dots, x_{n-1}, y_k\}_p, \dots, \theta y_n\}_p\\ &= \sum_{k=0}^{p} [\theta^2 y_1, \dots, \theta[\theta x_1, \dots, \theta x_p, \dots, x_{n-1}, y_k], \dots, \theta^2 y_p, \dots, \theta y_n]\\ &\quad + \sum_{k=0}^{p} [\theta^2 y_1, \dots, \theta^2 y_p, \dots, [\theta x_1, \dots, \theta x_p, \dots, x_{n-1}, y_k], \dots, \theta y_n]\\ &= \sum_{k=0}^{p} \theta^{2p+n-1} [y_1, \dots, [x_1, \dots, x_{n-1}, y_k], \dots, y_n]\\ &\quad + \sum_{k=0}^{p} \theta^{2p+n-1} [2y_1, \dots, [x_1, \dots, x_{n-1}, y_k], \dots, y_n]\\ &= \theta^{2p+n-1}(\sum_{k=0}^{n} [y_1, \dots, [x_1, \dots, x_{n-1}, y_k], \dots, y_n]).\end{aligned}$$

Therefore the Hom-Nambu identity with respect to the bracket $[\cdot, \dots, \cdot]$ leads to the Hom-Nambu identity for $\{\cdot, \dots, \cdot\}_l$. The skew-symmetry is proved by second assertion of Lemma 7.1.

7.2 Centroids and Derivations of n-ary Hom-Nambu Algebras

Let $(\mathcal{N}, [\cdot, \dots, \cdot], \alpha)$ be a multiplicative n-ary Hom-Nambu-Lie algebra.

Definition 7.3 For any $k \geq 1$, we call $D \in End(\mathcal{N})$ an *α^k-derivation* of the multiplicative n-ary Hom-Nambu-Lie $(\mathcal{N}, [\cdot, \dots, \cdot], \alpha)$ if D and α commute and we have

$$D[x_1,\dots,x_n]=\sum_{i=1}^{n}[\alpha^k(x_1),\dots,\alpha^k(x_{i-1}),D(x_i),\alpha^k(x_{i+1}),\dots,\alpha^k(x_n)]. \quad (62)$$

We denote by $Der_{\alpha^k}(\mathcal{N})$ the set of α^k-derivations.

For $x=(x_1,\dots,x_{n-1})\in\mathcal{N}^{n-1}$ satisfying $\alpha(x)=x$ and $k\geq 1$, we define the map $ad_k(x)\in End(\mathcal{N})$ by

$$ad_k(x)(y)=[x_1,\dots,x_{n-1},\alpha^k(y)]\ \forall y\in\mathcal{N}. \quad (63)$$

The map $ad_k(x)$ is an α^{k+1}-derivation, that we call inner α^{k+1}-derivation. We denote by $Inn_{\alpha^k}(\mathcal{N})$ the space generated by all inner α^{k+1}-derivations.

Set $Der(\mathcal{N})=\bigoplus_{k\geq -1}Der_{\alpha^k}(\mathcal{N})$ and $Inn(\mathcal{N})=\bigoplus_{k\geq -1}Inn_{\alpha^k}(\mathcal{N})$.

Lemma 7.2 *For $D\in Der_{\alpha^k}(\mathcal{N})$ and $D'\in Der_{\alpha^{k'}}(\mathcal{N})$, where $k+k'\geq -1$, we have $[D,D']\in Der_{\alpha^{k+k'}}(\mathcal{N})$, where the commutator $[D,D']$ is defined as usual.*

Now, we define a linear map $\varsigma : Der_{\alpha^k}(\mathcal{N})\to Der_{\alpha^{k+1}}(\mathcal{N})$ by $\varsigma(D)=\alpha\circ D$. Since the elements of $Der_{\alpha^k}(\mathcal{N})$ and α commute then ς is in the centroid of the Lie algebra $(Der(\mathcal{N}),[\cdot,\cdot])$.

Hence, using Proposition 7.1 we have

Proposition 7.2 *Let $(\mathcal{N},[\cdot,\dots,\cdot],\alpha)$ be a multiplicative n-ary Hom-Nambu-Lie algebra. The triple $(Der(\mathcal{N}),[\cdot,\cdot]_\varsigma,\varsigma)$, where the bracket is defined by $[\cdot,\cdot]_\varsigma=\varsigma\circ[\cdot,\cdot]$, is a Hom-Lie algebra.*

Proposition 7.3 *Let $(\mathcal{N},[\cdot,\dots,\cdot],\alpha)$ be a multiplicative n-ary Hom-Nambu-Lie algebra. If $D\in Der_{\alpha^k}(\mathcal{N})$ and $\theta\in Cent_{\alpha^{k'}}(\mathcal{N})$, then $\theta D\in Der_{\alpha^{k+k'}}(\mathcal{N})$.*

Proof Let $x_1,\dots,x_n\in\mathcal{N}$ then

$$\begin{aligned}\theta D([x_1,\dots,x_n])&=\sum_{i=1}^{n}\theta[\alpha^k(x_1),\dots,D(x_i),\dots,\alpha^k(x_n)]\\&=\sum_{i=1}^{n}[\alpha^{k+k'}(x_1),\dots,\theta D(x_i),\dots,\alpha^{k+k'}(x_n)].\end{aligned}$$

Thus θD is an α^k-derivation.

Now we define the notion of central derivation. Let $(\mathcal{N},[\cdot,\dots,\cdot],\alpha)$ be a multiplicative n-ary Hom-Nambu-Lie algebra. We set $Z(\mathcal{N})=\{x\in\mathcal{N}:[x,y_1,\dots,y_{n-1}]=0,\ \forall\, y_1,\dots,y_{n-1}\in\mathcal{N}\}$, the center of the n-ary Hom-Nambu-Lie algebra.

Definition 7.4 Let $\varphi \in End(\mathcal{N})$, then φ is said to be a central derivation if $\varphi(\mathcal{N}) \subset Z(\mathcal{N})$ and $\varphi([\mathcal{N}, \dots, \mathcal{N}]) = 0$.

The set of all central derivations of $\mathcal{N}$ is denoted by $C(\mathcal{N})$.

Notice that an α^k-derivation φ is a central derivation if $\varphi(\mathcal{N}) \subset Z(\mathcal{N})$.

Theorem 7.1 *Let $(\mathcal{N}, [\cdot, \dots, \cdot], \alpha)$ be a multiplicative n-ary Hom-Nambu-Lie algebra. Let D in $Der_{\alpha^k}(\mathcal{N})$ and θ in $Cent(\mathcal{N})$ such that $[\theta, \alpha] = 0$, then we have*

1. *$[D, \theta]$ is in the α^k-centroid of $\mathcal{N}$,*
2. *if $[D, \theta]$ is a central derivation then $D\theta$ is an α^k-derivation of $\mathcal{N}$.*

Proof (1) Let $D \in Der_{\alpha^k}(\mathcal{N})$, $\theta \in Cent(\mathcal{N})$ and $x_1, \dots, x_n \in \mathcal{N}$ we have

$$\begin{aligned} D\theta([x_1, \dots, x_n]) &= D([\theta x_1, \dots, x_n]) \\ &= [D\theta x_1, \dots, \alpha^k(x_n)] + \sum_{i=2}^{n} [\alpha^k(\theta x_1), \dots, D(x_i), \dots, \alpha^k(x_n)] \\ &= [D\theta x_1, \dots, \alpha^k(x_n)] + \sum_{i=2}^{n} [\theta\alpha^k(x_1), \dots, D(x_i), \dots, \alpha^k(x_n)] \\ &= [D\theta x_1, \dots, \alpha^k(x_n)] + \sum_{i=2}^{n} [\alpha^k(x_1), \dots, \theta D(x_i), \dots, \alpha^k(x_n)] \\ &= [D\theta x_1, \dots, \alpha^k(x_n)] + \theta D([x_1, \dots, x_n]) - [\theta D x_1, \dots, \alpha^k(x_n)]. \end{aligned}$$

Then

$$(D\theta - \theta D)([x_1, \dots, x_n]) = [(D\theta - \theta D)x_1, \alpha^k(x_2), \dots, \alpha^k(x_n)].$$

That is, $[D, \theta] = D\theta - \theta D \in Cent_k(\mathcal{N})$.

(2) Using Proposition 6.3, θD is an α^k-derivation and since $[D, \theta]$ is a α^k-derivation, then $D\theta = [D, \theta] + \theta D$ is also an α^k-derivation.

Let A be a $\mathbb{K}$-vector space, μ be an n-linear map on A and η be a linear map on A. Let (A, μ, η) be a multiplicative symmetric n-ary totally Hom-associative algebra. The η^k-centroid $Cent_{\eta^k}(A)$ of A is defined by

$$Cent_{\eta^k}(A) = \{f \in End(A) : f(\mu(a_1, \dots, a_n)) = \mu(f(a_1), \eta^k(a_2), \dots, \eta^k(a_n))\},$$

for all $a_i \in A$ and $i \in \{1, \dots, n\}$. The set of η^k-derivation, $Der_{\eta^k}(A)$, is a subset of $End(A)$ defined by $\varphi \in End(A)$ such that

$$\varphi(\mu(a_1, \dots, a_n)) = \sum_{i=1}^{n} \mu(\eta^k(a_1) \dots, \eta^k(a_{i-1}), \varphi(a_i), \eta^k(a_{i+1}), \dots, \eta^k(a_n))\},$$

for all $a_i \in A$.

Theorem 7.2 *Let (A, μ, η) be a multiplicative symmetric n-ary Hom-associative algebra and $(\mathcal{N}, [\cdot, \dots, \cdot], \alpha)$ be a multiplicative n-ary Hom-Nambu-Lie algebra, then we have the following assertion*

- *If $f \in Cent_{\eta^k}(A)$ and $\theta \in Cent_{\alpha^k}(\mathcal{N})$, then $f \otimes \theta$ is in the ζ^k-centroid, where $\zeta^k = \eta^k \otimes \alpha^k$, of the Hom-Nambu-Lie algebra $A \otimes \mathcal{N}$ defined in 2.3.*
- *If $f \in Cent_{\eta^k}(A)$ and $D \in Der_{\alpha^k}(\mathcal{N})$, then $f \otimes D$ is a ζ^k-derivation of the Hom-Nambu-Lie algebra $A \otimes \mathcal{N}$.*

Proof Let $a_i \in A$, $x_i \in \mathcal{N}$ where $i \in \{1, \dots, n\}$ and f be a η^k-centroid on A.
• If $\theta \in Cent_{\alpha^k}(\mathcal{N})$, then

$$\begin{aligned}(f \otimes \theta)([a_1 \otimes x_1, \dots, a_n \otimes x_n]) &= (f \otimes \theta)(\mu(a_1, \dots, a_n) \otimes [x_1, \dots, x_n]_{\mathcal{N}})\\ &= \mu(f(a_1), \eta^k(a_2), \dots, \eta^k(a_n)) \otimes [\theta(x_1), \alpha^k(x_2), \dots, \alpha^k(x_n)]_{\mathcal{N}}\\ &= [(f \otimes \theta)(a_1 \otimes x_1), \zeta^k(a_2 \otimes x_2), \dots, \zeta^k(a_n \otimes x_n)].\end{aligned}$$

Thus $f \otimes \theta$ is in the ζ^k-centroid of $A \otimes \mathcal{N}$.
• If $D \in Der_{\alpha^k}(\mathcal{N})$, then

$$\begin{aligned}(f \otimes D)([a_1 \otimes x_1, \dots, a_n \otimes x_n]) &= f \otimes D((a_1 \cdot \dots \cdot a_n) \otimes [x_1, \dots, x_n])\\ &= \mu(f(a_1), \eta^k(a_2), \dots, \eta^k(a_n)) \otimes \sum_{i=1}^{n} [\alpha^k(x_1), \dots, D(x_i), \dots, \alpha^k(x_n)]\\ &= \sum_{i=1}^{n} \mu(\eta^k(a_1), \dots, f(a_i), \dots, \eta^k(a_n)) \otimes [\alpha^k(x_1), \dots, D(x_i), \dots, \alpha^k(x_n)]_{\mathcal{N}}\\ &= \sum_{i=1}^{n} [\zeta^k(a_1 \otimes x_1), \dots, (f \otimes D)(a_i \otimes x_i), \dots, \zeta^k(a_n \otimes x_n)].\end{aligned}$$

Therefore $f \otimes D$ is a ζ^k-derivation of $A \otimes \mathcal{N}$.

7.3 Centroids and Quadratic n-ary Hom-Nambu Algebras

Let $\theta \in Cent(\mathcal{N})$ such that θ is invertible and symmetric with respect to B (i.e. $B(\theta x, y) = B(x, \theta y)$). We set

$$Cent_S(\mathcal{N}) = \{\theta \in Cent(\mathcal{N}) : \theta \text{ symmetric with respect to } B\}.$$

Theorem 7.3 *Let $(\mathcal{N}, [\cdot, \dots, \cdot], B)$ be a quadratic n-ary Nambu-Lie algebra and $\theta \in Cent_S(\mathcal{N})$ such that θ is invertible. We consider a bilinear form B_θ defined by*

$$\begin{aligned} B_\theta : \mathcal{N} \times \mathcal{N} &\longrightarrow \mathbb{K} \\ (x, y) &\longmapsto B(\theta x, y). \end{aligned}$$

Then, $(\mathcal{N}, \{\cdot, \ldots, \cdot\}_l, (\theta, \ldots, \theta), B_\theta)$ *is a quadratic n-ary Hom-Nambu-Lie algebra.*

Proof It easy to proof that B_θ is symmetric and nondegenerate.

We have also θ is symmetric with respect to B_θ, indeed

$$B_\theta(\theta x, y) = B(\theta^2 x, y) = B(\theta x, \theta y) = B_\theta(x, \theta y).$$

The invariance of B_θ is given by, set $l \in \{1, \ldots, n-1\}$

$$\begin{aligned} B_\theta(\{x_1, \ldots, x_{n-1}, y\}_l, z) &= B_\theta([\theta x_1, \ldots, \theta x_l, \ldots, x_{n-1}, y], z) \\ &= B(\theta[\theta x_1, \ldots, \theta x_l, \ldots, x_{n-1}, y], z) \\ &= B([\theta^2 x_1, \ldots, \theta x_l, \ldots, x_{n-1}, y], z) \\ &= -B(y, [\theta^2 x_1, \ldots, \theta x_l, \ldots, x_{n-1}, z]) \\ &= -B_\theta(y, [\theta x_1, \ldots, \theta x_l, \ldots, x_{n-1}, z]) \\ &= -B_\theta(y, \{x_1, \ldots, x_{n-1}, z\}_l). \end{aligned}$$

In the other hand, when $l = n$ we have $[\theta x_1, \ldots, \theta x_n] = [\theta^2 x_1, \ldots, \theta x_{n-1}, x_n]$ then it's a consequence of the previous calculations.

Therefore $(\mathcal{N}, \{\cdot, \ldots, \cdot\}_l, (\theta, \ldots, \theta), B_\theta)$ is a quadratic n-ary Hom-Nambu-Lie algebra.

Notice that B_θ is also an invariant scalar product of the n-ary Nambu-Lie algebra $\mathcal{N}$.

References

1. Ammar, F., Mabrouk, S., Makhlouf, A.: Representation and cohomology of multiplicative n-ary Hom-Nambu-Lie algebras. J. Geom. Phys. (2011). doi:10.1016/j.geomphys.2011.04.022
2. Ammar, F., Makhlouf, A.: Hom-Lie algebras and Hom-Lie admissible superalgebras. J. Algebra **324**(7), 1513–1528 (2010)
3. Ammar, F., Ejbehi, Z., Makhlouf, A.: Cohomology and deformations of Hom-algebras. J. Lie Theory **21**(4), 813–836 (2011)
4. Ammar, F., Makhlouf, A., Silvestrov, S.: Ternary q-Virasoro-Witt Hom-Nambu-Lie algebras. J. Phys. A Math. Theor. **43**, 265204 (2010)
5. Ataguema, H., Makhlouf A.: Deformations of ternary algebras. J. Gen. Lie Theory Appl. **1**, 41–45 (2007)
6. Ataguema, H., Makhlouf, A.: Notes on cohomologies of ternary algebras of associative type. J. Gen. Lie Theory Appl. **3**(3), 157–174 (2009)
7. Ataguema, H., Makhlouf, A., Silvestrov, S.: Generalization of n-ary Nambu algebras and beyond. J. Math. Phys. **50**, 1 (2009)
8. Arnlind, J., Makhlouf, A., Silvestrov, S.: Ternary Hom-Nambu-Lie algebras induced by Hom-Lie algebras. J. Math. Phys. **51**, 043515 (2010)

9. Ayadi, I., Benamor, H., Benayadi, S.: Lie superalgebras with some homogeneous structures (2010). arXiv:1011.2516v1
10. Ayadi, I., Benayadi, S.: Associative superalgebras with some homogeneous structures (2010). arXiv:1011.3002v1
11. Bagger, J., Lambert, N.: Gauge symmetry and supersymmetry of multiple M2-Branes. Phys. Rev. D **77**, 065008 (2008). arXiv:0711.0955v2 [hep-th]
12. Bai, C., An, H., Li, Z.: Centroid structures of n-Lie algebras. Linear Algebra Appl. **430**, 229–240 (2009)
13. Benayadi, S., Makhlouf, A.: Hom-Lie algebras with symmetric invariant nondegenerate bilinear forms. J. Geom. and Phys., **76**, 38–60 (2014). arXiv:0113.111
14. Bordemann, M.: Nondegenerate invariant bilinear forms in nonassociative algebras. Acta Math. Univ. Comenian. LXVI **2**, 151–201 (1997)
15. Daletskii, Y.L., Takhtajan, L.A.: Leibniz and Lie structures for Nambu algebra. Lett. Math. Phys. **39**, 127–141 (1997)
16. De Medeiros, P., Figueroa-O'Farrill, J., Escobar, M.: Lorentz Lie 3-algebras and their Bagger-Lambert moduli space (2008). arXiv:0805.4363v3
17. Figueroa-O'Farrill, J.: Lorentz Lie n-algebras (2008). arXiv:0805.4760v3
18. Figueroa-O'Farrill, J.: Metric Lie n-algebras and double extensions (2008). arXiv:0806.3534v1
19. Figueroa-O'Farrill, J.: Deformations of 3-algebras (2009). arXiv:0903.4871v2
20. Filippov, V.: n-Lie algebras. Sibirsk. Mat. Zh. **26**, 126–140 (1985) (English transl.: Siberian Math. J. **26**, 879–891 (1985))
21. Gautheron, P.: Some remarks concerning Nambu mechanics. Lett. Math. Phys. **37**, 103–116 (1996)
22. Kerner, R.: Ternary algebraic structures and their applications in physics. In: The Proceedings of BTLP 23rd International Colloquium on Group Theoretical Methods in Physics (2000). arXiv:math-ph/0011023
23. Kerner, R.: Z3-graded algebras and non-commutative gauge theories. In: Oziewicz, Z., Jancewicz, B., Borowiec, A. (eds.) Spinors, Twistors, Clifford Algebras and Quantum Deformations, pp. 349–357. Kluwer Academic Publishers, Dordrecht (1993)
24. Kerner, R.: The cubic chessboard : geometry and physics. Classical Quantum Gravity. **14**, A203–A225 (1997)
25. Larsson, D., Silvestrov, S.D.: Quasi-Hom-Lie algebras, central extensions and 2-cocycle-like identities. J. Algebra **288**, 321–344 (2005)
26. Ling, W.X.: On the structure of n-Lie algebras. Ph.D. thesis, Siegen (1993)
27. Loday, J.L., Pirashvili, T.: Universal enveloping algebras of Leibniz algebras and (co)homology. Math. Ann. **296**, 139–158 (1993)
28. Makhlouf, A.: Paradigm of nonassociative Hom-algebras and Hom-superalgebras. Carmona Tapia, J., Morales Campoy, A., Peralta Pereira, A.M., Ramirez Alvarez, M.I. (eds.) Proceedings of Jordan Structures in Algebra and Analysis Meeting, pp. 145–177. Publishing House: Circulo Rojo, Almeria (2010)
29. Makhlouf, A., Silvestrov, S.D.: Hom-algebra structures. J. Gen. Lie Theory Appl. **2**(2), 51–64 (2008)
30. Makhlouf, A., Silvestrov, S.D.: Hom-Lie admissible Hom-coalgebras and Hom-Hopf algebras. In: Silvestrov, S., Paal, E., Abramov, V., Stolin, A. (eds.) Generalized Lie Theory in Mathematics, Physics and Beyond. Published as Chapter 17, pp 189–206. Springer, Berlin (2008)
31. Makhlouf, A., Silvestrov, S.D.: Notes on formal deformations of Hom-associative and Hom-Lie algebras. Forum Mathematicum **22**(4), 715–739 (2010)
32. Makhlouf, A., Silvestrov S.D.: Hom-algebras and Hom-coalgebras. J. Algebra Appl. **9** (2010) doi:10.1142/S0219498810004117
33. Medina, A., Revoy, Ph: Algèbres de Lie et produit scalaire invariant. Ann. Sci. Ecole Norm. Sup. **4**(18), 553–561 (1985)
34. Nambu, Y.: Generalized Hamiltonian mechanics. Phys. Rev. **D7**, 2405–2412 (1973)
35. Okubo, S.: Triple products and Yang-Baxter equation (I): octonionic and quaternionic triple systems. J. Math. Phys. **34**, 3273–3291 (1993)

36. Sheng, Y.: Representations of hom-Lie algebras. Algebras Representation Theory 1–18 (2011). doi:10.1007/s10468-011-9280-8
37. Takhtajan, L.: On foundation of the generalized Nambu mechanics. Comm. Math. Phys. **160**, 295–315 (1994)
38. Takhtajan, L.: Leibniz and Lie algebra structures for Nambu algebra. Lett. Math. Phys. **39**, 127–141 (1997)
39. Yau, D.: Hom-algebras and homology. J. Lie Theory **19**(2), 409–421 (2009)
40. Yau, D.: On n-ary Hom-Nambu and Hom-Nambu-Lie algebras. J. Geom. Phys. **62**, 506–522 (2012)
41. Yau, D.: on n-ary Hom-Nambu and Hom-Maltsev algebras (2010). arXiv:1004.4795v1
42. Yau, D.: A Hom-associative analogue of n-ary Hom-Nambu algebras (2010). arXiv:1005.2373v1

A Comparison of Leibniz and Lie Cohomology and Deformations

Alice Fialowski

Abstract In this talk we compare Leibniz and Lie algebra cohomology and deformations of a given Lie algebra. We get some sufficient conditions for not getting more Leibniz deformations just the Lie ones. These conditions are easy to verify. As an example, we describe the universal infinitesimal versal Leibniz deformation of the 4-dimensional diamond algebra.

1 Introduction

Leibniz algebras were introduced in [10] as a non antisymmetric version of Lie algebras. Lie algebras are special Leibniz algebras, and Pirashvili introduced [16] a spectral sequence, that, when applied to Lie algebras, measures the difference between the Lie algebra cohomology and the Leibniz cohomology. Lie algebras have deformations as Leibniz algebras and those are piloted by the adjoint Leibniz 2-cocycles. In the present talk, we focus on the second Leibniz cohomology groups $HL^2(\mathfrak{g}, \mathfrak{g})$, $HL^2(\mathfrak{g}, \mathbb{C})$ with adjoint and trivial representations of a complex Lie algebra $\mathfrak{g}$. We adopt a very elementary approach, to compare $HL^2(\mathfrak{g}, \mathfrak{g})$ and $HL^2(\mathfrak{g}, \mathbb{C})$ to $H^2(\mathfrak{g}, \mathfrak{g})$ and $H^2(\mathfrak{g}, \mathbb{C})$ respectively. In both cases, HL^2 is the direct sum of 3 spaces: $H^2 \oplus ZL_0^2 \oplus \mathscr{C}$ wher H^2 is the Lie algebra cohomology group, ZL_0^2 is the space of symmetric Leibniz 2-cocycles and $\mathscr{C}$ is a space of coupled Leibniz 2-cocycles, the nonzero elements of which have the property that their symmetric and antisymmetric parts are not Leibniz cocycles. Our comparison gives some useful practical information about the structure of Lie and Leibniz cocycles. As an example, we analyse the

---

The research was partially supported by OTKA grant K77757.

A. Fialowski (✉)
Eötvös Loránd University Budapest, Pázmány Péter sétány 1/C, Budapest H-1117, Hungary
e-mail: fialowsk@cs.elte.hu

A. Makhlouf et al. (eds.), *Algebra, Geometry and Mathematical Physics*, 233
Springer Proceedings in Mathematics & Statistics 85, DOI: 10.1007/978-3-642-55361-5_13,

4-dimensional diamond algebra which is used to construct a Wess-Zumino-Witten model. We completely describe its universal infinitesimal Leibniz and Lie deformation by computing Massey products.

The talk is based on joint work with Mandal and Magnin [5].

2 Leibniz Cohomology and Deformations

Leibniz algebras were introduced by J.-L. Loday [10, 12]. Let $\mathbb{K}$ denote a field.

Definition 2.1 A Leibniz algebra is a $\mathbb{K}$-module L, equipped with a bracket operation that satisfies the Leibniz identity:

$$[x, [y, z]] = [[x, y], z] - [[x, z], y], \quad \text{for } x, y, z \in L.$$

Any Lie algebra is automatically a Leibniz algebra, as in the presence of antisymmetry, the Jacobi identity is equivalent to the Leibniz identity. More examples of Leibniz algebras were given in [10–12], and recently Leibniz algebras are intesively studied.

Let L be a Leibniz algebra and M a representation of L. By definition, M is a $\mathbb{K}$-module equipped with two actions (left and right) of L,

$$[-,-] : L \times M \longrightarrow M \ \text{ and } [-,-] : M \times L \longrightarrow M \ \text{ such that}$$

$$[x, [y, z]] = [[x, y], z] - [[x, z], y]$$

holds, whenever one of the variables is from M and the two others from L. Define $CL^n(L; M) := \mathrm{Hom}_{\mathbb{K}}(L^{\otimes n}, M),\ n \geq 0$. Let

$$\delta^n : CL^n(L; M) \longrightarrow CL^{n+1}(L; M)$$

be a $\mathbb{K}$-homomorphism defined by

$$\begin{aligned}
&\delta^n f(x_1, \dots, x_{n+1}) \\
&\quad := [x_1, f(x_2, \dots, x_{n+1})] + \sum_{i=2}^{n+1} (-1)^i [f(x_1, \dots, \hat{x}_i, \dots, x_{n+1}), x_i] \\
&\qquad + \sum_{1 \leq i < j \leq n+1} (-1)^{j+1} f(x_1, \dots, x_{i-1}, [x_i, x_j], x_{i+1}, \dots, \hat{x}_j, \dots, x_{n+1}).
\end{aligned}$$

Then $(CL^*(L; M), \delta)$ is a cochain complex, whose cohomology is called the cohomology of the Leibniz algebra L with coefficients in the representation M. The n-th cohomology is denoted by $HL^n(L; M)$. In particular, L is a representation of itself with the obvious action given by the bracket in L. The n-th cohomology of L with coefficients in itself is denoted by $HL^n(L; L)$.

Let S_n be the symmetric group. Recall that a permutation $\sigma \in S_{p+q}$ is called a (p, q)-shuffle, if $\sigma(1) < \sigma(2) < \cdots < \sigma(p)$, and $\sigma(p+1) < \sigma(p+2) < \cdots < \sigma(p+q)$. We denote the set of all (p, q)-shuffles in S_{p+q} by $Sh(p, q)$.

For $\alpha \in CL^{p+1}(L; L)$ and $\beta \in CL^{q+1}(L; L)$, define $\alpha \circ \beta \in CL^{p+q+1}(L; L)$ by

$$\begin{aligned}\alpha \circ \beta(x_1, \ldots, x_{p+q+1}) \\ = \sum_{k=1}^{p+1}(-1)^{q(k-1)}\{ \sum_{\sigma \in Sh(q,p-k+1)} & sgn(\sigma)\alpha(x_1, \ldots, x_{k-1}, \beta(x_k, x_{\sigma(k+1)}, \ldots, \\ & x_{\sigma(k+q)}), x_{\sigma(k+q+1)}, \ldots, x_{\sigma(p+q+1)})\}.\end{aligned}$$

The graded cochain module $CL^*(L; L) = \bigoplus_r CL^r(L; L)$ equipped with the bracket defined by

$$[\alpha, \beta] = \alpha \circ \beta + (-1)^{pq+1}\beta \circ \alpha \quad \text{for } \alpha \in CL^{p+1}(L; L) \text{ and } \beta \in CL^{q+1}(L; L)$$

and the differential map d by $d\alpha = (-1)^{|\alpha|}\delta\alpha$ for $\alpha \in CL^*(L; L)$ is a differential graded Lie algebra. (Here $|\alpha|$ denotes the degree of the cochain α.)

Let now $\mathbb{K}$ a field of zero characteristic and the tensor product over $\mathbb{K}$ will be denoted by $\otimes$. We recall the notion of deformation of a Lie (Leibniz) algebra $\mathfrak{g}$ (L) over a commutative algebra with identity base A with a fixed augmentation $\varepsilon : A \to \mathbb{K}$ and maximal ideal $\mathfrak{M}$. Assume $dim(\mathfrak{M}^k/\mathfrak{M}^{k+1}) < \infty$ for every k (see [6]).

Definition 2.2 A deformation λ of a Lie algebra $\mathfrak{g}$ (or a Leibniz algebra L) with base $(A, \mathfrak{M})$, or simply with base A is an A-Lie algebra (or an A-Leibniz algebra) structure on the tensor product $A \otimes \mathfrak{g}$ (or $A \otimes L$) with thc bracket $[,]_\lambda$ such that

$$\varepsilon \otimes id : A \otimes \mathfrak{g} \to \mathbb{K} \otimes \mathfrak{g} \quad (\text{or } \varepsilon \otimes id : A \otimes L \to \mathbb{K} \otimes L)$$

is an A-Lie algebra (A-Leibniz algebra) homomorphism.

A deformation of the Lie (Leibniz) algebra $\mathfrak{g}$ (L) with base A is called *infinitesimal*, or *first order*, if in addition to this, $\mathfrak{M}^2 = 0$. We call a deformation of *order k*, if $\mathfrak{M}^{k+1} = 0$. A deformation with base is called local if A is a local algebra over $\mathbb{K}$, which means A has a unique maximal ideal.

Suppose A is a complete local algebra ($A = \varprojlim_{n\to\infty} (A/\mathfrak{M}^n)$), where $\mathfrak{M}$ is the maximal ideal in A. Then a deformation of $\mathfrak{g}$ (L) with base A which is obtained as the projective limit of deformations of $\mathfrak{g}$ (L) with base $A/\mathfrak{M}^n$ is called a *formal deformation* of $\mathfrak{g}$ (L).

Definition 2.3 Suppose λ is a given deformation of L with base $(A, \mathfrak{M})$ and augmentation $\varepsilon : A \to \mathbb{K}$. Let A' be another commutative algebra with identity and a fixed augmentation $\varepsilon' : A' \to \mathbb{K}$. Suppose $\phi : A \to A'$ is an algebra homomorphism with $\phi(1) = 1$ and $\varepsilon' \circ \phi = \varepsilon$. Let $ker(\varepsilon') = \mathfrak{M}'$. Then the push-out $\phi_*\lambda$ is the deformation of L with base $(A', \mathfrak{M}')$ and bracket

$$[a_1{}' \otimes_A (a_1 \otimes l_1), a_2' \otimes_A (a_2 \otimes l_2)]_{\phi_* \lambda} = a_1' a_2' \otimes_A [a_1 \otimes l_1, a_2 \otimes l_2]_\lambda$$

where $a_1', a_2' \in A'$, $a_1, a_2 \in A$ and $l_1, l_2 \in L$. Here A' is considered as an A-module by the map $a' \cdot a = a'\phi(a)$ so that

$$A' \otimes L = (A' \otimes_A A) \otimes L = A' \otimes_A (A \otimes L).$$

Definition 2.4 (see [2]) Let C be a complete local algebra. A formal deformation η of a Lie algebra $\mathfrak{g}$ (Leibniz algebra L) with base C is called versal, if

(i) for any formal deformation λ of $\mathfrak{g}$ (L) with base A there exists a homomorphism $f : C \to A$ such that the deformation λ is equivalent to $f_*\eta$;
(ii) if A satisfies the condition $\mathfrak{M}^2 = 0$, then f is unique.

Theorem 2.1 *If $H^2(\mathfrak{g}; \mathfrak{g})$ is finite dimensional, then there exists a of $\mathfrak{g}$ (similarly for L).*

Proof Follows from the general theorem of Schlessinger [17], like it was shown for Lie algebras in [2].

In [3] a construction for a versal deformation of a Lie algebra was given and it was generalized to Leibniz algebras in [6]. The computation of a specific Leibniz algebra example is given in [4].

3 Comparison of the Cohomology Spaces HL^2 and H^2 for a Lie Algebra

In [16] the relation between Chevalley-Eilenberg and Leibniz homology with coefficients in a right module is considered via a spectral sequence. The statements are valid in the cohomological version as well. As a corollary, one deduces

Proposition 3.1 [16] *Let $\mathfrak{g}$ be a Lie algebra over a field $\mathbb{K}$ and M be a right $\mathfrak{g}$-module. If*

$$H_*(\mathfrak{g}, M) = 0, \textit{ then } HL_*(\mathfrak{g}, M) = 0.$$

As the similar statement is true for cohomologies, it implies that rigid Lie algebras are Leibniz rigid as well.

Now we describe the Leibniz 2-cohomology spaces with the help of Lie 2-cohomology space of a Lie algebra $\mathfrak{g}$.

Recall that a symmetric bilinear form $B \in S^2\mathfrak{g}^*$ is invariant, i.e. $B \in \left(S^2\mathfrak{g}^*\right)^{\mathfrak{g}}$ if and only if $B([Z, X], Y) = -B(X, [Z, Y])$ for every $X, Y, Z \in \mathfrak{g}$. The Koszul map [9] $\mathscr{I} : \left(S^2\mathfrak{g}^*\right)^{\mathfrak{g}} \to \left(\bigwedge^3 \mathfrak{g}^*\right)^{\mathfrak{g}} \subset Z^3(\mathfrak{g}, \mathbb{C})$ is defined by $\mathscr{I}(B) = I_B$, with $I_B(X, Y, Z) = B([X, Y], Z)$ for every $X, Y, Z \in \mathfrak{g}$. Since the projection $\pi : \mathfrak{g} \to \mathfrak{g}/\mathscr{C}^2\mathfrak{g}$ induces an isomorphism

$$\varpi : \ker \mathscr{I} \to S^2\left(\mathfrak{g}/\mathscr{C}^2\mathfrak{g}\right)^*,$$

(where $\mathscr{C}^2\mathfrak{g} = [\mathfrak{g}, \mathfrak{g}]$), $\dim\left(S^2\mathfrak{g}^*\right)^{\mathfrak{g}} = \frac{p(p+1)}{2} + \dim \operatorname{Im} \mathscr{I}$, with $p = \dim H^1(\mathfrak{g}, \mathbb{C})$. For reductive $\mathfrak{g}$, $\dim\left(S^2\mathfrak{g}^*\right)^{\mathfrak{g}} = \dim H^3(\mathfrak{g}, \mathbb{C})$. Note also that the restriction of $\delta_{\mathbb{C}}$ to $\left(S^2\mathfrak{g}^*\right)^{\mathfrak{g}}$ is $-\mathscr{I}$.

Definition 3.1 $\mathfrak{g}$ is said to be $\mathscr{I}$-null (resp. $\mathscr{I}$-exact) if $\mathscr{I} = 0$ (resp. $\operatorname{Im} \mathscr{I} \subset B^3(\mathfrak{g}, \mathbb{C})$).

Example 3.1 The $(2N + 1)$-dimensional complex Heisenberg Lie algebra $\mathscr{H}_N$ $(N \geqslant 1)$ with basis $(x_i)_{1\leqslant i\leqslant 2N+1}$ and nonzero commutation relations (with anticommutativity) $[x_i, x_{N+i}] = x_{2N+1}$ $(1 \leqslant i \leqslant N)$ is $\mathscr{I}$-null, for any $B \in \left(S^2\mathscr{H}_N{}^*\right)^{\mathscr{H}_N}$, $B(x_i, x_{2N+1}) = B(x_i, [x_i, x_{N+i}]) = -B([x_i, x_i], x_{N+i}) = 0$ (similarly with x_{N+i} instead of x_i) $(1 \leqslant i \leqslant N)$, and $B(x_{2N+1}, x_{2N+1}) = B(x_{2N+1}, [x_1, x_{N+1}]) = -B([x_1, x_{2N+1}], x_{N+1}) = 0$.

If $\mathfrak{c}$ denotes the center of $\mathfrak{g}$, then $\mathfrak{c} \otimes \left(S^2\mathfrak{g}^*\right)^{\mathfrak{g}}$ is the space of invariant $\mathfrak{c}$-valued symmetric bilinear maps and we denote $F = Id \otimes \mathscr{I} : \mathfrak{c} \otimes \left(S^2\mathfrak{g}^*\right)^{\mathfrak{g}} \to C^3(\mathfrak{g}, \mathfrak{g}) = \mathfrak{g} \otimes \bigwedge^3 \mathfrak{g}^*$. Then $\operatorname{Im} F = \mathfrak{c} \otimes \operatorname{Im} \mathscr{I}$.

Theorem 3.1 *Let* $\mathfrak{g}$ *be any finite dimensional complex Lie algebra and* $ZL_0^2(\mathfrak{g}, \mathfrak{g})$ *(resp.* $ZL_0^2(\mathfrak{g}, \mathbb{C})$*) the space of symmetric adjoint (resp. trivial) Leibniz 2-cocycles.*

(i) $ZL_0^2(\mathfrak{g}, \mathfrak{g}) = \mathfrak{c} \otimes \ker \mathscr{I}$. *In particular,* $\dim ZL_0^2(\mathfrak{g}, \mathfrak{g}) = c\,\frac{p(p+1)}{2}$ *where* $c = \dim \mathfrak{c}$ *and* $p = \dim \mathfrak{g}/\mathscr{C}^2\mathfrak{g} = \dim H^1(\mathfrak{g}, \mathbb{C})$.
(ii) $ZL^2(\mathfrak{g}, \mathfrak{g})/\left(Z^2(\mathfrak{g}, \mathfrak{g}) \oplus ZL_0^2(\mathfrak{g}, \mathfrak{g})\right) \cong (\mathfrak{c} \otimes \mathit{Im}\, \mathscr{I}) \cap B^3(\mathfrak{g}, \mathfrak{g})$.
(iii) $HL^2(\mathfrak{g}, \mathfrak{g}) \cong H^2(\mathfrak{g}, \mathfrak{g}) \oplus (\mathfrak{c} \otimes \ker \mathscr{I}) \oplus \left((\mathfrak{c} \otimes \mathit{Im}\, \mathscr{I}) \cap B^3(\mathfrak{g}, \mathfrak{g})\right)$.
(iv) $ZL_0^2(\mathfrak{g}, \mathbb{C}) = \ker \mathscr{I}$.
(v) $ZL^2(\mathfrak{g}, \mathbb{C})/\left(Z^2(\mathfrak{g}, \mathbb{C}) \oplus ZL_0^2(\mathfrak{g}, \mathbb{C})\right) \cong \mathit{Im} \mathscr{I} \cap B^3(\mathfrak{g}, \mathbb{C})$.
(vi) $HL^2(\mathfrak{g}, \mathbb{C}) \cong H^2(\mathfrak{g}, \mathbb{C}) \oplus \ker \mathscr{I} \oplus \left(\mathit{Im}\, \mathscr{I} \cap B^3(\mathfrak{g}, \mathbb{C})\right)$.

Proof (i) The Leibniz 2-cochain space $CL^2(\mathfrak{g}, \mathfrak{g}) = \mathfrak{g} \otimes (\mathfrak{g}^*)^{\otimes 2}$ decomposes as $\left(\mathfrak{g} \otimes \bigwedge^2 \mathfrak{g}^*\right) \oplus \left(\mathfrak{g} \otimes S^2\, \mathfrak{g}^*\right)$ with $\mathfrak{g} \otimes S^2\, \mathfrak{g}^*$ the space of symmetric elements in $CL^2(\mathfrak{g}, \mathfrak{g})$. By definition of the Leibniz coboundary δ, one has for $\psi \in CL^2(\mathfrak{g}, \mathfrak{g})$ and $X, Y, Z \in \mathfrak{g}$

$$(\delta\psi)(X, Y, Z) = u + v + w + r + s + t \tag{1}$$

with $u = [X, \psi(Y, Z)]$, $v = [\psi(X, Z), Y]$, $w = -[\psi(X, Y), Z]$, $r = -\psi([X, Y], Z)$, $s = \psi(X, [Y, Z])$, $t = \psi([X, Z], Y)$. δ coincides with the usual coboundary operator on $\mathfrak{g}\otimes \bigwedge^2 \mathfrak{g}^*$. Now, let $\psi = \psi_1 + \psi_0 \in CL^2(\mathfrak{g}, \mathfrak{g})$, $\psi_1 \in \mathfrak{g}\otimes \bigwedge^2 \mathfrak{g}^*$, $\psi_0 \in \mathfrak{g}\otimes S^2\, \mathfrak{g}^*$.

Suppose $\psi \in ZL^2(\mathfrak{g}, \mathfrak{g}) : \delta\psi = 0 = \delta\psi_1 + \delta\psi_0 = d\psi_1 + \delta\psi_0$. Then $\delta\psi_0 = -d\psi_1 \in \mathfrak{g}\otimes \bigwedge^3 \mathfrak{g}^*$ is antisymmetric. Then permuting X and Y in formula (1) for ψ_0 yields $(\delta\psi_0)(Y, X, Z) = -v - u + w - r + t + s$. As $\delta\psi_0$ is antisymmetric, we get

$$w + s + t = 0. \quad (2)$$

Now, the circular permutation (X, Y, Z) in (1) for ψ_0 yields $(\delta\psi_0)(Y, Z, X) = -v - w + u - s - t + r$. Again, by antisymmetry,

$$v + w + s + t = 0, \quad (3)$$

i.e. $(\delta\psi_0)(X, Y, Z) = u + r$.

From (2) and (3), $v = 0$. Applying twice the circular permutation (X, Y, Z) to v, we get first $w = 0$ and then $u = 0$. Hence $(\delta\psi_0)(X, Y, Z) = r = -\psi_0([X, Y], Z)$. Note first that $u = 0$ reads $[X, \psi_0(Y, Z)] = 0$. As X, Y, Z are arbitrary, ψ_0 is $\mathfrak{c}$-valued. Now the permutation of Y and Z changes r to $-t = s$ (from (3)). Again, by antisymmetry of $\delta\psi_0$, $r = t = -s$. As X, Y, Z are arbitrary, one gets $\psi_0 \in \mathfrak{c} \otimes \left(S^2\mathfrak{g}^*\right)^{\mathfrak{g}}$. Now $F(\psi_0) = -r = -\delta\psi_0 = d\psi_1 \in B^3(\mathfrak{g}, \mathfrak{g})$. Hence

$$\psi_0 \in ZL_0^2(\mathfrak{g}, \mathfrak{g}) \Leftrightarrow F(\psi_0) = 0 \Leftrightarrow \psi_1 \in Z^2(\mathfrak{g}, \mathfrak{g}) \Leftrightarrow \psi_0 \in \mathfrak{c} \otimes \ker \mathscr{I}.$$

Consider now the linear map $\Phi : ZL^2(\mathfrak{g}, \mathfrak{g}) \to F^{-1}(B^3(\mathfrak{g}, \mathfrak{g}))/\ker F$ defined by $\psi \mapsto [\psi_0] \pmod{\ker F}$. Φ is onto: for any $[\varphi_0] \in F^{-1}(B^3(\mathfrak{g}, \mathfrak{g}))/\ker F$, $\varphi_0 \in \mathfrak{c} \otimes \left(S^2\mathfrak{g}^*\right)^{\mathfrak{g}}$, one has $F(\varphi_0) \in B^3(\mathfrak{g}, \mathfrak{g})$, hence $F(\varphi_0) = d\varphi_1$, $\varphi_1 \in C^2(\mathfrak{g}, \mathfrak{g})$, and then $\varphi = \varphi_0 + \varphi_1$ is a Leibniz cocycle such that $\Phi(\varphi) = [\varphi_0]$. Now $\ker \Phi = Z^2(\mathfrak{g}, \mathfrak{g}) \oplus ZL_0^2(\mathfrak{g}, \mathfrak{g})$, since condition $[\psi_0] = [0]$ reads $\psi_0 \in \ker F$ which is equivalent to $\psi \in Z^2(\mathfrak{g}, \mathfrak{g}) \oplus ZL_0^2(\mathfrak{g}, \mathfrak{g})$. Hence Φ yields an isomorphism $ZL^2(\mathfrak{g}, \mathfrak{g})/\left(Z^2(\mathfrak{g}, \mathfrak{g}) \oplus ZL_0^2(\mathfrak{g}, \mathfrak{g})\right) \cong F^{-1}(B^3(\mathfrak{g}, \mathfrak{g}))/\ker F$. The latter is isomorphic to $\operatorname{Im} F \cap B^3(\mathfrak{g}, \mathfrak{g}) \cong (\mathfrak{c} \otimes \operatorname{Im} \mathscr{I}) \cap B^3(\mathfrak{g}, \mathfrak{g})$.

(ii) results from the invariance of $\psi_0 \in ZL_0^2(\mathfrak{g}, \mathfrak{g})$.

(iii) results immediately from (i) and (ii) since $BL^2(\mathfrak{g}, \mathfrak{g}) = B^2(\mathfrak{g}, \mathfrak{g})$ as the Leibniz differential on $CL^1(\mathfrak{g}, \mathfrak{g}) = \mathfrak{g}^* \otimes \mathfrak{g} = C^1(\mathfrak{g}, \mathfrak{g})$ coincides with the usual one.

(iv)-(vi) are similar.

Remark 3.1 Since $\ker \mathscr{I} \oplus \left(\operatorname{Im} \mathscr{I} \cap B^3(\mathfrak{g}, \mathbb{C})\right) \cong \ker h$ where h denotes $\mathscr{I}$ composed with the projection of $Z^3(\mathfrak{g}, \mathbb{C})$ onto $H^3(\mathfrak{g}, \mathbb{C})$, the result (vi) is the same as in [13].

Remark 3.2 Any supplementary subspace to $Z^2(\mathfrak{g}, \mathbb{C}) \oplus ZL_0^2(\mathfrak{g}, \mathbb{C})$ in $ZL^2(\mathfrak{g}, \mathbb{C})$ consists of coupled Leibniz 2-cocycles, i.e. the nonzero elements have the property that their symmetric and antisymmetric parts are not cocycles. To get such a supplementary subspace, pick any supplementary subspace W to $\ker \mathscr{I}$ in $\left(S^2\mathfrak{g}^*\right)^{\mathfrak{g}}$ and take $\mathscr{C} = \left\{B + \omega; B \in W \cap \mathscr{I}^{-1}(B^3(\mathfrak{g}, \mathbb{C})), I_B = d\omega\right\}$.

Definition 3.2 $\mathfrak{g}$ is said to be an adjoint (resp. trivial) ZL^2-uncoupling if

$$(\mathfrak{c} \otimes \operatorname{Im} \mathscr{I}) \cap B^3(\mathfrak{g}, \mathfrak{g}) = \{0\} \left(\text{resp. } \operatorname{Im} \mathscr{I} \cap B^3(\mathfrak{g}, \mathbb{C}) = \{0\}\right).$$

The class of adjoint ZL^2-uncoupling Lie algebras is rather extensive since it contains all zero-center Lie algebras and all $\mathscr{I}$-null Lie algebras. For non zero-center, adjoint ZL^2-uncoupling implies trivial ZL^2-uncoupling, since $\mathfrak{c} \otimes \left(\operatorname{Im} \mathscr{I} \cap B^3(\mathfrak{g}, \mathbb{C})\right) \subset (\mathfrak{c} \otimes \operatorname{Im} \mathscr{I}) \cap B^3(\mathfrak{g}, \mathfrak{g})$. The reciprocal holds obviously true for $\mathscr{I}$-exact Lie algebras. However we do not know if it holds true in general (e.g. we do not know of a nilpotent Lie algebra which is not $\mathscr{I}$-exact).

Corollary 3.1 (i) $HL^2(\mathfrak{g}, \mathfrak{g}) \cong H^2(\mathfrak{g}, \mathfrak{g}) \oplus (\mathfrak{c} \otimes \ker \mathscr{I})$ *if and only if* $\mathfrak{g}$ *is adjoint* ZL^2*-uncoupling.*

(ii) $HL^2(\mathfrak{g}, \mathbb{C}) \cong H^2(\mathfrak{g}, \mathbb{C}) \oplus \ker \mathscr{I}$ *if and only if* $\mathfrak{g}$ *is trivial* ZL^2*-uncoupling.*

Corollary 3.2 *For any Lie algebra* $\mathfrak{g}$ *with trivial center* $\mathfrak{c} = \{0\}$, $HL^2(\mathfrak{g}, \mathfrak{g}) = H^2(\mathfrak{g}, \mathfrak{g})$.

Remark 3.3 This fact also follows from the cohomological version of Theorem A in [16].

Proof Let $\mathfrak{g}$ be a Lie algebra and M be a right $\mathfrak{g}$-module. Consider the product map $m : \mathfrak{g} \otimes \Lambda^n \mathfrak{g} \longrightarrow \Lambda^{n+1}$ in the exterior algebra. This map yields an epimorphism of chain complexes

$$C_*(\mathfrak{g}, \mathfrak{g}) \longrightarrow C_*(\mathfrak{g}, \mathbb{K})[-1],$$

where $C_*(\mathfrak{g}, \mathbb{K})$ is the reduced chain complex:

$$C_0(\mathfrak{g}, \mathbb{K}) = 0, \quad C_i(\mathfrak{g}, \mathbb{K}) = C_i(\mathfrak{g}, \mathbb{K}) \text{ for } i > 0.$$

Define the reduced chain complex $CR_*(\mathfrak{g})$ such that $CR_*(\mathfrak{g}[1])$ is the kernel of the epimorphism $C_*(\mathfrak{g}, \mathfrak{g}) \longrightarrow C_*(\mathfrak{g}, \mathbb{K})[-1]$. Denote the cohomology of $CR_*(\mathfrak{g})$ by $HR_*(\mathfrak{g})$.

Let us recall Theorem A in [16]. It states that there exists a spectral sequence

$$E^2_{pq} = HR_p(\mathfrak{g} \otimes HL_q(\mathfrak{g}, M)) \Longrightarrow H^{rel}_{p+q}(\mathfrak{g}, M).$$

As the center of our Lie algebra is 0, it follows that $E^2_{00} = 0$, and so we get $H^{rel}_0(\mathfrak{g}, \mathfrak{g}) = 0$.

But then from the exact sequence in [16]

$$0 \leftarrow H_2(\mathfrak{g}, M) \leftarrow HL_2(\mathfrak{g}, M) \leftarrow H^{rel}_0(\mathfrak{g}, M) \leftarrow H_3(\mathfrak{g}, M) \leftarrow \ldots$$

we get

$$HL_2(\mathfrak{g}, M) = H_2(\mathfrak{g}, M).$$

Corollary 3.3 *For any reductive Lie algebra* $\mathfrak{g}$ *with center* $\mathfrak{c}$, $HL^2(\mathfrak{g}, \mathfrak{g}) = H^2(\mathfrak{g}, \mathfrak{g}) \oplus \left(\mathfrak{c} \otimes S^2\mathfrak{c}^*\right)$, *and* $\dim H^2(\mathfrak{g}, \mathfrak{g}) = \frac{c^2(c-1)}{2}$ *with* $c = \dim \mathfrak{c}$.

Proof $\mathfrak{g} = \mathfrak{s} \oplus \mathfrak{c}$ with $\mathfrak{s} = \mathscr{C}^2\mathfrak{g}$ semisimple. We first prove that $\mathfrak{g}$ is adjoint ZL^2-uncoupling. We have $\mathfrak{c} \otimes \left(S^2\mathfrak{g}^*\right)^{\mathfrak{g}} = \left(\mathfrak{c} \otimes \left(S^2\mathfrak{s}^*\right)^{\mathfrak{s}}\right) \oplus \left(\mathfrak{c} \otimes S^2\mathfrak{c}^*\right) = c\,\left(S^2\mathfrak{s}^*\right)^{\mathfrak{s}} \oplus c\,\left(S^2\mathfrak{c}^*\right)$. Suppose first $\mathfrak{s}$ simple. Then any bilinear symmetric invariant form on $\mathfrak{s}$ is some multiple of the Killing form K. Hence $\mathfrak{c} \otimes \left(S^2\mathfrak{g}^*\right)^{\mathfrak{g}} = c\,(\mathbb{C}K) \oplus c\,\left(S^2\mathfrak{c}^*\right)$. For any $\psi_0 \in \mathfrak{c} \otimes \left(S^2\mathfrak{g}^*\right)^{\mathfrak{g}}$, $F(\psi_0)$ is then some linear combination of copies of I_K. It is well-known, I_K is not a coboundary. Hence if we suppose that $F(\psi_0)$ is a coboundary, necessarily $F(\psi_0) = 0$. The Lie algebra $\mathfrak{g}$ is adjoint ZL^2-uncoupling when $\mathfrak{s}$ is simple. Now, if $\mathfrak{s}$ is not simple, $\mathfrak{s}$ can be decomposed as a direct sum $\mathfrak{s}_1 \oplus \cdots \oplus \mathfrak{s}_m$ of simple ideals of $\mathfrak{s}$. Then $\left(S^2\mathfrak{s}^*\right)^{\mathfrak{s}} = \bigoplus_{i=1}^m \left(S^2\mathfrak{s}_i{}^*\right)^{\mathfrak{s}_i} = \bigoplus_{i=1}^m \mathbb{C}K_i$ (K_i Killing form of $\mathfrak{s}_i$.) The same reasoning then applies and shows that $\mathfrak{g}$ is adjoint ZL^2-uncoupling. From (ii) in Theorem 3.1, we have $ZL_0^2(\mathfrak{g}, \mathfrak{g}) = \mathfrak{c} \otimes S^2\mathfrak{c}^*$. Now, $\mathfrak{g} = \mathfrak{s} \oplus \mathfrak{c}$ with $\mathfrak{s} = \mathscr{C}^2\mathfrak{g}$ semisimple. The subalgebra $\mathfrak{s}$ can be decomposed as a direct sum $\mathfrak{s}_1 \oplus \cdots \oplus \mathfrak{s}_m$ of ideals of $\mathfrak{s}$, hence of $\mathfrak{g}$. Then $H^2(\mathfrak{g}, \mathfrak{g}) = \bigoplus_{i=1}^m H^2(\mathfrak{g}, \mathfrak{s}_i) \oplus H^2(\mathfrak{g}, \mathfrak{c})$. As $\mathfrak{s}_i$ is a nontrivial $\mathfrak{g}$-module, $H^2(\mathfrak{g}, \mathfrak{s}_i) = \{0\}$ ([8], Prop. 11.4, page 154). So we get $H^2(\mathfrak{g}, \mathfrak{g}) = H^2(\mathfrak{g}, \mathfrak{c}) = c\,H^2(\mathfrak{g}, \mathbb{C})$. By the Künneth formula and Whitehead's lemmas,

$$\begin{aligned} H^2(\mathfrak{g}, \mathbb{C}) &= \left(H^2(\mathfrak{s}, \mathbb{C}) \otimes H^0(\mathfrak{c}, \mathbb{C})\right) \oplus \left(H^1(\mathfrak{s}, \mathbb{C})\right. \\ &\quad \left.\otimes H^1(\mathfrak{c}, \mathbb{C})\right) \oplus \left(H^0(\mathfrak{s}, \mathbb{C}) \otimes H^2(\mathfrak{c}, \mathbb{C})\right) \\ &= H^0(\mathfrak{s}, \mathbb{C}) \otimes H^2(\mathfrak{c}, \mathbb{C}) \\ &= \mathbb{C} \otimes H^2(\mathfrak{c}, \mathbb{C}). \end{aligned}$$

Hence

$$\dim H^2(\mathfrak{g}, \mathfrak{g}) = \frac{c^2(c-1)}{2}.$$

4 The Diamond Algebra

The 4-dimensional complex solvable "diamond" Lie algebra $\mathfrak{d}$ has basis (x_1, x_2, x_3, x_4) and nonzero commutation relations (with anticommutativity)

$$[x_1, x_2] = x_3,\ [x_1, x_3] = -x_2,\ [x_2, x_3] = x_4. \tag{4}$$

The relations show that $\mathfrak{d}$ is an extension of the one-dimensional abelian Lie algebra $\mathbb{C}x_1$ by the Heisenberg algebra $\mathfrak{n}_3$ with basis x_2, x_3, x_4. It is also known as the Nappi-Witten Lie algebra [14] or the central extension of the Poincaré Lie algebra in two dimensions. It is a solvable quadratic Lie algebra, as it admits a nondegenerate bilinear symmetric invariant form. Because of these properties, it plays an important role in conformal field theory.

We can use $\mathfrak{d}$ to construct a Wess-Zumino-Witten model, which describes a homogeneous four-dimensional Lorentz-signature space time [14].

It is easy to check that $\mathfrak{d}$ is $\mathscr{I}$-exact. In fact, one verifies that all other solvable 4-dimensional Lie algebras are $\mathscr{I}$-null (for a list, see e.g. [15]).

Consider $\mathfrak{d}$ as Leibniz algebra with basis $\{e_1, e_2, e_3, e_4\}$ over $\mathbb{C}$. Define a bilinear map $[\ ,\ ] : \mathfrak{d} \times \mathfrak{d} \longrightarrow \mathfrak{d}$ by $[e_2, e_3] = e_1$, $[e_3, e_2] = -e_1$, $[e_2, e_4] = e_2$, $[e_4, e_2] = -e_2$, $[e_3, e_4] = e_2 - e_3$ and $[e_4, e_3] = e_3 - e_2$, all other products of basis elements being 0.

We get a basis satisfying the usual commutation relations (4) by letting

$$x_1 = ie_4, \quad x_2 = e_3, \quad x_3 = i(-e_2 + e_3), \quad x_4 = ie_1. \tag{5}$$

One should mention that even though these two forms are equivalent over $\mathbb{C}$, they represent the two nonisomorphic real forms of the complex diamond algebra.

We found that by considering Leibniz algebra deformations of $\mathfrak{d}$ one gets more structures. Indeed it gives not only extra structures but also keeps the Lie structures obtained by considering Lie algebra deformations. To get the precise deformations we need to consider the cohomology groups.

We compute cohomologies necessary for our purpose. Let us use the simpler notation L for the diamond algebra. First consider the Leibniz cohomology space $HL^2(L; L)$. Our computation consists of the following steps:

(i) determine a basis of the space of cocycles $ZL^2(L; L)$,
(ii) determine a basis of the coboundary space $BL^2(L; L)$,
(iii) determine a basis of the quotient space $HL^2(L; L)$.

(i) Let $\psi \in ZL^2(L; L)$. Then $\psi : L \otimes L \longrightarrow L$ is a linear map and $\delta\psi = 0$, where

$$\begin{aligned}\delta\psi(e_i, e_j, e_k) = [e_i, \psi(e_j, e_k)] + [\psi(e_i, e_k), e_j] - [\psi(e_i, e_j), e_k] - \psi([e_i, e_j], e_k)\\ +\psi(e_i, [e_j, e_k]) + \psi([e_i, e_k], e_j) \text{ for } 0 \le i, j, k \le 4.\end{aligned}$$

Suppose $\psi(e_i, e_j) = \sum_{k=1}^{4} a_{i,j}^k e_k$ where $a_{i,j}^k \in \mathbb{C}$; for $1 \le i, j, k \le 4$. Since $\delta\psi = 0$, equating the coefficients of e_1, e_2, e_3 and e_4 in $\delta\psi(e_i, e_j, e_k)$ we get the following relations:

(i) $a_{1,1}^1 = a_{1,1}^2 = a_{1,1}^3 = a_{1,1}^4 = a_{1,2}^1 = a_{1,2}^3 = a_{1,2}^4 = 0$;
(ii) $a_{1,3}^4 = a_{1,4}^3 = a_{1,4}^4 = a_{2,1}^1 = a_{2,1}^3 = a_{2,1}^4 = a_{2,2}^1 = a_{2,2}^2 = a_{2,2}^3 = a_{2,2}^4 = 0$;
(iii) $a_{3,1}^4 = a_{3,3}^2 = a_{3,3}^3 = a_{3,3}^4 = a_{4,1}^3 = a_{4,1}^4 = a_{4,4}^2 = a_{4,4}^3 = a_{4,4}^4 = 0$;
(iv) $a_{1,2}^2 = -a_{2,1}^2 = a_{1,3}^2 = -a_{1,3}^3 = -a_{3,1}^2 = a_{3,1}^3$;
(v) $a_{1,3}^1 = -a_{3,1}^1 = a_{1,4}^2 = -a_{4,1}^2$;
(vi) $a_{2,3}^3 = -a_{3,2}^3 = -a_{2,4}^4 = a_{4,2}^4$; $a_{2,3}^4 = -a_{3,2}^4$; $a_{2,3}^2 = -a_{3,2}^2$;
(vi) $a_{2,4}^1 = -a_{4,2}^1$; $a_{2,4}^2 = -a_{4,2}^2$; $a_{2,4}^3 = -a_{4,2}^3$;

(vii) $a^1_{3,4} = -a^1_{4,3}$; $a^2_{3,4} = -a^2_{4,3}$; $a^3_{3,4} = -a^3_{4,3}$; $a^4_{3,4} = -a^4_{4,3}$

(ix) $a^3_{3,4} = (a^1_{14} - a^2_{24})$; $a^4_{3,4} = (a^2_{14} + a^2_{23})$

(x) $a^1_{33} = \frac{1}{2}(a^1_{23} + a^1_{32})$; $a^1_{41} = -(a^1_{14} + a^1_{23} + a^1_{32})$.

Therefore, in terms of the ordered basis $\{e_i \otimes e_j\}_{1 \le i,j \le 4}$ of $L \otimes L$ and $\{e_i\}_{1 \le i \le 4}$ of L, the transpose of the matrix corresponding to ψ is of the form

$$M^t = \begin{pmatrix} 0 & 0 & 0 & 0 \\ 0 & x_1 & 0 & 0 \\ x_2 & x_1 & -x_1 & 0 \\ x_3 & x_2 & 0 & 0 \\ 0 & -x_1 & 0 & 0 \\ 0 & 0 & 0 & 0 \\ x_4 & x_5 & x_6 & x_7 \\ x_8 & x_9 & x_{10} & -x_6 \\ -x_2 & -x_1 & x_1 & 0 \\ x_{11} & -x_5 & -x_6 & -x_7 \\ \frac{1}{2}(x_4 + x_{11}) & 0 & 0 & 0 \\ x_{12} & x_{13} & (x_3 - x_9) & (x_2 + x_5) \\ -(x_4 + x_3 + x_{11}) & -x_2 & 0 & 0 \\ -x_8 & -x_9 & -x_{10} & x_6 \\ -x_{12} & -x_{13} & -(x_3 - x_9) & -(x_2 + x_5) \\ x_{14} & 0 & 0 & 0 \end{pmatrix}.$$

where $x_1 = a^2_{1,2}$; $x_2 = a^1_{1,3}$; $x_3 = a^1_{1,4}$; $x_4 = a^1_{2,3}$; $x_5 = a^2_{2,3}$; $x_6 = a^3_{2,3}$; $x_7 = a^4_{2,3}$; $x_8 = a^1_{2,4}$; $x_9 = a^2_{2,4}$; $x_{10} = a^3_{2,4}$; $x_{11} = a^1_{3,2}$; $x_{12} = a^1_{3,4}$; $x_{13} = a^2_{3,4}$ and $x_{14} = a^1_{4,4}$

are in $\mathbb{C}$. Let $\phi_i \in ZL^2(L; L)$ for $1 \le i \le 14$, be the cocyle with $x_i = 1$ and $x_j = 0$ for $i \neq j$ in the above matrix of ψ. It is easy to check that $\{\phi_1, \dots, \phi_{14}\}$ forms a basis of $ZL^2(L; L)$.

(ii) Let $\psi_0 \in BL^2(L; L)$. We have $\psi_0 = \delta g$ for some 1-cochain $g \in CL^1(L; L) =$ Hom $(L; L)$. Suppose the matrix associated to ψ_0 is the same as the above matrix M.

Let $g(e_i) = a^1_i e_1 + a^2_i e_2 + a^3_i e_3 + a^4_i e_4$ for $i = 1, 2, 3, 4$. The matrix associated to g is given by

$$(a^j_i)_{i,j=1,\dots,4}$$

From the definition of the coboundary we get

$$\delta g(e_i, e_j) = [e_i, g(e_j)] + [g(e_i), e_j] - \psi([e_i, e_j])$$

for $0 \leq i, j \leq 4$. If we write out the transpose matrix of

$$\delta g,$$

and compare it with M (since $\psi_0 = \delta g$ is also a cocycle in $CL^2(L; L)$), we conclude that the transpose matrix of ψ_0 is of the form

$$M^t = \begin{pmatrix} 0 & 0 & 0 & 0 \\ 0 & x_1 & 0 & 0 \\ x_2 & x_1 & -x_1 & 0 \\ 0 & x_2 & 0 & 0 \\ 0 & -x_1 & 0 & 0 \\ 0 & 0 & 0 & 0 \\ x_4 & x_5 & x_6 & x_1 \\ x_8 & x_9 & x_{10} & -x_6 \\ -x_2 & -x_1 & x_1 & 0 \\ -x_4 & -x_5 & -x_6 & -x_1 \\ 0 & 0 & 0 & 0 \\ x_{12} & x_{13} & -x_9 & (x_2 + x_5) \\ 0 & -x_2 & 0 & 0 \\ -x_8 & -x_9 & -x_{10} & x_6 \\ -x_{12} & -x_{13} & x_9 & -(x_2 + x_5) \\ 0 & 0 & 0 & 0 \end{pmatrix}.$$

Let $\phi_i{}' \in BL^2(L; L)$ for $i = 1, 2, 4, 5, 6, 8, 9, 10, 12, 13$ be the coboundary with $x_i = 1$ and $x_j = 0$ for $i \neq j$ in the above matrix of ψ_0. It follows that $\{\phi'_1, \phi'_2, \phi'_4, \phi'_5, \phi'_6, \phi'_8, \phi'_9, \phi'_{10}, \phi'_{12}, \phi'_{13}\}$ forms a basis of the coboundary space $BL^2(L; L)$.

(iii) It is straightforward to check that

$$\{[\phi_3], [\phi_7], [\phi_{11}], [\phi_{14}]\}$$

span $HL^2(L; L)$ where $[\phi_i]$ denotes the cohomology class represented by the cocycle ϕ_i.

Thus $\dim(HL^2(L; L)) = 4$.

The representative cocycles of the cohomology classes forming a basis of $HL^2(L; L)$ are given explicitely as the following.

(1) $\phi_3 : \phi_3(e_1, e_4) = e_1,\ \phi_3(e_4, e_1) = -e_1;\ \phi_3(e_3, e_4) = e_3;\ \phi_3(e_4, e_3) = -e_3;$
(2) $\phi_7 : \phi_7(e_2, e_3) = e_4,\ \phi_7(e_3, e_2) = -e_4;$
(3) $\phi_{11} : \phi_{11}(e_3, e_2) = e_1,\ \phi_{11}(e_3, e_3) = \frac{1}{2}e_1,\ \phi_{11}(e_4, e_1) = -e_1;$
(4) $\phi_{14} : \phi_{14}(e_4, e_4) = e_1.$

Here ϕ_3 and ϕ_7 are skew-symmetric, so $\phi_i \in Hom(\Lambda^2 L; L) \subset Hom(L^{\otimes 2}; L)$ for $i = 3$ and 7.

Consider $\mu_i = \mu_0 + t\phi_i$ for $i = 3, 7, 11, 14$, where μ_0 denotes the original bracket in L.

This gives 4 non-equivalent infinitesimal deformations of the Leibniz bracket μ_0 with μ_3 and μ_7 giving the Lie algebra structure on the factor space $L[[t]]/ < t^2 >$.

Now we have to compute the nontrivial Massey brackets which give relations on the base of the miniversal deformation.

Let us start to compute the nonzero brackets $[\phi_i, \phi_i]$ which are the obstructions to extending infinitesimal deformations. We find

$$[\phi_3, \phi_3] = 0, \quad [\phi_7, \phi_7] = 0.$$

That means that these two infinitesimal Lie deformations can be extended. In fact, they can be extended to real Lie deformations as follows.

We give the new nonzero Lie brackets (and their anticommutative analogue).

The first deformation

$$\begin{aligned} [e_2, e_3]_t &= e_1 + te_4 \\ [e_2, e_4]_t &= e_2 \\ [e_3, e_4]_t &= e_2 - e_3 \end{aligned}$$

is isomorphic to $\mathfrak{sl}(2, \mathbb{C}) \oplus \mathbb{C}$ for every nonzero value of t, see [7].

The second deformation represents a 2-parameter projective family $d(\lambda, \mu)$, for which each projective parameter (λ, μ) defines a nonisomorphic Lie algebra (in fact, the diamond algebra is a member of this family with $(\lambda, \mu) = (1, -1)$):

$$\begin{aligned} [e_2, e_3]_{\lambda,\mu} &= e_1 \\ [e_2, e_4]_{\lambda,\mu} &= \lambda e_2 \\ [e_3, e_4]_{\lambda,\mu} &= e_2 + \mu e_3 \\ [e_1, e_4]_{\lambda,\mu} &= (\lambda + \mu)e_1. \end{aligned}$$

Furthermore, we also have $[\phi_{14}, \phi_{14}] = 0$ which means that ϕ_{14} defines a real Leibniz deformation:

$$\begin{aligned} [e_2, e_3]_t &= e_1 \\ [e_2, e_4]_t &= e_2 \\ [e_3, e_4]_t &= e_2 - e_3 \\ [e_4, e_4]_t &= te_1. \end{aligned}$$

We note that this Leibniz algebra is not nilpotent.

For the bracket $[\phi_{11}, \phi_{11}]$ we get a nonzero 3-cocycle, so the infinitesimal Leibniz deformation with infinitesimal part being ϕ_{11} can not be extended even to the next order. That means it gives a relation on the base of the versal deformation.

The nontrivial mixed brackets $[\phi_i, \phi_j]$ also determine relations on the base of the versal deformation.

Among the six possible cases $[\phi_3, \phi_{11}]$, $[\phi_3, \phi_{14}]$ and $[\phi_{11}, \phi_{14}]$ are nontrivial 3-cocycles, the others are represented by 3-coboundaries.

Thus we need to check the Massey 3-brackets which are defined, namely
$< \phi_3, \phi_3, \phi_7 >, < \phi_3, \phi_7, \phi_7 >, < \phi_7, \phi_7, \phi_{11} >,$
$< \phi_7, \phi_7, \phi_{14} >, < \phi_7, \phi_{14}, \phi_{14} >.$

In these five possible Massey 3-brackets, only $< \phi_3, \phi_3, \phi_7 >$ is represented by nontrivial cocycle.

So we now proceed to compute the possible Massey 4-brackets. We get that four of them are nontrivial:
$< \phi_3, \phi_7, \phi_7, \phi_{11} >, < \phi_3, \phi_7, \phi_7, \phi_{14} >,$
$< \phi_7, \phi_7, \phi_{14}, \phi_{11} >, < \phi_7, \phi_7, \phi_{14}, \phi_{14} >.$

At the next step, we get that all the Massey 5-brackets which are defined are trivial.

So we can write the universal infinitesimal Leibniz deformation of our Lie algebra:

$$
\begin{aligned}
&[e_1, e_2]_v = [e_2, e_1]_v = [e_1, e_3]_v = [e_3, e_1]_v = 0,\\
&[e_1, e_4]_v = te_1, \quad [e_4, e_1]_v = -(t+u)e_1,\\
&[e_2, e_3]_v = e_1 + se_4, \quad [e_3, e_2]_v = (u-1)e_1 - se_4,\\
&[e_2, e_4]_v = e_2, \quad [e_4, e_2] = -e_2,\\
&[e_3, e_4]_v = e_2 + (t-1)e_3, \quad [e_4, e_3]_v = -e_2 + (1-t)e_3,\\
&[e_1, e_1]_v = [e_2, e_2]_v = 0, \quad [e_3, e_3]_v = 1/2ue_1,\\
&[e_4, e_4]_v = we_1.
\end{aligned}
$$

With the nontrivial Massey brackets and the identification $t = \phi_3, s = \phi_7, u = \phi_{11}, w = \phi_{14}$, we get that the base of the infinitesimal deformation is

$$\mathbb{C}[[t, s, u, w]]/\{u^2, tu, tw, uw; t^2 s; ts^2 u, ts^2 w, s^2 uw, s^2 w^2\}.$$

References

1. Fialowski, A.: Deformations of Lie algebras, Mat. Sbornyik USSR, **127**(169), 476–482 (1985); English translation: Math. USSR-Sb., **55**(2), 467–473 (1986)
2. Fialowski, A.: An example of formal deformations of Lie algebras. In: Proceedings of NATO Conference on Deformation Theory of Algebras and Applications, Il Ciocco, Italy, 1986, pp. 375–401. Kluwer, Dordrecht (1988)
3. Fialowski, A., Fuchs, D.B.: Construction of miniversal deformation of lie algebras. J. Funct. Anal. **161**, 76 (1999)

4. Fialowski, A., Mandal, A.: Leibniz algebra deformations of a Lie algebra. J. Math. Phys. **49**, 093512, 10 (2008)
5. Fialowski, A., Mandal, A., Magnin, L.: About Leibniz cohomology and deformations of Lie algebras. J. Algebra **383**, 63–77 (2013)
6. Fialowski, A., Mandal, A., Mukherjee, G.: Versal deformations of leibniz algebras. J. K-Theor. **3**, 327–358 (2009). doi:10.1017/is008004027jkt049
7. Fialowski, A., Penkava, M.: Versal deformations of four dimensional lie algebras. Commun. Contemp. Math. **9**, 41–79 (2007)
8. Guichardet, A.: Cohomologie des groupes topologiques et des algèbres de Lie. Cedic/Fernand Nathan, Paris (1980)
9. Koszul, J.L.: Homologie et cohomologie des algèbres de lie. Bull. Soc. Math. France **78**, 67–127 (1950)
10. Loday, J.-L.: Une version non commutative des algèbres de lie: les algèbres de leibniz. Ens. Math. **39**, 269–293 (1993)
11. Loday, J.-L.: Overview on leibniz algebras, dialgebras and their homology. Fields Inst. Commun. **17**, 91 (1997)
12. Loday, J.-L., Pirashvili, T.: Universal enveloping algebras of leibniz algebras and (co)homology. Math. Ann. **296**, 139 (1993)
13. Hu, N., Pei, Y., Liu, D.: A cohomological characterization of leibniz central extensions of lie algebras. Proc. Amer. Math. Soc. **136**, 437–477 (2008)
14. Nappi, C.R., Witten, E.: Wess-zumino-witten model based on a nonsemisimple lie group. Phys. Rev. Lett. **71**, 3751 (1993)
15. Ovando, G.: Complex, symplectic and kähler structures on 4 dimensional lie groups. Rev. Un. Mat. Argentina **45**, 55–67 (2004)
16. Pirashvili, T.: On leibniz homology. Ann. Instit. Fourier **44**, 401–411 (1994)
17. Schlessinger, M.: Functors of artin rings. Trans. Am. Soc. **130**, 208 (1968)

Rigid Current Lie Algebras

Elisabeth Remm and Michel Goze

Abstract A current Lie algebra is constructed from a tensor product of a Lie algebra and a commutative associative algebra of dimension greater than 2. In this work we are interested in deformations of finite dimensional current Lie algebras and in the problem of rigidity. In particular we prove that a complex finite dimensional current Lie algebra with trivial center is rigid if it is isomorphic to a direct product $\mathfrak{g} \times \mathfrak{g} \times \cdots \times \mathfrak{g}$ where $\mathfrak{g}$ is a rigid Lie algebra.

1 Current Lie Algebras

If $\mathfrak{g}$ is a Lie algebra over a algebraically closed field $\mathbb{K}$ and $\mathscr{A}$ a $\mathbb{K}$-associative commutative algebra, then $\mathfrak{g} \otimes \mathscr{A}$, provided with the bracket

$$[X \otimes a, Y \otimes b] = [X, Y] \otimes ab$$

for every $X, Y \in \mathfrak{g}$ and $a, b \in \mathscr{A}$ is a Lie algebra. If $dim(\mathscr{A}) = 1$ such an algebra is isomorphic to $\mathfrak{g}$. If $dim(\mathscr{A}) > 1$ we will say that $\mathfrak{g} \otimes \mathscr{A}$ with the previous bracket is a current Lie algebra.

In [16] we have shown that if $\mathscr{P}$ is a quadratic operad, there is an associated quadratic operad, noted $\tilde{\mathscr{P}}$ such that the tensor product of a $\mathscr{P}$-algebra by a $\tilde{\mathscr{P}}$-algebra is a $\mathscr{P}$-algebra for the natural product. In particular, if the operad $\mathscr{P}$ is $\mathscr{L}ie$, then $\tilde{\mathscr{L}ie} = \mathscr{L}ie^! = \mathscr{C}om$ and a $\mathscr{C}om$-algebra is a commutative associative algebra. In this context we find again the notion of current Lie algebra.

E. Remm (✉) · M. Goze
Université de Haute Alsace, 68093 Mulhouse, France
e-mail: elisabeth.remm@uha.fr

M. Goze
e-mail: michel.goze@uha.fr

A. Makhlouf et al. (eds.), *Algebra, Geometry and Mathematical Physics*,
Springer Proceedings in Mathematics & Statistics 85, DOI: 10.1007/978-3-642-55361-5_14,

Remark In [3], the notion of duplication of algebras constructed by tensor product is presented. If $\mathfrak{g}$ is a Lie algebra, we define on $\mathfrak{g} \otimes \mathfrak{g}$ the product

$$\mu(X \otimes Y, X' \otimes Y') = [X, Y] \otimes [X', Y'].$$

But, in this case, $\mathfrak{g} \otimes \mathfrak{g}$ is not a Lie algebra, but is related with the notion of n-Lie algebras.

In this work we study the deformations of finite dimensional current Lie algebras and we study the rigidity. The notion of rigidity is related to the second group of the Chevalley-Eilenberg cohomology. For the current Lie algebras, this group is not well known. Recently some relations between $H^2(\mathfrak{g} \otimes \mathscr{A}, \mathfrak{g} \otimes \mathscr{A})$, $H^2(\mathfrak{g}, \mathfrak{g})$ and $H^2_H(\mathscr{A}, \mathscr{A})$ have been given in [18] but often when $\mathfrak{g}$ is abelian. Let us note also that the scalar cohomology has been studied in [15].

2 Determination of Rigid Current Lie Algebras

In all this work, Lie algebras or associative algebras are of finite dimension over the algebraically closed field $\mathbb{K}$.

2.1 On the Rigidity of Lie Algebras

Let us remind briefly some properties of the variety of Lie algebras (for more details, see [1]). Let $\mathfrak{g}$ be a n-dimensional $\mathbb{K}$-Lie algebra. Since the underlying vector space is isomorphic to $\mathbb{K}^n$, there exists a one-to-one correspondance between the set of Lie brackets of n-dimensional Lie algebras and the skew-symmetric bilinear maps $\mu : \mathbb{K}^n \times \mathbb{K}^n \rightarrow \mathbb{K}^n$ satisfying the Jacobi identity. We denote by $\mu_{\mathfrak{g}}$ this bilinear map corresponding to $\mathfrak{g}$. In this framework, we can identify $\mathfrak{g}$ with the pair $(\mathbb{K}^n, \mu_{\mathfrak{g}})$. Let us fix definitively a basis $\{X_1, \dots, X_n\}$ of $\mathbb{K}^n$. The structure constants (C_{ij}^k) of $\mu_{\mathfrak{g}}$ are given by

$$\mu_{\mathfrak{g}}(X_i, X_j) = \sum_{k=1}^{n} C_{ij}^k X_k$$

and we can identify $\mu_{\mathfrak{g}}$ with the N-tuple (C_{ij}^k) with $N = \frac{n^2(n-1)}{2}$. The Jacobi identity satisfied by $\mu_{\mathfrak{g}}$ is equivalent to the polynomial system :

$$\sum_{l=1,\dots,n} C_{ij}^l C_{lk}^s + C_{jk}^l C_{li}^s + C_{ki}^l C_{lj}^s = 0. \tag{1}$$

In this context, a Lie algebra is a point of $\mathbb{K}^N$ whose coordinates (C_{ij}^k) satisfy (1). The set of n-dimensional Lie algebras over $\mathbb{K}$ is identified with the algebraic variety L_n embedded into $\mathbb{K}^N$ and defined by the system of polynomial Eq. (1). We will always denote by μ a point of L_n. The algebraic group $GL(n, \mathbb{K})$ acts on L_n by:

$$(f, \mu) \in GL(n, \mathbb{K}) \times L_n \longrightarrow \mu_f \in L_n \tag{2}$$

where μ_f is given by $\mu_f(X, Y) = f^{-1}(\mu(f(X), f(Y))$ for every $X, Y \in \mathbb{K}^n$. The orbit $\mathscr{O}(\mu)$ of μ related to this action corresponds to the Lie algebras isomorphic to $\mathfrak{g} = (\mathbb{K}^n, \mu)$. We provide the algebraic variety L^n with the Zariski topology.

Definition 2.1 The Lie algebra $\mathfrak{g} = (\mathbb{K}^n, \mu)$ is rigid if the orbit $\mathscr{O}(\mu)$ is open in L_n.

A way of constructing rigid Lie algebras rests on the Nijenhuis-Richardson Theorem : Let $H^*(\mathfrak{g}, \mathfrak{g})$ be the Chevalley-Eilenberg cohomology of $\mathfrak{g}$. If $H^2(\mathfrak{g}, \mathfrak{g}) = 0$ then $\mathfrak{g}$ is rigid. Let us note that the converse is false, numerous examples are described in [1, 9] (in fact, a rigid Lie algebra whose cohomology $H^2(\mathfrak{g}, \mathfrak{g})$ is not trivial is such that the affine schema $\mathscr{L}_n$ given by the Jacobi ideal is not reduced to the point μ defining $\mathfrak{g}$.)

An intuitive way of defining the notion of rigidity is to consider a rigid algebra as not deformable, that is, any close algebra is isomorphic to it. A general definition of deformations was proposed in [12]. Let A be a commutative $\mathbb{K}$-algebra of valuation such that the residual field $A/\mathfrak{m}$ is isomorphic to $\mathbb{K}$ where $\mathfrak{m}$ is the maximal ideal of A. If $\mathfrak{g}$ is a $\mathbb{K}$-Lie algebra then the tensor product $\mathfrak{g} \otimes A$ is an A-algebra denoted by $\mathfrak{g}_A$.

Definition 2.2 A deformation of $\mathfrak{g}$ is an A-Lie algebra $\mathfrak{g}'_A$ such that the underlying A-module is $\mathfrak{g}_A$ and the brackets $[u, v]_{\mathfrak{g}'_A}$ and $[u, v]_{\mathfrak{g}_A}$ of $\mathfrak{g}'_A$ and $\mathfrak{g}_A$ satisfy

$$[u, v]_{\mathfrak{g}'_A} - [u, v]_{\mathfrak{g}_A} \in \mathfrak{g} \otimes \mathfrak{m}.$$

When $A = \mathbb{C}[[t]]$ we find the classical notion of deformation given by Gerstenhaber. When A is the ring of limited elements in a Robinson non archimedean extension of $\mathbb{C}$, we find the notion of perturbations [8]. If $\mathfrak{g}'_A$ is a deformation of $\mathfrak{g}$ then we have

$$[u, v]_{\mathfrak{g}'_A} - [u, v]_{\mathfrak{g}_A} = \sum_{i=1}^{k} \varepsilon_1 \varepsilon_2 \cdots \varepsilon_i \phi_i$$

where $\varepsilon_i \in \mathfrak{m}$ and $\{\phi_1, \ldots, \phi_k\}$ a family of independent skew symmetric bilinear maps on $\mathbb{K}^n \times \mathbb{K}^n$ with values in $\mathbb{K}^n$. In particular $\phi_1 \in Z^2(\mathfrak{g}, \mathfrak{g})$ and if $\mathfrak{g}'_A$ is isomorphic to $\mathfrak{g}_A$ this map belongs to $B^2(\mathfrak{g}, \mathfrak{g})$. We deduce that the deformations of $\mathfrak{g}$ are parameterized by $H^2(\mathfrak{g}, \mathfrak{g})$. In the following, we are going to determine the current Lie algebras which are rigid.

Remark In [4, 6], we find a similar definition of deformations, but without the hypothesis concerning the valuation. We assume that A is a commutative algebra over the field $\mathbb{K}$ which admits an augmentation $\varepsilon : A \to \mathbb{K}$. This says that ε is a $\mathbb{K}$-algebra homomorphism, e.g. $\varepsilon(1_A) = 1$. The ideal $\mathfrak{m}_\varepsilon := Ker(\varepsilon)$ is a maximal ideal of A (Let us note that any maximal ideal of A gives an augmentation). Let us consider a Lie algebra $\mathfrak{g}$ over $\mathbb{K}$, ε a fixed augmentation of A, and $\mathfrak{m} = Ker(\varepsilon)$ the associated maximal ideal. A global deformation λ of $\mathfrak{g}$ with base $(A, \mathfrak{m})$, is a Lie A-algebra structure on $\mathfrak{g} \otimes A$ with Lie bracket $[.,.]_\lambda$ such that for all $a, b \in A$ and $X, Y \in \mathfrak{g}$,

1. $[a \otimes X, b \otimes Y]_\lambda = (ab \otimes id)[1 \otimes X, 1 \otimes Y]_\lambda$,
2. $\varepsilon \otimes id([1 \otimes X, 1 \otimes Y]_\lambda) = 1 \otimes [X, Y]$.

2.2 The Manifold $L_{(p,q)}$

Let $\mathfrak{g} = \mathfrak{g} = \mathfrak{g}_p \otimes \mathscr{A}_q$ be a pq-dimensional current $\mathbb{K}$-Lie algebra where $\mathfrak{g}_p$ is a p-dimensional $\mathbb{K}$-Lie algebra and $\mathscr{A}_q$ a q-dimensional associative commutative $\mathbb{K}$-algebra. Let $\{X_1, \dots, X_p\}$ be a basis of $\mathfrak{g}_p$ and $\{e_1, \dots, e_q\}$ a basis of $\mathscr{A}_q$. If we denote by $\{C_{ij}^k\}$ and $\{D_{ab}^c\}$ the structure constants of $\mathfrak{g}_p$ and $\mathscr{A}_q$ with regards to these basis, then the Lie bracket $\mu_\mathfrak{g} = \mu_{\mathfrak{g}_p} \otimes \mu_{\mathscr{A}_q}$ of $\mathfrak{g}$ where $\mu_{\mathfrak{g}_p}$ is the multiplication of $\mathfrak{g}_p$ and $\mu_{\mathscr{A}_q}$ the multiplication of $\mathscr{A}_q$, satisfy:

$$\mu_\mathfrak{g}(X_i \otimes e_a, X_j \otimes e_b) = \sum_{k,c} C_{ij}^k D_{ab}^c X_k \otimes e_c,$$

and the structure constants of $\mathfrak{g}$ with respect to the basis $\{X_i \otimes e_a\}_{i=1,\dots,p;\ a=1,\dots,q}$ are $\{C_{ij}^k D_{ab}^c\}$. Thus, the Jacobi relations are written as

$$\sum_{l,r} C_{ij}^l C_{lk}^s D_{ab}^r D_{rc}^t + C_{jk}^l C_{li}^s D_{bc}^r D_{ra}^t + C_{ki}^l C_{lj}^s D_{ca}^r D_{rb}^t = 0$$

for any (s, t) in $\{\{1, \dots, p\} \times \{1, \dots, q\}\}$. These polynomial relations define a structure of algebraic variety denoted by $L_{(p,q)}$ and embedded in the vector space whose coordinates are the structure constants $\{C_{ij}^k D_{ab}^c\}$. It is a closed subvariety of L_{pq}. Let $G(p, q)$ be the algebraic group $G(p, q) = GL(p) \times GL(q)$. This group acts naturally on $L_{(p,q)}$ by

$$(f, g).(\mu_{\mathfrak{g}_p} \otimes \mu_{\mathscr{A}_q})(X \otimes a, Y \otimes b) = f^{-1}(\mu_{\mathfrak{g}_p}(f(X), f(Y))) \otimes g^{-1}(\mu_{\mathscr{A}_q}(g(a), g(b))).$$

We denote by $\mathscr{O}_{p,q}(\mathfrak{g}_p \otimes \mathscr{A}_q)$ the orbit in $L_{(p,q)}$ of $\mu_\mathfrak{g}$ corresponding to this action.

Thus, there are two types of deformations:

- The deformations of $\mathfrak{g}$ in the manifold L_{pq}. These deformations are parameterized by the second Chevalley-Eilenberg cohomology space $H_C^2(\mathfrak{g}, \mathfrak{g})$.

- The deformations of $\mathfrak{g}$ in the manifold $L_{(p,q)}$. They are parameterized by the space $H^2_C(\mathfrak{g}_p, \mathfrak{g}_p) \oplus H^2_H(\mathscr{A}_q, \mathscr{A}_q)$ where $H^2_H(\mathscr{A}_q, \mathscr{A}_q)$ is the Hochschild cohomology of the associative commutative algebra $\mathscr{A}_q$ [13, 14, 17].

Definition 2.3 The Lie algebra $\mathfrak{g}_p \otimes \mathscr{A}_q$ is rigid in $L_{(p,q)}$ if the orbit $\mathscr{O}_{p,q}(\mu_{\mathfrak{g}})$ is open (in the Zariski sense). It is rigid if the orbit $\mathscr{O}(\mu_{\mathfrak{g}})$ related to the action of $GL(pq)$ in L_{pq} is open.

It is clear that the rigidity implies the rigidity in $L_{(p,q)}$.

Proposition 2.1 *A current Lie algebra $\mathfrak{g} = \mathfrak{g}_p \otimes \mathscr{A}_q$ is rigid in $L_{(p,q)}$ if and only if $\mathfrak{g}_p$ is rigid in L_p and $\mathscr{A}_q$ is rigid in $\mathscr{C}om(q)$, the variety of q-dimensional associative commutative $\mathbb{K}$-algebras.*

In fact, if $\mathfrak{g}_p$ (respectively $\mathscr{A}_q$) is not rigid in L_p (respectively in $\mathscr{C}om(q)$), then we can find a non isomorphic deformation of $\mathfrak{g}_p$ (respectively $\mathscr{A}_q$), this gives a non isomorphic deformation of $\mathfrak{g}$. For the general notion of associative rigid algebras see [11].

The main part of this work is to describe rigid current algebras which are rigid (in L_{pq}, that is, rigid in the variety of pq-dimensional Lie algebras).

Example $p = 2, q = 2$ ($\mathbb{K} = \mathbb{C}$). There is, up to isomorphism, only one 2-dimensional rigid Lie algebra. It is defined by $[X_1, X_2] = X_2$. There is only one 2-dimensional associative commutative algebra. It is given by $e_1^2 = e_1, e_2^2 = e_2, e_1e_2 = 0$ and corresponds to the semi-simple algebra $A_1^2 = M_1(\mathbb{K}) \times M_1(\mathbb{K})$ where $M_n(\mathbb{K})$ is the algebra of n-matrices on $\mathbb{K}$. The Lie algebra $\mathfrak{g}_2 \otimes A_1^2$ is rigid in $L_{(2,2)}$. This algebra is isomorphic to $\mathfrak{g}_2 \times \mathfrak{g}_2$. It is also rigid in L_4.

2.3 Structure of Rigid Current Lie Algebras

Recall that a finite dimensional rigid $\mathbb{K}$-Lie algebra $\mathfrak{g}$ is algebraic (that is, isomorphic to a Lie algebra of an algebraic Lie group) and then admits the decomposition $\mathfrak{g} = \mathfrak{s} \oplus \mathfrak{t} \oplus \mathfrak{n}$ where $\mathfrak{t} \oplus \mathfrak{n}$ is the radical of $\mathfrak{g}$, $\mathfrak{t}$ is a maximal abelian subalgebra whose adjoint operators $ad X$, $X \in \mathfrak{t}$, are semi-simple and $\mathfrak{n}$ is the nilradical [5, 7]. If $\mathfrak{g} = \mathfrak{g}_p \otimes \mathscr{A}_q$ is rigid, then $\mathfrak{g}_p$ is rigid in L_p. If $\mathfrak{g}_p$ is solvable, then so is $\mathfrak{g}$ and we have

$$\mathfrak{g}_p = \mathfrak{t}_p \oplus \mathfrak{n}_p \quad \text{and} \quad \mathfrak{g} = \mathfrak{t} \oplus \mathfrak{n}.$$

Since $\mathfrak{n}_p \otimes \mathscr{A}_q$ is a nilpotent ideal of $\mathfrak{g}$, $\mathfrak{n}_p \otimes \mathscr{A}_p \subset \mathfrak{n}$.

Lemma 2.1 *If $\mathfrak{g} = \mathfrak{g}_p \otimes \mathscr{A}_q$ is rigid, then $\mathscr{A}_q$ has a non zero idempotent.*

Remark If $\mathscr{A}_q$ is a nilalgebra, then $\mathfrak{g}$ is nilpotent. In fact if $X \in \mathfrak{g}_p$ and $a \in \mathscr{A}_q$, we have $[ad(X \otimes a)]^m = (ad X)^m \otimes (L_a)^m$ where $L_a : \mathscr{A}_q \to \mathscr{A}_q$ is the left multiplication by a. Since $\mathscr{A}_q$ is a nilalgebra, every element is nilpotent and there exits m_0 such that $(L_a)^{m_0} = 0$. Thus $ad(X \otimes a)$ is a nilpotent operator for any X and a. This implies that

$\mathfrak{g}$ is nilpotent (this doesn't imply that $\mathfrak{g}_p$ is nilpotent). Let f be a derivation of $\mathfrak{g}_p$. Then $f \otimes Id$ is a derivation of $\mathfrak{g}$. Since $\mathfrak{g}_p$ is rigid, we can find a inner non trivial derivation $ad\,X$ which is diagonal. In this case $ad\,X \otimes Id$ is a non trivial diagonal derivation of $\mathfrak{g}$. By hypothesis $\mathfrak{g}$ is rigid. But any rigid nilpotent Lie algebra is characteristically nilpotent [9], that is, every derivation is nilpotent. We have a contradiction and $\mathscr{A}_p$ can not be a nilalgebra. Since it is finite dimensional, it admits a non zero idempotent.

Proposition 2.2 *If* $\mathfrak{g} = \mathfrak{g}_p \otimes \mathscr{A}_q$ *is rigid then* $\mathscr{A}_q$ *is an associative commutative rigid unitary algebra in* $\mathscr{C}om(q)$.

Remark Let $e \neq 0$ be in $\mathscr{A}_q$ and satisfying $e^2 = e$. The associated Pierce decomposition

$$\mathscr{A}_q = \mathscr{A}_q^{00} \oplus \mathscr{A}_q^{10} \oplus \mathscr{A}_q^{01} \oplus \mathscr{A}_q^{11}$$

where

$$\mathscr{A}_q^{ij} = \{x \in \mathscr{A}_q \text{ such that } e \cdot x = ix, x \cdot e = jx\}$$

reduces to $\mathscr{A}_q = \mathscr{A}_q^{11} \oplus \mathscr{A}_q^{00}$ because $\mathscr{A}_q$ is commutative and we have $\mathscr{A}_q^{11} \cdot \mathscr{A}_q^{00} = \{0\}$. Thus $\mathscr{A}_q$ is a direct sum of two commutative algebras. Since $\mathscr{A}_q$ is rigid, the algebras $\mathscr{A}_q^{11}$ and $\mathscr{A}_q^{00}$ are also rigid. The subalgebra $\mathscr{A}_q^{11}$ is unitary (e is the unit element). From the previous lemma $\mathscr{A}_q^{00}$ has an idempotent and admits a decomposition

$$\mathscr{A}_q^{00} = \mathscr{A}_q^{0011} \oplus \mathscr{A}_q^{0000}$$

with $\mathscr{A}_q^{0011} \neq \{0\}$. By induction we deduce that

$$\mathscr{A}_q = \mathscr{A}_q^{1} \oplus \ldots \oplus \mathscr{A}_q^{p}$$

with $\mathscr{A}_q^{i}$ with unit e_i and $\{e_1, \ldots, e_p\}$ is a system of pairwise orthogonal idempotents. Then $e_1 + \cdots + e_p$ is a unit of $\mathscr{A}_q$.

Theorem 2.1 *Let* $\mathfrak{g}_p$ *be a rigid Lie algebra with solvable non nilpotent radical such that* $Z(\mathfrak{g}_p) = \{0\}$. *Then* $\mathfrak{g} = \mathfrak{g}_p \otimes \mathscr{A}_q$ *is rigid if and only if* $\mathscr{A}_q = M_1^q(\mathbb{K})$ *is given by*

$$e_i^2 = e_i \,, i = 1, \ldots, q \text{ and } e_i \cdot e_j = 0 \text{ if } i \neq j.$$

Proof Since $\mathscr{A}_q$ is unitary, the radical of $\mathfrak{g}$ solvable and non nilpotent. Moreover $Z(\mathfrak{g}_p) = \{0\}$ implies that $Z(\mathfrak{g}) = \{0\}$. In fact if $U = \sum_{j,a} \alpha_{ja} X_j \otimes x_a$ is in the center of $\mathfrak{g}$, then $[U, X \otimes 1] = 0$ for each $X \in \mathfrak{g}_p$. Thus

$$\sum \alpha_{j,a}[X_j, X] \otimes x_a = 0.$$

We have $[\sum_j \alpha_{ja} X_j, X] = 0$ for each a and X. So $\sum_j \alpha_{ja} X_j \in Z(\mathfrak{g}_p)$ for any a. Therefore $\alpha_{ja} = 0$ for any a and $U = 0$.

Consequently, $\mathfrak{g}$ is a rigid Lie algebra with trivial center whose radical is non nilpotent. This implies that all derivations are inner. Let f be a non trivial derivation of $\mathscr{A}_q$. Since $\mathscr{A}_q$ is commutative, it is necessarily an outer derivation. Then $Id \otimes f$ is a derivation of $\mathfrak{g}$ and satisfies $(Id \otimes f)(X \otimes 1) = X \otimes f(1) = 0$ because $f(1 \cdot 1) = 2f(1) = f(1) = 0$. Suppose that $Id \otimes f \in Int(\mathfrak{g})$, that is $Id \otimes f = ad(\sum \alpha_{ij} X_i \otimes x_j)$. Thus $(Id \otimes f)(X \otimes 1) = \sum \alpha_{ij}[X_i, X] \otimes x_j = 0$ which implies $\sum \alpha_{ij}[X_i, X] = 0$ for any j and X. So $\sum \alpha_{ij} X_i \in Z(\mathfrak{g}_p)$ for any j. Since the center is trivial, then $\sum \alpha_{ij} X_j = 0$ for any j and $Id \otimes f \notin Int(\mathfrak{g})$. There is a contradicion. Therefore $\mathscr{A}_q$ is such that any external derivation is trivial. We deduce that $\mathscr{A}_q = M_1^q(\mathbb{K})$.

Remark 1. The current Lie algebra $\mathfrak{g}_p \otimes M_1^q(\mathbb{K})$ is isomorphic to $\mathfrak{g}_p \times \cdots \times \mathfrak{g}_p$ with q factors. If $\mathfrak{g}$ is a rigid current algebra with $Z(\mathfrak{g}_p)$ trivial, then it is isomorphic to $\mathfrak{g}_p \times \cdots \times \mathfrak{g}_p$.

2. In the theorem, we have a hypothesis concerning the center of $\mathfrak{g}_p$. This hypothesis is probably superfluous. In fact, since the orbit in L_n of a rigid n-dimensional Lie algebra is Zariski open, the Zariski closure of this orbit is an algebraic component of L_n. This assures that, for a fixed dimension, there exist only a finite number of non isomorphic rigid Lie algebras. But, for all the known examples of rigid Lie algebras, the center is trivial. We can naturally conjecture that any finite dimensional complex rigid Lie algebra has a trivial center.

3 Cohomology and Deformations

The Chevalley-Eilenberg cohomology of current Lie algebras was computed in [18] for the degrees 1 and 2. It is shown that the algebra of derivations of $\mathfrak{g} = \mathfrak{g}_p \otimes \mathscr{A}_q$ is equal to

$$\begin{aligned} \mathscr{D}er(\mathfrak{g}) = \mathscr{D}er(\mathfrak{g}_p) \otimes \mathscr{A}_q \oplus Hom_{\mathfrak{g}_p}(\mathfrak{g}_p, \mathfrak{g}_p) \otimes \mathscr{D}er(\mathscr{A}_q) \\ \oplus Hom(\mathfrak{g}_p/_{[\mathfrak{g}_p,\mathfrak{g}_p]}, Z(\mathfrak{g}_p))] \otimes \frac{End(\mathscr{A}_q)}{\mathscr{A}_q + \mathscr{D}er\mathscr{A}_q} \end{aligned}$$

and the first space of cohomology $H^1(\mathfrak{g}, \mathfrak{g})$ is

$$\begin{aligned} H^1(\mathfrak{g}, \mathfrak{g}) = H^1(\mathfrak{g}_p, \mathfrak{g}_p) \otimes \mathscr{A}_q \oplus Hom_{\mathfrak{g}_p}(\mathfrak{g}_p, \mathfrak{g}_p) \otimes \mathscr{D}er(\mathscr{A}_q) \\ \oplus Hom(\mathfrak{g}_p/_{[\mathfrak{g}_p,\mathfrak{g}_p]}, Z(\mathfrak{g}_p))] \otimes \frac{Hom(\mathscr{A}_q, \mathscr{A}_q)}{\mathscr{A}_q + \mathscr{D}er\mathscr{A}_q}. \end{aligned}$$

Assume that $\mathfrak{g} = \mathfrak{g}_p \otimes \mathscr{A}_q$ is a rigid current Lie algebra. Then $\mathfrak{g}_p$ is rigid. Assume also that $Z(\mathfrak{g}_p) = 0$. Then

$$H^1(\mathfrak{g}, \mathfrak{g}) = H^1(\mathfrak{g}_p, \mathfrak{g}_p) \otimes \mathscr{A}_q \oplus Hom_{\mathfrak{g}_p}(\mathfrak{g}_p, \mathfrak{g}_p) \otimes \mathscr{D}er(\mathscr{A}_q).$$

If $\mathfrak{g}_p$ is a rigid Lie algebra with non nilpotent radical (we do not know examples of rigid Lie algebras with a nilpotent radical), any derivation of $\mathfrak{g}_p$ is inner. This implies that $H^1(\mathfrak{g}_p, \mathfrak{g}_p) = 0$ and $H^1(\mathfrak{g}, \mathfrak{g}) = Hom_{\mathfrak{g}_p}(\mathfrak{g}_p, \mathfrak{g}_p) \otimes \mathscr{D}er(\mathscr{A}_q)$.

Proposition 3.1 *Let $\mathfrak{g} = \mathfrak{g}_p \otimes \mathscr{A}_q$ be a current Lie algebra such that $\mathfrak{g}_p$ is rigid with trivial center and a non nilpotent radical. Then $H^1(\mathfrak{g}, \mathfrak{g}) = 0$ if and only if $\mathscr{D}er(\mathscr{A}_q) = \{0\}$.*

Example Consider $\mathscr{A}_q = M_1^q(\mathbb{K})$. Let $\{e_i\}$ be a basis of $\mathscr{A}_q$ satisfying $e_i^2 = e_i, e_ie_j = 0$. Let f be in $\mathscr{D}er(\mathscr{A}_q)$. We have

$$f(e_i^2) = f(e_i) = 2e_if(e_i).$$

This induces $f(e_i) = 0$ and finally $f = 0$.

A Chevalley-Eilenberg 2-cochain of $\mathfrak{g} = \mathfrak{g}_p \otimes \mathscr{A}_q$ decomposes as a finite sum of bilinear forms of type:

$$\varphi = \psi_1 \otimes \varphi_2 + \varphi_3 \otimes \psi_4$$

with $\psi_1 \in \mathscr{C}^2(\mathfrak{g}_p, \mathfrak{g}_p)$, $\varphi_2 \in \mathscr{S}^2(\mathfrak{g}_p, \mathfrak{g}_p)$ and $\varphi_3 \in \mathscr{S}^2(\mathfrak{g}_p, \mathfrak{g}_p)$, $\psi_4 \in \mathscr{C}^2(\mathscr{A}_q, \mathscr{A}_q)$, where $\mathscr{C}^2(\mathfrak{g}_p, \mathfrak{g}_p)$ denotes the space of Chevalley-Eilenberg 2-cochains of $\mathfrak{g}_p$, $\mathscr{S}^2(\mathfrak{g}_p, \mathfrak{g}_p)$ the space of symmetric bilinear maps with values in $\mathfrak{g}_p$ and $\mathscr{C}^2(\mathscr{A}_q, \mathscr{A}_q)$ the space of 2-cochains of the Harrison cohomology of $\mathscr{A}_q$. We deduce using this decomposition that $H^2(\mathfrak{g}, \mathfrak{g}) = (H^2)' \oplus (H^2)''$. The first space is computed in ([18], proposition 3.1). We find

$$(H^2)' = H^2(\mathfrak{g}_p, \mathfrak{g}_p) \otimes \mathscr{A}_q \oplus \mathscr{B}(\mathfrak{g}_p, \mathfrak{g}_p) \otimes \frac{H_H^2(\mathscr{A}_q, \mathscr{A}_q)}{\mathscr{P}_+(\mathscr{A}_q, \mathscr{A}_q)} \oplus \chi(\mathfrak{g}_p, \mathfrak{g}_p) \otimes \frac{\mathscr{A}(\mathscr{A}_q, \mathscr{A}_q)}{\mathscr{P}_+(\mathscr{A}_q, \mathscr{A}_q)}$$

(see [18] for notations). But the second space was just computed when $\mathfrak{g}_p$ is abelian.

For example assume that we have a primitive infinitesimal deformation of $\mu_1 \otimes \mu_2$, that is, $\mu_1 \otimes \mu_2 + \epsilon(\psi_1 \otimes \varphi_2 + \varphi_3 \otimes \psi_4)$. The linear part of the Jacobi identity gives the expression of a 2-cocycle of Chevalley-Eilenberg cohomology of $\mu_1 \otimes \mu_2$. We find:

$$\begin{aligned}
&\delta_{\mu_1 \otimes \mu_2}(\psi_1 \otimes \varphi_2 + \varphi_3 \otimes \psi_4)(X_1, X_2, X_3, a_1, a_2, a_3)\\
&\quad = \Sigma \mu_1(\psi_1(X_1, X_2), X_3) \otimes \mu_2(\varphi_2(a_1, a_2), a_3)\\
&\qquad + \Sigma \mu_1(\varphi_3(X_1, X_2), X_3) \otimes \mu_2(\psi_4(a_1, a_2), a_3)\\
&\qquad + \Sigma \psi_1(\mu_1(X_1, X_2), X_3) \otimes \varphi_2(\mu_2(a_1, a_2), a_3)\\
&\qquad + \Sigma \varphi_3(\mu_1(X_1, X_2), X_3) \otimes \psi_4(\mu_2(a_1, a_2), a_3) = 0
\end{aligned}$$

for any $X_1, X_2, X_3 \in \mathfrak{g}_p$ and $a_1, a_2, a_3 \in \mathscr{A}_q$, and the sum is taken on the cyclic permutations of $(1, 2, 3)$. We deduce

Proposition 3.2 *If $\mathscr{A}_q$ is unitary then $\psi_1 \in Z^2(\mathfrak{g}_p, \mathfrak{g}_p)$ as soon as $\varphi_2(1,1) \neq 0$.*

If $X_1 = X_2 = X_3$, the above identity reduce to:

$$\mu_1(\varphi_3(X_1, X_1), X_1) \otimes \Sigma \mu_2(\psi_4(a_1, a_2), a_3) = 0.$$

Proposition 3.3 *If there exits $X \in \mathfrak{g}_p$ such that $\mu_1(\varphi_3(X_1, X_1), X_1) \neq 0$ then*

$$\mu_2 \bullet \psi_4 = 0$$

with

$$\mu_2 \bullet \psi_4(a_1, a_2, a_3) = \Sigma \mu_2(\psi_4(a_1, a_2), a_3).$$

Note that ψ_4 is a 2-cocyle for the Harrison cohomology of μ_2 so $\mu_2 \bullet \psi_4 = \psi_4 \bullet \mu_2$.
Suppose that $\mathfrak{g}$ is rigid solvable with trivial center. Then $\mathscr{A}_q$ is unitary and $\psi_1 \in Z^2(\mathfrak{g}_p, \mathfrak{g}_p)$ as soon as $\varphi_2(1,1) \neq 0$.

4 Application: Associative Commutative Real Rigid Algebras

4.1 Real Rigid Lie Algebras

The study of the rigid real Lie algebras was recently initiated in [2]. Let us point out the principal results. An external torus of derivations of $\mathfrak{n}$ is an abelian subalgebra $\mathfrak{t}$ of $\mathscr{D}er(\mathfrak{n})$, the Lie algebra of derivations of $\mathfrak{n}$, such as the elements are semi-simple. This means that complex derivations $f \otimes Id \in \mathfrak{t} \otimes \mathbb{C}$ are simultaneously diagonalizable. If $\mathfrak{t}$ is a maximal (with respect to inclusion) external torus of $\mathfrak{n}$ then $\mathfrak{t} \otimes \mathbb{C}$ is a maximal external torus of $\mathfrak{n} \otimes \mathbb{C}$. From a result of Malcev (see e.g. [10]), all the maximal tori of $\mathfrak{n} \otimes \mathbb{C}$ are conjugated with respect to $\mathrm{Aut}(\mathfrak{n} \otimes \mathbb{C})$ so they have the same dimension (thus a maximal exterior torus is sometimes called a Malcev torus). It is the same for the maximal tori $\mathfrak{t}$ of $\mathfrak{n}$. This dimension is called the rank of $\mathfrak{n}$. But contrary to the complex case, all the tori are not conjugated with respect to the group of automorphisms.

Definition 4.1 Let $\mathfrak{n}$ be a finite dimensional real nilpotent Lie algebra. We call a toroidal index of $\mathfrak{n}$ the number of conjugation classes of a maximal external torus with respect to the group of automorphisms $\mathrm{Aut}_{\mathbb{R}}(\mathfrak{n})$ of $\mathfrak{n}$.

Example The toroidal index of the real abelian Lie algebra $\mathfrak{a}_n$ of dimension n is equal to $[n/2] + 1$ where $[p]$ is the integer part of the rational number p. In fact, let $\{X_1, \ldots, X_n\}$ be a basis of $\mathfrak{a}_n$. Let us denote by f_i the derivation defined by $f_i(X_j) = \delta_i^j X_j$ and by $f_{1,2p}$ the derivation given by

$$\begin{cases} f_{1,2p}(X_{2p-1}) = X_{2p}, \\ f_{1,2p}(X_{2p}) = X_{2p-1}. \end{cases}$$

Up to conjugation, the maximal exterior tori are the subalgebras of $gl(n,\mathbb{R})$ generated by

$$\begin{array}{l} \mathfrak{t}_1 = \mathbb{R}\{f_1,\dots,f_n\} \\ \mathfrak{t}_2 = \mathbb{R}\{f_{1,2}, f_1+f_2, f_3\dots,f_n\} \\ \mathfrak{t}_3 = \mathbb{R}\{f_{1,2}, f_1+f_2, f_{1,4}, f_3+f_4, f_5,\dots,f_n\} \\ \dots \\ \mathfrak{t}_n = \mathbb{R}\{f_{1,2}, f_1+f_2, f_{1,4}, f_3+f_4,\dots,f_{1,n}, f_{n-1}+f_n\} \end{array}$$

if n is even, if not the last relation is replaced by

$$\mathfrak{t}_n = \mathbb{R}\{f_{1,2}, f_1+f_2, f_{1,4}, f_3+f_4,\dots,f_{1,n-1}, f_{n-2}+f_{n-1}, f_n\}.$$

4.2 Real Rigid Associative Commutative Algebras

Let $\mathfrak{r}_2$ be the real nonabelian 2-dimensional Lie algebra. There exists a basis $\{X_1, X_2\}$ with regard to which the bracket is given by $[X_1, X_2] = X_2$. Let $\mathscr{A}_n$ be a n-dimensional real rigid commutative associative algebra. Its complexification is isomorphic to $M_1^n(\mathbb{C})$. Thus the real current Lie algebra $\mathfrak{g} = \mathfrak{r}_2 \otimes \mathscr{A}_n$ is rigid. We deduce that its complexification is rigid and isomorphic to $\mathfrak{r}_2^n$. These remarks allow to write the following decomposition:

$$\mathfrak{g} = \mathfrak{r}_2 \otimes \mathscr{A}_n = \mathfrak{t}_n \oplus \mathfrak{a}_n$$

where $\mathfrak{a}_n$ is the n-dimensional abelian Lie algebra. We can deduce from this the structure of $\mathscr{A}_n$. In fact, if $\{Y_1, \dots, Y_n\}$ is a basis of $\mathfrak{t}_n$ corresponding to the derivations $f_{1,2}, f_1+f_2,\dots,f_{1,2s}, f_{2s-1}+f_{2s}, f_{2s+1},\dots,f_n\}$ described in the previous section, the Lie bracket of $\mathfrak{g}$ satisfies

$$\begin{cases} [Y_1, X_1] = -X_2,\ [Y_1, X_2] = X_1, \\ [Y_2, X_1] = X_1,\ [Y_2, X_2] = X_2, \\ \dots \\ [Y_{2s-1}, X_{2s-1}] = -X_{2s},\ [Y_{2s-1}, X_{2s}] = X_{2s-1}, \\ [Y_{2s}, X_{2s-1}] = X_{2s-1},\ [Y_{2s}, X_{2s}] = X_{2s}, \\ [Y_i, X_i] = X_i,\ i = 2s+1,\dots,n. \end{cases}$$

Let $\{e_1,\dots,e_n\}$ be a basis of $\mathscr{A}_n$ such that the isomorphism between $\mathfrak{r}_2 \otimes \mathscr{A}_n$ and $\mathfrak{t}_n \oplus \mathfrak{a}_n$ is given by $U_1 \otimes e_i = Y_i$ and $X_{2i} = U_2 \otimes e_{2i-1}$, $X_{2i-1} = U_2 \otimes e_{2i}$ for $i = 1,\dots,s$ and $X_j = U_2 \otimes e_j$ for $j = 2s+1, .., n$. The rigid associative algebra $\mathscr{A}_n$ is thus defined by

$$\begin{cases} e_{2i-1}^2 = e_{2i-1}, \quad i = 1, \ldots, s; \\ e_{2i-1}e_{2i} = e_{2i}e_{2i-1} = e_{2i}, \quad i = 1, \ldots, s; \\ e_{2i}^2 = -e_{2i-1}, \quad i = 1, \ldots, s; \\ e_j^2 = e_j, \quad j = 2s+1, \ldots, n. \end{cases}$$

Proposition 4.1 *Let $\mathscr{A}_n$ be a n-dimensional real rigid associative algebra. There exists an integer s, $1 \leq s \leq n$ and a basis $\{e_1, \ldots, e_n\}$ of $\mathscr{A}_n$ such that the multiplication of $\mathscr{A}_n$ is given by*

$$\begin{cases} e_{2i-1}^2 = e_{2i-1}, \quad i = 1, \ldots, s; \\ e_{2i-1}e_{2i} = e_{2i}e_{2i-1} = e_{2i}, \quad i = 1, \ldots, s; \\ e_{2i}^2 = -e_{2i-1}, \quad i = 1, \ldots, s; \\ e_j^2 = e_j, \quad j = 2s+1, \ldots, n. \end{cases}$$

References

1. Ancochea Bermudez J.M.: On the rigidity of solvable Lie algebras. Deformation theory of algebras and structures and applications (Il Ciocco, 1986). NATO Advanced Science Institutes Series C: Mathematical and Physical Sciences, vol. 247, pp. 403–445. Kluwer Academic Publisher, Dordrecht (1988)
2. Ancochea Bermudez J.M., Campoamor Stursberg R., Garcia Vergnolle L., Goze M.: Algèbres de Lie réelles résolubles algébriquement rigides. Monatsh. Math. **152**(3), 187–195 (2007)
3. Boers, A.H.: Duplication of algebras. Indag. Math. **44**(2), 121–125 (1982)
4. Doubek, M., Markl, M., Zima, P.: Deformation theory (lecture notes). Arch. Math. (Brno) **43**(5), 333–371 (2007)
5. Carles, R.: Sur la structure des algèbres de Lie rigides. Ann. Inst. Fourier **34**(3), 65–82 (1984)
6. Fialowski, A., Schlichenmaier, M.: Global deformations of the Witt algebra of Krichever-Novikov type. Commun. Contemp. Math. **5**(6), 921–945 (2003)
7. Goze, M.: Algèbres de Lie: classifications, déformations et rigidité, géométrie différentielle. In: Barbot, T., Belbachir, H., Mehdi, S. (eds.) Algèbre, dynamique et analyse pour la géométrie: aspects récents, pp. 39–98. Ellipse, Paris (2010)
8. Goze, M.: Perturbations of Lie algebra structures. Deformation theory of algebras and structures and applications (Il Ciocco, 1986). NATO Advanced Science Institutes Series C: Mathematical and Physical Sciences, vol. 247, pp. 265–355. Kluwer Academic Publishers, Dordrecht (1988)
9. Goze, M., Ancochea Bermudez, J.M.: On the classification of rigid Lie algebras. J. Algebra **245**(1), 68–91 (2001)
10. Goze, M., Khakimdjanov, Y.: Nilpotent and solvable Lie algebras. Handbook of algebra, vol. 2, pp. 615–663, North-Holland, Amsterdam (2000)
11. Goze, M., Makhlouf, A.: On the rigid complex associative algebras. Commun. Algebra **18**(12), 4031–4046 (1990)
12. Goze, M., Remm, E.: Valued deformations of algebras. J. Algebra Appl. **3**(4), 345–365 (2004)
13. Harrison, D.K.: Commutative algebras and cohomology. Trans. Am. Math. Soc. **104**, 191–204 (1962)
14. Hochschild, G.: On the cohomology theory for associative algebras. Ann. Math. **47**(2), 568–579 (1946)
15. Neeb, K.H., Wagemanm F.: The second cohomology of current algebras of general Lie algebras. Canad. J. Math. **60**(4), 892–922

16. Remm, E., Goze, M.: On algebras obtained by tensor product. J. Algebra **327**, 13–30 (2011)
17. Schaps, M.: Deformations of algebras and Hochschild cohomology. Perspectives in ring theory (Antwerp, 1987). NATO Advanced Science Institutes Series C: Mathematical and Physical Sciences, vol. 233, pp. 41–58. Kluwer Academic Publishers, Dordrecht (1988)
18. Zusmanovich, P.: Low-dimensional cohomology of current Lie algebras and analogs of the Riemann tensor for loop manifold. Linear Algebra Appl. **407**, 71–104 (2005)

Deformations of Diagrams

Arvid Siqveland

Abstract In this this paper we introduce entanglement among the points in a noncommutative scheme, in addition to the tangent directions. A diagram of A-modules is a pair $\underline{c} = (|\underline{c}|, \Gamma)$ where $|\underline{c}| = \{V_1, \ldots, V_r\}$ is a set of A-modules, and $\Gamma = \{\gamma_{ij}(l)\}$ is a set of A-module homomorphisms $\gamma_{ij}(l)\colon V_i \rightarrow V_j$, seen as the 0'th order tangent directions. We define the deformation theory for diagrams, making these the fundamental points in noncommutative algebraic geometry. Two simple examples of the theory are given: The space of a line through the origin and a point, which is a noncommutative but untangled example, and the space of a line through the origin and a point on the line, in which the condition of the point gives an entanglement between the point and the line.

1 Introduction

Throughout, k is an algebraically closed field of characteristic 0. Let A be a (not necessarily commutative) k-algebra, and let $V = \{V_1, \ldots, V_r\}$ be a set of A-modules. In the article [3], Laudal defines the noncommutative deformation functor $\mathrm{Def}_V : a_r \rightarrow \mathrm{Sets}$, see also Eriksen [1]. Here a_r is the category of r-pointed, Artinian k-algebras S, fitting into the diagram

$$\begin{array}{ccc} k^r & \longrightarrow & S \\ & \overset{\mathrm{Id}}{\searrow} & \downarrow \rho \\ & & k^r, \end{array}$$

A. Siqveland (✉)
Faculty of Technology, Buskerud Vestfold University College, Kongsberg, Norway
e-mail: arvid.siqveland@hbv.no

A. Makhlouf et al. (eds.), *Algebra, Geometry and Mathematical Physics*, 259
Springer Proceedings in Mathematics & Statistics 85, DOI: 10.1007/978-3-642-55361-5_15,

with $(\ker \rho)^n = \mathrm{rad}(S)^n = 0$. In [4], we define a noncommutative scheme theory, generalizing the commutative one in the geometric situation: Let M be a simple (right) A-module, and let $\mathfrak{m}_M$ be the corresponding (right) maximal ideal. A is called geometric if $0 = \mathrm{rad}(A)^\infty = \bigcap_{\substack{M \in \mathrm{Simp}(A) \\ n \geq 1}} \mathfrak{m}_M^n$. In [3], Laudal proves that a pro-representing hull for the noncommutative deformation functor of $V = \{V_1, \dots, V_r\}$ exists when V is a family of finite dimensional (right or left) A-modules. This is a k-algebra $\hat{H} = (\hat{H}_{ij})_{1 \leq i,j \leq r}$ in the pro-category $\hat{a}_r$ together with a pro-versal (also called mini-versal) family

$$A \xrightarrow{\iota} (\hat{H}_{ij} \otimes_k \mathrm{Hom}_k(V_i, V_j)) = \hat{\mathcal{O}}_V$$

satisfying the pro-versality conditions, see e.g. Schlessinger [4]. First of all, the property of A being geometric assures that the pro-versal morphism ι is injective. Secondly, $\hat{\mathcal{O}}_V \twoheadrightarrow \oplus_{i=1}^r \mathrm{End}_k(V_i, V_i)$, and it is known that this surjection implies that, as sets, $\mathrm{Simp}(\hat{\mathcal{O}}_V) = V$, see e.g. [4]. Thus the sub k-algebra $\mathcal{O}_V \subseteq \hat{\mathcal{O}}_V$ generated by the image of the generators of A and the inverses of the generated elements not in any corresponding maximal ideal is the localization of A in V: It is a fractional k-algebra of a finitely generated k-algebra, and the only simple modules are the modules in V (or equivalently, the only maximal ideals are the maximal ideals corresponding to the modules in V).

On the set Simp(A) we now pose the following saturated Zariski (or Jacobson) topology: First of all, the Zariski topology is the topology generated by the open base, over $f \in A$, $D(f) = \{V \in \mathrm{Simp}(A) | \rho(f) \in \mathrm{End}_k(V) \text{ is injective}\}$, where ρ is the structure morphism. We let the saturation relation be the equivalence relation generated by the condition that V_i and V_j are related if $\mathrm{Ext}_A^1(V_i, V_j) \neq 0$. This means that an open subset is saturated with all related points, and it is straight forward to prove that this gives a topology.

Just as in the commutative situation, we define a sheaf of rings, the structure sheaf, on Simp(A) by

$$\mathcal{O}(U) = \varprojlim_{\underline{c} \subseteq U} \mathcal{O}_{\underline{c}},$$

where the limit is taken over subsets of equivalence classes $\underline{c}$ with respect to the equivalence relation above. Writing out this definition, we see that it is a true generalization of the definition given in Hartshorne [2] for commutative schemes.

We need to study entangled systems. This means that the equivalence relation above should include a zero'th derivative, that is elements in $\mathrm{Hom}_A(V_i, V_j)$. So, we define a diagram as a pair $\underline{c} = (|\underline{c}|, \Gamma)$ where $|\underline{c}| = \{V_1, \dots, V_r\}$ is a set of A-modules, and $\Gamma = \{\gamma_{ij}(l)\}$ is a set of A-module homomorphisms $\gamma_{ij} : V_i \to V_j$.

Extending the equivalence relation demands a generalization of the category a_r and its deformation functor. This is the main result of the text.

2 Algebras Over k^r and Their Geometry

Let $r \in \mathbb{N}$ and let (d_{ij}) be an $r \times r$-matrix with entries $d_{ij} \in \mathbb{N}$. We recall the definition of the free $r \times r$ matrix polynomial algebra generated by the matrix variables $t_{ij}(l)$, $1 \le l \le d_{ij}$, in entry $1 \le i, j \le r$.

By the notation (S_{ij}) where S_{ii} is a k-algebra for each i, $1 \le i \le r$, and S_{ij} is a k-vector space, we mean the k^r-algebra generated by the matrices $M = (m_{ij})$ with $m_{ij} \in S_{ij}$, $1 \le i, j \le r$.

Definition 2.1 For a positive integer r, for each pair (i, j), $1 \le i, j \le r$, let $d_{ij} \in \mathbb{N}$. Then the free $r \times r$ matrix polynomial algebra in the matrix variables $t_{ij}(l)$, $1 \le i, j \le d_{ij}$, is the k^r-algebra S generated by the matrix elements in

$$\begin{pmatrix} k\langle t_{11}(1), \dots, t_{11}(d_{11})\rangle & \cdots & \sum_{v=1}^{d_{1r}} kt_{1r}(v) \\ \vdots & \ddots & \vdots \\ \sum_{v=1}^{d_{r1}} kt_{r1}(v) & \cdots & k\langle t_{rr}(1), \dots, t_{rr}(d_{rr})\rangle \end{pmatrix}.$$

Alternatively, we consider the k^r-module V generated by $t_{ij}(l)$, and let S be the tensor algebra

$$S = T_{k^r}(V).$$

Definition 2.2 For a positive integer r, a finitely generated $r \times r$ matrix polynomial algebra is a quotient of a free $r \times r$ matrix polynomial algebra.

Recall the following, proved in e.g. [4]:

Lemma 2.1 *Let R be a k-algebra, k algebraically closed, and let V be a finite dimensional R-module. Then V is simple if and only if the structure morphism*

$$\rho : R \to \operatorname{End}_k(V),$$

sending $r \in R$ to $\rho(r)(v) = rv$, is surjective.

Let $S_{ii} = k\langle t_{ii}(1), \dots, t_{ii}(d_{ii})\rangle$. Then there is a surjection $\rho_{ii}: S \to S_{ii}$ sending e_i to 1, $t_{ii}(l)$ to $t_{ii}(l)$ and all other generators to 0.

Lemma 2.2 *Simple S-modules are exactly the modules on the form $V_i = S/\rho_{ii}^{-1}(\mathfrak{m}_{ii})$ where $\mathfrak{m}_{ii} \subset S_{ii}$ is a maximal ideal for some i, $1 \le i \le r$.*

Proof For a maximal ideal $\mathfrak{m}_{ii} \subset S_{ii}$, we have an isomorphism

$$S/\rho_{ii}^{-1}(\mathfrak{m}_{ii}) \xrightarrow{\simeq} S_{ii}/\mathfrak{m}_{ii}.$$

This proves that V_i is a simple S-module. For the converse, assume $\mathfrak{m} \subset S$ is maximal. If $\rho_{ii}(\mathfrak{m}) = S_{ii}$ for all i, it follows that $1 = \sum e_i$ is in $\mathfrak{m}$ which is impossible. Thus there exists an i where $\rho_{ii}(\mathfrak{m}) \subseteq \mathfrak{m}_{ii}$ for a maximal ideal $\mathfrak{m}_{ii} \subset S_{ii}$.

Then $\mathfrak{m} \subseteq \rho_{ii}^{-1}(\rho_{ii}(\mathfrak{m})) \subseteq \rho_{ii}^{-1}(\mathfrak{m}_{ii}) \subsetneq S$. Then $\mathfrak{m} = \rho_{ii}^{-1}(\mathfrak{m}_{ii})$, by maximality, and the lemma is proved.

Let Simp(S) be the set of simple modules. We generalize the Zariski topology to the noncommutative case by as follows: For an element $s \in S$ we define the subset $D(s) = \{V \in \text{Simp}(S) | \rho(s) \colon V \to V \text{ is invertible}\}$. The noncommutative Zariski topology is the topology generated by the sets $D(s)$, $s \in S$.

We also generalize the tangent space. In the case of the free k^r-algebra S, every simple module is one-dimensional, and k is an S bimodule in the following way: $k \cong \text{Hom}_k(k, k)$, and for any two S-modules V_1, V_2, the left and right actions of $s \in S$ on the bimodule $\text{Hom}_k(V_1, V_2)$ is given respectively by the left and right skew morphism in the diagram

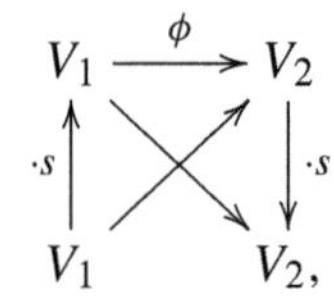

i.e. $(s\phi)(v) = \phi(vs)$ and $(\phi s)(v) = \phi(v)s$, (for right modules).

Definition 2.3 For a general k-algebra A, the tangent space between the two A-modules V_1 and V_2 is

$$T_{V_1,V_2} = \text{Ext}_A^1(V_1, V_2) \cong \text{HH}^1(A, \text{Hom}_k(V_1, V_2)) \\ \cong \text{Der}_k(A, \text{Hom}_k(V_1, V_2))/\,\text{Inner}\,.$$

The following is a straightforward computation:

Lemma 2.3 *Let S be the general, free, $r \times r$ matrix polynomial algebra, and let $V_i = V_{ii}(p_{ii})$ be the point p_{ii} in entry i, i. Then the tangent space from V_i to V_j is* $T_{V_i,V_j} = \text{Ext}_S^1(V_i, V_j) = \oplus_{l=1}^{d_{ij}} k\text{d}_{t_{ij}(l)}$.

3 Higher Order Derivatives: Generalized Matric Massey Products

In [1] Eriksen has given the description of the noncommutative deformation functor, in [5] we have defined the generalized matric Massey products. We recall the parts necessary to make the generalization:

Let A be a k-algebra. A deformation M_S of an A-module M to an Artinian local k-algebra S with residue field k, i.e. $S \in \mathrm{ob}(\ell)$, is an $S \otimes_k A$-module, flat over S, such that $k \otimes_S M_S \cong M$. Two deformations M_S and M'_S are equivalent if there exists an isomorphism $\phi: M_S \to M'_S$ commuting in the diagram

$$\begin{array}{ccc} M_S & \xrightarrow{\phi} & M'_S \\ & \searrow \quad \swarrow & \\ & M & \end{array} .$$

This gives the deformation functor $\mathrm{Def}_M : \ell \to$ Sets satisfying Schlessinger's well-known criteria for the existence of a pro-representing hull, see [4].

The flatness of $M_S \in \mathrm{Def}_M(S)$ over S is equivalent with the fact that as S-module, $M_S \cong S \otimes_k M$. For a small surjective morphism $0 \to I \to S \overset{\pi}{\twoheadrightarrow} R \to 0$, we use induction and linear algebra on the exact sequence $0 \to I \to S \to k \to 0$ to see this. So to give an A-module structure on M_S that is a lifting of the R-module structure, is equivalent to give a k-algebra homomorphism $\sigma_S: A \to \mathrm{End}_k(M_S)$ commuting in the diagram

$$\begin{array}{ccc} A & \xrightarrow{\sigma_S} & \mathrm{End}_k(M_S) \\ & \overset{\sigma}{\searrow} & \downarrow \\ & & \mathrm{End}_k(M_R). \end{array}$$

Using the fact that σ_S should commute with the action of S, that is, it should be S-linear, it is sufficient to define $\sigma_S(a): M \to S \otimes_k M$. For each $a \in A$, σ_S should be a lifting of σ_R, and so we choose the obvious lifting of σ_R. Then all properties but the associativity are fulfilled, and the associativity of σ_S says $\sigma_S(ab) - \sigma_S(a)\sigma_S(b) = 0$. So our obstructions for lifting M_R are the elements $\sigma_S(ab) - \sigma_S(a)\sigma_S(b) \in \mathrm{Hom}_k(M, M \otimes_k I) \cong \mathrm{End}_k(M, M) \otimes_k I$. As these are Hochschild two-cocycles, we have our obstructions

$$o(M_R, \pi) \in \mathrm{HH}^2(A, \mathrm{End}_k(M, M)) \otimes_k I,$$

with the property that M_R can be lifted to a M_S if and only if $o(M_R, \pi) = 0$.

Then we have the an alternative way of viewing this: Choose a free resolution of the A-module M,

$$0 \leftarrow M \overset{\varepsilon}{\leftarrow} L_0 \overset{d_0}{\leftarrow} L_1 \overset{d_1}{\leftarrow} L_2 \overset{d_2}{\leftarrow} \cdots .$$

We have proved that to give a lifting of M to S is equivalent to give a lifting of complexes

$$\begin{array}{ccccccccccc}
 & & 0 & & 0 & & 0 & & 0 & & \\
 & & \downarrow & & \downarrow & & \downarrow & & \downarrow & & \\
0 & \leftarrow & I\otimes_k M & \leftarrow & I\otimes_k L_0 & \leftarrow & I\otimes_k L_1 & \leftarrow & I\otimes_k L_2 & \leftarrow & \cdots \\
 & & \downarrow & & \downarrow & & \downarrow & & \downarrow & & \\
0 & \leftarrow & M_S & \leftarrow & S\otimes_k L_0 & \leftarrow & S\otimes_k L_1 & \leftarrow & S\otimes_k L_2 & \leftarrow & \cdots \\
 & & \downarrow & & \downarrow & & \downarrow & & \downarrow & & \\
0 & \leftarrow & M & \leftarrow & L_0 & \leftarrow & L_1 & \leftarrow & L_2 & \leftarrow & \cdots \\
 & & \downarrow & & \downarrow & & \downarrow & & \downarrow & & \\
 & & 0 & & 0 & & 0 & & 0. & &
\end{array}$$

For $k[\varepsilon] = k[x]/(x^2)$ as usual, i.e. $\varepsilon^2 = 0$, the tangent space of the deformation functor is

$$\mathrm{Def}_M(k[\varepsilon]) \cong \mathrm{Ext}^1_A(M, M) \cong \mathrm{HH}^1(A, \mathrm{End}_k(M)),$$

and likewise for the obstruction space. Using this, we find the correspondence $\mathrm{Ext}^1_A(M, M) \xrightarrow{\phi} \mathrm{HH}^1(A, \mathrm{End}_k(M))$ given as follows: Given a representative $\xi \in \mathrm{Hom}_A(L_1, M)$ for $\overline{\xi} \in \mathrm{Ext}^1_A(M, M)$. Choose a k-linear section $\sigma\colon V \to L_0$ and let $x \in L_1$ map to $\sigma(am) - a\sigma(m) \in L_0$. Then $\phi(\overline{\xi})(a)(m) = \xi(x)$.

So we can equally well work in the Yoneda complex, with homomorphisms $d_i : L_i \to L_{i-1}$ being matrices, as each L_i is assumed to be free. In the above diagram, choose the obvious liftings d_i^S to S. Then the obstruction for lifting M_R to M_S via π is represented by

$$o(M_R, \pi) = \{d^S_{i-1} d^S_i\} \in \mathrm{Hom}^2(L., L.) \otimes_k I$$

which is a 2-cocycle in the Yoneda complex. This theory can be generalized to the r-pointed situation:

Consider the category of r-pointed, Artinian k-algebras which we treated above. Let $\{V_1, \dots, V_r\}$ be a set of r A-modules and put $V = \oplus_{i=1}^r V_i$. Then a deformation V_S of V to S is a $S\otimes_k A$-module V_S, flat over S, such that $k^r \otimes_S V_S \cong V$ as A^r-module. As before, two deformations V_S and V'_S are equivalent, if there exists an isomorphism ϕ of $S \otimes_k A$-modules commuting in the diagram

$$\begin{array}{ccc}
V_S & \xrightarrow{\ \phi\ } & V'_S \\
 & \searrow \quad \swarrow & \\
 & V. &
\end{array}$$

As in the commutative situation, we let flatness (and we can prove that it in fact is) be equivalent to, as left S-module,

$$V_S \cong S \otimes_{k^r} V \cong (S_{ij} \otimes_k V_j)_{1 \le i,j \le r}.$$

Assume we have a small, surjective homomorphism $0 \to I \to S \xrightarrow{\pi} R \to 0$ in a_r. To give a lifting of V_R to S is to give an A^r-module structure on $(S_{ij} \otimes_k V_j)_{1 \le i,j \le r}$, lifting the action on V_R, which is a k-algebra homomorphism σ_S commuting in the diagram

$$\begin{array}{ccc} A^r & \xrightarrow{\sigma_S} & \mathrm{End}_k((S_{ij} \otimes_k V_j)) \\ & \searrow^{\sigma_R} & \downarrow \\ & & \mathrm{End}_k((R_{ij} \otimes_k V_j)). \end{array}$$

As the A-action is assumed to commute with the S and R-actions, by associativity, this is to give, for each $a \in A^r$, a k-linear homomorphism $\sigma_a \colon V \to (S_{ij} \otimes_k V_j)$. Also, as for each idempotent $e_i \in S$, $\sigma_a(e_i v) = e_i \sigma_a(v)$, this is equivalent to giving a k-linear homomorphism $\sigma_a \colon V_i \to S_{ij} \otimes_k V_j$ for each $a \in A$. Using this exactly as in the commutative situation, we get the natural k-linear lifting of σ_R to S. Everything is fulfilled but the associativity, and we get an obstruction

$$o(V_R, \pi) = (o_{ij}) \in (\mathrm{HH}^2(A, \mathrm{Hom}_k(V_i, V_j) \otimes_k I_{ij})),$$

where $I = (I_{ij})$ is the kernel of π, such that V_R can be lifted to V_S if and only if $o(V_R, \pi) = 0$.

We have to replace $k[\varepsilon]$ in the r-pointed situation. The new basic element in a_r is denoted *the test algebra*, and is not surprisingly given as

$$k[\varepsilon_{ij}] = \begin{pmatrix} k\langle t_{11}\rangle & \cdots & kt_{1r} \\ \vdots & \ddots & \vdots \\ kt_{r1} & \cdots & k\langle t_{rr}\rangle \end{pmatrix} / (t_{ij})^2.$$

The tangent space of the deformation functor is then $\mathrm{Def}_V(k[\varepsilon_{ij}])$, and again it can be seen that this is isomorphic to the matrix $(\mathrm{HH}^1(A, \mathrm{Hom}_k(V_i, V_j)))$.

To find the correspondence as above, we use free resolutions: For each V_i we choose free resolutions $0 \leftarrow V_i \leftarrow L.^i$ with differential $d.^i$, we put $L. = \oplus_{i=1}^r L.^i$, and think of this as a free resolution of V with differential $d. = \oplus_{i=1}^r d^i_.$.

Any morphism $\phi \colon : L_i \to L_{i-1}$ can be represented by a matrix $\phi = (\phi_{ij})^T$ where $\phi_{ij} : V_i \to V_j$. Note that multiplying from the left, we have to transpose the matrices. So in our case, we use "matrices of matrices". Then all computations, all choices of bases etc. can be done exactly as in the case with one-pointed algebras. The notation is somewhat more cumbersome because of the matrix expressions, but that is not a problem as one will see from the examples.

4 Incidence-Free Example: The Pair (Line Through Origin, Point)

We consider the plane $A = k[x, y]$, the x-axis $V_1 = k[x, y]/(y)$, and the origin $V_2 = k[x, y]/(x, y)$.

We put $V = V_1 \oplus V_2$ and construct the following resolution

$$0 \longleftarrow V \xleftarrow{\begin{pmatrix}1&0\\0&1\end{pmatrix}} A \oplus A \xleftarrow{\begin{pmatrix}(y)&0\\0&(x\ y)\end{pmatrix}} A \oplus A^2 \xleftarrow{\begin{pmatrix}(0)&0\\0&\begin{pmatrix}y\\-x\end{pmatrix}\end{pmatrix}} A \longleftarrow 0.$$

Lemma 4.1 *The tangent space of the deformation functor of V is the following:*

$$\operatorname{Ext}^1_A(V_1, V_1) = V_1,\ \operatorname{Ext}^1_A(V_1, V_2) = V_2 \cong k,$$
$$\operatorname{Ext}^1_A(V_2, V_1) \cong k,\ \operatorname{Ext}^1_A(V_2, V_2) = k^2.$$

Proof Taking the $\operatorname{Hom}(-, V)$ of the sequence, computing componentwise, we get:

$\operatorname{Ext}^1_A(V_1, V_1)$: $V_1 \xrightarrow{0} V_1 \longrightarrow 0 \Rightarrow \operatorname{Ext}^1_A(V_1, V_1) = V_1$,

$\operatorname{Ext}^1_A(V_1, V_2)$: $V_2 \xrightarrow{0} V_2 \longrightarrow 0 \Rightarrow \operatorname{Ext}^1_A(V_1, V_1) = V_2 \cong k$,

$\operatorname{Ext}^1_A(V_2, V_1)$: $V_1 \xrightarrow{\begin{pmatrix}x\\0\end{pmatrix}} V_1^2 \xrightarrow{(0\ -x)} V_1 \longrightarrow 0$. We give the straightforward computation:

$$(0\ -x)\begin{pmatrix}\overline{v}_1\\\overline{v}_2\end{pmatrix} = 0 \Leftrightarrow -x\overline{v}_2 = 0 \Leftrightarrow -xv_2 = hy \Leftrightarrow v_2 \in (y) \Leftrightarrow \overline{v}_2 = 0.$$

Thus the kernel is the set of elements of the form $(\overline{v}_1, 0)$, the image is the elements of the form $(xv, 0)$ so that $\operatorname{Ext}^1_A(V_2, V_1) = \langle(\alpha, 0)\rangle \cong k$.

$\operatorname{Ext}^1_A(V_2, V_2)$: $V_2 \xrightarrow{0} V_2^2 \xrightarrow{0} V_2 \Rightarrow \operatorname{Ext}^1_A(V_2, V_2) \cong V_2^2 \cong k^2$.

A line can be deformed flatly into any other curve passing through the origin. This is the result of $\operatorname{Ext}^1_A(V_1, V_1) = V_1$. In this example, we are interested in deformations of lines, thus we will only consider deformations of the line that are also lines. This equals to deformations of linear homogeneous curves, and we choose $x \in V_1 = \operatorname{Ext}^1_A(V_1, V_1)_{(1)}$ as our tangent direction. So for this example, our free noncommutative algebra with $H/\operatorname{rad}(H)^2 = S/\operatorname{rad}(S)^2$ is

$$S = \begin{pmatrix} k\langle t_{11}\rangle & t_{12} \\ t_{21} & k\langle t_{22}(1), t_{22}(2)\rangle \end{pmatrix}.$$

Now we will give the Yoneda representation of the tangent space. Recall that the Yoneda complex for a resolution $L.$ of M is given as

$$\mathrm{Hom}_A^p(L., L.) = \{\xi_i : L_{i+p} \to L_i\}_i,$$

with differential $d^p : \mathrm{Hom}^p(L., L.) \to \mathrm{Hom}^{p+1}(L., L.)$ given by

$$d^p(\{\xi\}) = \{\xi_i \circ d - (-1)^p d \circ \xi_{i+1}\}.$$

The tangent space is given by the following obvious diagrams:

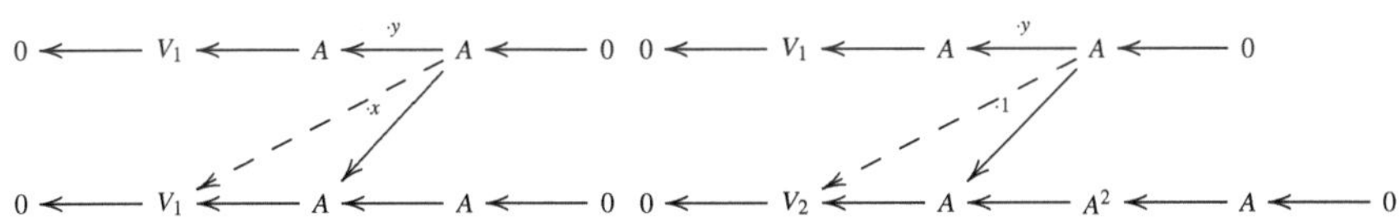

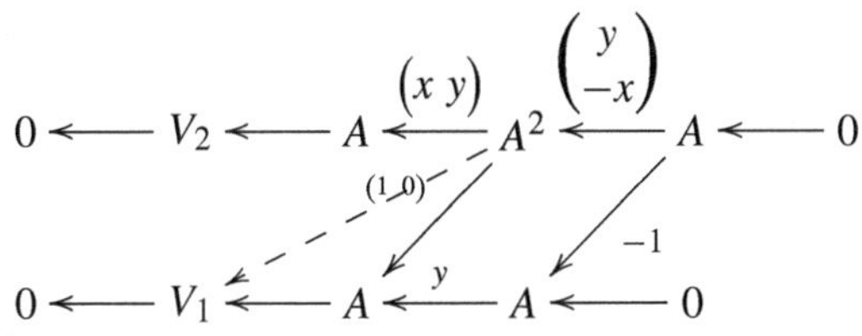

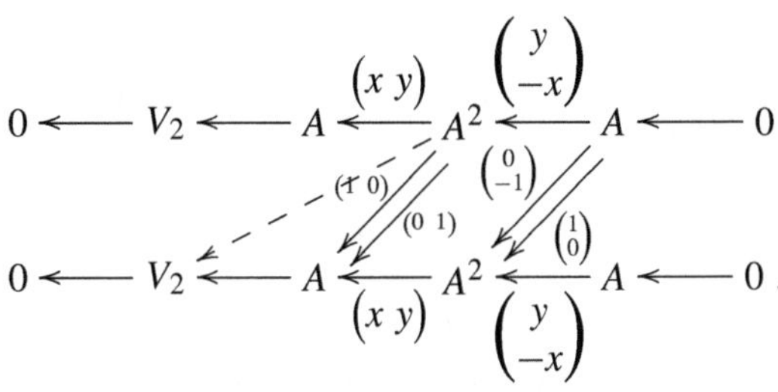

Given the Yoneda representation of the tangent space, we can compute the 2. Order Massey Products (the cup products). To make clear how morphisms are composed, notice the following illustrative way of thinking:

$$L_2 \to S \otimes_k L_1 \to S \otimes_k S \otimes_k L_0$$

sends l to the sequence:

$$l \mapsto \underline{t}_1 \otimes \alpha_{\underline{t}_1}(l) \mapsto \underline{t}_1(\underline{t}_2 \otimes \alpha_{\underline{t}_2}(\alpha_{\underline{t}_1}(l))) = \underline{t}_1\, \underline{t}_2 \otimes \alpha_{t_2}(\alpha_{\underline{t}_1}(l)).$$

We get the following results:

$$\begin{gathered}\langle t_{11}^2\rangle = 0,\ \langle t_{11}t_{12}\rangle = 0,\ \langle t_{22}(1)t_{21}\rangle = 0,\\ \langle t_{22}(1)^2\rangle = 0,\ \langle t_{22}(2)t_{22}(1)\rangle = 1,\ \langle t_{12}t_{21}\rangle = 0,\\ \langle t_{12}t_{22}(1)\rangle = 0,\ \langle t_{22}(2)t_{21}\rangle = 1,\ \langle t_{22}(2)^2\rangle = 0,\\ \langle t_{21}t_{12}\rangle = -1,\ \langle t_{11}^2\rangle = 0,\ \langle t_{12}t_{22}(2)\rangle = 0,\\ \langle t_{21}t_{11}\rangle = -x,\ \langle t_{22}(1)t_{22}(2)\rangle = -1,\ \langle t_{21}t_{11}(2)\rangle = 0.\end{gathered}$$

Then this is nearly as simple as it can be, every cup product is zero or a base element, as far as $\langle t_{21}t_{11}\rangle = -x = 0 \in \mathrm{Ext}^2_A(V_2, V_1)$, forcing us to choose the following 2. order defining system:

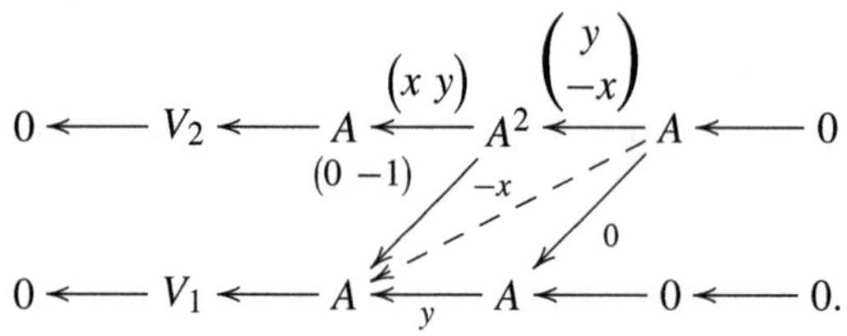

This means that $\alpha_{t_{21}t_{11}} = \begin{cases}\begin{pmatrix}0 & -1\end{pmatrix}\\ 0\end{cases}$, and all the rest of the 2. order defining systems can be chosen to be 0. The only 3. order Massey products to be computed are then:

$$\langle t_{21}t_{11}t_{11}\rangle = 0,\ \langle t_{22}(1)t_{21}t_{11}\rangle = 1,\ \langle t_{22}(2)t_{21}t_{11}\rangle = 0.$$

As the remaining differences are trivial cohomology classes, identically 0, we have the following result

Proposition 4.1 *The versal base space of a line through the origin and a point is*

$$H \simeq \frac{\begin{pmatrix}k[t_{11}] & t_{12}\\ t_{21} & k\langle t_{22}(1), t_{22}(2)\rangle\end{pmatrix}}{(t_{22}(2)t_{22}(1) - t_{22}(1)t_{22}(2) - t_{21}t_{12}, t_{22}(2)t_{21} + t_{22}(1)t_{21}t_{11})}.$$

We can interpret this result geometrically: Putting $t_{12} = t_{21} = 0$ we get families:
(1,1): $k[x, y]/(t_{11}x + y)$, $t_{11} \in k$: Lines with slope t_{11}.
(2,2): $k[x, y]/(x + t_{22}(1), y + t_{22}(2))$: Points $(-t_{22}(1), -t_{22}(2))$.
The variables t_{12}, t_{21} tell how the objects are related at tangent level,

$$t_{22}(2)t_{22}(1) - t_{22}(1)t_{22}(2) - t_{21}t_{12}$$

gives no forced tangent relations, it is just a description of the geometry. Then

$$t_{22}(2)t_{21} + t_{22}(1)t_{21}t_{11} \Rightarrow t_{21}(t_{11}t_{22}(1) + t_{22}(2)) = t_{21}(t_{11}x + y)$$

means that the constant ext-locus consists of points on the line (the line). So this is the moduli of all pairs of a point and a line through the origin. In particular, the Ext-dimension is correct for points on the line.

For the above, we give the computation:

$$\delta(t_{22}(2)t_{21} + t_{22}(1)t_{21}t_{11}) = 0 \Leftrightarrow t_{22}(2)\delta(t_{21}) + t_{22}(1)\delta(t_{21}t_{11}) = 0$$
$$\Leftrightarrow t_{22}(2)\delta(t_{21}) + t_{22}(1)\delta(t_{21})t_{11} = 0 \Leftrightarrow \delta(t_{21})(t_{22}(2) + t_{22}(1)t_{11}) = 0.$$

5 Deformations and Interactions

Definition 5.1 A *diagram* $\underline{c}$ of right A-modules consists of a family $|\underline{c}|$ of right A-modules, together with a set $\Gamma(V, W) \subseteq \mathrm{Hom}_A(V, W)$ of A-module homomorphisms for each pair of modules $V, W \in |\underline{c}|$.

If $A = k$, this is called a representation of the corresponding quiver, i.e. the quiver with $|\underline{c}|$ as set of nodes and $\Gamma = \bigcup_{V,W\in|\underline{c}|} \Gamma(V, W)$ as arrows. Throughout, we will just use the notation Γ for the corresponding quiver.

Let $k[\Gamma]$ denote the the quiver algebra. By definition, the quiver algebra is the k-algebra generated by all finite paths $\gamma_1\gamma_2\cdots\gamma_i\gamma_{i+1}\cdots\gamma_n$ such that the head of γ_i is the tail of γ_{i+1}. Note that e_i, the identity at the node i, is considered as a finite path. Thus $k[\Gamma]$ is isomorphic to $k^r[\Gamma]$.

Definition 5.2 The category a_Γ of pointed Artinian Γ-algebras is the category of $k[\Gamma]$-algebras fitting into the diagram

$$\begin{array}{ccc} k[\Gamma] & \longrightarrow & S \\ & \overset{\mathrm{Id}}{\searrow} & \downarrow \rho \\ & & k[\Gamma], \end{array}$$

such that $\mathrm{rad}(S)^n = (\ker \rho)^n = 0$. The morphisms of a_Γ are the commuting $k[\Gamma]$-homomorphisms.

Notice that when $\Gamma = \emptyset$, that is we have a diagram with a trivial quiver (no morphisms but the identities at each node), then $k[\Gamma] = k^r$ so that this definition is a generalization of the r-pointed algebras. Also notice that by Lemma 2.1, S has exactly r simple modules.

For the deformation theory, we recall that in the discrete situation, i.e. $\Gamma = \emptyset$, we considered $V = \bigoplus_{i=1}^{r} V_i$ as an A^r-module, and a lifting V_S of V to $S \in \mathrm{ob}(a_r)$ is an

$S \otimes_k A$-module satisfying $k^r \otimes_S V_S \cong V$. So we consider V as a $k^r \otimes_k A$-module. This makes perfectly sense also in the situation with nontrivial quivers:

Consider $\underline{c}$ as a $k[\Gamma] \otimes_k A$-module, for short, as an $A[\Gamma]$-module, by letting the elements in Γ act by right multiplication. That is, an element $v = (v_1, \ldots, v_r) \in V$ is given the action

$$v\gamma_{ij}(l) = \begin{pmatrix} v_1 & \cdots & v_r \end{pmatrix} \begin{pmatrix} & \vdots & \\ \cdots & \gamma_{ij}(l) & \cdots \\ & \vdots & \end{pmatrix} = \begin{pmatrix} \cdots & v_i\gamma_{ij}(l) & \cdots \end{pmatrix}$$

where we let $v_i\gamma_{ij}(l) = \gamma_{ij}(l)(v_i)$. So V is a right A, right $k[\Gamma]$-module, and because Γ consists of A-linear morphisms, these actions commute. Thus V is an $A[\Gamma]$-module.

We want to generalize the deformation functor $\mathrm{Def}_V : a_r \to \mathrm{Sets}$ to the category a_Γ. A deformation of the diagram $\underline{c}$ to an object S in a_Γ should be a deformation V_S which is a deformation of $V = |\,\underline{c}\,|$ to S as an object in a_r, but it should also lift the morphisms in the diagram, i.e. the quiver Γ of $\underline{c}$. Here, V_S a lifting of $V = |\,\underline{c}\,|$ to S as object in a_r, means the natural restriction to $S_r = S/\Gamma$.

Definition 5.3 Let $\underline{c}$ be a diagram of A-modules. We define $\mathrm{Def}_{\underline{c}} : a_\Gamma \to \mathrm{Sets}$ by letting a deformation, or lifting, of $\underline{c}$ to S be an $S \otimes_k A$-module V_S, flat over S, such that $k[\Gamma] \otimes_S V_S \cong \underline{c}$, as an $A[\Gamma]$-module.

Two deformations are equivalent, $V_S \sim V'_S$, if they are isomorphic over S, i.e. there exists an isomorphism $\iota : V_S \to V'_S$ commuting with the induced isomorphism $k[\Gamma] \otimes_S V_S \cong k[\Gamma] \otimes_S V'_S$.

Lemma 5.1 *$V_S \in \mathrm{Def}_V^\Gamma(S)$ is S-flat if and only if $V_S \cong S \otimes_{k[\Gamma]} V$ as S-module.*

Proof This follows exactly as in the discrete situation; for $R \in a_r$, $V_R \in \mathrm{Def}_V(R)$ is R-flat if and only if $V_R \cong R \otimes_{k^r} V$ as R-module.

Thus, a deformation, or lifting, of V to S is an A-module structure on $S \otimes_{k[\Gamma]} V$, commuting with the action of S (and then the induced action of $k[\Gamma]$). Following step by step the discrete situation, that is, for every $a \in A$ we give an action morphism $\sigma_a : V \to S \otimes_{k[\Gamma]} V$, commuting with the $k[\Gamma]$-action. There is a k^r-morphism $\kappa : S \otimes_{k[\Gamma]} V \twoheadrightarrow S \otimes_{k^r} V$ given by $\kappa(\gamma_{ij} \otimes v) = 1 \otimes \gamma_{ij}v$, e.g., the Γ-action on $S \otimes_{k^r} V$ is right Γ action on V. So this is equivalent to, for each $a \in A$, giving an action morphism $\sigma_a : V \to S \otimes_{k^r} V$ commuting with all $\gamma \in \Gamma$. The obstruction theory is then exactly as before, except that the cohomology controlling deformations is $\mathrm{Ext}_A^\Gamma(V_i, V_j)$, the left derived functor of $\mathrm{Hom}_A^\Gamma(V, -)$, where the superscript Γ denotes the subspace of morphisms commuting with all $\gamma \in \Gamma$. Notice in particular that the test-algebra in the incidence situation is

$$k^\Gamma[\varepsilon] = k[\Gamma] \otimes_{k^r} (t_{ij})/(t_{ij})^2.$$

As in the discrete situation, the obstruction calculus can be performed in the Hochschild cohomology; the σ's give a homomorphism

$$\sigma : A \to \operatorname{Hom}_A^{\Gamma}(A, \operatorname{Hom}_k(V_i, V_j)) \otimes_k I \subseteq \operatorname{Hom}_A(A, \operatorname{Hom}_k(V_i, V_j)) \otimes_k I,$$

where I is the kernel of a small morphism $\pi : S \twoheadrightarrow R$. All computations are identical, we should only be sure they respects the action of Γ.

Experience proves that, in some situations, it is easier to work with free resolutions of modules. The computations in the Hochschild cohomology then translates as follows:

Choose resolutions $0 \leftarrow V_i \to L^i_.$. We can lift Γ to the components in the respective projective resolutions, so that $L_i = \bigoplus_{j=1}^{r} L_i^j$ becomes an $A[\Gamma]$-module as well as $V = \bigoplus_{j=1}^{r} V_j$:

$$\begin{array}{ccccccccccc}
0 & \longleftarrow & V_i & \longleftarrow & L_0^i & \xleftarrow{d_0^i} & L_1^i & \xleftarrow{d_1^i} & L_2^i & \xleftarrow{d_2^i} & \cdots \\
 & & \downarrow{\scriptstyle \gamma_{ij}} & & \downarrow{\scriptstyle \gamma_{ij}} & & \downarrow{\scriptstyle \gamma_{ij}} & & \downarrow{\scriptstyle \gamma_{ij}} & & \\
0 & \longleftarrow & V_j & \longleftarrow & L_0^j & \xleftarrow{d_0^j} & L_1^j & \xleftarrow{d_1^j} & L_2^j & \xleftarrow{d_2^j} & \cdots .
\end{array}$$

In the discrete situation, we worked in the Yoneda complex $\operatorname{Hom}_A^{\cdot}(L., L.)$, using the quasi isomorphism

$$\iota : \operatorname{Hom}_A^{\cdot}(L., L.) \to \operatorname{Hom}_A(L., V).$$

Lifting the action of Γ as above, we get the natural action of Γ on $\operatorname{Hom}_A^{\cdot}(L., L.)$ and $\operatorname{Hom}_A(L., V)$, giving us the possibility to consider $\operatorname{Hom}_A^{\Gamma}(L., V)$ and $\operatorname{Hom}_A^{\cdot,\Gamma}$ $(L., L.)$, the morphisms that are invariant under Γ. There is no reason why these two are quasi-isomorphic in general, so we have to take invariant cycles, construct obstructions in $H^2(\operatorname{Hom}_A^{\cdot}(L., L.)) \cong H^2(\operatorname{Hom}(L., V)) \cong \operatorname{HH}^2(A, \operatorname{End}_k(V))$, knowing that the resulting class is an invariant, that is an element in

$$^{\Gamma}\operatorname{HH}^2(A, \operatorname{End}_k(V)) \subseteq \operatorname{Ext}_A^2(L., V).$$

We start by choosing bases $\{(t_{ij}^*(l))\} \subset {}^{\Gamma}\operatorname{Ext}_A^1(V, V)$, and let the test algebra be

$$S = k[\Gamma] \otimes_{k^r} (t_{ij}^*(l)),$$

and we do the computations exactly as before.

6 Example with Incidence: The Pair (Line Through the Origin, Point on the Line)

We will parameterize pairs (L, p) where L is a line through the origin in the plane and p is a point on the line L.

We consider the plane $k[x, y]$, the x-axis $V_1 = k[x, y]/(y)$, and the origin $V_2 = k[x, y]/(x, y)$.

Inside the moduli of the pairs (L, p), a line and a point, lies the moduli space of pairs (L, p) with p being a point on the line L. To say algebraically, there is a homomorphism $\gamma : A(L) \to k(p)$, where $A(L)$ denotes the affine ring of L. For this subspace of moduli, we use the corresponding notations:

$$A = k[x, y],\ V_1 = A/(y) \overset{\gamma_{12}}{\to} A/(x, y) = V_2.$$

Lift the quotient morphism, which is our incidence in this example, to the resolution $0 \leftarrow V = V_1 \oplus V_2 \leftarrow L. = L^1. \oplus L^2.$ of V according to the following

$$\begin{array}{ccccccccccc}
0 & \longleftarrow & V_1 & \longleftarrow & A & \overset{y}{\longleftarrow} & A & \longleftarrow & 0 & & \\
 & & \downarrow{\scriptstyle \gamma_{12}} & & \downarrow{\scriptstyle \mathrm{Id}} & & \downarrow{\scriptstyle \begin{pmatrix}0\\1\end{pmatrix}} & & & & \\
0 & \longleftarrow & V_2 & \longleftarrow & A & \underset{\begin{pmatrix}x & y\end{pmatrix}}{\longleftarrow} & A^2 & \longleftarrow & A & \longleftarrow & 0.
\end{array}$$

Then we start computing, taking the incidence into consideration. Let $\phi = \begin{pmatrix}\phi_{11} & \phi_{21} \\ \phi_{12} & \phi_{22}\end{pmatrix} \in \mathrm{Ker}(\mathrm{Hom}_A(L_1, V) \to \mathrm{Hom}_A(L_2, V))$. From the computations in the previous subsection, we then know

$$\begin{aligned}
\phi_{11} &= v \in V_1 \\
\phi_{21} &= (v, 0) \in V_1^2 \\
\phi_{12} &= \alpha \in V_2 \cong k \\
\phi_{22} &= (\alpha, \beta) \in V_2^2 \cong k^2.
\end{aligned}$$

For ϕ to be invariant under the action of Γ, i.e. $\phi \in \mathrm{Hom}_A^{\Gamma}(L_1, V)$, the diagram

$$\begin{array}{ccc}
L_1 & \overset{\gamma_{12}}{\longleftarrow} & L_1 \\
\downarrow{\scriptstyle \phi} & & \downarrow{\scriptstyle \phi} \\
V & \underset{\gamma_{12}}{\longleftarrow} & V
\end{array}$$

must be commutative. We get $\phi \circ \gamma_{12} = \gamma_{12} \circ \phi \Leftrightarrow$

$$\begin{pmatrix}\phi_{11} & \phi_{21}\\ \phi_{12} & \phi_{22}\end{pmatrix}\begin{pmatrix}0 & 0\\ \gamma_{12} & 0\end{pmatrix} = \begin{pmatrix}0 & 0\\ \gamma_{12} & 0\end{pmatrix}\begin{pmatrix}\phi_{11} & \phi_{21}\\ \phi_{12} & \phi_{22}\end{pmatrix} \Leftrightarrow \begin{pmatrix}\phi_{21}\circ\gamma_{12} & 0\\ \phi_{22}\circ\gamma_{12} & 0\end{pmatrix} = \begin{pmatrix}0 & 0\\ \gamma_{12}\circ\phi_{11} & \gamma_{12}\circ\phi_{12}\end{pmatrix}$$

which gives the equations

$$\begin{aligned}\phi_{21}\circ\gamma_{12} = 0 &\Leftrightarrow \begin{pmatrix}v & 0\end{pmatrix}\begin{pmatrix}0\\ 1\end{pmatrix} = 0,\\ \gamma_{12}\circ\phi_{21} = 0 &\Leftrightarrow \phi_{21} = (v, 0),\ v \in (x) \subset V_1,\\ \phi_{22}\circ\gamma_{12} = \gamma_{12}\circ\phi_{11} &\Leftrightarrow \begin{pmatrix}\alpha\beta\end{pmatrix}\begin{pmatrix}0\\ 1\end{pmatrix} = 0 \Leftrightarrow \phi_{22} = (\alpha, 0),\ \alpha \in k.\end{aligned}$$

We want to divide out by $\mathrm{Im}(\mathrm{Hom}_A^{\Gamma}(L_0, V) \to \mathrm{Hom}_A^{\Gamma}(L_1, V))$. For this, there is only one point of interest;

for $\psi = \begin{pmatrix}\psi_{11} & \psi_{21}\\ \psi_{12} & \psi_{22}\end{pmatrix}$, we have that $d\psi = \begin{pmatrix}\psi_{11} & \psi_{21}\\ \psi_{12} & \psi_{22}\end{pmatrix}\begin{pmatrix}d_1 & 0\\ 0 & d_2\end{pmatrix} = \begin{pmatrix}0 & (x,0)\psi_{21}\\ 0 & 0\end{pmatrix}$.

An element $\psi = \begin{pmatrix}\psi_{12} & \psi_{21}\\ \psi_{12} & \psi_{22}\end{pmatrix}$ is invariant if and only if $\psi\circ\gamma_{12} = \gamma_{12}\circ\psi \Leftrightarrow$

$$\begin{aligned}&\begin{pmatrix}\psi_{11} & \psi_{21}\\ \psi_{12} & \psi_{22}\end{pmatrix}\begin{pmatrix}0 & 0\\ \gamma_{12} & 0\end{pmatrix} = \begin{pmatrix}0 & 0\\ \gamma_{12} & 0\end{pmatrix}\begin{pmatrix}\psi_{11} & \psi_{21}\\ \psi_{12} & \psi_{22}\end{pmatrix}\\ \Leftrightarrow &\begin{pmatrix}\psi_{21} & 0\\ \psi_{22} & 0\end{pmatrix} = \begin{pmatrix}0 & 0\\ \gamma_{12}\circ\psi_{11} & \gamma_{12}\circ\psi_{21}\end{pmatrix} \Leftrightarrow \psi_{21} = 0,\end{aligned}$$

implying that the invariant image is the zero space.

This means that there are additional deformations in the case with incidences, and $\mathrm{Ext}_A^1(V_1, V_1)^{\Gamma}$ is infinite dimensional. Now, we choose a basis for the tangent space, contained in the case with incidences (notice that ϕ_{21} is killed by $\mathrm{Hom}_A(L_0, V)$ forgetting the incidences), i.e., we choose the following, invariant tangent space:

$$T_A^{\Gamma} = \{\phi \mid \phi = \begin{pmatrix}\alpha_1 x & 0\\ \alpha_2 & (\alpha_3\ 0)\end{pmatrix},\ \alpha_i \in k, 1 \le i \le 3\}.$$

The Yoneda representations are given by the following diagrams:

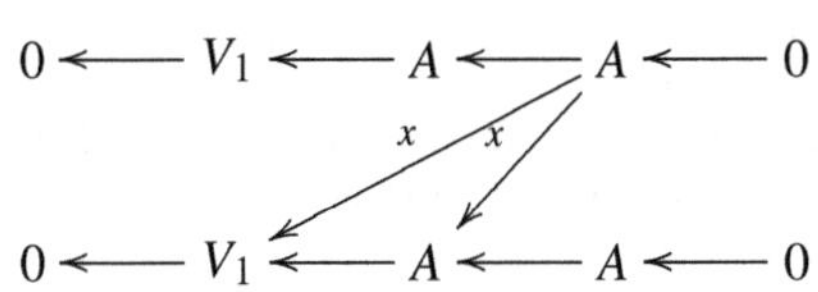

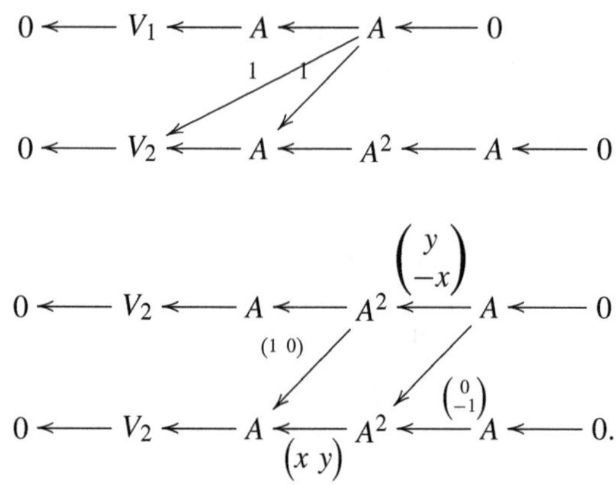

So the 2. order Massey products, the cup products, are

$$\langle t_{11}^2 \rangle = 0, \ \langle t_{12}t_{22} \rangle = \begin{pmatrix} 0 \\ -1 \end{pmatrix} : L_2^1 \to L_1^2 = A \in \mathrm{Hom}_A(L_2^1, V_1),$$
$$\langle t_{22}^2 \rangle = 0.$$

Because the only nonzero element is also necessarily nonzero in cohomology, this means that we end up with the following:

Proposition 6.1 *The moduli space of the pair (L, p) with p a point on the line L, inside the discrete moduli, is*

$$\begin{pmatrix} k[t_{11}] & t_{12} \\ 0 & k[t_{22}] \end{pmatrix} / (t_{12}t_{22}).$$

The geometric interpretation of this is the set of lines with slope t_{11}, and the point $(t_{22}, t_{11}t_{22})$. The relation just tells that the point has to move along the line.

References

1. Eriksen, E.: An Introduction to noncommutative deformations of modules. Lect. Notes Pure Appl. Math. **243**, 90–126 (2005)
2. Hartshorne, R.: Algebraic Geom. Grad. Texts Math. New York **52** (1977). ISBN 0-387-90244-9
3. Laudal, O.A.: Noncommutative algebraic geometry. Rev. Mat. Iberoamericana **19**(2), 509–580 (2003)
4. Schlessinger, M.: Functors of artin rings. Trans. Amer. Math. Soc. **130**, 208–222 (1968)
5. Siqveland, A.: Geometry of noncommutative k-algebras. J. Gen. Lie Theory Appl. **5**, 1–12 (2011)

Noncommutative Algebraic Varieties

Arvid Siqveland

Abstract For a natural number r, we define the free $r \times r$ matrix polynomial algebras and their quotients. We define algebraic sets and tangent spaces between different points. We then study their naive geometry by deformation theory, and prove that this defines noncommutative varieties in a natural way.

1 Introduction

Algebraic geometry has a long tradition, and in fact comes from a natural place. Then after making algebraic geometry to a categorical theme, it is possible to define noncommutative algebraic geometry. In this text we try to take noncommutative algebraic geometry back to the natives. We will use deformation theory to define higher order derivatives between points, and then use this to construct a noncommutative variety. Our main commutative reference is Hartshorne's classical book [2].

Through this notes, k is an algebraically closed field of characteristic 0.

2 Polynomial Matrix Algebras

Let $r \in \mathbb{N}$ and let (d_{ij}) be an $r \times r$-matrix with entries $d_{ij} \in \mathbb{N}$. We start by defining the free $r \times r$ matrix polynomial algebra generated by the matrix variables $t_{ij}(l)$, $1 \leq l \leq d_{ij}$, in entry $1 \leq i, j \leq r$. To get into the language, consider the following (in which $r = 2$ and $d_{ij} = 1$, $1 \leq i, j \leq 2$):

A. Siqveland (✉)
Faculty of Technology, Buskerud Vestfold University College, Kongsberg, Norway
e-mail: arvid.siqveland@hbv.no

A. Makhlouf et al. (eds.), *Algebra, Geometry and Mathematical Physics*, Springer Proceedings in Mathematics & Statistics 85, DOI: 10.1007/978-3-642-55361-5_16,

Example 2.1 Let the matrices

$$X = \begin{pmatrix} x & 0 \\ 0 & 0 \end{pmatrix}, Y = \begin{pmatrix} 0 & y \\ 0 & 0 \end{pmatrix}, Z = \begin{pmatrix} 0 & 0 \\ z & 0 \end{pmatrix}, \text{ and } W = \begin{pmatrix} 0 & 0 \\ 0 & w \end{pmatrix}$$

be given. Together with the idempotents $e_1 = \begin{pmatrix} 1 & 0 \\ 0 & 0 \end{pmatrix}$ and $e_2 = \begin{pmatrix} 0 & 0 \\ 0 & 1 \end{pmatrix}$ these matrix variables generates a k^2-algebra which is denoted

$$S = \begin{pmatrix} k\langle x\rangle & ky \\ kz & k\langle w\rangle \end{pmatrix}.$$

By the notation (S_{ij}) where S_{ii} is a k-algebra for each i, $1 \leq i \leq r$, and S_{ij} is a k-vector space, we mean the k^r-algebra generated by the matrices $M = (m_{ij})$ with $m_{ij} \in S_{ij}$, $1 \leq i, j \leq r$.

Definition 2.1 For a positive integer r, for each pair (i, j), $1 \leq i, j \leq r$, let $d_{ij} \in \mathbb{N}$. Then the free polynomial algebra in the matrix variables $t_{ij}(l)$, $1 \leq i, j \leq d_{ij}$, is the k^r-algebra generated by the matrix elements in

$$\begin{pmatrix} k\langle t_{11}(1), \ldots, t_{11}(d_{11})\rangle & \cdots & \sum_{v=1}^{d_{1r}} kt_{1r}(v) \\ \vdots & \ddots & \vdots \\ \sum_{v=1}^{d_{r1}} kt_{r1}(v) & \cdots & k\langle t_{rr}(1), \ldots, t_{rr}(d_{rr})\rangle \end{pmatrix}.$$

Alternatively, we consider the k^r-module V generated by $t_{ij}(l)$, and let S be the tensor algebra

$$S = T_{k^r}(V).$$

Definition 2.2 For a positive integer r, a finitely generated $r \times r$ matrix polynomial algebra is a quotient of a free $r \times r$ matrix polynomial algebra.

Lemma 2.1 *Let $\pi : R \twoheadrightarrow S$ be a surjective k-algebra homomorphism sending non-units to non-units, and let $\mathfrak{m} \subset R$ be a maximal ideal. Then $\pi(\mathfrak{m})$ is maximal in S.*

Proof First of all, as $s\pi(m) = \pi(r)\pi(m) = \pi(rm) \in \pi(\mathfrak{m})$ for some $r \in R$, $\pi(\mathfrak{m})$ is an ideal. Assume $\pi(\mathfrak{m}) \subsetneq \mathfrak{a}$ and let $a = \pi(r) \in \mathfrak{a} \setminus \pi(\mathfrak{m})$. Then $r \in \pi^{-1}(\mathfrak{a}) \setminus \mathfrak{m}$ so that $\mathfrak{m} \subsetneq \pi^{-1}(\mathfrak{a})$ so that $\pi^{-1}(\mathfrak{a}) = R$. But then $1 = \pi(1) \in \mathfrak{a}$ implying $\mathfrak{a} = S$ and we conclude that $\pi(\mathfrak{m})$ is maximal.

Remark 2.1 We could have simplified by working only with commutative polynomial algebras on the diagonal. However, for obvious reasons, we choose to be as general as reasonable.

Lemma 2.2 *The maximal (right or left or both) ideals, corresponding to one-dimensional simple modules, of the free noncommutative k-algebra $k\langle t_1, \ldots, t_d\rangle$ are the ideals generated by $(t_1 - a_1, \ldots, t_d - a_d)$ with $a_1, \ldots, a_d \in k$.*

Proof Let $\mathfrak{m}$ be a maximal ideal in $S = k\langle t_1, \dots, t_d \rangle$. We have a surjection $\pi_0 : S \twoheadrightarrow k[t_1, \dots, t_d]$. We let $\mathfrak{m}_0 = \pi_0(\mathfrak{m})$ which is a maximal ideal, and because k is algebraically closed, $k[t_1, \dots, t_d]/\mathfrak{m}_0 \simeq k$. Letting $\pi : S \to k$ be the canonical homomorphism and letting $\pi(t_i) = a_i$, we find $\mathfrak{m} = \ker \pi = (t_1 - a_1, \dots, t_d - a_d)$.

Lemma 2.3 *For each $i \leq r$, let $S_{ii} = k\langle t_{ii}(1), \dots, t_{ii}(d_{ii}) \rangle$, and let $\pi_{ii} : S \twoheadrightarrow S_{ii}$ be the natural morphism. Then the maximal ideals of the free matrix algebra S are the ideals $\mathfrak{m}_{ii} = \pi_{ii}^{-1}(\mathfrak{m})$ where $\mathfrak{m} \subset S_{ii}$ is a maximal ideal. This means that the maximal ideals of S are the maximal ideals on the diagonal.*

Proof For a maximal ideal $\mathfrak{m}_{ii} \subset S_{ii}$, we have an isomorphism

$$S/\pi_{ii}^{-1}(\mathfrak{m}_{ii}) \overset{\simeq}{\to} S_{ii}/\mathfrak{m}_{ii} \ .$$

This proves that $\pi_{ii}^{-1}(\mathfrak{m}_{ii})$ is a maximal ideal. For the converse, assume $\mathfrak{m} \subset S$ is maximal. If $\pi_{ii}(\mathfrak{m}) = S_{ii}$ for all i, it follows that $1 = \sum e_i$ is in $\mathfrak{m}$ which is impossible. Thus there exists an i where $\pi_{ii}(\mathfrak{m}) \subseteq \mathfrak{m}_{ii}$ for a maximal ideal $\mathfrak{m}_{ii} \subset S_{ii}$. Then $\mathfrak{m} \subseteq \pi_{ii}^{-1}(\pi_{ii}(\mathfrak{m})) \subseteq \pi_{ii}^{-1}(\mathfrak{m}_{ii}) \subsetneq S$. Then $\mathfrak{m} = \pi_{ii}^{-1}(\mathfrak{m}_{ii})$ by maximality, and the lemma is proved.

3 Algebraic Spaces and Matrix Coordinate Algebras

For ordinary polynomial algebras, the evaluation in points of affine space is clear. We give the definition for the $r \times r$ polynomial matrix algebras. Let $S = (S_{ij})$ be an $r \times r$ matrix polynomial algebra. Let as before $S_{ii} = k\langle t_{ii}(1), \dots, t_{ii}(d_{ii}) \rangle$ and let $\pi_{ii} : S \to S_{ii}$ be the morphism defined by sending $t_{ii}(l) \in S$ to $t_{ii}(l) \in S_{ii}$, and all other generators to 0. We have seen that the maximal ideals in S are in one to one correspondence with the collection of maximal ideals in the k-algebras S_{ii}, $1 \leq i \leq r$.

Definition 3.1 The affine $r \times r$-space $\mathbb{A}_S^{r \times r}$ is the set of points (maximal ideals) in the free $r \times r$ matrix polynomial algebra S. (Together with the additional structure given by S to be defined in the next section).

We define the evaluation of $f \in S$ in the point $p = m_{ii} \in S_{ii}$ as $f(p) = \overline{\pi_{ii}(f)}$, the class of $\pi_{ii}(f) \in S_{ii}/\mathfrak{m}_{ii}$. So, in the situation with polynomial matrix algebras, we have the following naive definition.

Definition 3.2 Let $S = (S_{ij})$ be a free $r \times r$ matrix polynomial algebra, and let $I = (I_{ij}) \subseteq S$ be an ideal. Then an algebraic set is a set on the form $Z(I) = \{p \in \mathbb{A}_S^{r \times r} : f(p) = 0, \forall f \in I\}$. Conversely, let $V \subseteq \mathbb{A}_S^{r \times r}$. Then the ideal of V is $I(V) = \{f \in S : f(p) = 0, \forall p \in V\}$, and the affine matrix ring coordinate ring is defined as $S(V) = S/I(V)$.

4 Tangent Spaces for Finitely Generated Matrix Algebras

Speaking differential geometric, for an affine variety $V = Z(I) \subseteq \mathbb{A}^n$, $I \subseteq k[x_1, \ldots, x_d]$ an ideal, the tangent directions are the directions along which we can differentiate, so that the total differential is a sum of the differentials along the directions. Even better, the k-vector space of derivations has a basis indexed over the tangent directions. Translated to algebraic geometry, for a point $\mathfrak{m} \in V$, we consider the $A(V)$-module $A(V)/\mathfrak{m} \simeq k$ and find a basis for the vector space of k-derivations $\mathrm{Der}_k(A(V), k)$ indexed over what we could call tangent directions, spanning the tangent space. So we just call $\mathrm{Der}_k(A(V), k)$ the tangent space. To recognize this in other textbooks, e.g. Hartshorne [2], we notice the following:

Lemma 4.1 *For a general vector space W, letting W^* denote the dual vector space, we have that*

$$\mathrm{Der}_k(A(V), k) \simeq (\mathfrak{m}/\mathfrak{m}^2)^* \ .$$

Proof As $A(V)$ is generated in degree one by $\mathfrak{m}$, a derivation is determined by its value on the generators on $\mathfrak{m}$. In addition, as the target module is $k = A(V)/\mathfrak{m}$, any derivation δ satisfies $\delta(\mathfrak{m}^2) = 0$ giving a linear transformation $\delta : \mathfrak{m}/\mathfrak{m}^2 \to k$. Also, given such a linear transformation δ with $\delta(\mathfrak{m}^2) = 0$, δ defines a derivation.

Now, we generalize this to the noncommutative situation, that is to the finitely generated matrix polynomial algebras. For any two points in a variety V, that is for any two maximal ideals $\mathfrak{m}_1$ and $\mathfrak{m}_2$, put $V_1 = S(V)/\mathfrak{m}_1$ and $V_2 = S(V)/\mathfrak{m}_2$. Then we have proved above that $S(V)/\mathfrak{m}_i \simeq S_{jj}/\mathfrak{m}'_i$ for $i = 1, 2$ and some j's, so we can consider $\mathrm{Hom}_k(V_1, V_2)$ as an S-bimodule by defining $(s\phi)(v) = \phi(sv)$ and $(\phi s)(v) = s\phi(v)$, with the given multiplication by s. We then define the tangent space between two closed points as

$$\begin{aligned} T_{V_1,V_2} &= \mathrm{Ext}^1_{S(V)}(V_1, V_2) = \mathrm{HH}^1(S(V), \mathrm{Hom}_k(V_1, V_2)) \\ &= \mathrm{Der}_k(S(V), \mathrm{Hom}_k(V_1, V_2))/\,\mathrm{Inner} \ . \end{aligned}$$

In the commutative situation, for a commutative k-algebra A, and two different simple A-modules $V_1 = A/\mathfrak{m}_1$, $V_2 = A/\mathfrak{m}_2$, it is well known that $\mathrm{Ext}^1_A(V_1, V_2) \cong \mathrm{Der}_k(A, \mathrm{Hom}_k(V_1, V_2))/\,\mathrm{Inner} = 0$. In the noncommutative case however, this is is different. The noncommutative information is contained in the different tangent spaces and higher order derivations between the different points. For simplicity, we give the following definition in all generality, even if it makes sense only for noncommutative k-algebras.

Definition 4.1 Let S be any k-algebra. The tangent space between two S-modules M_1 and M_2 is

$$\mathrm{Ext}^1_S(M_1, M_2) \cong \mathrm{HH}^1(S, \mathrm{Hom}_k(M_1, M_2))$$

where $\mathrm{HH}^{\cdot}$ is the Hochschild cohomology.

Example 4.1 Let $S = \begin{pmatrix} k[t_{11}] & kt_{12} \\ kt_{21} & k[t_{22}] \end{pmatrix}$ and consider two general points

$$V_1 = k[t_{11}]/(t_{11} - a)\ ,\ V_2 = k[t_{22}]/(t_{22} - b)\ .$$

First, we compute

$$\mathrm{Ext}^1_S(V_i, V_j) \cong \mathrm{Der}_k(S, \mathrm{Hom}_k(V_i, V_j))/\,\mathrm{Inner}$$

by derivations:
$\mathrm{Ext}^1_S(V_1, V_1)$: Let $\delta \in \mathrm{Der}_k(S, \mathrm{End}_k(V_1))$. Then

$$\delta(e_i) = \delta(e_i^2) = 2\delta(e_i) \Rightarrow \delta(e_i) = 0, i = 1, 2\ .$$

$$\delta(t_{12}) = \delta(t_{12}e_2) = \delta(t_{12})e_2 = 0\ ,$$

$$\delta(t_{21}) = \delta(e_2 t_{21}) = e_2\delta(t_{21}) = 0\ ,$$

$$\delta(t_{22}) = \delta(t_{22})e_2 = 0\ ,$$

and finally

$$\delta(t_{11}) = \alpha\ .$$

As all inner derivations are zero (easily seen from the computation above), we find that $\mathrm{Ext}^1_S(V_1, V_1)$ is generated by the derivation sending t_{11} to α, and all other generators to 0.
$\mathrm{Ext}^1_S(V_1, V_2)$:

For $\delta \in \mathrm{Der}_k(S, \mathrm{End}_k(V_1, V_2))$ things are slightly different. $\delta(e_1) = \delta(e_1^2) = e_1\delta(e_1) + \delta(e_1)e_1 = \delta(e_1)$, that is, the above trick doesn't work quite the same way. However, as $\delta(1) = \delta(e_1 + e_2) = 0$, for every derivation $\delta : S \to \mathrm{End}_k(V_1, V_2)$, we find $\delta(e_1) = \alpha$, $\delta(e_2) = -\alpha$,

$$\begin{aligned} \delta(e_1) = \alpha,\ \delta(e_2) &= -\alpha, \\ \delta(t_{11}) = \delta(t_{11}e_1) = \delta(t_{11})e_1 + t_{11}\delta(e_1) &= a\alpha, \\ \delta(t_{21}) = \delta(t_{21}e_1) = \delta(t_{21})e_1 &= 0, \\ \delta(t_{22}) = \delta(e_2 t_{22}) = \delta(e_2)t_{22} &= -b\alpha, \\ \delta(t_{12}) &= \rho\ . \end{aligned}$$

So a general derivation can be written, the * denoting the dual,

$$\delta = \alpha e_1^* - \alpha e_2^* + a\alpha t_{11}^* - b\alpha t_{22}^* + \rho t_{12}^*\ .$$

For the inner derivations, we compute

$$\begin{aligned} \mathrm{ad}_{\beta}(e_1) &= \beta e_1 - e_1\beta = -\beta \ , \\ \mathrm{ad}_{\beta}(e_2) &= \beta e_2 - e_2\beta = \beta \ , \\ \mathrm{ad}_{\beta}(t_{11}) &= -\beta a \ , \\ \mathrm{ad}_{\beta}(t_{22}) &= \beta b \ , \end{aligned}$$

saying that

$$\mathrm{ad}_{\beta} = \gamma e_1^* - \gamma e_2^* + a\gamma t_{11}^* - b\gamma t_{22}^*, \text{ where we have put } \gamma = -\beta \ .$$

So as $\mathrm{ad}_{\beta}(t_{12}) = 0$, and there are no conditions on $\delta(t_{12})$, we get

$$\mathrm{Ext}_S^1(V_1, V_2) = kt_{12}^* = kd_{t_{12}} \ .$$

The cases $\mathrm{Ext}_S^1(V_2, V_1)$ and $\mathrm{Ext}_S^1(V_2, V_2)$ are exactly similar.

Generalizing the computation in the above example, we have proved the following:

Lemma 4.2 *Let S be a general free $r \times r$ matrix polynomial algebra, and let $V_i = V_{ii}(p_{ii})$ be the point p_{ii} in entry i, i. Then the tangent space from V_i to V_j is* $\mathrm{Ext}_S^1(V_i, V_j) = \oplus_{l=1}^{d_{ij}} k\mathrm{d}_{t_{ij}(l)}$.

Now, we will explain what happens in the case with relations, that is, quotients of a matrix polynomial algebra.

Example 4.2 We let $R = \begin{pmatrix} k[t_{11}] & kt_{12} \\ kt_{21} & k[t_{22}] \end{pmatrix} / (t_{11}t_{12} - t_{12}t_{22})$. The polynomial in the ideal is really in the entry $(1, 2)$, but there is no ambiguity writing it like this. The points are still the simple modules along the diagonal, but a derivation $\delta \in \mathrm{Der}_k(R, \mathrm{Hom}_k(V_{ii}(p_{ii}), V_{jj}(p_{jj})))$, must this time respect the quotient;

$$\delta(t_{11}t_{12} - t_{12}t_{22}) = 0 \ .$$

This says

$$\delta(t_{11}t_{12} - t_{12}t_{22}) = t_{11}\delta(t_{12}) + \delta(t_{11})t_{12} - t_{12}\delta(t_{22}) - \delta(t_{12})t_{22} = 0 \ ,$$

and is fulfilled for any $\delta \in \mathrm{Ext}_R^1(V_i, V_j)$, $(i, j) \neq (1, 2)$. When $\delta \in \mathrm{Ext}_R^1(V_1, V_2)$, we get that the above equation is equivalent to

$$t_{11}\delta(t_{12}) - \delta(t_{12})t_{22} = \delta(t_{12})(t_{11} - t_{22}) = 0 \ .$$

Thus in the case that $p_{11} \neq p_{22}$ the tangent direction is annihilated: This quotient has no tangent direction from $V_1(p_1)$ to $V_2(p_2)$ unless $p_1 = p_2$.

This example illustrates the geometry of matrix polynomial algebras, and is of course nothing else than the obvious generalization of the the ordinary tangent space:

Lemma 4.3 *Let S be a finitely generated $r \times r$ matrix polynomial algebra with residue $\rho : S \to k^r$ and radical $\mathfrak{m} = \ker \rho$. Let p_1, p_2 be two points on the diagonal of S with respective quotients $V_1 \cong V_2 \cong k$. Then $T_{p_1,p_2} = \mathrm{Ext}^1_S(V_1, V_2) = \mathrm{Hom}_k(\mathfrak{m}/\mathfrak{m}^2, k)$ where the action on $k \cong \mathrm{Hom}_k(V_1, V_2)$ is the left-right action defined by $(s\phi)(v) = \phi(vs)$, $(\phi s)(v) = \phi(v)s$ (for right modules).*

The tangent space is not enough to reconstruct the algebra, not even in the commutative situation. As always, to get the full geometric picture we also need the higher order derivatives. Even if we cannot reconstruct the algebra in all cases, we get an algebra that is geometrically equivalent (Morita equivalent), and that suffices in construction of moduli.

5 Noncommutative Deformation Theory

For ordinary, commutative, varieties V, for each closed point $\mathfrak{m}$, we have the ring of local regular functions. For noncommutative k-algebras, there are serious challenges with localizing. These challenges are already present when it comes to finitely generated matrix polynomial algebras, and as the noncommutative deformation theory is the solution, we need to go through the basics of this. However, the constructive proof of existence of a local formal moduli is found in the classical works of Laudal [3], also formulated by Eriksen in [1].

Definition 5.1 The objects in the category a_r are the k-algebras S with morphisms commuting in the diagram

$$\begin{array}{ccc} k^r & \xrightarrow{\iota} & S \\ & \underset{\mathrm{Id}}{\searrow} & \downarrow{\scriptstyle \rho} \\ & & k^r \end{array},$$

such that $\ker(\rho)^n = 0$. We call $\ker(\rho) = \mathrm{rad}(S)$ the radical, the morphisms are the morphisms commuting with ι and ρ. The category a_r is called the category of r-pointed Artinian k-algebras. The notation $\hat{a}_r$ denotes the procategory of a_r, the category of objects that are projective limits of objects in a_r.

Definition 5.2 Let A be a k-algebra, let $V = \{V_1, \dots, V_r\}$ be A-modules. The noncommutative deformation functor $\mathrm{Def}_V : a_r \to \mathrm{Sets}$ is given by:

$$\mathrm{Def}_V(S) = \{S \otimes_k A\text{-Mod } V_S, \text{ flat over} S : k^r \otimes_S V_S \simeq V\}/\cong$$

where the equivalence of M_S and M'_S is given as an isomorphism

$$\begin{array}{ccc} M_S & \xrightarrow{\simeq} & M'_S \\ & \searrow \quad \swarrow & \\ & M & \end{array} .$$

Lemma 5.1 *(Yoneda). Consider a covariant functor* $F : C \to$ Sets. *Then there is an isomorphism*

$$\psi : F(R) \to \mathrm{Hom}(\mathrm{Hom}(R, -), F)$$

given by $\psi(\xi)(\eta) = F(\eta)(\xi)$, *for* $\xi \in F(R)$ *and* $\eta : R \to R'$ *any morphism.*

Definition 5.3 In the above situation, $(\hat{H}, \hat{\xi})$ is said to prorepresent $\mathrm{Def}_V : \hat{a}_r \to$ Sets if $\psi(\hat{\xi})$ is an isomorphism. If $\psi(\hat{\xi})$ is smooth and an isomorphism for the $r \times r$ matrix polynomial algebra R in the variables ε_{ij}, $1 \leq i, j \leq r$, $(\varepsilon_{ij})^2 = 0$, we call $(\hat{H}, \hat{\xi})$ a prorepresenting hull, or a local formal moduli.

Theorem 5.1 *There exists a local formal moduli* $(\hat{H}_V, \hat{\xi}_V)$ *for the noncommutative deformation functor* Def_V. *There is a homomorphism*

$$\iota : A \to (H_{ij}) \otimes_{k^r} \mathrm{Hom}_k(V_i, V_j) .$$

Its kernel is given by $\ker \iota = \bigcap_{i,n} \mathfrak{a}_i^n$ *where* $\mathfrak{a}_i = \ker \rho_i : A \to \mathrm{End}_k(V_i)$.

Proof The proof is given by Laudal in [3].

In our situation, what we need is the following:

Corollary 5.1 *For* $V = \{V_1, \ldots, V_r\}$ *a collection of simple* S*-modules where* S *is a finitely generated matrix polynomial algebra, there exists an injection*

$$\iota : S \hookrightarrow \hat{H}_V$$

such that $\iota(f)$ *is a unit if* $f \in S \setminus \cup_{i=1}^r \mathfrak{m}_i$, *where* $\mathfrak{m}_i$, $1 \leq i \leq r$, *is the maximal ideal corresponding to* V_i.

We notice that this holds also in the ordinary commutative situation, allowing us to replace a localization with the image of S.

Definition 5.4 For a finite family of simple modules $V = \{V_1, \ldots, V_r\}$, the localization of S in V is the k-algebra S_V generated by the image of ι in $\hat{H}_V$, together with the inverses of the images of elements not contained in any of the maximal ideals.

6 Definition of Noncommutative Varieties

In this final section, we make the direct translation of the general theory in [4] to the affine varieties. As in the commutative situation, we let $\mathbb{A}(S)$ denote the set of maximal ideals in S. We define a topology on $\mathbb{A}(S)$ by letting the closed sets be the algebraic sets $Z(I)$ where $I \subseteq S$ is an ideal. Alternatively, the sets $D(f)$, $f \in S$, given by $D(f) = \{\mathfrak{m} : f \notin \mathfrak{m}\}$, is a generating set for the topology.

For any set U, let $\mathrm{Pf}(U)$ denote the set of finite subsets of U. We define a sheaf of rings on the topological space: For an open U we let

$$\mathcal{O}_S(U) = \{f : \mathrm{Pf(U)} \rightarrow \coprod_{\underline{c} \in \mathrm{Pf(U)}} S_{\underline{c}}\}$$

such that f is locally regular: For each $\underline{c} \in \mathrm{Pf(U)}$ there exists an open subset $V \subseteq U$ containing $\underline{c}$ and elements $f, g \in S$ with g not in the unions of the corresponding maximal ideals of any of the subsets $\underline{c}' \in \mathrm{Pf}(V)$.

Then all theorems from the commutative situations are prolonged, and we have the category of noncommutative varieties

$$(\mathbb{A}(S), \mathcal{O}_S) .$$

References

1. Eriksen, E.: An introduction to non-commutative deformations of modules. Lect. Notes Pure Appl. Math. **243**, 90–126 (2005)
2. Hartshorne, R.: Algebraic Geometry Graduate Texts in Mathematics, No. 52. New York. ISBN 0-387-90244-9 (1977)
3. Laudal, O.A.: Non-commutative algebraic geometry. Rev. Mat. Iberoamericana **19(2)**, 509–580 (2003)
4. Siqveland, A.: Geometry of noncommutative k-algebras. J. Gen. Lie Theory Appl. **5**, 1–12 (2011)

Computing Noncommutative Deformations

Eivind Eriksen

Abstract Let M be a right module over an associative k-algebra A, where k is a field. We show how to compute noncommutative deformations of M in concrete terms, using an obstruction calculus based on free resolutions.

1 Introduction

Let A be an associative k-algebra, where k is a field. For any right A-module M, there is a noncommutative deformation functor $\mathsf{Def}_M : \mathsf{a}_1 \to \mathsf{Sets}$, introduced in Laudal [2], defined on the category a_1 of local Artinian k-algebras with residue field k. The noncommutative deformation functor extends the classical deformation functor $\mathsf{Def}_M^{\mathsf{cl}} : \mathbf{l} \to \mathsf{Sets}$, defined on the category $\mathbf{l}$ of local commutative Artinian k-algebras with residue field k.

In this paper, we show how to compute noncommutative deformations of M in concrete terms, using an obstruction calculus based on free resolutions. We show the computations explicitly in the example with $A = k[x, y]$ and $M = A/(x^2, y)$, which is obstructed. We also compare the result with the classical deformations of M.

2 Noncommutative Deformations of Modules

Let M be a right module over an associative k-algebra A, where k is a field. Then there is a classical deformation functor $\mathsf{Def}_M^{\mathsf{cl}} : \mathbf{l} \to \mathsf{Sets}$, where $\mathbf{l}$ is the category of commutative Artinian local k-algebras with residue field k. We fix a free resolution $(L_\bullet, d_\bullet)$ of M. For any algebra R in $\mathbf{l}$, a lifting of complexes from $(L_\bullet, d_\bullet)$ to R is a

E. Eriksen (✉)
Department of Economics, BI Norwegian Business School, N-0442, Oslo, Norway
e-mail: eivind.eriksen@bi.no

A. Makhlouf et al. (eds.), *Algebra, Geometry and Mathematical Physics*, Springer Proceedings in Mathematics & Statistics 85, DOI: 10.1007/978-3-642-55361-5_17,

complex $(L^R_\bullet, d^R_\bullet)$ of R-A bimodules, with $L^R_m = R \otimes_k L_m$, such that the following diagram commutes:

$$\begin{array}{ccccccc} L^R_0 & \xleftarrow{d^R_0} & L^R_1 & \xleftarrow{d^R_1} & L^R_2 & \xleftarrow{d^R_2} & \cdots \\ \downarrow & & \downarrow & & \downarrow & & \\ L_0 & \xleftarrow[d_0]{} & L_1 & \xleftarrow[d_1]{} & L_2 & \xleftarrow[d_2]{} & \cdots \end{array}$$

It is well-known that $\mathsf{Def}^{\mathsf{cl}}_M(R)$ can be identified with the set of equivalence classes of liftings of $(L_\bullet, d_\bullet)$ to R, and that $\mathsf{Def}^{\mathsf{cl}}_M$ has tangent space $\mathsf{t}(\mathsf{Def}^{\mathsf{cl}}_M) \cong \mathrm{Ext}^1_A(M, M)$ and an obstruction theory with cohomology $\{\mathrm{Ext}^p_A(M, M)\}$. When $d = \dim_k \mathrm{Ext}^1_A(M, M)$ and $r = \dim_k \mathrm{Ext}^2_A(M, M)$ are finite, there is an obstruction morphism

$$o^{\mathsf{cl}} : k[[s_1, s_2, \ldots, s_r]] \to k[[t_1, t_2, \ldots, t_d]]$$

such that $H^{\mathsf{cl}} = k[[t_1, \ldots, t_d]]/(f^{\mathsf{cl}}_1, \ldots, f^{\mathsf{cl}}_r)$ is a pro-representing hull of $\mathsf{Def}^{\mathsf{cl}}_M$, with $f^{\mathsf{cl}}_i = o^{\mathsf{cl}}(s_i)$ for $1 \le i \le r$. Its versal family is given by a lifting of complexes of $(L_\bullet, d_\bullet)$ to H^{cl}.

There is an extension of the classical deformation functor $\mathsf{Def}^{\mathsf{cl}}_M$ of M to a noncommutative deformation functor $\mathsf{Def}_M : \mathsf{a}_1 \to \mathsf{Sets}$, where a_1 is the category of local Artinian k-algebras with residue field k. This extension is due to Laudal [2]; see also Eriksen [1] for details. We remark that $\mathsf{Def}_M(R)$ can be identified with the set of equivalence classes of liftings of $(L_\bullet, d_\bullet)$ to R. When $d = \dim_k \mathrm{Ext}^1_A(M, M)$ and $r = \dim_k \mathrm{Ext}^2_A(M, M)$ are finite, there is an obstruction morphism

$$o : k\langle\langle s_1, s_2, \ldots, s_r \rangle\rangle \to k\langle\langle t_1, t_2, \ldots, t_d \rangle\rangle$$

such that $H = k\langle\langle t_1, t_2, \ldots, t_d \rangle\rangle/(f_1, \ldots, f_r)$ is a pro-representing hull of Def_M, with $f_i = o(s_i)$ for $1 \le i \le r$. Its versal family is given by a lifting of complexes of $(L_\bullet, d_\bullet)$ to H.

The relationship between classical and noncommutative deformations are given by the following commutative diagram

$$\begin{array}{ccc} k\langle\langle s_1, s_2, \ldots, s_r \rangle\rangle & \xrightarrow{o} & k\langle\langle t_1, t_2, \ldots, t_d \rangle\rangle \\ \downarrow & & \downarrow \\ k[[s_1, s_2, \ldots, s_r]] & \xrightarrow{o^{\mathsf{cl}}} & k[[t_1, t_2, \ldots, t_d]] \end{array}$$

where the vertical maps are the natural commutativization homomorphisms given by $A \to A^{\mathsf{cl}} = A/(xy - yx : x, y \in A)$. In particular, f^{cl}_i is the image of f_i in $k[[t_1, \ldots, t_d]]$ for $1 \le i \le r$.

3 A Concrete Description of Lifting of Complexes

Let R be an algebra in a_1, and choose a k-linear base $\{r_i : 0 \le i \le l\}$ of R with $r_0 = 1$. Then a lifting of complexes of $(L_\bullet, d_\bullet)$ to R is given by R-A linear maps $d_m^R : L_{m+1}^R \to L_m^R$ for $m \ge 0$, and d_m^R is determined by its value on elements of the form $1 \otimes f$ in $L_{m+1}^R = R \otimes_k L_{m+1}$. Therefore the differential d_m^R can be considered as an element in $\mathrm{Hom}_A(L_{m+1}, R \otimes_k L_m) \cong R \otimes_k \mathrm{Hom}_A(L_{m+1}, L_m)$, described in concrete terms as

$$d_m^R = 1 \otimes d_m + \sum_{i=1}^{l} r_i \otimes \alpha(r_i)_m$$

where $\underline{\alpha} = \{\alpha(r_i)_m : m \ge 0,\ 0 \le i \le l\}$ is a family of A-linear homomorphisms $\alpha(r_i)_m : L_{m+1} \to L_m$ with $\alpha(1)_m = d_m$. Conversely, such a family $\underline{\alpha}$ of A-linear homomorphisms represents a lifting of complexes of $(L_\bullet, d_\bullet)$ to R if and only if $d_m^R \circ d_{m+1}^R = 0$ for all $m \ge 0$. This condition can be expressed in terms of $\underline{\alpha}$ as

$$\sum_{1 \le i \le l} r_i \otimes (\alpha(r_i)_m\, d_{m+1} + d_m\, \alpha(r_i)_{m+1}) + \sum_{1 \le i,j \le l} r_j r_i \otimes \alpha(r_i)_m\, \alpha(r_j)_{m+1} = 0$$

Notice that on the tangent level, where $r_j r_i = 0$ for $1 \le i, j \le l$, $\underline{\alpha}$ determines a lifting of complexes to R if and only if $\alpha(r_i)$ is a 1-cocycle in the Yoneda complex $\mathsf{YC}^\bullet(L_\bullet, L_\bullet)$. We recall that the Yoneda complex $\mathsf{YC}^\bullet(L_\bullet, L_\bullet)$ is defined by

$$\mathsf{YC}^n(L_\bullet, L_\bullet) = \prod_{m \ge 0} \mathrm{Hom}_A(L_{m+n}, L_m)$$

for all $n \ge 0$, and with differential $d^n : \mathsf{YC}^n(L_\bullet, L_\bullet) \to \mathsf{YC}^{n+1}(L_\bullet, L_\bullet)$ given by

$$d^n(\phi)_m = \phi_m\, d_{n+m} + (-1)^{n+1} d_m\, \phi_{m+1} \quad \text{for } m \ge 0$$

for all $\phi = (\phi_m)_{m \ge 0} \in \mathsf{YC}^n(L_\bullet, L_\bullet)$. It is well-known that the cohomology of the Yoneda complex is $\mathsf{YH}^p(M, M) = \mathsf{H}^p(\mathsf{YC}^\bullet(L_\bullet, L_\bullet)) \cong \mathrm{Ext}_A^p(M, M)$.

4 Computing Noncommutative Deformations in an Example

Let $A = k[x, y]$, and let M be the right A-module $M = A/(x^2, y)$ with free resolution $(L_\bullet, d_\bullet)$ given by

$$0 \leftarrow M \leftarrow A \xleftarrow{(x^2\ y)\cdot} A^2 \xleftarrow{\begin{pmatrix} y \\ -x^2 \end{pmatrix}\cdot} A \leftarrow 0$$

To compute $\operatorname{Ext}^p_A(M, M)$ for $p = 1$ (the tangent space) and $p = 2$ (the obstruction space), we consider the complex $\operatorname{Hom}_A(L_\bullet, M)$:

$$M \xrightarrow{\cdot (x^2 \; y)} M^2 \xrightarrow{\begin{pmatrix} y \\ -x^2 \end{pmatrix}\cdot} M \to 0$$

Note that the differentials in this complex are zero. Since $M = k[x, y]/(x^2, y) \simeq k + kx$ has dimension two, we see that

$$\operatorname{Ext}^p_A(M, M) = \begin{cases} (k + kx)^2 \cong k^4, & p = 1 \\ k + kx \cong k^2, & p = 2 \end{cases}$$

Hence there are noncommutative power series $f_1, f_2 \in k\langle\langle t_1, t_2, t_3, t_4\rangle\rangle$ determined by the obstruction morphism such that $H = k\langle\langle t_1, t_2, t_3, t_4\rangle\rangle/(f_1, f_2)$ is a pro-representing hull of the noncommutative deformation functor Def_M. We shall compute f_1 and f_2 in concrete terms.

At the tangent level, $H_2 = k\langle\langle t_1, t_2, t_3, t_4\rangle\rangle/(t_1, t_2, t_3, t_4)^2$, and the versal family $\xi_2 \in \mathsf{Def}_M(H_2)$ is given by a lifting of complexes of $(L_\bullet, d_\bullet)$ to H_2. In concrete terms, the differential in $H_2 \otimes_k \operatorname{Hom}_A(L_{m+1}, L_m)$ is given by

$$d_m^{H_2} = 1 \otimes d_m + \sum_{1 \le i \le 4} t_i \otimes \alpha(t_i)_m$$

for all $m \ge 0$. We let $t_1^* = (1, 0),\ t_2^* = (x, 0),\ t_3^* = (0, 1),\ t_4^* = (0, x)$ such that $\{t_1^*, t_2^*, t_3^*, t_4^*\}$ is a k-linear base for $\mathsf{t}(\mathsf{Def}_M)$, and let $\alpha(t_i)$ be a 1-cocycle in the Yoneda complex $\mathsf{YC}^\bullet(L_\bullet, L_\bullet)$ that represents $t_i^* \in \mathsf{YH}^1(M, M) \cong \operatorname{Ext}^1_A(M, M)$. Note that a 1-cocyle $\phi \in \mathsf{YC}^1(L_\bullet, L_\bullet)$ is a pair (ϕ_0, ϕ_1) of A-linear maps $\phi_i : L_{i+1} \to L_i$ such that $d_0\phi_1 + \phi_0 d_1 = 0$ since $L_i = 0$ for $i > 2$. We may therefore choose

$$\alpha(t_1) = \left\{ (1\ 0)\cdot,\ \begin{pmatrix} 0 \\ -1 \end{pmatrix}\cdot \right\} \qquad \alpha(t_3) = \left\{ (0\ 1)\cdot,\ \begin{pmatrix} 1 \\ 0 \end{pmatrix}\cdot \right\}$$

$$\alpha(t_2) = \left\{ (x\ 0)\cdot,\ \begin{pmatrix} 0 \\ -x \end{pmatrix}\cdot \right\} \qquad \alpha(t_4) = \left\{ (0\ x)\cdot,\ \begin{pmatrix} x \\ 0 \end{pmatrix}\cdot \right\}$$

Then the differential $d^{H_2} = (d_0^{H_2}, d_1^{H_2})$ is explicitly given by

$$d_0^{H_2} = d_0 + \sum_{1 \le i \le 4} t_i\, \alpha(t_i)_0 = \left(x^2 + t_1 + t_2 x \quad y + t_3 + t_4 x\right)\cdot$$

$$d_1^{H_2} = d_1 + \sum_{1 \le i \le 4} t_i\, \alpha(t_i)_1 = \begin{pmatrix} y + t_3 + t_4 x \\ -x^2 - t_1 - t_2 x \end{pmatrix}\cdot$$

By construction, $d_0^{H_2} \circ d_1^{H_2} = 0$ in $H_2 \otimes_k \mathrm{Hom}_A(L_2, L_0)$ and we may check that this is the case:

$$\begin{aligned} d_0^{H_2} \circ d_1^{H_2} &= \begin{pmatrix} x^2 + t_1 + t_2 x & y + t_3 + t_4 x \end{pmatrix} \cdot \begin{pmatrix} y + t_3 + t_4 x \\ -x^2 - t_1 - t_2 x \end{pmatrix} \\ &= (x^2 + t_1 + t_2 x)(y + t_3 + t_4 x) + (y + t_3 + t_4 x)(-x^2 - t_1 - t_2 x) \\ &= [t_3, t_1] + [t_4, t_1]x + [t_3, t_2]x + [t_4, t_2]x^2 \end{aligned}$$

Since $(t_1, \dots, t_4)^2 = 0$ in H_2, this obstruction vansishes. The obstruction space $\mathrm{Ext}^2_A(M, M) \cong \mathsf{YH}^2(M, M) = k + kx$ has a k-linear base $\{s_1^* = 1, s_2^* = x\}$, and we can write the obstruction as

$$[t_3, t_1]s_1^* + ([t_4, t_1] + [t_3, t_2])s_2^* + [t_4, t_2]x^2 s_1^*$$

Since $s_1^*, s_2^* \neq 0$ while $x^2 s_1^* = 0$ in $\mathsf{YH}^2(M, M)$, it follows that $H_3 = k\langle\langle t_1, t_2\rangle\rangle / a_3$, where $a_3 = (f_1^2, f_2^2) + (t_1, t_2)^3$ and $f_1^2 = [t_3, t_1]$, $f_2^2 = [t_4, t_1] + [t_3, t_2]$ are the second order approximations of f_1 and f_2. To lift ξ_2 to H_3, we choose $\alpha(t_4 t_2)$ and $\alpha(t_2 t_4)$ such that $d^1\alpha(t_4 t_2) = -x^2 s_1^*$ and $d^1\alpha(t_2 t_4) = x^2 s_1^*$, and find that

$$\begin{aligned} d^1\left(\left\{\begin{pmatrix} 0 & 1 \end{pmatrix}\cdot, \begin{pmatrix} 0 \\ 0 \end{pmatrix}\cdot\right\}\right) = -x^2 s_1^* \quad &\Rightarrow \quad \alpha(t_4 t_2) = \left\{\begin{pmatrix} 0 & 1 \end{pmatrix}\cdot, \begin{pmatrix} 0 \\ 0 \end{pmatrix}\cdot\right\} \\ &\Rightarrow \quad \alpha(t_2 t_4) = \left\{\begin{pmatrix} 0 & -1 \end{pmatrix}\cdot, \begin{pmatrix} 0 \\ 0 \end{pmatrix}\cdot\right\} \end{aligned}$$

Explicitly, the lifting ξ_3 is represented by the differential d^{H_3}, given by

$$\begin{aligned} d_0^{H_3} &= \begin{pmatrix} x^2 + t_1 + t_2 x & y + t_3 + t_4 x + [t_4, t_2] \end{pmatrix} \cdot \\ d_1^{H_3} &= \begin{pmatrix} y + t_3 + t_4 x \\ -x^2 - t_1 - t_2 x \end{pmatrix} \cdot \end{aligned}$$

Again, we compute the obstruction given by $d_0^{H_3} d_1^{H_3}$, and find that

$$\begin{aligned} d_0^{H_3} \circ d_1^{H_3} &= \begin{pmatrix} x^2 + t_1 + t_2 x & y + t_3 + t_4 x + [t_4, t_2] \end{pmatrix} \cdot \begin{pmatrix} y + t_3 + t_4 x \\ -x^2 - t_1 - t_2 x \end{pmatrix} \\ &= [t_3, t_1] + ([t_4, t_1] + [t_3, t_2])x - t_1[t_4, t_2] - t_2[t_4, t_2]x \\ &= ([t_3, t_1] - t_1[t_4, t_2])s_1^* + ([t_4, t_1] + [t_3, t_2] - t_2[t_4, t_2])s_2^*. \end{aligned}$$

This implies that $H_4 = k\langle\langle t_1, t_2, t_3, t_4\rangle\rangle / a_4$, where $a_4 = (f_1^3, f_2^3) + (t_1, t_2, t_3, t_4)^4$ and $f_1^3 = [t_3, t_1] - t_1[t_4, t_2]$, $f_2^3 = [t_4, t_1] + [t_3, t_2] - t_2[t_4, t_2]$ are the third order approximations of f_1 and f_2. We see that ξ_3 can be lifted to

$$H = k\langle\langle t_1, t_2, t_3, t_4\rangle\rangle / ([t_3, t_1] - t_1[t_4, t_2], [t_4, t_1] + [t_3, t_2] - t_2[t_4, t_2])$$

and this implies that H is the pro-representing hull of the noncommutative deformation functor Def_M (with $f_1 = [t_3, t_1] - t_1[t_4, t_2]$ and $f_2 = [t_4, t_1] + [t_3, t_2] - t_2[t_4, t_2]$), and that Def_M is obstructed. The versal family $\xi \in \mathsf{Def}_M(H)$ is given by a lifting of complexes of $(L_\bullet, d_\bullet)$ to H. In concrete terms, the differential $d^H = (d^H_m)$ is given by

$$d^H_0 = \begin{pmatrix} x^2 + t_1 + t_2 x & y + t_3 + t_4 x + [t_4, t_2] \end{pmatrix} \cdot \quad \text{and} \quad d^H_1 = \begin{pmatrix} y + t_3 + t_4 x \\ -x^2 - t_1 - t_2 x \end{pmatrix} \cdot$$

and the versal family $\xi \in \mathsf{Def}_M(H)$ is the H-A bimodule $M_H = \operatorname{coker}(d^H_0)$.

5 Comparison with Classical Deformations

From the computations above, it follows that the classical deformation functor $\mathsf{Def}^{\mathsf{cl}}_M$ has a pro-representing hull $H^{\mathsf{cl}} = k[[t_1, \ldots, t_4]]$ since $f^{\mathsf{cl}}_1 = f^{\mathsf{cl}}_2 = 0$, and that $\mathsf{Def}^{\mathsf{cl}}_M$ is unobstructed. Its versal family is given by the differential

$$d^H_0 = \begin{pmatrix} x^2 + t_1 + t_2 x & y + t_3 + t_4 x \end{pmatrix} \cdot \quad \text{and} \quad d^H_1 = \begin{pmatrix} y + t_3 + t_4 x \\ -x^2 - t_1 - t_2 x \end{pmatrix} \cdot$$

since $[t_4, t_2] = 0$ in H^{cl}. Hence the versal family is the H^{cl}-A bimodule $M_{H^{\mathsf{cl}}}$ given by

$$M_{H^{\mathsf{cl}}} = k[[t_1, \ldots, t_4]][x, y]/(x^2 + t_1 + t_2 x,\ y + t_3 + t_4 x)$$

We see that there is an algebraization of H^{cl} and its versal family, given by the algebra $\mathfrak{H}^{\mathsf{cl}} = k[t_1, t_2, t_3, t_4]$ and the versal family

$$M_{\mathfrak{H}^{\mathsf{cl}}} = k[t_1, \ldots, t_4][x, y]/(x^2 + t_1 + t_2 x,\ y + t_3 + t_4 x)$$

The corresponding family of classical deformations of M, parameterized by the closed points of spec $\mathfrak{H}^{\mathsf{cl}} = \mathbb{A}^4$, is $\{M_{\mathfrak{H}^{\mathsf{cl}}}(\tau) : \tau = (\tau_1, \ldots, \tau_4) \in \mathbb{A}^4\}$ with

$$M_{\mathfrak{H}^{\mathsf{cl}}}(\tau) \cong k[x, y]/(x^2 + \tau_1 + \tau_2 x,\ y + \tau_3 + \tau_4 x)$$

This is a family of right A-modules of length 2.

References

1. Eivind, E.: An Introduction to Noncommutative Deformations of Modules, Noncommutative Algebra and Geometry, Lecture Notes in Pure and Applied Mathematics, vol. 243, pp. 90–125. CRC, Boca Raton. MR2189988 (2006m:16037) (2006)
2. Laudal, O.A.: Noncommutative deformations of modules, Homol. Homotopy Appl. **4**(2), part 2, 357–396 (electronic), The Roos Festschrift volume, 2. MR1918517 (2003e:16005) (2002)

Geometric Classification of 4-Dimensional Superalgebras

Aaron Armour and Yinhuo Zhang

Abstract In this paper, we give a geometric classification of 4-dimensional superalgebras over an algebraic closed field. It turns out that the number of irreducible components of the variety of 4-dimensional superalgebras Salg_4 under the Zariski topology is between 20 and 22. One of the significant differences between the variety Alg_n and the variety Salg_n is that Salg_n is disconnected while Alg_n is connected. Under certain conditions on n, one can show that the variety Salg_n is the disjoint union of n connected subvrieties. We shall present the degeneration diagrams of the 4 disjoint connected subvarieties Salg_4^i of Salg_4.

1 Introduction

The algebraic and geometric classification of finite dimensional algebras over an algebraic closed field k was initiated by Gabriel in [5], and has been being one of the interesting topics in the study of geometric methods in representation theory of algebras for the last three decades. In [5], Gabriel gave a complete list of nonisomorphic 4-dimensional algebras over an algebraic closed field k with characteristic not equal to 2. The number of irreducible components of the variety Alg_4 is 5. The classification of 5-dimensional k-algebras was done by Mazzola in [10]. The number of irreducible components of the variety Alg_5 was showed to be 10. Further studies on the classification of low dimensional (rigid) algebras can be found in [4, 6, 8, 9, 13]. With the dimension n increasing, both algebraic and geometric

A. Armour (✉)
School of Mathematics, Statistics and Computer Science, Victoria University of Wellington, PO Box 600, Wellington, New Zealand
e-mail: aaron.armour@mcs.vuw.ac.nz

Y. Zhang
Department of Mathematics and Statistics, University of Hasselt, B-3590 Diepenbeek, Belgium
e-mail: yinhuo.zhang@uhasselt.be

A. Makhlouf et al. (eds.), *Algebra, Geometry and Mathematical Physics*,
Springer Proceedings in Mathematics & Statistics 85, DOI: 10.1007/978-3-642-55361-5_18,

classifications of n-dimensional k-algebras become more and more difficult. However, A lower and a upper bound for the number of irreducible components of Alg_n can be given (see [11]). Now let V_n be an n-dimensional vector space over k with a basis $\{e_1, e_2, \dots, e_n\}$. An algebra structure on V_n is determined by a set of structure constants c_{ij}^h, where $e_i \cdot e_j = \sum_h c_{ij}^h e_h$. Requiring the algebra structure to be associative and unitary gives rise to a subvariety Alg_n of k^{n^3}. Base changes in V_n result in the natural transport of structure action on Alg_n, namely the action of $\mathrm{GL}_n(k)$ on Alg_n. Thus isomorphism classes of n-dimensional algebras are in one-to-one correspondence with the orbits of the action of $\mathrm{GL}_n(k)$ on Alg_n. The decomposition of Alg_n into its irreducible components under the Zariski topology is called the geometric classification of n-dimensional algebras.

Our main interest is to give a geometric classification of 4-dimensional superalgebras, i.e. $\mathbb{Z}_2$-graded algebras. We notice that a $\mathbb{Z}_2$-graded algebra is the same as a pair (A, σ) consisting of an algebra A and an algebra involution σ. This enables us to define the variety Salg_n—the variety of n-dimensional superalgebras—as a subvariety of $k^{n^3+n^2}$. One of the significant differences between the variety Alg_n and the variety Salg_n is that Salg_n is disconnected while Alg_n is connected. Under certain assumptions on n and $\mathrm{ch}(k)$ it can be shown that Salg_n is the disjoint union of n connected subvarieties, for example, when $n \leq 6$ or $ch(k) = 0$.

The paper is organized as follows. In Sect. 2, we define the variety Salg_n of n-dimensional superalgebras, a closed subvariety of SA_n of n-dimensional algebras. Salg_n is a disjoint union of the subsets Salg_n^i, $i = 1, 2, \dots, n$. When $n \leq 6$ or $ch(k) = 0$, they are closed in Salg_n and form connected components of Salg_n.

In Sect. 3, we compute the dimensions of the orbits in Salg_4, which will help us to determine the degenerations of the superalgebras. So we need to recall the algebraic group G_n and the transport of its structure action on Salg_n. Since G_n is connected, every irreducible component of Salg_n is the closure of either a single orbit or an infinite family of orbits. In Sect. 4, we will use the ring properties of superalgebras to determine some closed sets of Salg_n. For instance, the set of superalgebras A with $A_1^2 = 0$ is a closed subset. Similarly the set of superalgebras with A_0 being commutative is also a closed subset. The closed subsets can help us to determine some superalgebras that can not degenerate to other superalgebras.

In the last section, we give the degeneration diagrams of Salg_4^i, where $i = 2, 3$. The degeneration diagram of $\mathrm{Salg}_4^4 = \mathrm{Alg}_4$ has been given by Gabriel, and Salg_4^1 has only one orbit. In total, we have found 20 irreducible components of Salg_4. However, Salg_4 may posses up to 22 irreducible components.

To end the introduction, let us recall from [1] the algebraic classification of 4-dimensional superalgebras over an algebraic closed field k as the geometric classification must be made on the basis of the algebraic classification. Throughout, k is an algebraic closed field with $ch(k) \neq 2$. All the algebras without other specified are over k.

Theorem 1.1 *[5] The following algebras are pairwise non-isomorphic except pairs within the family* $(18; \lambda|0)$ *where* $(18; \lambda_1|0) \cong (18; \lambda_2|0)$ *if and only if* $\lambda_1 = \lambda_2$ *or* $\lambda_1\lambda_2 = 1$.

(1|0) $k \times k \times k \times k$,
(2|0) $k \times k \times k[X]/(X^2)$,
(3|0) $k[X]/(X^2) \times k[Y]/(Y^2)$,
(4|0) $k \times k[X]/(X^3)$,
(5|0) $k[X]/(X^4)$,
(6|0) $k \times k[X, Y]/(X, Y)^2$,
(7|0) $k[X, Y]/(X^2, Y^2)$,
(8|0) $k[X, Y]/(X^3, XY, Y^2)$,
(9|0) $k[X, Y, Z]/(X, Y, Z)^2$,
(10|0) M_2,
(11|0) $\left\{ \begin{pmatrix} a & 0 & 0 & 0 \\ 0 & a & 0 & d \\ c & 0 & b & 0 \\ 0 & 0 & 0 & b \end{pmatrix} \middle| \, a, b, c, d \in k \right\}$,
(12|0) $\wedge k^2$,
(13|0) $k \times \begin{pmatrix} k & k \\ 0 & k \end{pmatrix} = \left\{ \left(a, \begin{pmatrix} b & c \\ 0 & d \end{pmatrix}\right) \middle| \, a, b, c, d \in k \right\}$,
(14|0) $\left\{ \begin{pmatrix} a & 0 & 0 \\ c & a & 0 \\ d & 0 & b \end{pmatrix} \middle| \, a, b, c, d \in k \right\}$,
(15|0) $\left\{ \begin{pmatrix} a & c & d \\ 0 & a & 0 \\ 0 & 0 & b \end{pmatrix} \middle| \, a, b, c, d \in k \right\}$,
(16|0) $k\langle X, Y\rangle/(X^2, Y^2, YX)$,
(17|0) $\left\{ \begin{pmatrix} a & 0 & 0 \\ 0 & a & 0 \\ c & d & b \end{pmatrix} \middle| \, a, b, c, d \in k \right\}$,
(18; λ|0) $k\langle X, Y\rangle/(X^2, Y^2, YX - \lambda XY)$, $\lambda \neq -1, 0, 1$,
(19|0) $k\langle X, Y\rangle/(Y^2, X^2 + YX, XY + YX)$.

Theorem 1.2 *[1, Thm 3.1] Suppose A is a superalgebra with* $\dim A_0 = 3$ *and* $\dim A_1 = 1$. *Then A is isomorphic to one of the superalgebras in the following pairwise non-isomorphic families:*

(1|1) $k \times k \times k \times k$:
$(1|1)_0 = k(1, 1, 1, 1) \oplus k(1, 0, 0, 0) \oplus k(0, 0, 1, 1)$, $(1|1)_1 = k(0, 0, 1, -1)$,
(2|1) $k \times k \times k[X]/(X^2)$:
$(2|1)_0 = k(1, 1, 1) \oplus k(1, 0, 0) \oplus k(0, 1, 0)$, $(2|1)_1 = k(0, 0, X)$,
(2|2) $(2|2)_0 = k(1, 1, 1) \oplus k(1, 1, 0) \oplus k(0, 0, X)$, $(2|2)_1 = k(1, -1, 0)$,
(3|1) $k[X]/(X^2) \times k[Y]/(Y^2)$:
$(3|1)_0 = k(1, 1) \oplus k(1, 0) \oplus k(X, 0)$, $(3|1)_1 = k(0, Y)$,
(4|1) $k \times k[X]/(X^3)$:
$(4|1)_0 = k(1, 1) \oplus k(1, 0) \oplus k(0, X^2)$, $(4|1)_1 = k(0, X)$,
(6|1) $k \times k[X, Y]/(X, Y)^2$:
$(6|1)_0 = k(1, 1) \oplus k(1, 0) \oplus k(0, X)$, $(6|1)_1 = k(0, Y)$,
(7|1) $k[X, Y]/(X^2, Y^2)$:
$(7|1)_0 = k1 \oplus k(X + Y) \oplus kXY$, $(7|1)_1 = k(X - Y)$,
(8|1) $k[X, Y]/(X^3, XY, Y^2)$:
$(8|1)_0 = k1 \oplus kX \oplus kX^2$, $(8|1)_1 = kY$,
(8|2) $(8|2)_0 = k1 \oplus kX^2 \oplus kY$, $(8|2)_1 = kX$,
(9|1) $k[X, Y, Z]/(X, Y, Z)^2$:
$(9|1)_0 = k1 \oplus kX \oplus kY$, $(9|1)_1 = kZ$,
(11|1) $\left\{ \begin{pmatrix} a & 0 & 0 & 0 \\ 0 & a & 0 & d \\ c & 0 & b & 0 \\ 0 & 0 & 0 & b \end{pmatrix} \middle| \, a, b, c, d \in k \right\}$:

$$(11|1)_0 = k\begin{pmatrix} 1 & 0 & 0 & 0 \\ 0 & 1 & 0 & 0 \\ 0 & 0 & 1 & 0 \\ 0 & 0 & 0 & 1 \end{pmatrix} \oplus k\begin{pmatrix} 1 & 0 & 0 & 0 \\ 0 & 1 & 0 & 0 \\ 0 & 0 & 0 & 0 \\ 0 & 0 & 0 & 0 \end{pmatrix} \oplus k\begin{pmatrix} 0 & 0 & 0 & 0 \\ 0 & 0 & 0 & 1 \\ 0 & 0 & 0 & 0 \\ 0 & 0 & 0 & 0 \end{pmatrix}$$

$$(11|1)_1 = k\begin{pmatrix} 0 & 0 & 0 & 0 \\ 0 & 0 & 0 & 0 \\ 1 & 0 & 0 & 0 \\ 0 & 0 & 0 & 0 \end{pmatrix},$$

(13|1) $k \times \begin{pmatrix} k & k \\ 0 & k \end{pmatrix} = \left\{ \left(a, \begin{pmatrix} b & c \\ 0 & d \end{pmatrix} \right) \middle| a, b, c, d \in k \right\}$:

$(13|1)_0 = k\left(1, \begin{pmatrix} 1 & 0 \\ 0 & 1 \end{pmatrix}\right) \oplus k\left(0, \begin{pmatrix} 1 & 0 \\ 0 & 0 \end{pmatrix}\right) \oplus k\left(0, \begin{pmatrix} 0 & 0 \\ 0 & 1 \end{pmatrix}\right)$

$(13|1)_1 = k\left(0, \begin{pmatrix} 0 & 1 \\ 0 & 0 \end{pmatrix}\right)$,

(14|1) $\left\{ \begin{pmatrix} a & 0 & 0 \\ c & a & 0 \\ d & 0 & b \end{pmatrix} \middle| a, b, c, d \in k \right\}$:

$(14|1)_0 = k\begin{pmatrix} 1 & 0 & 0 \\ 0 & 1 & 0 \\ 0 & 0 & 1 \end{pmatrix} \oplus k\begin{pmatrix} 1 & 0 & 0 \\ 0 & 1 & 0 \\ 0 & 0 & 0 \end{pmatrix} \oplus k\begin{pmatrix} 0 & 0 & 0 \\ 0 & 0 & 0 \\ 1 & 0 & 0 \end{pmatrix}$

$(14|1)_1 = k\begin{pmatrix} 0 & 0 & 0 \\ 1 & 0 & 0 \\ 0 & 0 & 0 \end{pmatrix}$,

(14|2) $(14|2)_0 = k\begin{pmatrix} 1 & 0 & 0 \\ 0 & 1 & 0 \\ 0 & 0 & 1 \end{pmatrix} \oplus k\begin{pmatrix} 1 & 0 & 0 \\ 0 & 1 & 0 \\ 0 & 0 & 0 \end{pmatrix} \oplus k\begin{pmatrix} 0 & 0 & 0 \\ 1 & 0 & 0 \\ 0 & 0 & 0 \end{pmatrix}$

$(14|2)_1 = k\begin{pmatrix} 0 & 0 & 0 \\ 0 & 0 & 0 \\ 1 & 0 & 0 \end{pmatrix}$,

(15|1) $\left\{ \begin{pmatrix} a & c & d \\ 0 & a & 0 \\ 0 & 0 & b \end{pmatrix} \middle| a, b, c, d \in k \right\}$:

$(15|1)_0 = k\begin{pmatrix} 1 & 0 & 0 \\ 0 & 1 & 0 \\ 0 & 0 & 1 \end{pmatrix} \oplus k\begin{pmatrix} 1 & 0 & 0 \\ 0 & 1 & 0 \\ 0 & 0 & 0 \end{pmatrix} \oplus k\begin{pmatrix} 0 & 0 & 1 \\ 0 & 0 & 0 \\ 0 & 0 & 0 \end{pmatrix}$

$(15|1)_1 = k\begin{pmatrix} 0 & 1 & 0 \\ 0 & 0 & 0 \\ 0 & 0 & 0 \end{pmatrix}$,

(15|2) $(15|2)_0 = k\begin{pmatrix} 1 & 0 & 0 \\ 0 & 1 & 0 \\ 0 & 0 & 1 \end{pmatrix} \oplus k\begin{pmatrix} 1 & 0 & 0 \\ 0 & 1 & 0 \\ 0 & 0 & 0 \end{pmatrix} \oplus k\begin{pmatrix} 0 & 1 & 0 \\ 0 & 0 & 0 \\ 0 & 0 & 0 \end{pmatrix}$

$(15|2)_1 = k\begin{pmatrix} 0 & 0 & 1 \\ 0 & 0 & 0 \\ 0 & 0 & 0 \end{pmatrix}$,

(17|1) $\left\{ \begin{pmatrix} a & 0 & 0 \\ 0 & a & 0 \\ c & d & b \end{pmatrix} \middle| a, b, c, d \in k \right\}$:

$(17|1)_0 = k\begin{pmatrix} 1 & 0 & 0 \\ 0 & 1 & 0 \\ 0 & 0 & 1 \end{pmatrix} \oplus k\begin{pmatrix} 1 & 0 & 0 \\ 0 & 1 & 0 \\ 0 & 0 & 0 \end{pmatrix} \oplus k\begin{pmatrix} 0 & 0 & 0 \\ 0 & 0 & 0 \\ 1 & 0 & 0 \end{pmatrix}$

$(17|1)_1 = k\begin{pmatrix} 0 & 0 & 0 \\ 0 & 0 & 0 \\ 0 & 1 & 0 \end{pmatrix}$.

Theorem 1.3 *[1, Thm 4.1] Suppose A is a superalgebra with* $\dim A_0 = \dim A_1 = 2$. *Then A is isomorphic to one of the superalgebras in the following pairwise non-isomorphic families:*

(1|2) $k \times k \times k \times k$:

$(1|2)_0 = k(1, 1, 1, 1) \oplus k(1, 1, 0, 0)$ and $(1|2)_1 = k(1, -1, 0, 0)$ $\oplus k(0, 0, 1, -1)$,

(2|3) $k \times k \times k[X]/(X^2)$:

$(2|3)_0 = k(1, 1, 1) \oplus k(1, 1, 0)$ and $(2|3)_1 = k(1, -1, 0) \oplus k(0, 0, X)$,

(3|2) $k[X]/(X^2) \times k[Y]/(Y^2)$:

$(3|2)_0 = k(1, 1) \oplus k(1, 0)$ and $(3|2)_1 = k(X, 0) \oplus k(0, Y)$,

(3|3) $(3|3)_0 = k(1, 1) \oplus k(X, Y)$ and $(3|3)_1 = k(1, -1) \oplus k(X, -Y)$,

(5|1) $k[X]/(X^4)$:

$(5|1)_0 = k1 \oplus kX^2$ and $(5|1)_1 = kX \oplus kX^3$,

(6|2) $k \times k[X, Y]/(X, Y)^2$:

$(6|2)_0 = k(1,1) \oplus k(1,0)$ and $(6|2)_1 = k(0,X) \oplus k(0,Y)$,

(7|2) $k[X,Y]/(X^2,Y^2)$:

$(7|2)_0 = k1 \oplus kX$ and $(7|2)_1 = kY \oplus kXY$,

(7|3) $(7|3)_0 = k1 + kXY$ and $(7|3)_1 = kX + kY$.

(8|3) $k[X,Y]/(X^3,XY,Y^2)$:

$(8|3)_0 = k1 + kX^2$ and $(8|3)_1 = kX + kY$.

(9|2) $k[X,Y,Z]/(X,Y,Z)^2$,

$(9|2)_0 = k1 \oplus kX$ and $(9|2)_1 = kY \oplus kZ$,

(10|1) M_2 :

$(10|1)_0 = k\begin{pmatrix}1&0\\0&1\end{pmatrix} \oplus k\begin{pmatrix}1&0\\0&0\end{pmatrix}$ and $(10|1)_1 = k\begin{pmatrix}0&1\\0&0\end{pmatrix} \oplus k\begin{pmatrix}0&0\\1&0\end{pmatrix}$,

(11|2) $\left\{ \left.\begin{pmatrix}a&0&0&0\\0&a&0&d\\c&0&b&0\\0&0&0&b\end{pmatrix} \right| a,b,c,d \in k \right\}$:

$$(11|2)_0 = k\begin{pmatrix}1&0&0&0\\0&1&0&0\\0&0&1&0\\0&0&0&1\end{pmatrix} \oplus k\begin{pmatrix}1&0&0&0\\0&1&0&0\\0&0&0&0\\0&0&0&0\end{pmatrix} \text{ and}$$

$$(11|2)_1 = k\begin{pmatrix}0&0&0&0\\0&0&0&1\\0&0&0&0\\0&0&0&0\end{pmatrix} \oplus k\begin{pmatrix}0&0&0&0\\0&0&0&0\\1&0&0&0\\0&0&0&0\end{pmatrix},$$

(11|3) $$(11|3)_0 = k\begin{pmatrix}1&0&0&0\\0&1&0&0\\0&0&1&0\\0&0&0&1\end{pmatrix} \oplus k\begin{pmatrix}0&0&0&0\\0&0&0&1\\1&0&0&0\\0&0&0&0\end{pmatrix} \text{ and}$$

$$(11|3)_1 = k\begin{pmatrix}1&0&0&0\\0&1&0&0\\0&0&-1&0\\0&0&0&-1\end{pmatrix} \oplus k\begin{pmatrix}0&0&0&0\\0&0&0&-1\\1&0&0&0\\0&0&0&0\end{pmatrix},$$

(12|1) $\wedge k^2 \cong k\langle X,Y\rangle/(X^2,Y^2,XY+YX)$:

$(12|1)_0 = k1 \oplus kX$ and $(12|1)_1 = kY \oplus kXY$,

(12|2) $(12|2)_0 = k1 + kXY$ and $(12|2)_1 = kX + kY$.

(14|3) $\left\{ \left.\begin{pmatrix}a&0&0\\c&a&0\\d&0&b\end{pmatrix} \right| a,b,c,d \in k \right\}$:

$$(14|3)_0 = k\begin{pmatrix}1&0&0\\0&1&0\\0&0&1\end{pmatrix} \oplus k\begin{pmatrix}1&0&0\\0&1&0\\0&0&0\end{pmatrix} \text{ and}$$

$$(14|3)_1 = k\begin{pmatrix}0&0&0\\1&0&0\\0&0&0\end{pmatrix} \oplus k\begin{pmatrix}0&0&0\\0&0&0\\1&0&0\end{pmatrix},$$

(15|3) $\left\{ \left.\begin{pmatrix}a&c&d\\0&a&0\\0&0&b\end{pmatrix} \right| a,b,c,d \in k \right\}$:

$$(15|3)_0 = k\begin{pmatrix}1&0&0\\0&1&0\\0&0&1\end{pmatrix} \oplus k\begin{pmatrix}1&0&0\\0&1&0\\0&0&0\end{pmatrix} \text{ and}$$

$$(15|3)_1 = k\begin{pmatrix}0&1&0\\0&0&0\\0&0&0\end{pmatrix} \oplus k\begin{pmatrix}0&0&1\\0&0&0\\0&0&0\end{pmatrix},$$

(16|1) $k\langle X,Y\rangle/(X^2,Y^2,YX)$:

$(16|1)_0 = k1 \oplus kX$ and $(16|1)_1 = kY \oplus kXY$,

(16|2) $(16|2)_0 = k1 \oplus kY$ and $(16|2)_1 = kX \oplus kXY$,

(16|3) $(16|3)_0 = k1 + kXY$ and $(16|3)_1 = kX + kY$.

(17|2) $\left\{\begin{pmatrix} a & 0 & 0 \\ 0 & a & 0 \\ c & d & b \end{pmatrix} \middle| a, b, c, d \in k\right\}$:

$(17|2)_0 = k\begin{pmatrix} 1 & 0 & 0 \\ 0 & 1 & 0 \\ 0 & 0 & 1 \end{pmatrix} \oplus k\begin{pmatrix} 1 & 0 & 0 \\ 0 & 1 & 0 \\ 0 & 0 & 0 \end{pmatrix}$ and

$(17|2)_1 = k\begin{pmatrix} 0 & 0 & 0 \\ 0 & 0 & 0 \\ 1 & 0 & 0 \end{pmatrix} \oplus k\begin{pmatrix} 0 & 0 & 0 \\ 0 & 0 & 0 \\ 0 & 1 & 0 \end{pmatrix}$,

(18; λ|1) $k\langle X, Y\rangle/(X^2, Y^2, YX - \lambda XY)$, where $\lambda \in k$ *with* $\lambda \neq -1, 0, 1$:
$(18; \lambda|1)_0 = k1 \oplus kX$ and $(18; \lambda|1)_1 = kY \oplus kXY$.

(18; λ|2) $(18; \lambda|2)_0 = k1 + kXY$ and $(18; \lambda|2)_1 = kX + kY$.

(19|1) $k\langle X, Y\rangle/(Y^2, X^2 + YX, YX + XY)$:
$(19|1)_0 = k1 + kXY$ and $(19|1)_1 = kX + kY$.

Moreover, $(18; \lambda|2) \cong (18; \lambda'|2)$ *if and only if* $\lambda' = \lambda$ *or* $\lambda\lambda' = 1$.

There exists only one 4-dimensional superalgebra $A = k \oplus I$ with $A_0 = k$ and $A_1 = I^2 = 0$. We denote it by (9|3) as its underlying algebra is isomorphic to (9).

Theorem 1.4 (Algebraic classification of 4-dimensional graded algebras) *Assume that k is an algebraically closed field and that* $\mathrm{ch}(k) \neq 2$. *Let A be a 4-dimensional superalgebra. Then A is isomorphic to one of the following superalgebras. Moreover each pair of listed superalgebras is non-isomorphic except the superalgebras within the same family* $(18; \lambda|i)$, *where* $(18; \lambda|i) \cong (18; \lambda'|i)$ *if and only if* $\lambda' = \lambda$ *or* $\lambda\lambda' = 1$, $i = 0, 1, 2$.

(1) : (1|0), (1|1), (1|2),
(2) : (2|0), (2|1), (2|2), (2|3),
(3) : (3|0), (3|1), (3|2), (3|3),
(4) : (4|0), (4|1),
(5) : (5|0), (5|1),
(6) : (6|0), (6|1), (6|2),
(7) : (7|0), (7|1), (7|2), (7|3),
(8) : (8|0), (8|1), (8|2), (8|3),
(9) : (9|0), (9|1), (9|2), (9|3),
(10) : (10|0), (10|1),
(11) : (11|0), (11|1), (11|2), (11|3),
(12) : (12|0), (12|1), (12|2),
(13) : (13|0), (13|1),
(14) : (14|0), (14|1), (14|2), (14|3),
(15) : (15|0), (15|1), (15|2), (15|3),
(16) : (16|0), (16|1), (16|2), (16|3),
(17) : (17|0), (17|1), (17|2),
(18; λ) : (18; λ|0), (18; λ|1), (18; λ|2), *where* $\lambda \in k$ *with* $\lambda \neq 1, 0, -1$,
(19) : (19|0), (19|1).

2 The Variety Salg_n and Its Properties

In this section we introduce the variety Salg_n of n-dimensional superalgebras. Let $A = A_0 \oplus A_1$ be a superalgebra $A = A_0 \oplus A_1$. The $\mathbb{Z}_2$-grading of A induces an involution given by $\sigma(a_0 + a_1) = a_0 - a_1$ where $a_i \in A_i$. Conversely, any algebra involution σ of A induces a $\mathbb{Z}_2$-grading on A, that is, $A = A_0 \oplus A_1$ with $A_0 = \{a \in A \mid \sigma(a) = a\}$ and $A_1 = \{a \in A \mid \sigma(a) = -a\}$. Thus we can identify a superalgebra A with an algebra A with an involution σ, denoted (A, σ).

Let (A, σ) be an n-dimensional superalgebra and $\{e_1, e_2, \ldots, e_n\}$ be a basis of A. The (unitary associative) algebra structure on vector space A gives rise to a set of structure constants $(\alpha_{ij}^k) \in \mathbb{A}^{n^3}$ determined by the multiplication of basis vectors so that

$$e_i e_j = \sum_{k=1}^{n} \alpha_{ij}^k e_k.$$

The involution σ on A may be also described by a set of constants $(\gamma_i^j) \in \mathbb{A}^{n^2}$ satisfying $\sigma(e_i) = \sum_{j=1}^n \gamma_i^j e_j$. It follows that to each superalgebra, (A, σ), we can associate a set of augmented *structure constants* $(\alpha_{ij}^k, \gamma_i^j) \in \mathbb{A}^{n^3+n^2}$, where (α_{ij}^k) are the structure constants determined by the algebra structure of A and (γ_i^j) the constants determined by the $\mathbb{Z}_2$-grading in the above manner. However it is not true that an arbitrary set of augmented structure constants can give rise to a superalgebra. The structure constants must obey certain relations to reflect how we define a superalgebra.

As a superalgebra (A, σ) must in particular be a unitary associative algebra, we have a multiplicative identity which we always take to be the first element of our basis, e_1. Then to be a unitary associative algebra we have the following conditions:

$$\begin{aligned} e_1 e_i &= e_i \\ e_i e_1 &= e_i \\ (e_i e_j) e_k &= e_i (e_j e_k) \end{aligned}$$

which translate into the following relations amongst the structure constants:

$$\alpha_{1i}^j - \delta_i^j = 0 \tag{1}$$

$$\alpha_{i1}^j - \delta_i^j = 0 \tag{2}$$

$$\sum_{l=1}^n (\alpha_{ij}^l \alpha_{lk}^m - \alpha_{il}^m \alpha_{jk}^l) = 0 \tag{3}$$

For σ to be an algebra involution means that:

$$\sigma(e_1) = e_1$$
$$\sigma(e_i e_j) = \sigma(e_i)\sigma(e_j)$$
$$\sigma^2(e_i) = e_i$$

These become the following relations in terms of the structure constants:

$$\gamma_1^j - \delta_1^j = 0 \tag{4}$$
$$\sum_{k=1}^n \alpha_{ij}^k \gamma_k^m - \sum_{k,l=1}^n \gamma_i^k \gamma_j^l \alpha_{kl}^m = 0 \tag{5}$$
$$\sum_{j=1}^n \gamma_i^j \gamma_j^k - \delta_i^k = 0 \tag{6}$$

It is precisely those structure constants obeying the relations (1)–(6) given above which give rise to superalgebras.

Definition 2.1 The Eqs. (1)–(6) given above cut out a variety in $\mathbb{A}^{n^3+n^2}$ which we shall call Salg_n—the variety of n-dimensional superalgebras.

In the rest of this paper we will study the geometry of Salg_n. The geometry of Salg_n is influenced by that of Alg_n, but Salg_n has a richer geometrical structure.

Definition 2.2 We define SA_n—the variety of n-dimensional superalgebras not requiring existence of a unit—to be the subvariety of $\mathbb{A}^{n^3+n^2}$ cut out by Eqs. (3), (5) and (6).

One checks that if A is a unitary algebra and $\sigma : A \to A$ satisfies $\sigma(xy) = \sigma(x)\sigma(y)$ and $\sigma^2 = \mathrm{id}_A$ then $\sigma(1_A) = 1_A$ (This follows from the more general fact that any invertible homomorphism $\sigma : A \to B$ between rings with unit must map the identity to the identity, i.e. $\sigma(1_A) = 1_B$), which after a little thought shows that $\mathrm{Salg}_n = \mathrm{SA}_n \cap V(\{\alpha_{1i}^j - \delta_i^j, \alpha_{i1}^j - \delta_i^j\})$. So we obtain the following result:

Lemma 2.1 *Salg_n is a closed subvariety of SA_n.*

It is important to notice the way that we have defined Salg_n—requiring the identity to be fixed—is analogous to the way Alg_n is defined in [2], but is not analogous to the way Alg_n was defined in [5]. We may define Salg'_n as an analogue to Alg_n in [5], which is the subset of SA_n consisting of superalgebras with unit, but not necessarily requiring the unit to be the first element of the basis or even in the basis. A similar proof to the one given in [3] shows that Salg'_n is an open affine subvariety of SA_n.

Similarly to the situation remarked in [2], since for our definition of Salg_n we require that the identity be the first element in the basis of any superalgebra, a subgroup G_n of GL_n acts on Salg_n (not the full group GL_n as one may expect). This action is induced by considering what happens to the structure constants when one makes a basis change. As the identity must be the first element in the basis, this means that the first column of the matrix describing the basis change must be $\begin{pmatrix}1 & 0 & \dots & 0\end{pmatrix}^T$ (identifying the given basis $\{e_1 = 1, e_2, \dots, e_n\}$ with the standard basis vectors for k^n). Hence we can describe G_n for $n \geq 2$ as follows: $G_n = \left\{\begin{pmatrix}1 & b^T \\ 0 & \Sigma\end{pmatrix} : \Sigma \in \mathrm{GL}_{n-1}, b \in k^{n-1}\right\}$. Thus the algebraic group G_n is of dimension $n^2 - n$.

Remark 2.1 If one so desired, our methods could be modified to study Salg'_n with the action of GL_n. However, one would hope that the geometry of both spaces are very similar—in particular we would like the degeneration partial orders induced in each space to coincide (the degeneration partial order will be introduced in Sect. 3). We would hope that such properties are intrinsic to the superalgebras and thus not depend on the way in which they are represented by a particular variety. We have not investigated this thoroughly, although in [2], it is remarked that this is the case for the degeneration partial orders in Alg_n and Alg'_n.

Let $\Lambda = (\lambda_i^j) \in G_n$ and $(\nu_i^j) = \Lambda^{-1}$. Then we can describe the action of G_n on Salg_n as follows:

$$\Lambda \cdot (\alpha_{ij}^k, \gamma_i^j) = \left(\sum_{l,p,q=1}^{n} \nu_l^k \alpha_{pq}^l \lambda_i^p \lambda_j^q, \sum_{k,l=1}^{n} \nu_k^j \gamma_l^k \lambda_i^l \right) = (\alpha_{ij}^{\prime k}, \gamma_i^{\prime j})$$

Firstly, recall that the formula for the inverse of a matrix means that we can express the entries ν_i^j of the matrix Λ^{-1} as a polynomial in the entries λ_i^j of the matrix Λ and $1/\det(\Lambda)$. Then the above formula expresses the new structure constants $\alpha_{ij}^{\prime k}, \gamma_i^{\prime j}$ in Salg_n as a polynomial in the old structure constants $\alpha_{ij}^k, \gamma_i^j$, the entries of the matrix $\Lambda \in G_n$ and $1/\det(\Lambda)$ which has non-vanishing denominator. Hence the action gives us a morphism $G_n \times \mathrm{Salg}_n \to \mathrm{Salg}_n$. The same reasoning also shows that the transport of structure action on Alg_n gives a morphism $G_n \times \mathrm{Alg}_n \to \mathrm{Alg}_n$.

We may refer to the above action of G_n on Salg_n as the **transport of structure action**. However as it is the only action of G_n on Salg_n considered here, we shall often simply refer to it as the action of G_n on Salg_n. It is clear that the orbits of Salg_n under the action of G_n can be identified with the isomorphism classes of n-dimensional superalgebras.

For an n-dimensional superalgebra A, we will sometimes use $G_n \cdot A$ to represent the orbit in Salg_n which the isomorphism class of A can be identified with. If in some basis the superalgebra A has structure constants $(\alpha_{ij}^k, \gamma_i^j)$, then $G_n \cdot A = G_n \cdot (\alpha_{ij}^k, \gamma_i^j)$.

There are two interesting morphisms between Salg_n and Alg_n. They arise from the following observations: any n-dimensional superalgebra may be regarded as an n-dimensional algebra and any n-dimensional algebra can be endowed with the trivial $\mathbb{Z}_2$-grading making it into an n-dimensional superalgebra.

The first morphism: $U : \mathrm{Salg}_n \to \mathrm{Alg}_n$ is defined by $(\alpha_{ij}^k, \gamma_i^j) \mapsto (\alpha_{ij}^k)$ is the forgetful map, which forgets the superalgebra structure on A and only remembers the algebra structure on A.

The second morphism: $I : \mathrm{Alg}_n \to \mathrm{Salg}_n$ is defined by $(\alpha_{ij}^k) \mapsto (\alpha_{ij}^k, \delta_i^j)$ where δ_i^j is the Kronecker delta function. This takes an algebra structure on A and endows it with the trivial $\mathbb{Z}_2$-grading making it a superalgebra on A.

Notice that the subset of Salg_n consisting of superalgebras with the trivial $\mathbb{Z}_2$-grading is a closed subset of Salg_n and is given by $V(\{\gamma_i^j - \delta_i^j\}) \cap \mathrm{Salg}_n$. The morphism I above identifies Alg_n with this subset. This result is a part of the following proposition.

Proposition 2.1 *The morphisms U and I described above are continuous closed maps. Moreover I provides an isomorphism of* Alg_n *with the closed subset of* Salg_n *consisting of the superalgebras with the trivial $\mathbb{Z}_2$-grading.*

We point out that both morphisms U and I are G_n-equivariant. That is, for $\Lambda \in G_n$ and $(\alpha_{ij}^k, \gamma_i^j) \in \mathrm{Salg}_n$, we have $U(\Lambda\cdot(\alpha_{ij}^k, \gamma_i^j)) = \Lambda\cdot U((\alpha_{ij}^k, \gamma_i^j))$ and $I(\Lambda\cdot(a_{ij}^k)) = \Lambda \cdot I((\alpha_{ij}^k))$. As a consequence of the G_n-equivariance of U, we obtain the following:

Corollary 2.1 $U\left(\overline{G_n \cdot (\alpha_{ij}^k, \gamma_i^j)}\right) = \overline{G_n \cdot (\alpha_{ij}^k)}.$

Suppose that one has a superalgebra A with $\dim A_0 = i$ and $\mathbb{Z}_2$-grading given by the algebra involution σ. Now change to a homogeneous basis (say by a linear map represented by the matrix Λ), which clearly has $\mathbb{Z}_2$-grading σ' given by the linear map represented by the diagonal matrix with 1 for the first i entries and -1 for the last $n-i$ entries. From the above we have $\sigma' = \Lambda\sigma\Lambda^{-1}$, so $\sigma = \Lambda^{-1}\sigma'\Lambda$. Thus $\mathrm{tr}(\sigma) = \mathrm{tr}(\Lambda^{-1}\sigma'\Lambda) = \mathrm{tr}(\sigma'\Lambda\Lambda^{-1}) = \mathrm{tr}(\sigma') = i - (n-i) = 2i - n$ and $\det(\sigma) = \det(\Lambda^{-1}\sigma'\Lambda) = \det(\Lambda^{-1})\det(\sigma')\det(\Lambda) = \det(\sigma') = (-1)^{n-i}$.

We now define Salg_n^i to be the subset of Salg_n consisting of the superalgebras A with $\dim A_0 = i$. Obviously we have $\mathrm{Salg}_n = \bigcup_{i=1}^n \mathrm{Salg}_n^i$. Hence, from above, **the trace and determinant are constant on** Salg_n^i. It is clear that these subsets must be disjoint. We are interested in when these subsets are also closed. The following lemma gives some sufficient conditions for this to be the case.

Before stating the next couple of results we mention how vital the assumption that $\mathrm{ch}(k) \neq 2$ is to Lemma 2.2 and Proposition 2.2. These are very basic results about the geometry of Salg_n—the study of Salg_n over an algebraically closed field k with $\mathrm{ch}(k) = 2$ would require new techniques as the proofs of these two results do not work in the case $\mathrm{ch}(k) = 2$.

Lemma 2.2 *The sets* Salg_n^i *are closed subsets of* Salg_n *in the following situations:*

(a) $\mathrm{ch}(k) = p$ *and* $n \le 2p$
(b) $\mathrm{ch}(k) = 0$ *(with no restriction on n in this case)*
(c) $n \le 6$ *(for any algebraically closed field k with* $\mathrm{ch}(k) \neq 2$*)*

Proof Define $S_n^i = V(\{\sum_{j=1}^n \gamma_j^j - (2i-n), \sum_\pi \operatorname{sgn}(\pi)\gamma_1^{\pi(1)} \dots \gamma_n^{\pi(n)} - (-1)^{n-i}\}) \cap \mathrm{Salg}_n$ for $i \in \{1, \dots, n\}$, (where $\operatorname{sgn}(\pi)$ denotes the signature of the permutation π, and the sum is taken over all permutations of $\{1, \dots, n\}$). Thus the S_n^i are closed subsets of Salg_n. From the statements above, it is clear that $\mathrm{Salg}_n^i \subseteq S_n^i$. The first polynomial $\sum_{j=1}^n \gamma_j^j$ represents the trace of the $\mathbb{Z}_2$-grading and the second $\sum_\pi \operatorname{sgn}(\pi)\gamma_1^{\pi(1)} \dots \gamma_n^{\pi(n)}$ represesnts its determinant.

For the proof of part (a), consider the following. Let $i, j \in \{1, \dots, n\}, i \neq j$. If i and j differ by $2p$ then both the traces and the determinants for Salg_n^i and S_n^j will agree, so $\mathrm{Salg}_n^i \subseteq S_n^j$. If i and j differ by less than $2p$, then the traces of Salg_n^i and S_n^j will differ unless i and j differ by p, in which case, since p is odd (remember we are excluding the case $\mathrm{ch}(k) = 2$ throughout this paper) the determinants will differ. Thus Salg_n^i and S_n^j are disjoint. From these comments one can see that we have the equality $\mathrm{Salg}_n^i = S_n^i$ for all $i \in \{1, \dots, n\}$ if and only if there are no two distinct integers $i, j \in \{1, \dots, n\}$ which differ by $2p$. One can always be sure that this is condition is met when $n \leq 2p$. This completes the proof of (a).

For part (b), we have $\mathrm{ch}(k) = 0$. Here one simply needs to consider the traces on Salg_n^i and S_n^j, which must differ unless $i = j$, showing that the subsets Salg_n^i and S_n^j are disjoint unless $i = j$, that is $\mathrm{Salg}_n^i = S_n^i$.

Finally, for part (c) we combine the results of (a) and (b). In the case of positive characteristic p, then as $p \geq 3$, from part (a) we know that these subsets are disjoint and closed for $n \leq 6$, while in the case of zero characteristic from part (b) we know that these subsets are disjoint and closed for any n. Combine these statements to see that regardless of the characteristic of the field k, the subsets Salg_n^i are all closed subsets when $n \leq 6$.

Remark 2.2 Lemma 2.2 is likely to be general enough for us to use in all cases where determining irreducible components of Salg_n is currently practical. The irreducible components of Alg_n have so far only been described for $n \leq 5$ (with some special—"rigid"—components described in the case $n = 6$), and finding these irreducible components is a more basic question than finding the irreducible components of Salg_n. However, it is of theoretical interest to determine whether the subsets Salg_n^i are in fact closed subsets of Salg_n for all n and any field k with $\mathrm{ch}(k) \neq 2$, or if there is some field k of prime characteristic, p, and some integer, n, such that the variety Salg_n over the field k has one of its subsets Salg_n^i which is not closed. As we shall see, when the Salg_n^i are closed they form the connected components of Salg_n. Thus it would be interesting to know if the geometry of Salg_n can change in this manner for some integer, n, and field, k, of prime characteristic, p.

Using the notation from the proof of Lemma 2.2 we have the following situation for the variety Salg_7 over an algebraically closed field of characteristic 3. $S_7^1 = S_7^7 = V(\{\sum_{j=1}^n \gamma_j^j - 1, \sum_\pi \operatorname{sgn}(\pi)\gamma_1^{\pi(1)} \dots \gamma_n^{\pi(n)} - 1\}) \cap \mathrm{Salg}_7$. This is the smallest example of where the above lemma may not be applied. While it is clear that Salg_7^1 and Salg_7^7 are disjoint, it may be possible that $\overline{\mathrm{Salg}_7^1}$ and Salg_7^7 have some point in common. (Recall that we remarked earlier that Salg_n^n is closed—so $\overline{\mathrm{Salg}_n^n} = \mathrm{Salg}_n^n$ and thus we do know that $\overline{\mathrm{Salg}_7^7} = \mathrm{Salg}_7^7$ and Salg_7^1 are disjoint).

Proposition 2.2 Salg_n *is disconnected for* $n \geq 2$.

Proof By the comments above Lemma 2.2, for each superalgebra, the determinant of the $\mathbb{Z}_2$-grading is either -1 or 1. Since $\mathrm{ch}(k) \neq 2$, -1 and 1 are distinct elements of k, hence $X_{-1} = V(\{\sum_\pi \mathrm{sgn}(\pi)\gamma_1^{\pi(1)} \dots \gamma_n^{\pi(n)} - (-1)\}) \cap \mathrm{Salg}_n$ and $X_1 = V(\{\sum_\pi \mathrm{sgn}(\pi)\gamma_1^{\pi(1)} \dots \gamma_n^{\pi(n)} - 1\}) \cap \mathrm{Salg}_n$ are disjoint closed subsets whose union is Salg_n. But $X_{-1} = \mathrm{Salg}_n \setminus X_1$ and $X_1 = \mathrm{Salg}_n \setminus X_{-1}$, hence both are open sets too. Thus Salg_n is a union of two disjoint open subsets. Both subsets are non-empty for $n \geq 2$. Thus for $n \geq 2$, Salg_n is disconnected.

From here onwards, we make **the assumption that** Salg_n^i **are closed subsets of** Salg_n. The main examples which we are interested in are Salg_n for $n = 2, 3, 4$, and in these cases this assumption is satisfied by Lemma 2.2.

Since some algebras and superalgebras will arise frequently, we shall name them for convenience. Define C_n to be the algebra $k[X_1, \dots, X_{n-1}]/(X_1, \dots, X_{n-1})^2$ and for $i = 1, \dots, n$, let $C_n(i)$ be the superalgebra which has C_n as its underlying algebra and the $\mathbb{Z}_2$-grading is given by $C_n(i)_0 = \mathrm{span}\{1, X_1, \dots, X_{i-1}\}$, $C_n(i)_1 = \mathrm{span}\{X_i, \dots, X_{n-1}\}$. It is clear that algebra C_n and the superalgebras $C_n(i)$ for $i = 1, \dots, n$ all have dimension n.

The following lemma shows that each superalgebra structure on C_n is isomorphic to one of the $C_n(i)$.

Lemma 2.3 *Consider the algebra* C_n. *There are n distinct isomorphism classes of superalgebras on this algebra, which are* $C_n(1), \dots, C_n(n)$.

Proof Let $B = B_0 \oplus B_1$ be a superalgebra structure on C_n where $\dim B_0 = i + 1$ with $0 \leq i \leq n-1$ (so $\dim B_1 = n-i-1$). Suppose B_0 has basis $\{1, u_1, \dots, u_i\}$ and B_1 has basis $\{u_{i+1}, \dots, u_{n-1}\}$. There must be scalars such that for $1 \leq j \leq n-1$, $u_j = \alpha_{j1}1 + \alpha_{j2}X_1 + \dots + \alpha_{jn}X_{n-1}$.

Now let $u'_j = u_j - \alpha_{j1}1 = \alpha_{j2}X_1 + \dots + \alpha_{jn}X_{n-1}$. Then $\{1, u'_1, \dots, u'_i\}$ is also a basis for B_0.

If $\alpha_{j1} \neq 0$ for any $i+1 \leq j \leq n-1$ then $u_j = \alpha_{j1}1 + \sum_{i=1}^{n-1} \alpha_{ji+1}X_i$, so $u_j^2 = \alpha_{j1}^2 1 + 2\sum_{i=1}^{n-1} \alpha_{ji+1}X_i$. Since $u_j^2 \in B_0$ we must have $\sum_{i=1}^{n-1} \alpha_{ji+1}X_i \in B_0$, say $\sum_{i=1}^{n-1} \alpha_{ji+1}X_i = \beta_1 1 + \sum_{k=1}^{i} \beta_{k+1}u_k$ then $(\beta_1 + \alpha_{j1})1 + \sum_{k=1}^{i} \beta_{k+1}u_k - u_j = 0$, which contradicts the linear independence of the basis. So $\alpha_{j1} = 0$ for all $i+1 \leq j \leq n-1$.

It is easy to check that any two of $u'_1, \dots, u'_i, u_{i+1}, \dots, u_{n-1}$ have product zero (including a product involving two of the same terms). So we can define a map $\phi : B \to C_n(i+1)$ by $1 \mapsto 1, u'_1 \mapsto X_1, \dots, u'_i \mapsto X_i, u_{i+1} \mapsto X_{i+1}, \dots, u_{n-1} \mapsto X_{n-1}$. It is easy to see that this is a bijection, which preserves the algebra structure and $\mathbb{Z}_2$-grading, hence is an isomorphism of superalgebras. Thus a superalgebra structure on C_n must be isomorphic to one of those described in the lemma.

To conclude the proof, we note that the n superalgebra structures given in the lemma are clearly mutually non-isomorphic.

So for each i there is a unique (up to isomorphism) superalgebra structure A on $k[X_1, \dots, X_{n-1}]/(X_1, \dots, X_{n-1})^2$ which has $\dim A_0 = i$.

In the case of n-dimensional algebras, Gabriel showed that the closed orbit consists of algebras isomorphic to C_n. The closed orbits in Salg_n consist of superalgebras isomorphic to one of the superalgebras $C_n(i)$, as the following Proposition shows.

Proposition 2.3 *There are n closed orbits in* Salg_n. *They are all disjoint,* $C_n(i)$ *being the closed orbit in* Salg_n^i.

Proof Suppose $G_n \cdot A$ is a closed orbit, i.e. $\overline{G_n \cdot A} = G_n \cdot A$. As $U(A)$ is an n-dimensional algebra, $G_n \cdot U(A)$ is an orbit in Alg_n. Now by Corollary 2.1 $\overline{G_n \cdot U(A)} = U(\overline{G_n \cdot A}) = U(G_n \cdot A) = G_n \cdot U(A)$. Thus the orbit $G_n \cdot U(A)$ is closed in Alg_n but then, by the results of [5], $U(A)$ must be isomorphic to C_n. That is, A must be isomorphic to a superalgebra structure on C_n.

It remains to show that the orbits, $G_n \cdot C_n(i)$, corresponding to the isomorphism classes of the superalgebras $C_n(i)$ are, in fact, closed. Notice that $C_n = U(C_n(i))$ is the algebra structure whose isomorphism class corresponds to the closed orbit in Alg_n. That is, the orbit $G_n \cdot C_n$ is closed in Alg_n and thus $U^{-1}(G_n \cdot C_n)$ is closed in Salg_n. Now, by assumption, Salg_n^i are closed disjoint subsets, thus $U^{-1}(G_n \cdot C_n) \cap \mathrm{Salg}_n^i$ is closed. However this set is the orbit $G_n \cdot C_n(i)$ (since Lemma 2.3 above showed that all superalgebra structures on algebra C_n with the degree zero component having dimension i are all isomorphic). The result follows.

Lemma 2.4 *Suppose that* Salg_n^i *are closed subsets. Let A be a superalgebra with* $\dim A_0 = i$. *Assume that there is only one isomorphism class of superalgebras on* $U(A)$ *which has* $\dim_0 = i$. *If the orbit* $G_n \cdot U(A)$ *is open in* Alg_n *then the orbit* $G_n \cdot A$ *is open in* Salg_n.

Proof Since Salg_n^i are all disjoint closed subsets by assumption, they are also each open. Now $U^{-1}(G_n \cdot U(A)))$ is the collection of superalgebra structures on $U(A)$. Since $G_n \cdot U(A)$ is open, so too must be $U^{-1}(G_n \cdot U(A))$, by the continuity of U. Now by the assumptions made $G_n \cdot A = U^{-1}(G_n \cdot U(A)) \cap \mathrm{Salg}_n^i$. Thus $G_n \cdot A$ is the intersection of two open sets, so it is open itself.

Example 2.1 *This is indeed the case for several orbits in* Salg_4. *Using this result and the fact that the orbits of* (1) *and* (10) *are open in* Alg_4 *we discover that the orbits* (1|0), (1|1), (1|2), (10|0) *and* (10|1) *are open in* Salg_4.

3 Algebraic Groups and Their Actions

Recall that an algebraic group G is an algebraic variety which additionally has the structure of a group. That is, the multiplication $\mu : G \times G \to G$ given by $\mu(x, y) = xy$ and inversion $\text{ß} : G \to G$ given by $\text{ß}(x) = x^{-1}$ are morphisms of varieties. An algebraic group is said to be connected if it is irreducible as a variety. The algebra group G_n and GL_n are connected with dimensions $n^2 - n$ and n^2 respectively.

The following result is well-known, for example see [3].

Lemma 3.1 *Let G be a connected algebraic group acting on a variety X, then:*

(a) *Each orbit $G \cdot x$ is locally closed (i.e. $G \cdot x$ is open in $\overline{G \cdot x}$) and irreducible*
(b) $\dim G \cdot x = \dim G - \dim \mathrm{Stab}_G(x)$
(c) $\overline{G \cdot x} \backslash G \cdot x$ *is a union of orbits of dimension* $< \dim G \cdot x$

Note that in the case of the G_n-action on Salg_n, the stabiliser subgroup of a point $(\alpha_{ij}^k, \gamma_i^j)$ of Salg_n is the automorphism group of the superalgebra given by the point $(\alpha_{ij}^k, \gamma_i^j)$.

Whenever we have a connected algebraic group G acting on a variety X, we have the idea of degeneration. The action of G on X partitions the variety into equivalence classes under the equivalence relation $x \equiv y \Leftrightarrow \exists\, g \in G$ such that $y = g \cdot x$. The equivalence classes are the G-orbits. Because of this, we shall use the notation $[x] = G \cdot x$ for brevity, while stating and proving results about this more general notion of degeneration.

Definition 3.1 We say that $[x]$ **degenerates** to $[y]$ if $y \in \overline{G \cdot x}$ and will write $[x] \to [y]$.

It is not difficult to see that $[x] \to [y]$ if and only if $G \cdot y \subseteq \overline{G \cdot x}$. The latter provides a useful way to visualize the notion of degeneration—that an orbit is contained in the closure of some other orbit.

By appealing to Lemma 3.1 we can show that this idea of degeneration is not only well-defined on the G-orbits of X, but it also gives rise to a partial order on the G-orbits in X. We define $[y] \leq_{degr} [x]$ if and only if $[x]$ degenerates to $[y]$. (Note that in some places the degeneration partial order is defined to be the opposite to this. This happens for example in [14]).

Our main interest is in the degeneration of superalgebras and the degeneration partial order on the isomorphism classes of n-dimensional superalgebras.

For n-dimensional superalgebras A and B, if $(\alpha_{ij}^k, \gamma_i^j) \in G_n \cdot B$ and $(a_{ij}^k, \gamma_i^j) \in \overline{G_n \cdot A}$, then A **degenerates to** B and denote this by $A \to B$. In some places the terminology A **dominates** B is used instead of A degenerates to B. Clearly, whenever $(\alpha_{ij}^k, \gamma_i^j) \in G_n \cdot A$, then we also have $(\alpha_{ij}^k, \gamma_i^j) \in \overline{G_n \cdot A}$ since $G_n \cdot A \subseteq \overline{G_n \cdot A}$. A degeneration of this form is referred to as a **trivial degeneration**, any degeneration which is not of this form is called a **non-trivial degeneration.**

Intuitively, if the superalgebra A degenerates to the superalgebra B (where $B \ncong A$ that is, this is a proper degeneration) then we think of the orbit $G_n \cdot B$ as consisting of some of those points outside the orbit $G_n \cdot A$, but which are "close to" some of the points in the orbit $G_n \cdot A$. This is supported by observing that the orbit $G_n \cdot B$ belongs to the boundary of $G_n \cdot A$ (i.e. the set $\overline{G_n \cdot A} \backslash G_n \cdot A$) as we shall see in the next section. Another observation supporting this intuition is that some degenerations may be obtained by taking a sequence of points in the orbit $G_n \cdot A$ whose "limit" lies in the orbit $G_n \cdot B$ (see Corollary 4.1).

It is well-known that when G is a connected algebraic group acting on a variety X, the irreducible components of X are stable under the action of G. Thus we have the following.

Corollary 3.1 *When G is a connected algebraic group acting on a variety, the irreducible components are closures of a single orbit or closures of an infinite family of orbits.*

Proof We know that irreducible components are G-stable. We also know that components are closed, hence each component can be taken to be the closure of a union of orbits. If there are only finitely many orbits in the union, then by using $\overline{A \cup B} = \overline{A} \cup \overline{B}$ we see that the component is not irreducible unless it is the closure of a single orbit. This gives the required statement.

In the case of the G_n transport of structure action on Alg_n Flanigan goes further, and in [4] proves a result describing algebraic properties of algebras belonging to some infinite family, whose orbits give rise to an irreducible component as described above.

In the following we shall abuse the terminology, and refer to the situation when some structure is contained in the closure of the union of the orbits of an infinite family of orbits, as a degeneration. We see an example of this in Alg_4 in the results of Gabriel, where the structure (19) is contained in the closure of the union of orbits of the family of structures $(18; \lambda)$. It is important to notice, however, that this is not a degeneration as defined earlier. Similarly, when an infinite family of orbits is contained in another infinite family of structures, we may also wish to refer to this as a degeneration too. We have an example of this given by Mazzola's work on Alg_5 in [10], where the orbits of the infinite family of structures $(35; \lambda)$ is contained in the closure of the union of the orbits in the infinite family of structures $(13; \lambda)$. Finally, one may wish to refer to the case where an infinite family of structures is contained in the closure of a single orbit as a degeneration. This idea is less of an abuse of terminology than the others mentioned above, however, since we could consider it to be an infinite family of degenerations (in the original sense), one to each of the orbits in the infinite family. Although an abuse of terminology, it is useful to extend the notion of degeneration in this way, as it helps with determining the irreducible components.

Corollary 3.2 *When G is a connected algebraic group acting on a variety X, we have the following statements regarding the notions of degeneration and irreducible components:*

(a) If $[x] \to [y]$ then $[y]$ belongs to all the irreducible components to which $[x]$ belongs (and possibly more too).
(b) If there is no degeneration to $[x]$, then its closure is an irreducible component.
(c) If $\cup_\lambda [x(\lambda)]$ is irreducible and there is no degeneration to $\cup_\lambda [x(\lambda)]$ then its closure is an irreducible component.

Proof For part (a) $G \cdot y \subseteq \overline{G \cdot x}$, so that any irreducible component containing $G \cdot x$ must also contain $G \cdot y$.

For parts (b) and (c), consider what happens if $\overline{G \cdot x}$ (respectively $\overline{\cup_\lambda G \cdot x(\lambda)}$) is not an irreducible component. Then, as an irreducible set, it must be contained in some irreducible component implying that $[x]$ (respectively $\cup_\lambda [x(\lambda)]$) is contained in the closure of an orbit, or in the closure of the union of an infinite family of orbits. This means that there is a degeneration to $[x]$ (respectively $\cup_\lambda [x(\lambda)]$), contrary to our assumption.

Remark 3.1 The above to wonder when a union of a family of orbits is irreducible, so that we may apply part (c) of the above. This might not be true for arbitrary actions of algebraic groups on a variety. However the infinite families which arise in Alg_4 and Alg_5 can be shown to be irreducible. We illustrate this idea using the superalgebras $(18; \lambda|i)$. Firstly fix i as either 0, 1 or 2. Use the basis $e_1 = 1, e_2 = X, e_3 = Y, e_4 = XY$ of $(18; \lambda|i)$. Then for the member of the family with parameter value $\lambda \neq -1$ we have that the structure constant, $\alpha_{23}^4 = \lambda$. Hence, using this basis, we obtain a set of points in Salg_4. Call this set S—one point from each orbit corresponding to a member of the family $(18; \lambda|i)$. This set of points can be identified with $k\backslash\{-1\}$ which is irreducible in $\mathbb{A}^1$ (being the distinguished open $D(x+1)$ of $\mathbb{A}^1$), thus the set of points, S, is also irreducible. Now denote by $\phi : G_n \times \mathrm{Salg}_n \to \mathrm{Salg}_n$ the morphism arising from the transport of structure action of G_n on Salg_n. The union of the orbits of $(18; \lambda|i)$ is given by $\phi(G_n \times S)$, which, exactly as remarked proceeding to Corollary 3.2, is seen to be irreducible. So we have shown that the union of orbits of superalgebras $(18; \lambda|i)$ for $i = 0, 1, 2$ are irreducible. The infinite families in Alg_5 can be shown to be irreducible in a similar manner.

Corollary 3.2 tells us that the irreducible components are the orbits or infinite families of orbits, which no other orbit or infinite family of orbits degenerates to. So if one knows all degenerations between orbits and infinite families of orbits, then it is a trivial matter to determine the irreducible components. Unfortunately, the problem of determining all these degenerations is usually difficult. The problem of determining the irreducible components is somewhat easier, but can still be difficult too.

Definition 3.2 An n-dimensional superalgebra A (respectively, a family of superalgebras $A(\lambda)$) is called **generic**, if the closure of its orbit in Salg_n—$\overline{G_n \cdot A}$ (respectively, the closure of the union of the family of orbits—$\overline{\bigcup_\lambda G_n \cdot A(\lambda)}$), is an irreducible component of Salg_n.

Remark 3.2 A superalgebra A, whose orbit is open, is always generic. Since it must lie in some irreducible component (being an irreducible set by part (a) of Lemma 3.1) and, as an open subset of any irreducible set is dense, we must have that $\overline{G_n \cdot A}$ is the entire component.

However the observations in Corollary 3.2 applies more generally and can also aid us in finding the irreducible components. For example, after finding that no algebras degenerate to (17) in Alg_4, by applying the closed continuous map U, we discover that no superalgebras can degenerate to any of $(17|i)$ for $i = 0, 1, 2$ in Salg_4. Then, by using the observations given in Corollary 3.2, we see that $(17|i)$ for $i = 0, 1, 2$ give rise to irreducible components of Salg_4, hence these algebras are also generic.

The next two lemmas of this section are concerned with calculating the dimensions of the orbits in Salg_n. We explain how to read these tables now. Each row corresponds to a different algebra structure and the columns of the table are for different $\mathbb{Z}_2$-gradings on that given underlying algebra structure. Thus the underlying algebra structure of the superalgebra determines which row you look in, and which particular $\mathbb{Z}_2$-grading is used to obtain the given superalgebra structure determines which column you look under. We illustrate this by using an example. To find the dimension of the stabilizer of a point in the orbit of $(3|2)$ we look in the row labelled $(3|\cdot)$ and then look under the column labelled 2 to see that the dimension of the required stabilizer is 2.

Lemma 3.2 *The following gives the dimensions of the stabilizers of points in the orbits in* Salg_4*:*

Stabilizer dimensions				
$\cdot$	0	1	2	3
$(1\|\cdot)$	0	0	0	
$(2\|\cdot)$	1	1	1	1
$(3\|\cdot)$	2	2	2	1
$(4\|\cdot)$	2	1		
$(5\|\cdot)$	3	2		
$(6\|\cdot)$	4	2	4	
$(7\|\cdot)$	4	2	3	2
$(8\|\cdot)$	5	3	3	3
$(9\|\cdot)$	9	5	5	9
$(10\|\cdot)$	3	1		
$(11\|\cdot)$	4	3	2	2
$(12\|\cdot)$	6	3	4	
$(13\|\cdot)$	2	1		
$(14\|\cdot)$	3	3	2	2
$(15\|\cdot)$	3	3	2	2
$(16\|\cdot)$	4	3	3	2
$(17\|\cdot)$	6	3	4	
$(18; \lambda\|\cdot)$	4	3	2	
$(19\|\cdot)$	4	2		

Proof If the point $(\alpha_{ij}^k, \gamma_i^j)$ is in the orbit, $G_4 \cdot A$, which is identified with the isomorphism class of superalgebra A, then $\mathrm{Stab}_{G_4}((\alpha_{ij}^k, \gamma_i^j)) \cong \mathrm{Aut}(A)$ where $\mathrm{Aut}(A)$ is the group of automorphisms of the superalgebra A as mentioned in the paragraph below Lemma 3.1.

We remark that $\dim \mathrm{PGL}_n(k) = n^2 - 1$, so that $\dim \mathrm{PGL}_2(k) = 2^2 - 1 = 3$ (see for example [7])

Orbit dimensions				
·	0	1	2	3
(1|·)	12	12	12	
(2|·)	11	11	11	11
(3|·)	10	10	10	11
(4|·)	10	11		
(5|·)	9	10		
(6|·)	8	10	8	
(7|·)	8	10	9	10
(8|·)	7	9	9	9
(9|·)	3	7	7	3
(10|·)	9	11		
(11|·)	8	9	10	10
(12|·)	6	9	8	
(13|·)	10	11		
(14|·)	9	9	10	10
(15|·)	9	9	10	10
(16|·)	8	9	9	10
(17|·)	6	9	8	
(18; λ|·)	8	9	10	
(19|·)	8	10		

Proposition 3.1 *The following gives the dimensions of the orbits in* Salg_4*:*

Proof We have calculated the dimensions of the automorphism groups, or equivalently, the dimensions of stabilizers of any point in each orbit in Lemma 3.2 above. We know that the dimension of G_4 is 12. By using part (b) of Lemma 3.1, we can calculate the dimension of the orbit $G_4 \cdot (\alpha_{ij}^k, \gamma_i^j)$ by subtracting the dimension of the stabilizer, $\mathrm{Stab}_{G_4}((\alpha_{ij}^k, \gamma_i^j))$, from the dimension of G_4 which is 12.

Remark 3.3 We remark that to calculate the dimensions of the orbits in the case where we don't require the identity to be fixed (i.e. the orbits in Salg_4' and in which case GL_4 acts on this variety) we can subtract the dimensions of the stabilizers found in Lemma 3.2 from 16. If we then compare the dimensions of the orbits of the trivially $\mathbb{Z}_2$-graded superalgebras $(i|0)$ for $i = 1, \ldots, 18; \lambda, 19$, thus calculated, with those given by Gabriel in [5], we find that the two sets of numbers do not agree. In fact the orbit dimensions that Gabriel gives are exactly one less than the orbit dimensions we calculate in each case. This is strange. Since Gabriel did not give the proof of these facts in [5] it is difficult to find an explanation for this difference. However in Mazzola's paper [10] on classifying algebras of dimension five, the orbit dimensions are calculated by subtracting the dimension of the automorphism groups from 25 (25 being the dimension of GL_5)—this would tend to suggest that our methodology for calculating orbit dimensions is correct.

4 Degenerations in Salg_n

In this section we concern ourselves with conditions determining when a degeneration of superalgebras in Salg_n can or cannot exist. When looking for conditions for the non-existence of degenerations between a given pair of superalgebras, it would be helpful to have some invariants of the superalgebra which are "rigid" in the sense that if there is a degeneration of superalgebras from A to B, then the superalgebras A and B must have the same value for the invariant. Unfortunately, the only such invariant that we know of is $\dim_0$, the dimension of the trivial degree part. The next best thing is a property of a superalgebra which any degeneration of this superalgebra must inherit, or some property which cannot increase or decrease upon degeneration. Such properties are analogous to those described in [5, Proposition 2.7], which states, for example, the fact that the dimension of the radical cannot decrease upon degeneration. Later in the section we determine several properties from which any degeneration of a given superalgebra must share.

Lemma 4.1 *Let $\Omega : k \to \mathrm{Salg}_n$ be a polynomial function and $U \subseteq \mathrm{Salg}_n$. If there are infinitely many points of $\Omega(k)$ in U then $\Omega(k) \subseteq \overline{U}$.*

Proof First, note that we think of Ω as describing a curve in Salg_n. $\overline{U}$ is defined to be the intersection of all closed sets containing U. A closed set is the vanishing set of polynomials (intersected with Salg_n), so it is enough to show that any polynomial vanishing on U must also vanish on all of $\Omega(k)$. By applying the appropriate projections to Ω, we may write $\alpha_{ij}^k = a_{ij}^k(t)$ and $\gamma_i^j = g_i^j(t)$ (letting the indeterminate be t), to describe the coordinates of this curve.

It is standard that $\Omega^{-1}(U) = \{t \in k : \Omega(t) \in U\}$, but notice that this set gives the t values such that the curve Ω lies inside the set U. We consider a polynomial function in $(\alpha_{ij}^k, \gamma_i^j)$, which vanishes on U, $f(\alpha_{ij}^k, \gamma_i^j) = 0$. Since f vanishes on U it must vanish at the points of $\Omega(k)$ lying inside U. So we have $t \in \Omega^{-1}(U) \Rightarrow f(a_{ij}^k(t), g_i^j(t)) = 0$. Note that $f(a_{ij}^k(t), g_i^j(t))$ is a polynomial in t. Suppose the degree $\deg(f(a_{ij}^k(t), g_i^j(t))) = d$.

If $d \geq 1$, then $f(a_{ij}^k(t), g_i^j(t)) = 0$ has at most d zeros, which contradicts the fact that we assumed to vanish on all of $\Omega(k) \cap U$, which has infinitely many points. Thus $d = 0$, hence $f(a_{ij}^k(t), g_i^j(t))$ must be a constant. The only way that $f(a_{ij}^k(t), g_i^j(t)) = 0$ is satisfied for points in $\Omega^{-1}(U)$ is if $f(a_{ij}^k(t), g_i^j(t))$ is the zero polynomial, in which case $f(a_{ij}^k(t), g_i^j(t)) = 0$ is satisfied for all $t \in k$. This completes the proof.

Now we consider a practical method for computing degeneration of superalgebras, called a specialization of superalgebras. This method was first introduced by Gabriel in [5]. We formulate it in the form of superalgebras.

Definition 4.1 If A and B are n-dimensional superalgebras, a **specialization** of A to B is the following situation: one makes a change of basis in A to a "variable" basis, i.e. one involving some unknown t, such that the point of Salg_n obtained by structural transport is given by some polynomial functions in t and lies in the orbit of A for $t \neq 0$, yet at $t = 0$ lies in the orbit B. We think of B as being obtained by a formal limit of the basis change in A.

A specialization of superalgebras A to B is a more restrictive notion than a specialization of algebras, since not only must there be a specialization of the underlying algebras, but also must this occur in such a way that under the specialization. The $\mathbb{Z}_2$-grading on A also tends to the $\mathbb{Z}_2$-grading on B. This is usually a non-trivial constraint. So some specializations between algebras may not give rise to specializations of superalgebras on these algebras. Or perhaps one must use different specializations for different superalgebra structures on the same underlying algebra.

With this idea of specialization we obtain a useful corollary of the above lemma.

Corollary 4.1 *A specialization of A to B implies that A degenerates to B.*

Proof Clearly the specialization gives us a curve $\Omega : k \to \mathrm{Salg}_n$. We let the set U in Lemma 4.1 be the orbit $G_n \cdot A$. Now, as k is algebraically closed, it has infinitely many elements. Thus so does k^*. Then $\Omega(k^*) \subseteq G_n \cdot A$, so $G_n \cdot A$ contains infinitely many elements of $\Omega(k)$. Thus we may apply Lemma 4.1. Now note that $\Omega(0)$ gives structure constants for a point in the orbit $G_n \cdot B$. Hence, by Lemma 4.1 the point in the orbit $G_n \cdot B$ given by $\Omega(0)$ lies in the closure of the orbit of A—this means that A degenerates to B.

Remark 4.1 Let A be a superalgebra with $\dim A_0 = i$, in other words $A \in \mathrm{Salg}_n^i$. Suppose the bases of A_0 and A_1 are given by $\{1, e_2, \ldots, e_i\}$ and $\{e_{i+1}, \ldots, e_n\}$ respectively. The specialization described by Gabriel in [5] given by $1 \mapsto 1, e_2 \mapsto te_2, \ldots, e_n \mapsto te_n$ and letting $t \to 0$ implies that any algebra degenerates to the algebra C_n. This specialization does not alter the $\mathbb{Z}_2$-grading, which implies (by Corollary 4.1) any superalgebra in Salg_n^i degenerates to the superalgebra $C_n(i)$ in Salg_n^i. Stated another way, the closure of any orbit in Salg_n^i contains the orbit of the superalgebra $C_n(i)$ in Salg_n^i (which is the closed orbit in Salg_n^i).

Earlier in Remark 2.2 we mentioned that Salg_n^i are the connected components of Salg_n. Using Corollary 4.1 above, we can now prove this to be the case.

We know that $\mathbb{A}^m$ is a Noetherian space and we have assumed that Salg_n^i is a closed subset of $\mathbb{A}^m$ (for $m = n^3 + n^2$). Thus Salg_n^i is a union of a finite number of irreducible components. However, irreducible components are closed and they must all contain the orbit of the superalgebra $C_n(i)$ by the above remark. Hence the irreducible components have a non-empty intersection. Thus Salg_n^i is a finite union of its irreducible components which are connected and have non-empty intersection. Thus we have showed the following.

Proposition 4.1 *The set $\{\mathrm{Salg}_n^i\}_{i=1}^n$ are the connected components of* Salg_n.

Note that we needed to assume that $\{\mathrm{Salg}_n^i\}_{i=1}^n$ are closed subsets of Salg_n in order to prove Proposition 4.1. In fact one can actually see that $\{\mathrm{Salg}_n^i\}_{i=1}^n$ are the connected components of Salg_n if and only if $\{\mathrm{Salg}_n^i\}_{i=1}^n$ are closed subsets. Proposition 4.1 shows one of the directions, and for the converse we note that connected components are closed (a fact from General Topology).

Given n-dimensional superalgebras A and B. To show that A can not degenerate to B, it is sufficient to exhibit a closed set in Salg_n containing the orbit $G_n \cdot A$ which is disjoint from $G_n \cdot B$. Note that if there are two disjoint closed sets in Salg_n one containing the orbit $G_n \cdot A$ and the other containing the orbit of $G_n \cdot B$, then there cannot be any degenerations between A and B. We now look for some necessary conditions for a degeneration of superalgebras to exist.

Remark 4.2 Suppose that A and B are n-dimensional superalgebras. The following conditions are necessary for a degeneration.

(a) If A doesn't degenerate to B as algebras, then A cannot degenerate to B as superalgebras. This condition becomes sufficient in case the $\mathbb{Z}_2$-gradings of the superalgebras are trivial.
(b) There is no degeneration from A to B unless $\dim A_0 = \dim B_0$.
(c) When $n \geq 3$, Salg_n^1 consists only of the closed orbit of the superalgebra $C_n(1)$. In this case, there is no degenerations in Salg_n^1.

The above facts follow from considering either the algebra structure or the $\mathbb{Z}_2$-grading in isolation. For some more necessary conditions for the existence of a degeneration we must exploit both the algebra structure and the $\mathbb{Z}_2$-grading simultaneously.

Now we look for closed G_n-stable subsets defined by some superalgebraic properties. We need the notion of a *upper semicontinuous function*. One may find it, for example in [3]. Given two topological space X. A function $f : X \to \mathbb{Z}$ is said to be upper semicontinuous if the set $\{x \in X : f(x) \geq n\}$ is closed in X for all $n \in \mathbb{Z}$.

Lemma 4.2 ([12, Chapter 1 §8 Corollary 3]) *If $f : X \to Y$ is a morphism of varieties, then the function $x \mapsto \dim_x f^{-1}(f(x))$ is upper semicontinuous.*

If V is a vector space and W a subset of V, then W is called a *cone in V* if W contains the zero vector and is closed under scalar multiplication. The following lemma can be found in [3].

Lemma 4.3 *Suppose X is a variety, V a vector space and we are given subsets $V_x \subseteq V$ for all $x \in X$. Suppose that*

(a) each V_x is a cone in V,
(b) $\{(x, v) : v \in V_x\}$ is closed in $X \times V$.

Then the map $x \mapsto \dim V_x$ is upper semicontinuous.

We have the following facts about dimension.

Lemma 4.4 *(a) For an algebraic set X, the dimension of X is equal to the Krull dimension of its coordinate ring $A(X)$.*
(b) The dimension of $\mathbb{A}^n$ is n.
(c) If $U \neq \emptyset$ is open in an irreducible variety X, then $\dim U = \dim X$.

(d) *If* $X = \bigcup_{i=1}^{n} U_i$ *with the* U_i *irreducible, then* $\dim X = \max_{i\in\{1,\dots,n\}}\{\dim U_i\}$.
(e) *If* $X \subseteq Y$ *then* $\dim X \leq \dim Y$*, moreover if* X *is closed and* Y *is irreducible, then* $X \subset Y$ *implies* $\dim X < \dim Y$.

Lemma 4.5 *The following sets are closed in* Salg_n*:*

(a) $\{A \in \mathrm{Salg}_n : A_1^2 = \{0\}\}$.
(b) $\{A \in \mathrm{Salg}_n : A_0$ *is commutative* $\}$.

Proof Recall that we defined superalgebra structures on an n-dimensional vector space V with a basis $\{e_1, \dots, e_n\}$.

For the set in part (a) we assign to a superalgebra A the following subset $W_A = \{v \otimes w : v, w \in A_1, vw = 0\}$ of $V \otimes V$. For the set in part (b) we assign to a superalgebra A the following subset $W'_A = \{v \otimes w : v, w \in A_0, vw = wv\}$ of $V \otimes V$. It is straightforward to check that these are both cones in $V \otimes V$.

Then we may write $v = \sum_{i=1}^{n} c_i e_i$ and $w = \sum_{i=1}^{n} d_i e_i$. Now from $v \otimes w \neq 0$ it is possible to recover v and w up to scalar multiple. This fact shall cause us no problems, however, since W_A and W'_A are cones in $V \otimes V$.

We show now that $\{(A, v \otimes w) : v, w \in A_1, vw = 0\}$ is closed in $\mathrm{Salg}_n \times (V \otimes V)$. If $v \otimes w = 0$ then either $v = 0$ or $w = 0$, in which case $c_i = 0$ for $i = 1, \dots, n$ or $d_i = 0$ for $i = 1, \dots, n$. So for $v \otimes w \neq 0$, $v \in A_1 \Leftrightarrow \sum_{i=1}^{n} c_i \gamma_i^j + c_j = 0$ for $j = 1, \dots, n$; $w \in A_1 \Leftrightarrow \sum_{i=1}^{n} d_i \gamma_i^j + d_j = 0$ for $j = 1, \dots, n$; and $vw = 0 \Leftrightarrow \sum_{i,j=1}^{n} c_i d_j \alpha_{ij}^k = 0$ for $1 \leq i, j \leq n$. We remark that if coordinates of v and w with respect to the given basis, i.e. (c_i), (d_i), satisfy these equations, then so too must (λc_i), (μd_i) for any $\lambda, \mu \in k$. Thus it does not matter that we can only obtain v and w up to scalar multiple. Thus $\{(A, v \otimes w) : v, w \in A_1, vw = 0\} = V(\{c_i\}) \cup V(\{d_i\}) \cup V(\{\sum_{i=1}^{n} c_i \gamma_i^j + c_j, \sum_{i=1}^{n} d_i \gamma_i^j + d_j, \sum_{i,j=1}^{n} c_i d_j \alpha_{ij}^k\})$, which is closed in $\mathrm{Salg}_n \times (V \otimes V)$.

We show next that $\{(A, v \otimes w) : v, w \in A_0, vw = wv\}$ is closed in $\mathrm{Salg}_n \times (V \otimes V)$. If $v \otimes w = 0$ then either $v = 0$ or $w = 0$, in which case $c_i = 0$ for $i = 1, \dots, n$ or $d_i = 0$ for $i = 1, \dots, n$. So for $v \otimes w \neq 0$, $v \in A_0 \Leftrightarrow \sum_{i=1}^{n} c_i \gamma_i^j - c_j = 0$ for $j = 1, \dots, n$; $w \in A_0 \Leftrightarrow \sum_{i=1}^{n} d_i \gamma_i^j - d_j = 0$ for $j = 1, \dots, n$; and $vw = wv \Leftrightarrow \sum_{i,j=1}^{n} c_i d_j (\alpha_{ij}^k - \alpha_{ji}^k) = 0$ for $1 \leq i, j \leq n$. Thus $\{(A, v \otimes w) : v, w \in A_0, vw = wv\} = V(\{c_i\}) \cup V(\{d_i\}) \cup V(\{\sum_{i=1}^{n} c_i \gamma_i^j - c_j, \sum_{i=1}^{n} d_i \gamma_i^j - d_j, \sum_{i,j=1}^{n} c_i d_j (\alpha_{ij}^k - \alpha_{ji}^k)\})$, which is closed in $\mathrm{Salg}_n \times (V \otimes V)$.

It follows from Lemma 4.3 that the maps $A \mapsto \dim W_A$ and $A \mapsto \dim W'_A$ are upper semicontinuous. Now since Salg_n^i are closed subsets of Salg_n it suffices to show that the sets mentioned in the lemma intersected with Salg_n^i are closed in Salg_n^i for each $i = 1, \dots, n$. That is, we may assume $\dim A_0 = i$. We note that $W_A \subseteq A_1 \otimes A_1$. Now if $A_1^2 = 0$, then $W_A = A_1 \otimes A_1$ which has dimension $(n-i)^2$. If $A_1^2 \neq \{0\}$, then $W_A \subset A_1 \otimes A_1$. We can see from the above, that for a given superalgebra A, W_A is closed in $V \otimes V$, and we note that $A_1 \otimes A_1$ is irreducible and has dimension $(n-i)^2$ (as a variety) since it is isomorphic to the $(n-i)^2$-dimensional affine space $\mathbb{A}^{(n-i)^2}$. Thus $\dim W_A < (n-i)^2$ by Lemma 4.4.

Therefore the set $\{A \in \mathrm{Salg}_n^i : A_1^2 = \{0\}\} = \{A \in \mathrm{Salg}_n^i : \dim W_A \geq (n-i)^2\}$ is a closed set by the upper semicontinuity. This proves part (a).

Similarly $W'_A \subseteq A_0 \otimes A_0$, and if A_0 is commutative then $W'_A = A_0 \otimes A_0$ which has dimension i^2. If A_0 is not commutative then $W'_A \subset A_0 \otimes A_0$ and so similarly as above $\dim W'_A < i^2$ (we just need to note that W'_A is closed and $A_0 \otimes A_0$ is irreducible). Thus the set $\{A \in \mathrm{Salg}_n^i : A_0$ is commutative $\} = \{A \in \mathrm{Salg}_n^i : \dim W'_A \geq i^2\}$ is a closed set by the upper semicontinuity. This proves part (b).

For Salg_n^2 we have other closed subsets. Since $\dim A_0 = 2$, $J(A_0) = \{x \in A_0 : x^2 = 0\}$, notice that this is a vector subspace of A_0.

Lemma 4.6 *The following are closed sets in* Salg_n^2*:*

(a) $\{A \in \mathrm{Salg}_n^2 : \dim J(A_0) = 1\}$.
(b) $\{A \in \mathrm{Salg}_n^2 : \dim J(A_0) = 1, J(A_0)A_1 = \{0\}\}$.
(c) $\{A \in \mathrm{Salg}_n^2 : \dim J(A_0) = 1, A_1 J(A_0) = \{0\}\}$.

Proof We give the proof for subset (b), since the proof for subset (c) is very similar and the proof for subset (a) follows by just simplifying this proof.

For subset (b) we assign to a superalgebra A the subset $W_A = \{v \otimes w : v \in A_0, w \in A_1, v^2 = 0, vw = 0\}$ of $V \otimes V$. This is clearly a cone. We also note $W_A \subseteq J(A_0) \otimes A_1$. Suppose $v = \sum_{i=1}^n c_i e_i$, $w = \sum_{i=1}^n d_i e_i$. We discover $\{(A, v \otimes w) : v \otimes w \in W_A\} = V(\{c_i\}) \cup V(\{d_i\}) \cup V(\{\sum_{i=1}^n c_i \gamma_i^j - c_j, \sum_{i=1}^n c_i c_j \alpha_{ij}^k, \sum_{i=1}^n d_i \gamma_i^j + d_j, \sum_{i,j=1}^n c_i d_j \alpha_{ij}^k\})$. Which is closed in $\mathrm{Salg}_n \times (V \otimes V)$.

So by Lemma 4.3, $A \mapsto \dim W_A$ is an upper semicontinuous map. Now, if $A \in \{A \in \mathrm{Salg}_n^2 : \dim J(A_0) = 1, J(A_0)A_1 = \{0\}\}$ then $\dim W_A = n-2$. If $A \notin \{A \in \mathrm{Salg}_n^2 : \dim J(A_0) = 1, J(A_0)A_1 = \{0\}\}$ then either $\dim J(A_0) = 0$ in which case $W_A = \{0\}$ and $\dim W_A = 0$ or $\dim J(A_0) = 1$ and $J(A_0)A_1 \neq \{0\}$ in which case $W_A \subset J(A_0) \otimes A_1$. In this case $\dim W_A < n-2$ since W_A is closed, and $J(A_0) \otimes A_1 \cong A_1 \cong \mathbb{A}^{n-2}$ as vector spaces, so $J(A_0) \otimes A_1$ is an irreducible subset of dimension $n-2$ as an $(n-2)$-dimensional vector subspace W of $\mathbb{A}^n$ with $n > r$ is isomorphic as a variety to $\mathbb{A}^r$. In particular this means that W is irreducible and as a variety has dimension r.

Hence $\{A \in \mathrm{Salg}_n^2 : \dim J(A_0) = 1, J(A_0)A_1 = \{0\}\} = \{A \in \mathrm{Salg}_n^2 : \dim W_A \geq n-2\}$ which is closed by the upper semi-continuity.

One can quickly check that if a superalgebra belongs to one of the closed sets described in Lemma 4.5, or Lemma 4.6, then any isomorphic superalgebra must also belong to the same set. Thus these closed sets are stable under the action of G_n.

5 Degenerations in Salg_4

In this section we are interested in determining when 4-dimensional superalgebra structures do or do not degenerate to one another. Here we use the results derived in the previous section to help us.

The results of this section give us most of the degenerations in Salg_4. Before giving the degeneration diagrams we shall first explain how to interpret them. We follow this by giving a partial classification theorem for Salg_4—we determine twenty irreducible components. There are, however, two other structures which may or may not give rise to irreducible components, and finally we give the details of the degenerations or the non-existence of degenerations, which were shown in the degeneration diagram.

As we shall soon see, there can be no degenerations amongst 4-dimensional superalgebras A and B with $\dim A_0 \neq \dim B_0$. Thus we can give the degeneration diagram for Salg_4 by giving the degeneration diagrams for each of the connected components Salg_4^i for $i = 1, 2, 3, 4$ separately. However we shall omit the diagram for Salg_4^1 since this consists of the solitary orbit of $(9|3)$.

Before giving these diagrams we shall explain the notations that we use in these diagrams.

We represent the orbits of isomorphism classes of superalgebras, by using the $(i|j)$ notation from [1]; $(i|j)$ shall be used to denote the orbit $G_4 \cdot (i|j)$ in Salg_4.

The families of superalgebras $(18; \lambda|i)$, $i = 0, 1, 2$ consist of those superalgebras for all values of λ except -1, which in particular includes the values $\lambda = 0$ and $\lambda = 1$. In these cases these orbits coincide with some of the other orbits. This is because, as superalgebras, we have the following equalities or isomorphisms: $(18; 0|0) = (16|0)$, $(18; 0|1) = (16|1)$, $(18; 0|2) = (16|3)$, $(18; 1|0) \cong (7|0)$, $(18; 1|1) \cong (7|2)$, $(18; 1|2) \cong (7|3)$.

In the degeneration diagrams Figs. 1 and 3, we use a dashed line to indicate a "degeneration" by a family of superalgebra structures; that is, when an orbit lies in the closure of the union of a family of orbits. This explains the use of the dashed lines through the families $(18; \lambda|i)$, $i = 0, 1, 2$. The fact that we use an arrow from $(18; \lambda|0)$ to $(8|0)$ and from $(18; \lambda|2)$ to $(8|3)$ is because there is a genuine degeneration from each of the orbits in these families to the orbits $(8|0)$ or $(8|3)$.

The dotted arrows (or dotted lines in the case of degenerations by a family of structures), are used to indicate those degenerations which we are unsure of—there may or may not be a degeneration between the indicated superalgebras.

From this we get the following (partial) result classifying 4-dimensional superalgebras:

Theorem 5.1 (Partial Geometric Classification of 4-dimensional Superalgebras)
In Salg_4 *there are at least twenty irreducible components. The following structures (or families of structures) are known to be generic:*

In Salg_4^4: $(1|0)$, $(10|0)$, $(13|0)$, $(17|0)$, $(18; \lambda|0)$.
In Salg_4^3: $(1|1)$, $(11|1)$, $(13|1)$, $(14|1)$, $(15|1)$, $(17|1)$.
In Salg_4^2: $(1|2)$, $(10|1)$, $(11|3)$, $(14|3)$, $(15|3)$, $(17|2)$, $(18; \lambda|1)$, $(18; \lambda|2)$.
In Salg_4^1: $(9|3)$.

Proof This follows from the degeneration diagrams Figs. 1, 2 and 3 and Corollary 3.2 which gives the relationship between the degeneration partial order and the irreducible components.

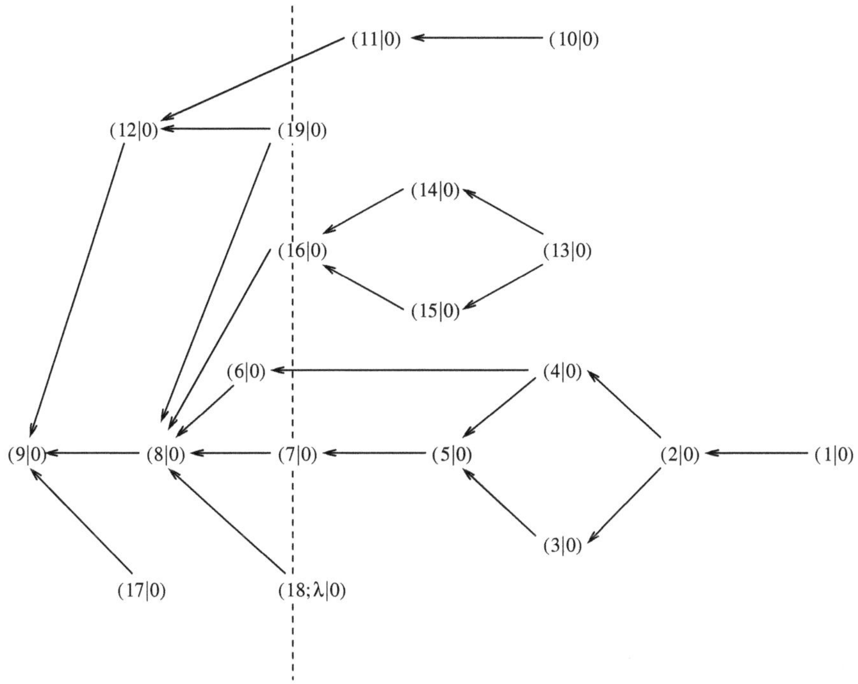

Fig. 1 Degenerations in the component Salg_4^4

Remark 5.1 The result above guarantees the existence of twenty irreducible components, however there could be up to two more irreducible components as well. It is the connected component Salg_4^2 in which we are unsure if we have found all of the irreducible components. It is not known whether the following two structures in Salg_4^2 are generic or not: $(6|2)$, $(19|1)$—so Salg_4^2 could have as few as eight irreducible components or as many as ten.

We are unsure if $(18;\lambda|2)$ degenerates to $(19|1)$ or not. This is why the dashed line through $(18;\lambda|2)$ changes to a dotted line after passing through $(16|3)$. We point this out to the reader to ensure that this important detail is not missed.

Remark 5.2 Porposition 3.1 gives the dimensions of these orbits, which for the generic structures gives the dimensions of the components too. However, for the generic families $(18;\lambda|i)$ for $i = 0, 1, 2$, the dimension of the component must be at least one larger than the dimension of any single orbit in this family. Since the family depends on one parameter λ, we would suspect that the dimensions of these components of the generic families are exactly one larger than the dimension of any single orbit in this family. However, we have not proved this. To prove that this is indeed the case, it would suffice to show that there can be no closed irreducible set Y lying properly between $\overline{G_n \cdot (18;\lambda|i)}$ and $\overline{\bigcup_\lambda G_n \cdot (18;\lambda|i)}$, i.e. that it is impossible to have $\overline{G_n \cdot (18;\lambda|i)} \subset Y \subset \overline{\bigcup_\lambda G_n \cdot (18;\lambda|i)}$ when Y is closed and irreducible.

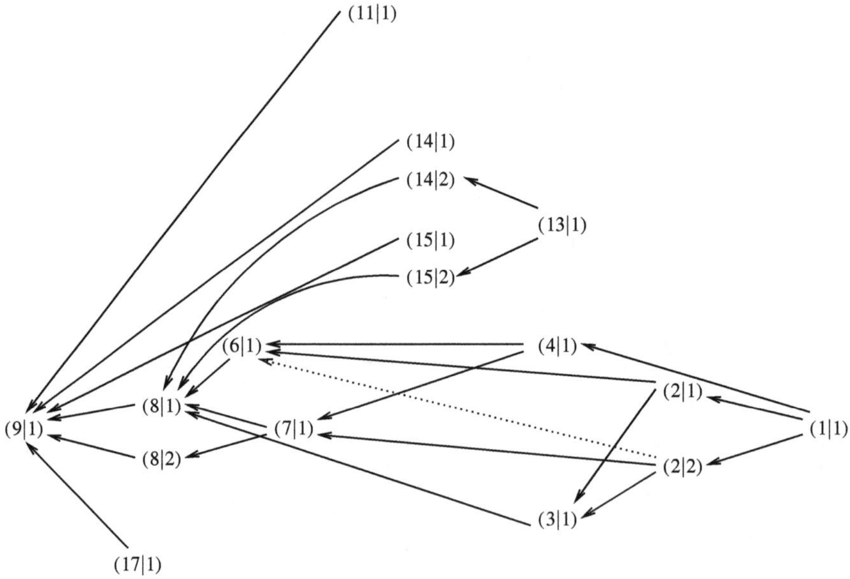

Fig. 2 Degenerations in the component Salg_4^3

We now provide the details which were used to obtain the degeneration diagrams Figs 1, 2 and 3 :

We apply the following useful facts mentioned in Remark 4.2 in the previous section which shall help us here. Since $n = 4$ we may appeal Lemma 2.2 to see that Salg_4^i for $i = 1, 2, 3, 4$ are all closed disjoint subsets (and in fact by Proposition 4.1 are the connected components of Salg_4). Thus by part (c) of Remark 4.2 there cannot be a degeneration from A to B unless $\dim_0 A = \dim_0 B$. Thus we need only to look at the degenerations amongst superalgebras belonging to the same subset Salg_4^i.

Another remark made in part (a) of Remark 4.2 is the following: If $U(A)$ doesn't degenerate to $U(B)$ as algebras, then A cannot degenerate to B as superalgebras. So we simply focus on degenerations from A to B, when there is a degeneration from $U(A)$ to $U(B)$ of underlying algebras. These two remarks represent large simplifications for us, as they greatly reduce the number of degenerations we must consider. Since two different superalgebras on the same underlying algebra have a trivial degeneration of the underlying algebra, we must however check to see if there are degenerations between different superalgebras on the same underlying algebra.

We also recall, any superalgebra in Salg_4^i degenerates to the superalgebra structure on $k[X, Y, Z]/(X, Y, Z)^2$ in Salg_4^i for $i = 1, 2, 3, 4$. The orbit of this superalgebra is the closed orbit in Salg_4^i. We will not mention this degeneration further since it always exists. We gave the specialization giving rise to this degeneration in Remark 4.1.

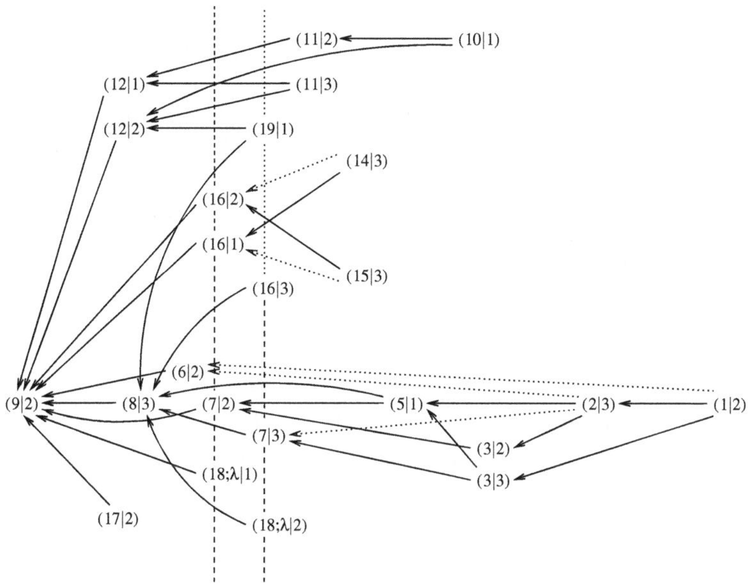

Fig. 3 Degenerations in the component Salg_4^2

By Corollary 4.1, to show the existence of a degeneration, it suffices to exhibit a specialization. In this section to show the existence of degenerations we shall do this, except in one instance where we shall appeal to Lemma 4.1 directly.

We mention that all the specializations given in this section are "homogeneous", that is, the basis changes replace degree zero terms by degree zero terms, and similarly replace degree one terms by degree one terms. Corollary 4.1 applies equally well to non-homogeneous specializations, however, such specializations are more difficult to determine. In fact, there are some superalgebras which we haven't determined whether there is or is not a degeneration between (e.g. does (1|2) degenerate to (6|2)?), but if the degeneration was to be obtained by a specialization it would necessarily have to be non-homogeneous. For an example of a degeneration obtained by a non-homogeneous specialization we have the following in the dimension 2 case, where each superalgebra is given the non-trivial $\mathbb{Z}_2$-grading:

$k \times k \to k[X]/(X^2)$ by $e_1 = (1, 1), e_2 = (1, -1), e_1' = e_1, e_2' = te_1 + te_2$ let $t \to 0$

To show the non-existence of a degeneration we list the method which we use. There are several different methods. We give the name and a brief explanation for each below.

- By Lemma 3.1 part (c) the orbit dimension must strictly decrease upon proper degeneration. So a superalgebra cannot degenerate to another superalgebra of the same or greater dimension. We abbreviate this method by (OD). Note however that it is possible for a family of structures of a given dimension to "degenerate" to a structure of the same dimension. As an example of this, each orbit in (18; $\lambda|0$) has dimension 8 as does the orbit (19|0), yet the family (18; $\lambda|0$) "degenerates" to (19|0).

- For the other methods we use the closed G_n-stable subsets found in the previous section. If A belongs to one of these subsets, and B does not, then A cannot degenerate to B. We shall refer to this set of methods by which of the closed G_n-stable subsets we apply. The abbreviation we give to the method by applying one of the closed sets is listed below.
  - (A) $\{A \in \mathrm{Salg}_n : A_1^2 = \{0\}\}$.
  - (B) $\{A \in \mathrm{Salg}_n : A_0$ is commutative $\}$.
  - (C) $\{A \in \mathrm{Salg}_4^2 : \dim J(A_0) = 1\}$.
  - (D) $\{A \in \mathrm{Salg}_4^2 : \dim J(A_0) = 1, J(A_0)A_1 = \{0\}\}$.
  - (E) $\{A \in \mathrm{Salg}_4^2 : \dim J(A_0) = 1, A_1 J(A_0) = \{0\}\}$.

In the following, when $\alpha \neq 0$, we will use the shorthand, $\sqrt{\alpha}$ to denote *some element*, x, of k^*, such that $x^2 = \alpha$. (Such an element x always exists as k is algebraically closed. Moreover, if x is such an element, then so too is $-x$).

Case $\dim_0 = 4$:

Applying part (b) in Remark 4.2 from the previous section, we notice that the degeneration diagram of Salg_4^4 corresponds exactly to the degeneration diagram of Alg_4. These degenerations have been completely described by Gabriel in [5], where he gives the degeneration diagram.

Case $\dim_0 = 3$:

Existence of Degenerations:

$(1|1) \to (2|1)$: $e_1 = (1,1,1,1), e_2 = (0,1,0,0), e_3 = (0,0,1,1), e_4 = (0,0,1,-1), e_1' = e_1, e_2' = e_2, e_3' = e_3, e_4' = te_4$ let $t \to 0$.

$(1|1) \to (2|2)$: $e_1 = (1,1,1,1), e_2 = (0,0,1,1), e_3 = (1,-1,0,0), e_4 = (0,0,1,-1), e_1' = e_1, e_2' = e_2, e_3' = te_3, e_4' = e_4$ let $t \to 0$.

$(1|1) \to (4|1)$: $e_1 = (1,1,1,1), e_2 = (1,0,0,0), e_3 = (0,0,1,1), e_4 = (0,0,1,-1), e_1' = e_1, e_2' = e_2, e_3' = t^2e_3, e_4' = te_4$ let $t \to 0$.

$(2|1) \to (3|1)$: $e_1 = (1,1,1), e_2 = (1,1,0), e_3 = (1,-1,0), e_4 = (0,0,X), e_1' = e_1, e_2' = e_2, e_3' = te_3, e_4' = e_4$ let $t \to 0$.

$(2|1) \to (6|1)$: $e_1 = (1,1,1), e_2 = (1,0,0), e_3 = (0,-1,1), e_4 = (0,0,X), e_1' = e_1, e_2' = e_2, e_3' = te_3, e_4' = e_4$ let $t \to 0$

$(2|2) \to (3|1)$: $e_1 = (1,1,1), e_2 = (1,1,0), e_3 = (0,0,X), e_4 = (1,-1,0), e_1' = e_1, e_2' = e_2, e_3' = e_3, e_4' = te_4$ let $t \to 0$.

$(2|2) \to (7|1)$: $e_1 = (1,1,1), e_2 = (1,1,0), e_3 = (0,0,X), e_4 = (1,-1,0), e_1' = e_1, e_2' = \sqrt{2}te_2 + e_3, e_3' = t^2e_2, e_4' = \sqrt{-2}te_4$ let $t \to 0$.

$(3|1) \to (8|1)$: $e_1 = (1,1), e_2 = (1,0), e_3 = (X,0), e_4 = (0,Y), e_1' = e_1, e_2' = te_2 + e_3, e_3' = te_3, e_4' = e_4$ let $t \to 0$.

$(4|1) \to (6|1)$: $e_1 = (1,1), e_2 = (1,0), e_3 = (0,X^2), e_4 = (0,X), e_1' = e_1, e_2' = e_2, e_3' = e_3, e_4' = te_4$ let $t \to 0$.

$(4|1) \to (7|1) : e_1 = (1,1), e_2 = (-1,1), e_3 = (0, X^2), e_4 = (0, X), e_1' = e_1, e_2' = t^2 e_2 + e_3, e_3' = t^2 e_3, e_4' = \sqrt{-2} t e_4$ let $t \to 0$.

$(6|1) \to (8|1) : e_1 = (1,1), e_2 = (-1,1), e_3 = (0, X), e_4 = (0, Y), e_1' = e_1, e_2' = te_2 + e_3, e_3' = 2te_3, e_4' = e_4$ let $t \to 0$.

$(7|1) \to (8|1) : e_1 = 1, e_2 = X + Y, e_3 = XY, e_4 = X - Y, e_1' = e_1, e_2' = e_2, e_3' = 2e_3, e_4' = te_4$ let $t \to 0$.

$(7|1) \to (8|2) : e_1 = 1, e_2 = X + Y, e_3 = XY, e_4 = X - Y, e_1' = e_1, e_2' = -2e_3, e_3' = te_2, e_4' = e_4$ let $t \to 0$.

$(13|1) \to (14|2) : e_1 = \left(1, \begin{pmatrix} 1 & 0 \\ 0 & 1 \end{pmatrix}\right), e_2 = \left(1, \begin{pmatrix} 0 & 0 \\ 0 & 1 \end{pmatrix}\right), e_3 = \left(1, \begin{pmatrix} 0 & 0 \\ 0 & 0 \end{pmatrix}\right), e_4 = \left(0, \begin{pmatrix} 0 & 1 \\ 0 & 0 \end{pmatrix}\right), e_1' = e_1, e_2' = e_2, e_3' = te_3, e_4' = e_4$ let $t \to 0$.

$(13|1) \to (15|2) : e_1 = \left(1, \begin{pmatrix} 1 & 0 \\ 0 & 1 \end{pmatrix}\right), e_2 = \left(1, \begin{pmatrix} 1 & 0 \\ 0 & 0 \end{pmatrix}\right), e_3 = \left(1, \begin{pmatrix} 0 & 0 \\ 0 & 0 \end{pmatrix}\right), e_4 = \left(0, \begin{pmatrix} 0 & 1 \\ 0 & 0 \end{pmatrix}\right), e_1' = e_1, e_2' = e_2, e_3' = te_3, e_4' = e_4$ let $t \to 0$.

$(14|2) \to (8|1) : e_1 = \begin{pmatrix} 1 & 0 & 0 \\ 0 & 1 & 0 \\ 0 & 0 & 1 \end{pmatrix}, e_2 = \begin{pmatrix} 1 & 0 & 0 \\ 0 & 1 & 0 \\ 0 & 0 & -1 \end{pmatrix}, e_3 = \begin{pmatrix} 0 & 0 & 0 \\ 1 & 0 & 0 \\ 0 & 0 & 0 \end{pmatrix}, e_4 = \begin{pmatrix} 0 & 0 & 0 \\ 0 & 0 & 0 \\ 1 & 0 & 0 \end{pmatrix}, e_1' = e_1, e_2' = te_2 + e_3, e_3' = 2te_3, e_4' = e_4$ let $t \to 0$.

$(15|2) \to (8|1) : e_1 = \begin{pmatrix} 1 & 0 & 0 \\ 0 & 1 & 0 \\ 0 & 0 & 1 \end{pmatrix}, e_2 = \begin{pmatrix} 1 & 0 & 0 \\ 0 & 1 & 0 \\ 0 & 0 & -1 \end{pmatrix}, e_3 = \begin{pmatrix} 0 & 1 & 0 \\ 0 & 0 & 0 \\ 0 & 0 & 0 \end{pmatrix}, e_4 = \begin{pmatrix} 0 & 0 & 1 \\ 0 & 0 & 0 \\ 0 & 0 & 0 \end{pmatrix}, e_1' = e_1, e_2' = te_2 + e_3, e_3' = 2te_3, e_4' = e_4$ let $t \to 0$.

Non-existence of Degenerations:

$(2|1) \not\to (2|2)$ (OD);
$(2|1) \not\to (4|1)$ (OD);
$(2|1) \not\to (7|1)$ (A);
$(2|1) \not\to (8|2)$ (A);
$(2|2) \not\to (2|1)$ (OD);
$(2|2) \not\to (4|1)$ (OD);
$(3|1) \not\to (7|1)$ (OD);
$(3|1) \not\to (8|2)$ (A);
$(6|1) \not\to (8|2)$ (A);
$(8|1) \not\to (8|2)$ (OD);
$(8|2) \not\to (8|1)$ (OD);
$(13|1) \not\to (8|2)$ (A);
$(13|1) \not\to (14|1)$ (B);
$(13|1) \not\to (15|1)$ (B);
$(14|1) \not\to (8|1)$ (OD);
$(14|1) \not\to (8|2)$ (OD);
$(14|1) \not\to (14|2)$ (OD);
$(14|2) \not\to (8|2)$ (A);
$(14|2) \not\to (14|1)$ (B);
$(15|1) \not\to (8|1)$ (OD);
$(15|1) \not\to (8|2)$ (OD);
$(15|1) \not\to (15|2)$ (OD);

$(15|2) \not\to (8|2)$ (A);
$(15|2) \not\to (15|1)$ (B).

Undetermined Degeneration:

$(2|2) \overset{?}{\to} (6|1)$.

Case $\dim_0 = 2$:

Existence of Degenerations:

$(1|2) \to (2|3)$: $e_1 = (1,1,1,1), e_2 = (0,0,1,1), e_3 = (1,-1,0,0), e_4 = (0,0,1,-1), e_1' = e_1, e_2' = e_2, e_3' = e_3, e_4' = te_4$ let $t \to 0$.

$(1|2) \to (3|3)$: $e_1 = (1,1,1,1), e_2 = (1,1,0,0), e_3 = (1,-1,1,-1), e_4 = (1,-1,0,0), e_1' = e_1, e_2' = te_2, e_3' = e_3, e_4' = te_4$ let $t \to 0$.

$(2|3) \to (3|2)$: $e_1 = (1,1,1), e_2 = (1,1,0), e_3 = (1,-1,0), e_4 = (0,0,X), e_1' = e_1, e_2' = e_2, e_3' = te_3, e_4' = e_4$ let $t \to 0$.

$(2|3) \to (5|1)$: $e_1 = (1,1,1), e_2 = (1,1,0), e_3 = (1,-1,0), e_4 = (0,0,X), e_1' = e_1, e_2' = t^2e_2, e_3' = te_3 + e_4, e_4' = t^3e_3$ let $t \to 0$.

$(3|2) \to (7|2)$: $e_1 = (1,1), e_2 = (1,-1), e_3 = (X,Y), e_4 = (X,-Y), e_1' = e_1, e_2' = te_2, e_3' = e_3, e_4' = te_4$ let $t \to 0$.

$(3|3) \to (5|1)$: $e_1 = (1,1), e_2 = (X,Y), e_3 = (1,-1), e_4 = (X,-Y), e_1' = e_1, e_2' = 2te_2, e_3' = te_3 + e_4, e_4' = 2t^2e_4$ let $t \to 0$.

$(3|3) \to (7|3)$: $e_1 = (1,1), e_2 = (X,Y), e_3 = (1,-1), e_4 = (X,-Y), e_1' = e_1, e_2' = te_2, e_3' = te_3, e_4' = e_4$ let $t \to 0$.

$(5|1) \to (7|2)$: $e_1 = 1, e_2 = X^2, e_3 = X, e_4 = X^3, e_1' = e_1, e_2' = e_2, e_3' = te_3, e_4' = te_4$ let $t \to 0$.

$(5|1) \to (8|3)$: $e_1 = 1, e_2 = X^2, e_3 = X, e_4 = X^3, e_1' = e_1, e_2' = t^2e_2, e_3' = te_3, e_4' = e_4$ let $t \to 0$

$(7|3) \to (8|3)$: $e_1 = 1, e_2 = XY, e_3 = X+Y, e_4 = X-Y, e_1' = e_1, e_2' = 2e_2, e_3' = e_3, e_4' = te_4$ let $t \to 0$.

$(10|1) \to (11|2)$: $e_1 = \begin{pmatrix}1&0\\0&1\end{pmatrix}, e_2 = \begin{pmatrix}1&0\\0&-1\end{pmatrix}, e_3 = \begin{pmatrix}0&1\\0&0\end{pmatrix}, e_4 = \begin{pmatrix}0&0\\1&0\end{pmatrix}, e_1' = e_1, e_2' = e_2, e_3' = te_3, e_4' = te_4$ let $t \to 0$.

$(10|1) \to (12|2)$: $e_1 = \begin{pmatrix}1&0\\0&1\end{pmatrix}, e_2 = \begin{pmatrix}1&0\\0&-1\end{pmatrix}, e_3 = \begin{pmatrix}0&1\\-1&0\end{pmatrix}, e_4 = \begin{pmatrix}0&1\\1&0\end{pmatrix}, e_1' = e_1, e_2' = t^2e_2, e_3' = te_3, e_4' = te_4$ let $t \to 0$.

$(11|2) \to (12|1)$: $e_1 = \begin{pmatrix}1&0&0&0\\0&1&0&0\\0&0&1&0\\0&0&0&1\end{pmatrix}, e_2 = \begin{pmatrix}1&0&0&0\\0&1&0&0\\0&0&-1&0\\0&0&0&-1\end{pmatrix}, e_3 = \begin{pmatrix}0&0&0&0\\0&0&0&1\\1&0&0&0\\0&0&0&0\end{pmatrix},$
$e_4 = \begin{pmatrix}0&0&0&0\\0&0&0&1\\-1&0&0&0\\0&0&0&0\end{pmatrix}, e_1' = e_1, e_2' = te_2, e_3' = e_3, e_4' = te_4$ let $t \to 0$.

$(11|3) \to (12|1)$: $e_1 = \begin{pmatrix}1&0&0&0\\0&1&0&0\\0&0&1&0\\0&0&0&1\end{pmatrix}, e_2 = \begin{pmatrix}0&0&0&0\\0&0&0&1\\1&0&0&0\\0&0&0&0\end{pmatrix}, e_3 = \begin{pmatrix}1&0&0&0\\0&1&0&0\\0&0&-1&0\\0&0&0&-1\end{pmatrix},$
$e_4 = \begin{pmatrix}0&0&0&0\\0&0&0&-1\\1&0&0&0\\0&0&0&0\end{pmatrix}, e_1' = e_1, e_2' = e_2, e_3' = te_3, e_4' = te_4$ let $t \to 0$.

$(11|3) \to (12|2) : e_1 = \begin{pmatrix} 1&0&0&0\\0&1&0&0\\0&0&1&0\\0&0&0&1 \end{pmatrix}, e_2 = \begin{pmatrix} 0&0&0&0\\0&0&0&1\\1&0&0&0\\0&0&0&0 \end{pmatrix}, e_3 = \begin{pmatrix} 1&0&0&0\\0&1&0&0\\0&0&-1&0\\0&0&0&-1 \end{pmatrix},$
$e_4 = \begin{pmatrix} 0&0&0&0\\0&0&0&1\\-1&0&0&0\\0&0&0&0 \end{pmatrix}, e_1' = e_1, e_2' = te_2, e_3' = te_3, e_4' = e_4$ let $t \to 0$.

$(14|3) \to (16|1) : e_1 = \begin{pmatrix} 1&0&0\\0&1&0\\0&0&1 \end{pmatrix}, e_2 = \begin{pmatrix} 1&0&0\\0&1&0\\0&0&-1 \end{pmatrix}, e_3 = \begin{pmatrix} 0&0&0\\1&0&0\\1&0&0 \end{pmatrix}, e_4 =$
$\begin{pmatrix} 0&0&0\\1&0&0\\-1&0&0 \end{pmatrix}, e_1' = e_1, e_2' = te_2, e_3' = e_3, e_4' = te_4$ let $t \to 0$.

$(15|3) \to (16|2) : e_1 = \begin{pmatrix} 1&0&0\\0&1&0\\0&0&1 \end{pmatrix}, e_2 = \begin{pmatrix} 1&0&0\\0&1&0\\0&0&-1 \end{pmatrix}, e_3 = \begin{pmatrix} 0&1&1\\0&0&0\\0&0&0 \end{pmatrix}, e_4 =$
$\begin{pmatrix} 0&1&-1\\0&0&0\\0&0&0 \end{pmatrix}, e_1' = e_1, e_2' = te_2, e_3' = e_3, e_4' = te_4$ let $t \to 0$.

$(16|3) \to (8|3) : e_1 = 1, e_2 = XY, e_3 = X + Y, e_4 = X - Y, e_1' = e_1, e_2' = e_2, e_3' = e_3, e_4' = te_4$ let $t \to 0$.

$(18; \lambda|1) \to (7|2), (16|1)$: Since the orbits of $(7|2)$ and $(16|1)$ coincide with the orbits of $(18; 1|1)$ and $(18; 0|1)$ respectively, $(7|2)$ and $(16|1)$ are included in the closure of the union of the family of orbits $(18; \lambda|1)$.

$(18; \lambda|1) \to (16|2)$: Also $(16|2)$ is included in the closure of the union of the family of orbits $(18; \lambda|1)$. To see this, we look at the structure constants of $(18; t^{-1}|1)$ in the basis $e_1 = 1, e_2 = X, e_3 = Y, e_4 = YX$. This gives us a curve in Salg_4 which lies in the family of orbits of $(18; \lambda|1)$ for $t \neq 0$, yet lies in the orbit of $(16|2)$ when $t = 0$. By appealing to Lemma 4.1 directly the result follows.

$(18; \lambda|2) \to (7|3), (16|3)$: Similarly the orbits of $(7|3)$ and $(16|3)$ are included in the closure of the union of the family of orbits $(18; \lambda|2)$.

$(18; \lambda|2) \to (8|3) : e_1 = 1, e_2 = XY, e_3 = X + Y, e_4 = X - Y, e_1' = e_1, e_2' = (1 + \lambda)e_2, e_3' = e_3, e_4' = te_4$ let $t \to 0$.

$(19|1) \to (8|3) : e_1 = 1, e_2 = XY, e_3 = X + Y, e_4 = X - Y, e_1' = e_1, e_2' = e_2, e_3' = e_3, e_4' = te_4$ let $t \to 0$.

$(19|1) \to (12|2) : e_1 = 1, e_2 = XY, e_3 = X, e_4 = Y, e_1' = e_1, e_2' = te_2, e_3' = te_3, e_4' = e_4$ let $t \to 0$.

Non-existence of Degenerations:

$(2|3) \not\to (3|3)$ (OD);
$(3|2) \not\to (3|3)$ (OD);
$(3|2) \not\to (5|1)$ (OD);
$(3|2) \not\to (7|3)$ (OD);
$(3|2) \not\to (8|3)$ (A);
$(3|3) \not\to (3|2)$ (C);
$(5|1) \not\to (7|3)$ (OD);
$(6|2) \not\to (8|3)$ (OD);
$(7|2) \not\to (7|3)$ (OD);
$(7|2) \not\to (8|3)$ (OD);
$(7|3) \not\to (7|2)$ (D);

(10|1) $\nrightarrow$ (11|3) (OD);
(11|2) $\nrightarrow$ (11|3) (OD);
(11|2) $\nrightarrow$ (12|2) (A);
(11|3) $\nrightarrow$ (11|2) (C);
(12|1) $\nrightarrow$ (12|2) (A);
(12|2) $\nrightarrow$ (12|1) (OD);
(14|3) $\nrightarrow$ (16|3) (OD);
(14|3) $\nrightarrow$ (8|3) (A);
(15|3) $\nrightarrow$ (16|3) (OD);
(15|3) $\nrightarrow$ (8|3) (A);
(16|1) $\nrightarrow$ (16|2) (OD);
(16|1) $\nrightarrow$ (16|3) (OD);
(16|1) $\nrightarrow$ (8|3) (OD);
(16|2) $\nrightarrow$ (16|1) (OD);
(16|2) $\nrightarrow$ (16|3) (OD);
(16|2) $\nrightarrow$ (8|3) (OD);
(16|3) $\nrightarrow$ (16|1) (D);
(16|3) $\nrightarrow$ (16|2) (E);
(18; λ|1) $\nrightarrow$ (7|3), (16|3), (18; λ|2), (19|1) (A);
(18; λ|1) $\nrightarrow$ (8|3) (A);
(18; λ|2) $\nrightarrow$ (7|2), (16|1), (16|2), (18; λ|1) (D), (E);
(19|1) $\nrightarrow$ (12|1) (D).

Undetermined Degenerations:

(1|2) $\xrightarrow{?}$ (6|2);
(2|3) $\xrightarrow{?}$ (6|2);
(18; λ|2) $\xrightarrow{?}$ (19|1);
(2|3) $\xrightarrow{?}$ (7|3);
(14|3) $\xrightarrow{?}$ (16|2);
(15|3) $\xrightarrow{?}$ (16|1).

The first three of these undetermined degenerations are related to discovering whether (6|2) or (19|1) give rise to irreducible components in Salg_4^2.

Remark 5.3 We close with the remark that in Salg_4 no two superalgebra structures A and B on the same underlying algebra can degenerate to each other, even if $\dim_0 A = \dim_0 B$. We have seen this from brute force checking of each case. Is it a general result that there can be no degeneration from a superalgebra to any other superalgebra having the same underlying algebra?

References

1. Armour, A., Chen, H., Zhang, Y.: Classification of 4-dimensional graded algebras. Comm. Algebra **37**, 3697–3728 (2009)
2. Bruyn, L.L., Reichstein, Z.: Smoothness in algebraic geography, Proc. London Math. Soc. (3) **79**(1), 158–190 (1999)
3. Crawley-Boevey, W.: Geometry of Representations of Algebras, Lecture Notes. Oxford University, Oxford (1993)
4. Flanigan, F.J.: Algebraic geography: varieties of structure constants. Pacific J. Math. **27**, 71–79 (1968)
5. Gabriel, P.: Finite representation type is open. Lect. Notes Math. **488**, 132–155 (1975)
6. Goze, M., Makhlouf, A.: Classification of Rigid Associative Algebras in Low Dimensions, Lois d'algèbres et variétés algébriques (Colmar, 1991), Travaux en Cours, vol. 50, pp. 5–22. Hermann, Paris (1996)
7. Harris, J.: Algebraic Geometry: A First Course, Graduate Texts in Mathematics, p. 133. Springer, New York (1992)
8. Makhlouf, A.: Algèbres associatives et calcul formel. Theor. Comput. Sci. **187**(1–2), 123–145 (1997)
9. Makhlouf, A.: Degeneration, rigidity and irreducible components of Hopf algebras. Algebra Colloq. **12**(2), 241–254 (2005)
10. Mazzola, G.: The algebraic and geometric classification of associative algebras of dimension five. Manuscripta Math. **27**(1), 81–101 (1979); Comment. Math. Helv. **55**(2), 267–293 (1980)
11. Mazzola, G.: How to count the number of irreducible components of the schemes of finite-dimensional algebra structures. J. Algebra **78**(2), 292–302 (1982)
12. Mumford, D.: The Red Book of Varieties and Schemes, Lecture Notes in Mathematics, vol. 1358. Springer, Berlin (1988)
13. Riedtmann, C: Degenerations for representations of quivers with relations. Ann. Sci. École Norm. Sup. (4) **19**(2), 275–301 (1986)
14. Zwara, G.: Degenerations of finite-dimensional modules are given by extensions. Compositio Math. **121**(2), 205–218 (2000)

Distributivity in Quandles and Quasigroups

Mohamed Elhamdadi

Abstract Distributivity in algebraic structures appeared in many contexts such as in quasigroup theory, semigroup theory and algebraic knot theory. In this paper we give a survey of distributivity in quasigroup theory and in quandle theory.

1 Introduction

Quandles are in general non-associative structures whose axioms correspond to the algebraic distillation of the three Reidemeister moves in knot theory. Quandles appeared in the literature with many different names. If one restricts himself to the most important axiom of a quandle which is the self-distributivity axiom (see definition below), then one can trace this back to 1880 in the work of Pierce [52] where one can read the following comments, *"These are other cases of the distributive principle ....These formulae, which have hitherto escaped notice, are not without interest."* Another early work fully devoted to self-distributivity appeared in 1929 by Burstin and Mayer [7] where normal subquasigroups are studied and an attempt is made to show that every minimal subquasigroup of a finite distributive quasigroup is normal. This is considered as the starting point for the investigation of normality problems in distributive quasigroups. In 1942 Mituhisa Takasaki [57] introduced the notion of kei (involutive quandle in Joyce's terminology [37]) as an abstraction of the notion of symmetric transformation. The earliest known work on racks (see definition below) is contained in the 1959 correspondence between John Conway and Gavin Wraith who studied racks in the context of the conjugation operation in a group. Around 1982, Joyce [37] (used the term quandle) and Matveev [40] (who call them distributive groupoids) introduced independently the notion of a quandle. Joyce and Matveev associated to each oriented knot K a quandle $Q(K)$ called the knot

M. Elhamdadi (✉)
University of South Florida, Tampa, FL, USA
e-mail: emohamed@math.usf.edu

A. Makhlouf et al. (eds.), *Algebra, Geometry and Mathematical Physics*, Springer Proceedings in Mathematics & Statistics 85, DOI: 10.1007/978-3-642-55361-5_19,

quandle. The knot quandle is a complete invariant up to orientation. Since then quandles and racks have been investigated by topologists in order to construct knot and link invariants and their higher analogues (see for example [20] and references therein). In 1986, Brieskorn [5] introduced the concept of automorphic sets to describe a set Δ with a binary operation $*$ such that all left multiplications $b \mapsto a * b$ are automorphisms of Δ. He considered the action of the braid group B_n on the Cartesian product Δ^n and introduced invariants of the orbit; for example, monodromy groups. In 1991, Kauffman introduced a similar notion called crystal ([38], p. 186) as a generalization of the fundamental group of a knot in the sense that the crystal has more information than the fundamental group alone. In 1992, Fenn and Rourke [29] showed that any codimension-two link has a fundamental rack which contains more information than the fundamental group. They gave some examples of computable link invariants derived from the fundamental rack and explained the connection of the theory of racks with that of braids. In 2003, Fenn, Rourke and Sanderson [30] introduced rack homology. This (co)homology was modified in 1999 by Carter et al. [18] to give a cohomology theory for quandles. This cohomology was used to define state-sum invariant for knots in three space and knotted surfaces in four space. A nice survey paper on quandle ideas is a paper by Scott Carter [8] showing the applications of quandle cocycle invariants.

In this paper, we give a survey of distributivity in quasigroup theory and in quandle theory.

In Sect. 2, we review the basics of quandles and give examples. Section 3 deals with the problem of classification of quandles. In Sect. 4 we relate quandles to quasigroups and Moufang loops. Section 5 deals with the quandle cohomology and cocycle knot invariants.

2 Basics of Quandles

We start by reviewing the basics of quandles and give some examples.

Definition 2.1 [37] A *quandle*, X, is a set with a binary operation $(a, b) \mapsto a * b$ such that

(1) For any $a \in X$, $a * a = a$.
(2) For any $a, b \in X$, there is a unique $c \in X$ such that $a = c * b$.
(3) For any $a, b, c \in X$, we have $(a * b) * c = (a * c) * (b * c)$.

Axiom (2) states that for each $u \in X$, the map $R_u : X \to X$ with $R_u(x) := x * u$ is a bijection. The axioms for a quandle correspond respectively to the Reidemeister moves of type I, II, and III as can be seen from Fig. 1.

Quandles have been used to study colorings of knots and links and to define some of their invariants, see for example [17].

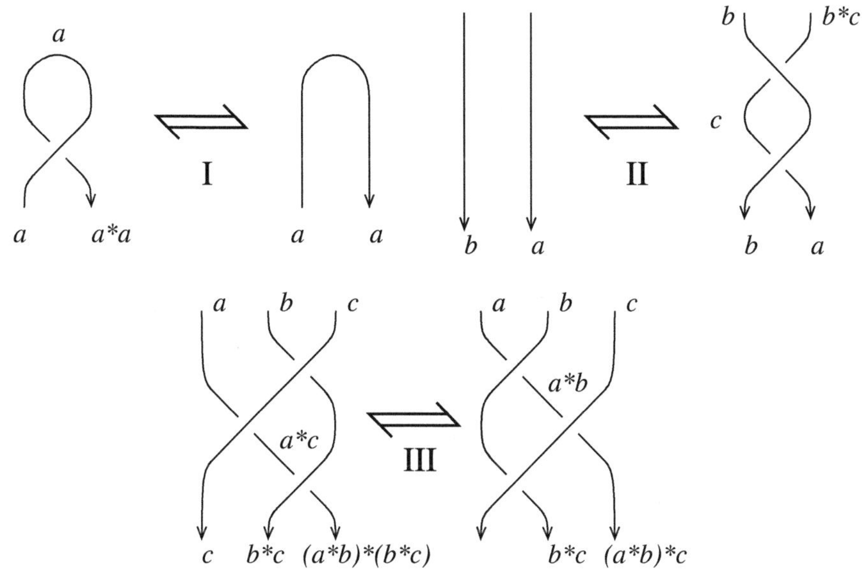

Fig. 1 Reidemeister moves and quandle axioms

Here are some examples of quandles:

- Any set X with the operation $x * y = x$ for all $x, y \in X$, is a quandle called the *trivial* quandle.
- Any group $X = G$ with conjugation $a * b = bab^{-1}$ is a quandle.
- Let n be a positive integer. For elements $i, j \in \mathbb{Z}_n$ (integers modulo n), define $i * j \equiv 2j - i \pmod{n}$. Then $*$ defines a quandle structure called the *dihedral quandle*, R_n. This set can be identified with the set of reflections of a regular n-gon with conjugation as the quandle operation. If we denote the group of symmetry of a regular n-gon by $D_n =< u, v \mid u^n = 1, v^2 = 1, vuv = u^{-1} >$, then conjugation on reflections is given by $(u^i v)*(u^j v) = u^j v u^i v (u^j v)^{-1} = u^j u^{-i} v u^{-j} = u^{2j-i} v$.
- A group $X = G$ with operation $x * y = yx^{-1}y$ is called the *core* quandle of G, denoted $Core(G)$.
- For any abelian group M and automorphism t of M define a quandle structure on M by $x * y = t(x - y) + y$. This is called an *Alexander quandle*.
- A generalization of the last example is, let G be a group and ϕ be an automorphism of G, then define a quandle structure on G by $x * y = \phi(xy^{-1})y$. Further, let H be a subgroup of G such that $\phi(h) = h$, for all $h \in H$. Then G/H is a a quandle with operation $Hx * Hy = H\phi(xy^{-1})y$. It is called the *homogeneous* quandle (G, H, ϕ).
- Let $< \, , \, >: \mathbb{R}^n \times \mathbb{R}^n \to \mathbb{R}$ be a symmetric bilinear form on $\mathbb{R}^n$. Let X be the subset of $\mathbb{R}^n$ consisting of vectors x such that $< x, x > \neq 0$. Then the operation

$$x * y = \frac{2 < x, y >}{< x, x >} y - x$$

defines a quandle structure on X. Note that, $x * y$ is the image of x under the reflection in y. This quandle is called a *Coxeter* quandle.

A function $\phi : (X, *) \to (Y, \rhd)$ is a quandle *homomorphism* if $\phi(a * b) = \phi(a) \rhd \phi(b)$ for any $a, b \in X$. Axiom (3) of definition 2.1 state that for each $u \in X$, the map R_u is a quandle homomorphism. Let Aut(X) denotes the automorphism group of X. The subgroup of Aut(X) generated by the permutations R_x is called the *inner* automorphism group of X and denoted by Inn(X). By axiom (3) of definition 2.1, the map $R : X \to \text{Inn}(X)$, sending u to R_u, satisfies the equation $R_z\ R_y = R_{y*z}\ R_z$, which can be written as $R_z\ R_y\ {R_z}^{-1} = R_{y*z}$, for all $y, z \in X$. Thus, if the group Inn(X) is considered as a quandle with conjugation then the map R becomes a quandle homomorphism. The subgroup of Aut(X) generated by $R_x R_y^{-1}$, for all $x, y \in X$, is called the *transvection* group of X denoted by $Transv(X)$. It is well known (see for example [37]) that the *transvection* group is a normal subgroup of the inner group and the later group is normal subgroup of the automorphism group of X. The quotient group Inn(X)/Transv(X) is a cyclic group (see [37]). For each $u \in X$, let us denote the left multiplication by u by the map $L_u : X \to X$ with $L_u(x) := u * x$. We list some properties and some definitions of quandles below.

- A quandle X is *involutory*, or a *kei*, if the right translations are involutions: $R_a^2 = \text{id}$, for all $a \in X$.
- A quandle is *faithful* if the mapping $a \mapsto R_a$ is an injection from X to Inn(X).
- A quandle is *connected* if Inn(X) acts transitively on X.
- A *Latin quandle* is a quandle such that for each $a \in X$, the left translation L_a is a bijection. That is, the multiplication table of the quandle is a Latin square.
- A quandle X is *medial* if $(a * b) * (c * d) = (a * c) * (b * d)$ for all $a, b, c, d \in X$. It is well known that a quandle is medial if and only if its transvection group is abelian, that is why it is also called *abelian*. It is known and easily seen that every Alexander quandle is medial.
- A quandle X is called *simple* if the only surjective quandle homomorphisms on X have trivial image or are bijective.

3 The Problem of Classification of Quandles

The problem of classification of quandles and racks was attempted by many authors mainly because computable invariants of knots such as, the quandle cocycle invariant of Carter et al. [18, 20], and enhancement of counting homomorphisms from the knot quandle to a fixed quandle of Nelson et al. [45, 46] can be defined from quandles. Racks and quandles are used in the classification of pointed Hopf algebras [1] since they help in the understanding of Yetter-Drinfeld modules over groups. Below, we give a survey of the classification of finite quandles.

In 2003, Nelson gave a classification of finite Alexander quandles proving the following

Theorem 3.1 [47] *Two finite Alexander quandles M and N of the same cardinality are isomorphic as quandles if and only if $(1-t)M$ and $(1-t)N$ are isomorphic as $\mathbb{Z}[t, t^{-1}]$-modules.*

As a consequence of this theorem Ho and Nelson [34] computed isomorphism classes of quandles up to order 5 and their automorphism groups. Quandles of order 6, 7 and 8 were given by Henderson, Macedo and Nelson in [33] but isomorphism class representatives were not determined. In 2006, Nelson and Wong [48] obtained the orbit decomposition of finite quandles: A subset A of a quandle X is said to be X-complemented if the complement of A in X is a subquandle of X. They proved the following

Theorem 3.2 [48] *Up to isomorphism, every finite quandle has a unique decomposition into subquandles $A_1, A_2, \ldots, A_n$ such that every A_j is X-complemented and no proper subquandle of any A_j is X-complemented.*

Independently around the same time Yetter et al. [26] obtained a similar decomposition theorem for quandles in terms of an operation of "semidisjoint union", showing that all finite quandles canonically decompose via iterated semidisjoint unions into connected subquandles. Murillo and Nelson [43, 44] proved in 2006 that there are 24 isomorphism classes of Alexander quandles of order 16. In [28] quandles up to order 9 were classified, automorphism groups of quandles (with orders up to 7) were determined and the automorphism group of the dihedral quandle R_n was proven to be isomorphic to the affine group of $\mathbb{Z}_n$. The number of isomorphism classes of quandles of order 3, 4, 5, 6, 7, 8 and 9 are respectively 3, 7, 22, 73, 298, 1581, 11079. The list of isomorphism classes can be found in https://sites.google.com/a/exactas.udea.edu.co/restrepo/quandles. Independently the same classification result was obtained in [41] by McCarron.

In [35] it was shown first that the isomorphism class of an Alexander quandle $(M, *)$ is determined by the isomorphism type of the Λ-module $(1-t)M$ and the cardinality of the quotient A/K, where A is the annihilator of $(1-t)$ in M, $K = A \cap (1-t)M$ and $\Lambda = \mathbb{Z}[t, t^{-1}]$. This recovers a result of Sam Nelson [47]. The structure of the automorphism group of a general Alexander quandle $(M, *)$ is completely determined (see [35] for more details). Enumeration of Alexander quandles has been much improved. Edwin Clark computed the number of Alexander quandles of orders up to 255 (see http://oeis.org/A193024, for more details) based on results from [36] which contains other interesting enumeration results concerning Alexander quandles. More sequences related to quandles can be found on http://oeis.org.

Example 3.1 One way of describing a finite quandle is by the Cayley table. Since by the second axiom of a quandle right multiplication by a fixed i, $R_i : j \mapsto j * i$ is a permutation. We then can describe each quandle by writing each column R_i of the Cayley table as a product of disjoint cycles. Here we include the list of quandles of order 4. The notation (1) in the table means that the permutation is the identity

Table 1 Quandles of order 4 in terms of disjoint cycles of columns

Quandle	Disjoint cycle notation for the columns of the quandle
Q_1	(1), (1), (1), (1)
Q_2	(1), (1), (1), (23)
Q_3	(1), (1), (1), (123)
Q_4	(1), (1), (12), (12)
Q_5	(1), (34), (24), (23)
Q_6	(34), (34), (12), (12)
Q_7	(234), (143), (124), (132)

permutation. For example the quandle Q_5 is the set $\{1, 2, 3, 4\}$ where R_1 is the identity permutation, R_2 is the transposition sending 3 to 4, R_3 is the transposition sending 2 to 4 and R_4 is the transposition sending 2 to 3 (Table 1).

Using computers the search space in general becomes too large to obtain the computation of all quandles up to isomorphism for higher cardinality. Clearly, this depends on the algorithm used to find quandles. However if one restricts himself to the subclass of connected quandles then classification becomes more accessible to calculation in a somehow comparable way to the classification of finite groups. In [25], Clauwens studied connected quandles and proved the following

Proposition 3.1 [25] *If $f : Q \to P$ is a surjective quandle homomorphism and P is connected then for all $x, y \in P$, there is a bijection between $f^{-1}(x)$ and $f^{-1}(y)$. In particular the cardinality of P divides the cardinality of Q.*

This allowed him to obtain isomorphism classes of connected quandles up to order 14, in particular he showed that there is no connected quandle of order 14. In [59], Vendramin extended Clauwens results to the list of all connected quandles of orders less than 36. He used the classification of transitive groups and the program described in [26] based mainly on the following

Theorem 3.3 [59] *Let X be a connected quandle of cardinality n. Let $x_0 \in X$ and $z = R_{x_0}$ be the right multiplication by x_0, $G = \text{Inn}(X)$ and $H = Stab_G(x_0) = \{g \in G, gx_0 = x_0\}$. Then (1) G is a transitive group of order n, (2) z is central element of H and (3) X is isomorphic to the homogeneous quandle (G, H, I_z), where I_z is the conjugation by z.*

A complete list of isomorphism classes of quandles with up to 6 elements appeared in the appendix [20].

4 Quandles and Quasigroups

In this section we will discuss the relation between left and right distributive quasigroups and the following types of quandles: Alexander, Latin and medial quandles. Two connections between quasigroups and quandles were established in [54].

Self-distributivity appeared in 1929 by Burstin and Mayer [7] where they studied quasigroups which are left- and right-distributive. They stated that there are none of orders 2 and 6, observed that the group of automorphisms is transitive, and showed that such a quasigroup is idempotent.

Definition 4.1 [6] (1) A quasigroup is a set Q with a binary operation $*$ such that for all $u \in Q$ the right translation R_u and left translation L_u by u are both permutations. (2) If the operation $*$ has an identity element e in Q then the quasigroup is called a *loop* and denoted $(Q, *, e)$.

Quasigroups differ from groups in the sense that they satisfy identities which usually conflict with associativity. Distributive quasigroups have transitive groups of automorphisms but the only group with this property is the trivial group. In [56] it is shown that there are no right-distributive quasigroups whose order is twice an odd number. Right-distributive quasigroups are intimately connected with the binary operation of a conjugation in a group since in a right-distributive quasigroup it holds that $R_{y*z} = R_z R_y R_z^{-1}$ and the mapping $x \mapsto R_x$ is injective. We will see below that distributive quasigroups relate to Moufang loops.

Definition 4.2 [6] Let $(M, *)$ be a set with a binary operation. It is called a *Moufang loop* if it is a loop such that the binary operation satisfies one of the following equivalent identities:

$$x * (y * (x * z)) = ((x * y) * x) * z, \tag{1}$$

$$z * (x * (y * x)) = ((z * x) * y) * x, \tag{2}$$

$$(x * y) * (z * x) = (x * (y * z)) * x. \tag{3}$$

As the name suggests, the Moufang identity is named for Ruth Moufang who discovered it in some geometrical investigations in the first half of this century [42]. Moufang loops differ from groups in that they need not be associative. A Moufang loop that is associative is a group. The Moufang identities may be viewed as weaker forms of associativity. The typical examples include groups and the set of nonzero octonions which gives a nonassociative Moufang loop.

Theorem 4.1 (Moufang's Theorem) *Let a, b, c be three elements in a commutative Moufang loop (abbreviated CML) M for which the relation $(a * b) * c = a * (b * c)$ holds. Then the subloop generated by them is associative and hence is an Abelian group.*

A consequence of this theorem is that every two elements in CML generate an Abelian subgroup. Let $(X, *)$ be a right-distributive quasigroup. Then $(x * x) * x = (x * x) * (x * x)$ which implies that each element is idempotent and $(X, *)$ is then a Latin quandle. Fix $a \in X$ and define the following operation, denoted $+$, on X by $x + y := R_a{}^{-1}(x) * L_a{}^{-1}(y)$. Then $a + y = y$ and $y + a = y$. Thus $(X, +, a)$ is a loop. Therefore any right-distributive quasigroup satisfying one of the Moufang identities (1), (2) and (3) is a Moufang loop. Note that $R_a(x) + L_a(y) = x * y$. The Moufang loop is commutative if and only if

$$(u * v) * (w * z) = (u * w) * (v * z). \tag{4}$$

Recall that a *magma* is a set with a binary operation. A magma $(X, *)$ that satisfies Eq. (4) is said to be *medial* (Belousov [3]) or *abelian* (Joyce [37]). The Bruck-Toyoda theorem gives the following characterization of medial quasigroup. Given an Abelian group M, two commuting automorphisms f and g of M and a fixed element a of M, define an operation $*$ on M by $x * y = f(x) + g(y) + a$. This quasigroup is called *affine* quasigroup. It's clear that $(M, *)$ is a medial quasigroup. The Bruck-Toyoda theorem states that every medial quasigroup is of this form, i.e. is isomorphic to a quasigroup defined from an abelian group in this way. Belousov gave the connection between distributive quasigroups and Moufang loops in the following

Theorem 4.2 [3] *If $(X, *)$ be a distributive quasigroup then for all $a \in X$, $(X, +, a)$ is a commutative Moufang loop.*

Now let $(X, *)$ be a Latin quandle (that is right-distributive quasigroup), then the automorphism $\phi = R_a$ satisfies $2\phi(a) = a$. If the order of a is odd then one can write $\phi(a) = \frac{1}{2}a$. The map $x \mapsto 2x$ being a homomorphism is equivalent to $(x + y) + (x + y) = (x + x) + (y + y)$, (mediality property).

We have the following question: do the following three properties imply associativity for a finite magma $(X, +)$?

1. $(X, +)$ is a commutative loop with identity element 0.
2. For all x, y in X we have the identity $(x + y) + (z + z) = (x + z) + (y + z)$.
3. There is an automorphism f of $(X, +)$ satisfying $f(x) + f(x) = x$ for all x. (in other words, the map $x \mapsto 2x$ is onto and $(x + x) + (y + y) = (x + y) + (x + y)$.

In fact, if $(X, +)$ is a loop satisfying condition 2, then $(X, +)$ is a commutative Moufang loop, necessarily satisfying the other conditions. There exist nonassociative commutative Moufang loops. The smallest order at which such loops occur is 81, and there are, in fact, two such loops of that order. The easier to describe of the two commutative Moufang loops of order 81 is the one of exponent 3. Special thanks to Michael Kinyon and David Stanovsky for telling us about the following example and some other results about quasigroups. Let $F = \mathbb{Z}_3$ and on F^4, define

$$\begin{aligned}(x_0, x_1, x_2, x_3) + (y_0, y_1, y_2, y_3) \\ = (x_0 + y_0 + (x_1 - y_1)(x_2 y_3 - x_3 y_2), x_1 + y_1, x_2 + y_2, x_3 + y_3),\end{aligned}$$

This is very first known example, published by Bol, who attributed it to Zassenhaus [4].

The construction from loops to quandles requires the maps $x \mapsto 2x$ to be bijections as well as a homomorphisms. Is this guaranteed for commutative Moufang loops? Every abelian group is a commutative Moufang loop, so squaring is not always a bijection, of course. For the two examples we mentioned above (loops of order 81), the answer is yes. Any commutative Moufang loop modulo its center will have exponent 3. If you have a commutative Moufang loop which is indecomposable in the sense that it is not a direct product of smaller loops, then it will have order a power of 3. Nonassociativity starts showing up at order 81. Classification of commutative Moufang loops of higher order has not been worked out in detail because of the computational difficulties. Much literature has been about free commutative Moufang loops of exponent 3, because they turn out to be finite and of order 3^n. Quandles which are also quasigroups correspond to a class of loops known as Bruck loops. Commutative Moufang loops have been investigated in detail by Bruck and Salby.

Theorem 4.3 [6] *If $(X, +)$ is a commutative Moufang loop then $X = A \times B$ is a direct product of an abelian group A with order coprime to 3 and a commutative Moufang loop of order 3^k.*

Latin quandles are right distributive quasigroups and left-distributive Latin quandles are distributive quasigroups. Belousov's theorem tells us that if $(X, *)$ is left-distributive Latin quandle then $(X, +)$ is a commutative Moufang loop and then Bruck-Slaby theorem tells us that $(X, *)$ is affine over a commutative Moufang loop, and then medial. The smallest Latin quandle that is not left distributive is of order 15 and was found by David Stanovsky (see [55], p 29) using an automatic model builder SEM for all quasigroups satisfying left distributivity, but not mediality. This motivated Jan Vlachy [60] to look for a more theoretical argument that would explain the nonexistence of any smaller quasigroups of this kind and proved that there are exactly two non-isomorphic types of these smallest non-right-distributive left-distributive quasigroups with 15 elements. He constructed them explicitly using the Galkin's representation [32]. In the survey paper [31], page 950, Galkin states that nonmedial quasigroups of order less than 27 appear only in orders 15 and 21 and are given by the following construction: Define a binary operation on $\mathbb{Z}_3 \times \mathbb{Z}_p$ by

$$(x, a) * (y, b) = (2y - x, -a + \mu(x - y)b + \tau(x - y)) \quad x, y \in \mathbb{Z}_3, \ a, b \in \mathbb{Z}_p,$$

where $\mu(0) = 2$, $\mu(1) = \mu(2) = -1$, and $\tau : \mathbb{Z}_3 \to \mathbb{Z}_p$ is such that $\tau(0) = 0$. This construction was generalized by replacing $\mathbb{Z}_p$ by any abelian group A in [24]. Let A be an abelian group, also regarded naturally as a $\mathbb{Z}$-module. Let $\mu : \mathbb{Z}_3 \to \mathbb{Z}$, $\tau : \mathbb{Z}_3 \to A$ be functions. These functions μ and τ need not be homomorphisms. Define a binary operation on $\mathbb{Z}_3 \times A$ by

$$(x, a) * (y, b) = (2y - x, -a + \mu(x - y)b + \tau(x - y)) \quad x, y \in \mathbb{Z}_3, \ a, b \in A.$$

Proposition 4.1 [24] *For any abelian group A, the above operation $*$ defines a quandle structure on $\mathbb{Z}_3 \times A$ if $\mu(0) = 2$, $\mu(1) = \mu(2) = -1$, and $\tau(0) = 0$.*

This quandle $(\mathbb{Z}_3 \times A, *)$ is called the *Galkin quandle* and denoted by $G(A, \tau)$.

Lemma 4.1 [24] *For any abelian group A and $c_1, c_2 \in A$, $G(A, c_1, c_2)$ and $G(A, 0, c_2 - c_1)$ are isomorphic.*

Various properties of Galkin quandles were studied in [24] and their classification in terms of pointed abelian groups was given. We mention a few properties. Each $G(A, c)$ is connected but not Latin unless A has odd order, $G(A, c)$ is non-medial unless $3A = 0$.

We conclude with the following properties relating distributivity and mediality to quandles [24]: Alexander quandles are left-distributive and medial. It is easy to check that for a finite Alexander quandle (M, T) with $T \in \mathrm{Aut}(M)$, the following are equivalent: (1) (M, T) is connected, (2) $(1-T)$ is an automorphism of M, and (3) (M, T) is Latin. It was also proved by Toyoda [58] that a Latin quandle is Alexander if and only if it is medial. As noted by Galkin, $G(\mathbb{Z}_5, 0)$ and $G(\mathbb{Z}_5, 1)$ are the smallest non-medial Latin quandles and hence the smallest non-Alexander Latin quandles. We note that medial quandles are left-distributive (by idempotency). It is proved in [24] that any left-distributive connected quandle is Latin. This implies, by Toyoda's theorem, that every medial connected quandle is Alexander and Latin. The smallest Latin quandles that are not left-distributive are the Galkin quandles of order 15. It is known that the smallest left-distributive Latin quandle that is not Alexander is of order 81. This is due to V. D. Belousov.

5 Quandle Cohomology and Cocycle Invariant of Knots

In the classical theory of knots and links in 3-space, one utilizes projections of knots and links and applies to them the Reidemeister moves, a sequence of which will take one from any one projection of a given knot or link to any other projection of that knot or link. The Reidemeister moves have played an essential role in the development of a wide variety of invariants for knots and links, since any quantity that remains unchanged by the three moves is an invariant for knots and links. In 1999, Carter et al. [18] used quandle cohomology to define combinatorial "state-sum" invariants for classical knots and knotted surfaces called quandle cocycle invariant (see definition below). Here we mention some interesting results on surfaces in 4-space they obtained: (1) constructing an example of a sphere that is knotted in 4-dimensional space [18], (2) giving obstructions to ribbon concordance for knotted surfaces [22], and (3) detecting non-invertibility of knotted surfaces [18]. This was extended to some other examples [14, 17].

In order to define quandle homology and the cocycle knot invariant we need to define coloring of knots by a quandle. A *coloring* of an oriented classical knot K is a function $\mathscr{C} : R \to X$, where X is a fixed quandle and R is the set of over-arcs in a fixed diagram of K, satisfying the condition depicted in the top of Fig. 2. This definition of colorings on knot diagrams has been known, see [29] for example. In the bottom of Fig. 2, the relation between Reidemeister type III move and a quandle axiom (self-distributivity) is indicated. In particular, the colors of the bottom right segments before and after the move correspond to the self-distributivity. By assigning a weight $\phi(x, y)$ at each crossing of a knot diagram (as in the top Fig. 2) we obtain a 2-cocycle condition which can be generalized to a homology of cohomology theory which we describe now.

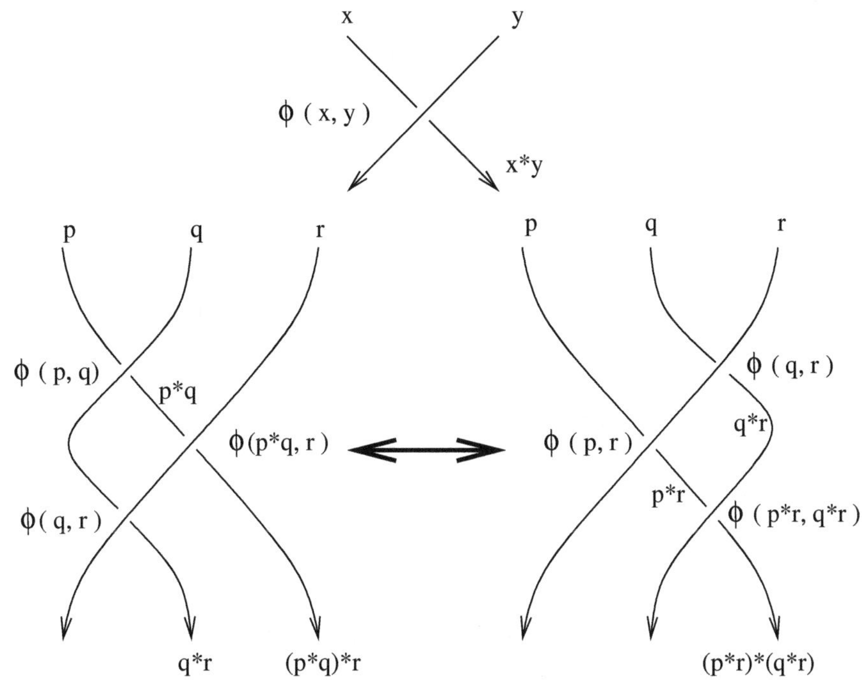

Fig. 2 2-cocycle condition and Reidemeister move III

Let $C_n(X)$ be the free abelian group generated by n-tuples $(x_1, \ldots, x_n)$ of elements of a quandle X. Define a homomorphism $\partial_n : C_n(X) \to C_{n-1}(X)$ by

$$\begin{aligned} &\partial_n(x_1, x_2, \ldots, x_n) \\ &\quad = \sum_{i=2}^{n} (-1)^i \left[(x_1, x_2, \ldots, x_{i-1}, x_{i+1}, \ldots, x_n) \right. \\ &\qquad \left. - (x_1 * x_i, x_2 * x_i, \ldots, x_{i-1} * x_i, x_{i+1}, \ldots, x_n) \right] \end{aligned} \tag{5}$$

for $n \geq 2$ and $\partial_n = 0$ for $n \leq 1$. Then $C_*(X) = \{C_n(X), \partial_n\}$ is a chain complex. The nth *quandle homology group* and the nth *quandle cohomology group* [18] of a quandle X with coefficient in a group A can be defined. One can consider cohomology also and for example:

A *2-cocycle* is a function $\phi : X \times X \to A$ such that $\phi(x, y) + \phi(x * y, z) = \phi(x, z) + \phi(x * z, y * z)$, and for all x, $\phi(x, x) = 0$.

A *3-cocycle* is a function $\psi : X \times X \times X \to A$ such that

$$\psi(x, y, z) + \psi(x, z, w) + \psi(x * z, y * z, w) = \psi(x * y, z, w) + \psi(x * w, y * w, z * w) + \psi(x, y, w),$$

and and for all x, y, $\psi(x, x, y) = \psi(x, y, y) = 0$.

Let $\mathscr{C}$ denote a coloring of a knot K by a quandle X and choose a quandle 2-cocycle ϕ, Then define a *(Boltzmann) weight*, $B(\tau, \mathscr{C})$, at a crossing τ, by $B(\tau, \mathscr{C}) = \phi(x, y)^{\epsilon(\tau)}$, where $\epsilon(\tau) = 1$ or -1, if the sign of τ is positive or negative, respectively. The *partition function* (called also *state-sum*) is the expression

$$\Phi_\phi(K) := \sum_{\mathscr{C}} \prod_{\tau} B(\tau, \mathscr{C}).$$

The product is taken over all crossings of the given diagram, and the sum is taken over all possible colorings. The values of the partition function are taken to be in the group ring $\mathbb{Z}[A]$ where A is the coefficient group written multiplicatively.

Theorem 5.1 [18] *The state sum $\Phi_\phi(K)$ does not depend on the choice of a diagram of a knot K, so that it is a knot invariant.*

This knot invariant is also called quandle cocycle invariant associated with the quandle 2-cocycle ϕ.

Example 5.1 see [19] p 52, Let $X = \mathbb{Z}_2[T, T^{-1}]/(T^2 + T + 1)$, $A = \mathbb{Z}_2$, and cocycle $\Phi = \prod \chi_{(a,b)}$ where $a,\ b \in \{0, 1, T + 1\}$ and $a \neq b$.

For knots K (up to nine crossings, see [23] for diagrams and other information) the Invariants $\Phi(K)$ are:

- $4(1 + 3T)$ for $3_1, 4_1, 7_2, 7_3, 8_1, 8_4, 8_{11}, 8_{13}, 9_1, 9_6, 9_{12}, 9_{13}, 9_{14}, 9_{21}, 9_{23}, 9_{35}, 9_{37}$,
- $16(1 + 3T)$ for 8_{18}, and 9_{40}
- 16 for $8_5, 8_{10}, 8_{15}, 8_{19} - 8_{21}, 9_{16}, 9_{22}, 9_{24}, 9_{25}, 9_{28} - 9_{30}, 9_{36}, 9_{38}, 9_{39}, 9_{41} - 9_{45}, 9_{49}$
- 4 otherwise.

Generalizations, variations, and applications of the cocycle knot invariants have been discovered; for example, see [2, 9–15, 17]. Quandle homology has also been investigated in [49–51, 53].

5.1 Extensions of Quandles

Quandle extension theory was developed in [16] by analogy with group extensions defined for low dimensional group cocycles. Let X be a quandle, A be an abelian group and given a 2-cocycle $\phi \in Z^2_Q(X; A)$, the quandle operation in extension is defined on $E = A \times X$ by $(a_1, x_1) * (a_2, x_2) = (a_1 + \phi(x_1, x_2),\ x_1 * x_2)$. The following lemma is the converse of the fact proved in [21] that $E(X, A, \phi)$ is a quandle.

Lemma 5.1 [16] *Let X, E be finite quandles, and A be a finite abelian group written multiplicatively. Suppose there exists a bijection $f : E \to A \times X$ with the following* *property. There exists a function* $\phi : X \times X \to A$ *such that for any* $e_i \in E$ $(i = 1, 2)$, *if* $f(e_i) = (a_i, x_i)$, *then* $f(e_1 * e_2) = (a_1\phi(x_1, x_2), x_1 * x_2)$. *Then* $\phi \in Z^2_Q(X; A)$.

The following two theorems produce examples of extensions of quandles.

Theorem 5.2 [16] *For any positive integers* q *and* m, $E = \mathbb{Z}_{q^{m+1}}[T, T^{-1}]/(T - 1 + q)$ *is an abelian extension* $E = E(\mathbb{Z}_{q^m}[T, T^{-1}]/(T - 1 + q), \mathbb{Z}_q, \phi)$ *of* $X = \mathbb{Z}_{q^m}[T, T^{-1}]/(T - 1 + q)$ *for some cocycle* $\phi \in Z^2_Q(X; \mathbb{Z}_q)$.

Theorem 5.3 [16] *For any positive integer* q *and* m, *the quandle* $E = \mathbb{Z}_q[T, T^{-1}]/(1 - T)^{m+1}$ *is an abelian extension of* $X = \mathbb{Z}_q[T, T^{-1}]/(1 - T)^m$ *over* $\mathbb{Z}_q$: $E = E(X, \mathbb{Z}_q, \phi)$, *for some* $\phi \in Z^2_Q(X; \mathbb{Z}_q)$.

Below are some explicit examples of extensions.

Example 5.2 [16] For any positive integer q and m, the quandle $E = \mathbb{Z}_q[T, T^{-1}]/(1 - T)^{m+1}$ is an abelian extension of $X = \mathbb{Z}_q[T, T^{-1}]/(1 - T)^m$ over $\mathbb{Z}_q$: $E = E(X, \mathbb{Z}_q, \phi)$, for some $\phi \in Z^2_Q(X; \mathbb{Z}_q)$.

Example 5.3 [16] Consider the case $q = 2$, $m = 2$ in Example 5.2. In this case

$$\mathbb{Z}_4[T, T^{-1}]/(T + 1) = R_4, \quad \text{and}$$

$$\mathbb{Z}_8[T, T^{-1}]/(T + 1) = R_8 = E(R_4, \mathbb{Z}_2, \phi)$$

for some $\phi \in Z^2_Q(R_4; \mathbb{Z}_2)$. We obtain an explicit formula for this cocycle ϕ by computation:

$$\phi = \chi_{0,2} + \chi_{0,3} + \chi_{1,0} + \chi_{1,3} + \chi_{2,0} + \chi_{2,3} + \chi_{3,0} + \chi_{3,1},$$

where

$$\chi_{a,b}(x, y) = \begin{cases} 1 \text{ if } (x, y) = (a, b), \\ 0 \text{ if } (x, y) \neq (a, b) \end{cases}$$

denotes the characteristic function.

Other extensions of quandles have been considered by some authors; see, for example, in [1] where a more general homology theory is developed and in [27] where algebraic covering theory of quandle is established.

Dynamical cocycles [1] Let X be a quandle and S be a non-empty set. Let $\alpha : X \times X \to \text{Fun}(S \times S, S) = S^{S \times S}$ be a function, so that for $x, y \in X$ and $a, b \in S$ we have $\alpha_{x,y}(a, b) \in S$.

Then it is checked by computations that $S \times X$ is a quandle by the operation $(a, x) * (b, y) = (\alpha_{x,y}(a, b), x * y)$, where $x * y$ denotes the quandle product in X, if and only if α satisfies the following conditions:

1. $\alpha_{x,x}(a, a) = a$ for all $x \in X$ and $a \in S$;
2. $\alpha_{x,y}(-, b) : S \to S$ is a bijection for all $x, y \in X$ and for all $b \in S$;
3. $\alpha_{x*y,z}(\alpha_{x,y}(a, b), c) = \alpha_{x*z,y*z}(\alpha_{x,z}(a, c), \alpha_{y,z}(b, c))$, $\forall x, y, z \in X$ and $\forall a, b, c \in S$.

Such a function α is called a *dynamical quandle cocycle* [1]. The quandle constructed above is denoted by $S \times_\alpha X$, and is called the *extension* of X by a dynamical cocycle α. The construction is general, as Andruskiewitsch and Graña show:

Lemma 5.2 [1] *Let $p : Y \to X$ be a surjective quandle homomorphism between finite quandles such that the cardinality of $p^{-1}(x)$ is a constant for all $x \in X$. Then Y is isomorphic to an extension $S \times_\alpha X$ of X by some dynamical cocycle on the set S such that $|S| = |p^{-1}(x)|$.*

Acknowledgments I would like to thank Scott Carter, Edwin Clark and Masahico Saito for commenting on an early version and for fruitful suggestions. Thanks also to the referee for all comments.

References

1. Andruskiewitsch, N., Graña, M.: From racks to pointed hopf algebras. Adv. Math. **178**(2), 177–243 (2003)
2. Ameur, K., Elhamdadi, M., Rose, T., Saito, M., Smudde, C.: Tangle embeddings and quandle cocycle invariants. Experiment. Math. **17**(4), 487–497 (2008)
3. Belousov, V.D.: The structure of distributive quasigroups (in Russian). Mat. Sb. (N.S.) **50**(92), 267–298 (1960)
4. Bol, G.: Gewebe und gruppen. Math. Ann. **114**(1), 414–431 (1937)
5. Brieskorn, E.: Automorphic sets and singularities. Contemp. math. **78**, 45–115 (1988)
6. Bruck, H.: A Survey of Binary Systems, Ergeb. Math. Grenzgeb. Neue Folge, Heft 20. Reihe: Gruppentheorie. Springer, Berlin (1958)
7. Burstin, C., Mayer, W.: Distributive gruppen. J. Reine Angew. Math. **160**, 111–130 (1929)
8. Carter, J.S.: A Survey of Quandle Ideas, Introductory Lectures on Knot Theory, Ser. Knots Everything, vol. 46, 22–53. World Scientific Publishing, Hackensack (2012)
9. Carter, S., Elhamdadi, M., Saito, M., Silver, D., Williams, S.: Virtual knot invariants from group biquandles and their cocycles. J. Knot Theor. Ramifications **18**(7), 957–972 (2009)
10. Carter, S., Crans, A., Elhamdadi, M., Karadayi, E., Saito, M.: Cohomology of frobenius algebras and the yang-baxter equation. Commun. Contemp. Math. **10**(suppl. 1), 791–814 (2008)
11. Carter, S., Crans, A., Elhamdadi, M., Saito, M.: Cohomology of categorical self-distributivity. J. Homotopy Relat. Struct. **3**(1), 13–63 (2008)
12. Carter, S., Crans, A., Elhamdadi, M., Saito, M.: Cohomology of the adjoint of hopf algebras. J. Gen. Lie Theor. Appl. **2**(1), 19–34 (2008)
13. Carter, S., Elhamdadi, M., Saito, M., Satoh, S.: A lower bound for the number of reidemeister moves of type iii. Topology Appl. **153**(15), 2788–2794 (2006)
14. Carter, J.S., Elhamdadi, M., Graña, M., Saito, M.: Cocycle knot invariants from quandle modules and generalized quandle homology. Osaka J. Math. **42**, 499–541 (2005)
15. Carter, J.S., Elhamdadi, M., Saito, M.: Homology theory for the set-theoretic yang-baxter equation and knot invariants from generalizations of quandles. Fund. Math. **184**, 31–54 (2004)
16. Carter, S., Elhamdadi, M., Nikifourou, M., Saito, M.: Extensions of quandles and cocycle knot invariants. J. Knot Theor. Ramifications **12**(6), 725–738 (2003)
17. Carter, S., Elhamdadi, M., Saito, M.: Twisted quandles homology and cocycle knot invariants. Algebr. Geom. Topol. **2**, 95–135 (2002)

18. Carter, J.S., Jelsovsky, D., Kamada, S., Langford, L., Saito, M.: Quandle cohomology and state-sum invariants of knotted curves and surfaces. Trans. Amer. Math. Soc. **355**, 3947–3989 (2003)
19. Carter, J.S., Jelsovsky, D., Kamada, S., Saito, M.: Computations of quandle cocycle invariants of knotted curves and surfaces. Adv. Math. **157**(1), 36–94 (2001)
20. Carter, S., Kamada, S., Saito, M.: Surfaces in 4-space, Encyclopaedia of Mathematical Sciences: Low-Dimensional Topology. III, vol. 142. Springer, Berlin (2004)
21. Carter, S., Kamada, S., Saito, M.: Diagrammatic Computations for Quandles and Cocycle Knot Invariants, Diagrammatic Morphisms and Applications (San Francisco, CA, 2000), pp. 51–74. Contemp. Math., 318, Amer. Math. Soc., Providence, RI (2003)
22. Carter, S., Saito, M., Satoh, S.: Ribbon concordance of surface-knots via quandle cocycle invariants. J. Aust. Math. Soc. **80**(1), 131–147 (2006)
23. Cha, J.C., Livingston, C.: KnotInfo: table of knot invariants. http://www.indiana.edu/knotinfo
24. Clark, W., Elhamdadi, M., Hou, X., Saito, M., Yetman, T.: Connected quandles associated with pointed Abelian groups. Pac. J. Math. **264**(1), 31–60 (2013)
25. Clauwens, F.J.B.J.: Small connected quandles, preprint (2010). arXiv:1011.2456
26. Ehrman, G., Gurpinar, A., Thibault, M., Yetter, D.N.: Toward a classification of finite quandles. J. Knot Theor. Ramifications **17**(4), 511–520 (2008)
27. Eisermann, M.: Quandle coverings and their Galois correspondence. arXiv:math/0612459
28. Elhamdadi, M., MacQuarrie, J., Restrepo, R.: Automorphism groups of quandles. J. Algebra Appl. **11**(1), 1250008 (9 pages) (2012)
29. Fenn, R., Rourke, C.: Racks and links in codimension two. J. Knot Theor. Ramifications **1**, 343–406 (1992)
30. Fenn, R., Rourke, C., Sanderson, B.: 1992. The rack space. Trans. Amer. Math. Soc. **359**(2), 701–740 (2007)
31. Galkin, V.M.: Quasigroups, Itogi Nauki i Tekhniki, Algebra. Topology. Geometry, vol. 26 (Russian), pp. 3–44, 162, Akad. Nauk SSSR, Vsesoyuz. Inst. Nauchn. i Tekhn. Inform., Moscow (1988). (Translated in J. Soviet Math. 49 (1990), no. 3, 941–967)
32. Galkin, V.M.: Left distributive finite order quasigroups, (Russian) quasigroups and loops. Mat. Issled. **51**, 43–54, 163 (1979)
33. Henderson, R., Macedo, T., Nelson, S.: Symbolic computation with finite quandles. J. Symbolic Comput. **41**(7), 811–817 (2006)
34. Ho, B., Nelson, S.: Matrices and finite quandles. Homology Homotopy Appl. **7**(1), 197–208 (2005)
35. Hou, X.: Automorphism groups of alexander quandles. J. Algebra **344**, 373–385 (2011)
36. Hou, X.: Finite modules over $\mathbb{Z}[t, t^{-1}]$ J. Knot Theory Ramifications **21**(8), 1250079, 28 (2012)
37. Joyce, D.: A classifying invariant of knots, the knot quandle. J. Pure Appl. Alg. **23**, 37–65 (1982)
38. Kauffman, L.H.: Knots and Physics, World Scientific, Series on Knots and Everything, vol. 1. World Scientific, Singapore (1991)
39. Lopes, P., Roseman, D.: On finite racks and quandles. Comm. Algebra **34**(1), 371–406 (2006)
40. Matveev, S.: Distributive groupoids in knot theory, (Russian). Mat. Sb. (N.S.) **119**(161), no. 1, 78–88, 160 (1982)
41. McCarron, J.: The on-line Encyclopedia of integer sequences. http://oeis.org/A181769
42. Moufang, R.: Alternativkorper und der Satz vom vollstandigen Vierseit (D9). Abh. Math. Sem. Univ. Hamburg **9**, 207–222 (1933)
43. Murillo, G., Nelson, S.: Erratum: Alexander quandles of order 16. J. Knot Theor. Ramifications **18**(5), 727 (2009)
44. Murillo, G., Nelson, S.: Alexander quandles of order 16. J. Knot Theor. Ramifications **17**(3), 273–278 (2008)
45. Navas, E.A., Nelson, S.: On symplectic quandles. Osaka J. Math. **45**(4), 973–985 (2008)
46. Nelson, S.: A polynomial invariant of finite quandles. J. Algebra Appl. **7**(2), 263–273 (2008)
47. Nelson, S.: Classification of finite Alexander quandles, proceedings of the spring topology and dynamical systems conference. Topology Proc. **27**(1), 245–258 (2003)

48. Nelson, S., Wong, C.: On the orbit decomposition of finite quandles. J. Knot Theor. Ramifications **15**(6), 761–772 (2006)
49. Niebrzydowski, M., Przytycki, J.H.: The second quandle homology of the takasaki quandle of an odd abelian group is an exterior square of the group. J. Knot Theor. Ramifications **20**(1), 171–177 (2011)
50. Niebrzydowski, M., Przytycki, J.H.: Homology of dihedral quandles. J. Pure Appl. Algebra **213**(5), 742–755 (2009)
51. Niebrzydowski, M., Przytycki, J.H.: Homology operations on homology of quandles. J. Algebra **324**(7), 1529–1548 (2010)
52. Pierce, C.S.: On the algebra of logic. Amer. J. Math. **3**, 15–57 (1880)
53. Przytycki, J.H.: Distributivity versus associativity in the homology theory of algebraic structures. Demonstratio Mathematica **44**(4), 823–869 (2011)
54. Smith, J.D.H.: Quasigroups and quandles. Discrete Math. **109**(1–3), 277–282 (1992)
55. Stanovsky, D.: Left distributive left quasigroups. Ph.D. thesis, Charles University in Prague (2004)
56. Stein, S.: On the foundations of quasigroups. Trans. Amer. Math. Soc. **85**, 228–256 (1957)
57. Takasaki, M.: Abstraction of symmetric transformation. Tohoku Math. J. **49**, 145–207 (1942/3). (in Japanese)
58. Toyoda, K.: On axioms of linear functions. Proc. Imperial Acad. **17**(7), 221–227 (1941)
59. Vendramin, L.: On the classification of quandles of low order, J. Knot Theor. Ramifications **21**(9), 1250088 (10 pages) (2012)
60. Vlach'y, J.: Small left distributive quasigroups. Thesis (2010)

Part II
Geometry

Torsors and Ternary Moufang Loops Arising in Projective Geometry

Wolfgang Bertram and Michael Kinyon

Abstract A projective space gives rise to an affine space V_a by taking out a hyperplane a. We define a natural ternary product on the set $U_{ab} = V_a \cap V_b$, for any pair (a, b) of hyperplanes. If the space is Desarguesian, we show that this ternary product is para-associative and that it coincides with the torsor structures considered in preceding work by the authors. Compared with that work, it is remarkable that—in the case of a projective space—the torsor structure can be expressed solely in terms of the lattice structure of the geometry. For general projective planes, our construction is closely related to the classical construction of ternary rings associated to such planes. In particular, for Moufang planes we show that U_{ab} is a ternary Moufang loop.

1 The Geometric Construction

In this first section, we describe the general construction of torsors and of ternary loops associated to projective spaces; proofs and computational descriptions are given in the two following sections. We assume that $\mathscr{X}$ is a projective space of dimension at least two. For projective subspaces a, b of $\mathscr{X}$, let as usual $a \wedge b$ be the meet (intersection) and $a \vee b$ be the join (smallest subspace containing a and b). In the following, the letters a and b will denote two hyperplanes of $\mathscr{X}$ (the case $a = b$ is not excluded), and the set-theoretic complement of $a \cup b$ is denoted by

W. Bertram (✉)
CNRS, INRIA, Institut Élie Cartan de Lorraine, Université de Lorraine,
Boulevard des Aiguillettes, BP 239, 54506 Vandœuvre-lès-Nancy, France
e-mail: wolfgang.bertram@univ-lorraine.fr

M. Kinyon
Department of Mathematics, University of Denver, 2360 South Gaylord Street,
Denver, CO 80208, USA
e-mail: mkinyon@du.edu

A. Makhlouf et al. (eds.), *Algebra, Geometry and Mathematical Physics*,
Springer Proceedings in Mathematics & Statistics 85, DOI: 10.1007/978-3-642-55361-5_20,

$$U_{ab} := \mathscr{X} \setminus (a \cup b) = V_a \cap V_b. \tag{1}$$

It is well-known that $V_a := U_{aa}$ is an affine space. Something similar is true for any U_{ab}: consider a triple of points (x, y, z) from U_{ab}. We will define a fourth point w in U_{ab}, depending on these data, so we write $w = (xyz)_{ab}$.

1.1 The Generic Case

By "generic" we mean that x, y, z are not collinear.

Definition 1.1 Notation being as above, we define w to be the intersection point of

- the parallel of the line $x \vee y$ through z in the affine space $V_a := \mathscr{X} \setminus a$, with
- the parallel of the line $z \vee y$ through x in the affine space $V_b = \mathscr{X} \setminus b$; that is:

$$w = (xyz)_{ab} = \Big(\big((x \vee y) \wedge a\big) \vee z\Big) \wedge \Big(\big((z \vee y) \wedge b\big) \vee x\Big).$$

Note that this point of intersection exists since all lines belong to the projective plane spanned by x, y, z. For $a = b$, this is the usual "parallelogram definition" of vector addition in the affine space V_a with origin y, that is, $(xyz)_{aa} = x + z$ in this case. Hence, for $a \neq b$, $(xyz)_{ab}$ may be seen as a kind of "deformation of vector addition": we have a sort of "fake parallelogram" with vertices y, x, z, w, as shown in the following illustration:

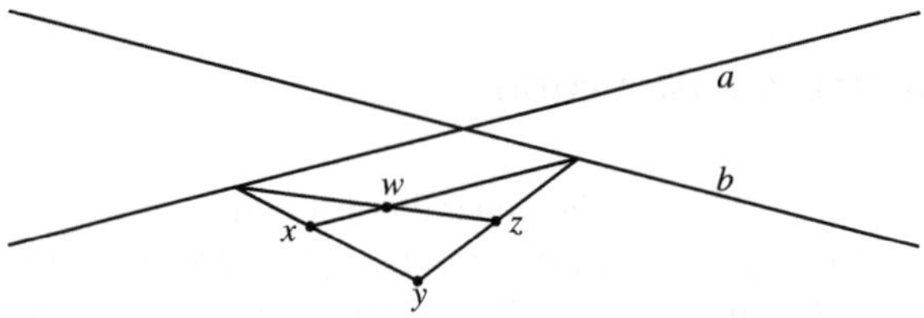

As for "usual" parallelograms, it is easily seen that, with $(xyz) := (xyz)_{ab}$ for fixed (a, b), the conditions

$$w = (xyz), \qquad y = (zwx), \qquad z = (yxw), \qquad x = (wzy) \tag{2}$$

are all equivalent. Note also that the following *symmetry relation* is obvious from the definition:

$$(xyz)_{ba} = (zyx)_{ab}. \tag{3}$$

If we choose a as "line at infinity" of our drawing plane, and draw b horizontally, we get the following image:

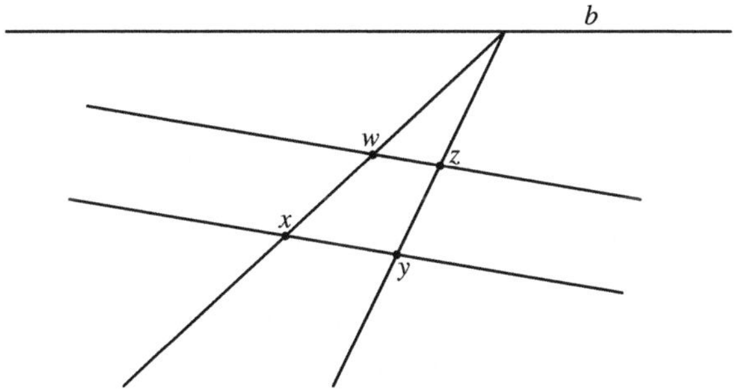

These images admit a spacial interpretation: we may imagine the observer placed in affine space $\mathbb{R}^3$ inside a plane B which is vizualized only by its "horizon", the line b; then we think of the line $y \vee z$ as lying in a plane B' parallel to B, and of the line $x \vee w$ as lying in another such plane B''; the other two lines $w \vee z$ and $x \vee y$ lie in planes that are parallel to the drawing plane P. This interpretation is not symmetric in x and z: the point z lies "behind" (or "in front of") y, whereas x is considered to be "on the same level" as y.

The product $xz := (xyz)_{ab}$ is in general not commutative, but it is *associative*: we show that, if $\mathscr{X}$ is Desarguesian, then, for any fixed origin y, the binary map $(x, z) \mapsto xz$ gives rise to a *group law* on U_{ab}. More generally and more conceptually, we show that the ternary law $(x, y, z) \mapsto (xyz)_{ab}$ defines a *torsor structure on* U_{ab} (Theorem 2.1). Here, we use the term "torsor" in the sense of "group without distinguished origin":

Definition 1.2 A set G with a map $G^3 \to G$, $(x, y, z) \mapsto (xyz)$ is called a *torsor* if

$$(xxy) = y = (yxx) \tag{T0}$$

$$(xy(zuv)) = (x(uzy)v) = ((xyz)uv) \tag{T1}$$

(There are other terms in use for this concept, such as *groud, heap, flock, principal homogeneous space*—see [2] for some remarks on the terminology we use.) Naturally, the question arises what we can say for general, non-Desarguesian projective planes, or for still more general lattices. The most prominent class of non-Desarguesian projective planes are the *Moufang planes*: we show that in this case we get a kind of "alternative version of a torsor" which we call a *ternary Moufang loop* (Theorem 3.1). For $a = b$, these ternary Moufang loops contract to the abelian vector group of an affine plane. For very general projective planes (which need not be "translation planes") it remains an interesting open problem to relate this new algebraic structure to those traditionally considered in the literature: indeed, our definition is closely related to the more traditional ways of coordinatizing projective planes by *ternary rings*. This is related to the following item.

1.2 The Collinear Case

We have not yet defined what $(xyz)_{ab}$ should mean if x, y, z are *collinear*. If $\mathcal{X}$ is a *topological* projective plane, then one would like to complete our definition simply "by continuity", e.g., by taking the limit of $(xyz)_{ab}$ as y, not lying on the line $x \vee z$, converges to a point on $x \vee z$. This is indeed what happens in the classical planes over the division algebras $\mathbb{R}$, $\mathbb{C}$, $\mathbb{H}$, $\mathbb{O}$. (We encourage the reader to vizualise this, for the real plane, by using some dynamical geometry software, such as `geogebra`!) Since we do not know whether in very general cases such a "limit" exists, we restrict ourselves here to the Moufang case, and leave the general case for later work.

Definition 1.3 Assume that $\mathcal{X}$ is a Moufang plane or a projective space of dimension bigger than 2. Consider a pair (a, b) of hyperplanes and a *collinear* triple (x, y, z) of points, none of them in a or b.

(1) If $x = y = z$, let $(xyz)_{ab} := x$.
(2) If $x \neq y$, then let $L := x \vee y$ and choose a point u not belonging to L or to a, and we let $w := (xyz)_{ab} :=$

$$(x \vee y) \wedge \Big[\big((z \vee u) \wedge b \big) \vee \Big(\big[((x \vee y) \wedge a) \vee u \big] \wedge \big[((u \vee y) \wedge b) \vee x \big] \Big) \Big].$$

(It will be shown below that w does not depend on the choice of u.)
(3) If $z \neq y$, then we let $L := z \vee y$ and define

$$w := (xyz)_{ab} := (zyx)_{ba},$$

where the right hand side is defined by the preceding case.

This definition can be interpreted from two different viewpoints:

(A) *Algebraic* In the Desarguesian case, the expression in (2) is derived from our definition in the generic case by using (T0) and (T1)

$$((xyu)uz) = (xy(uuz)) = (xyz),$$

where now the left hand side can be expressed by using twice Definition 1.1, giving (2) (see Theorem 2.2). This is indeed in keeping with idea explained above of "taking a limit" (imagine u tending towards a point on the line L). The argument still goes through in the Moufang case since one does not need for it full (T1), but just a special case which remains valid precisely in the Moufang case (but it breaks down as soon as one wants to go further).

(B) *Geometric* The formula in (2) corresponds to classical "constructions of the field associated to a plane". It is known that in the Moufang case the field does not depend on the "off-line" point u. More specifically, we distinguish two cases in (2):

- the generic case corresponds to the *product* of the field: if the points $L \wedge a$ and $L \wedge b$ are different, then $(xyz)_{ab}$ is the product $zy^{-1}x$ on the vector line L with "point at infinity" $L \wedge a$ and "zero point" $L \wedge b$ (in the illustration below, a is the line at infinity and b the horizontal line; in usual textbook drawings, the inverse choice is made. We have marked the points $p = (xyu)$ and $q = (puz) = w$.)

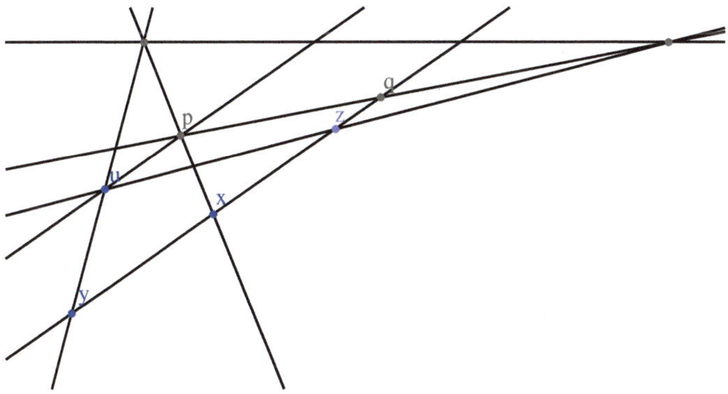

- a special case corresponds to the *addition* of the field: if the line $L := x \vee y$ intersects $a \wedge b$, then $(xyz)_{ab}$ is the "ternary sum" $x - y + z$ in the affine line L (with $L \wedge (a \wedge b)$ as "point at infinity", see the following illustration, which is the limit case of the preceding one, as L becomes parallell to b).

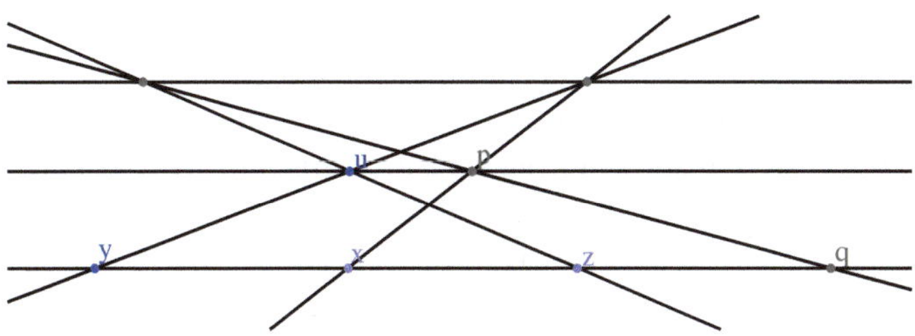

The main result of the present work can now be stated as follows:

Definition 1.4 A set G with a map $G^3 \to G$, $(x, y, z) \mapsto (xyz)$ is called a *ternary Moufang loop* if it satisfies (T0) and

$$(uv(xyx)) = ((uvx)yx) \tag{MT1}$$

$$(xy(xyz)) = ((xyx)yz) \tag{MT2}$$

Theorem 1.1 *If $\mathscr{X}$ is a projective space of dimension bigger than one over a skew-field (i.e., a Desarguesian space), then the preceding constructions define a torsor law on U_{ab}. If $\mathscr{X}$ is a Moufang projective plane, then the constructions define a ternary Moufang loop.*

1.3 Generalized Cross-Ratios, and Associative Geometries

In the Desarguesian case, a very general theory describing torsors of the kind of U_{ab} has been developed in [2]. Comparing with the approach presented here, one may ask for what kinds of lattices there are similar theories—we will, in subsequent work, investigate in more depth the case of Moufang spaces, related to *alternative* algebras, triple systems and pairs. Returning to the Desarguesian case and to classical projective geometry, the link between the lattice and the structure defined in [2] is surprisingly close; however, one should not forget that for *projective lines* the lattice structure is completely useless, whereas the structures from [2] are at least as strong as the classical *cross-ratio*, and hence are much stronger than the lattice structure. Let us briefly explain this. Given a unital ring $\mathbb{K}$ and $\mathcal{X} := \mathcal{X}(\Omega)$, the full Grassmannian geometry of some $\mathbb{K}$-module Ω (set of all submodules of Ω), we have associated in [2] to any 5-tuple $(x, a, y, b, z) \in \mathcal{X}^5$ another element of $\mathcal{X}$ by

$$\Gamma(x,a,y,b,z) := \left\{ \omega \in \Omega \,\middle|\, \begin{array}{c} \exists \xi \in x, \exists \alpha \in a, \exists \eta \in y, \exists \beta \in b, \exists \zeta \in z : \\ \omega = \zeta + \alpha = \alpha + \eta + \beta = \xi + \beta \end{array} \right\}. \qquad (4)$$

In [2], Theorem 2.4, it is shown that the lattice structure is recovered via

$$x \wedge a = \Gamma(x,a,y,x,a), \qquad b \vee a = \Gamma(a,a,y,b,b) \qquad (5)$$

for any $y \in \mathcal{X}$. On the other hand, in the present work we prove (Theorem 2.2) that, if $\mathbb{K}$ is a field, if a, b are hyperplanes and x, y, z one-dimensional subspaces, then $\Gamma(x, a, y, b, z)$ can be recovered from the lattice structure via

$$\Gamma(x,a,y,b,z) = (xyz)_{ab}. \qquad (6)$$

Thus, roughly speaking, for Desarguesian projective spaces of dimension bigger than one, Γ and the lattice structure are essentially equivalent data. Summing up, there are two major approaches to our object: the algebraic approach [2], based on associative algebras and -pairs and on an underlying group structure of the "background" Ω (cf. [1]), and the lattice theoretic approach from the present work, keeping close to classical geometric language, and paving the way to incorporate exceptional geometries into the picture.

2 The Desarguesian Case

Theorem 2.1 *Assume $\mathcal{X}$ is a Desarguesian projective space of dimension bigger than one, and fix a pair (a, b) of hyperplanes. Then U_{ab}, together with the ternary product $(xyz) := (xyz)_{ab}$ defined above, is a torsor. In particular, if we fix an* *"origin" $y \in U_{ab}$, then U_{ab} with product $xz = (xyz)_{ab}$ and origin y becomes a group. If $a \neq b$, then this is group is not commutative.*

Proof We give three different proofs: "axiomatic", "computational", and "algebraic".

(a) The first proof is in terms of axiomatic geometry. The idempotency law $(xzz) = x$, for $x \neq z$, is a fairly direct consequence of item (2) in definition 1.3, and the other law $(zzx) = x$ then follows immediately from (3). Next, let us show, by using Desargues' Theorem, that $((xyz)uv) = (xy(zuv))$ in the non-collinear case. We construct first the point $((xyz)uv)$. This is best visualized by choosing a as line at infinity of our drawing plane, and we may draw the lines $y \vee x$ and $z \vee (xyz)$ as vertical lines. Then $((xyz)uv)$ is the point q in the illustration given below. Next, construct the point $(xy(zuv))$ and observe that the triangles $u, z, (xyz)$ and $v, (zuv), q$ are in a Desargues configuration, and conclude that the line $q \vee (zuv)$ is parallel to $z \vee (xyz)$, i.e., it is vertical. But then the triangles $y, z, (zuv)$ and $x, (xyz), q$ are also in Desargues configuration, i.e., the intersection points of corresponding sides lie on a common line, which must be b. It follows that $(x \vee q) \wedge b = (y \vee (zuv)) \wedge b$, from which the desired equality follows.

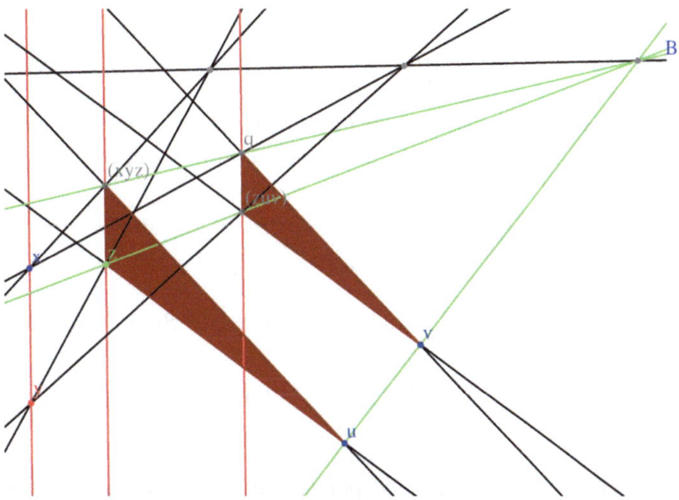

The identity $(x(yzu)v) = (xu(zyv))$, in the non-collinear case, is proved in a similar way. Using this, as explained in the introduction, the definition in the collinear case reads $(xyz) = ((xyu)uz)$, proving that the result does not depend on the choice of u. By purely algebraic computations it follows now that para-associativity also holds in the collinear case.

(b) A computational proof. Let $\mathbb{K}$ be the (skew)field of $\mathcal{X}$, and work in the affine space $V := V_a$. If $a = b$, then (as mentioned above), $(xyz)_{aa} = x - y + z$ is the torsor law of the affine space V_a, and the claim is obviously true. If $a \neq b$, fix some arbitrary origin o in the affine hyperplane $V_a \cap b$. There is a linear form $\beta : V \to \mathbb{K}$ such that $b \cap V = \ker(\beta)$, so that $U_{ab} = \{x \in V_a \mid \beta(x) \in \mathbb{K}^\times\}$.

Lemma 2.1 *For all $x, y, z \in U_{ab}$, in the vector space (V_a, o), we have*

$$(xyz)_{ab} = \beta(z)\beta(y)^{-1}(x - y) + z.$$

Proof Assume first that x, y, z are not collinear. The parallel of $x \vee y$ through z in V_a is

$$\big((x \vee y) \wedge a\big) \vee z = \{z + s(x - y) \mid s \in \mathbb{K}\}.$$

We determine the point $(y \vee z) \wedge b$. If $y \vee z$ is parallel to b, then we get easily from the definition that $(xyz)_{ab} = x - y + z$ is the usual sum, which is in keeping with our claim. Assume that $y \vee z$ is not parallel to b. Then the intersection point $(y \vee z) \wedge$ is obtaining by solving $\beta\big((1-t)y + tz\big) = 0$, whence $t = \beta(y)(\beta(y) - \beta(z))^{-1}$, whence $1 - t = -\beta(z)(\beta(y) - \beta(z))^{-1}$ and

$$(y \vee z) \wedge b = -\beta(z)(\beta(y) - \beta(z))^{-1}y + \beta(y)(\beta(y) - \beta(z))^{-1}.$$

The intersection of $\big((z \vee y) \wedge b\big) \vee x$ and $\big((x \vee y) \wedge a\big) \vee z = z + \mathbb{K}(x - y)$ is determined by $r, s \in \mathbb{K}$ such that

$$(1 - r)x + r\big(-\beta(z)(\beta(y) - \beta(z))^{-1}y + \beta(y)(\beta(y) - \beta(z))^{-1}z\big) = sx - sy + z.$$

Since both sides are barycentric combinations of x, y, z, we may consider y as new origin. Then, if x and z are linearly independent with respect to this origin, this condition is equivalent to

$$1 - r = s, \qquad r\big(\beta(y)\big(\beta(y) - \beta(z)\big)^{-1}\big) = 1$$

whence $r = \big(\beta(y) - \beta(z)\big)\beta(y)^{-1}$ and $s = 1 - r = \beta(z)\beta(y)^{-1}$, and finally

$$(xyz)_{ab} = s(x - y) + z = \beta(z)\beta(y)^{-1}\big(x - y\big) + z,$$

proving our claim in the non-collinear case.

Now consider the collinear case. As pointed out after Definition 1.3, in this case the definition of $(xyz)_{ab}$ amounts to the geometric definition of the field operations. If the line L spanned by x, y, z is parallel to b, then $\beta(z) = \beta(y)$, and the formula from the lemma gives the additive torsor law $x - y + x$, as required. Else, choose $o := L \wedge b$ as origin, let $u \in L$ with $\beta(u) = 1$ and write $x = \xi u$, $y = \eta u$, $z = \zeta u$ with $\xi, \eta, \zeta \in \mathbb{K}^\times$, and then the formula from the lemma gives $(xyz)_{ab} = \zeta\eta^{-1}(\xi u - \eta u) + \zeta u = \zeta\eta^{-1}\xi u$, which again corresponds to the definition given in this case. Thus the claim holds in all cases.

Using the lemma, we now prove the torsor laws: first of all, we have

$$\beta\big((xyz)\big) = \beta\Big(\beta(z)\beta(y)^{-1}\big(x - y\big) + z\Big) = \beta(z)\beta(y)^{-1}\beta(x), \tag{7}$$

showing that $U_{ab} = V_a \setminus \ker(\beta)$ is stable under the ternary law. The idempotent laws follow by an easy computation from the lemma. For para-associativity, using (7), a straightforward computation shows that both $((xyz)uv)$ and $(x(uzy)v)$ are given by

$$\beta(v)\beta(u)^{-1}\beta(z)\beta(y)^{-1}(x-y)+\beta(v)\beta(u)^{-1}(z-u)+v. \tag{8}$$

(b') *Remark* there is a slightly different version of (b), having the advantage that the cases $a=b$ and $a \neq b$ can be treated simultaneously, and the drawback that the dependence on y is not visible: choose $o := y$ as origin in $V = V_a$, and a linear form $\beta : V \to \mathbb{K}$ such that $b \cap V = \{x \in V \mid \beta(x) = 1\}$. The case $a = b$ then corresponds to $\beta = 0$. A computation similar as above yields

$$xz = (xyz)_{ab} = (1-\beta(z))x + z = x - \beta(z)x + z \tag{9}$$

from which associativity of the product xz follows easily. Note that Formula (9) is a special case of the formulae given in Sect. 1.4 of [2].

(c) A third and algebraic proof goes by first establishing that our lattice theoretic definition of $(xyz)_{ab}$ coincides with the algebraic definition of $\Gamma(x,a,y,b,z)$ by Eq. (4) (see items (1) and (3) of the following theorem), and then using Theorem 2.3 in [2], saying that the map Γ defines a torsor structure on U_{ab}:

Theorem 2.2 *Let $\mathbb{K}$ be a unital ring and $\mathcal{X} = \mathcal{X}(\Omega)$ be the full Grassmannian geometry of some $\mathbb{K}$-module Ω (set of all submodules of Ω), and define, for a 5-tuple $(x,a,y,b,z) \in \mathcal{X}^5$, the submodule $\Gamma(x,a,y,b,z)$ by Eq.* (4).

(1) *Assume that the triple (x,y,z) is in* general position, *that is,*

$$x \wedge (y \vee z) = 0, \quad \textit{or} \quad y \wedge (x \vee z) = 0, \quad \textit{or} \quad z \wedge (x \vee y) = 0.$$

Then we have the following equality of submodules of Ω:

$$\Gamma(x,a,y,b,z) = \Big(\big((x \vee y) \wedge a\big) \vee z\Big) \wedge \Big(\big((z \vee y) \wedge b\big) \vee x\Big).$$

(2) *Assume that z is contained in $x \vee y$, i.e., $z \wedge (x \vee y) = z$. Then, for any choice of $u \in U_{ab}$ satisfying $u \wedge (x \vee y) = 0$, we have*

$$\begin{aligned}\Gamma(x,a,y,b,z) = {} & ([([\Big(\big((x \vee y) \wedge a\big) \vee u\Big) \wedge \Big(\big((u \vee y) \wedge b\big) \vee x\Big)] \vee u) \wedge a] \vee z) \\ & \wedge\ [((z \vee u) \wedge b) \vee \Big(\big((x \vee y) \wedge a\big) \vee u\Big) \wedge \Big(\big((u \vee y) \wedge b\big) \vee x\Big)]\end{aligned}$$

(3) *Let a, b be hyperplanes in a vector space and x, y, z lines. Retain assumptions from the preceding item and assume that $x \neq y$. Then the expression given there simplifies to*

$$\Gamma(x,a,y,b,z) = (x \vee y) \wedge [((z \vee u) \wedge b) \vee \Big(\big((x \vee y) \wedge a\big) \vee u\Big) \wedge \Big(\big((u \vee y) \wedge b\big) \vee x\Big)].$$

Proof (1) We prove first the inclusion "$\subset$" (which holds in fact for all triples (x, y, z)): on the one hand, the set $\big(((x \vee y) \wedge a) \vee z\big) \wedge \big(((z \vee y) \wedge b) \vee x\big)$ is the set of all $\omega \in \Omega$ such that we can write

$$\omega = \alpha + \zeta, \qquad \omega = \beta + \xi$$

with $\alpha \in a$, $\beta \in b$, which in turn can be written

$$\alpha = \xi' + \eta, \qquad \beta = \zeta' + \eta'$$

with $\xi' \in x$, etc. This gives us a system (S) of 4 equations.

On the other hand, by definition, $\Gamma(x, a, y, b, z)$ is the set of all $\omega \in \Omega$ such that

$$\exists \xi \in x, \exists \alpha \in a, \exists \eta \in y, \exists \beta \in b, \exists \zeta \in z : \quad \omega = \zeta + \alpha = \alpha + \eta + \beta = \xi + \beta$$

There are several equivalent versions of this system (R) of three equations—see [1], Lemma 2.3., from which it is read off that the four conditions from (S) are satisfied for $\omega \in \Gamma(x, a, y, b, z)$ if we choose $\xi' = \xi$, $\eta' = \eta$, $\zeta' = \zeta$. Thus the inclusion "$\subset$" holds always.

The other inclusion does not always hold, but the theorem gives a sufficient condition: indeed, if ω belongs to the set on the right hand side, then (S) implies

$$\omega = \xi' + \eta + \zeta = \zeta' + \eta' + \xi,$$

whence $\xi - \xi' \in y \vee z$. If $x \wedge (y \vee z) = 0$, this implies that $\xi = \xi'$, and three of the four equations from (S) are equivalent to (R). If $y \wedge (x \vee z) = 0$ or $z \wedge (x \vee y) = 0$, then the same argument applies (with respect to another choice of three from the four equations of (S)). In all cases, it follows that $\omega \in \Gamma(x, a, y, b, z)$.

(2) From [2], Theorem 2.3, we know that Γ is para-associative and satisfies the idempotent law:

$$\Gamma\big(\Gamma(x, a, y, b, u), a, u, b, z\big) = \Gamma(x, a, y, b, \Gamma(u, a, u, b, z)) = \Gamma(x, a, y, b, z).$$

By assumption, the triple (x, y, u) is in general position, and from this it follows that the triple $((xyu), u, z)$ is also in general position; therefore the left-hand side may be expressed in terms of the lattice structure by applying twice part (1), which leads to the expression from the claim: in a first step, we get

$$(xyu) = \big(((x \vee y) \wedge a) \vee u\big) \wedge \big(((u \vee y) \wedge b) \vee x\big),$$

and in a second step

$$((xyu)uz) = (([(\big(((x \vee y) \wedge a) \vee u\big) \wedge \big(((u \vee y) \wedge b) \vee x\big)] \vee u) \wedge a] \vee z)$$

$$\wedge\ [((z \vee u) \wedge b) \vee \Big(\big((x \vee y) \wedge a\big) \vee u\Big) \wedge \Big(\big((u \vee y) \wedge b\big) \vee x\Big)].$$

(3) Under the given assumptions, the first term on the right hand side in (2) reduces to the line $x \vee y$, and hence the claim follows directly from (2).

Remarks (a) Not all possible relative positions of (x, y, z) are covered by Theorem 2.2, that is, the lattice theoretic formula for $\Gamma(x, a, y, b, z)$ does not hold for all triples of submodules of Ω. For instance, if $\Omega = \mathbb{K}^{2n}$ and x, y, z are of dimension n, then they cannot be in general position, and in general no u as in (2) exists. This case illustrates the special rôle of "generalized projective lines" (cf. [2]) with respect to lattice approaches.

(b) Both for the definitions given here and in [2], it is not strictly necessary that x, y, z belong to U_{ab}: they may belong to V_a, or to V_b, or (in [2]) be completely arbitrary. We will not enter here into a discussion of the relation of both definitions if x, y or z does not belong to U_{ab}.

(c) Both approaches lead to their own notions of *morphisms*. In the situation of Part (3) of Theorem 2.2, both of these notions must lead to the same result: this is precisely the famous "second fundamental theorem of projective geometry". Indeed, the "construction of the field" is contained in our approach, and hence morphisms in the lattice theoretic sense must induce morphisms of the field and the corresponding semi-linear mapping.

3 The Moufang Case

Theorem 3.1 *Assume $\mathcal{X}$ is a Moufang projective plane and (a, b) a pair of lines. Then U_{ab}, together with the ternary product $(xyz)_{ab}$ defined in the first section, is a* ternary Moufang loop. *In particular, if we fix an element $y \in U_{ab}$ as origin, then U_{ab} with product $xz = (xyz)_{ab}$ and origin y becomes a (binary) Moufang loop.*

Before proving the theorem, we recall the relevant definitions (cf., e.g., [10]):

Definition 3.1 A projective plane $\mathcal{X}$ is a *Moufang plane* if it satisfies one of the following equivalent conditions

1. The group of automorphisms fixing all points of any given line acts transitively on the points not on the line.
2. The group of automorphisms acts transitively on quadrangles.
3. Any two ternary rings of the plane are isomorphic.
4. Some ternary ring of the plane is an alternative division algebra, i.e., it is a division algebra satisfying the following identities:

$$x(xy) = (xx)y, \qquad (yx)x = y(xx), \qquad (xy)x = x(yx).$$

5. $\mathcal{X}$ is isomorphic to the projective plane over an alternative division ring.
6. The "small Desargues theorem" holds in all affine parts of $\mathcal{X}$.

The set of invertible elements in alternative algebra forms a *Moufang loop*. A basic reference for loops in general and Moufang loops in particular is [4].

Definition 3.2 A *loop* $(Q, \cdot)$ is a set Q with a binary operation $Q^2 \to Q; (x, y) \mapsto xy$ such that for each x, the maps $y \mapsto xy$ and $y \mapsto yx$ are bijections of Q, and having an element e such that $ex = xe = x$ for all $x \in Q$. A *Moufang loop* is a loop Q that satisfies any, and hence all of the following equivalent identities (the *Moufang identities*):

$$z(x(zy)) = ((zx)z)y \tag{M1}$$

$$x(z(yz)) = ((xz)y)z \tag{M2}$$

$$(zx)(yz) = (z(xy))z \tag{N1}$$

$$(zx)(yz) = z((xy)z) \tag{N2}$$

The *left* and *right multiplication maps* (sometimes called translations) in a loop are defined, respectively by $L_x y := xy =: R_y x$. The Moufang identities can be written in terms of the left and right multiplication maps. For instance, the first two identities state that

$$L_z L_x L_z = L_{zxz} \quad \text{and} \qquad R_z R_y R_z = R_{zyz}.$$

Moufang's Theorem implies that Moufang loops are *diassociative*, that is, for any a, b, the subloop $\langle a, b\rangle$ generated by a, b is a group. This can be seen as a loop theoretic analog of Artin's Theorem for alternative algebras. Two particular instances of diassociativity are the *left* and *right inverse properties*

$$x^{-1}(xy) = y \tag{LIP}$$

$$(xy)y^{-1} = x, \tag{RIP}$$

where x^{-1} is the unique element satisfying $xx^{-1} = x^{-1}x = e$. The following lemma gives the Moufang analog of the well-known relation between torsors and groups:

Lemma 3.1 *Let Q be a Moufang loop, and define a ternary operation $(\cdots)$: $Q^3 \to Q$ by $(xyz) := (xy^{-1})z$. Then the following three identities hold:*

$$(xxy) = y = (yxx) \tag{MT0}$$

$$(uv(xyx)) = ((uvx)yx) \tag{MT1}$$

$$(xy(xyz)) = ((xyx)yz) \tag{MT2}$$

Conversely, if M is a set with a ternary operation $(\cdots) : M^3 \to M$ satisfying (MT0), (MT1) and (MT2), then, for every choice of "origin" $e \in M$, the binary *operation* $x \cdot y := (xey)$ *and the unary operation* $x^{-1} := (exe)$ *define the structure of a Moufang loop on* M *with neutral element* e.

Proof First assume Q is a Moufang loop. The leftmost identity in (MT0) is trivial while the rightmost follows immediately from (RIP). For (MT1), we compute

$$\begin{aligned}(uv(xyx)) &= (uv^{-1})((xy^{-1})x) \\ &= (uv^{-1})(x(y^{-1}x)) && (\langle x, y\rangle \text{ is a group}) \\ &= (((uv^{-1})x)y^{-1})x && (M2) \\ &= ((uvx)yx).\end{aligned}$$

For (MT2),

$$\begin{aligned}(xy(xyz)) &= (xy^{-1})((xy^{-1})z) \\ &= ((xy^{-1})(xy^{-1}))z && (\langle xy^{-1}, z\rangle \text{ is a group}) \\ &= (((xy^{-1})x)y^{-1})z && (\langle x, y\rangle \text{ is a group}) \\ &= (((xyx)y)z).\end{aligned}$$

Conversely, suppose M is a set with a ternary operation $(\cdots) : M^3 \to M$ satisfying (MT0), (MT1) and (MT2). Fix $e \in M$ and define $x \cdot y := (xey)$ and $x^{-1} := (exe)$ for all $x, y \in M$. By (MT0), we see that e is neutral element for the binary operation.

First we establish the following identities:

$$x \cdot y^{-1} = (xye), \tag{10}$$

$$(x \cdot y^{-1}) \cdot z^{-1} = (xyz^{-1}), \tag{11}$$

$$(x^{-1})^{-1} \cdot x = e, \tag{12}$$

$$((x^{-1})^{-1}xy^{-1}) = y. \tag{13}$$

For (10) we compute $x \cdot y^{-1} = (xe(eye)) = ((xee)ye) = (xye)$ using (MT1) in the second equality and (MT0) in the third. For (11), we have $(x \cdot y^{-1}) \cdot z^{-1} = ((xye)ze) = (xy(eze)) = (xyz^{-1})$, using (10) (twice) and (MT1). For (12), $(x^{-1})^{-1} \cdot x = ((x^{-1})^{-1}xe) = ((ex^{-1}e)xe) = (ex^{-1}(exe)) = (ex^{-1}x^{-1}) = e$, using (10) in the first equality, (MT1) in the third and (MT0) in the fourth. Finally, for (13), $((x^{-1})^{-1}xy^{-1}) = ((x^{-1})^{-1} \cdot x^{-1}) \cdot y^{-1} = e \cdot y^{-1} = y^{-1}$, using (11) followed by (12).

Next we prove

$$(xy(y^{-1})^{-1}) = x. \tag{14}$$

Indeed,

$$(xy(y^{-1})^{-1}) = ((x(y^{-1})^{-1}(y^{-1})^{-1})y(y^{-1})^{-1}) \quad = (x(y^{-1})^{-1}((y^{-1})^{-1}y(y^{-1})^{-1}))$$
$$= (x(y^{-1})^{-1}(y^{-1})y^{-1}) = x,$$

where we have used (MT0), (MT1), (13) and (MT0).

Taking $y = x$ in (14) and applying (MT0), we obtain

$$(x^{-1})^{-1} = x. \tag{15}$$

From this it follows that $e^{-1} = e$, since $e^{-1} = e \cdot e^{-1} = (e^{-1})^{-1} \cdot e^{-1} = e$.

Now in (11), replace z with z^{-1} and use (15) to obtain

$$(x \cdot y^{-1}) \cdot z = (xyz). \tag{16}$$

Replacing y with y^{-1} and then setting $z = y^{-1}$ in (16), we obtain $(x \cdot y) \cdot y^{-1} = (xy^{-1}y^{-1}) = x$ using (MT0). Thus the right inverse property (RIP) holds.

Next we almost obtain the Moufang identity (M2) as follows:

$$\begin{aligned}((x \cdot y) \cdot z) \cdot y &= (((x \cdot e) \cdot y) \cdot z) \cdot y \\ &= ((xey)z^{-1}y) \\ &= (xe(yz^{-1}y)) \\ &= (x \cdot e) \cdot ((y \cdot z) \cdot y),\end{aligned}$$

using (16), (MT1) and (16) again. In loop theory, this is known as the *right Bol identity*.

We also have the left alternative law:

$$\begin{aligned}(x \cdot x) \cdot y &= (((x \cdot e) \cdot x) \cdot e) \cdot y \\ &= ((xex)ez) \\ &= (xe(xez)) \\ &= (x \cdot e) \cdot ((x \cdot e) \cdot y) \\ &= x \cdot (x \cdot y),\end{aligned}$$

using (16), (MT2) and (16) again.

The rest of the argument is standard. A magma satisfying the right Bol identity and (RIP) is a loop, called a *right Bol loop* (see, e.g., [5], Theorem 3.11, suitably dualized). A right Bol loop satisfying the left alternative law is a Moufang loop [9].

Definition 3.3 A set M with a map $(\cdot\cdot\cdot) : M^3 \to M$ satisfying the three identities from Lemma 3.1 will be called a *ternary Moufang loop*.

Remarks (1) The axioms (MT1) and (MT2) for ternary Moufang loops are precisely the identities (AP2) and (AP3) in Loos' axiomatization of an alternative pair [8].

(2) For an associative torsor $(\cdots) : M^3 \to M$, the groups determined by different choices of "origin," that is, fixed middle slot, are all isomorphic. The analog of this does not hold for ternary Moufang loops. Instead the different Moufang loops are *isotopic* [4]. In fact, it is straightforward to show that for a Moufang loop Q, each isotope of Q is isomorphic to an isotope with multiplication given by $x \circ z = (xy^{-1})z$ for some $y \in Q$. Thus just as alternative triple systems encode all homotopes of an alternative algebra into a single structure, so do ternary Moufang loops encode all isotopes of Moufang loops.

(3) Though we did not bother to state this in the lemma, it is clear from the proof that if we start with a Moufang loop Q with neutral element e, construct the corresponding ternary operation $(\cdots)$ and then construct the binary and unary operations induced by $(\cdots)$ with origin e, we recover the original loop operations. Similarly, if we start with a ternary Moufang loop M, construct the binary and unary operations with origin e and then construct the corresponding ternary operation induced by the loop structure, we recover the original ternary Moufang loop.

Proof (Proof of Theorem 3.1.) In principle, the first two strategies of proof of Theorem 2.1 carry over:

(a) A proof in the framework of axiomatic geometry. Instead of the full Desargues theorem we now can only use the Little Desargues theorem. The drawings will become more complicated than above since one has to introduce auxiliary points. We will not pursue this proof here.

(b) A computational proof. Let $\mathbb{K}$ be the alternative division ring belonging to the plane. Then the affine space $V := V_a$ is isomorphic to $\mathbb{K}^2$, and affine lines can be described as in the Desarguesian case, eg. $x \vee y = \{(1-t)x + ty \mid t \in \mathbb{K}\}$, where multiplication by "scalars" in $\mathbb{K}^2$ is componentwise. If $a = b$, then $(xyz)_{aa} = x - y + z$ is the torsor law of the abelian group $V_a \cong (\mathbb{K}^2, +)$, and the claim is true. If $a \neq b$, fix some arbitrary origin o in the affine hyperplane $V_a \cap b$. There is a linear form $\beta : V \to \mathbb{K}$ such that $b \cap V = \ker(\beta)$, so that $U_{ab} = \{x \in V_a \mid \beta(x) \in \mathbb{K}^\times\}$. (To fix things, one may choose coordinates such that $\beta = \mathrm{pr}_1$ is the projection onto the first coordinate of $\mathbb{K}^2$, so b is the vertical axis.)

Lemma 3.2 *Let notation be as above. Then, for all $x, y, z \in U_{ab}$, we have*

$$(xyz)_{ab} = (\beta(z)\beta(y)^{-1}) \cdot (x - y) + z.$$

Proof The proof of Lemma 2.1 carries over without any changes—associativity of the ring has not been used there, only some elementary properties of inverses which are direct consequences of the left and right inverse properties (LIP) and (RIP).

From the lemma we get, as before, the formula

$$\beta\big((xyz)\big) = [\beta(z)\beta(y)^{-1}]\beta(x) = \big(\beta(z)\beta(y)\beta(x)\big) \tag{17}$$

which means that β induces a homomorphism from U_{ab} to the ternary Moufang loop $\mathbb{K}^\times$.

This formula is crucial in the proof of the alternative laws of U_{ab}: essentially, it implies that identities holding in $\mathbb{K}^\times$ will carry over to U_{ab}; but the unit loop of $\mathbb{K}$ is a ternary Moufang loop, and hence so will be U_{ab}. For instance, for the proof of (MT1), $(uv(xyx)) = ((uvx)yx)$, write both sides, using the lemma: one sees that equality holds iff, for the vector $w := u - v \in \mathbb{K}^2$ and for all $x, y, v \in \mathbb{K}^2 \setminus \ker(\beta)$, we have

$$(((\beta(x)\beta(y)^{-1})\beta(x))\beta(v)^{-1})w = (\beta(x)\beta(y)^{-1})((\beta(x)\beta(v)^{-1})w)$$

But this amounts to an identity in $\mathbb{K}$ (or, if one prefers, two identities, one for each component of w), of the same form as the one we want to prove; this identity holds since $\mathbb{K}$ is an alternative algebra.

4 Prospects

In subsequent work, we will investigate more thoroughly the geometry corresponding to alternative algebras and alternative pairs (cf. [8]): "alternative geometries" correspond to such algebras in a similar way as the associative geometries from [2] correspond to associative algebras and associative pairs. They play a key rôle in the construction of exceptional spaces corresponding to Jordan algebas and Jordan pairs. In the following, we briefly mention some topics to be discussed in this context.

4.1 Structure of the Torsors and Ternary Moufang Loops

First of all, it is easy to understand the structure of the groups U_{ab} in the Desarguesian case: for $a = b$, $U_{ab} = V_a$ is a vector group (this is true even in the Moufang case), and for $a \neq b$, U_{ab} is isomorphic to the *dilation* or *ax+b-group*

$$\mathrm{Dil}(E) := \{f : E \to E \mid f(x) = ax + b,\ b \in E, a \in \mathbb{K}^\times\} \tag{18}$$

of the affine space $E = a_b = a \setminus b$ (where $a \cap b$ is considered as hyperplane at infinity of a). This dilation group, in turn, is a semidirect product of $\mathbb{K}^\times$ with the translation group of E. The resulting homomorphism $U_{ab} \to \mathbb{K}^\times$ can be described in a purely geometric way (cf. [1], Theorem 7.4 for the case of very general Grassmannians). For Moufang planes, partial analogs of this hold: there is a split exact sequence of ternary Moufang loops

$$a_b \to U_{ab} \to \mathbb{K}^\times ,$$

where any line L in $\mathcal{X}$ which intersects $a \cup b$ in exactly two different points provides a splitting. But, if the plane is not Desarguesian, the set $\mathrm{Dil}(E)$ defined by (18) is then no longer a group, nor is it contained in the automorphism group of the plane.

However, it remains true in the Moufang case that one obtains a *symmetric plane* (defined in [7]; in the rough classification of symmetric planes by H. Löwe [6], our spaces appear among the *split symmetric planes*.)

4.2 Duality

Carrying out our geometric construction from Chap. "Poincaré Duality for Koszul Algebras" in the *dual projective space*, by general duality principles of projective geometry, we get again torsors, respectively ternary Moufang loops. Remarkably, the description of the torsors in the Desarguesian case by Eq. (4) does not change, except for a switch in a and b. In other words, up to this switch, the map Γ is "self-dual", which is in keeping with results on anti-automorphisms from [3]. For the moment, it is an open problem whether a similar "self-dual description" exists also in the Moufang case.

4.3 General Projective Planes

Our definition of $(xyz)_{ab}$ in the generic case (Definition 1.1) makes sense for any projective plane (and even for any lattice if we admit 0 as possible result). What, then, are its properties? In particular, what is its relation with the "ternary field" associated to a quadruple of points in the plane? Put differently, how do we have to modify the definition in the collinear case (Definition 1.3)? Does the "split exact sequence" $a_b \to U_{ab} \to \mathbb{K}^\times$ survive in some suitable algebraic category? Can one re-interprete the classical Lenz-Barlotti types of projective planes (cf., e.g., [10], p. 142) in terms of $(xyz)_{ab}$?

4.4 Perspective Drawing

Our construction also has aspects that should be interesting for applied sciences: as already pointed out, our two-dimensional drawings have a "spacial interpretation". This can be explained by observing that the torsors U_{AB} living in a three-dimensional space $\mathbb{KP}^3$ can be mapped homomorphically onto torsors U_{ab} living in a projective plane P (by choosing $P \subset \mathbb{KP}^3$ intersecting $A \wedge B$ in a single point and projecting from a point $q \in A \wedge B, q \notin P$, onto P; then let $a := P \wedge A$ and $b := P \wedge B$). A careful look shows that the torsor structure thus represented on P is quite often implicitly used in two-dimensional "perspective representations" of three-dimensional space; however, to our knowledge, the underlying algebraic structure has so far not yet been clearly recognized.

References

1. Bertram, W.: The projective geometry of a group, preprint 2012. http://arxiv.org/abs/1201.6201
2. Bertram, W., Kinyon, M.: Associative geometries. I: torsors, linear relations and Grassmannians. J. Lie Theory **20**(2), 215–252 (2010). arXiv:0903.5441 [math.RA]
3. Bertram, W., Kinyon, M.: Associative geometries. II: involutions, the classical torsors, and their homotopes. J. Lie Theory **20**(2), 253–282 (2010). arXiv:0909.4438
4. Bruck, R.H.: A Survey of Binary Systems. Springer, Berlin (1958)
5. Kiechle, H.: Theory of K-loops, Lecture Notes in Mathematics, vol. 1778. Springer, Berlin (2002)
6. Löwe, H.: A rough classification of symmetric planes. Adv. Geom. **1**, 1–21 (2001)
7. Löwen, R.: Symmetric planes. Pacific J. Math. **84**(2), 367–390 (1979)
8. Loos, O.: Jordan Pairs, Lecture Notes in Mathematics, vol. 460. Springer, New York (1975)
9. Robinson, D.A.: Bol loops. Trans. Amer. Math. Soc. **123**, 341–354 (1966)
10. Salzmann, H., Betten, D., Grundhöfer, T., Hähl, H., Löwen, R., Stroppel, M.: Compact Projective Planes. De Gruyter, Berlin (1995)

Graded q-Differential Polynomial Algebra of Connection Form

Viktor Abramov and Olga Liivapuu

Abstract Given a graded associative unital algebra we construct a graded q-differential algebra, where q is a primitive Nth root of unity and prove that the generalized cohomologies of the corresponding N-complex are trivial. We construct a graded q-differential algebra of polynomials and introduce a notion of connection form. We find explicit formula for the curvature of connection form and prove Bianchi identity.

1 Introduction

An idea to generalize the concept of a differential module and to elaborate the corresponding algebraic structures by giving the basic property of differential $d^2 = 0$ a more general form $d^N = 0$, $N \geq 2$ seems to be very natural. Taking the equation $d^N = 0$ as a starting point one should choose a space where a calculus with $d^N = 0$ will be constructed. As a calculus with $d^N = 0$ may be considered as a generalization of $d^2 = 0$ and taking into account that there is an exterior calculus of differential forms with exterior differential $d^2 = 0$ on a smooth manifold one way to construct $d^N = 0$ is to take a smooth manifold and to consider objects on this manifold more general than the differentials forms. Our approach is based on q-deformed structures such as graded q-Leibniz rule, graded q-commutator, graded inner q-derivation, where q is a primitive Nth root of unity [1–6].

A notion of graded q-differential algebra was introduced in [7] (see also in [8–10]) and it may be viewed as a generalization of a graded differential algebra. Let us

V. Abramov (✉)
Institute of Mathematics, University of Tartu, Liivi 2, 50409 Tartu, Estonia
e-mail: viktor.abramov@ut.ee

O. Liivapuu
Institute of Technology, Estonian University of Life Sciences, Kreutzwaldi 56, Tartu, Estonia
e-mail: olga.liivapuu@emu.ee

A. Makhlouf et al. (eds.), *Algebra, Geometry and Mathematical Physics*, Springer Proceedings in Mathematics & Statistics 85, DOI: 10.1007/978-3-642-55361-5_21,

mention that a concept of graded q-differential algebra is closely related to the monoidal structure introduced in [11] for the category of N-complexes and it is proved in [12] that the monoids of the category of N-complexes can be identified as the graded q-differential algebras. It is well known that a connection and its curvature are basic elements of the theory of fiber bundles and they play an important role not only in a modern differential geometry but also in theoretical physics namely in a gauge field theory. A basic algebraic structure used in the theory of connections on modules is a graded differential algebra. A graded differential algebra is an algebraic model for the de Rham algebra of differential forms on a smooth manifold. Consequently considering a concept of graded q-differential algebra which is more general structure than a graded differential algebra we can develop a generalization of the theory of connections on modules. One of the aims of this paper is to present and study algebraic structures based on the relation $d^N = 0$ and to generalize a concept of connection and its curvature applying a concept of graded q-differential algebra to the theory of connections on modules.

In Sect. 2 we prove Theorem 2.1 which is very useful in the sense that we can construct various cochain N-complexes by means of this theorem. Theorem 2.1 asserts if their exist an element v of grading one of a graded associative unital algebra $\mathscr{A}$ which satisfies $v^N \in \mathscr{Z}(\mathscr{A})$, where $\mathscr{Z}(\mathscr{A})$ is the graded center of $\mathscr{A}$, then the inner graded q-derivation ad_v^q is N-differential. Next we prove that the generalized cohomologies of cochain N-complex of Theorem 2.1 are trivial. In Sect. 3 we give the definition of a graded q-differential algebra. We introduce the algebra of polynomials and endow it with the structure of graded q-differential algebra. We introduce two operators D, ∇ and the polynomials f_k, which are defined with the help of recurrent relation. We prove the Theorem 3.2 which give explicit power expansion formulae for the operator D and the polynomials f_k.

2 N-Complexes and Cohomologies

A concept of cohomology of a differential module or of a cochain complex with coboundary operator d is based on the quadratic nilpotency condition $d^2 = 0$. It is obvious that one can construct a generalization of a concept of cohomology of a cochain complex if the quadratic nilpotency $d^2 = 0$ is replaced by a more general nilpotency condition $d^N = 0$, where N is an integer satisfying $N \geq 2$.

Let $\mathscr{A} = \oplus_{k \in \mathbb{Z}_N} \mathscr{A}^k = \mathscr{A}^0 \oplus \mathscr{A}^1 \oplus \ldots \oplus \mathscr{A}^{N-1}$ be a $\mathbb{Z}_N$-graded associative unital $\mathbb{C}$-algebra whose identity element is denoted by $\mathbb{1}$. The subspace $\mathscr{A}^0 \subset \mathscr{A}$ of elements of grading zero is the subalgebra of an algebra $\mathscr{A}$. Since this subalgebra plays an important role in several structures related to a graded algebra $\mathscr{A}$ we will denote it by $\mathfrak{A}$, i.e. $\mathfrak{A} \equiv \mathscr{A}^0$. It is easy to see that each subspace $\mathscr{A}^i \subset \mathscr{A}$ of homogeneous elements of grading i is the $\mathfrak{A}$-bimodule. Hence in the case of a graded algebra $\mathscr{A}$ we have the set of $\mathfrak{A}$-bimodules $\mathscr{A}^0, \mathscr{A}^1, \mathscr{A}^2, \ldots, \mathscr{A}^{N-1}$. The graded subspace $\mathscr{Z}(\mathscr{A}) \subset \mathscr{A}$ generated by homogeneous elements $u \in \mathscr{A}^k$, which for any $v \in \mathscr{A}^l$ satisfy $uv = (-1)^{kl} vu$, is called a *graded center* of a graded algebra $\mathscr{A}$.

The derivation of degree m induced by an element $v \in \mathscr{A}^m$ will be denoted by

$$\mathrm{ad}_v(u) = [v, u] = v\,u - (-1)^{ml} uv, \tag{1}$$

where $u \in \mathscr{A}^l$. The graded derivation ad_v is called an *inner graded derivation* of an algebra $\mathscr{A}$.

The notions of graded commutator and graded derivation of a graded algebra can be generalized within the framework of noncommutative geometry and the theory of quantum groups with the help of q-deformations. Let q be a primitive Nth root of unity. The *graded q-commutator* $[\,,\,]_q : \mathscr{A}^k \otimes \mathscr{A}^l \to \mathscr{A}^{k+l}$ is defined by

$$[u, v]_q = uv - q^{kl} vu. \tag{2}$$

A graded q-derivation of degree m of a graded algebra $\mathscr{A}$ is a linear mapping $\delta : \mathscr{A} \to \mathscr{A}$ of degree m with respect to graded structure of $\mathscr{A}$, i.e. $\delta : \mathscr{A}^i \to \mathscr{A}^{i+m}$, which satisfies the graded q-Leibniz rule

$$\delta(uv) = \delta(u)\, v + q^{ml} u\, \delta(v), \tag{3}$$

where u is a homogeneous element of grading l, i.e. $u \in \mathscr{A}^l$. In analogy with an inner graded derivation one defines an inner graded q-derivation of degree m of a graded algebra $\mathscr{A}$ associated to an element $v \in \mathscr{A}^m$ by the formula

$$\mathrm{ad}_v^q(u) = [v, u]_q = vu - q^{ml} uv, \tag{4}$$

where $u \in \mathscr{A}^l$.

A left K-module E is said to be an *N-differential module* if it is equipped with an endomorphism $d : E \to E$ which satisfies $d^N = 0$. An N-differential module E with N-differential d is said to be a *cochain N-complex of modules* or simply *N-complex* if E is a graded module $E = \oplus_{k\in\mathbb{Z}} E^k$ and its N-differential d has degree 1 with respect to a graded structure of E, i.e. $d : E^k \to E^{k+1}$.

We prove the following theorem which can be used to construct a cochain N-complex for a certain class of graded associative unital algebras (see also [8], p. 394).

Theorem 2.1 *Let $\mathscr{A} = \oplus_{k\in\mathbb{Z}_N} \mathscr{A}^k$ be a graded associative unital algebra and q be a primitive Nth root of unity. If there exists an element $v \in \mathscr{A}^1$ of grading one which satisfies the condition $v^N \in \mathscr{Z}(\mathscr{A})$ then the inner graded q-derivation $d = \mathrm{ad}_v^q$ of degree 1 is an N-differential and the sequence*

$$\mathscr{A}^0 \xrightarrow{d} \mathscr{A}^1 \xrightarrow{d} \mathscr{A}^2 \xrightarrow{d} \cdots \xrightarrow{d} \mathscr{A}^{N-1} \tag{5}$$

is the cochain N-complex.

Proof We begin the proof with a power expansion of d^k, where $1 \le k \le N$. Let u be a homogeneous element of an algebra $\mathscr{A}$ whose grading will be denoted by $|u|$. For the first values of $k = 1, 2, 3$ a straightforward computation gives

$$\begin{aligned}
du &= [v, u]_q = vu - q^{|u|}uv, \\
d^2u &= [v, [v, u]_q]_q = v^2u - q^{|u|}[2]_q vuv + q^{2|u|+1}uv^2, \\
d^3u &= v^3u - q^{|u|}[3]_q v^2uv + q^{2|u|+1}[3]_q vuv^2 - q^{3|u|+3}uv^3.
\end{aligned}$$

We state that for any $k \in \{1, 2, \dots, N\}$ and any homogeneous $u \in \mathscr{A}$ a power expansion of d^k has the form

$$d^k u = \sum_{i=0}^{k} (-1)^i p_i \begin{bmatrix} k \\ i \end{bmatrix}_q v^{k-i}\, uv^i, \tag{6}$$

where $p_i = q^{i|u|+\sigma(i)}$ and $\sigma(i) = \frac{i(i-1)}{2}$. We proof this statement by means of mathematical induction assuming that the above power expansion (6) for d^k is true and then showing that it has the same form for $k+1$. Indeed we have

$$\begin{aligned}
d^{k+1}u &= d(d^k u) = d\Big(\sum_{i=0}^{k} (-1)^i p_i \begin{bmatrix} k \\ i \end{bmatrix}_q v^{k-i}uv^i\Big) \\
&= \sum_{i=0}^{k} (-1)^i p_i \begin{bmatrix} k \\ i \end{bmatrix}_q (v^{k+1-i}uv^i - q^{|u|+k}v^{k-i}uv^{i+1}) \\
&= \sum_{i=0}^{k} (-1)^i p_i \begin{bmatrix} k \\ i \end{bmatrix}_q v^{k+1-i}uv^i - \sum_{i=0}^{k} (-1)^i q^{|u|+k} p_i \begin{bmatrix} k \\ i \end{bmatrix}_q v^{k-i}uv^{i+1} \\
&= v^{k+1}u + \sum_{i=1}^{k} (-1)^i p_i \begin{bmatrix} k \\ i \end{bmatrix}_q v^{k+1-i}uv^i \\
&\quad - \sum_{i=0}^{k-1} (-1)^i q^{|u|+k} p_i \begin{bmatrix} k \\ i \end{bmatrix}_q v^{k-i}uv^{i+1} - (-1)^k q^{|u|+k} p_k\, uv^{k+1} \\
&= v^{k+1}u + \sum_{i=1}^{k} (-1)^i p_i \begin{bmatrix} k \\ i \end{bmatrix}_q v^{k+1-i}uv^i \\
&\quad + \sum_{i=1}^{k} (-1)^i q^{|u|+k} p_{i-1} \begin{bmatrix} k \\ i-1 \end{bmatrix}_q v^{k-i}uv^{i+1} + (-1)^{k+1} q^{|u|+k} p_k\, uv^{k+1} \\
&= v^{k+1}u + \sum_{i=1}^{k} (-1)^i \Big(p_i \begin{bmatrix} k \\ i \end{bmatrix}_q + q^{|u|+k} p_{i-1} \begin{bmatrix} k \\ i-1 \end{bmatrix}_q\Big) v^{k+1-i}uv^i \\
&\quad + (-1)^{k+1} q^{|u|+k} p_k\, uv^{k+1}.
\end{aligned}$$

Now the coefficients in the last expansion we can write as follows

$$p_i \begin{bmatrix} k \\ i \end{bmatrix}_q + q^{|u|+k} p_{i-1} \begin{bmatrix} k \\ i-1 \end{bmatrix}_q = p_i \Big(\begin{bmatrix} k \\ i \end{bmatrix}_q + q^{k+\sigma(i-1)-\sigma(i)} \begin{bmatrix} k \\ i-1 \end{bmatrix}_q \Big),$$

and making use of

$$\sigma(i-1) - \sigma(i) = \frac{(i-1)(i-2)}{2} - \frac{i(i-1)}{2} = 1 - i$$

and making use of well known recurrent relation for q-binomial coefficients we get

$$\begin{bmatrix} k \\ i \end{bmatrix}_q + q^{k+1-i} \begin{bmatrix} k \\ i-1 \end{bmatrix}_q = \begin{bmatrix} k+1 \\ i \end{bmatrix}_q.$$

As $p_{k+1} = q^{|u|+k} p_k$ we finally obtain

$$d^{k+1} u = \sum_{i=0}^{k+1} (-1)^i p_i \begin{bmatrix} k+1 \\ i \end{bmatrix}_q v^{k-i} u v^i,$$

and this ends the proof of the formula for power expansion of d^k.

Now our aim is to show that the power expansion (6) implies $d^N u = 0$ for any $u \in \mathscr{A}$. Indeed making use of (6) we can express the Nth power of d as follows

$$d^N u = \sum_{i=0}^{N} (-1)^i p_i \begin{bmatrix} N \\ i \end{bmatrix}_q v^{k-i} u v^i. \tag{7}$$

Taking into account that q is a primitive Nth root of unity we get

$$\begin{bmatrix} N \\ i \end{bmatrix}_q = 0, \quad i \in \{1, 2, \ldots, N-1\}.$$

Hence the terms in (7), which are numbered with $i = 1, 2, \ldots, N-1$, vanish, and we are left with two terms

$$d^N u = v^N u + (-1)^N q^{\sigma(N)} u v^N.$$

As v^N is the element of grading zero (modulo N) of the graded center $\mathscr{Z}(\mathscr{A})$ we can rewrite the above formula as follows

$$d^N u = (1 + (-1)^N q^{\sigma(N)})\, u v^N, \quad \sigma(N) = \frac{N(N-1)}{2}.$$

In order to show that the multiplier in the above formula vanish for any $N \geq 2$ we consider separately two cases for N to be an odd or even positive integer. If N is an odd positive integer then the multiplier $1 + (-1)^N q^{\sigma(N)}$ vanish because in this case

$$1 + (-1)^N q^{\sigma(N)} = 1 - (q^N)^{\frac{N-1}{2}} = 0.$$

If N is an even positive integer then

$$1 + (-1)^N q^{\sigma(N)} = 1 + (q^{\frac{N}{2}})^{N-1} = 1 + (-1)^{N-1} = 0.$$

Hence for any $N \geq 2$ we have $d^N = 0$, and this ends the proof of the theorem. □

Let us fix a positive integer $m \in \{1, 2, \ldots, N-1\}$ and split up the Nth power of N-differential d as follows $d^N = d^m \circ d^{N-m}$. Then the nilpotency condition for N-differential can be written in the form $d^N = d^m \circ d^{N-m} = 0$ and this leads to possible generalization of a concept of cohomology. For each integer $1 \leq m \leq N-1$ one can define the submodules

$$Z_m(E) = \{x \in E : d^m x = 0\} \subset E, \tag{8}$$

$$B_m(E) = \{x \in E : \exists y \in E,\ x = d^{N-m} y\} \subset E. \tag{9}$$

From $d^N = 0$ it follows that $B_m(E) \subset Z_m(E)$. For each $m \in \{1, 2, \ldots, N-1\}$ the quotient module $H_m(E) := Z_m(E)/B_m(E)$ is said to be a *generalized homology of order m* of N-differential module E. The following lemma which is proved in [8] gives a very useful criteria for the triviality of the generalized cohomologies of an N-differential module.

Lemma 2.1 *Let E be an N-differential module over a ring* $\mathbf{k}$, *$N \geq 2$ be an integer and q be an element of* $\mathbf{k}$ *satisfying the conditions $[N]_q = 0$ and $[n]_q$ is invertible for any integer $1 \leq n \leq N-1$. If there is a module-endomorphism $h : E \to E$ satisfying $h \circ d - q\, d \circ h = Id_E$ then the generalized cohomologies of an N-differential module E are trivial, i.e. for any integer $1 \leq n \leq N-1$ it holds $H_n(E) = 0$.*

Based on this lemma we can prove that the generalized cohomologies of the cochain N-complex described in Theorem 2.1 are trivial. It is worth mentioning that the same argument is used in [10] to show that the generalized cohomologies of the N-differential module constructed by means of the algebra of $N \times N$-matrices $M_N(\mathbf{k})$ are trivial.

Theorem 2.2 *Let q be a primitive Nth root of unity, $\mathscr{A} = \oplus_{i \in \mathbb{Z}_N} \mathscr{A}^i$ be a graded associative unital algebra with an element $v \in \mathscr{A}^1$ satisfying $v^N = \lambda \mathbb{1}$, where $\lambda \neq 0$. Then the generalized cohomologies $H_n(\mathscr{A})$ of the cochain N-complex of Theorem 2.1*

$$\mathscr{A}^0 \xrightarrow{d} \mathscr{A}^1 \xrightarrow{d} \mathscr{A}^2 \xrightarrow{d} \cdots \xrightarrow{d} \mathscr{A}^{N-1} \tag{10}$$

with N-differential $d = \mathrm{ad}_v^q$, induced by an element v, are trivial, i.e. for any $n \in \{1, 2, \ldots, N-1\}$ we have $H_n(\mathscr{A}) = 0$.

Proof Let us define the endomorphism h of the vector space of $\mathscr{A}$ as follows

$$h(u)=\frac{1}{(1-q)\,\lambda}\,v^{N-1}\,u,$$

where u is an element of an algebra $\mathscr{A}$. If u is a homogeneous element of a graded algebra $\mathscr{A}$ then $|h(u)|=|u|+N-1$, where $|u|$ is the grading of an element u. For any homogeneous $u\in\mathscr{A}$ we have

$$\begin{aligned}(h\circ d-q\,d\circ h)(u)&=h(du)-q\,d(h(u))\\&=h(\mathrm{ad}_v^q(u))-\frac{q}{(1-q)\,\lambda}\,\mathrm{ad}_v^q(v^{N-1}u)\\&=h([v,u]_q)-\frac{q}{(1-q)\,\lambda}\,[v,v^{N-1}]_q\\&=h\,(v\,u-q^{|u|}uv)-\frac{q}{(1-q)\,\lambda}\,(v^N u-q^{|u|+N-1}v^{N-1}u\,v)\\&=\frac{1}{(1-q)\,\lambda}\,v^N u-\frac{q^{|u|}}{(1-q)\,\lambda}v^{N-1}u\,v\\&\quad-\frac{q}{(1-q)\,\lambda}\,v^N u+\frac{q^{|u|}}{(1-q)\,\lambda}v^{N-1}u\,v\\&=\frac{(1-q)\,\lambda}{(1-q)\,\lambda}\,u=\mathrm{Id}_{\mathscr{A}}(u).\end{aligned}$$

The endomorphism $h:\mathscr{A}\to\mathscr{A}$ of the vector space of an algebra $\mathscr{A}$ satisfies $h\circ d-q\,d\circ h=\mathrm{Id}_{\mathscr{A}}$ and it follows from Lemma 2.1 that the generalized cohomology of the cochain N-complex

$$\mathscr{A}^0\xrightarrow{d}\mathscr{A}^1\xrightarrow{d}\mathscr{A}^2\xrightarrow{d}\ldots\xrightarrow{d}\mathscr{A}^{N-1}$$

are trivial. □

3 Graded q-Differential Algebras

In this section we use the cochain N-complex described in the Theorem 2.1 to construct a graded q-differential algebra which can be viewed as a natural generalization of the notion of graded differential. Then we will describe a graded q-differential polynomial algebra which arises in relation with a connection form which can be viewed as analog of connection form in a graded differential algebra introduced by Quillen in [13].

Definition 3.1 *A graded q-differential algebra* is a graded associative unital algebra $\mathscr{A}=\oplus_{k\in\mathbb{Z}}\mathscr{A}^k$ endowed with a linear mapping d of degree one such that the sequence

$$\cdots \xrightarrow{d} \mathscr{A}^{k-1} \xrightarrow{d} \mathscr{A}^{k} \xrightarrow{d} \mathscr{A}^{k+1} \xrightarrow{d} \cdots$$

is an N-complex with N-differential d satisfying the graded q-Leibniz rule

$$d(u\,v) = d(u)v + q^k u d(v), \tag{11}$$

where $u \in \mathscr{A}^k, v \in \mathscr{A}$.

It follows from Theorem 2.1

Theorem 3.1 *Let $\mathscr{A}$ be a graded associative unital algebra $\mathscr{A} = \oplus_k \mathscr{A}^k$, and q be a primitive Nth root of unity. If there exists an element of grading one $v \in \mathscr{A}^1$ which satisfies the condition $v^N \in \mathscr{Z}(\mathscr{A})$, where $\mathscr{Z}(\mathscr{A})$ is the graded center of $\mathscr{A}$, then the graded algebra $\mathscr{A}$ endowed with the inner graded q-derivation $d = \mathrm{ad}_v^q$ is a graded q-differential algebra (d is its N-differential).*

Indeed we can prove this theorem by taking into account that an inner graded q-derivation satisfies the graded q-Leibniz rule (3) and the inner graded q-derivation $d = \mathrm{ad}_v^q$, induced by an element of grading one $v \in \mathscr{A}^1$ such that $v^N \in \mathscr{Z}(\mathscr{A})$, is the N-differential of the cochain complex (Theorem 2.1)

$$\cdots \xrightarrow{d} \mathscr{A}^{k-1} \xrightarrow{d} \mathscr{A}^{k} \xrightarrow{d} \mathscr{A}^{k+1} \xrightarrow{d} \cdots.$$

Now we introduce a graded q-differential algebra of polynomials which arises in relation with an algebraic model of a connection form and this algebraic model is based on exterior calculus with differential satisfying $d^N = 0$. This algebra will be used in the next section in order to calculate the curvature of a connection form.

Let $\mathbb{N}_1 = \{i \in \mathbb{Z} : i \geq 1\}$ be the set of integers greater than or equal to one and $\{\mathfrak{d}, a_i\}_{i\in\mathbb{N}_1}$ be a set of variables. We consider the algebra of noncommutative polynomials $\mathfrak{P}_q[\mathfrak{d}, a]$ over $\mathbb{C}$ generated by the variables $\{\mathfrak{d}, a_i\}_{i\in\mathbb{N}_1}$ which are subjected to the commutation relations

$$\mathfrak{d} a_i = q^i\, a_i \mathfrak{d} + a_{i+1}, \qquad \forall i \in \mathbb{N}_1 \tag{12}$$

where q is any complex number different from zero. We denote the identity element of this algebra by $\mathbb{1}$. Obviously we can split up the set of variables of the algebra $\mathfrak{P}_q[\mathfrak{d}, a]$ into two subsets $\{\mathfrak{d}\}$, $\{a_i\}_{i\in\mathbb{N}_1}$ which generate respectively the subalgebras $\mathfrak{P}_q[\mathfrak{d}] \subset \mathfrak{P}_q[\mathfrak{d}, a]$ and $\mathfrak{P}_q[a] \subset \mathfrak{P}_q[\mathfrak{d}, a]$. Hence the subalgebra $\mathfrak{P}_q[\mathfrak{d}]$ is generated by a single variable $\mathfrak{d}$, and the subalgebra $\mathfrak{P}_q[a]$ is freely generated by the variables $\{a_i\}_{i\in\mathbb{N}_1}$ because we do not assume any relation between variables a_i.

Now our aim is to equip the algebra of polynomials $\mathfrak{P}_q[\mathfrak{d}, a]$ with a graded structure so that $\mathfrak{P}_q[\mathfrak{d}, a]$ will become a graded algebra. This can be done as follows: we assign grading zero to the identity element $\mathbb{1}$ of the algebra $\mathfrak{P}_q[\mathfrak{d}, a]$, grading one to the generator $\mathfrak{d}$ and grading i to a generator a_i, where $i \in \mathbb{N}_1$. Thus making use of previously defined notations we can describe the graded structure of generators of $\mathfrak{P}_q[\mathfrak{d}, a]$ by the formulae

$$|\mathbb{1}| = 0, \quad |\mathfrak{d}| = |a_1| = 1, \quad |a_i| = i, \quad i \geq 2. \tag{13}$$

As usual we extend this graded structure to the whole algebra $\mathfrak{P}_q[\mathfrak{d}, a]$ by defining the grading of any product of variables $\{\mathfrak{d}, a_i\}_{i\in\mathbb{N}_1}$ as the sum of gradings of its factors. It is easy to see that the algebra of polynomials $\mathfrak{P}_q[\mathfrak{d}, a]$ becomes the positively graded algebra. Hence we can write

$$\mathfrak{P}_q[\mathfrak{d}, a] = \oplus_{k\in\mathbb{N}}\mathfrak{P}_q^k[\mathfrak{d}, a],$$

where $\mathfrak{P}_q^k[\mathfrak{d}, a]$ is the subspace of homogeneous polynomials of grading k. It should be mentioned that the graded structure of $\mathfrak{P}[\mathfrak{d}, a]$ induces the graded structures of the subalgebras $\mathfrak{P}_q[\mathfrak{d}]$, $\mathfrak{P}_q[a]$ which are positively graded algebras as well. Clearly the positively graded algebra $\mathfrak{P}_q[\mathfrak{d}, a]$ becomes the $\mathbb{Z}_N$-graded algebra, where N any integer greater than 1, if we slightly modify the above described gradation by taking all gradings modulo N. Let us denote by $\mathrm{Lin}\,\mathfrak{P}_q[a]$ the algebra of $\mathbb{C}$-endomorphisms of vector space of $\mathfrak{P}_q[a]$. Obviously $\mathrm{Lin}\,\mathfrak{P}_q[a]$ is a graded algebra with gradation induced by the gradation of $\mathfrak{P}_q[a]$. Having defined the positively graded structure of the algebra $\mathfrak{P}_q[\mathfrak{d}, a]$ we can apply the notions of graded commutator and inner graded q-derivation described in the previous chapter to study the structure of $\mathfrak{P}_q[\mathfrak{d}, a]$. First of all we observe that the commutation relations (12) can be written by means of graded commutator and inner graded q-derivation in the form

$$[\mathfrak{d}, a_i]_q = a_{i+1}, \qquad \text{or} \qquad \mathrm{ad}^q_{\mathfrak{d}}(a_i) = a_{i+1}, \tag{14}$$

where $i \in \mathbb{N}_1$. This form of commutation relations suggests us to consider the inner graded q-derivation $\mathrm{ad}^q_{\mathfrak{d}}$ of the algebra $\mathfrak{P}_q[\mathfrak{d}, a]$ associated with a variable $\mathfrak{d}$. If we restrict $\mathrm{ad}^q_{\mathfrak{d}}$ to the subalgebra $\mathfrak{P}_q[a]$ we get the graded q-derivation of subalgebra $\mathfrak{P}_q[a]$ which we will denote by d, i.e.

$$d := \mathrm{ad}^q_{\mathfrak{d}}\,|_{\mathfrak{P}_q[a]}, \qquad d : \mathfrak{P}_q[a] \to \mathfrak{P}_q[a]. \tag{15}$$

Obviously d is a graded q-derivation of grading one of the $\mathbb{Z}_N$-graded algebra $\mathfrak{P}_q[a]$. From the commutation relations (14) it follows that

$$d(\mathbb{1}) = 0, \quad d(a_i) = a_{i+1},$$

for any $i \geq 1$. Let us define $D, \nabla \in \mathrm{Lin}\,\mathfrak{P}_q[a]$ of grading one and the polynomials $f_k \in \mathfrak{P}_q[a]$, where k is an integer greater than or equal to zero, by the formulae

$$D(P) = d(P) + a_1\,P, \tag{16}$$
$$\nabla(P) = d(P) + [a_1, P]_q, \tag{17}$$
$$f_0 = \mathbb{1},$$
$$f_1 = a_1,$$

$$f_k = D(f_{k-1}), \tag{18}$$

where $P \in \mathfrak{P}_q[a]$ is a homogeneous polynomial. We can write the linear mapping ∇ in the form $\nabla = \mathrm{ad}^q_{\mathfrak{d}+a_1}$ which clearly shows that ∇ is a inner graded q-derivation of the algebra $\mathfrak{P}_q[a]$. Hence for any polynomials $P, Q \in \mathfrak{P}_q[a]$, where P is homogeneous, it holds

$$D(PQ) = D(P)\,Q + q^{|P|}P\,d(Q), \tag{19}$$

$$\nabla(PQ) = \nabla(P)\; + q^{|P|}P\,\nabla(Q). \tag{20}$$

For the first values of k we calculate by means of the recurrent relation (18)

$$\begin{aligned} f_2 &= a_2 + a_1^2, \\ f_3 &= a_3 + a_2\,a_1 + [2]_q\,a_1\,a_2 + a_1^3, \\ f_4 &= a_4 + a_3\,a_1 + [3]_q\,a_1a_3 + [3]_q\,a_2^2 \\ &\quad + a_2a_1^2 + [3]_q\,a_1^2a_2 + [2]_q\,a_1a_2a_1 + a_1^4, \\ f_5 &= a_5 + a_4a_1 + [4]_q\,a_1a_4 + [4]_q\,a_3a_2 \\ &\quad + \begin{bmatrix} 4 \\ 2 \end{bmatrix}_q a_2a_3 + a_3a_1^2 + [3]_q\,a_2^2a_1 + [4]_q\,a_2a_1a_2 \\ &\quad + [2]_q\,[4]_q\,a_1a_2^2 + \begin{bmatrix} 4 \\ 2 \end{bmatrix}_q a_1^2a_3 + [3]_q\,a_1a_3a_1 \\ &\quad + [2]_q\,a_1a_2a_1^2 + [3]_q\,a_1^2a_2a_1 + a_2a_1^3 + [4]_q\,a_1^3a_2 + a_1^5. \end{aligned} \tag{21}$$

Getting a bit ahead we would like to point out that the polynomials f_k may be interpreted as the curvature of a connection if we view the generator a_1 as an algebraic model for a connection one form. Let us remind that if k is a positive integer then a composition of k is a representation of k as the sum of a sequence of strictly positive integers, and two sequences that differ in the order of their terms give different compositions of their sum while they define the same partition of k. For example if $k = 3$ then there are 4 compositions

$$3 = 3,\;\; 3 = 2 + 1,\;\; 3 = 1 + 2,\;\; 3 = 1 + 1 + 1.$$

Let Ψ_k be the set of all compositions of an integer k. We will write a composition of an integer k in the form of a sequence of strictly positive integers $\sigma = (i_1, i_2, \ldots, i_r)$, where $i_1 + i_2 + \cdots + i_r = k$. Let us denote

$$\begin{aligned} k_1 &= i_1, \\ k_2 &= i_1 + i_2, \\ k_3 &= i_1 + i_2 + i_3, \\ &\;\;\ldots \end{aligned}$$

$$k_{r-1} = i_1 + i_2 + \cdots + i_{r-1}.$$

It can be proved [14] that the number of all possible compositions of a positive integer k is 2^{k-1}, i.e. the set Ψ_k contains 2^{k-1} elements. The following theorem gives an explicit formula for the polynomials f_k:

Theorem 3.2 *For any integer $k \geq 2$ we have the following expansion of power of the operator D and the expansion of a polynomial f_k in terms of generators a_i:*

$$D^k = \sum_{i=0}^{k} \begin{bmatrix} k \\ i \end{bmatrix}_q f_i\, d^{k-i},$$

$$f_k = \sum_{\sigma \in \Psi_k} \begin{bmatrix} k_2 - 1 \\ k_1 \end{bmatrix}_q \begin{bmatrix} k_3 - 1 \\ k_2 \end{bmatrix}_q \cdots \begin{bmatrix} k - 1 \\ k_{r-1} \end{bmatrix}_q a_{i_1} a_{i_2} \ldots a_{i_r},$$

where $\sigma = (i_1, i_2, \ldots, i_r)$ is a composition of an integer k.

Proof We will prove the expansion formulae of this theorem by the method of mathematical induction. In order to prove the expansion of power of the operator D by means of mathematical induction we begin with the base case and show that this formula holds when k is equal to 1. This is true because

$$D = \begin{bmatrix} 1 \\ 0 \end{bmatrix}_q f_0\, d + \begin{bmatrix} 1 \\ 1 \end{bmatrix}_q f_1 = d + a_1.$$

Next step in the proof is an inductive step, i.e. we assume that the expansion formula holds for some integer $k > 1$ and show that it also holds when $k + 1$ is substituted for k. Indeed we have

$$\begin{aligned}
D^{k+1} &= D(D^k) = D\Big(\sum_{i=0}^{k} \begin{bmatrix} k \\ i \end{bmatrix}_q f_i\, d^{k-i}\Big) \\
&= \sum_{i=0}^{k} \begin{bmatrix} k \\ i \end{bmatrix}_q \Big(D(f_i)\, d^{k-i} + q^i f_i d^{k+1-i}\Big) \\
&= \sum_{i=0}^{k} \begin{bmatrix} k \\ i \end{bmatrix}_q \Big(f_{i+1}\, d^{k-i} + q^i f_i d^{k+1-i}\Big) \\
&= f_{k+1} + \sum_{i=0}^{k-1} \begin{bmatrix} k \\ i \end{bmatrix}_q f_{i+1}\, d^{k-i} + q^i \sum_{i=1}^{k} \begin{bmatrix} k \\ i \end{bmatrix}_q f_i\, d^{k+1-i} + d^{k+1} \\
&= f_{k+1} + \sum_{i=1}^{k} \begin{bmatrix} k \\ i-1 \end{bmatrix}_q f_i\, d^{k+1-i} + q^i \sum_{i=1}^{k} \begin{bmatrix} k \\ i \end{bmatrix}_q f_i\, d^{k+1-i} + d^{k+1}
\end{aligned}$$

$$
\begin{aligned}
&= f_{k+1} + \sum_{i=1}^{k} \left(\begin{bmatrix} k \\ i-1 \end{bmatrix}_q + q^i \begin{bmatrix} k \\ i \end{bmatrix}_q \right) f_i \, d^{k+1-i} + d^{k+1} \\
&= f_{k+1} + \sum_{i=1}^{k} \begin{bmatrix} k+1 \\ i \end{bmatrix}_q f_i \, d^{k+1-i} + d^{k+1} \\
&= \sum_{i=0}^{k+1} \begin{bmatrix} k+1 \\ i \end{bmatrix}_q f_i \, d^{k+1-i}.
\end{aligned}
$$

Thus the expansion of power of the operator D is proved. Now if we apply the both sides of the proved formula to a_1 we obtain

$$
f_{k+1} = \sum_{i=0}^{k} \begin{bmatrix} k \\ i \end{bmatrix}_q f_i \, a_{k+1-i}, \tag{22}
$$

and this is the recurrent formula for the polynomials f_k which we will use in the second part of the present proof in order to prove the expansion formula for f_k.

We start the proof of the expansion formula for a polynomial f_k with the base case when $k = 2$. In this case there are two compositions $2 = 2,\ 2 = 1 + 1$. Hence we have

$$
f_2 = \begin{bmatrix} 1 \\ 0 \end{bmatrix}_q a_2 + \begin{bmatrix} 1 \\ 0 \end{bmatrix}_q \begin{bmatrix} 1 \\ 1 \end{bmatrix}_q a_1^2 = a_2 + a_1^2.
$$

Comparing this result with the first formula in (21) we see that in the case when $k = 2$ the expansion formula for f_k is correct. The next step is an inductive step, i.e. we assume that the expansion formula holds for some positive integer $k > 2$ and show that it also holds when $k + 1$ is substituted for k. Let us consider the sum

$$
\sum_{\sigma \in \Psi_{k+1}} \begin{bmatrix} k_2 - 1 \\ k_1 \end{bmatrix}_q \begin{bmatrix} k_3 - 1 \\ k_2 \end{bmatrix}_q \cdots \begin{bmatrix} k \\ k_r \end{bmatrix}_q a_{i_1} a_{i_2} \ldots a_{i_{r+1}}, \tag{23}
$$

where $\sigma = (i_1, i_2, \ldots, i_r, i_{r+1})$ is a composition of an integer $k+1$. Hence $i_1 + \cdots + i_r + i_{r+1} = k+1$. Our aim is to show that this sum is equal to the polynomial f_{k+1}. Let us fix an integer $i \in \{0, 1, \ldots, k\}$ and a generator a_{k+1-i}. It is clear that if we select the compositions of an integer $k+1$ which have the form $(i_1, i_2, \ldots, i_r, k+1-i)$, i.e. the last integer of each composition is previously fixed integer $k + 1 - i$, and we remove in each composition the last integer then the set of compositions $(i_1, i_2, \ldots, i_r)$ is the set of all compositions of an integer i, i.e. $\{(i_1, i_2, \ldots, i_r)\} = \Psi_i$. Indeed we have

$$
i_1 + i_2 + \cdots + i_r + k + 1 - i = k + 1,
$$

which implies $i_1 + i_2 + \cdots + i_r = i$. Consequently if we select in the sum (23) all terms with $i_{r+1} = k+1-i$ (i.e. containing a generator a_{k+1-i} at the end of a product of generators) then we get the sum

$$\sum_{\sigma\in\Psi_{k+1}} \begin{bmatrix} k_2-1 \\ k_1 \end{bmatrix}_q \begin{bmatrix} k_3-1 \\ k_2 \end{bmatrix}_q \cdots \begin{bmatrix} k \\ i \end{bmatrix}_q a_{i_1}a_{i_2}\ldots a_{i_r}a_{k+1-i}, \tag{24}$$

where the sum is taken over the compositions of integer $k+1$ which have the form $\sigma=(i_1,i_2,\ldots,i_r,k+1-i)\in\Psi_{k+1}$. We would like to point out that the product of binomial coefficients of each term in this sum contains the factor

$$\begin{bmatrix} k \\ i \end{bmatrix}_q .$$

Hence we can write the sum (24) as follows

$$\begin{bmatrix} k \\ i \end{bmatrix}_q \Big(\sum_{\tau\in\Psi_i} \begin{bmatrix} k_2-1 \\ k_1 \end{bmatrix}_q \begin{bmatrix} k_3-1 \\ k_2 \end{bmatrix}_q \cdots \begin{bmatrix} i-1 \\ k_{r-1} \end{bmatrix}_q a_{i_1}a_{i_2}\ldots a_{i_r}\Big)\, a_{k+1-i},$$

where $\tau=(i_1,i_2,\ldots,i_r)\in\Psi_i$ and the sum is taken over all compositions of integer i. Now we make use of the assumption of an inductive step that the expansion formula for a polynomial f_m holds for each integer $m\in\{1,2,\ldots,k\}$. Hence the sum in the previous formula is equal to f_i, i.e

$$\sum_{\tau\in\Psi_i} \begin{bmatrix} k_2-1 \\ k_1 \end{bmatrix}_q \begin{bmatrix} k_3-1 \\ k_2 \end{bmatrix}_q \cdots \begin{bmatrix} i-1 \\ k_{r-1} \end{bmatrix}_q a_{i_1}a_{i_2}\ldots a_{i_r} = f_i .$$

Thus the sum (24) is equal to

$$\begin{bmatrix} k \\ i \end{bmatrix}_q f_i\, a_{k+1-i},$$

and summing up all these terms with respect to i we get the sum (23). Consequently the sum (23) we started with is equal to the sum

$$\sum_{i=0}^{k} \begin{bmatrix} k \\ i \end{bmatrix}_q f_i\, a_{k+1-i},$$

which in turn is equal to f_{k+1} (see the recurrent relation (22)). This ends the proof. □

We remind a reader that the parameter q which plays an important role in the structure of the algebra $\mathfrak{P}_q[\mathfrak{d},a]$ is any complex number different from zero. Now we will study the structure of the algebra of polynomials $\mathfrak{P}_q[\mathfrak{d},a]$ at a primitive Nth root of unity, i.e. we assume q to be a primitive Nth root of unity. We may expect that in this case the infinite set of variables $\{\mathfrak{d},a_1,a_2,\ldots\}$ is "cut off" and we get an algebra whose vector space is finite dimensional. Indeed we can prove the following proposition:

Proposition 3.1 *Let $\mathfrak{P}_q[\eth, a]$ be the algebra of polynomials generated by the set of variables $\{\eth, a_i\}_{i\in\mathbb{N}_1}$ which obey the commutation relations (12). If we assume that q is a primitive Nth root of unity and the variable $\eth$ is subjected to the additional relation $\eth^N = \lambda \cdot \mathbb{1}$, where λ is a complex number, then for any integer $k > N$ a variable a_k vanishes, i.e. the algebra $\mathfrak{P}_q[\eth, a]$ is generated by the finite set of variables $\{\eth, a_k\}_{k=1}^N$ which obey the relations*

$$\begin{aligned} \eth a_1 &= q\, a_1 \eth + a_2, \\ \eth a_2 &= q^2\, a_2 \eth + a_3, \\ &\dots \\ \eth a_{N-1} &= q^{N-1}\, a_{N-1} \eth + a_N, \\ \eth a_N &= a_N \eth, \\ \eth^N &= \lambda \cdot \mathbb{1}. \end{aligned} \tag{25}$$

The graded q-derivation $d = \mathrm{ad}^q_{\eth} : \mathfrak{P}_q[a] \to \mathfrak{P}_q[a]$ associated to variable $\eth$ is an N-differential, i.e. $d^N = 0$, and the sequence

$$\cdots \xrightarrow{d} \mathfrak{P}^{i-1}_q[a] \xrightarrow{d} \mathfrak{P}^{i}_q[a] \xrightarrow{d} \mathfrak{P}^{i+1}_q[a] \xrightarrow{d} \cdots$$

is a cochain N-complex. The graded algebra $\mathfrak{P}_q[a]$ equipped with the N-differential d is a graded q-differential algebra.

Proof We suppose that the algebra of polynomials is equipped with the $\mathbb{Z}_N$-gradation as it was explained earlier (13). It easily follows from the commutation relations of the algebra $\mathfrak{P}_q[\eth, a]$ that for any integer $k \geq 2$ we have

$$a_{k+1} = d^k(a_1),$$

where $d = \mathrm{ad}^q_{\eth}$ is the graded q-derivation associated with a variable $\eth$. Making use of the expansion of power of graded q-derivation used in the proof of Theorem 2.1 we obtain

$$a_{k+1} = d^k(a_1) = (\mathrm{ad}^q_{\eth})^k(a_1) = \sum_{i=0}^{k} (-1)^i\, p_i \begin{bmatrix} k \\ i \end{bmatrix}_q \eth^{k-i}\, u\, \eth^i.$$

Consequently if q is a primitive Nth root of unity, $\eth$ satisfies $\eth^N = \lambda \cdot \mathbb{1}$ and $k = N$ then making use of the same arguments as in the proof of Theorem 2.1) we conclude that all terms of the sum at the right-hand side of the above expansion formula vanish. Consequently we have $a_{N+1} = a_{N+2} = \ldots = 0$ and this ends the proof. □

It is well known that locally a connection of a vector bundle can be described with the help of matrix-valued 1-form. From an algebraic point of view this matrix-valued 1-form is an element of degree one of differential algebra of matrix-valued differential forms, where differential is identified with exterior differential and graduation is induced by degree of differential form. Hence an algebraic model for a connection can be constructed if we take a differential algebra $\mathscr{A}$ (over $\mathbb{C}$) and consider an element of degree one of this algebra A calling it connection form. Then a covariant differential induced by this connection form is the operator $\nabla = d + A$, and the curvature is the element of degree 2 given by $F = dA + A^2 = dA + \frac{1}{2}[A, A]$, where $[,]$ is the graded commutator of $\mathscr{A}$. This approach was proposed by Quillen in [13]. Following this approach we introduce a notion of N-connection form which particularly gives a connection form if $N = 2$. Let us denote by $\mathfrak{P}_q[\mathfrak{d}, a]$ the finite dimensional graded algebra generated by $\{\mathfrak{d}, a_k\}_{k=1}^N$ which obey relations (26) and by $\mathfrak{P}_q[a]$ the graded q-differential algebra generated by $\{a_k\}_{k=1}^N$ with N-differential d. Now we give the following definition:

Definition 3.2 The generator a_1 of $\mathbb{Z}_N$-graded q-differential algebra $\mathfrak{P}_q[a]$ will be referred to as an N-connection form and the algebra $\mathfrak{P}_q[a]$ will be referred to as an algebra of N-connection form. The operator $D = d + a_1 : \mathfrak{P}_q[a] \to \mathfrak{P}_q[a]$ will be called a covariant N-differential, and the polynomial f_N, whose explicit power expansion formula given in (3.2), will be called the curvature of N-connection form a_1.

Proposition 3.2 *If $\mathfrak{P}_q[a]$ is the algebra of N-connection form and d is its N-differential then the Nth power of the covariant N-differential D is the operator of multiplication by the curvature of N-connection form f_N.*

Proof The proof of this proposition is based on the first expansion formula proved in the Theorem 3.2. Indeed we can expand an Nth power of the covariant N-differential D into the sum of products of polynomials f_i and the powers of the N-differential d as follows

$$D^N = \sum_{i=0}^{N} \begin{bmatrix} N \\ i \end{bmatrix}_q f_i\, d^{N-i}.$$

As q is a primitive Nth root of unity this expansion can be essentially simplified in the case $k = N$ if we take into account that all q-binomial coefficients with $i \in \{1, 2, \ldots, N-1\}$ vanish. The first term of this expansion also vanishes because d is the N-differential. Hence for any polynomial $P \in \mathfrak{P}_q[a]$ we have

$$D^N(P) = f_N \cdot P,$$

and this ends the proof. □

Proposition 3.3 *If $\mathfrak{P}_q[a]$ is the algebra of connection form and f_N is the curvature of connection form then the curvature satisfies the identity*

$$\nabla(f_N) = 0. \tag{26}$$

Proof Let us remind a reader that $\nabla = d + \mathrm{ad}^q_{a_1}$. We prove this proposition by means of the recurrent relation for polynomials f_k

$$f_{k+1} = \sum_{i=0}^{k} \begin{bmatrix} k \\ i \end{bmatrix}_q f_i\, a_{k+1-i}.$$

Substituting N for k in the above relation we obtain

$$f_{N+1} = \sum_{i=0}^{N} \begin{bmatrix} N \\ i \end{bmatrix}_q f_i\, a_{N+1-i}. \tag{27}$$

As q is a primitive Nth root of unity we have

$$\begin{bmatrix} N \\ i \end{bmatrix}_q = 0,$$

for any integer $i \in \{1, 2, \ldots, N-1\}$. Consequently there are only two terms with non-zero q-binomial coefficients (labeled by $i = 0, N$) at the right-hand side of the relation (27) and

$$f_{N+1} = f_0\, a_{N+1} + f_N\, a_1.$$

The first term at the right-hand side of the above formula is also zero because of $a_{N+1} = 0$ (Proposition 3.1). Hence

$$\begin{aligned} 0 &= f_{N+1} - f_N\, a_1 = D(f_N) - f_N\, a_1 \\ &= d(f_N) + a_1 f_N - f_N\, a_1 = d(f_N) + [a_1, f_N]_q = (d + \mathrm{ad}^q_{a_1})(f_N) = \nabla(f_N). \end{aligned}$$

□

The identity (26) is an analogue of Bianchi identity for the curvature of N-connection form. It is worth mentioning that we can write the Bianchi identity for the curvature of N-connection form (26) in a different way if we consider the covariant N-differential D and the curvature f_N as the linear operators $D, f_N : \mathfrak{P}_q[a] \to \mathfrak{P}_q[a]$, i.e. $D, f_N \in \mathrm{Lin}\, \mathfrak{P}_q[a]$, where f_N is the operator of multiplication by f_N (we denote it by the same symbol as the curvature f_N in order not to make the notations very complicated). Then the Bianchi identity may be written in the form

$$[D, f_N]_q = 0.$$

Indeed

$$\begin{aligned}[D, f_N]_q &= D \circ f_N - f_N \circ D \\ &= d(f_N) + f_N \circ d + a_1 f_N - f_N \circ d - f_N\, a_1 \\ &= d(f_N) + [a_1, f_N]_q = \nabla(f_N) = 0.\end{aligned}$$

Acknowledgments The authors are grateful to Michel Dubois-Violette of the University Paris XI for valuable discussions and suggestions that improved the manuscript. The authors also gratefully acknowledge the financial support of the Estonian Science Foundation under the research grant ETF9328, target finance grant SF0180039s08 and ESF DoRa programme.

References

1. Abramov, V.: Generalization of superconnection in non-commutative geometry. Proc. Est. Acad. Sci. Phys. Math. **55**(1), 3–15 (2006)
2. Abramov, V.: On a graded q-differential algebra. J. Nonlinear Math. Phys. **13**(Supplement), 1–8 (2006)
3. Abramov, V.: Graded q-differential algebra approach to q-connection. In: Silvestrov, S., Paal, E., Abramov, V., Stolin, A. (eds.) Generalized Lie Theory in Mathematics, Physics and Beyond, pp. 71–79. Springer, Berlin (2009)
4. Abramov, V., Liivapuu, O.: Generalization of connection based on a concept of graded q-differential algebra. Proc. Est. Acad. Sci. Phys. Math. **59**(4), 256–264 (2010)
5. Abramov, V., Liivapuu, O.: N-complex, graded q-differential algebra and N-connection on modules. J. Math. Sci. (to appear)
6. Liivapuu, O.: Graded q-differential algebras and algebraic models in noncommutative geometry, Dissertationes Mathematicae Universitatid Tartuensis, Tartu University Press (2011)
7. Dubois-Violette, M., Kerner, R.: Universal q-differential calculus and q-analog of homological algebra. Acta Math. Univ. Comenianae **LXV**, 175–188 (1996)
8. Dubois-Violette, M.: $d^N = 0$. K Theor. **14**, 371–404 (1998)
9. Dubois-Violette, M.: Generalized homologies for $d^N = 0$ and graded q-differential algebras. In: Henneaux, M., Krasilshchik, J., Vinogradov, A. (eds.) Contemporary Mathematics, vol. 219, pp. 69–79. American Mathematical Society (1998)
10. Dubois-Violette, M.: Lectures on differentials, generalized differentials and some examples related to theoretical physics. In: Coquereaux, R., Garcia, A., Trinchero, R. (eds.) Contemporary Mathematics, pp. 59–94. American Mathematical Society (2002)
11. Kapranov, M.: On the q-analog of homological algebra. Preprint Cornell University, q-alg/961005
12. Dubois-Violette, M.: Tensor product of N-complexes and generalization of graded differential algebras. Bulg. J. Phys. **36**, 227–236 (2009)
13. Quillen, D.: Chern-Simons Forms and Cyclic Cohomology. In: Quillen, D.G., Segal, G.B., Tsou, S.T. (eds.) The Interface of Mathematics and Particle Physics, pp. 117–134. Clarendon Press, Oxford (1990)
14. Heubach, S., Mansour, T.: Combinatorics of Compositions and Words. CRC Press, Boca Raton (2009)

Double Category Related to Path Space Parallel Transport and Representations of Lie 2 Groups

Saikat Chatterjee, Amitabha Lahiri and Ambar N. Sengupta

Abstract From a differential geometric approach parallel transport on path spaces has been addressed. Integral version establishes the category theoretic frame work. We have glossed over the topic of representations of categorical groups and its relation with categorical framework of path space parallel transport.

1 Introduction

A growing body of literature [1–5] has been dovoted to the study of parallel transport of surfaces. Perhaps most interesting aspect of the surface parallel transport is the observation that a single group is not adequate for an 'well defined' Non-Abelian description. [1, 2] etc. address the problem from a purely category theoretic argumets. On the other hand there is also gerbe theoretic approach to the problem [6–8]. In [4] starting from a differential geometry, a category theoretic frame work for the parallel transport of surfaces has been established. Our approach in [4] mainly focussed on differential geometry on the path space. In Sect. 2 we first review few important

S. Chatterjee (✉)
School of Mathematics, Tata Institute of Fundamental Research,
Mumbai 400005, Maharashtra, India
e-mail: saikat.chat01@gmail.com

S. Chatterjee
Institut des Hautes Études Scientifiques, 35 Route de Chartres, Bures Sur Yvette 91440, France

A. Lahiri
S. N. Bose National Centre for Basic Sciences, Block JD, Sector III, Salt Lake,
Kolkata 700098, West Bengal, India
e-mail: amitabha@bose.res.in

A. N. Sengupta
Department of Mathematics, Louisiana State University Baton Rouge, Louisiana 70803, USA
e-mail: ambarnsg@gmail.com

A. Makhlouf et al. (eds.), *Algebra, Geometry and Mathematical Physics*, 379
Springer Proceedings in Mathematics & Statistics 85, DOI: 10.1007/978-3-642-55361-5_22,

results of [4]. We choose a principal G bundle (Π, P, M). Let $\bar{A}$ be a connection on this G-bundle. We construct another G-bundle Π, $\mathscr{P}_{\bar{A}}P$, $\mathscr{P}M$, where $\mathscr{P}M$ is the path space over manifold M and $\mathscr{P}_{\bar{A}}P$ is the space of $\bar{A}$ horizontal paths over P. Keeping in mind that a single group is insufficient to describe the surface parallel transport, we introduce another group H related to G. We construct a connection $\mathscr{A}$ on the G-bundle Π, $\mathscr{P}_{\bar{A}}P$, $\mathscr{P}M$. Differential geometry with this connection $\mathscr{A}$ leads to a natural construction of integrated picture or category structure of the surface parallel transport. Section 3 provides a glimpse of [9]. We first define *representations of categorical groups*. Then we show that it defines a double category over a category of some vector spaces. Then we touch upon the relation between category theoretic frame work of path space parallel transport (discussed in Sect. 2) and representations of categorical groups [9].

2 Connections on Path Spaces

We define the path space $\mathscr{P}M$ of a given manifold M as the space of all parametrized smooth paths in M,

$$\gamma : I \to M \qquad I = [0, 1]$$

.i.e. if $\gamma \in \mathscr{P}M$, then $\gamma(t) \in M$, where $t \in [0, 1]$. The tangent space of the path space is defined as follows. For $\gamma \in \mathscr{P}M$, a vector $X \in T_\gamma(\mathscr{P}M)$ is given by a vector field $X(t) \in T_{\gamma(t)}(M)$ [3].

Let ev be the general evaluation map. i.e.

$$ev : \mathscr{P}M \times I \longrightarrow M, \qquad ev_t \stackrel{\text{def}}{=} ev(\cdot, t) \tag{1}$$

Then

$$\begin{aligned} ev_t :\ & \mathscr{P}M \to M \\ & \gamma \mapsto \gamma(t). \end{aligned}$$

We denote corresponding pull back operator as ev_t^*.

Let us consider a principal G-bundle (Π, P, M)

$$\Pi : P \to M$$

with the usual right action of the Lie group G on P

$$P \times G \to P : (p, g) \to pg$$

If $\bar{A}$ is a connection on this bundle, we can construct the space of $\bar{A}$ horizontal paths in P. Then we have a natural projection map $\Pi : \mathscr{P}_{\bar{A}}P \to \mathscr{P}M$ given by

$$\Pi(\tilde{\gamma})(t) = \gamma(t),$$

where $\tilde{\gamma}$ is a lift of γ by connection $\bar{A}$, and a natural G action $G \times \mathscr{P}_{\bar{A}}P \to \mathscr{P}_{\bar{A}}P$ given by

$$(\tilde{\gamma}.g)(t) = \tilde{\gamma}(t).g$$

Note if $\tilde{\gamma}$ is a $\bar{A}$-horizontal path, then $\tilde{\gamma}.g$ is also a $\bar{A}$-horizontal path. Given any $u \in LG$ we can a construct a *vertical vector* field on $\mathscr{P}_{\bar{A}}P$ by

$$X^u(\tilde{\gamma})(t) := \frac{d}{ds}|_{s=0}\tilde{\gamma}(t)exp(su) \tag{2}$$

Naturally, but informally we treat $\mathscr{P}_{\bar{A}}P \to \mathscr{P}M$ as a principal G-bundle. For details see [3, 4]. It can be shown (Proposition 2.1 in [4]) that if $\tilde{\Gamma} : [0, 1] \times [0, 1] \to P : (t, s) \to \tilde{\Gamma}(t, s) = \tilde{\Gamma}_s(t)$ is a smooth map and $\tilde{X}_s(t) = \partial_s\tilde{\Gamma}(t, s)$, then *each transverse path* $\tilde{\Gamma}_s : [0, 1] \to P$ *is* $\bar{A}$*-horizontal* if and only if *the initial path* $\tilde{\Gamma}_0$ *is* $\bar{A}$*-horizontal, and the tangency condition*

$$\frac{\partial\bar{A}(\tilde{X}_s(t))}{\partial t} = F^{\bar{A}}\left(\partial_t\tilde{\Gamma}(t, s), \tilde{X}_s(t)\right) \tag{3}$$

holds. It is often conveniently written in integral form as

$$ev_T^*\bar{A} - ev_0^*\bar{A} = \int_0^T F^{\bar{A}} \tag{4}$$

The right hand side is a Chen integral [10, 11] in the interval $[0, T]$. We define the tangent space $T_{\tilde{\gamma}}\mathscr{P}_{\bar{A}}P$ at a point $\tilde{\gamma}$ of $\mathscr{P}_{\bar{A}}P$ to be space of all vector fields $t \to \tilde{X}(t) \in T_{\tilde{\gamma}(t)}P$ along $\tilde{\gamma}$ for which (3) holds, i.e.

$$\frac{\partial\bar{A}(\tilde{X}(t))}{\partial t} = F^{\bar{A}}\left(\tilde{\gamma}'(t), \tilde{X}(t)\right) \tag{5}$$

for all $t \in [0, 1]$. The vertical subspace of $T_{\tilde{\gamma}}\mathscr{P}_{\bar{A}}P$ is the collection of all vectors $\tilde{X}$ for which $\tilde{X}(t)$ is vertical, a more detailed discussion can be found in [4].

2.1 Parallel Transport on Path Space

A description of parallel transport on path space with a single gauge leads to a serious inconsistency, which is quite obvious from the following category theoretic argument and forces us to introduce two different gauge groups for the parallel transport on the path space. In order to have a consistent parallel transport along a surface, the

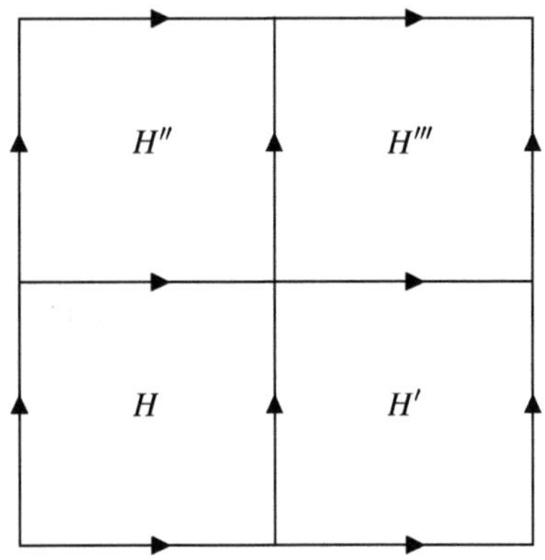

Fig. 1 No go theorem!

following equation must be satisfied

$$(H' \bullet H) \times (H''' \bullet H'') = (H' \times H''') \bullet (H \times H'') \tag{6}$$

here H, H', H'', H''' are 'surface parallel transport' operators in the Fig. 1 and $\times$ and $\bullet$ denote vertical and horizontal composition of surfaces respectively.

So it is obvious that if we take surface parallel transport operator to be a group element and assign the same composition law (the group product) for both horizontal and vertical compositions, the group must be Abelian. It is a 'No Go theorem' [1]! To avoid that we introduce two Lie groups G and H to describe surface parallel transport and define different composition laws for the 'horizontal' and 'vertical' compositions. Basic goal here is to arrive at a double category structure starting from a gauge theoretic argument and the Mac Lanes consistency condition [12]. The necessary framework for the purpose, which we discuss below, is known as *Lie 2-group* [4, 13].

A Lie 2-group is given by two Lie groups G and H, along with a smooth homomorphism $\tau : H \to G$ and a smooth map for $\alpha : G \to H$, such that for any $g \in G$ and $h, h' \in H$ following identities hold:

$$\tau(\alpha(g)h) = g\tau(h)g^{-1} \tag{7}$$

$$\alpha(\tau(h))h' = hh'h^{-1} \tag{8}$$

For simplicity we will denote the mapping $\alpha'(e) : LG \to LH$ and $\tau'(e) : LH \to LG$ as α and τ respectively, here LG and LH are Lie algebras of G and H respectively.

2.1.1 A Connection Form on the Path Space Bundle

Suppose we have a connection A on the bundle P and an LH valued α- equivariant (under the right action of G) 2—form B on P, which vanishes on vertical vectors, i.e.

$$B(X, Y) = 0, \quad \text{if } X \text{ or } Y \text{ is vertical} \tag{9}$$

and

$$R_g^* B = \alpha(g^{-1})B, \quad \text{for all } g \in G. \tag{10}$$

Here $R_g : P \to P : p \mapsto pg$ and according to our convention $\alpha(g^{-1})B = d\alpha(g^{-1})|_e B$.

Proposition 2.2 of [4] states that

$$\mathscr{A} = ev_1^* A + \tau \int_0^1 B \tag{11}$$

is a connection on the principal G-bundle $\Pi : \mathscr{P}_{\bar{A}} P \to \mathscr{P} M$, where the integration on the right hand side is a first order Chen integral. A sketchy description of the proof might be pertinent here. In order to prove that $\mathscr{A}$ is a connection we check that $\mathscr{A}$ satisfies following two conditions:

1. $\mathscr{A}(R_{g*}\tilde{v}) = Ad(g^{-1})\mathscr{A}(\tilde{v})$
2. $\mathscr{A}(X^u) = u$,

for every $g \in G$ and $\tilde{v} \in T\mathscr{P}_{\bar{A}} P$ and X^u is as defined in (2). The first condition is met due the fact that A is a connection on principal G-bundle $\Pi : P \to M$ and B is α equivariant (10). As B is zero on vertical vectors (9) second condition is also satisfied. Thus $\mathscr{A}$ is indeed a connection on $\Pi : \mathscr{P}_{\bar{A}} P \to \mathscr{P} M$.

The first step to describe the parallel transport of a path by the connection $\mathscr{A}$ would be to give a prescription for lifting a given vector field $X : I \to TM$, along $\gamma \in \mathscr{P} M$, to a vector field $\tilde{X}$ along a $\tilde{\gamma}$, such that it is (i)$\mathscr{A}$ horizontal and (ii) is actually a vector in $T_{\tilde{\gamma}} \mathscr{P}_{\bar{A}} P$. That is $\tilde{X}$ must satisfies following equations:

$$A(\tilde{X}(1)) + \tau \int_0^1 B(\tilde{\gamma}'(t), \tilde{X}(t))dt = 0 \tag{12}$$

$$\frac{\partial \bar{A}(\tilde{X}(t))}{\partial t} = F^{\bar{A}}\left(\tilde{\gamma}'(t), \tilde{X}(t)\right) \tag{13}$$

(12) is for the $\mathscr{A}$-horizontality and (13) is to ensure (5). Proposition 2.3 [4] shows that given a $X \in T_\gamma M$ conditions (12), (13) uniquely determine the lifted vector $\tilde{X} \in T_{\tilde{\gamma}} \mathscr{P}_{\bar{A}} P$. In order to precisely determine the lifted vector field $\tilde{X}$. We employ following technique. First let us decompose the lifted vector field along $\tilde{\gamma}$ into horizontal and vertical parts with respect to the connection $\bar{A}$, $\tilde{X}(t) = \tilde{X}^h_{\bar{A}}(t) + \tilde{X}^V(t)$. Note as B is zero on the vertical vectors, (13) gives

$$A(\tilde{X}(1)) + \tau \int_0^1 B(\tilde{\gamma}'(t), \tilde{X}^h_{\bar{A}}(t))dt = 0 \tag{14}$$

Fig. 2 A plaquette

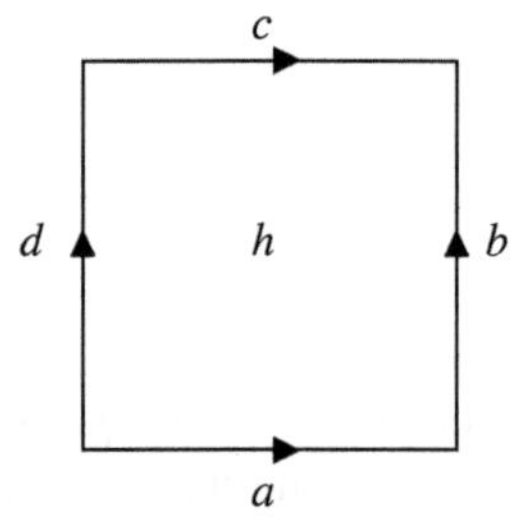

and $\tilde{X}^h_{\bar{A}}(t)$ is completely determined by the connection $\bar{A}$. Thus (14) determines $A(\tilde{X}(1))$. Also (13) leads to

$$\frac{\partial \bar{A}(\tilde{X}^V(t))}{\partial t} = F^{\bar{A}}\left(\tilde{\gamma}'(t), \tilde{X}^h_{\bar{A}}(t)\right). \tag{15}$$

Next we split $\tilde{X}(1)$ into A horizontal and corresponding vertical part. (15) is a first order differential equation. With a tediuos but straightforward calculation we obtain following result (see (2.22)–(2.24) in [4]):

$$\tilde{X}(t) = \tilde{X}^h_{\bar{A}}(t) + X^u(\tilde{\gamma})(t), \tag{16}$$

where

$$u = \bar{A}(\tilde{v}(1)) - \int_t^1 F^{\bar{A}}(\tilde{\gamma}'(s), \tilde{v}^h_{\bar{A}}(s))ds,$$

$$\tilde{v}(1) = \tilde{v}^h_{\bar{A}}(1) + X^w(1), \qquad w = -\int_0^1 \tau(B(\tilde{\gamma}'(s), \tilde{v}^h_{\bar{A}}(s))ds$$

and $X^u(\tilde{\gamma})$ is as defined in (2). The integrated version of the above construction leads to the description of the parallel transport of paths. The pivotal part of the description is that (12) and (13) specify 'parallel transport' of the 'right endpoint' $\tilde{\gamma}(1)$ and then (13) specifies the parallel transport of the entire path $\tilde{\gamma}$. Theorem 2.4 in [4] provides the explicit expression for the parallely transported path by connection $\mathscr{A}$. Our main focus will be (2.46) in [4]. A categorical description for the parallel transport of paths directly follows from that equation follows.

2.2 Categorical Picture

Let $\Gamma : [0, 1] \to \mathscr{P}M$ and denote $(\Gamma(s))(t) := \Gamma(s, t) = \Gamma_s(t) = \Gamma^t(s)$.i.e. we denote 'longitudinal' paths as Γ^t and 'transverse' paths as Γ_s. Let $a \in G$ and $c \in G$ are $\bar{A}$-parallel transports along Γ_0 and Γ_1 respectively. $d \in G$ and $b \in G$ are A-parallel transports along Γ^0 and Γ^1 respectively. The $h \in H$ is a term dependent on the 'surface' Γ, given by the solution of

$$\frac{dh(s)}{ds}h(s)^{-1} = -\int_0^1 B'(\partial_t\Gamma(s,t), \partial_s\Gamma(s,t))dt$$

at $s = 1$ with the initial condition $h(0) = e$. Here, B' is a LH valued 2-form on M determined by $A, \bar{A}, B$. The explicit expression for B is given in (2.47) of [4]. However for our current purpose we do not need to bother about the exact expression of B'. Now consider a plaquette as in the Fig. 2, whose edges are labeled with the elements of the group G, $a \in G$, $b \in G$, $c \in G$ and $d \in G$ and the 'surface' is labeled by $h \in H$. As a consequence of Eq. (2.46) in [4] we have following relation

$$\tau(h) = a^{-1} \cdot b^{-1} \cdot c \cdot d \tag{17}$$

As a consequence of the above equation we have a 2-categorical picture for parallel transport of paths, where the set of objects for both of the categories is the group G and the set of morphisms is $G^4 \times H$. We call these two categories involved as **Vert** (vertical category) and **Horz** (horizontal category) respectively. We denote a morphism associated with a plaquette as in Fig. 2 by $(a, b, c, d; h)$. The source and target maps for **Vert** and **Horz** are respectively given as follows

$$s_{\mathbf{Vert}}(a, b, c, d; h) = a$$
$$t_{\mathbf{Vert}}(a, b, c, d; h) = c$$
$$s_{\mathbf{Horz}}(a, b, c, d; h) = d$$
$$t_{\mathbf{Horz}}(a, b, c, d; h) = b$$

Using the (17) we define composition law for **Vert** as

$$(a, b, c, d; h) \times (c, b', d, d'; h') = (a, b' \cdot b, d, d' \cdot d; h(\alpha(d)h')) \tag{18}$$

and that of **Horz** is given by

$$(a, b, c, d; h) \bullet (a', f, c', b; h') = (a' \cdot a, f, c' \cdot c, d; (\alpha(d^{-1})h')h) \tag{19}$$

It is easy to check that the identity morphism $a \to a$ for **Vert** is $(a, e, a, e; e)$, on the other hand for the **Horz** the identity morphism $d \to d$ is $(e, d, e, d; e)$, where e denotes the the identity element for both G and H. Associative property of those two categories is obvious. It should be noted here that all morphisms (in both **Vert** and **Horz**) are isomorphisms. For instance, inverse of $(a, b, c, d; h)$ in **Vert** is $(c, b^{-1}, a, d^{-1}; \alpha(d)h^{-1})$ and that in **Horz** is $(a^{-1}, d, c^{-1}, b; \alpha(a)h^{-1})$. Now it can be checked that whenever it is well defined following identity holds,

$$(F_1 \times F_2) \bullet (F_1' \times F_2') = (F_1 \bullet F_1') \times (F_2 \bullet F_2'), \tag{20}$$

where F_1, F_2 etc. are elements of $G^4 \times H$. Thus we have a consistent 'window diagram' as in Fig. 1 and as demanded in (6).

In ordinary gauge theory a parallel transport operator $\mathscr{H}(\gamma, 0, 1)$ between $\gamma(0)$ and $\gamma(1)$ along the path γ transforms homogeneously as $U(\gamma(1))\mathscr{H}(\gamma, 0, 1)$ $U(\gamma(0))^{-1}$, here $U(\gamma(0))$ and $U(\gamma(1))$ are two elements of the gauge group associated with the end points of the path and $\mathscr{H}(\gamma, 0, 1)$ is also an element of the same group. Now consider a plaquette like Fig. 2, here instead of a group valued parallel transport operator we have a morphism like $(a, b, c, d; h)$ and have two end paths rather than two end points. So in the same spirit we define gauge transformation of a surface parallel transport operator as

$$(\bar{a}, \bar{b}, \bar{c}, \bar{d}; \bar{h}) \stackrel{\text{def}}{=} (c, \tilde{U}(1), c, \tilde{U}(0); \tilde{W}) \times (a, b, c, d; h) \times (a, U(1), a, U(0); W)^{-1} \tag{21}$$

Here $U(0), U(1) \in G$ are group elements associated with the left and right end points of the initial path in Fig. 2 respectively, $\tilde{U}(0), \tilde{U}(1) \in G$ are those of the final path, and $W, \tilde{W} \in H$ are path ordered exponentials of some LH-valued one form λ over the initial and the final path respectively. As we have already defined the vertical composition in (18), from (21) we have following transformations

$$\begin{aligned}
\bar{a} &= U(1) \cdot a \cdot \tau(W) \cdot U(0)^{-1} \\
\bar{b} &= \tilde{U}(0) \cdot b \cdot U(0)^{-1} \\
\bar{c} &= \tilde{U}(1) \cdot c \cdot \tau(\tilde{W}) \cdot \tilde{U}(0)^{-1} \\
\bar{d} &= \tilde{U}(1) \cdot b \cdot U(1)^{-1} \\
\bar{h} &= (\alpha(U(0))(W^{-1}.h.((\alpha(d^{-1})\tilde{W}))
\end{aligned}$$

3 Representations of Categorical Groups

This section is based on an ongoing work [9]. Here we extend the notion of representations for the categorical groups and discuss some consequences. We will farther explore few aspects which will be interesting from the categorical frame work we described at the end of preceding section.

A *categorical group* is a monoidal category, where both object and arrow sets form groups under the monoidal product. For a detail exposition on the subject of categorical groups [14] or [15] may be consulted. We will denote monoidal or group product between a and b as ab both at the level of objects and morphisms. Our primary interest would be a specific type of categorical groups, whose object and morphism sets are both Lie groups. We will call such categorical groups *Lie 2-groups*. Let $\mathscr{G}$ be a categorical group and group G and group $\mathscr{K}$ be object and morphism sets respectively. We denote an arrow (morphism) $k \in \mathscr{K}$ from $g_1 \in G$ to $g_2 \in G$ as

$$g_1 \xrightarrow{k} g_2.$$

Suppose $g_1 \xrightarrow{k_1} g_2$ and $g_1' \xrightarrow{k_2} g_2'$, then monoidality implies

$$g_1 g_1' \xrightarrow{k_1 k_2} g_2 g_2'.$$

Also if $k_1, k_2, k_3, k_4 \in \mathscr{K}$ then we have the following 'exchange law':

$$(k_1 k_2) * (k_3 k_4) = (k_1 * k_3)(k_2 * k_4), \tag{22}$$

whenever left hand side is well defined, where $*$ is the composition law in $\mathscr{G}$.

Given r number of n-dimensional vector spaces V_i, $i = [1, r]$, we define a category $\mathbf{V}[n]$, whose object set is $\{V_i\}$ and a morphism from $V_1 \in \mathrm{Obj}(\mathbf{V}[n])$ to $V_2 \in \mathrm{Obj}(\mathbf{V}[n])$ is given by a linear map $f : V_1 \to V_2$. Obviously, identity morphism associated with $V \in \mathrm{Obj}(\mathbf{V}[n])$ is given by the idenity element I_V of $\mathrm{Aut}(V)$ and the composition law in category $\mathbf{V}[n]$ is the composition of set maps.

Motivated by the definition of representation of ordinary groups into a vector space, we define representation of a categorical group $\mathscr{G}$ on to the category $\mathbf{V}[n]$ as follows. Let $\mathscr{G}$ be a categorical group, G and $\mathscr{K}$ are object and morphism set respectively. A represenation of $\mathscr{G}$ is given by a functor

$$\rho : \mathscr{G} \times \mathbf{V}[n] \to \mathbf{V}[n], \tag{23}$$

such that for any $k_1, k_2 \in \mathrm{Mor}(\mathscr{G}) = \mathscr{K}$ and $f \in \mathrm{Mor}(\mathbf{V}[n])$ following identity holds;

$$\rho(k_2, \rho(k_1, f)) = \rho(k_2 k_1, f). \tag{24}$$

From the condition in (24) we observe that

$$\rho(1_e, f) = f, \tag{25}$$

where $e \in G$ is the identity element in G, and $1_e \in \mathscr{K}$ is the corresponding identity morphism.

Let $\mathbf{C}$ be a category. Recall that a double category, over the base category $\mathbf{C}$, is a category which has another set of 'morphisms between morphisms in the category $\mathbf{C}$'. The second level of morphisms are equipped with two types of composition laws, namely *horizontal composition* and *vertical composition*. Also horizontal and vertical composition together satisfy certian 'exchange law'. Here we keep the definition vague and to a bare essentials to avoid lengthening the paper. More rigorous definition of a double category is provided in numerous literature, such as [1, 2, 13]. A double category over $\mathbf{V}[n]$ is a category whose base category is $\mathbf{V}[n]$.

Proposition 3.1 *Representation of a Lie 2-group on* $\mathbf{V}[n]$ *defines a double category over* $\mathbf{V}[n]$.

Proof Let the object set of $\mathscr{G}$ be a Lie group G and the morphism set be the Lie group $\mathscr{K}$ and ρ is a representation of $\mathscr{G}$

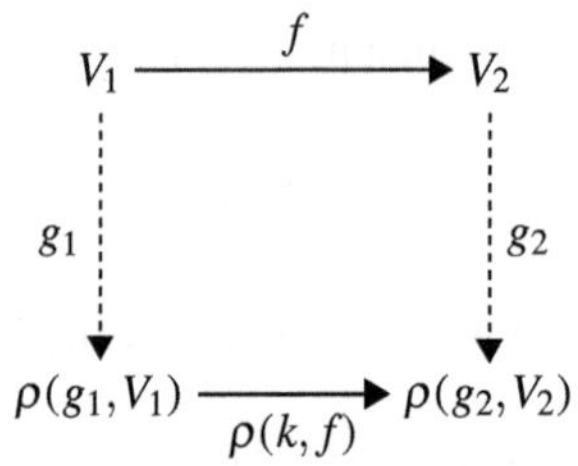

Fig. 3 Representation functor

$$\rho : \mathscr{G} \times \mathbf{V}[n] \to \mathbf{V}[n]$$

Thus we have following maps

$$\begin{aligned} \rho : G \times \text{Obj}\big(\mathbf{V}[n]\big) &\to \text{Obj}\big(\mathbf{V}[n]\big) \\ (g, V) &\mapsto \rho(g, V) \in \text{Obj}\big(\mathbf{V}[n]\big) \end{aligned} \tag{26}$$

and

$$\begin{aligned} \rho : \mathscr{K} \times \text{Mor}\big(\mathbf{V}[n]\big) &\to \text{Mor}\big(\mathbf{V}[n]\big) \\ (k, f) &\mapsto \rho(k, f) : \rho(g_1, V_1) \to \rho(g_2, V_2), \end{aligned} \tag{27}$$

where $g_1 \xrightarrow{k} g_2$ and $V_1 \xrightarrow{f} V_2$. Hence we have the following diagram

For brevity let us write $\rho(g_1, V_1)$ as $g_1 V_1$ etc. and $\rho(k_1, f_1)$ etc. as $k_1 f_1$ etc. Let $\text{Hom}(V_1, V_2)$ be the set of morphisms from V_1 to V_2. Thus given a representation ρ and $k_1 \in \mathscr{K}$ we have

$$(k_1, \rho) : \text{Hom}(V_1, V_2) \to \text{Hom}(g_1 V_1, g_2 V_2)$$

given by

$$(k_1, \rho)(f_1) = \rho(k_1, f_1) = k_1 f_1.$$

Now suppose k_2 is another element of $\mathscr{K}$, then

$$(k_2, \rho) : \text{Hom}(g_1 V_1, g_2 V_2) \to \text{Hom}(g_1' g_1 V_1, g_2' g_2 V_2).$$

So if $(k_2, \rho) : k_1 f_1 \mapsto k_2 k_1 f_1$, using (24) we have a composition law

$$(k_2, \rho) \circ (k_1, \rho) := (k_2 k_1, \rho). \tag{28}$$

Also as $k_2 k_1 : g_1' g_1 \to g_2' g_2$, we have diagram in Fig. 3 for the above composition Under this composition law associativity follows naturally, and we define $1_f := (1_e, \rho) : f \to f$. So we have a category here. Let us call this composition law *vertical composition.*

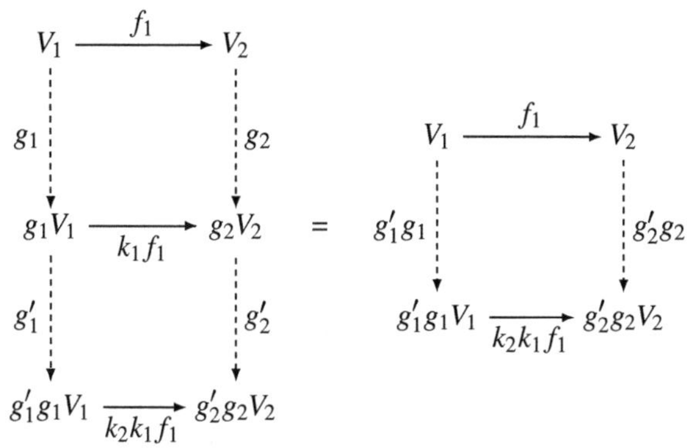

Fig. 4 Vertical composition for the representation

Now suppose $k_1, k_1' \in \mathscr{K}$ are composable as morphism and so are $f_1, f_1' \in \mathrm{Mor}\big(\mathbf{V}[n]\big)$. Let

$$s(k_1) = g_1$$
$$s(k_1') = g_2 = t(k_1)$$
$$t(k_1') = g_3$$

and

$$s(f_1) = V_1$$
$$s(f_1') = V_2 = t(f_1)$$
$$t(f_1') = V_3,$$

where s, t denote the source and target maps respectively. Functoriality of ρ implies

$$\begin{aligned} &\rho((k_1, f_1) * (k_1', f_1')) \\ &= \rho(k_1 * k_1', f_1 * f_1') \\ &= \rho(k_1, f_1) * \rho(k_1', f_1'). \end{aligned} \tag{29}$$

Hence

$$(k_1, \rho)(f_1) * (k_1', \rho)(f_1') = (k_1 * k_1', \rho)(f_1' * f_1) \tag{30}$$

where * denotes various compositions of morphisms. (30) defines a horizontal composition. Following digram explains the composition

So horizontal composition naturally follows from the morphism composition of $\mathscr{G}$.

Our next task is to check the consistency 'window diagram'. With the aid of (22) and definitions of horizontal and vertical composition in (30), (28) respectively it can be easily shown that

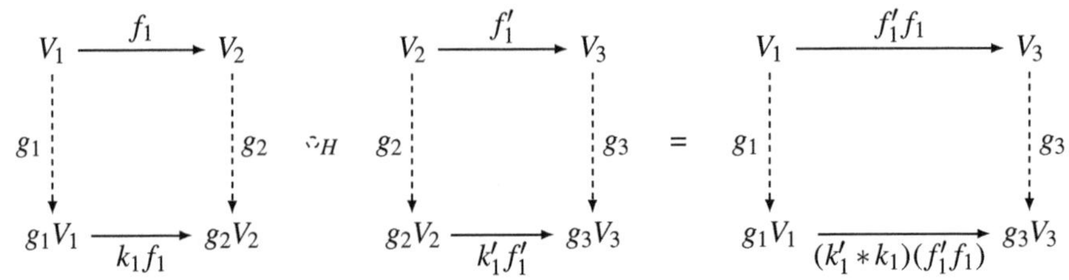

Fig. 5 Horizontal composition for the representation

$$\begin{aligned}&\big((k_1, \rho) \circ_H (k_1', \rho)\big) \circ \big((k_2, \rho) \circ_H (k_2', \rho)\big)\\ &= ((k_1, \rho) \circ (k_2, \rho)) \circ_H \big((k_1', \rho) \circ (k_2', \rho)\big),\end{aligned} \tag{31}$$

whenever left and right sides are well defined. Thus a representation of a categorical group defines a double category.

Here we will give a cursory description of relation between the representation of a categorical group and the categorical frame work described in Sect. 2.2. A more in depth and serious study shall be pursued in [9].

Recall the principal G bundle Π, P, M. Let us consider a category Σ, whose Object set $\mathrm{Obj}(\Sigma) := \{T_pP|p \in P\}$ be the set of all tangent vector spaces on P and a morphism from T_pP to T_qP is given by the parallel transport along a path connecting p to q. Suppose (G, H, τ, α) be a Lie crossed module, as described in (7)–(8). It can be shown that a Lie crossed module defines a categorical group (see [1] or Theorem 4.1 [15]). Let us consider the Lie 2-group $\mathscr{G}^{para}$ given by the Lie-crossed module (G, H, τ, α). $\mathscr{G}^{para}$ has the object group G and a morphism is given by the ordered pair $(g, h) \in G \times H$, with source being $s(g, h) = g \in G$ and the target being $t(g, h) = \tau(h)g \in G$. Suppose $\mathscr{R}$ be a representation of the Lie 2-group $\mathscr{G}^{para}$ into Σ as defined in (23). Then according to Proposition 3.1 $\mathscr{R}$ defines a double category. If we compare the Plaquette in Fig. 2 with the diagram in Fig. 3, we realize that the Plaquette obtained by the parallel transport on the principal G bundle Π, $\mathscr{P}_{\bar{A}}P$, $\mathscr{P}M$ can be viewed as a 'second level' of morphism given by a representation of $\mathscr{G}^{para}$ on the category Σ. Let us now compare (18) and (19) with diagrams in (4) and (5) respectively. We readily see that categories **Vert** and **Horz** of Sect. 2.2 can be respectively viewed as (28) and (30) defined by the representation $\mathscr{R}$ of the Lie 2-group $\mathscr{G}^{para}$. It is now obvious that (31) ensures the consistent window diagram of (20).

Acknowledgments Authors thank the anonymous referee for making many useful comments and suggestions. S. Chatterjee acknowledges travel grant provided by the Department of Science and Technology, Government of India. S. Chatterjee is grateful to the organizers of AGMP-2011, Mulhouse, France for their kind hospitality and other means of support.

References

1. Baez, J.: Higher Yang-Mills theory (2002). arXiv:math/0206130
2. Baez, J., Schreiber, U.: Higher gauge theory II: 2-connections on 2-bundles (2004). arXiv:hep-th/0412325
3. Cattaneo, A.S., Cotta-Ramusino, P., Rinaldi, M.: Loop and path spaces and four-dimensional bf theories: connections, holonomies and observables. Com. Math. Phys. **204**, 493–524 (1999)
4. Chatterjee, S., Lahiri, A., Sengupta, A.N.: Parallel transport over path spaces. Rev. Math. Phys. **9**, 1033–1059 (2010)
5. Girelli, F., Pfeiffer, H.: Higher gauge theory—differential versus integral formulation. J. Math. Phys. **45**, 3949–3971 (2004)
6. Breen, L., Messing, W.: Differential geometry of gerbes (2001). arXiv:math/0106083v3
7. Murray, M.: Bundle gerbes. J. London. Math. Soc. **54**, 403–416 (1994)
8. Schreiber, U. Waldorf, K.: Connections on non-abelian gerbes and their holonomy (2008). arXiv:0808.1923v1
9. Chatterjee, S., Lahiri, A., Sengupta, A.N.: Twisted actions of categorical groups (2014). arXiv:1404.6365
10. Chen, K-T.: Algebras of Iterated path integrals and fundamental groups. Trans. Amer. Math. Soc. **156**, 359–379 (1971)
11. Chen, K-T.: Iterated integrals of differential forms and loop space homology. Ann. Math., 2nd Ser. **97**(2) 217–246 (1973)
12. Mac lane, S.: Categories for the Working Mathematician. Springer, New York (1971)
13. Baez, J., Schreiber, U.:Higher gauge theory (2005). http://math.ucr.edu/home/baez/higher.pdf
14. Barrett, J.W., Mackaay, M.: Categorical representations of categorical groups. Theor. Appl. Categories **16**(20), 529–557 (2006)
15. Chatterjee, S., Lahiri, A., Sengupta, A.N.: Path space connections and categorical geometry. J. Geom. Phys. **75**, (2014) arXiv:1207.5488

Differential Geometry of Microlinear Frölicher Spaces IV-1

Hirokazu Nishimura

Abstract The fourth paper of our series of papers entitled "Differential Geometry of Microlinear Frölicher Spaces" is concerned with jet bundles. We present three distinct approaches together with transmogrifications of the first into the second and of the second to the third. The affine bundle theorem and the equivalence of the three approaches with coordinates are relegated to a subsequent paper.

1 Introduction

As the fourth of our series of papers entitled "Differential Geometry of Microlinear Frölicher Spaces" [14–16], this paper will discuss jet bundles. Since the paper has become somewhat too long as a single paper, we have decided to divide it into two parts. In this first part we will present three distinct approaches to jet bundles in the general context of Weil exponentiable and microlinear Frölicher spaces. In the subsequent part [17], we will establish the affine bundle theorem in the second and the third approaches, and we will show that the three approaches are equivalent, as far as coordinates are available (i.e., in the classical context).

This part consisits of 7 sections. The first section is this introduction, while the second section is devoted to some preliminaries. We will present three distinct approaches to jet bundles in Sects. 3, 4 and 5. In Sect. 6 we will show how to translate the first approach into the second, while Sect. 7 is devoted to the transmogrification of the second approach into the third.

We have already discussed these three approaches to jet bundles in the context of synthetic differential geometry, for which the reader is referred to our previous work [8–13]. Now we have emancipated them to the real world of Frölicher spaces.

H. Nishimura (✉)
Institute of Mathematics, University of Tsukuba, Tsukuba, Ibaraki 305-8571, Japan
e-mail: logic@math.tsukuba.ac.jp

A. Makhlouf et al. (eds.), *Algebra, Geometry and Mathematical Physics*, Springer Proceedings in Mathematics & Statistics 85, DOI: 10.1007/978-3-642-55361-5_23,

2 Preliminaries

2.1 Frölicher Spaces

Frölicher and his followers have vigorously and consistently developed a general theory of smooth spaces, often called *Frölicher spaces* for his celebrity, which were intended to be the *maximal class of* spaces where smooth structures can live. A Frölicher space is an underlying set endowed with a class of real-valued functions on it (simply called *structure functions*) and a class of mappings from the set $\mathbb{R}$ of real numbers to the underlying set (simply called *structure curves*) subject to the condition that structure curves and structure functions should compose so as to yield smooth mappings from $\mathbb{R}$ to itself. It is required that the class of structure functions and that of structure curves should determine each other so that each of the two classes is maximal with respect to the other as far as they abide by the above condition. What is most important among many nice properties about the category **FS** of Frölicher spaces and smooth mappings is that it is cartesian closed, while neither the category of finite-dimensional smooth manifolds nor that of infinite-dimensional smooth manifolds modelled after any infinite-dimensional vector spaces such as Hilbert spaces, Banach spaces, Fréchet spaces or the like is so at all. For a standard reference on Frölicher spaces, the reader is referred to [2].

2.2 Weil Algebras and Infinitesimal Objects

2.2.1 The Category of Weil Algebras and the Category of Infinitesimal Objects

The notion of a Weil algebra was introduced by Weil himself in [18]. We denote by **W** the category of Weil algebras, which is well known to be left exact. Roughly speaking, each Weil algebra corresponds to an infinitesimal object in the shade. By way of example, the Weil algebra $\mathbb{R}[X]/(X^2)$ (=the quotient ring of the polynomial ring $\mathbb{R}[X]$ of an indeterminate X over $\mathbb{R}$ modulo the ideal (X^2) generated by X^2) corresponds to the infinitesimal object of first-order nilpotent infinitesimals, while the Weil algebra $\mathbb{R}[X]/(X^3)$ corresponds to the infinitesimal object of second-order nilpotent infinitesimals. Although an infinitesimal object is undoubtedly imaginary in the real world, as has harassed both mathematicians and philosophers of the 17th and the 18th centuries such as philosopher Berkley (because mathematicians at that time preferred to talk infinitesimal objects as if they were real entities), each Weil algebra yields its corresponding *Weil functor* or *Weil prolongation* on the category of smooth manifolds of some kind to itself, which is no doubt a real entity. By way of example, the Weil algebra $\mathbb{R}[X]/(X^2)$ yields the tangent bundle functor as its corresponding Weil functor. Intuitively speaking, the Weil functor corresponding to a Weil algebra stands for the exponentiation by the infinitesimal object corresponding to the Weil algebra at issue. For Weil functors on the category of finite-dimensional smooth manifolds, the reader is referred to §35 of [5], while the reader can find a readable treatment of Weil functors on the category of smooth manifolds modeled on convenient vector spaces in §31 of [6]. In [14] we have discussed how to assign, to each pair (X, W) of a Frölicher space X and a Weil algebra W, another Frölicher space $X \otimes W$ called the Weil prolongation *Weil prolongation of X with respect to W*, which is naturally extended to a bifunctor $\mathbf{FS} \times \mathbf{W} \to \mathbf{FS}$. And we have shown that, given a Weil algebra W, the functor assigning $X \otimes W$ to each object X in **FS** and $f \otimes \mathrm{id}_W$ to each morphism f in **FS**, namely, the Weil functor on **FS** corresponding to W is product-preserving. The proof can easily be strengthened to

Theorem 2.1 *The Weil functor on the category **FS** corresponding to any Weil algebra is left exact.*

There is a canonical projection $\pi : X \otimes W \to X$. Given $x \in X$, we write $(X \otimes W)_x$ for the inverse image of x under the mapping π. We denote by $\mathbf{S}_n$ the symmetric group of the set $\{1, ..., n\}$, which is well known to be generated by $n-1$ transpositions $\langle i, i+1 \rangle$ exchanging i and $i+1 (1 \leq i \leq n-1)$ while keeping the other elements fixed. Given $\sigma \in \mathbf{S}_n$ and $\gamma \in X \otimes \mathcal{W}_{D^n}$, we define $\gamma^\sigma \in X \otimes \mathcal{W}_{D^n}$ to be

$$\gamma^\sigma = \left(\mathrm{id}_X \otimes \mathcal{W}_{(d_1,...,d_n) \in D^n \mapsto (d_{\sigma(1)},...,d_{\sigma(n)}) \in D^n}\right)(\gamma)$$

Given $\alpha \in \mathbb{R}$ and $\gamma \in X \otimes \mathcal{W}_{D^n}$, we define $\alpha \underset{i}{\cdot} \gamma \in \gamma \in X \otimes \mathcal{W}_{D^n}$ $(1 \leq i \leq n)$ to be

$$\alpha \underset{i}{\cdot} \gamma = \left(\mathrm{id}_X \otimes \mathcal{W}_{(d_1,...,d_n) \in D^n \mapsto (d_1,...,d_{i-1},\alpha d_i,d_{i+1},...,d_n) \in D^n}\right)(\gamma)$$

Given $\alpha \in \mathbb{R}$ and $\gamma \in X \otimes \mathcal{W}_{D_n}$, we define $\alpha\gamma \in X \otimes \mathcal{W}_{D_n}$ $(1 \leq i \leq n)$ to be

$$\alpha\gamma = \left(\mathrm{id}_X \otimes \mathcal{W}_{d \in D_n \mapsto \alpha d \in D_n}\right)(\gamma)$$

for any $d \in D_n$. The restriction mapping $\gamma \in \mathbf{T}_x^{D_{n+1}}(M) \mapsto \gamma|_{D_n} \in \mathbf{T}_x^{D_n}(M)$ is often denoted by $\pi_{n+1,n}$.

Between $X \otimes \mathcal{W}_{D^n}$ and $X \otimes \mathcal{W}_{D^{n+1}}$ there are $2n+2$ canonical mappings:

$$X \otimes \mathcal{W}_{D^{n+1}} \underset{\mathbf{s}_i}{\overset{\mathbf{d}_i}{\rightleftarrows}} X \otimes \mathcal{W}_{D^n} \quad (1 \leq i \leq n+1)$$

For any $\gamma \in X \otimes \mathcal{W}_{D^n}$, we define $\mathbf{s}_i(\gamma) \in X \otimes \mathcal{W}_{D^{n+1}}$ to be

$$\mathbf{s}_i(\gamma) = \left(\mathrm{id}_X \otimes \mathcal{W}_{(d_1,...,d_{n+1}) \in D^{n+1} \mapsto (d_1,...,d_{i-1},d_{i+1},...,d_{n+1}) \in D^n}\right)(\gamma)$$

For any $\gamma \in X \otimes \mathcal{W}_{D^{n+1}}$, we define $\mathbf{d}_i(\gamma) \in X \otimes \mathcal{W}_{D^n}$ to be

$$\mathbf{d}_i(\gamma) = \left(\mathrm{id}_X \otimes \mathcal{W}_{(d_1,...,d_n) \in D^n \mapsto (d_1,...,d_{i-1},0,d_i,...,d_n) \in D^{n+1}}\right)(\gamma)$$

These operations satisfy the so-called simplicial identities (cf. Goerss and Jardine [3]), so that the family of $X \otimes \mathcal{W}_{D^n}$'s together with mappings $\mathbf{s}_i$'s and $\mathbf{d}_i$'s form a so-called simplicial set.

Synthetic differential geometry (usually abbreviated to SDG), which is a kind of differential geometry with a cornucopia of nilpotent infinitesimals, was forced to invent its models, in which nilpotent infinitesimals were visible. For a standard textbook on SDG, the reader is referred to [7], while he or she is referred to [4] for the model theory of SDG constructed vigorously by Dubuc [1] and others. Although we do not get involved in SDG herein, we will exploit locutions in terms of infinitesimal objects so as to make the paper highly readable. Thus we prefer to write $\mathcal{W}_D$ and $\mathcal{W}_{D_2}$ in place of $\mathbb{R}[X]/(X^2)$ and $\mathbb{R}[X]/(X^3)$ respectively, where D stands for the infinitesimal object of first-order nilpotent infinitesimals, and D_2 stands for the infinitesimal object of second-order nilpotent infinitesimals. To Newton and Leibniz, D stood for

$$\{d \in \mathbb{R} \mid d^2 = 0\}$$

while D_2 stood for

$$\{d \in \mathbb{R} \mid d^3 = 0\}$$

More generally, given a natural number n, we denote by D_n the set

$$\{d \in \mathbb{R} | d^{n+1} = 0\},$$

which stands for the infinitesimal object corresponding to the Weil algebra $\mathbb{R}[X]/(X^{n+1})$. Even more generally, given natural numbers m, n, we denote by $D(m)_n$ the infinitesimal object

$$\{(d_1, ..., d_m) \in \mathbb{R}^m | d_{i_1} ... d_{i_{n+1}} = 0\},$$

where $i_1, ..., i_{n+1}$ shall range over natural numbers between 1 and m including both ends. It corresponds to the Weil algebra $\mathbb{R}[X_1, ..., X_m]/I$, where I is the ideal generated by $X_{i_1} ... X_{i_{n+1}}$'s. Therefore we have

$$D(1)_n = D_n$$
$$D(m)_1 = D(m)$$

Trivially we have

$$D(m)_n \subseteq D(m)_{n+1}$$

It is easy to see that

$$D(m_1)_n \times D(m_2)_1 \subseteq D(m_1 + m_2)_{n+1}$$
$$D(m_1 + m_2)_n \subseteq D(m_1)_n \times D(m_2)_n$$

By convention, we have

$$D^0 = D_0 = \{0\} = 1$$

A polynomial ρ of $d \in D_n$ is called a *simple* polynomial of $d \in D_n$ if every coefficient of ρ is either 1 or 0, and if the constant term is 0. A simple polynomial ρ of $d \in D_n$ is said to be of dimension m, in notation $\dim(\rho) = m$, provided that m is the least integer with $\rho^{m+1} = 0$. By way of example, letting $d \in D_3$, we have

$$\dim(d) = \dim(d + d^2) = \dim(d + d^3) = 3$$
$$\dim(d^2) = \dim(d^3) = \dim(d^2 + d^3) = 1$$

We will write $\mathscr{W}_{d \in D_2 \mapsto d^2 \in D}$ for the homomorphism of Weil algebras $\mathbb{R}[X]/(X^2) \to \mathbb{R}[X]/(X^3)$ induced by the homomorphism $X \to X^2$ of the polynomial ring $\mathbb{R}[X]$ to itself. Such locutions are justifiable, because the category **W** of Weil algebras in the real world and the category **D** of infinitesimal objects in the shade are dual to each other in a sense. Thus we have a contravariant functor $\mathscr{W}$ from the category of infinitesimal objects in the shade to the category of Weil algebras in the real world. Its inverse contravariant functor from the category of Weil algebras in the real world to the category of infinitesimal objects in the shade is denoted by $\mathscr{D}$. By way of example, $\mathscr{D}_{\mathbb{R}[X]/(X^2)}$ and $\mathscr{D}_{\mathbb{R}[X]/(X^3)}$ stand for D and D_2, respectively. Since the category **W** is left exact, the category **D** is right exact, in which we write $\mathbb{D} \oplus \mathbb{D}'$ for the coproduct of infinitesimal objects $\mathbb{D}$ and $\mathbb{D}'$. For any two infinitesimal objects $\mathbb{D}, \mathbb{D}'$ with $\mathbb{D} \subseteq \mathbb{D}'$, we write i or $i_{\mathbb{D} \to \mathbb{D}'}$ for its natural injection of $\mathbb{D}$ into $\mathbb{D}'$. We write $\mathbf{m}$ or $\mathbf{m}_{D_n \times D_m \to D_n}$ for the mapping $(d, d') \in D_n \times D_m \mapsto dd' \in D_n$. Given $\alpha \in \mathbb{R}$, we write $\left(\underset{i}{\alpha \cdot}\right)_{D^n}$ for the mapping

$$(d_1, ..., d_n) \in D^n \mapsto (d_1, ...d_{i-1}, \alpha d_i, d_{i+1}, ..., d_n) \in D^n$$

To familiarize himself or herself with such locutions, the reader is strongly encouraged to read the first two chapters of [7], even if he or she is not interested in SDG at all.

2.2.2 Simplicial Infinitesimal Objects

Definition 2.1 1. Simplicial infinitesimal spaces are objects of the form

$$D\{m; \mathscr{S}\} = \{(d_1, ..., d_m) \in D^m | d_{i_1} ... d_{i_k} = 0 \text{ for any } (i_1, ..., i_k) \in \mathscr{S}\},$$

where $\mathscr{S}$ is a finite set of sequences $(i_1, ..., i_k)$ of natural numbers with $1 \leq i_1 < \cdots < i_k \leq m$.

2. A simplicial infinitesimal object $D\{m; \mathscr{S}\}$ is said to be *symmetric* if $(d_1, ..., d_m) \in D\{m; \mathscr{S}\}$ and $\sigma \in \mathbf{S}_m$ always imply $(d_{\sigma(1)}, ..., d_{\sigma(m)}) \in D\{m; \mathscr{S}\}$.

To give examples of simplicial infinitesimal spaces, we have

$$D(2) = D\{2; (1,2)\}$$
$$D(3) = D\{3; (1,2), (1,3), (2,3)\},$$

which are all symmetric.

Definition 2.2 1. The number m is called the *degree* of $D\{m; \mathcal{S}\}$, in notation: $m = \deg D\{m; \mathcal{S}\}$.

2. The maximum number n such that there exists a sequence $(i_1, ..., i_n)$ of natural numbers of length n with $1 \leq i_1 < \cdots < i_n \leq m$ containing no subsequence in $\mathcal{S}$ is called the *dimension* of $D\{m; \mathcal{S}\}$, in notation: $n = \dim D\{m; \mathcal{S}\}$.

By way of example, we have

$$\deg D(3) = \deg D\{3; (1,2)\} = \deg D\{3; (1,2), (1,3)\} = \deg D^3 = 3$$
$$\dim D(3) = 1$$
$$\dim D\{3; (1,2)\} = \dim D\{3; (1,2), (1,3)\} = 2$$
$$\dim D^3 = 3$$

It is easy to see that

Proposition 2.1 *if* $n = \dim D\{m; \mathcal{S}\}$, *then*

$$d_1 + \cdots + d_m \in D_n$$

for any $(d_1, ..., d_m) \in D\{m; \mathcal{S}\}$, *so that we have the mapping*

$$+_{D\{m;\mathcal{S}\}\to D_n} : D\{m; \mathcal{S}\} \to D_n$$

Definition 2.3 Infinitesimal objects of the form D^m are called basic infinitesimal objects.

Definition 2.4 Given two simplicial infinitesimal objects $D\{m; \mathcal{S}\}$ and $D\left\{m'; \mathcal{S}'\right\}$, a mapping

$$\varphi = (\varphi_1, ..., \varphi_{m'}) : D\{m; \mathcal{S}\} \to D\left\{m'; \mathcal{S}'\right\}$$

is called a *monomial mapping* if every φ_j is a monomial in $d_1, ..., d_m$ with coefficient 1.

Notation 2.2 *We denote by* $D\{m\}_n$ *the infinitesimal object*

$$\{(d_1, ..., d_m) \in D^m | d_{i_1} ... d_{i_{n+1}} = 0\},$$

where $i_1, ..., i_{n+1}$ *shall range over natural numbers between* 1 *and* m *including both ends.*

2.2.3 Quasi-Colimit Diagrams

Definition 2.5 A diagram in the category **D** is called a quasi-colimit diagram if its dually corresponding diagram in the category **W** is a limit diagram.

Theorem 2.3 (The Fundamental Theorem on Simplicial Infinitesimal Objects) *Any simplicial infinitesimal object $\mathbb{D}$ of dimension n is the quasi-colimit of a finite diagram whose objects are of the form D^k's $(0 \leq k \leq n)$ and whose arrows are natural injections.*

Proof Let $\mathbb{D} = \mathscr{D}(m; \mathscr{S})$. For any maximal sequence $1 \leq i_1 < \cdots < i_k \leq m$ of natural numbers containing no subsequence in $\mathscr{S}$ (maximal in the sense that it is not a proper subsequence of such a sequence), we have a natural injection of D^k into $\mathbb{D}$. By collecting all such D^k's together with their natural injections into $\mathbb{D}$, we have an overlapping representation of $\mathbb{D}$ in terms of basic infinitesimal spaces. This representation is completed into a quasi-colimit representation of $\mathbb{D}$ by taking D^l together with its natural injections into D^{k_1} and D^{k_2} for any two basic infinitesimal spaces D^{k_1} and D^{k_2} in the overlapping representation of $\mathbb{D}$, where if D^{k_1} and D^{k_2} come from the sequences $1 \leq i_1 < \cdots < i_{k_1} \leq m$ and $1 \leq \bar{i}_1 < \cdots < \bar{i}_{k_2} \leq m$ in the above manner, then D^l together with its natural injections into D^{k_1} and D^{k_2} comes from the maximal common subsequence $1 \leq \tilde{i}_1 < \cdots < \tilde{i}_l \leq m$ of both the preceding sequences of natural numbers in the above manner. By way of example, the above method leads to the following quasi-colimit representation of $\mathbb{D}{=}D\,\{3\}_2$:

$$
\begin{array}{ccccc}
 & & D^2 & & \\
 & i_1 \nearrow & & \nwarrow i_2 & \\
D & & \downarrow i_{12} & & D \\
i_1 \downarrow & & D(3)_2 & & \downarrow i_1 \\
D^2 & i_{13} \nearrow & & \nwarrow i_{23} & D^2 \\
 & i_2 \nwarrow & & \nearrow i_2 & \\
 & & D & &
\end{array}
$$

In the above representation i_{jk}'s and i_j's are as follows:

1. the j-th and k-th components of $i_{jk}(d_1, d_2) \in D(3)_2$ are d_1 and d_2, respectively, while the remaining component is 0;
2. the j-th component of $i_j(d) \in D^2$ is d, while the other component is 0.

Definition 2.6 The quasi-colimit representation of $\mathbb{D}$ depicted in the proof of the above theorem is called *standard*.

Remark 2.1 Generally speaking, there are multiple ways of quasi-colimit representation of a given simplicial infinitesimal space. By way of example, two quasi-colimit representations of $D\,\{3; (1, 3), (2, 3)\}\,(= (D \times D) \oplus D)$ were given in Lavendhomme [7, pp. 92–93] (§3.4, pp. 92–93), only the second one being standard.

2.3 *Weil-Exponentiability and Microlinearity*

2.3.1 Weil-Exponentiability

We have no reason to hold that all Frölicher spaces credit Weil prolongations as exponentiations by infinitesimal objects in the shade. Therefore we need a notion which distinguishes Frölicher spaces that do so from those that do not.

Definition 2.7 A Frölicher space X is called *Weil exponentiable* if

$$(X \otimes (W_1 \otimes_\infty W_2))^Y = (X \otimes W_1)^Y \otimes W_2 \tag{1}$$

holds naturally for any Frölicher space Y and any Weil algebras W_1 and W_2.

If $Y = 1$, then (1) degenerates into

$$X \otimes (W_1 \otimes_\infty W_2) = (X \otimes W_1) \otimes W_2$$

If $W_1 = \mathbb{R}$, then (1) degenerates into

$$(X \otimes W_2)^Y = X^Y \otimes W_2$$

The following three propositions have been established in our previous paper [14].

Proposition 2.2 *Convenient vector spaces are Weil exponentiable.*

Corollary 2.1 *C^∞-manifolds in the sense of* [6] *(cf. Section 27) are Weil exponentiable.*

Proposition 2.3 *If X is a Weil exponentiable Frölicher space, then so is $X \otimes W$ for any Weil algebra W.*

Proposition 2.4 *If X and Y are Weil exponentiable Frölicher spaces, then so is $X \times Y$.*

The last proposition can be strengthened to

Proposition 2.5 *The limit of a diagram in* **FS** *whose objects are all Weil-exponentiable is also Weil-exponentiable.*

Proof Let Γ be a diagram in **FS**. Given a Weil algebra W, we write $\Gamma \otimes W$ for the diagram obtained from Γ by putting $\otimes W$ to the right of every object in Γ and $\otimes \mathrm{id}_W$ to the right of every morphism in Γ.We have

$$
\begin{aligned}
&((\mathrm{Lim}\ \Gamma) \otimes (W_1 \otimes_\infty W_2))^Y \\
&= (\mathrm{Lim}\ (\Gamma \otimes (W_1 \otimes_\infty W_2)))^Y \\
&= \mathrm{Lim}\ (\Gamma \otimes (W_1 \otimes_\infty W_2))^Y \\
&= \mathrm{Lim}\ \left((\Gamma \otimes W_1)^Y \otimes W_2\right) \\
&= \left(\mathrm{Lim}\ (\Gamma \otimes W_1)^Y\right) \otimes W_2 \\
&= (\mathrm{Lim}\ (\Gamma \otimes W_1))^Y \otimes W_2 \\
&= ((\mathrm{Lim}\ \Gamma) \otimes W_1)^Y \otimes W_2
\end{aligned}
$$

so that we have the coveted result.

We have already established the following proposition and theorem in in our previous paper [14].

Proposition 2.6 *If X is a Weil exponentiable Frölicher space, then so is X^Y for any Frölicher space Y.*

Theorem 2.4 *Weil exponentiable Frölicher spaces, together with smooth mappings among them, form a Cartesian closed subcategory* $\mathbf{FS}_{\mathbf{WE}}$ *of the category* **FS**.

2.3.2 Microlinearity

The central object of study in SDG is *microlinear* spaces. Although the notion of a manifold (=a pasting of copies of a certain linear space) is defined on the local level, the notion of microlinearity is defined on the genuinely infinitesimal level. For the historical account of microlinearity, the reader is referred to §§2.4 of [7] or Appendix D of [4]. To get an adequately restricted cartesian closed subcategory of Frölicher spaces, we have emancipated microlinearity from within a well-adapted model of SDG to Frölicher spaces in the real world in [15]. Recall that

Definition 2.8 A Frölicher space X is called *microlinear* providing that any finite limit diagram Γ in **W** yields a limit diagram $X \otimes \Gamma$ in **FS**, where $X \otimes \Gamma$ is obtained from Γ by putting $X\otimes$ to the left of every object in Γ and $\mathrm{id}_X\otimes$ to the left of every morphism in Γ.

Generally speaking, limits in the category **FS** are bamboozling. The notion of limit in **FS** should be elaborated geometrically.

Definition 2.9 A finite cone Γ in **FS** is called a *transversal limit diagram* providing that $\Gamma \otimes W$ is a limit diagram in **FS** for any Weil algebra W, where the diagram $\Gamma \otimes W$ is obtained from Γ by putting $\otimes W$ to the right of every object in Γ and $\otimes\mathrm{id}_W$ to the right of every morphism in Γ. The limit of a finite diagram of Frölicher spaces is said to be *transversal* providing that its limit diagram is a transversal limit diagram.

Remark 2.2 By taking $W = \mathbb{R}$, we see that a transversal limit diagram in **FS** is always a limit diagram in **FS**.

We have already established the following two propositions in [15].

Proposition 2.7 *If Γ is a transversal limit diagram in* **FS** *whose objects are all Weil exponentiable, then Γ^X is also a transversal limit diagram for any Frölicher space X, where Γ^X is obtained from Γ by putting X as the exponential over every object in Γ and over every morphism in Γ.*

Proposition 2.8 *If Γ is a transversal limit diagram in* **FS** *whose objects are all Weil exponentiable, then $\Gamma \otimes W$ is also a transversal limit diagram for any Weil algebra W.*

The following results have been established in [15].

Proposition 2.9 *Convenient vector spaces are microlinear.*

Corollary 2.2 *C^∞-manifolds in the sense of* [6] *(cf. Section 27) are microlinear.*

Proposition 2.10 *If X is a Weil exponentiable and microlinear Frölicher space, then so is $X \otimes W$ for any Weil algebra W.*

Proposition 2.11 *The class of microlinear Frölicher spaces is closed under transversal limits.*

Corollary 2.3 *Direct products are transversal limits, so that if X and Y are microlinear Frölicher spaces, then so is $X \times Y$.*

Proposition 2.12 *If X is a Weil exponentiable and microlinear Frölicher space, then so is X^Y for any Frölicher space Y.*

Proposition 2.13 *If a Weil exponentiable Frölicher space X is microlinear, then any finite limit diagram Γ in* **W** *yields a transversal limit diagram $X \otimes \Gamma$ in* **FS**.

Theorem 2.5 *Weil exponentiable and microlinear Frölicher spaces, together with smooth mappings among them, form a cartesian closed subcategory* $\mathbf{FS}_{\mathbf{WE},\mathbf{ML}}$ *of the category* **FS**.

2.4 Convention

Unless stated to the contrary, every Frölicher space occurring in the sequel is assumed to be microlinear and Weil exponentiable. We will fix a smooth mapping $\pi: E \to M$ arbitrarily. In this paper we will naively speak of *bundles* simply as smooth mappings of microlinear and Weil exponentiable Frölicher spaces, for which we will develop three theories of jet bundles. We say that $t \in M \otimes \mathcal{W}_D$ is *degenerate* providing that

$$t = \left(i_{\{x\}\to M} \otimes \mathrm{id}_{\mathcal{W}_D}\right)\left(t'\right)$$

for some $x \in M$ and some $t' \in \{x\} \otimes \mathcal{W}_D$. We say that $t \in E \otimes \mathcal{W}_D$ is *vertical* provided that $\left(\pi \otimes \mathrm{id}_{\mathcal{W}_D}\right)(t)$ is degenerate. We write $(E \otimes \mathcal{W}_D)^{\perp}$ for the totality of vertical $t \in E \otimes \mathcal{W}_D$.

3 The First Approach to Jets

Definition 3.1 A 1-*tangential over* the bundle $\pi : E \to M$ *at* $x \in E$ is a mapping $\nabla_x : (M \otimes \mathcal{W}_D)_{\pi(x)} \to (E \otimes \mathcal{W}_D)_x$ subject to the following three conditions:

1. We have
$$\left(\pi \otimes \mathrm{id}_{\mathcal{W}_D}\right)(\nabla_x(t)) = t$$
for any $t \in (M \otimes \mathcal{W}_D)_{\pi(x)}$.
2. We have
$$\nabla_x(\alpha t) = \alpha \nabla_x(t)$$
for any $t \in (M \otimes \mathcal{W}_D)_{\pi(x)}$ and any $\alpha \in \mathbb{R}$.
3. The diagram
$$\begin{array}{ccc} (M \otimes \mathcal{W}_D)_{\pi(x)} & \xrightarrow{\mathrm{id}_M \otimes \mathcal{W}_{(d,e)\in D\times D_m \mapsto ed \in D}} & (M \otimes \mathcal{W}_D)_{\pi(x)} \otimes \mathcal{W}_{D_m} \\ \nabla_x \downarrow & & \downarrow \nabla_x \otimes \mathrm{id}_{\mathcal{W}_{D_m}} \\ (E \otimes \mathcal{W}_D)_x & \xrightarrow[\mathrm{id}_E \otimes \mathcal{W}_{(d,e)\in D\times D_m \mapsto ed \in D}]{} & (E \otimes \mathcal{W}_D)_x \otimes \mathcal{W}_{D_m} \end{array}$$
is commutative, where m is an arbitrary natural number.

We note in passing that condition (1.2) implies that ∇_x is linear by dint of Proposition 10 in §1.2 of [7].

Notation 3.1 *We denote by* $\mathbf{J}^1_x(\pi)$ *the totality of* 1*-tangentials* ∇_x *over the bundle* $\pi : E \to M$ *at* $x \in E$. *We denote by* $\mathbf{J}^1(\pi)$ *the set-theoretic union of* $\mathbf{J}^1_x(\pi)$*'s for all* $x \in E$. *The canonical projection* $\mathbf{J}^1(\pi) \to E$ *is denoted by* $\pi_{1,0}$ *with*

$$\pi_1 = \left(\pi \otimes \mathrm{id}_{\mathcal{W}_D}\right) \circ \pi_{1,0}.$$

Definition 3.2 Let F be a morphism of bundles over M from π to π' over the same base space M. We say that a 1 *-tangential* ∇_x over π at a point x of E is F-related *to* a 1-*tangential* $\nabla_{F(x)}$ over π' at $F(x)$ of E' provided that

$$\left(F \otimes \mathrm{id}_{\mathcal{W}_D}\right)(\nabla_x(t)) = \nabla_{F(x)}(t)$$

for any $t \in (M \otimes \mathcal{W}_D)_{\pi(x)}$.

Notation 3.2 *By convention, we let*

$$\tilde{\mathbf{J}}^0(\pi) = \hat{\mathbf{J}}^0(\pi) = \mathbf{J}^0(\pi) = E$$

with

$$\tilde{\pi}_{0,0} = \hat{\pi}_{0,0} = \pi_{0,0} = \mathrm{id}_E$$

and

$$\tilde{\pi}_0 = \hat{\pi}_0 = \pi_0 = \pi$$

We let

$$\tilde{\mathbf{J}}^1(\pi) = \hat{\mathbf{J}}^1(\pi) = \mathbf{J}^1(\pi)$$

with

$$\tilde{\pi}_{1,0} = \hat{\pi}_{1,0} = \pi_{1,0}$$

and

$$\tilde{\pi}_1 = \hat{\pi}_1 = \pi_1$$

Notation 3.3 *Now we are going to define $\tilde{\mathbf{J}}^{k+1}(\pi)$, $\hat{\mathbf{J}}^{k+1}(\pi)$ and $\mathbf{J}^{k+1}(\pi)$ together with mappings $\tilde{\pi}_{k+1,k}$: $\tilde{\mathbf{J}}^{k+1}(\pi) \to \tilde{\mathbf{J}}^k(\pi)$, $\hat{\pi}_{k+1,k}$: $\hat{\mathbf{J}}^{k+1}(\pi) \to \hat{\mathbf{J}}^k(\pi)$ and $\pi_{k+1,k}$: $\mathbf{J}^{k+1}(\pi) \to \mathbf{J}^k(\pi)$ by induction on $k \geq 1$. Intuitively speaking, these are intended for non-holonomic, semi-holonomic and holonomic jet bundles in order. We let $\tilde{\pi}_{k+1} = \tilde{\pi}_k \circ \tilde{\pi}_{k+1,k}$, $\hat{\pi}_{k+1} = \hat{\pi}_k \circ \hat{\pi}_{k+1,k}$ and $\pi_{k+1} = \pi_k \circ \pi_{k+1,k}$.*

1. *First we deal with $\tilde{\mathbf{J}}^{k+1}(\pi)$, which is defined to be $\mathbf{J}^1(\tilde{\pi}_k)$ with $\tilde{\pi}_{k+1,k} = (\tilde{\pi}_k)_{1,0}$.*
2. *Next we deal with $\hat{\mathbf{J}}^{k+1}(\pi)$, which is defined to be the subspace of $\mathbf{J}^1(\hat{\pi}_k)$ consisting of ∇_x's with $x = \nabla_y \in \hat{\mathbf{J}}^k(\pi)$ abiding by the condition that ∇_x is $\hat{\pi}_{k,k-1}$-related to ∇_y.*
3. *Finally we deal with $\mathbf{J}^{k+1}(\pi)$, which is defined to be the subspace of $\mathbf{J}^1(\pi_k)$ consisting of ∇_x's with $x = \nabla_y \in \mathbf{J}^k(\pi)$ abiding by the conditions that ∇_x is $\pi_{k,k-1}$-related to ∇_y and that the composition of mappings*

$$\begin{aligned}
&\left(M \otimes \mathcal{W}_{D^2}\right)_{\pi_k(x)} \\
&\xrightarrow{\left\langle \mathrm{id}_M \otimes \mathcal{W}_{d\in D \mapsto (d,0)\in D^2},\, \mathrm{id}_M \otimes \mathcal{W}_{(d_1.d_2)\in D^2 \mapsto (d_2.d_1)\in D^2}\right\rangle} \\
&\left((M \otimes \mathcal{W}_D) \times_{M\otimes\mathcal{W}_D} \left(M \otimes \mathcal{W}_{D^2}\right)\right)_{\pi_k(x)} \\
&\xrightarrow{\nabla_x \times \mathrm{id}_{M\otimes\mathcal{W}_{D^2}}} \\
&\left(\left(\mathbf{J}^k(\pi) \otimes \mathcal{W}_D\right) \times_{M\otimes\mathcal{W}_D} \left(M \otimes \mathcal{W}_{D^2}\right)\right)_{\pi_k(x)} \\
&= \left(\left(\mathbf{J}^k(\pi) \otimes \mathcal{W}_D\right) \times_{M\otimes\mathcal{W}_D} \left((M \otimes \mathcal{W}_D) \otimes \mathcal{W}_D\right)\right)_{\pi_k(x)} \\
&= \left(\left(\mathbf{J}^k(\pi) \times_M (M \otimes \mathcal{W}_D)\right) \otimes \mathcal{W}_D\right)_{\pi_k(x)}
\end{aligned}$$

$$\begin{array}{l}
\xrightarrow{\left(\left(\nabla,t\right)\in\mathbf{J}^{k}\left(\pi\right)\times_{M}\left(M\otimes\mathcal{W}_{D}\right)\mapsto\nabla\left(t\right)\in\left(\mathbf{J}^{k-1}\left(\pi\right)\otimes\mathcal{W}_{D}\right)\right)\otimes\mathrm{id}_{\mathcal{W}_{D}}} \\
\left(\mathbf{J}^{k-1}\left(\pi\right)\otimes\mathcal{W}_{D}\right)\otimes\mathcal{W}_{D} \\
=\mathbf{J}^{k-1}\left(\pi\right)\otimes\mathcal{W}_{D^{2}}
\end{array}$$

is equal to the composition of mappings

$$\begin{array}{l}
\left(M\otimes\mathcal{W}_{D^{2}}\right)_{\pi_{k}(x)} \\
\xrightarrow{\left\langle\mathrm{id}_{M}\otimes\mathcal{W}_{d\in D\mapsto(0,d)\in D^{2}},\,\mathrm{id}_{M\otimes\mathcal{W}_{D^{2}}}\right\rangle} \\
\left(\left(M\otimes\mathcal{W}_{D}\right)\times_{M\otimes\mathcal{W}_{D}}\left(M\otimes\mathcal{W}_{D^{2}}\right)\right)_{\pi_{k}(x)} \\
\xrightarrow{\nabla_{x}\times\mathrm{id}_{M\otimes\mathcal{W}_{D^{2}}}} \\
\left(\left(\mathbf{J}^{k}\left(\pi\right)\otimes\mathcal{W}_{D}\right)\times_{M\otimes\mathcal{W}_{D}}\left(M\otimes\mathcal{W}_{D^{2}}\right)\right)_{\pi_{k}(x)} \\
=\left(\left(\mathbf{J}^{k}\left(\pi\right)\otimes\mathcal{W}_{D}\right)\times_{M\otimes\mathcal{W}_{D}}\left(\left(M\otimes\mathcal{W}_{D}\right)\otimes\mathcal{W}_{D}\right)\right)_{\pi_{k}(x)} \\
=\left(\left(\mathbf{J}^{k}\left(\pi\right)\times_{M}\left(M\otimes\mathcal{W}_{D}\right)\right)\otimes\mathcal{W}_{D}\right)_{\pi_{k}(x)} \\
\xrightarrow{\left(\left(\nabla,t\right)\in\mathbf{J}^{k}\left(\pi\right)\times_{M}\left(M\otimes\mathcal{W}_{D}\right)\mapsto\nabla\left(t\right)\in\left(\mathbf{J}^{k-1}\left(\pi\right)\otimes\mathcal{W}_{D}\right)\right)\otimes\mathrm{id}_{\mathcal{W}_{D}}} \\
\left(\mathbf{J}^{k-1}\left(\pi\right)\otimes\mathcal{W}_{D}\right)\otimes\mathcal{W}_{D} \\
=\mathbf{J}^{k-1}\left(\pi\right)\otimes\mathcal{W}_{D^{2}} \\
\xrightarrow{\mathrm{id}_{\mathbf{J}^{k-1}(\pi)}\otimes\mathcal{W}_{(d_{1},d_{2})\in D^{2}\mapsto(d_{2},d_{1})\in D^{2}}} \\
\mathbf{J}^{k-1}\left(\pi\right)\otimes\mathcal{W}_{D^{2}}
\end{array}$$

Definition 3.3 Elements of $\tilde{\mathbf{J}}^{n}(\pi)$ are called n-subtangentials, while elements of $\hat{\mathbf{J}}^{n}(\pi)$ are called n-quasitangentials. Elements of $\mathbf{J}^{n}(\pi)$ are called n-tangentials.

4 The Second Approach to Jets

Definition 4.1 Let n be a natural number. A D^n-pseudotangential *over* the bundle $\pi: E \to M$ *at* $x \in E$ is a mapping $\nabla_x: (M \otimes \mathcal{W}_{D^n})_{\pi(x)} \to (E \otimes \mathcal{W}_{D^n})_x$ abiding by the following conditions:

1. We have

$$\left(\pi\otimes\mathrm{id}_{\mathcal{W}_{D^{n}}}\right)\left(\nabla_{x}(\gamma)\right)=\gamma$$

for any $\gamma \in (M \otimes \mathcal{W}_{D^n})_{\pi(x)}$.

2. We have

$$\nabla_x(\alpha \underset{i}{\cdot} \gamma) = \alpha \underset{i}{\cdot} \nabla_x(\gamma) \qquad (1 \leq i \leq n)$$

for any $\gamma \in (M \otimes \mathcal{W}_{D^n})_{\pi(x)}$ and any $\alpha \in \mathbb{R}$.

3. The diagram

$$\begin{array}{ccc} (M \otimes \mathcal{W}_{D^n})_{\pi(x)} & \rightarrow & (M \otimes \mathcal{W}_{D^n})_{\pi(x)} \otimes \mathcal{W}_{D_m} \\ \nabla_x \downarrow & & \downarrow \nabla_x \otimes \mathrm{id}_{\mathcal{W}_{D_m}} \\ (E \otimes \mathcal{W}_{D^n})_x & \rightarrow & (E \otimes \mathcal{W}_{D^n})_x \otimes \mathcal{W}_{D_m} \end{array}$$

is commutative, where m is an arbitrary natural number, the upper horizontal arrow is

$$\mathrm{id}_M \otimes \mathcal{W}_{(d_1,\ldots,d_n,e) \in D^n \times D_m \mapsto (d_1,\ldots,d_{i-1},ed_i,d_{i+1},\ldots d_n) \in D^n},$$

and the lower horizontal arrow is

$$\mathrm{id}_E \otimes \mathcal{W}_{(d_1,\ldots,d_n,e) \in D^n \times D_m \mapsto (d_1,\ldots,d_{i-1},ed_i,d_{i+1},\ldots d_n) \in D^n}.$$

4. We have

$$\nabla_x(\gamma^\sigma) = (\nabla_x(\gamma))^\sigma$$

for any $\gamma \in (M \otimes \mathcal{W}_{D^n})_{\pi(x)}$ and for any $\sigma \in \mathbf{S}_n$.

Remark 4.1 The third condition in the above definition claims what is called infinitesimal multilinearity, while the second claims what is authentic multilinearity.

Notation 4.1 *We denote by* $\hat{\mathbb{J}}_x^{D^n}(\pi)$ *the totality of* D^n*-pseudotangentials* ∇_x *over the bundle* $\pi : E \to M$ *at* $x \in E$. *We denote by* $\hat{\mathbb{J}}^{D^n}(\pi)$ *the set-theoretic union of* $\hat{\mathbb{J}}_x^{D^n}(\pi)$*'s for all* $x \in E$. *In particular,* $\hat{\mathbb{J}}^{D^0}(\pi) = E$ *by convention.*

Lemma 4.1 *The diagram*

$$E \otimes \mathcal{W}_{D^n} \xrightarrow{\mathrm{id}_E \otimes \mathcal{W}_{(d_1,\ldots,d_n,d_{n+1}) \in D^{n+1} \mapsto (d_1,\ldots,d_n) \in D^n}} E \otimes \mathcal{W}_{D^{n+1}}$$

$$\underset{\mathrm{id}_E \otimes \mathcal{W}_{(d_1,\ldots,d_n,d_{n+1}) \in D^{n+1} \mapsto (d_1,\ldots,d_n,0) \in D^{n+1}}}{\overset{\mathrm{id}_{E \otimes \mathcal{W}_{D^{n+1}}}}{\rightrightarrows}} E \otimes \mathcal{W}_{D^{n+1}}$$

is an equalizer.

Proof It is well known that the diagram

$$\mathcal{W}_{D^n} \xrightarrow{\mathcal{W}_{(d_1,\ldots,d_n,d_{n+1}) \in D^{n+1} \mapsto (d_1,\ldots,d_n) \in D^n}} \mathcal{W}_{D^{n+1}} \underset{\mathcal{W}_{(d_1,\ldots,d_n,d_{n+1}) \in D^{n+1} \mapsto (d_1,\ldots,d_n,0) \in D^{n+1}}}{\overset{\mathrm{id}_{\mathcal{W}_{D^{n+1}}}}{\rightrightarrows}} \mathcal{W}_{D^{n+1}}$$

is an equalizer in the category of Weil algebras, so that the desired result follows from the microlinearity of E.

Corollary 4.1 *$\gamma \in E \otimes \mathcal{W}_{D^{n+1}}$ is in the equalizer of*

$$E \otimes \mathcal{W}_{D^{n+1}} \underset{\mathrm{id}_E \otimes \mathcal{W}_{(d_1,\ldots,d_n,d_{n+1}) \in D^{n+1} \mapsto (d_1,\ldots,d_n,0) \in D^{n+1}}}{\overset{\mathrm{id}_{E \otimes \mathcal{W}_{D^{n+1}}}}{\rightrightarrows}} E \otimes \mathcal{W}_{D^{n+1}}$$

iff

$$\gamma = (\mathbf{s}_{n+1} \circ \mathbf{d}_{n+1})(\gamma)$$

Proof This follows simply from

$$\mathbf{s}_{n+1} \circ \mathbf{d}_{n+1} = \mathrm{id}_E \otimes \mathcal{W}_{(d_1,\ldots,d_n,d_{n+1}) \in D^{n+1} \mapsto (d_1,\ldots,d_n,0) \in D^{n+1}}$$

Proposition 4.1 *Let ∇_x be a D^{n+1}-pseudotangential over the bundle $\pi : E \to M$ at $x \in E$. Let $\gamma \in (M \otimes \mathcal{W}_{D^n})_{\pi(x)}$. Then we have*

$$\nabla_x(\mathbf{s}_{n+1}(\gamma)) = \left(\mathrm{id}_E \otimes \mathcal{W}_{(d_1,\ldots,d_n,d_{n+1}) \in D^{n+1} \mapsto (d_1,\ldots,d_n,0) \in D^{n+1}}\right)(\nabla_x(\mathbf{s}_{n+1}(\gamma)))$$

so that

$$\nabla_x(\mathbf{s}_{n+1}(\gamma)) = (\mathbf{s}_{n+1} \circ \mathbf{d}_{n+1})(\nabla_x(\mathbf{s}_{n+1}(\gamma)))$$

Proof For any $\alpha \in \mathbb{R}$, we have

$$\begin{aligned} &\alpha \underset{n+1}{\cdot} (\nabla_x(\mathbf{s}_{n+1}(\gamma))) \\ &= \nabla_x(\alpha \underset{n+1}{\cdot} (\mathbf{s}_{n+1}(\gamma))) \\ &= \nabla_x(\mathbf{s}_{n+1}(\gamma)) \end{aligned}$$

Therefore we have the desired result by letting $\alpha = 0$ in the above calculation.

Corollary 4.2 *The assignment*

$$\gamma \in (M \otimes \mathcal{W}_{D^n})_{\pi(x)} \longmapsto \mathbf{d}_{n+1}(\nabla_x(\mathbf{s}_{n+1}(\gamma))) \in (E \otimes \mathcal{W}_{D^n})_x$$

is an n-pseudotangential over the bundle $\pi : E \to M$ at x.

Notation 4.2 *By this Corollary, we have canonical projections $\widehat{\pi}_{n+1,n}: \hat{\mathbb{J}}^{D^{n+1}}(\pi) \to \hat{\mathbb{J}}^{D^n}(\pi)$. By assigning $\pi(x) \in M$ to each n-pseudotangential ∇_x over the bundle $\pi: E \to M$ at $x \in E$, we have the canonical projections $\widehat{\pi}_n : \hat{\mathbb{J}}^{D^n}(\pi) \to M$. Note that $\widehat{\pi}_n \circ \widehat{\pi}_{n+1,n} = \widehat{\pi}_{n+1}$ For any natural numbers n, m with $m \leq n$, we define $\widehat{\pi}_{n,m}$: $\hat{\mathbb{J}}^{D^n}(\pi) \to \hat{\mathbb{J}}^{D^m}(\pi)$ to be $\widehat{\pi}_{m+1,m} \circ \ldots \circ \widehat{\pi}_{n,n-1}$.*

Now we are going to show that

Proposition 4.2 *Let* $\nabla_x \in \hat{\mathbb{J}}^{D^{n+1}}(\pi)$. *Then the following diagrams are commutative:*

$$\begin{array}{ccc} \left(M \otimes \mathcal{W}_{D^{n+1}}\right)_{\pi(x)} & \xrightarrow{\nabla_x} & \left(E \otimes \mathcal{W}_{D^{n+1}}\right)_x \\ \mathbf{s}_i \uparrow & & \uparrow \mathbf{s}_i \\ (M \otimes \mathcal{W}_{D^n})_{\pi(x)} & \xrightarrow[\widehat{\pi}_{n+1,n}(\nabla_x)]{} & (E \otimes \mathcal{W}_{D^n})_x \end{array}$$

$$\begin{array}{ccc} \left(M \otimes \mathcal{W}_{D^{n+1}}\right)_{\pi(x)} & \xrightarrow{\nabla_x} & \left(E \otimes \mathcal{W}_{D^{n+1}}\right)_x \\ \mathbf{d}_i \downarrow & & \downarrow \mathbf{d}_i \\ (M \otimes \mathcal{W}_{D^n})_{\pi(x)} & \xrightarrow[\widehat{\pi}_{n+1,n}(\nabla_x)]{} & (E \otimes \mathcal{W}_{D^n})_x \end{array}$$

Proof By the very definition of $\widehat{\pi}_{n+1,n}$, we have

$$\mathbf{s}_{n+1}(\widehat{\pi}_{n+1}(\nabla_x)(\gamma)) = \nabla_x(\mathbf{s}_{n+1}(\gamma))$$

for any $\gamma \in (M \otimes \mathcal{W}_{D^n})_{\pi(x)}$. For $i \neq n+1$, we have

$$\begin{aligned} & \mathbf{s}_i(\widehat{\pi}_{n+1,n}(\nabla_x)(\gamma)) \\ & = \left(\left(\mathbf{s}_{n+1}(\widehat{\pi}_{n+1,n}(\nabla_x)(\gamma))\right)^{\langle i,n+1\rangle}\right)^{\langle i+1,i+2,\ldots,n,n+1\rangle} \\ & = \left((\nabla_x(\mathbf{s}_{n+1}(\gamma)))^{\langle i,n+1\rangle}\right)^{\langle i+1,i+2,\ldots,n,n+1\rangle} \\ & = \left(\nabla_x\left((\mathbf{s}_{n+1}(\gamma))^{\langle i,n+1\rangle}\right)\right)^{\langle i+1,i+2,\ldots,n,n+1\rangle} \\ & = \nabla_x\left(\left((\mathbf{s}_{n+1}(\gamma))^{\langle i,n+1\rangle}\right)^{\langle i+1,i+2,\ldots,n,n+1\rangle}\right) \\ & = \nabla_x\left(\mathbf{s}_i\left(\gamma\right)\right) \end{aligned}$$

Now we are going to show that

$$\mathbf{d}_i(\nabla_x(\gamma)) = (\widehat{\pi}_{n+1,n}(\nabla_x))(\mathbf{d}_i(\gamma))$$

for any $\gamma \in \left(M \otimes \mathcal{W}_{D^{n+1}}\right)_{\pi(x)}$. First we deal with the case of $i = n+1$. We have

$$\begin{aligned} & \mathbf{d}_{n+1}(\nabla_x(\gamma)) \\ & = \mathbf{d}_{n+1}(0 \underset{n+1}{\cdot} \nabla_x(\gamma)) \\ & = \mathbf{d}_{n+1}(\nabla_x(0 \underset{n+1}{\cdot} \gamma)) \\ & = \mathbf{d}_{n+1}(\nabla_x(\mathbf{s}_{n+1}(\mathbf{d}_{n+1}(\gamma)))) \\ & = (\widehat{\pi}_{n+1,n}(\nabla_x))(\mathbf{d}_{n+1}(\gamma)) \end{aligned}$$

For $i \neq n+1$, we have

$$\begin{aligned}
&\mathbf{d}_i(\nabla_x(\gamma)) \\
&= \left(\mathbf{d}_{n+1}\left((\nabla_x(\gamma))^{\langle i,n+1\rangle}\right)\right)^{\langle n,n-1,\ldots,i+1,i\rangle} \\
&= \left(\mathbf{d}_{n+1}(\nabla_x(\gamma^{\langle i,n+1\rangle}))\right)^{\langle n,n-1,\ldots,i+1,i\rangle} \\
&= (((\widehat{\pi}_{n+1,n}(\nabla_x))\left(\mathbf{d}_{n+1}(\gamma^{\langle i,n+1\rangle}))\right)^{\langle n,n-1,\ldots,i+1,i\rangle} \\
&= (\widehat{\pi}_{n+1,n}(\nabla_x))\left(\left(\mathbf{d}_{n+1}(\gamma^{\langle i,n+1\rangle})\right)^{\langle n,n-1,\ldots,i+1,i\rangle}\right) \\
&= (\widehat{\pi}_{n+1,n}(\nabla_x))\,(\mathbf{d}_i(\gamma))
\end{aligned}$$

Thus we are done through.

Corollary 4.3 *Let* $\nabla_x^+, \nabla_x^- \in \hat{\mathbb{J}}^{D^{n+1}}(\pi)$ *with*

$$\widehat{\pi}_{n+1,n}(\nabla_x^+) = \widehat{\pi}_{n+1,n}(\nabla_x^-)$$

Then

$$\left(\mathrm{id}_E \otimes \mathscr{W}_{i_{D\{n+1\}_n \to D^{n+1}}}\right)(\nabla_x^+(\gamma)) = \left(\mathrm{id}_E \otimes \mathscr{W}_{i_{D\{n+1\}_n \to D^{n+1}}}\right)(\nabla_x^-(\gamma))$$

for any $\gamma \in \left(M \otimes \mathscr{W}_{D^{n+1}}\right)_{\pi(x)}$.

Definition 4.2 The notion of a D^n-tangential *over* the bundle $\pi: E \to M$ *at* x is defined by induction on n. The notion of a D-tangential *over* the bundle $\pi: E \to M$ *at* x shall be identical with that of a D-pseudotangential *over* the bundle $\pi: E \to M$ *at* x . Now we proceed inductively. A D^{n+1}-pseudotangential

$$\nabla_x : \left(M \otimes \mathscr{W}_{D^{n+1}}\right)_{\pi(x)} \to \left(E \otimes \mathscr{W}_{D^{n+1}}\right)_x$$

over the bundle $\pi: E \to M$ at $x \in E$ is called a D^{n+1}-tangential *over* the bundle $\pi: E \to M$ *at* x if it acquiesces in the following two conditions:

1. $\widehat{\pi}_{n+1,n}(\nabla_x)$ is a D^n-tangential *over* the bundle $\pi: E \to M$ *at* x.
2. For any $\gamma \in (M \otimes \mathscr{W}_{D^n})_{\pi(x)}$, we have

$$\begin{aligned}
&\nabla_x\left(\left(\mathrm{id}_M \otimes \mathscr{W}_{(d_1,\ldots,d_n,d_{n+1})\in D^{n+1}\mapsto(d_1,\ldots,d_n d_{n+1})\in D^n}\right)(\gamma)\right) \\
&= \left(\mathrm{id}_E \otimes \mathscr{W}_{(d_1,\ldots,d_n,d_{n+1})\in D^{n+1}\mapsto(d_1,\ldots,d_n d_{n+1})\in D^{n+1}}\right)\left((\widehat{\pi}_{n+1,n}(\nabla_x))\,(\gamma)\right)
\end{aligned}$$

Notation 4.3 *We denote by* $\mathbb{J}_x^{D^n}(\pi)$ *the totality of* D^n*-tangentials* ∇_x *over the bundle* $\pi: E \to M$ *at* $x \in E$*. We denote by* $\mathbb{J}^{D^n}(\pi)$ *the set-theoretic union of* $\mathbb{J}_x^{D^n}(\pi)$*'s for all* $x \in E$*. In particular,* $\mathbb{J}^{D^0}(\pi) = \hat{\mathbb{J}}^{D^0}(\pi) = E$ *by convention and* $\mathbb{J}^D(\pi) = \hat{\mathbb{J}}^D(\pi)$ *by definition. By the very definition of* D^n*-tangential, the projections* $\widehat{\pi}_{n+1,n}$*:*

$\hat{\mathbb{J}}^{D^{n+1}}(\pi) \to \hat{\mathbb{J}}^{D^n}(\pi)$ *are naturally restricted to mappings* $\pi_{n+1,n}\colon \mathbb{J}^{D^{n+1}}(\pi) \to \mathbb{J}^{D^n}(\pi)$. *Similarly for* $\pi_n\colon \mathbb{J}^{D^n}(\pi) \to M$ *and* $\pi_{n,m}\colon \mathbb{J}^{D^n}(\pi) \to \mathbb{J}^{D^m}(\pi)$ *with* $m \leq n$.

It is easy to see that

Proposition 4.3 *Let* m, n *be natural numbers with* $m \leq n$. *Let* $k_1, \ldots, k_m$ *be positive integers with* $k_1 + \cdots + k_m = n$. *For any* $\nabla_x \in \mathbb{J}^{D^n}(\pi)$, *any* $\gamma \in (M \otimes \mathcal{W}_{D^m})_{\pi(x)}$ *and any* $\sigma \in \mathbf{S}_n$, *we have*

$$\nabla_x \left(\left(\mathrm{id}_M \otimes \mathcal{W}_{(d_1,\ldots,d_n)\in D^n \mapsto \left(d_{\sigma(1)}\ldots d_{\sigma(k_1)}, d_{\sigma(k_1+1)}\ldots d_{\sigma(k_1+k_2)}, \ldots, d_{\sigma(k_1+\ldots+k_{m-1}+1)}\ldots d_{\sigma(n)}\right)} \right)(\gamma) \right)$$
$$= \left(\mathrm{id}_E \otimes \mathcal{W}_{(d_1,\ldots,d_n)\in D^n \mapsto \left(d_{\sigma(1)}\ldots d_{\sigma(k_1)}, d_{\sigma(k_1+1)}\ldots d_{\sigma(k_1+k_2)}, \ldots, d_{\sigma(k_1+\ldots+k_{m-1}+1)}\ldots d_{\sigma(n)}\right)} \right)$$
$$\left(\left(\pi_{n,m}(\nabla_x)\right)(\gamma) \right)$$

Interestingly enough, any D^n-pseudotangential naturally gives rise to what might be called a $\mathbb{D}$-*pseudotangential* for any simplicial infinitesimal space $\mathbb{D}$ of dimension less than or equal to n.

Theorem 4.4 *Let* n *be a natural number. Let* $\mathbb{D}$ *be a simplicial infinitesimal space of dimension less than or equal to* n. *Any* D^n-*pseudotangential* ∇_x *over the bundle* $\pi\colon E \to M$ *at* $x \in E$ *naturally induces a mapping* $\nabla_x^{\mathbb{D}}\colon (M \otimes \mathcal{W}_{\mathbb{D}})_{\pi(x)} \to (E \otimes \mathcal{W}_{\mathbb{D}})_x$ *abiding by the following three conditions:*

1. *We have*
$$\left(\pi \otimes \mathrm{id}_{\mathcal{W}_{\mathbb{D}}}\right)\left(\nabla_x^{\mathbb{D}}(\gamma)\right) = \gamma$$
for any $\gamma \in (M \otimes \mathcal{W}_{\mathbb{D}})_{\pi(x)}$.
2. *We have*
$$\nabla_x^{\mathbb{D}}(\alpha \underset{i}{\cdot} \gamma) = \alpha \underset{i}{\cdot} \left(\nabla_x^{\mathbb{D}}(\gamma)\right)$$
for any $\alpha \in \mathbb{R}$ *and any* $\gamma \in (M \otimes \mathcal{W}_{\mathbb{D}})_{\pi(x)}$, *where* i *is a natural number with* $1 \leq i \leq \deg \mathbb{D}$.
3. *The diagram*
$$\begin{array}{ccc} (M \otimes \mathcal{W}_{\mathbb{D}})_{\pi(x)} & \to & (M \otimes \mathcal{W}_{\mathbb{D}})_{\pi(x)} \otimes \mathcal{W}_{D_m} \\ \nabla_x \downarrow & & \downarrow \nabla_x \otimes \mathrm{id}_{\mathcal{W}_{D_m}} \\ (E \otimes \mathcal{W}_{\mathbb{D}})_x & \to & (E \otimes \mathcal{W}_{\mathbb{D}})_x \otimes \mathcal{W}_{D_m} \end{array}$$
is commutative, where m *is an arbitrary natural number, the upper horizontal arrow is*
$$\mathrm{id}_M \otimes \mathcal{W}_{(d_1,\ldots,d_k,e)\in \mathbb{D}\times D_m \mapsto (d_1,\ldots,d_{i-1},ed_i,d_{i+1},\ldots d_k)\in \mathbb{D}},$$
and the lower horizontal arrow is
$$\mathrm{id}_E \otimes \mathcal{W}_{(d_1,\ldots,d_k,e)\in \mathbb{D}\times D_m \mapsto (d_1,\ldots,d_{i-1},ed_i,d_{i+1},\ldots d_k)\in \mathbb{D}}$$

with $k = \deg \mathbb{D}$ and $1 \leq i \leq k$.

If the simplicial infinitesimal space $\mathbb{D}$ is symmetric, the induced mapping $\nabla_x^{\mathbb{D}}$: $(M \otimes \mathcal{W}_{\mathbb{D}})_{\pi(x)} \to (E \otimes \mathcal{W}_{\mathbb{D}})_x$ acquiesces in the following condition of symmetry besides the above ones:

- *We have*

$$\nabla_x^{\mathbb{D}}(\gamma^\sigma) = (\nabla_x^{\mathbb{D}}(\gamma))^\sigma$$

for any $\sigma \in \mathbf{S}_k$ and any $\gamma \in (M \otimes \mathcal{W}_{\mathbb{D}})_{\pi(x)}$.

Proof For the sake of simplicity in description, we deal, by way of example, with the case that $n = 3$ and $\mathbb{D} = D\{3\}_2$, for which the standard quasi-colimit representation was given in the proof of Theorem 2.3. Therefore, giving $\gamma \in \left(M \otimes \mathcal{W}_{D\{3\}_2}\right)_{\pi(x)}$ is equivalent to giving $\gamma_{12}, \gamma_{13}, \gamma_{23} \in \left(M \otimes \mathcal{W}_{D^2}\right)_{\pi(x)}$ with $\mathbf{d}_2(\gamma_{12}) = \mathbf{d}_2(\gamma_{13})$, $\mathbf{d}_1(\gamma_{12}) = \mathbf{d}_2(\gamma_{23})$ and $\mathbf{d}_1(\gamma_{13}) = \mathbf{d}_1(\gamma_{23})$. By Proposition 4.2, we have

$$\begin{aligned}
\mathbf{d}_2(\widehat{\pi}_{3,2}(\nabla_x)(\gamma_{12})) &= \widehat{\pi}_{3,2}(\nabla_x)(\mathbf{d}_2(\gamma_{12})) = \widehat{\pi}_{3,2}(\nabla_x)(\mathbf{d}_2(\gamma_{13})) = \mathbf{d}_2(\widehat{\pi}_{3,2}(\nabla_x)(\gamma_{13}))\\
\mathbf{d}_1(\widehat{\pi}_{3,2}(\nabla_x)(\gamma_{12})) &= \widehat{\pi}_{3,2}(\nabla_x)(\mathbf{d}_1(\gamma_{12})) = \widehat{\pi}_{3,2}(\nabla_x)(\mathbf{d}_2(\gamma_{23})) = \mathbf{d}_2(\widehat{\pi}_{3,2}(\nabla_x)(\gamma_{23}))\\
\mathbf{d}_1(\widehat{\pi}_{3,2}(\nabla_x)(\gamma_{13})) &= \widehat{\pi}_{3,2}(\nabla_x)(\mathbf{d}_1(\gamma_{13})) = \widehat{\pi}_{3,2}(\nabla_x)(\mathbf{d}_1(\gamma_{23})) = \mathbf{d}_1(\widehat{\pi}_{3,2}(\nabla_x)(\gamma_{23})),
\end{aligned}$$

which determines a unique $\nabla_x^{D\{3\}_2}(\gamma) \in \left(E \otimes \mathcal{W}_{D\{3\}_2}\right)_x$ with

$$\begin{aligned}
\mathbf{d}_1(\nabla_x^{D\{3\}_2}(\gamma)) &= \widehat{\pi}_{3,2}(\nabla_x)(\gamma_{23})\\
\mathbf{d}_2(\nabla_x^{D\{3\}_2}(\gamma)) &= \widehat{\pi}_{3,2}(\nabla_x)(\gamma_{13})\\
\mathbf{d}_3(\nabla_x^{D\{3\}_2}(\gamma)) &= \widehat{\pi}_{3,2}(\nabla_x)(\gamma_{12}).
\end{aligned}$$

The proof that $\nabla_x^{D\{3\}_2}$: $\left(M \otimes \mathcal{W}_{D\{3\}_2}\right)_{\pi(x)} \to \left(E \otimes \mathcal{W}_{D\{3\}_2}\right)_x$ acquiesces in the desired four properties is safely left to the reader.

Remark 4.2 The reader should note that the induced mapping $\nabla_x^{\mathbb{D}}$ is defined in terms of the standard quasi-colimit representation of $\mathbb{D}$. The concluding corollary of this subsection will show that the induced mapping $\nabla_x^{\mathbb{D}}$ is independent of our choice of a quasi-colimit representation of $\mathbb{D}$ to a large extent, whether it is standard or not, as long as ∇ is not only a D^n-pseudotangential but also a D^n-*tangential*. We note in passing that $\hat{\pi}_{n,m}(\nabla)$ with $m \leq n$ is no other than $\nabla_x^{D^m}$.

Proposition 4.4 *Let π': $P \to E$ be another bundle with $x \in P$. If $\nabla_{\pi'(x)}$ is a n-tangential$_2$ over the bundle π: $E \to M$ at $\pi'(x) \in E$ and ∇_x is a n-tangential$_2$ over the bundle π': $P \to E$ at $x \in E$, then the composition $\nabla_x \circ \nabla_{\pi'(x)}$ is a n-tangential$_2$ over the bundle $\pi \circ \pi'$: $P \to M$ at $x \in E$, and $\pi_{n,n-1}(\nabla_x \circ \nabla_{\pi'(x)}) = \pi_{n,n-1}(\nabla_x) \circ \pi_{n,n-1}(\nabla_{\pi'(x)})$ provided that $n \geq 1$.*

Proof In case of $n = 0$, there is nothing to prove. It is easy to see that if $\nabla_{\pi'(x)}$ is a n-tangential$_2$ over the bundle π: $E \to M$ *at* $\pi'(x) \in E$ and ∇_x is a n-tangential$_2$

over the bundle $\pi'\colon P \to E$ *at* $x \in E$, then the composition $\nabla_x \circ \nabla_{\pi'(x)}$ is an n-pseudoconnection *over* the bundle $\pi\colon E \to M$ *at* $x \in P$. If $\nabla_{\pi'(x)}$ is a $(n+1)$-tangential$_2$ over the bundle $\pi\colon E \to M$ *at* $\pi'(x) \in E$ and ∇_x is a $(n+1)$-tangential$_2$ over the bundle $\pi'\colon P \to E$ *at* $x \in P$, then we have

$$\begin{aligned}\pi_{n+1,n}(\nabla_x \circ \nabla_{\pi'(x)}) &= \mathbf{d}_{n+1} \circ \nabla_x \circ \nabla_{\pi'(x)} \circ \mathbf{s}_{n+1} \\ &= \mathbf{d}_{n+1} \circ \nabla_x \circ \mathbf{s}_{n+1} \circ \mathbf{d}_{n+1} \circ \nabla_{\pi'(x)} \circ \mathbf{s}_{n+1} \\ &\text{[By Proposition 4.1]} \\ &= \pi_{n+1,n}(\nabla_x) \circ \pi_{n+1,n}(\nabla_{\pi'(x)})\end{aligned}$$

Therefore we have

$$\begin{aligned}&\nabla_x \circ \nabla_{\pi'(x)}((\mathrm{id}_M \otimes \mathscr{W}_{(d_1,\ldots,d_n,d_{n+1})\in D^{n+1}\mapsto(d_1,\ldots,d_nd_{n+1})\in D^n})(\gamma)) \\ &= \nabla_x\left(\nabla_{\pi'(x)}\left((\mathrm{id}_M \otimes \mathscr{W}_{(d_1,\ldots,d_n,d_{n+1})\in D^{n+1}\mapsto(d_1,\ldots,d_nd_{n+1})\in D^n})(\gamma)\right)\right) \\ &= \nabla_x\left((\mathrm{id}_E \otimes \mathscr{W}_{(d_1,\ldots,d_n,d_{n+1})\in D^{n+1}\mapsto(d_1,\ldots,d_nd_{n+1})\in D^n})(\pi_{n+1,n}(\nabla_{\pi'(x)})(\gamma))\right) \\ &= (\mathrm{id}_P \otimes \mathscr{W}_{(d_1,\ldots,d_n,d_{n+1})\in D^{n+1}\mapsto(d_1,\ldots,d_nd_{n+1})\in D^n})\left(\pi_{n+1,n}(\nabla_x)\left(\pi_{n+1,n}(\nabla_{\pi'(x)})(\gamma)\right)\right) \\ &= (\mathrm{id}_P \otimes \mathscr{W}_{(d_1,\ldots,d_n,d_{n+1})\in D^{n+1}\mapsto(d_1,\ldots,d_nd_{n+1})\in D^n})\left(\pi_{n+1,n}(\nabla_x \circ \nabla_{\pi'(x)})\right)\end{aligned}$$

Thus we can prove by induction on n that if $\nabla_{\pi'(x)}$ is a n-tangential$_2$ over the bundle $\pi\colon E \to M$ *at* $\pi'(x) \in E$ and ∇_x is a n-tangential$_2$ *over* the bundle $\pi'\colon P \to E$ *at* $x \in E$, then the composition $\nabla_x \circ \nabla_{\pi'(x)}$ is a n-tangential$_2$ *over* the bundle $\pi \circ \pi'\colon P \to M$ *at* $x \in E$.

Theorem 4.5 *Let* ∇ *be a* D^n*-tangential over the bundle* $\pi\colon E \to M$ *at* $x \in E$. *Let* $\mathbb{D}$ *and* $\mathbb{D}'$ *be simplicial infinitesimal spaces of dimension less than or equal to* n. *Let* χ *be a monomial mapping from* $\mathbb{D}$ *to* $\mathbb{D}'$. *Let* $\gamma \in \mathbf{T}_x^{\mathbb{D}'}(M)$. *Then we have*

$$\nabla_{\mathbb{D}}((\mathrm{id}_M \otimes \mathscr{W}_\chi)(\gamma)) = (\mathrm{id}_E \otimes \mathscr{W}_\chi)(\nabla_{\mathbb{D}'}(\gamma))$$

Remark 4.3 The reader should note that the above far-flung generalization of Proposition 4.3 subsumes Proposition 4.2.

Proof In place of giving a general proof with formidable notation, we satisfy ourselves with an illustration. Here we deal only with the case that $\mathbb{D} = D^3$, $\mathbb{D}' = D(3)$ and χ is

$$\chi(d_1, d_2, d_3) = (d_1d_2, d_1d_3, d_2d_3)$$

for any $(d_1, d_2, d_3) \in D^3$. We assume that $n \geq 3$. We note first that the monomial mapping $\chi\colon D^3 \to D(3)$ is the composition of two monomial mappings

$$\begin{aligned}\chi_1 &: D^3 \to D\{6; (1,2), (3,4), (5,6)\} \\ \chi_2 &: D\{6; (1,2), (3,4), (5,6)\} \to D(3)\end{aligned}$$

with

$$\chi_1(d_1, d_2, d_3) = (d_1, d_1, d_2, d_2, d_3, d_3)$$

for any $(d_1, d_2, d_3) \in D^3$ and

$$\chi_2(d_1, d_2, d_3, d_4, d_5, d_6) = (d_1 d_3, d_2 d_5, d_4 d_6)$$

for any $(d_1, d_2, d_3, d_4, d_5, d_6) \in D\{6; (1,2), (3,4), (5,6)\}$, while the former monomial mapping $\chi_1 \colon D^3 \to D\{6; (1,2), (3,4), (5,6)\}$ is in turn the composition of three monomial mappings

$$\begin{aligned} \chi_1^1 &: D^3 \to D\{4; (1,2)\} \\ \chi_1^2 &: D\{4; (1,2)\} \to D\{5; (1,2), (3,4)\} \\ \chi_1^3 &: D\{5; (1,2), (3,4)\} \to D\{6; (1,2), (3,4), (5,6)\} \end{aligned}$$

with

$$\chi_1^1(d_1, d_2, d_3) = (d_1, d_1, d_2, d_3)$$

for any $(d_1, d_2, d_3) \in D^3$,

$$\chi_1^2(d_1, d_2, d_3, d_4) = (d_1, d_2, d_3, d_3, d_4)$$

for any $(d_1, d_2, d_3, d_4) \in D\{4; (1,2)\}$ and

$$\chi_1^3(d_1, d_2, d_3, d_4, d_5) = (d_1, d_2, d_3, d_4, d_5, d_5)$$

for any $(d_1, d_2, d_3, d_4, d_5) \in D\{5; (1,2), (3,4)\}$. Therefore it suffices to prove that

$$\nabla\left(\left(\mathrm{id}_M \otimes \mathcal{W}_{\chi_1^1}\right)(\gamma')\right) = \left(\mathrm{id}_E \otimes \mathcal{W}_{\chi_1^1}\right)\left(\nabla_{D\{4;(1,2)\}}(\gamma')\right) \tag{2}$$

for any $\gamma' \in \left(M \otimes \mathcal{W}_{D\{4;(1,2)\}}\right)_{\pi(x)}$, that

$$\nabla_{D\{4;(1,2)\}}\left(\left(\mathrm{id}_M \otimes \mathcal{W}_{\chi_1^2}\right)(\gamma'')\right) = \left(\mathrm{id}_E \otimes \mathcal{W}_{\chi_1^2}\right)\left(\nabla_{D\{5;(1,2),(3,4)\}}(\gamma'')\right) \tag{3}$$

for any $\gamma'' \in \left(M \otimes \mathcal{W}_{D\{5;(1,2),(3,4)\}}\right)_{\pi(x)}$, that

$$\nabla_{D\{5;(1,2),(3,4)\}}\left(\left(\mathrm{id}_M \otimes \mathcal{W}_{\chi_1^3}\right)(\gamma''')\right) = \left(\mathrm{id}_E \otimes \mathcal{W}_{\chi_1^3}\right)\left(\nabla_{D\{6;(1,2),(3,4),(5,6)\}}(\gamma''')\right) \tag{4}$$

for any $\gamma''' \in \left(M \otimes \mathcal{W}_{D\{6;(1,2),(3,4),(5,6)\}}\right)_{\pi(x)}$, and that

$$\nabla_{D\{6;(1,2),(3,4),(5,6)\}}\left(\left(\mathrm{id}_M \otimes \mathcal{W}_{\chi_2}\right)(\gamma'''')\right) = \left(\mathrm{id}_E \otimes \mathcal{W}_{\chi_2}\right)\left(\nabla_{D(3)}(\gamma'''')\right) \tag{5}$$

for any $\gamma'''' \in \left(M \otimes \mathcal{W}_{D(3)}\right)_{\pi(x)} \mathbf{T}_x^{D(3)}(M)$. Since $D\{4;(1,2)\} = D(2) \times D^2$, it is easy to see that

$$\nabla\left(\left(\mathrm{id}_M \otimes \mathcal{W}_{\chi_1^1}\right)(\gamma')\right) = \nabla(\gamma_1' \underset{1}{+} \gamma_2') = \nabla(\gamma_1') + \nabla(\gamma_2')$$

where $\gamma_1' = \gamma' \circ (i_1 \times \mathrm{id}_{D^2})$ and $\gamma_2' = \gamma' \circ (i_2 \times \mathrm{id}_{D^2})$ with $i_1(d) = (d, 0) \in D(2)$ and $i_2(d) = (0, d) \in D(2)$ for any $d \in D$. On the other hand, we have

$$\left(\mathrm{id}_E \otimes \mathcal{W}_{\chi_1^1}\right)\left(\nabla_{D(4;(1,2))}(\gamma')\right) = \left(\mathrm{id}_E \otimes \mathcal{W}_{\chi_1^1}\right)\left(\mathbf{l}_{(\nabla(\gamma_1'),\nabla(\gamma_2'))}\right) = \nabla(\gamma_1') + \nabla(\gamma_2')$$

where $\mathbf{l}_{(\nabla(\gamma_1'),\nabla(\gamma_2'))}$ is the unique element of $E \otimes \mathcal{W}_{D(2)\times D^2}$ with

$$\left(\mathrm{id}_E \otimes \mathcal{W}_{i_1 \times \mathrm{id}_{D^2}}\right)\left(\mathbf{l}_{(\nabla(\gamma_1'),\nabla(\gamma_2'))}\right) = \nabla(\gamma_1')$$

and

$$\left(\mathrm{id}_E \otimes \mathcal{W}_{i_2 \times \mathrm{id}_{D^2}}\right)\left(\mathbf{l}_{(\nabla(\gamma_1'),\nabla(\gamma_2'))}\right) = \nabla(\gamma_2')$$

Thus we have established (2). By the same token, we can establish (3) and (4). In order to prove (5), it suffices to note that

$$\begin{aligned}&\left(\mathrm{id}_E \otimes \mathcal{W}_{i_{135}}\right)\left(\nabla_{D\{6;(1,2),(3,4),(5,6)\}}\left(\left(\mathrm{id}_M \otimes \mathcal{W}_{\chi_2}\right)(\gamma'''')\right)\right)\\ &= \left(\mathrm{id}_E \otimes \mathcal{W}_{\chi_2 \circ i_{135}}\right)\left(\nabla_{D(3)}(\gamma'''')\right)\end{aligned}$$

together with the seven similar identities obtained from the above by replacing i_{135} by seven other $i_{jkl}\colon D^3 \to D\{6;(1,2),(3,4),(5,6)\}$ in the standard quasi-colimit representation of $D\{6;(1,2),(3,4),(5,6)\}$, where $i_{jkl}\colon D^3 \to D\{6;(1,2),(3,4),(5,6)\}$ $(1 \le j < k < l \le 6)$ is a mapping with $i_{jkl}(d_1, d_2, d_3) = (\ldots, \underset{j}{d_1}, \ldots, \underset{k}{d_2}, \ldots, \underset{l}{d_3}, \ldots)$ (d_1, d_2 and d_3 are inserted at the j-th, k-th and l-th positions respectively, while the other components are fixed at 0). Its proof goes as follows. Since

$$\begin{aligned}&\left(\mathrm{id}_E \otimes \mathcal{W}_{i_{135}}\right)\left(\nabla_{D\{6;(1,2),(3,4),(5,6)\}}\left(\left(\mathrm{id}_M \otimes \mathcal{W}_{\chi_2}\right)(\gamma'''')\right)\right)\\ &= \nabla\left(\left(\mathrm{id}_M \otimes \mathcal{W}_{\chi_2 \circ i_{135}}\right)(\gamma'''')\right),\end{aligned}$$

it suffices to show that

$$\nabla\left(\left(\mathrm{id}_M \otimes \mathcal{W}_{\chi_2 \circ i_{135}}\right)(\gamma'''')\right) = \left(\mathrm{id}_E \otimes \mathcal{W}_{\chi_2 \circ i_{135}}\right)\nabla_{D(3)}(\gamma'''')$$

However the last identity follows at once by simply observing that the mapping $\chi_2 \circ i_{135}\colon D^3 \to D(3)$ is the mapping

$$(d_1, d_2, d_3) \in D^3 \longmapsto (d_1 d_2, 0, 0) \in D(3),$$

which is the successive composition of the following three mappings:

$$\begin{aligned}(d_1, d_2, d_3) \in D^3 &\longmapsto (d_1, d_2) \in D^2 \\ (d_1, d_2) \in D^2 &\longmapsto d_1 d_2 \in D \\ d \in D &\longmapsto (d, 0, 0) \in D(3).\end{aligned}$$

Corollary 4.4 *Let ∇ be a D^n-tangential over the bundle $\pi : E \to M$ at $x \in E$. Let $\mathbb{D}$ be a simplicially infinitesimal spaces of dimension less than or equal to n. Any nonstandard quasi-colimit representation of $\mathbb{D}$, if any mapping into $\mathbb{D}$ in the representation is monomial, induces the same mapping as $\nabla_{\mathbb{D}}$ (induced by the standard quasi-colimit representation of $\mathbb{D}$) by the method in the proof of Theorem 4.4.*

Proof It suffices to note that

$$\nabla_{D^m}\left(\left(\mathrm{id}_M \otimes \mathcal{W}_\chi\right)(\gamma)\right) = \left(\mathrm{id}_E \otimes \mathcal{W}_\chi\right)\left(\nabla_{\mathbb{D}}(\gamma)\right)$$

for any mapping $\chi : D^m \to \mathbb{D}$ in the given nonstandard quasi-colimit representation of $\mathbb{D}$, which follows directly from the above theorem.

5 The Third Approach to Jets

Definition 5.1 Let n be a natural number. A *D_n-pseudotangential over* the bundle $\pi : E \to M$ *at* $x \in E$ is a mapping

$$\nabla_x : \left(M \otimes \mathcal{W}_{D_n}\right)_{\pi(x)} \to \left(E \otimes \mathcal{W}_{D_n}\right)_x$$

abiding by the following two conditions:

1. We have

$$\left(\pi \otimes \mathrm{id}_{\mathcal{W}_{D_n}}\right)(\nabla_x(\gamma)) = \gamma$$

for any $\gamma \in \left(M \otimes \mathcal{W}_{D_n}\right)_{\pi(x)}$.

2. For any $\gamma \in \left(E \otimes \mathcal{W}_{D_n}\right)_x$ and any $\alpha \in \mathbb{R}$, we have

$$\nabla_x(\alpha\gamma) = \alpha\nabla_x(\gamma)$$

3. The diagram

$$\begin{array}{ccc} \left(M \otimes \mathcal{W}_{D_n}\right)_{\pi(x)} & \xrightarrow{\mathrm{id}_M \otimes \mathcal{W}_{(d_1,d_2)\in D_n \times D_m \mapsto d_1 d_2 \in D_n}} & \left(M \otimes \mathcal{W}_{D_n}\right)_{\pi(x)} \otimes \mathcal{W}_{D_m} \\ \nabla_x \downarrow & & \downarrow \nabla_x \otimes \mathrm{id}_{\mathcal{W}_{D_m}} \\ \left(E \otimes \mathcal{W}_{D_n}\right)_x & \xrightarrow{\mathrm{id}_E \otimes \mathcal{W}_{(d_1,d_2)\in D_n \times D_m \mapsto d_1 d_2 \in D_n}} & \left(E \otimes \mathcal{W}_{D_n}\right)_x \otimes \mathcal{W}_{D_m} \end{array}$$

commutes, where m is an arbitrary natural number.

Remark 5.1 The third condition in the above definition claims what is called infinitesimal linearity.

Notation 5.1 *We denote by* $\widehat{\mathbb{J}}_x^{D_n}(\pi)$ *the totality of* D_n*-pseudotangentials over the bundle* $\pi : E \to M$ *at* $x \in E$*. We denote by* $\widehat{\mathbb{J}}^{D_n}(\pi)$ *the set-theoretic union of* $\widehat{\mathbb{J}}_x^{D_n}(\pi)$*'s for all* $x \in E$*.*

It is easy to see that

Lemma 5.1 *The following diagram is an equalizer in the category of Weil algebras:*

$$\mathcal{W}_{D_n} \xrightarrow{\mathcal{W}_{(d_1,d_2)\in D_{n+1}\times D_n \mapsto d_1d_2\in D_n}} \mathcal{W}_{D_{n+1}\times D_n} \underset{\mathcal{W}_{(d_1,d_2,d_3)\in D_{n+1}\times D_{n+1}\times D_n \mapsto (d_1,d_2d_3)\in D_{n+1}\times D_n}}{\overset{\mathcal{W}_{(d_1,d_2,d_3)\in D_{n+1}\times D_{n+1}\times D_n \mapsto (d_1d_2,d_3)\in D_{n+1}\times D_n}}{\rightrightarrows}} \mathcal{W}_{D_{n+1}\times D_{n+1}\times D_n}$$

Proposition 5.1 *Let* ∇_x *be a* D_{n+1}*-pseudotangential overthe bundle* $\pi : E \to M$ *at* $x \in E$ *and* $\gamma \in \left(M \otimes \mathcal{W}_{D_n}\right)_{\pi(x)}$*. Then th ere exists a unique* $\gamma' \in \left(E \otimes \mathcal{W}_{D_n}\right)_x$ *such that the composition of mappings*

$$\left(M \otimes \mathcal{W}_{D_n}\right)_{\pi(x)} \xrightarrow{\mathrm{id}_M \otimes \mathcal{W}_{(d_1,d_2)\in D_{n+1}\times D_n \mapsto d_1d_2\in D_n}} \left(M \otimes \mathcal{W}_{D_{n+1}}\right)_{\pi(x)} \otimes \mathcal{W}_{D_n}$$
$$\xrightarrow{\nabla_x \otimes \mathrm{id}_{\mathcal{W}_{D_n}}} \left(E \otimes \mathcal{W}_{D_{n+1}}\right)_x \otimes \mathcal{W}_{D_n} \tag{6}$$

applied to γ *results in*

$$\left(\mathrm{id}_E \otimes \mathcal{W}_{(d_1,d_2)\in D_{n+1}\times D_n \mapsto d_1d_2\in D_n}\right)\left(\gamma'\right) \tag{7}$$

Proof By dint of Lemma 5.1, it suffices to show that the composition of mappings

$$\left(M \otimes \mathcal{W}_{D_n}\right)_{\pi(x)} \xrightarrow{\mathrm{id}_M \otimes \mathcal{W}_{(d_1,d_2)\in D_{n+1}\times D_n \mapsto d_1d_2\in D_n}} \left(M \otimes \mathcal{W}_{D_{n+1}}\right)_{\pi(x)} \otimes \mathcal{W}_{D_n}$$
$$\xrightarrow{\nabla_x \otimes \mathrm{id}_{\mathcal{W}_{D_n}}} \left(E \otimes \mathcal{W}_{D_{n+1}}\right)_x \otimes \mathcal{W}_{D_n}$$
$$\xrightarrow{\mathrm{id}_E \otimes \mathcal{W}_{(d_1,d_2,d_3)\in D_{n+1}\times D_{n+1}\times D_n \mapsto (d_1,d_2d_3)\in D_{n+1}\times D_n}}$$
$$\left(E \otimes \mathcal{W}_{D_{n+1}}\right)_x \otimes \mathcal{W}_{D_{n+1}\times D_n} \tag{8}$$

is equal to the composition of mappings

$$\left(M \otimes \mathcal{W}_{D_n}\right)_{\pi(x)} \xrightarrow{\mathrm{id}_M \otimes \mathcal{W}_{(d_1,d_2)\in D_{n+1}\times D_n \mapsto d_1 d_2 \in D_n}} \left(M \otimes \mathcal{W}_{D_{n+1}}\right)_{\pi(x)} \otimes \mathcal{W}_{D_n}$$
$$\xrightarrow{\nabla_x \otimes \mathrm{id}_{\mathcal{W}_{D_n}}} \left(E \otimes \mathcal{W}_{D_{n+1}}\right)_x \otimes \mathcal{W}_{D_n}$$
$$\xrightarrow{\mathrm{id}_E \otimes \mathcal{W}_{(d_1,d_2,d_3)\in D_{n+1}\times D_{n+1}\times D_n \mapsto (d_1 d_2, d_3)\in D_{n+1}\times D_n}}$$
$$\left(E \otimes \mathcal{W}_{D_{n+1}}\right)_x \otimes \mathcal{W}_{D_{n+1}\times D_n} \tag{9}$$

Since $\otimes$ is a bifunctor, the diagram

$$\begin{array}{ccc} \left(M \otimes \mathcal{W}_{D_{n+1}}\right)_{\pi(x)} \otimes \mathcal{W}_{D_n} & \rightarrow & \left(M \otimes \mathcal{W}_{D_{n+1}}\right)_{\pi(x)} \otimes \mathcal{W}_{D_{n+1}\times D_n} \\ \nabla_x \otimes \mathrm{id}_{\mathcal{W}_{D_n}} \downarrow & & \downarrow \nabla_x \otimes \mathrm{id}_{\mathcal{W}_{D_{n+1}\times D_n}} \\ \left(E \otimes \mathcal{W}_{D_{n+1}}\right)_x \otimes \mathcal{W}_{D_n} & \rightarrow & \left(E \otimes \mathcal{W}_{D_{n+1}}\right)_x \otimes \mathcal{W}_{D_{n+1}\times D_n} \end{array}$$

commutes, where the upper horizontal arrow is

$$\mathrm{id}_M \otimes \mathcal{W}_{(d_1,d_2,d_3)\in D_{n+1}\times D_{n+1}\times D_n \mapsto (d_1, d_2 d_3)\in D_{n+1}\times D_n},$$

while the lower horizontal arrow is

$$\mathrm{id}_E \otimes \mathcal{W}_{(d_1,d_2,d_3)\in D_{n+1}\times D_{n+1}\times D_n \mapsto (d_1, d_2 d_3)\in D_{n+1}\times D_n}.$$

Therefore the composition of mappings in (8) is equal to the composition of mappings

$$\left(M \otimes \mathcal{W}_{D_n}\right)_{\pi(x)} \xrightarrow{\mathrm{id}_M \otimes \mathcal{W}_{(d_1,d_2)\in D_{n+1}\times D_n \mapsto d_1 d_2 \in D_n}} \left(M \otimes \mathcal{W}_{D_{n+1}}\right)_{\pi(x)} \otimes \mathcal{W}_{D_n}$$
$$\xrightarrow{\mathrm{id}_M \otimes \mathcal{W}_{(d_1,d_2,d_3)\in D_{n+1}\times D_{n+1}\times D_n \mapsto (d_1, d_2 d_3)\in D_{n+1}\times D_n}} \left(M \otimes \mathcal{W}_{D_{n+1}}\right)_{\pi(x)} \otimes \mathcal{W}_{D_{n+1}\times D_n}$$
$$\xrightarrow{\nabla_x \otimes \mathrm{id}_{\mathcal{W}_{D_{n+1}\times D_n}}} \left(E \otimes \mathcal{W}_{D_{n+1}}\right)_x \otimes \mathcal{W}_{D_{n+1}\times D_n} \tag{10}$$

Since the composition of mappings

$$M \otimes \mathcal{W}_{D_n} \xrightarrow{\mathrm{id}_M \otimes \mathcal{W}_{(d_1,d_2)\in D_{n+1}\times D_n \mapsto d_1 d_2 \in D_n}} M \otimes \mathcal{W}_{D_{n+1}\times D_n}$$
$$\xrightarrow{\mathrm{id}_M \otimes \mathcal{W}_{(d_1,d_2,d_3)\in D_{n+1}\times D_{n+1}\times D_n \mapsto (d_1, d_2 d_3)\in D_{n+1}\times D_n}} M \otimes \mathcal{W}_{D_{n+1}\times D_{n+1}\times D_n}$$

is trivially equal to the composition of mappings

$$M \otimes \mathcal{W}_{D_n} \xrightarrow{\mathrm{id}_M \otimes \mathcal{W}_{(d_1,d_2)\in D_{n+1}\times D_n \mapsto d_1 d_2 \in D_n}} M \otimes \mathcal{W}_{D_{n+1}\times D_n}$$
$$\xrightarrow{\mathrm{id}_M \otimes \mathcal{W}_{(d_1,d_2,d_3)\in D_{n+1}\times D_{n+1}\times D_n \mapsto (d_1 d_2, d_3)\in D_{n+1}\times D_n}} M \otimes \mathcal{W}_{D_{n+1}\times D_{n+1}\times D_n},$$

the composition of mappings in (10) is equal to the composition of mappings

$$\left(M\otimes\mathcal{W}_{D_n}\right)_{\pi(x)}\xrightarrow{\mathrm{id}_M\otimes\mathcal{W}_{(d_1,d_2)\in D_{n+1}\times D_n\mapsto d_1d_2\in D_n}}\left(M\otimes\mathcal{W}_{D_{n+1}}\right)_{\pi(x)}\otimes\mathcal{W}_{D_n}$$
$$\xrightarrow{\mathrm{id}_M\otimes\mathcal{W}_{(d_1,d_2,d_3)\in D_{n+1}\times D_{n+1}\times D_n\mapsto(d_1d_2,d_3)\in D_{n+1}\times D_n}}\left(M\otimes\mathcal{W}_{D_{n+1}}\right)_{\pi(x)}\otimes\mathcal{W}_{D_{n+1}\times D_n}$$
$$\xrightarrow{\nabla_x\otimes\mathrm{id}_{\mathcal{W}_{D_{n+1}\times D_n}}}\left(E\otimes\mathcal{W}_{D_{n+1}}\right)_x\otimes\mathcal{W}_{D_{n+1}\times D_n} \tag{11}$$

By dint of the third condition in Definition 5.1.1, the diagram

$$\begin{array}{ccc}\left(M\otimes\mathcal{W}_{D_{n+1}}\right)_{\pi(x)}\otimes\mathcal{W}_{D_n} & \rightarrow & \left(M\otimes\mathcal{W}_{D_{n+1}}\right)_{\pi(x)}\otimes\mathcal{W}_{D_{n+1}\times D_n}\\ \nabla_x\otimes\mathrm{id}_{\mathcal{W}_{D_n}}\downarrow & & \downarrow\nabla_x\otimes\mathrm{id}_{\mathcal{W}_{D_{n+1}\times D_n}}\\ \left(E\otimes\mathcal{W}_{D_{n+1}}\right)_x\otimes\mathcal{W}_{D_n} & \rightarrow & \left(E\otimes\mathcal{W}_{D_{n+1}}\right)_x\otimes\mathcal{W}_{D_{n+1}\times D_n}\end{array}$$

commutes, where the upper horizontal arrow is

$$\mathrm{id}_M\otimes\mathcal{W}_{(d_1,d_2,d_3)\in D_{n+1}\times D_{n+1}\times D_n\mapsto(d_1d_2,d_3)\in D_{n+1}\times D_n},$$

and the lower horizontal arrow is

$$\mathrm{id}_E\otimes\mathcal{W}_{(d_1,d_2,d_3)\in D_{n+1}\times D_{n+1}\times D_n\mapsto(d_1d_2,d_3)\in D_{n+1}\times D_n}.$$

Therefore the composition of mappings in (11) is equal to the composition of mappings in (9), which completes the proof.

It is not difficult to see that

Proposition 5.2 *Given a D_{n+1}-pseudotangential ∇_x overthe bundle $\pi: E\rightarrow M$ at $x\in E$, the assignment $\gamma\in\left(M\otimes\mathcal{W}_{D_n}\right)_{\pi(x)}\mapsto\gamma'\in\left(E\otimes\mathcal{W}_{D_n}\right)_x$ in the above proposition, denoted by $\hat{\pi}_{n+1,n}(\nabla_x)$, is a D_n-pseudotangential over the bundle $\pi: E\rightarrow M$ at $x\in E$.*

Proof We have to verify the three conditions in Definition 5.1 concerning the mapping $\hat{\pi}_{n+1,n}(\nabla_x)$: $\left(M\otimes\mathcal{W}_{D_n}\right)_{\pi(x)}\rightarrow\left(E\otimes\mathcal{W}_{D_n}\right)_x$.

1. To see the first condition, it suffices to show that

$$\left(\mathrm{id}_M\otimes\mathcal{W}_{(d_1,d_2)\in D_{n+1}\times D_n\mapsto d_1d_2\in D_n}\right)\circ\left(\pi\otimes\mathrm{id}_{\mathcal{W}_{D_n}}\right)\left(\left(\hat{\pi}_{n+1,n}(\nabla_x)\right)(\gamma)\right)=\gamma,$$

which is equivalent to

$$\left(\pi\otimes\mathrm{id}_{\mathcal{W}_{D_{n+1}\times D_n}}\right)\circ\left(\mathrm{id}_E\otimes\mathcal{W}_{(d_1,d_2)\in D_{n+1}\times D_n\mapsto d_1d_2\in D_n}\right)\left(\left(\hat{\pi}_{n+1,n}(\nabla_x)\right)(\gamma)\right)=\gamma,$$

since $\otimes$ is a bifunctor. Therefore it suffices to show that the composition of mappings

$$\left(M\otimes\mathcal{W}_{D_n}\right)_{\pi(x)}\xrightarrow{\mathrm{id}_M\otimes\mathcal{W}_{(d_1,d_2)\in D_{n+1}\times D_n\mapsto d_1d_2\in D_n}}\left(M\otimes\mathcal{W}_{D_{n+1}}\right)_{\pi(x)}\otimes\mathcal{W}_{D_n}$$
$$\xrightarrow{\nabla_x\otimes\mathrm{id}_{\mathcal{W}_{D_n}}}\left(E\otimes\mathcal{W}_{D_{n+1}}\right)_x\otimes\mathcal{W}_{D_n}\xrightarrow{\pi\otimes\mathrm{id}_{\mathcal{W}_{D_{n+1}\times D_n}}}\left(M\otimes\mathcal{W}_{D_{n+1}}\right)_x\otimes\mathcal{W}_{D_n}$$

applied to γ results in

$$\left(\mathrm{id}_M\otimes\mathcal{W}_{(d_1,d_2)\in D_{n+1}\times D_n\mapsto d_1d_2\in D_n}\right)(\gamma),$$

which follows directly from the first condition in Definition 5.1.

2. To see the second, let us note first that the composition of mappings

$$\left(M\otimes\mathcal{W}_{D_n}\right)_{\pi(x)}\xrightarrow{\mathrm{id}_M\otimes\mathcal{W}_{d\in D_n\mapsto\alpha d\in D_n}}\left(M\otimes\mathcal{W}_{D_n}\right)_{\pi(x)}$$
$$\xrightarrow{\mathrm{id}_M\otimes\mathcal{W}_{(d_1,d_2)\in D_{n+1}\times D_n\mapsto d_1d_2\in D_n}}\left(M\otimes\mathcal{W}_{D_{n+1}}\right)_{\pi(x)}\otimes\mathcal{W}_{D_n}$$

is equal to the composition of mappings

$$\left(M\otimes\mathcal{W}_{D_n}\right)_{\pi(x)}\xrightarrow{\mathrm{id}_M\otimes\mathcal{W}_{(d_1,d_2)\in D_{n+1}\times D_n\mapsto d_1d_2\in D_n}}\left(M\otimes\mathcal{W}_{D_{n+1}}\right)_{\pi(x)}\otimes\mathcal{W}_{D_n}$$
$$\xrightarrow{\mathrm{id}_M\otimes\mathcal{W}_{(d_1,d_2)\in D_{n+1}\times D_n\mapsto(\alpha d_1,d_2)\in D_{n+1}\times D_n}}\left(M\otimes\mathcal{W}_{D_{n+1}}\right)_{\pi(x)}\otimes\mathcal{W}_{D_n}$$

Since ∇_x is a D_{n+1}-pseudotangential over the bundle $\pi\colon E\to M$ *at* $x\in E$, the diagram

$$\begin{array}{ccc}\left(M\otimes\mathcal{W}_{D_{n+1}}\right)_{\pi(x)}\otimes\mathcal{W}_{D_n} & \to & \left(M\otimes\mathcal{W}_{D_{n+1}}\right)_{\pi(x)}\otimes\mathcal{W}_{D_n}\\ \nabla_x\otimes\mathrm{id}_{\mathcal{W}_{D_n}}\downarrow & & \downarrow\nabla_x\otimes\mathrm{id}_{\mathcal{W}_{D_n}}\\ \left(E\otimes\mathcal{W}_{D_{n+1}}\right)_x\otimes\mathcal{W}_{D_n} & \to & \left(E\otimes\mathcal{W}_{D_{n+1}}\right)_x\otimes\mathcal{W}_{D_n}\end{array}$$

commutes, where the upper horizontal arrow is

$$\mathrm{id}_M\otimes\mathcal{W}_{(d_1,d_2)\in D_{n+1}\times D_n\mapsto(\alpha d_1,d_2)\in D_{n+1}\times D_n},$$

while the lower horizontal arrow is

$$\mathrm{id}_E\otimes\mathcal{W}_{(d_1,d_2)\in D_{n+1}\times D_n\mapsto(\alpha d_1,d_2)\in D_{n+1}\times D_n}.$$

Therefore the composition of mappings

$$\left(M\otimes\mathcal{W}_{D_n}\right)_{\pi(x)}\xrightarrow{\mathrm{id}_M\otimes\mathcal{W}_{d\in D_n\mapsto\alpha d\in D_n}}\left(M\otimes\mathcal{W}_{D_n}\right)_{\pi(x)}$$
$$\xrightarrow{\mathrm{id}_M\otimes\mathcal{W}_{(d_1,d_2)\in D_{n+1}\times D_n\mapsto d_1d_2\in D_n}}\left(M\otimes\mathcal{W}_{D_{n+1}}\right)_{\pi(x)}\otimes\mathcal{W}_{D_n}$$
$$\xrightarrow{\nabla_x\otimes\mathrm{id}_{\mathcal{W}_{D_n}}}\left(E\otimes\mathcal{W}_{D_{n+1}}\right)_x\otimes\mathcal{W}_{D_n}$$

is equal to the composition of mappings

$$\left(M\otimes\mathcal{W}_{D_n}\right)_{\pi(x)} \xrightarrow{\mathrm{id}_M\otimes\mathcal{W}_{(d_1,d_2)\in D_{n+1}\times D_n\mapsto d_1d_2\in D_n}} \left(M\otimes\mathcal{W}_{D_{n+1}}\right)_{\pi(x)}\otimes\mathcal{W}_{D_n}$$
$$\xrightarrow{\nabla_x\otimes\mathrm{id}_{\mathcal{W}_{D_n}}} \left(E\otimes\mathcal{W}_{D_{n+1}}\right)_x\otimes\mathcal{W}_{D_n} \xrightarrow{\mathrm{id}_E\otimes\mathcal{W}_{(d_1,d_2)\in D_{n+1}\times D_n\mapsto(\alpha d_1,d_2)\in D_{n+1}\times D_n}}$$
$$\left(E\otimes\mathcal{W}_{D_{n+1}}\right)_x\otimes\mathcal{W}_{D_n}$$

The former composition of mappings applied to $\gamma\in\left(M\otimes\mathcal{W}_{D_n}\right)_{\pi(x)}$ results in

$$\left(\mathrm{id}_E\otimes\mathcal{W}_{(d_1,d_2)\in D_{n+1}\times D_n\mapsto d_1d_2\in D_n}\right)\left(\hat{\pi}_{n+1,n}(\nabla_x)(\alpha\gamma)\right),$$

while the latter composition of mappings applied to γ results in

$$\begin{aligned}&\left(\mathrm{id}_E\otimes\mathcal{W}_{(d_1,d_2)\in D_{n+1}\times D_n\mapsto(\alpha d_1,d_2)\in D_{n+1}\times D_n}\right)\circ\\&\left(\mathrm{id}_E\otimes\mathcal{W}_{(d_1,d_2)\in D_{n+1}\times D_n\mapsto d_1d_2\in D_n}\right)\left(\hat{\pi}_{n+1,n}(\nabla_x)(\gamma)\right)\\&=\left(\mathrm{id}_E\otimes\mathcal{W}_{(d_1,d_2)\in D_{n+1}\times D_n\mapsto d_1d_2\in D_n}\right)\left(\alpha\left(\hat{\pi}_{n+1,n}(\nabla_x)(\gamma)\right)\right).\end{aligned}$$

Therefore we have

$$\hat{\pi}_{n+1,n}(\nabla_x)(\alpha\gamma)=\alpha\left(\hat{\pi}_{n+1,n}(\nabla_x)(\gamma)\right)$$

3. To see the third, we have to show that the diagram

$$\begin{array}{ccc}\left(M\otimes\mathcal{W}_{D_n}\right)_{\pi(x)} & \xrightarrow{\mathrm{id}_M\otimes\mathcal{W}_{\mathbf{m}_{D_n\times D_m\to D_n}}} & \left(M\otimes\mathcal{W}_{D_n}\right)_{\pi(x)}\otimes\mathcal{W}_{D_m}\\ \hat{\pi}_{n+1,n}(\nabla_x)\downarrow & & \downarrow\hat{\pi}_{n+1,n}(\nabla_x)\otimes\mathrm{id}_{\mathcal{W}_{D_m}}\\ \left(E\otimes\mathcal{W}_{D_n}\right)_x & \xrightarrow{\mathrm{id}_E\otimes\mathcal{W}_{\mathbf{m}_{D_n\times D_m\to D_n}}} & \left(E\otimes\mathcal{W}_{D_n}\right)_x\otimes\mathcal{W}_{D_m}\end{array}\qquad(12)$$

commutes, where m is an arbitrary natural number. Since the lower square of the diagram

$$\begin{array}{ccc}\left(M\otimes\mathcal{W}_{D_n}\right)_{\pi(x)} & \xrightarrow{\mathrm{id}_M\otimes\mathcal{W}_{\mathbf{m}_{D_n\times D_m\to D_n}}} & \left(M\otimes\mathcal{W}_{D_n}\right)_{\pi(x)}\otimes\mathcal{W}_{D_m}\\ \hat{\pi}_{n+1,n}(\nabla_x)\downarrow & & \downarrow\hat{\pi}_{n+1,n}(\nabla_x)\otimes\mathrm{id}_{\mathcal{W}_{D_n}}\\ \left(E\otimes\mathcal{W}_{D_n}\right)_x & \xrightarrow{\mathrm{id}_E\otimes\mathcal{W}_{\mathbf{m}_{D_n\times D_m\to D_n}}} & \left(E\otimes\mathcal{W}_{D_n}\right)_x\otimes\mathcal{W}_{D_m}\\ \mathrm{id}_E\otimes\mathcal{W}_{\mathbf{m}_{D_{n+1}\times D_n\to D_n}}\downarrow & & \downarrow\mathrm{id}_E\otimes\mathcal{W}_{\mathbf{m}_{D_{n+1}\times D_n\to D_n}\times\mathrm{id}_{D_m}}\\ \left(E\otimes\mathcal{W}_{D_{n+1}}\right)_x\otimes\mathcal{W}_{D_n} & \xrightarrow{\mathrm{id}_E\otimes\mathcal{W}_{\mathrm{id}_{D_{n+1}}\times\mathbf{m}_{D_n\times D_m\to D_n}}} & \left(E\otimes\mathcal{W}_{D_{n+1}}\right)_x\otimes\mathcal{W}_{D_n\times D_m}\end{array}$$
(13)

commutes, so that the commutativity of the diagram in (12) is equivalent to the commutativity of the outer square of the diagram in (13). .The composition of mappings

$$\left(M\otimes\mathcal{W}_{D_n}\right)_{\pi(x)}\xrightarrow{\hat{\pi}_{n+1,n}(\nabla_x)}\left(E\otimes\mathcal{W}_{D_n}\right)_x\xrightarrow{\mathrm{id}_E\otimes\mathcal{W}_{\mathbf{m}_{D_{n+1}\times D_n\to D_n}}}\left(E\otimes\mathcal{W}_{D_{n+1}}\right)_x\otimes\mathcal{W}_{D_n}$$

is equal to the composition of mappings

$$\left(M\otimes\mathcal{W}_{D_n}\right)_{\pi(x)}\underrightarrow{\mathrm{id}_M\otimes\mathcal{W}_{\mathbf{m}_{D_{n+1}\times D_n\to D_n}}}\left(M\otimes\mathcal{W}_{D_{n+1}}\right)_{\pi(x)}\otimes\mathcal{W}_{D_n}$$
$$\underrightarrow{\nabla_x\otimes\mathrm{id}_{\mathcal{W}_{D_n}}}\left(E\otimes\mathcal{W}_{D_{n+1}}\right)_x\otimes\mathcal{W}_{D_n},$$

while the composition of mappings

$$\left(M\otimes\mathcal{W}_{D_n}\right)_{\pi(x)}\otimes\mathcal{W}_{D_m}\underrightarrow{\hat{\pi}_{n+1,n}(\nabla_x)\otimes\mathrm{id}_{\mathcal{W}_{D_m}}}\left(E\otimes\mathcal{W}_{D_n}\right)_x\otimes\mathcal{W}_{D_m}$$
$$\underrightarrow{\mathrm{id}_E\otimes\mathcal{W}_{\mathbf{m}_{D_{n+1}\times D_n\to D_n}\times\mathrm{id}_{D_m}}}\left(E\otimes\mathcal{W}_{D_{n+1}}\right)_x\otimes\mathcal{W}_{D_n\times D_m}$$

is equal to the composition of mappings

$$\left(M\otimes\mathcal{W}_{D_n}\right)_{\pi(x)}\otimes\mathcal{W}_{D_m}\underrightarrow{\mathrm{id}_M\otimes\mathcal{W}_{\mathbf{m}_{D_{n+1}\times D_n\to D_n}\times\mathrm{id}_{D_m}}}\left(M\otimes\mathcal{W}_{D_{n+1}}\right)_{\pi(x)}$$
$$\otimes\mathcal{W}_{D_n\times D_m}\underrightarrow{\nabla_x\otimes\mathrm{id}_{\mathcal{W}_{D_n\times D_m}}}\left(E\otimes\mathcal{W}_{D_{n+1}}\right)_x\otimes\mathcal{W}_{D_n\times D_m}$$

It is easy to see that the diagram

$$\begin{array}{ccc}
\left(M\otimes\mathcal{W}_{D_n}\right)_{\pi(x)} & \underrightarrow{\mathrm{id}_M\otimes\mathcal{W}_{\mathbf{m}_{D_n\times D_m\to D_n}}} & \left(M\otimes\mathcal{W}_{D_n}\right)_{\pi(x)}\otimes\mathcal{W}_{D_m} \\
\mathrm{id}_M\otimes\mathcal{W}_{\mathbf{m}_{D_{n+1}\times D_n\to D_n}}\downarrow & & \downarrow\mathrm{id}_M\otimes\mathcal{W}_{\mathbf{m}_{D_{n+1}\times D_n\to D_n}\times\mathrm{id}_{D_m}} \\
\left(M\otimes\mathcal{W}_{D_{n+1}}\right)_{\pi(x)}\otimes\mathcal{W}_{D_n} & \underrightarrow{\mathrm{id}_M\otimes\mathcal{W}_{\mathrm{id}_{D_{n+1}}\times\mathbf{m}_{D_n\times D_m\to D_n}}} & \left(M\otimes\mathcal{W}_{D_{n+1}}\right)_{\pi(x)}\otimes\mathcal{W}_{D_n\times D_m} \\
\nabla_x\otimes\mathrm{id}_{\mathcal{W}_{D_n}}\downarrow & & \downarrow\nabla_x\otimes\mathrm{id}_{\mathcal{W}_{D_n\times D_m}} \\
\left(E\otimes\mathcal{W}_{D_{n+1}}\right)_x\otimes\mathcal{W}_{D_n} & \underrightarrow{\mathrm{id}_E\otimes\mathcal{W}_{\mathrm{id}_{D_{n+1}}\times\mathbf{m}_{D_n\times D_m\to D_n}}} & \left(E\otimes\mathcal{W}_{D_{n+1}}\right)_x\otimes\mathcal{W}_{D_n\times D_m}
\end{array}$$

commutes, which implies that the outer square of the diagram in (13) commutes. This completes the proof.

Notation 5.2 *By the above proposition, we have the canonical projection* $\hat{\pi}_{n+1,n}$*:* $\hat{\mathbb{J}}^{D_{n+1}}(\pi)\to\hat{\mathbb{J}}^{D_n}(\pi)$ *so that, given* $\nabla_x\in\hat{\mathbb{J}}_x^{D_{n+1}}(\pi)$ *and* $\gamma\in\left(M\otimes\mathcal{W}_{D_n}\right)_{\pi(x)}$*, the composition of mappings in (6) applied to* γ *results in*

$$\left(\mathrm{id}_E\otimes\mathcal{W}_{(d_1,d_2)\in D_{n+1}\times D_n\mapsto d_1d_2\in D_n}\right)\left(\hat{\pi}_{n+1,n}(\nabla_x)(\gamma)\right)$$

For any natural numbers n*,* m *with* $m\leq n$*, we define* $\hat{\pi}_{n,m}$*:* $\hat{\mathbb{J}}^{D_n}(\pi)\to\hat{\mathbb{J}}^{D_m}(\pi)$ *to be* $\hat{\pi}_{m+1,m}\circ\ldots\circ\hat{\pi}_{n,n-1}$*.*

Proposition 5.3 *Let* ∇_x *be a* D_{n+1}*-pseudotangential over the bundle* $\pi:E\to M$ *at* $x\in E$*. Then the diagram*

$$\begin{array}{ccc}
\left(M\otimes\mathcal{W}_{D_{n+1}}\right)_{\pi(x)} & \xrightarrow{\nabla_x} & \left(E\otimes\mathcal{W}_{D_{n+1}}\right)_x \\
\widehat{\pi}_{n+1,n}\downarrow & & \downarrow\widehat{\pi}_{n+1,n} \\
\left(M\otimes\mathcal{W}_{D_n}\right)_{\pi(x)} & \xrightarrow[\widehat{\pi}_{n+1,n}(\nabla_x)]{} & \left(E\otimes\mathcal{W}_{D_n}\right)_x
\end{array}$$

is commutative.

Proof It is easy to see that the following four diagrams are commutative:

$$\begin{array}{ccc}
M\otimes\mathcal{W}_{D_{n+1}} & \xrightarrow{\mathrm{id}_M\otimes\mathcal{W}_{(d_1,d_2)\in D_{n+1}\times D_{n+1}\mapsto d_1d_2\in D_{n+1}}} & M\otimes\mathcal{W}_{D_{n+1}\times D_{n+1}} \\
\mathrm{id}_M\otimes\mathcal{W}_{\mathbf{i}_{D_n\subseteq D_{n+1}}}\downarrow & & \downarrow\mathrm{id}_M\otimes\mathcal{W}_{\mathbf{i}_{D_{n+1}\times D_n\subseteq D_{n+1}\times D_{n+1}}} \\
M\otimes\mathcal{W}_{D_n} & \xrightarrow[\mathrm{id}_M\otimes\mathcal{W}_{(d_1,d_2)\in D_{n+1}\times D_n\mapsto d_1d_2\in D_n}]{} & M\otimes\mathcal{W}_{D_{n+1}\times D_n}
\end{array}$$

$$\begin{array}{ccc}
M\otimes\mathcal{W}_{D_{n+1}\times D_{n+1}} & \xrightarrow{\nabla_x\otimes\mathrm{id}_{\mathcal{W}_{D_{n+1}}}} & E\otimes\mathcal{W}_{D_{n+1}\times D_{n+1}} \\
\mathrm{id}_M\otimes\mathcal{W}_{\mathbf{i}_{D_{n+1}\times D_n\subseteq D_{n+1}\times D_{n+1}}}\downarrow & & \downarrow\mathrm{id}_E\otimes\mathcal{W}_{\mathbf{i}_{D_{n+1}\times D_n\subseteq D_{n+1}\times D_{n+1}}} \\
M\otimes\mathcal{W}_{D_{n+1}\times D_n} & \xrightarrow[\nabla_x\otimes\mathrm{id}_{\mathcal{W}_{D_n}}]{} & E\otimes\mathcal{W}_{D_{n+1}\times D_n}
\end{array}$$

$$\begin{array}{ccc}
M\otimes\mathcal{W}_{D_{n+1}} & \xrightarrow{\mathrm{id}_M\otimes\mathcal{W}_{(d_1,d_2)\in D_{n+1}\times D_{n+1}\mapsto d_1d_2\in D_{n+1}}} & M\otimes\mathcal{W}_{D_{n+1}\times D_{n+1}} \\
\nabla_x\downarrow & & \downarrow\nabla_x\otimes\mathrm{id}_{\mathcal{W}_{D_{n+1}}} \\
E\otimes\mathcal{W}_{D_{n+1}} & \xrightarrow[\mathrm{id}_E\otimes\mathcal{W}_{(d_1,d_2)\in D_{n+1}\times D_{n+1}\mapsto d_1d_2\in D_{n+1}}]{} & E\otimes\mathcal{W}_{D_{n+1}\times D_{n+1}}
\end{array}$$

[By the second condition in Definition 5.1]

$$\begin{array}{ccc}
E\otimes\mathcal{W}_{D_{n+1}} & \xrightarrow{\mathrm{id}_E\otimes\mathcal{W}_{(d_1,d_2)\in D_{n+1}\times D_{n+1}\mapsto d_1d_2\in D_{n+1}}} & E\otimes\mathcal{W}_{D_{n+1}\times D_{n+1}} \\
\mathrm{id}_E\otimes\mathcal{W}_{\mathbf{i}_{D_n\subseteq D_{n+1}}}\downarrow & & \downarrow\mathrm{id}_E\otimes\mathcal{W}_{\mathbf{i}_{D_{n+1}\times D_n\subseteq D_{n+1}\times D_{n+1}}} \\
E\otimes\mathcal{W}_{D_n} & \xrightarrow[\mathrm{id}_E\otimes\mathcal{W}_{(d_1,d_2)\in D_{n+1}\times D_n\mapsto d_1d_2\in D_n}]{} & E\otimes\mathcal{W}_{D_{n+1}\times D_n}
\end{array}$$

Therefore the composition of mappings

$$\begin{aligned}
&M\otimes\mathcal{W}_{D_{n+1}}\xrightarrow{\mathrm{id}_M\otimes\mathcal{W}_{\mathbf{i}_{D_n\subseteq D_{n+1}}}}M\otimes\mathcal{W}_{D_n}\\
&\xrightarrow{\mathrm{id}_M\otimes\mathcal{W}_{(d_1,d_2)\in D_{n+1}\times D_n\mapsto d_1d_2\in D_n}}M\otimes\mathcal{W}_{D_{n+1}\times D_n}\\
&=\left(M\otimes\mathcal{W}_{D_{n+1}}\right)\otimes\mathcal{W}_{D_n}\xrightarrow{\nabla_x\otimes\mathrm{id}_{\mathcal{W}_{D_n}}}\left(E\otimes\mathcal{W}_{D_{n+1}}\right)\otimes\mathcal{W}_{D_n}\\
&=E\otimes\mathcal{W}_{D_{n+1}\times D_n}
\end{aligned}$$

is equal to the composition of mappings

$$\begin{aligned}
&M\otimes\mathcal{W}_{D_{n+1}}\xrightarrow{\nabla_x}E\otimes\mathcal{W}_{D_{n+1}}\xrightarrow{\mathrm{id}_E\otimes\mathcal{W}_{\mathbf{i}_{D_n\to D_{n+1}}}}E\otimes\mathcal{W}_{D_n}\\
&\xrightarrow{\mathrm{id}_E\otimes\mathcal{W}_{(d_1,d_2)\in D_{n+1}\times D_n\mapsto d_1d_2\in D_n}}E\otimes\mathcal{W}_{D_{n+1}\times D_n}
\end{aligned}$$

which yields the coveted result.

Corollary 5.1 *Let ∇_x be a D_{n+1}-pseudotangential over the bundle $\pi: E \to M$ at $x \in E$. For any $\gamma, \gamma' \in \left(M \otimes \mathcal{W}_{D_{n+1}}\right)_{\pi(x)}$, if*

$$\pi_{n+1,n}(\gamma) = \pi_{n+1,n}\left(\gamma'\right)$$

then

$$\pi_{n+1,n}\left(\nabla_x(\gamma)\right) = \pi_{n+1,n}\left(\nabla_x(\gamma')\right)$$

Proof By the above proposition, we have

$$\begin{aligned}\pi_{n+1,n}(\nabla_x(\gamma)) &= \hat{\pi}_{n+1,n}(\nabla_x)(\pi_{n+1,n}(\gamma)) \\ &= \hat{\pi}_{n+1,n}(\nabla_x)(\pi_{n+1,n}(\gamma')) = \pi_{n+1,n}(\nabla_x(\gamma')),\end{aligned}$$

which establishes the coveted proposition.

Definition 5.2 The notion of a *D_n-tangential over* the bundle $\pi: E \to M$ *at* $x \in E$ is defined inductively on n. The notion of a *D_0-tangential over* the bundle $\pi: E \to M$ *at* $x \in E$ and that of a *D_1-tangential over* the bundle $\pi: E \to M$ *at* $x \in E$ shall be identical with that of a D_0-pseudotangential *over* the bundle $\pi: E \to M$ *at* $x \in E$ and that of a D_1-pseudotangential *over* the bundle $\pi: E \to M$ *at* $x \in E$ respectively. Now we proceed by induction on n. A D_{n+1}-pseudotangential ∇_x: $\left(M \otimes \mathcal{W}_{D_{n+1}}\right)_{\pi(x)} \to \left(E \otimes \mathcal{W}_{D_{n+1}}\right)_x$ *over* the bundle $\pi: E \to M$ *at* $x \in E$ is called a *D_{n+1}-tangential over* the bundle $\pi: E \to M$ *at* $x \in E$ if it acquiesces in the following two conditions:

1. $\hat{\pi}_{n+1,n}(\nabla_x)$ is a D_n-tangential *over* the bundle $\pi: E \to M$ *at* $x \in E$.
2. For any simple polynomial ρ of $d \in D_{n+1}$ with $l = \dim \rho$ and any $\gamma \in \left(M \otimes \mathcal{W}_{D_l}\right)_{\pi(x)}$, we have

$$\nabla_x(\gamma \circ \rho) = (\pi_{n+1,l}(\nabla_x)(\gamma)) \circ \rho$$

Notation 5.3 *We denote by $\mathbb{J}_x^{D_n}(\pi)$ the totality of D_n-tangentials over the bundle π: $E \to M$ at $x \in E$, while we denote by $\mathbb{J}^{D_n}(\pi)$ the totality of D_n-tangentials over the bundle $\pi: E \to M$. By the very definition of a D_n-tangential, the projection $\hat{\pi}_{n+1,n}$: $\widehat{\mathbb{J}}^{D_{n+1}}(\pi) \to \widehat{\mathbb{J}}^{D_n}(\pi)$ is naturally restricted to a mapping $\pi_{n+1,n}$: $\mathbb{J}^{D_{n+1}}(\pi) \to \mathbb{J}^{D_n}(\pi)$. Similarly for $\pi_{n,m}$: $\mathbb{J}^{D_n}(\pi) \to \mathbb{J}^{D_m}(\pi)$ with $m \leq n$.*

6 From the First Approach to the Second

Definition 6.1 Mappings φ_n: $\mathbf{J}^n(\pi) \to \mathbb{J}^{D^n}(\pi)$ $(n = 0, 1)$ shall be the identity mappings. We are going to define φ_n: $\mathbf{J}^n(\pi) \to \mathbb{J}^{D^n}(\pi)$ for any natural number n by induction on n. Let $x_n = \nabla_{x_{n-1}} \in \mathbf{J}^n(\pi)$ and $\nabla_{x_n} \in \mathbf{J}^{n+1}(\pi)$. We define $\varphi_{n+1}(\nabla_{x_n})$ as the composition of mappings

$$\begin{aligned}
&\left(M \otimes \mathcal{W}_{D^{n+1}}\right)_{\pi(x_n)} \\
&= ((M \otimes \mathcal{W}_{D^n}) \otimes \mathcal{W}_D)_{(M \otimes \mathcal{W}_{D^n})_{\pi(x_n)}} \\
&\xrightarrow{\left\langle \pi_M^{M \otimes \mathcal{W}_{D^n}} \otimes \mathrm{id}_{\mathcal{W}_D}, \mathrm{id}_{(M \otimes \mathcal{W}_{D^n}) \otimes \mathcal{W}_D} \right\rangle} \\
&(M \otimes \mathcal{W}_D)_{\pi(x_n)} \underset{M \otimes \mathcal{W}_D}{\times} ((M \otimes \mathcal{W}_{D^n}) \otimes \mathcal{W}_D)_{(M \otimes \mathcal{W}_{D^n})_{\pi(x_n)}} \\
&\xrightarrow{\nabla_{x_n} \times \mathrm{id}_{(M \otimes \mathcal{W}_{D^n}) \otimes \mathcal{W}_D}} \\
&\left(\mathbf{J}^n(\pi) \otimes \mathcal{W}_D\right)_{x_n} \underset{M \otimes \mathcal{W}_D}{\times} ((M \otimes \mathcal{W}_{D^n}) \otimes \mathcal{W}_D)_{(M \otimes \mathcal{W}_{D^n})_{\pi(x_n)}} \\
&\xrightarrow{\left(\varphi_n \otimes \mathrm{id}_{\mathcal{W}_D}\right) \times \mathrm{id}_{(M \otimes \mathcal{W}_{D^n}) \otimes \mathcal{W}_D}} \\
&\left(\mathbb{J}^{D^n}(\pi) \otimes \mathcal{W}_D\right)_{\varphi_n(x_n)} \underset{M \otimes \mathcal{W}_D}{\times} ((M \otimes \mathcal{W}_{D^n}) \otimes \mathcal{W}_D)_{(M \otimes \mathcal{W}_{D^n})_{\pi(x_n)}} \\
&= \left(\left(\mathbb{J}^{D^n}(\pi) \underset{M}{\times} (M \otimes \mathcal{W}_{D^n})\right) \otimes \mathcal{W}_D\right)_{\varphi_n(x_n) \times (M \otimes \mathcal{W}_{D^n})_{\pi(x_n)}} \\
&\xrightarrow{\left((\nabla, \gamma) \in \mathbb{J}^{D^n}(\pi) \underset{M}{\times} (M \otimes \mathcal{W}_{D^n}) \mapsto \nabla(\gamma) \in E \otimes \mathcal{W}_{D^n}\right) \otimes \mathrm{id}_{\mathcal{W}_D}} \\
&((E \otimes \mathcal{W}_{D^n}) \otimes \mathcal{W}_D)_{(E \otimes \mathcal{W}_{D^n})_{\pi_0(x_n)}} \\
&= \left(E \otimes \mathcal{W}_{D^{n+1}}\right)_{\pi_0(x_n)}
\end{aligned}$$

Surely we have to show that

Lemma 6.1 *We have*

$$\varphi_{n+1}(\nabla_{x_n}) \in \hat{\mathbb{J}}^{n+1}(\pi)$$

Proof We have to show that for any $\gamma \in \mathbf{T}^{n+1}_{\pi_n(x_n)}(M)$, any $\alpha \in \mathbb{R}$ and any $\sigma \in \mathbf{S}_{n+1}$, we have

$$\gamma = \left(\pi \otimes \mathrm{id}_{\mathcal{W}_{D^{n+1}}}\right) \circ \left(\varphi_{n+1}(\nabla_{x_n})\right)(\gamma) \tag{14}$$

$$\varphi_{n+1}(\nabla_{x_n})(\alpha \underset{i}{\cdot} \gamma) = \alpha \underset{i}{\cdot} \varphi_{n+1}(\nabla_{x_n})(\gamma) \quad (1 \leq i \leq n+1) \tag{15}$$

$$\varphi_{n+1}(\nabla_{x_n})(\gamma^\sigma) = (\varphi_{n+1}(\nabla_{x_n})(\gamma))^\sigma \tag{16}$$

We proceed by induction on n.

1. First we deal with (14). The mapping

$$\left(\pi \otimes \mathrm{id}_{\mathcal{W}_{D^{n+1}}}\right) \left(\varphi_{n+1}(\nabla_{x_n})\right)$$

is the composition of mappings

$$
\begin{aligned}
&\left(M\otimes\mathcal{W}_{D^{n+1}}\right)_{\pi(x_n)} \\
&= ((M\otimes\mathcal{W}_{D^n})\otimes\mathcal{W}_D)_{(M\otimes\mathcal{W}_{D^n})_{\pi(x_n)}} \\
&\xrightarrow{\left\langle\pi_M^{M\otimes\mathcal{W}_{D^n}}\otimes\mathrm{id}_{\mathcal{W}_D},\,\mathrm{id}_{(M\otimes\mathcal{W}_{D^n})\otimes\mathcal{W}_D}\right\rangle} \\
&(M\otimes\mathcal{W}_D)_{\pi(x_n)}\underset{M\otimes\mathcal{W}_D}{\times}((M\otimes\mathcal{W}_{D^n})\otimes\mathcal{W}_D)_{(M\otimes\mathcal{W}_{D^n})_{\pi(x_n)}} \\
&\xrightarrow{\nabla_{x_n}\times\mathrm{id}_{(M\otimes\mathcal{W}_{D^n})\otimes\mathcal{W}_D}} \\
&\left(\mathbf{J}^n(\pi)\otimes\mathcal{W}_D\right)_{x_n}\underset{M\otimes\mathcal{W}_D}{\times}((M\otimes\mathcal{W}_{D^n})\otimes\mathcal{W}_D)_{(M\otimes\mathcal{W}_{D^n})_{\pi(x_n)}} \\
&\xrightarrow{\left(\varphi_n\otimes\mathrm{id}_{\mathcal{W}_D}\right)\times\mathrm{id}_{(M\otimes\mathcal{W}_{D^n})\otimes\mathcal{W}_D}} \\
&\left(\mathbb{J}^{D^n}(\pi)\otimes\mathcal{W}_D\right)_{\varphi_n(x_n)}\underset{M\otimes\mathcal{W}_D}{\times}((M\otimes\mathcal{W}_{D^n})\otimes\mathcal{W}_D)_{(M\otimes\mathcal{W}_{D^n})_{\pi(x_n)}} \\
&=\left(\left(\mathbb{J}^{D^n}(\pi)\underset{M}{\times}(M\otimes\mathcal{W}_{D^n})\right)\otimes\mathcal{W}_D\right)_{\varphi_n(x_n)\times(M\otimes\mathcal{W}_{D^n})_{\pi(x_n)}} \\
&\xrightarrow{\left((\nabla,\gamma)\in\mathbb{J}^{D^n}(\pi)\underset{M}{\times}(M\otimes\mathcal{W}_{D^n})\mapsto\nabla(\gamma)\in E\otimes\mathcal{W}_{D^n}\right)\otimes\mathrm{id}_{\mathcal{W}_D}} \\
&((E\otimes\mathcal{W}_{D^n})\otimes\mathcal{W}_D)_{(E\otimes\mathcal{W}_{D^n})_{\pi_0(x_n)}} \\
&=\left(E\otimes\mathcal{W}_{D^{n+1}}\right)_{\pi_0(x_n)}\xrightarrow{\pi\otimes\mathrm{id}_{\mathcal{W}_{D^{n+1}}}}\left(M\otimes\mathcal{W}_{D^{n+1}}\right)_{\pi(x_n)}
\end{aligned}
$$

It is easy to see that the composition of mappings

$$
\begin{aligned}
&\left(\mathbf{J}^n(\pi)\otimes\mathcal{W}_D\right)_{x_n}\underset{M\otimes\mathcal{W}_D}{\times}((M\otimes\mathcal{W}_{D^n})\otimes\mathcal{W}_D)_{(M\otimes\mathcal{W}_{D^n})_{\pi(x_n)}} \\
&\xrightarrow{\left(\varphi_n\otimes\mathrm{id}_{\mathcal{W}_D}\right)\times\mathrm{id}_{(M\otimes\mathcal{W}_{D^n})\otimes\mathcal{W}_D}} \\
&\left(\mathbb{J}^{D^n}(\pi)\otimes\mathcal{W}_D\right)_{\varphi_n(x_n)}\underset{M\otimes\mathcal{W}_D}{\times}((M\otimes\mathcal{W}_{D^n})\otimes\mathcal{W}_D)_{(M\otimes\mathcal{W}_{D^n})_{\pi(x_n)}} \\
&=\left(\left(\mathbb{J}^{D^n}(\pi)\underset{M}{\times}(M\otimes\mathcal{W}_{D^n})\right)\otimes\mathcal{W}_D\right)_{\{\varphi_n(x_n)\}\times(M\otimes\mathcal{W}_{D^n})_{\pi(x_n)}} \\
&\xrightarrow{\left((\nabla,\gamma)\in\mathbb{J}^{D^n}(\pi)\underset{M}{\times}(M\otimes\mathcal{W}_{D^n})\mapsto\nabla(\gamma)\in E\otimes\mathcal{W}_{D^n}\right)\otimes\mathrm{id}_{\mathcal{W}_D}} \\
&((E\otimes\mathcal{W}_{D^n})\otimes\mathcal{W}_D)_{(E\otimes\mathcal{W}_{D^n})_{\pi_0(x_n)}} \\
&=\left(E\otimes\mathcal{W}_{D^{n+1}}\right)_{\pi_0(x_n)}\xrightarrow{\pi\otimes\mathrm{id}_{\mathcal{W}_{D^{n+1}}}}\left(M\otimes\mathcal{W}_{D^{n+1}}\right)_{\pi(x_n)}
\end{aligned}
$$

is no other than the canonical projection of

$$\left(\mathbf{J}^n(\pi)\otimes\mathcal{W}_D\right)_{x_n}\underset{M\otimes\mathcal{W}_D}{\times}((M\otimes\mathcal{W}_{D^n})\otimes\mathcal{W}_D)_{(M\otimes\mathcal{W}_{D^n})_{\pi(x_n)}}$$

to the second factor $((M\otimes\mathcal{W}_{D^n})\otimes\mathcal{W}_D)_{(M\otimes\mathcal{W}_{D^n})_{\pi(x_n)}}$. It is also easy to see that the composition of mappings

$$\begin{aligned}
&((M\otimes\mathcal{W}_{D^n})\otimes\mathcal{W}_D)_{(M\otimes\mathcal{W}_{D^n})_{\pi(x_n)}}\\
&\xrightarrow{\left\langle\pi_M^{M\otimes\mathcal{W}_{D^n}}\otimes\mathrm{id}_{\mathcal{W}_D},\,\mathrm{id}_{(M\otimes\mathcal{W}_{D^n})\otimes\mathcal{W}_D}\right\rangle}\\
&(M\otimes\mathcal{W}_D)_{\pi(x_n)}\underset{M\otimes\mathcal{W}_D}{\times}((M\otimes\mathcal{W}_{D^n})\otimes\mathcal{W}_D)_{(M\otimes\mathcal{W}_{D^n})_{\pi(x_n)}}\\
&\xrightarrow{\nabla_{x_n}\times\mathrm{id}_{(M\otimes\mathcal{W}_{D^n})\otimes\mathcal{W}_D}}\\
&\left(\mathbf{J}^n(\pi)\otimes\mathcal{W}_D\right)_{x_n}\underset{M\otimes\mathcal{W}_D}{\times}((M\otimes\mathcal{W}_{D^n})\otimes\mathcal{W}_D)_{(M\otimes\mathcal{W}_{D^n})_{\pi(x_n)}}\\
&\xrightarrow{\left(\varphi_n\otimes\mathrm{id}_{\mathcal{W}_D}\right)\times\mathrm{id}_{(M\otimes\mathcal{W}_{D^n})\otimes\mathcal{W}_D}}\\
&\left(\mathbb{J}^{D^n}(\pi)\otimes\mathcal{W}_D\right)_{\varphi_n(x_n)}\underset{M\otimes\mathcal{W}_D}{\times}((M\otimes\mathcal{W}_{D^n})\otimes\mathcal{W}_D)_{(M\otimes\mathcal{W}_{D^n})_{\pi(x_n)}}
\end{aligned}$$

is

$$\begin{aligned}
&((M\otimes\mathcal{W}_{D^n})\otimes\mathcal{W}_D)_{(M\otimes\mathcal{W}_{D^n})_{\pi(x_n)}}\\
&\xrightarrow{\left\langle\left(\varphi_n\otimes\mathrm{id}_{\mathcal{W}_D}\right)\circ\nabla_{x_n}\circ\left(\pi_M^{M\otimes\mathcal{W}_{D^n}}\otimes\mathrm{id}_{\mathcal{W}_D}\right),\,\mathrm{id}_{(M\otimes\mathcal{W}_{D^n})\otimes\mathcal{W}_D}\right\rangle}\\
&\left(\mathbb{J}^{D^n}(\pi)\otimes\mathcal{W}_D\right)_{\varphi_n(x_n)}\underset{M\otimes\mathcal{W}_D}{\times}((M\otimes\mathcal{W}_{D^n})\otimes\mathcal{W}_D)_{(M\otimes\mathcal{W}_{D^n})_{\pi(x_n)}}.
\end{aligned}$$

Therefore (14) follows at once.

2. Now we deal with (15), the treatment of which is divided into two cases, namely, $i\leq n$ and $i=n+1$. Since both of them are almost trivial, they can safely be left to the reader.
3. Finally we must deal with (16), for which it suffices to consider only transpositions $\sigma=\langle i,i+1\rangle$ $(1\leq i\leq n)$. Here we deal only with the most difficult case of $\sigma=\langle n,n+1\rangle$. We consider the composition of mappings

$$\begin{aligned}
&\left(M\otimes\mathcal{W}_{D^{n+1}}\right)_{\pi(x_n)}\xrightarrow{\gamma\in\left(M\otimes\mathcal{W}_{D^{n+1}}\right)_{\pi(x_n)}\mapsto\gamma^{\langle n,n+1\rangle}\in\left(M\otimes\mathcal{W}_{D^{n+1}}\right)_{\pi(x_n)}}\\
&\left(M\otimes\mathcal{W}_{D^{n+1}}\right)_{\pi(x_n)}\\
&=((M\otimes\mathcal{W}_{D^n})\otimes\mathcal{W}_D)_{(M\otimes\mathcal{W}_{D^n})_{\pi(x_n)}}
\end{aligned}$$

$$
\xrightarrow{\left\langle \pi_M^{M\otimes\mathcal{W}_{D^n}} \otimes \mathrm{id}_{\mathcal{W}_D}, \mathrm{id}_{(M\otimes\mathcal{W}_{D^n})\otimes\mathcal{W}_D} \right\rangle}
$$

$$
(M \otimes \mathcal{W}_D)_{\pi(x_n)} \underset{M\otimes\mathcal{W}_D}{\times} ((M \otimes \mathcal{W}_{D^n}) \otimes \mathcal{W}_D)_{(M\otimes\mathcal{W}_{D^n})_{\pi(x_n)}}
$$

$$
\xrightarrow{\nabla_{x_n} \times \mathrm{id}_{(M\otimes\mathcal{W}_{D^n})\otimes\mathcal{W}_D}}
$$

$$
\left(\mathbf{J}^n(\pi) \otimes \mathcal{W}_D\right)_{x_n} \underset{M\otimes\mathcal{W}_D}{\times} ((M \otimes \mathcal{W}_{D^n}) \otimes \mathcal{W}_D)_{(M\otimes\mathcal{W}_{D^n})_{\pi(x_n)}}
$$

$$
\xrightarrow{\left(\varphi_n \otimes \mathrm{id}_{\mathcal{W}_D}\right) \times \mathrm{id}_{(M\otimes\mathcal{W}_{D^n})\otimes\mathcal{W}_D}}
$$

$$
\left(\mathbb{J}^{D^n}(\pi) \otimes \mathcal{W}_D\right)_{\varphi_n(x_n)} \underset{M\otimes\mathcal{W}_D}{\times} ((M \otimes \mathcal{W}_{D^n}) \otimes \mathcal{W}_D)_{(M\otimes\mathcal{W}_{D^n})_{\pi(x_n)}}
$$

$$
= \left(\left(\mathbb{J}^{D^n}(\pi) \underset{M}{\times} (M \otimes \mathcal{W}_{D^n})\right) \otimes \mathcal{W}_D\right)_{\varphi_n(x_n)\times(M\otimes\mathcal{W}_{D^n})_{\pi(x_n)}}
$$

$$
\xrightarrow{\left((\nabla, \gamma) \in \mathbb{J}^{D^n}(\pi) \underset{M}{\times} (M \otimes \mathcal{W}_{D^n}) \mapsto \nabla(\gamma) \in E \otimes \mathcal{W}_{D^n}\right) \otimes \mathrm{id}_{\mathcal{W}_D}}
$$

$$
((E \otimes \mathcal{W}_{D^n}) \otimes \mathcal{W}_D)_{(E\otimes\mathcal{W}_{D^n})_{\pi_0(x_n)}}
$$

$$
= \left(E \otimes \mathcal{W}_{D^{n+1}}\right)_{\pi_0(x_n)} \tag{17}
$$

By the very definition of φ_n, the composition of mappings

$$
\left(\mathbf{J}^n(\pi) \otimes \mathcal{W}_D\right)_{x_n} \underset{M\otimes\mathcal{W}_D}{\times} ((M \otimes \mathcal{W}_{D^n}) \otimes \mathcal{W}_D)_{(M\otimes\mathcal{W}_{D^n})_{\pi(x_n)}}
$$

$$
\xrightarrow{\left(\varphi_n \otimes \mathrm{id}_{\mathcal{W}_D}\right) \times \mathrm{id}_{(M\otimes\mathcal{W}_{D^n})\otimes\mathcal{W}_D}}
$$

$$
\left(\mathbb{J}^{D^n}(\pi) \otimes \mathcal{W}_D\right)_{\varphi_n(x_n)} \underset{M\otimes\mathcal{W}_D}{\times} ((M \otimes \mathcal{W}_{D^n}) \otimes \mathcal{W}_D)_{(M\otimes\mathcal{W}_{D^n})_{\pi(x_n)}}
$$

$$
= \left(\left(\mathbb{J}^{D^n}(\pi) \underset{M}{\times} (M \otimes \mathcal{W}_{D^n})\right) \otimes \mathcal{W}_D\right)_{\varphi_n(x_n)\times(M\otimes\mathcal{W}_{D^n})_{\pi(x_n)}}
$$

$$
\xrightarrow{\left((\nabla, \gamma) \in \mathbb{J}^{D^n}(\pi) \underset{M}{\times} (M \otimes \mathcal{W}_{D^n}) \mapsto \nabla(\gamma) \in E \otimes \mathcal{W}_{D^n}\right) \otimes \mathrm{id}_{\mathcal{W}_D}}
$$

$$
((E \otimes \mathcal{W}_{D^n}) \otimes \mathcal{W}_D)_{(E\otimes\mathcal{W}_{D^n})_{\pi_0(x_n)}}
$$

$$
= \left(E \otimes \mathcal{W}_{D^{n+1}}\right)_{\pi_0(x_n)}
$$

is equivalent to the composition of mappings

$$
\left(\mathbf{J}^n(\pi) \otimes \mathcal{W}_D\right)_{x_n} \underset{M\otimes\mathcal{W}_D}{\times} ((M \otimes \mathcal{W}_{D^n}) \otimes \mathcal{W}_D)_{(M\otimes\mathcal{W}_{D^n})_{\pi(x_n)}}
$$

$$
\begin{aligned}
&= \left(\mathbf{J}^n(\pi) \otimes \mathcal{W}_D\right)_{x_n} \underset{M \otimes \mathcal{W}_D}{\times} \\
&\left(\left(\left(M \otimes \mathcal{W}_{D^{n-1}}\right) \otimes \mathcal{W}_D\right) \otimes \mathcal{W}_D\right)_{\left(\left(M \otimes \mathcal{W}_{D^{n-1}}\right) \otimes \mathcal{W}_D\right)_{\left(M \otimes \mathcal{W}_{D^{n-1}}\right)_{\pi(x_n)}}} \\
&= \left(\left(\mathbf{J}^n(\pi) \underset{M}{\times} \left(\left(M \otimes \mathcal{W}_{D^{n-1}}\right) \otimes \mathcal{W}_D\right)\right) \otimes \mathcal{W}_D\right)_* \\
&[* = x_n \times \left(\left(M \otimes \mathcal{W}_{D^{n-1}}\right) \otimes \mathcal{W}_D\right)_{\left(M \otimes \mathcal{W}_{D^{n-1}}\right)_{\pi(x_n)}}] \\
&\xrightarrow{\left(\mathrm{id}_{\mathbf{J}^n(\pi)} \times \left\langle \pi_M^{M \otimes \mathcal{W}_{D^{n-1}}} \otimes \mathrm{id}_{\mathcal{W}_D}, \mathrm{id}_{\left(M \otimes \mathcal{W}_{D^{n-1}}\right) \otimes \mathcal{W}_D} \right\rangle\right) \otimes \mathrm{id}_{\mathcal{W}_D}} \\
&\left(\left(\mathbf{J}^n(\pi) \underset{M}{\times} (M \otimes \mathcal{W}_D) \underset{M}{\times} \left(\left(M \otimes \mathcal{W}_{D^{n-1}}\right) \otimes \mathcal{W}_D\right)\right) \otimes \mathcal{W}_D\right)_* \\
&[* = x_n \times \pi(x_n) \times \left(\left(M \otimes \mathcal{W}_{D^{n-1}}\right) \otimes \mathcal{W}_D\right)_{\left(M \otimes \mathcal{W}_{D^{n-1}}\right)_{\pi(x_n)}}] \\
&\xrightarrow{\left(\left(\begin{array}{c} (\nabla, t) \in \mathbf{J}^n(\pi) \times (M \otimes \mathcal{W}_D) \mapsto \\ \nabla(t) \in \mathbf{J}^{n-1}(\pi) \otimes \mathcal{W}_D \end{array}\right) \times \mathrm{id}_{\left(\left(M \otimes \mathcal{W}_{D^{n-1}}\right) \otimes \mathcal{W}_D\right)}\right) \otimes \mathrm{id}_{\mathcal{W}_D}} \\
&\left(\left(\left(\mathbf{J}^{n-1}(\pi) \otimes \mathcal{W}_D\right) \underset{M}{\times} \left(\left(M \otimes \mathcal{W}_{D^{n-1}}\right) \otimes \mathcal{W}_D\right)\right) \otimes \mathcal{W}_D\right)_* \\
&[* = \left(\mathbf{J}^{n-1}(\pi) \otimes \mathcal{W}_D\right)_{\pi_{n-1}(x_n)} \times \left(\left(M \otimes \mathcal{W}_{D^{n-1}}\right) \otimes \mathcal{W}_D\right)_{\left(M \otimes \mathcal{W}_{D^{n-1}}\right)_{\pi(x_n)}}] \\
&= \left(\left(\mathbf{J}^{n-1}(\pi) \underset{M}{\times} \left(M \otimes \mathcal{W}_{D^{n-1}}\right)\right) \otimes \mathcal{W}_{D^2}\right)_{\pi_{n-1}(x_n) \times \left(M \otimes \mathcal{W}_{D^{n-1}}\right)_{\pi(x_n)}} \\
&\xrightarrow{\varphi_{n-1} \times \mathrm{id}_{\left(M \otimes \mathcal{W}_{D^n}\right) \otimes \mathcal{W}_D}} \\
&\left(\left(\mathbb{J}^{D^{n-1}}(\pi) \underset{M}{\times} \left(M \otimes \mathcal{W}_{D^{n-1}}\right)\right) \otimes \mathcal{W}_{D^2}\right)_{\pi_0(x_n) \times \left(M \otimes \mathcal{W}_{D^{n-1}}\right)_{\pi(x_n)}} \\
&\xrightarrow{\left((\nabla, \gamma) \in \mathbb{J}^{D^{n-1}}(\pi) \times \left(M \otimes \mathcal{W}_{D^{n-1}}\right) \mapsto \nabla(\gamma) \in E \otimes \mathcal{W}_{D^{n-1}}\right) \otimes \mathrm{id}_{\mathcal{W}_{D^2}}} \\
&\left(\left(E \otimes \mathcal{W}_{D^{n-1}}\right) \otimes \mathcal{W}_{D^2}\right)_{\left(E \otimes \mathcal{W}_{D^{n-1}}\right)_{\pi_0(x_n)}} \\
&= \left(E \otimes \mathcal{W}_{D^{n+1}}\right)_{\pi_0(x_n)}
\end{aligned}
$$

Therefore (17) is no other than the composition of mappings

$$
\begin{aligned}
&\left(M \otimes \mathcal{W}_{D^{n+1}}\right)_{\pi(x_n)} \\
&\xrightarrow{\gamma \in \left(M \otimes \mathcal{W}_{D^{n+1}}\right)_{\pi(x_n)} \mapsto \gamma^{\langle n, n+1 \rangle} \in \left(M \otimes \mathcal{W}_{D^{n+1}}\right)_{\pi(x_n)}} \\
&\left(M \otimes \mathcal{W}_{D^{n+1}}\right)_{\pi(x_n)} \\
&= \left(\left(M \otimes \mathcal{W}_{D^n}\right) \otimes \mathcal{W}_D\right)_{\left(M \otimes \mathcal{W}_{D^n}\right)_{\pi(x_n)}}
\end{aligned}
$$

$$\xrightarrow{\left\langle \pi_M^{M\otimes\mathcal{W}_{D^n}} \otimes \mathrm{id}_{\mathcal{W}_D}, \mathrm{id}_{(M\otimes\mathcal{W}_{D^n})\otimes\mathcal{W}_D} \right\rangle}$$

$$(M\otimes\mathcal{W}_D)_{\pi(x_n)} \underset{M\otimes\mathcal{W}_D}{\times} ((M\otimes\mathcal{W}_{D^n})\otimes\mathcal{W}_D)_{(M\otimes\mathcal{W}_{D^n})_{\pi(x_n)}}$$

$$\xrightarrow{\nabla_{x_n} \times \mathrm{id}_{(M\otimes\mathcal{W}_{D^n})\otimes\mathcal{W}_D}}$$

$$\left(\mathbf{J}^n(\pi)\otimes\mathcal{W}_D\right)_{x_n} \underset{M\otimes\mathcal{W}_D}{\times} ((M\otimes\mathcal{W}_{D^n})\otimes\mathcal{W}_D)_{(M\otimes\mathcal{W}_{D^n})_{\pi(x_n)}}$$

$$= \left(\mathbf{J}^n(\pi)\otimes\mathcal{W}_D\right)_{x_n} \underset{M\otimes\mathcal{W}_D}{\times}$$

$$\left(\left(\left(M\otimes\mathcal{W}_{D^{n-1}}\right)\otimes\mathcal{W}_D\right)\otimes\mathcal{W}_D\right)_{\left(\left(M\otimes\mathcal{W}_{D^{n-1}}\right)\otimes\mathcal{W}_D\right)_{\left(M\otimes\mathcal{W}_{D^{n-1}}\right)_{\pi(x_n)}}}$$

$$= \left(\left(\mathbf{J}^n(\pi) \underset{M}{\times} \left(\left(M\otimes\mathcal{W}_{D^{n-1}}\right)\otimes\mathcal{W}_D\right)\right)\otimes\mathcal{W}_D\right)_*$$

$$[* = x_n \times \left(\left(M\otimes\mathcal{W}_{D^{n-1}}\right)\otimes\mathcal{W}_D\right)_{\left(M\otimes\mathcal{W}_{D^{n-1}}\right)_{\pi(x_n)}}]$$

$$\xrightarrow{\left(\mathrm{id}_{\mathbf{J}^n(\pi)} \times \left\langle \pi_M^{M\otimes\mathcal{W}_{D^{n-1}}} \otimes \mathrm{id}_{\mathcal{W}_D}, \mathrm{id}_{\left(M\otimes\mathcal{W}_{D^{n-1}}\right)\otimes\mathcal{W}_D}\right\rangle\right)\otimes\mathrm{id}_{\mathcal{W}_D}}$$

$$\left(\left(\mathbf{J}^n(\pi) \underset{M}{\times} (M\otimes\mathcal{W}_D) \underset{M}{\times} \left(\left(M\otimes\mathcal{W}_{D^{n-1}}\right)\otimes\mathcal{W}_D\right)\right)\otimes\mathcal{W}_D\right)_*$$

$$[* = x_n \times \pi(x_n) \times \left(\left(M\otimes\mathcal{W}_{D^{n-1}}\right)\otimes\mathcal{W}_D\right)_{\left(M\otimes\mathcal{W}_{D^{n-1}}\right)_{\pi(x_n)}}]$$

$$\xrightarrow{\left(\left(\begin{array}{c}(\nabla,t)\in\mathbf{J}^n(\pi)\times(M\otimes\mathcal{W}_D)\mapsto\\ \nabla(t)\in\mathbf{J}^{n-1}(\pi)\otimes\mathcal{W}_D\end{array}\right)\times\mathrm{id}_{\left(\left(M\otimes\mathcal{W}_{D^{n-1}}\right)\otimes\mathcal{W}_D\right)}\right)\otimes\mathrm{id}_{\mathcal{W}_D}}$$

$$\left(\left(\left(\mathbf{J}^{n-1}(\pi)\otimes\mathcal{W}_D\right) \underset{M}{\times} \left(\left(M\otimes\mathcal{W}_{D^{n-1}}\right)\otimes\mathcal{W}_D\right)\right)\otimes\mathcal{W}_D\right)_*$$

$$[* = \left(\mathbf{J}^{n-1}(\pi)\otimes\mathcal{W}_D\right)_{\pi_{n-1}(x_n)} \times \left(\left(M\otimes\mathcal{W}_{D^{n-1}}\right)\otimes\mathcal{W}_D\right)_{\left(M\otimes\mathcal{W}_{D^{n-1}}\right)_{\pi(x_n)}}]$$

$$= \left(\left(\mathbf{J}^{n-1}(\pi)\times\left(M\otimes\mathcal{W}_{D^{n-1}}\right)\right)\otimes\mathcal{W}_{D^2}\right)_{\pi_{n-1}(x_n)\times\left(M\otimes\mathcal{W}_{D^{n-1}}\right)_{\pi(x_n)}}$$

$$\xrightarrow{\varphi_{n-1}\times\mathrm{id}_{(M\otimes\mathcal{W}_{D^n})\otimes\mathcal{W}_D}}$$

$$\left(\left(\mathbb{J}^{D^{n-1}}(\pi) \underset{M}{\times} \left(M\otimes\mathcal{W}_{D^{n-1}}\right)\right)\otimes\mathcal{W}_{D^2}\right)_{\pi_0(x_n)\times\left(M\otimes\mathcal{W}_{D^{n-1}}\right)_{\pi(x_n)}}$$

$$\xrightarrow{\left((\nabla,\gamma)\in\mathbb{J}^{D^{n-1}}(\pi)\times\left(M\otimes\mathcal{W}_{D^{n-1}}\right)\mapsto\nabla(\gamma)\in E\otimes\mathcal{W}_{D^{n-1}}\right)\otimes\mathrm{id}_{\mathcal{W}_{D^2}}}$$

$$\left(\left(E\otimes\mathcal{W}_{D^{n-1}}\right)\otimes\mathcal{W}_{D^2}\right)_{\left(E\otimes\mathcal{W}_{D^{n-1}}\right)_{\pi_0(x_n)}}$$

$$= \left(E\otimes\mathcal{W}_{D^{n+1}}\right)_{\pi_0(x_n)}$$

On the other hand, the composition of mappings

$$
\begin{aligned}
&\left(M \otimes \mathcal{W}_{D^{n+1}}\right)_{\pi(x_n)} \\
&= ((M \otimes \mathcal{W}_{D^n}) \otimes \mathcal{W}_D)_{(M \otimes \mathcal{W}_{D^n})_{\pi(x_n)}} \\
&\xrightarrow{\left\langle \pi_M^{M \otimes \mathcal{W}_{D^n}} \otimes \mathrm{id}_{\mathcal{W}_D}, \mathrm{id}_{(M \otimes \mathcal{W}_{D^n}) \otimes \mathcal{W}_D} \right\rangle} \\
&(M \otimes \mathcal{W}_D)_{\pi(x_n)} \underset{M \otimes \mathcal{W}_D}{\times} ((M \otimes \mathcal{W}_{D^n}) \otimes \mathcal{W}_D)_{(M \otimes \mathcal{W}_{D^n})_{\pi(x_n)}} \\
&\xrightarrow{\nabla_{x_n} \times \mathrm{id}_{(M \otimes \mathcal{W}_{D^n}) \otimes \mathcal{W}_D}} \\
&\left(\mathbf{J}^n(\pi) \otimes \mathcal{W}_D\right)_{x_n} \underset{M \otimes \mathcal{W}_D}{\times} ((M \otimes \mathcal{W}_{D^n}) \otimes \mathcal{W}_D)_{(M \otimes \mathcal{W}_{D^n})_{\pi(x_n)}} \\
&\xrightarrow{\left(\varphi_n \otimes \mathrm{id}_{\mathcal{W}_D}\right) \times \mathrm{id}_{(M \otimes \mathcal{W}_{D^n}) \otimes \mathcal{W}_D}} \\
&\left(\mathbb{J}^{D^n}(\pi) \otimes \mathcal{W}_D\right)_{\varphi_n(x_n)} \underset{M \otimes \mathcal{W}_D}{\times} ((M \otimes \mathcal{W}_{D^n}) \otimes \mathcal{W}_D)_{(M \otimes \mathcal{W}_{D^n})_{\pi(x_n)}} \\
&= \left(\left(\mathbb{J}^{D^n}(\pi) \underset{M}{\times} (M \otimes \mathcal{W}_{D^n})\right) \otimes \mathcal{W}_D\right)_{\varphi_n(x_n) \times (M \otimes \mathcal{W}_{D^n})_{\pi(x_n)}} \\
&\xrightarrow{\left((\nabla, \gamma) \in \mathbb{J}^{D^n}(\pi) \underset{M}{\times} (M \otimes \mathcal{W}_{D^n}) \mapsto \nabla(\gamma) \in E \otimes \mathcal{W}_{D^n}\right) \otimes \mathrm{id}_{\mathcal{W}_D}} \\
&((E \otimes \mathcal{W}_{D^n}) \otimes \mathcal{W}_D)_{(E \otimes \mathcal{W}_{D^n})_{\pi_0(x_n)}} \\
&= \left(E \otimes \mathcal{W}_{D^{n+1}}\right)_{\pi_0(x_n)} \\
&\xrightarrow{\gamma \in E \otimes \mathcal{W}_{D^{n+1}} \mapsto \gamma^{\langle n, n+1 \rangle} \in E \otimes \mathcal{W}_{D^{n+1}}} \left(E \otimes \mathcal{W}_{D^{n+1}}\right)_{\pi_0(x_n)}
\end{aligned}
$$

is the composition of mappings

$$
\begin{aligned}
&\left(M \otimes \mathcal{W}_{D^{n+1}}\right)_{\pi(x_n)} \\
&= ((M \otimes \mathcal{W}_{D^n}) \otimes \mathcal{W}_D)_{(M \otimes \mathcal{W}_{D^n})_{\pi(x_n)}} \\
&\xrightarrow{\left\langle \pi_M^{M \otimes \mathcal{W}_{D^n}} \otimes \mathrm{id}_{\mathcal{W}_D}, \mathrm{id}_{(M \otimes \mathcal{W}_{D^n}) \otimes \mathcal{W}_D} \right\rangle} \\
&(M \otimes \mathcal{W}_D)_{\pi(x_n)} \underset{M \otimes \mathcal{W}_D}{\times} ((M \otimes \mathcal{W}_{D^n}) \otimes \mathcal{W}_D)_{(M \otimes \mathcal{W}_{D^n})_{\pi(x_n)}} \\
&\xrightarrow{\nabla_{x_n} \times \mathrm{id}_{(M \otimes \mathcal{W}_{D^n}) \otimes \mathcal{W}_D}} \\
&\left(\mathbf{J}^n(\pi) \otimes \mathcal{W}_D\right)_{x_n} \underset{M \otimes \mathcal{W}_D}{\times} ((M \otimes \mathcal{W}_{D^n}) \otimes \mathcal{W}_D)_{(M \otimes \mathcal{W}_{D^n})_{\pi(x_n)}} \\
&= \left(\mathbf{J}^n(\pi) \otimes \mathcal{W}_D\right)_{x_n} \underset{M \otimes \mathcal{W}_D}{\times}
\end{aligned}
$$

$$\left(\left(\left(M \otimes \mathcal{W}_{D^{n-1}}\right) \otimes \mathcal{W}_D\right) \otimes \mathcal{W}_D\right)_{\left(\left(M \otimes \mathcal{W}_{D^{n-1}}\right) \otimes \mathcal{W}_D\right)_{\left(M \otimes \mathcal{W}_{D^{n-1}}\right)_{\pi(x_n)}}}$$

$$= \left(\left(\mathbf{J}^n(\pi) \times \left(\left(M \otimes \mathcal{W}_{D^{n-1}}\right) \otimes \mathcal{W}_D\right)\right) \otimes \mathcal{W}_D\right)_*$$

$$[* = x_n \times \left(\left(M \otimes \mathcal{W}_{D^{n-1}}\right) \otimes \mathcal{W}_D\right)_{\left(M \otimes \mathcal{W}_{D^{n-1}}\right)_{\pi(x_n)}}]$$

$$\xrightarrow{\left(\mathrm{id}_{\mathbf{J}^n(\pi)} \times \left\langle \pi_M^{M \otimes \mathcal{W}_{D^{n-1}}} \otimes \mathrm{id}_{\mathcal{W}_D}, \mathrm{id}_{\left(M \otimes \mathcal{W}_{D^{n-1}}\right) \otimes \mathcal{W}_D}\right\rangle\right) \otimes \mathrm{id}_{\mathcal{W}_D}}$$

$$\left(\left(\mathbf{J}^n(\pi) \underset{M}{\times} (M \otimes \mathcal{W}_D) \underset{M}{\times} \left(\left(M \otimes \mathcal{W}_{D^{n-1}}\right) \otimes \mathcal{W}_D\right)\right) \otimes \mathcal{W}_D\right)_*$$

$$[* = x_n \times \pi(x_n) \times \left(\left(M \otimes \mathcal{W}_{D^{n-1}}\right) \otimes \mathcal{W}_D\right)_{\left(M \otimes \mathcal{W}_{D^{n-1}}\right)_{\pi(x_n)}}]$$

$$\xrightarrow{\left(\left(\begin{array}{c}(\nabla, t) \in \mathbf{J}^n(\pi) \times (M \otimes \mathcal{W}_D) \mapsto \\ \nabla(t) \in \mathbf{J}^{n-1}(\pi) \otimes \mathcal{W}_D\end{array}\right) \times \mathrm{id}_{\left(\left(M \otimes \mathcal{W}_{D^{n-1}}\right) \otimes \mathcal{W}_D\right)}\right) \otimes \mathrm{id}_{\mathcal{W}_D}}$$

$$\left(\left(\left(\mathbf{J}^{n-1}(\pi) \otimes \mathcal{W}_D\right) \underset{M \otimes \mathcal{W}_D}{\times} \left(\left(M \otimes \mathcal{W}_{D^{n-1}}\right) \otimes \mathcal{W}_D\right)\right) \otimes \mathcal{W}_D\right)_*$$

$$[* = \left(\mathbf{J}^{n-1}(\pi) \otimes \mathcal{W}_D\right)_{\pi_{n-1}(x_n)} \underset{M \otimes \mathcal{W}_D}{\times} \left(\left(M \otimes \mathcal{W}_{D^{n-1}}\right) \otimes \mathcal{W}_D\right)_{\left(M \otimes \mathcal{W}_{D^{n-1}}\right)_{\pi(x_n)}}]$$

$$= \left(\left(\mathbf{J}^{n-1}(\pi) \times \left(M \otimes \mathcal{W}_{D^{n-1}}\right)\right) \otimes \mathcal{W}_{D^2}\right)_{\pi_{n-1}(x_n) \times \left(M \otimes \mathcal{W}_{D^{n-1}}\right)_{\pi(x_n)}}$$

followed by the composition of mappings

$$\left(\left(\mathbf{J}^{n-1}(\pi) \times \left(M \otimes \mathcal{W}_{D^{n-1}}\right)\right) \otimes \mathcal{W}_{D^2}\right)_{\pi_{n-1}(x_n) \times \left(M \otimes \mathcal{W}_{D^{n-1}}\right)_{\pi(x_n)}}$$

$$\xrightarrow{\varphi_{n-1} \times \mathrm{id}_{\left(M \otimes \mathcal{W}_{D^n}\right) \otimes \mathcal{W}_D}}$$

$$\left(\left(\mathbb{J}^{D^{n-1}}(\pi) \underset{M}{\times} \left(M \otimes \mathcal{W}_{D^{n-1}}\right)\right) \otimes \mathcal{W}_{D^2}\right)_{\pi_0(x_n) \times \left(M \otimes \mathcal{W}_{D^{n-1}}\right)_{\pi(x_n)}}$$

$$\xrightarrow{\left((\nabla, \gamma) \in \mathbb{J}^{n-1}(\pi) \times \left(M \otimes \mathcal{W}_{D^{n-1}}\right) \mapsto \nabla(\gamma) \in E \otimes \mathcal{W}_{D^{n-1}}\right) \otimes \mathrm{id}_{\mathcal{W}_{D^2}}}$$

$$\left(E \otimes \mathcal{W}_{D^{n-1}}\right) \otimes \mathcal{W}_{D^2} = \left(E \otimes \mathcal{W}_{D^{n+1}}\right)_{\pi_0(x_n)}$$

$$\xrightarrow{\gamma \in E \otimes \mathcal{W}_{D^{n+1}} \mapsto \gamma^{\langle n, n+1 \rangle} \in E \otimes \mathcal{W}_{D^{n+1}}} \left(E \otimes \mathcal{W}_{D^{n+1}}\right)_{\pi_0(x_n)},$$

which is easily seen to be equivalent to the composition of mappings

$$\left(\left(\mathbf{J}^{n-1}(\pi) \times \left(M \otimes \mathcal{W}_{D^{n-1}}\right)\right) \otimes \mathcal{W}_{D^2}\right)_{\pi_{n-1}(x_n) \times \left(M \otimes \mathcal{W}_{D^{n-1}}\right)_{\pi(x_n)}}$$

$$\xrightarrow{\mathrm{id}_{\mathbf{J}^{n-1}(\pi)\times(M\otimes\mathcal{W}_{D^{n-1}})}\otimes\mathcal{W}_{(d_1.d_2)\in D^2\mapsto(d_2.d_1)\in D^2}}$$

$$\left(\left(\mathbf{J}^{n-1}(\pi)\times\left(M\otimes\mathcal{W}_{D^{n-1}}\right)\right)\otimes\mathcal{W}_{D^2}\right)_{\pi_{n-1}(x_n)\times\left(M\otimes\mathcal{W}_{D^{n-1}}\right)_{\pi(x_n)}}$$

$$\xrightarrow{\varphi_{n-1}\times\mathrm{id}_{(M\otimes\mathcal{W}_{D^n})\otimes\mathcal{W}_D}}$$

$$\left(\left(\mathbb{J}^{D^{n-1}}(\pi)\underset{M}{\times}\left(M\otimes\mathcal{W}_{D^{n-1}}\right)\right)\otimes\mathcal{W}_{D^2}\right)_{\pi_0(x_n)\times\left(M\otimes\mathcal{W}_{D^{n-1}}\right)_{\pi(x_n)}}$$

$$\xrightarrow{\left((\nabla,\gamma)\in\mathbb{J}^{D^{n-1}}(\pi)\times\left(M\otimes\mathcal{W}_{D^{n-1}}\right)\mapsto\nabla(\gamma)\in E\otimes\mathcal{W}_{D^{n-1}}\right)\otimes\mathrm{id}_{\mathcal{W}_{D^2}}}$$

$$\left(\left(E\otimes\mathcal{W}_{D^{n-1}}\right)\otimes\mathcal{W}_{D^2}\right)_{\left(E\otimes\mathcal{W}_{D^{n-1}}\right)_{\pi_0(x_n)}}$$

$$=\left(E\otimes\mathcal{W}_{D^{n+1}}\right)_{\pi_0(x_n)}$$

Therefore the desired result follows from the second condition in the item 3 of Notation 3.3.

Lemma 6.2 *The diagram*

$$\begin{array}{ccc}\mathbf{J}^{n+1}(\pi) & \xrightarrow{\varphi_{n+1}} & \widehat{\mathbb{J}}^{D^{n+1}}(\pi)\\ \pi_{n+1,n}\downarrow & & \downarrow\widehat{\pi}_{n+1,n}\\ \mathbf{J}^{n}(\pi) & \xrightarrow{\varphi_{n}} & \widehat{\mathbb{J}}^{D^{n}}(\pi)\end{array}$$

is commutative.

Proof Given $\nabla_{x_n}\in\mathbf{J}^{n+1}(\pi)$, $\left(\widehat{\pi}_{n+1,n}\circ\varphi_{n+1}\right)\left(\nabla_{x_n}\right)$ is, by the very definition of $\widehat{\pi}_{n+1,n}$, the composition of mappings

$$\left(M\otimes\mathcal{W}_{D^n}\right)_{\pi(x_n)}\xrightarrow{\mathbf{s}_{n+1}}\left(M\otimes\mathcal{W}_{D^{n+1}}\right)_{\pi(x_n)}\xrightarrow{\varphi_{n+1}(\nabla_{x_n})}$$
$$\left(E\otimes\mathcal{W}_{D^{n+1}}\right)_{\pi_0(x_n)}\xrightarrow{\mathbf{d}_{n+1}}\left(E\otimes\mathcal{W}_{D^n}\right)_{\pi_0(x_n)}$$

which is equivalent, by the very definition of $\varphi_{n+1}(\nabla_{x_n})$, to the composition of mappings

$$\left(M\otimes\mathcal{W}_{D^n}\right)_{\pi(x_n)}\xrightarrow{\mathbf{s}_{n+1}}\left(M\otimes\mathcal{W}_{D^{n+1}}\right)_{\pi(x_n)}$$

$$=\left(\left(M\otimes\mathcal{W}_{D^n}\right)\otimes\mathcal{W}_D\right)_{\left(M\otimes\mathcal{W}_{D^n}\right)_{\pi(x_n)}}$$

$$\xrightarrow{\left\langle\pi_M^{M\otimes\mathcal{W}_{D^n}}\otimes\mathrm{id}_{\mathcal{W}_D},\mathrm{id}_{(M\otimes\mathcal{W}_{D^n})\otimes\mathcal{W}_D}\right\rangle}$$

$$\left(\left(M\otimes\mathcal{W}_D\right)\underset{M\otimes\mathcal{W}_D}{\times}\left(\left(M\otimes\mathcal{W}_{D^n}\right)\otimes\mathcal{W}_D\right)\right)_{\{\pi(x_n)\}\times\left(M\otimes\mathcal{W}_{D^n}\right)_{\pi(x_n)}}$$

$$\xrightarrow{\nabla_{x_n}\times\mathrm{id}_{(M\otimes\mathcal{W}_{D^n})\otimes\mathcal{W}_D}}$$

$$\left((\mathbf{J}^n(\pi) \otimes \mathcal{W}_D) \underset{M \otimes \mathcal{W}_D}{\times} ((M \otimes \mathcal{W}_{D^n}) \otimes \mathcal{W}_D) \right)_{\{\pi(x_n)\} \times (M \otimes \mathcal{W}_{D^n})_{\pi(x_n)}}$$

$$\xrightarrow{\left(\varphi_n \otimes \mathrm{id}_{\mathcal{W}_D}\right) \times \mathrm{id}_{(M \otimes \mathcal{W}_{D^n}) \otimes \mathcal{W}_D}}$$

$$\left(\left(\mathbb{J}^{D^n}(\pi) \otimes \mathcal{W}_D\right) \underset{M \otimes \mathcal{W}_D}{\times} ((M \otimes \mathcal{W}_{D^n}) \otimes \mathcal{W}_D) \right)_{\{\pi(x_n)\} \times (M \otimes \mathcal{W}_{D^n})_{\pi(x_n)}}$$

$$= \left(\left(\mathbb{J}^{D^n}(\pi) \underset{M}{\times} (M \otimes \mathcal{W}_{D^n}) \right) \otimes \mathcal{W}_D \right)_{\{\pi(x_n)\} \times (M \otimes \mathcal{W}_{D^n})_{\pi(x_n)}}$$

$$\xrightarrow{\left((\nabla, \gamma) \in \mathbb{J}^{D^n}(\pi) \times (M \otimes \mathcal{W}_{D^n}) \mapsto \nabla(\gamma) \in E \otimes \mathcal{W}_{D^n}\right) \otimes \mathrm{id}_{\mathcal{W}_D}}$$

$$((E \otimes \mathcal{W}_{D^n}) \otimes \mathcal{W}_D)_{(E \otimes \mathcal{W}_{D^n})_{\pi_0(x_n)}}$$

$$= \left(E \otimes \mathcal{W}_{D^{n+1}}\right)_{\pi_0(x_n)} \xrightarrow{\mathbf{d}_{n+1}} (E \otimes \mathcal{W}_{D^n})_{\pi_0(x_n)}$$

This is easily seen to be equivalent to $\varphi_n(\pi_{n+1,n}(\nabla_{x_n}))$, which completes the proof.

Lemma 6.1 can be strengthened as follows:

Lemma 6.3 *We have*

$$\varphi_{n+1}(\nabla_{x_n}) \in \mathbb{J}^{n+1}(\pi)$$

Proof With due regard to Lemmas 6.1 and 6.2, we have only to show that

$$\begin{aligned}
&\left(\varphi_{n+1}(\nabla_{x_n})\right) \circ \left(\mathrm{id}_M \otimes \mathcal{W}_{(d_1,\ldots,d_n,d_{n+1}) \in D^{n+1} \mapsto (d_1,\ldots,d_n d_{n+1}) \in D^n}\right) \\
&= \left(\mathrm{id}_E \otimes \mathcal{W}_{(d_1,\ldots,d_n,d_{n+1}) \in D^{n+1} \mapsto (d_1,\ldots,d_n d_{n+1}) \in D^{n+1}}\right) \circ \\
&\left(\widehat{\pi}_{n+1,n}(\varphi_{n+1}(\nabla_{x_n}))\right) \qquad (18)
\end{aligned}$$

For $n = 0$, there is nothing to prove. We proceed by induction on n. By the very definition of φ_{n+1}, the left-hand side of (18) is the composition of mappings

$$(M \otimes \mathcal{W}_{D^n})_{\pi(x_n)}$$

$$\xrightarrow{\mathrm{id}_M \otimes \mathcal{W}_{(d_1,\ldots,d_n,d_{n+1}) \in D^{n+1} \mapsto (d_1,\ldots,d_n d_{n+1}) \in D^n}}$$

$$\left(M \otimes \mathcal{W}_{D^{n+1}}\right)_{\pi(x_n)}$$

$$= ((M \otimes \mathcal{W}_{D^n}) \otimes \mathcal{W}_D)_{(M \otimes \mathcal{W}_{D^n})_{\pi(x_n)}}$$

$$\xrightarrow{\left\langle \pi_M^{M \otimes \mathcal{W}_{D^n}} \otimes \mathrm{id}_{\mathcal{W}_D}, \mathrm{id}_{(M \otimes \mathcal{W}_{D^n}) \otimes \mathcal{W}_D} \right\rangle}$$

$$(M \otimes \mathcal{W}_D)_{\pi(x_n)} \underset{M \otimes \mathcal{W}_D}{\times} ((M \otimes \mathcal{W}_{D^n}) \otimes \mathcal{W}_D)_{(M \otimes \mathcal{W}_{D^n})_{\pi(x_n)}}$$

$$\xrightarrow{\nabla_{x_n} \times \mathrm{id}_{(M \otimes \mathcal{W}_{D^n}) \otimes \mathcal{W}_D}}$$

$$
\begin{aligned}
&\left(\mathbf{J}^n(\pi)\otimes\mathcal{W}_D\right)_{\pi(x_n)} \underset{M\otimes\mathcal{W}_D}{\times} ((M\otimes\mathcal{W}_{D^n})\otimes\mathcal{W}_D)_{(M\otimes\mathcal{W}_{D^n})_{\pi(x_n)}} \\
&\xrightarrow{\left(\varphi_n\otimes\mathrm{id}_{\mathcal{W}_D}\right)\times\mathrm{id}_{(M\otimes\mathcal{W}_{D^n})\otimes\mathcal{W}_D}} \\
&\left(\mathbb{J}^{D^n}(\pi)\otimes\mathcal{W}_D\right)_{\pi(x_n)} \underset{M\otimes\mathcal{W}_D}{\times} ((M\otimes\mathcal{W}_{D^n})\otimes\mathcal{W}_D)_{(M\otimes\mathcal{W}_{D^n})_{\pi(x_n)}} \\
&=\left(\left(\mathbb{J}^{D^n}(\pi)\underset{M}{\times}(M\otimes\mathcal{W}_{D^n})\right)\otimes\mathcal{W}_D\right)_{\{\pi(x_n)\}\times(M\otimes\mathcal{W}_{D^n})_{\pi(x_n)}} \\
&\xrightarrow{\left((\nabla,\gamma)\in\mathbb{J}^{D^n}(\pi)\times(M\otimes\mathcal{W}_{D^n})\mapsto\nabla(\gamma)\in E\otimes\mathcal{W}_{D^n}\right)\otimes\mathrm{id}_{\mathcal{W}_D}} \\
&((E\otimes\mathcal{W}_{D^n})\otimes\mathcal{W}_D)_{(E\otimes\mathcal{W}_{D^n})_{\pi_0(x_n)}} \\
&=\left(E\otimes\mathcal{W}_{D^{n+1}}\right)_{\pi_0(x_n)}
\end{aligned}
$$

which is easily seen, by dint of Lemma 6.1, to be equivalent to the right-hand side of (18).

Thus we have established the mappings $\varphi_n : \mathbf{J}^n(\pi) \to \mathbb{J}^{D^n}(\pi)$.

7 From the Second Approach to the Third

The principal objective in this section is to define a mapping $\psi_n\colon \mathbb{J}^{D^n}(\pi) \to \mathbb{J}^{D_n}(\pi)$. Let us begin with

Proposition 7.1 *Let ∇_x be a D^n-pseudotangential over the bundle $\pi\colon E \to M$ at $x \in E$ and $\gamma \in \left(M \otimes \mathcal{W}_{D_n}\right)_{\pi(x)}$. Then there exists a unique $\gamma' \in \left(E \otimes \mathcal{W}_{D_n}\right)_x$ such that*

$$
\begin{aligned}
&\nabla_x(\left(\mathrm{id}_M \otimes \mathcal{W}_{(d_1,\ldots,d_n)\in D^n \longmapsto (d_1+\ldots+d_n)\in D_n}\right)(\gamma)) \\
&= \left(\mathrm{id}_E \otimes \mathcal{W}_{(d_1,\ldots,d_n)\in D^n \longmapsto (d_1+\ldots+d_n)\in D_n}\right)\left(\gamma'\right)
\end{aligned}
$$

Proof This stems easily from the following simple lemma.

Lemma 7.1 *The diagram*

$$
\mathcal{W}_{D_n} \xrightarrow{\mathcal{W}_{(d_1,\ldots,d_n)\in D^n \longmapsto (d_1+\ldots+d_n)\in D_n}} \mathcal{W}_{D^n} \;
\begin{array}{c}
\xrightarrow{\mathcal{W}_{\tau_1}} \\
\vdots \\
\xrightarrow{\mathcal{W}_{\tau_i}} \\
\vdots \\
\xrightarrow{\mathcal{W}_{\tau_{n-1}}}
\end{array}
\; \mathcal{W}_{D^n}
$$

is a limit diagram in the category of Weil algebras, where $\tau_i \colon D^n \to D^n$ is the mapping permuting the i-th and $(i+1)$-th components of D^n while fixing the other components.

Notation 7.1 *We will denote by $\widehat{\psi}_n(\nabla_x)(\gamma)$ the unique γ' in the above proposition, thereby getting a function $\widehat{\psi}_n(\nabla_x) \colon \left(M \otimes \mathcal{W}_{D_n}\right)_{\pi(x)} \to \left(E \otimes \mathcal{W}_{D_n}\right)_x$.*

Proposition 7.2 *For any $\nabla_x \in \widehat{\mathbb{J}}_x^{D^n}(\pi)$, we have $\widehat{\psi}_n(\nabla_x) \in \widehat{\mathbb{J}}_x^{D_n}(\pi)$.*

Proof We have to verify the three conditions in Definition 5.1 concerning the mapping $\widehat{\psi}_n(\nabla_x) \colon \left(M \otimes \mathcal{W}_{D_n}\right)_{\pi(x)} \to \left(E \otimes \mathcal{W}_{D_n}\right)_x$.

1. To see the first condition, it suffices to show that

$$\left(\mathrm{id}_M \otimes \mathcal{W}_{(d_1,\ldots,d_n)\in D^n \longmapsto (d_1+\ldots+d_n)\in D_n}\right)(\gamma)$$
$$= \left(\mathrm{id}_E \otimes \mathcal{W}_{(d_1,\ldots,d_n)\in D^n \longmapsto (d_1+\ldots+d_n)\in D_n}\right)\left(\left(\pi \otimes \mathrm{id}_{\mathcal{W}_{D_n}}\right)\left(\widehat{\psi}_n(\nabla_x)(\gamma)\right)\right),$$

which follows from

$$\left(\mathrm{id}_M \otimes \mathcal{W}_{(d_1,\ldots,d_n)\in D^n \longmapsto (d_1+\ldots+d_n)\in D_n}\right)\left(\left(\pi \otimes \mathrm{id}_{\mathcal{W}_{D_n}}\right)\left(\widehat{\psi}_n(\nabla_x)(\gamma)\right)\right)$$
$$= \left(\pi \otimes \mathrm{id}_{\mathcal{W}_{D^n}}\right)\left(\left(\mathrm{id}_E \otimes \mathcal{W}_{(d_1,\ldots,d_n)\in D^n \longmapsto (d_1+\ldots+d_n)\in D_n}\right)\left(\widehat{\psi}_n(\nabla_x)(\gamma)\right)\right)$$
[By the bifunctionality of $\otimes$]
$$= \left(\pi \otimes \mathrm{id}_{\mathcal{W}_{D^n}}\right)\left(\nabla_x\left(\left(\mathrm{id}_M \otimes \mathcal{W}_{(d_1,\ldots,d_n)\in D^n \longmapsto (d_1+\ldots+d_n)\in D_n}\right)(\gamma)\right)\right)$$
[By the very definition of $\widehat{\psi}_n(\nabla_x)$]
$$= \left(\mathrm{id}_M \otimes \mathcal{W}_{(d_1,\ldots,d_n)\in D^n \longmapsto (d_1+\ldots+d_n)\in D_n}\right)(\gamma)$$

2. Now we are going to deal with the second condition. It is easy to see that the composition of mappings

$$\left(M \otimes \mathcal{W}_{D_n}\right)_{\pi(x)} \xrightarrow{\mathrm{id}_M \otimes \mathcal{W}_{(\alpha\cdot)_{D_n}}} \left(M \otimes \mathcal{W}_{D_n}\right)_{\pi(x)} \xrightarrow{\mathcal{W}_{(d_1,\ldots,d_n)\in D^n \longmapsto (d_1+\ldots+d_n)\in D_n}} (M \otimes \mathcal{W}_{D^n})_{\pi(x)}$$

is equivalent to the composition of mappings

$$\left(M \otimes \mathcal{W}_{D_n}\right)_{\pi(x)} \xrightarrow{\mathcal{W}_{(d_1,\ldots,d_n)\in D^n \longmapsto (d_1+\ldots+d_n)\in D_n}} (M \otimes \mathcal{W}_{D^n})_{\pi(x)} \xrightarrow{\mathrm{id}_M \otimes \mathcal{W}_{\left(\underset{1}{\alpha\cdot}\right)_{D^n}}} (M \otimes \mathcal{W}_{D^n})_{\pi(x)} \ldots \xrightarrow{\mathrm{id}_M \otimes \mathcal{W}_{\left(\underset{n}{\alpha\cdot}\right)_{D^n}}} (M \otimes \mathcal{W}_{D^n})_{\pi(x)},$$

while the composition of mappings

$$(M \otimes \mathcal{W}_{D^n})_{\pi(x)} \underrightarrow{\mathrm{id}_M \otimes \mathcal{W}_{\left(\alpha \cdot_1\right)_{D^n}}} (M \otimes \mathcal{W}_{D^n})_{\pi(x)} \ldots \underrightarrow{\mathrm{id}_M \otimes \mathcal{W}_{\left(\alpha \cdot_n\right)_{D^n}}}$$

$$(M \otimes \mathcal{W}_{D^n})_{\pi(x)} \underrightarrow{\nabla_x} (E \otimes \mathcal{W}_{D^n})_x$$

is equivalent to the composition of mappings

$$(M \otimes \mathcal{W}_{D^n})_{\pi(x)} \underrightarrow{\nabla_x} (E \otimes \mathcal{W}_{D^n})_x \underrightarrow{\mathrm{id}_E \otimes \mathcal{W}_{\left(\alpha \cdot_1\right)_{D^n}}} (E \otimes \mathcal{W}_{D^n})_x \ldots$$

$$\underrightarrow{\mathrm{id}_E \otimes \mathcal{W}_{\left(\alpha \cdot_n\right)_{D^n}}} (E \otimes \mathcal{W}_{D^n})_x$$

Therefore the composition of mappings

$$\left(M \otimes \mathcal{W}_{D_n}\right)_{\pi(x)} \underrightarrow{\mathrm{id}_M \otimes \mathcal{W}_{(\alpha \cdot)_{D_n}}} \left(M \otimes \mathcal{W}_{D_n}\right)_{\pi(x)}$$

$$\underrightarrow{\mathcal{W}_{(d_1,\ldots,d_n)\in D^n \longmapsto (d_1+\ldots+d_n)\in D_n}} (M \otimes \mathcal{W}_{D^n})_{\pi(x)} \underrightarrow{\nabla_x} (E \otimes \mathcal{W}_{D^n})_x$$

is equivalent to the composition of mappings

$$\left(M \otimes \mathcal{W}_{D_n}\right)_{\pi(x)} \underrightarrow{\mathcal{W}_{(d_1,\ldots,d_n)\in D^n \longmapsto (d_1+\ldots+d_n)\in D_n}} (M \otimes \mathcal{W}_{D^n})_{\pi(x)} \underrightarrow{\nabla_x}$$

$$(E \otimes \mathcal{W}_{D^n})_x \underrightarrow{\mathrm{id}_E \otimes \mathcal{W}_{\left(\alpha \cdot_1\right)_{D^n}}} (E \otimes \mathcal{W}_{D^n})_x \ldots \underrightarrow{\mathrm{id}_E \otimes \mathcal{W}_{\left(\alpha \cdot_n\right)_{D^n}}} (E \otimes \mathcal{W}_{D^n})_x ,$$

which should be equivalent in turn to

$$\left(M \otimes \mathcal{W}_{D_n}\right)_{\pi(x)} \underrightarrow{\widehat{\Psi}_n(\nabla_x)} \left(E \otimes \mathcal{W}_{D_n}\right)_x \underrightarrow{\mathrm{id}_E \otimes \mathcal{W}_{(d_1,\ldots,d_n)\in D^n \longmapsto (d_1+\ldots+d_n)\in D_n}}$$

$$(E \otimes \mathcal{W}_{D^n})_x \underrightarrow{\mathrm{id}_E \otimes \mathcal{W}_{\left(\alpha \cdot_1\right)_{D^n}}} (E \otimes \mathcal{W}_{D^n})_x \ldots \underrightarrow{\mathrm{id}_E \otimes \mathcal{W}_{\left(\alpha \cdot_n\right)_{D^n}}} (E \otimes \mathcal{W}_{D^n})_x$$

Since the composition of mappings

$$\left(E \otimes \mathcal{W}_{D_n}\right)_x \underrightarrow{\mathcal{W}_{(d_1,\ldots,d_n)\in D^n \longmapsto (d_1+\ldots+d_n)\in D_n}} (E \otimes \mathcal{W}_{D^n})_x \underrightarrow{\mathrm{id}_E \otimes \mathcal{W}_{\left(\alpha \cdot_1\right)_{D^n}}}$$

$$(E \otimes \mathcal{W}_{D^n})_x \ldots \underrightarrow{\mathrm{id}_E \otimes \mathcal{W}_{\left(\alpha \cdot_n\right)_{D^n}}} (E \otimes \mathcal{W}_{D^n})_x$$

is equivalent to the composition of mappings

$$\left(E \otimes \mathcal{W}_{D_n}\right)_x \xrightarrow{\mathrm{id}_E \otimes \mathcal{W}_{(\alpha\cdot)_{D_n}}} \left(E \otimes \mathcal{W}_{D_n}\right)_x \xrightarrow{\mathrm{id}_E \otimes \mathcal{W}_{(d_1,\ldots,d_n)\in D^n \longmapsto (d_1+\ldots+d_n)\in D_n}}$$
$$(E \otimes \mathcal{W}_{D^n})_x ,$$

the coveted result follows.

3. We are going to deal with the third condition. We have to show that the diagram

$$\begin{array}{ccc} \left(M \otimes \mathcal{W}_{D_n}\right)_{\pi(x)} & \xrightarrow{\mathrm{id}_M \otimes \mathcal{W}_{\mathbf{m}_{D_n \times D_m \to D_n}}} & \left(M \otimes \mathcal{W}_{D_n}\right)_{\pi(x)} \otimes \mathcal{W}_{D_m} \\ \widehat{\psi}_n(\nabla_x) \downarrow & & \downarrow \widehat{\psi}_n(\nabla_x) \otimes \mathrm{id}_{\mathcal{W}_{D_m}} \\ \left(E \otimes \mathcal{W}_{D_n}\right)_x & \xrightarrow{\mathrm{id}_E \otimes \mathcal{W}_{\mathbf{m}_{D_n \times D_m \to D_n}}} & \left(E \otimes \mathcal{W}_{D_n}\right)_x \otimes \mathcal{W}_{D_m} \end{array} \tag{19}$$

commutes. It is easy to see that the diagram

$$\begin{array}{ccc} \left(E \otimes \mathcal{W}_{D_n}\right)_x & \xrightarrow{\mathrm{id}_E \otimes \mathcal{W}_{+_{D^n \to D_n}}} & (E \otimes \mathcal{W}_{D^n})_x \\ \mathrm{id}_E \otimes \mathcal{W}_{\mathbf{m}_{D_n \times D_m \to D_n}} \downarrow & & \downarrow \mathrm{id}_E \otimes \mathcal{W}_\eta \\ \left(E \otimes \mathcal{W}_{D_n}\right)_x \otimes \mathcal{W}_{D_m} & \xrightarrow{\mathrm{id}_E \otimes \mathcal{W}_{+_{D^n \to D_n}} \times \mathrm{id}_{D_m}} & (E \otimes \mathcal{W}_{D^n})_x \otimes \mathcal{W}_{D_m} \end{array}$$

commutes, where η stands for

$$(d_1, \ldots, d_n, e) \in D^n \times D_m \longmapsto (d_1 e, \ldots, d_n e) \in D^n$$

so that the commutativity of the diagram in (19) is equivalent to the commutativity of the outer square of the diagram

$$\begin{array}{ccc} \left(M \otimes \mathcal{W}_{D_n}\right)_{\pi(x)} & \xrightarrow{\mathrm{id}_M \otimes \mathcal{W}_{\mathbf{m}_{D_n \times D_m \to D_n}}} & \left(M \otimes \mathcal{W}_{D_n}\right)_{\pi(x)} \otimes \mathcal{W}_{D_m} \\ \widehat{\psi}_n(\nabla_x) \downarrow & & \downarrow \widehat{\psi}_n(\nabla_x) \otimes \mathrm{id}_{\mathcal{W}_{D_m}} \\ \left(E \otimes \mathcal{W}_{D_n}\right)_x & \xrightarrow{\mathrm{id}_E \otimes \mathcal{W}_{\mathbf{m}_{D_n \times D_m \to D_n}}} & \left(E \otimes \mathcal{W}_{D_n}\right)_x \otimes \mathcal{W}_{D_m} \\ \mathrm{id}_E \otimes \mathcal{W}_{+_{D^n \to D_n}} \downarrow & & \downarrow \mathrm{id}_E \otimes \mathcal{W}_{+_{D^n \to D_n}} \times \mathrm{id}_{D_m} \\ (E \otimes \mathcal{W}_{D^n})_x & \xrightarrow{\mathrm{id}_E \otimes \mathcal{W}_\eta} & (E \otimes \mathcal{W}_{D^n})_x \otimes \mathcal{W}_{D_m} \end{array} \tag{20}$$

where $+_{D^n \to D_n}$ stands for

$$(d_1, \ldots, d_n) \in D^n \longmapsto (d_1 + \ldots + d_n) \in D_n$$

The composition of mappings

$$\left(M \otimes \mathcal{W}_{D_n}\right)_{\pi(x)} \xrightarrow{\widehat{\psi}_n(\nabla_x)} \left(E \otimes \mathcal{W}_{D_n}\right)_x \xrightarrow{\mathrm{id}_E \otimes \mathcal{W}_{+_{D^n \to D_n}}} (E \otimes \mathcal{W}_{D^n})_x$$

is equal to the composition of mappings

$$\left(M \otimes \mathcal{W}_{D_n}\right)_{\pi(x)} \xrightarrow{\mathrm{id}_M \otimes \mathcal{W}_{+_{D^n \to D_n}}} (M \otimes \mathcal{W}_{D^n})_{\pi(x)} \xrightarrow{\nabla_x} (E \otimes \mathcal{W}_{D^n})_x$$

while the composition of mappings

$$\left(M\otimes\mathcal{W}_{D_n}\right)_{\pi(x)}\otimes\mathcal{W}_{D_m}\xrightarrow{\widehat{\psi}_n(\nabla_x)\otimes\mathrm{id}_{\mathcal{W}_{D_m}}}\left(E\otimes\mathcal{W}_{D_n}\right)_x\otimes\mathcal{W}_{D_m}$$
$$\xrightarrow{\mathrm{id}_E\otimes\mathcal{W}_{+_{D^n\to D_n}}\times\mathrm{id}_{D_m}}(E\otimes\mathcal{W}_{D^n})_x\otimes\mathcal{W}_{D_m}$$

is equal to the composition of mappings

$$\left(M\otimes\mathcal{W}_{D_n}\right)_{\pi(x)}\xrightarrow{\mathrm{id}_M\otimes\mathcal{W}_{+_{D^n\to D_n}}\times\mathrm{id}_{D_m}}(M\otimes\mathcal{W}_{D^n})_{\pi(x)}\xrightarrow{\nabla_x\otimes\mathrm{id}_{\mathcal{W}_{D_m}}}$$
$$(E\otimes\mathcal{W}_{D^n})_x\otimes\mathcal{W}_{D_m}$$

Since the diagram

$$\begin{array}{ccc}\left(M\otimes\mathcal{W}_{D_n}\right)_{\pi(x)} & \xrightarrow{\mathrm{id}_M\otimes\mathcal{W}_{\mathbf{m}_{D_n\times D_m\to D_n}}} & \left(M\otimes\mathcal{W}_{D_n}\right)_{\pi(x)}\otimes\mathcal{W}_{D_m}\\ \mathrm{id}_M\otimes\mathcal{W}_{+_{D^n\to D_n}}\downarrow & & \downarrow\mathrm{id}_M\otimes\mathcal{W}_{+_{D^n\to D_n}}\times\mathrm{id}_{D_m}\\ (M\otimes\mathcal{W}_{D^n})_{\pi(x)} & \xrightarrow{\mathrm{id}_M\otimes\mathcal{W}_\eta} & (M\otimes\mathcal{W}_{D^n})_{\pi(x)}\otimes\mathcal{W}_{D_m}\\ \nabla_x\downarrow & & \downarrow\nabla_x\otimes\mathrm{id}_{\mathcal{W}_{D_m}}\\ (E\otimes\mathcal{W}_{D^n})_x & \xrightarrow{\mathrm{id}_E\otimes\mathcal{W}_\eta} & (E\otimes\mathcal{W}_{D^n})_x\otimes\mathcal{W}_{D_m}\end{array}$$

commutes, the outer square of the diagram in (20) commutes. This completes the proof.

Proposition 7.3 *The diagram*

$$\begin{array}{ccc}\hat{\mathbb{J}}_x^{D^{n+1}}(\pi) & \xrightarrow{\widehat{\psi}_{n+1}} & \hat{\mathbb{J}}_x^{D_{n+1}}(\pi)\\ \widehat{\pi}_{n+1,n}\downarrow & & \downarrow\widehat{\pi}_{n+1,n}\\ \hat{\mathbb{J}}_x^{D^n}(\pi) & \overrightarrow{\widehat{\psi}_n} & \hat{\mathbb{J}}_x^{D_n}(\pi)\end{array}$$

commutes.

Proof Given $\nabla_x\in\hat{\mathbb{J}}_x^{D^{n+1}}(\pi)$, the composition of mappings

$$\left(M\otimes\mathcal{W}_{D_n}\right)_{\pi(x)}\xrightarrow{\widehat{\pi}_{n+1,n}\left(\widehat{\psi}_{n+1}(\nabla_x)\right)}\left(E\otimes\mathcal{W}_{D_n}\right)_x\xrightarrow{\mathrm{id}_E\otimes\mathcal{W}_{\mathbf{m}_{D_n\times D_n\to D_n}}}$$
$$\left(E\otimes\mathcal{W}_{D_n}\right)_x\otimes\mathcal{W}_{D_n}\xrightarrow{\mathrm{id}_E\otimes\mathcal{W}_{+_{D^n\to D_n}}\times\mathrm{id}_{D_n}}(E\otimes\mathcal{W}_{D^n})_x\otimes\mathcal{W}_{D_n}\qquad(21)$$

is equivalent to the composition of mappings

$$\left(M\otimes\mathcal{W}_{D_n}\right)_{\pi(x)}\xrightarrow{\widehat{\pi}_{n+1,n}\left(\widehat{\psi}_{n+1}(\nabla_x)\right)}\left(E\otimes\mathcal{W}_{D_n}\right)_x\xrightarrow{\mathrm{id}_E\otimes\mathcal{W}_{\mathbf{m}_{D_{n+1}\times D_n\to D_n}}}$$
$$\left(E\otimes\mathcal{W}_{D_{n+1}}\right)_x\otimes\mathcal{W}_{D_n}\xrightarrow{\mathrm{id}_E\otimes\mathcal{W}_{+_{D^{n+1}\to D_{n+1}}}\times\mathrm{id}_{D_n}}\left(E\otimes\mathcal{W}_{D^{n+1}}\right)_x$$

$$\otimes \mathcal{W}_{D_n} \xrightarrow{\mathbf{d}_{n+1} \otimes \mathrm{id}_{\mathcal{W}_{D_n}}}$$
$$(E \otimes \mathcal{W}_{D^n})_x \otimes \mathcal{W}_{D_n}$$

which is in turn equivalent to the composition of mappings

$$\left(M \otimes \mathcal{W}_{D_n}\right)_{\pi(x)} \xrightarrow{\mathrm{id}_M \otimes \mathcal{W}_{\mathbf{m}_{D_{n+1} \times D_n \to D_n}}} \left(M \otimes \mathcal{W}_{D_{n+1}}\right)_{\pi(x)}$$
$$\otimes \mathcal{W}_{D_n} \xrightarrow{\widehat{\psi}_{n+1}(\nabla_x) \otimes \mathcal{W}_{\mathrm{id}_{\mathcal{W}_{D_n}}}} \left(E \otimes \mathcal{W}_{D_{n+1}}\right)_x \otimes \mathcal{W}_{D_n} \xrightarrow{\mathrm{id}_E \otimes \mathcal{W}_{+_{D^{n+1} \to D_{n+1}}} \times \mathrm{id}_{D_n}}$$
$$\left(E \otimes \mathcal{W}_{D^{n+1}}\right)_x \otimes \mathcal{W}_{D_n} \xrightarrow{\mathbf{d}_{n+1} \otimes \mathrm{id}_{\mathcal{W}_{D_n}}} (E \otimes \mathcal{W}_{D^n})_x \otimes \mathcal{W}_{D_n}$$

This is to be supplanted by the composition of mappings

$$\left(M \otimes \mathcal{W}_{D_n}\right)_{\pi(x)} \xrightarrow{\mathrm{id}_M \otimes \mathcal{W}_{\mathbf{m}_{D_{n+1} \times D_n \to D_n}}} \left(M \otimes \mathcal{W}_{D_{n+1}}\right)_{\pi(x)} \otimes \mathcal{W}_{D_n}$$
$$\xrightarrow{\mathrm{id}_M \otimes \mathcal{W}_{+_{D^{n+1} \to D_{n+1}}} \times \mathrm{id}_{D_n}} \left(M \otimes \mathcal{W}_{D^{n+1}}\right)_{\pi(x)} \otimes \mathcal{W}_{D_n} \xrightarrow{\nabla_x \otimes \mathrm{id}_{\mathcal{W}_{D_n}}}$$
$$\left(E \otimes \mathcal{W}_{D^{n+1}}\right)_x \otimes \mathcal{W}_{D_n} \xrightarrow{\mathbf{d}_{n+1} \otimes \mathrm{id}_{\mathcal{W}_{D_n}}} (E \otimes \mathcal{W}_{D^n})_x \otimes \mathcal{W}_{D_n},$$

which is in turn equivalent to the composition of mappings

$$\left(M \otimes \mathcal{W}_{D_n}\right)_{\pi(x)} \xrightarrow{\mathrm{id}_M \otimes \mathcal{W}_{\mathbf{m}_{D_{n+1} \times D_n \to D_n}}} \left(M \otimes \mathcal{W}_{D_{n+1}}\right)_{\pi(x)}$$
$$\otimes \mathcal{W}_{D_n} \xrightarrow{\mathrm{id}_M \otimes \mathcal{W}_{+_{D^{n+1} \to D_{n+1}}} \times \mathrm{id}_{D_n}} \left(M \otimes \mathcal{W}_{D^{n+1}}\right)_{\pi(x)} \otimes \mathcal{W}_{D_n} \xrightarrow{\mathbf{d}_{n+1} \otimes \mathrm{id}_{\mathcal{W}_{D_n}}}$$
$$(M \otimes \mathcal{W}_{D^n})_{\pi(x)} \otimes \mathcal{W}_{D_n} \xrightarrow{\widehat{\pi}_{n+1,n}(\nabla_x) \otimes \mathrm{id}_{\mathcal{W}_{D_n}}} (E \otimes \mathcal{W}_{D^n})_x \otimes \mathcal{W}_{D_n}$$

by Proposition 4.2. This is to be supplanted by the composition of mappings

$$\left(M \otimes \mathcal{W}_{D_n}\right)_{\pi(x)} \xrightarrow{\mathrm{id}_M \otimes \mathcal{W}_{\mathbf{m}_{D_n \times D_n \to D_n}}} \left(M \otimes \mathcal{W}_{D_n}\right)_{\pi(x)}$$
$$\otimes \mathcal{W}_{D_n} \xrightarrow{\mathrm{id}_M \otimes \mathcal{W}_{+_{D^n \to D_n}} \times \mathrm{id}_{D_n}} (M \otimes \mathcal{W}_{D^n})_{\pi(x)} \otimes \mathcal{W}_{D_n} \xrightarrow{\widehat{\pi}_{n+1,n}(\nabla_x) \otimes \mathrm{id}_{\mathcal{W}_{D_n}}}$$
$$(E \otimes \mathcal{W}_{D^n})_x \otimes \mathcal{W}_{D_n},$$

which is equivalent to the composition of mappings

$$\left(M \otimes \mathcal{W}_{D_n}\right)_{\pi(x)} \xrightarrow{\mathrm{id}_M \otimes \mathcal{W}_{\mathbf{m}_{D_n \times D_n \to D_n}}} \left(M \otimes \mathcal{W}_{D_n}\right)_{\pi(x)}$$
$$\otimes \mathcal{W}_{D_n} \xrightarrow{\widehat{\psi}_n\left(\widehat{\pi}_{n+1,n}(\nabla_x)\right) \otimes \mathrm{id}_{\mathcal{W}_{D_n}}} \left(E \otimes \mathcal{W}_{D_n}\right)_x \otimes \mathcal{W}_{D_n} \xrightarrow{\mathrm{id}_E \otimes \mathcal{W}_{+_{D^n \to D_n}} \times \mathrm{id}_{D_n}}$$
$$(E \otimes \mathcal{W}_{D^n})_x \otimes \mathcal{W}_{D_n}$$

This is really equivalent to the composition of mappings

$$\left(M\otimes\mathcal{W}_{D_n}\right)_{\pi(x)}\underrightarrow{\widehat{\psi}_n\left(\widehat{\pi}_{n+1,n}\left(\nabla_x\right)\right)}\left(E\otimes\mathcal{W}_{D_n}\right)_x\underrightarrow{\mathrm{id}_E\otimes\mathcal{W}_{\mathbf{m}_{D_n\times D_n\to D_n}}}$$
$$\left(E\otimes\mathcal{W}_{D_n}\right)_x\otimes\mathcal{W}_{D_n}\underrightarrow{\mathrm{id}_E\otimes\mathcal{W}_{+_{D^n\to D_n}\times\mathrm{id}_{D_n}}}E\otimes\mathcal{W}_{D^n\times D_n} \tag{22}$$

This just established fact that the composition of mappings in (21) and that in (22) are equivalent implies the coveted result at once. This completes the proof.

Proposition 7.4 *Let $\mathbb{D}$ be a simplicial infinitesimal space of dimension n and degree m. Let ∇_x be a D^n-pseudotangential over the bundle $\pi\colon E\to M$ at $x\in E$ and $\gamma\in\left(M\otimes\mathcal{W}_{D_n}\right)_{\pi(x)}$. Then the composition of mappings*

$$\left(M\otimes\mathcal{W}_{D_n}\right)_{\pi(x)}\underrightarrow{\mathrm{id}_M\otimes\mathcal{W}_{+_{\mathbb{D}\to D_n}}}\left(M\otimes\mathcal{W}_{\mathbb{D}}\right)_{\pi(x)}\underrightarrow{\nabla_x^{\mathbb{D}}}\left(E\otimes\mathcal{W}_{\mathbb{D}}\right)_x$$

is equivalent to the composition of mappings

$$\left(M\otimes\mathcal{W}_{D_n}\right)_{\pi(x)}\underrightarrow{\widehat{\psi}_n(\nabla_x)}\left(E\otimes\mathcal{W}_{D_n}\right)_x\underrightarrow{\mathrm{id}_E\otimes\mathcal{W}_{+_{\mathbb{D}\to D_n}}}\left(E\otimes\mathcal{W}_{\mathbb{D}}\right)_x$$

Proof Let $i\colon D^k\to\mathbb{D}$ be any mapping in the standard quasi-colimit representation of $\mathbb{D}$. The composition of mappings

$$\left(M\otimes\mathcal{W}_{D_n}\right)_{\pi(x)}\underrightarrow{\mathrm{id}_M\otimes\mathcal{W}_{+_{\mathbb{D}\to D_n}}}\left(M\otimes\mathcal{W}_{\mathbb{D}}\right)_{\pi(x)}\underrightarrow{\nabla_x^{\mathbb{D}}}\left(E\otimes\mathcal{W}_{\mathbb{D}}\right)_x$$
$$\underrightarrow{\mathrm{id}_E\otimes\mathcal{W}_i}\left(E\otimes\mathcal{W}_{D^k}\right)_x \tag{23}$$

is equivalent, by dint of Theorem 4.5, to the composition of mappings

$$\left(M\otimes\mathcal{W}_{D_n}\right)_{\pi(x)}\underrightarrow{\mathrm{id}_M\otimes\mathcal{W}_{\mathbf{i}_{D_k\to D_n}}}\left(M\otimes\mathcal{W}_{D_k}\right)_{\pi(x)}\underrightarrow{\mathrm{id}_M\otimes\mathcal{W}_{+_{D^k\to D_k}}}$$
$$\left(M\otimes\mathcal{W}_{D^k}\right)_{\pi(x)}\underrightarrow{\nabla_x^{D^k}}\left(E\otimes\mathcal{W}_{D^k}\right)_x,$$

which is in turn equivalent, by the very definition of $\widehat{\psi}_k$, to the composition of mappings

$$\left(M\otimes\mathcal{W}_{D_n}\right)_{\pi(x)}\underrightarrow{\mathrm{id}_M\otimes\mathcal{W}_{\mathbf{i}_{D_k\to D_n}}}\left(M\otimes\mathcal{W}_{D_k}\right)_{\pi(x)}\underrightarrow{\widehat{\psi}_k\left(\nabla_x^{D^k}\right)}\left(E\otimes\mathcal{W}_{D_k}\right)_x$$
$$\underrightarrow{\mathrm{id}_E\otimes\mathcal{W}_{+_{D^k\to D_k}}}\left(E\otimes\mathcal{W}_{D^k}\right)_x.$$

This is indeed equivalent, by dint of Proposition 7.3, to the composition of mappings

$$\left(M\otimes\mathcal{W}_{D_n}\right)_{\pi(x)}\xrightarrow{\widehat{\psi}_n(\nabla_x)}\left(E\otimes\mathcal{W}_{D_n}\right)_x\xrightarrow{\mathrm{id}_E\otimes\mathcal{W}_{\mathbf{i}_{D_k\to D_n}}}\left(E\otimes\mathcal{W}_{D_k}\right)_x$$
$$\xrightarrow{\mathrm{id}_E\otimes\mathcal{W}_{+_{D^k\to D_k}}}\left(E\otimes\mathcal{W}_{D^k}\right)_x,$$

which is in turn equivalent to the composition of mappings

$$\left(M\otimes\mathcal{W}_{D_n}\right)_{\pi(x)}\xrightarrow{\widehat{\psi}_n(\nabla_x)}\left(E\otimes\mathcal{W}_{D_n}\right)_x\xrightarrow{\mathrm{id}_E\otimes\mathcal{W}_{+_{\mathbb{D}\to D_n}}}(E\otimes\mathcal{W}_{\mathbb{D}})_x$$
$$\xrightarrow{\mathrm{id}_E\otimes\mathcal{W}_i}\left(E\otimes\mathcal{W}_{D^k}\right)_x \tag{24}$$

The just established fact that the composition of mappings in (23) and that in (24) are equivalent implies the coveted result at once. This completes the proof.

Theorem 7.1 *For any* $\nabla_x\in\mathbb{J}_x^{D^n}(\pi)$, *we have* $\widehat{\psi}_n(\nabla_x)\in\mathbb{J}_x^{D_n}(\pi)$.

Proof In view of Proposition 7.2, it suffices to show that $\widehat{\psi}_n(\nabla_x)$ satisfies second the condition in Definition 5.2. Here we deal only with the case that $n=3$ and the simple polynomial ρ at issue is $d\in D_3\longmapsto d^2\in D$, leaving the general case safely to the reader. Since

$$(d_1+d_2+d_3)^2=2(d_1d_2+d_1d_3+d_2d_3)$$

for any $(d_1,d_2,d_3)\in D^3$, we have the commutative diagram

$$\begin{array}{ccc} D^3 & \overset{\chi}{\to} & D(6)\\ +_{D^3\to D_3}\downarrow & & \downarrow +_{D(6)\to D}\\ D_3 & \underset{\rho}{\to} & D \end{array} \tag{25}$$

where χ stands for the mapping

$$(d_1,d_2,d_3)\in D^3\mapsto(d_1d_2,d_1d_3,d_2d_3,d_1d_2,d_1d_3,d_2d_3)\in D(6)$$

Then the composition of mappings

$$(M\otimes\mathcal{W}_D)_{\pi(x)}\xrightarrow{\mathrm{id}_M\otimes\mathcal{W}_\rho}\left(M\otimes\mathcal{W}_{D_3}\right)_{\pi(x)}\xrightarrow{\widehat{\psi}_3(\nabla_x)}\left(E\otimes\mathcal{W}_{D_3}\right)_x$$
$$\xrightarrow{\mathrm{id}_E\otimes\mathcal{W}_{+_{D^3\to D_3}}}\left(E\otimes\mathcal{W}_{D^3}\right)_x$$

is equivalent, by the very definition of $\widehat{\psi}_3$, to the composition of mappings

$$(M\otimes\mathcal{W}_D)_{\pi(x)}\xrightarrow{\mathrm{id}_M\otimes\mathcal{W}_\rho}\left(M\otimes\mathcal{W}_{D_3}\right)_{\pi(x)}\xrightarrow{\mathrm{id}_M\otimes\mathcal{W}_{+_{D^3\to D_3}}}\left(M\otimes\mathcal{W}_{D^3}\right)_{\pi(x)}$$
$$\xrightarrow{\nabla_x}\left(E\otimes\mathcal{W}_{D^3}\right)_x$$

which is in turn equivalent to the composition of mappings

$$(M \otimes \mathcal{W}_D)_{\pi(x)} \xrightarrow{\mathrm{id}_M \otimes \mathcal{W}_{+_{D(6)\to D}}} (M \otimes \mathcal{W}_{D(6)})_{\pi(x)} \xrightarrow{\mathrm{id}_M \otimes \mathcal{W}_\chi} (M \otimes \mathcal{W}_{D^3})_{\pi(x)} \xrightarrow{\nabla_x} (E \otimes \mathcal{W}_{D^3})_x$$

with due regard to the commutative diagram in (25). By Theorem 4.5, this is equivalent to the composition of mappings

$$(M \otimes \mathcal{W}_D)_{\pi(x)} \xrightarrow{\mathrm{id}_M \otimes \mathcal{W}_{+_{D(6)\to D}}} (M \otimes \mathcal{W}_{D(6)})_{\pi(x)} \xrightarrow{\nabla_x^{D(6)}} (E \otimes \mathcal{W}_{D(6)})_x \xrightarrow{\mathrm{id}_E \otimes \mathcal{W}_\chi} (E \otimes \mathcal{W}_{D^3})_x$$

which is in turn equivalent by Proposition 7.4 to the composition of mappings

$$(M \otimes \mathcal{W}_D)_{\pi(x)} \xrightarrow{\widehat{\psi}_1(\pi_{3,1}(\nabla_x))} (E \otimes \mathcal{W}_D)_x \xrightarrow{\mathrm{id}_E \otimes \mathcal{W}_{+_{D(6)\to D}}} (E \otimes \mathcal{W}_{D(6)})_x \xrightarrow{\mathrm{id}_E \otimes \mathcal{W}_\chi} (E \otimes \mathcal{W}_{D^3})_x$$

Since

$$\widehat{\psi}_1(\widehat{\pi}_{3,1}(\nabla_x)) = \widehat{\pi}_{3,1}(\widehat{\psi}_3(\nabla_x))$$

by Proposition 7.3 and the commutativity of the diagram (25), this is equivalent to the composition of mappings

$$(M \otimes \mathcal{W}_D)_{\pi(x)} \xrightarrow{\pi_{3,1}(\widehat{\psi}_3(\nabla_x))} (E \otimes \mathcal{W}_D)_x \xrightarrow{\mathrm{id}_E \otimes \mathcal{W}_\rho} (E \otimes \mathcal{W}_{D_3})_x \xrightarrow{\mathrm{id}_E \otimes \mathcal{W}_{+_{D^3\to D_3}}} (E \otimes \mathcal{W}_{D^3})_x,$$

which completes the proof.

Notation 7.3 *Thus the mapping* $\widehat{\psi}_n\colon \widehat{\mathbb{J}}^{D^n}(\pi) \to \widehat{\mathbb{J}}^{D_n}(\pi)$ *is naturally restricted to a mapping* $\psi_n\colon \mathbb{J}^{D^n}(\pi) \to \mathbb{J}^{D_n}(\pi)$.

References

1. Dubuc, E.: Sur les modèles de la géométrie différentielle synthétique. Cahiers de Topologie et Géométrie Différentielle **20**, 231–279 (1979)
2. Frölicher, A., Kriegl, A.: Linear Spaces and Differentiation Theory. Wiley, Chichester (1988)
3. Goerss, P.G., Jardine, J.F.: Simplicial Homotopy Theory, Progress in Mathematics, vol. 174. Birkhäuser, Basel (1999)
4. Kock, A.: Synthetic Differential Geometry, London Mathematical Society Lecture Note Series, vol. 333, 2nd edn. Cambridge University Press, Cambridge (2006)

5. Kolář, I., Michor, P.W., Slovák, J.: Natural Operations in Differential Geometry. Springer, Berlin (1993)
6. Kriegl, A., Michor, P.W.: The Convenient Setting of Global Analysis. American Mathematical Society, Rhode Island (1997)
7. Lavendhomme, R.: Basic Concepts of Synthetic Differential Geometry. Kluwer, Dordrecht (1996)
8. Nishimura, H.: Synthetic differential geometry of jet bundles. Bull. Belg. Math. Soc. Simon Stevin **8**, 639–650 (2001)
9. Nishimura, H.: Corrigenda to "Synthetic differential geometry of jet bundles". Bull. Belg. Math. Soc. Simon Stevin **9**, 473 (2002)
10. Nishimura, H.: Holonomicity in synthetic differential geometry of jet bundles. Beiträge zur Algebra und Geometrie **44**, 471–481 (2003)
11. Nishimura, H.: Higher-order preconnections in synthetic differential geometry of jet bundles. Beiträge zur Algebra und Geometrie **45**, 677–696 (2004)
12. Nishimura, H.: Synthetic differential geometry of higher-order total differentials. Cahiers de Topologie et Géométrie Différentielle Catégoriques **47**, 129–154, 207–232 (2006)
13. Nishimura, H.: The affince bundle theorem in synthetic differential geometry of jet bundles. Beiträge zur Algebra und Geometrie **48**, 351–365 (2007)
14. Nishimura, H.: A much larger class of Frölicher spaces than that of convenient vector spaces may embed into the Cahiers topos. Far East J. Math. Sci. **35**, 211–223 (2009)
15. Nishimura, H.: Microlinearity in Frölicher spaces -beyond the regnant philosophy of manifolds-. Int. J. Pure Appl. Math. **60**, 15–24 (2010)
16. Nishimura, H.: Differential geometry of microlinear Frölicher spaces I. Int. J. Pure Appl. Math. **64**, 43–83 (2010)
17. Nishimura, H.: Differential geometry of microlinear Frölicher spaces IV-2, in preparation. http://zbmath.org/?q=an:06160038
18. Weil, A.: Théorie des points proches sur les variétés différentiables, pp. 111–117. Colloques Internationaux du Centre National de la Reserche Scientifique, Strassbourg (1953)

A Note on Jet and Geometric Approach to Higher Order Connections

Maïdo Rahula and Petr Vašík

Abstract We compare two ways of interpreting higher order connections. The geometric approach lies in the decomposition of higher order tangent space into the horizontal and vertical structures while the jet-like approach considers a higher order connection as the section of a jet prolongation of a fibered manifold. Particularly, we use the Ehresmann prolongation of a general connection and study the result from the point of view of geometric theory. We pay attention to linear connections, too.

1 Introduction

Several models of real objects are given as a smooth manifold and one or more linear connections, e.g. material elasticity, see [1]. To obtain a manifold with just one characterization, one has to consider a concept of a higher order connection. In this paper, we recall the basic concepts of higher order connections from both geometric and jet–like point of view, Sects. 2 and 4. Let us note that the original ideas are those of Ehresmann, i.e. the definition of a connection by means of a horizontal distribution in a tangent space, the double fibered manifolds and holonomic and nonholonomic jets of fibered mappings. The first idea can be found in [2], the second one in [3]. The second idea was used for the case of vector bundles by Pradines, [4]. Finally, the concept of holonomic and nonholonomic jets is widely studied in [5–9]. The first idea was extended in [10], where the main formulae of higher order objects in multiple tangent spaces are derived, see also [11]. In this paper we compare the jet–like and

M. Rahula (✉)
Institute of Mathematics, University of Tartu, J.Liivi 2, 50409 Tartu, Estonia
e-mail: rahula@ut.ee

P. Vašík
Faculty of Mechanical Engineering, Institute of Mathematics, Brno University of Technology, Technická 2, 61669 Brno, Czech Republic
e-mail: vasik@fme.vutbr.cz

A. Makhlouf et al. (eds.), *Algebra, Geometry and Mathematical Physics*, Springer Proceedings in Mathematics & Statistics 85, DOI: 10.1007/978-3-642-55361-5_24,

geometric approach. We also recall a product of general connections which leads to the so called Ehresmann prolongation and show the reason why this operation is outstanding, especially concerning semiholonomic connections, Sect. 6.1. We study Ehresmann prolongation of a connection from both points of view and show the analogues in both approaches.

2 Jet Prolongation of a Fibered Manifold

Classical theory reads that r-th holonomic prolongation $J^r Y$ of $Y \to M$ is the space of r–jets of local sections $M \to Y$. The nonholonomic prolongation $\widetilde{J}^r Y$ of $Y \to M$ is defined by the following iteration:

1. $\widetilde{J}^1 Y = J^1 Y$, i.e. $\widetilde{J}^1 Y$ is a space of 1-jets of sections $M \to Y$ over the target space Y.
2. $\widetilde{J}^r Y = J^1(\widetilde{J}^{r-1} Y \to M)$.

Clearly, we have an inclusion $J^r Y \subset \widetilde{J}^r Y$ given by $j^r_x \gamma \mapsto j^1_x(j^{r-1}\gamma)$. Further, r-th semiholonomic prolongation $\overline{J}^r Y \subset \widetilde{J}^r Y$ is defined by the following induction. First, by $\beta_1 = \beta_Y$ we denote the projection $J^1 Y \to Y$ and by $\beta_r = \beta_{\widetilde{J}^{r-1} Y}$ the projection $\widetilde{J}^r Y = J^1 \widetilde{J}^{r-1} Y \to \widetilde{J}^{r-1} Y$, $r = 2, 3, \ldots$. If we set $\overline{J}^1 Y = J^1 Y$ and assume we have $\overline{J}^{r-1} Y \subset \widetilde{J}^{r-1} Y$ such that the restriction of the projection $\beta_{r-1} : \widetilde{J}^{r-1} Y \to \widetilde{J}^{r-2} Y$ maps $\overline{J}^{r-1} Y$ into $\overline{J}^{r-2} Y$, we can construct $J^1 \beta_{r-1} : J^1 \overline{J}^{r-1} Y \to J^1 \overline{J}^{r-2} Y$ and define

$$\overline{J}^r Y = \{A \in J^1 \overline{J}^{r-1} Y;\ \beta_r(A) = J^1 \beta_{r-1}(A) \in \overline{J}^{r-1} Y\}.$$

If we denote by $\mathcal{FM}_{m,n}$ the category with objects composed of fibered manifolds with m-dimensional bases and n-dimensional fibres and morphisms formed by locally invertible fiber-preserving mappings, then, obviously, J^r, $\overline{J}^r$ and $\widetilde{J}^r$ are bundle functors on $\mathcal{FM}_{m,n}$.

Alternatively, one can define the r-th order semiholonomic prolongation $\overline{J}^r Y$ by means of natural target projections of nonholonomic jets, see [9]. For $r \geq q \geq 0$ let us denote by π^r_q the target surjection $\pi^r_q : \widetilde{J}^r Y \to \widetilde{J}^q Y$ with π^r_r being the identity on $\widetilde{J}^r Y$. We note that the restriction of these projections to the subspace of semiholonomic jet prolongations will be denoted by the same symbol. By applying the functor J^k we have also the surjections $J^k \pi^{r-k}_{q-k} : \widetilde{J}^r Y \to \widetilde{J}^q Y$ and, consequently, the element $X \in \widetilde{J}^r Y$ is semiholonomic if and only if

$$(J^k \pi^{r-k}_{q-k})(X) = \pi^r_q(X) \text{ for any integers } 1 \leq k \leq q \leq r. \tag{1}$$

In [9], the proof of this property can be found and the author finds it useful when handling semiholonomic connections and their prolongations.

Now let us recall local coordinates on higher order jet prolongations of a fibered manifold $Y \to M$. Let us denote by x^i, $i = 1, \ldots, m$ the local coordinates on M and y^p, $p = 1, \ldots, n$ the fiber coordinates of $Y \to M$. We recall that the induced coordinates on the holonomic prolongation $J^r Y$ are given by (x^i, y^p_α), where α is a multiindex of range m satisfying $|\alpha| \leq r$. Clearly, the coordinates y^p_α on $J^r Y$ are characterized by the complete symmetry in the indices of α. Having the nonholonomic prolongation $\widetilde{J}^r Y$ constructed by the iteration, we define the local coordinates inductively as follows:

(1) Suppose that the induced coordinates on $\widetilde{J}^{r-1} Y$ are of the form

$$(x^i, y^p_{k_1 \ldots k_{r-1}}),\ k_1, \ldots, k_{r-1} = 0, 1, \ldots, m.$$

(2) We define the induced coordinates on $\widetilde{J}^r Y$ by

$$(x^i, y^p_{k_1 \ldots k_{r-1} 0} = y^p_{k_1 \ldots k_{r-1}}, y^p_{k_1 \ldots k_{r-1} i} = \frac{\partial}{\partial x^i} y^p_{k_1 \ldots k_{r-1}}),$$

i.e. induced coordinates are partial derivatives are obtained as partial derivatives of fiber coordinates with respect to the base coordinates.

It remains to describe coordinates on the semiholonomic prolongation $\overline{J}^r Y$. Let $(k_1, \ldots, k_r)$, $k_1, \ldots, k_r = 0, 1, \ldots, m$ be a sequence of indices and denote by $\langle k_1, \ldots, k_s \rangle$, $s \leq r$ the sequence of non-zero indices in $(k_1, \ldots, k_r)$ respecting the order. Then the definition of $\overline{J}^r Y$ reads that the point $(x^i, y^p_{k_1 \ldots k_r}) \in \widetilde{J}^r Y$ belongs to $\overline{J}^r Y$ if and only if $y^p_{k_1 \ldots k_r} = y^p_{l_1 \ldots l_r}$ whenever $\langle k_1, \ldots, k_r \rangle = \langle l_1, \ldots, l_r \rangle$

3 Iterated Tangents

Another concept, in this paper called geometric, of a connection rises from the theory of iterated tangent spaces. Let us recall that the bundle $T^k M \to T^{k-1} M$ is equipped with the structure of a k-fold vector bundle. Particularly, $T^k M$ admits k different projections to $T^{k-1} M$,

$$\rho_s := T^{k-s} \pi_s : T^k M \to T^{k-1} M,$$

where π_s is the natural projection $T^s M \to T^{s-1} M$, $s = 1, 2, \ldots, k$. Each projection defines a vector bundle with basis $T^{k-1} M$ and the total space is composed of $2^{k-1} n$-dimensional vector spaces as fibers. The local coordinates on the neighborhoods

$$T^s U \subset T^s M, \quad \text{where} \quad T^{s-1} U = \pi_s(T^s U),\ s = 1, 2, \ldots, k,$$

are derived from coordinates, or coordinate mappings, (u^i), which are given on the neighborhood $U \subset M$:

U: (u^i), $i = 1, 2, \ldots, n$,
TU: (u^i, u^i_1), where $u^i := u^i \circ \pi_1$, $u^i_1 := du^i$,
T^2U: $(u^i, u^i_1, u^i_2, u^i_{12})$,
where $u^i := u^i \circ \pi_1\pi_2$, $u^i_1 := du^i \circ \pi_2$, $u^i_2 := d(u^i \circ \pi_1)$, $u^i_{12} := d^2u^i$,
etc.

Proposition 3.1 *Coordinate mappings given on the neighborhood $T^{s-1}U$ induce coordinate mappings on the neighborhood T^sU with respect to the projection π_s by adding the differentials of these mappings.*

Local coordinates are obtained by the following principle:
to the coordinates of a point of a manifold we attach the coordinates of the vector tangent to the manifold at that point. We use the following notation: the coordinates of a neighborhood T^kU consist of two copies of local coordinates on $T^{k-1}U$ where the second copy is equipped with an additional subscript k. This principle is suitable in the sense that the coordinates with index s are recognized as the fiber coordinates for projections ρ_s, $s = 1, 2, \ldots, k$, i.e. the coordinates with index s disappear after the application of projection ρ_s.

The coordinate form of the three projections $\rho_s : T^3U \to T^2U$, $s = 1, 2, 3$, is given by the following diagram:

$$
\begin{array}{ccc}
 & (u^i, u^i_1, u^i_2, u^i_{12}, u^i_3, u^i_{13}, u^i_{23}, u^i_{123}) & \\
\rho_1 \swarrow & \rho_2 \downarrow & \searrow \rho_3 \\
(u^i, u^i_2, u^i_3, u^i_{23}) & (u^i, u^i_1, u^i_3, u^i_{13}) & (u^i, u^i_1, u^i_2, u^i_{12}).
\end{array}
$$

Remark 3.1 Let us note that the semiholonomity condition is connected to the notion of the osculating bundle, see [11], and can be defined as the equalizer of all possible projections, which corresponds to (1).

4 Connections

We start with the jet–like approach to connections. This rather structural description is quite suitable for determining natural operators on connections, for details see [5].

Definition 4.1 A general connection on the fibered manifold $Y \to M$ is a section $\Gamma : Y \to J^1Y$ of the first jet prolongation $J^1Y \to Y$.

Further generalization of this idea leads us to the definition of r-th order connection, which is a section of r-th order jet prolongation of a fibered manifold. According to the character of the target space we distinguish holonomic, semiholonomic and nonholonomic general connections. The coordinate form of a second order nonholonomic connection $\Delta : Y \to \widetilde{J}^2 Y$ is given by

$$y_i^p = F_i^p(x, y), \quad y_{0i}^p = G_i^p(x, y), \quad y_{ij}^p = H_{ij}^p(x, y),$$

where F, G, H are arbitrary smooth functions. In case of linear connections all functions are linear in fiber coordinates.

Let us now recall the geometric concept of a connection and its extension to higher order connections. The following section is based on the paper [11].

Definition 4.2 A connection on bundle $\pi : M_1 \to M$ is defined by the structure $\Delta_h \oplus \Delta_v$ on a manifold M_1 where $\Delta_v = \ker T\pi$ is *vertical distribution* tangent to the fibers and Δ_h is *horizontal distribution* complementary to the distribution Δ_v. The transport of the fibers along the path $\gamma \subset M$ is realized by the horizontal lifts given by the distribution Δ_h on the surface $\pi^{-1}(\gamma)$. If the bundle is a vector one and the transport of fibers along an arbitrary path is linear, then the connection is called linear.

We will assume that the base manifold M is of dimension n and the fibers are of dimension r. Then

$$\dim \Delta_h = n\,, \quad \dim \Delta_v = r\,.$$

On the neighborhood $U \subset M_1$, let us consider local base and fiber coordinates:

$$(u^i, u^\alpha)\,,\ i = 1, 2, \ldots, n\,;\ \alpha = n+1, \ldots, n+r.$$

Base coordinates (u^i) are determined by the projection π and the coordinates $(\bar{u}^i)$ on a neighborhood $\bar{U} = \pi(U)$, $u^i = \bar{u}^i \circ \pi$.

Definition 4.3 On a neighborhood $U \subset M_1$ we define a local (adapted) basis of the structure $\Delta_h \oplus \Delta_v$,

$$(X_i\ X_\alpha) = \begin{pmatrix} \dfrac{\partial}{\partial u^j} & \dfrac{\partial}{\partial u^\beta} \end{pmatrix} \cdot \begin{pmatrix} \delta_i^j & 0 \\ \Gamma_i^\beta & \delta_\alpha^\beta \end{pmatrix}, \quad \begin{pmatrix} \omega^i \\ \omega^\alpha \end{pmatrix} = \begin{pmatrix} \delta_j^i & 0 \\ -\Gamma_j^\alpha & \delta_\beta^\alpha \end{pmatrix} \cdot \begin{pmatrix} du^j \\ du^\beta \end{pmatrix}.$$

The horizontal distribution Δ_h is the linear span of the vector fields (X_i) and the annihilator of the forms (ω^α),

$$X_i = \partial_i + \Gamma_i^\beta \partial_\beta, \quad \omega^\alpha = du^\alpha - \Gamma_i^\alpha du^i.$$

Definition 4.4 A classical affine connection on manifold M is seen as a linear connection on the bundle $\pi_1 : TM \to M$. On the tangent bundle $TM \to M$ one can define the structure $\Delta_h \oplus \Delta_v$. The indices in the formulas are denoted by Latin letters all of them ranging from 1 to n. The functions Γ_i^α, X_i, ω^α are of the form (in Γ_i^α the sign is changed to comply with the classical theory):

$$
\begin{aligned}
\Gamma_i^{\alpha} \quad &\rightsquigarrow \quad -\Gamma^i_{jk} u^k_1 \,, \\
X_i = \partial_i + \Gamma_i^{\alpha}\partial_\alpha \quad &\rightsquigarrow \quad X_i = \partial_i - \Gamma^k_{ij} u^i_1 \partial^1_k \,, \\
\omega^{\alpha} = du^{\alpha} - \Gamma_i^{\alpha} du^i \quad &\rightsquigarrow \quad U^i_{12} = u^i_{12} + \Gamma^i_{jk} u^k_1 u^j_2 \,.
\end{aligned}
$$

Definition 4.5 Higher order connections are defined as follows: on tangent bundle *TM* the structure $\Delta \oplus \Delta_1$ is defined where $\ker T\rho_1 = \Delta_1$, on $T(TM)$ the structure $\Delta \oplus \Delta_1 \oplus \Delta_2 \oplus \Delta_{12}$ is defined where $\ker T\rho_s = \Delta_s \oplus \Delta_{12}$, $s = 1, 2$, etc.

5 Connections on Two-Fold Fibered Manifolds

More generally, one can define a second order connection by means of a two-fold fibered manifold. Note that the Definition 4.5 is a special case of the following. A two-fold fibered manifold is a commutative diagram

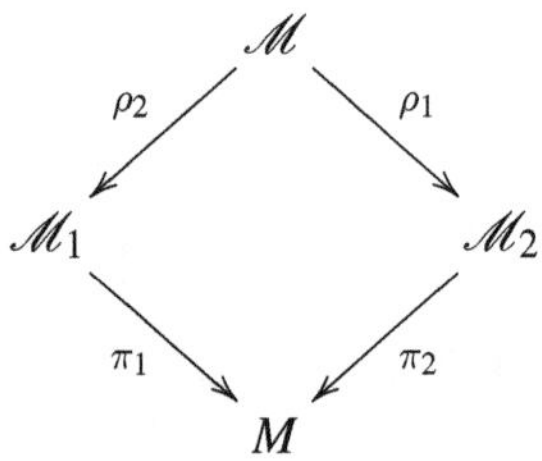

where ρ_1, ρ_2 and π_1, π_2—four fibered manifolds
$\dim M = n$, $\dim \mathcal{M}_1 = n + r_1$, $\dim \mathcal{M}_2 = n + r_2$, $\dim \mathcal{M} = n + r_1 + r_2 + r_{12}$.
The double projection

$$\pi = \pi_1 \circ \rho_2 = \pi_2 \circ \rho_1 : \mathcal{M} \to M$$

divides a manifold $\mathcal{M}$ to n-parameter family of fibers of dimensions $(r_1 + r_2 + r_{12})$. Each fiber carries structure of another two fibers of dimensions $r_1 + r_{12}$ and $r_2 + r_{12}$ and these two fibers have the common intersection of dimension r_{12}.

A two-fold fibered manifold is called a vector bundle if both fibrations π_1, π_2, ρ_1 and ρ_2—form vector bundles.

An example of a two-fold fibered manifold is the second order tangent bundle T^2M of a manifold M. In this case $n = r_1 = r_2 = r_{12}$.

Definition 5.1 A connection on a two-fold fibered manifold is defined by a structure on a manifold $\mathcal{M}$:

$$\Delta \otimes \Delta_1 \otimes \Delta_2 \otimes \Delta_{12} \,, \tag{2}$$

$$\dim \Delta = n, \quad \dim \Delta_1 = r_1 \,, \quad \dim \Delta_2 = r_2 \,, \quad \dim \Delta_{12} = r_{12} \,,$$

$$\mathrm{Ker}T\rho_2 = \Delta_2 \oplus \Delta_{12}, \qquad \mathrm{Ker}T\rho_1 = \Delta_1 \oplus \Delta_{12}$$

$$T\rho_2(\Delta \oplus \Delta_1) = T\mathcal{M}_1, \quad T\rho_1(\Delta \oplus \Delta_2) = T\mathcal{M}_2,$$

$$T\pi\Delta = TM.$$

Remark 5.1 A connection on a two-fold vector fibered manifold is called linear if the structure (2) induces on the manifolds π_1, π_2, ρ_1 and ρ_2 linear connections.

Remark 5.2 Similarly, one can define a connection on a k–fold fibered manifold. In such case the commutative diagram would be represented by a k–dimensional cube. These manifolds would correspond to the k–th tangent bundle $T^k M$ of a manifold M.

On the neighborhoods

$$\mathcal{U} \subset \mathcal{M}, \ \mathcal{U}_1 = \rho_2(\mathcal{U}) \subset \mathcal{M}_1, \ \mathcal{U}_2 = \rho_1(\mathcal{U}) \subset \mathcal{M}_2, \ U = \pi(\mathcal{U}) \subset M$$

we have the coordinate systems
$(u^i, u^{\alpha_1}, u^{\alpha_2}, u^{\alpha_{12}})$, (u^i, u^{α_1}), (u^i, u^{α_1}), (u^i).

The transformation of coordinates on the neighborhoods $\mathcal{U}$,

$$(u^i, u^{\alpha_1}, u^{\alpha_2}, u^{\alpha_{12}}) \rightsquigarrow (\tilde{u}^i, \tilde{u}^{\alpha_1}, \tilde{u}^{\alpha_2}, \tilde{u}^{\alpha_{12}}) = (a^i, a^{\alpha_1}, a^{\alpha_2}, a^{\alpha_{12}}),$$

gives a Jacobi matrix:

$$\begin{pmatrix} a^i_j & 0 & 0 & 0 \\ a^{\alpha_1}_j & a^{\alpha_1}_{\beta_1} & 0 & 0 \\ a^{\alpha_2}_j & 0 & a^{\alpha_2}_{\beta_2} & 0 \\ a^{\alpha_{12}}_j & a^{\alpha_{12}}_{\beta_1} & a^{\alpha_{12}}_{\beta_2} & a^{\alpha_{12}}_{\beta_{12}} \end{pmatrix}.$$

See [10, 12]. Let us mention that the local (adapted) basis of such decomposition is represented by a matrix of the form

$$\begin{pmatrix} \delta^i_j & 0 & 0 & 0 \\ \Gamma^{\alpha_1}_j & \delta^{\alpha_1}_{\beta_1} & 0 & 0 \\ \Gamma^{\alpha_2}_j & 0 & \delta^{\alpha_2}_{\beta_2} & 0 \\ \Gamma^{\alpha_{12}}_j & \Gamma^{\alpha_{12}}_{\beta_1} & \Gamma^{\alpha_{12}}_{\beta_2} & \delta^{\alpha_{12}}_{\beta_{12}} \end{pmatrix}. \tag{3}$$

The dual frame is given by the system of 1–forms:

$$\begin{aligned} \omega^i &= du^i, \\ \omega^{\alpha_1} &= du^{\alpha_1} - \Gamma^{\alpha_1}_i du^i, \\ \omega^{\alpha_2} &= du^{\alpha_2} - \Gamma^{\alpha_2}_i du^i, \\ \omega^{\alpha_{12}} &= du^{\alpha_{12}} - \Gamma^{\alpha_{12}}_{\alpha_1} du^{\alpha_1} - \Gamma^{\alpha_{12}}_{\alpha_2} du^{\alpha_2} - \bar{\Gamma}^{\alpha_{12}}_i du^i, \end{aligned}$$

$$\text{where} \quad \Gamma_i^{\alpha_{12}} - \bar{\Gamma}_i^{\alpha_{12}} = \Gamma_{\beta_1}^{\alpha_{12}} \Gamma_i^{\beta_1} + \Gamma_{\beta_2}^{\alpha_{12}} \Gamma_i^{\beta_2}.$$

In case of linear connection the elements of the matrix (3) are of the form

$$\Gamma_j^{\alpha_1} = \Gamma_{j\beta_1}^{\alpha_1} u^{\beta_1}, \quad \Gamma_j^{\alpha_2} = \Gamma_{j\beta_2}^{\alpha_2} u^{\beta_2},$$
$$\Gamma_{\beta_1}^{\alpha_{12}} = \Gamma_{\beta_1\beta_2}^{\alpha_{12}} u^{\beta_2}, \quad \Gamma_{\beta_2}^{\alpha_{12}} = \Gamma_{\beta_2\beta_1}^{\alpha_{12}} u^{\beta_1},$$
$$\Gamma_j^{\alpha_{12}} = \Gamma_{j\beta_1\beta_2}^{\alpha_{12}} u^{\beta_1} u^{\beta_2} + \Gamma_{j\beta_{12}}^{\alpha_{12}} u^{\beta_{12}}, \quad \bar{\Gamma}_j^{\alpha_{12}} = \bar{\Gamma}_{j\beta_1\beta_2}^{\alpha_{12}} u^{\beta_1} u^{\beta_2} + \bar{\Gamma}_{j\beta_{12}}^{\alpha_{12}} u^{\beta_{12}},$$

$$\Gamma_{j\beta_1\beta_2}^{\alpha_{12}} - \bar{\Gamma}_{j\beta_1\beta_2}^{\alpha_{12}} = \Gamma_{\gamma_2\beta_1}^{\alpha_{12}} \Gamma_{j\beta_2}^{\gamma_2},$$

where the coefficients depend on the base coordinates u^i only.

6 Ehresmann Prolongation

First, let us now recall a concept of a product of two connections.

Given two higher order connections $\Gamma : Y \to \widetilde{J}^r Y$ and $\overline{\Gamma} : Y \to \widetilde{J}^s Y$, the product of Γ and $\overline{\Gamma}$ is the $(r+s)$-th order connection $\Gamma * \overline{\Gamma} : Y \to \widetilde{J}^{r+s} Y$ defined by

$$\Gamma * \overline{\Gamma} = \widetilde{J}^s \Gamma \circ \overline{\Gamma}.$$

Particularly, if both Γ and $\overline{\Gamma}$ are of the first order, then $\Gamma * \overline{\Gamma} : Y \to \widetilde{J}^2 Y$ is semiholonomic if and only if $\Gamma = \overline{\Gamma}$ and $\Gamma * \overline{\Gamma}$ is holonomic if and only if Γ is curvature-free, [9, 13].

As an example we show the coordinate expression of an arbitrary nonholonomic second order connection and of the product of two first order connections. The coordinate form of $\Delta : Y \to \widetilde{J}^2 Y$ is

$$y_i^p = F_i^p(x, y), \quad y_{0i}^p = G_i^p(x, y), \quad y_{ij}^p = H_{ij}^p(x, y),$$

where F, G, H are arbitrary smooth functions. Further, if the coordinate expressions of two first order connections $\Gamma, \overline{\Gamma} : Y \to J^1 Y$ are

$$\Gamma : \quad y_i^p = F_i^p(x, y), \qquad \overline{\Gamma} : \quad y_i^p = G_i^p(x, y), \tag{4}$$

then the second order connection $\Gamma * \overline{\Gamma} : Y \to \widetilde{J}^2 Y$ has equations

$$y_i^p = F_i^p, \quad y_{0i}^p = G_i^p, \quad y_{ij}^p = \frac{\partial F_i^p}{\partial x^j} + \frac{\partial F_i^p}{\partial y^q} G_j^q.$$

For linear connections, the coordinate form would be obtained by substitution

$$F_i^p = F_{iq}^p y^q,$$
$$G_i^p = G_{iq}^p y^q$$

in the Eq. (4), where F_{iq}^p and G_{iq}^p are functions of the base manifold coordinates x_i. For order three see [8].

In the above process, if $\Gamma = \overline{\Gamma}$, the connection $\Gamma * \Gamma$ is called the Ehresmann prolongation of Γ, iteratively we obtain the r–th Ehresmann prolongation of Γ. We show that Ehresmann prolongation plays an important role in determining all natural operators transforming first order connections into higher order connections. Let us note that also natural transformations of semiholonomic jet prolongation functor $\overline{J}^r$ are involved. To find the details about this topic we refer to [5–7]. For our purposes, it is enough to consider $r = 2$. We use the notation of [5], where the map $e : \overline{J}^2 Y \to \overline{J}^2 Y$ is obtained from the natural exchange map $e_\Lambda : J^1 J^1 Y \to J^1 J^1 Y$ as a restriction to the subbundle $\overline{J}^2 Y \subset J^1 J^1 Y$. Note that while e_Λ depends on the linear connection Λ on M, its restriction e is independent of any auxiliary connections. We remark, that originally the map e_Λ was introduced by M. Modugno. We also recall that J. Pradines introduced a natural map $\overline{J}^2 Y \to \overline{J}^2 Y$ with the same coordinate expression.

Now we are ready to recall the following assertion, see [7] for the proof.

Proposition 6.1 *All natural operators transforming first order connection $\Gamma : Y \to J^1 Y$ into second order semiholonomic connection $Y \to \overline{J}^2 Y$ form a one parameter family*

$$\Gamma \mapsto k \cdot (\Gamma * \Gamma) + (1 - k) \cdot e(\Gamma * \Gamma), \qquad k \in \mathbb{R}.$$

This shows the importance of Ehresmann prolongation in the theory of prolongations of connections.

7 Tangent Functor and Ehresmann Prolongation

If we apply the tangent functor T two times on a projection $\pi : E \to M$ and a section $\sigma : M \to E$ we obtain

$$T\pi : TE \to TM\,,\ \ T^2\pi : T^2E \to T^2M,$$

$$T\sigma : TM \to TE,\ T^2\sigma : T^2M \to T^2E,$$

respectively. The mappings σ, $T\sigma$ and $T^2\sigma$ define the sections of fibered manifolds π, $T\pi$ and $T^2\pi$.

Let us consider local coordinates on the following manifolds in the form

$$\text{on } \ M,\ TM,\ T^2M\ :\ (x^i),\ (x^i, x_1^i),\ (x^i, x_1^i, x_2^i, x_{12}^i),$$

$$\text{and on } \ E,\ TE,\ T^2E\ :\ (y^p),\ (y^p, y_1^p),\ (y^p, y_1^p, y_2^p, y_{12}^p).$$

Let us also consider for a function f defined on a manifold M, its following differentials on T^2M in local coordinate form:

$$f_1 \doteq f_i x_1^i, \quad f_2 \doteq f_i x_2^i, \quad f_{12} \doteq f_{ij} x_1^i x_2^j + f_i x_{12}^i, \quad \text{where } f_i = \frac{\partial f}{\partial x^i}, \; f_{ij} = \frac{\partial^2 f}{\partial x^i \partial x^j}.$$

Furthermore, $f_1 = df \circ \rho_1$, $f_2 = df \circ \rho_2$, $f_{12} = d^2 f$. We use these notations in the formulae bellow.

If the section σ is defined by local functions Γ^p, then the sections $T\sigma$ and $T^2\sigma$ are defined by its differentials Γ_1^p, Γ_2^p and Γ_{12}^p,

$$\begin{aligned}
&\sigma : \; x^i \rightsquigarrow y^p = \Gamma^p, \\
&T\sigma : \; (x^i, x_1^i) \rightsquigarrow (y^p, y_1^p) = (\Gamma^p, \Gamma_1^p), \\
&T^2\sigma : \; (x^i, x_1^i, x_2^i, x_{12}^i) \rightsquigarrow (y^p, y_1^p, y_2^p, y_{12}^p) = (\Gamma^p, \Gamma_1^p, \Gamma_2^p, \Gamma_{12}^p), \\
&\text{where } \; \Gamma_1^p = \Gamma_i^p x_1^i, \quad \Gamma_2^p = \Gamma_i^p x_2^i, \quad \Gamma_{12}^p = \Gamma_{ij}^p x_1^i x_2^j + \Gamma_i^p x_{12}^i. \qquad (5)
\end{aligned}$$

The case when the coefficients Γ_i^p, Γ_{ij}^p in (5) are arbitrary functions, corresponds to a nonholonomic connection on the fibered manifold π.

The case when $\Gamma_{ij}^p = \dfrac{\partial \Gamma_i^p}{\partial x^j}$, where Γ_i^p are arbitrary functions corresponds to a semiholonomic connection on the fibered manifold π.

The case when $\Gamma_1^p = d\Gamma^p \circ \rho_1$, $\Gamma_2^p = d\Gamma^p \circ \rho_2$, $\Gamma_{12}^p = d^2\Gamma^p$, corresponds to a holonomic connection on the fibered manifold π.

The functions Γ_i^p, Γ_{ij}^p define nonholonomic, semiholonomic or holonomic Ehresmann prolongation of a connection, respectively.

Remark 7.1 Nonholonomic prolongation induces a connection on a double fibered manifold

$$J \to E \to M : \; y_i^p \rightsquigarrow y^p \rightsquigarrow x^i.$$

On the fibered manifold $E \to M$ the fiber transformations are given by the Pfaff system

$$\omega^p \equiv dy^p - \Gamma_i^p dx^i = 0,$$

more precisely, along a curve $x^i(t)$ – by the system of first order ODEs

$$\dot{y}^p = \Gamma_i^p \dot{x}^i. \qquad (6)$$

In case $(\Gamma_{12}^p, x_1^i, x_2^j, x_{12}^i) \rightsquigarrow (\ddot{y}^p, \dot{x}^i, \dot{x}^j, \ddot{x}^i)$ we obtain the system of second order ODEs:

$$\Gamma_{12}^p = \Gamma_{ij}^p x_1^i x_2^j + \Gamma_i^p x_{12}^i \rightsquigarrow \ddot{y}^p = \Gamma_{ij}^p \dot{x}^i \dot{x}^j + \Gamma_i^p \ddot{x}^i.$$

Considering the system (6), we obtain for fiber coordinates y^α, y_i^α system of first order ODEs

$$\begin{cases} \dot{y}^p = \Gamma_i^p \dot{x}^i, \\ \dot{y}_i^p = \Gamma_{ij}^p \dot{x}^j. \end{cases}$$

The sections of fibers along a curve $x^i(t)$ are given.
The horizontal distribution Δ_h is n-dimensional and described by the vector field

$$X_i = \partial_i + \Gamma_i^p \partial_p + \Gamma_{ij}^p \partial_p^j, \quad \text{where } \partial_i = \frac{\partial}{\partial x^i}, \ \partial_p = \frac{\partial}{\partial y^p}, \ \partial_p^j = \frac{\partial}{\partial y_j^p}.$$

Acknowledgments The first author was supported by the grant of Tartu University TMTMM 0039, the second author by the project NETME CENTRE PLUS (LO1202). The results of the project NETME CENTRE PLUS (LO1202) were co-funded by the Ministry of Education, Youth and Sports within the support programme "National Sustainability Programme I".

References

1. Epstein, M.: The Geometrical Language of Continuum Mechanics. Cambridge University Press, New York (2010)
2. Ehresmann, C.: Les connexions infinitésimales dans un espace fibré différentiable. Coll. de Topologie, 29–55 (1950)
3. Ehresmann, C.: Catégories doubles et catégories structurées. C.R. Acad. Sci. **256**, 1198–1201 (1958)
4. Pradines, J.: Suites exactes vectorielles doubles et connexions. C.R. Acad. Sci. **278**, 1587–1590 (1974)
5. Kolář, I., Michor, P.W., Slovák, J.: Natural Operations in Differential Geometry. Springer, Berlin (1993)
6. Kolář, I., Modugno, M.: Natural maps on the iterated jet prolongation of a fibered manifold. Annali di Matematica CLVIII, 151–165 (1991)
7. Vašík, P.: On the Ehresmann Prolongation. Ann. Univ. Mariae Curie Sklodowska Sectio A **LXI**, 145–153(2007)
8. Vašík, P.: Transformations of semiholonomic 2- and 3-jets and semiholonomic prolongation of connections. Proc. Est. Acad. Sci. **59**, 375–380 (2010)
9. Virsik, G.: On the holonomity of higher order connections. Cahiers Topol. Géom. Diff. **12**, 197–212 (1971)
10. Rahula, M.: New Problems in Differential Geometry. WSP, London (1993)
11. Rahula, M., Vašík, P., Voicu, N.: Tangent structures: sector-forms, jets and connections. J. Phys. Conf. Ser. **346**, 012023 (2012)
12. Atanasiu, G., Balan, V., Brînzei, N., Rahula, M.: Differential Geometric Structures: Tangent Bundles, Connections in Bundles, Exponential Law in the Jet Space (in Russian). Librokom, Moscow (2010)
13. Kolář, I.: On the torsion of spaces with connections. Czechoslovak Math. J. **21**, 124–136 (1971)

Classification of Principal Connections on $\widetilde{W}^2PE$

Jan Vondra

Abstract We assume a vector bundle $E \to M$ and the principal bundle PE of frames of E. Let K be a general linear connection on E and Λ be a linear connection on M. We classify all connections on $\widetilde{W}^2PE$ naturally given by K and Λ.

1 Introduction

We study principal connections on principal prolongation $\widetilde{W}^rP$ of a principal bundle P. Some authors, i.e. Kolář, Janyška, Doupovec, Mikulski, deal with holonomic case of this problem, see [1–4]. Classification of all principal connections on $\widetilde{W}^rP$ naturally given by principal connection Γ on P depends on auxiliary linear connection on the base. Moreover solution of this problem is depends essentially on the structure group of principal bundles. In this paper we give the full classification of principal connections on $\widetilde{W}^2PE$, i.e. for the linear gauge group $GL(n)$ and order two.

We use the terminology of natural bundles theory in sense of [5–7] and of gauge natural bundles theory in sense of [8, 9]. Let as recall that some other view is in [10].

Results of this paper are based on reduction theorems by Janyška [11, 12].

We denote by $\mathscr{P}\mathscr{B}_m(G)$ the category of principal G-bundles with m-dimensional bases and principal bundle morphisms over diffeomorphisms of bases.

All manifolds and maps are assumed to be smooth.

J. Vondra (✉)
Department of Mathematics and Statistics, Masaryk University,
Kotlářská 2, 611 37 Brno, Czech Republic
e-mail: vondra@math.muni.cz

A. Makhlouf et al. (eds.), *Algebra, Geometry and Mathematical Physics*, Springer Proceedings in Mathematics & Statistics 85, DOI: 10.1007/978-3-642-55361-5_25,

2 Principal Bundles and Principal Connections

2.1 Principal Bundles and Principal Connections

We consider a principal bandle $P = (P, M, \pi; G)$ with a structure group G. We denote by (x^λ, z^a) fibered coordinates on P, $\lambda = 1, \dots, \dim M$, $a = 1, \dots, \dim G$.

By $\mathrm{ad}(P)$ we denote the vector bundle associated to P with respect to the adjoint action of G on its Lie algebra $\mathfrak{g}$. We denote by (B_a) a base of left invariant vector fields on $\mathfrak{g}$ and we denote by (x^λ, u^a) the induced fiber linear local coordinates on $\mathrm{ad}(P)$. Let us denote by c^a_{bd} the related structure constants, i.e. $[B_b, B_d] = c^a_{bd} B_a$.

For the right action $r_g \colon P \to P$ given by an element $g \in G$ we consider the tangent mapping $Tr_g \colon\ TP \to TP$. Let Ξ be a vector field on P. We say that Ξ is right invariant if $\Xi(pg) = Tr_g \Xi(p)$ for all $p \in P$ and $g \in G$. In coordinates we have

$$\Xi = \xi^\lambda(x)\partial_\lambda + \Xi^a(x)\widetilde{B}_a, \tag{1}$$

where $(\widetilde{B}_a)$ is the base of vertical right invariant vector fields on P which are induced by (B_a). So Ξ are sections of the bundle $TP/G \to M$.

Definition 2.1 For each principal bundle $P = (P, M, \pi; G)$ and for each integer r we can define the principal bundle $\widetilde{W}^r P = \widetilde{P}^r M \times_M \widetilde{J}^r P \equiv (\widetilde{W}^r P, M, p;\ \widetilde{W}^r_m G)$. The structure group is the semidirect product $\widetilde{W}^r_m G = \widetilde{G}^r_m \rtimes \widetilde{T}^r_m G$.

The group $\widetilde{W}^r_m G$ is the group of nonholonomic r-jets at $(0, e)$ of all automorphisms $\varphi \colon \mathbb{R}^m \times G \to \mathbb{R}^m \times G$ with $\underline{\varphi}(0) = 0$, where the multiplication μ is defined by the composition of jets,

$$\mu(\widetilde{j}^r \varphi(0, e), \widetilde{j}^r \psi(0, e)) = \widetilde{j}^r(\psi \circ \varphi)(0, e). \tag{2}$$

Remark 2.1 Let us recall that elements of $\widetilde{T}^r_m G = \widetilde{J}^r_0(\mathbb{R}^m, G)$ are said to be the m-dimensional nonholonomic velocities of order r on G.

In the semiholonomic case we hawe $\overline{W}^r_m G = \overline{G}^r_m \rtimes \overline{T}^r_m G$.

In order one coincide holonomic, semiholonomic and nonholonomic jets, i.e. $\widetilde{W}^1_m G = \overline{W}^1_m G = W^1_m G$.

In order two, we have canonical group homomorphisms

$$\pi^2_{10} \colon \widetilde{W}^2_m G \to \widetilde{W}^1_m G, \qquad \pi^2_{01} \colon \widetilde{W}^2_m G \to \widetilde{W}^1_m G, \qquad p_1 \colon \widetilde{W}^2_m G \to \widetilde{G}^2_m, \tag{3}$$

which induce homomorphisms of Lie algebras

$$\pi^2_{10} \colon \widetilde{\mathfrak{w}}^2_{\mathfrak{m}}\mathfrak{g} \to \widetilde{\mathfrak{w}}^1_{\mathfrak{m}}\mathfrak{g}, \qquad \pi^2_{01} \colon \widetilde{\mathfrak{w}}^2_{\mathfrak{m}}\mathfrak{g} \to \widetilde{\mathfrak{w}}^1_{\mathfrak{m}}\mathfrak{g}, \qquad p_1 \colon \widetilde{\mathfrak{w}}^2_{\mathfrak{m}}\mathfrak{g} \to \widetilde{\mathfrak{g}}^2_{\mathfrak{m}}, \tag{4}$$

principle bundle morphisms, over the identity of M,

$$\pi^2_{10}\colon \widetilde{W}^2 P \to \widetilde{W}^1 P, \qquad \pi^2_{01}\colon \widetilde{W}^2 P \to \widetilde{W}^1 P, \qquad p_1\colon \widetilde{W}^2 P \to \widetilde{P}^2 M, \tag{5}$$

and homomorphisms of associated vector bundles

$$\pi^2_{10}\colon \mathrm{ad}(\widetilde{W}^2 P) \to \mathrm{ad}(\widetilde{W}^1 P), \qquad \pi^2_{01}\colon \mathrm{ad}(\widetilde{W}^2 P) \to \mathrm{ad}(\widetilde{W}^1 P), \tag{6}$$
$$p_1\colon \mathrm{ad}(\widetilde{W}^2 P) \to \mathrm{ad}(\widetilde{P}^2 M).$$

2.2 Principal Connection on *P* and on $\widetilde{W}^2 P$

A principal connection on P is defined as lifting linear mapping

$$\Gamma\colon TM \to TP/G.$$

In coordinates

$$\Gamma = \mathrm{d}^\lambda \otimes \left(\partial_\lambda + \Gamma^a{}_\lambda(x)\widetilde{B}_a\right), \tag{7}$$

where $\Gamma^a{}_\lambda(x)$ are functions on M. If we identify Γ with the functions $\Gamma^a{}_\lambda(x)$ then Γ can be considered as a section of the bundle $QP \to M$ of principal connections on P.

Moreover, we have [9].

Proposition 2.1 *Let $P \to M$ be a principal bundle and $QP \to M$ is the bundle of principal connections on P. Then QP is a 1-order G-gauge-natural affine bundle associated with the vector bundle* $\mathrm{ad}(P) \otimes T^*M \to M$*, which implies that the standard fiber of the functor Q is* $\mathfrak{g} \otimes \mathbb{R}^{m*}$.

Example 2.1 Let us recall that $\widetilde{P}^2 M$ is the 2nd order nonholonomic frame bundle of M. A principal connection Λ_2 on $\widetilde{P}^2 M$ has form

$$\Lambda_2 = \mathrm{d}^\lambda \otimes \left(\partial_\lambda + \Lambda^\nu_{\mu 0 \lambda}\widetilde{B}^{\mu 0}_\nu + \Lambda^\nu_{0\mu\lambda}\widetilde{B}^{0\mu}_\nu + \Lambda^\nu_{\mu_1\mu_2\lambda}\widetilde{B}^{\mu_1\mu_2}_\nu\right). \tag{8}$$

We remark that principal connections on $\widetilde{P}^1 M = P^1 M$ are in the bijection with classical connections on M (linear connections on TM).

Let Γ_2 be a principal connection on $\widetilde{W}^2 P$ given in coordinates by

$$\Gamma_2 = \mathrm{d}^\lambda \otimes \Big(\partial_\lambda + \Lambda^\nu_{\mu 0 \lambda}\widetilde{B}^{\mu 0}_\nu + \Lambda^\nu_{0\mu\lambda}\widetilde{B}^{0\mu}_\nu + \Lambda^\nu_{\mu_1\mu_2\lambda}\widetilde{B}^{\mu_1\mu_2}_\nu \tag{9}$$
$$+ \Gamma^a{}_\lambda\widetilde{B}_a + \Gamma^a{}_{\mu 0\lambda}\widetilde{B}^{\mu 0}_a + \Gamma^a{}_{0\mu\lambda}\widetilde{B}^{0\mu}_a + \Gamma^a{}_{\mu_1\mu_2\lambda}\widetilde{B}^{\mu_1\mu_2}_a\Big).$$

The projections (5) of $\widetilde{W}^2 P$ and the functor Q induce projections

$$\pi^2_{10}\colon Q\widetilde{W}^2P \to Q\widetilde{W}^1P, \qquad \pi^2_{01}\colon Q\widetilde{W}^2P \to Q\widetilde{W}^1P, \qquad (10)$$
$$p_1\colon Q\widetilde{W}^2P \to Q\widetilde{P}^2M,$$

so any principal connection Γ_2 on $\widetilde{W}^2P$ projects on the principal connection Λ_2 on $\widetilde{P}^2M$, see (8), and on the principal connections Γ_{10} and Γ_{01} on $\widetilde{W}^1P$.

By Proposition 2.1 $Q\widetilde{W}^2P \to M$ is the affine bundle modeled over the vector bundle $\mathrm{ad}(\widetilde{W}^2P) \otimes T^*M \to M$.

Now, if we consider two principal connections Γ_2 and $\bar{\Gamma}_2$ on $\widetilde{W}^2P$ such that they are over the same Λ_2 on $\widetilde{P}^2M$ and over the same Γ_{10} and Γ_{01} on $\widetilde{W}^1P$, then the difference $\Gamma_2 - \bar{\Gamma}_2$ is in the intersection of the kernels of projections

$$\pi^2_{10} \otimes id_{T^*M}\colon \mathrm{ad}(\widetilde{W}^2P) \otimes T^*M \to \mathrm{ad}(\widetilde{W}^1P) \otimes T^*M,$$

$$\pi^2_{01} \otimes id_{T^*M}\colon \mathrm{ad}(\widetilde{W}^2P) \otimes T^*M \to \mathrm{ad}(\widetilde{W}^1P) \otimes T^*M$$

and

$$p_1 \otimes id_{T^*M}\colon \mathrm{ad}(\widetilde{W}^2P) \otimes T^*M \to \mathrm{ad}(\widetilde{P}^2M) \otimes T^*M.$$

Let ξ be a vector field on M, Γ be a principal connection on P and Λ be a principal connection on P^1M. Let $h^\Gamma(\xi)$ denote the horizontal lift of ξ with respect to Γ. Let us denote by $Fl_t(h^\Gamma(\xi))$ the flow of $h^\Gamma(\xi)$. Then the expresion

$$\widetilde{W}^2(Fl_t(h^\Gamma(\xi))) = (\widetilde{P}^2(Fl_t(\xi)), \widetilde{J}^2(Fl_t(h^\Gamma(\xi)))) = Fl_t(h^{\widetilde{\mathcal{W}}^2\Gamma}(\xi))$$

gives principal connection $\widetilde{\mathcal{W}}^2\Gamma$ on $\widetilde{W}^2P$ which depends on Γ in order 2 and on Λ in order 1. So $\widetilde{\mathcal{W}}^2\Gamma$ is a natural operator

$$\widetilde{\mathcal{W}}^2\Gamma\colon J^1QP^1M \times_M J^2QP \to Q\widetilde{W}^2P$$

called the *flow prolongation* of Γ with respect to Λ.

The general problem is the classification of all principal connections on $\widetilde{W}^rP$ which are naturally given by a principal connection Γ on P and by a classical connection Λ on the base M. In general this problem is still open. There are only some particular results, but in holonomic case only. Namely, the classification of principal connections on W^1P for a torsion free Λ and any Γ is given in [4] and the full classification for the linear gauge group $GL(n)$ and order one is given in [3]. Finally, the full classification for the linear gauge group $GL(n)$ and order two in holonomic case is given in [13].

Clearly, the solution of this problem depends essentially on the structure group of principal bundles. In this paper we give the full classification of principal connections for the linear gauge group $GL(n)$ and order two.

2.3 Connections on Frame Bundle PE

Let $E \to M$ be a vector bundle with m-dimensional base and n-dimensional fibers. Let us denote by (x^λ, y^i) local linear fiber coordinate charts on E. Let $PE \to M$ be the frame bundle of E, i.e. PE is the principal bundle with the structure group $GL(n)$ and the induced fiber coordinates (x^λ, x^i_j).

Principal connections on PE are in bijection with general linear connections on E. The coordinate expression of a linear connection K on E is of the type

$$K = \mathrm{d}^\lambda \otimes (\partial_\lambda + K^i_{j\lambda}(x) y^j \partial_i)$$

and, if we consider K as a principal connection on PE,

$$K = \mathrm{d}^\lambda \otimes (\partial_\lambda + K^i_{p\lambda}(x) x^p_j \partial^j_i) = \mathrm{d}^\lambda \otimes (\partial_\lambda + K^i_{j\lambda}(x) \widetilde{B}^j_i).$$

Applying the functor $\widetilde{W}^r$ on PE we obtain the principle prolongation $\widetilde{W}^r PE$ with the structure group $\widetilde{W}^r_m GL(n) = \widetilde{G}^r_m \rtimes \widetilde{T}^r_m GL(n)$. For example, in order 2 we have the structure group $\widetilde{W}^2_m GL(n)$ with coordinates $(a^\lambda_{\mu 0}, a^\lambda_{0\mu}, a^\lambda_{\mu\nu}, a^i_j, a^i_{j\mu 0}, a^i_{j0\nu}, a^i_{j\mu\nu})$. If we denote by $(X^\lambda_{\mu 0}, X^\lambda_{0\mu}, X^\lambda_{\mu\nu}, X^i_j, X^i_{j\mu 0}, X^i_{j0\nu}, X^i_{j\mu\nu})$ the induced coordinates on the Lie algebra $\widetilde{\mathfrak{w}}^2_m\mathfrak{gl}(n)$ we can compute easily the adjoint action of the group $\widetilde{W}^2_m GL(n)$ on $\mathfrak{w}^2_m\mathfrak{gl}(n)$. In coordinates if $\bar{X} = ad(g)(X)$ we have

$$\begin{aligned}
\bar{X}^\lambda_{\mu 0} &= a^\lambda_\rho X^\rho_{\tau 0} \tilde{a}^\tau_\mu \,, \qquad \bar{X}^\lambda_{0\mu} = a^\lambda_\rho X^\rho_{0\tau} \tilde{a}^\tau_\mu \,,\\
\bar{X}^\lambda_{\mu\nu} &= a^\lambda_\rho X^\rho_{\alpha\beta} \tilde{a}^\alpha_\mu \tilde{a}^\beta_\nu + pol(a, X) \qquad \bar{X}^i_j = a^i_p X^p_q \tilde{a}^q_j \,,\\
\bar{X}^i_{j\rho 0} &= a^i_p X^p_{q\alpha 0} \tilde{a}^\alpha_\rho \tilde{a}^q_j + pol(a, X) \,, \qquad \bar{X}^i_{j0\rho} = a^i_p X^p_{q0\alpha} \tilde{a}^\alpha_\rho \tilde{a}^q_j + pol(a, X),\\
\bar{X}^i_{j\rho\sigma} &= a^i_{q\tau} X^\tau_{\alpha\beta} \tilde{a}^\alpha_\rho \tilde{a}^\beta_\sigma \tilde{a}^q_j + a^i_p X^p_{q\alpha\beta} \tilde{a}^\alpha_\rho \tilde{a}^\beta_\sigma \tilde{a}^q_j + pol(a, X),
\end{aligned}$$

where $pol(X, a)$ is a polynome on $\widetilde{W}^2_m GL(n) \times \widetilde{\mathfrak{w}}^1_m\mathfrak{gl}(n)$ such that any monome contains exactly one coordinate of $\pi^2_{10}(X)$ or $\pi^2_{01}(X)$ of orders less then the leading terms.

Proposition 2.2 *1. The restriction of the adjoint action of the group $\widetilde{W}^2_m Gl(n)$ on the intersection of the kernels of the projections $\pi^2_{10} \colon \widetilde{\mathfrak{w}}^2_m\mathfrak{gl}(n) \to \widetilde{\mathfrak{w}}^1_m\mathfrak{gl}(n)$ and $\pi^2_{01} \colon \widetilde{\mathfrak{w}}^2_m\mathfrak{gl}(n) \to \widetilde{\mathfrak{w}}^1_m\mathfrak{gl}(n)$ have the form*

$$\begin{aligned}
\bar{X}^\lambda_{\mu\nu} &= a^\lambda_\rho X^\rho_{\alpha\beta} \tilde{a}^\alpha_\mu \tilde{a}^\beta_\nu \,,\\
\bar{X}^i_{j\rho\sigma} &= a^i_{q\tau} X^\tau_{\alpha\beta} \tilde{a}^\alpha_\rho \tilde{a}^\beta_\sigma \tilde{a}^q_j + a^i_p X^p_{q\alpha\beta} \tilde{a}^\alpha_\rho \tilde{a}^\beta_\sigma \tilde{a}^q_j \,.
\end{aligned}$$

2. *The restriction of the adjoint action of the group $\widetilde{W}^2_m Gl(n)$ on the kernel of the projection $p_1 \colon \widetilde{\mathfrak{w}}^2_m\mathfrak{gl}(n) \to \widetilde{\mathfrak{g}}^2_m$ have the form*

$$\bar{X}^i_j = a^i_p X^p_q \tilde{a}^q_j \,,$$
$$\bar{X}^i_{j\rho 0} = a^i_p X^p_{q\alpha 0} \tilde{a}^\alpha_\rho \tilde{a}^q_j + pol(a,X)\,, \qquad \bar{X}^i_{j0\rho} = a^i_p X^p_{q0\alpha} \tilde{a}^\alpha_\rho \tilde{a}^q_j + pol(a,X),$$
$$\bar{X}^i_{j\rho\sigma} = a^i_p X^p_{q\alpha\beta} \tilde{a}^\alpha_\rho \tilde{a}^\beta_\sigma \tilde{a}^q_j + pol(a,X),$$

where $pol(X,a)$ is a polynom on $\widetilde{W}^2_m GL(n) \times \widetilde{\mathfrak{t}}^1_{\mathfrak{m}}\mathfrak{gl}(n)$ such that any monom contains exactly one coordinate of orders less then the leading terms.

Proof 1. The intersection of the kernels of the projections $\pi^2_{10} \colon \widetilde{\mathfrak{w}}^2_{\mathfrak{m}}\mathfrak{gl}(n) \to \widetilde{\mathfrak{w}}^1_{\mathfrak{m}}\mathfrak{gl}(n)$ and $\pi^2_{01} \colon \widetilde{\mathfrak{w}}^2_{\mathfrak{m}}\mathfrak{gl}(n) \to \widetilde{\mathfrak{w}}^1_{\mathfrak{m}}\mathfrak{gl}(n)$ is given by $X^\lambda_{\mu 0} = 0, X^\lambda_{0\mu} = 0, X^i_j = 0$, $X^i_{j\mu 0} = 0$ and $X^i_{j0\nu} = 0$, i.e. all $pol(a,X) = 0$.
2. The kernel of the projection $p_1 \colon \widetilde{\mathfrak{w}}^2_{\mathfrak{m}}\mathfrak{gl}(n) \to \widetilde{\mathfrak{g}}^2_{\mathfrak{m}}$ is given by $X^\lambda_{\mu 0} = 0, X^\lambda_{0\mu} = 0$, and $X^\lambda_{\mu\nu} = 0$. □

Now, as a direct consequence of Proposition 2.2, we have.

Theorem 2.1 *The restriction of the adjoint action of the group $\widetilde{W}^2_m Gl(n)$ on the intersection of the kernels of the projections $\pi^2_{10} \colon \widetilde{\mathfrak{w}}^2_{\mathfrak{m}}\mathfrak{gl}(n) \to \widetilde{\mathfrak{g}}^2_{\mathfrak{m}}$, $\pi^2_{01} \colon \widetilde{\mathfrak{w}}^2_{\mathfrak{m}}\mathfrak{gl}(n) \to \widetilde{\mathfrak{g}}^2_{\mathfrak{m}}$ and $p_1 \colon \widetilde{\mathfrak{w}}^2_{\mathfrak{m}}\mathfrak{gl}(n) \to \widetilde{\mathfrak{g}}^2_{\mathfrak{m}}$ is given by*

$$\bar{X}^i_{j\rho\sigma} = a^i_p X^p_{q\alpha\beta} \tilde{a}^\alpha_\rho \tilde{a}^\beta_\sigma \tilde{a}^q_j \,,$$

i.e. $\operatorname{Ker}\pi^2_{10} \cap \operatorname{Ker}\pi^2_{01} \cap \operatorname{Ker} p_1$ *is* $\mathfrak{gl}(n) \otimes \otimes^2\mathbb{R}^{m*}$ *with the action of the group $G^1_m \times GL(n)$ given as the tensor product of the adjoint action of $GL(n)$ on its Lie algebra $\mathfrak{gl}(n)$ and the tensor action of the group G^1_m on $\otimes^2\mathbb{R}^{m*}$.*

Corollary 2.1 *The intersection of the kernels of the projections* $\pi^2_{10} \colon \operatorname{ad}(\widetilde{W}^2_m PE) \to \operatorname{ad}(\widetilde{W}^1_m PE)$, $\pi^2_{01} \colon \operatorname{ad}(\widetilde{W}^2_m PE) \to \operatorname{ad}(\widetilde{W}^1_m PE)$ *and* $p_1 \colon \operatorname{ad}(\widetilde{W}^2_m PE) \to \operatorname{ad}(\widetilde{P}^2 M)$ *is the vector bundle*

$$\operatorname{ad}(PE) \otimes \otimes^2 T^*M \to M.$$

Next we will describe the flow lift (see Sect. 2.1) of a principal connection K on PE and a classical connection Λ on M on the principle bundle $\widetilde{W}^2 PE$. We have the induced fibered coordinates on $\widetilde{W}^2 PE$ denoted by $(x^\lambda_{\mu 0}, x^\lambda_{0\mu}, x^\lambda_{\mu\nu}, x^i_j, x^i_{j\mu 0}, x^i_{j0\nu}, x^i_{j\mu\nu})$.

The flow lift on $\widetilde{W}^2 PE$ of a right invariant vector field Ξ on PE is then

$$\widetilde{\mathscr{W}^2}\Xi = \widetilde{\mathscr{P}^2}\xi +_\xi \widetilde{\mathscr{J}^2}\Xi,$$

where $\widetilde{\mathscr{P}^2}\xi$ is the flow lift of the vector field ξ on $\widetilde{P}^2 M$ and $\widetilde{\mathscr{J}^2}(\Xi)$ is the 2nd order nonholonomic jet lift of Ξ on $\widetilde{J}^2 PE$, see [5, 9].

3 Classification of Connections on $\widetilde{W}^2PE$

Let as recall that any principal connection on $\widetilde{W}^2PE$ projects by (10) on principal connection Λ_2 on $\widetilde{P}^2M$ and principal connections Γ_{10} and Γ_{01} on $\widetilde{W}^1PE$. So first we will classify principal connections on $\widetilde{P}^2M$ naturally given by a classical connection Λ and a general linear connection K (considered as principal connections on the corresponding frame bundle).

3.1 Natural Connections on the 2nd Order Frame Bundle

Any natural principal connection Λ_2 on $\widetilde{P}^2M$ given by Λ projects on principal connections Λ_{10} and Λ_{01} on P^1M given by Λ. So, we have to classify first all natural connections Λ_1 on P^1M given by Λ. This result is known and we have.

Proposition 3.1 [9, p. 220] *All natural operators transforming a principal connection Λ on P^1M into principal connections Λ_1 on P^1M form the 3-parameter family*

$$\Lambda_1(\Lambda) = \Lambda + \Phi_1\,,$$

where Φ_1 is a $(1,2)$-tensor field of the form

$$\Phi_1 = a_1\, T + a_2\, Id_{TM} \otimes \hat{T} + a_3\, \hat{T} \otimes Id_{TM}, \qquad a_i \in \mathbb{R}\,,$$

where T denote the torsion tensor of Λ and $\hat{T}$ denote the contraction.

Theorem 3.1 *Let $\Lambda_2(\Lambda)$ on $\widetilde{P}^2M$ be a natural connection given by Λ projectable on $\Lambda_1(\Lambda)$ then the difference*

$$\Lambda_2(\Lambda) - \mathscr{P}^2\Lambda_1(\Lambda)\colon M \to TM \otimes \otimes^3 T^*M.$$

Proof As a corollary of Theorem 2.1 we obtain that the intersection of the kernels of the projections $\pi^2_{10} : \widetilde{\mathfrak{g}}^2_m \to \mathfrak{g}^1_m$ and $\pi^2_{01} : \widetilde{\mathfrak{g}}^2_m \to \mathfrak{g}^1_m$ is $\mathfrak{g}^1_m \otimes \mathbb{R}^{m*}$ which implies that the intersection of kernels of projections $\pi^2_{10}\mathrm{ad}(\widetilde{P}^2M) \to \mathrm{ad}(P^1M)$ and $\pi^2_{01}\mathrm{ad}(\widetilde{P}^2M) \to \mathrm{ad}(P^1M)$ is the vector bundle $\mathrm{ad}(P^1M) \otimes T^*M$. The difference of two principle connections on $\widetilde{P}^2M$ over the same Λ_{10} and Λ_{01} on P^1M is then in the $\mathrm{ad}(P^1M) \otimes T^*M \otimes T^*M$. Moreover, $\mathrm{ad}(P^1M) = TM \otimes T^*M$. □

Lemma 3.1 *Any natural tensor field $\Phi_2(\Lambda)\colon M \to TM \otimes \otimes^3 T^*M$ naturally given by Λ on M and by general linear connection K on E is of maximal order one, is of the form*

$$\Psi(j^1\Lambda, j^1K) = \bar{\Psi}(R[\widetilde{\Lambda}], R[K], T, \widetilde{\nabla}T)$$

and all $\Phi_2(\Lambda, K)$ form 32-parameter family.

Proof A non-symmetric connection Λ can be decomposed as the sum of the classical symmetric connection $\widetilde{\Lambda}$ (obtained by the symmetrization of Λ) and the torsion tensor T. From reduction theorems for classical non-symmetric connections [12] we obtain that $\Phi_2(\Lambda)$ is given as a zero order operator on the curvature tensor of $\widetilde{\Lambda}$ and its covariant differentials (with respect to $\widetilde{\Lambda}$) and the torsion tensor and its covariant differentials (with respect to $\widetilde{\Lambda}$). From the homogeneous function theorem [9, p. 213] it follows, that a natural (1, 3)-tensor field can be constructed only from the curvature tensors of $\widetilde{\Lambda}$ and K, torsion T and the first covariant differential of T. □

Theorem 3.2 *All natural principal connections $\Lambda_2(\Lambda)$ on $\widetilde{P}^2M$ naturally given by a classical connection Λ on M and by general linear connection K on E form a* 38*-parameter family.*

Proof This theorem is the corollary of Proposition 3.1 (2 times 3 parameters) and Lemma 3.1 (32 parameters). □

3.2 Classification of Natural Connections on $\widetilde{W}^2PE$

Proposition 3.2 *Any principal connection Γ_2 on $\widetilde{W}^2PE$ naturally given by Λ and K is in the correspondence*

$$\Gamma_2 \approx (\Lambda_2, \Gamma_{10}, \Gamma_{01}, \Psi_2),$$

where Λ_2 is principal connection on $\widetilde{P}^2M$, Γ_{10} and Γ_{01} are principal connections on $\widetilde{W}^1PE$ and Ψ_2 is natural tensor field $M \to E \otimes E^* \otimes \otimes^3 T^*M$ all naturally given by Λ and K.

Proof From coordinate expression of Γ_2 and projections (10) we have the correspondence with Γ_{10}, Γ_{01} and Λ_2. The difference of two connection over the same connections Γ_{10}, Γ_{01} and Λ_2 is by the Corollary 2.1 exactly natural tensor field of Ψ_2. □

From the previous follows that for the classification of all Γ_2 we need to classify all Λ_2, all Γ_1 and all natural tensor fields Ψ_2.

Now, we have to classify all natural connections Γ_1 on $\widetilde{W}^1PE$ given by Λ and K. Let as recall that we have correspondance $\Gamma_1 \approx (\Lambda_1, \Gamma_0, \Psi_1)$. Λ_1 is described by the Proposition 3.1.

Lemma 3.2 *All natural operators transforming a principal connection K on PE into principal connections Γ_0 on PE naturally given by Λ and K form a* 1*-parameter family*

$$\Gamma_0 = K + \Psi_0 = K + mId_E \otimes c_1^1 T, \quad m \in \mathbb{R}.$$

Proposition 3.3 [3, p. 107] *All natural operators transforming a classical connection Λ on M and a linear connection K on PE into principal connections Γ_1 on W^1PE form the* 14*-parameter family.*

Proof We have 3 parameters from Λ_1, 1 parameter from Γ_0 and 10 parameters come from Ψ_1. Ψ_1 is given by the curvature tensor of K, by the tensor product of identity of E and by the contracted curvature tensor of Λ and of K, by contracted tensor product of T (two times) and by the contracted covariant differentials of T For more details see [3]. □

Lemma 3.3 *All natural tensor fields $\omega\colon M \to \otimes^3 T^*M$ of type* $(0, 3)$ *naturally given by Λ and K are of maximal order two and form* 72*-parameter family.*

Proof From reduction theorems for classical non-symmetric connections [12] we obtain that ω is given by the curvature tensor of $\widetilde{\Lambda}$ and its covariant differentials (with respect to $\widetilde{\Lambda}$) and the torsion tensor and its covariant differentials (with respect to $\widetilde{\Lambda}$). From the homogeneous function theorem [9, p. 213] it follows, that a natural $(0, 3)$-tensor field is of maximal order two. First part of ω has order two and is given by contracted covariant differential of the curvature tensor of K, contracted covariant differential of the curvature tensor of $\widetilde{\Lambda}$ and contracted second order covariant differential of the torsion of Λ. From Bianchi and Ricci identities we obtain a 12-parameter family. The second part has order one and is given by the contracted curvature tensor of K and the torsion tensor T, by the contracted curvature tensor of $\widetilde{\Lambda}$ and the torsion tensor T and by the contracted covariant differential of the torsion tensor and torsion tensor. In this way we obtain a 33-parameter family. The last part has order zero and is given by the tensor product of torsion (three times) only and form a 27-parameter family. □

Lemma 3.4 *All natural tensor fields $\Psi_2(\Lambda, K)\colon M \to E \otimes E^* \otimes \otimes^3 T^*M$ naturally given by Λ and K form* 86*-parameter family.*

Proof A non-symmetric connection Λ can be decomposed as the sum of the classical symmetric connection $\widetilde{\Lambda}$ (obtained by the symmetrization of Λ) and the torsion tensor T. From reduction theorems for general linear connections [11] we obtain that $\Psi_2(\Lambda, K)$ is given by the curvature tensor of $\widetilde{\Lambda}$ and its covariant differentials (with respect to $\widetilde{\Lambda}$) and the torsion tensor and its covariant differentials and by the curvature tensor of K and its covariant differentials.

From the homogeneous function theorem it follows, that $\Psi_2(\Lambda, K)$ has the decomposable part, which is given by the covariant differential of curvature tensor of K and by curvature tensor of K and torsion of Λ. The second part is of form $Id_E \otimes \omega$, where ω is tensor of type $(0, 3)$ described in Lemma 3.3. In all we have 86-parameter family. □

Theorem 3.3 *Natural principal connections on $\widetilde{W}^2PE$ given by a classical connection Λ on M and by a general linear connection K on E form* 145*-parameter family.*

Proof By Proposition 3.2 we have $\Gamma_2 \approx (\Lambda_2, \Gamma_{10}, \Gamma_{01}, \Psi_2) \approx (\Lambda_2, \Gamma_0, \Psi_{10}, \Psi_{01}, \Psi_2)$. Λ_2 form by Theorem 3.2 38-parameter family. Γ_0 form by Lemma 3.2 1-parameter family. Ψ_{10} and Ψ_{01} form two 10-parameter families (see proof of Proposition 3.3). And finally Ψ_2 form by Lemma 3.4 86-parameter family. In sum we obtain 145-parameter family. □

Corollary 3.1 *Natural principal connections on* $\overline{W}^2 PE$ *given by a classical connection* Λ *on* M *and by a general linear connection* K *on* E *form* 132*-parameter family.*

Proof In the semiholonomic case we have $\Psi_{10} = \Psi_{01}$ and $\Phi_{10} = \Phi_{01}$. □

Acknowledgments This research has been supported by the Grant agency of the Czech Republic under the project 201/09/0981.

References

1. Doupovec, M., Mikulski, W.M.: Reduction theorems for principal and classical connections. Acta Math. Sini. **26**(1), 169–184 (2010)
2. Doupovec, M., Mikulski, W.M.: Holonomic extensions of connections and symmetrization of jets. Rep. Math. Phys. **60**, 299–316 (2007)
3. Kolář, I.: Some natural operators in differential geometry, Differential Geometry and Its Applications, Proceedings, pp. 91–110. D. Reidel (1987)
4. Kolář, I., Virsik, G.: Connections in first principal prolongations. Suppl. ai Rendiconti del Circolo Matematico di Palermo, Serie **2**(43), 163–171 (1996)
5. Krupka, D., Janyška, J.: Lectures on differential invariants. Folia Fac. Sci. Nat. Univ. Purkynianae Brunensis, Brno (1990)
6. Nijenhuis, A.: Natural bundles and their general properties. Differential Geometry, In honour of K. Yano, Kinokuniya, Tokyo, pp. 317 (1972)
7. Terng, C.L.: Natural vector bundles and natural differential operators. Am. J. Math. **100**, 775 (1978)
8. Eck, D.E.: Gauge-natural bundles and generalized gauge theories. Mem. Amer. Math. Soc. **33**(247), (1981)
9. Kolář, I., Michor, P.W., Slovák, J.: Natural Operations in Differential Geometry. Springer, Berlin (1993)
10. Fatibene, L., Francaviglia, M.: Natural and Gauge Natural Formalism for Classical Field Theories. Kluwer Academic Publishers, London (2003)
11. Janyška, J.: Reduction theorems for general linear connections. Diff. Geom. Appl. **20**, 177 (2004)
12. Janyška, J.: Higher order valued reduction theorems for classical connections. Cent. Eur. J. Math. **3**, 294–308 (2005)
13. Vondra, J.: Classification of principal connections naturally induced on W^2PE. Arch. Math. (Brno) **44**(5), 535–547 (2008)

The Generic Rank of A-Planar Structures

Jaroslav Hrdina

Abstract The paper mostly collects material on generic rank of A-modules with respect to differential geometric applications. Our research was motivated by geometry of A structures. In particular, we discuss the case of A being an unitary associative algebra not necessarily with inversion. Some of the examples are studied in detail.

1 Motivation

Let us say a few words about our geometric motivation. Various concepts generalizing geodesics have been studied for almost quaternionic and similar geometries. Also various structures on manifolds are defined as smooth distribution in the vector bundle $T^*M \otimes TM$ of all endomorphisms of the tangent bundle. Very well known are two examples: almost complex and almost quaternionic structures. Let us extract some formal properties from these examples. Unless otherwise stated, all manifolds are smooth and they have the dimension m. Let ∇ be a linear connection and let $c : \mathbb{R} \to M$ be a curve on M. Then there is a vector field $\dot{c} := \frac{dc(t)}{dt} : \mathbb{R} \to TM$ along the curve c. Classically, a curve c is a geodesic if and only if its tangent vectors $\dot{c}(t)$ are parallely transported along $c(t)$. Let M be a smooth manifold equipped with a linear connection ∇ and let F be an affinor on M. A curve c is called F-planar curve if there is its parametrization $c : \mathbb{R} \to M$ satisfying the condition

$$\nabla_{\dot{c}}\dot{c} \in \langle \dot{c}, F(\dot{c}) \rangle.$$

It is easy to see that geodesics are F-planar curves for all affinors F, because of $\nabla_{\dot{c}}\dot{c} \in \langle \dot{c} \rangle \subset \langle \dot{c}, F(\dot{c}) \rangle$. The best known example is an almost complex structure.

J. Hrdina (✉)
Faculty of Mechanical Engineering, Institute of Mathematics, Brno University of Technology, Brno, Czech Republic
e-mail: hrdina@fme.vutbr.cz

A. Makhlouf et al. (eds.), *Algebra, Geometry and Mathematical Physics*, Springer Proceedings in Mathematics & Statistics 85, DOI: 10.1007/978-3-642-55361-5_26,

We have to be careful about the dimension of M. Let M be a manifold of dimension two and let I be a complex structure. A curve c is F-planar for $F = I$ if and only if c is satisfying the identity $\nabla_{\dot{c}}\dot{c} \in \langle \dot{c}, I\dot{c} \rangle \cong \mathbb{R}^2$, and any curve c satisfy the identity $\nabla_{\dot{c}}\dot{c} \in \mathbb{R}^2$. In other words any curve c is F-planar on the manifold of dimension two. The concept of F-planar curves makes sense for dimension at least four. Consider almost hypercomplex structure (I, J, K). The curve $c : \mathbb{R} \to M$ such that $\nabla_{\dot{c}}\dot{c} \in \langle \dot{c}, I(\dot{c}), J(\dot{c}), K(\dot{c}) \rangle$ is called 4-planar. It is easy to see that all geodesics are 4-planar curves, because of $\nabla_{\dot{c}}\dot{c} \in \langle \dot{c} \rangle \subset \langle \dot{c}, I(\dot{c}), J(\dot{c}), K(\dot{c}) \rangle$ and also all F-planar curves are 4-planar, for $F \in \langle I, J, K, E \rangle$. This simple consequence of standard behavior of the generators of a vector subspace suggests the generalization of the planarity concept below.

Definition 1.1 Let M be a smooth manifold of dimension m. Let A be a smooth ℓ-rank ($\ell < m$) vector subbundle in $T^*M \otimes TM$, such that the identity affinor $E = id_{TM}$ restricted to T_xM belongs to $A_x \subset T_x^*M \otimes T_xM$ at each point $x \in M$. We say that M is equipped by ℓ-rank A-structure.

In Definition 1.1, the dimension of M is higher than the rank of A. This is not a restriction, because there are no A-structures of rank ℓ higher than m. The possibility $\ell = m$ is not interesting, because in this case every curve is A-planar.

Definition 1.2 For any tangent vector $X \in T_xM$ we shall write $A(X)$ for the vector subspace

$$A(X) = \{F(X) | F \in A_xM\} \subset T_xM$$

and we call $A(X)$ the A-hull of the vector X. Similarly, the A-hull of vector field will be a subbundle in TM obtained pointwise. To find more details about A-structures we refer to [1–3].

2 The Generic Rank

For every smooth parameterized curve $c : \mathbb{R} \to M$ we write $\dot{c}$ and $A(\dot{c})$ for the tangent vector field and its A-hull along the curve c.

Definition 2.1 Let M be a smooth manifold equipped with an A-structure and a linear connection ∇. A smooth curve $c : \mathbb{R} \to M$ is told to be A-planar if

$$\nabla_{\dot{c}}\dot{c} \in A(\dot{c}).$$

Clearly, A-planarity means that the parallel transport of any vector tangent to c has to stay within the A-hull $A(\dot{c})$ of the tangent vector field $\dot{c}$ along the curve.

Definition 2.2 Let (M, A) be a smooth manifold M equipped with an ℓ-rank A-structure. We say that the A-structure has

1. generic rank ℓ if for each $x \in M$ the subset of vectors $(X, Y) \in T_xM \oplus T_xM$, such that the A-hulls $A(X)$ and $A(Y)$ generate a vector subspace $A(X) \oplus A(Y)$ of dimension 2ℓ is open and dense in $T_xM \oplus T_xM$.
2. weak generic rank ℓ if for each $x \in M$ the subset of vectors

$$\mathscr{V} := \{X \in T_xM \,|\, \dim A(X) = \ell\}$$

is open and dense in T_xM.

One immediately checks that any A-structure which has generic rank ℓ has weak generic rank ℓ. Indeed, if $U \subset T_xM$ is an open subset of vectors X with $A(X)$ of dimension lower than ℓ, then $U \times U$ is an open subset with the dimension, too.

Lemma 2.1 *Let M be a smooth manifold of dimension at least two and F be an affinor such that $F \neq q \cdot E$, $q \in \mathbb{R}$. Then the A-structure, where $A = \langle E, F \rangle$, has weak generic rank* 2.

Proof Consider A-structure $A = \langle E, F \rangle$. The complement of $\mathscr{V}$ consists of vectors $X \in T_xM$ such that:

$$X + aF(X) = 0, \; a \in \mathbb{R},$$

i.e. eigenspace of F. Dimension of A is two and F is not multiple of the identity. Thus, the union of eigenspaces of F is closed or trivial vector subspace of T_xM. Thus, the complement $\mathscr{V}$ is open and nontrivial, i.e. open and dense.

There is only one possibility for the A-structures in the lowest dimension one $A = \langle E \rangle$. The algebra $\langle E \rangle$ is an algebra with inversion, such that $E \cdot E = E$. For every $X \in T_xM$, $A(X)$ is the straight line containing A.

Finally, let us recall two important examples. The pair (M, F), where M is a smooth manifold and F is an affinor on M, is called a complex structure if and only if $F^2 = -E = -id_{TM}$. An almost complex structure has generic rank two on all manifolds of dimension at least four, because of Lemma 2.1. The pair (M, F) is called a product structure on M if and only if $F^2 = E$ and $F \neq E$. An almost product structure has generic rank two on all manifolds of dimension at least four, because of Lemma 2.1.

3 The Case of A Being an Algebra

Lemma 3.1 *Every A-structure (M, A) on a manifold M,* $\dim M \geq \dim A$, *where A is an algebra with inversion, has weak generic rank* $\dim A$.

Proof Consider X such that $X \notin \mathscr{V}$, therefore $\exists F \in A = \langle E, G \rangle$, $FX = 0$, and $F^{-1}FX = 0$ implies $X = 0$.

Theorem 3.1 *Let M be an A-structure and let $X_1, \ldots, X_m$ be a basis of $V := T_x M$, i.e. V is an A-module. Let A be an n-dimensional $\mathbb{R}$-algebra, where $n < m$. If there exists $X \in V$ such that $\dim(A(X)) = n$ then the A-structure has weak generic rank n.*

Proof We prove equivalent statement, A-module V does not have a generic rank ℓ if and only if there is a vector $X \in V$, such that for any vector $Y \in V$ there is an affinor G_Y, such that $G_Y(X - \varepsilon Y) = 0$, for small ε. Therefore, for any vector $Y \in V$ there is an affinor G_Y such that

$$G_Y(X) = \varepsilon G_Y(Y)$$

for small ε. Hence, the affinor $\frac{1}{\sqrt{\varepsilon}} G_Y$ maps $\frac{1}{\sqrt{\varepsilon}} X$ to a vector $G_Y(Y)$ and therefore, for any vector $Y \in V$ there is an affinor H_Y such that $H_Y(\frac{1}{\sqrt{\varepsilon}} X) = H_Y(Y)$. In particular, there is an affinor S such that $S(\frac{1}{\sqrt{\varepsilon}} X) = S(Y + \frac{1}{\sqrt{\varepsilon}} X)$ and therefore, for any $Y \in V$ there is an affinor S_Y such that $S_Y(Y) = 0$.

Theorem 3.2 *Let (M, A) be a smooth manifold of dimension m equipped with an A-structure of rank ℓ, such that $2\ell \leq m$. If A_x is an algebra (i.e. for all $f, g \in A_x$, $fg := f \circ g \in A_x$) for all $x \in M$, and A has weak generic rank ℓ then the structure has generic rank ℓ.*

Proof Since the A-structure has a weak generic rank ℓ, there is an open and dense subset $\mathcal{V} \subset T_x M$ such that $\dim A(X) = \ell$ for all $X \in \mathcal{V}$.

Because A is an algebra, for any $X, Z \in T_x M$, $Z \in A(X)$ implies also $A(Z) \subset A(X)$, and moreover $A(Z) = A(X)$ for all $X, Z \in \mathcal{V}$ because of the dimension. Thus, whenever there is a non-trivial vector $0 \neq Z \in A(X) \cap A(Y)$, the entire subspaces coincide, i.e. $A(X) = A(Y)$.

In particular, whenever $X, Y \in \mathcal{V}$ and the dimension of $A(X) \oplus A(Y)$ is less then 2ℓ, we know $A(X) = A(Y)$.

Let us consider a couple of vectors $(Y, Z) \in A(X) \oplus A(X)$ for some $X \in \mathcal{V}$. Consider a vector $W \notin A(X)$. An open neighbourhood $\mathcal{U}$ of Y has to include $(Y + aW, Y)$ for all sufficiently small $a \in \mathbb{R}$. But if $Y + aW \in A(X)$ for some $a \neq 0$ then $W \in A(X)$ and this is not true. Thus, for every couple of vectors in $A(X) \oplus A(X)$ and for its every open neighbourhood, we have found another couple $(Y' = Y + aW, Z)$ for which the dimension of $A(Y') + A(Z)$ is 2ℓ. This proves the density of the set of couples of vectors generating the maximal dimension 2ℓ.

Of course, the requirement on the maximal dimension is an open condition and the theorem is proved.

Corollary 3.1 *Let (M, A) be a smooth manifold with A-structure of rank ℓ, such that $2\ell \leq \dim M$. If $A_x \subset T_x^* M \otimes T_x M$ is an algebra with inversion then A has weak generic rank. Moreover, if $\dim M \geq 2\ell$ than A has generic rank ℓ.*

Corollary 3.2 *Let (M, A) be a smooth manifold with A-structure of rank ℓ, such that $2\ell \leq \dim M$ and let A be an algebra. If there exists $X \in T_x M$ such that $\dim(A(X)) = n$ then the A-structure has generic rank n.*

4 Remark on Frobenious Algebras

Let A be an algebra over $\mathbb{R}$ with basis $\{F_i\}$, where $i = 1, \dots, n$, $F_1 := E$, with structure constants

$$C_{ij}^k \in \mathbb{R} \text{ (i.e. } F_i F_j = C_{ij}^s F_s).$$

In particular, one can easily see that

$$F_i = F_1 F_i = C_{1i}^s F_s = \delta_i^s F_s, \text{ i.e. } C_{1i}^s = \delta_i^s.$$

We introduce matrices

$$\hat{C}_i = (C_{ji}^k), \quad \hat{C}_i^* = (C_{ik}^j),$$

where j is a number of rows. Then the associativity condition can be written as

$$\hat{C}_j \hat{C}_k = C_{jk}^s \hat{C}_s \quad \text{or} \quad \hat{C}_j^* \hat{C}_k^* = C_{jk}^s \hat{C}_s^*$$

and the unity can be written as $\hat{C}_1 = E$. A linear functional $\varepsilon : A \to \mathbb{R}$ is determined by the choice of a n-dimensional vector $\lambda = (\lambda_1, \dots, \lambda_n)$.

Now, for $\varepsilon(F_i) = \lambda_i$ and for $F = \sum_{i=1}^n a_i F_i \in A$ we can see immediately that $\varepsilon(F) = \sum_{i=1}^n a_i \lambda_i$ and finally $F_i F_j = C_{ij}^s F_s$. If $\lambda \in \mathbb{R}^n$ is a vector such that the matrix $G = (g_{ij})$ is regular, where $g_{ij} := C_{ij}^s \lambda_s$, we prove that the functional $\varepsilon : A \to \mathbb{R}$, such that

$$\varepsilon : \sum_{i=1}^n a_i F_i \mapsto \sum_{i=1}^n a_i \lambda_i,$$

is a Frobenius form.

The formula for generic rank n from the Theorem 3.1 reads that if there exists $X \in V$ such that $\{F_i X\}$ is linearly independent then V has a generic rank. On the other hand, if there exists $\lambda \in \mathbb{R}^n$ such that $\{\hat{C}_i \lambda\}$ are linearly independent then A is a Frobenius algebra. This indicates that these properties lead to similar conditions.

In other words, if A is an algebra over $\mathbb{R}$ and the matrices $\hat{C}_i$ are structural matrices of A, then there is a B-module $\mathbb{R}^n$, where $B = \langle \hat{C}_1, \dots, \hat{C}_n \rangle$. Therefore, the algebra A is a Frobenius algebra if and only if the B-module $\mathbb{R}^n$ has generic rank n.

5 Examples

One can apply these results to two big groups of geometric structures, Clifford algebras and distributions.

5.1 Almost Cliffordian Manifolds

Almost Clifford and almost Cliffordian manifolds are G-structures based on the definitionition of Clifford algebras. An almost Clifford manifold based on $\mathcal{C}l(s,t)$ is given by a reduction of the structure group $GL(km,\mathbb{R})$ to $GL(m,\mathcal{O})$, where $k = 2^{s+t}$, $m \in \mathbb{N}$ and $\mathcal{O}$ is an arbitrary Clifford algebra. An almost Cliffordian manifold is given by a reduction of the structure group to $GL(m,\mathcal{O})GL(1,\mathcal{O})$. It is easy to see that an almost Cliffordian structure is an A-structure, where A is a Clifford algebra $\mathcal{O}$ because the affinors in the form of $F_0,\dots,F_\ell \in A$ can be defined only locally. In [4] authors prove the following theorem.

Theorem 5.1 *Let $F_0, \dots F_k$ denote the $k+1$ elements of the matrix representation of Clifford algebra $\mathcal{C}l(s,t)$. Then there exists a real vector X such that the dimension of a linear span $\langle F_i X | i = 1,\dots,k\rangle$ equals to $k+1$.*

Finally, let M be a Cliffordian manifold, i.e. let (M, A) be a smooth manifold with $A = \mathcal{C}l(s,t)$, such that $2^{s+t+1} \leq \dim M$, then Cliffordian manifold has generic rank 2^{s+t}. For more information about almost Cliffordian structures see papers [4, 6].

5.2 Distributions

If $D, \overline{D}$ form a complete system of distributions (i.e. they are disjoint and $D + \overline{D} = TM$) then there are two affinors $P, \overline{P}$ associated with them such that

$$P^2 = P, \quad \overline{P}^2 = \overline{P}, \quad P\overline{P} = \overline{P}P = 0 \quad \text{and} \quad P + \overline{P} = E,$$

where rank $P = r$ and rank $\overline{P} = \overline{r}$.

The representation of distributions by affinors can by extended to any complete system D_i such that the affinors P_i satisfy the properties

$$P_i^2 = P_i, \quad P_i P_j = 0 \ \text{ for } \ i \neq j, \quad \text{and} \quad \sum_i P_i = E.$$

Considering the element $P = a_1 P_1 + \cdots + a_n P_n \in A$, the matrix

$$\begin{pmatrix} EP \\ P_1 P \\ \vdots \\ P_n P \end{pmatrix}$$

is the following

$$\begin{pmatrix} a_1 & 0 & \cdots & 0 \\ 0 & a_2 & \cdots & 0 \\ \vdots & \vdots & \cdots & \vdots \\ 0 & 0 \cdots & 0 & a_n \end{pmatrix}$$

and therefore $\langle P_1, \ldots, P_n \rangle$ has weak generic rank n by Lemma 3.1. Finally, let M be a manifold with complete system of distributions $D_1, \ldots, D_n$, i.e. let (M, A) be a smooth manifold with $A = \langle P_1, \ldots, P_n \rangle$, such that $2n \leq \dim M$, then the A-structure has generic rank n. See [5].

Acknowledgments This work was supported by the project NETME CENTRE PLUS (LO1202). The results of the project NETME CENTRE PLUS (LO1202) were co-funded by the Ministry of Education, Youth and Sports within the support programme "National Sustainability Programme I".

References

1. Hrdina, J., Slovák, J.: Generalized planar curves and quaternionic geometry. Glob. Anal. Geom. **29**, 349–360 (2006)
2. Hrdina, J.: Remarks on F-planar curves and their generalizations, vol. 93, pp. 241–249. Banach Center Publications (2011)
3. Hrdina, J.: Notes on connections attached to A-structures. Differ. Geom. Appl. **29**(Supl. 1), 91–97 (2011)
4. Hrdina, J., Vašík, P.: Generalized geodesics on some almost Cliffordian geometries. Balkan J. Geom. Appl. **17**(2), 41–48 (2012)
5. Hrdina, J.: Geometric properties of Frobenius algebras. J. Phys. Conf. Ser. **1**, 1–5 (2012)
6. Hrdina, J., Vašík, P.: Geometry of almost Cliffordian manifolds: classes of subordinated connections. Turkish J. Math. **38**(1), 179–190 (2014)

Geometric Approach to Ghost Fields

Viktor Abramov and Jaan Vajakas

Abstract An infinite dimensional Grassmann algebra on a compact Riemannian manifold is constructed by means of rigged Hilbert spaces of differential forms. We give a notion of p-form of order α on a product manifold and define a wedge product of these forms. The set of involutive generators of infinite dimensional Grassmann algebra which can be used for geometric approach to ghost fields appearing in quantized gauge theory is introduced. We extend our approach to vector bundles and construct an infinite dimensional Grassmann algebra with generators by means of the rigged Hilbert spaces of sections of a vector bundle.

1 Introduction

The quantization of Yang-Mills field theory based on a functional integral approach is the most suitable scheme for quantization of gauge field theories because the principle of gauge invariance could be expressed in terms of this approach very easily: one should integrate not over the space of all field configurations, but only over the space of gauge-equivalent classes of field configurations. However this method leads to the well known problem of non-local functional which appears in the functional integral for S-matrix. This problem is solved if one introduces the auxiliary fields $c^a(x), \bar{c}^b(x)$, where x is a point of a manifold and the superscripts a, b run from one to the dimension of a gauge group, and then uses them to write the determinant of a differential operator in a form of the Berezin integral over the infinite dimensional Grassmann algebra. From an algebraic point of view the auxiliary fields $c^a(x), \bar{c}^b(x)$, which are called Faddeev-Popov ghost fields, are the generators of an infinite dimensional Grassmann algebra. Hence they are subjected to the relations

V. Abramov (✉) · J. Vajakas
University of Tartu, Liivi 2, 50409 Tartu, Estonia
e-mail: viktor.abramov@ut.ee

J. Vajakas
e-mail: jv@ut.ee

A. Makhlouf et al. (eds.), *Algebra, Geometry and Mathematical Physics*,
Springer Proceedings in Mathematics & Statistics 85, DOI: 10.1007/978-3-642-55361-5_27,

$$c^a(x)\, c^b(y) = -c^b(y)\, c^a(x), \tag{1}$$

where x and y are points of a base manifold M. It is very important that from these relations it follows that the ghost fields anticommute not only with respect to superscripts a and b, but also with respect to a point x of a base manifold M. Therefore one can multiply the ghost fields even in the case of different points $x_1, x_2, \ldots, x_N$ of a base manifold and the product $c^{a_1}(x_1)c^{a_2}(x_2)\ldots c^{a_N}(x_N)$ can be considered as an element of an infinite dimensional Grassmann algebra. The geometric interpretation of ghost fields in terms of differential forms proposed in [1, 2] does not cover this property of ghost fields because it is well known that one can multiply differential forms pointwise and the product of two differential forms has no sense if they are taken at different points of a manifold. In this paper we develop a geometric approach to ghost fields proposed in [3] which is based on a notion of an infinite dimensional Grassmann algebra. In this approach ghost fields are the generators of an infinite dimensional Grassmann algebra which is constructed with the help of rigged Hilbert spaces of differential forms of a manifold and the rigged Hilbert spaces of sections of a vector bundle.

2 Infinite Dimensional Grassmann Algebra

Let us remind that a finite dimensional Grassmann algebra is an associative unital algebra generated by a finite set of variables $x_1, x_2, \ldots, x_k$ which are subjected to the relations

$$x_i x_j + x_j x_i = 0. \tag{2}$$

Grassmann algebra is graded algebra if one assigns grading zero to the unit element, grading one to each variable $x_1, x_2, \ldots, x_k$ and defines the grading of a product as the sum of gradings of its factors. An infinite dimensional Grassmann algebra with generators was introduced by Berezin in order to describe generating functionals of quantum field theory in Fermi case [4]. The Gauss integral on infinite dimensional Grassmann algebra with generators was used by Faddeev and Popov in quantization of gauge field theory and this has led to appearance of ghost fields in quantized gauge field theory [5].

Let $\mathbb{N}$ be the set of non-negative integers. A graded linear space is a family of topological linear spaces $\Psi = \{\Psi^\alpha\}_{\alpha\in\mathbb{N}}$ indexed by non-negative integers with natural linear operations, i.e. for any elements of graded linear space

$$\xi = (\xi^0, \xi^1, \ldots) = (\xi^\alpha) \in \Psi, \quad \eta = (\eta^0, \eta^1, \ldots) = (\eta^\alpha) \in \Psi,$$

where $\xi^\alpha, \eta^\alpha \in \Psi^\alpha$, and any complex number $a \in \mathbb{C}$ we have

$$\xi + \eta = (\xi^\alpha + \eta^\alpha) \in \Psi, \quad a\,\xi = (a\,\xi^\alpha) \in \Psi.$$

As usual the elements of Ψ^α will be called homogeneous elements of graded linear space Ψ, and the degree of homogeneous element ξ will be denoted by $|\xi|$. Hence if $\xi \in \Psi^\alpha$ then $|\xi| = \alpha$. We will also assume that if $\xi_{(\mu)}$ is a sequence of elements of graded linear space Ψ, where $\mu \geq 1$ is an integer and

$$\xi_{(\mu)} = (\xi^0_{(\mu)}, \xi^1_{(\mu)}, \ldots, \xi^\alpha_{(\mu)}, \ldots), \quad \xi^\alpha_{(\mu)} \in \Psi^\alpha,$$

such that $\lim_\mu \xi^\alpha_{(\mu)} = \xi^\alpha \in \Psi^\alpha$ in the topology of linear topological space Ψ^α then $\lim_\mu \xi_{(\mu)} = \xi$, where $\xi = (\xi^\alpha) \in \Psi$.

A graded topological algebra is a graded linear space $\Psi = \{\Psi^\alpha\}_{\alpha\in\mathbb{N}}$ endowed with a multiplication $\Psi^\alpha \otimes \Psi^\beta \to \Psi^{\alpha+\beta}$ that is continuous and associative with the unit element in Ψ^0. A graded topological algebra Ψ is said to be graded commutative if for any homogeneous elements $\xi, \eta \in \Psi$ we have $\xi\eta = (-1)^{|\xi||\eta|}\eta\xi$. A graded commutative topological algebra $\Psi = \oplus_{\alpha\in\mathbb{N}}\Psi^\alpha$ is said to be a Grassmann algebra if it satisfies

1. Ψ^0 is the one-dimensional space generated by the unit element of Ψ,
2. the subspace of all finite linear combinations of products of homogeneous elements $\xi\eta$, where $\xi \in \Psi^\alpha$, $\eta \in \Psi^\beta$, is dense in $\Psi^{\alpha+\beta}$,
3. all products $\xi_{i_1}\xi_{i_2}\ldots\xi_{i_\alpha}$, where $1 \leq i_1 < i_2 < \ldots < i_n$ and $\xi_1, \xi_2, \ldots, \xi_N, \ldots$ are linearly independent elements of Ψ^1, are linearly independent elements of Ψ^α.

The one-dimensional space of elements of degree zero Ψ^0 will be identified with complex numbers, i.e. $\Psi^0 \equiv \mathbb{C}$. Obviously Ψ is a finite-dimensional Grassmann algebra (2) generated by a basis for a linear space Ψ^1 if this linear space is finite-dimensional. A Grassmann algebra is called infinite dimensional Grassmann algebra if Ψ^1 is an infinite-dimensional topological linear space. As our aim in this paper is to construct and study infinite dimensional Grassmann algebras we will assume that Ψ^1 is an infinite-dimensional linear space and Grassmann algebra will mean an infinite-dimensional Grassmann algebra.

A Grassmann $*$-algebra is a Grassmann algebra $\Psi = \{\Psi^\alpha\}_{\alpha\in\mathbb{N}}$ with involution $* : \Psi \to \Psi$ which satisfies

1. $* : \xi \in \Psi^\alpha \to \xi^* \in \Psi^\alpha$ is a continuous antilinear isomorphism, i.e. $(a\xi + \eta)^* = \bar{a}\xi^* + \eta^*$,
2. $(\xi\eta)^* = \eta^*\xi^*$,
3. $(\xi^*)^* = \xi$,
4. Ψ^1 is a direct sum of linear subspaces Φ_+, Φ_-, i.e. $\Psi^1 = \Phi_+ \oplus \Phi_-$.

A Grassmann $*$-algebra $\Psi = \{\Psi^\alpha\}_{\alpha\in\mathbb{N}}$ is said to be a Grassmann $*$-algebra with an inner product if each subspace of homogeneous elements of degree α has a structure of rigged Hilbert space, i.e. the subspace of homogeneous elements of degree α is a triple $(\tilde{\Psi}^\alpha, H^\alpha, \Psi^\alpha)$, where

1. $\tilde{\Psi}^\alpha$ is a nuclear space equipped with an inner product, i.e. for any homogeneous degree α elements $\xi, \eta \in \tilde{\Psi}^\alpha$ the inner product $<\xi, \eta>$ is a continuous

(with respect to each argument) positively definite Hermitian functional such that $\lim_\mu <\xi_{(\mu)}, \eta> = <\xi, \eta>$ for any sequence $(\xi_{(\mu)})$ of elements of $\tilde{\Psi}^\alpha$ satisfying $\lim_\mu \xi_{(\mu)} = \xi \in \tilde{\Psi}^\alpha$,

2. H^α is a Hilbert space that is a completion of $\tilde{\Psi}^\alpha$ by the inner product $<\, ,\, >$, $\tilde{\Psi}^\alpha$ is dense in H^α and this determines a continuous natural inclusion map $\tilde{\Psi}^\alpha \subset H^\alpha$,
3. Ψ^α is a dual space of $\tilde{\Psi}^\alpha$,
4. if $\xi_1, \xi_2, \ldots, \xi_\alpha, \ldots$ is an orthonormal basis for H^1 then $\xi_{i_1}\xi_{i_2}\ldots\xi_{i_\alpha}$, where $1 \leq i_1 < i_2 < \ldots i_\alpha$, is the orthonormal basis for H^α,
5. $\Phi_+ \cap H^1$ is orthogonal to $\Phi_- \cap H^1$ and for any elements ξ, η such that $<\xi, \eta>$ is defined it holds $<\xi^*, \eta^*> = <\eta, \xi>$.

3 Spaces of Differential Forms of a Manifold

Let M be a Riemannian n-dimensional manifold with a metric g. Rather than considering the spaces of smooth differential forms with compact support we will assume that M is a compact manifold. We will also assume that M is an orientable manifold. Let $\Omega(M) = \oplus_p \Omega^p(M)$ be the algebra of smooth complex valued p-forms of a manifold M, where $\Omega^p(M)$ is a space of p-forms. It is convenient to use multi-indices to write a differential form locally that is if $x^1, x^2, \ldots, x^n$ are local coordinates of M then locally any p-form $\omega \in \Omega^p(M)$ can be written as

$$\omega = \omega_I \, dx^I,$$

where $I = \{i_1, i_2, \ldots, i_p\}$ is a subset of the set $\mathbb{N}_n = \{1, 2, \ldots, n\}$, $\omega_I = \omega_{i_1 i_2,\ldots i_p}$ and $dx^I = dx^{i_1} \wedge dx^{i_2} \wedge \ldots \wedge dx^{i_p}$. We will denote by $|I|$ the number of elements of I. It is well known that given two smooth p-forms ω, θ one can define the Hermitian inner product of these p-forms

$$(\omega, \theta) = \int_M \omega \cdot \theta \, d\mu, \tag{3}$$

where locally $\omega = \omega_I \, dx^I, \theta = \theta_J \, dx^J, d\mu = |\det(g_{ij})|^{1/2} dx^1 dx^2 \ldots dx^n$ is the density relative to g,

$$\omega \cdot \theta = \omega^I \bar{\theta}_I, \quad \omega^I = \omega^{i_1 i_2 \ldots i_p} = g^{i_1 k_1} \ldots g^{i_p k_p} \omega_{k_1 k_2 \ldots k_p}.$$

Supposing that orientation of a manifold M is determined by $dx^1 \wedge dx^2 \wedge \ldots \wedge dx^n$ and making use of the Hodge operator $\star : \Omega^p(M) \to \Omega^{n-p}(M)$ one can write the inner product (3) in the form

$$(\omega, \theta) = \int_M \omega \wedge \star\theta, \tag{4}$$

where locally

$$(\star\theta)_{i_{p+1}\dots i_n} = |\det(g_{ij})|^{1/2}\,\theta^{i_1\dots i_p}\varepsilon_{i_1\dots i_p i_{p+1}\dots i_n}.$$

The completion of the space of smooth complex valued p-forms $\Omega^p(M)$ by the inner product (4) is the Hilbert space which will be denoted by $H^p(M)$. Let us denote by $\Omega^{*p}(M)$ the space of all continuous linear functionals on the space of p-forms $\Omega^p(M)$. We will follow the terminology proposed by De Rham in [6] and call a continuous linear functional on $\Omega^p(M)$ an p-dimensional current.

At each point $x \in M$ of a manifold M we have the finite dimensional Grassmann algebra $\bigwedge T_x^*M$ which is the vector space spanned by $\{dx^I\}$, where I runs over all subsets of the set $\mathbb{N}_n$. Let us mention that the Grassmann algebra $\bigwedge T_x^*M$ can be viewed as the fiber of the vector bundle $\bigwedge T^*M$ at a point $x \in M$. Then any differential form on M can be considered as a smooth section of $\bigwedge T^*M$. Consider the product manifold $M^{(2)} = M \times M$ equipped with the Riemannian metric $g^{(2)} = g \oplus g$. Just as in the case of a manifold M at each point (x, y) of the product manifold $M^{(2)}$ we have the finite dimensional Grassmann algebra $\bigwedge T^*_{(x,y)}M^{(2)}$. If $U \times V$ is a local chart in a neighborhood of (x, y) with local coordinates $x^1, x^2, \dots, x^n$ on $U \subset M$ and local coordinates $y^1, y^2, \dots, y^n$ on $V \subset M$ then $\bigwedge T^*_{(x,y)}M^{(2)}$ is spanned by the wedge products of differentials $dx^I \wedge dy^J$, where I, J are subsets of $\mathbb{N}_n$. Evidently if $|I| = p$, $|J| = q$ then $dx^I \wedge dy^J = (-1)^{pq} dy^J \wedge dx^I$.

For infinite dimensional Grassmann algebra which will be described in the next section we will need a wedge product of differentials dx^I, dy^J which is slightly different from $dx^I \wedge dy^J$. We introduce a notion of a double p-form on the product manifold $M^{(2)}$ giving it locally by an expression

$$\omega(x, y) = \sum_{|I|=|J|=p} \omega_{I,J}(x, y)\, dx^I \bar{\wedge} dy^J, \tag{5}$$

where the multiplication $\bar{\wedge}$ is subjected to the relations

$$dx^I \bar{\wedge} dy^J = -dy^J \bar{\wedge} dx^I. \tag{6}$$

It should be noted that in our approach to a notion of double p-form wedge products $dx^I = dx^{i_1} \wedge \dots \wedge dx^{i_p}$ anticommute with wedge products $dy^J = dy^{j_1} \wedge \dots \wedge dy^{j_p}$ while in approach proposed by De Rham in [6] they commute. A double p-form ω is said to be symmetric if $\omega(x, y) = \omega(y, x)$. Easy calculation shows

$$\begin{aligned}\omega(x, y) - \omega(y, x) &= \omega_{I,J}(x, y) dx^I \bar{\wedge} dy^J - \omega_{J,I}(y, x) dy^J \bar{\wedge} dx^I \\ &= (\omega_{I,J}(x, y) + \omega_{J,I}(y, x)) dx^I \bar{\wedge} dy^J = 0,\end{aligned}$$

that a double p-form (5) is symmetric if and only if $\omega_{I,J}(x, y) = -\omega_{J,I}(y, x)$. Given two p-forms $\omega(x) = \omega_I(x) dx^I, \theta(y) = \theta_J(y) dy^J$ we construct the symmetric double p-form as follows $\sigma(x, y) = \omega(x) \bar{\wedge} \theta(y) + \theta(y) \bar{\wedge} \omega(x)$. If we write $\sigma(x, y) = \sigma_{I,J}(x, y) dx^I \bar{\wedge} dy^J$ then

$$\sigma_{I,J}(x,y) = \omega_I(x)\theta_J(y) - \omega_J(y)\theta_I(x).$$

Let us denote the space of smooth complex valued symmetric double p-forms by $\Omega^p(M^{(2)})$. In analogy with (3) we define an inner product of symmetric double p-forms ω, θ by

$$(\omega,\theta) = \int_{M^{(2)}} \omega(x,y)\cdot\theta(x,y)\, d\mu^{(2)}, \tag{7}$$

where locally $\omega(x,y) = \omega_{I,J}(x,y)dx^I\bar{\wedge}dy^J$, $\theta(x,y) = \theta_{I,J}(x,y)dx^I\bar{\wedge}dy^J$, $d\mu^{(2)}$ is the density relative to metric $g^{(2)} = g \oplus g$,

$$\omega(x,y)\cdot\theta(x,y) = \omega^{I,J}(x,y)\bar{\theta}_{I,J}(x,y),$$

and

$$\begin{aligned}\omega^{I,J}(x,y) &= \omega^{i_1i_2\ldots i_p,j_1j_2\ldots j_p}(x,y)\\ &= g^{i_1k_1}(x)g^{i_2k_2}(x)\ldots g^{i_pk_p}(x)g^{j_1l_1}(y)g^{j_2l_2}(y)\ldots\\ &\quad g^{j_pl_p}(y)\omega_{k_1k_2\ldots k_p,l_1l_2\ldots l_p}(x,y)\end{aligned}$$

The completion of the space $\Omega^p(M^{(2)})$ with respect to the inner product (7) is the Hilbert space which will be denoted by $H^p(M^{(2)})$. The space of continuous linear functionals on the space of symmetric double p-forms $\Omega^p(M^{(2)})$ will be denoted by $\Omega^{*p}(M^{(2)})$.

Now in analogy with the space of symmetric double p-forms $\Omega^p(M^{(2)})$ for any integer $\alpha > 2$ we can construct a space of symmetric p-forms of order α. Consider the product manifold $M^{(\alpha)} = M \times M \times \ldots \times M$ (α times) equipped with the Riemannian metric $g^{(\alpha)} = g \oplus g \oplus \ldots \oplus g$ (α times). In order to introduce a notion of p-form of order α we will assume that the multiplication (6) is associative, i.e.

$$(dx^I\bar{\wedge}dy^J)\bar{\wedge}dz^K = dx^I\bar{\wedge}(dy^J\bar{\wedge}dz^K).$$

We will call ω an p-form of order α on the manifold $M^{(\alpha)}$ if in each local chart of $M^{(\alpha)}$ it is given by an expression

$$\omega(x_1,x_2,\ldots,x_\alpha) = \omega_{I_1I_2\ldots I_\alpha}(x_1,x_2,\ldots,x_\alpha)\, dx_1^{I_1}\bar{\wedge}dx_2^{I_2}\bar{\wedge}\ldots\bar{\wedge}dx_\alpha^{I_\alpha},$$

where $(x_1,x_2,\ldots,x_\alpha) \in M^{(\alpha)}$, $I_1, I_2, \ldots, I_\alpha$ are subsets of $\mathbb{N}_n$ each consisting of p elements, i.e. for any $k = 1,2,\ldots,\alpha$ we have $|I_k| = p$. In the case of the product manifold $M^{(\alpha)}$ the local expressions for p-forms of αth order and related formulae have a cumbersome appearance and in order to write them in a more compact way it is useful to combine a multi-index $I = (i_1i_2\ldots i_p) \subset \mathbb{N}_n$ and a point x of a manifold M into a single symbol $\mathscr{I} = (I,x)$ setting

$$\omega_{\mathscr{I}_1\mathscr{I}_2\ldots\mathscr{I}_\alpha} = \omega_{I_1I_2\ldots I_\alpha}(x_1, x_2, \ldots, x_\alpha),$$

where $\mathscr{I}_k = (I_k, x_k)$. It is also useful to denote a point $(x_1, x_2, \ldots, x_\alpha)$ of the product manifold $M^{(\alpha)}$ by $x^{(\alpha)}$. Making use of these notations one can write a local expression of p-form of αth order as follows

$$\omega(x^{(\alpha)}) = \omega_{\mathscr{I}_1\mathscr{I}_2\ldots\mathscr{I}_\alpha}\, dx_1^{I_1}\bar{\wedge} dx_2^{I_2}\bar{\wedge}\ldots\bar{\wedge} dx_\alpha^{I_\alpha}$$

An p-form of order α is said to be symmetric if for any permutation $r : \mathbb{N}_\alpha \to \mathbb{N}_\alpha$ it holds

$$\omega(x_{r(1)}, x_{r(2)}, \ldots, x_{r(\alpha)}) = \omega(x_1, x_2, \ldots, x_\alpha).$$

It is easy to show that an p-form of αth order $\omega(x_1, x_2, \ldots, x_\alpha)$ is symmetric if and only if for each local chart of $M^{(\alpha)}$ and for any permutation $r : \mathbb{N}_\alpha \to \mathbb{N}_\alpha$ its coefficient functions of local expression obey the relations

$$\omega_{\mathscr{I}_{r(1)}\mathscr{I}_{r(2)}\ldots\mathscr{I}_{r(\alpha)}} = (-1)^{\mathfrak{p}(r)}\omega_{\mathscr{I}_1\mathscr{I}_2\ldots\mathscr{I}_\alpha},$$

where $\mathfrak{p}(r)$ is a parity of a permutation r. Let us denote by $\Omega^p(M^{(\alpha)})$ the space of smooth symmetric p-forms of αth order on the product manifold $M^{(\alpha)}$.

As $M^{(\alpha)}$ is a Riemannian manifold we can raise and lower the indexes, i.e. given a coefficient $\omega_{I_1I_2\ldots I_\alpha}(x_1, x_2, \ldots, x_\alpha)$ of an p-form of αth order ω we raise the subscripts in each multi-index $I_k = (i_1^{(k)}, i_2^{(k)}, \ldots, i_p^{(k)})$ by means of the metric $g^{j_l^{(k)} i_l^{(k)}}(x_k)$. Now in analogy with the inner product for symmetric double p-forms we define an inner product of symmetric p-forms of order α by

$$(\omega, \theta) = \int_{M^{(\alpha)}} \omega(x^{(\alpha)}) \cdot \theta(x^{(\alpha)})\, d\mu^{(\alpha)}, \tag{8}$$

where $d\mu^{(\alpha)}$ is the density on the product manifold relative to the Riemannian metric $g^{(\alpha)}$,

$$\omega(x^{(\alpha)}) = \omega_{\mathscr{I}_1\mathscr{I}_2\ldots\mathscr{I}_\alpha}\, dx_1^{I_1}\bar{\wedge} dx_2^{I_2}\bar{\wedge}\ldots\bar{\wedge} dx_\alpha^{I_\alpha},$$
$$\theta(x^{(\alpha)}) = \theta_{\mathscr{J}_1\mathscr{J}_2\ldots\mathscr{J}_\alpha}\, dx_1^{J_1}\bar{\wedge} dx_2^{J_2}\bar{\wedge}\ldots\bar{\wedge} dx_\alpha^{J_\alpha},$$

and

$$\omega(x^{(\alpha)}) \cdot \theta(x^{(\alpha)}) = \omega^{I_1\ldots I_\alpha}(x^{(\alpha)})\bar{\theta}_{I_1\ldots I_\alpha}(x^{(\alpha)}).$$

The completion of $\Omega^p(M^{(\alpha)})$ with respect to the inner product (8) is the Hilbert space which we denote by $H^p(M^{(\alpha)})$. The space of continuous linear functionals on the space $\Omega^p(M^{(\alpha)})$ will be denoted by $\Omega^{*p}(M^{(\alpha)})$.

Given a symmetric p-form of αth order ω and a symmetric p-form of βth order θ we can define their product $\omega\bar{\wedge}\theta$ which is the symmetric p-form of order $\alpha+\beta$. Indeed let locally

$$\omega(x^{(\alpha)}) = \omega_{\mathscr{I}_1\mathscr{I}_2\ldots\mathscr{I}_\alpha}\, dx_1^{I_1}\bar{\wedge}dx_2^{I_2}\bar{\wedge}\ldots\bar{\wedge}dx_\alpha^{I_\alpha},$$
$$\theta(x^{(\beta)}) = \theta_{\mathscr{J}_1\mathscr{J}_2\ldots\mathscr{J}_\beta}\, dx_1^{J_1}\bar{\wedge}dx_2^{J_2}\bar{\wedge}\ldots\bar{\wedge}dx_\beta^{J_\beta}.$$

Then we identify

$$(x^{(\alpha)}, x^{(\beta)}) \in M^{(\alpha)} \times M^{(\beta)} \to (x_1, x_2, \ldots, x_\alpha, x_{\alpha+1}, \ldots, x_{\alpha+\beta}) \in M^{(\alpha+\beta)}$$

and define the product of forms ω, θ by

$$\omega\bar{\wedge}\theta = \sigma_{\mathscr{I}_1\mathscr{I}_2\ldots\mathscr{I}_{\alpha+\beta}} dx_1^{I_1}\bar{\wedge}dx_2^{I_2}\bar{\wedge}\ldots\bar{\wedge}dx_{\alpha+\beta}^{I\alpha+\beta}, \tag{9}$$

where

$$\sigma_{\mathscr{I}_1\mathscr{I}_2\ldots\mathscr{I}_{\alpha+\beta}} = \sum (-1)^{\mathfrak{p}(r)} \omega_{\mathscr{I}_{r(1)}\ldots\mathscr{I}_{r(\alpha)}} \theta_{\mathscr{I}_{r(\alpha+1)}\ldots\mathscr{I}_{r(\alpha+\beta)}}, \tag{10}$$

and $\mathfrak{p}(r)$ is the parity of a permutation r. It can be shown that

$$\omega\bar{\wedge}\theta = (-1)^{\alpha\beta}\theta\bar{\wedge}\omega. \tag{11}$$

4 Generators of Grassmann Algebra on a Manifold

In this section our aim is to construct an infinite dimensional Grassmann algebra with inner product described in Sect. 2 on an oriented compact Riemannian manifold M with Riemannian metric g. We will introduce the generators of this infinite dimensional Grassmann algebra.

First of all we identify $\Psi^0 \equiv \mathbb{C}$. If $\alpha \geq 1$ then let us remind that according to the description of the structure of Grassmann algebra given in Sect. 2 the space of elements of degree α of this algebra has the structure of rigged Hilbert space, i.e. it is the triple $\tilde{\Psi}^\alpha \subset H^\alpha \subset \Psi^\alpha$, where H^α is the Hilbert space and Ψ^α is dual space to $\tilde{\Psi}^\alpha$. We begin our construction of infinite dimensional Grassmann algebra on a Riemannian manifold M by fixing a sequence of Hilbert spaces $\{H^\alpha\}_{\alpha\geq 1}$. We consider the realization of a Hilbert space H^α of elements of degree α by the elements of the Hilbert space $H^p(M^{(\alpha)})$ of p-forms of order α on the product manifold $M^{(\alpha)}$. By other words we suppose that each Hilbert space H^α of a sequence of Hilbert spaces $\{H^\alpha\}_{\alpha\geq 1}$ is isomorphic to the Hilbert space $H^p(M^{(\alpha)})$ of p-forms of order α of the product manifold $M^{(\alpha)}$. Let us denote this isomorphism by $\psi_{(\alpha)} : H^p(M^{(\alpha)}) \to H^\alpha$. If $\omega, \theta \in H^p(M^{(\alpha)})$ and $\psi_{(\alpha)}(\omega) = \xi$, $\psi_{(\alpha)}(\theta) = \eta$, where $\xi, \eta \in H^\alpha$, then

$$<\xi, \eta> = (\psi_{(\alpha)}(\omega), \psi_{(\alpha)}(\theta)) = (\omega, \theta) = \int_{M^{(\alpha)}} \omega(x^{(\alpha)}) \cdot \theta(x^{(\alpha)})\, d\mu^{(\alpha)}. \quad (12)$$

So a sequence of Hilbert spaces $\{H^\alpha\}_{\alpha \geq 1}$ is constructed. Our next step is to construct the sequence of spaces $\{\tilde{\Psi}^\alpha\}_{\alpha \geq 1}$ such that $\tilde{\Psi}^\alpha \subset H^\alpha$ and H^α is a completion of $\tilde{\Psi}^\alpha$. For this purpose we restrict an isomorphism $\psi_{(\alpha)} : H^p(M^{(\alpha)}) \to H^\alpha$ to the subspace $\Omega^p(M^{(\alpha)})$ of smooth p-forms of αth order and set $\tilde{\Psi}^\alpha = \psi_{(\alpha)}(\Omega^p(M^{(\alpha)}))$. Evidently $\tilde{\Psi}^\alpha$ satisfies the conditions 1,2 of the definition of Grassmann algebra with inner product, i.e. $\tilde{\Psi}^\alpha$ is a nuclear space equipped with an inner product (12), H^α is a completion of $\tilde{\Psi}^\alpha$ by the inner product $<,>$ and $\tilde{\Psi}^\alpha$ is dense in H^α. Hence we get the realization of space $\tilde{\Psi}^\alpha$ by the space of smooth complex valued p-forms of order α which is induced by the realization $\psi_{(\alpha)} : H^p(M^{(\alpha)}) \to H^\alpha$. Each smooth p-form of order α defines the continuous linear functional on $\Omega^p(M^{(\alpha)})$ by means of the inner product (8). Hence we can endow the space $\Omega^p(M^{(\alpha)})$ with the topology of space of continuous linear functionals and completion of $\Omega^p(M^{(\alpha)})$ by this topology is the dual space $\Omega^{*p}(M^{(\alpha)})$. Now we can extend $\psi_{(\alpha)} : \Omega^p(M^{(\alpha)}) \to \tilde{\Psi}^\alpha$ to the dual space $\Omega^{*p}(M^{(\alpha)})$ by continuity. Taking into account that Ψ^α is dual space to $\tilde{\Psi}^\alpha$ we obtain $\psi_{(\alpha)} : \Omega^{*p}(M^{(\alpha)}) \to \Psi^\alpha$. Thus the sequence of spaces $\Psi = \{\Psi^\alpha\}_{\alpha \geq 1}$, where each space Ψ^α has a structure of rigged Hilbert space, is constructed. We will refer to the space $\Omega^p(M^{(\alpha)})$ as the space of basic forms of Ψ and to the space $\Omega^{*p}(M^{(\alpha)})$ as the space of generalized forms or currents of Ψ.

Given two elements $\xi \in \Psi^\alpha$, $\eta \in \Psi^\beta$ of Ψ we define their product $\xi\,\eta$ by

$$\psi_{(\alpha+\beta)}(\xi\,\eta) = \psi_{(\alpha)}(\xi) \bar{\wedge} \psi_{(\beta)}(\eta). \quad (13)$$

It is evident that $\xi\,\eta \in \Psi^{\alpha+\beta}$, i.e. the multiplication (13) determines the mapping $\Psi^\alpha \otimes \Psi^\beta \to \Psi^{\alpha+\beta}$ which is continuous, and it follows from (11) that $\xi\,\eta = (-1)^{\alpha\beta}\eta\xi$. Consequently $\Psi = \{\Psi^\alpha\}_{\alpha \in \mathbb{N}}$ equipped with the multiplication (13) is the graded commutative algebra.

Now our aim is to define the generators of Grassmann algebra Ψ by means of the realization of Hilbert space H^1 by the Hilbert space of p-forms of a manifold M. Let $\xi \in H^1$ be an element of degree one of Grassmann algebra Ψ and $\psi(\xi) = \omega$, where $\omega \in H^{1,p}(M)$ is an p-form which locally can be written as

$$\omega = \frac{1}{p!}\omega_{i_1 i_2 \ldots i_p} dx^{i_1} \wedge \ldots \wedge dx^{i_p}.$$

In every local chart of M we introduce the set of symbols $\{\psi^{i_1 i_2 \ldots i_p}, \overset{*}{\psi}{}^{i_1 i_2 \ldots i_p}\}$ by writing an element of degree one $\xi \in H^1$ and its conjugate $\xi^* \in H^1$ in symbolic form which is used in the theory of generalized functions

$$\xi = \frac{1}{p!}\int_M \omega_{i_1 i_2 \ldots i_p}\, \psi^{i_1 i_2 \ldots i_p}\, d\mu, \quad \xi^* = \frac{1}{p!}\int_M \bar{\omega}_{i_1 i_2 \ldots i_p}\, \overset{*}{\psi}{}^{i_1 i_2 \ldots i_p}\, d\mu. \quad (14)$$

We also assume that the set of symbols $\{\psi^{i_1 i_2 \ldots i_p}, \overset{*}{\psi}{}^{i_1 i_2 \ldots i_p}\}$ is defined at each point $x \in M$ of a manifold M and under a change of local coordinates $x^{i'} = x^{i'}(x^i)$ they behave as a totally antisymmetric contravariant p-tensor, i.e.

$$\psi^{i'_1 i'_2 \ldots i'_p} = \frac{\partial x^{i'_1}}{\partial x^{i_1}} \frac{\partial x^{i'_2}}{\partial x^{i_2}} \cdots \frac{\partial x^{i'_p}}{\partial x^{i_p}} \psi^{i_1 i_2 \ldots i_p}, \quad \overset{*}{\psi}{}^{i'_1 i'_2 \ldots i'_p} = \frac{\partial x^{i'_1}}{\partial x^{i_1}} \frac{\partial x^{i'_2}}{\partial x^{i_2}} \cdots \frac{\partial x^{i'_p}}{\partial x^{i_p}} \overset{*}{\psi}{}^{i_1 i_2 \ldots i_p}.$$

We will refer to the symbols $\{\psi^{i_1 i_2 \ldots i_p}, \overset{*}{\psi}{}^{i_1 i_2 \ldots i_p}\}$ as the involutive generators of Grassmann algebra Ψ on a Riemannian manifold M.

It is useful to write the involutive generators of a Grassmann algebra Ψ in a covariant form

$$\psi_{i_1 i_2 \ldots i_p} = g_{i_1 j_1} g_{i_2 j_2} \cdots g_{i_p j_p} \psi^{j_1 j_2 \ldots j_p},$$
$$\overset{*}{\psi}{}_{i_1 i_2 \ldots i_p} = g_{i_1 j_1} g_{i_2 j_2} \cdots g_{i_p j_p} \overset{*}{\psi}{}^{j_1 j_2 \ldots j_p},$$

which can be used in order to combine the involutive generators of Grassmann algebra into the formal p-forms

$$\psi = \frac{1}{p!} \psi_{i_1 i_2 \ldots i_p} dx^{i_1} \wedge dx^{i_2} \wedge \ldots \wedge dx^{i_p},$$
$$\overset{*}{\psi} = \frac{1}{p!} \overset{*}{\psi}{}_{i_1 i_2 \ldots i_p} dx^{i_1} \wedge dx^{i_2} \wedge \ldots \wedge dx^{i_p}.$$

It is worth mentioning that $\psi, \overset{*}{\psi}$ are not ordinary differential p-forms on a manifold M, but they are formal p-forms whose coefficients are the involutive generators of infinite dimensional Grassmann algebra Ψ on a manifold M. We will refer to these formal p-forms as p-forms of involutive generators of Grassmann algebra.

We can extend the Hodge operator to the p-form of involutive generators ψ by means of the formula

$$\star\psi = \frac{1}{(n-p)!} (\star\psi)_{i_1 \ldots i_{n-p}} dx^{i_1} \wedge dx^{i_2} \wedge \ldots \wedge dx^{i_{n-p}},$$

where

$$(\star\psi)_{i_{p+1} \ldots i_n} = \frac{1}{(n-p)!} |\det(g_{ij})|^{1/2} \psi^{i_1 \ldots i_p} \varepsilon_{i_1 \ldots i_p i_{p+1} \ldots i_n},$$

and similarly in the case of the p-form $\overset{*}{\psi}$. Now the formulae (14) can be written in the form

$$\xi = \int_M \omega \wedge \star\psi, \quad \xi^* = \int_M \bar{\omega} \wedge \star\overset{*}{\psi},$$

where $\bar{\omega} = \frac{1}{p!} \bar{\omega}_{i_1 i_2 \ldots i_p} dx^{i_1} \wedge dx^{i_2} \wedge \ldots dx^{i_p}$.

5 Infinite-Dimensional Grassmann Algebra of Sections of a Vector Bundle

In this section we will construct an infinite dimensional Grassmann algebra whose elements of first order are sections of a vector bundle. As we shall see, this will generalize the construction in the previous sections.

Let $\pi_1 : E_1 \to M_1$ and $\pi_2 : E_2 \to M_2$ be two vector bundles over $\mathbb{C}$, where M_1 and M_2 are two manifolds. Let $\operatorname{rank} E_i =: n_i$ and $(e^i_1, \dots, e^i_{n_i})$ be a basis of $\mathbb{C}^{n_i}$, for $i \in \{1, 2\}$. Then we define their tensor product $E_1 \bar{\otimes} E_2$ to be the vector bundle with base manifold $M_1 \times M_2$ and fiber $\mathbb{C}^{n_1} \otimes \mathbb{C}^{n_2} = \mathbb{C}^{n_1 n_2}$ which is obtained by equipping the set

$$E_1 \bar{\otimes} E_2 := \bigsqcup_{\substack{x_1 \in M_1 \\ x_2 \in M_2}} \pi_1^{-1}(x_1) \otimes \pi_2^{-1}(x_2)$$

with projection

$$\pi : E_1 \bar{\otimes} E_2 \to M_1 \times M_2, \quad u \mapsto (x_1, x_2) \text{ such that } u \in \pi_1^{-1}(x_1) \otimes \pi_2^{-1}(x_2)$$

and local trivializations $(U_1 \times U_2, \ \phi_1 \bar{\otimes} \phi_2)$ where (U_1, ϕ_1) and (U_2, ϕ_2) are local trivializations of E_1 and E_2, correspondingly (i. e. for each $i \in \{1, 2\}$, $U_i \subset M_i$ is an open set and $\phi_i : U_i \times \mathbb{C}^{n_i} \to \pi_i^{-1}(U_i)$ is a diffeomorphism whose restrictions to fibers are isomorphisms), and

$$(\phi_1 \bar{\otimes} \phi_2)\left((x_1, x_2),\ a^{jk} e^1_j \otimes e^2_k\right) := a^{jk} \phi_1(x_1, e^1_j) \otimes \phi_2(x_2, e^2_k).$$

It is easy to see that this is a correctly defined vector bundle. We use the symbol $\bar{\otimes}$ to distinguish, in the case $M_1 = M_2$, this tensor product bundle from the usual tensor product bundle $E_1 \otimes E_2$ with base manifold M_1.

Analogously, if $\alpha \in \mathbb{N}$ and $E_1, \dots, E_\alpha$ are vector bundles respectively over manifolds $M_1, \dots, M_\alpha$, we define the tensor product $E_1 \bar{\otimes} \dots \bar{\otimes} E_\alpha$ over the product manifold $M_1 \times \dots \times M_\alpha$.

Now, let $\pi : E \to M$ be a vector bundle over a manifold M. Let us denote $E^{\bar{\otimes}(\alpha)} := \underbrace{E \bar{\otimes} \dots \bar{\otimes} E}_{\alpha}$. Let S_α denote the symmetric group on $\mathbb{N}_\alpha$, i. e. the set of all permutations $r : \mathbb{N}_\alpha \to \mathbb{N}_\alpha$. For each permutation $r \in S_\alpha$, let us define a map

$$s_r : M^{(\alpha)} \to M^{(\alpha)}, \quad (x_1, \dots, x_\alpha) \mapsto (x_{r^{-1}(1)}, \dots, x_{r^{-1}(\alpha)}).$$

Let us also define a map

$$\hat{s}_r : E^{\bar{\otimes}(\alpha)} \to E^{\bar{\otimes}(\alpha)}$$

that satisfies

$$\hat{s}_r(u_1 \otimes \ldots \otimes u_\alpha) = (-1)^{\mathfrak{p}(r)} u_{r^{-1}(1)} \otimes \ldots \otimes u_{r^{-1}(\alpha)}$$

for all points $x_1, \ldots, x_\alpha \in M$ and vectors $u_1 \in \pi_1^{-1}(x_1) \ldots, u_\alpha \in \pi_\alpha^{-1}(x_\alpha)$, and is linear on each fiber of $E^{\bar{\otimes}(\alpha)}$. It is easy to see that for each r, there exists exactly one such map $\hat{s}_r$, and in a sense, it is an automorphism of the vector bundle (it is a diffeomorphism on $E^{\bar{\otimes}(\alpha)}$ and for each point $x = (x_1, \ldots, x_\alpha) \in M^{(\alpha)}$, the restriction of $\hat{s}_r$ to the fiber of $E^{\bar{\otimes}(\alpha)}$ at x is a vector space isomorphism from the fiber at x to the fiber at $s_r(x)$). As $\hat{s}_{r_2 \circ r_1} = \hat{s}_{r_2} \circ \hat{s}_{r_1}$ and $\hat{s}_{\mathrm{id}_{\mathbb{N}_\alpha}} = \mathrm{id}_{E^{\bar{\otimes}(\alpha)}}$ where id_X denotes the identity transform on a set X, we have a faithful left action of S_α on $E^{\bar{\otimes}(\alpha)}$.

Let us define the set of *smooth totally antisymmetric sections of E of order* α as

$$\Gamma^{(\alpha)}(E) := \left\{ \sigma \in \Gamma(E^{\bar{\otimes}(\alpha)}) \,\middle|\, \sigma \circ s_r = \hat{s}_r \circ \sigma \quad \forall r \in S_\alpha \right\}.$$

From the definition we see that $\Gamma^{(0)}(E) = \mathbb{C}$ (since $M^{(0)}$ consists of only one point, $E^{\bar{\otimes}(0)} = \mathbb{C}$ and the only permutation on $\mathbb{N}_0 = \emptyset$ is the identity map) and $\Gamma^{(1)}(E) = \Gamma(E)$.

For every two sections of E, say, $\sigma \in \Gamma(E^{\bar{\otimes}(\alpha)})$ and $\tau \in \Gamma(E^{\bar{\otimes}(\beta)})$, we define their wedge product $\sigma \bar{\wedge} \tau \in \Gamma^{(\alpha+\beta)}(E)$ the same way we did for differential forms in formulae (9)–(10). First, notice that if E_1 and E_2 are two vector bundles and $\sigma_1 \in \Gamma(E_1)$ and $\sigma_2 \in \Gamma(E_2)$ then we can form a tensor product $\sigma_1 \bar{\otimes} \sigma_2 \in \Gamma(E_1 \bar{\otimes} E_2)$ by taking tensor products of the values of the two sections pointwise. Now, we identify $M^{(\alpha)} \times M^{(\beta)} = M^{(\alpha+\beta)}$ and $E^{\bar{\otimes}(\alpha)} \bar{\otimes} E^{\bar{\otimes}(\beta)} = E^{\bar{\otimes}(\alpha+\beta)}$ and define

$$(\sigma \bar{\wedge} \tau)|_{(x_1, \ldots x_{\alpha+\beta})} := \sum_{r \in S_\alpha} \hat{s}_r \left(\sigma|_{(x_{r(1)}, \ldots, x_{r(\alpha)})} \otimes \tau|_{(x_{r(\alpha+1)}, \ldots, x_{r(\alpha+\beta)})} \right). \tag{15}$$

For example, let $\alpha = 1$ and $\beta = 2$ and let $x_1, x_2, x_3 \in M$ be three points. At each point x_i choose a basis $e_1^i, \ldots, e_n^i$ of the fiber $\pi^{-1}(x_i)$. Then, if we express

$$\sigma|_{x_k} =: \sigma_k^i e_i^k, \quad \tau|_{(x_k, x_l)} =: \tau_{kl}^{ij} e_i^k \otimes e_j^l,$$

(where summation is only over i and j, not k and l), the formula (15) yields

$$(\sigma \bar{\wedge} \tau)|_{(x_1, x_2, x_3)} = (\sigma_1^i \tau_{23}^{jk} - \sigma_1^i \tau_{32}^{kj} + \sigma_2^j \tau_{31}^{ki} - \sigma_2^j \tau_{13}^{ik} + \sigma_3^k \tau_{12}^{ij} - \sigma_3^k \tau_{21}^{ji}) e_i^1 \otimes e_j^2 \otimes e_k^3.$$

As a special case, we can apply this approach to p-forms on a manifold M by taking $E = \bigwedge^p T^*M$. Then the set of smooth complex-valued double p-forms can be identified with $\Gamma(E^{\bar{\otimes}(2)})$: if $x_0, y_0 \in M$ are two points and we have local charts near each of these points with local coordinates $x^1, \ldots, x^n$ in the neighborhood of x_0 and local coordinates $y^1, \ldots, y^n$ near y_0, then, for multi-indices I and J such that $|I| = |J| = p$, we may identify $dx^I \bar{\wedge} dy^J$ in formula (5) with the vector $dx^I \otimes dy^J$ belonging to the fiber of $E^{\bar{\otimes}(2)}$ at (x_0, y_0). Under this identification, if $\omega, \theta \in \Omega^p(M)$ are two forms of order one, then the notions of product $\omega \bar{\wedge} \theta$, given for p-forms by (9) and for sections by (15), coincide. Let us identify every element $u \in E^{\bar{\otimes}(2)}$ with $\hat{s}_{21}(u)$, where 21 denotes the non-identity element of S_2, i. e. let us consider the quotient space of $E^{\bar{\otimes}(2)}$ under this identification. Then the vector $dx^I \otimes dy^J$, belonging to the fiber of $E^{\bar{\otimes}(2)}$ at point (x_0, y_0), is identified with $-dy^J \otimes dx^I$, belonging to the fiber of $E^{\bar{\otimes}(2)}$ at (y_0, x_0); in this sense $\otimes : E \times E \to E^{\bar{\otimes}(2)}$ is anticommutative and we can see that smooth complex-valued symmetric p-forms of order 2 correspond to exactly those sections $\sigma \in \Gamma(E^{\bar{\otimes}(\alpha)})$ that satisfy $\sigma \circ s_{21} = \hat{s}_{21} \circ \sigma$, i.e. $\Omega^p(M^{(2)}) = \Gamma^{(2)}(E)$.

More generally, we can identify the set of smooth complex-valued p-forms of order α with $\Gamma(E^{\bar{\otimes}(\alpha)})$ and by identifying $u \equiv \hat{s}_r(u)$ for all $u \in E^{\bar{\otimes}(\alpha)}$ and $r \in S_\alpha$ (i.e. instead of $E^{\bar{\otimes}(\alpha)}$ considering the orbit space of $E^{\bar{\otimes}(\alpha)}$ under the action $\hat{s}_r$ of S_α on $E^{\bar{\otimes}(\alpha)}$) we see that $\Gamma^{(\alpha)}(E) = \Omega^p(M^{(\alpha)})$ and the notion $\bar{\wedge}$, as defined by (15), coincides with the product of two p-forms of order α and β defined by formula (9).

Next, suppose furthermore, that M is a compact Riemannian manifold and $\pi : E \to M$ is a Hermitian vector bundle. Let $M^{(\alpha)}$ be equipped with the product metric, as before, and let each fiber of $E^{\bar{\otimes}(\alpha)}$ be equipped with the tensor product Hermitian metric, i.e. the Hermitian metric on the fiber of $E^{\bar{\otimes}(\alpha)}$ at each point $(x_1, \ldots, x_\alpha)$ satisfies the formula

$$\langle u_1 \otimes \ldots \otimes u_\alpha,\ v_1 \otimes \ldots \otimes v_\alpha \rangle = \langle u_1, v_1 \rangle \cdot \ldots \cdot \langle u_\alpha, v_\alpha \rangle$$

for all choices of vectors $u_i, v_i \in \pi^{-1}(x_i)$. Then we define an inner product of two sections $\sigma, \tau \in \Gamma^{(\alpha)}(E)$ using the formula

$$\langle \sigma,\ \tau \rangle = \int_{M^{(\alpha)}} \langle \sigma(x^{(\alpha)}),\ \tau(x^{(\alpha)}) \rangle \, d\mu^{(\alpha)}. \tag{16}$$

Analogously to what was done in the preceding section, an infinite-dimensional Grassmann algebra can now be constructed by realizing each $\tilde{\Psi}^\alpha$ as $\Gamma^{(\alpha)}(E)$, each Hilbert space H^α as the completion of $\Gamma^{(\alpha)}(E)$ with respect to the inner product above and each Ψ^α as the space of continuous linear functionals $\Gamma^{(\alpha)}(E)$, equipped with $*$-weak topology. The completion of $\Gamma^{(\alpha)}(E)$ with respect to the inner product is actually the space of totally antisymmetric square-integrable sections of $E^{\bar{\otimes}(\alpha)}$: indeed, using the similar proposition for functions on Euclidean spaces, and a partition of unity on $M^{(\alpha)}$, one can show that $\Gamma(E^{\bar{\otimes}(\alpha)})$ is dense in the space of all square-integrable sections of $E^{\bar{\otimes}(\alpha)}$, and hence also $\Gamma^{(\alpha)}(E)$ is dense in the space of totally antisymmetric square-integrable sections of $E^{\bar{\otimes}(\alpha)}$, since if $(\sigma_n) \in \Gamma(E^{\bar{\otimes}(\alpha)})$ is a sequence in $\Gamma(E^{\bar{\otimes}(\alpha)})$ that converges to a totally antisymmetric square-integrable section σ then $(\frac{1}{\alpha!} \sum_{r \in S_\alpha} \hat{s}_{r^{-1}} \circ \sigma_n \circ s_r)$ is a sequence in $\Gamma^{(\alpha)}(E)$ which converges to σ, too. Again we see that the Grassmann algebra of p-forms appears as a special case of the Grassmann algebra of sections of a vector bundle.

Acknowledgments The authors gratefully acknowledge the financial support of the Estonian Science Foundation under the research grant ETF9328, target finance grant SF0180039s08 and Estonian Doctoral School in Mathematics and Statistics.

References

1. Thierry-Mieg, J.: Geometrical reinterpretation of Faddeev-Popov ghost particles and BRS transformations. J. Math. Phys. **21**(12), 2834–2838 (1980)
2. Thierry-Mieg, J.: Explicit classical construction of the Faddeev-Popov ghost field. Nuovo Cimento A(11), **56**(4), 396–404 (1980)
3. Abramov, V., Lumiste, Ü.: Superspace with underlying Banach bundle of connections and supersymmetry of effective action (Russian). Izv. Vyssh. Uchebn. Zaved., Mat. **1**, 3–12 (1986)
4. Berezin, F.A.: The Method of Second Quantization. Academic Press, New York (1966)
5. Faddeev, D.D., Slavnov, A.A.: Gauge Fields: Introduction to Quantum Theory. Benjamin/Cummings Publishing Company, New York (1980)
6. De Rham, G.: Differentable Manifolds. Springer, Berlin (1984)

On Sinyukov's Equations in Their Relation to a Curvature Operator of Second Kind

Irena Hinterleitner, Josef Mikeš and Elena Stepanova

Abstract Many authors have studied Riemannian manifolds admitting a geodesic mapping. Fundamental results of the theory of geodesic mapping were settled by Sinyukov. In the present paper we analyze the Sinykov equations of the geodesic mappings of Riemannian manifolds by using the curvature operator of the second kind. This approach to the study of geodesic mapping is essentially new.

1 Introduction

In a Riemannian manifold, the Riemannian curvature tensor R defines two kinds of curvature operators: the operator $\mathring{R}$ of first kind, acting on 2-forms, and the operator $\mathring{R}$ of second kind, acting on symmetric 2-tensors. In our paper we analyze the Sinyukov equations of geodesic mappings of Riemannian manifolds by using the curvature operator of the second kind.

I. Hinterleitner (✉)
Brno University of Technology, Brno, Czech Republic
e-mail: hinterleitner.irena@seznam.cz

J. Mikeš · E. Stepanova
Palacky University Olomouc, Olomouc, Czech Republic
e-mail: josef.mikes@upol.cz

E. Stepanova
e-mail: stephelena@list.ru

A. Makhlouf et al. (eds.), *Algebra, Geometry and Mathematical Physics*, 489
Springer Proceedings in Mathematics & Statistics 85, DOI: 10.1007/978-3-642-55361-5_28,

2 Equation Systems of Geodesic Mappings and Einstein Manifolds

The condition, for which an n-dimensional ($n \geq 2$) Riemannian manifold (M, g) admits a geodesic mapping onto another n-dimensional Riemannian manifold $(\bar{M}, \bar{g})$, has the following form of differential equations of Cauchy type in covariant derivatives

$$\nabla_k a_{ij} = \lambda_i g_{jk} + \lambda_j g_{ik}, \tag{1}$$

$$n \nabla_j \lambda_i = \mu g_{ij} - a_{ik} R^k_j + a^{kl} R_{ikjl}, \tag{2}$$

$$(n-1)\nabla_i \mu = -2(n+1)\lambda_k R^k_i + a_{kl}\left(\nabla_i R^{kl} - 2g^{kj}\nabla_j R^l_i\right). \tag{3}$$

These equations were obtained by Sinyukov more than fifty years ago (see [1–4]).

A geodesic mapping is *non trivial* (or *non affine*) if $\lambda \not\equiv$ const.

Here $a = (a_{ij})$ is a regular symmetric 2-tensor, $Ric = (R_{ij})$ is the Ricci tensor whose components are given by $R_{jl} = g^{ik} R_{ijkl}$ for local components of the Riemannian curvature tensor $R = (R_{ijkl})$; λ_i and μ are defined in the following way

$$\lambda_i = \frac{1}{2}\nabla_i (g^{kl} a_{kl}); \tag{4}$$

$$\mu = g^{kl} \nabla_k \lambda_l. \tag{5}$$

With respect to the above Eqs. (4) and (5) Eq. (2) can be rewritten as

$$n\, \nabla_i \nabla_j \lambda = \Delta\lambda \cdot g_{ij} - a_{ik} R^k_j + a^{kl} R_{ikjl}, \tag{6}$$

where

$$\Delta\lambda = g^{ij} \nabla_i \nabla_j \lambda$$

is the Laplace operator acting on the scalar function $\lambda = \frac{1}{2}\, g^{kl} a_{kl}$.

From (6) follows (see [4, p. 138]):

$$a_{ik} R^k_j = R^k_i a_{kj}. \tag{7}$$

If we suppose that the manifold (M, g) is an Einstein manifold, then from Eq. (3) we conclude the following

$$(n-1)\nabla_i \Delta\lambda = -2\,\frac{n+1}{n}\, S\, \nabla_i \lambda \tag{8}$$

for the scalar curvature S (= const). After multiplying the left and right sides of (8) by $\nabla^i \lambda$ and after integration over the compact manifold we have

$$(n-1)\int_M (\Delta\lambda)^2 dv = 2\,\frac{n+1}{n}\,S\int_M \left(\nabla_i\lambda\cdot\nabla^i\lambda\right)dv, \tag{9}$$

because $\nabla_i\left(\Delta\lambda\cdot\nabla^i\lambda\right)=\nabla_i\Delta\lambda\cdot\nabla^i\lambda+(\Delta\lambda)^2$. From (9) we conclude that $S>0$. This is in accordance with the paper Couty [5].

3 An Algebraic Operator Associated with the Curvature Tensor

We will consider the space of symmetric 2-forms S^2M over the Riemannian manifold (M,g). In particular, the tensor $a=(a_{ij})$ is a smooth cross-section of S^2M. The space S^2M (see [6]) has the pointwise orthogonal decomposition

$$S^2M = C^\infty M\cdot g\oplus S_0^2M,$$

where $C^\infty M$ is a space C^∞-functions on M and S_0^2M is a subspace of the space S^2M, which contains symmetric 2-forms with zero traces.

We introduce (see [7, 8]) a curvature operator of second kind $\mathring{R}\colon S^2M\to S^2M$ with components

$$R^{ij}{}_{kl}=\frac{1}{2}\left(g^{im}R^j_{kml}+g^{jm}R^i_{kml}\right)$$

for the curvature tensor $R=(R^i_{jkl})$, whose actions are defined by the formulas $\mathring{R}(b_{ij})=R_{ikjl}b^{kl}$ for any smooth cross-section $b=(b_{ij})$ of S^2M. On the basis of the curvature operator of second kind (see [9]) we can define a linear symmetric operator $B_2\colon S^2M\to S_0^2M$ with components

$$B^{ij}_{kl}=\frac{1}{2}\left(g^{im}R^j_{kml}+g^{jm}R^i_{kml}\right)+\frac{1}{4}\left(\delta^i_kR^j_l+\delta^j_kR^i_l+\delta^i_lR^j_k+\delta^j_lR^i_k\right) \tag{10}$$

for the Ricci operator $Ric^*=(g^{im}R_{mj})$. From (10) we have

$$B_2(b_{ij})=R_{ikjl}b^{kl}-\frac{1}{2}\left(R^m_ib_{mj}+R^m_jb_{mi}\right) \tag{11}$$

for any smooth section $b=(b_{ij})$ on S^2M. The operator $B_2\colon S^2M\to S_0^2M$ is a linear and symmetric operator. Then there exists a pointwise orthogonal decomposition $S^2M=ImB_2\oplus KerB_2$ of the space S^2M of symmetric 2-tensors on M. It is obvious that $C^\infty M\cdot g\subset KerB_2$ and $ImB_2\subset S_0^2M$. The following theorem holds.

Theorem 3.1 *If a complete Riemannian manifold (M,g) of dimension $n\geq 2$ admits geodesic mapping is a non affine and the tensor $a=(a_{ij})$ belongs to the kernel of the symmetric linear operator $B_2\colon S^2M\to S_0^2M$ then (M,g) is conformal to a sphere S^n in $(n+1)$ dimensional Euclidean space.*

Proof According to (11) we can rewrite Eq. (6) in the following form

$$\nabla_i \nabla_j \lambda - \frac{1}{n} \Delta\lambda g_{ij} = -B_2(a_{ij}). \tag{12}$$

If we assume that $B_2(a_{ij}) = 0$ in (12), then we have the following

$$\nabla_i \nabla_j \lambda = \frac{1}{n} \Delta\lambda g_{ij}. \tag{13}$$

We note that (see [10]) the complete Riemannian manifold (M, g) of dimension $n \geq 2$ is conformal to a sphere S^n in $(n + 1)$-dimensional Euclidean space if on (M, g) exists a non-constant function $\lambda \in C^\infty M$ satisfying Eq. (13). Therefore a complete Riemannian manifold (M, g) of dimension $n \geq 2$ admitting geodesic mappings onto another n-dimensional Riemannian manifold $(\bar{M}, \bar{g})$ is conformal to a sphere S^n in $(n + 1)$-dimensional Euclidean space if the tensor $a = (a_{ij})$ from the equations of geodesic mappings (1–3) belongs to the kernel $Ker B_2$ of the linear operator B_2: $S^2M \to S_0^2M$.

As a corollary to the Theorem 3.1, we can deduce the following theorem.

Theorem 3.2 *If a compact Riemannian manifold (M, g) of dimension $n \geq 2$ admits geodesic mapping is a non affine and the tensor $a = (a_{ij})$ belongs to the kernel of the symmetric linear operator B_2: $S^2M \to S_o^2M$ then (M, g) is isometric to a sphere S^n in $(n + 1)$-dimensional Euclidean space.*

Proof It is know (see [11]) that if a compact Riemannian manifold (M, g) of n-dimension of $n \geq 2$ admits an infinitesimal conformal transformation which is not an isometry:

$$L_X g_{ij} = 2\rho g_{ij} \tag{14}$$

for $\rho \neq 0$, and if the vector field X is a gradient of a scalar function then (M, g) is isometric to a Euclidean n-sphere (see [11]). Here L_X is the operator of the Lie derivation with respect to X. If we assume that $X = \text{grad}\,\lambda$ then (14) we can rewritten as $\nabla_i \nabla_j \lambda = \frac{1}{n}\ \Delta\lambda\ g_{ij}$. Then our Theorem 3.2 follows from Theorem 3.1 and the above result by Lichnerowicz.

Remark The Lichnerowicz Laplacian (see [6], p. 54) acting on symmetric covariant 2-tensor is $\Delta_L = \bar{\Delta} - 2B_2$ where we denote by $\bar{\Delta}$ the rough Bochner Laplacian (see [6], p. 52). Then $B_2(b) = 0$ if and only if $\Delta_L b = \bar{\Delta} b$ for a symmetric covariant 2-tensor b.

4 Principle Directions of the Ricci Tensor in the Case of Degenerate Geodesic Mappings

From Eq. (7) we conclude that the Ricci tensor $Ric = (R_{ij})$ can be diagonalised in any point $x \in M$ in the same orthonormal basis $\{e_1, \dots, e_n\}$ as the symmetric non-degenerate tensor $a = (a_{ij})$. Therefore in any point $x \in M$ vectors of the orthonormal basis $\{e_1, \dots, e_n\}$ of the tangent space $T_x M$ define principle directions not only of the tensor $a = (a_{ij})$, but also principle directions of the Ricci tensor (see [12, § 34]). In this case the basis $\{e_1, \dots, e_n\}$ gets an invariant meaning for the manifold (M, g), independent of the tensor $a = (a_{ij})$.

The following theorem holds.

Theorem 4.1 *Let the Riemannian manifold (M, g) of dimension $n \geq 2$ admit a geodesic mapping and the tensor $a = (a_{ij})$ belong to the kernel of the symmetric linear operator $B_2 : S^2M \to S_0^2M$. If in each point $x \in M$ the sectional curvature $K(e_i, e_j) > 0$ (or $K(e_i, e_j) < 0$) to the direction $e_i \wedge e_j$ for the ortonormal basis $\{e_1, \dots, e_n\}$ of the vectors of principle directions of the Ricci tensor then the geodesic mapping is an affine mapping.*

Proof The quadratic form $\Phi_2(b_{ij}) = g(B_2(b_{ij}), b_{ij})$ can be written in the following from (see [6, § 16.9]).

$$\Phi_2(b_{ij}) = -\sum_{i<j} K(e_i, e_j)(b_i - b_j)^2,$$

where $K(e_i, e_j)$ is the sectional curvature to the direction $e_i \wedge e_j$ for any vectors of the orthonormal basis $\{e_1, \dots, e_n\}$ of eigenvectors of the tensor $b = (b_{ij})$ of the space $T_x M$ in any point $x \in M$ i.e. $b(e_i, e_j) = b_i \delta_{ij}$, where δ_{ij} is the Kroneker symbol. Then from the condition $B_2(a_{ij}) = 0$ follows $\Phi_2(a_{ij}) = 0$ and under the condition $K(e_i, e_j) > 0$ (or $K(e_i, e_j) < 0$) from the equality $\Phi_2(g_{ij}) = 0$ follows that $a(e_i, e_j) = a\delta_{ij}$ for the corresponding ortonormal basis $\{e_1, \dots, e_n\}$ of $T_x M$ in any point $x \in M$. This means that the geodesic mappings is an affine mapping (see [3, p. 93]).

Acknowledgments The paper was supported by the project FAST-S-11-47 of the Brno University of Technology, by grant P201/11/0356 of The Czech Science Foundation, and by the project POST-UP CZ 1.07/2.3.00/30.0004.

References

1. Hinterleitner, I., Mikeš, J.: Geodesic Mappings and Einstein Spaces. In: Kielanowski, p. et al. (eds.) Geometric Methods in Physics, pp. 331–336. Birkhäuser, Basel (2013)
2. Mikeš, J.: Geodesic mappings of affine-connected and Riemannian spaces. J. Math. Sci. **78**, 311–333 (1996)

3. Mikeš, J., Vanžurová, A., Hinterleitner, I.: Geodesic mappings and some generalizations. Palacky University Press, Olomouc (2009)
4. Sinyukov, N.S.: Geodesic Mappings of Riemannian Spaces. Nauka, Moscow (1979)
5. Couty, R.: Transformations infinitésimales projectives. C. R. Acad. Sci. **247**, 804–806 (1958)
6. Besse, A.L.: Einstein Manifolds. Springer, Berlin (1987)
7. Ogiue, K., Tachibana, S.: Les variétés riemanniennes dont l'opérateur de courbure restreint est positif sont des sphères d'homologie réelle. C. R. Acad. Sci. Paris **289**, A29–A30 (1979)
8. Sampson, J.H.: On a theorem of chern. Trans. Amer. Math. Soc. **177**, 141–153 (1973)
9. Stepanov, S.E.: Fields of symmetric tensors on a compact Riemannian manifold. Math. Notes **52**, 1048–1050 (1992) (Translation from Mat. Zametki **52**, 85–88 (1992))
10. Tashiro, Y.: Complete riemannian manifolds and some vector fields. Trans. Amer. Math. Soc. **117**, 251–275 (1965)
11. Lichnerowicz, A.: Sur les transformations conformes d'une variété riemannienne compacte. C.R. Acad. Sci. Paris **259**, 697–700 (1964)
12. Eisenhart, L.P.: Riemannian Geometry. Princeton University Press, Princeton (1926)

On Dimensions of Vector Spaces of Conformal Killing Forms

Sergey E. Stepanov, Marek Jukl and Josef Mikeš

Abstract In this article there are found precise upper bounds of dimension of vector spaces of conformal Killing forms, closed and coclosed conformal Killing r-forms $(1 \leq r \leq n-1)$ on an n-dimensional manifold. It is proved that, in the case of n-dimensional closed Riemannian manifold, the vector space of conformal Killing r-forms is an orthogonal sum of the subspace of Killing forms and of the subspace of exact conformal Killing r-forms. This is a generalization of related local result of Tachibana and Kashiwada on pointwise decomposition of conformal Killing r-forms on a Riemannian manifold with constant curvature. It is shown that the following well known proposition may be derived as a consequence of our result: any closed Riemannian manifold having zero Betti number and admitting group of conformal mappings, which is non equal to the group of motions, is conformal equivalent to a hypersphere of Euclidean space.

1 Introduction

1.1 The history of conformal Killing forms has started almost a half of century ago by works of Tachibana and Kashiwada [7, 26]. During this long time, this topic has given rise to an active interest (see for example [8, 16, 17, 22]) because of great number of its applications (see e.g. [4, 10, 20]). This paper is devoted to the study of dimensions of vector spaces of Killing forms (see [17]).

S. E. Stepanov (✉)
Finance University Under the Government of the Russian Federation,
49-55, Leningradsky Prospect, Moscow, Russian Federation 125468
e-mail: s.e.stepanov@mail.ru

M. Jukl · J. Mikeš
Palacky University, 17. Listopadu 12, 77146 Olomouc, Czech Republic
e-mail: marek.jukl@upol.cz

J. Mikeš
e-mail: josef.mikes@upol.cz

A. Makhlouf et al. (eds.), *Algebra, Geometry and Mathematical Physics*, 495
Springer Proceedings in Mathematics & Statistics 85, DOI: 10.1007/978-3-642-55361-5_29,

1.2 In the second section, we investigate *conformal Killing differential r-forms* $(1 \leq r \leq n-1)$ on local coordinates of an arbitrary neighbourhood U of n-dimensional Riemannian manifold (M, g). We consider a vector space T^r of these forms and deal with two subspaces—the subspace P^r of *planar r-forms* (i.e. *closed conformal Killing forms)* and subspace K^r of *Killing r-forms (coclosed conformal Killing r-form)*. For $r = 1$ we obtain the following three vector spaces of 1-forms: dual to conformal Killing vector field, concircular vector field and Killing vector field. In this section there are found dimensions t_r, k_r and p_r of these free "local" vector spaces on the manifold (M, g) with constant curvature.

1.3 In the third section there are studied "complete" conformal Killing r-forms $(1 \leq r \leq n-1)$ on n-dimensional closed Riemannian manifold (M, g), vector space $T^r(M, \mathbb{R})$ of these forms and two subspaces of it $K^r(M, \mathbb{R})$ of Killing forms and $P^r(M, \mathbb{R})$ of planar r-forms; dimensions of these spaces are denoted by $t_r(M)$, $k_r(M)$ and $p_r(M)$, respectively. In the case of closed manifold (M, g) with zero Betti numbers $b_r(M) = 0$, we show the orthogonal decomposition of T^r in the form $T^r(M, \mathbb{R}) = K^r(M, \mathbb{R}) + P^r(M, \mathbb{R})$, which implies the relation $t_r(M) = k_r(M) + p_r(M)$. In the case $b_1(M) = 0$ and $t_1(M) \neq k_1(M)$, we will prove that (M, g) is globally conformal to the n-dimensional sphere S^n of the Euclidean space R^{n+1}.

2 Definitions and Notations

2.1 Let us consider n-dimensional Riemannian manifold (M, g) with Levi-Civita connection. Denote by $C^\infty(M)$ a vector space of C^∞-function on M and by $\Omega^r(M)$ a vector space of differentiable r-forms on M. Taking local orientation of M we introduce the Hodge operator $*$ defining an isomorphism $* : \Omega^r \to \Omega^{n-r}$ such that $g(\omega, *\Theta) = (-1)^{r(n-r)} g(*\omega, \Theta)$, for any $\omega \in \Omega^r(M)$, $\Theta \in \Omega^{n-r}(M)$, and $*^2 = (-1)^{r(n-r)} Id_{\Omega^r(M)}$ (see [2, definition 1.51], [13, p. 203]).

For the exterior differential operator $\mathrm{d} : C^\infty \Lambda^r(M) \to C^\infty \Lambda^{r+1}(M)$ there exists a formal adjoint operator $\delta : \Omega^{r+1}(M) \to \Omega^r(M)$ which is called *codifferential operator* (see [2, definition 1.56], [13, p. 203, 204], [15, § 25]) and it is defined by

$$\delta = (-1)^{(n-r)(r+1)} * \mathrm{d} * \tag{1}$$

2.2 Let us remind well known definitions of three types of Killing vector fields in Riemannian geometry.

A vector field Z on a Riemannian manifold (M, g) is called *infinitesimal conformal transformations* or *conformal Killing vector field* (see [5, § 69], [11, p. 120]) if $L_Z g = 2\sigma g$ for Lie derivative L_Z with respect to the vector field Z and some $\sigma \in C^\infty M$.

Defining for a vector field Z a dual 1-form ω by the relation $\omega = g(Z, \cdot)$ we introduce a denotation $\omega^\# = Z$ (see [2, denotation 1.38]). Now, the identity $L_Z g =$

$2\sigma \cdot g$, by which an infinitesimal conformal transformation is defined, may be written by

$$(L_Z g)(X, Y) \stackrel{\text{def}}{=} (\nabla_X)Y + (\nabla_Y \omega)X = -\frac{2}{n}(\delta\omega)g(X, Y) \quad (2)$$

or, equivalently, by

$$\nabla\omega = -\frac{1}{2}\mathrm{d}\omega - \frac{1}{n}g \cdot \delta\omega = 0. \quad (3)$$

Any vector field Z with $L_Z g = 0$ on a Riemannian manifold (M, g) is called *infinitesimal isometry* or Killing vector field.

Clearly, if $\sigma = 0$ then every infinitesimal conformal transformation Z is an infinitesimal isometry. Because $\sigma = n^{-1}(-\delta\omega)$, for 1-form ω dual to the vector field $Z = \omega^{\#}$, we see that any infinitesimal isometry may be considered as coclosed infinitesimal conformal transformation.

Let us remind that a vector field Z is called *concircular* (see [30]) if $\nabla Z = \rho\, Id_M$ for $\rho \in C^{\infty}M$. In this case, for every 1-form ω with $\omega^{\#} = Z$ we have $\nabla\omega = n^{-1}(-\delta\omega)g$. Therefore, a concircular vector field may be defined as closed infinitesimal conformal transformation.

2.3 Let us deal with a generalization of three types of Killing vector field defined above.

Let an n-dimensional Riemannian manifold (M, g) be given. A form $\omega \in \Omega^r(M)$ is called *conformal Killing r-form* (see [7]) if there exists a form $\Theta \in \Omega^{r-1}(M)$ with

$$(\nabla_Y\omega)(X, X_2, \ldots, X_r) + (\nabla_X\omega)(Y, X_2, \ldots, X_r) = 2g(X, Y)\Theta(X_2, \ldots, X_r)$$
$$- \sum_{a=2}^{r}(-1)^a \Big(g(Y, X_a)\Theta(X, X_2, \ldots, \hat{X}_a, \ldots, X_r + g(X, X_a)\Theta(Y, Y_2, \ldots, \hat{X}_a, \ldots, X_r)\Big) \quad (4)$$

for any vector fields $Y, X, X_2, \ldots, X_r \in C^{\infty}(TM)$, where $\hat{X}_a$ means that X_a is omitted. The form $\Theta \in \Omega^{r-1}(M)$ is called an *associated form* of the conformal Killing form $\omega \in \Omega^r(M)$. Moreover, the identity (see [7])

$$\delta\omega = -(n - r + 1)\Theta \quad (5)$$

holds as the corollary of (3).

Equation (4) are a natural generalization of Eq. (2). Equation (4) may be written in the form (see [20, 22])

$$\nabla\omega = (r + 1)^{-1}\mathrm{d}\omega + (n - r + 1)^{-1}g \wedge \delta\omega. \quad (6)$$

These relations constitute necessary and sufficient conditions that an r-form ω is a conformal Killing form $(1 \le r \le n - 1)$.

The set of conformal Killing forms on a Riemannian manifold (T, g) forms a vector space T^r (with real coefficients) (see [19]).

A form $\omega \in \Omega^r(M)$ on an n-dimensional Riemannian manifold (M, g) is called a *Killing r-form* if it is a *coclosed conformal Killing r-form*. Such form $\omega \in \Omega^r(M)$ fulfils its definition equations (see [14] and [10], Definition 31.3.1)

$$(\nabla_Y \omega)(X, X_2, \dots, X_r) + (\nabla_X \omega)(Y, X_2, \dots, X_r) = 0 \tag{7}$$

or equivalent equations

$$\nabla \omega = (r+1)^{-1} \mathrm{d}\omega. \tag{8}$$

This form ω is a generalization of an 1-form $\omega \in \Omega^{-1}(M)$, which is dual to the Killing vector field $Z = \omega^{\#}$. The set of all Killing r-forms constitutes a vector space $K^r \subset T^r$ (see [19]).

A form $\omega \in C^\infty \Lambda^r(M)$ on an n-dimensional Riemannian manifold (M, g) is called a *planar r-form* if it is a *closed conformal Killing r-form* (see [19]). Such form $\omega \in C^\infty \Lambda^r(M)$ fulfils its definition equations

$$\nabla \omega = (n - r + 1)^{-1} g \wedge \delta\omega, \tag{9}$$

for $1 \le r \le n-1$.

This form ω is a generalization of an 1-form $\omega \in \Omega^1(M)$, which is dual to the concircular vector field $Z = \omega^{\#}$. The set of all such r-forms constitutes a vector space $P^r \subset T^r$ (see [19]).

3 Dimensions of Vector Spaces of Killing Forms and Vector Fields on Non-compact Riemannian Manifold

3.1 Let (M, g) be an n-dimensional connected Riemannian manifold. Let us remind some well known facts on dimensions of three types spaces of Killing vector fields on (M, g). It is known that the dimension of a Lie algebra of a group $C(M, g)$ of infinitesimal conformal transformations of connected Riemannian n-dimensional manifold (M, g) is not greater than $\frac{1}{2}(n+1)(n+2)$ and this algebra is a vector space of conformal Killing vector fields (see [11, p. 120]). The equality is obtained in the case of conformally flat Riemannian manifold (M, g).

The dimension of a Lie algebra of a subgroup $I(M, g)$ of infinitesimal transformations is not greater than $\frac{1}{2}(n+1)n$ and this algebra is a vector space of Killing vector fields (see [11, p. 101]). The equality is obtained in the case of Riemannian manifold (M, g) with constant curvature.

The dimension of a vector space of concircular vector fields on connected n-dimensional manifold (M, g) is not greater than $n+1$ and the equality is obtained for manifold with constant curvature (see [6]).

3.2 To generalize facts presented above we will find precise upper bounds of dimensions of vector spaces of three types of Killing r-forms ($1 \le r \le n-1$). Let us investigate a connected n-dimensional Riemannian manifold (M, g). To form $\omega \in C^{\infty}\Lambda^r(M)$ the condition of integrability of arbitrary equation of a systems of Eqs. (5), (7) or (8) is the *Ricci identity* ([5, § 11]). This identity has in any local coordinate system $x^1, \ldots, x^n$ of a manifold (M, g) the following expression:

$$\nabla_j \nabla_k \omega_{i_1 i_2 \ldots i_r} - \nabla_k \nabla_j \omega_{i_1 i_2 \ldots i_r} = -\sum_{\alpha=1}^{r} \omega_{i_1 i_2 \ldots i_{\alpha-1} l i_{\alpha+1} \ldots i_r} R^l_{i_\alpha jk}, \tag{10}$$

where $\omega_{i_1, i_2 \ldots i_p} = \omega(X_{i_1}, X_{i_2}, \ldots, X_{i_p})$ and $R^i_{jkl} X_i = R(X_k, X_l) X_j$ are local components of a conformal Killing r-form and curvature tensor R for $X_k = \frac{\partial}{\partial x^k}$ and $\nabla_k = \nabla_{X_k}$.

Ricci identity (10) establishes restrictions not only the choice of components of r-form ω but also on the curvature tensor R of the manifold (M, g). We have the following theorem.

Theorem 3.1 *On an n-dimensional connected Riemannian manifold* (M, g), *the dimensions* t_r, k_r *and* p_r *of vector spaces of conformal Killing r-form* T^r, *co-closed confomal Killing (Killing) r-form* K^r *and closed conformal Killing r-form* P^r, $(1 \le r \le n-1)$, *respectively have the following upper bounds*

$$t_r \le \frac{(n+2)!}{(r+1)!(n-r+1)!}, \quad k_r \le \frac{(n+1)!}{(r+1)!(n-r)!}, \quad p_r \le \frac{(n+1)!}{r!(n-r+1)!}.$$

The equalities are obtained in the case of Riemannian manifold (M, g) *with constant nonzero curvature.*

Proof The case when (M, g) is locally flat manifold is trivial; in [19, 20] on the basis of (10) there are defined components $\omega_{i_1 \ldots i_r} = A_{ki_1 \ldots i_r} x^k + B_{i_1 \ldots i_r}$ of Killing r-form ω, for an arbitrary local Cartesian coordinate system $x^1, \ldots, x^n$. Here, $A_{ki_1 \ldots i_r}$, $B_{i_1 \ldots i_r}$ are local components of constant skew-symmetric $(r+1)$-forms and r-forms, respectively.

With respect to this result, in [18] there for some special coordinate system $x^1, \ldots, x^n$ of manifold (M, g) with constant sectional curvature $C \neq 0$ were found components $\omega_{i_1 \ldots i_r} = \mathrm{e}^{(r+1)\varphi}(A_{ki_1 \ldots i_r} x^k + B_{i_1 \ldots i_r})$, $\varphi = \frac{1}{2(n+1)} \ln(\det g)$, of Killing r-form ω. Therefore the dimension k_r of a space K^r of coclosed conformal Killing r-forms on Riemannian manifold with constant curvature is equal to

$$k_r = \binom{n}{r+1} + \binom{n}{r} = \binom{n+1}{r+1} = \frac{(n+1)!}{(r+1)!(n-r)!}.$$

In the case of an arbitrary connected manifold (M, g), it is evident that the dimension k_r of a space K^r is not greater then this number, i.e. $k_r \le \frac{(n+1)!}{(r+1)!(n-r)!}$.

It is known (see [8, 19, 20]) that there exists an isomorphism $* : K^{n-r} \to P^r$. It gives the possibility to count the dimension p_r of a space P^r of closed conformal Killing r-forms on a manifold (M, g) with nonzero constant sectional curvature

$C \neq 0$; this dimension is equal to $p_r = \frac{(n+1)!}{r!(n-r+1)!}$. For arbitrary connected manifold (M, g), it is evident that the dimension p_r of a space P^r is not greater then this number, i.e. $p_r \leq \frac{(n+1)!}{r!(n-r+1)!}$.

For manifold (M, g) with constant sectional curvature $C \neq 0$, in [7, 26] there by direct calculation was obtained decomposition of any conformal Killing r-form ω into direct sum $\omega = \omega_1 + \omega_2$ of a coclosed conformal Killing (Killing) r-form ω_1 and of closed conformal Killing (planar) r-form $\omega_2 = \nabla\Theta$ for Killing $(r-1)$-form Θ.

Clearly, the arbitrariness of choice of an r-form $\omega_2 = \nabla\Theta$ is given by the number of parameters on which a Killing $(r-1)$-form depends. This number is equal to $\binom{n}{r} + \binom{n}{n-1} = \binom{n+1}{r}$. In the case of an arbitrary connected manifold (M, g), it is obvious that the arbitrariness of determination of an exact conformal Killing r-form $\omega_2 = \nabla\Theta$ does not be greater then this number.

Based on a pointwise decomposition $\omega = \omega_1 + \omega_2$ the expression of an arbitrary conformal Killing r-form ω in some special local coordinate system $x^1, \ldots, x^n$ on a manifold (M, g) with nonzero constant curvature $C \neq 0$

$$\omega_{i_1\ldots i_r} = \mathrm{e}^{(r+1)\varphi}\left(A_{ki_1\ldots i_r}x^k + B_{i_1\ldots i_r} - \frac{1}{C}\left(\varphi_{[i_1}C_{ki_2\ldots i_r]}x^k + \varphi_{[i_1}D_{i_2\ldots i_r]} + \frac{1}{r}C_{i_1\ldots i_r}\right)\right),$$

where $A_{k,i_1\ldots i_r}$, $B_{i_1\ldots i_r}$, $C_{i_1 i_2\ldots i_r}$ and $D_{i_1\ldots i_{r-1}}$ are local components of constant skew-symmetric forms. Now, we may compute the dimension t_r of a space T^r of conformal Killing r-forms on a manifold with constant curvature $C \neq 0$ which is equal to the following summ

$$t_r = \binom{n+1}{r+1} + \binom{n+1}{r} = \binom{n+2}{r+1} = \frac{(n+2)!}{(r+1)!(n-r+1)!}.$$

In the case of an arbitrary connected manifold (M, g), it is obvious that the dimension of a space T^r is not greater then this number, i. e.

$$t_r \leq \frac{(n+2)!}{(r+1)!(n-r+1)!}.$$

We have proved the theorem.

Considering the upper bounds of dimension of vector space of conformal Killing forms T^r and of space of Killing r-forms K^r as found in Theorem 3.1 we for $r = 1$ obtain the following well known proposition.

Corollary 3.1 *The dimensions of vector spaces of conformal Killing vector field T^1, Killing vector fields K^1 and concircular vector fields P^1 on connected Riemannian manifold (M, g) do not be greater then $\frac{1}{2}(n+1)(n+2)$, $\frac{1}{2}(n+1)n$ and $n+1$, respectively. The equalities are obtained in the case of manifold with constant nonzero section curvature.*

3.3 Formulated in the theorem and its corollary results on dimensions of vector spaces of Killing and conformal Killing forms and of vector fields are *substantially local.*

As an example, let us consider an n-dimensional ($n \geq 2$) hyperbolic space which is a Riemannian manifold with constant negative curvature. As we have proved above, in this space the dimension k_r of the space of coclosed conformal Killing (Killing) r-forms ($1 \leq r \leq n-1$) and, especially, the dimension k_1 of Killing vector spaces is equal to $\frac{(n+1)!}{(r+1)!(n-r)!}$ and $\frac{1}{2}(n+1)n$, respectively. Factorizing hyperbolic space according to a suitable discrete group of motions (see [29, § 2.4]) we obtain a compact manifold with constant curvature. On this manifold there exists "generally" no nonzero Killing r-form (see [32, § 1 of Chap. VI]). Therefore, our result on the dimension of a space of Killing r-forms as well as well known result on the dimension of a space of Killing vector field deals with "local dimensions" k_r and k_1, especially.

An analogous conclusion may be obtained for dimensions of spaces of conformal Killing forms and vector fields and also for closed conformal Killing forms and concircular vector field.

4 Dimensions of Spaces of Conformal Killing Forms on a Closed Riemannian Manifold

4.1 Let (M, g) be a closed (i.e. compact without the border ∂M) oriented Riemannian manifold. Let us denote by $\langle \cdot, \cdot \rangle$ the global inner product

$$\langle \omega, \omega' \rangle = \int_M \frac{1}{r!} g(\omega, \omega') dv \tag{11}$$

for arbitrary compact carriers $\omega, \omega' \in \Omega^r(M)$ of r-form and volume element dv. Then the exterior differential operator d: $\Omega^r(M) \to \Omega^{r+1}(M)$ and the adjoin codifferential operator δ: $\Omega^{r+1}(M) \to \Omega^r(M)$ are connected by the following equality (see [13, p. 204])

$$\langle \mathrm{d}\omega, \Theta \rangle = \langle \omega, \delta\Theta \rangle \tag{12}$$

for any $\omega \in \Omega^r(M)$ and $\Theta \in \Omega^{r+1}(M)$.

On a closed manifold (M, g), it holds the following Hodge-de Ram orthogonal decomposition with respect to the global inner product (11) (see [12]):

$$\Omega^r(M) = \operatorname{Im} \mathrm{d} \oplus \operatorname{Im} \delta \oplus \operatorname{Ker} \Delta, \tag{13}$$

where $\Delta = \mathrm{d}\delta + \delta\mathrm{d}$ is Laplace operator with $\operatorname{Ker} \Delta = \operatorname{Im} \mathrm{d} \cap \operatorname{Ker} \delta$. In addition, there are following orthogonal decompositions (see [12])

$$\operatorname{Ker} \delta = \operatorname{Im} \delta \oplus \operatorname{Ker} \Delta, \tag{14}$$

$$\operatorname{Ker} \mathrm{d} = \operatorname{Im} \mathrm{d} \oplus \operatorname{Ker} \Delta. \tag{15}$$

The kernel of Laplace operator Δ on (M, g) is a finite dimension vector space $H^r(M, \mathbb{R}) = \{\omega \in \Omega^r(M) | \Delta\omega = 0\}$ of harmonic r-forms, for $r = 1, \dots, n-1$ (see [15, § 25], [12]). The dimension of $H^r(M, \mathbb{R})$ is equal to the Betti number $b_r(M)$ of a manifold (M, g), i.e. $b_r(M) = \dim_{\mathbb{R}} \operatorname{Ker} \Delta$. It is known (see [12]), that Betti numbers of a manifold (M, g) are dual in the sense of $b_r(M) = b_{n-r}(M)$ and for $n = 2r$ Betti numbers are invariant with respect to the conformal transformation of metric $\overline{g} = e^{2F} g$, because in this case $\overline{\delta} = \delta$ (see [2, Corollary 1.162]).

4.2 Let an n-dimensional closed manifold (M, g) be given and let us consider a natural with respect to isometric dipheomorphisms differential operator of the first order D: $\Omega^r(M) \to \Omega^1(M) \otimes \Omega^1(M)$ being define by the following (see [3, pp. 312–313], [16, 20])

$$\mathrm{D} = \nabla - (r+1)^{-1}\mathrm{d} - (n-r+1)^{-1} g \wedge \delta, \tag{16}$$

where $\wedge$ denotes multiplication of an $(r-1)$-form $\delta\omega$ by a metric tensor which is defined by the following rule

$$(g \wedge \delta\omega)(X_0, X_1, \dots, X_r) = \sum_{a=1}^{r} (-1)^a g(X_0, X_a)(\delta\omega)(X_1, \dots, \widehat{X_a}, \dots, x_r)$$

for arbitrary $(X_0, \dots, X_r) \in C^\infty(TM)$.

Then the condition $\omega \in \operatorname{Ker} \mathrm{D}$, which is equal to the identity $\nabla\omega = (r+1)^{-1}\mathrm{d}\omega + (\mathrm{n}-\mathrm{r}+1)^{-1} g \wedge \delta\omega$, is a necessary and sufficient condition that an r-form ω is a conformal Killing form, $r = 1, \dots, n-1$ (see [16, 19, 20]).

Especially, it follows from this that for a conformal Killing vector field $Z = \omega^\#$ any 1-form ω belongs to the kernel of differential operator of the first order

$$\mathrm{D} := \nabla\omega - \frac{1}{2}\mathrm{d}\omega + \frac{1}{\mathrm{n}} g \cdot \delta\omega,$$

which is called Ahlfors operator (see [14]).

In the [23] there is for operator D, defined by (16), found an adjoint operator D^*. Moreover, there is also the rough Laplacian constructed by (see [2, Definition 1.135], [3, pp. 316–317])

$$\mathrm{D}^*\mathrm{D} = \frac{1}{r(r+1)} \left(\nabla^*\nabla - \frac{1}{r+1}\delta\mathrm{d} - \frac{1}{n-r+1}\mathrm{d}\delta \right), \tag{17}$$

where $\nabla^*\nabla$ is the *rough Bochner Laplacian* ([21]).

It follows from general theory that rough Laplacian on a closed Riemannian manifold (M, g) is positive and elliptic (see [3, pp. 316–317]). Because it is an elliptic operator, its kernel is a finite dimensional vector subspace. Therefore, the kernel of a rough Laplacian $\operatorname{Ker} \mathrm{D}^*\mathrm{D}$ is a finite dimensional vector space $T^r(M, \mathbb{R}) = \{\omega \in$

$\Omega^r(M) \mid \mathrm{D}^*\mathrm{D}\omega = 0\}$. Using the identity $\langle \mathrm{D}^*\mathrm{D}\omega, \omega\rangle = \langle \mathrm{D}\omega, \mathrm{D}\omega\rangle$, we have that this kernel consists of conformal Killing r-forms, for all $r = 1, \dots, n-1$ (see [16]).

In [17] and [24], the dimension $t_r(M) = \dim_{\mathbb{R}} \mathrm{Ker}\, \mathrm{D}^*\mathrm{D}$, for all $r=1, \dots, n-1$ is called *Tachibana number* of closed Riemannian manifold (M, g) as an analogy to the Betti number $b_r(M) = \dim_{\mathbb{R}} \mathrm{Ker}\, \Delta$. Tachibana numbers as well as Betti numbers are dual in the sense of $t_r(M) = t_{n-r}(M)$ (see [17, 22, 24].) Evidently,

$$t_r(M) \leq \frac{(n+2)!}{(r+1)!(n-r+1)!}.$$

One of the most important properties of conformal Killing forms is their conformal invariance (see [4]), i.e. for an arbitrary conformal Killing r-form ω the form $\bar{\omega} = e^{(r+1)f}\omega$ is a conformal Killing form with respect to the conformally equivalent metric $\bar{g} = \mathrm{e}^{2\mathrm{f}}\mathrm{g}$. This implies that Tachibana numbers $t_r(M)$ for $r = 1, \dots, n-1$ are conformal scalar invariants of a closed Riemannian manifold (see [17, 24]).

Coclosed conformal Killing (Killing) r-forms $(1 \leq r \leq n-1)$ form the vector space $K^r(M, \mathbb{R}) = \{\omega \in \Omega^r(M) \mid \mathrm{D}^*\mathrm{D}\omega = \delta\omega = 0\}$. The dimension $k_r(M) = \dim_{\mathbb{R}}(\mathrm{Ker}\, \mathrm{D}^*\mathrm{D} \cap \mathrm{Ker}\, \delta)$ was in [17] and [24] called *Killing number* of closed Riemannian manifold (M, g). Evidently,

$$k_r(M) \leq \frac{(n+1)!}{(r+1)!(n-r)!}.$$

Closed conformal Killing (planar) r-forms $(1 \leq r \leq n-1)$ form the vector space $P^r(M, \mathbb{R}) = \{\omega \in \Omega^r(M) \mid \mathrm{D}^*\mathrm{D}\omega = d\omega = 0\}$. Its dimension $p_r(M) = \dim_{\mathbb{R}}(\mathrm{Ker}\, \mathrm{D}^*\mathrm{D} \cap \mathrm{Ker}\, \mathrm{d})$ was in [17] and [24] called *planar number* of closed Riemannian manifold (M, g). Evidently,

$$p_r(M) \leq \frac{(n+1)!}{r!(n-r+1)!}.$$

One of the most important properties of Killing and planar r-forms is their conformal invariance (see [4]), i.e. for an arbitrary conformal Killing (or planar) r-form ω the form

$$\bar{\omega} = \mathrm{e}^{-(r+1)f}\omega \;\text{ for }\; f = (n+1)^{-1}\ln\sqrt{\frac{\det g}{\det \bar{g}}}$$

is a Killing (respectively, planar) r-form with respect to the projectively equivalent metric $\bar{g}$ (see [24]).This implies that the Killing numbers $k_r(M)$ and planar numbers $p_r(M)$ for $r = 1, \dots, n-1$ are projective scalar invariants of the Riemannian manifold (M, g).

To summarize, we may formulate

Proposition 4.1 *On an n-dimensional closed Riemannian manifold (M, g) the following hold for all r, $r=1, \dots, n-1$*

1. *Tachibana numbers $t_r(M)$ are conformal scalar invariants, they are dual in the sense $t_r(M) = t_{n-r}(M)$ and they fulfill the relation*

$$t_r(M) \leq \frac{(n+2)!}{(r+1)!(n-r+1)!};$$

2. *Killing numbers $t_r(M)$ and planar numbers $p_r(M)$ are projective scalar invariants, they are dual in the sense $k_r(M) = p_{n-r}(M)$ and they fulfill the relations*

$$k_r(M) \leq \frac{(n+1)!}{(r+1)!(n-r)!} \quad \text{and} \quad p_r(M) \leq \frac{(n+1)!}{r!(n-r+1)!}.$$

4.3 We establish a connection between Betti numbers and Tachibana numbers. Two following theorems hold.

Theorem 4.1 *If Ricci tensor Ric of an n-dimensional compact and oriented conformal planar Riemannian manifold (M, g), $n \geq 2$, is definite, then Tachibana $t_k(M)$ and Betti $b_l(M)$ numbers cannot be different from zero for arbitrary pair of indices $k, l = 1, \ldots, n-1$.*

Proof Let Ricci tensor Ric be definite on a compact manifold (M, g), i.e. quadratic form $Ric(X, X)$ is definite for arbitrary nonzero vector field $X \in C^\infty TM$. Let us suppose that quadratic form $Ric(X, X)$ is positive definite and manifold (M, g) be compact and oriented conformal flat manifold. Then $b_1(M) = \cdots = b_{n-1}(M) = 0$, in accordance with [31].

Further, to get a new expression of the rough Laplacian (17), let us use the classical Bochner-Weitzenböck formula [2], which gives $\Delta = \nabla^*\nabla + F_r$, where F_r may be a algebraically (even linearly) expressed in terms of the curvature tensor R of manifold (M, g). Now, the rough Laplacian may be written in the form

$$\mathrm{D}^*\mathrm{D} = \frac{1}{r(r+1)}\left(\mathrm{d}^*\mathrm{d} - \frac{n-r}{n-r+1}\mathrm{d}\mathrm{d}^* - F_r\right).$$

Then any conformal Killing r-form ω must fulfill the following equation

$$\int_M g(F_r(\omega), \omega)dv = \frac{r}{r+1}\langle \mathrm{d}\omega, \mathrm{d}\omega\rangle + \frac{n-r}{n-r+1}\langle \mathrm{d}^*\omega, \mathrm{d}^*\omega\rangle.$$

If we suppose that quadratic form $Ric(X, X)$ is negative definite, then on a compact conformal flat manifold we have the following inequality [31]

$$g(F_r(\Theta), \Theta) \leq -\frac{n-r}{n-1}\lambda \cdot g(\Theta, \Theta),$$

for the greatest (negative) eigenvalue $-\lambda$ of the matrix $\|Ric\|$, for all $1 \leq r \leq n-1$ and any nonzero form $\Theta \in \Omega^r(M)$. Consequently, any conformal Killing r-form ω must fulfill the inequality

$$\frac{r}{(n-r)(r+1)}\langle \mathrm{d}\omega, \mathrm{d}\omega\rangle + \frac{1}{n-r+1}\langle \mathrm{d}^*\omega, \mathrm{d}^*\omega\rangle \leq -\frac{n-r}{n-1}\lambda\langle \omega, \omega\rangle.$$

This is possible only if a conformal Killing r-form vanishes at any point of manifold (M, g) and then $t_1(M) = \cdots = t_{n-1}(M) = 0$. The theorem is proved.

Theorem 4.2 *If on an n-dimensional closed Riemannian manifold (M, g) Betti number $b_r(M) = 0$, for $1 \leq r \leq n-1$, and Tachibana number $t_r(M) > k_r(M) \neq 0$, then for Killing numbers $k_r(M)$ and planar numbers $p_r(M)$ it holds $t_r(M) = k_r(M) + p_r(M)$.*

Proof Let on an n-dimensional closed Riemannian manifold (M, g) Betti number $b_r(M) = 0$, for $1 \leq r \leq n-1$. Then in decompositions (13–15) there is $\mathrm{Ker}\,\Delta = 0$ and therefore we have the following orthogonal decomposition

$$\Omega^r(M) = \mathrm{Im}\,\mathrm{d} \oplus \mathrm{Im}\,\delta \tag{18}$$

and besides $\mathrm{Ker}\,\mathrm{d} = \mathrm{Im}\,\mathrm{d}$ and $\mathrm{Ker}\,\delta = \mathrm{Im}\,\delta$. Since $t_r(M) > k_r(M) \neq 0$ we obtain for Killing number $k_r(M) = \dim_{\mathbb{R}}(\mathrm{Ker}\,\mathrm{D}^*\mathrm{D} \cap \mathrm{Im}\,\mathrm{d}^*)$ and for planar number $p_r(M) = \dim_{\mathbb{R}}(\mathrm{Ker}\,\mathrm{D}^*\mathrm{D} \cap \mathrm{Im}\,\mathrm{d}) \neq 0$. Consequently, it is clear that decomposition (18) implies the following orthogonal decomposition

$$\mathrm{Ker}\,\mathrm{D}^*\mathrm{D} = (\mathrm{Ker}\,\mathrm{D}^*\mathrm{D} \cap \mathrm{Im}\,\delta) \oplus (\mathrm{Ker}\,\mathrm{D}^*\mathrm{D} \cap \mathrm{Im}\,\mathrm{d}). \tag{19}$$

It may be rewritten in the form $T^r(M, \mathbb{R}) = K^r(\mathbb{R}, M) \oplus P^r(M, \mathbb{R})$, $1 \leq r \leq n-1$, where in accordance with (14) a space $K^r(M, \mathbb{R})$ consists of co-exact conformal Killing r-forms and in accordance with (15) a space $P^r(M, \mathbb{R})$ consists of exact ones. It follows from this the equality $t_r(M) = k_r(M) + p_r(M)$. The summands on the right side represent dimensions of namely these vector spaces. Let us remark that in the case $t_r(M) = 0$ the equality which may be proved turns into an identity.

The proof is finished.

Remark 4.1 In the article [25], an analogical orthogonal decomposition $T^r(M, \mathbb{R}) = K^r(M, \mathbb{R}) + P^r(M, \mathbb{R})$ was established on an $2r$-dimensional closed conformal flat Riemannian manifold (M, g) with constant positive scalar curvature. Let us remark, that $b_r(M) = 0$ is fulfilled in such manifold in accordance with [31].

In [17] this decomposition is established for a closed manifold with positive curvature operator, where $b_1(M) = \cdots = b_{n-1}(M) = 0$ in accordance with [13]. It is important to say that in the both decompositions the space $P^r(M, \mathbb{R})$ consists of exact conformal Killing r-forms, $1 \leq r \leq n-1$, and the space $K^r(M, \mathbb{R})$ consists of coexact ones. These facts does not contained in the cited article.

Especially, for $r = 1$ we have the following corollary.

Corollary 4.1 *If for an n-dimensional closed Riemannian the first Betti number $b_1(M) = 0$ and at the same time $t_1(M) \neq 0$, $t_1(M) \neq k_1(M)$, then (M, g) is globally conformal to an n-dimensional sphere S^n of Euclidean space $\mathbb{R}^{n+1}$. If $s = const$,*

$s > 0$, additionally, then a manifold (M, g) is globally isometric to the sphere S^n. If for an n-dimensional closed Riemannian manifold $s = const$ and $s = 0$ or $s = const$ and $s < 0$, then $t_1(M) = k_1(M)$.

Proof For an n-dimensional closed Riemannian manifold (M, g) with $b_1(M) = 0$, $t_1(M) \neq 0$, $t_1(M) \neq k_1(M)$ we have the following orthogonal decomposition with respect to the global inner product

$$T^1(M, \mathbb{R}) = K^1(M, \mathbb{R}) + P^1(M, \mathbb{R}), \tag{20}$$

where $P^1(M, \mathbb{R})$ contains at least one exact conformal Killing 1-form, i.e. a form which may be expressed as $\omega = \text{gradf}$ with $\nabla\nabla f = \rho g$, where $\rho = -\frac{1}{n}\Delta f$. In this case, manifold (M, g) is globally conformal to an Euclidean sphere S^n with the standard metric $\bar{g}_{can}$ (see [27]). It is known (see [32]), that the presence of orthogonal decomposition (20) on closed Riemannian manifold (M, g) with an additional condition $s = const$ implies that (M, g) is globally isometric to the Euclidean sphere S^n. Let us remark, that for a compact Riemannian manifold (M, g) with constant negative or zero curvature it holds $t_1(M) = k_1(M)$, because in this case any conformal Killing vector field is a Killing field (see [9]).

Remark 4.2 Conformal Killing and closed conformal Killing vector field is dual to conformal Killing and planar 1-form, respectively. At the end of the past and beginning of this century, these vector fields have been objects of an intensive interest in connection with the study of groups of infinitesimal conformal transformations of search criteria for conformity and isometry Riemannian manifold to the Euclidean sphere (see [1, 6, 23, 28] and others.) Therefore, there exists a great number of propositions, which are analogical to our proved corollary. Let us mention one of them (see [23]). It says that any compact Riemannian manifold having finite fundamental group $\pi_1(M)$ and admitting a closed conformal Killing vector field, which is not an infinitesimal isometry, is diffeomorphic to the Euclidean sphere. Adding the condition of constancy of the scalar curvature we obtain that such manifold must be isometric to the Euclidean sphere (see [28]). Le us add, the finiteness of a fundamental group $\pi_1(M)$ implies automatically that first Betti number $b_1(M)$ is equal to zero and closed conformal Killing vector field is the gradient at the same time.

Acknowledgments The paper was supported by grant P201/11/0356 of The Czech Science Foundation.

References

1. Alodan, H.: Conformal gradient vector fields. Differential Geometry—Dynamical System **12**, 1–3 (2010)
2. Besse, A.L.: Einstein Manifolds. Springer, Berlin (1987)
3. Besse, A.L.: Géométrie Riemannienne en dimension 4. Cedic-Fernand Nathan, Paris (1981)

4. Benn, I.M., Chalton, P.: Dirac symmetry operators from conformal Killing-Yano tensors. Class. Quantum Gravity **14**(5), 1037–1042 (1997)
5. Eisenhart, L.P.: Riemannian Geometry. Princeton University Press, Princeton (1926)
6. Kühnel, W., Rademacher, H.-B.: Conformal transformations of pseudo-Riemannian manifolds. Recent developments in pseudo-Riemannian geometry. Zürich Eur. Math. Soc. 261–298 (2008)
7. Kashiwada, T.: On conformal Killing tensor. Nat. Sci. Rep. Ochanomizu Univ **19**(2), 67–74 (1968)
8. Kora, M.: On conformal Killing forms and the proper space of for p-forms. Math. J. Okayama Univ. **22**, 195–204 (1980)
9. Lichnerowicz, A.: Géométrie des groupes de transformations. Dunod, Paris (1958)
10. Maccallum, M., Herlt, E., Schmutzer, E.E.: Exact Solutions of the Einstein Field Equations. Deutscher Verlag der Wissenschaften, Berlin (1980)
11. Mikeš, J., Vanžurová, A., Hinterleitner, I.: Geodesic Mappings and Some Generalizations. Palacký University Press, Olomouc (2009)
12. Novikov, S.P.: Topology I. Encycl. Math. Sci. **12**, 1–310 (1996) (translation from Itogi Nauki Tekh., Ser. Sovrem. Probl. Mat., Fundam. Napravleniya **12**, 5–252 (1986))
13. Petersen, P.: Riemannian Geometry. Springer, New York (1997)
14. Pierzchalski, A.: Ricci curvature and quasiconformal deformations of a Riemannian manifold. Manuscripta Math. **66**, 113–127 (1989)
15. De Rham, G.: Variétés Différentiables. Hermann, Paris (1955)
16. Stepanov, S.E.: A new strong Laplacian on differential forms. Math. Notes **76**(3), 420–425 (2004)
17. Stepanov, S.E.: Curvature and Tachibana numbers. Sbornik Math. **202**(7), 135–146 (2011)
18. Stepanov, S.E.: The Killing-Yano tensor. Theoret. Math. Phys. **134**(3), 333–338 (2003)
19. Stepanov, S.E.: The vector space of conformal Killing forms on a Riemannian manifold. J. Math. Sci. **110**(4), 2892–2906 (2002)
20. Stepanov, S.E.: On conformal Killing 2-form of the electromagnetic field. J. Geom. Phys. **33**, 191–209 (2000)
21. Stepanov, S.E.: Vanishing theorems in affine, Riemannian and Lorenz geometries. J. Math. Sci. **141**(1), 929–964 (2007)
22. Semmelmann, U.: Conformal Killing forms on Riemannian manifolds. Math. Z. **245**(3), 503–627 (2003)
23. Suyama, Y., Tsukamoto, Y.: Riemannian manifolds admitting a certain conformal transformation group. J. Diff. Geom. **5**, 415–426 (1971)
24. Stepanov, S.E.: Some conformal and projective scalar invariants of Riemannian manifolds. Math. Notes **80**(6), 848–852 (2006)
25. Tachibana, S.: On the proper space of for m-forms im 2m dimensional conformal flat Riemannian manifolds. Nat. Sci. Rep. Ochanomizu Univ. **28**, 111–115 (1978)
26. Tachibana, S.: On conformal Killing tensor in a Riemannian space. Tohoku Math. J. **21**, 56–64 (1969)
27. Tashiro, Y.: Complete Riemannian manifolds and some vector fields. Trans. Am. Math. Soc. **117**, 251–275 (1965)
28. Tanno, S., Weber, W.C.: Closed conformal vector fields. J. Diff. Geom. **3**, 361–366 (1969)
29. Wolf, J.A.: Space of Constant Curvature. California University Press, Berkley (1972)
30. Yano, K.: Concircular geometry. Proc. Imp. Acad. Tokyo **16**, 195–200, 354–360, 442–448, 505–511 (1940)
31. Yano, K., Bochner, C.: Curvature and Betti Numbers. Princeton University Press, Princeton (1953)
32. Yano, K., Sawaki, S.: Riemannian manifolds admitting a conformal transformation group. J. Diff. Geom. **2**, 161–184 (1968)

Part III
Dynamical Symmetries and Conservation Laws

Causality from Dynamical Symmetry: An Example from Local Scale-Invariance

Malte Henkel

Abstract Physical ageing phenomena far from equilibrium naturally lead to dynamical scaling. It has been proposed to consider the consequences of an extension to a larger Lie algebra of local scale-transformation. The best-tested applications of this are explicitly computed co-variant two-point functions which have been compared to non-equilibrium response functions in a large variety of statistical mechanics models. It is shown that the extension of the Schrödinger Lie algebra $\mathfrak{sch}(d)$ to a maximal parabolic sub-algebra, when combined with a dualisation approach, is sufficient to derive the causality condition required for the interpretation of two-point functions as physical response functions. The proof is presented for the recent logarithmic extension of the differential operator representation of the Schrödinger algebra.

1 Motivation and Background

Physicists have valued since a long time the important rôle of symmetries, be it for their usefulness in simplifying practical calculations, be it for making progress in issues of conceptual understanding. Arguably the most famous instance of this is *relativistic covariance* in mechanics and electrodynamics,[1] formally described by the Lie group of Lorentz transformations which has been introduced almost exactly a century ago [10, 40]. Almost three quarters of a century later, it has been realised that by the inclusion of scale-invariance and the subsequent extension

[1] Physicists carefully distinguish between *co*-variance and *in*variance: for example, a scalar is invariant under rotations, while a vector or a tensor transforms covariantly. Since the equations of mechanics or electrodynamics are in general vector or tensor equations, it is appropriate to speak of relativistic co-variance.

M. Henkel (✉)
Groupe de Physique Statistique, DP2M, Institut Jean Lamour (CNRS UMR 7198),
Université de Lorraine Nancy, B.P. 70239, F-54506 Vandœuvre-lès-Nancy Cedex, France
e-mail: malte.henkel@univ-lorraine.fr

A. Makhlouf et al. (eds.), *Algebra, Geometry and Mathematical Physics*,
Springer Proceedings in Mathematics & Statistics 85, DOI: 10.1007/978-3-642-55361-5_30,

of the Lorentz group to the *conformal group* considerable advances can be made, simultaneously in cooperative phenomena in statistical mechanics as well as in string theory. A special rôle is herein played by the case of two dimensions, where the infinite-dimensional Lie algebra of conformal transformations is centrally extended to the Virasoro algebra, in order to be able to take the physical effects of either thermal or quantum fluctuations into account [3].

Here, we shall consider a different example of covariance under a certain class of space-time transformations. Historically, these were found by considering the dynamical symmetries of what in physics is called by an abuse of language the 'non-relativistic limit' of mechanics where the speed of light $c \to \infty$. Specifically, we shall be interested in the transformations of the *Schrödinger group* $Sch(d)$ which is defined by the following transformation on space-time coordinates $(t, r) \in \mathbb{R} \times \mathbb{R}^d$:

$$t \mapsto t' := \frac{\alpha t + \beta}{\gamma t + \delta} \ , \ r \mapsto r' := \frac{\mathscr{R} r + vt + a}{\gamma t + \delta} \ ; \ \alpha\delta - \beta\gamma = 1 \tag{1}$$

with $\mathscr{R} \in SO(d)$, $a, v \in \mathbb{R}^d$ and $\alpha, \beta, \gamma, \delta \in \mathbb{R}$. Indeed, it has been known to mathematicians since a long time that free-particle motion (be it classical, quantum mechanical or probabilistic) is invariant under the Schrödinger group in the sense that a solution of the equation of motion is mapped onto a different solution of the same equation of motion in the transformed coordinates [36, 39]. During the past century, this has been re-discovered a couple of times, both in mathematics and physics, see e.g. [29] and references therein. It is often convenient to study instead the Lie algebra $\mathfrak{sch}(d) = \mathrm{Lie}(Sch(d)) = \left\langle X_{0,\pm 1}, Y^{(j)}_{\pm 1/2}, M_0, R^{(jk)}_0 \right\rangle_{j,k=1,\dots d}$ with the explicit generators (where $\partial_j := \partial/\partial r_j$ and $\nabla_r = (\partial_1, \dots, \partial_d)^{\mathrm{T}}$)

$$\begin{aligned}
X_n &= -t^{n+1}\partial_t - \frac{n+1}{2} t^n r \cdot \nabla_r - \frac{\mathscr{M}}{2}(n+1)nt^{n-1}r^2 - \frac{n+1}{2} x t^n \\
Y^{(j)}_m &= -t^{m+1/2}\partial_j - \left(m + \frac{1}{2}\right) t^{m-1/2} \mathscr{M} r_j \\
M_n &= -t^n \mathscr{M} \\
R^{(jk)}_n &= -t^n \left(r_j \partial_k - r_k \partial_j\right) \ = \ -R^{(kj)}_n
\end{aligned} \tag{2}$$

Herein, the non-derivative terms (characterised by a dimensionful constant $\mathscr{M}$ ('mass') and a scaling dimension x) describe how the solution of a Schrödinger/diffusion equation will transform under the action of $\mathfrak{sch}(d)$. One has the non-vanishing commutation relations

$$\begin{aligned}
&[X_n, X_{n'}] = (n - n')X_{n+n'} \ , && [X_n, Y^{(j)}_m] = \left(\frac{n}{2} - m\right) Y^{(j)}_{n+m} \\
&[X_n, M_{n'}] = -n' M_{n+n'} \ , && [X_n, R^{(jk)}_{n'}] = -n' R^{(jk)}_{n+n'} \\
&[Y^{(j)}_m, Y^{(k)}_{m'}] = \delta^{j,k} \left(m - m'\right) M_{m+m'} \ , && [R^{(jk)}_n, Y^{(\ell)}_m] = \delta^{j,\ell}\, Y^{(k)}_{n+m} - \delta^{k,\ell}\, Y^{(j)}_{n+m}
\end{aligned} \tag{3}$$

up to the commutators of $\mathfrak{so}(d)$, involving the $R_0^{(jk)}$, which are not spelled out. The Schrödinger algebra is also the Lie symmetry algebra of non-linear (systems of) equations. Probably one of the best-known examples of this kind are the Euler equations of a compressible fluid of velocity $u = u(t, r)$ and density $\rho = \rho(t, r)$

$$\partial_t \rho + \nabla \cdot \big(\rho u\big) = 0 \ , \ \ \rho\big(\partial_t + (u \cdot \nabla)\big)u + \nabla P = 0 \tag{4}$$

together with the polytropic equation of state $P = \rho^{1+2/d}$. This has been known to russian and ukrainian mathematicians at least since the 1960s [15, 49] and was re-discovered by european physicists around the turn of the century [20, 48]. Many more Schrödinger-invariant non-linear equations and systems exist, see [14–16, 55]. Analogously to conformal invariance in $2D$, an infinite-dimensional extension of $\mathfrak{sch}(d)$ is the *Schrödinger-Virasoro algebra* $\mathfrak{sv}(d) = \left\langle X_n, Y_m^{(j)}, M_n, R_n^{(jk)} \right\rangle_{n\in\mathbb{Z}, m\in\mathbb{Z}+\frac{1}{2}, j,k\in\{1,\ldots,d\}}$, with an explicit representation in (2) and an immediate extension of the commutators (3) [22]. The mathematical properties of $\mathfrak{sv}$ are studied in detail in [57, 61], the geometry in [9] and physical applications are reviewed in [29].

Contrary to a widespread belief, when taking the non-relativistic limit of the conformal algebra, one does *not* obtain the Schrödinger algebra, but a different Lie algebra, which by now is usually called the *conformal Galilean algebra* $\mathrm{CGA}(d) = \langle X_{\pm 1,0}, Y_{\pm 1,0}^{(j)}, R_0^{(jk)} \rangle_{j,k=1,\ldots,d}$ [1, 21, 23, 26, 42, 47, 64]. Its most general known differential operator representation is [7]

$$\begin{aligned}
X_n &= -t^{n+1}\partial_t - (n+1)t^n r \cdot \nabla_r - n(n+1)t^{n-1}\gamma \cdot r - x(n+1)t^n \\
Y_n^{(j)} &= -t^{n+1}\partial_j - (n+1)t^n \gamma_j \\
R_n^{(jk)} &= -t^n\big(r_j\partial_k - r_k\partial_j\big) - t^n\big(\gamma_j\partial_{\gamma_k} - \gamma_k\partial_{\gamma_j}\big) \ = \ -R_n^{(kj)}
\end{aligned} \tag{5}$$

where $\gamma = (\gamma_1, \ldots, \gamma_d)$ is a vector of dimensionful constants and x is again a scaling dimension. Its non-vanishing commutators read, again up to those of $\mathfrak{so}(d)$

$$\begin{aligned}
&[X_n, X_{n'}] = (n - n')X_{n+n'} \ , \ \ [X_n, Y_m^{(j)}] \ = \ (n - m)\, Y_{n+m}^{(j)} \\
&[X_n, R_{n'}^{(jk)}] = -n' R_{n+n'}^{(jk)} \ \ \ \ , \ \ [R_n^{(jk)}, Y_m^{(\ell)}] \ = \ \delta^{j,\ell} Y_{n+m}^{(k)} - \delta^{k,\ell} Y_{n+m}^{(j)}
\end{aligned} \tag{6}$$

The non-linear systems for which $\mathrm{CGA}(d)$ arises as a (conditional) dynamical symmetry are distinct from (4) [7, 62]. As before, the systematic organisation of the generators allows for an immediate infinite-dimensional extension $\mathfrak{av}(d) := \left\langle X_n, Y_n^{(j)}, R_n^{(jk)} \right\rangle_{n\in\mathbb{Z}, j,k=1,\ldots,d}$ [23, 50] (*'altern-Virasoro algebra'*).

In $d = 2$ spatial dimensions, it was recently shown [41] that the conformal Galilean algebra admits a so-called 'exotic' central extension. This is achieved by adding to the commutator relations (6) the following commutator

$$\big[Y_n^{(1)}, Y_m^{(2)}\big] \ = \ \delta_{n+m,0}\,\big(3\delta_{n,0} - 2\big)\, \Theta, \quad n, m \in \{\pm 1, 0\}, \tag{7}$$

where the new central generator Θ is needed for this central extension. Physicists usually call this central extension of CGA(2) the *exotic Galilean conformal algebra*, and we shall denote it by ECGA $=$ CGA$(2) + \mathbb{C}\Theta$.[2] A differential operator representation of ECGA reads [7, 42] (ε_{jk} is the totally antisymmetric tensor)

$$
\begin{aligned}
X_n &= -t^{n+1}\partial_t - (n+1)t^n r\cdot\nabla_r - \lambda(n+1)t^n - (n+1)nt^{n-1}\gamma\cdot r - (n+1)nh\cdot r \\
Y_n^{(j)} &= -t^{n+1}\partial_j - (n+1)t^n\gamma_j - (n+1)t^n h_j - (n+1)n\,(\varepsilon_{jk}r_k)\,\theta \\
R_0^{(12)} &= -\left(r_1\partial_2 - r_2\partial_1\right) - \left(\gamma_1\partial_{\gamma_2} - \gamma_2\partial_{\gamma_1}\right) - \frac{1}{2\theta}h\cdot h
\end{aligned}
\tag{8}
$$

where $n \in \{\pm1, 0\}$ and $j, k \in \{1, 2\}$.[3] Because of Schur's lemma, the central generator Θ can be replaced by its eigenvalue $\theta \neq 0$. The components of the vector-operator $h = (h_1, h_2)$ are connected by the commutator $[h_1, h_2] = \Theta$. For illustration, we quote the following non-linear system which has ECGA as a Lie symmetry [7]

$$
\nabla\wedge u = 0 \ , \quad \partial_t u + (u\cdot\nabla)\,u + \frac{1}{2}\,(u\wedge\nabla)\wedge u = q\nabla\wedge\omega \tag{9}
$$

where q is a constant, $u = u(t, r) = (u_1, u_2, 0)^{\mathrm{T}}$ is a planar vector embedded into $\mathbb{R}^3$ (and similarly for ∇) and $\omega = (0, 0, w)^{\mathrm{T}}$ is constructed from the coordinate dual to the central charge according to $\Theta = \partial_w$. Clearly, (9) is very different from (4).

Remark In analogy to the Virasoro algebra of $2D$ conformal invariance, it is natural to ask if the full definition of algebras such as $\mathfrak{sv}(d)$ or $\mathfrak{av}(d)$ may include central extensions. For the Schrödinger-Virasoro algebra $\mathfrak{sv}(1)$, one merely has the central Virasoro-like extension of $[X_n, X_m]$ [22, 57, 61]. On the other hand, if in $\mathfrak{sv}(1)$ one considers the generators Y_n with *integer* indices $n \in \mathbb{Z}$, then three distinct central extensions are possible [57], [61, Theorem 7.4]. For the 'altern-Virasoro algebra' or the infinite-dimensional extension of CGA(1) one has the central extensions [28, 50]

$$
\begin{aligned}
[X_n, X_{n'}] &= (n-n')X_{n+n'} + \frac{c_X}{12}\delta_{n+n',0}\left(n^3 - n\right) \\
[X_n, Y_{n'}] &= (n-n')X_{n+n'} + \frac{c_Y}{12}\delta_{n+n',0}\left(n^3 - n\right)
\end{aligned}
\tag{10}
$$

with two independent central charges. The independence of the two central charges $c_{X,Y}$ can be illustrated by the following example: let L_n and L_n' with $n \in \mathbb{Z}$ stand for the generators of two commuting Virasoro algebras with central charges c and c'. Then the generators

$$
X_n := \begin{pmatrix} L_n + L_n' & 0 \\ 0 & L_n + L_n' \end{pmatrix} , \ Y_n := \begin{pmatrix} 0 & L_n \\ 0 & 0 \end{pmatrix} , \ K_X := \begin{pmatrix} 1 & 0 \\ 0 & 1 \end{pmatrix} , \ K_Y := \begin{pmatrix} 0 & 1 \\ 0 & 0 \end{pmatrix} \tag{11}
$$

---

[2] All Lie algebras are complex, unless explicitly stated otherwise.

[3] An infinite-dimensional extension of ECGA does not appear to be possible.

obey the commutators (10), with $c_X = (c+c')K_X$ and $c_Y = cK_Y$ [28] [29, Exercise 5.5].

In statistical physics, many situations are known and well-understood where the usual space-time symmetries of temporal and spatial translation-invariance and rotation-invariance are supplemented by dilatation (or scale-) invariance.[4] The paradigmatic examples are provided by various phase transitions—often-mentioned examples include the liquid-gas transition, the ferromagnetic-paramagnetic transition, the transition between normal conductivity and superconductivity, the electroweak phase transition in the early universe and so on. Here, we shall be interested in instances of *dynamical* scaling, which involves the space-time rescaling $t \mapsto b^z t$, $r \mapsto br$ and is characterised by a constant, the *dynamical exponent* z. It arises naturally in various many-body systems far from equilibrium, often without having to fine-tune external parameters. Paradigmatic examples are *ageing phenomena*, which may arise in systems quenched, from some initial state, either (i) into a coexistence phase with more than one stable equilibrium state or else (ii) onto a critical point of the stationary state, see [4, 8, 29] for reviews. Phenomenologically, ageing can be characterised through three (symmetry) properties: namely [29]

1. slow, non-exponential relaxation,
2. breaking of time-translation-invariance
3. dynamical scaling.

For equilibrium critical phenomena, it was believed for a long time that under relatively weak conditions scale-invariance could be extended to conformal invariance. Recent work has considerably clarified that this conclusion cannot always be drawn so readily [56], although there exist many theoretical models which are indeed both scale- and conformally invariant, with many important consequences [3, 53]. Drawing on this analogy, we look for situations when dynamical scaling can be extended to a larger group, such as the Schrödinger group when $z = 2$. Quite analogously with respect to conformal invariance, one is looking for co-variant two-point functions, such that the co-variance under Schrödinger transformations leads to a set of differential equations for the said two-point function. However, in contrast to conformal invariance, it has turned out that this kind of co-variance condition is *not* satisfied by correlation functions but rather by the so-called *response functions*. As an example, we quote the basic prediction of Schrödinger-invariance for the linear two-time auto-response function [24–27]

$$\begin{aligned} R(t,s) &= \left.\frac{\delta\langle\phi(t,r)\rangle}{\delta h(s,r)}\right|_{h=0} = \langle\phi(t,r)\widetilde{\phi}(s,r)\rangle = s^{-1-a} f_R\left(\frac{t}{s}\right), \\ f_R(y) &= f_0 y^{1+a'-\lambda_R/z}(y-1)^{-1-a'}\Theta(y-1) \end{aligned} \tag{12}$$

which measures the linear response of the order-parameter $\phi(t,r)$ with respect to its canonically conjugated external field $h(s,r)$. In stochastic field-theory using the

[4] In the physicists terminology: at an equilibrium critical point, the partition function is *in*variant under dilatations, whereas correlators of physical observables transform *co*-variantly.

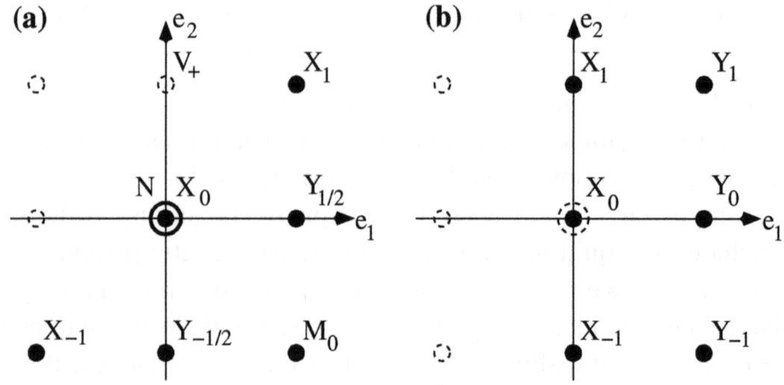

Fig. 1 Root diagrammes of some sub-algebras of the complex Lie algebra B_2. The roots of B_2 are indicated by the full and *broken dots*, those of the sub-algebras by the *full dots* only. **a** Schrödinger algebra $\mathfrak{sch}(1) = \langle X_{\pm 1,0}, Y_{\pm 1/2}, M_0 \rangle$ and the maximal parabolic sub-algebra $\widetilde{\mathfrak{sch}}(1) = \mathfrak{sch}(1) + \mathbb{C}N$. **b** Conformal Galilean algebra $\text{CGA}(1) = \langle X_{\pm 1,0}, Y_{\pm 1,0} \rangle$

Janssen-de Dominicis formalism, see e.g. [8, 29], it can be shown that response functions can be written as a correlator between the order-parameter ϕ and an associated *'response field'* $\widetilde{\phi}$.[5] The auto-response exponent λ_R and the ageing exponents a, a' are universal non-equilibrium exponents.[6] This prediction has been tested extensively, and the computation of correlators can be understood along different lines, as reviewed in [29].

The main distinction of response functions with respect to correlation functions is the *causality condition* $t > s$, which is spelt out in (12) through the Heaviside Θ-function. Here, we shall show *how the origin of this causality condition can be understood from an algebraic symmetry hypothesis*. The central observation is that there exists a natural way to imbed the Schrödinger algebra $\mathfrak{sch}(d)$ into a (semi-simple) conformal Lie algebra in $d + 2$ dimensions [5, 26]. This opens the route to introduce a powerful mathematical concept, namely the parabolic sub-algebras of that conformal Lie algebra. By definition, a (standard) *parabolic sub-algebra* is made up by the Cartan sub-algebra and a selected set of positive roots [37]. It turns out that *a sufficient condition for deriving a causality condition for the co-variant two-point functions as in* (12) is the co-variance under a maximal parabolic sub-algebra dualised in such a way that translation-invariance in the dual variable becomes part of the algebra. For example, rather than requiring Schrödinger-covariance under the algebra $\mathfrak{sch}(d)$, one considers an extended co-variance under the *maximal parabolic sub-algebra* $\widetilde{\mathfrak{sch}}(d) = \mathfrak{sch}(d) + \mathbb{C}N$, with a single extra generator N, to be specified below [26]. In Fig. 1a, we illustrate the inclusion, for the $d = 1$ case, $\widetilde{\mathfrak{sch}}(1) = \mathfrak{sch}(1) + \mathbb{C}N \subset B_2$ to the simple complex Lie algebra B_2, isomorphic to the conformal algebra

[5] The example of the free field equations of motion already shows that while the order-parameter ϕ has a positive 'mass' $\mathcal{M} > 0$, the 'mass' associated to the response field is negative $\widetilde{\mathcal{M}} = -\mathcal{M} < 0$.

[6] In magnets, the temperature is rapidly lowered ('quenched') from a very high initial value to a finite value T. Mean-field theory suggests that usually $a = a'$ for low final temperatures $T < T_c$ and $a \neq a'$ for critical quenches at $T = T_c$, where T_c is the equilibrium critical temperature [29].

$B_2 \cong (\mathfrak{conf}(3))_{\mathbb{C}}$ in three dimensions. Similarly, Fig. 1b illustrates the inclusion $\mathrm{CGA}(1) \subset B_2$ and an extension by the second independent generator in the Cartan sub-algebra would give an inclusion $\widetilde{\mathrm{CGA}}(1) \subset B_2$. Maximal parabolic sub-algebras of B_2 are distinguished in that the addition of any further generator produces the entire conformal algebra. Furthermore, in view of may important physical applications (some of them to be mentioned briefly below), we shall see that the same kind of causality condition is also obtained for the novel logarithmic extensions of the Schrödinger and/or conformal Galilean algebras [30, 32–34, 58].

This paper is organised as follows. The first sections recall basic facts on the ingredients required. In Sect. 2, we recall briefly those elements of *logarithmic* conformal invariance as required here and quote the corresponding logarithmic extensions of $\mathfrak{sch}(d)$- and $\mathrm{CGA}(d)$-invariance. In Sect. 3, specialising to $d = 1$ for brevity, we describe the inclusion of the Schrödinger algebra into B_2 by a canonical dualisation procedure and its extension to the logarithmic case. In Sect. 4, the shapes of the dual logarithmic Schrödinger-covariant two-point functions will be derived and we shall see that Schrödinger-covariance alone is *not* enough to derive a causality condition. In Sect. 5 we finally derive our main result, namely that $\widetilde{\mathfrak{sch}}(1)$-covariant two-point functions automatically must obey causality. In this way, a combination of dualisation with an extended dynamical co-variance requirement allows to derive the causality condition algebraically.

2 Logarithmic Conformal Invariance

In various physical situations presenting an equilibrium phase transition, for example disordered systems [6], percolation [12, 43] or sand-pile models [52], it has been useful to consider degenerate vacuum states. Formally, this can be implemented [19, 54] by replacing the order parameter ϕ by a vector $\begin{pmatrix} \psi \\ \phi \end{pmatrix}$ and the scaling dimension x by a Jordan matrix $\begin{pmatrix} x & 1 \\ 0 & x \end{pmatrix}$. For reviews, see [11, 17].

Here, we consider an analogous extension of the representations of the Schrödinger and conformal Galilean algebras. Consider the two-point functions[7]

$$F := \langle \phi_1(t_1, r_1)\phi_2(t_2, r_2)\rangle \ , \quad G := \langle \phi_1(t_1, r_1)\psi_2(t_2, r_2)\rangle \ , \\ H := \langle \psi_1(t_1, r_1)\psi_2(t_2, r_2)\rangle \tag{13}$$

Temporal and spatial translation-invariance imply that the are functions $F = F(t, r)$, $G = G(t, r)$ and $H = H(t, r)$ with $t = t_1 - t_2$ and $r = r_1 - r_2$. Since we shall explain the method in more detail below, we now simply quote the results and generalise them immediately to an arbitrary space dimension d. Co-variance under the logarithmic extension of either $\mathfrak{sch}(d)$ or $\mathrm{CGA}(d)$ implies $x_1 = x_2 =: x$ and

[7] Here and later, $\langle \cdot \rangle$ refers to an average over the thermal noise.

$F = 0$. For logarithmic Schrödinger invariance [32]

$$G = G_0|t|^{-x}\exp\left[-\frac{\mathcal{M}}{2}\frac{r^2}{t}\right] \;,\; H = \big(H_0 - G_0\ln|t|\big)\,|t|^{-x}\exp\left[-\frac{\mathcal{M}}{2}\frac{r^2}{t}\right] \quad (14)$$

subject to the constraint [2] $\mathcal{M} := \mathcal{M}_1 = -\mathcal{M}_2$.[8] For the case of logarithmic conformal Galilean invariance [30]

$$G = G_0|t|^{-2x}\exp\left[-2\frac{\gamma\cdot r}{t}\right] \;,\; H = \big(H_0 - 2G_0\ln|t|\big)\,|t|^{-2x}\exp\left[-2\frac{\gamma\cdot r}{t}\right] \quad (15)$$

together with the constraint $\gamma := \gamma_1 = \gamma_2$. Here, G_0, H_0 are normalisation constants. The presence of the logarithmic terms explain the name of 'logarithmic extension'.

3 Extension to Maximal Parabolic Sub-Algebras

Clearly, the results (14, 15) do not contain any information on causality. In order to write down the required extension of the symmetry algebras, we first *consider the 'mass' parameter $\mathcal{M}$ as a further variable* (for the moment for the scalar case) and write [18]

$$\widehat{\phi}(\zeta, t, r) := \frac{1}{\sqrt{2\pi}}\int_{\mathbb{R}} \mathrm{d}\mathcal{M}\; e^{\mathrm{i}\mathcal{M}\zeta}\,\phi_{\mathcal{M}}(t, r) \quad (16)$$

which defines the coordinate ζ dual to $\mathcal{M}$ which we shall consider as a '$(-1)^{\mathrm{st}}$' coordinate.[9] From now on, we concentrate on the case $d = 1$ for simplicity. The generators of $\mathfrak{sch}(1)$ become

[8] In order to keep the physical convention of non-negative masses $\mathcal{M} \geq 0$, one may introduce a 'complex conjugate' ϕ^* to the scaling field ϕ, with $\mathcal{M}^* = -\mathcal{M}$. In dynamics, co-variant two-point functions are interpreted as response functions, written as $R(t, s) = \langle\phi(t)\widetilde{\phi}(s)\rangle$ in the context of Janssen-de Dominicis theory, where the response field $\widetilde{\phi}$ has a mass $\widetilde{\mathcal{M}} = -\mathcal{M}$, see e.g. [8, 29] for details.
Furthermore, the physical relevant equations are *stochastic* Langevin equations, whose noise terms do break any interesting extended dynamical scale-invariance. However, one may identify a 'deterministic part' which may be Schrödinger-invariant, such that the predictions (14) remain valid even in the presence of noise [51]. This was rediscovered recently under name of 'time-dependent deformation of Schrödinger geometry' [46].

[9] In string theory and non-relativistic versions of the celebrated AdS/CFT correspondence [44], an analogous construction is used [38, 45, 59], with interesting applications to cold atoms [13].

$$
\begin{aligned}
X_n &= \frac{\mathrm{i}}{2}(n+1)nt^{n-1}r^2\partial_\zeta - t^{n+1}\partial_t - \frac{n+1}{2}t^n r\partial_r - \frac{n+1}{2}xt^n \\
Y_m &= \mathrm{i}\left(m+\frac{1}{2}\right)t^{m-1/2}r\partial_\zeta - t^{m+1/2}\partial_j \\
M_n &= \mathrm{i}t^n\partial_\zeta
\end{aligned}
\tag{17}
$$

The extension to the maximal parabolic sub-algebra $\widetilde{\mathfrak{sch}}(1) = \mathfrak{sch}(1) + \mathbb{C}N$ is achieved by including the generator

$$
N := \zeta\partial_\zeta - t\partial_t + \xi\,. \tag{18}
$$

In order to understand the origin of the constant term ξ, which in what follows will turn out to play the rôle of a second scaling dimension, consider another representation of the conformal Galilean algebra $\mathrm{CGA}(1) = \langle X_1, Y_{\pm 1/2}, M_0, V_+, 2X_0 - N\rangle$, see Fig. 1a. Herein, the generator X_1 takes a slightly generalised form[10]

$$
X_1 = \mathrm{i}r^2\partial_\zeta - t^2\partial_t - tr\partial_r - (x+\xi)\,t \tag{19}
$$

along with the new generator

$$
V_+ = -\zeta r\partial_\zeta - tr\partial_t - \left(\mathrm{i}\zeta t + \frac{r^2}{2}\right)\partial_r - (x+\xi)\,r \tag{20}
$$

All other generators are as in (17). One readily verifies that $[V_+, Y_{-1/2}] = 2X_0 - N$, with the explicitly given forms and this explains the presence of the constant ξ in (18).

The chosen normalisation of the generators is clarified by the commutator $[V_+, Y_{1/2}] = X_1$ and the remaining commutators of $\mathrm{CGA}(1)$ are promptly verified. *These generators act as a dynamical symmetry of the Schrödinger equation*

$$
\mathscr{S}\widehat{\phi} = 0\ ,\quad \mathscr{S} = -2\mathrm{i}\partial_\zeta\partial_t - \partial_r^2 - 2\mathrm{i}\left(x+\xi-\frac{1}{2}\right)t^{-1}\partial_\zeta \tag{21}
$$

in the sense that the generators of $\mathrm{CGA}(1)$ *map solutions of* $\mathscr{S}\widehat{\phi} = 0$ *onto another solution.*

Proof To check this, it suffices to verify the commutators

$$
\begin{aligned}
&[\mathscr{S}, V_+] = -2r\mathscr{S},\quad [\mathscr{S}, X_1] = -2t\mathscr{S},\quad [\mathscr{S}, X_0] = -\mathscr{S} \\
&[\mathscr{S}, N] = [\mathscr{S}, Y_{-1/2}] = [\mathscr{S}, M_0] = 0
\end{aligned}
$$

[10] The same form of X_1 also arises in the ageing sub-algebra $\mathfrak{age}(1) = \langle X_{1,0}, Y_{\pm 1/2}, M_0\rangle \subset \mathfrak{sch}(1)$. Physically, the presence of ξ, together with the absence of the time-translations $X_{-1} = -\partial_t$, leads to distinct exponents a and a' in (12).

and to recall that $X\widehat{\phi}$ with $X \in \text{CGA}(1)$ generates an infinitesimal transformation on the solution $\widehat{\phi}$. □

In general, a *standard parabolic sub-algebra* of a simple complex Lie algebra is spanned by the Cartan sub-algebra $\mathfrak{h}$ and a set of 'positive' generators [37]. We illustrate this for the example B_2, using Fig. 1a. The separation between positive and non-positive generators can be introduced by drawing a straight line through the Cartan sub-algebra $\mathfrak{h}$, indicated by the double point in the centre and then defining all generators who are represented by a dot to the right of this line as 'positive'. It is well-known that the Weyl group (which acts on the root diagramme) maps isomorphic sub-algebras onto each other. Hence, it is enough to consider the cases when the straight line mentioned above has a slope between unity and infinity. Then one finds the following classification of the non-isomorphic maximal standard parabolic sub-algebras of B_2 [26]: (i) if the slope is unity, one has $\widetilde{\mathfrak{sch}}(1)$, (ii) for a finite slope larger than unity, one has $\widetilde{\mathfrak{age}}(1) = \langle X_{0,1}, Y_{\pm 1/2}, M_0, N\rangle$ and (iii) if the slope is infinite, one has $\widetilde{\text{CGA}}(1)$.

4 Dual Logarithmic Schrödinger-Invariance

We now describe the consequences of logarithmic Schrödinger-invariance for the 'dual' formulation introduced in the previous section. This representation is constructed from (17) by the formal substitution $x \to \begin{pmatrix} x & x' \\ 0 & x \end{pmatrix}$, where we explicitly keep the two possibilities $x' = 0$ and $x' = 1$. Only the generators $X_{0,1}$ are modified and now read

$$
\begin{aligned}
X_0 &= -t\partial_t - \frac{1}{2}r\partial_r - \frac{1}{2}\begin{pmatrix} x & x' \\ 0 & x \end{pmatrix} \\
X_1 &= \frac{\mathrm{i}}{2}r^2\partial_\zeta - t^2\partial_t - tr\partial_r - t\begin{pmatrix} x & x' \\ 0 & x \end{pmatrix}
\end{aligned}
\tag{22}
$$

The co-variant two-point functions, built from quasi-primary scaling operators $\begin{pmatrix} \phi_i \\ \psi_i \end{pmatrix}$ which are characterised by the values of x_i and x_i', to be studied are

$$
\begin{aligned}
\widehat{F}(\zeta,t,r) &:= \langle \widehat{\phi}_1(\zeta_1,t_1,r_1)\widehat{\phi}_2(\zeta_2,t_2,r_2)\rangle \\
\widehat{G}_{12}(\zeta,t,r) &:= \langle \widehat{\phi}_1(\zeta_1,t_1,r_1)\widehat{\psi}_2(\zeta_2,t_2,r_2)\rangle \\
\widehat{G}_{21}(\zeta,t,r) &:= \langle \widehat{\psi}_1(\zeta_1,t_1,r_1)\widehat{\phi}_2(\zeta_2,t_2,r_2)\rangle \\
\widehat{H}(\zeta,t,r) &:= \langle \widehat{\psi}_1(\zeta_1,t_1,r_1)\widehat{\psi}_2(\zeta_2,t_2,r_2)\rangle
\end{aligned}
\tag{23}
$$

where $\zeta = \zeta_1 - \zeta_2, t = t_1 - t_2$ and $r = r_1 - r_2$. This form already takes translation-invariance in the three variables ζ, t, r into account which in turn follow from the covariance under $M_0, Y_{-1/2}, X_{-1}$, respectively.[11] Next, we consider the consequences of co-variance under the Galilei-transformations generated by $Y_{1/2}$. For the first of the two-point functions (23) this implies the differential equation (called 'projective Ward identity' in physics[12])

$$\left(\mathrm{i}(r_1 - r_2)\partial_\zeta - (t_1 - t_2)\partial_r\right) \widehat{F} = 0 \tag{24}$$

whose general solution (and similarly for the other two-point functions) is

$$\widehat{F} = \widehat{F}(t,u) \ , \ \widehat{G}_{12} = \widehat{G}_{12}(t,u) \ , \ \widehat{G}_{21} = \widehat{G}_{21}(t,u) \ , \ \widehat{H} = \widehat{H}(t,u) \ ; \quad u := 2\zeta t + \mathrm{i}r^2 \tag{25}$$

The new specific information of the logarithmic representations becomes first evident from dilatation-covariance, generated by X_0. When taking the previous results (25) into account, the projective Ward identities become, for the four distinct functions in (23)

$$\begin{aligned}
\left(-t\partial_t - u\partial_u - \frac{1}{2}(x_1 + x_2)\right) \widehat{F}(t,u) &= 0 \\
\left(-t\partial_t - u\partial_u - \frac{1}{2}(x_1 + x_2)\right) \widehat{G}_{12}(t,u) &= \frac{x_2'}{2}\widehat{F}(t,u) \\
\left(-t\partial_t - u\partial_u - \frac{1}{2}(x_1 + x_2)\right) \widehat{G}_{21}(t,u) &= \frac{x_1'}{2}\widehat{F}(t,u) \\
\left(-t\partial_t - u\partial_u - \frac{1}{2}(x_1 + x_2)\right) \widehat{H}(t,u) &= \frac{x_1'}{2}\widehat{G}_{12}(t,u) + \frac{x_2'}{2}\widehat{G}_{21}(t,u)
\end{aligned} \tag{26}$$

Rather than solving this directly, it is more efficient to use first the information coming from the special Schrödinger transformations generated by X_1. Applied to the first two-point function $\widehat{F}$, the use of (24, 26) gives

$$\left(\frac{\mathrm{i}}{2}r^2\partial_\zeta - t^2\partial_t - tr\partial_r - tx_1\right) \widehat{F}(t,u) = 0 \tag{27}$$

Applying again (26), we have the system

$$\left.\begin{aligned} (-t\partial_t - u\partial_u - x_1)\,\widehat{F} &= 0 \\ (-t\partial_t - u\partial_u - (x_1 + x_2)/2)\,\widehat{F} &= 0 \end{aligned}\right\} \Longrightarrow (x_1 - x_2)\,\widehat{F} = 0 \tag{28}$$

[11] Since the kinetic term of the invariant Schrödinger Eq. (21) reduces to a Laplace operator in a convenient basis, the calculations are analogous to those of logarithmic conformal invariance.

[12] We prefer to include the terms describing the transformation of the physical scaling operators right into the generators, while many authors only include them into the projective Ward identities. The end result is the same, the difference corresponds to the distinction between active and passive transformations.

and we have proven the following

Proposition 1 *If* $\begin{pmatrix}\phi\\ \psi\end{pmatrix}$ *is a quasi-primary scaling operator of logarithmic Schrödinger-invariance with generators* (17, 22), it the two-point function $\widehat{F} = \langle\widehat{\phi}_1\widehat{\phi}_2\rangle$ *satisfies one of the following conditions: (i)* $x_1 = x_2$*, (ii)* $\widehat{F} = 0$.

Now, we consider the mixed two-point functions $\widehat{G}_{12}$ and $\widehat{G}_{21}$. In complete analogy with the above calculations, we find

$$\left.\begin{array}{r}(-t\partial_t - u\partial_u - x_1)\,\widehat{G}_{12} = 0\\ (-t\partial_t - u\partial_u - (x_1+x_2)/2)\,\widehat{G}_{12} - \frac{1}{2}x_2'\widehat{F} = 0\end{array}\right\} \Longrightarrow (x_1-x_2)\,\widehat{G}_{12} = x_2'\widehat{F} \tag{29}$$

$$\left.\begin{array}{r}(-t\partial_t - u\partial_u - x_1)\,\widehat{G}_{21} = 0\\ (-t\partial_t - u\partial_u - (x_1+x_2)/2)\,\widehat{G}_{21} - \frac{1}{2}x_1'\widehat{F} = 0\end{array}\right\} \Longrightarrow (x_1-x_2)\,\widehat{G}_{21} = x_1'\widehat{F} \tag{30}$$

Proposition 2 *If either* $x_2' \neq 0$ *and* $\widehat{G}_{12} \neq 0$ *or else* $x_1' \neq 0$ *and* $\widehat{G}_{21} \neq 0$*, then both (i)* $x := x_1 = x_2$ *and (ii)* $\widehat{F} = 0$ *hold true.*

Obviously, at least one of $\widehat{G}_{12}$ or $\widehat{G}_{21}$ must be non-zero in order to a have non-trivial answer. More information is obtained from the last two-point function $\widehat{H}$, for which covariance under the generators $X_{0,1}$ implies, using also that $x_1 = x_2$

$$\left.\begin{array}{r}(-t\partial_t - u\partial_u - x_1)\,\widehat{H} - x_1'\widehat{G}_{12} = 0\\ (-t\partial_t - u\partial_u - (x_1+x_2)/2)\,\widehat{H} - \frac{1}{2}x_1'\widehat{G}_{12} - \frac{1}{2}x_2'\widehat{G}_{21} = 0\end{array}\right\} \Longrightarrow x_1'\widehat{G}_{12} = x_2'\widehat{G}_{21} \tag{31}$$

Consequently, one must distinguish two essentially distinct cases:

$\boxed{x_1' = x_2' = 1}$. We shall refer to this situation as the **symmetric case**. The scaling operators $\begin{pmatrix}\widehat{\phi}_1\\ \widehat{\psi}_1\end{pmatrix}$ and $\begin{pmatrix}\widehat{\phi}_2\\ \widehat{\psi}_2\end{pmatrix}$ are identical. Since under the exchange of the two operators, one has $t \mapsto -t$ and $u \mapsto u$, it follows that $\widehat{G}_{12} = \widehat{G}(t,u)$ and $\widehat{G}_{21} = \widehat{G}(-t,u)$. Because of (31), the function $\widehat{G}(t,u) = \widehat{G}(-t,u)$ is symmetric. Solving the differential Eq. (29), we have

$$\widehat{G}(t,u) = |t|^{-x}\,\widehat{g}\left(u|t|^{-1}\right) \tag{32}$$

where $\widehat{g}$ is a differentiable scaling function. Inserting into (31) and integrating

$$\widehat{H}(t,u) = |t|^{-x}\left(\widehat{h}\left(u|t|^{-1}\right) - \ln|t|\,\widehat{g}\left(u|t|^{-1}\right)\right) \tag{33}$$

Finally, we return to the formulation with fixed masses $\mathscr{M}_{1,2}$, which gives

Proposition 3 *The co-variant two-point functions of the logarithmic representation* (17, 22) *of* $\mathfrak{sch}(1)$ *are, with* $x := x_1 = x_2$

$$
\begin{aligned}
F(t,r) &= 0 \\
G(t,r) &= \delta(\mathscr{M}_1+\mathscr{M}_2)\,|t|^{-x}\,\exp\left[-\frac{\mathscr{M}_1}{2}\frac{r^2}{t}\right] g_0\,(\mathrm{sign}\,(t),\mathscr{M}_1) \\
H(t,r) &= \delta(\mathscr{M}_1+\mathscr{M}_2)\,|t|^{-x}\,\exp\left[-\frac{\mathscr{M}_1}{2}\frac{r^2}{t}\right] \\
&\qquad \times \left(h_0\,(\mathrm{sign}\,(t),\mathscr{M}_1) - \ln|t|\,g_0\,(\mathrm{sign}\,(t),\mathscr{M}_1)\right)
\end{aligned}
\tag{34}
$$

where g_0 and h_0 *are unspecified functions and* $\delta(\mathscr{M})$ *is the Dirac distribution.*

Comparing with the prediction (14), we can identify $G_0 = g_0$ and $H_0 = h_0$. Notice: logarithmic Schrödinger-invariance did *not* produce the causality constraint $t > 0$!

Proof We illustrate the proof of (34) for $G(t,r)$. Using $\zeta = \zeta_1 - \zeta_2$, $\eta = \zeta_1 + \zeta_2$, we have

$$
\begin{aligned}
G(t,r) &= \frac{1}{2\pi}\int_{\mathbb{R}^2} \mathrm{d}\zeta_1 \mathrm{d}\zeta_2\, e^{-\mathrm{i}\mathscr{M}_1\zeta_1 - \mathrm{i}\mathscr{M}_2\zeta_2}\,|t|^{-x}\,\widehat{g}\left(\frac{2(\zeta_1-\zeta_2)t + \mathrm{i}r^2}{|t|}\right) \\
&= \frac{1}{4\pi}|t|^{-x}\int_{\mathbb{R}}\mathrm{d}\eta\, e^{-\mathrm{i}(\mathscr{M}_1+\mathscr{M}_2)\eta/2}\int_{\mathbb{R}}\mathrm{d}\zeta\, e^{-\mathrm{i}(\mathscr{M}_1-\mathscr{M}_2)\zeta/2}\,\widehat{g}\left(2\mathrm{sign}\,(t)\left(\zeta + \frac{\mathrm{i}}{2}\frac{r^2}{\mathrm{sign}\,(t)\,|t|}\right)\right) \\
&= \delta(\mathscr{M}_1+\mathscr{M}_2)|t|^{-x}\int_{\mathbb{R}}\mathrm{d}\zeta\, e^{-\mathrm{i}\mathscr{M}_1\zeta}\,\widehat{g}\left(2\,\mathrm{sign}\,(t)\left(\zeta+\frac{\mathrm{i}}{2}\frac{r^2}{t}\right)\right) \\
&= \delta(\mathscr{M}_1+\mathscr{M}_2)|t|^{-x}\exp\left[-\frac{\mathscr{M}_1}{2}\frac{r^2}{t}\right]\underbrace{\int_{\mathbb{R}}\mathrm{d}\zeta\, e^{-\mathrm{i}\mathscr{M}_1\zeta}\,\widehat{g}\,(2\,\mathrm{sign}\,(t)\zeta)}_{=g_0(\mathrm{sign}\,(t),\mathscr{M}_1)}
\end{aligned}
$$

with a change of variables in the last line and we have also assumed that $\widehat{g}$ has no singularity 'near to' the real axis which could prevent shifting the contour. H is derived similarly. □

$\boxed{\begin{array}{l} x'_1 = 0 \\ x'_2 = 1 \end{array}}$. This is called the **asymmetric case**. The mirror situation $x'_1 = 1, x'_2 = 0$ is analogous. Now, from (31) we have $G_{21} = 0$. Inserting into and solving (29, 31), we have

$$
\widehat{G}_{12}(t,u) = t^{-x}\,\widehat{g}\left(ut^{-1}\right) \quad , \quad \widehat{H}(t,u) = t^{-x}\,\widehat{h}\left(ut^{-1}\right) \tag{35}
$$

without any logarithmic term ! Again, no causality condition is produced.

5 Causality in Maximal Parabolic Sub-Algebras

In the previous section we had seen that $\mathfrak{sch}(1)$-covariance alone is not strong enough to derive the causality condition $t > 0$ for the two-point function. We now show that indeed *causality is implied if covariance under the maximal parabolic sub-algebra* $\widetilde{\mathfrak{sch}}(1)$ *is required.* In what follows, it will be essential that $M_0 = \mathrm{i}\partial_\zeta$ generates translations in the dual coordinate. In consequence, the M_0-covariant two-point functions merely depend on $\zeta = \zeta_1 - \zeta_2$.

We begin by extending N to a logarithmic representation by replacing the second scaling dimension ξ by a matrix $\Xi = \begin{pmatrix} \xi & \xi' \\ \xi'' & \xi \end{pmatrix}$ and write

$$N = \zeta\partial_\zeta - t\partial_t + \begin{pmatrix} \xi & \xi' \\ \xi'' & \xi \end{pmatrix}. \tag{36}$$

Proposition 4 *One can always arrange in (36) for $\xi'' = 0$.*

Proof Since both X_0 and N are in the Cartan sub-algebra of B_2, see Fig. 1a, we must have $[X_0, N] = \frac{1}{2}x'\xi'' \begin{pmatrix} 1 & 0 \\ 0 & -1 \end{pmatrix} = 0$, hence $x'\xi'' = 0$. If $x' = 0$, one asks whether Ξ can be diagonalised. If that is so, one has the non-interesting case of a pair of non-logarithmic quasi-primary operators. If Ξ cannot be diagonalised, it can be brought to a Jordan form and one can always arrange for $\xi'' = 0$. Therefore, we can set $\xi'' = 0$ in (36) without restriction of the generality. One can check the commutators of $\widetilde{\mathfrak{sch}}(1)$, notably $[X_1, N] = X_1$. □

Using the results of Sect. 4, co-variance under N yields

$$N\widehat{G}_{12}(t,u) = (-t\partial_t + \xi_1 + \xi_2)\,\widehat{G}_{12}(t,u) = 0 \tag{37}$$

Solving this first for $t > 0$, this implies $\widehat{G}_{12}(t,u) = t^{\xi_1+\xi_2}\widehat{\gamma}(u)$. Comparison with the scaling form (32) leads to $\widehat{G}_{12} = \widehat{g}_0 t^{\xi_1+\xi_2}u^{-x-\xi_1-\xi_2}$. Together with the results of Sect. 4, and setting $v = u/t$, we have the scaling function

$$\widehat{g}(v) = \widehat{g}_0\, v^{-x-\xi_1-\xi_2} \tag{38}$$

where $\widehat{g}_0$ is a normalisation constant. The last two-point function $\widehat{H}$ can be found from

$$N\widehat{H}(t,u) = (-t\partial_t + \xi_1 + \xi_2)\,\widehat{H}(t,u) + \xi_1'\widehat{G}_{12}(t,u) + \xi_2'\widehat{G}_{21}(t,u) = 0\,. \tag{39}$$

We now look at the two cases defined in Sect. 4.

5.1 Symmetric Case

A straightforward calculation gives, using (32, 33, 38, 39)

$$\widehat{G}(\zeta,t,r) = \widehat{g}_0\,|t|^{-x}\left(\frac{2\zeta t + \mathrm{i}r^2}{|t|}\right)^{-x-\xi_1-\xi_2}$$

$$\widehat{H}(\zeta,t,r) = |t|^{-x}\left(\frac{2\zeta t + \mathrm{i}r^2}{|t|}\right)^{-x-\xi_1-\xi_2} \times \left(\widehat{h}_0 + \widehat{g}_0(1+\xi_1'+\xi_2')\ln\left(\frac{2\zeta t + \mathrm{i}r^2}{|t|}\right) - \widehat{g}_0 \ln|t|\right) \tag{40}$$

where $\widehat{g}_0$ and $\widehat{h}_0$ are normalisation constants. We can now state the main result.

Theorem *Quasi-primary scaling operators* $\begin{pmatrix}\phi_i\\ \psi_i\end{pmatrix}$*, which are scalars under spatial rotations and transform co-variantly under a logarithmic representation of the parabolic sub-algebra* $\widetilde{\mathfrak{sch}}(d)$*, are characterised by the simultaneous Jordan matrices* $\begin{pmatrix}x_i & x_i'\\ 0 & x_i\end{pmatrix}$ *and* $\begin{pmatrix}\xi_i & \xi_i'\\ 0 & \xi_i\end{pmatrix}$ *and the masses* $\mathscr{M}_i$*. Assume that* $\mathscr{M}_1 > 0$ *and furthermore that* $\frac{1}{2}(x_1+x_2)+\xi_1+\xi_2 > 0$*. If* $x_1' = x_2' = 1$ *, the co-variant two-point functions (13) have the following causal forms*

$$\begin{aligned}
F(t,r) &= 0\\
G(t,r) &= \delta(\mathscr{M}_1+\mathscr{M}_2)\,\delta_{x_1,x_2}\,\Theta(t)\,t^{-x_1}\exp\left[-\frac{\mathscr{M}_1}{2}\frac{r^2}{t}\right]G_0\\
H(t,r) &= \delta(\mathscr{M}_1+\mathscr{M}_2)\,\delta_{x_1,x_2}\,\Theta(t)\,t^{-x_1}\exp\left[-\frac{\mathscr{M}_1}{2}\frac{r^2}{t}\right](H_0 - G_0\ln t)
\end{aligned} \tag{41}$$

where G_0 *and* H_0 *are normalisation constants,* $\Theta(t)$ *is the Heaviside function and* $\delta_{a,b} = 1$ *if* $a=b$ *and zero otherwise.*

Here, we are mainly interested in the causality statement which is essentially contained in the following

Proposition 5 *Let* $x > 0$*,* n *be a non-negative integer and consider the integrals, in the limit* $\varepsilon \to 0+$

$$I_{\pm}^{(n)}(x) := \int_{\mathbb{R}\pm\mathrm{i}\varepsilon} \mathrm{d}\zeta\; e^{-\mathrm{i}\zeta}\,\zeta^{-x}\ln^n\zeta \tag{42}$$

Then $I_{-}^{(n)}(x) = 0$*. There is no simple known expression for* $I_{+}^{(n)}(x)$*.*

Proof To prove the proposition, consider the contour integrals

$$J_{\pm} := \oint_{C_\pm} \mathrm{d}\zeta\; e^{-\mathrm{i}\zeta}\,\zeta^{-x}\ln^n\zeta$$

where the contours $C_\pm$ correspond to $t > 0$ and $t < 0$, respectively, as we shall see below and are indicated in Fig. 2. For $x > 0$, the only singularity is the cut along the negative real axis, hence $J_\pm = 0$. We now estimate the contribution of the lower half-circle, $J_{-,\mathrm{inf}}$. Setting $\zeta = Re^{\mathrm{i}\theta}$ such that $\ln R > 1$, one has

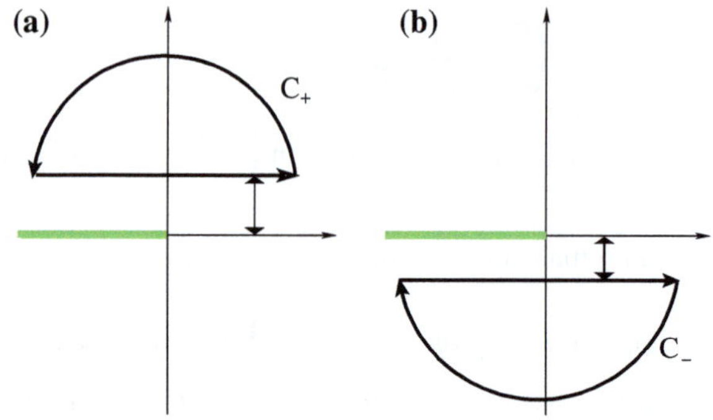

Fig. 2 Integration contours **a** C_+ for $t > 0$ and **b** C_- for $t < 0$. The cut is indicated by the *thick line*

$$J_{-,\inf} = \frac{1}{\mathrm{i}} \int_0^{\pi} \mathrm{d}\theta\; R^{1-x} e^{\mathrm{i}\theta(x-1) - \mathrm{i}R\cos\theta} e^{-R\sin\theta} \left(\ln R e^{-\mathrm{i}\theta}\right)^n$$

Computing the complex logarithm via the binomial theorem, one has the estimate

$$\begin{aligned}
\left|J_{-,\inf}\right| &\le \int_0^{\pi} \mathrm{d}\theta\; R^{1-x} e^{-R\sin\theta} \sum_{k=0}^{n} \binom{n}{k} \ln^{n-k} R\, \theta^k \\
&\le \sum_{k=0}^{n} \binom{n}{k} R^{1-x} (\ln R)^{n-k}\, \pi^k \underbrace{\int_0^{\pi} \mathrm{d}\theta\; e^{-R\sin\theta}}_{\le \pi R^{-1}} \\
&\le \pi R^{-x} (\pi + \ln R)^n \to 0
\end{aligned}$$

as $R \to \infty$. Hence, since $J_- = I_-^{(n)}(x) + J_{-,\inf} = 0$, the assertion follows. □

Proof (of the Theorem) In order to prove the theorem, recall first that for quasi-primary operators which are scalars under rotations, one can always reduce to the case $d = 1$. Hence the spatial dependence in (41) is a direct consequence of (34). Writing $\xi := \xi_1 + \xi_2$, we use the physical convention of positive masses $\mathscr{M}_1 > 0$ and have along the lines of the proof of proposition 3

$$\begin{aligned}
G &= \delta(\mathscr{M}_1 + \mathscr{M}_2)|t|^{-x} \widehat{g}_0 \int_{\mathbb{R}} \mathrm{d}\zeta\; e^{-\mathrm{i}\mathscr{M}_1\zeta} \left(2\mathrm{sign}\,(t)\right)^{-x-\xi} \left(\zeta + \frac{\mathrm{i}r^2}{2\mathrm{sign}\,(t)|t|}\right)^{-x-\xi} \\
&= \delta(\mathscr{M}_1 + \mathscr{M}_2) \left(2\mathrm{sign}\,(t)\right)^{-x-\xi} \mathscr{M}_1^{x+\xi-1} |t|^{-x} \widehat{g}_0 \underbrace{\int_{\mathbb{R} + \frac{\mathrm{i}\mathscr{M}_1}{2}\frac{r^2}{t}} \mathrm{d}\zeta\; e^{-\mathrm{i}\zeta} \zeta^{-x-\xi}}_{I_{\pm}^{(0)}(x+\xi)}\; e^{-\frac{\mathscr{M}_1}{2}\frac{r^2}{t}} \\
&= \delta(\mathscr{M}_1 + \mathscr{M}_2)|t|^{-x} \underbrace{2^{-x-\xi} \mathscr{M}_1^{x+\xi-1} \widehat{g}_0 I_+^{(0)}(x+\xi)}_{=: G_0}\; e^{-\frac{\mathscr{M}_1}{2}\frac{r^2}{t}} \Theta(t)
\end{aligned}$$

where in the second line we see that for $t > 0$ ($t < 0$) the contour is slightly above (below) the real axis and we need $I_+^{(0)}$ ($I_-^{(0)}$). In the last line, we used the statement $I_-^{(0)}(x+\xi) = 0$ of proposition 5 and expressed this by the Heaviside function. Similarly, for H we use (41) and obtain along the same lines

$$H = \delta(\mathscr{M}_1 + \mathscr{M}_2)|t|^{-x} e^{-\frac{\mathscr{M}_1}{2}\frac{r^2}{t}} 2^{-x-\xi} \mathscr{M}_1^{x+\xi-1} \Big[-\widehat{g}_0 \ln|t| I_\pm^{(0)}(x+\xi)$$
$$+ \left(\widehat{h}_0 + \widehat{g}_0(1+\xi_1'+\xi_2')\ln(2\mathrm{sign}\,(t)/\mathscr{M}_1)\right) I_\pm^{(0)}(x+\xi) + \widehat{g}_0(1+\xi_1'+\xi_2') I_\pm^{(1)}(x+\xi) \Big]$$

and by proposition 5 and defining H_0 from the constants in the second line, the announced causal form follows. □

Remarks and Generalisations:

(a) Equation (41) reproduces the known form (14) [32] of logarithmic Schrödinger-covariance, but adds the causality condition $t > 0$ described by the extra factor $\Theta(t)$. Our derivation generalises earlier causality proofs for the non-logarithmic case and under the more strong condition $x > 0$ [26].

The extension to $d > 1$ dimensions is immediate.

(b) For physical applications, recall the form (12) of the response function $R = \langle \phi\widetilde{\phi} \rangle$ with a positive mass $\mathscr{M}_\phi > 0$ and a negative mass $\mathscr{M}_{\widetilde{\phi}} = -\mathscr{M}_\phi < 0$ such that the 'mass conservation' following from galilean invariance is accounted for. The response field $\widetilde{\phi}$ is associated with the complex conjugate ϕ^* in (16).

(c) Since the generator of time-translations $X_{-1} \in \mathfrak{sch}(1)$, the proven scaling forms (41) correspond to $a = a'$ in (12). However, the specific form (18, 36) of the generator N is already compatible with the more general representations (or equivalently the Ward identities) required for the maximal parabolic extension of the ageing algebra, $\widetilde{\mathfrak{age}}(1)$ [27, 30]. Hence the causality arguments presented here explicitly for Schrödinger-invariance can be directly generalised to ageing-invariance, including the logarithmic extension. Hence our present results also provide a mathematical justification for the successful empirical comparison of numerical data of response functions from critical directed percolation [30] and the $1D$ KPZ Eq. [31] with the co-variant two-point function of logarithmic ageing-invariance.

(d) Galilei-covariance is an essential assumption. While it seems to be well confirmed in many numerical tests of specific models, see [29] and references therein, it is very difficult to prove formally. Finding such an argument remains an important open problem.[13]

(e) The second essential ingredient is the dualisation with respect to the mass $\mathscr{M}$, and that co-variance under the corresponding generator $M_0 = \mathrm{i}\partial_\zeta$ takes the form of translation-invariance in the dual coordinate ζ. For illustration of this ingredient, consider the (non-logarithmic) representation $\widetilde{\mathrm{CGA}}(1) = \langle X_1, Y_{\pm 1/2}, D, M_0, V_+, N \rangle$ from Sect. 3, with the dilatations $D = 2X_0 - N = -\zeta\partial_\zeta - t\partial_t - r\partial_r - (x+\xi)$.

[13] At present, the nearest one might come to a formal proof is to consider the models in the dualised form as introduced in Sect. 4. Then, one trades the phase changes of the usual solution of the 'Schrödinger equation' $\mathscr{S}\phi = 0$ for a transformation in the dual coordinate ζ. This seriously modifies the equation under study, but galilean co-variance can be checked [60].

Hence, the only effective scaling dimension appearing is $x + \xi$, hence the dual CGA(1)-covariant two-point function can be read from the literature [26, 28]

$$\left\langle\widehat{\phi}_1(\zeta_1,t_1,r_1)\widehat{\phi}_2(\zeta_2,t_2,r_2)\right\rangle = (t_1-t_2)^{-\frac{1}{2}(x_1+\xi_1+x_2+\xi_2)}\left(\frac{t_1}{t_2}\right)^{\frac{1}{2}(x_2+\xi_2-x_1-\xi_1)} \times f\left(\zeta_1-\zeta_2+\frac{\mathrm{i}}{2}\frac{(r_1-r_2)^2}{t_1-t_2}\right) \tag{43}$$

Requiring the co-variance $N\left\langle\widehat{\phi}_1\widehat{\phi}_2\right\rangle = 0$, with N given by (18), leads as before to $f(u) = \widehat{f}_0 u^{-(x_1+3\xi_1+x_2+3\xi_2)/2}$ and transforming back, we recover the causality condition $t_1 - t_2 > 0$, provided only that $x_1 + 3\xi_1 + x_2 + 3\xi_2 > 0$.
(f) M_0 plays the rôle of a central extension in the *Schrödinger* algebra. Such a central extension does not exist for CGA(d) with $d \neq 2$, but we expect that an argument similar to the one used here should apply to the exotic central generator Θ in the ECGA, after dualisation. This should allow, after the identification of the corresponding parabolic sub-algebra, to derive causality conditions in this case as well. We hope to return to this question in the future.

5.2 Asymmetric Case

Applying the conditions (37, 39) to the previously derived scaling forms (35), we promptly have

$$\widehat{g}(v) = \widehat{g}_0 v^{-x-\xi_1-\xi_2} \quad , \quad \widehat{h}(v) = v^{-x-\xi_1-\xi_2}\left(\widehat{h}_0 - \xi_1'\widehat{g}_0 \ln v\right) \tag{44}$$

Transforming back as before to the situation with fixed masses, we obtain under the same conditions as for the main theorem, but now with $x_1' = 0$ and $x_2' = 1$, that $F(t,r) = G_{21}(t,r) = 0$ and the causal, but non-logarithmic forms

$$G_{12}(t,r) = G_0\delta(\mathscr{M}_1+\mathscr{M}_2)\,\delta_{x_1,x_2}\,\Theta(t)\,t^{-x_1}\exp\left[-\frac{\mathscr{M}_1}{2}\frac{r^2}{t}\right]$$
$$H(t,r) = H_0\delta(\mathscr{M}_1+\mathscr{M}_2)\,\delta_{x_1,x_2}\,\Theta(t)\,t^{-x_1}\exp\left[-\frac{\mathscr{M}_1}{2}\frac{r^2}{t}\right] \tag{45}$$

Note added in proof: For the representation (5) of the non-exotic CGA, an analogous dualisation and parabolic extension rather shows that $<\phi_1(t)\phi_2(s)> = <\phi_1(s)\phi_2(t)>$ is fully symmetric [63].

Acknowledgments It is a pleasure to thank the organisers of the 7th AGMP conference in Mulhouse and especially R. Cherniha for their kind invitation. This work was partly supported by the Collège doctoral franco-allemand Nancy-Leipzig-Coventry (Systèmes complexes à l'équilibre et hors équilibre) of UFA-DFH.

References

1. Bagchi, A., Mandal, I.: On representations and correlation functions of Galilean conformal algebra. Phys. Lett. B **675**, 393 (2009). arXiv:0903.0580
2. Bargman, V.: Unitary ray representations of continuous groups. Ann. Math. **56**, 1 (1954)
3. Belavin, A.A., Polyakov, A.M., Zamolodchikov, A.B.: Infinite conformal symmetry in two-dimensional quantum field-theory. Nucl. Phys. **B241**, 333 (1984)
4. Bray, A.J.: Theory of phase-ordering. Adv. Phys. **43**, 357 (1994)
5. Burdet, G., Perrin, M., Sorba, P.: About the non-relativistic structure of the conformal algebra. Comm. Math. Phys. **34**, 85 (1973)
6. Caux, J.-S., Kogan, I.I., Tsvelik, A.M.: Logarithmic operators and hidden continuous symmetry in critical disordered models. Nucl. Phys. **B466**, 444 (1996). [hep-th/9511134]
7. Cherniha, R., Henkel, M.: The exotic conformal Galilei algebra and non-linear partial differential equations. J. Math. Anal. Appl. **369**, 120 (2010). arXiv:0910.4822
8. Cugliandolo, L.F.: Dynamics of glassy systems. In: Barrat, J.-L., Dalibard, J., Kurchan, J., Feigel'man, M.V., (eds.) Slow Relaxation and Non Equilibrium Dynamics in Condensed Matter, Les Houches Session 77 July 2002. Springer, Heidelberg (2003) [cond-mat/0210312]
9. Duval, C., Horváthy, P.A.: Non-relativistic conformal symmetries and Newton-Cartan structures. J. Phys. A: Math. Theor. **42**, 465206 (2009). arXiv:0904.0531
10. Einstein, A.: Zur Elektrodynamik bewegter Körper. Ann der Physik **17**, 891 (1905)
11. Flohr, M.: Bits and pieces in logarithmic conformal field-theory. Int. J. Mod. Phys. **A18**, 4497 (2003). [hep-th/0111228]
12. Flohr, M., Müller-Lohmann, A.: Proposal for a CFT interpretation of Watts' differential equation for percolation. J. Stat. Mech. P12004 (2005). [hep-th/0507211]
13. Fuertes, C.A., Moroz, S.: Correlation functions in the non-relativistic AdS/CFT correspondence, Phys. Rev. D**79**, 106004 (2009). arXiv:0903.1844
14. Fushchych, W.I., Cherniha, R.: Galilei-invariant nonlinear equations of Schrödinger-type and their exact solutions I. Ukrainian Math. J. **41**, 1161 (1989)
15. Fushchych, W.I., Shtelen, W.M., Serov, M.I.: Symmetry Analysis and Exact Solutions of Equations of Nonlinear Mathematical Physics. Kluwer, Dordrecht (1993)
16. Fushchych, W.I., Cherniha, R.: Galilei-invariant systems of nonlinear systems of evolution equations. J. Phys. **A28**, 5569 (1995)
17. Gaberdiel, M.R.: An algebraic approach to logarithmic conformal field theory. Int. J. Mod. Phys. **A18**, 4593 (2003). [hep-th/0111260]
18. Giulini, D.: On Galilei-invariance in quantum mechanics and the Bargmann superselection rule. Ann. Phys. **249**, 222 (1996). [quant-ph/9508002]
19. Gurarie, V.: Logarithmic operators in conformal field-theory. Nucl. Phys. **B410**, 535 (1993). [hep-th/9303160]
20. Hassaïne, M., Horváthy, P.A.: Field-dependent symmetries of a non-relativistic fluid model. Ann. Phys. **282**, 218 (2000) [math-ph/9904022]; Field-dependent symmetries of a non-relativistic fluid model. Phys. Lett. A**279**, 215 (2001). [hep-th/0009092]
21. Havas, P., Plebanski, J.: Conformal extensions of the Galilei group and their relation to the schrödinger group. J. Math. Phys. **19**, 482 (1978)
22. Henkel, M.: Schrödinger-invariance and strongly anisotropic critical systems. J. Stat. Phys. **75**, 1023 (1994). [hep-th/9310081]
23. Henkel, M.: Extended scale-invariance in strongly anisotropic equilibrium critical systems. Phys. Rev. Lett. **78**, 1940 (1997). [cond-mat/9610174]
24. Henkel, M., Pleimling, M., Godrèche, C., Luck, J.-M.: Ageing, phase ordering and conformal invariance. Phys. Rev. Lett. **87**, 265701 (2001). [hep-th/0107122]
25. Henkel, M.: Phenomenology of local scale invariance: from conformal invariance to dynamical scaling. Nucl. Phys. **B641**, 405 (2002). [hep-th/0205256]
26. Henkel, M., Unterberger, J.: Schrödinger invariance and space-time symmetries. Nucl. Phys. **B660**, 407 (2003). [hep-th/0302187]

27. Henkel, M., Enss, T., Pleimling, M.: On the identification of quasiprimary operators in local scale-invariance. J. Phys. A Math. Gen. **39**, L589 (2006). [cond-mat/0605211]
28. Henkel, M., Schott, R., Stoimenov, S., Unterberger, J.: The Poincaré algebra in the context of ageing systems: Lie structure, representations, Appell systems and coherent states. Confluentes Mathematici **4**, 125006 (2012). [math-ph/0601028]
29. Henkel, M., Pleimling, M.: Non-equilibrium Phase Transitions, vol. 2: Ageing and Dynamical Scaling Far from Equilibrium, Springer, Heidelberg (2010)
30. Henkel, M.: On logarithmic extensions of local scale-invariance. Nucl. Phys. B **864** [FS], 282 (2013). arXiv:1009.4139
31. Henkel, M., Noh, J.D., Pleimling, M.: Phenomenology of ageing in the Kardar-Parisi-Zhang equation. Phys. Rev. E **85**, 030102(R) (2012). 1109.5022
32. Hosseiny, A., Rouhani, S.: Logarithmic correlators in non-relativistic conformal field-theory. J. Math. Phys. **51** 102303 (2010). arXiv:1001.1036
33. Hosseiny, A., Naseh, A.: On holographic realization of logarithmic Galilean conformal algebra. J. Math. Phys. **52** 092501 (2011). arXiv:1101.2126
34. Hyun, S., Jeong, J., Kim, B.S.: Aging logarithmic conformal field theory : A holographic view. J. High Energy Phys. **1301** 141 (2013). arXiv:1209.2417
35. Hyun, S., Jeong, J., Kim, B.S.: Aging logarithmic galilean field-theories. Nucl. Phys. B **874**, 358 (2013). arXiv:1304.0007
36. Jacobi, C.G.: Vorlesungen über Dynamik (1842/1843), 4. Vorlesung. In: Clebsch, A., Lottner, E., (eds.) Gesammelte Werke, Akademie der Wissenschaften, Berlin 1866/1884
37. Knapp, A.W.: Representation theory of semisimple groups: An overview based on examples. Princeton University Press, Princeton (1986).
38. Leigh, R.G., Hoang, N.N.: Real-time correlators and non-relativistic holography. J. High-energy Phys. **0911**, 010 (2009) arXiv:0904.4270; Fermions and the Sch/nrCFT Correspondence. J. High-energy Phys. **1003**, 027 (2010) arXiv:0909.1883
39. Lie, S.: Über die Integration durch bestimmte Integrale von einer Klasse linearer partieller Differentialgleichungen. Arch. for Mathematik og Naturvidenskab **6**, 328 (1881)
40. Lorentz, H.A.: Electromagnetic phenomena in a system moving with any velocity smaller than that of light. Proc. Acad. Science Amsterdam **6**, 809 (1904)
41. Lukierski, J., Stichel, P.C., Zakrewski, W.J.: Exotic galilean conformal symmetry and its dynamical realisations, Phys. Lett. **A357**, 1 (2006) [hep-th/0511259]; Accelaration-extended galilean symmetries with central charges and their dynamical realisations. Phys. Lett. **B650**, 203 (2007). [hep-th/0702179]
42. Martelli, D., Tachikawa, Y.: Comments on Galiean conformal field-theories and their geometric realisation. J. High-energy Phys. **1005**, 091 (2010). arXiv:0903.5184
43. Mathieu, P., Ridout, D.: From percolation to logarithmic conformal field theory. Phys. Lett. B **657**, 120 (2007) arXiv:0708.0802; Logarithmic M(2, p) Minimal Models, their Logarithmic Couplings, and Duality. Nucl. Phys. B **801**, 268 (2008). arXiv:0711.3541
44. Maldacena, J.M.: The large-n limit of superconformal field-theories and super-gravity. Adv. Theor. Math. Phys. **2**, 231 (1998). [hep-th/9711200]
45. Minic, D., Pleimling, M.: Correspondence between nonrelativistic anti-de Sitter space and conformal field theory, and ageing-gravity duality. Phys. Rev. E **78**, 061108 (2008) arXiv:0807.3665; The Jarzynski Identity and the AdS/CFT Duality. Phys. Lett. B **700**, 277 (2011). arXiv:1007.3970
46. Nakayama, Y.: Universal time-dependent deformations of Schrödinger geometry. J. High-energy Phys. **04**, 102 (2010). arXiv:1002.0615
47. Negro, J., del Olmo, M.A., Rodríguez-Marco, A.: Nonrelativistic conformal groups. J. Math. Phys. **38** 3786–3809 (1997)
48. O'Raifeartaigh, L., Sreedhar, V.V.: The maximal kinematical invariance group of fluid dynamics and explosion-implosion duality. Ann. of Phys. **293**, 215 (2001)
49. Ovsiannikov, L.V.: The Group Analysis of Differential Equations. Academic Press, London (1980)

50. Ovsienko, V., Roger, C.: Generalisations of Virasoro group and Virasoro algebras through extensions by modules of tensor-densities on S^1. Indag. Math. **9**, 277 (1998)
51. Picone, A., Henkel, M.: Local scale-invariance and ageing in noisy systems. Nucl. Phys. B **688** 217 (2004). [cond-mat/0402196]
52. Poghosyan, V.S., Grigorev, S.Y., Priezzhev, V.B., Ruelle, P.: Pair correlations in sandpile model: A check of logarithmic conformal field theory. Phys. Lett. B **659**, 768 (2008). arXiv:0710.3051; Logarithmic two-point correlators in the Abelian sandpile model. J. Stat. Mech. P07025 (2010). arXiv:1005.2088
53. Polyakov, A.M.: Conformal symmetry of critical fluctuations. Sov. Phys. JETP Lett. **12**, 381 (1970)
54. Rahimi Tabar, M.R., Aghamohammadi, A., Khorrami, M.: The logarithmic conformal field theories. Nucl. Phys. B **497**, 555 (1997). [hep-th/9610168]
55. Rideau, G., Winternitz, P.: Evolution equations invariant under two-dimensional space-time Schrödinger group. J. Math. Phys. **34**, 558 (1993)
56. Riva, V., Cardy, J.L.: Scale and conformal invariance in field theory: a physical counterexample. Phys. Lett. **B622**, 339 (2005). [hep-th/0504197]
57. Roger, C., Unterberger, J.: The Schrödinger-Virasoro Lie group and algebra: From geometry to representation theory. Ann. Inst. H. Poincaré **7**, 1477 (2006). [math-ph/0601050]
58. Setare, M.R., Kamali, V.: Galilean conformal algebra in semi-infinite space. Int. J. Mod. Phys. A **27**, 1250044 (2012). arXiv:1101.2339; Anti-de Sitter/ boundary conformal field theory correspondence in the non-relativistic limit, Eur. Phys. J. **C72**, 2115 (2012). arXiv:1202.4917
59. Son, D.T.: Towards an AdS/cold atom correspondence: A geometric realisation of the Schrödinger symmetry. Phys. Rev. D **78**, 106005 (2008). arXiv:0804.3972
60. Stoimenov, S., Henkel, M.: Dynamical symmetries of semi-linear Schrödinger and diffusion equations. Nucl. Phys. B **723**, 205 (2005). [math-ph/0504028]
61. Unterberger, J., Roger, C.: The Schrödinger-Virasoro algebra. Springer, Heidelberg (2011)
62. Zhang, P.-M., Horváthy, P.A.: Non-relativistic conformal symmetries in fluid mechanics. Eur. Phys. J. C **65**, 607 (2010). arXiv:0906.3594
63. Henkel, M., Stoimenov, S.: Physical ageing and new representations of some Liealgebras of local scale-invarience, Proc. LT-10 Varna, Bulgariea (2014) arXiv:1401.6086
64. Negro, J., del Olmo, M.A., Rodríguez-Marco, A.: Nonrelativistic conformal groups II. J. Math. Phys. **38** 3810–3831 (1997)

Reaction-Diffusion Systems with Constant Diffusivities: Conditional Symmetries and Form-Preserving Transformations

Roman Cherniha and Vasyl' Davydovych

Abstract Q-conditional symmetries (nonclassical symmetries) for a general class of two-component reaction-diffusion systems with constant diffusivities are studied. Using the recently introduced notion of Q-conditional symmetries of the first type (R. Cherniha J. Phys. A: Math. Theor., 2010. vol. 43., 405207), an exhaustive list of reaction-diffusion systems admitting such symmetry is derived. The form-preserving transformations for this class of systems are constructed and it is shown that this list contains only non-equivalent systems. The obtained symmetries permit to reduce the reaction-diffusion systems under study to two-dimensional systems of ordinary differential equations and to find exact solutions. As a non-trivial example, multiparameter families of exact solutions are explicitly constructed for two nonlinear reaction-diffusion systems. A possible interpretation to a biologically motivated model is presented.

1 Introduction

The paper is devoted to the investigation of the two-component reaction-diffusion (RD) systems of the form

$$\begin{aligned} u_t &= d_1 u_{xx} + F(u, v), \\ v_t &= d_2 v_{xx} + G(u, v). \end{aligned} \tag{1}$$

R. Cherniha (✉) · V. Davydovych
Institute of Mathematics, Ukrainian National Academy of Sciences, Tereshchenkivs'ka Street 3, Kyiv 01601, Ukraine
e-mail: cherniha@gmail.com

V. Davydovych
e-mail: davydovych@imath.kiev.ua

R. Cherniha
School of Mathematical Sciences, University of Nottingham, Nottingham 2RD NG7, UK

A. Makhlouf et al. (eds.), *Algebra, Geometry and Mathematical Physics*,
Springer Proceedings in Mathematics & Statistics 85, DOI: 10.1007/978-3-642-55361-5_31,

where $u = u(t, x)$ and $v = v(t, x)$ are two unknown functions representing the densities of populations (cells), the concentrations of chemicals, the pressures in thin films, etc. F and G are the given smooth functions describing interaction between them and environment, d_1 and d_2 are diffusivities assumed to be positive constants. The subscripts t and x denote differentiation with respect to these variables. The class of RD systems (1) generalizes many well-known nonlinear second-order models and is used to describe various processes in physics, biology, chemistry and ecology (see, e.g., the well-known books [1–5]).

Nevertheless the search for Lie symmetries of the class of RD systems (1) was initiated about 30 years ago [6], this problem was completely solved only during the last decade because of its complexity. Now one can claim that all possible Lie symmetries of (1) were completely described in [7–9].

The time is therefore ripe for a complete description of non-Lie symmetries for the class of the RD systems (1). However, it seems to be extremely difficult task because, firstly, several definitions of non-Lie symmetries have been introduced (nonclassical symmetry [1, 10], conditional symmetry [11, 12], generalized conditional symmetry [13, 14] etc.), secondly, the complete description of non-Lie symmetries needs to solve the corresponding system of determining equations, which is *non-linear* and can fully be solved only in exceptional cases.

Hereafter we use the most common notion among non-Lie symmetries, non-classical symmetry, which we continuously call the Q-conditional symmetry following the well-known book [11] and our previous papers [15, 16]. It is well-known that the notion of Q-conditional symmetry plays an important role in investigation of the nonlinear evolution equations because, having such symmetries in the explicit form, one may construct new exact solutions, which are not obtainable by the classical Lie machinery. However, for a complete description of such symmetries, one needs to solve the corresponding non-linear system of determining equations that usually is very difficult task. Thus, to solve the Q-conditional symmetry classification problem for the class of RD systems (1), one should look for new constructive approaches helping to solve the relevant nonlinear system of determining equations. A possible approach was recently proposed in [17] and is used in this paper.

It can be noted that the diffusion coefficient d_1 in system (1) can be omitted without losing of generality because the simple substitution

$$t \to t/d_1, F \to -d_1 C^1, G \to -d_2 C^2$$

reduces the system to the form

$$\begin{aligned} u_{xx} &= u_t + C^1(u, v), \\ v_{xx} &= d v_t + C^2(u, v), \end{aligned} \tag{2}$$

where $d = \frac{d_1}{d_2}$. Thus, we consider system (2) in what follows.

The paper is organized as follows. In Sect. 2, two different definitions of Q-conditional invariance for the class of RD systems (2) are presented and the system of determining equations is derived. The theorem giving the complete description of Q-conditional symmetries of the first type is proved. In Sect. 3, the form-preserving transformations for the class of RD systems (2) are constructed and applied to the RD systems derived in Sect. 2. In Sect. 4, the Q-conditional symmetry obtained for reducing of the RD systems to the ODE systems are applied. Examples of finding exact solutions are presented together with a possible interpretation for population dynamics. Finally, we summarize and discuss the results obtained in the Sect. 5.

2 Conditional Symmetries of the RD Systems

Here we use the definition of Q-conditional symmetry of the first type for the RD systems (see [17] for details). It is well-known that to find Lie invariance operators, one needs to consider system (2) as the manifold $\mathcal{M} = \{S_1 = 0,\ S_2 = 0\}$ where

$$\begin{aligned} S_1 &\equiv u_{xx} - u_t - C^1(u,v), \\ S_2 &\equiv v_{xx} - dv_t - C^2(u,v), \end{aligned}$$

in the prolonged space of the variables: $t, x, u, v, u_t, v_t, u_x, v_x, u_{xx}, v_{xx}, u_{xt}, v_{xt}, u_{tt}, v_{tt}$. According to the definition, system (2) is invariant under the transformations generated by the infinitesimal operator

$$Q = \xi^0(t,x,u,v)\partial_t + \xi^1(t,x,u,v)\partial_x + \eta^1(t,x,u,v)\partial_u + \eta^2(t,x,u,v)\partial_v, \tag{3}$$

if the following invariance conditions are satisfied:

$$\begin{aligned} \underset{2}{Q} S_1 &\equiv \underset{2}{Q}, \left(u_{xx} - u_t - C^1(u,v)\right)\Big|_{\mathcal{M}} = 0, \\ \underset{2}{Q} S_2 &\equiv \underset{2}{Q} \left(v_{xx} - dv_t - C^2(u,v)\right)\Big|_{\mathcal{M}} = 0. \end{aligned}$$

The operator $\underset{2}{Q}$ is the second prolongation of the operator Q, i.e.

$$\underset{2}{Q} = Q + \rho_t^1 \frac{\partial}{\partial u_t} + \rho_t^2 \frac{\partial}{\partial v_t} + \rho_x^1 \frac{\partial}{\partial u_x} + \rho_x^2 \frac{\partial}{\partial v_x} + \sigma_{xx}^1 \frac{\partial}{\partial u_{xx}} + \sigma_{xx}^2 \frac{\partial}{\partial v_{xx}},$$

where the coefficients ρ and σ with relevant subscripts are expressed via the functions ξ^0, ξ^1, η^1 and η^2 by well-known formulae (see, e.g., [11, 18, 19]).

Hereafter the listed above differential operators act on functions and differential expressions in a natural way, particularly $Q(u) = \xi^0 u_t + \xi^1 u_x - \eta^1$ and $Q(v) = \xi^0 v_t + \xi^1 v_x - \eta^2$.

Definition 1 ([17]) Operator (3) is called the Q-conditional symmetry of the first type for the RD system (2) if the following invariance conditions are satisfied:

$$\underset{2}{Q}\, S_1 \equiv \underset{2}{Q}\left(u_{xx} - u_t - C^1(u, v)\right)\Big|_{\mathscr{M}_1} = 0,$$
$$\underset{2}{Q}\, S_2 \equiv \underset{2}{Q}\left(v_{xx} - dv_t - C^2(u, v)\right)\Big|_{\mathscr{M}_1} = 0,$$

where the manifold $\mathscr{M}_1$ is either $\{S_1 = 0,\ S_2 = 0,\ Q\,(u) = 0\}$ or $\{S_1 = 0,\ S_2 = 0,\ Q\,(v) = 0\}$.

Definition 2 Operator (3) is called the Q-conditional symmetry of the second type, i.e., the standard non-classical symmetry for the RD system (2) if the following invariance conditions are satisfied:

$$\underset{2}{Q}\, S_1 \equiv \underset{2}{Q}\left(u_{xx} - u_t - C^1(u, v)\right)\Big|_{\mathscr{M}_2} = 0,$$
$$\underset{2}{Q}\, S_2 \equiv \underset{2}{Q}\left(v_{xx} - dv_t - C^2(u, v)\right)\Big|_{\mathscr{M}_2} = 0,$$

where the manifold $\mathscr{M}_2 = \{S_1 = 0,\ S_2 = 0,\ Q\,(u) = 0,\ Q\,(v) = 0\}$.

Remark 1 It is easily seen that $\mathscr{M}_2 \subset \mathscr{M}_1 \subset \mathscr{M}$, hence, each Lie symmetry is automatically the Q-conditional symmetry of the first and second type, while each Q-conditional symmetry of the first type is one of the second type (non-classical symmetry).

Remark 2 To the best of our knowledge, there are not many paper devoted to search of Q-conditional symmetries for *the systems of PDEs* [20–24]. One may easily check that Definition 2 was only used in all these papers.

Statement. Let us assume that

$$X = (h_1(t, x)v + h_0(t, x))\partial_v \tag{4}$$

(hereafter $h_1(t, x)$ and $h_0(t, x)$ are the given functions) is the Lie symmetry operator of the RD system (2) while Q_1 is the known Q-conditional symmetry of the first type, which was found using the manifold $\mathscr{M}_1 = \{S_1 = 0, S_2 = 0, Q\,(u) = 0\}$. Then any linear combination $C_1Q_1 + C_2X$ (C_1 and $C_2 \neq 0$ are arbitrary constants) produces new Q-conditional symmetry of the first type.

Application of Definition 2 for finding Q-conditional symmetry (non-classical symmetry) operators of the RD system (2) leads to a complicated system of determining equations (DEs) (see system 19 in [17]), which seems to be extremely difficult for exact solving.

It turns out that application of definition 1 leads to essentially simpler system of DEs, which can be fully integrated. Here we present the result under the restrictions $\xi^0 \neq 0$ and $d \neq 1$ (the cases $\xi^0 = 0$ and $d = 1$ must be investigated separately). Thus, the system of DEs corresponding to the manifold $\mathscr{M}_1 = \{S_1 = 0,\ S_2 = 0,\ Q\,(u) = 0\}$ takes the form

$$\xi^0_x = \xi^0_u = \xi^0_v = \xi^1_u = \xi^1_v = 0, \tag{5}$$

$$\eta^1_v = \eta^1_{uu} = \eta^2_{uu} = \eta^2_{vv} = \eta^2_{uv} = 0, \tag{6}$$

$$2\xi^0 \eta^2_{xu} + (d-1)\xi^1 \eta^2_u = 0, \tag{7}$$

$$2\eta^1_{xu} + \xi^1_t = 0, \tag{8}$$

$$2\eta^2_{xv} + d\xi^1_t = 0, \tag{9}$$

$$2\xi^1_x - \xi^0_t = 0, \tag{10}$$

$$\eta^1 C^1_u + \eta^2 C^1_v + (2\xi^1_x - \eta^1_u)C^1 = \eta^1_{xx} - \eta^1_t, \tag{11}$$

$$\eta^1 C^2_u + \eta^2 C^2_v + (2\xi^1_x - \eta^2_v)C^2 = \eta^2_u C^1 + (1-d)\frac{\eta^1}{\xi^0}\eta^2_u + \eta^2_{xx} - d\eta^2_t. \tag{12}$$

Note that there is no any need to solve the similar system of DEs corresponding to the manifold $\mathcal{M}^*_1 = \{S_1 = 0, S_2 = 0, Q(v) = 0\}$ because the discrete transformations $u \to v,\ v \to u$ transform each symmetry found using $\mathcal{M}_1$ to one corresponding to the manifold $\mathcal{M}^*_1$.

It should be also noted that we find purely conditional symmetry operators, i.e., exclude all such operators, which are equivalent to Lie symmetry operators described in [7, 8]. Having this aim, we use the system DEs for search Lie symmetry operators (see [16] for details):

$$\xi^0_x = \xi^0_u = \xi^0_v = \xi^1_u = \xi^1_v = 0, \tag{13}$$

$$\eta^1_v = \eta^2_u = \eta^1_{uu} = \eta^2_{vv} = 0, \tag{14}$$

$$2\xi^1_x - \xi^0_t = 0, \tag{15}$$

$$2\eta^1_{xu} + \xi^1_t = 0, \tag{16}$$

$$2\eta^2_{xv} + d\xi^1_t = 0, \tag{17}$$

$$\eta^1 C^1_u + \eta^2 C^1_v + (2\xi^1_x - \eta^1_u)C^1 = \eta^1_{xx} - \eta^1_t, \tag{18}$$

$$\eta^1 C^2_u + \eta^2 C^2_v + (2\xi^1_x - \eta^2_v)C^2 = \eta^2_{xx} - d\eta^2_t. \tag{19}$$

Comparing DEs (5)–(12) with (13)–(19) one concludes that $\eta^2_u \neq 0$ is the necessary and sufficient condition, which guarantees this property.

Now we need to solve the nonlinear system (5)–(12). Obviously Eqs. (5) and (6) can be easily integrated:

$$\begin{aligned} &\xi^0 = \xi^0(t),\ \xi^1 = \xi^1(t,x),\\ &\eta^1 = r^1(t,x)u + p^1(t,x),\ \eta^2 = q(t,x)u + r^2(t,x)v + p^2(t,x), \end{aligned} \tag{20}$$

where $\xi^0(t)$, $\xi^1(t,x)$, $q(t,x)$, $r^k(t,x)$, $p^k(t,x)$ $(k = 1,2)$ are to-be-determined functions. Thus, substituting (20) into (7)–(12), one obtains the nonlinear system of PDEs:

$$2\xi^0 q_x + \xi^1(d-1)q = 0, \tag{21}$$

$$2r^1_x + \xi^1_t = 0, \tag{22}$$

$$2r^2_x + d\xi^1_t = 0, \tag{23}$$

$$2\xi^1_x - \xi^0_t = 0, \tag{24}$$

$$(r^1u+p^1)C^1_u + (qu+r^2v+p^2)C^1_v + (2\xi^1_x - r^1)C^1 = (r^1_{xx} - r^1_t)u + p^1_{xx} - p^1_t, \tag{25}$$

$$\begin{aligned} (r^1u+p^1)C^2_u + (qu+r^2v+p^2)C^2_v + (2\xi^1_x - r^2)C^2 = qC^1 + \frac{r^1u+p^1}{\xi^0}q(1-d)\\ + (r^2_{xx} - dr^2_t)v + (q_{xx} - dq_t)u + p^2_{xx} - dp^2_t, \end{aligned} \tag{26}$$

to find the functions $\xi^0(t)$, $\xi^1(t,x)$, $q(t,x) \neq 0$, $r^k(t,x)$, $p^k(t,x)$. In other words, all possible Q-conditional symmetries of the first type are easily constructed provided the general solution of system (21)–(26) is known.

Theorem 2.1 *The nonlinear RD system* (2) *with* $d \neq 1$ *is invariant under the* Q*-conditional operator of the first type* (3) *if and only if one and the corresponding operator have the forms listed in* Table 1. *Any other RD system admitting such* Q*-conditional operator is reduced to one of those from* Table 1 *by the local transformations*

$$\begin{aligned} &t \to C_1t + C_2,\\ &x \to C_3x + C_4,\\ &u \to C_5e^{C_6t}u + C_7t + C_8,\\ &v \to C_9e^{C_{10}t}v + C_{11}t^2 + C_{12}t + C_{13}, \end{aligned} \tag{27}$$

with correctly-specified constants $C_l, l = 1,\dots,13$ *and/or by adding a Lie symmetry operator of the form* (4).

Sketch of proof To prove the theorem one needs to solve the nonlinear PDE system (21)–(26) with restriction $q(t,x) \neq 0$. We remind the reader that C^1 and C^2 should be treated as unknown functions. As follows from the preliminary analysis (see Eqs. (25) and (26) involving the functions C^1 and C^2), we should examine six cases:

(1) $r^1 = r^2 = p^1 = 0,$
(2) $r^1 = r^2 = 0,\ p^1 \neq 0,$
(3) $r^1 = p^1 = 0,\ r^2 \neq 0,$
(4) $r^1 = 0,\ p^1 \neq 0,\ r^2 \neq 0,$
(5) $r^2 = 0,\ r^1 \neq 0,$
(6) $r^1 \neq 0,\ r^2 \neq 0.$

Solving system (21)–(26) in each case one obtains the list of Q-conditional symmetries of the first type together with the correctly-specified functions C^1 and C^2. Note that the symmetry operators have the different structures depending on the case.

Let us consider case (1) in details. Equations (25) and (26) take the form

$$\begin{aligned} &(qu+p^2)C^1_v + 2\xi^1_x C^1 = 0,\\ &(qu+p^2)C^2_v + 2\xi^1_x C^2 = qC^1 + (q_{xx} - dq_t)u + p^2_{xx} - dp^2_t. \end{aligned} \tag{28}$$

Differentiating the first equation of (28) with respect to x, one arrives at the equation $(q_x u + p^2_x)C^1_v = 0$, which lead to the requirement $C^1_v = 0$. In fact, if $q_x \neq 0$ then immediately $C^1_v = 0$. If $q_x = 0$ then Eq. (21) produces $\xi^1 = 0$, hence, $C^1_v = 0$. Thus, the first equation of system (28) takes the form $\xi^1_x C^1 = 0$ and two subcases $\xi^1_x \neq 0$ and $\xi^1_x = 0$ should be examined.

The general solution of (28) with $\xi^1_x \neq 0$ is

$$C^1 = 0,\ C^2 = \exp\left(-\frac{2\xi^1_x}{qu+p^2}v\right)g(u) + \frac{q_{xx} - dq_t}{2\xi^1_x}u + \frac{p^2_{xx} - dp^2_t}{2\xi^1_x}, \tag{29}$$

where $g(u)$ is an arbitrary (at the moment) function. Because the function C^2 doesn't depend on t and x, Eq. (29) with $g(u) \neq 0$ immediately produces the restrictions $q = \alpha_1 \xi^1_x$, $p^2 = \alpha_2 \xi^1_x$, where α_1 and α_2 are arbitrary constants. Differentiating Eq. (24) with respect to x, one obtains $\xi^1_{xx} = 0$. So $q_x \equiv \alpha_1 \xi^1_{xx} = 0$, however, this contradicts to the assumption $\xi^1_x \neq 0$. The remaining possibility $g(u) = 0$ leads to the linear RD system (2).

Now we examine the subcase $\xi^1_x = 0$, i.e., $\xi^1 = \lambda_1 = const$. The general solution of (28) takes the form

$$C^1 = f(u),\ C^2 = \frac{qf(u) + (q_{xx} - dq_t)u + p^2_{xx} - dp^2_t}{qu+p^2}v + g(u), \tag{30}$$

where $f(u)$ and $g(u)$ are arbitrary (at the moment) functions.

If $f(u)$ is an arbitrary function then we obtain $p^2 = \beta q$ ($\beta = const$) Hence $C^2 = \frac{f(u)}{u+\beta}v + \alpha v + g(u)$, where $\alpha = \frac{q_{xx} - dq_t}{q}$. Having this, we use renaming $\frac{f(u)}{u+\beta} \to f(u)$ and solve the overdetermined system

$$\frac{q_{xx}-dq_t}{q} = \alpha,$$
$$2q_x + \lambda_1(d-1)q = 0.$$

Thus, the system of DEs (21)–(26) is completely solved (under above listed restrictions!) and we obtain the conditional symmetry operator

$$Q = \partial_t + \lambda_1\partial_x + \lambda_2 \exp\left(\frac{\lambda_1(1-d)}{2}x + \frac{\lambda_1^2(1-d)^2 - 4\alpha}{4d}t\right)(u+\beta)\partial_v,$$

where λ_1 and $\lambda_2 \neq 0$ are arbitrary constants, of the RD system

$$\begin{aligned} u_{xx} &= u_t + (u+\beta)f(u), \\ v_{xx} &= dv_t + f(u)v + \alpha v + g(u). \end{aligned} \tag{31}$$

Finally, using the simple transformation

$$u \to u - \beta, \tag{32}$$

one sees that it is exactly case 6 of Table 1.

To complete the examination of case (1) we look for the correctly-specified function $f(u)$, which satisfies (30) without the restriction $p^2 = \beta q$. Indeed, if one finds the differential consequences of the second order (see equation for C^2) then $C^2_{vx} = 0,\ C^2_{vt} = 0$ and two algebraic equation to find the function $f(u)$ are obtained:

$$\begin{aligned}(q_tp^2 - qp_t^2)f &= ((q_{xx} - dq_t)u + p^2_{xx} - dp^2_t)(q_tu + p^2_t) \\ &-((q_{xx} - dq_t)_tu + (p^2_{xx} - dp^2_t)_t)(qu + p^2), \\ (q_xp^2 - qp_x^2)f &= ((q_{xx} - dq_t)u + p^2_{xx} - dp^2_t)(q_xu + p^2_x) \\ &-((q_{xx} - dq_t)_xu + (p^2_{xx} - dp^2_t)_x)(qu + p^2),\end{aligned}$$

Thus, $f(u) = \alpha_1 + \alpha_2 u + \alpha_3 u^2$ provided $p^2 \neq \beta q$. Substituting this expression into (30) and making the standard routine, one arrives at case 8 of Table 1 if $\alpha_3 \neq 0$ and case 9 if $\alpha_3 = 0$.

Cases (2)–(6) were treated in the similar way and the results are listed in Table 1. It should be noted that several local transformations (32 is the simplest example) were used to reduce the number of cases and simplify structures of the relevant RD systems. These transformations can be presented in the general form (27).

The sketch of proof is now completed. □

In Table 1, the function p^2 is the general solution of the equations

$$\begin{aligned} p^2_{xx} &= dp^2_t + \alpha_3 p^2 - \alpha_1 q, \\ p^2_{xx} &= dp^2_t + \alpha_3 p^2 + \lambda_2, \\ p^2_{xx} &= dp^2_t + \alpha_3 p^2 + \lambda_1, \\ p^2_{xx} &= dp^2_t + \alpha_3 p^2 + (d-1)qp^1 - \alpha_1 q, \end{aligned}$$

Table 1 Q-conditional symmetry operators of the first type of the RD system (2) with $d \neq 1$

	$C^1(u, v)$ in (2)	$C^2(u, v)$ in (2)	The operators Q and restrictions
1	$uf(\omega), \omega = u^{-k}(v-u)$	$u\big(f(\omega) + \alpha(1-d)\big) + u^k g(\omega)$	$\partial_t + \alpha u\partial_u + \alpha\big((1-k)u + kv\big)\partial_v,\ \alpha \neq 0,\ k \neq 1$
2	$uf(\omega), \omega = u\exp(-\frac{v}{u})$	$u(g(\omega) + f(\omega)\ln u + \alpha(1-d)\ln u)$	$\partial_t + \alpha u\partial_u + \alpha(u+v)\partial_v,\ \alpha \neq 0$
3	$uf(\omega), \omega = u\exp(u-v)$	$u(f(\omega) + \alpha(1-d)) + g(\omega)$	$\partial_t + \alpha u\partial_u + \alpha(u+1)\partial_v,\ \alpha \neq 0$
4	$f(\omega), \omega = e^{-u}(u+v)$	$e^u g(\omega) - \alpha(1-d) - f(\omega)$	$\partial_t + \alpha\partial_u + \alpha(u+v-1)\partial_v,\ \alpha \neq 0$
5	$f(\omega),\quad \omega = u^2 - 2v$	$uf(\omega) + g(\omega) + (1-d)u$	$\partial_t + \partial_u + u\partial_v$
6	$uf(u)$	$vf(u) + g(u) + \alpha v$	$\partial_t + \lambda_1\partial_x + qu\partial_v,\ q = \lambda_2\exp\Big(\frac{\lambda_1(1-d)}{2}x + \frac{\lambda_1^2(1-d)^2-4\alpha}{4d}t\Big),\ \lambda_2 \neq 0$
7	$f(u)$	$(u+v)\big(\alpha\ln(u+v) + g(u)\big) - f(u)$	$\partial_t + \lambda\exp(-\frac{\alpha}{d}t)(u+v)\partial_v,\ \lambda \neq 0$
8	$\alpha_1 + \alpha_2 u + u^2$	$g(u) + uv$	$\partial_t + \lambda_1\partial_x + (qu+p^2)\partial_v,\ p^2 = \Big(\frac{\lambda_1^2(1-d)^2+4\alpha_2}{4}\varphi_1 - d\dot{\varphi}_1\Big)\exp\big(\frac{\lambda_1(1-d)}{2}x\big),$ $q = \varphi_1(t)\exp\big(\frac{\lambda_1(1-d)}{2}x\big),\ \varphi_1 \neq 0$
9	$\alpha_1 + \alpha_2 u$	$g(u) + \alpha_3 v$	$\partial_t + \lambda_1\partial_x + (qu+p^2)\partial_v,\ q = \lambda_2\exp\Big(\frac{\lambda_1(1-d)}{2}x + \frac{\lambda_1^2(1-d)^2+4(\alpha_2-\alpha_3)}{4d}t\Big),\ \lambda_2 \neq 0$
10	$\alpha_1 + \alpha_2 u + \alpha_4\ln(u+v)$	$\alpha_3 v + (\alpha_3 - \alpha_2)u - \alpha_4\ln(u+v)$	$\partial_t + (\psi(x)\exp\big(\frac{\alpha_1\alpha_2}{\alpha_4(1-d)}t\big) - \frac{\alpha_1}{1-d})(\partial_u - \partial_v) + \frac{\alpha_1\alpha_2}{\alpha_4(1-d)}(u+v)\partial_v,\ \alpha_1\alpha_2\alpha_4 \neq 0$
11	$\alpha_2 u + v$	$\alpha_3 v + \frac{1}{2}(1+d)\frac{v^2}{u} + \alpha_1 u\ln u + \alpha_4 u$	$\partial_t + \varphi_2(t)(u\partial_u + v\partial_v) - \dot{\varphi}_2(t)u\partial_v,\ \alpha_2\dot{\varphi}_2(t) \neq 0$
12	$\alpha_2 u$	$\alpha_3 v + \alpha_4 u + u^k$	$\partial_t + \lambda_1 u\partial_u + (\varphi_3(t)u + \lambda_1 kv)\partial_v, \alpha_4\lambda_1\varphi_3 \neq 0$
13	$\alpha_1 u\ln u$	$\alpha_3 v + \alpha_1 v\ln u + \alpha_2 u^{\frac{1}{d}}$	$\partial_t + \lambda_1\partial_x + \lambda_2 e^{-\alpha_1 t}u\partial_u + (qu + \frac{\lambda_2}{d}e^{-\alpha_1 t}v)\partial_v,$ $q = \lambda_3\exp\Big(\frac{\lambda_1(1-d)}{2}x + \frac{\lambda_1^2(1-d)^2-4\alpha_3}{4d}t + \frac{d-1}{\alpha_1 d}\lambda_2 e^{-\alpha_1 t}\Big),\ \alpha_1\lambda_2\lambda_3 \neq 0$
14	$\alpha_1 u\ln u$	$\alpha_3 v + \alpha_1 v\ln u + \alpha_4 u$	$\partial_t + \lambda_2 e^{-\alpha_1 t}u\partial_u + (\varphi_4(t)u + (\frac{\lambda_2}{d}e^{-\alpha_1 t} + \lambda_1)v)\partial_v,\ \alpha_1\lambda_2\varphi_4 \neq 0$
15	$\alpha_1 u\ln u$	$\alpha_3 v + \alpha_1 v\ln u + \alpha_4 u + \alpha_2 u^{\frac{1}{d}}$	$\partial_t + \lambda_2 e^{-\alpha_1 t}u\partial_u + (\varphi_5(t)u + \frac{\lambda_2}{d}e^{-\alpha_1 t}v)\partial_v,\ \alpha_1\alpha_2\alpha_4\lambda_2\varphi_5 \neq 0$

(continued)

Table 1 (continued)

	$C^1(u,v)$ in (2)	$C^2(u,v)$ in (2)	The operators Q and restrictions
16	$\alpha_1 u \ln u$	$\alpha_3 v + \alpha_1 d v \ln u + \alpha_1(1-d) u \ln u - \alpha_3 u$	$\partial_t + \lambda_2 e^{-\alpha_1 t} u \partial_u + (\alpha_1 u + (\lambda_2 e^{-\alpha_1 t} - \alpha_1) v)\partial_v,\ \alpha_1\lambda_2 \neq 0$
17	0	$\alpha_3 v + \ln u$	$\partial_t + \lambda_1 \partial_x + \lambda_2 u \partial_u + (qu + p^2)\partial_v,\ q = \lambda_3$ $\exp\left(\frac{\lambda_1(1-d)}{2} x + \frac{\lambda_1^2(1-d)^2 + 4(-\alpha_3 + \lambda_2(1-d))}{4d} t\right),\ \lambda_2\lambda_3 \neq 0$
18	$\alpha_2 u$	$\alpha_3 v + \ln u + \alpha_4 u$	$\partial_t + \lambda_1 u \partial_u + (\varphi_6(t) u + p^2)\partial_v,\ \alpha_4\lambda_1\varphi_6 \neq 0$
19	$\alpha_2 u$	$\alpha_3 v + u \ln u$	$\partial_t + \lambda_1 u \partial_u + (\varphi_7(t) u + \lambda_1 v)\partial_v,\ \lambda_1\varphi_7 \neq 0$
20	$\alpha_1 + \alpha_2 u + u^2$	$uv + \alpha_3$	$\partial_t + \lambda\big(v + \varphi_8(t) u + \alpha_2\varphi_8(t) - d\dot{\varphi}_8(t)\big)\partial_v,\ \lambda\varphi_8 \neq 0$
21	$\alpha_1 + \alpha_2 u$	$\alpha_3 v + u^2$	$\partial_t + \lambda_1 \partial_x + p^1 \partial_u + (qu + p^2)\partial_v,\ p^1 = \frac{1}{2}\Big(\frac{\lambda_1^2(1-d)^2 + 4(\alpha_2 - \alpha_3)}{4}\varphi_9 - d\dot{\varphi}_9\Big)$ $\exp\left(\frac{\lambda_1(1-d)}{2} x\right),\ q = \varphi_9(t)\exp\left(\frac{\lambda_1(1-d)}{2} x\right),\ \varphi_9 \neq 0$
22	0	$\alpha_3 v + e^u$	$\partial_t + \lambda_1 \partial_x + \lambda_2 \partial_u + (qu + \lambda_2 v + p^2)\partial_v,\ q = \lambda_3$ $\exp\left(\frac{\lambda_1(1-d)}{2} x + \frac{\lambda_1^2(1-d)^2 - 4\alpha_3}{4d} t\right),\ \lambda_3 \neq 0$
23	α_1	$\alpha_3 v + e^u + \alpha_4 u$	$\partial_t + \lambda_1 \partial_u + (\varphi_{10}(t) u + \lambda_1 v + p^2)\partial_v,\ \alpha_4\lambda_1\varphi_{10} \neq 0$
24	0	$\alpha_1(u+v)\ln(u+v) + \alpha_2$	$\partial_t + \frac{\alpha_2}{d-1}(\partial_u - \partial_v) + \lambda\exp(-\frac{\alpha_1}{d} t)(u+v)\partial_v,\ \alpha_1\alpha_2\lambda \neq 0$
25	0	u^2	$2t\partial_t + x\partial_x + \frac{2t - x^2}{4\sqrt{t^5}}\exp(-\frac{x^2}{4t})\partial_u + (\frac{1}{\sqrt{t}}\exp(-\frac{x^2}{4t})u + p^2 + 2v)\partial_v,\ d = 3$
26	0	u^5	$t^2\partial_t + tx\partial_x - \frac{x^2 + 2t}{4} u\partial_u + (\lambda\exp(-\frac{x^2}{t})u - \frac{5x^2 + 2t}{4} v + p^2)\partial_v,\ d = 5,\ \lambda \neq 0$

$$p_{xx}^2 = dp_t^2 + \alpha_3 p^2 - \lambda_2(1-d)q,$$
$$p_{xx}^2 = dp_t^2 + \alpha_3 p^2 - (\lambda_2(1-d)+\alpha_1)q + \alpha_4\lambda_1,$$
$$p_{xx}^2 = 3p_t^2 + \frac{2t-x^2}{4t^4}\exp(-\frac{x^2}{2t}),$$
$$p_{xx}^2 = 5p_t^2$$

of the cases 9, 17, 18, 21, 22, 23, 25 and 26, respectively.

The functions $\psi(x)$, $\varphi_1(t)$, $\varphi_2(t)$, $\varphi_4(t)$, $\varphi_5(t)$, $\varphi_8(t)$, and $\varphi_9(t)$ are the general solutions of the equations

$$\psi'' - \left(\alpha_1 + \frac{\alpha_1\alpha_2}{1-d}\right)\psi = 0,$$

$$d\ddot{\varphi}_1 - \frac{\lambda_1^2(1-d)^2 + 2\alpha_2}{2}\dot{\varphi}_1 + \frac{\lambda_1^4(1-d)^4 + 4\alpha_2\lambda_1^2(1-d)^2 + 16\alpha_1}{16d}\varphi_1 = 0,$$

$$d\ddot{\varphi}_2 - (\alpha_2 - \alpha_3 + (1-d)\varphi_2)\dot{\varphi}_2 - \alpha_3\varphi_2 = 0,$$

$$d\dot{\varphi}_4 + (\alpha_3 + \lambda_2(d-1)e^{-\alpha_1 t})\varphi_4 - \alpha_4\left(\lambda_1 + \lambda_2\frac{1-d}{d}e^{-\alpha_1 t}\right) = 0,$$

$$d\dot{\varphi}_5 + (\alpha_3 + \lambda_2(d-1)e^{-\alpha_1 t})\varphi_5 - \alpha_4\lambda_2\frac{1-d}{d}e^{-\alpha_1 t} = 0,$$

$$d\ddot{\varphi}_8 - \alpha_2\dot{\varphi}_8 + \frac{\alpha_1}{d}\varphi_8 + \frac{\alpha_3}{d^2} = 0,$$

and

$$d\ddot{\varphi}_9 - \frac{\lambda_1^2(1+d)(1-d)^2 + 4(\alpha_2(1-d)-\alpha_3)}{4}\dot{\varphi}_9 + \frac{\lambda_1^4(1-d)^4 - 4\alpha_3\lambda_1^2(1-d)^2 - 16\alpha_2(\alpha_2-\alpha_3)}{16}\varphi_9 = 0,$$

respectively. The functions

$$\varphi_3(t) = \begin{cases} \lambda_2\exp\left(\frac{\alpha_2-\alpha_3+\lambda_1(1-d)}{d}t\right) + \frac{\alpha_4\lambda_1(1-k)}{\alpha_2-\alpha_3+\lambda_1(1-d)}, & \text{if } \alpha_2 \neq \alpha_3 - \lambda_1(1-d), \\ -\frac{\alpha_4\lambda_1(1-k)}{d}t + \lambda_2, & \text{if } \alpha_2 = \alpha_3 - \lambda_1(1-d); \end{cases}$$

$$\varphi_{10}(t) = \begin{cases} \lambda_2\exp\left(-\frac{\alpha_3}{d}t\right) + \frac{\alpha_4\lambda_1}{\alpha_3}, & \text{if } \alpha_3 \neq 0, \\ \frac{\alpha_4\lambda_1}{d}t, & \text{if } \alpha_3 = 0. \end{cases}$$

Finally, the function $\varphi_6(t) = \varphi_3(t)$ at $k = 0$, while $\varphi_7(t) = \varphi_3(t)$ at $k = 0$ and $\alpha_4 = 1$. Hereafter the upper dot index denotes differentiation with respect to the variable t.

3 Form-Preserving Transformations of the RD Systems

A natural question is: Can we claim that 26 systems listed in Table 1 are inequivalent up to any local substitutions (not only of the form 27!)? It turns out that the answer is positive. To present the rigorous proof of this, we used the notion of the set of form-preserving point transformations introduced in [25] and now extensively used for Lie symmetry classification problems (see, e.g. [26, 27]). Note that finding these transformations for systems of PDEs is a difficult problem because of technical problems occurring in computations and there is no many results for systems. To the best of our knowledge, the recent paper [28] is the first one presenting an explicit result for a class of PDE systems.

The form-preserving transformations present the most general and correctly-specified form of local substitutions, which can map *some equations* from a given class to other those belonging to the same class. They contain as particular cases the well-known equivalence transformations and discrete transformations, which maps *each equation* from the class to another one from this class, used in the well-known Ovsiannikov method of Lie symmetry classification. Here we construct such transformations with the aim to show that Table 1 cannot be shortened.

Theorem 3.1 *An arbitrary RD system of the form* (2) *with $d \neq 1$ can be reduced to another system of the same form*

$$\begin{aligned} w_{yy} &= w_\tau + F^1(w, z), \\ z_{yy} &= \lambda z_\tau + F^2(w, z), \end{aligned} \tag{33}$$

by the local non-degenerate transformation

$$\tau = a(t, x, u, v), \ \ y = b(t, x, u, v), \tag{34}$$

$$w = \varphi(t, x, u, v), \ \ z = \psi(t, x, u, v), \tag{35}$$

if and only if the smooth functions a, b, φ and ψ are:

$$\begin{aligned} a &= \alpha(t), \ b = \beta(t)x + \gamma(t), \\ \varphi &= f(t) \exp\Big(- \tfrac{1}{4\beta}(\dot\beta x^2 + 2\dot\gamma x)\Big) u + P(t, x), \\ \psi &= g(t) \exp\Big(- \tfrac{d}{4\beta}(\dot\beta x^2 + 2\dot\gamma x)\Big) v + Q(t, x), \end{aligned} \tag{36}$$

where the functions $\alpha(t)$, $\beta(t)$, $f(t)$, $g(t)$, $\gamma(t)$, $P(t,x)$, *and* $Q(t,x)$ *are such that the equalities*

$$\dot{\alpha} = \beta^2, \quad \lambda = d, \tag{37}$$

$$\beta^2 F^1(\varphi, \psi) = \varphi_u C^1(u,v) + \varphi_{xx} - \varphi_t - 2\frac{\varphi_x}{\varphi_u}\varphi_{xu}, \tag{38}$$

$$\beta^2 F^2(\varphi, \psi) = \psi_v C^2(u,v) + \psi_{xx} - d\psi_t - 2\frac{\psi_x}{\psi_v}\psi_{xv} \tag{39}$$

take place provided $\beta f g \neq 0$.

Proof First of the all we note that each non-degenerate transformation (34)–(35) should satisfy the conditions

$$\Delta_1 = \begin{vmatrix} a_x & a_t \\ b_x & b_t \end{vmatrix} \neq 0, \qquad \Delta_2 = \begin{vmatrix} \varphi_u & \varphi_v \\ \psi_u & \psi_v \end{vmatrix} \neq 0, \tag{40}$$

which are used to prove the theorem.

Let us choose an arbitrary RD system of the form (2). The main idea of the proof is based on substituting the expressions for u_{xx}, v_{xx}, u_t, v_t using the formulae (34) and (35) into this system and on analysis conditions when the system obtained is equivalent to system (33). The expressions for the first-order derivatives have the form

$$u_x = \frac{\begin{vmatrix} \varphi_x - a_x w_\tau - b_x w_y & a_v w_\tau + b_v w_y - \varphi_v \\ \psi_x - a_x z_\tau - b_x z_y & a_v z_\tau + b_v z_y - \psi_v \end{vmatrix}}{\begin{vmatrix} a_u w_\tau + b_u w_y - \varphi_u & a_v w_\tau + b_v w_y - \varphi_v \\ a_u z_\tau + b_u z_y - \psi_u & a_v z_\tau + b_v z_y - \psi_v \end{vmatrix}},$$

$$u_t = \frac{\begin{vmatrix} \varphi_t - a_t w_\tau - b_t w_y & a_v w_\tau + b_v w_y - \varphi_v \\ \psi_t - a_t z_\tau - b_t z_y & a_v z_\tau + b_v z_y - \psi_v \end{vmatrix}}{\begin{vmatrix} a_u w_\tau + b_u w_y - \varphi_u & a_v w_\iota + b_v w_y - \varphi_v \\ a_u z_\tau + b_u z_y - \psi_u & a_v z_\tau + b_v z_y - \psi_v \end{vmatrix}}.$$

The expressions for the second-order derivatives are very cumbersome, however, it can be noted that they contain the derivative $w_{\tau\tau}$ and $w_{\tau y}$. Because τ is a new time-variable we conclude that the coefficient next to $w_{\tau\tau}$ and $w_{\tau y}$ must vanish otherwise system (33) are not obtainable. These coefficients vanish if and only if the equalities take place:

$$a_x = a_u = a_v = b_u = b_v = 0 \Rightarrow a = \alpha(t), \ b = b(t,x). \tag{41}$$

Moreover, taking into account (40), the restriction $\dot{\alpha}\, b_x \neq 0$ is also obtained.

Having the set of equalities (41), the expressions for u_{xx} and u_t can be essentially simplified, namely:

$$
\begin{aligned}
u_{xx} = {} & \tfrac{\psi_v b_x^2}{\Delta_2} w_{yy} - \tfrac{\varphi_v b_x^2}{\Delta_2} z_{yy} + \tfrac{(\psi_v b_x)_x \Delta_2 - (\Delta_2)_x \psi_v b_x}{\Delta_2^2} w_y - \tfrac{(\varphi_v b_x)_x \Delta_2 - (\Delta_2)_x \varphi_v b_x}{\Delta_2^2} z_y \\
& + \tfrac{(\psi_x \varphi_v - \psi_v \varphi_x)_x \Delta_2 - (\Delta_2)_x (\psi_x \varphi_v - \psi_v \varphi_x)}{\Delta_2^2}, \\
u_t = {} & \tfrac{1}{\Delta_2}\big(\psi_v(\dot{\alpha} w_\tau + b_t w_y - \varphi_t) - \varphi_v(\dot{\alpha} z_\tau + b_t z_y - \psi_t)\big).
\end{aligned} \tag{42}
$$

Substituting (42) into the first equation of (2). Omitting the full expression of the equation obtained, we note that one contains the terms

$$
(i) - \frac{\varphi_v b_x^2}{\Delta_2}\left(z_{yy} - \frac{\dot{\alpha}}{b_x^2} z_\tau\right), \qquad (ii)\ \frac{\psi_v b_x^2}{\Delta_2}\left(w_{yy} - \frac{\dot{\alpha}}{b_x^2} w_\tau\right),
$$

while other terms don't depend on z_{yy}, w_{yy}, z_τ, and w_τ.

Now there is two possibilities. If the first equation of (2) is transformed into the first one of (33) then we immediately obtain

$$
\dot{\alpha} = b_x^2, \ \varphi_v = 0. \tag{43}
$$

If the first equation of (2) is transformed into the second one then the conditions

$$
\dot{\alpha} = \lambda b_x^2, \ \psi_v = 0 \tag{44}
$$

must be satisfied. It turns out that conditions (44) lead to the result, which is obtainable from (36) by the discrete transformations $u \to v$ and $v \to u$.

Let us consider conditions (43). Taking into account (40) and $\varphi_v = 0$, the restriction $\varphi_u \psi_v \neq 0$ springs up.

On the other hand, $b(t,x) = \beta(t)x + \gamma(t)$, follows from (43) where β and γ are arbitrary smooth functions. Thus, the first equation from (37) is derived.

Substituting (43) into expressions for u_{xx} and u_t (see formulae 42), one obtains

$$
\begin{aligned}
u_{xx} &= \tfrac{\beta^2}{\varphi_u} w_{yy} - 2\tfrac{\beta \varphi_{xu}}{\varphi_u^2} w_y + \tfrac{2\varphi_x \varphi_{xu} - \varphi_u \varphi_{xx}}{\varphi_u^2} - \tfrac{\varphi_{uu}}{\varphi_u^3}(\beta w_y - \varphi_x)^2, \\
u_t &= \tfrac{1}{\varphi_u}(\dot{\alpha} w_\tau + (\dot{\beta} x + \dot{\gamma}) w_y - \varphi_t).
\end{aligned} \tag{45}
$$

Since the first equation of system (33) does not contain the terms w_y and w_y^2, we should vanish the relevant coefficient, namely:

$$
\begin{aligned}
& 2\tfrac{\beta \varphi_{xu}}{\varphi_u^2} + \tfrac{\dot{\beta} x + \dot{\gamma}}{\varphi_u} = 0, \\
& \varphi_{uu} = 0.
\end{aligned}
$$

The general solution of this system can be easily constructed so that obtains

$$
\varphi = f(t) \exp\Big(-\frac{1}{4\beta}(\dot{\beta} x^2 + 2\dot{\gamma} x)\Big) u + P(t,x), \tag{46}
$$

where $f(t) \neq 0$ and $P(t,x)$ are arbitrary functions at the moment. Thus, the first, second and third equations from (36) are derived. Moreover, substituting (45) and (46) into the first equation of system (2), we arrive at the equation

$$w_{yy} = w_\tau + \frac{\varphi_u}{\beta^2}\Big(C^1(u,v) - \frac{\varphi_t}{\varphi_u} - \frac{2\varphi_x\varphi_{xu} - \varphi_u\varphi_{xx}}{\varphi_u^2}\Big). \quad (47)$$

Now one realizes that (47) coincides with the first equation of system (33) iff condition (38) takes place.

The analogous routine involving the second equation of system (2) leads to the condition $\dot{\alpha} = \frac{\lambda}{d}\beta^2 \Rightarrow \lambda = d$ (see 43), the function ψ of the form (36) and Eq. (39).

The proof is now completed. □

Consequence 1 The set of transformations (27) arising in theorem 1 is a subset of form-preserving transformations (36).

Consequence 2 If the nonlinear RD system of the form (1) is transformed to another one from this class, say, to the system

$$\begin{aligned} u^*_{t^*} &= d_1^* u^*_{x^*x^*} + F^*(u^*, v^*), \\ v^*_{t^*} &= d_2^* v^*_{x^*x^*} + G^*(u^*, v^*) \end{aligned}$$

by a local substitution then they have the proportional diffusivities. Moreover, there are two linear combinations for the reaction terms F and F^*, and for G and G^* resulting $\alpha_1 u + \alpha_2$ and $\alpha_3 v + \alpha_4$, respectively (here α_k, $k = 1, \ldots, 4$ are correctly-specified constants).

Roughly speaking, consequence 2 says that the locally-equivalent RD systems have the same structure up to additive terms $\alpha_1 u + \alpha_2$ and $\alpha_3 v + \alpha_4$. At the first sight, there are some systems in Table 1 satisfying this consequence, for example in cases 17 and 18. However, according to consequence 2, the term $\alpha_4 u$ arising in the second equation of the RD system (see case 18) cannot be removed by any local substitution. We have carefully checked all cases listed in Table 1 and concluded that there aren't any locally-equivalent systems therein.

Thus, we have shown that the list of RD systems presented in Table 1 cannot be reduced (shortened) by any local substitution.

4 New Exact Solutions and Their Interpretation

It is well-known that using the known Q-conditional symmetry (non-classical symmetry), one reduces the given system of PDEs to a system of ODEs via the same procedure as for classical Lie symmetries. Since each Q-conditional symmetry of the first type is automatically one of the second type, i.e., non-classical symmetry, we apply this procedure for finding exact solutions. Thus, to construct an ansatz corresponding to the given operator Q, the system of the linear first-order PDEs

$$Q(u) = 0, \quad Q(v) = 0 \tag{48}$$

should be solved. Substituting the ansatz obtained into the RD system with correctly-specified coefficients, one obtains the reduced system of ODEs.

Let us construct exact solutions of the non-linear RD system listed in the case 1 of Table 1, when the system and the corresponding symmetry operator have the form

$$\begin{aligned} u_t &= u_{xx} - uf(\omega), \\ dv_t &= v_{xx} - u^k g(\omega) - u(f(\omega) + \alpha(1-d)), \ \omega = u^{-k}(v-u) \end{aligned} \tag{49}$$

and

$$Q = \partial_t + \alpha u \partial_u + \alpha\big((1-k)u + kv\big)\partial_v, \tag{50}$$

In this case system (48) takes the form

$$\begin{aligned} u_t &= \alpha u, \\ v_t &= \alpha(1-k)u + \alpha k v \end{aligned} \tag{51}$$

and its general solution produces the ansatz (the functions u and v depend on two variables t and x):

$$\begin{aligned} u &= \varphi(x)e^{\alpha t}, \\ v &= \psi(x)e^{k\alpha t} + \varphi(x)e^{\alpha t}, \end{aligned} \tag{52}$$

where $\varphi(x)$ and $\psi(x)$ are new unknown functions. Substituting ansatz (52) into (49), one obtains so called reduced system of ODEs

$$\begin{aligned} \varphi'' &= \varphi\big(\alpha + f(\omega)\big), \\ \psi'' &= \varphi^k g(\omega) + \alpha k d\psi, \ \omega = \psi\varphi^{-k}. \end{aligned} \tag{53}$$

Because system (53) is non-linear (excepting, of course, some special cases) it can be integrated only for the correctly-specified functions f and g. We specify f and g in a such way, when the RD system in question will be still non-linear (otherwise the result will be rather trivial). Thus, setting $f(\omega) = \gamma\omega^{\frac{1}{k}} - \alpha, \ g(\omega) = \beta\omega$, β and γ are arbitrary non-zero constants, the RD system takes the form

$$\begin{aligned} u_t &= u_{xx} - \gamma(v-u)^{\frac{1}{k}} + \alpha u, \\ dv_t &= v_{xx} - \gamma(v-u)^{\frac{1}{k}} - \beta v + (\beta + \alpha d)u, \end{aligned} \tag{54}$$

while the corresponding reduced system is

$$\begin{aligned} \varphi'' &= \gamma\psi^{\frac{1}{k}}, \\ \psi'' &= (\beta + \alpha k d)\psi. \end{aligned} \tag{55}$$

The general solution of (55) can be easily constructed:

$$\varphi(x) = \gamma \int \Big(\int \psi^{\frac{1}{k}}(x)dx \Big) dx + c_3 x + c_4, \tag{56}$$

$$\psi(x) = \begin{cases} c_1 \exp(\mu x) + c_2 \exp(-\mu x), & \text{if } \mu^2 = \beta + \alpha k d > 0, \\ c_1 \cos(\nu x) + c_2 \sin(\nu x), & \text{if } \nu^2 = -(\beta + \alpha k d) > 0, \\ c_1 x + c_2, & \text{if } \beta + \alpha k d = 0. \end{cases} \tag{57}$$

Thus, substituting (56) and (57) into (52), the 4-parameter family of solutions for the non-linear RD system (54) is constructed.

Hereafter we highlight the solutions satisfying the zero Neumann boundary conditions, which widely arise in biologically motivated boundary-value problems. Hence, setting $c_3 = c_4 = 0,\ k = \frac{1}{3},\ \psi = c_1 \cos(\nu x)$, one obtains the solution

$$\begin{aligned} u &= -\gamma \frac{c_1^3}{9\nu^2}(\cos^2(\nu x) + 6)\cos(\nu x)e^{\alpha t}, \\ v &= c_1 \cos(\nu x) e^{\frac{1}{3}\alpha t} + u. \end{aligned}$$

It can be noted that this solution satisfies the zero Neumann boundary conditions

$$u_x|_{x=0} = 0,\ v_x|_{x=0} = 0,\ u_x|_{x=j\frac{\pi}{\nu}} = 0, v_x|_{x=j\frac{\pi}{\nu}} = 0$$

on the interval $[0, j\frac{\pi}{\nu}]$, where $j \in \mathbb{N}$.

Let us set $f(\omega) = -(a_1 + b\omega)$, $g(\omega) = (\alpha(1-d) - a_1)\omega$ (hereafter α, a_1 and b are arbitrary non-zero constants) in (49), hence, it takes the form

$$\begin{aligned} u_t &= u_{xx} + a_1 u - bu^{2-k} + bvu^{1-k}, \\ dv_t &= v_{xx} - a_2 v - bu^{2-k} + bvu^{1-k}, \end{aligned} \tag{58}$$

where $a_2 = \alpha(1-d) - a_1$. The corresponding reduced system of ODEs is

$$\begin{aligned} &\varphi'' + b\psi\varphi^{1-k} + (a_1 - \alpha)\varphi = 0, \\ &\psi'' = (\alpha k d + a_2)\psi. \end{aligned} \tag{59}$$

Nevertheless we have not constructed the general solution of system (59), its particular solution was found by setting $\psi = -\delta,\ \delta \neq 0$. In this case, the first-order ODE

$$\varphi' = \pm\sqrt{(\alpha - a_1)\varphi^2 + \frac{2b\delta}{2-k}\varphi^{2-k} + c_1},\ \ \alpha = \frac{a_1}{d(k-1)+1},\ \ k \neq 2 \tag{60}$$

for the function φ is obtained (the value $k = 2$ is special and leads to ODE $\varphi' = \pm\sqrt{(\alpha - a_1)\varphi^2 + 2b\delta \ln\varphi + c_1}$).

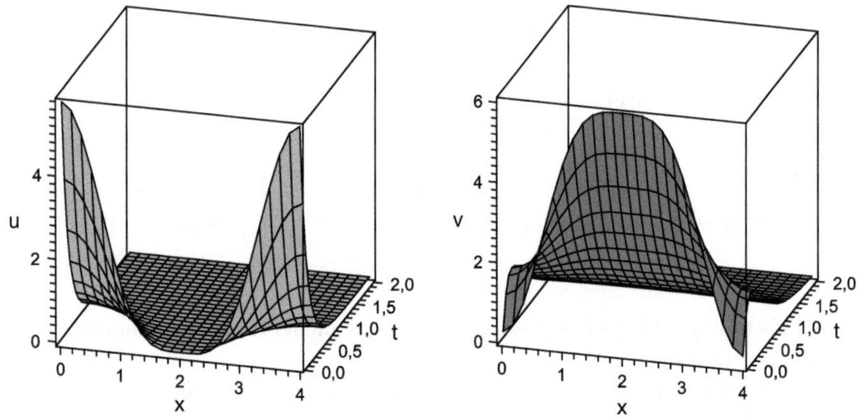

Fig. 1 Exact solution (63) with $k = 0.5,\ \delta = 6,\ d = 4,\ a_1 = 5,\ b = 3$

If $c_1 \neq 0$ then the general solution of (60) can be expressed via hypergeometric functions. Here we present the solution for (60) with $c_1 = 0,\ k \neq 0$:

$$\varphi(x) = \begin{cases} \left(\beta\left(\tan^2\left(\frac{k\sqrt{a_1-\alpha}}{2}(x \pm c_2)\right) + 1\right)\right)^{-\frac{1}{k}}, & a_1 > \alpha, \\ \left(-\beta\left(\tanh^2\left(\frac{k\sqrt{\alpha-a_1}}{2}(x \pm c_2)\right) - 1\right)\right)^{-\frac{1}{k}}, & a_1 < \alpha, \end{cases} \tag{61}$$

($\beta = \frac{(a_1-\alpha)(2-k)}{2b\delta}$), which seems to be the most interesting. Note that another arbitrary constant can be removed by the trivial substitution $x \pm c_2 \to x$.

Now we rewrite system (58) setting $v \to -v$ with the aim to obtain a biologically motivated model. So the system takes form

$$\begin{aligned} u_t &= u_{xx} + u(a_1 - bu^{1-k}) - bvu^{1-k}, \\ dv_t &= v_{xx} + v(-a_2 + bu^{1-k}) + bu^{2-k}, \end{aligned} \tag{62}$$

where all coefficients (excepting k) should be positive. Equation (62) can be treated as a prey-predator model for the population dynamics. In fact, the species u is prey and described by the first equation. Its population decreases proportionally to the predator density v. The natural birth-dead rule for the prey is $u(a_1 - bu^{1-k})$ and can be treated as a generalization of the standard logistic rule $u(a_1 - bu)$ (see, e.g., [2]). The similar arguments are also valid for the second equation. The model should involve also the zero Neumann boundary conditions (zero-flux on the boundaries), which indicate that both species cannot widespread over the globe but occupy a bounded domain.

Using (52) with $v \to -v$, $\psi = -\delta$ and (61) with $a_1 > \alpha$ we construct the exact solution

$$
\begin{aligned}
u &= \left(\beta\left(\tan^2\left(\tfrac{k\sqrt{a_1-\alpha}}{2}x\right)+1\right)\right)^{-\frac{1}{k}} e^{\alpha t},\\
v &= \delta e^{\alpha k t} - \left(\beta\left(\tan^2\left(\tfrac{k\sqrt{a_1-\alpha}}{2}x\right)+1\right)\right)^{-\frac{1}{k}} e^{\alpha t}
\end{aligned}
\tag{63}
$$

of (62). It turns out that solution can describe interaction between prey and predator on the space interval $[0, l]$, (here $l = \frac{2\pi j}{k\sqrt{a_1-\alpha}}$, $j \in \mathbb{N}$) provided

$$
0 < k < 1 - \frac{1}{d},\ 0 < \delta \le \left(\frac{(2-k)(a_1-\alpha)}{2b}\right)^{\frac{1}{1-k}},\ \alpha = \frac{a_1}{d(k-1)+1} < 0. \tag{64}
$$

One easily checks that solution (63) is non-negative, bounded in the domain $\Omega = \{(t, x) \in (0, +\infty) \times (0, l)\}$ and satisfy the given zero Neumann boundary conditions, i.e.

$$
u_x|_{x=0} = 0,\ v_x|_{x=0} = 0,\ u_x|_{x=l} = 0,\ v_x|_{x=l} = 0.
$$

As example we present this solution (63) with the parameters satisfying the restrictions (64) in Fig. 1. This solution can describe such type of the interaction between the species u and v when both of them eventually die, i.e. $(u, v) \to (0, 0)$ if $t \to +\infty$.

5 Conclusions

In this paper, Q-conditional symmetries for the class of RD systems (2) (that is equivalent to the class of systems (1)) and their application for finding exact solutions are studied. Following the recent paper [17], the notion of Q-conditional symmetry of the first type was used for these purposes. The main result is presented in theorem 1 giving the exhaustive list of RD systems of the form (1) with $d_1 \neq d_2$ (the case $d_1 = d_2$ should be analyzed separately), which admit such symmetry. It turns out that there are exactly 26 locally-inequivalent RD systems admitting the Q-conditional symmetry operators of the first type of the form (3) with $\xi^0 \neq 0$ (the case $\xi^0 = 0$ should be analyzed separately). To show local non-equivalence of the systems listed in Table 1, we proved theorem 2 describing the set of form-preserving point transformations for the class of RD systems (2). Note that all the operators found are inequivalent to the Lie symmetry operators presented in [7, 8] because the necessary and sufficient condition, which guarantees this property, was used.

The Q-conditional operator listed in case 1 of Table 1 was used to construct the non-Lie ansatz and to reduce two nonlinear RD systems to the corresponding ODE systems. Solving these ODE systems, the two-parameter families of exact solutions were explicitly constructed for the RD systems in question. Moreover, application of the exact solutions for solving the prey-predator system (62) was presented. It turns out that the relevant boundary value problem with the zero Neumann conditions can be exactly solved and the solution can describe the densities of two interacting species.

The work is in progress to construct conditional symmetries for *multicomponent* RD systems. In particular case, a wide list of the Q-conditional symmetries of the first type for the three-component diffusive Lotka-Volterra system is presented in [29].

Finally, we point out that this paper is a natural continuation of the recent paper [16], where RD systems with non-constant diffusivities were examined.

Acknowledgments R.Ch. thanks the Organizing Committee of the 7th Workshop 'Algebra, Geometry, and Mathematical Physics' (Mulhouse, 24-26 October 2011) for the financial support.

References

1. Ames, W.F.: Nonlinear Partial Differential Equations in Engineering. Academic, New York (1972)
2. Murray, J.D.: Mathematical Biology. Springer, Berlin (1989)
3. Murray, J.D.: Mathematical Biology II: Spatial Models and Biomedical Applications. Springer, Berlin (2003)
4. Aris, R.: The Mathematical Theory of Diffusion and Reaction in Permeable Catalysts. Clarendon, Oxford (1975)
5. Okubo, A., Levin, S.A.: Diffusion and Ecological Problems. Modern Perspectives, 2nd edn. Springer, Berlin (2001)
6. Zulehner, W., Ames, W.F.: Group analysis of a semilinear vector diffusion equation. Nonlinear Analysis **7**, 945–969 (1983)
7. Cherniha, R., King, J.R.: Lie symmetries of nonlinear multidimensional reaction-diffusion systems: I. J. Phys. A: Math. Gen. **33**(267–82), 7839–7841 (2000)
8. Cherniha, R., King, J.R.: Lie symmetries of nonlinear multidimensional reaction-diffusion systems: II. J. Phys. A: Math. Gen. **36**, 405–425 (2003)
9. Nikitin, A.G.: Group classification of systems of non-linear reaction-diffusion equations. Ukrainian Math. Bull. **2**, 153–204 (2005)
10. Bluman, G.W., Cole, J.D.: The general similarity solution of the heat equation. J. Math. Mech. **18**, 1025–1042 (1969)
11. Fushchych, W.I., Shtelen, W.M., Serov, M.I.: Symmetry Analysis and Exact Solutions of Equations of Nonlinear Mathematical Physics. Kluwer, Dordrecht (1993)
12. Cherniha, R., Henkel, M.: On nonlinear partial differential equations with an infinite-dimensional conditional symmetry. J. Math. Anal. Appl. **298**, 487–500 (2004)
13. Fokas, A.S., Liu, Q.M.: Nonlinear interaction of traveling waves of nonintegrable equations. Phys. Rev. Lett. **72**, 3293–3296 (1994)
14. Liu, Q.M., Fokas, A.S.: Exact interaction of solitary waves for certain nonintegrable equations. J. Math. Phys. **37**, 324–345 (1996)
15. Cherniha, R., Davydovych, V.: Conditional symmetries and exact solutions of the diffusive Lotka-Volterra system. Math. Comput. Modelling. **54**, 1238–1251 (2011)
16. Cherniha, R., Davydovych, V.: Conditional symmetries and exact solutions of nonlinear reaction-diffusion systems with non-constant diffusivities. Commun. Nonlinear. Sci. Numer. Simulat. **17**, 3177–3188 (2012)
17. Cherniha, R.: Conditional symmetries for systems of PDEs: new definition and its application for reaction-diffusion systems. J. Phys. A: Math. Theor. **43**, 405207, 13 (2010)
18. Olver, P.J.: Applications of Lie Groups to Differential Equations. Springer, Berlin (1986)
19. Bluman, G.W., Kumei, S.: Symmetries and Differential Equations. Springer, Berlin (1989)
20. Cherniha, R., Pliukhin, O.: New conditional symmetries and exact solutions of reaction-diffusion systems with power diffusivities. J. Phys. A: Math. Theor. **41**, 185208, 15 (2008)

21. Barannyk, T.: Symmetry and exact solutions for systems of nonlinear reaction-diffusion equations. Proc. Inst. Math. NAS Ukraine **43**, 80–85 (2002)
22. Murata, S.: Non-classical symmetry and Riemann invariants. Int. J. Non-Lin. Mech. **41**, 242–246 (2006)
23. Arrigo, D.J., Ekrut, D.A., Fliss, J.R.: Long, Le: Nonclassical symmetries of a class of Burgers' systems. J. Math. Anal. Appl. **371**, 813–820 (2010)
24. Cherniha, R., Serov, M.: Nonlinear systems of the Burgers-type equations: Lie and Q-conditional symmetries, ansätze and solutions. J. Math. Anal. Appl. **282**, 305–328 (2003)
25. Kingston, J.G.: On point transformations of evolution equations. J. Phys. A. **24**, 769–774 (1991)
26. Cherniha, R., Serov, M., Rassokha, I.: Lie symmetries and form-preserving transformations of reaction-diffusion-convection equations. J. Math. Anal. Appl. **342**, 1363–1379 (2008)
27. Popovych, R., Sophocleous, C., Vaneeva, O.: Enhanced group analysis and exact solutions of variable coefficient semilinear diffusion equations with a power source. Acta Appl. Math. **106**, 1–46 (2009)
28. Cherniha, R., Myroniuk, L.: Lie symmetries and exact solutions of a class of thin film equations. J. Phys. Math. **2**, 1–19 (2010)
29. Cherniha, R., Davydovych, V.: Lie and Conditional symmetries of the 3-D diffusive Lotka-Volterra system. J. Phys. A: Math. Theor. **46**, 185204, 14 (2013)

Solution to the Inverse Problem of Reconstructing Permittivity of an n-Sectional Diaphragm in a Rectangular Waveguide

Yuri G. Smirnov, Yuri V. Shestopalov and Ekaterina D. Derevyanchuk

Abstract We have developed a numerical-analytical method of solution to the inverse problem of reconstructing permittivities of n-sectional diaphragms in a waveguide of rectangular cross section. For a one-sectional diaphragm, a solution in the closed form is obtained and the uniqueness is proved.

1 Introduction

Determination of electromagnetic parameters of dielectric bodies that have complicated geometry or structure is an urgent problem arising e.g. when nanocomposite or artificial materials and media are used as elements of various devices. However, as a rule, these parameters cannot be directly measured (because of composite character of the material and small size of samples), which leads to the necessity of applying methods of mathematical modeling and numerical solution of the corresponding forward and inverse electromagnetic problems [21]. It is especially important to develop the solution techniques when the inverse problem for bodies of complicated shape are considered in the resonance frequency range, which is the case when permittivity of nanocomposite materials must be reconstructed [19, 20].

One of possible applications of composites is the creation of radio absorbing materials that can be used in systems that provide electromagnetic compatibility of modern electronic devices and in 'Stealth'-type systems aimed at damping and decreasing

Y. G. Smirnov (✉) · E. D. Derevyanchuk
Penza State University, Penza, Russia
e-mail: smirnovyug@mail.ru

E. D. Derevyanchuk
e-mail: catherinderevyanchuk@mail.ru

Y. V. Shestopalov
University of Gävle, Gävle, Sweden
e-mail: youri.shestopalov@kau.se

A. Makhlouf et al. (eds.), *Algebra, Geometry and Mathematical Physics*, 555
Springer Proceedings in Mathematics & Statistics 85, DOI: 10.1007/978-3-642-55361-5_32,

reflectivity of microwave electromagnetic radiation from objects to be detected [7, 17]. When calculating reflection and absorption characteristics of electromagnetic microwave radiation of radio absorbing materials, researchers use models employing the data on the material constants (permittivity, permeability, conductivity) of these materials in the microwave range. Such composites often contain carbon particles, short carbon fibers, carbon nanofibers, and multilayer carbon nanotubes as fillers for polymer dielectric matrices [7, 13, 17]. The use of carbon nanotubes enables one to achieve a significant (up to 10 dB) absorption of microwave electromagnetic radiation at relatively thin composite layers and low volume fractions of nanotubes and hence a small weight, in a broad frequency range (up to 5 GHz). Such characteristics are caused by both the geometrical sizes of individual nanotubes and their electrophysical properties; among the most important parameters here are permittivity and electric conductivity (which can vary over very wide ranges).

It is important to determine permittivity and conductivity not only of a composite as a solid body (as in [7, 17]), but also of its components, e.g., nanotubes, whose physical characteristics can vary substantially in the process of composite formation.

The forward scattering problem for a diaphragm in a parallel-plane waveguide was considered in [14]. In papers [1, 3, 8, 9, 16, 23–25] the inverse problem of reconstructing complex permittivity was analyzed from the measurements of the transmission coefficient; in [8, 9, 15] the artificial neural networks method was applied.

Several techniques for the permittivity determination of homogeneous materials loaded in a waveguide are reported [1, 3, 6]. The permittivity reconstruction of inhomogeneous structures are not as widely investigated and only a few studies exist for multilayered materials [2, 10]. Note a recently developed advanced approach [5] that can be also applied to numerical solution of this inverse problem.

However, the solution in closed form to the inverse problem of permittivity determination of materials loaded in a waveguide is not available in the literature, to the best of our knowledge, even for the simplest configuration of a parallel-plane dielectric insert in a guide of rectangular cross section. This fact dictates the aim of this work: to develop a method of solution to the inverse problem of reconstructing effective permittivity of layered dielectrics in the form of diaphragms in a waveguide of rectangular cross section that would enable both obtaining solution in a closed form for benchmark problems and efficient numerical implementation. We note that the corresponding forward problem for a one-sectional diaphragm is considered in [11] and [22].

2 Statement of the Problem

Assume that a waveguide $P = \{x : 0 < x_1 < a, 0 < x_2 < b, -\infty < x_3 < \infty\}$ with the perfectly conducting boundary surface ∂P is given in Cartesian coordinate system. A three-dimensional body Q ($Q \subset P$)

$$Q = \{x : 0 < x_1 < a, 0 < x_2 < b, 0 < x_3 < l\}$$

is placed in the waveguide; the body has the form of a diaphragm (an insert), namely, a parallelepiped separated into n sections adjacent to the waveguide walls. Domain $P\backslash\bar{Q}$ is filled with an isotropic and homogeneous layered medium having constant permeability $(\mu_0 > 0)$ in whole waveguide P, the sections of the diaphragm

$$\begin{aligned} Q_0 &= \{x : 0 < x_1 < a, 0 < x_2 < b, -\infty < x_3 < 0\} \\ Q_j &= \left\{x : 0 < x_1 < a, 0 < x_2 < b, l_{j-1} < x_3 < l_j\right\}, j = 1, \ldots, n \\ Q_{n+1} &= \{x : 0 < x_1 < a, 0 < x_2 < b, l < x_3 < +\infty\} \end{aligned}$$

are filled each with a medium having constant permittivity $\varepsilon_j > 0$; $l_0 := 0$, $l_n := l$.

The electromagnetic field inside and outside of the object in the waveguide is governed by Maxwell's equation:

$$\begin{aligned} \text{rot } \mathbf{H} &= -i\omega\varepsilon\mathbf{E} + \mathbf{j}_E^0 \\ \text{rot } \mathbf{E} &= i\omega\mu_0\mathbf{H}, \end{aligned} \tag{1}$$

where $\mathbf{E}$ and $\mathbf{H}$ are the vectors of the electric and magnetic field intensity, $\mathbf{j}$ is the electric polarization current, and ω is the circular frequency.

Assume that $\pi/a < k_0 < \pi/b$, where k_0 is the wavenumber, $k_0^2 = \omega^2\varepsilon_0\mu_0$ [12]. In this case, only one wave H_{10} propagates in the waveguide without attenuation (we have a single-mode waveguide [12]).

The incident electrical field is

$$\mathbf{E}^0 = \mathbf{e}_2 A \sin\left(\frac{\pi x_1}{a}\right) e^{-i\gamma_0 x_3} \tag{2}$$

with a known A and $\gamma_0 = \sqrt{k_0^2 - \pi^2/a^2}$.

Solving the forward problem for Maxwell's equations with the aid of (1) and the propagation scheme in Fig. 1, we obtain explicit expressions for the field inside every section of diaphragm Q and outside the diaphragm:

$$E_{(0)} = \sin\left(\frac{\pi x_1}{a}\right)(Ae^{-i\gamma_0 x_3} + Be^{i\gamma_0 x_3}), \quad x \in Q_0, \tag{3}$$

$$\begin{aligned} E_{(j)} &= \sin\left(\frac{\pi x_1}{a}\right)(C_j e^{-i\gamma_j x_3} + D_j e^{i\gamma_j x_3}), \\ j &= 1, \ldots, n+1; \qquad D_{n+1} = 0, \quad x \in Q_j, \end{aligned} \tag{4}$$

where $\gamma_j = \sqrt{k_j^2 - \pi^2/a^2}$ and $k_j^2 = \omega^2\varepsilon_j\mu_0$.

From the transmission conditions on the boundary surfaces of the diaphragm sections

$$[E_{(j)}] = [E_{(j+1)}] = 0; \quad \frac{\partial[E_{(j)}]}{\partial x_3} = \frac{\partial[E_{(j+1)}]}{\partial x_3} = 0, \ j = 0, \ldots, n+1. \tag{5}$$

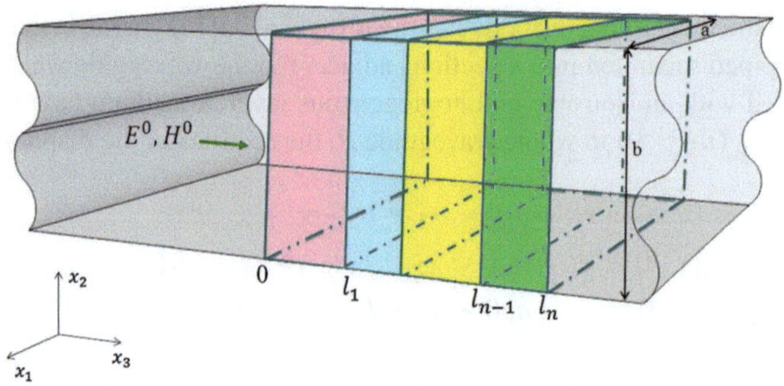

Fig. 1 Multilayered diaphragms in a waveguide

applied to (3) and (4) we obtain using conditions (5) a system of equations for the unknown coefficients

$$\begin{cases} A + B = C_1 + D_1 \\ \gamma_0 (B - A) = \gamma_1 (D_1 - C_1) \\ C_j e^{-i\gamma_j l_j} + D_j e^{i\gamma_j l_j} = C_{j+1} e^{-i\gamma_{j+1} l_j} + D_{j+1} e^{i\gamma_{j+1} l_j} \\ \gamma_j (D_j e^{i\gamma_j l_j} - C_j e^{-i\gamma_j l_j}) = \gamma_{j+1} (D_{j+1} e^{i\gamma_{j+1} l_j} - C_{j+1} e^{-i\gamma_{j+1} l_j}), \ \ j = 1, \ldots, n. \end{cases} \tag{6}$$

where $C_{n+1} = F$, $D_{n+1} = 0$. In system (6) coefficients A, B, C_j, D_j, ε_j, ($j = 1, \ldots, n$) are supposed to be complex.

We can express C_j, D_j from C_{j+1}, D_{j+1} in order to obtain a recurrent formula that couples amplitudes A and F.

We prove that this recurrent formula has the form

$$A = \frac{1}{2 \prod\limits_{j=0}^{n} \gamma_j} (\gamma_n p_{n+1} + \gamma_0 q_{n+1}) F e^{-i\gamma_0 l_n}, \tag{7}$$

where

$$p_{j+1} = \gamma_{j-1} p_j \cos \alpha_j + \gamma_j q_j i \sin \alpha_j; \quad p_1 := 1, \tag{8}$$

$$q_{j+1} = \gamma_{j-1} p_j i \sin \alpha_j + \gamma_j q_j \cos \alpha_j; \quad q_1 := 1. \tag{9}$$

Here $\alpha_j = \gamma_j (l_j - l_{j-1}), j = 2, \ldots, n$. Note that similar formulas are obtained in classical monographs dealing with wave propagation in layered media, e.g, in [4].

3 Inverse Problem for Multisectional Diaphragm

Formulate the inverse problem for a multisectional diaphragm that will be addressed in this work.

Inverse problem P: *find (complex) permittivity ε_j of each section from the known amplitude A of the incident wave and amplitude F of the transmitted wave at different frequencies.*

It is reasonable to consider the right-hand side of (7) as a complex-valued function with respect to n variables ε_j. For n sections we must know amplitudes A and F for each of n frequency values to have a consistent system of n equations with respect to n unknown permittivity values ε_j. This system is then solved to obtain the sought-for permittivity values.

Let us rewrite Eq. (7) in the form

$$G(h) = H, \quad H := \frac{2A\gamma_0 e^{i\gamma_0 l_n}}{F}, \tag{10}$$

where

$$G(h) := \frac{1}{\prod\limits_{j=1}^{n} \gamma_j} (\gamma_n p_{n+1} + \gamma_0 q_{n+1}), \tag{11}$$

and $h := (\varepsilon_1, \dots, \varepsilon_n)$.

We will consider (11) as a complex function of n complex variables. It follows from (8) and (9) that

$$\begin{pmatrix} p_{j+1} \\ q_{j+1} \end{pmatrix} = \begin{pmatrix} \cos\alpha_j & i\sin\alpha_j \\ i\sin\alpha_j & \cos\alpha_j \end{pmatrix} \begin{pmatrix} \gamma_{j-1} & 0 \\ 0 & \gamma_j \end{pmatrix} \begin{pmatrix} p_j \\ q_j \end{pmatrix} \tag{12}$$

$(j = 1, \dots, n)$. Thus we can represent p_{n+1}, q_{n+1} via finite multiplication of matrices by formula (12). From representation (12) we select, for every fixed j, only the matrices depending on γ_j. Finally we obtain

$$\begin{pmatrix} \gamma_j & 0 \\ 0 & \gamma_{j+1} \end{pmatrix} \begin{pmatrix} \cos\alpha_j & i\sin\alpha_j \\ i\sin\alpha_j & \cos\alpha_j \end{pmatrix} \begin{pmatrix} \gamma_{j-1} & 0 \\ 0 & \gamma_j \end{pmatrix} = \begin{pmatrix} \gamma_j\gamma_{j-1}\cos\alpha_j & i\gamma_j^2\sin\alpha_j \\ i\gamma_{j+1}\gamma_{j-1}\sin\alpha_j & \gamma_j\gamma_{j+1}\cos\alpha_j \end{pmatrix}. \tag{13}$$

Dividing matrix (13) by γ_j we have

$$\begin{pmatrix} \gamma_{j-1}\cos\alpha_j & i\gamma_j\sin\alpha_j \\ i\gamma_{j+1}\gamma_{j-1}\sin\alpha_j/\gamma_j & \gamma_{j+1}\cos\alpha_j \end{pmatrix}. \tag{14}$$

Taking into account Taylor series for functions $\sin\alpha_j$ and $\cos\alpha_j$ and that $\alpha_j = \gamma_j(l_j - l_{j-1})$ (14) we see that each coefficient of this matrix depends on γ_j^2. Since $\gamma_j^2 = \varepsilon_j\mu_0\omega^2 - \pi^2/a^2$ we have that each coefficient of matrix (14) is an analytical function w.r.t. ε_j. Hence function $G(h)$ depends on ε_j analytically for every j, $(j = 1, \dots, n)$.

Using Hartogs' theorem [18] we obtain the following statement

Theorem 3.1 *$G(h)$ is holomorphic on $\mathbf{C}^n$ as a function of n complex variables.*

Let us formulate inverse problem P for n-sectional diaphragm in the following form. Consider n different frequencies $\Omega = (\omega_1, \dots, \omega_n)$ and functions $G_j(h) := G(h, \omega_j), j = 1, \dots, n$. It is necessary find a solution to the (nonlinear) system of n equations w.r.t. n variables $\varepsilon_1, \dots, \varepsilon_n$:

$$G_j(h) = H_j, \quad H_j = H(\omega_j), \quad j = 1, \dots, n. \tag{15}$$

Theorem 3.1 implies [18].

Theorem 3.2 *If Jacobian $\frac{\partial(G_1,\dots,G_n)}{\partial(h_1,\dots,h_n)} \neq 0$ at the point h^* then function $G(h)$ is locally invertible in a vicinity of h^* and inverse problem P has unique solution for every h from that vicinity.*

Below we present an example of numerical solutions to inverse problem P for a three-sectional diaphragm. The table shows the test results of numerical solution to the inverse problem of reconstructing permittivities of a three-section diaphragm at three frequencies. The test values of the transmission coefficient are taken from the solution to the forward problem.

$F(\omega_{1,2,3})$	Calculated $\varepsilon_{1,2,3}$	True $\varepsilon_{1,2,3}$
$0.012 + i \cdot 0.036$	$-1.713 + i \cdot 0.078$,	-1.7
$0.025 - i \cdot 0.012$	$1.523 - 0.085$	1.5
$0.029 + i \cdot 0.014$	$4.01 + i \cdot 0.13$	4
$0.073 - i \cdot 0.177$	$1.702 - i \cdot 0.0004$,	1.7
$-0.269 - i \cdot 0.197$	$-1.499 + i \cdot 0.006$	-1.5
$-0.052 - i \cdot 0.22$	$3.996 + i \cdot 0.003$	4
$0.106 + i \cdot 0.061$	$1.716 - i \cdot 0.0004$,	1.7
$0.096 - i \cdot 0.023$	$1.48 - i \cdot 0.013$	1.5
$0.102 + i \cdot 0.046$	$-3.928 + i \cdot 0.037$	-4
$-0.0004 - i \cdot 0.00376$	$-1.708 - i \cdot 0.008$,	-1.7
$-0.0045 + i \cdot 0.00368$	$1.5 + i \cdot 0.0007$	1.5
$-0.0057 - i \cdot 0.00134$	$-3.982 + i \cdot 0.017$	-4
$-0.00005 - i \cdot 0.0005$	$-1.708 + i \cdot 0.03$,	-1.7
$-0.002 + i \cdot 0.001$	$-1.499 - i \cdot 0.03$	-1.5
$-0.002 - i \cdot 0.0004$	$-3.982 - i \cdot 0.044$	-4

Parameters of the three-section diaphragm are $a = 2$, $b = 1$, $c = 2$, $A = 1$, $l_1 = 1$, $l_2 = 1.5$; the excitation frequencies $\omega_1 = 2.5$, $\omega_2 = 1.7$, and $\omega_3 = 2$. The first, second, and third columns of the table shows, respectively, the values of transmission coefficient F, calculated values of permittivity of a section, and true values of (real) permittivity of a section.

We see that in all examples the error of computations does not exceed 3 % which proves high efficiency of the method.

4 One-Sectional Diaphragm: Explicit Solution to the Inverse Problem

From (7) for a one-sectional diaphragm we have

$$\frac{Ae^{i\gamma_0 l_1}}{F} = g(z),$$
$$g(z) = \cos z + i\left(\frac{z}{2\gamma_0 l_1} + \frac{\gamma_0 l_1}{2z}\right)\sin z, \tag{16}$$
$$z = \gamma_1 l_1 = l_1\sqrt{k_1^2 - \frac{\pi^2}{a^2}},$$

where z is generally a complex variable. From (16) we obtain a relation for the transmission coefficient

$$F = \frac{Ae^{i\gamma_0 l_1}}{g(z)}, \tag{17}$$

which, together with formulas (3) and (4), gives an explicit solution to the forward problem under study.

When the inverse problem is solved, ε_1 is considered as an unknown quantity that should be determined from Eq. (16) in terms of F.

List the most important properties of $g(z)$ which easily follows from its explicit representation:

(i) $g(z)$ is an entire function.
(ii) $g(z)$ has neither real zeros nor poles. This fact is in line with physical requirements that the transmission coefficient does not vanish and is a bounded quantity at real frequencies.
(iii) $g(z)$, also considered as a function of real τ, is not invertible locally at the origin because it is easy to check that $g'(0) = 0$. Next, the inverse of $g(z)$ is a multi-valued function. In fact, the inverse function does not exist globally according to the statement in Remark concerning violation of uniqueness.
(iv) $g(z)$ is not a fractional-linear function; therefore $g(z)$ performs one-to-one conformal mappings only of certain regions of the complex plane onto regions of the complex plane.
(v) It is easy to check up that $g'(\tau) \neq 0$ for (real) $\tau \neq 0$. Hence, $g(z)$ is invertible locally at the real point $\tau \neq 0$.

Assuming that ε_1 is real it is reasonable to introduce a real variable

$$\tau = \gamma_1 l_1 = l_1\sqrt{k_1^2 - \frac{\pi^2}{a^2}} > 0 \tag{18}$$

which may be used for parametrization. Extract the real and imaginary part of $g(\tau)$, denoting them by x and y,

$$\begin{cases} x = \cos\tau, \\ y = h(\tau)\sin\tau, \end{cases} \quad \text{where } h(\tau) = \frac{\tau}{2C} + \frac{C}{2\tau}, \quad C = \gamma_0 l_1. \tag{19}$$

Equation (16) is equivalent to the system

$$\begin{cases} \cos\tau = p, & p = \operatorname{Re}\left(\frac{Ae^{-i\gamma_0 l_1}}{F}\right), \\ h(\tau)\sin(\tau) = q, & q = \operatorname{Im}\left(\frac{Ae^{-i\gamma_0 l_1}}{F}\right), \end{cases} \tag{20}$$

where p and q are known values. Using the results of Appendix I we finally obtain from (20) an explicit formula for the sought (real) permittivity

$$\varepsilon_1 = \frac{1}{\omega^2\mu_0}\left(\left(\frac{\pi}{a}\right)^2 + \left(\frac{\tau}{l_1}\right)^2\right), \tag{21}$$

here

$$\tau = \tau_1 = C\left(\frac{|q| + \sqrt{p^2+q^2-1}}{\sqrt{1-p^2}}\right) \tag{22}$$

when $\varepsilon_1 > \varepsilon_0$ and

$$\tau = \tau_2 = C\left(\frac{\sqrt{1-p^2}}{|q| + \sqrt{p^2+q^2-1}}\right) \tag{23}$$

when $\frac{\pi^2}{a^2\omega^2\mu_0} < \varepsilon_1 < \varepsilon_0$.

Formulas (21)–(23) constitute explicit solution of inverse problem P under study.

Using the reasoning and results of Appendix I we prove the following result stating the existence and uniqueness of solution to the inverse problem of finding permittivity of a one-sectional diaphragm in a waveguide of rectangular cross-section.

Theorem 4.1 *Assume that* $|p| < 1$ *and* $p^2+q^2 \geq 1$. *Then inverse problem P has only one solution expressed by* (22) *if* $\frac{\tau_1}{C} > 1$, $\cos\tau_1 = p$, *and* $sign(q) = sign(\sin(\tau_1))$. *If* $\frac{\tau_2}{C} < 1$, $\cos\tau_2 = p$, *and* $sign(q) = sign(\sin(\tau_2))$, *inverse problem P has only one solution expressed by* (23). *Otherwise, inverse problem P has no solution.*

Remark 4.1 If $p = 1$, then q must be equal zero and $\tau = 2\pi n$, $n \in Z$. If $p = -1$, then q must be equal to zero and $\tau = \pi + 2\pi n$, $n \in Z$. In these cases inverse problem P has has infinitely many solutions; therefore they are excluded from Theorem.

5 Conclusion

We have developed a numerical-analytical method of solution to the inverse problem of reconstructing permittivities of n-sectional diaphragms in a waveguide of rectangular cross-section. For a one-sectional diaphragm, a solution in the closed form is obtained and the uniqueness is proved. These results make it possible to use the case of a one-sectional diaphragm in a waveguide of rectangular cross-section as a benchmark test problem and perform a complete analysis of the inverse scattering problem for arbitrary n-sectional diaphragms.

Acknowledgments This work is partially supported by Russian Foundation of Basis Research 11-07-00330-a and Visby Program of the Swedish Institute.

Appendix 1

Reduce Eq. (16) to a quadratic equation. From (18) it follows that (on the domain of all the functions involved):

$$p^2 + \frac{q^2}{h^2(\tau)} = 1, \quad h(\tau) > 0.$$

From (20) we obtain:

$$h^2(\tau) = \frac{q^2}{1-p^2}, \quad |p| < 1. \tag{24}$$

Then

$$h(\tau) = Q, \quad h(\tau) := \frac{\tau}{2C} + \frac{C}{2\tau}, \quad Q := \frac{|q|}{\sqrt{1-p^2}} > 0, \tag{25}$$

and we obtain a quadratic equation

$$\tau^2 - 2CQ\tau + C^2 = 0 \tag{26}$$

which has the roots

$$\tau_1 = C(Q + \sqrt{Q^2-1}), \quad \tau_2 = \frac{C}{Q + \sqrt{Q^2-1}}. \tag{27}$$

$\tau_{1,2}$ are real if $Q \geq 1$; therefore,

$$p^2 + q^2 \geq 1. \tag{28}$$

Inequality (28) constitutes the existence condition for the solution of equation (16). Since $\tau = \gamma_1 l_1$ and $C = \gamma_0 l_1$, we have

$$\frac{\tau}{C} = \frac{\gamma_1}{\gamma_0} = \frac{\sqrt{\omega^2\mu_0\varepsilon_1 - \frac{\pi^2}{a^2}}}{\sqrt{\omega^2\mu_0\varepsilon_0 - \frac{\pi^2}{a^2}}},$$

so that, in view of the assumption $\varepsilon_1 > \varepsilon_0$,

$$\frac{\tau}{C} > 1.$$

Similarly, for $\frac{\pi^2}{a^2\omega^2\mu_0} < \varepsilon_1 < \varepsilon_0$,

$$\frac{\tau}{C} < 1.$$

Thus, for $\varepsilon_1 > \varepsilon_0$ we obtain

$$\frac{\tau_1}{C} = Q + \sqrt{Q^2 - 1} \quad (> 1).$$

For $\frac{\pi^2}{a^2\omega^2\mu_0} < \varepsilon_1 < \varepsilon_0$

$$\frac{\tau_2}{C} = \frac{1}{Q + \sqrt{Q^2 - 1}} \quad (< 1).$$

Thus, when $\varepsilon_1 > \varepsilon_0$ Eq. (25) has only one root (24) τ_1. Similarly, Eq. (27) has the only one root (25) τ_2 for $\frac{\pi^2}{a^2\omega^2\mu_0} < \varepsilon_1 < \varepsilon_0$.

It should be noted that reduction of (16) to quadratic equation (26) is not an equivalent transformation. It is necessary to complement (26) with one of the equations of system (19), for example, with the first, and take into accounts the signs of p and q. As a result, (16) will be equivalent to the system

$$\begin{cases} \cos\tau = p, \ sign(q) = sign(\sin(\tau)), \\ p^2 + \frac{q^2}{h^2(\tau)} = 1. \end{cases}$$

References

1. Akleman, F.: Reconstruction of complex permittivity of a longitudinally inhomogeneous material loaded in a rectangular waveguide. IEEE Microw. Wirel. Compon. Lett. **18**(3), 158–160 (2008)
2. Baginski, M.E., Faircloth, D.L., Deshpande, M.D.: Comparison of two optimization techniques for the estimation of complex complex permittivities of multilayered structures using waveguide measurements. IEEE Trans. Microw. Theory Tech. **53**(10), 3251–3259 (2005)
3. Baker-Jarvis, J., Vanzura, E.J.: Improved technique for determining complex permittivity with the transmission-reflection method. IEEE Trans. Microw. Theory Tech. **38**(8), 1096–1103 (1990)
4. Brekhovskih, L.: Waves in Layered Media. Academic Press, New York (1980)
5. Beilina, L., Klibanov, M.: Approximate Global Convergence and Adaptivity for Coefficient Inverse Problems. Springer, New York (2012)
6. Dediu, S., McLaughlin, J.R.: Recovering inhomogeneities in a waveguide using eigensystem decomposition. Inverse Prob. **22**(4), 1227–1246 (2006)
7. De Rosa, I.M., Dinescu, A., Sarasini, F., Sarto, M.S., Tamburrano, A.: Effect of short carbon fibers and MWCNTs on microwave absorbing properties of polyester composites containing nickel-coated carbon fibers. Compos. Sci. Tech. **70**, 102–109 (2010)
8. Eves, E.E., Kopyt, P., Yakovlev, V.V.: Determination of complex permittivity with neural networks and FDTD modeling. Microw. Opt. Tech. Let. **40**(3), 183–188 (2004)
9. Eves, E.E., Kopyt, P., Yakovlev, V.V.: Practical aspects of complex permittivity reconstruction with neural-network-controlled FDTD modeling of a two-port fixture. J. Microw. Power Electromagn. Energy **41**(4), 81–94 (2007)
10. Faircloth, D.L., Baginski, M.E., Wentworth, S.M.: Complex permittivity and permeability extraction for multilayered samples using S-parameter waguide measurements. IEEE Trans. Microw. Theory Tech. **54**(3), 1201–1209 (2006)
11. Grishina, E.E., Derevyanchuk, E.D., Medvedik, M.Y., Smirnov, Y.G.: Numerical and analytical solutions of electromagnetic field diffraction on the two-sectional body with different permittivity in the rectangular waveguide. Izv. Vyssh. Uchebn. Zaved. Povolzh. Region Fiz.-Mat. Nauki No. **4**, 73–81 (2010)
12. Jackson, J.D.: Classical Electromagnetics. Wiley, New York (1967)
13. Kelly, J.M., Stenoien, J.O., Isbell, D.E.: Wave-guide measurements in the microwave region on metal powders suspended in paraffin wax. J. Appl. Phys. **24**(3), 258–262 (1953)
14. Lurie, K.A., Yakovlev, V.V.: Optimization of electric field in rectangular waveguide with lossy layer. IEEE Trans. Magn. **36**, 1094–1097 (2000)
15. Lurie, K.A., Yakovlev, V.V.: Control over electric field in traveling wave applicators. J. Eng. Math. **44**(2), 107–123 (2002)
16. Outifa, L., Delmotte, M., Jullien, H.: Dielectric and geometric dependence of electric field and power distribution in a waveguide heterogeneously filled with lossy dielectrics. IEEE Trans. Microw. Theor. Tech. **45**(Suppl 1), 1154–1161 (1997)
17. Saib, A., Bednarz, L., Daussin, R., Bailly, C., Lou, X., Thomassin, J.-M., Pagnoulle, C., Detrembleur, C., Jerome, R., Huynen, I.: Carbon nanotube composites for broadband microwave absorbing materials. IEEE Trans. Microw. Theor. Techn. **54**(6), 2745–2754 (2006)
18. Shabat, B.: Introduction to Complex Analysis. Part II: Functions of Several Variables. American Mathematical Society, Town (2003)
19. Shestopalov, YuV, Smirnov, YuG, Yakovlev, V.V.: Volume singular integral equations method for determination of effective permittivity of meta- and nanomaterials. In: Proceedings of Progress in Electromagnetics Research Symposium, pp. 291–292. Cambridge, USA (2008)
20. Shestopalov, Yu.V., Smirnov, Yu.G., Yakovlev, V.V.: Development of mathematical methods for reconstructing complex permittivity of a scatterer in a waveguide. In: Proceedings of 5th International Workshop on Electromagnetic Wave Scattering, Antalya, Turkey (2008)

21. Smirnov, Y.G., Medvedik, M.Y., Derevyanchuk, E.D.: Collocation method of solving volume singular integral equation for diffraction by dielectric body in rectangular waveguide. In: Proceedings of Progsess in Electromagnetics Research Symposium, Moscow, Russia (2009)
22. Solymar, L., Shamonina, E.: Waves in Metamaterials. Oxford University Press Inc., New York (2009)
23. Usanov, D.A., Skripal, A.V., Abramov, A.V., Bogolyubov, A.S.: Determination of the metal nanometer layer thicknesss and semiconductor conductivity in metal-semiconductor structures from electromagnetic reflection and transmission spectra. Tech. Phys. **51**(5), 644–649 (2006)
24. Usanov, D.A., Skripal, A.V., Abramov, A.V., Bogolyubov, A.S.: Complex permittivity of composites based on dielectric matrices with carbon nanotubes. Tech. Phys. **56**(1), 102–106 (2011)
25. Yakovlev, V.V., Murphy, E.K., Eves, E.E.: Neural networks for FDTD-backed permittivity reconstruction. COMPEL **33**(1), 291–304 (2005)

Some Conservation Laws for a Class of Hamilton-Jacobi-Bellman Equations

Maria Luz Gandarias, Maria V. Redondo and Maria S. Bruzón

Abstract The idea of a conservation law has its origin in mechanics and physics. Since a large number of physical theories, including some of the 'laws of nature', are usually expressed as systems of nonlinear differential equations, it follows that conservation laws are useful in both general theory and the analysis of concrete systems. In [3] one of the present authors has introduced the concept of weak self-adjoint equations. This definition generalizes the concept of self-adjoint and quasi self-adjoint equations that were introduced by Ibragimov in [8]. Recently [4] we found a class of weak self-adjoint Hamilton-Jacobi-Bellman equations which are neither self-adjoint nor quasi self-adjoint. In this paper, by using a general theorem on conservation laws proved in [7] and the new concept of weak self-adjointness [3] we find conservation laws for some of these partial differential equations.

1 Introduction

The class of Hamilton-Jacobi-Bellman (HJB) equations arises in the sphere of stochastic control theory [12]. In [13] the properties of one of these HJB equations were investigated from the viewpoint of the Lie symmetry analysis. It was pointed out in [13] that the HJB equation

$$u_t + au_x + \frac{b^2}{2}u_{xx} - \frac{1}{2}(u_x)^2 + \left(\frac{c}{x}\right)^2 = 0 \tag{1}$$

M. L. Gandarias (✉) · M. Redondo · M. S. Bruzón
Departamento de Matemáticas, Universidad de Cádiz, Puerto Real (Cádiz), Spain
e-mail: marialuz.gandarias@uca.es

M. S. Bruzón
e-mail: maria.redondo@uca.es

M. Redondo
e-mail: m.bruzon@uca.es

A. Makhlouf et al. (eds.), *Algebra, Geometry and Mathematical Physics*,
Springer Proceedings in Mathematics & Statistics 85, DOI: 10.1007/978-3-642-55361-5_33,

with the terminal condition $u(T,x) = 0$ is not an obvious relative of the Black-Scholes or heat equations and is presented by Heath, Platin and Schweizer [5] as an equation for mean-variance hedging. The authors observe that c in the ultimate term of (1) needs not be a constant. In fact it is better to write the HJB equation as

$$u_t + au_x + \frac{b^2}{2}u_{xx} - \frac{1}{2}(u_x)^2 + c(x) = 0 \tag{2}$$

to allow for the possible dependence of c upon x.

The idea of a conservation law, or more particularly, of a conserved quantity, has its origin in mechanics and physics. Since a large number of physical theories, including some of the 'laws of nature', are usually expressed as systems of nonlinear differential equations, it follows that conservation laws are useful in both general theory and the analysis of concrete systems [14]. In [1] Anco and Bluman gave a general treatment of a direct conservation law method for partial differential equations expressed in a standard Cauchy-Kovaleskaya form

$$u_t = G(x, u, u_x, u_{xx}, \ldots, u_{nx}).$$

In [11] Kara and Mahomed showed to construct conservation laws of Euler-Lagrange type equations via Noether type symmetry operators associated with partial Lagrangians.

In [7] (see also [6]) a general theorem on conservation laws for arbitrary differential equations which do not require the existence of Lagrangians has been proved. This new theorem is based on the concept of adjoint equations for non-linear equations. There are many equations with physical significance which are not self-adjoint. Therefore one cannot eliminate the nonlocal variables from the conservation laws of these equations. In [8] Ibragimov generalized the concept of by introducing the definition of quasi self-adjoint equations.

It happens that many equations having remarkable symmetry properties, such as the HJB equation, are neither self-adjoint nor quasi self-adjoint.

In [3] one of the present authors has generalized the concept of quasi-self-adjoint equations by introducing the concept of weak self-adjoint equations. In [9] has generalized this concept and has introduced the concept of nonlinear self-adjointness. By using these two recent developments Freire and Sampaio [2] have deter mined the nonlinear self-adjoint class of a generalized fifth order equation and by using Ibragimov theorem [6] the authors have established some local conservation laws. In [10] Johnpillai and Khalique have studied study the conservation laws of some special forms of the nonlinear scalar evolution equation, the modified Korteweg-De Vries (mKdV) equation with time dependent variable coefficients of damping and dispersion

$$u_t + u^2u_x + a(t)u + b(t)u_{xxx} = 0.$$

The authors use the new conservation theorem (Ibragimov [6]) and the partial Lagrangian approach (Kara and Mahomed [11]).

In [3] one of the present authors has generalized the concept of quasi self-adjoint equations by introducing the concept of.

In a previous work we have determined, for Eq. (2), the subclasses of equations which are self-adjoint, quasi self-adjoint and weak self-adjoint. The aim of this work is to determine, by using the notation and techniques of [7], some nontrivial conservation laws for a class of Eq. (2).

1.1 The Class of Self-Adjoint and Quasi Self-Adjoint Equations

Definition 1 Consider an sth-order partial differential equation

$$F(x, u, u_{(1)}, \dots, u_{(s)}) = 0 \tag{3}$$

with independent variables $x = (x^1, \dots, x^n)$ and a dependent variable u, where $u_{(1)} - \{u_i\}$, $u_{(2)} = \{u_{ij}\}, \dots$ denote the sets of the partial derivatives of the first, second, etc. orders, $u_i = \partial u/\partial x^i$, $u_{ij} = \partial^2 u/\partial x^i \partial x^j$. The adjoint equation to (12) is

$$F^*(x, u, v, u_{(1)}, v_{(1)}, \dots, u_{(s)}, v_{(s)}) = 0, \tag{4}$$

with

$$F^*(x, u, v, u_{(1)}, v_{(1)}, \dots, u_{(s)}, v_{(s)}) = \frac{\delta(v F)}{\delta u}, \tag{5}$$

where

$$\frac{\delta}{\delta u} = \frac{\partial}{\partial u} + \sum_{s=1}^{\infty}(-1)^s D_{i_1} \dots D_{i_s} \frac{\partial}{\partial u_{i_1 \dots i_s}} \tag{6}$$

denotes the variational derivative (the Euler-Lagrange operator), and v is a new dependent variable. Here

$$D_i = \frac{\partial}{\partial x^i} + u_i \frac{\partial}{\partial u} + u_{ij} \frac{\partial}{\partial u_j} + \cdots$$

are the total differentiations.

Definition 2 Equation (12) is said to be if the equation obtained from the adjoint Eq. (13) by the substitution $v = u$:

$$F^*(x, u, u, u_{(1)}, u_{(1)}, \dots, u_{(s)}, u_{(s)}) = 0,$$

is identical to the original Eq. (12). In other words, if

$$F^*(x, u, u_{(1)}, u_{(1)}, \dots, u_{(s)}, u_{(s)}) = \lambda(x, u, u_{(1)}, \dots) \, F(x, u, u_{(1)}, \dots, u_{(s)}). \tag{7}$$

Definition 3 Equation (12) is said to be if the equation obtained from the adjoint Eq. (13) by the substitution $v = h(u)$ with a certain function $h(u)$ such that $h'(u) \neq 0$ is identical to the original Eq. (12).

The following statement was proved in [4].

Theorem 11 *Equation* (2) *is quasi self-adjoint if and only if* $c(x) = 0$.

1.2 The Class of Weak Self-Adjoint Equations

In [3] the following definition has been introduced.

Definition 4 Eq. (12) is said to be if the equation obtained from the adjoint Eq. (13) by the substitution $v = h(x, t, u)$ with $h_x(x, t, u) \neq 0$ or $h_t(x, t, u) \neq 0$ and $h_u(x, t, u) \neq 0$ is identical to the original Eq. (12).

In [4] was given the following statement.

Theorem 12 *Equation* (2) *is weak self-adjoint for any arbitrary function* $c(x)$ *upon the substitution* $v = \alpha(x)e^{-\frac{u}{b^2}}$, *where* $\alpha(x)$ *must satisfy the following condition*

$$b^4\alpha_{xx} - 2ab^2\alpha_x - 2\,c\,\alpha = 0. \tag{8}$$

1.3 General Theorem on Conservation Laws

We use the following theorem on proved in [7].

Theorem 13 *Any, Lie-Bäcklund or non-local symmetry*

$$X = \xi^i(x, u, u_{(1)}, \ldots)\frac{\partial}{\partial x^i} + \eta(x, u, u_{(1)}, \ldots)\frac{\partial}{\partial u} \tag{9}$$

of Eq. (12) *provides a conservation law* $D_i(C^i) = 0$ *for the simultaneous system* (12), (13). *The conserved vector is given by*

$$\begin{aligned} C^i = \xi^i \mathscr{L} + W\left[\frac{\partial \mathscr{L}}{\partial u_i} - D_j\left(\frac{\partial \mathscr{L}}{\partial u_{ij}}\right) + D_j D_k\left(\frac{\partial \mathscr{L}}{\partial u_{ijk}}\right) - \cdots\right] \\ + D_j(W)\left[\frac{\partial \mathscr{L}}{\partial u_{ij}} - D_k\left(\frac{\partial \mathscr{L}}{\partial u_{ijk}}\right) + \cdots\right] + D_j D_k(W)\left[\frac{\partial \mathscr{L}}{\partial u_{ijk}} - \cdots\right] + \cdots, \end{aligned} \tag{10}$$

where W *and* $\mathscr{L}$ *are defined as follows:*

$$W = \eta - \xi^j u_j, \quad \mathscr{L} = v\,F\left(x, u, u_{(1)}, \ldots, u_{(s)}\right). \tag{11}$$

The proof is based on the following operator identity:

$$X + D_i(\xi^i) = W\frac{\delta}{\delta u} + D_i \mathcal{N}^i, \tag{12}$$

where X is operator (19) taken in the prolonged form:

$$X = \xi^i \frac{\partial}{\partial x^i} + \eta \frac{\partial}{\partial u} + \zeta_i \frac{\partial}{\partial u_i} + \zeta_{i_1 i_2} \frac{\partial}{\partial u_{i_1 i_2}} + \cdots,$$

$$\zeta_i = D_i(\eta) - u_j D_i(\xi^j), \quad \zeta_{i_1 i_2} = D_{i_2}(\zeta_{i_1}) - u_{j i_1} D_{i_2}(\xi^j), \ \ldots .$$

For the expression of operator $\mathcal{N}^i$ and a discussion of the identity (22) in the general case of several dependent variables, we refer the reader to [7], Sect. 8.4.4.

We will write the generators of a point transformation group admitted by Eq. (2) in the form

$$X = \xi^1 \frac{\partial}{\partial t} + \xi^2 \frac{\partial}{\partial x} + \eta \frac{\partial}{\partial u} \tag{13}$$

by setting $t = x^1, \ x = x^2$. The conservation law will be written

$$D_t(C^1) + D_x(C^2) = 0. \tag{14}$$

1.4 Conservation Laws for a Class of Quasi Self-Adjoint Equations

Let us apply the general Theorem on conservation laws to the quasi self-adjoint equation:

$$u_t + a u_x + \frac{b^2}{2} u_{xx} - \frac{1}{2}(u_x)^2 = 0. \tag{15}$$

In this case we have

$$\mathcal{L} = \left(u_t + a u_x + \frac{b^2}{2} u_{xx} - \frac{1}{2}(u_x)^2\right) v, \qquad v = e^{-\frac{u}{b^2}} \tag{16}$$

Let us find the conservation law provided by the following symmetry of Eq. (2):

$$X = t\frac{\partial}{\partial x} - (x - at)\frac{\partial}{\partial u}. \tag{17}$$

This symmetry was derived by Naicker, Andriopulos, and Leach in [13]. In this case we have $W = -(x - at) - t u_x$ and Eq. (20) yield conservation law (23) with

$$C^1 = e^{-\frac{u}{b^2}}\,(x - a\,t) + D_x\left(b^2\,t\,e^{-\frac{u}{b^2}}\right),$$
$$C^2 = \frac{e^{-\frac{u}{b^2}}\,\left(u_x\,x - 2\,a\,x - a\,t\,u_x + 2\,a^2\,t + b^2\right)}{2} - D_t\left(b^2\,t\,e^{-\frac{u}{b^2}}\right).$$

We simplify the conserved vector by transferring the terms of the form $D_x(\ldots)$ from C^1 to C^2 and obtain

$$C^1 = -e^{-\frac{u}{b^2}}\,(x - a\,t),$$
$$C^2 = \frac{e^{-\frac{u}{b^2}}\,\left(u_x\,x - 2\,a\,x - a\,t\,u_x + 2\,a^2\,t + b^2\right)}{2}.$$

1.5 Conservation Laws for a Subclass of Weak Self-Adjoint Equations

In [13] it was pointed out that $c(x)$ plays the role of a potential. The "potential" $c(x) = \dfrac{\mu^2}{2x^2}$ has a rich history dating back to the time of Newton. The symmetries for Eq. (2) were derived by Naicker, Andriopulos and Leach in [13]. In [4] we have applied the general Theorem on conservation laws to the weak self-adjoint equation:

$$u_t + au_x + \frac{b^2}{2}u_{xx} - \frac{1}{2}(u_x)^2 + \frac{\mu^2}{2x^2} = 0. \tag{18}$$

We will now find conservation laws of Eq. (2) with $c(x) = constant$ provided by the symmetries of this equation derived by Naicker, Andriopulos and Leach in [13].

Let us apply the general Theorem on conservation laws to the weak self-adjoint equation:

$$u_t + au_x + \frac{b^2}{2}u_{xx} - \frac{1}{2}(u_x)^2 + c = 0.$$

For $c = -\frac{a^2}{2}$ the solution of Eq. (8) is

$$\alpha(x) = (k_2\,x + k_1)\,e^{\frac{a\,x}{b^2}}.$$

In this case we have

$$\mathscr{L} = \left(u_t + au_x + \frac{b^2}{2}u_{xx} - \frac{1}{2}(u_x)^2 + c\right)v. \tag{19}$$

We will write generators of point transformation group admitted by Eq. (15) in the form (13) by setting $t = x^1,\ x = x^2$. The conservation law will be written (23).

1. Let us find the conservation law provided by the following symmetry of Eq. (2) where $c(x) = -\frac{a^2}{2}$:

$$X = t\frac{\partial}{\partial x} - (x - at)\frac{\partial}{\partial u}. \tag{20}$$

In this case we have $W = -(x - at) - tu_x$ and Eq. (20) yield the conservation law (23) with

$$C^1 = -x\, e^{\frac{a\,x}{b^2} - \frac{u}{b^2}} + D_x\left(b^2\, t\, e^{\frac{a\,x}{b^2} - \frac{u}{b^2}}\right),$$
$$C^2 = \frac{\left(u_x\, x - a\, x + 2\, c\, t + a^2\, t + b^2\right)\, e^{\frac{a\,x}{b^2} - \frac{u}{b^2}}}{2} - D_t\left(b^2\, t\, e^{\frac{a\,x}{b^2} - \frac{u}{b^2}}\right).$$

We simplify the conserved vector by transferring the terms of the form $D_x(\ldots)$ from C^1 to C^2 and obtain

$$C^1 = -x\, e^{\frac{a\,x}{b^2} - \frac{u}{b^2}},$$
$$C^2 = \frac{\left(u_x\, x - a\, x + 2\, c\, t + a^2\, t + b^2\right)\, e^{\frac{a\,x}{b^2} - \frac{u}{b^2}}}{2}.$$

2. Let us find the conservation law provided by the following o symmetry of Eq. (2) where $c(x) = -\frac{a^2}{2}$:

$$X = t\frac{\partial}{\partial t} + \frac{x + at}{2}\frac{\partial}{\partial x} + \frac{a^2 t}{2}\frac{\partial}{\partial u}. \tag{21}$$

In this case we have $W = \dfrac{a^2 t}{2} - tu_t - \dfrac{x + at}{2}u_x$ and Eq. (20) yield the conservation law (23) with

$$C^1 = -\frac{\left(a\, x + b^2\right)\, e^{\frac{a\,x}{b^2} - \frac{u}{b^2}}}{2} + D_x\left(\frac{b^2\, (x + t\, u_x)\, e^{\frac{a\,x}{b^2} - \frac{u}{b^2}}}{2}\right),$$
$$C^2 = \frac{\left(a\, u_x\, x - a^2\, x + b^2\, u_x\right)\, e^{\frac{a\,x}{b^2} - \frac{u}{b^2}}}{4} - D_t\left(\frac{b^2\, (x + t\, u_x)\, e^{\frac{a\,x}{b^2} - \frac{u}{b^2}}}{2}\right).$$

We simplify the conserved vector by transferring the terms of the form $D_x(\ldots)$ from C^1 to C^2 and obtain

$$C^1 = -\frac{\left(a\, x + b^2\right)\, e^{\frac{a\,x}{b^2} - \frac{u}{b^2}}}{2},$$
$$C^2 = \frac{\left(a\, u_x\, x - a^2\, x + b^2\, u_x\right)\, e^{\frac{a\,x}{b^2} - \frac{u}{b^2}}}{4}.$$

3. Let us find the conservation law provided by the following symmetry of Eq. (2) where $c(x) = -\frac{a^2}{2}$:

$$X = tx\frac{\partial}{\partial x} + t^2\frac{\partial}{\partial x} + \frac{1}{2}(b^2t + a^2t^2 - (x - at)^2)\frac{\partial}{\partial u}. \tag{22}$$

In this case we have $W = -\frac{1}{2}(b^2t + a^2t^2 - (x - at)^2) - txu_x - t^2u_t$ and Eq. (20) yield the conservation law (23) with

$$C^1 = -\frac{\left(x^2 + b^2\,t\right)\, e^{\frac{a\,x}{b^2} - \frac{u}{b^2}}}{2} + D_x(B)$$
$$C^2 = \frac{\left(u_x\,x^2 - a\,x^2 + 2\,b^2\,x + b^2\,t\,u_x - a\,b^2\,t\right)\, e^{\frac{a\,x}{b^2} - \frac{u}{b^2}}}{4} - D_t(B)$$

where

$$B = \left(\frac{b^2\,t\,\left(2\,x + t\,u_x - a\,t\right)\, e^{\frac{a\,x}{b^2} - \frac{u}{b^2}}}{2}\right)$$

We simplify the conserved vector by transferring the terms of the form $D_x(\ldots)$ from C^1 to C^2 and obtain

$$C^1 = -\frac{\left(x^2 + b^2\,t\right)\, e^{\frac{a\,x}{b^2} - \frac{u}{b^2}}}{2}$$
$$C^2 = \frac{\left(u_x\,x^2 - a\,x^2 + 2\,b^2\,x + b^2\,t\,u_x - a\,b^2\,t\right)\, e^{\frac{a\,x}{b^2} - \frac{u}{b^2}}}{4}$$

4. Let us find the conservation law provided by the following symmetry of Eq. (2) where $c(x) = -\frac{a^2}{2}$:

$$X = \frac{\partial}{\partial u}. \tag{23}$$

In this case we have $W = 1$ and Eq. (20) yield conservation law (23) with

$$C^1 = \alpha\, e^{-\frac{u}{b^2}},$$
$$C^2 = -\frac{e^{-\frac{u}{b^2}}\,\left(\alpha\,u_x + \alpha_x\,b^2 - 2\,a\,\alpha\right)}{2}.$$

5. Let us find the conservation law provided by the following symmetry of Eq. (2) where $c(x) = -\frac{a^2}{2}$:

$$X = f(x,t)\exp\left(\frac{u}{b^2}\right)\frac{\partial}{\partial u}, \tag{24}$$

where $f(x,t)$ satisfies the linear equation

$$b^4\, f_{x\,x} + 2\,a\,b^2\, f_x + 2\,b^2\, f_t + a^2\, f = 0. \tag{25}$$

In this case we have $W = f(x,t)\exp\left(\frac{u}{b^2}\right)$ and Eq. (20) yield conservation law (23) with

$$C^1 = \alpha\, f,$$
$$C^2 = \frac{\alpha\, b^2\, f_x}{2} - \frac{\alpha_x\, b^2\, f}{2} + a\,\alpha\, f.$$

Substituting the expression of u_t from Eq. (2) into (23) we get the following expression

$$\alpha\, b^2\, f_{x\,x} + 2\,a\,\alpha\, f_x + 2\,\alpha\, f_t - \alpha_{x\,x}\, b^2\, f + 2\,a\,\alpha_x\, f = 0.$$

This condition is satisfied whenever $\alpha(x)$ satisfies the condition of weak self-adjointness (8) and $f(x,t)$ satisfies the linear Eq. (25).

2 Conclusions

The concepts of self-adjoint and quasi self-adjoint equations were introduced by NH Ibragimov in [8]. In [3] one of the present authors has generalized the concept of self-adjoint and quasi self-adjoint equations by introducing the definition of weak self-adjoint equations. In this paper we prove that the only class of quasi self-adjoint equations is such that $c(x) = 0$. Nevertheless, for any $c(x)$ arbitrary we find a class of weak self-adjoint Hamilton-Jacobi-Bellman equations. This new property of weak self-adjointness allow us, by using a general theorem on conservation laws, to find some conservation laws for a nonlinear HJB equation which arises in the modelling of mean-variance hedging. We point out that in physical systems, many conservation laws that arise can usually be identified with a physical quantity, like energy or linear momentum, being conserved. It would be of interest to identify what conserved economic or financial phenomena can be associated with the conservation laws identified in this paper. Finally, we remark that the search for conservation laws is also useful to determine potential symmetries. These symmetries will allow us to find new solutions for the HJB equations.

Acknowledgments The support of DGICYT project MTM2009-11875, Junta de Andalucía group FQM-201 are gratefully acknowledged.

References

1. Anco, S.C., Bluman, G.: Direct construction method for conservation laws for partial differential equations part II: general treatment em Euro. J. Appl. Math. **41**, 567–585 (2002)

2. Freire, I.L., Sampaio, J.C.S.: Nonlinear self-adjointness of a generalized fifth-order KdV equation. J. Phys. A: Math. Theor. **45**, 032001 (2012)
3. Gandarias, M.L.: Weak self-adjoint differential equations. J. Phys. A.: Math. Theor. **44**, 262–301 (2011)
4. Gandarias, M.L., Redondo, M., Bruzon, M.S.: Some weak self-adjoint Hamilton-Jacobi-Bellman equations arising in financial mathematics. Nonlinear Anal. Real World Appl. **13**, 340–347 (2012)
5. Heath, D., Platin, E., Schweizer, M.: Numerical comparison of local riskminimisation and mean-variance hedging. In: Jouini, E., Cvitanić, J., Musiela, M. (eds.) Option Pricing, Interest Rates and Risk Management. Elyès, CUP, Cambridge (2001)
6. Ibragimov, N.H.: A new conservation theorem. J. Math. Anal. Appl. **333**, 311–328 (2007)
7. Ibragimov, N.H.: Quasi self-adjoint differential equations. Prepr. Arch. ALGA **4**, 55–60 (2007)
8. Ibragimov, N.H.: The answer to the question put to me by L.V. Ovsyannikov 33 years ago. Arch. ALGA **3**, 80 (2006)
9. Ibragimov, N.H.: Nonlinear self-adjointness and conservation laws. J. Phys. A Math. Theor. **44**, 432002 (2011)
10. Johnpillai, A.G., Khalique, C.M.: Variational approaches to conservation laws for a nonlinear evolution equation with time dependent coefficients. Quaestiones Mathematicae. **34**, 235–245 (2011)
11. Kara, A.H., Mahomed, F.M.: Noether-type symmetries and conservation laws via partial lagrangians. Nonlinear Dyn. **45**, 367–383 (2006)
12. Kushner, H.J.: Introduction to Stochastic Control. Holt, New York (1971)
13. Naicker, V., Andriopulos, K., Leach, P.G.L.: Symmetry reductions of a Hamilton-Jacobi-Bellman equation arising in financial mathematics. J. Nonlinear Mathe. Phys. **12**, 268–283 (2005)
14. Vinogradov, A.M.: Local symmetries and conservation laws. Acta Appl. Math. **2**, 21–78 (1984)

Self-Adjointness and Conservation Laws for a Generalized Dullin-Gottwald-Holm Equation

Maria S. Bruzón and Maria Luz Gandarias

Abstract We consider the problem of group classification and conservation laws of some Generalized Dullin-Gottwald-Holm equations. We obtain the subclasses of these general equations which are self-adjoint. By using the recent Ibragimov's Theorem on conservation laws, we establish some conservation laws of the self-adjoint equations.

1 Introduction

Dullin, Gottwald and Holm derived a new equation describing unidirectional propagation of surface waves on a shallow layer of water which is at rest at infinity [4].

$$m_t + 2\omega u_x + 2m u_x + u m_x = -\gamma u_{xxx}, \quad t > 0,\ x \in \mathscr{R} \tag{1}$$

where $m = u - \alpha^2 u_{xx}$, $u(x,t)$ stands for the fluid velocity, $x \in \mathscr{R}$ and $t > 0$. The constants α^2 and $\frac{\gamma}{c_0}$ are squares of length scales, $c_0 = \sqrt{gh}$ is the linear wave speed for undisturbed water at rest at spatial infinity.

Equation (1) is completely integrable and its traveling wave solutions contains both the Korteweg-de Vries solitons and the Camassa-Holm peakons as limiting cases [4]. When $\alpha \to 0$, this equation becomes the Korteweg-de Vries equation

$$u_t + 2\omega u_x + 3u u_x = -\gamma u_{xxx}$$

M. S. Bruzón (✉) · M. L. Gandarias
Department of Mathematics, University of Cádiz, P.O. BOX 40,
11510 Puerto Real, (Cádiz), Spain
e-mail: m.bruzon@uca.es

M. L. Gandarias
e-mail: marialuz.gandarias@uca.es

A. Makhlouf et al. (eds.), *Algebra, Geometry and Mathematical Physics*,
Springer Proceedings in Mathematics & Statistics 85, DOI: 10.1007/978-3-642-55361-5_34,

which for $\omega = 0$, has the famous smooth soliton solution $u(x,t) = u_0 \mathrm{sech}^2(x - ct)\sqrt{u_0\gamma/2}$. Instead taking $\gamma -> 0$ in the Eq. (1), it turns out to be the Camassa-Holm equation

$$u_t - \alpha^2 u_{xxt} + 2\omega u_x + 3uu_x = \alpha^2(2u_x u_{xx} + uu_{xxx}).$$

Tian, Fang and Gui, applying Kato's semigroup approach, obtained the well-posedness of the equation and showed the existence of global smooth solutions. The authors proved that the equation has solutions that exist for indefinite times as well as solutions that blow up in finite time, [13].

Biswas and Kara [1] obtained the 1-soliton solution by the aid of solitary wave ansatz. The conserved quantities were obtained by utilising the interplay between the multipliers and underlying Lie point symmetry generators of the equation.

In [10] Liu and Yin established the local well-posedness by using Kato's theory for the generalized Dullin-Gottwald-Holm equation

$$u_t - u_{txx} + (h(u))_x + bu_{xxx} = a\left(\frac{g'(u)}{2}u_x^2 + g(u)u_{xx}\right)_x. \tag{2}$$

They proved the orbital stability of the peaked solitary waves.

Symmetry groups have several different applications in the context of nonlinear differential equations. For example, they are used to obtain exact solutions and conservation laws of partial differential equations (PDEs) [3, 5]. The classical method for finding symmetry groups of PDEs is the Lie group method [2, 6, 11, 12].

In this work, we study Eq. (2) with $a, b \neq 0$ from the point of view of the theory of symmetry group transformations in PDEs. We determine the subclasses of equations which are self-adjoint. We also determine, by using the notation and techniques of the work [8, 9], some nontrivial conservation laws for Eq. (2). The paper is organized as follows. In Sect. 2 we give the Lie symmetries of (2) equation. In Sect. 3 we determine the subclasses of equations of (2) which is self-adjoint. In Sect. 4 we obtain some nontrivial conservation laws for Eq. (2). Finally, in Sect. 5 we give conclusions.

2 Classical Symmetries

To apply the Lie classical method to Eq. (2) we consider the one-parameter Lie group of infinitesimal transformations in (x, t, u) given by

$$x^* = x + \varepsilon\xi(x,t,u) + O(\varepsilon^2), \tag{3}$$

$$t^* = t + \varepsilon\tau(x,t,u) + O(\varepsilon^2), \tag{4}$$

$$u^* = u + \varepsilon\eta(x,t,u) + O(\varepsilon^2), \tag{5}$$

where ε is the group parameter. We require that this transformation leaves invariant the set of solutions of Eq. (2). This yields to an overdetermined, linear system of equations for the infinitesimals $\xi(x,t,u)$, $\tau(x,t,u)$ and $\eta(x,t,u)$. The associated Lie algebra of infinitesimal symmetries is the set of vector fields of the form

$$\mathbf{v} = \xi(x,t,u)\frac{\partial}{\partial x} + \tau(x,t,u)\frac{\partial}{\partial t} + \eta(x,t,u)\frac{\partial}{\partial u}. \tag{6}$$

Having determined the infinitesimals, the symmetry variables are found by solving the characteristic equation which is equivalent to solving the invariant surface condition

$$\eta(x,t,u) - \xi(x,t,u)\frac{\partial u}{\partial x} - \tau(x,t,u)\frac{\partial u}{\partial t} = 0. \tag{7}$$

The set of solutions of Eq. (2) is invariant under the transformation (3)–(5) provided that

$$\mathrm{pr}^{(3)}\mathbf{v}(\Delta) = 0 \quad \text{when} \quad \Delta = 0,$$

where $\mathrm{pr}^{(3)}\mathbf{v}$ is the third prolongation of the vector field (6) given by

$$\mathrm{pr}^{(3)}\mathbf{v} = \mathbf{v} + \sum_J \eta^J(x,t,u^{(3)})\frac{\partial}{\partial u_J}$$

where

$$\eta^J(x,t,u^{(3)}) = D_J(\eta - \xi u_x - \tau u_t) + \xi u_{Jx} + \eta u_{Jt},$$

with $J = (j_1, \ldots, j_k)$, $1 \le j_k \le 2$ y $1 \le k \le 3$. Hence we obtain the following 13 determining equations for the infinitesimals:

$$\begin{aligned}
&\tau_u = 0,\\
&\tau_x = 0,\\
&\xi_u = 0,\\
&\eta_{uu} = 0,\\
&2\eta_{ux} - \xi_{xx} = 0,\\
&\eta_{uxx} - 2\xi_x = 0,\\
&3g_{uu}\eta_x + 8g_u\eta_{ux} - 4\xi_{xx}g_u = 0,\\
&g_u\eta_u + g_{uu}\eta + \tau_t g_u - \xi_x g_u = 0,\\
&2g_{uu}\eta_u + g_{uuu}\eta + \tau_t g_{uu} - \xi_x g_{uu} = 0,\\
&-ag\eta_{xxx} + b\eta_{xxx} + h_u\eta_x - \eta_{txx} + \eta_t = 0,\\
&ag_u\eta + a\tau_t g - a\xi_x g - b\tau_t + b\xi_x - \xi_t = 0,\\
&2ag_u\eta_x + 3ag\eta_{ux} - 3b\eta_{ux} + \eta_{tu} - 3a\xi_{xx}g + 3b\xi_{xx} - 2\xi_{tx} = 0,\\
&2ag_u\eta_{xx} + 3ag\eta_{uxx} - 3b\eta_{uxx} + 2\eta_{tux} - h_{uu}\eta - \tau_t h_u - \xi_x h_u - a\xi_{xxx}g\\
&+b\xi_{xxx} - \xi_{txx} + \xi_t = 0.
\end{aligned} \tag{8}$$

From (8) we obtain $g, h, \xi = \xi(x,t), \tau = \tau(t), \phi = \frac{(\delta+\xi_x)u}{2} + \nu$ with $\delta = \delta(t)$ and $\nu = \nu(x,t)$ where ξ, τ, δ and ν are related by the following conditions:

$$\begin{aligned}
&\xi_{xxx} - 4\xi_x = 0,\\
&g_{uu}\left(\xi_{xx}u + 2\nu_x\right) = 0,\\
&2a\xi_{xx}g_u u + \delta_t + 4ag_u\nu_x - 3a\xi_{xx}g + 3b\xi_{xx} - 3\xi_{tx} = 0,\\
&\left(ag_u\delta + a\xi_x g_u\right)u + 2ag_u\nu + \left(2a\tau_t - 2a\xi_x\right)g - 2b\tau_t + 2b\xi_x - 2\xi_t = 0,\\
&\left(g_{uu}\delta + \xi_x g_{uu}\right)u + g_u\delta + 2g_{uu}\nu + \left(2\tau_t - \xi_x\right)g_u = 0,\\
&\left(g_{uuu}\delta + \xi_x g_{uuu}\right)u + 2g_{uu}\delta + 2g_{uuu}\nu + 2\tau_t g_{uu} = 0,\\
&\left(h_{uu}\delta + \xi_x h_{uu} - 2a\xi_{xxx}g_u\right)u - 4ag_u\nu_{xx}\\
&+2h_{uu}\nu + \left(2\tau_t + 2\xi_x\right)h_u - a\xi_{xxx}g + b\xi_{xxx} - 2\xi_t = 0,\\
&\left(\delta_t + \xi_{xx}h_u - a\xi_{xxxx}g + b\xi_{xxxx} - \xi_{txxx} + \xi_{tx}\right)u + (2b - 2ag)\,\nu_{xxx}\\
&+2h\nu_x - 2\nu_{txx} + 2\nu_t = 0.
\end{aligned} \tag{9}$$

Solving system (9) we obtain that if g and h are arbitrary functions the only symmetries admitted by (2) are

$$\xi = k_1, \quad \tau = k_2, \quad \eta = 0. \tag{10}$$

The generators are $\mathbf{X}_1 = \frac{\partial}{\partial x}$ (corresponding to space translational invariance) and $\mathbf{X}_2 = \frac{\partial}{\partial t}$ (time translational invariance). In the following cases Eq. (2) have extra symmetries:

Case 1: If $g = (a_1u + a_2)^n + a_3$ and $h = \left(\frac{b_1}{a_1}(a_1u + a_2)\right)^{n+1} + (aa_3 - b)u, n \neq 1$, $a_1 \neq 0$,

$$\xi = (b - aa_3)k_1 n\, t + k_2, \quad \tau = -k_1 n\, t + k_3, \quad \eta = \frac{k_1}{a_1}(a_1u + a_2).$$

The generators are: $\mathbf{X_1}, \mathbf{X_2}$ and $\mathbf{X_3^1} = \frac{(b - aa_4)n}{2} t\frac{\partial}{\partial x} - \frac{n}{2}t\frac{\partial}{\partial t} + \frac{1}{2a_1}(a_1u + a_2)\frac{\partial}{\partial u}$.

Case 2: If $g = a_1u + a_2$ and $h = \frac{b_1}{2}u^2 + b_2u, a_1 \neq 0, b_1 \neq aa_1$,

$$\xi = c_1k_1t + k_3, \quad \tau = k_1t + k_2, \quad \eta = -k_1(u + c_2).$$

We have $\mathbf{X_1}, \mathbf{X_2}$ and $\mathbf{X_3^2} = c_1t\frac{\partial}{\partial x} + t\frac{\partial}{\partial t} - (c_2 + u)\frac{\partial}{\partial u}$, where $c_1 = -\frac{aa_1b_2 + bb_1 - aa_2b_1}{b_1 - aa_1}$ and $c_2 = \frac{b_2 + b - aa_2}{b_1 - aa_1}$.

Case 3: If $g = a_1u + a_2$ and $h = \frac{aa_1}{2}u^2 + (aa_2 - b)u, a_1 \neq 0$

$$\xi = k_1t + k_2, \quad \tau = k_3t + k_4, \quad \eta = -k_3u + \frac{(b - aa_2)k_3 + k_1}{aa_1}.$$

The generators are $\mathbf{X_1}, \mathbf{X_2}$ and $\mathbf{X_3^3} = t\frac{\partial}{\partial x} + \frac{1}{aa_1}\frac{\partial}{\partial u}, \mathbf{X_4^3} = t\frac{\partial}{\partial t} - (u+c_1)\frac{\partial}{\partial u}$, where $c_1 = \frac{b-aa_2}{aa_1}$.

Case 4: If $g = c$=constant and $h = ku$ with $k \neq ac - b$,

$$\xi = k_1, \quad \tau = k_2, \quad \eta = k_3 u + \alpha(x,t),$$

where

$$(ac - b)\alpha_{xxx} - k\alpha_x + \alpha_{txx} - \alpha_t = 0. \tag{11}$$

In this case besides $\mathbf{X_1}$ and $\mathbf{X_2}$ we obtain the generators $\mathbf{X_3^4} = u\frac{\partial}{\partial u}$ and $\mathbf{X_\infty} = \alpha(x,t)\frac{\partial}{\partial u}$.

Case 5: If $g = c$ and $h = (ac - b)u$ with $c \neq \frac{b}{a}$,

$$\xi = k_1 e^{2x+2(b-ac)t} + k_3 e^{2(ac-b)t-2x} + (ac-b)\beta(t) + k_2, \quad \tau = \beta(t),$$

$$\eta = u\left(k_1 e^{2x+2(b-ac)t} - k_3 e^{2(ac-b)t-2x} + k_5\right) + \alpha(x,t),$$

where α satisfies Eq. (11) with $k = ac - b$.

We obtain the generators: $\mathbf{X_1}$, $\mathbf{X_2}$, $\mathbf{X_\infty}$, $\mathbf{X_3^5} = \left(\frac{\partial}{\partial x} + u\frac{\partial}{\partial u}\right) e^{2[x+(b-ac)t]}$, $\mathbf{X_4^5} = \left(\frac{\partial}{\partial x} - u\frac{\partial}{\partial u}\right) e^{-2[x+(b-ac)t]}$.

Case 6: If $g = a_1\, e^{a_2\, u} + a_3$ and $h_u = k\, e^{a_2\, u} - b + a\, a_3$

$$\xi = a_2\,(b - a\,a_3)\; k_2\, t + k_1, \quad \tau = k_3 - a_2\, k_2\, t, \quad \eta = k_2.$$

The generators are: $\mathbf{X_1}$, $\mathbf{X_2}$ and $\mathbf{X_3^6} = a_2\,(b - a\,a_3)\; t\frac{\partial}{\partial x} - a_2 t\frac{\partial}{\partial t} + \frac{\partial}{\partial u}$.

Case 7: If $g = a_2 \ln(a_1\, u + b_1) + b_2$ and $h_u = aa_2\, \ln(a_1 u + b_1) + b_3$,

$$\xi = k_3 t + k_1, \quad \tau = k_2, \quad \eta = \frac{k_3}{aa_1a_2}(a_1 u + b_1).$$

The generators are $\mathbf{X_1}, \mathbf{X_2}$ and $\mathbf{X_3^7} = t\frac{\partial}{\partial x} + \frac{a_1 u + b_1}{aa_1a_2}\frac{\partial}{\partial u}$.

3 Determination of Self-Adjoint Equations

In [8] Ibragimov introduced a new theorem on conservation laws. The theorem is valid for any system of differential equations where the number of equations is equal to the number of dependent variables. The new theorem does not require existence of a Lagrangian and this theorem is based on a concept of an adjoint equation for nonlinear equations.

Consider an sth-order partial differential equation

$$F(x, u, u_{(1)}, \dots, u_{(s)}) = 0 \tag{12}$$

with independent variables $x = (x^1, \dots, x^n)$ and a dependent variable u, where $u_{(1)} = \{u_i\}$, $u_{(2)} = \{u_{ij}\}, \dots$ denote the sets of the partial derivatives of the first, second, etc. orders, $u_i = \partial u/\partial x^i$, $u_{ij} = \partial^2 u/\partial x^i \partial x^j$. The adjoint equation to (12) is

$$F^*(x, u, v, u_{(1)}, v_{(1)}, \dots, u_{(s)}, v_{(s)}) = 0, \tag{13}$$

with

$$F^*(x, u, v, u_{(1)}, v_{(1)}, \dots, u_{(s)}, v_{(s)}) = \frac{\delta(vF)}{\delta u}, \tag{14}$$

where

$$\frac{\delta}{\delta u} = \frac{\partial}{\partial u} + \sum_{s=1}^{\infty} (-1)^s D_{i_1} \cdots D_{i_s} \frac{\partial}{\partial u_{i_1 \cdots i_s}} \tag{15}$$

denotes the variational derivative (the Euler-Lagrange operator), and v is a new dependent variable. Here

$$D_i = \frac{\partial}{\partial x^i} + u_i \frac{\partial}{\partial u} + u_{ij} \frac{\partial}{\partial u_j} + \cdots$$

are the total differentiations.

Equation (12) is said to be *self-adjoint* if the equation obtained from the adjoint Eq. (13) by the substitution $v = u$:

$$F^*(x, u, u, u_{(1)}, u_{(1)}, \dots, u_{(s)}, u_{(s)}) = 0,$$

is identical to the original Eq. (12). In other words, if

$$F^*(x, u, u_{(1)}, u_{(1)}, \dots, u_{(s)}, u_{(s)}) = \lambda(x, u, u_{(1)}, \dots) F(x, u, u_{(1)}, \dots, u_{(s)}). \tag{16}$$

Let us single out self-adjoint equations from the equation of the form (2). Equation (14) yields

$$\begin{aligned} F^* \equiv\, & ag v_{xxx} - b v_{xxx} + a g_u u_x v_{xx} + a g_u u_{xx} v_x - a g_{uu} (u_x)^2 v_x \\ & - h_u v_x + v_{txx} - v_t - 3 a g_{uu} u_x u_{xx} v - \tfrac{3}{2} a g_{uuu} (u_x)^3 v. \end{aligned} \tag{17}$$

By substituting $v = u$ into (17) we obtain

$$\begin{aligned} F^* \equiv\, & ag u_{xxx} - b u_{xxx} - 3 a g_{uu} u u_x u_{xx} + 2 a g_u u_x u_{xx} - \tfrac{3}{2} a g_{uuu} u (u_x)^3 \\ & - a g_{uu} (u_x)^3 - h_u u_x + u_{txx} - u_t. \end{aligned} \tag{18}$$

Comparing F^* with F we obtain the following result:

Proposition Equation $F \equiv u_t - u_{txx} + (h(u))_x + b u_{xxx} - a \left(\frac{g'(u)}{2} u_x^2 + g(u) u_{xx} \right)_x = 0$ is self-adjoint if g and h are arbitrary functions.

4 General Theorem on Conservation Laws

We use the following theorem on conservation laws proved in [8]. Any Lie point, Lie-Bäcklund or non-local symmetry

$$X = \xi^i (x, u, u_{(1)}, \ldots) \frac{\partial}{\partial x^i} + \eta (x, u, u_{(1)}, \ldots) \frac{\partial}{\partial u} \tag{19}$$

of Eq. (12) provides a conservation law $D_i(C^i) = 0$ for the simultaneous system (12), (13). The conserved vector is given by

$$\begin{aligned} C^i = \xi^i \mathscr{L} + W & \left[\frac{\partial \mathscr{L}}{\partial u_i} - D_j \left(\frac{\partial \mathscr{L}}{\partial u_{ij}} \right) + D_j D_k \left(\frac{\partial \mathscr{L}}{\partial u_{ijk}} \right) - \cdots \right] \\ + D_j(W) & \left[\frac{\partial \mathscr{L}}{\partial u_{ij}} - D_k \left(\frac{\partial \mathscr{L}}{\partial u_{ijk}} \right) + \cdots \right] + D_j D_k(W) \left[\frac{\partial \mathscr{L}}{\partial u_{ijk}} - \cdots \right] + \cdots, \end{aligned} \tag{20}$$

where W and $\mathscr{L}$ are defined as follows:

$$W = \eta - \xi^j u_j, \quad \mathscr{L} = v\, F\left(x, u, u_{(1)}, \ldots, u_{(s)}\right). \tag{21}$$

The proof is based on the following operator identity (N.H. Ibragimov 1979):

$$X + D_i(\xi^i) = W \frac{\delta}{\delta u} + D_i \mathscr{N}^i, \tag{22}$$

where X is operator (19) taken in the prolonged form:

$$X = \xi^i \frac{\partial}{\partial x^i} + \eta \frac{\partial}{\partial u} + \zeta_i \frac{\partial}{\partial u_i} + \zeta_{i_1 i_2} \frac{\partial}{\partial u_{i_1 i_2}} + \cdots,$$

$$\zeta_i = D_i(\eta) - u_j D_i(\xi^j), \quad \zeta_{i_1 i_2} = D_{i_2}(\zeta_{i_1}) - u_{j i_1} D_{i_2}(\xi^j), \; \ldots .$$

For the expression of operator $\mathscr{N}^i$ and a discussion of the identity (22) in the general case of several dependent variables to see [7] (Sect. 8.4.4).

We will write the generators of a point transformation group admitted by Eq. (2) in the form

$$X = \xi^1 \frac{\partial}{\partial t} + \xi^2 \frac{\partial}{\partial x} + \eta \frac{\partial}{\partial u}$$

by setting $t = x^1,\ x = x^2$. The conservation law will be written

$$D_t(C^1) + D_x(C^2) = 0. \tag{23}$$

Now we use the Ibragimov's Theorem on conservation laws to establish some conservation laws of Eq. (2). We obtain trivial conservation laws for $g = a_1 u + a_2$, from generators $\mathbf{X}_1$ and $\mathbf{X}_2$.
For **Case 2**, $g = a_1 u + a_2$, $h = |\frac{b_1}{2} + b_2 u$, from generator $\mathbf{X}_2^3$ we obtain the conserved vector associated,

$$C^1 = -(u_x)^2 - u^2 - \tfrac{b_2 + b - a a_2}{b_1 - a a_1}\, u,$$

$$\begin{aligned} C^2 = \tfrac{1}{6(b_1 - a a_1)} \big[&\left(\left(12 a a_1 b_1 - 12 a^2 {a_1}^2\right) u^2 \right.\\ &+ \left(6 a a_1 b_2 + (12 a a_2 - 12 b)\, b_1 + 18 a a_1 b - 18 a^2 a_1 a_2\right) u \\ &+ (6 a a_2 - 6 b)\, b_2 - 6 b^2 + 12 a a_2 b - 6 a^2 {a_2}^2\big)\, u_{xx} \\ &+ \left(3 a a_1 b_2 + (6 b - 6 a a_2)\, b_1 - 3 a a_1 b + 3 a^2 a_1 a_2\right) (u_x)^2 \\ &+ \left((12 b_1 - 12 a a_1)\, u + 6 b_2 + 6 b - 6 a a_2\right) u_{tx} + \left(4 a a_1 b_1 - 4 {b_1}^2\right) u^3 \\ &+ \left((6 a a_1 - 9 b_1)\, b_2 + (3 a a_2 - 3 b)\, b_1\right) u^2 + \left((6 a a_2 - 6 b)\, b_2 - 6 {b_2}^2\right) u \big]. \end{aligned}$$

We use the symmetry of the **Case 3** of Eq. (2) for $g = a_1 u + a_2$ and $h = \frac{a a_1}{2} u^2 + (a a_2 - b) u$ with $a_1 \neq 1$. Proceeding as before we obtain the conserved vector associated with the following symmetries.
For $\mathbf{X_3^3}$:

$$C^1 = \frac{1}{a a_1} u,$$

$$C^2 = -u u_{xx} + \frac{b}{a a_1} u_{xx} - \frac{a_2}{a_1} u_{xx} - \frac{1}{2} (u_x)^2 - \frac{1}{a a_1} u_{tx} + \frac{1}{2} u^2 - \frac{b}{a a_1} u + \frac{a_2}{a_1} u. \tag{24}$$

For $\mathbf{X_4^3}$:

$$C^1 = -(u_x)^2 - u^2 + \left(\frac{b}{a a_1} - \frac{a_2}{a_1}\right) u,$$

$$C^2 = 2aa_1u^2u_{xx} + 3(aa_2 - b)uu_{xx} + \frac{(b-aa_2)^2}{aa_1}u_{xx} + \frac{b-aa_2}{2}(u_x)^2 + 2uu_{tx} + \frac{a_2a-b}{a_1}u_{tx} - \frac{2}{3}aa_1u^3 + \frac{3}{2}(b-aa_2)u^2 - \frac{(b-aa_2)^2}{aa_1}u. \tag{25}$$

For **Case 4**, if $g = c$ and $h = ku$, from generator $\mathbf{X}_\infty = \alpha(x,t)$, where α satisfies Eq. (11), the normal form for this group is $W = \alpha(x,t)$. By applying (20) the vector components are

$$\begin{aligned} C^1 &= -\tfrac{1}{3}\alpha v_{xx} + \tfrac{1}{3}\alpha_x v_x - \tfrac{1}{3}\alpha_{xx}v + \alpha v = 0. \\ C^2 &= -a\alpha c v_{xx} + \alpha b v_{xx} + a\alpha_x c v_x - \alpha_x b v_x + \tfrac{1}{3}\alpha_t v_x - \tfrac{2}{3}\alpha v_{tx} + \tfrac{1}{3}\alpha_x v_t \\ &\quad + \alpha k v - a\alpha_{xx}cv + \alpha_{xx}bv - \tfrac{2}{3}\alpha_{tx}v. \end{aligned} \tag{26}$$

Setting $v = u$ in (26)

$$\begin{aligned} C^1 &= -\tfrac{\alpha u_{xx}}{3} + \tfrac{\alpha_x u_x}{3} - \tfrac{\alpha_{xx}u}{3} + \alpha u. \\ C^2 &= -a\alpha c u_{xx} + \alpha b u_{xx} + a\alpha_x c u_x - \alpha_x b u_x + \tfrac{1}{3}\alpha_t u_x - \tfrac{2}{3}\alpha u_{tx} + \tfrac{1}{3}\alpha_x u_t \\ &\quad + \alpha k u - a\alpha_{xx}cu + \alpha_{xx}bu - \tfrac{2}{3}\alpha_{tx}u. \end{aligned} \tag{27}$$

We simplify the conserved vector (27) by transferring the terms of the form $D_x(\ldots)$ from C^1 to C^2 and obtain

$$\begin{aligned} C^1 &= (\alpha - \alpha_{xx})u. \\ C^2 &= \alpha(b-ac)u_{xx} + \alpha_x(ac-b)u_x - \alpha u_{tx} + \alpha_x u_t + \alpha k u - \alpha_{xx}(ac-b)u. \end{aligned}$$

For **Case 5**, if $g = c$ and $h = (ac-b)u$, from generators $\mathbf{X}_3^5 = \frac{\partial}{\partial x} + u\frac{\partial}{\partial u}$ and $\mathbf{X}_4^5 = \frac{\partial}{\partial x} - u\frac{\partial}{\partial u}$, proceeding as in the Case 1 we obtain the conserved vector associated,

$$\begin{aligned} C^1 &= n\left((u_x)^2 + u^2\right), \\ C^2 &= 2(b-ac)\,nuu_{xx} + (ac-b)\,n\,(u_x)^2 - 2nuu_{tx} + (ac-b)\,nu^2, \end{aligned}$$

where $n = \pm 1$.

5 Conclusions

In this work we have considered the generalized Dullin-Gottwald-Holm Eq. (2). We have derived the Lie classical symmetries. We have determined the subclasses of Eq. (2) which are self-adjoint. By using a general theorem on conservation laws proved by Nail Ibragimov we found conservation laws for some of these partial differential equations without classical Lagrangians. We point out that in physical systems, many conservation laws that arise can usually be identified with a physical quantity, like energy or linear momentum, being conserved. Finally, we remark that the search for conservation laws is also useful to determine potential symmetries.

Acknowledgments The support of DGICYT project MTM2009-11875, Junta de Andalucía group FQM-201 are gratefully acknowledged.

References

1. Biswas, A., Kara, A.H.: 1-soliton solution and conservation laws of the generalized dullin-gottwald-holm equation. Appl. Math. Comput. **217**, 929–932 (2010)
2. Bluman, G.W., Kumei, S.: Symmetries and Differential Equations. Springer, New York (1989)
3. Bruzón, M.S., Gandarias, M.L., Ibragimov, N.H.: Self-adjoint sub-classes of generalized thin film equations. J. Math. Anal. Appl. **357**, 307–313 (2009)
4. Dullin, H.R., Gottwald, G.A., Holm, D.D.: An integrable shallow water equation with linear and nonlinear dispersion. Phys. Rev. Lett. **87**, 1945–1948 (2001)
5. Gandarias, M.L., Redondo, M., Bruzon, M.S.: Some weak self-adjoint Hamilton-Jacobi-Bellman equations arising in financial mathematics. Nonl. Anal.: Real World Appl. **13**, 340–347 (2011)
6. Ibragimov, N.H.: Transformation Groups Applied to Mathematical Physics. Reidel-Dordrecht, Dordrecht (1985)
7. Ibragimov, N.H.: Elementary Lie Group Analysis and Ordinary Differential Equations. Wiley, Chichester (1999)
8. Ibragimov, N.H.: A new conservation theorem. J. Math. Anal. Appl. **333**, 311–328 (2007)
9. Ibragimov, N.H.: Quasi self-adjoint differential equations. Archives of ALGA **4**, 55–60 (2007)
10. Liu, X., Yin, Z.: Local well-posedness and stability of peakons for a generalized dullin-gottwald-holm equation. Nonlinear Anal. **74**, 2497–2507 (2011)
11. Olver, P.J.: Applications of Lie Groups to Differential Equations. Springer, New York (1986)
12. Ovsyannikov, L.V.: Group Analysis of Differential Equations. Academic, New York (1982)
13. Tian, L., Fang, G., Gui, G.: Well-posedness and blowup for an integrable shallowwater equation with strong dispersive term. Int. J. Nonlinear Sci. **1**, 3–13 (2006)

Semi Analytical Solutions of a Heat-Mass Transfer Problem Via Group Theoretic Approach

Mina B. Abd-el-Malek and Medhat M. Helal

Abstract Problem of heat-mass transfer in non-Newtonian power law, two-dimensional, laminar, boundary layer flow of a viscous incompressible fluid over an inclined plate has been studied by applying the method of group theoretic approach. The governing system of nonlinear partial differential equation describing the flow and heat transfer problem are transformed into a system of nonlinear ordinary differential equation which has been solved semi analytically. Exact solutions for the dimensionless temperature and concentration profiles, are presented graphically for different physical parameters and for the different power law exponents $n \in (0, 0.5)$ and for $n > 0.5$. Also the effect of n, the Prandtl number, and the heat generation parameter on both the temperature and the concentration of the fluid inside the boundary layer have been studied.

1 Introduction

The flow of non-Newtonian fluids, including the power-law model, has attracted the interest of many researchers and scientists in the recent time due to its several applications in food, polymer, petrol-chemical, geothermal, rubber, paint and biological

M. B. Abd-el-Malek (✉)
Department of Engineering Mathematics and Physics, Faculty of Engineering,
Alexandria University, Alexandria 21544, Egypt
e-mail: minab@aucegypt.edu

M. B. Abd-el-Malek
Department of Mathematics and Actuarial Sciences, The American University in Cairo,
Cairo, Egypt

M. M. Helal
Department of Engineering Mathematics and Physics, Faculty of Engineering,
Zagazig University, Zagazig, Egypt
e-mail: mmhelal@aucegypt.edu

A. Makhlouf et al. (eds.), *Algebra, Geometry and Mathematical Physics*,
Springer Proceedings in Mathematics & Statistics 85, DOI: 10.1007/978-3-642-55361-5_35,

industries, as mentioned in [1]. The main difficulty in the analytical consideration of the power-law fluid flow is the non-linearity of governing differential equations.

Similarity analysis is convenient methods of reducing system of partial differential equations into system of ordinary differential equations. Group theoretic method, is a class of methods which lead to a reduction of the number of independent variables, which were first introduced by Birkhoff [2] in 1948. In 1966 and then in 1968, Moran and Gaggioli [3, 4] presented a theory which has led to improvements over earlier similarity methods. Similarity analysis has been applied intensively by Gabbert [5] in 1967. For more discussions see Ames [6, 7], Bluman and Cole [8], Boisvert et al. [9], Gaggioli and Moran [10, 11]. Throughout the history of similarity analysis, a variety of problems in science and engineering have been solved. Many physical applications are carried out by Abd-el-Malek et al. [12, 13].

In 2006, Sivasankaran et al. [14], have applied the Lie group analysis to study the same problem but without considering the power-law fluid in the momentum equation, the heat generation in the energy equation, and the thermophoretic velocity in the diffusion equation.

In the present work, we provide semi analytical solution and qualitative discussion for the laminar boundary-layer flow of non-Newtonian power law fluids using group theoretic approach. Under the application of one-parameter group, the governing system of partial differential equations and the associated boundary conditions are reduced to a system of ordinary differential equations with the corresponding boundary conditions, which are solved semi analytically.

2 The Mathematical Formulation

Following system [15], by Olajuwon in 2009, we study the following system:

$$\frac{\partial u}{\partial x}+\frac{\partial v}{\partial y}, \tag{1}$$

$$u\frac{\partial u}{\partial x}+v\frac{\partial u}{\partial y}=-\nu\frac{\partial}{\partial y}\Big(-\frac{\partial u}{\partial y}\Big)^{n}+g\beta(T-T_{\infty})\cos\alpha+g\beta^{*}(C-C_{\infty})\cos\alpha \tag{2}$$

$$u\frac{\partial T}{\partial x}+v\frac{\partial T}{\partial y}=\frac{k}{\rho c_{\rho}}\frac{\partial^{2}T}{\partial y^{2}}+\frac{Q}{\rho c}(T-T_{\infty}) \tag{3}$$

$$u\frac{\partial C}{\partial x}+v\frac{\partial C}{\partial y}=D\frac{\partial^{2}C}{\partial y^{2}}-\frac{\partial}{\partial y}\big(V_{T}(C-C_{\infty})\big), \tag{4}$$

where Eqs. (1)–(4) represent respectively, continuity, momentum, energy, and diffusion equations. This system is subjected to the following boundary conditions

$$u=v=0,\ T=T_{W},\ C=C_{W}\ \text{at}\ y=0,\ \text{ while, } u=0,\ T=T_{\infty},\ C=C_{\infty}\ \text{as}\ y\to\infty, \tag{5}$$

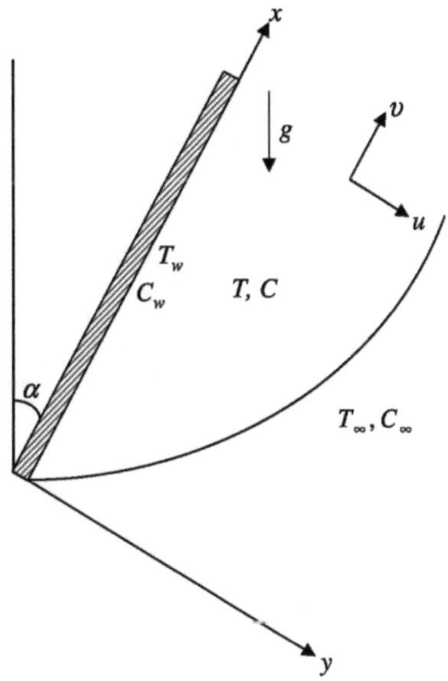

Fig. 1 The physical problem

where "u" and "v" are velocity components; "x" and "y" are space coordinates as illustrated in Fig. 1; T, T_w, and T_∞ are the temperature of the fluid inside the boundary layer, the plate and the fluid temperature in the free stream, respectively. The plate is maintained at a temperature T_w, and the free stream air is at a temperature T_∞, where $T_\infty > T_w$ to a cold surface. C, C_w, C_∞ are the concentration of the fluid inside the boundary layer, beside the plate, and the fluid concentration in the free stream, respectively; ν is the kinematic viscosity of the fluid; g is the acceleration due to gravity; β is the coefficient of thermal expansion; β^* is the coefficient of expansion with concentration; k is the thermal conductivity of fluid; ρ is the density of the fluid; c_ρ is the specific heat of the fluid; Q is the heat generation constant; D is the diffusion coefficient; α is the angle of inclination, n is the non-Newtonian parameter (power index). The thermophoretic velocity V_T, has the form:

$$V_T = -\frac{k\nu}{T_r}\frac{\partial T}{\partial y} \tag{6}$$

where T_r, is some reference temperature and ν is the thermo-phoretic coefficient.

The non-dimensional variables are:

$$\bar{x} = \frac{xU_\infty}{\nu},\ \bar{y} = \frac{yU_\infty}{\nu},\ \bar{u} = \frac{u}{U_\infty}\ \bar{v} = \frac{v}{V_\infty},\ \theta(x, y) = \frac{T(x, y) - T_\infty}{T_w - T_\infty},\ \phi(x, y) = \frac{C(x, y) - C_\infty}{C_w - C_\infty}. \tag{7}$$

The stream function formulation is:

$$\bar{u} = \frac{\partial\psi}{\partial y},\ \bar{v} = -\frac{\partial\psi}{\partial x} \tag{8}$$

By this assumption, the governing differential Eqs. (1)–(4) transform to:

$$\frac{\partial \psi}{\partial y}\frac{\partial^2 \psi}{\partial x \partial y} - \frac{\partial \psi}{\partial x}\frac{\partial^2 \psi}{\partial y^2} = -n\nu U_\infty^{n-2}\left(-\frac{\partial^2 \psi}{\partial y^2}\right)^{n-1}\frac{\partial^3 \psi}{\partial y^3}$$

$$+\frac{g\cos\alpha}{U_\infty^2}\left[\beta\theta(T_w - T_\infty)q_1(x) + \beta^*\phi(C_w - C_\infty)q_2(x)\right] \quad (9)$$

$$\frac{\partial \psi}{\partial y}\frac{\partial \theta}{\partial x} - \frac{\partial \psi}{\partial x}\frac{\partial \theta}{\partial y} = \frac{k}{\rho c_\rho U_\infty}q_3(x)\frac{\partial^2 \theta}{\partial y^2} + \frac{Q}{\rho c_\rho U_\infty}q_4(x)\theta \quad (10)$$

$$\frac{\partial \psi}{\partial y}\frac{\partial \theta}{\partial x} - \frac{\partial \psi}{\partial x}\frac{\partial \theta}{\partial y} = \frac{D}{U_\infty}q_5(x)\frac{\partial^2 \phi}{\partial y^2} + \frac{k\nu}{U_\infty T_r}(T_w - T_\infty)q_6(x)\left(\phi\frac{\partial^2 \theta}{\partial y^2} + \frac{\partial \theta}{\partial y}\frac{\partial \phi}{\partial y}\right). \quad (11)$$

The corresponding boundary and initial conditions become,

$$\frac{\partial \psi}{\partial x} = \frac{\partial \psi}{\partial y} = 0,\ \theta = 1,\ \phi = 1 \text{ at } y = 0, \text{ and } \frac{\partial \psi}{\partial y} = 0,\ \theta = 0,\ \phi = 0, \text{ as } y \to \infty, \quad (12)$$

where the functions $q_i(x)$; $i = 1, 2, ..., 6$ are auxiliary functions in x that are useful in applying the group theoretic approach and must vanish at the end.

3 Solution of the Problem

Solution depends on the application of a one-parameter group transformation to the system of partial differential Eqs. (9)–(11), where the two independent variables will be reduced by one and the equations will be transformed to a system of ordinary differential equations in only one independent variable, which is the similarity variable.

3.1 The Group Systematic Formulation

The procedure is initiated with the group G, a class of transformation of one parameter "a" of the form

$$G : \bar{S} = K^s(a)S + P^s(a), \quad (13)$$

where, "S" stands for $x,\ y;\ \psi,\ \theta,\ \phi,\ q_1(x),\ q_2(x),\ q_3(x),\ q_4(x),\ q_5(x),$ and $q_6(x)$ and the K's and P's are real-valued functions and at least differentiable in the real argument "a".

3.2 The Invariance Analysis

To transform the differential equation, transformations of the derivatives are obtained from G via chain—rule operations:

$$\bar{S}_i = \left(\frac{K^s}{K^i}\right) S_i,\ \bar{S}_{ij} = \left(\frac{K^s}{K^i K^j}\right) S_i,\ i = x, y;\ j = x, y, \tag{14}$$

where, "S" stands for $x,\ y;\ \psi,\ \theta,\ \phi,\ q_1(x),\ q_2(x),\ q_3(x),\ q_4(x),\ q_5(x),$ and $q_6(x)$.

Equations (9)–(11) are said to be invariantly transformed whenever

$$\begin{aligned}
&\frac{\partial \bar{\psi}}{\partial \bar{y}}\frac{\partial^2 \bar{\psi}}{\partial \bar{x}\partial \bar{y}} - \frac{\partial \bar{\psi}}{\partial \bar{x}}\frac{\partial^2 \bar{\psi}}{\partial \bar{y}^2} + n\nu U_\infty^{n-2}\Big(-\frac{\partial^2 \bar{\psi}}{\partial \bar{y}^2}\Big)^{n-1}\frac{\partial^3 \bar{\psi}}{\partial \bar{y}^3} - \frac{g\cos\alpha}{U_\infty^2}\big[\beta\bar{\theta}(T_w - T_\infty)\bar{q}_1(\bar{x}) \\
&-\beta^*\bar{\phi}(C_w - C_\infty)\bar{q}_2(\bar{x})\big] = H_1(a)\Big[\frac{\partial \psi}{\partial y}\frac{\partial^2 \psi}{\partial x\partial y} - \frac{\partial \psi}{\partial x}\frac{\partial^2 \psi}{\partial y^2} + n\nu U_\infty^{n-2}\Big(-\frac{\partial^2 \psi}{\partial y^2}\Big)^{n-1}\frac{\partial^3 \psi}{\partial y^3} \\
&\qquad - \frac{g\cos\alpha}{U_\infty^2}\big[\beta\theta(T_w - T_\infty)q_1(x) - \beta^*\phi(C_w - C_\infty)q_2(x)\big]\Big]
\end{aligned} \tag{15}$$

$$\begin{aligned}
&\frac{\partial \bar{\psi}}{\partial \bar{y}}\frac{\partial \bar{\theta}}{\partial \bar{x}} - \frac{\partial \bar{\psi}}{\partial \bar{x}}\frac{\partial \bar{\theta}}{\partial \bar{y}} - \frac{k}{\rho c_\rho U_\infty}\bar{q}_3(\bar{x})\frac{\partial^2 \bar{\theta}}{\partial \bar{y}^2} - \frac{Q}{\rho c_\rho U_\infty}\bar{q}_4(\bar{x})\bar{\theta} \\
&\quad = H_2(a)\Big[\frac{\partial \psi}{\partial y}\frac{\partial \theta}{\partial x} - \frac{\partial \psi}{\partial x}\frac{\partial \theta}{\partial y} - \frac{k}{\rho c_\rho U_\infty}q_3(x)\frac{\partial^2 \theta}{\partial y^2} - \frac{Q}{\rho c_\rho U_\infty}q_4(x)\theta\Big]
\end{aligned} \tag{16}$$

$$\begin{aligned}
&\frac{\partial \bar{\psi}}{\partial \bar{y}}\frac{\partial \bar{\theta}}{\partial \bar{x}} - \frac{\partial \bar{\psi}}{\partial \bar{x}}\frac{\partial \bar{\theta}}{\partial \bar{y}} - \frac{D}{U_\infty}\bar{q}_5(\bar{x})\frac{\partial^2 \bar{\phi}}{\partial \bar{y}^2} - \frac{k\nu}{U_\infty T_r}(T_w - T_\infty)\bar{q}_6(\bar{x})\Big(\bar{\phi}\frac{\partial^2 \bar{\theta}}{\partial \bar{y}^2} + \frac{\partial \bar{\theta}}{\partial \bar{y}}\frac{\partial \bar{\phi}}{\partial \bar{y}}\Big) \\
&\quad = H_3(a)\Big[\frac{\partial \psi}{\partial y}\frac{\partial \theta}{\partial x} - \frac{\partial \psi}{\partial x}\frac{\partial \theta}{\partial y} - \frac{D}{U_\infty}q_5(x)\frac{\partial^2 \phi}{\partial y^2} - \frac{k\nu}{U_\infty T_r}(T_w - T_\infty)q_6(x) \\
&\qquad \Big(\phi\frac{\partial^2 \theta}{\partial y^2} + \frac{\partial \theta}{\partial y}\frac{\partial \phi}{\partial y}\Big)\Big]
\end{aligned} \tag{17}$$

for some functions $H_1(a),\ H_2(a),$ and $H_3(a),$ which may be constants.

Substitution from (13) into Eqs. (15)–(17) for the independent variables, the functions and their derivatives yield

$$\begin{aligned}
&\frac{(K^\psi)^2}{K^x(K^y)^2}\Big(\frac{\partial \psi}{\partial y}\frac{\partial^2 \psi}{\partial x\partial y} - \frac{\partial \psi}{\partial x}\frac{\partial^2 \psi}{\partial y^2}\Big) + n\nu U_\infty^{n-2}\Big(\frac{K^\psi}{(K^y)^2}\Big)^{n-1}\frac{K^\psi}{(K^y)^3}\Big(-\frac{\partial^2 \psi}{\partial y^2}\Big)^{n-1}\frac{\partial^3 \psi}{\partial y^3} \\
&-\frac{g\cos\alpha}{U_\infty^2}\big[K^{q_1}K^\theta\beta\theta(T_w - T_\infty)q_1(x) - \beta^*\phi(C_w - C_\infty)K^{q_2}K^\phi q_2(x)\big] + R_1(a) \\
&\quad = H_1(a)\Big[\frac{\partial \psi}{\partial y}\frac{\partial^2 \psi}{\partial x\partial y} - \frac{\partial \psi}{\partial x}\frac{\partial^2 \psi}{\partial y^2} + n\nu U_\infty^{n-2}\Big(-\frac{\partial^2 \psi}{\partial y^2}\Big)^{n-1}\frac{\partial^3 \psi}{\partial y^3}
\end{aligned}$$

$$-\frac{g\cos\alpha}{U_\infty^2}\big[\beta\theta(T_w-T_\infty)q_1(x)-\beta^*\phi(C_w-C_\infty)q_2(x)\big]\Big] \tag{18}$$

$$\frac{K^\psi K^\phi}{K^x K^y}\Big(\frac{\partial\psi}{\partial y}\frac{\partial\theta}{\partial x}-\frac{\partial\psi}{\partial x}\frac{\partial\theta}{\partial y}\Big)-\frac{K^\theta K^{q_3}}{(K^y)^2}\frac{k}{\rho c_\rho U_\infty}q_3(x)\frac{\partial^2\theta}{\partial y^2}-\frac{QK^\theta K^{q_4}}{\rho c_\rho U_\infty}q_4(x)\theta+R_2(a)$$

$$=H_2(a)\Big[\frac{\partial\psi}{\partial y}\frac{\partial\theta}{\partial x}-\frac{\partial\psi}{\partial x}\frac{\partial\theta}{\partial y}-\frac{k}{\rho c_\rho U_\infty}q_3(x)\frac{\partial^2\theta}{\partial y^2}-\frac{Q}{\rho c_\rho U_\infty}q_4(x)\theta\Big] \tag{19}$$

and

$$\frac{K^\psi K^\phi}{K^x K^y}\Big(\frac{\partial\psi}{\partial y}\frac{\partial\theta}{\partial x}-\frac{\partial\psi}{\partial x}\frac{\partial\theta}{\partial y}\Big)-\frac{K^\phi K^{q_5}}{(K^y)^2}\frac{D}{U_\infty}q_5(x)\frac{\partial^2\phi}{\partial y^2}$$

$$-\frac{k\nu}{U_\infty T_r}(T_w-T_\infty)\frac{K^{q_6}K^\theta K^\phi}{(K^y)^2}q_6(x)\Big(\phi\frac{\partial^2\theta}{\partial y^2}+\frac{\partial\theta}{\partial y}\frac{\partial\phi}{\partial y}\Big)+R_3(a)$$

$$=H_3(a)\Big[\frac{\partial\psi}{\partial y}\frac{\partial\theta}{\partial x}-\frac{\partial\psi}{\partial x}\frac{\partial\theta}{\partial y}-\frac{D}{U_\infty}q_5(x)\frac{\partial^2\phi}{\partial y^2}-\frac{k\nu}{U_\infty T_r}(T_w-T_\infty)q_6(x)\Big(\phi\frac{\partial^2\theta}{\partial y^2}+\frac{\partial\theta}{\partial y}\frac{\partial\phi}{\partial y}\Big)\Big] \tag{20}$$

$$R_1(a)=-\frac{g\cos\alpha}{U_\infty^2}\big[\beta\theta(T_w-T_\infty)(K^{q_1}P^\theta q_1+K^\theta P^{q_1}\theta+P^\theta P^{q_1})$$

$$-\beta^*\phi(C_w-C_\infty)(K^{q_2}P^\phi q_2+K^\phi P^{q_1}\theta+P^\phi P^{q_2})\big] \tag{21}$$

$$R_2(a)=-\frac{k}{\rho c_\rho U_\infty}\frac{P^{q_3}K^\theta}{(K^y)^2}\frac{\partial^2\theta}{\partial y^2}-\frac{Q}{\rho c_\rho U_\infty}(K^{q_4}P^\theta q_4+K^\theta P^{q_4}\theta+P^{q_4}P^\theta) \tag{22}$$

and

$$R_3(a)=-\frac{D}{U_\infty}\frac{P^{q_5}K^\phi}{(K^y)^2}-\frac{k\nu}{U_\infty T_r}(T_w-T_\infty)P^{q_6}\frac{K^\theta K^\phi}{(K^y)^2}\Big(\phi\frac{\partial^2\theta}{\partial y^2}+\frac{\partial\theta}{\partial y}\frac{\partial\phi}{\partial y}\Big) \tag{23}$$

The invariance of Eqs. (18)–(20) follows. This is satisfied by putting

$$P^\phi=P^\theta=P^{q_i}=0;\ i=1,2,...,6 \tag{24}$$

and

$$\frac{(K^\psi)^2}{K^x(K^y)^2}=\frac{(K^\psi)^n}{(K^y)^{2n+1}}=K^\theta K^{q_1}=K^{q_2}K^\phi=H_1(a), \tag{25}$$

$$\frac{K^\psi K^\phi}{K^x K^y}=\frac{K^{q_3}K^\theta}{(K^y)^2}=K^\theta K^{q_4}=H_2(a), \tag{26}$$

$$\frac{K^\psi K^\phi}{K^x K^y}=\frac{K^{q_5}K^\phi}{(K^y)^2}=\frac{K^{q_6}K^\theta K^\phi}{(K^y)^2}=H_3(a), \tag{27}$$

Moreover, the boundary conditions (12) are also invariant, that implies

$$K^{\theta}K^{\phi} = 1. \tag{28}$$

Equations (24)–(28) yield

$$K^{\psi} = \frac{(K^{y})^{\frac{2n-1}{n-2}}}{(K^{x})^{\frac{1}{n-2}}},\ K^{q_1}K^{q_2} = \frac{(K^{y})^{\frac{2}{n-2}}}{(K^{x})^{\frac{n}{n-2}}},\ K^{q_4} = \frac{(K^{y})^{\frac{n+1}{n-2}}}{(K^{x})^{\frac{n-1}{n-2}}},\ K^{q_3} = K^{q_5} = K^{q_6} = \frac{(K^{y})^{\frac{3n-3}{n-2}}}{(K^{x})^{\frac{n-1}{n-2}}} \tag{29}$$

Finally, we get the one-parameter group G which transforms invariantly, the differential Eqs. (9)–(11) and the boundary conditions (12). The group G is of the form

$$G: \begin{cases} \bar{x} = K^{x} + P^{x},\ \bar{y} = K^{y}y + P^{y},\ \bar{\Psi} = \dfrac{(K^{y})^{\frac{n+1}{n-2}}}{(K^{x})^{\frac{n-1}{n-2}}}\psi + P^{\psi},\ \bar{\theta} = \theta,\ \bar{\phi} = \phi, \\ \bar{q_1} = \bar{q_2} = \dfrac{(K^{y})^{\frac{2}{n-2}}}{(K^{x})^{\frac{n}{n-2}}},\ \bar{q_3} = \dfrac{(K^{y})^{\frac{3n-3}{n-2}}}{(K^{x})^{\frac{n-1}{n-2}}}q_3,\ \bar{q_4} = \dfrac{(K^{y})^{\frac{n+1}{n-2}}}{(K^{x})^{\frac{n-1}{n-2}}}q_4,\ \bar{q_5} = \dfrac{(K^{y})^{\frac{3n-3}{n-2}}}{(K^{x})^{\frac{n-1}{n-2}}}q_5, \\ \bar{q_6} = \dfrac{(K^{y})^{\frac{3n-3}{n-2}}}{(K^{x})^{\frac{n-1}{n-2}}}q_6 \end{cases} \tag{30}$$

3.3 The Complete Set of Absolute Invariants

Our aim is to make use of group theoretic method to represent the problem in the form of an ordinary differential equation (similarity representation) in a single independent variable (similarity variable). Then we have to proceed in our analysis to obtain a complete set of absolute invariants. In addition to the absolute invariant of the independent variable, there are nine absolute invariants of the dependent variables: $\psi,\ \theta,\ \phi,\ q_i(x),\ i = 1, 2, ..., 6$. If $\eta = \eta(x, y)$ is the absolute invariant of the independent variables, then

$$g_j\big(x, y;\ \psi, \phi, q_i(x),\ i = 1, 2, ..., 6\big) = F_j[\eta(x, y)],\ j = 1, 2, ..., 9 \tag{31}$$

which are the nine absolute invariants corresponding to $\psi,\ \theta,\ q_i,\ i = 1, 2, ..., 6$. A function $g_j\big(x, y;\ \psi, \phi, q_i(x),\ i = 1, 2, ..., 6\big)\ j = 1, 2, ..., 9$ is an absolute invariant of a one parameter group if it satisfies the following first-order linear differential equation:

$$\sum_{i=1}^{11}(\alpha_i S_i + \beta_i)\frac{\partial g}{\partial S_i} = 0, \tag{32}$$

where S_i stands for $x, y;\ \psi, \phi, q_i(x),\ i = 1, 2, ..., 6,$ and;

$$\alpha_i = \frac{\partial K^{S_i}}{\partial a}(a^0),\ \beta_i = \frac{\partial P^{S_i}}{\partial a}(a^0),\ i = 1, 2, ..., 11, \tag{33}$$

where a^0 denotes the value of "a" which yields the identity element of the group.

From group (30) and (33), we get: $\beta_i = 0,\ i = 3, 4, ..., 11$.

At first, we seek the absolute invariant of the independent variables. Owing to (32), $\eta(x, y)$ is an absolute invariant in condition the following first-order partial differential equation is satisfied:

$$(\alpha_1 x + \beta_1)\frac{\partial \eta}{\partial x} + (\alpha_2 y + \beta_2)\frac{\partial \eta}{\partial y} = 0. \tag{34}$$

Case(1): $\gamma_1 = \frac{\beta_1}{\alpha_1} \neq 0,\ \gamma_2 = \frac{\beta_2}{\alpha_2} \neq 0$, and $\gamma = \frac{\alpha_2}{\alpha_1} \neq 0$. Equation (34), has a solution of the form

$$\eta(x, y) = \frac{y + \gamma_2}{(x + \gamma_1)^{\gamma}}. \tag{35}$$

Case(2): $\alpha_1 = \alpha_2 = 0$, and $\beta = \beta_2/\beta_1$. Equation (34) gives the solution

$$\eta(x, y) = y + \beta x. \tag{36}$$

Case(3): $\alpha_2 = \beta_2 = 0$. Equation (34), has solution in the form

$$\eta(x, y) = y. \tag{37}$$

Case(4): $\alpha_1 = \beta_1 = 0$. Equation (34), has solution in the form

$$\eta(x, y) = x. \tag{38}$$

Case(5): $\alpha_1 = \beta_2 = 0$. Equation (34) has a solution

$$\eta(x, y) = y e^{\frac{\alpha_2}{\beta_1} x} \tag{39}$$

Case(6): $\alpha_2 = \beta_1 = 0$. Equation (34) has a solution

$$\eta(x, y) = x e^{\frac{\beta_2}{\alpha_1} y} \tag{40}$$

Similarly, analysis the absolute invariants of the dependent variables: $\psi,\ \theta$, and ϕ are:

$$\psi(x, y) = \Gamma(x)\Psi(\eta),\ \theta(x, y) = \Theta(\eta),\ \phi(x, y) = \Phi(\eta). \tag{41}$$

3.4 The Reduction to Ordinary Differential Equation

As the general analysis proceeds, the established forms of the dependent and independent absolute invariant are used to obtain ordinary differential equation. Generally, the absolute invariant $\eta(x,y)$ has the form given in (35).

Case(1): $\gamma_1 = \frac{\beta_1}{\alpha_1} \neq 0,\ \gamma_2 = \frac{\beta_2}{\alpha_2} \neq 0,$ and $\gamma = \frac{\alpha_2}{\alpha_1} \neq 0.$

Substituting from (35) and (41) into (9) yields

$$\frac{\Gamma(x)\Gamma'(x)\Psi'^2}{(x+\gamma_1)^{2\gamma}} - \frac{\gamma\Gamma^2(x)\Psi'^2}{(x+\gamma_1)^{2\gamma+1}} - \frac{\Gamma(x)\Gamma'(x)\Psi''\Psi}{(x+\gamma_1)^{2\gamma}} + n\nu(U_\infty)^{n-2}\frac{\Psi'''(-\Psi'')^{n-1}}{(x+\gamma_1)^{(2n+1)\gamma}}$$

$$-\frac{g\beta(T_w-T_\infty)\cos\alpha}{(U_\infty)^2}q_1(x)\Theta - \frac{g\beta(C_w-C_\infty)\cos\alpha}{(U_\infty)^2}q_2(x)\Phi = 0. \tag{42}$$

For (42) to be reduced to an expression in a single independent variable η, its coefficients should be constants or functions of η only. Thus,

$$2\gamma + 1 = (2n+1)\gamma, \tag{43}$$

$$\Gamma(x) = 1. \tag{44}$$

From which we get $\gamma = \frac{1}{2n-1}$. Therefore, the absolute invariant $\eta(x,y)$ takes the form

$$\eta(x,y) = \frac{y+\gamma_2}{(x+\gamma_1)^{\frac{1}{2n-1}}}, \tag{45}$$

consequently, Eq. (42) reduces to

$$-\frac{\Psi'^2}{2n-1} + n\nu(U_\infty)^{n-2}\Psi'''(|\Psi'|)^{n-1} - \frac{g\beta(T_w-T_\infty)}{(U_\infty)^2}\cos\alpha(x+\gamma_1)^{\frac{2n+1}{2n-1}}\Theta$$

$$-\frac{g\beta(C_w-C_\infty)}{(U_\infty)^2}\cos\alpha(x+\gamma_1)^{\frac{2n+1}{2n-1}}\Phi = 0. \tag{46}$$

But since the coefficients in Eq. (46) are not functions in only or constants, then case (1) should be rejected. Case(2): $\alpha_1 = \alpha_2 = 0$, and $\beta = \beta_2/\beta_1$. Substituting from (36) and (41) into (9)–(11), we obtain:

$$na\nu(U_\infty)^{n-2}\Psi'''(|\Psi'|)^{n-1} - Gr\cos\alpha\Theta - Gc\cos\alpha\Phi = 0, \tag{47}$$

$$(2n-1)\Theta'' + PrHe\Theta = 0, \tag{48}$$

$$\Phi'' - \tau Sc(\Phi\Theta'' + \Phi'\Theta') = 0, \tag{49}$$

with the following appropriate corresponding conditions

$$\Psi' = 0,\ \Theta = 1,\ \text{and}\ \Phi = 1,\ \text{at}\ \eta = 0;\ \Psi' = 0,\ \Theta = 0,\ \text{and}\ \Phi = 0,\ \text{as}\ \eta \to \infty. \tag{50}$$

where the non-dimensional parameters introduced in the Eqs. (47)–(49) are defined as follows:

- thermal Grashof number: $Gr = \frac{g\nu\beta(T_w - T_\infty)}{b^2(U_\infty)^3}$
- solutal Grashof number: $Gc = \frac{g\nu\beta(C_w - C_\infty)}{b^2(U_\infty)^3}$
- Prandlt number: $Pr = \frac{\nu^2\rho c_\rho}{kbU_\infty}$
- heat generation parameter: $He = \frac{U_\infty Q(2n-1)}{b\nu^2\rho c_\rho}$
- Schmidt number: $Sc = \nu/D$
- Thermo—phoretic parameter: $\tau = -\frac{k}{T_r}(T_w - T_\infty)$

where a and b are positive constants.

It is clear from Eq. (48), we have different cases for n, namely $n \in (0, 0.5)$, $n = 0.5$ and $n > 0.5$.
Case(3): $\alpha_2 = \beta_2 = 0$. Its results are similar to case(2).
Case(4): $\alpha_1 = \beta_1 = 0$. Gives a trivial solution.
Case(5): $\alpha_1 = \beta_2 = 0$. Has no solution.
Case(6): $\alpha_2 = \beta_1 = 0$. Has no solution.

4 Numerical Results

The solution of Eq. (48) under its appropriate boundary conditions takes the form

$$\Theta(\eta) = \begin{cases} \cos\left(\sqrt{\dfrac{PrHe}{2n-1}}\eta\right), & n > 0.5, \\ e^{-\sqrt{\frac{PrHe}{1-2n}}\eta}, & n < 0.5. \end{cases} \tag{51}$$

where η_∞ is a sufficiently large number which makes $\Theta \to 0$ at the appropriate distance η_∞, for the first case of (51), $\eta_\infty = \frac{\pi}{2}\sqrt{\frac{2n-1}{PrHe}}$. Double successive integrals to Eq. (49) using its boundary

$$\Phi(\eta) = \left(\frac{I(\eta) - I_\infty}{I_0 - I_\infty}\right)e^{Sc\tau\Theta(\eta)}, \tag{52}$$

where $I(\eta) = \displaystyle\int e^{-Sc\tau\Theta(\eta)}d\eta$, and $I_0,\ I_\infty$ are the values of $I(\eta)$ at $\eta = 0$ and η_∞, respectively. After computing $\Theta(\eta)$ and $\Phi(\eta)$ we can calculate $\Psi(\eta)$ numerically

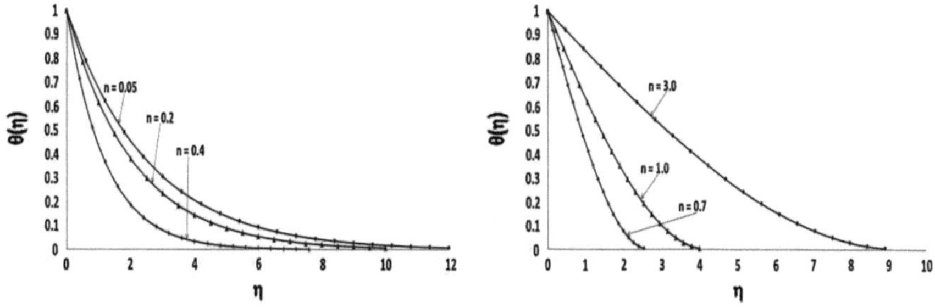

Fig. 2 Effect of n on the dimensionless temperature profiles across the boundary layer ($Pr = 1,\ He = 0.2,\ Sc = 0.22,\ \tau = 1,\ \nu = 1,\ U_\infty = 1,\ Gr = 0.9,\ Gc = 1,$ and $\alpha = 30°$)

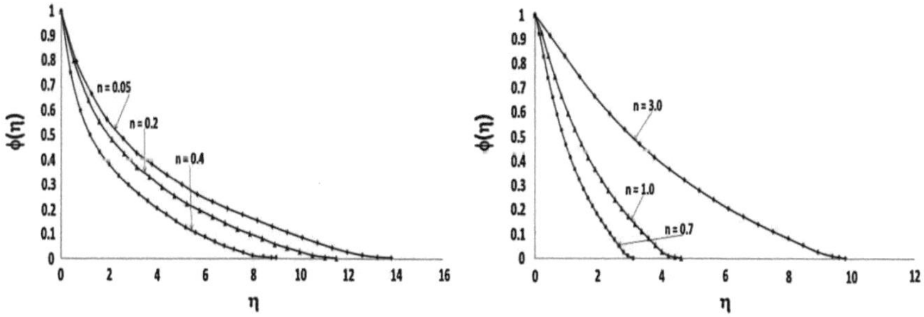

Fig. 3 Effect of n on the dimensionless concentration profiles across the boundary layer ($Pr = 1,\ He = 0.2,\ Sc = 0.22,\ \tau = 1,\ \nu = 1,\ U_\infty = 1,\ Gr = 0.9,\ Gc = 1,$ and $\alpha = 30°$)

from Eq. (47). The variation of dimensionless temperature profiles for different values of n; with $He = 0.2$ and $Pr = 1.0$ is illustrated in Fig. 2. It is interesting to note that the value of the minimum temperature decreases with increase in $n \in (0, 0.5)$ and also the value of the maximum temperature increases with increase in the value of $n > 0.5$. It is interesting to note from Fig. 2 and from (48) that the fluid flow will experience zero temperature for $n = 0.5$.

The variation of dimensionless concentration profiles for different values of n; with $He = 0.2,\ Pr = 1.0,\ Sc = 0.22$ and $\tau = 1$ is presented in Fig. 3. It is observed that boundary shows zero concentration at $n = 0.5$. The boundary layer thickness starts to increase again for $n > 0.5$.

Figures 4 and 5 display the effects of Pr on the temperature and concentration profiles, respectively. Physically speaking, Pr is an important parameter in heat transfer processes as it characterizes the ratio of thicknesses of the viscous and thermal boundary layers. Increasing the value of Pr causes the fluid temperature and its boundary layer thickness to decrease significantly as it is clear Fig. 4. This decrease in temperature produces a net reduction of the thermal buoyancy effect in the momentum equation which results in less induced flow along the plate and consequently, the fluid velocity decreases. In addition, it is clear that the concentration

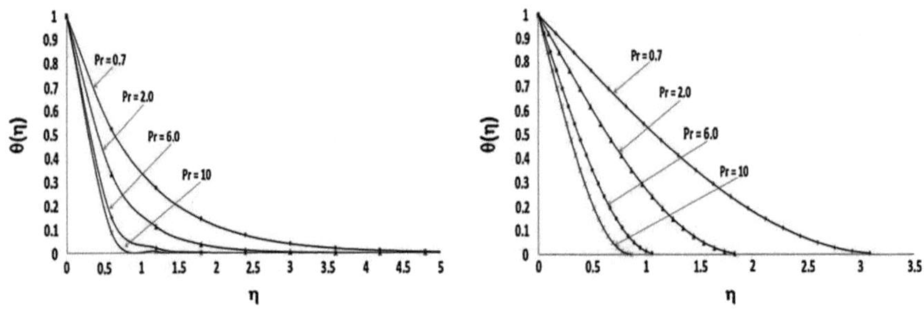

Fig. 4 Effect of Pr on the dimensionless temperature profiles across the boundary layer ($He = 1,\ Sc = 0.22,\ \tau = 1,\ \nu = 1, U_\infty = 1,\ Gr = 0.9,\ Gc = 1,$ and $\alpha = 30°$)

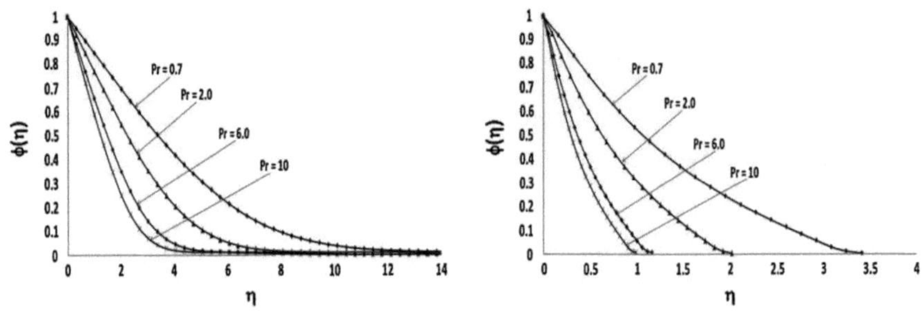

Fig. 5 Effect of Pr on the dimensionless concentration profiles across the boundary layer ($He = 1,\ Sc = 0.22,\ \tau = 1,\ \nu = 1,\ U_\infty = 1,\ Gr = 0.9,\ Gc = 1,$ and $\alpha = 30°$)

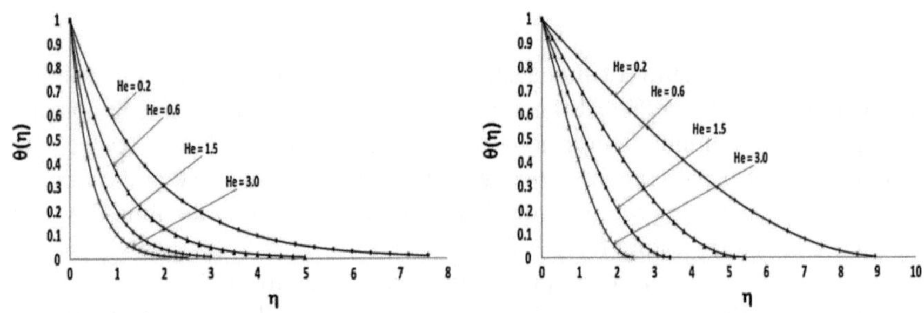

Fig. 6 Effect of He on the dimensionless temperature profiles across the boundary layer ($Pr = 0.7,\ Sc = 1,\ \tau = 1,\ \nu = 1,\ U_\infty = 1,\ Gr = 0.9,\ Gc = 1,$ and $\alpha = 30°$)

distribution inside the boundary layer also decreases. These behaviors are illustrated in Fig. 5.

Figure 6 shows the temperature profile for various values of He; with $n = 0.3$ and 3.0. It is observed that the fluid temperature decreases with increase in He and the rate at which the temperature approaches zero is fast for high value of He. It is observed that the fluid temperature tends to minimum for the fluids with $n < 0.5$ and experience a maximum for the fluids with $n > 0.5$.

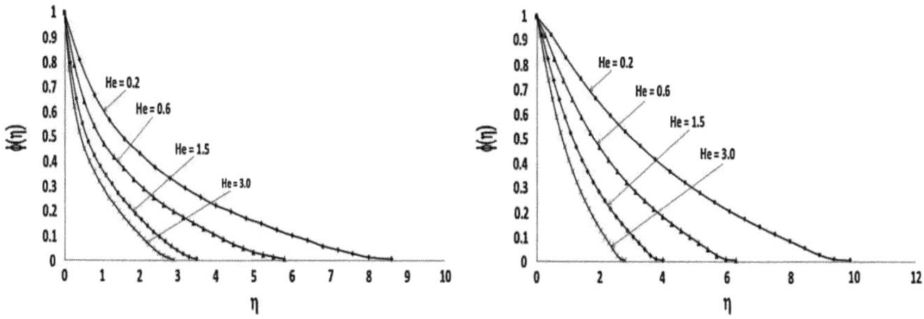

Fig. 7 Effect of *He* on the dimensionless concentration profiles across the boundary layer ($Pr = 0.7,\ Sc = 1,\ \tau = 1,\ \nu = 1,\ U_\infty = 1,\ Gr = 0.9,\ Gc = 1,$ and $\alpha = 30°$)

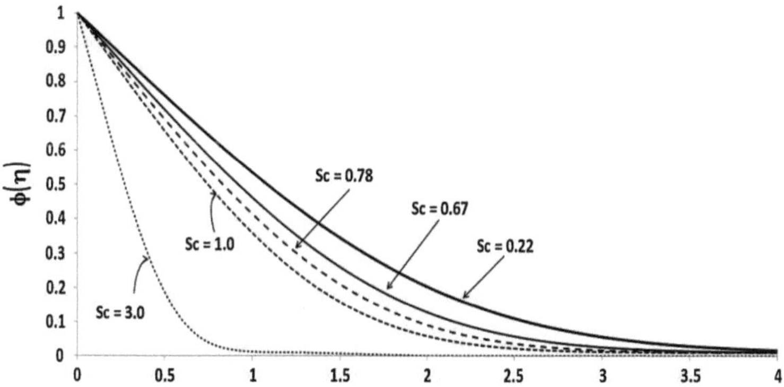

Fig. 8 Effect of *Sc* on the dimensionless concentration profiles across the boundary layer ($Pr = 1,\ He = 0.2,\ \tau = 1,\ \nu = 1,\ U_\infty = 1,\ Gr = 0.9,\ Gc = 1,$ and $\alpha = 30°$)

Figure 7 shows the concentration profile for various values of *He*; with $n = 0.3$ and 3.0. It is observed that the fluid concentration approaches its minimum faster with a high value of *He*. The concentration profile converges vertically as $He \to \infty$.

Figure 8 illustrates the influence of *Sc* on the concentration profiles. By analogy with the *Pr*, the *Sc* is an important parameter in mass transfer processes as it characterizes the ratio of thicknesses of the viscous and concentration boundary layers. Its effect on the species concentration boundary-layer thickness has similarities to the *Pr* effect on the thermal boundary-layer thickness, i.e., increases in the values of *Sc* cause the species concentration boundary layer thickness to decrease significantly.

5 Conclusion and Comments

Group theoretic approach is a powerful tool for solving the two-dimensional boundary-layer flow of non-Newtonian power-law fluids and for obtaining the velocity profiles. Referring to the numerical results and the illustrated figures it is observed that: (a) For all values of n, the value of the minimum temperature decreases with increase in $n \in (0, 0.5)$ and also the value of the maximum temperature increase with increase in the value of $n > 0.5$. The boundary layer thickness decreases as $n \in (0, 0.5)$, thus the fluid flow shows zero concentration at $n = 0.5$. The boundary layer thickness starts to increase again for $n > 0.5$. (b) Increasing the value of Pr causes the fluid temperature and its boundary layer thickness to decrease significantly and consequently, the fluid velocity decreases. In addition, the concentration distribution inside the boundary layer also decreases. (c) The fluid temperature decreases with the increase in He, and the rate at which the temperature reaches zero is fast with a low value of He. (d) The fluid concentration reaches its maximum faster with a high value of He. (e) Increasing in Sc forces the species concentration boundary layer thickness to decrease significantly.

Acknowledgments The authors would like to express their gratitude and appreciations to the professional reviewers for their valuable comments that improved the results and contents of the paper to the present form.

References

1. Djukic, D.S., Vujanovic, B.D.: A variational principle for the two-dimensional boundary-layer flow of non-Newtonian power-law fluids
2. Birkhoff, G.: Mathematics for engineers. Elect. Eng. **67**, 1185 (1948)
3. Gaggioli, R.A., Moran, M.J.: Group Theoretic techniques for the similarity solution of systems of partial differential equations with auxiliary conditions. Math. Res. Cent. (Univ. of Wisconsin) Tech. Summery Rep. No. 693 (1966)
4. Moran, M.J., Gaggioli, R.A.: A new systematic formalism for similarity analysis. J. Engg. Math. **3**, 151–162 (1969)
5. Gabbert, C.H.: Similarity for unsteady compressible boundary layers. AIAA J. **5**(6) (1967)
6. Ames, W.F.: Similarity for nonlinear diffusion equation i. Ec Fund. **4**, 72–76 (1965)
7. Ames, W.F., Nucci, M.C.: Analysis of fluid equations by group methods. J. Engg. Math. **20**, 181–187 (1985)
8. Bluman, G.W., Cole, J.D.: Similarity methods of differential equations. Springer, New York (1974)
9. Boisvert, R.F., Ames, W.F., Srivastava, U.N.: Group properties and new solutions of navier-stokes equations. J. Engg. Math. **17**, 203–221 (1983)
10. Gaggioli, R.A., Moran, M.J.: Similarity analysis via group theory. AIAA. J. **6**, 2014–2016 (1968)
11. Gaggioli, R.A., Moran, M.J.: Similarity analysis of compressible boundary layer flows via group theory. Math. Res. Center (Univ. of Wisconsin) Tech. Summery Rep. No. 838 (1967)
12. Abd-el-Malek, M.B., Helal, M.M.: Steady flow of an electrically conducting incompressible viscoelastic fluid over a heated plate. Z. Naturforsch. **60a**, 29–36 (2005)
13. Abd-el-Malek, M.B., Boutros, Y.Z., Badran, N.A.: Group method analysis of unsteady free-convective boundary-layer flow on a nonisothermal vertical flat plate. J. Eng. Math. **24**(4), 343–368 (1990)
14. Sivasankaran, S., Bhuvaneswari, M., Kandaswamy, P., Ramasami, E.K.: Lie group analysis of natural convection heat and mass transfer in an inclined surface. Nonlinear Anal. Model. Control **11**(1), 201–212 (2006)
15. Olajuwon, B.I.: Flow and natural convection heat transfer in a power law fluid past a vertical plate with heat generation. Int. J. Nonlinear Sci. **7**(1), 50–56 (2009)

Part IV
Mathematical Physics and Applications

Realizations of Affine Lie Algebra $A_1^{(1)}$ at Negative Levels

Jilan Dong and Naihuan Jing

Abstract A realization of the affine Lie algebra $A_1^{(1)}$ and the relevant Z-algebra at negative level $-k$ is given in terms of parafermions. This generalizes the recent work on realization of the affine Lie algebra at the critical level.

1 Introduction

Since the sixties the theory of affine Lie algebras has been one of the popular subjects in mathematical physics. Its physical applications and mathematical properties usually depend on whether one can use the matrix method to give a concrete realization or representation. This approach has been used in dual resonance models, infinitesimal Bäcklund transformations in soliton theory etc. The first concrete realization of the affine Lie algebra $\widehat{sl}(2)$ was Lepowsky-Wilson's vertex operator representation at level one [18] and was then generalized to arbitrary types by Kac et al. [15]. Later the homogeneous realization of simply laced affine Lie algebras at level one was given by Frenkel and Kac [12] and Segal [19]. Fermionic realizations were also constructed by Frenkel [10] and Kac and Peterson [16], and were generalized to arbitrary types by Feingold and Frenkel [9].

Representations of the affine Lie algebras at other levels also have attracted a lot of attention [5–7, 14, 17]. Wakimoto [20] derived a general scheme to realize the

J. Dong
School of Sciences, South China University of Technology, Guangzhou 510640, Guangdong, China
e-mail: dongjilan.11@163.com

N. Jing (✉)
Department of Mathematics, North Carolina State University, Raleigh, NC 27695, USA
e-mail: jing@math.ncsu.edu

A. Makhlouf et al. (eds.), *Algebra, Geometry and Mathematical Physics*, Springer Proceedings in Mathematics & Statistics 85, DOI: 10.1007/978-3-642-55361-5_36,

affine Lie algebra of type $A_1^{(1)}$ and this was generalized to higher rank by Feigin and Frenkel [8]. Generally speaking, highest weight representations of the affine Lie algebras with integral levels are built from the theory of vertex operators in terms of bosonic or fermionic operators. Recently Adamović [1] used vertex superalgebras to study critical modules $\widehat{sl}_2(\mathbb{C})$. Dunbar et al. [7] also gave a new representation of at the critical level using some technique similar to semi-infinite wedge products. This paper is a generalization of their work to arbitrary negative integral level using the theory of parafermions. Parafermions are introduced in statistical mechanics and conformal field theory, they are also related to Majorana fermions, fractional superstring, mirror symmetry, and have close connections with exclusion statistics, quantum computations, Bose-Einstein condensates etc. [2–4]. In particular, the parafermions, sometimes regarded as Z-algebras proposed in [22] contribute to various extensions of the Ising model and 3-state Potts model, all of which are basically relevant to $A_1^{(1)}$. These works show that the Z-algebra at a positive integral level is identical with that of $A_1^{(1)}$-parafermions.

In the monograph [5], Dong and Lepowsky constructed canonical generalized vertex operator algebras for $\widehat{\mathfrak{g}}$ (of simply laced types $\hat{A}$, $\hat{D}$ or $\hat{E}$) and pointed out that the corresponding quotient space for the vacuum space of any positive integer level k standard $\hat{\mathfrak{G}}$-module is a module of the generalized vertex operator algebra. Furthermore, as an illustration, they showed in details the construction for $A_1^{(1)}$. They used the vacuum space of $L(k, 0)$ ($k \in \mathbb{N}$) in terms of a natural Heisenberg subalgebra of $A_1^{(1)}$ to define a quotient spaces of this vacuum space by the action of an infinite cyclic group, and then realized the parafermion algebra as the canonically modified Z-algebra acting on certain quotient spaces.

In [7] the affine Lie algebra $\widehat{sl}_2(\mathbb{C})$ at the critical level -2 was realized using the generalized Clifford algebra. This shows that the case of negative level can be treated by parafermions as well. In this paper we generalize this result and realize the affine Lie algebra $\widehat{sl}_2$ at negative levels by parafermions. Although many results at negative levels are quite similar to the positive integral levels, we still give a complete treatment of the realization with the hope that this may be useful to understand Lusztig's theory of the relationship between quantum groups and affine Lie algebras. For completeness we include all necessary computation of operator product expansions of parafermions and also provide the detailed verifications of the Z-algebra relations.

The paper is organized as follows. In section two we first recall the basic definitions. The later part of section two reviews some basic results of parafermions based on [7], and briefly explains the physicists' approach to parafermion fields with respect to each form of current algebras, operator product expansions and so on, and also studies in detail the generalized commutation relations and in particular modifications needed in the paper. In section three, a parafermionic representation of $A_1^{(1)}$ at level $-k$ ($k \in \mathbb{N}$) is constructed and corresponding results for the associated Z-algebra are given.

2 Basic Definitions

2.1 The Affine Lie Algebra $\widehat{sl}_2$

Let $\widehat{sl}_2(\mathbb{C})$ be the affine Lie algebra of type $A_1^{(1)}$, which is generated by a 1-dimensional central c, a degree derivation $d = 1 \otimes t\partial t$ and elements $a(m) = a \otimes t^m \in sl_2(\mathbb{C}) \otimes \mathbb{C}[t, t^{-1}]$, where $\mathbb{C}[t, t^{-1}]$ is the algebra of Laurent polynomials in the indeterminate t. The Lie bracket operation is defined by

$$\begin{aligned} [c, \widehat{sl}_2(\mathbb{C})], \qquad [d, a(m)] = ma(m) \\ [a(m), b(n)] = [a, b](m+n) + Tr(ab)mc\delta_{m+n,0} \end{aligned} \tag{1}$$

for all $m, n \in \mathbb{Z}$, $a, b \in sl_2(\mathbb{C})$. The Chevalley basis of sl_2 consists of X, Y, H:

$$H = \begin{pmatrix} 1 & 0 \\ 0 & -1 \end{pmatrix}, X = \begin{pmatrix} 0 & 1 \\ 0 & 0 \end{pmatrix}, Y = \begin{pmatrix} 0 & 0 \\ 1 & 0 \end{pmatrix}$$

with brackets

$$[H, X] = 2X, [H, Y] = -2Y, [X, Y] = H \tag{2}$$

Besides the presentation of $A_1^{(1)}$ with the basis $\{H(p), X(m), Y(n), c \mid p, m, n \in \mathbb{Z}\}$, satisfying the commutation relations (1), there is also the Kac-Moody definition by the Chevalley generators $\{h_i, e_j, f_k | i, j, k \in \{0, 1\}\}$, subject to the conditions

$$[h_i, e_j] = A_{ij} e_j, [h_i, f_j] = -A_{ij} f_j, [e_i, f_j] = \delta_{ij} h_j, \tag{3}$$

where $A = (A_{ij}) = \begin{pmatrix} 2 & -2 \\ -2 & 2 \end{pmatrix}$ is the generalized Cartan matrix.

The two equivalent descriptions of $A_1^{(1)}$ are related under the following correspondence

$$\begin{aligned} e_0 \leftrightarrow Y(1), f_0 \leftrightarrow X(-1), h_0 = -H(0) + c, \\ e_1 \leftrightarrow X(0), f_1 \leftrightarrow Y(0), h_1 = H(0) \end{aligned}$$

Recall that the weight space $V_\mu = \{v \in V \mid h \cdot v = \mu(h)v, \forall h \in \eta\}$, where $\eta = (\oplus_{n \in \mathbb{Z}} H(n)) \oplus \mathbb{C}c \oplus \mathbb{C}d$ is the Cartan subalgebra of $\widehat{sl}_2(\mathbb{C})$. A highest weight module $V(\lambda) = \oplus_{\mu \leq \lambda} V_\mu$, or the highest weight representation , is the space generated by a highest weight vector v_λ of weight λ such that $e_i v_\lambda = 0$, $h_i v_\lambda = \lambda(h_i) v_\lambda$. The central element c acts on $V(\lambda)$ as a scalar k, which will be called the level of the module.

The elements of Heisenberg subalgebra $\mathfrak{h}' = (\oplus_{n \neq 0} H(n)) \oplus \mathbb{C}c$ of $A_1^{(1)}$ obey the following relations which are special cases of Eq. (1):

$$[H(m), H(n)] = 2mc\delta_{m+n,0} \tag{4}$$

Given level $-k$, any negative integer. Let $S(\mathfrak{h}'^-)$ the space of symmetric polynomials generated by elements in $\mathfrak{h}'^- = \oplus_{n<0} H(n)$. Then there is a canonical representation of the Heisenberg algebra $\mathfrak{h}'$ on $S(\mathfrak{h}'^-)$ via the actions in accordance with Eq. (1):

$$\begin{aligned} &c \cdot 1 = k, H(0) \cdot v = 0, \\ &H(m) \cdot v = H(m)v, m < 0 \\ &H(m) \cdot v = -2mk\partial_{H(m)}(v), m > 0 \end{aligned} \tag{5}$$

In fact, this can be verified pretty straightforward, one needs only to observe that $[H(m), H(n)] \cdot v = -2mk\delta_{m+n,0}v$ is valid under bracket relations.

We denote by $a(z) = \sum_{m\in\mathbb{Z}} a(m)z^{-m}$ the power formal series. Here and later, z, w mean any formal variables. In this form, $A_1^{(1)}$ is usually called a current algebra. To write commutation relations in formal series, we need to introduce the formal δ-function $\delta(\frac{w}{z}) = \sum_{m\in\mathbb{Z}}(\frac{w}{z})^n$, which possesses the fundamental property: for any $f(w,z) \in End(V)[[w, w^{-1}, z, z^{-1}]]$ such that

$$lim_{w\to z} f(w,z) = f(z,z), \quad f(w,z)\delta(\frac{w}{z}) = f(z,z)\delta(\frac{w}{z})$$

exists. More information on delta functions can be found in [11].

The commutation relations of the affine Lie algebra can now be given as follows.

$$\begin{aligned} &[H(z), X(w)] = 2X(w)\delta(\tfrac{w}{z}) \\ &[H(z), Y(w)] = -2X(w)\delta(\tfrac{w}{z}) \\ &[X(z), Y(w)] = H(w)\delta(\tfrac{w}{z}) - kw\partial_w\delta(\tfrac{w}{z}) \end{aligned} \tag{6}$$

2.2 Parafermions

We now discuss the parafermion theory [21]. Let Φ be the root system of the simple Lie algebra $\mathfrak{g}$ and let M (or $M \; mod \; kM_L$) denote the root lattice spanned by Φ, where $-k$ is identified with the level in the corresponding affine Lie algebra $\hat{g}$ and M_L is the long root sublattice. Let E_α be the root vector of $\mathfrak{g}$, and we normalize the Chevalley basis of $\mathfrak{g}$ via $[E_\alpha, E_\beta] = \varepsilon_{\alpha\beta}E_{\alpha+\beta}$ if $\alpha+\beta \in \Phi$. It is well-known that $\varepsilon_{\alpha\beta} \in \mathbb{Z}$.

General parafermions are defined for elements of M, but we will focus on parafermionic fields $\psi_\alpha(z), \psi_\beta(w)$ for roots $\alpha, \beta \in \Phi$ [13]. For two such parafermions the radial ordered product is defined as a multivalued function owning to the mutually semilocal property between them (cf. [21]). Instead of (anti-)commtativity the key relation is

$$R(\psi_\alpha(z)\psi_\beta(w)) = (-1)^{\frac{\alpha\beta}{-k}} R(\psi_\beta(w)\psi_\alpha(z)). \tag{7}$$

For simplicity we will drop the symbol R. For $\alpha, \beta \in \Phi$ the operator product expansion for two parafermions can be formulated as (cf. [2])

$$\psi_\alpha(z)\psi_\beta(w)(z-w)^{-\frac{\alpha\beta}{k}} = N\left(\psi_\alpha(z)\psi_\beta(w)\right) + \frac{\varepsilon_{\alpha,\beta}}{z-w}\psi_{\alpha+\beta}(w) + \frac{\delta_{\alpha+\beta,0} I_{\psi_\alpha(z)\psi_\beta(w)}}{(z-w)^2} \quad (8)$$

in which $N\left(\psi_\alpha(z)\psi_\beta(w)\right)$ can be seen as infinitesimal of higher order in $z-w$, namely inside $\mathcal{O}(z-w)$. Note that the regular part of the expression in parentheses satisfy

$$N\left(\psi_\alpha(z)\psi_\beta(w)\right) = N\left(\psi_\beta(w)\psi_\alpha(z)\right),$$

and

$$\varepsilon_{\alpha,\beta} = \begin{cases} \varepsilon_{\alpha\beta}/\sqrt{-k}, & \text{if } \alpha+\beta \in \Phi \\ 0, & \text{otherwise} \end{cases},$$

where $I_{\psi_\alpha(z)\psi_\beta(w)}$ are some constants to be fixed later.

According to parafermion theory, the conformal dimension of $\psi_l(z)$, $l \in \Phi$ is defined by $\Delta_l = \frac{l^2}{2k} + n(l)$ [13], where $n(l)$ is the minimal number of roots α_i in Φ by which l can be composed, $\alpha = \sum_{i=1}^{n(l)} \alpha_i$. Note that Eq. (8) can be equivalently written as

$$\psi_\alpha(z)\psi_\beta(w) = (z-w)^{\Delta_{\alpha+\beta}-\Delta_\alpha-\Delta_\beta}[\delta_{\alpha+\beta,0} I_{\psi_\alpha(z)\psi_\beta(w)} + \varepsilon_{\alpha,\beta}\psi_{\alpha+\beta}(w) + \cdots]. \quad (9)$$

In this case $\Delta_\alpha = \Delta_{-\alpha} = 1$ and $\Delta_0 = 0$, therefore Eq. (9) is simply

$$\begin{aligned}&\psi_\alpha(z)\psi_\beta(w)(z-w)^{-\frac{\alpha\beta}{k}} \\ &\quad = (z-w)^{n(\alpha+\beta)-2}[\delta_{\alpha+\beta,0} I_{\psi_\alpha(z)\psi_\beta(w)} + \varepsilon_{\alpha,\beta}\psi_{\alpha+\beta} + \mathcal{O}(z-w)].\end{aligned}$$

For $\psi_{\pm\alpha}(z)$ associated to Lie algebra $sl_2(\mathbb{C})$, we find that $I_{\psi_\alpha(z)\psi_\beta(w)} = -kzw$ for $\alpha = -\beta$, and it is 1 otherwise. We define the normal ordered product

$$:\psi_\alpha(z)\psi_\beta(w):(z-w)^{-\frac{\alpha\beta}{k}} = N\left(\psi_\alpha(z)\psi_\beta(w)\right). \quad (10)$$

We define the contraction function by

$$\underbrace{\psi_\alpha(z)\psi_\beta(w)}(z-w)^{-\frac{\alpha\beta}{k}} = \psi_\alpha(z)\psi_\beta(w)(z-w)^{-\frac{\alpha\beta}{k}} - :\psi_\alpha(z)\psi_\beta(w):(z-w)^{-\frac{\alpha\beta}{k}} \quad (11)$$

Then Eq. (8) can be written by

$$\psi_\alpha(z)\psi_\beta(w)(z-w)^{-\frac{\alpha\beta}{k}} = \begin{cases} :\psi_\alpha(z)\psi_\alpha(w):(z-w)^{-\frac{2}{k}}, & if\, \alpha = \beta \\ :\psi_\alpha(z)\psi_{-\alpha}(w):(z-w)^{\frac{2}{k}} + \frac{-kzw}{(z-w)^2}, & if\, \alpha = -\beta \end{cases} \quad (12)$$

Now for the affine Lie algebra $\widehat{sl}_2$ we define the Z-algebra operators $A_\alpha(z)$, $A^*_{-\alpha}(z)$ for $\psi_\alpha(z)$, $\psi_{-\alpha}(z)$. Using Eq. (12), we get the following equations:

$$\begin{aligned}
\underbrace{A_\alpha(z)A_\alpha(w)}(z-w)^{-\frac{2}{k}} &= \underbrace{A^*_{-\alpha}(z)A^*_{-\alpha}(w)}(z-w)^{-\frac{2}{k}} = 0,\\
\underbrace{A_\alpha(z)A^*_{-\alpha}(w)}(z-w)^{\frac{2}{k}} &= \tfrac{-kzw}{(z-w)^2},\\
\underbrace{A^*_{-\alpha}(w)A_\alpha(z)}(w-z)^{\frac{2}{k}} &= \tfrac{-kzw}{(w-z)^2}.
\end{aligned} \tag{13}$$

The operator A_α (or A^*_α) acts on the field operator $\Phi_{\lambda,\bar\lambda}(w,\bar w)$ with charge $(\lambda,\bar\lambda)$ (cf. [2, 4, 13, 22]) as follows.

$$A_\alpha(z)\Phi_{\lambda,\bar\lambda}(w,\bar w) = \sum_{m=\infty}^{\infty}(z-w)^{-m-1+\frac{\alpha\lambda}{k}}A_m^{\alpha,\lambda}\Phi_{\lambda,\bar\lambda}(w,\bar w), \tag{14}$$

then the component operator $A_m^{\alpha,\lambda}$ acts on $\Phi_{\lambda,\bar\lambda}$ via

$$A_m^{\alpha,\lambda}\Phi_{\lambda,\bar\lambda}(w,\bar w) = \int_w \frac{dz}{2\pi i}(z-w)^{m-\frac{\alpha\lambda}{2k}}A_\alpha(z)\Phi_{\lambda,\bar\lambda}(w,\bar w).$$

We are interested only in parafermions ψ_α, carrying charge $(\alpha,0)$ with $\Phi_{\alpha,0}(w,\bar w)$ [13]:

$$A_\alpha(z)\Phi_{\alpha,0}(w,\bar w) = \sum_{m=\infty}^{\infty}(z-w)^{-m-1+\frac{2}{k}}A_m^{\alpha,\alpha}\Phi_{\alpha,0}(w,\bar w). \tag{15}$$

2.3 *Action of the Group Algebra*

The group algebra $\mathbb{C}(\mathbb{Z}\alpha)$ is the associative algebra generated by $e^{n\alpha}$ ($n\in\mathbb{Z}$) under the multiplication

$$e^0 = 1, \qquad e^{m\alpha}\cdot e^{n\alpha} = e^{(m+n)\alpha}. \tag{16}$$

where $m,n\in\mathbb{Z}$. The group algebra $\mathbb{C}(\mathbb{Z}\alpha)$ acts on itself via multiplication, and we also introduce the operator $h(0)$ ($h\in\mathfrak{h}$) which acts on $\mathbb{C}(\mathbb{Z}\alpha)$ by

$$\begin{aligned}
h(0):\mathbb{C}(\mathbb{Z}\alpha) &\to \mathbb{C}(\mathbb{Z}\alpha)\\
e^\alpha &\mapsto \langle h,\alpha\rangle e^\alpha
\end{aligned}$$

so we get $[h(0),e^\alpha] = \langle h,\alpha\rangle e^\alpha$. Using the operator $h(0)$ we naturally define the operator $z^h\in(End\mathbb{C}(\mathbb{Z}\alpha))\{z\}$(can be seen as $z^{h(0)}$) for $h\in\mathbb{Z}\alpha$ by

$$z^h \cdot e^\alpha = z^{\langle h,\alpha\rangle} e^\alpha. \tag{17}$$

Then we get

$$\begin{aligned} [\alpha(0), z^\beta] &= 0, \\ z^\alpha e^\beta = z^{\langle\alpha,\beta\rangle} e^\beta z^\alpha &= e^\beta z^{\alpha+\langle\alpha,\beta\rangle}. \end{aligned} \tag{18}$$

3 Construction of the Parafermion Representations of $A_1^{(1)}$ and Z-Algebra

3.1 Action of Heisenberg Subalgebra

We define the following exponential operators on the space $S(\mathfrak{h}'^-)$ and their properties are given in Proposition 3.1.

$$E_+^\pm(z) = exp(\mp \sum_{n>0} \frac{H(-n)}{kn} z^n)$$
$$E_-^\pm(z) = exp(\pm \sum_{n>0} \frac{H(n)}{kn} z^{-n})$$

Proposition 3.1 *On the space $S(\mathfrak{h}'^-)$ we have*

$$\begin{aligned} &E_\pm^+(z)E_\pm^-(z) = E_\pm^-(z)E_\pm^+(z) = 1 \\ &E_+^+(z)E_+^\mp(z) = E_+^\mp(z)E_+^+(z) \\ &E_+^-(z)E_+^\mp(z) = E_+^\mp(z)E_+^-(z) \\ &\partial_z(E_+^\pm(z)E_-^\pm(z)) = \mp E_-^\pm(z)E_+^\pm(z) \sum_{n\neq 0} \frac{H(n)}{k} z^{-n-1} \end{aligned} \tag{19}$$

$$\begin{aligned} E_\pm^+(z)E_\pm^+(w) &= E_\pm^+(w)E_\pm^+(z) \\ E_\pm^-(z)E_\pm^-(w) &= E_\pm^-(w)E_\pm^-(z) \end{aligned} \tag{20}$$

$$\begin{aligned} E_\pm^+(z)E_\pm^-(w) &= E_\pm^-(w)E_\pm^+(z) \\ E_+^\pm(z)E_-^\mp(w) &= E_-^\mp(w)E_+^\pm(z)(1-\tfrac{z}{w})^{-\frac{2}{k}} \\ E_+^\pm(z)E_-^\pm(w) &= E_-^\pm(w)E_+^\pm(z)(1-\tfrac{z}{w})^{\frac{2}{k}} \end{aligned} \tag{21}$$

Proof These identities are proved by the Campbell-Hausdorf-Witt theorem. The commutativity relations are easy consequence of the fact that $H(m)$ and $H(n)$ commute if $m \neq -n$. For the other identities we compute that

$$\partial_z(E_+^-(z)E_-^-(z)) = \partial_z(exp(-\sum_{n\neq 0} \frac{H(n)}{kn} z^{-n})) = E_+^-(z)E_-^-(z) \sum_{n\neq 0} \frac{H(n)}{k} z^{-n-1}.$$

For the last two relations in Eq. (21), we use the identity $e^{x_1}e^{x_2} = e^{x_2}e^{x_1}e^{[x_1,x_2]}$ if x_1, x_2 commute with $[x_1, x_2]$:

$$\begin{aligned} E_+^+(z)E_-^-(w) &= E_+^-(w)E_+^+(z)exp([\sum_{m>0} \frac{-H(-m)}{km} z^m, -\sum_{n>0} \frac{H(n)}{kn} w^{-n}]) \\ &= E_+^-(w)E_+^+(z)exp(-\sum_{m,n>0} \frac{2mc\delta_{m-n,0}}{k^2mn} z^m w^{-n}) \\ &= E_+^-(w)E_+^+(z)(1-\tfrac{z}{w})^{-\frac{2}{k}}. \end{aligned}$$

□

3.2 The Realization

Let $V = S(\mathfrak{h}'^-) \otimes < \Phi_{\alpha,0}(\omega,\overline{\omega}) > \otimes \mathbb{C}(\mathbb{Z}\alpha)$, we define the map $\pi : \widehat{sl}_2(\mathbb{C}) \to End(V)\{z\}$ as follows:

$$\begin{aligned} X(z) &\longmapsto E_+^+(z)E_-^+(z) \otimes A_\alpha(z)e^\alpha z^{-\frac{\alpha}{k}} \\ Y(z) &\longmapsto E_+^-(z)E_-^-(z) \otimes A_{-\alpha}^*(z)e^{-\alpha} z^{\frac{\alpha}{k}} \\ H(z) &\longmapsto H(z) \otimes 1 \\ c &\longmapsto -k \\ d &\longmapsto deg. \end{aligned}$$

Theorem 3.1 *(π, V) defines a representation of $A_1^{(1)}$.*

Proof For convenience, we just check the relations in Eq. (6).

$$\begin{aligned} X(z)X(w) \mapsto\ & E_+^+(z)E_-^+(z)E_+^+(w)E_-^+(w) \otimes A_\alpha(z)e^\alpha z^{-\frac{\alpha}{k}} A_\alpha(w)e^\alpha w^{-\frac{\alpha}{k}} \\ =\ & E_+^+(z)E_+^+(w)E_-^+(z)E_-^+(w)(1-\frac{w}{z})^{-\frac{2}{k}} \otimes A_\alpha(z)A_\alpha(w)e^{2\alpha} z^{-\frac{\alpha}{k}-\frac{2}{k}} w^{-\frac{\alpha}{k}} \\ =\ & E_+^+(z)E_+^+(w)E_-^+(z)E_-^+(w) \otimes :A_\alpha(z)A_\alpha(w): (z-w)^{-\frac{2}{k}} e^{2\alpha}(zw)^{-\frac{\alpha}{k}} \\ & + E_+^+(z)E_+^+(w)E_-^+(z)E_-^+(w) \otimes \underbrace{A_\alpha(z)A_\alpha(w)}(z-w)^{-\frac{2}{k}} e^{2\alpha}(zw)^{-\frac{\alpha}{k}} \\ =\ & E_+^+(z)E_+^+(w)E_-^+(z)E_-^+(w) \otimes :A_\alpha(z)A_\alpha(w): (z-w)^{-\frac{2}{k}} e^{2\alpha}(zw)^{-\frac{\alpha}{k}} \end{aligned}$$

Hence,

$$\begin{aligned} [X(z), X(w)] &= X(z)X(w) - X(w)X(z) \\ &= E_+^+(z)E_+^+(w)E_-^+(z)E_-^+(w) \otimes \Big(:A_\alpha(z)A_\alpha(w): (z-w)^{-\frac{2}{k}} \\ &\quad - :A_\alpha(w)A_\alpha(z): (w-z)^{-\frac{2}{k}}\Big) e^{2\alpha}(zw)^{-\frac{\alpha}{k}} \\ &= 0 \end{aligned}$$

By similar method we get $[Y(z)Y(w)] = 0$. Next notice that

$$
\begin{aligned}
X(z)Y(w) &\mapsto E_+^+(z)E_-^+(z)E_+^-(w)E_-^-(w) \otimes A_\alpha(z)e^\alpha z^{-\frac{\alpha}{k}} A^*_{-\alpha}(w)e^{-\alpha} w^{\frac{\alpha}{k}} \\
&= E_+^+(z)E_+^-(w)E_-^+(z)E_-^-(w)(1-\frac{w}{z})^{\frac{2}{k}} \otimes (\,:A_\alpha(z)A^*_{-\alpha}(w):+\underbrace{A_\alpha(z)A^*_{-\alpha}(w)}\,) \\
&\quad e^{\alpha-\alpha} z^{-\frac{\alpha}{k}+\frac{2}{k}} w^{\frac{\alpha}{k}} \\
&= E_+^+(z)E_+^-(w)E_-^+(z)E_-^-(w) \otimes :A_\alpha(z)A^*_{-\alpha}(w):(z-w)^{\frac{2}{k}} z^{-\frac{\alpha}{k}} w^{\frac{\alpha}{k}} \\
&\quad + E_+^+(z)E_+^-(w)E_-^+(z)E_-^-(w) \otimes \frac{-kzw}{(z-w)^2} z^{-\frac{\alpha}{k}} w^{\frac{\alpha}{k}}
\end{aligned}
$$

Thus we get the computation as expected

$$
\begin{aligned}
[X(z), Y(w)] &= X(z)Y(w) - Y(w)X(z) \\
&= E_+^+(z)E_+^-(w)E_-^+(z)E_-^-(w) \otimes \big(\,:A_\alpha(z)A^*_{-\alpha}(w):(z-w)^{\frac{2}{k}} - :A^*_{-\alpha}(w)A_\alpha(z) \\
&\quad :(w-z)^{\frac{2}{k}}\big) z^{-\frac{\alpha}{k}} w^{\frac{\alpha}{k}} \\
&\quad + E_+^+(z)E_+^-(w)E_-^+(z)E_-^-(w) \otimes \big(\frac{-kzw}{(z-w)^2} - \frac{-kzw}{(w-z)^2}\big) z^{-\frac{\alpha}{k}} w^{\frac{\alpha}{k}} \\
&= -kE_+^+(z)E_+^-(w)E_-^+(z)E_-^-(w) z^{-\frac{\alpha}{k}} w^{\frac{\alpha}{k}} w\partial_w \delta(\frac{w}{z}) \\
&= -kw\partial_w \big(E_+^+(z)E_+^-(w)E_-^+(z)E_-^-(w) z^{-\frac{\alpha}{k}} w^{\frac{\alpha}{k}} \delta(\frac{w}{z})\big) \\
&\quad + kw\partial_w (E_+^+(z)E_+^-(w)E_-^+(z)E_-^-(w)) z^{-\frac{\alpha}{k}} w^{\frac{\alpha}{k}} \delta(\frac{w}{z}) \\
&\quad + kwE_+^+(z)E_+^-(w)E_-^+(z)E_-^-(w)\partial_w (z^{-\frac{\alpha}{k}} w^{\frac{\alpha}{k}})\delta(\frac{w}{z}) \\
&= -kw\partial_w \delta(\frac{w}{z}) + kw \sum_{n\neq 0} \frac{H(n)}{k} w^{n-1} w^{\frac{\alpha}{k}} z^{-\frac{\alpha}{k}} \delta(\frac{w}{z}) + \alpha w^{\frac{\alpha}{k}} z^{-\frac{\alpha}{k}} \delta(\frac{w}{z}) \\
&= -kw\partial_w \delta(\frac{w}{z}) + \sum_{n\neq 0} H(n) w^{-n} \delta(\frac{w}{z}) + H(0)\delta(\frac{w}{z}) \\
&= H(w)\delta(\frac{w}{z}) - kw\partial_w \delta(\frac{w}{z}).
\end{aligned}
$$

It is easy to compute that

$$
\begin{aligned}
[H(z), E_+^-(w)] &= \sum_{m\in\mathbb{Z}} [H(m), e^{\sum_{n>0} \frac{H(-n)}{kn} w^n}] z^{-m} = E_+^-(w) \sum_{\substack{m\in\mathbb{Z}\\ n>0}} \frac{[H(m), H(-n)]}{kn} z^{-m} w^n \\
&= E_+^-(w) \sum_{\substack{m\in\mathbb{Z}\\ n>0}} \frac{2mc\delta_{m-n,0}}{kn} z^{-m} w^n = E_+^-(w) \sum_{n>0} -2z^{-n} w^n.
\end{aligned}
$$

A similar calculation for $[H(z), E_+^+(w)]$, $[H(z), E_-^-(w)]$, $[H(z), E_-^+(w)]$ yields

$$\begin{aligned}
&[H(z), X(w)] = [H(z), E_+^+(w)]E_-^+(w) \otimes A_\alpha(w)e^\alpha w^{-\frac{\alpha}{k}} \\
&\quad + E_+^+(w)[H(z), E_-^+(w)] \otimes A_\alpha(w)e^\alpha w^{-\frac{\alpha}{k}} + E_+^+(w)E_-^+(w) \otimes A_\alpha(w)[H(0), e^\alpha]w^{-\frac{\alpha}{k}} \\
&= 2E_+^+(w)E_-^+(w)\Big(\sum_{n>0} z^{-n}w^n + \sum_{n<0} z^{-n}w^n + 1\Big) \otimes A_\alpha(w)e^\alpha w^{-\frac{\alpha}{k}} \\
&= 2E_+^+(w)E_-^+(w) \otimes A_\alpha(w)e^\alpha w^{-\frac{\alpha}{k}} = 2X(w)\delta(\frac{w}{z}).
\end{aligned}$$

It is immediate that $[H(z), Y(w)] = -2Y(w)\delta(\frac{w}{z})$. □

The action of c shows that the representation of $A_1^{(1)}$ just obtained has the level $-k$.

3.3 The Z-Algebra

Furthermore we can get the representation of Z-algebra as in [7]. We remark that this Z-algebra is fundamentally different from the Z-algebra in [21]. Taken the same definition of formal power series $Z^\pm(z)$, $x(\phi_1, z)$, $x(\phi_2, z)$:

$$Z^+(z) = Z(\alpha, z) = E_+^-(z)X(z)E_-^-(z),\ Z^-(z) = Z(-\alpha, z) = E_+^+(z)Y(z)E_-^+(z),$$

and the generalized commutator brackets

$$[\![x(\phi_1, z), x(\phi_2, z)]\!] = x(\phi_1, z)x(\phi_2, z)(1 - \frac{w}{z})^{\frac{(\phi_1, \phi_2)}{c}} - x(\phi_2, z)x(\phi_1, z)(1 - \frac{w}{z})^{\frac{(\phi_1, \phi_2)}{c}},$$

for $\phi_1, \phi_2 = \pm\alpha$, we can check that the lemmas given in paper [7] still hold. We state them here in Lemma 3.1 and Lemma 3.2.

Lemma 3.1 *Let Z-operators* $Z(z) = Z^+(z), Z^-(z)$, *we have that*

$$[E_+^\pm(z), Z(w)] = 0, \qquad [E_-^\pm(z), Z(w)] = 0 \tag{22}$$

Proof For $n \neq 0$, simple calculation yields

$$[H(n), X(w)] = \sum_{m\in\mathbb{Z}} [H(n), X(m)]w^{-m} = \sum_{m\in\mathbb{Z}} 2X(m+n)w^{-m} = 2X(w)w^n$$

and write $x_1 = \partial_s(e^{sx_1})|_{s=0}$, the following equations follow from $[x_1, e^{x_2}] = e^{x_2}[x_1, x_2]$

$$\begin{aligned}[H(n), E_-^-(w)] &= E_-^-(w)[H(n), \sum_{m>0} \tfrac{H(m)}{km} w^{-m}] \\ &= E_-^-(w) \sum_{m>0} \tfrac{2nc\delta_{m+n,0}}{km} w^{-m} = -2E_-^-(w) w^n \delta_{-m,n<0}, \\ [H(n), E_+^-(w)] &= -2E_+^-(w) w^n \delta_{m,n>0}.\end{aligned}$$

Note that

$$\begin{aligned}&[H(n), Z^+(w)] = [H(n), E_+^-(w)X(w)E_-^-(w)] \\ &= [H(n), E_+^-(w)]X(w)E_-^-(w) + E_+^-(w)[H(n), X(w)]E_-^-(w) \\ &\quad + E_+^-(w)X(w)[H(n), E_-^-(w)] \\ &= \big(-2w^n\delta_{m,n>0} + 2w^n + 2w^n\delta_{-m,n<0}\big)E_+^-(w)X(w)E_-^-(w) = 0.\end{aligned}$$

Similiarly, $[H(n), Z^-(w)] = [H(-n), Z^{\pm}(w)] = 0$, so

$$[E_-^+(z), Z^+(w)] = E_-^+(z)[\sum_{n>0} \frac{H(n)}{kn} z^{-n}, Z^+(w)] = 0$$

The calculation for the other brackets is similar. □

Lemma 3.2 *One has*

$$\begin{aligned}&[\![Z^{\pm}(z), Z^{\pm}(w)]\!] = 0 \\ &[\![Z^+(z), Z^-(w)]\!] = H(0)\delta(\tfrac{w}{z}) - kw\partial_w\delta(\tfrac{w}{z})\end{aligned} \tag{23}$$

Proof We calculate the product using Proposition 3.1 together with Lemma 3.1

$$\begin{aligned}Z^-(z)Z^-(w) &= Z^-(z)E_+^+(w)Y(w)E_-^+(w)E_+^+(w)Z^-(z)Y(w)E_-^+(w) \\ &= E_+^+(w)E_+^+(z)Y(z)\left(E_+^-(w)E_+^+(w)\right)E_-^+(z)Y(w)E^+(w) \\ &= E_+^+(w)E_+^+(z)Y(z)E_+^-(w)E_-^+(z)E_+^+(w)(1-\frac{w}{z})^{\frac{2}{k}}Y(w)E_-^+(w) \\ &= E_+^+(w)E_+^+(z)Y(z)E_+^-(w)\left(E_-^+(z)Z^-(w)\right)(1-\frac{w}{z})^{\frac{2}{k}} \\ &= E_+^+(w)E_+^+(z)Y(z)E_+^-(w)E_+^+(w)Y(w)E_-^+(w)E_-^+(z)(1-\frac{w}{z})^{\frac{2}{k}} \\ &= E_+^+(w)E_+^+(z)Y(z)Y(w)E_-^+(w)E_-^+(z)(1-\frac{w}{z})^{\frac{2}{k}}\end{aligned}$$

Consequently,

$$\begin{aligned}[\![Z^-(z)Z^-(w)]\!] &= Z^-(z)Z^-(w)(1-\frac{w}{z})^{-\frac{2}{k}} - Z^-(w)Z^-(z)(1-\frac{z}{w})^{-\frac{2}{k}} \\ &= E_+^+(z)E_+^+(w)[Y(z), Y(w)]E_-^+(z)E_-^+(w) = 0\end{aligned}$$

The calculation for $[\![Z^+(z), Z^+(w)]\!]$ is similar.
Also

$$
\begin{aligned}
Z^+(z)Z^-(w) &= E_+^+(w)E_+^-(z)X(z)Y(w)E_-^+(w)E_-^-(z)(1-\tfrac{w}{z})^{-\frac{2}{k}}\\
Z^-(w)Z^+(z) &= E_+^-(z)E_+^+(w)Y(w)X(z)E_-^-(z)E_-^+(w)(1-\tfrac{z}{w})^{-\frac{2}{k}}
\end{aligned}
$$

$$
\begin{aligned}
&[\![Z^+(z)Z^-(w)]\!] = Z^+(z)Z^-(w)(1-\frac{w}{z})^{\frac{2}{k}} - Z^-(w)Z^+(z)(1-\frac{z}{w})^{\frac{2}{k}}\\
&= E_+^+(w)E_+^-(z)[X(z)Y(w)]E_-^+(w)E_-^-(z)\\
&= E_+^-(z)E_+^+(w)H(w)\delta(\frac{w}{z})E_-^-(z)E_-^+(w) - E_+^-(z)E_+^+(w)kw\partial_w\delta(\frac{w}{z})E_-^-(z)E_-^+(w)\\
&= H(w)\delta(\frac{w}{z}) - kw\left(\partial_w(E_+^-(z)E_+^+(w)E_-^-(z)E_-^+(w)\delta(\frac{w}{z}))\right.\\
&\qquad\qquad \left.- \partial_w\big(E_+^-(z)E_+^+(w)E_-^-(z)E_-^+(w)\big)\delta(\frac{w}{z})\right)\\
&= \sum_{m\in\mathbb{Z}} H(m)w^{-m}\delta(\tfrac{w}{z}) - kw\partial_w\delta(\tfrac{w}{z}) + kw\sum_{m\neq 0}\tfrac{H(m)}{k}w^{-m-1}\delta(\tfrac{w}{z})\\
&= H(0)\delta(\frac{w}{z}) - kw\partial_w\delta(\frac{w}{z})
\end{aligned}
$$

□

For $A_1^{(1)}$-module $V = S(\mathfrak{h}'^-)\otimes\langle\Phi_{\alpha,0}(\omega,\overline{\omega})\rangle\otimes\mathbb{C}(\mathbb{Z}\alpha)$ in Theorem 3.1, we define the vacuum space $\Omega(V)$ of V by

$$\Omega(V) = \{v\in V, \eta'^+ = \oplus_{n>0}H(n) \mid \eta'^+\cdot v = 0\}.$$

Observe that we can decompose V by $V = S(\mathfrak{h}'^-)\otimes\Omega(V)$, then we get $\Omega(V) = \Phi_{\alpha,0}(\omega,\overline{\omega})\otimes\mathbb{C}(\mathbb{Z}\alpha)$ and furthermore.

Theorem 3.2 *The map $\pi_\Omega : Z \to gl(\Omega(V))$ gives a representation of Z-algebra on the vacuum space $\Omega(V)$ at level $-k$ via the action:*

$$
\begin{aligned}
Z^+(z) &\longmapsto A_\alpha(z)e^\alpha z^{-\frac{\alpha}{k}}\\
Z^-(z) &\longmapsto A^*_{-\alpha}(z)e^{-\alpha}z^{\frac{\alpha}{k}}
\end{aligned}
$$

.

Proof Under the map π we have

$$
\begin{aligned}
Z^+(z)Z^+(w) &\mapsto A_\alpha(z)e^\alpha z^{-\frac{\alpha}{k}}A_\alpha(w)e^\alpha w^{-\frac{\alpha}{k}}\\
&= A_\alpha(z)A_\alpha(w)e^{2\alpha}z^{-\frac{2}{k}}(zw)^{-\frac{\alpha}{k}}
\end{aligned}
$$

Therefore,

$$
\begin{aligned}
[\![Z^+(z),Z^+(w)]\!] &= Z^+(z)Z^+(w)(1-\frac{w}{z})^{-\frac{2}{k}} - Z^+(w)Z^+(z)(1-\frac{z}{w})^{-\frac{2}{k}}\\
&\mapsto \big(A_\alpha(z)A_\alpha(w)(z-w)^{-\frac{2}{k}} - A_\alpha(w)A_\alpha(z)(w-z)^{-\frac{2}{k}}\big)e^{2\alpha}(zw)^{-\frac{\alpha}{k}}
\end{aligned}
$$

$$= \big(: A_\alpha(z)A_\alpha(w) : (z-w)^{-\frac{2}{k}} - : A_\alpha(w)A_\alpha(z) : (w-z)^{-\frac{2}{k}}\big) e^{2\alpha}(zw)^{-\frac{\alpha}{k}}$$
$$+\big(\underbrace{A_\alpha(z)A_\alpha(w)}(z-w)^{-\frac{2}{k}} - \underbrace{A_\alpha(w)A_\alpha(z)}(w-z)^{-\frac{2}{k}}\big) e^{2\alpha}(zw)^{-\frac{\alpha}{k}} = 0$$

Similar calculations produce $[\![Z^-(z), Z^-(w) = 0]\!]$. Note that

$$Z^+(z)Z^-(w) \mapsto A_\alpha(z)e^\alpha z^{-\frac{\alpha}{k}} A^*_{-\alpha}(w)e^{-\alpha} w^{\frac{\alpha}{k}}$$
$$= A_\alpha(z)A^*_{-\alpha}(w) z^{\frac{2}{k}} z^{-\frac{\alpha}{k}} w^{\frac{\alpha}{k}}$$
$$Z^-(w)Z^+(z) \mapsto A^*_{-\alpha}(w)A_\alpha(z) w^{\frac{2}{k}} z^{-\frac{\alpha}{k}} w^{\frac{\alpha}{k}}$$

Then

$$[\![Z^+(z), Z^-(w)]\!] = Z^+(z)Z^-(w)(1-\frac{w}{z})^{\frac{2}{k}} - Z^-(w)Z^+(z)(1-\frac{z}{w})^{\frac{2}{k}}$$
$$\mapsto \big(A_\alpha(z)A^*{}_\alpha(w)(z-w)^{\frac{2}{k}} - A^*_{-\alpha}(w)A_\alpha(z)(w-z)^{\frac{2}{k}}\big) z^{-\frac{\alpha}{k}} w^{\frac{\alpha}{k}}$$
$$= \big(: A_\alpha(z)A^*_{-\alpha}(w) : (z-w)^{\frac{2}{k}} - : A^*_{-\alpha}(w)A_\alpha(z) : (w-z)^{\frac{2}{k}}\big) z^{-\frac{\alpha}{k}} w^{\frac{\alpha}{k}}$$
$$+\big(\underbrace{A_\alpha(z)A^*_{-\alpha}}(z-w)^{\frac{2}{k}} - \underbrace{A^*_{-\alpha}(w)A_\alpha(z)}(w-z)^{\frac{2}{k}}\big) z^{-\frac{\alpha}{k}} w^{\frac{\alpha}{k}}$$
$$= \big(\frac{-kzw}{(z-w)^2} - \frac{-kzw}{(w-z)^2}\big) z^{-\frac{\alpha}{k}} w^{\frac{\alpha}{k}} = -kw\partial_w \delta(\frac{w}{z}) z^{-\frac{\alpha}{k}} w^{\frac{\alpha}{k}}$$
$$= -kw\partial_w\big(\delta(\frac{w}{z}) z^{-\frac{\alpha}{k}} w^{\frac{\alpha}{k}}\big) + kw\delta(\frac{w}{z})\partial_w(z^{\frac{\alpha}{k}} w^{-\frac{\alpha}{k}})$$
$$= -kw\partial_w\delta(\frac{w}{z}) + az^{\frac{\alpha}{k}} w^{-\frac{\alpha}{k}}\delta(\frac{w}{z}) = H(0)\delta(\frac{w}{z}) - kw\partial_w\delta(\frac{w}{z}),$$

from which the theorem follows. □

References

1. Adamović, A.: Lie superalgebras and irreducibility of $A_1^{(1)}$-modules at the critical level. Commun. Math. Phys. **270**, 141–161 (2000)
2. Ding, X.M., Fan, H., Shi, K.J.: W_3-algebra constructed from the $SU(3)$ parafermion. Nucl. Phys. B **422**, 307–328 (1994)
3. Ding, X.M., Gould, M.D., Zhang, Y.Z.: Twisted parafermions. Phys. Lett. B **530**, 197–201 (2002)
4. Ding, X.M., Gould, M.D., Zhang, Y.Z.: $a_2^{(2)}$ parafermions: a new conformal field. Nucl. Phys. B **636**, 549–567 (2002)
5. Dong, C.Y., Lepowsky, J.: Generalized vertex algebras and relative vertex operators. Prog. Math, **112** (1993) (Birkhäuser, Boston)
6. Dong, C.Y., Wang, Q.: Parafermion vertex operator algebras. Front. Math. China **6**(4), 567–579 (2011)
7. Dunbar, J., Jing, N., Misra, K.: Realization of $\widehat{sl}_2(\mathbb{C})$ at the critical level and the Z-algebra. Commun. Contemp. Math. 16(2), 145006–145019 (2014)

8. Feigin, B.L., Frenkel, E.V.: Representations of affine Kac-Moody algebras and Bosonization. In: Brink, L., Freidan, D., Polyakov, A.M. (eds.) Physics and Mathematics of Strings, pp. 271–316. World Scientific, Singapore (1990)
9. Feingold, A., Frenkel, I.B.: Classical affine lie algebras. Adv. Math. **56**, 117–172 (1985)
10. Frenkel, I.B.: Spinor representations of affine lie algebras. Proc. Nat. Acad. Sci. U.S.A. **77**, 6303–6306 (1980)
11. Frenkel, I.B., Lepowsky, J., Meurman, A.: Vertex operator algebras and the monster. Pure Appl. Math. **134**, (1988) (Academic Press, Boston)
12. Frenkel, I.B., Kac, V.G.: Basic representations of affine lie algebras and dual resonance models. Invent. Math. **62**, 23–66 (1980)
13. Gepner, D.: New conformal filed theories associated with lie algebras and their partions functions. Nucl. Phys. B **290**, 10–24 (1987)
14. Jing, N.: Higher level representations of the quantum affine algebra $u_q(\widehat{sl}_2(\mathbb{C}))$. J. Alg. **182**, 448–468 (1996)
15. Kac, V.G., Kazhdan, D., Lepowsky, J., Wilson, R.L.: Realization of the basic representation of the euclidean lie algebras. Adv. Math. **42**, 83–112 (1981)
16. Kac, V.G., Peterson, D.: Spin and wedge representations of infinite-dimensional lie algebras and groups. Proc. Nat'l. Acad. Sci. U.S.A. **78**, 3308–3312 (1981)
17. ten Kroode, F., van de Leur, J.: Level one representations of the affine algebra $b_n^{(1)}$. Acta Appl. Math. **31**, 1–73 (1993)
18. Lepowsky, J., Wilson, R.L.: Construction of the affine lie algebra $a_1^{(1)}$. Commun. Math. Phys. **62**, 43–53 (1978)
19. Segal, G.: Unitary representations of some infinite dimensional groups. Commun. Math. Phys. **80**, 301–342 (1981)
20. Wakimoto, M.: Fock representations of the affine lie algebra $a_1^{(1)}$. Commun. Math. Phys. **104**, 605–609 (1986)
21. Wang, P., Zhu, C.Y.: Z-algebra construction of W-algebra in the parafermion model. Commun. Theor. Phys. **21**, 471–478 (1994)
22. Zamolodchikov, A.B., Fateev, V.A. : Nonlocal (parafermion) currents in two-dimensional conformal quantum field theory and self-dual critical points in Z_N-symmetric statistical systems. Sovet Phys. JETP **62**, 215–225 (1985)

Invariance and Symmetries of Cubic and Ternary Algebras

Richard Kerner

Abstract A class of Z_3-graded associative algebras with cubic Z_3-invariant constitutive relations is investigated. The invariant forms on finite algebras of this type are given in the low dimensional cases with two or three generators. We show how the Lorentz symmetry represented by the $SL(2, \mathbf{C})$ group emerges naturally without any notion of Minkowskian metric, just as the invariance group of the Z_3-graded cubic algebra and its constitutive relations. Its representation is found in terms of Pauli matrices. The relationship of this construction with the operators defining quark states is also considered.

1 Introduction

In most of the textbooks introducing the Lorentz-Poincaré group the accent is put on the transformation properties of space and time coordinates, and the invariance of the Minkowskian metric tensor $g_{\mu\nu} = diag(+, -, -, -)$. But neither the components of $g_{\mu\nu}$, nor the space-time coordinates of an observed event can be given an intrinsic physical meaning; they are not related to any conserved or directly observable quantities.

A more reliable physical content of Lorentz transformations is revealed when they are applied to the observable and measurable quantities such as electric charges and currents, or frequencies and wavelengths of electromagnetic waves. The Lorentz transformations apply directly to the four-current $j^\mu = [\rho c, \mathbf{j}]$ and to the four-vector $k^\mu = [\omega/c, \mathbf{k}]$. Two galilean observers comparing these quantities will arrive at the transformation property which is in agreement with charge conservation and the

R. Kerner (✉)
Laboratoire de Physique Théorique de la Matière Condensée, Université Pierre et Marie Curie (Paris-VI)-CNRS UMR 7600, Tour 23-13, B.C. 121, 4 Place Jussieu, Paris 75005, France
e-mail: richard.kerner@upmc.fr

A. Makhlouf et al. (eds.), *Algebra, Geometry and Mathematical Physics*, Springer Proceedings in Mathematics & Statistics 85, DOI: 10.1007/978-3-642-55361-5_37,

(relativistic) Doppler effect: in the appropriate units, one has

$$j^{\mu'} = \Lambda^{\mu'}_{\nu}\, j^{\nu}, \quad k^{\mu'} = \Lambda^{\mu'}_{\nu}\, k^{\nu}, \tag{1}$$

The differential form of the Lorentz force,

$$\frac{d\mathbf{p}}{dt} = q\,\mathbf{E} + q\,\frac{\mathbf{v}}{c} \wedge \mathbf{B} \tag{2}$$

combined with the energy conservation of a charged particle under the influence of electromagnetic field

$$\frac{d\mathscr{E}}{dt} = q\,\mathbf{E} \cdot \mathbf{v} \tag{3}$$

is also Lorentz-invariant:

$$dp^{\mu} = \frac{q}{mc}\, F^{\mu}_{\nu}\, p^{\nu} ds, \tag{4}$$

where $p^{\mu} = [p^0, \mathbf{p}]$ is the four-momentum and $F^{\mu}_{\ \nu}$ is the Maxwell-Faraday tensor. Reliable experimental confirmations of the validity of Lorentz transformations concern measurable quantities such as charges, currents, energies (frequencies) and momenta (wave vectors) much more than the less intrinsic quantities which are the *differentials* of the space-time variables. In principle, the Lorentz transformations could have been established by very precise observations of the Doppler effect alone.

It is often said that the vector space to which belong wave four-vectors k^{μ} is *dual* to the space-time in which we all live. But from the purely experimental point of view, the first vector space accessible for observation is the space of conserved wave vectors k^{μ}, and our space-time is its dual space. It should be also reminded that the Lorentz transformations first concerned the electromagnetic field and the forces it imposed on charged point-like masses, according to the formula (2).

Our questioning about the sources of Lorentz-Poincaré symmetry should not stop at the stage of *forces*, which are but expressions of effects of countless fundamental interactions, just like the thermodynamical pressure is in fact an averaged result of countless atomic collisions. On a classical level, when theory permits, the symbolical force can be replaced by a more explicit expression in which fields responsible for the forces do appear, like in the case of the Lorentz force (4).

But the fields acting on a test particle are usually generated by more or less distant charges and currents, according to the formula giving the retarded four-potential $A_{\mu}(x^{\lambda})$:

$$A_{\mu}(\mathbf{r}, t) = \frac{1}{4\pi c} \int\!\!\int\!\!\int \frac{j_{\mu}(\mathbf{r}', t - \frac{|\mathbf{r}-\mathbf{r}'|}{c})}{|\ \mathbf{r} - \mathbf{r}'\ |} d^3\mathbf{r}'. \tag{5}$$

Then we get the field tensor given by

$$F_{\mu\nu} = \partial_{\mu} A_{\nu} - \partial_{\nu} A_{\mu}.$$

The macroscopic currents are generated by electrons' collective motion. A single electron whose wave function is a bi-spinor gives rise to the Dirac current

$$j^{\mu} = \psi^{\dagger} \gamma^{\mu} \psi, \tag{6}$$

with

$$\psi^{\dagger} = \bar{\psi}^{T} \gamma^{5}, \quad \text{where} \quad \gamma^{5} = \gamma^{0}\gamma^{1}\gamma^{2}\gamma^{3} = \begin{pmatrix} I_2 & 0 \\ 0 & -I_2 \end{pmatrix}, \quad I_2 = \begin{pmatrix} 1 & 0 \\ 0 & 1 \end{pmatrix}.$$

In fact, the four-component complex function ψ is composed of two two-component spinors, ξ_{α} and $\chi_{\dot{\beta}}$ [1],

$$\psi = \begin{pmatrix} \xi \\ \chi \end{pmatrix},$$

which are supposed to transform under two non-equivalent representations of the $SL(2, \mathbb{C})$ group:

$$\xi_{\alpha'} = S^{\alpha}_{\alpha'} \xi_{\alpha}, \quad \chi_{\dot{\beta}'} = S^{\dot{\beta}}_{\dot{\beta}'} \chi_{\dot{\beta}'}, \tag{7}$$

The electric charge conservation is equivalent to the annulation of the four-divergence of j^{μ}:

$$\partial_{\mu} j^{\mu} = \left(\partial_{\mu} \psi^{\dagger} \gamma^{\mu}\right) \psi + \psi^{\dagger} \left(\gamma^{\mu} \partial_{\mu} \psi\right) = 0, \tag{8}$$

from which we infer that this condition will be satisfied if we have

$$\partial_{\mu} \psi^{\dagger} \gamma^{\mu} = -m \psi^{\dagger} \quad \text{and} \quad \gamma^{\mu} \partial_{\mu} \psi = m \psi, \tag{9}$$

which is the Dirac equation. In terms of the spinorial components ξ and χ the Dirac equation can be seen as a pair of two coupled equations which can be written in terms of Pauli's σ-matrices:

$$\left(-i\hbar \frac{1}{c} \frac{\partial}{\partial t} + mc\right) \xi = i\hbar \boldsymbol{\sigma} \cdot \nabla \chi,$$

$$\left(-i\hbar \frac{1}{c} \frac{\partial}{\partial t} - mc\right) \chi = i\hbar \boldsymbol{\sigma} \cdot \nabla \xi. \tag{10}$$

The relativistic invariance imposed on this equation is usually presented as follows: under a Lorentz transformation Λ the 4-current j^{μ} undergoes the following change:

$$j^{\mu} \rightarrow j^{\mu'} = \Lambda^{\mu'}_{\mu} j^{\mu}. \tag{11}$$

This means that the matrices γ^{μ} must transform as components of a 4-vector, too. Parallelly, the components of the bi-spinor ψ must be transformed in a way such as to leave the form of the Eq. (9) unchanged: writing symbolically the transformation of $\mid \psi >$ as $\mid \psi' >= S \mid \psi >$, and $< \psi' \mid = < \psi \mid S^{-1}$, we should have

$$j^{\mu'} = < \psi' \mid \gamma^{\mu'} \mid \psi' > = < \psi \mid S^{-1} \gamma^{\mu'} S \mid \psi > = \Lambda^{\mu'}_{\mu} j^{\mu} = \Lambda^{\mu'}_{\mu} < \psi \mid \gamma^{\mu} \mid \psi > \tag{12}$$

from which we infer the transformation rules for gamma-matrices:

$$S^{-1} \gamma^{\mu'} S = \Lambda^{\mu'}_{\mu} \gamma^{\mu}. \tag{13}$$

The relativistic quantum mechanics combines the electron and positron states in a single Dirac bi-spinor ψ comprising the two aforementioned two-component spinors, and transforming under a 4×4 representation of $SL(2, \mathbf{C})$ group.

The usual way of presenting the joint effect of a Lorentz transformation Λ on the coordinates and the wave function is as follows:

$$x^{\mu} \rightarrow x^{\mu'} = \Lambda^{\mu'}_{\nu} x^{\nu}, \quad \psi'(x^{\mu'}) = \psi'(\Lambda^{\mu'}_{\nu} x^{\nu}) = S(\Lambda) \psi(x^{\nu}). \tag{14}$$

This formula suggests that the transformation S of the states in the Hilbert space is imposed by the Lorentz transformation acting on the space-time coordinates, but in fact, decrypted from transformation properties of classical macro-objects such as wave vectors or the 4-momentum.

In view of the analysis of the causal chain, it seems more appropriate to write the same transformations with Λ depending on S:

$$\psi'(x^{\mu'}) = \psi'(\Lambda^{\mu'}_{\nu}(S) x^{\nu}) = S \psi(x^{\nu}) \tag{15}$$

This form of the same relation suggests that the transition form one quantum state to another, represented by the unitary transformation S is the primary cause that implies the transformation of observed quantities such as the electric 4-current, and as a final consequence, the apparent transformations of time and space intervals measured with physical devices.

Although mathematically the two formulations are equivalent, it seems more plausible that the Lorentz group resulting from the averaging of the action of the $SL(2, \mathbf{C})$ in the Hilbert space of states contains less information than the original double-valued representation, than the other way round.

In what follows, we shall draw physical consequences from this approach.

2 Pauli's Exclusion Principle and the $SL(2, \mathbf{C})$ Group

Our Universe, such as we know it now, could never exist without fermionic matter. If bosons were only present, they would all condensate in one common fundamental state, and only small fluctuations around that state would constitute the history of the Universe. On the contrary, fermions tend to occupy different states whenever possible, thus enlarging the overall configuration space.

The Pauli exclusion principle [2], according to which two electrons cannot be in the same state characterized by identical quantum numbers, is one of the most important cornerstones of quantum physics. This principle not only explains the structure of atoms and therefore the entire content of the periodic table of elements, but it also guarantees the stability of matter preventing its collapse, as suggested by Ehrenfest [3], and proved later by Dyson [3, 4]. The relationship between the exclusion principle and particle's spin, known under the name of the "spin-and-statistic theorem", represents one of the deepest results in quantum field theory [1].

In purely algebraical terms Pauli's exclusion principle amounts to the anti-symmetry of wave functions describing two coexisting particle states. The easiest way to see how the principle works is to apply Dirac's formalism in which wave functions of particles in given state are obtained as products between the "bra" and "ket" vectors [5].

Consider the probability amplitude to find a particle in the state $\mid x >$,

$$\Phi(x) =< \psi \mid x > . \tag{16}$$

The wave function of a two-particle state of which one is in the state $\mid x >$ and another in the state $\mid y >$ is represented by a superposition

$$\mid \psi >= \sum \Phi(x, y) \, (\mid x > \otimes \mid y >). \tag{17}$$

It is clear that if the wave function $\Phi(x, y)$ is anti-symmetric, i.e. if it satisfies

$$\Phi(x, y) = -\Phi(y, x), \tag{18}$$

then $\Phi(x, x) = 0$ and such states have vanishing both their wave function and probability. It is easy to prove using the superposition principle, that this condition is not only sufficient, but also necessary. Let us suppose that $\Phi(x, x)$ vanish. This should remain valid in any basis provided the new basis $\mid x' >, \mid y' >$ was obtained from the former one via an unitary transformation. Let us form an arbitrary state being a linear combination of $\mid x >$ and $\mid y >$,

$$\mid z >= \alpha \mid x > +\beta \mid y >, \quad \alpha, \beta \in \mathbf{C},$$

and let us form the wave function of a tensor product of such a state with itself:

$$\Phi(z, z) =< \psi \mid (\alpha \mid x > +\beta \mid y >) \otimes (\alpha \mid x > +\beta \mid y >), \tag{19}$$

which develops as follows:

$$\begin{aligned} &\alpha^2 < \psi \mid (x, x) > +\alpha\beta < \psi \mid (x, y) > +\beta\alpha < \psi \mid (y, x) > +\beta^2 < \psi \mid (y, y) >= \\ &= \Phi(x, y) = \alpha^2 \, \Phi(x, x) + \alpha\beta \, \Phi(x, y) + \beta\alpha \, \Phi(y, x) + \beta^2 \, \Phi(y, y). \end{aligned} \tag{20}$$

Now, as $\Phi(x, x) = 0$ and $\Phi(y, y) = 0$, the sum of remaining two terms will vanish if and only if (18) is satisfied, i.e. if $\Phi(x, y)$ is anti-symmetric in its two arguments.

After second quantization, when the states are obtained with creation and annihilation operators acting on the vacuum, the anti-symmetry is encoded in the anti-commutation relations

$$\psi(x)\psi(y) + \psi(y)\psi(x) = 0 \tag{21}$$

where $\psi(x) \mid 0 >=\mid \psi >$.

Now, according to the experiment, electrons having identical energy and momenta can still display two different states; in fact, this is the only possibility for two electrons to occupy otherwise identical states. This is why for a given principal quantum number n there are only $2n^2$ possible electron states in the corresponding electron shell. Therefore, if these states (which are just two opposite directions of spin) are labeled $\mid 1 >$ and $\mid 2 >$, their tensor product should contain only the anti-symmetric sector,

$$\mid 1 > \otimes \mid 2 >= - \mid 2 > \otimes \mid 1 >,$$

This property can be also expressed by admitting the existence of an anti-symmetric two-form in the Hilbert space of two-electron states, which can be normalized to 1 as follows:

$$\varepsilon_{\alpha\beta} = -\varepsilon_{\beta\alpha}, \quad \alpha, \beta = 1, 2; \quad \varepsilon^{12} = -\varepsilon^{21} = 1, \quad \varepsilon^{11} = 0, \quad \varepsilon^{22} = 0.$$

According to the superposition principle, another basis in the Hilbert space of two-electron states can be chosen; however, the Pauli principle should hold independently of such transformation. One should have then, after a linear transformation

$$\mid \psi_\alpha > \rightarrow \mid \psi_{\alpha'} >= S^{\alpha}_{\alpha'} \mid \psi_\alpha >,$$

the same anti-symmetric 2-form:

$$\varepsilon_{\alpha'\beta'} = S^{\alpha}_{\alpha'} S^{\beta}_{\beta'} \varepsilon_{\alpha\beta} \tag{22}$$

Requiring the invariance of the form $\varepsilon^{\alpha\beta}$, i.e. postulating that $\varepsilon^{\alpha'\beta'}$ has the same components as before,

$$\varepsilon_{1'2'} = -\varepsilon_{2'1'} = 1, \quad \varepsilon_{1'1'} = 0, \quad \varepsilon_{2'2'} = 0$$

leads to the unique condition on the components of the complex 2×2 matrix S, namely

$$\det S = 1, \tag{23}$$

which determines the group $SL(2,\mathbf{C})$ as the invariance group of tensor products of electron states.

The existence of the positron implies the existence of a different, although similar sector of Hilbert space of states. The complex conjugate matrices of the $SL(2,\mathbf{C})$ group yield another representation, which is not equivalent. It acts on the complex conjugate spinors, whose indeces are labeled with dots, to make the difference clearly visible [1]. We have therefore a skew-symmetric two-form

$$\varepsilon_{\dot{\alpha}\dot{\beta}} = -\varepsilon_{\dot{\beta}\dot{\alpha}}, \quad \dot{\alpha},\dot{\beta} = 1,2; \qquad \varepsilon^{\dot{1}\dot{2}} = -\varepsilon^{\dot{2}\dot{1}} = 1, \quad \varepsilon^{\dot{1}\dot{1}} = 0, \quad \varepsilon^{\dot{2}\dot{2}} = 0.$$

3 Ternary Generalization of Pauli's Principle

The electrons and positrons satisfying Dirac's equation are considered as elementary particles not only because their propagation is well described by the solutions of this equation and because they satisfy other physical predictions like the gyromagnetic ratio equal to 2, but also because there is no experimental evidence of any internal structure.

The situation is quite different when protons and neutrons are being considered. Although at first approximation their behavior can be also described quite successfully by the same Dirac equation, only with different mass parameter (and zero electric charge in the case of the neutron), their physical parameters other than spin do not display values imposed by the Dirac equation. The magnetic momentum is different and does not have the required gyromagnetic ratio. Moreover, high energy experiments known as the *deep inelastic scattering* show that the gamma-photons with very high energy penetrate inside the nucleon and are scattered by almost point-like entities, whose characteristic dimensions must be at least three orders smaller than that of the proton itself: about 10^{-16} cm versus 10^{-13} cm. These conctituents of nucleons are called *quarks*, and their characterstic dimensions are close to the size of the electron. Apparently, the Lorentz symmetry is valid also for quarks; however, like in the case of the electrons, one may think that it is not imposed by what is happening in the macroscopic world, but is a result of the action of certain form of $SL(2,\mathbf{C})$ in the Hilbert space of quantum states of quarks.

In the formalism called "Quantum Chromo-Dynamics" (QCD) quarks are considered as fermions, endowed with spin $\frac{1}{2}$. Free quarks are inaccessible for direct observation, only *three* quarks or anti-quarks can coexist inside a fermionic baryon (respectively, anti-baryon), and a pair quark-antiquark can form a meson with integer spin. Besides, they have to belong to different *colors*, also a three-valued set. There are two quarks in the first generation, u and d ("up" and "down"), which may be considered as two states of a more general object, just like proton and neutron are regarded upon as two isospin component of a doublet called "nucleon". With this in mind we see that in the same bound state there is place for *two* quarks in the same u-state or d-state, but not three.

This suggests that a convenient generalization of Pauli's exclusion principle would be the statement that no *three* quarks in the same state can be present in a nucleon. Let us require then the vanishing of wave functions corresponding to the tensor product of *three* (but not necessarily two) identical states. That is, we require that $\Phi(x,x,x) = 0$ for any state $| x >$. As in the former case, it is easy to prove that the necessary symmetry condition for that to be true in any basis is

$$\Phi(x,y,z) + \Phi(y,z,x) + \Phi(z,x,y) = 0. \tag{24}$$

Let us consider an arbitrary superposition of three different states, $| x >$, $| y >$ and $| z >$,

$$| w >= \alpha \, | x > + \beta \, | y > + \gamma \, | z >$$

and apply the same criterion, $\Phi(w,w,w) = 0$. We get then, after developing the tensor products,

$$\begin{aligned}\Phi(w,w,w) &= \alpha^3 \Phi(x,x,x) + \beta^3 \Phi(y,y,y) + \gamma^3 \Phi(z,z,z)\\ &+ \alpha^2\beta[\Phi(x,x,y) + \Phi(x,y,x) + \Phi(y,x,x)] + \gamma\alpha^2[\Phi(x,x,z) + \Phi(x,z,x) + \Phi(z,x,x)]\\ &+ \alpha\beta^2[\Phi(y,y,x) + \Phi(y,x,y) + \Phi(x,y,y)] + \beta^2\gamma[\Phi(y,y,z) + \Phi(y,z,y) + \Phi(z,y,y)]\\ &+ \beta\gamma^2[\Phi(y,z,z) + \Phi(z,z,y) + \Phi(z,y,z)] + \gamma^2\alpha[\Phi(z,z,x) + \Phi(z,x,z) + \Phi(x,z,z)]\\ &+ \alpha\beta\gamma[\Phi(x,y,z) + \Phi(y,z,x) + \Phi(z,x,y) + \Phi(z,y,x) + \Phi(y,x,z) + \Phi(x,z,y)] = 0.\end{aligned} \tag{25}$$

The three diagonal expressions $\Phi(x,x,x)$, $\Phi(y,y,y)$ and $\Phi(z,z,z)$ vanish by virtue of the original assumption; in what remains, every combination preceded by an independent powers of three independent numerical coefficients α, β and γ, must vanish separately.

This can be achieved if the following Z_3 symmetry is imposed on the wave functions of three arguments:

$$\Phi(x,y,z) = j\,\Phi(y,z,x) = j^2\,\Phi(z,x,y), \quad \text{with} \;\; j = e^{\frac{2\pi i}{3}} \tag{26}$$

where $j = e^{\frac{2\pi i}{3}}$ is the cubic root of unity, satisfying $j^3 = 1, \quad j + j^2 + 1 = 0$ Note that the complex conjugates of functions $\Psi(x,y,z)$ transform under cyclic permutations of their arguments with $j^2 = \bar{j}$ replacing j in the formula (26):

$$\Phi(x,y,z) = j^2\,\Phi(y,z,x) = j\,\Phi(z,x,y). \tag{27}$$

In terms of operators acting on vacuum state producing states with definite number of quarks (or antiquarks) this property will be translated into the following *cubic commutation relation* generalizing Pauli's principle in the Z_3-graded case [6, 7]:

$$\theta^A\theta^B\theta^C = j\,\theta^B\theta^C\theta^A = j^2\,\theta^C\theta^A\theta^B, \tag{28}$$

with $j = e^{2i\pi/3}$, the primitive root of 1. We have $1 + j + j^2 = 0$ and $\bar{j} = j^2$.

We shall also introduce a similar set of *conjugate* generators, $\bar{\theta}^{\dot{A}}$, $\dot{A}, \dot{B}, ... = 1, 2, ..., N$, satisfying similar condition with j^2 replacing j:

$$\bar{\theta}^{\dot{A}}\bar{\theta}^{\dot{B}}\bar{\theta}^{\dot{C}} = j^2\,\bar{\theta}^{\dot{B}}\bar{\theta}^{\dot{C}}\bar{\theta}^{\dot{A}} = j\,\bar{\theta}^{\dot{C}}\bar{\theta}^{\dot{A}}\bar{\theta}^{\dot{B}}, \tag{29}$$

A direct consequence of these constitutive relations is the impossibility of forming products of more than three quark or anti-quark operators. The proof is straightforward, using the associativity:

$$\theta^A\theta^B\theta^C\theta^D = j\,\theta^B\theta^C\theta^A\theta^D = j^2\,\theta^B\theta^A\theta^D\theta^C = j^3\,\theta^A\theta^D\theta^B\theta^C = j^4\,\theta^A\theta^B\theta^C\theta^D,$$

and because $j^4 = j \neq 1$, the only solution is $\theta^A\theta^B\theta^C\theta^D = 0$.

In order to make the constitutive relations for the set of operators θ^A and $\bar{\theta}^{\dot{B}}$ complete, we have to impose *binary* commutation relations between the "quark" and "anti-quark" generators. The two sets can be united in a common algebra if we decide which commutation relations should be imposed on the binary products $\theta^A\bar{\theta}^{\dot{B}}$ and $\bar{\theta}^{\dot{B}}\theta^A$.

Looking at the cubic commutation relations (28) and (29) we see that the product of tho θ's behave as an element of grade 2; however, we would like to make a clear distinction between such a product and a conjugate element $\bar{\theta}^{\dot{B}}$. Luckily enough, there exists such a possibility, namely, to require the following commutation relations:

$$\theta^A\bar{\theta}^{\dot{B}} = -j\,\bar{\theta}^{\dot{B}}\theta^A, \quad \bar{\theta}^{\dot{B}}\theta^A = -j^2\,\theta^A\bar{\theta}^{\dot{B}}, \tag{30}$$

as proposed in [8].

4 The Covariance Principle Applied to Finite Groups

We shall concentrate our attention on two simplest permutation groups, S_2 and S_3. The S_2 group contains only two elements, the identity keeping two items unchanged, and the only non-trivial permutation of two items, $(ab) \to (ba)$. This permutation is cyclic, so the S_2 group coincides with its Z_2 subgroup.

The simplest representations of the Z_2 group are realised via its actions on the complex numbers, $\mathbf{C}^1$. Three different inversions can be introduced, each of them generating a different representation of Z_2 in the complex plane $\mathbf{C}^1$:

(i) the sign inversion, $z \to -z$;
(ii) complex conjugation, $z \to \bar{z}$;
(iii) the combination of both, $z \to -\bar{z}$.

One should not forget about the fourth possibility, the trivial representation attributing the identity transformation to the two elements of the group, including the non-trivial one:

(iv) the identity transformation, $z \to z$;

Let us recall once again the principle of covariance:

Any meaningful quantity described by a set of functions $\psi^A(x^\mu)$, $A, B, ... = 1, 2, ..., N$, $\mu, \nu, ... = 0, 1, 2, 3$ defined on the Minkowskian space-time must be a representation of the Lorentz group, i.e. it should transform following one of its representations:

$$\psi^{A'}(x^{\mu'}) = \psi^{A'}(\Lambda^{\mu'}_{\rho}\, x^\rho) = S^{A'}_B(\Lambda^{\mu'}_{\rho})\,\psi^B(x^\rho). \tag{31}$$

which can be written even more concisely,

$$\psi(x') = S(\Lambda)(\psi(x)). \tag{32}$$

The important assumption here being the representation property of the linear transformations $S(\Lambda)$:

$$S(\Lambda_1)S(\Lambda_2) = S(\Lambda_1\Lambda_2). \tag{33}$$

The same principle can be applied in the discrete case, when continuous variables are replaced by indeces, and the group of continuous transformations by permutations.

In the case of the S_2 group, instead of a set of functions defined on the space-time, we consider the mapping of two indices into the complex numbers, i.e. a matrix or a two-valenced complex-valued tensor. Under the non-trivial permutation π of indices its value should change according to one of the possible representations of Z_2 in the complex plane. This leads to the following four possibilities:

(i) The trivial representation defines the symmetric tensors:

$$S_{\pi(AB)} = S_{BA} = S_{AB},$$

(ii) The sign inversion defines the anti-symmetric tensors:

$$A_{\pi(CD)} = A_{DC} = -A_{CD},$$

(iii) The complex conjugation defines the hermitian tensors:

$$H_{\pi(AB)} = H_{BA} = \bar{H}_{AB},$$

(iv) $(-1)\times$ complex conjugation defines the anti-hermitian tensors.

$$T_{\pi(AB)} = T_{BA} = -\bar{T}_{AB},$$

Table 1 The multiplication table for the S_3 symmetric group

	1	j	j^2	$-$	$\wedge$	$*$
1	1	j	j^2	$-$	$\wedge$	$*$
j	j	j^2	1	$*$	$-$	$\wedge$
j^2	j^2	1	j	$\wedge$	$*$	$-$
$-$	$-$	$\wedge$	$*$	1	j	j^2
$\wedge$	$\wedge$	$*$	$-$	j^2	1	j
$*$	$*$	$-$	$\wedge$	j	j^2	1

Consider now the symmetric group S_3.

The group S_3 containing all permutations of three different elements is a special case among all symmetry groups S_N. It is exceptional because it is the first in the row to be non-abelian, and the last one that possesses a faithful representation in the complex place $\mathbf{C}^1$.

It contains six elements, and can be generated with only two elements, corresponding to one cyclic and one odd permutation, e.g. $(abc) \to (bca)$, and $(abc) \to (cba)$. All permutations can be represented as different operations on complex numbers as follows.

Let us denote the primitive third root of unity by $j = e^{2\pi i/3}$.

The cyclic abelian subgroup Z_3 contains three elements corresponding to the three cyclic permutations, which can be represented via multiplication by j, j^2 and $j^3 = 1$ (the identity).

$$\begin{pmatrix} ABC \\ ABC \end{pmatrix} \to \mathbf{1}, \quad \begin{pmatrix} ABC \\ BCA \end{pmatrix} \to \mathbf{j}, \quad \begin{pmatrix} ABC \\ CAB \end{pmatrix} \to \mathbf{j}^2, \tag{34}$$

Odd permutations must be represented by idempotents, i.e. by operations whose square is the identity operation. We can make the following choice:

$$\begin{pmatrix} ABC \\ CBA \end{pmatrix} \to (\mathbf{z} \to \bar{\mathbf{z}}), \quad \begin{pmatrix} ABC \\ BAC \end{pmatrix} \to (\mathbf{z} \to \hat{\mathbf{z}}), \quad \begin{pmatrix} ABC \\ CBA \end{pmatrix} \to (\mathbf{z} \to \mathbf{z}^*), \tag{35}$$

Here the bar $(\mathbf{z} \to \bar{\mathbf{z}})$ denotes the complex conjugation, i.e. the reflection in the real line, the hat $\mathbf{z} \to \hat{\mathbf{z}}$ denotes the reflection in the root j^2, and the star $\mathbf{z} \to \mathbf{z}^*$ the reflection in the root j. The six operations close in a non-abelian group with six elements, and the corresponding multiplication rules is shown in Table 1:

As in the Z_2 case, one can define the Z_3-irreducible three-forms.

There are three possibilities of an action of Z_3 being represented by multiplication by a complex number: the trivial one (multiplication by 1), and the two other representations, the multiplication by $j = \mathrm{e}^{2\pi i/3}$ or by its complex conjugate,

$$j^2 = \bar{j} = \mathrm{e}^{4\pi i/3}$$

$$T \in \mathcal{T}: \quad T_{ABC} = T_{BCA} = T_{CAB}, \tag{36}$$

(totally symmetric)

$$S \in \mathscr{S}: \quad S_{ABC} = j\, S_{BCA} = j^2\, S_{CAB}, \tag{37}$$

(j-skew-symmetric)

$$\bar{S} \in \bar{\mathscr{S}}; \quad \bar{S}_{ABC} = j^2\, \bar{S}_{BCA} = j\, \bar{S}_{CAB}, \tag{38}$$

(j^2-skew-symmetric).

The space of all tri-linear forms is the sum of three irreducible subspaces,

$$\Theta_3 = \mathscr{T} \oplus \mathscr{S} \oplus \bar{\mathscr{S}}$$

the corresponding dimensions being, respectively, $(N^3 + 2N)/3$ for $\mathscr{T}$ and $(N^3 - N)/3$ for $\mathscr{S}$ and for $\bar{\mathscr{S}}$.

Any three-form W^{α}_{ABC} maps $\mathscr{A} \otimes \mathscr{A} \otimes \mathscr{A}$ into a vector space $\mathscr{X}$ of dimension k, $\alpha, \beta = 1, 2, ...k$, so that $X^{\alpha} = W^{\alpha}_{ABC}\, \theta^A \theta^B \theta^C$ can be represented as a linear combination of forms with specific symmetry properties,

$$W^{\alpha}_{ABC} = T^{\alpha}_{ABC} + S^{\alpha}_{ABC} + \bar{S}^{\alpha}_{ABC},$$

$$T^{\alpha}_{ABC} := \frac{1}{3}\,(W^{\alpha}_{ABC} + W^{\alpha}_{BCA} + W^{\alpha}_{CAB}), \tag{39}$$

$$S^{\alpha}_{ABC} := \frac{1}{3}\,(W^{\alpha}_{ABC} + j\, W^{\alpha}_{BCA} + j^2\, W^{\alpha}_{CAB}), \tag{40}$$

$$\bar{S}^{\alpha}_{ABC} := \frac{1}{3}\,(W^{\alpha}_{ABC} + j^2\, W^{\alpha}_{BCA} + j\, W^{\alpha}_{CAB}), \tag{41}$$

As in the Z_2 case, the three symmetries above define irreducible 3-forms.

5 General Definition of Invariant Forms

Let us consider multilinear forms defined on the algebra $\mathscr{A} \otimes \bar{\mathscr{A}}$. Because only cubic relations are imposed on products in $\mathscr{A}$ and in $\bar{\mathscr{A}}$, and the binary relations on the products of ordinary and conjugate elements, we shall fix our attention on tri-linear and bi-linear forms, conceived as mappings of $\mathscr{A} \otimes \bar{\mathscr{A}}$ into certain linear spaces over complex numbers.

Let us consider a tri-linear form ρ^{α}_{ABC}. We shall call this form Z_3-invariant if we can write:

$$\rho^{\alpha}_{ABC}\,\theta^A\theta^B\theta^C = \frac{1}{3}\left[\rho^{\alpha}_{ABC}\,\theta^A\theta^B\theta^C + \rho^{\alpha}_{BCA}\,\theta^B\theta^C\theta^A + \rho^{\alpha}_{CAB}\,\theta^C\theta^A\theta^B\right] =$$
$$= \frac{1}{3}\left[\rho^{\alpha}_{ABC}\,\theta^A\theta^B\theta^C + \rho^{\alpha}_{BCA}\,(j^2\,\theta^A\theta^B\theta^C) + \rho^{\alpha}_{CAB}\,j\,(\theta^A\theta^B\theta^C)\right], \tag{42}$$

by virtue of the commutation relations (28).

From this it follows that we should have

$$\rho^{\alpha}_{ABC}\;\theta^A\theta^B\theta^C = \frac{1}{3}\left[\rho^{\alpha}_{ABC} + j^2\,\rho^{\alpha}_{BCA} + j\,\rho^{\alpha}_{CAB}\right]\theta^A\theta^B\theta^C, \tag{43}$$

from which we get the following properties of the ρ-cubic matrices:

$$\rho^{\alpha}_{ABC} = j^2\,\rho^{\alpha}_{BCA} = j\,\rho^{\alpha}_{CAB}. \tag{44}$$

Even in this minimal and discrete case, there are covariant and contravariant indices: the lower and the upper indices display the inverse transformation property. If a given cyclic permutation is represented by a multiplication by j for the upper indices, the same permutation performed on the lower indices is represented by multiplication by the inverse, i.e. j^2, so that they compensate each other [9].

Similar reasoning leads to the definition of the conjugate forms $\bar{\rho}^{\dot{\alpha}}_{\dot{C}\dot{B}\dot{A}}$ satisfying the relations similar to (44) with j replaced be its conjugate, j^2:

$$\bar{\rho}^{\dot{\alpha}}_{\dot{A}\dot{B}\dot{C}} = j\,\bar{\rho}^{\dot{\alpha}}_{\dot{B}\dot{C}\dot{A}} = j^2\,\bar{\rho}^{\dot{\alpha}}_{\dot{C}\dot{A}\dot{B}} \tag{45}$$

In the simplest case of two generators, the j-skew-invariant forms have only two independent components:

$$\rho^1_{121} = j\,\rho^1_{211} = j^2\,\rho^1_{112},$$
$$\rho^2_{212} = j\,\rho^2_{122} = j^2\,\rho^2_{221},$$

and we can set

$$\rho^1_{121} = 1,\;\; \rho^1_{211} = j^2,\;\; \rho^1_{112} = j,$$
$$\rho^2_{212} = 1,\;\; \rho^2_{122} = j^2,\;\; \rho^2_{221} = j.$$

The anti-symmetric tensor ε^{AB} and its inverse ε_{BC} enable us to define the dual ρ-matrices:

$$\rho^{ABC}_{\beta} = \varepsilon_{\alpha\beta}\rho^{\alpha}_{DEF}\;\varepsilon^{AD}\varepsilon^{BE}\varepsilon^{CF}. \tag{46}$$

It is easy to check that the dual cubic matrices ρ^{ABC}_{α} have exactly the same properties as the original ones, ρ^{β}_{DEF}, with indeces 1 and 2 interchanged.

6 The Symmetry Group of Invariant 3-Forms

The idea of covariance can be applied now to cubic algebraic structures introduced in the Sect. 2.

The constitutive cubic relations between the generators of the Z_3 graded algebra can be considered as intrinsic if they are conserved after linear transformations with commuting (pure number) coefficients, i.e. if they are independent of the choice of the basis.

Let $U^{A'}_A$ denote a non-singular $N \times N$ matrix, transforming the generators θ^A into another set of generators, $\theta^{B'} = U^{B'}_B \, \theta^B$.

We are looking for the solution of the covariance condition for the ρ-matrices:

$$\Lambda^{\alpha'}_{\beta} \, \rho^{\beta}_{ABC} = U^{A'}_A \, U^{B'}_B \, U^{C'}_C \, \rho^{\alpha'}_{A'B'C'}. \tag{47}$$

Now, $\rho^1_{121} = 1$, and we have two equations corresponding to the choice of values of the index α' equal to 1 or 2. For $\alpha' = 1'$ the ρ-matrix on the right-hand side is $\rho^{1'}_{A'B'C'}$, which has only three components,

$$\rho^{1'}_{1'2'1'} = 1, \quad \rho^{1'}_{2'1'1'} = j^2, \quad \rho^{1'}_{1'1'2'} = j,$$

which leads to the following equation:

$$\Lambda^{1'}_1 = U^{1'}_1 \, U^{2'}_2 \, U^{1'}_1 + j^2 \, U^{2'}_1 \, U^{1'}_2 \, U^{1'}_1 + j \, U^{1'}_1 \, U^{1'}_2 \, U^{2'}_1 = U^{1'}_1 \, (U^{2'}_2 \, U^{1'}_1 - U^{2'}_1 \, U^{1'}_2), \tag{48}$$

because $j^2 + j = -1$. For the alternative choice $\alpha' = 2'$ the ρ-matrix on the right-hand side is $\rho^{2'}_{A'B'C'}$, whose three non-vanishing components are

$$\rho^{2'}_{2'1'2'} = 1, \quad \rho^{2'}_{1'2'2'} = j^2, \quad \rho^{2'}_{2'2'1'} = j.$$

The corresponding equation becomes now:

$$\Lambda^{2'}_1 = U^{2'}_1 \, U^{1'}_2 \, U^{2'}_1 + j^2 \, U^{1'}_1 \, U^{2'}_2 \, U^{2'}_1 + j \, U^{2'}_1 \, U^{2'}_2 \, U^{1'}_1 = U^{2'}_1 \, (U^{1'}_2 \, U^{2'}_1 - U^{1'}_1 \, U^{2'}_2), \tag{49}$$

The two remaining equations are obtained in a similar manner. We choose now the three lower indices on the left-hand side equal to another independent combination, (212). Then the ρ-matrix on the left hand side must be ρ^2 whose component ρ^2_{212} is equal to 1. This leads to the following equation when $\alpha' = 1'$:

$$\Lambda^{1'}_2 = U^{1'}_2 \, U^{2'}_1 \, U^{1'}_2 + j^2 \, U^{2'}_2 \, U^{1'}_1 \, U^{1'}_2 + j \, U^{1'}_2 \, U^{1'}_1 \, U^{2'}_2 = U^{1'}_2 \, (U^{1'}_2 \, U^{2'}_1 - U^{1'}_1 \, U^{2'}_2), \tag{50}$$

and the fourth equation corresponding to $\alpha' = 2'$ is:

$$\Lambda_2^{2'} = U_2^{2'} U_1^{1'} U_2^{2'} + j^2 U_2^{1'} U_1^{2'} U_2^{2'} + j U_2^{2'} U_1^{2'} U_2^{1'} = U_2^{2'} (U_1^{1'} U_2^{2'} - U_1^{2'} U_2^{1'}). \tag{51}$$

$$\Lambda_1^{2'} = -U_1^{2'} [\det(U)], \tag{52}$$

The remaining two equations are obtained in a similar manner, resulting in the following:

$$\Lambda_2^{1'} = -U_2^{1'} [det(U)], \quad \Lambda_2^{2'} = U_2^{2'} [\det(U)]. \tag{53}$$

The determinant of the 2×2 complex matrix $U_B^{A'}$ appears everywhere on the right-hand side. Taking the determinant of the matrix $\Lambda_\beta^{\alpha'}$ one gets immediately

$$\det(\Lambda) = [\det(U)]^3. \tag{54}$$

However, the U-matrices on the right-hand side are defined only up to the phase, which due to the cubic character of the covariance relations and they can take on three different values: 1, j or j^2, i.e. the matrices $j\, U_B^{A'}$ or $j^2\, U_B^{A'}$ satisfy the same relations as the matrices $U_B^{A'}$ defined above. The determinant of U can take on the values 1, j, or j^2 if $\det(\Lambda) = 1$.

Up to this point, there is no reason yet to impose the unitarity condition. It can be derived if we require the same behavior for the duals, ρ_β^{DEF}. This extra condition amounts to the invariance of the anti-symmetric tensor ε^{AB}, and this is possible only if the determinant of U-matrices is 1 (or j or j^2, because only cubic combinations of these matrices appear in the tranformation law for ρ-forms.

A similar covariance requirement can be formulated with respect to the set of 2-forms mapping the quadratic quark-anti-quark combinations into a four-dimensional linear real space. As we already saw, the symmetry (30) imposed on these expressions reduces their number to four. Let us define two quadratic forms, $\pi^\mu_{A\dot{B}}$ and its conjugate $\bar{\pi}^\mu_{\dot{B}A}$

$$\pi^\mu_{A\dot{B}}\, \theta^A \bar{\theta}^{\dot{B}} \quad \text{and} \quad \bar{\pi}^\mu_{\dot{B}A}\, \bar{\theta}^{\dot{B}} \theta^A. \tag{55}$$

The Greek indices μ, ν... take on four values, and we shall label them 0, 1, 2, 3.

The four tensors $\pi^\mu_{A\dot{B}}$ and their hermitian conjugates $\bar{\pi}^\mu_{\dot{B}A}$ define a bi-linear mapping from the product of quark and anti-quark cubic algebras into a linear four-dimensional vector space, whose structure is not yet defined.

Let us impose the following invariance condition:

$$\pi^\mu_{A\dot{B}}\, \theta^A \bar{\theta}^{\dot{B}} = \bar{\pi}^\mu_{\dot{B}A} \bar{\theta}^{\dot{B}} \theta^A. \tag{56}$$

It follows immediately from (30) that

$$\pi^{\mu}_{A\dot{B}} = -j^2\, \bar{\pi}^{\mu}_{\dot{B}A}. \tag{57}$$

Such matrices are non-hermitian, and they can be realized by the following substitution:

$$\pi^{\mu}_{A\dot{B}} = j^2\, i\, \sigma^{\mu}_{A\dot{B}}, \quad \bar{\pi}^{\mu}_{\dot{B}A} = -j\, i\, \sigma^{\mu}_{\dot{B}A} \tag{58}$$

where $\sigma^{\mu}_{A\dot{B}}$ are the unit 2 matrix for $\mu = 0$, and the three hermitian Pauli matrices for $\mu = 1, 2, 3$.

Again, we want to get the same form of these four matrices in another basis. Knowing that the lower indices A and $\dot{B}$ undergo the transformation with matrices $U^{A'}_{B}$ and $\bar{U}^{\dot{A}'}_{\dot{B}}$, we demand that there exist some 4×4 matrices $\Lambda^{\mu'}_{\nu}$ representing the transformation of lower indices by the matrices U and $\bar{U}$:

$$\Lambda^{\mu'}_{\nu}\, \pi^{\nu}_{A\dot{B}} = U^{A'}_{A}\, \bar{U}^{\dot{B}'}_{\dot{B}}\, \pi^{\mu'}_{A'\dot{B}'}, \tag{59}$$

It is clear that we can replace the matrices $\pi^{\nu}_{A\dot{B}}$ by the corresponding matrices $\sigma^{\nu}_{A\dot{B}}$, and this defines the vector (4×4) representation of the Lorentz group.

It can be checked that now det $(\Lambda) = [\det U]^2\, \left[\det \bar{U}\right]^2$.

The group of transformations thus defined is $SL(2, \mathbf{C})$, which is the covering group of the Lorentz group.

With the invariant "spinorial metric" in two complex dimensions, ε^{AB} and $\varepsilon^{\dot{A}\dot{B}}$ such that $\varepsilon^{12} = -\varepsilon^{21} = 1$ and $\varepsilon^{\dot{1}\dot{2}} = -\varepsilon^{\dot{2}\dot{1}}$, we can define the contravariant components $\pi^{\nu\, A\dot{B}}$. It is easy to show that the Minkowskian space-time metric, invariant under the Lorentz transformations, can be defined as

$$g^{\mu\nu} = \frac{1}{2}\left[\pi^{\mu}_{A\dot{B}}\, \pi^{\nu\, A\dot{B}}\right] = diag(+, -, -, -) \tag{60}$$

Together with the anti-commuting spinors ψ^{α} the four real coefficients defining a Lorentz vector, $x_{\mu}\, \pi^{\mu}_{A\dot{B}}$, can generate now the supersymmetry via standard definitions of super-derivations. Let us then choose the matrices $\Lambda^{\alpha'}_{\beta}$ to be the usual spinor representation of the $SL(2, \mathbf{C})$ group, while the matrices $U^{A'}_{B}$ will be defined as follows:

$$U^{1'}_{1} = j\Lambda^{1'}_{1}, \quad U^{1'}_{2} = -j\Lambda^{1'}_{2}, \quad U^{2'}_{1} = -j\Lambda^{2'}_{1}, \quad U^{2'}_{2} = j\Lambda^{2'}_{2}, \tag{61}$$

the determinant of U being equal to j^2. Obviously, the same reasoning leads to the conjugate cubic representation of $SL(2, \mathbf{C})$ if we require the covariance of the conjugate tensor

$$\bar{\rho}^{\dot{\beta}}_{\dot{D}\dot{E}\dot{F}} = j\, \bar{\rho}^{\dot{\beta}}_{\dot{E}\dot{F}\dot{D}} = j^2\, \bar{\rho}^{\dot{\beta}}_{\dot{F}\dot{D}\dot{E}},$$

by imposing the equation similar to (47)

$$\Lambda^{\dot{\alpha}'}_{\dot{\beta}}\, \bar{\rho}^{\dot{\beta}}_{\dot{A}\dot{B}\dot{C}} = \bar{\rho}^{\dot{\alpha}'}_{\dot{A}'\dot{B}'\dot{C}'} \bar{U}^{\dot{A}'}_{\dot{A}}\, \bar{U}^{\dot{B}'}_{\dot{B}}\, \bar{U}^{\dot{C}'}_{\dot{C}}\,. \tag{62}$$

The matrix $\bar{U}$ is the complex conjugate of the matrix U, with determinant equal to j.

Moreover, the two-component entities obtained as images of cubic combinations of quarks, $\psi^{\alpha} = \rho^{\alpha}_{ABC}\theta^A\theta^B\theta^C$ and $\bar{\psi}^{\dot{\beta}} = \bar{\rho}^{\dot{\beta}}_{\dot{D}\dot{E}\dot{F}}\bar{\theta}^{\dot{D}}\bar{\theta}^{\dot{E}}\bar{\theta}^{\dot{F}}$ should anti-commute, because their arguments do so, by virtue of (30):

$$(\theta^A\theta^B\theta^C)(\bar{\theta}^{\dot{D}}\bar{\theta}^{\dot{E}}\bar{\theta}^{\dot{F}}) = -(\bar{\theta}^{\dot{D}}\bar{\theta}^{\dot{E}}\bar{\theta}^{\dot{F}})(\theta^A\theta^B\theta^C)$$

We have found the way to derive the covering group of the Lorentz group acting on spinors via the usual spinorial representation. Spinors are obtained as a homomorphic image of tri-linear combinations of three quarks (or anti-quarks). The quarks transform with matrices U (or $\bar{U}$ for the anti-quarks), but these matrices are not unitary: their determinants are equal to j^2 or j, respectively.

So, quarks cannot be put on the same footing as classical spinors; they transform under the $Z_3 \times SL(2, \mathbf{C})$ group. There are strong reasons to believe that their wave functions in the Schroedinger picture should not obey exactly the same equations as the electrons; a modified version of Dirac's equation should be found to explain why they do not propagate as ordinary solutions do, while their tri-linear combinations can propagate if extra selection rules (only combinations with three different "colors") display behavior similar to that of the ordinary spin-one-half particles.

7 A Z_3 Generalization of Dirac's Equation

Lets us first underline the Z_2 symmetry of Maxwell and Dirac equations, which implies the hyperbolic character of both systems, and therefore makes the propagation possible. Maxwell's equations *in vacuo* can be written as follows:

$$\frac{1}{c}\frac{\partial \mathbf{E}}{\partial t} = \nabla \wedge \mathbf{B}, \quad -\frac{1}{c}\frac{\partial \mathbf{B}}{\partial t} = \nabla \wedge \mathbf{E}. \tag{63}$$

These equations can be decoupled by applying the time derivation twice, which in vacuum, where $div\mathbf{E} = 0$ (and $div\mathbf{B} = 0$ which holds always) leads to the d'Alembert equation satisfied by both components separately:

$$\frac{1}{c^2}\frac{\partial^2 \mathbf{E}}{\partial t^2} - \nabla^2 \mathbf{E} = 0, \quad \frac{1}{c^2}\frac{\partial^2 \mathbf{B}}{\partial t^2} - \nabla^2 \mathbf{B} = 0.$$

Nevertheless, neither of the components of the Maxwell tensor, be it $\mathbf{E}$ or $\mathbf{B}$, can propagate separately alone. It is also remarkable that although each of the fields $\mathbf{E}$

and **B** satisfies a second-order propagation equation, due to the coupled system (63) there exists a quadratic combination satisfying the first-order equation, the Poynting four-vector:

$$P^{\mu} = \left[P^0, \mathbf{P}\right], \quad P^0 = \frac{1}{2}\left(\mathbf{E}^2 + \mathbf{B}^2\right), \quad \mathbf{P} = \mathbf{E} \wedge \mathbf{B}, \quad \partial_{\mu} P^{\mu} = 0. \tag{64}$$

The Dirac equation for the electron displays a similar Z_2 symmetry, with two coupled equations which can be put in the following form:

$$\begin{aligned} i\hbar\frac{\partial}{\partial t}\psi_+ - mc^2\psi_+ &= i\hbar\boldsymbol{\sigma}\cdot\nabla\psi_-, \\ -i\hbar\frac{\partial}{\partial t}\psi_- - mc^2\psi_- &= -i\hbar\boldsymbol{\sigma}\cdot\nabla\psi_+, \end{aligned} \tag{65}$$

where ψ_+ and ψ_- are the positive and negative energy components of the Dirac equation; this is visible even better in the momentum representation:

$$\begin{aligned} \left[E - mc^2\right]\psi_+ &= c\boldsymbol{\sigma}\cdot\mathbf{p}\psi_-, \\ \left[-E - mc^2\right]\psi_- &= -c\boldsymbol{\sigma}\cdot\mathbf{p}\psi_+. \end{aligned} \tag{66}$$

Note that the same effect (negative energy states) can be obtained by changing the direction of time, and putting the minus sign in front of the time derivative, as suggested by Feynman [10]. Each of the components satisfies the Klein-Gordon equation, easily obtained by successive application of two operators and diagonalization:

$$\left[\frac{1}{c^2}\frac{\partial^2}{\partial t^2} - \nabla^2 - m^2\right]\psi_{\pm} = 0$$

As in the case of the electromagnetic waves, neither of the components of this complex entity can propagate by itself; only all the components can [11].

As it follows from the experiment, the two types of quarks, u and d, cannot propagate freely, but can form a freely propagating particle perceived as a fermion, but only under an extra condition: they must belong to three *different* species called *colors*; short of this they do not form a freely propagating entity. Therefore, quarks should be described by *three fields* satisfying a set of coupled linear equations, with the Z_3-symmetry playing a similar role that the Z_2-symmetry plays in the case of Maxwell's and Dirac's equations. Instead of the "–" sign multiplying the time derivative, we should use the cubic root of unity j and its complex conjugate j^2 according to the following scheme:

$$\frac{\partial}{\partial t} \mid \psi >= \hat{H}_{12} \mid \phi >, \quad j\frac{\partial}{\partial t} \mid \phi >= \hat{H}_{23} \mid \chi >, \quad j^2\frac{\partial}{\partial t} \mid \chi >= \hat{H}_{31} \mid \psi >, \tag{67}$$

We do not specify yet the number of components in each state vector, nor the character of the hamiltonian operators on the right-hand side; the three fields $\mid \psi >$, $\mid \phi >$ and $\mid \chi >$ should represent the three colors, none of which can propagate by itself. The quarks being endowed with mass, we can suppose that one of the main terms in the hamiltonians is the mass operator $\hat{m}$; and let us suppose that the remaining parts are the same in all three hamiltonians. This will lead to the following three equations:

$$\begin{aligned} \frac{\partial}{\partial t} \mid \psi > - \hat{m} \mid \psi > &= \hat{H} \mid \phi >, \\ j \frac{\partial}{\partial t} \mid \phi > - \hat{m} \mid \phi > &= \hat{H} \mid \chi >, \\ j^2 \frac{\partial}{\partial t} \mid \chi > - \hat{m} \mid \chi > &= \hat{H} \mid \psi >, \end{aligned} \tag{68}$$

Supposing that the mass operator commutes with time derivation, by applying three times the left-hand side operators, each of the components satisfies the same common *third order* equation:

$$\left[\frac{\partial^3}{\partial t^3} - \hat{m}^3 \right] \mid \psi > = \hat{H}^3 \mid \psi > . \tag{69}$$

The anti-quarks should satisfy a similar equation with the negative sign for the Hamiltonian operator. The fact that there exist two types of quarks in each nucleon suggests that the state vectors $\mid \psi >$, $\mid \phi >$ and $\mid \chi >$ should have two components each. When combined together, the two postulates lead to the conclusion that we must have three two-component functions and their three conjugates:

$$\begin{pmatrix} \psi_1 \\ \psi_2 \end{pmatrix}, \begin{pmatrix} \bar{\psi}_{\dot{1}} \\ \bar{\psi}_{\dot{2}} \end{pmatrix}, \begin{pmatrix} \varphi_1 \\ \varphi_2 \end{pmatrix}, \begin{pmatrix} \bar{\varphi}_{\dot{1}} \\ \bar{\varphi}_{\dot{2}} \end{pmatrix}, \begin{pmatrix} \chi_1 \\ \chi_2 \end{pmatrix}, \begin{pmatrix} \bar{\chi}_{\dot{1}} \\ \bar{\chi}_{\dot{2}} \end{pmatrix}.$$

which may represent three colors, two quark states (e.g. "up" and "down"), and two anti-quark states (with anti-colors, respectively) [12, 13].

Finally, in order to be able to implement the action of the $SL(2, \mathbf{C})$ group via its 2×2 matrix representation defined in the previous section, we choose the Hamiltonian $\hat{H}$ equal to the operator $\sigma \cdot \nabla$, the same as in the usual Dirac equation. The action of the Z_3 symmetry is represented by factors j and j^2, while the Z_2 symmetry between particles and anti-particles is represented by the "−" sign in front of the time derivative. An additional Z_2-symmetry comes from the presence of two components in each wave function, so that the final symmetry is $Z_3 \times Z_2 \times Z_2$, resulting in *twelve-component* wave functions. The differential system that satisfies all these assumptions is as follows:

$$-i\hbar\frac{\partial}{\partial t}\psi_+ - mc^2\psi_+ = -i\hbar c\sigma\cdot\nabla\varphi_-,$$
$$i\hbar\frac{\partial}{\partial t}\varphi_- - jmc^2\varphi_- = -ji\hbar c\sigma\cdot\nabla\chi_+$$
$$-i\hbar\frac{\partial}{\partial t}\chi_+ - j^2mc^2\chi_+ = -j^2i\hbar c\sigma\cdot\nabla\psi_-,$$
$$i\hbar\frac{\partial}{\partial t}\psi_- - mc^2\psi_- = -i\hbar c\sigma\cdot\nabla\varphi_+,$$
$$-i\hbar\frac{\partial}{\partial t}\varphi_+ - jmc^2\varphi_+ = -ji\hbar c\sigma\cdot\nabla\chi_-,$$
$$i\hbar\frac{\partial}{\partial t}\chi_- - j^2mc^2\bar{\chi} = -j^2i\hbar c\sigma\cdot\nabla\psi_+, \qquad (70)$$

Here we made a simplifying assumption that the mass operator is just proportional to the identity matrix, and therefore commutes with the operator $\sigma\cdot\nabla$.

The functions ψ, φ and χ are related to their conjugates via the following third-order equations:

$$-i\frac{\partial^3}{\partial t^3}\psi = \left[\frac{m^3c^6}{\hbar^3} - i(\sigma\cdot\nabla)^3\right]\bar{\psi} = \left[\frac{m^3c^6}{\hbar^3} - i\sigma\cdot\nabla\right](\Delta\bar{\psi}),$$
$$i\frac{\partial^3}{\partial t^3}\bar{\psi} = \left[\frac{m^3c^6}{\hbar^3} - i(\sigma\cdot\nabla)^3\right]\psi = \left[\frac{m^3c^6}{\hbar^3} - i\sigma\cdot\nabla\right](\Delta\psi), \qquad (71)$$

and the same, of course, for the remaining wave functions φ and χ.

This equation can be solved by separation of variables; the time-dependent and the space-dependent factors have the same structure:

$$A_1\,\mathrm{e}^{\omega t} + A_2\,\mathrm{e}^{j\,\omega t} + A_3\mathrm{e}^{j^2\,\omega t},\; B_1\,\mathrm{e}^{\mathbf{k}.\mathbf{r}} + B_2\,\mathrm{e}^{j\,\mathbf{k}.\mathbf{r}} + B_3\,\mathrm{e}^{j^2\,\mathbf{k}.\mathbf{r}}$$

The nine complex solutions can be displayed in a 3×3 matrix as follows:

$$\begin{pmatrix} e^{\omega t-\mathbf{k}\cdot\mathbf{r}} & e^{\omega t-j\mathbf{k}\cdot\mathbf{r}} & e^{\omega t-j^2\mathbf{k}\cdot\mathbf{r}} \\ e^{j\omega t-\mathbf{k}\cdot\mathbf{r}} & e^{j\omega t-j\mathbf{k}\cdot\mathbf{r}} & e^{j\omega t-j^2\mathbf{k}\cdot\mathbf{r}} \\ e^{j^2\omega t-\mathbf{k}\cdot\mathbf{r}} & e^{j^2\omega t-\mathbf{k}\cdot\mathbf{r}} & e^{j^2\omega t-j^2\mathbf{k}\cdot\mathbf{r}} \end{pmatrix} \qquad (72)$$

and their nine independent entries can be represented in a basis of real functions as

$$\begin{pmatrix} A_{11}\,e^{\omega t-\mathbf{k}\cdot\mathbf{r}} & A_{12}\,e^{\omega t+\frac{\mathbf{k}\cdot\mathbf{r}}{2}}\cos\mathbf{k}\cdot\xi & A_{13}\,e^{\omega t+\frac{\mathbf{k}\cdot\mathbf{r}}{2}}\sin\mathbf{k}\cdot\xi \\ A_{21}\,e^{-\frac{\omega t}{2}-\mathbf{k}\cdot\mathbf{r}}\cos\omega\tau & A_{22}\,e^{-\frac{\omega t}{2}+\frac{\mathbf{k}\cdot\mathbf{r}}{2}}\cos(\omega\tau-\mathbf{k}\cdot\xi) & A_{23}\,e^{-\frac{\omega t}{2}+\frac{\mathbf{k}\cdot\mathbf{r}}{2}}\cos(\omega\tau+\mathbf{k}\cdot\xi) \\ A_{31}\,e^{-\frac{\omega t}{2}-\mathbf{k}\cdot\mathbf{r}}\sin\omega\tau & A_{32}\,e^{-\frac{\omega t}{2}+\frac{\mathbf{k}\cdot\mathbf{r}}{2}}\sin(\omega\tau+\mathbf{k}\cdot\xi) & A_{33}\,e^{-\frac{\omega t}{2}+\frac{\mathbf{k}\cdot\mathbf{r}}{2}}\sin(\omega\tau-\mathbf{k}\cdot\xi) \end{pmatrix} \qquad (73)$$

where $\tau = \frac{\sqrt{3}}{2} t$ and $\xi = \frac{\sqrt{3}}{2}\mathbf{kr}$. The parameters ω, $\mathbf{k}$ and m must satisfy the cubic dispersion relation:

$$\omega^3 = k_x^3 + k_y^3 + k_z^3 - 3\, k_x k_y k_z + m^3 \tag{74}$$

This relation is invariant under the simultaneous change of sign of ω, $\mathbf{k}$ and m, which suggests the introduction of another set of solutions constructed in the same manner, but with minus sign in front of ω and $\mathbf{k}$, which we shall call *conjugate* solutions.

Its solutions can be readily found in the exponential form :

$$\Psi \sim e^{k_\mu x^\mu} = e^{\omega t - \mathbf{k}\cdot\mathbf{r}}, \quad \text{with} \quad \frac{\omega^3}{c^3} = m^3 + k_x^3 + k_y^3 + k_z^3 - 3k_x k_y k_z. \tag{75}$$

The functions displayed in the matrix do not represent a wave; however, one can produce a propagating solution by forming certain cubic combinations, e.g.

$$e^{\omega t - \mathbf{k}\cdot\mathbf{r}}\, e^{-\frac{\omega t}{2} + \frac{\mathbf{k}\cdot\mathbf{r}}{2}} \cos(\omega\tau - \mathbf{k}\cdot\xi)\, e^{-\frac{\omega t}{2} + \frac{\mathbf{k}\cdot\mathbf{r}}{2}} \sin(\omega\tau - \mathbf{k}\cdot\xi) = \frac{1}{2}\sin(2\omega\tau - 2\mathbf{k}\cdot\xi)$$

This model can explain why a single quark cannot propagate, while three quarks can form a freely propagating state.

References

1. Streater, R.F., Wightman, A.S.: PCT, Spin Ans Statistics and All That. W.A. Benjamin Inc, Amsterdam (1964)
2. Pauli, W.: The connection between spin and statistics. Phys. Rev. **58**, 716–722 (1940)
3. Dyson, F.J.: Ground-state energy of a finite system of charged particles. J. Math. Phys. **8**, 1538–1545 (1967)
4. Dyson, F.J., Lenard, A.: Stability of matter, parts I and II. J. Math. Phys. **8**, 423–434 (1967); J. Math. Phys. **9**, 698–711 (1968)
5. Dirac, P.A.M.: The Principles of Quantim Mechanics, 4th edn. Oxford University Press, New York (1958)
6. Kerner, R.: Graduation Z_3 et la racine cubique de léquation de Dirac. C.R. Acad. Sci. Paris **312**(série II), 191–196, (1991)
7. Kerner, R.: Z(3) graded algebras and the cubic root of the supersymmetry translations. Journ. Math. Phys. **33**, 403–411 (1992)
8. Abramov, V., Kerner, R., Le Roy, B.: Hypersymmetry: a Z_3-graded generalization of supersymmetry. J. Math. Phys. **38**, 1650–1669 (1997)
9. Kerner, R.: The cubic chessboard: geometry and physics. Class. Quantum Gravity **14**, A203–A225 (1997)
10. Feynman, R. P.: Space-Time Approach to Quantum Electrodynamics. Phys. Rev. **76**, 69–789 (1949)
11. Bjorken, J.D., Drell, S.D.: Relativistic Quantum Mechanics. McGraw-Hill, London (1964)
12. Lipatov, L.N., Rausch de Traubenberg, M., Volkov, G.G.: On the ternary complex analysis and its applications. J. Math. Phys. **49**, 013502 (2008)
13. Campoamor-Stursberg, R., Rausch de Traubenberg, M.: Kinematical superalgebras and Lie algebras of order 3. J. Math. Phys. **49**, 063506 (2008)

Calculation of Decay Times for Simple Modules

Olav Gravir Imenes

Abstract How can the physical processes of decay be modeled in a mathematically sensible way? One possible model, due to Laudal, see [3, 4, 6], is to start with an algebra $A(\sigma)$ of observables, assumed to contain all parameters of the states, represented by generators of the sub-algebra A, together with the parameters parametrising the dynamics of the phenomena we are studying. The geometry of the moduli space of isomorphism classes, $\text{Simp}(A(\sigma))$, of simple modules, $\rho : A(\sigma) \to \text{End}_k(V)$, will reflect how the system behave. A metric can be assumed to model time, and the canonical Dirac derivation δ of $A(\sigma)$ generalizes the equation of motion. Any simple module corresponds to a point of a "world curve". As time goes by, this point may leave the space $\text{Simp}(A(\sigma))$. This can be interpreted as a decay process, and the semi-simple representation, corresponding to the non-simple limit-representation, is said to be a system that have experienced decay. We shall consider the case of the harmonic oscillator, and show how to calculate decay and life-times of specific phenomena.

1 The Space of Simple Modules

Let k be a field of characteristic zero and let A be a finitely generated associative k-algebra. Consider left modules $\rho : A \to \text{End}_k(V)$ where V is a k-vector space. Define the space of simple modules,

$$\text{Simp}_n(A) := \{\rho : A \to \text{End}_k(V) | \rho \text{ simple}, \dim V = n\}/\sim, \tag{1}$$

where two modules ρ and ρ' are identified if they are isomorphic. For $n = 1$, $\underline{A} := \text{Simp}_1(A)$ is the space of k-points of A. For each n, it is possible to construct

O. G. Imenes (✉)
Buskerud University College, Postboks 7053, 3007 Drammen, Norway
e-mail: olav.gravir.imenes@hibu.no

A. Makhlouf et al. (eds.), *Algebra, Geometry and Mathematical Physics*, 639
Springer Proceedings in Mathematics & Statistics 85, DOI: 10.1007/978-3-642-55361-5_38,

a finitely generated k-algebra $C(n)$, and a family, $\tilde{\rho} : A \to \mathcal{M}_n(C(n))$, such that any homomorphism $p : C(n) \to k(p)$, i.e. any k-point of $C(n)$, corresponds to a module $\rho : A \to \mathrm{End}_k(V) = \mathcal{M}_n(k)$ via the diagram,

$$\begin{array}{ccc} A & \xrightarrow{\tilde{\rho}} & \mathcal{M}_n(C(n)) \\ & \searrow^{\rho} & \downarrow p \\ & & \mathcal{M}_n(k(p)) \end{array}$$

The existence of such families has been proved by Laudal [6]. Not all points in $\mathrm{Simp}_1 C(n)$ correspond to simple modules. However, Laudal has proved that if k is algebraically closed, there exists a Zariski open subspace $U(n) \subset \mathrm{Simp}_1(C(n))$ corresponding to simple modules, see Laudal [6], Chapter 3, Theorem 3.4.8. The correspondence between $U(n)$ and simple n-dimensional A-modules is not necessarily one-to-one. A finite number of points in $U(n)$ may correspond to isomorphic module structures, thus the same point in $\mathrm{Simp}_n(A)$.

The tangent space of a point $\rho \in \mathrm{Simp}_n(A)$ is

$$T_{\mathrm{Simp}_n(A),\rho} = \mathrm{Ext}^1_A(V, V) = \mathrm{Der}_k(A, \mathrm{End}_k(V))/\mathrm{Triv}(A, \mathrm{End}_k(V)) \ , \qquad (2)$$

where $\mathrm{Triv}(A, \mathrm{End}_k(V))$ denotes the set of trivial or inner derivations, defined by $\xi \in \mathrm{Triv}(A, \mathrm{End}_k(V))$, if there exist a $\phi \in \mathrm{End}_k(V)$, such that $\xi(a) = [\phi, \rho(a)] := \phi\rho(a) - \rho(a)\phi$.

2 Dynamics on the Space of Modules

For every point $\rho \in \mathrm{Simp}_n(A)$ and every derivation $\delta \in \mathrm{Der}_k(A, A)$, we have a tangent vector of $\mathrm{Simp}_n(A)$ at ρ. These tangent vectors provides us with a 1-dimensional distribution, and in good cases, with a vector field $[\delta]$, on $\mathrm{Simp}_1(C(n))$. We have the following theorem due to Laudal.

Theorem 2.1 *For any $\delta \in Der_k(A, A)$ and every $p \in Simp_1(C(n))$, with local ring $\widehat{C(n)}_{\mathfrak{m}_p}$, there exists a derivation $[\delta] \in Der_k(\widehat{C(n)}_{\mathfrak{m}_p}, \widehat{C(n)}_{\mathfrak{m}_p})$, and an element $Q \in End_{\widehat{C(n)}_{\mathfrak{m}_p}}(V \otimes_k \widehat{C(n)}_{\mathfrak{m}_p})$ such that, as operators on $V \otimes_k \widehat{C(n)}_{\mathfrak{m}_p}$, we must have*

$$\tilde{\rho}(\delta(a)) = [\delta](\tilde{\rho}(a)) + [Q, \tilde{\rho}(a)] \ . \qquad (3)$$

Proof See Laudal [6], Chapter 4, Theorem 4.2.1. □

The element Q corresponds to the Hamiltonian. Since $\mathrm{Simp}_1(C(n))$ is not a fine moduli space, we needed to restrict to the formal case. However, doing this at every point $p \in \mathrm{Simp}_1(C(n))$, leave us with a vector field $[\delta]$ on $\mathrm{Simp}_1(C(n))$ given by δ.

Laudal [4] introduces time as a metric on the moduli space, in this case $\mathrm{Simp}_1(C(n))$ or $\mathrm{Simp}_n(A)$. A deformation or change of state of the system, as defined by a simple representation, corresponds to a change of representation. Since time, according to our sensations, is a measure of change, it seems reasonable to let time be the measure of change, the metric, in this space of representations. An integral curve of $[\delta]$ will be a history of the system, and the arc-length of this curve is a measure of the time passed.

When we have chosen a point, ρ, corresponding to a module, we are interested in the evolution of the module along the integral curve, given by $[\delta]$, through this point. The metric g on $\mathrm{Simp}_1(C(n))$, or on $\mathrm{Simp}_n(A)$, gives the arc-length of a given integral curve, and is therefore the natural measure of change, time.

3 The Notions of Interaction and Decay

In physics, a system can be specified by a Lagrangian $\mathcal{L}$, from which an Hamiltonian H acting on a Hilbert space $\mathcal{H}$ is found. The states of the system are vectors $\Psi \in \mathcal{H}$.

If the Hamiltonian H depends upon *time*, we can view the changing of the Hamiltonian H as a deformation in the following way: Let $\mathcal{H}$ be a Hilbert space and let $\rho_t : A := k\langle x_0, .., x_3, dx_0, .., dx_3\rangle \to \mathrm{End}_k(\mathcal{H})$ be a homomorphism of algebras parametrised by t, i.e. a family of A-modules, with structure maps defined by $\rho_t(x_i) = q_i$, $\rho_t(dx_i) = \pi_i(H(t))$, where $\pi_i(H(t))$ is the canonical momentum. Note that in this setting we have considered π_i as a function of the Hamiltonian H instead of the Lagrangian $\mathcal{L}$. As time pass, the module ρ_{t_1} is deformed into the module ρ_{t_2}. A state $\Psi(t) \in \mathcal{H}$ describe the particle content of the system as time passes.

We refer to Laudal [6], Theorem 4.2.2, for a discussion of the possibility to use the time-dependent Hamiltonian as the propulsor.

To model interactions, physicists split, at some fixed time, the Hamiltonian: $H = H_0 + I$. The 'unperturbed', or 'free', system that we understand, is represented by H_0, while I is the interaction term, see for example Huang [1] p. 139. Both H_0 and I are time-dependent. The starting configuration of the system is given by the structure map ρ_0, given by H_0, and an initial state Ψ_i. The end configuration is given by ρ_0 and a final state Ψ_f. When the interaction part, I, is *turned on*, say at $t = t_1$, ρ_0 is deformed into ρ_{t_1}, which is deformed further as time passes, but deformed back again to ρ_0 when the interaction is *turned off*. The module itself has not changed, only the state. The module, $(\rho_0, \mathcal{H})$, determines the particle type, for example what the physicists call leptons, while the state determine the specifics, for example may Ψ_i be an electron and a neutrino. Since the module is unchanged, the final state Ψ_f still describes leptons, but now it may represent a family of different types of leptons. This happens if the interaction process include a gauge group action on the Hilbert space. Under the action of a gauge group, usually a Lie algebra, $\mathfrak{g}$, the Hilbert space $\mathcal{H}$ may have a natural splitting $\mathcal{H} = \oplus_{j=1}^{\infty}\mathcal{H}_j$. An in-state $\Psi_i = \sum_{l\in L} \Psi_i(l) \in \oplus_{l\in L}\mathcal{H}_l$, where L is an index set, is interpreted as a collection of certain particles, indexed by the elements of L. Now the out-state Ψ_f may have the form $\Psi_f = \sum_{p \in P} \Psi_f(p) \in \oplus_{p \in P} \mathscr{H}_p$, representing quite different particles. Thus Ψ_i and Ψ_f may correspond to the same particle type, but to very different collections of particles.

Physicists talk about decay as a special case of an interaction in the sense that the particles given by Ψ_i have decayed into the particles described by Ψ_f, see for example Weinberg [8], Sects. 3.2 and 3.4. The representations parametrised by the time parameter t are usually simple modules, and therefore both the decayed system and the original system is in a sense the same.

Decay in Laudal's model is different; we start with a simple module $\rho : A \to \mathrm{End}_k(V)$, and let a derivation $\delta \in \mathrm{Der}_k(A, A)$ guide the deformation of ρ, by using Theorem 2.1, into a non-simple module. This will be interpreted as if ρ has experienced decay. Thus a condition for decay in our sense is that $[\delta] \neq 0$. This is not a requirement in physics.

In Laudal's model, each simple module correspond to a 'time point' in a moduli space consisting of n-dimensional simple modules, thus we start with a module, and end up with time, which is the measure of change on the moduli space. This is in opposition to the above description, where we start with time as a parameter, and obtain a certain family of modules. For both cases, an interaction is described by a deformation.

To explain our sense of decay, we interpret the family $\tilde{\rho}$ as the particle type, for example what physicists call a lepton. A state $\tilde{\psi} \in C(n) \otimes_k V$ specifies certain parts of the system, for example that we have a lepton which is an electron with spin up, a neutrino, etc. A point $p : C(n) \to k(p)$ specifies a test particle, or starting configuration. If the particle type is stable, i.e. if $[\delta] = 0$, then after a time τ, we will still have a lepton, because ρ_0 and ρ_τ correspond to isomorphic modules, but we may have an electron with spin down, etc. If the derivation $[\delta]$ guide the particle into another point in $\mathrm{Simp}_n(A)$, the particle may be unstable, and we no longer necessarily have a lepton. We may for example have two different particle types, and this is what we interpret as 'decay'. See also [2] for a discussion of how to model electroweak interactions as stable points, i.e. points where $[\delta] = 0$.

4 Harmonic Oscillator

The harmonic oscillator is a system in which the second order rate of change of displacement from its equilibrium, is proportional to the displacement. We consider the one-dimensional harmonic oscillator, which is given by the k-algebra of observables $H(\sigma) = k\langle x, dx \rangle$, the free non-commutative k-algebra on two symbols. This k-algebra has a natural derivation $\delta \in \mathrm{Der}_k(H(\sigma), H(\sigma))$, given by the harmonic oscillator ideal,

$$\delta(x) = dx, \delta(dx) = -\omega^2 x . \tag{4}$$

The above equation is a force law.

Remark 4.1 Using the phase space construction of Laudal [5], it can be written

$$H(\sigma) := Ph^{\infty}(k[x])/(d^2x + \omega^2 x), \quad \omega \in k\ . \tag{5}$$

We will let $\omega = 1$, i.e. $\delta^2(x) = -x$. We consider the space of 2-dimensional representations of the harmonic oscillator.

Theorem 4.1 *There exists a $C(2) \cong k[t_1, .., t_5]$, and a versal family of 2-dimensional $H(\sigma)$-modules given by*

$$\begin{aligned} \tilde{\rho} : H(\sigma) &\longrightarrow \mathcal{M}_2(C(2)) \\ x &\longmapsto \begin{pmatrix} 0 & 1+t_3 \\ t_5 & t_4 \end{pmatrix} \\ dx &\longmapsto \begin{pmatrix} t_1 & t_2 \\ 1+t_3 & 0 \end{pmatrix}\ . \end{aligned} \tag{6}$$

There exists a Zariski-open subset $U(2)$ of $Simp_1(C(2))$ such that each point in $U(2)$ corresponds to a simple module.

Proof See Laudal [6], Chapter 3, Example 3.3. □

The polynomial

$$f := \det(\tilde{\rho}([x, y])) = ((1+t_3)^2 - t_2t_5)^2 + (t_1(1+t_3) + t_2t_4)(t_4(1+t_3) + t_1t_5)\ , \tag{7}$$

is called the Formanek center. The set of zeroes of f, $Z(f)$, corresponds to non-simple modules.

Let a general point $(t_1, ..., t_5) \in \mathrm{Simp}_1(C(2))$ provide a preparation or a startconfiguration of the system. We use the natural vector field $[\delta]$, defined by Eq. (4), to find how the starting point evolves. The derivation came from the fact that we have an harmonic oscillator.

Laudal [6], Example 4.4, calculates integral curves for the harmonic oscillator given by $\delta^2(x) = x$, in the moduli space consisting of two-dimensional representations of $Ph(k[x])$.

In this paper we shall calculate integral curves of the vector field $[\delta]$ for an harmonic oscillator given by $\delta^2(x) = -x$. For this harmonic oscillator, it is possible for a system described by a simple module to decay into a system described by a non-simple module. We want to calculate the amount of time until this happens. We do this by considering a metric g on $\mathrm{Simp}_2(H(\sigma))$, the time, to find the arc-length of the integral curve from the starting point until decay. For visualization and physical purposes we will let $k = \mathbf{R}$.

The *"formal flow"* of an observable $a \in H(\sigma)$ with respect to δ is $\exp(\tau\delta)(a)$, which acts on the completion of $H(\sigma)$. In this example we can use the local formal case of Theorem 2.1 and Eq. (3) to obtain a vector field $[\delta]$ on $\mathrm{Simp}_1(C(n))$ which governs the evolution of the system, and whose formal flow, parametrised by τ, is

$$\tilde{\rho}_\tau(a) := \exp(\tau[\delta])(\tilde{\rho}(a)) = \tilde{\rho}(\exp(\tau\delta)(a)) \,. \tag{8}$$

We expand the right hand side of Eq. (8), and use Eq. (4) with $\omega = 1$, to obtain expressions for $\tilde{\rho}_\tau(x)$ and $\tilde{\rho}_\tau(dx)$ for the case of the harmonic oscillator. For each τ, we need to write the structure map $\tilde{\rho}_\tau$ as a versal family of the form of Eq. (6). During calculations, this is accomplished by finding an $U(\tau)$ such that

$$\begin{aligned}\tilde{\rho}_\tau(x) &= U(\tau)(\tilde{\rho}(x)\cos\tau + \tilde{\rho}(dx)\sin\tau)U^{-1}(\tau)\\ \tilde{\rho}_\tau(dx) &= U(\tau)(\tilde{\rho}(x)\sin\tau + \tilde{\rho}(dx)\cos\tau)U^{-1}(\tau) \,.\end{aligned} \tag{9}$$

This is an analytical expression. The first problem is to find where the modules along the integral curves degenerate, i.e. for which τ's are $\tilde{\rho}_\tau$ a non-simple module. To determine if the module given by Eq. (9) is a non-simple module, we can, in the case of 2-dimensional representations, solve the equation $0 = \det(\tilde{\rho}_\tau([dx, x]))$. For the force-law $\delta^2(x) = -x$ we find that for the most general points $\underline{t} \in \mathrm{Simp}_1(C(2))$, the above equation holds if $\tau = \pm\pi/4$. Going forward in time, we have found a maximum time for decay. Note that the derivation $[\delta]$ provide us with a direction as well as an integral curve. For $\tau = \pi/4$ we find from Eq. (9) that the eigenvalues of $\tilde{\rho}(x)$ and $\tilde{\rho}(dx)$ are the same, namely

$$\lambda_{\tilde{\rho}(x)} = \lambda_{\tilde{\rho}(dx)} = \frac{\sqrt{2}}{4}(t_1 + t_4) \pm \lambda_0 \,, \tag{10}$$

where $\lambda_0 := \frac{1}{4}\sqrt{2t_4^2 - 4t_1t_4 + 2t_1^2 + 8t_5 + 8 + 16t_3 + 8t_3^2 + 8t_5t_2 + 8t_2 + 8t_3t_5 + 8t_2t_3}$. In Fig. 1 we have visualized the evolution, and subsequent decay. Suppose our starting point is $(t_1, t_2, t_3, t_4, t_5) = (0, 2.25, -0.5, 0, 1.5)$, a point corresponding to a simple module. The derivation δ, given by Eq. (4), gives four integral curves in the space $\mathrm{Simp}_1(C(2))$, two of which are shown in Fig. 1, corresponding to the same curve in $\mathrm{Simp}_2(H(\sigma))$. The particle represented by the starting point is said to have experienced decay when the integral curves reach a point corresponding to a non-simple module, in the figure represented by the curved surface, the zeros of the polynomial f of Eq. (7).

This far we have only showed that the derivation $[\delta]$ gives integral curves, which, if we start with a simple module, will lead us to a non-simple module, we have not yet introduced clocks. Now choose a metric g on $\mathrm{Simp}_2(H(\sigma))$. Integrating along the integral curve will give us an arc-length, which we interpret as the amount of time until decay. We express the chosen metric g in local coordinates on $\mathrm{Simp}_2(H(\sigma))$ using the trace ring of invariants of the module given by Eq. (6),

$$\begin{aligned} u_1 &= \mathrm{tr}(dx) &&= t_1\\ u_2 &= -\det(dx) &&= t_2(1+t_3)\\ u_3 &= \mathrm{tr}(xdx) &&= (1+t_3)^2 + t_2t_5\\ u_4 &= \mathrm{tr}(x) &&= t_4\\ u_5 &= -\det(x) &&= t_5(1+t_3) \,.\end{aligned} \tag{11}$$

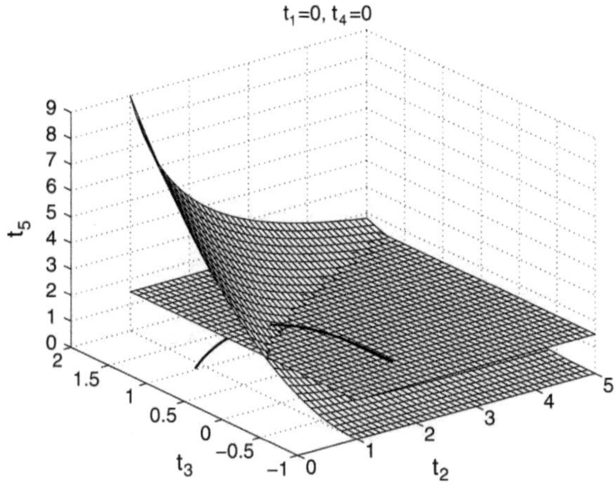

Fig. 1 A subset of $\mathrm{Simp}_1(C(2))$ where $t_1 = t_4 = 0$. The *curved surface* is the subset consisting of non-simple modules. Each point not on this surface correspond to a simple module. The *plane surface* is the subset where $t_5 = 1.5$. One of the curves start at the point $(0, 2.25, -0.5, 0, 1.5)$ and is drawn until it reaches a non-simple module. The other curve, *below* the curved surface, represents the same isomorphism classes of modules

Thus we let time be a metric g expressed on $\mathrm{Simp}_1(k[u_1, ..., u_5])$. For a general starting point $(t_1, t_2, t_3, t_4, t_5) \in \mathrm{Simp}_1(C(2))$, the evolution of the system is given as the following parametrised curve in $\mathrm{Simp}_1(k[u_1, ..., u_5])$,

$$\begin{aligned} u_1' &= u_1 \cos\tau + u_4 \sin\tau \\ u_2' &= -u_1 u_4 \sin\tau\cos\tau + u_2 \cos^2\tau + u_3 \sin\tau\cos\tau + u_5 \sin^2\tau \\ u_3' &= (u_1^2 + u_4^2)\sin\tau\cos\tau + 2(u_5 + u_2)\sin\tau\cos\tau + u_3 \\ u_4' &= u_1 \sin\tau + u_4 \cos\tau \\ u_5' &= -u_1 u_4 \sin\tau\cos\tau + u_2 \sin^2\tau + u_3 \sin\tau\cos\tau + u_5 \cos^2\tau\ , \end{aligned} \tag{12}$$

where $u_1, .., u_5$ are given by Eq. (11). As defined above, time passed is measured as the arc-length of this curve. We have a non-simple module if the polynomial of Eq. (7) is zero. By using Eq. (11) this can be written as

$$f = (u_3')^2 - 4u_2'u_5' + u_1'u_3'u_4' + (u_1')^2 u_5' + (u_4')^2 u_2' = 0\ . \tag{13}$$

If we use Eq. (11) we can find the corresponding points for a point in $\mathrm{Simp}_1(C(2))$, which resulted in the two corresponding integral curves of Fig. 1.

The new local coordinates of Eq. (11) means that the image in Fig. 1 becomes the image of Fig. 2. The two curves of Fig. 1 correspond to the curve of Fig. 2. The following theorem show that if $k = \mathbf{C}$, then each point in $\mathrm{Simp}_2(H(\sigma))$ corresponds to four points in $\mathrm{Simp}_1(C(2))$.

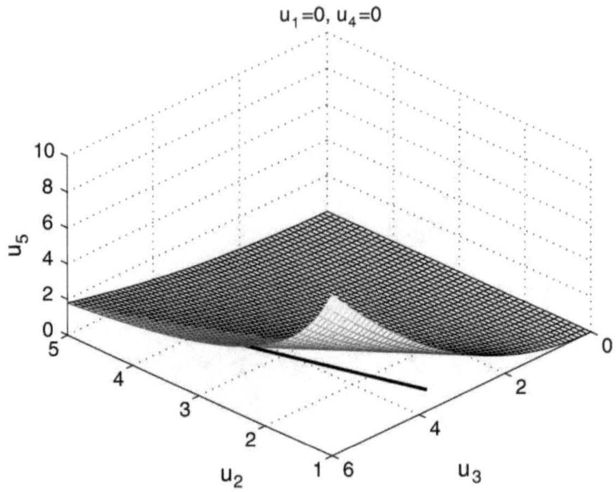

Fig. 2 A visualization of a subset of $\mathrm{Simp}_2(H(\sigma))$ parametrised by the coordinates of Eq. (11) where $u_1 = u_4 = 0$. The *curved surface* is the subset consisting of non-simple modules. Each point not on this surface correspond to a simple module. The curve corresponds to the curves in Fig. 1. The arc-length of the curve is the time until decay

Theorem 4.2 *Let $k = \mathbf{C}$ and consider the versal family given by Eq. (6). Let $(t_1, t_2, t_3, t_4, t_5)$ be a point in $Simp_1(C(2))$ such that $t_2 \neq 0, t_5 \neq 0$. The following four points in $Simp_1(C(2))$ corresponds to isomorphic modules,*

$$(t_1, t_2, t_3, t_4, t_5)\left(t_1, \frac{(1+t_3)\sqrt{t_2}}{\sqrt{t_5}}, \sqrt{t_2 t_5} - 1, t_4, \frac{(1+t_3)\sqrt{t_5}}{\sqrt{t_2}}\right)$$
$$(t_1, -t_2, -2-t_3, t_4, -t_5)\left(t_1, -\frac{(1+t_3)\sqrt{t_2}}{\sqrt{t_5}}, -\sqrt{t_2 t_5} - 1, t_4, -\frac{(1+t_3)\sqrt{t_5}}{\sqrt{t_2}}\right). \tag{14}$$

These are the only points corresponding to this isomorphism class of modules.

Proof Let $(t_1', t_2', t_3', t_4', t_5') \in \mathrm{Simp}_1(C(2))$ be a point corresponding to the same module. The invariants of Eq. (11) gives, if $1 + t_3' \neq 0$, that

$$((1+t_3')^2)^2 - u_3(1+t_3')^2 + u_2 u_5 = 0\,. \tag{15}$$

Solving the above equation gives

$$(1+t_3')^2 = (1+t_3)^2 \text{ or } (1+t_3')^2 = t_2 t_5\,. \tag{16}$$

For $1 + t_3' = 0$, we obtain from Eq. (11) that either $1 + t_3 = 0$ or $t_2 = t_5 = 0$, which means that Eq. (16) also holds for $1 + t_3' = 0$. Now using Eqs. (16) and (11) gives the four points of Eq. (14). We have now shown that there cannot be more than these four points. To show that these four points corresponds to isomorphic modules, i.e. that there exist U such that

$$U\rho(x)U^{-1} = \rho'(x), \quad U\rho(dx)U^{-1} = \rho'(dx) \,, \tag{17}$$

we compute the following U's;

$$U = \begin{pmatrix} 1 & 0 \\ 0 & -1 \end{pmatrix}, \; U = \begin{pmatrix} \frac{-t_1 t_5 \sqrt{t_2} - t_4(1+t_3)\sqrt{t_2}}{(1+t_3)^2 - t_2 t_5} & \sqrt{t_2} \\ \sqrt{t_5} & \frac{-t_1(1+t_3)\sqrt{t_5} - t_2 t_4 \sqrt{t_5}}{(1+t_3)^2 - t_2 t_5} \end{pmatrix}. \tag{18}$$

Combining the two U's give the transformation to the last point. □

5 Calculating the Arc-length

We want to calculate the arc-length of the parametrised curves given by Eq. (12) between the starting point at $\tau = 0$ and the point where decay have occurred. This may be interpreted as the time until decay.

Let $g = \sum_{i,j} g_{ij} du_i du_j$ be a metric on $\mathrm{Simp}_2(H(\sigma))$ and let γ be a path parametrized by τ, and let the coordinates of γ be given by functions $u_i(\tau)$. The arc length s of γ is then

$$s = \int_{\tau_0}^{\tau_1} \sqrt{\sum_{i,j} g_{ij}(u(\tau)) \frac{du_i}{d\tau} \frac{du_j}{d\tau}} d\tau \,. \tag{19}$$

The interesting thing is to calculate the arc-length between a starting point τ_0, corresponding to a simple module (initial point), and a point τ_1, corresponding to a non-simple module, where the system has experienced decay.

To calculate analytically the arc length for curves given by Eq. (12), is a tedious task, and we have therefore opted to calculate the path length numerically using MATLAB [7]. In addition we numerically search for semi-simple modules along the path, as we know that $\tau = \pi/4$ only provides us with a maximum value. To show the path lengths visually, we have restricted ourselves to starting points in a two-dimensional subset of $\mathrm{Simp}_1(C(2))$ where $t_1 = 0, 0 < t_2 < 5, -1 < t_3 < 4, t_4 = 0, t_5 = 1.5$. This corresponds to the plane $t_5 = 1.5$ of Fig. 1. If we let the metric on $\mathrm{Simp}_2(H(\sigma))$, corresponding to the notion of decay, be given by $g_{ij} = \delta_{ij}$, the time until decay for these starting points are shown in Fig. 3. The slopes of the graph of Fig. 3 near the Formanek center are due to the numerical calculations. If done analytically, the non-simple modules would probably have been cut out from otherwise gently sloped graphs.

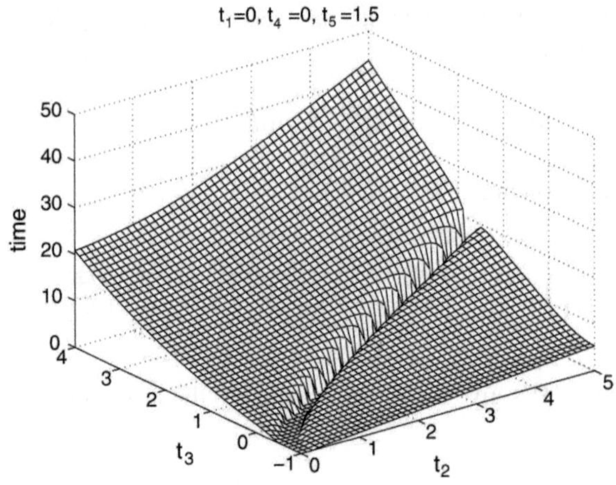

Fig. 3 Time until decay: The *vertical* axis measures the time it takes for the Dirac derivation to push a point in the subset $t_1 = 0, 0 < t_2 < 5, -1 < t_3 < 4, t_4 = 0, t_5 = 1.5$ into decay

6 Back to the Physics

In physics, we would have a device for producing particles with a given probability for specific measurements of the observables. The eigenvalues of the observables would be these measurements. In quantum mechanics, the commutation relation between position and momentum observables is fixed. However, when we consider finite dimensional representations, which is the case in this paper, this relation have to be changed. We therefore obtain an extra degree of freedom.

For the harmonic oscillator given by Eq. (5), consider observables with two different eigenvalues since we use two-dimensional representations. The eigenvalues of the observables of Eq. (6) are

$$\begin{aligned} \lambda_x &= \frac{1}{2}t_4 \pm \frac{1}{2}\sqrt{t_4^2 + 4t_5(1+t_3)} = \frac{1}{2}u_4 \pm \frac{1}{2}\sqrt{u_4^2 + 4u_3} \\ \lambda_{dx} &= \frac{1}{2}t_1 \pm \frac{1}{2}\sqrt{t_1^2 + 4t_2(1+t_3)} = \frac{1}{2}u_1 \pm \frac{1}{2}\sqrt{u_1^2 + 4u_3} . \end{aligned} \tag{20}$$

If we have a device producing particles with $\lambda_x = \pm\alpha, \lambda_{dx} = \pm\beta$, the possible starting points are $(0, t_2, t_3, 0, t_5) \in \mathrm{Simp}_1(C(2))$ such that

$$\alpha^2 = t_5(1+t_3), \quad \beta^2 = t_2(1+t_3) , \tag{21}$$

where $t_2 \neq 0, t_3 \neq -1, t_5 \neq 0$. We have that certain one-dimensional subsets of points in $\mathrm{Simp}_1(C(2))$ satisfies Eq. (21) and we can assign a probability for each of the points in this subset to be the starting point. The different points have different decay times, and therefore we may give probabilities to the decay times, i.e. we

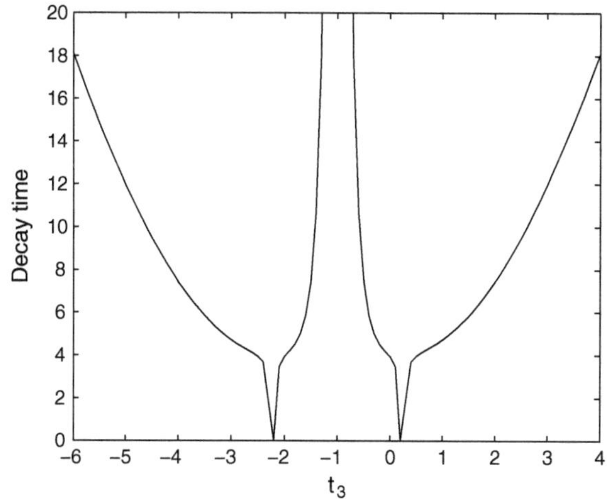

Fig. 4 Decay time as a function on the subset of $\mathrm{Simp}_1(C(2))$ consisting of the t_3-axis, except the point $t_3 = -1$

cannot say that a product of the particle producing device has a determined decay time.

We may for example consider the subset consisting of the t_3-axis in the space $\mathrm{Simp}_1(C(2))$ except for the point $t_3 = -1$. Then the possible start points are parametrised by t_3 such that we consider $\left(0, \beta^2/(1+t_3), t_3, 0, \alpha^2/(1+t_3)\right)$. In Fig. 4 we plot decay times as a function of t_3 for $\alpha = 1$, $\beta = 1.5$. Note that this function is not defined for $t_3 = -1$. Also note the zero values for $t_3 = -1 \pm \sqrt{1.5}$. However, the sloping of the graph near these zero values are due to the numerical calculations.

We are now able to talk about an average of decay times. Also note that the extra degree of freedom in this example was due to the fact that we did not specify the commutation relation between position and momentum. To choose t_3 is in effect to choose this commutation relation when the possible eigenvalues for position and momentum are given.

References

1. Huang, K.: Quantum Field Theory: From Operators to Path Integrals. Wiley, New York (1998)
2. Imenes, O.G.: The Concept of Charge in a Model based on Non-commutative Algebraic Geometry. Series of dissertations submitted to the Faculty of Mathematics and Natural Sciences, University of Oslo No. 1103. Unipub, Oslo (2011)
3. Laudal, O.A.: Noncommutative algebraic geometry. Rev. Mat. Iberoamericana **19**, 509–580 (2003)
4. Laudal, O.A.: Time-space and space-times. In: Fuchs, J. et al. (eds.) Noncommutative Geometry and Representation Theory in Mathematical Physics, 5–10 July 2004. AMS, Contemporary Mathematics, vol. 391, pp. 249–280, Providence, Rhode Island (2005)

5. Laudal, O.A.: Phase spaces & deformation theory. Acta Applicandae Mathematicae **101**, 191–204 (2008)
6. Laudal, O.A.: Geometry of Time-Spaces. World Scientific, Singapore (2011)
7. MATLAB version 7.10.0., The MathWorks Inc., Natick, Massachusetts (2010)
8. Weinberg, S.: The Quantum Theory of Fields I. Cambridge University Press, Cambridge (1995)

On the Detection of Permutation Polynomials

Mazen Gharibah

Abstract Multivariate Public key cryptosystems are widely spread and ever evolving domain. This study aims to find new techniques to characterize and detect permutation polynomials over finite fields, which enable us to find trapdoor, one way, functions that are essential to build robust cryptosystems. Let f be a polynomial over F_q, a finite field of order q, where $q = p^m$, p is a prime number. If f induces a bijective mapping, one-to-one mapping, of F_q, we call f a permutation polynomial over F_q. In order to detect these polynomials, we constructed a program implementing multiple algorithms based on Galois field arithmetic. As a result, we have the number of all possible permutation polynomials in the fields F_4, F_8 and F_{16}.

1 Introduction

The entire cryptographic schemes and systems are results of a long time mathematical researches and studies. Every cryptographic algorithm is based upon mathematical keystones, such as modular arithmetic [1], finite fields [9], and permutation polynomials [8].

Most of today's public key cryptosystems [5] are based on permutation polynomials, two common examples are RSA [7] and Dickson scheme [11].

Current cryptography algorithms were believed to be unbreakable by cryptanalysis attacks, until Peter Shor presented his algorithm that can factorize an integer N in polynomial time [13]. Shor's algorithm relies on what is called Quantum Computers [2], the new generation of computers based on Richard P. Feynman's proposal in 1984. This new era of quantum computers along with their quantum algorithms raise a big challenge for modern cryptographers.

M. Gharibah (✉)
Brno University of Technology, Brno, Czech Rep.
e-mail: mazen.gharibah@gmail.com

A. Makhlouf et al. (eds.), *Algebra, Geometry and Mathematical Physics*, Springer Proceedings in Mathematics & Statistics 85, DOI: 10.1007/978-3-642-55361-5_39,

Many scientists believe it is time to be prepared for the *quantum challenge* and it is time to focus on other classes of cryptography, like post quantum cryptography [2], which are not vulnerable to Shor's algorithm or any other quantum-based attacks.

Thus, a successful attempt to improve some of the public key algorithms should start with one of the most essential building blocks, i.e. Permutation Polynomials (PP). Over the last few decades, the interest for studying and analyzing these polynomials has been gradually increased considering their applications in key exchanges schemes, encryption and decryption, for an example see [6].

Unfortunately, a few techniques has been presented to test whether a given polynomial is a permutation polynomial or not and to state how many of these polynomials are there, most of these techniques are complicated, theoretical and without real world implementations, for example consider [10] and [15].

In this paper, a new application for determining if a given polynomial is a permutation polynomial and for listing the accurate number of such polynomials in a given finite field is proposed, along with a detailed analysis about these polynomials, their definition, their features and their number in the fields F_4, F_8, and F_{16}. (See Sect. 4).

Due to the excessive mathematical aspects behind this research, some concepts are briefly reviewed. In Sect. 2, modular arithmetic and finite fields, their definitions and some of their applications have been re-examined.

In Sect. 3, a general view of polynomials over finite fields applications are shortly studied, particularly Multivariate Quadratic Polynomials (MQP) in public key cryptography [17].

2 Modular Arithmetic

Modular arithmetic is a symbolic transformation; it can be viewed as a mapping between the integers domain and the integers *modulo* p domain, where p is also an integer.

This mapping makes the arithmetic operations easier, thus saves computational time. Once the solution in the second domain is obtained, it can be converted back to the original domain.

Definition 1 Let $m \in Z$ be a non-zero integer. For each $a \in Z$, there is a unique b, with $a \equiv b \ mod \ m$ and $0 \leq b < |m|$

We can use b as a new representation of a in $mod \ m$ domain, for example: We can re-represent 2 by 7 in mod-5: $7 \equiv 2 \ (mod \ 5)$.

Definition 2 a is congruent to b *modulo* m, if and only if $b = a + mk$, where $k \in Z$. see the proof in [16].

Definition 3 Modulo m maps all integers into the set $Z_m = \{0, 1, 2, ..., m - 1\}$. Proof is in [16].

The set from Definition 3. can be represented using different notations: Z_m, $Z/(m)$, Z/mZ. But what is Z_m? In order to answer this question, some notions have to be considered.

2.1 Rings and Fields

A ring is a triple $(S, +, .)$, S is a set. $+, .$ (Addition and multiplication) are binary operators satisfying the following axioms: $for\ all\ a, b, c \in S$.

1. Associativity: $(a + b) + c = a + (b + c)$, $(a \cdot b) \cdot c = a \cdot (b \cdot c)$.
2. Distributivity of Multiplication over addition: $a \cdot (b + c) = (a \cdot b) + (a \cdot c)$.
3. Existence of additive and multiplicative identities: $0 + a = a$, $1 \cdot a = a$.
4. Existence of additive inverse: there exist $-a \in S$ such that $a+(-a) = (-a)+a = 0$.

When the commutativity condition is satisfied, i.e. $a + b = b + a$, $a \cdot b = b \cdot a$, then the ring is called a commutative ring.

Proposition 1 *Z_m is a commutative ring. Why? See* [3].

2.2 Finite Fields

A field is a commutative ring with at least two elements, satisfying the existence of multiplication inverse axiom.

There exists an additive inverse in Z_m, but a multiplicative inverse for every element of Z_m does not exist, as a result Z_m is not a field.

In abstract algebra, a finite field or Galois field (after the French scientist Galois) is a field that contains finite number of elements.

Theorem 1 *For prime m, every element $a \in Z$ will be relatively prime to m. That implies that there exists a multiplicative inverse for every $a \in Z$ for prime m. for a proof see* [3].

Proposition 2 (from Theorem 1.) *The Commutative ring Z/mZ is finite field if and only if m is a prime.*

Definition 4 If p is a prime number, then the integers $mod\ p$ form a field denoted $F_p = \{0, 1, 2, p - 1\}$, its elements are the congruence classes of integers $mod\ p$, under $mod\ p$ addition and multiplication.

Proposition 3 *If p is any prime and m is a positive integer, we have the finite field F_{p^m} with p^m elements; this is an extension field of the finite field $F_p = Z/pZ$ with p elements. See* [8].

Theorem 2 *(prime field existence): For any prime power p^m , a field of order p^m exists. The proof is in* [14].

Theorem 3 *(prime field uniqueness): for any prime integer p and any integer m, $m > 0$ there is a unique field of p^m elements denoted F_{p^m}. See* [14] *for the proof.*

2.3 Polynomials in Finite Fields

Let $F[X]$ denotes the polynomials over F_p in one indeterminate X.

Proposition 4 *A polynomial $l(X) \in F_p[X]$ if and only if all its coefficients are elements over F_p. Obviously!*

Definition 5 We call $f(x)$ an irreducible polynomial over the field F_p , if $f(x)$ cannot be expressed as a product of two polynomials both in F_p, and both of them of a degree lower than $f(x)$ and greater than zero. Irreducible polynomials have no roots over their defined fields.

For example, considering the field F_{2^3} or F_8 , $f(x) = x^3 + x + 1$ is an irreducible polynomial over the field F_8.

Obviously, in all computer related calculations, we use the field F_{2^m} and all the operations over the field will be in $mod\ f(x)$. We can represent the elements of the field F_{2^3} using eight distinct polynomial representations:

$$0, 1, x, x + 1, x^2, x^2 + 1, x^2 + x, x^2 + x + 1.$$

3 Polynomials Over Finite Fields Applications

By generalizing the idea of polynomials in one indeterminate to polynomials with n variables, multivariate polynomials will be acquired, so the multivariate polynomials over the finite field will be denoted as $F_q[X_1, X_2, ..., X_n]$ or simply $F_q[X]$ where $X = (X_1, ..., X_n)$.

The most efficient and robust post quantum cryptosystems are based on multivariate polynomials, multivariate quadratic polynomials cryptography, see [2], presents a good example of this idea. The safety and security of these systems partially rely on the difficulty of solving polynomial equations with multiple variables, among other reasons which are beyond the scope of this paper.

3.1 Multivariate Quadratic Polynomials Cryptography (MQC)

As mentioned earlier, these kinds of polynomials are widely used in post quantum cryptography applications. For more information on MQC and its applications and variations see [17] and [2]. All public key schemes based on these polynomials have the following general form:

We have the polynomial vector $P = (p_1, p_2, ..., p_m)$

$$p_i = \sum_{1 \le i \le k \le m} \gamma_{i,j,k} x_j x_k + \sum_{j=1}^{n} \beta_{i,j} x_j + \alpha_i$$

$i \le i \le m$ and $\gamma_{i,j,k}, \beta_{i,j}, \alpha_i \in F$

3.1.1 Private Key

The private key is the triple $(S, P', T) : S \in Aff^{-1}(F^n), T \in Aff^{-1}(F^m)$ are affine transformations.

$P' \in MQ(F^n, F^m)$ is a polynomial vector.

$P' := (p'_1, p'_2, ..., p'_m)$, m polynomials each one depends on the input variables $(x'_1, x'_2, ..., x'_n)$.

3.1.2 Public Key

The public key is the composition:

$$S \circ P' \circ T$$

The output, $y = S(P'(T(x)))$

P' called the central mapping, the building block for all Multivariate Quadratic Cryptosystems (MQC), it varies from one application to another, depending on the used technique.

$$S(x) = M_s x + \upsilon_s \text{ where } M_s \in F^{(n \times n)} \text{ and } \upsilon_s \in F^n.$$

$$T(x) = M_T x + \upsilon_T \text{ where } M_T \in F^{(m \times m)} \text{ and } \upsilon_T \in F^m.$$

4 Permutation Polynomials

The most important kind of polynomials over the field $F_q[X]$ is the permutation polynomials (PP). Most of finite field based cryptosystems use these polynomials in Encryption and Decryption methods.

Very few algorithms exist to test whether a given polynomial is a permutation polynomial or not.

The general idea of these polynomials is simple. Consider the following set:

$$(0, 1, 2, 3, 4, 5, 6, 7)$$

If a function f was applied and the output was, for instance:

$$(2, 1, 3, 0, 7, 5, 6, 4)$$

It is simply said, f is a *permutation* function, because it permutes the input.

4.1 Definition and Properties

Considering the finite field F_q, where $q = p^m$ elements; p is a prime.

We call the polynomial $L(X) \in F_q[X]$ a *permutation polynomial*, if it induces a bijective, one-to-one, mapping from F_q to itself.

More specifically, a permutation polynomial $L(X) \in F_q[X]$, permutes the elements of F_q, under addition and multiplication. And has the form:

$$L(X) = \sum_{i=0}^{k} a_i X^{p^i}$$

Hermite's Criterion : $L(X)$ is a permutation polynomial in F_q, if and only if :

1. $L(X)$ has exactly one root in F_q.
2. For each integer t, with $1 \leq t \leq q-2$ and $t \equiv 0(mod\ p)$ the reduction $[L(X)]^t$ $mod(x^q - x)$ has a degree less than or equal to $q-2$ and greater than zero. [4].

Corollary 1 *If $L(X)$ is a permutation polynomial in F_q of degree $n \geq 1$, then $n|(q-1)$.*

4.2 Algorithm for Detecting Permutation Polynomials

This algorithm aims to decide if a polynomial given as input is a permutation polynomial and to count how many of these polynomials exist in a chosen finite field.

Suppose two polynomials L and K of the second degree are given, each one of them has two variables x and y.

$L = a_1x + b_1y + c_1xy + d_1x^2 + e_1y^2$

$K = a_2x + b_2y + c_2xy + d_2x^2 + e_2y^2$

Where: $a_1, b_1, c_1, d_1, e_1, a_2, b_2, c_2, d_2, e_2 \in F_q$ and $x, y \in F_q[X][Y]$.

A special library for finite fields operations has been used, it is called Galois Field Arithmetic library, an open source library, for its original specifications and description see [12]. This library has been adapted to serve the purpose of our algorithm, by developing it, so it can deal with multiple operations at once, using parallel programming techniques and it has been modified so it can work with polynomials with more than one variable instead of polynomials with one indeterminate as it was with the original library.

The algorithm consists of three main steps. First step: In this step, a two dimensional array will be initiated. It will be called *mat*, the first dimension of this array is for handling the values and the variations of the first polynomial, the second dimension, is for the second polynomial.

Second step: All the possible cases of the two polynomials coefficients will be visited through a main loop. This main loop includes sub loops for each coefficient of each polynomial, in every one of these coefficients sub loops, all the values of x and y will be considered.

The result of each of these nested loops will be stored inside the two dimensional array *mat*, where the first dimension will held the value of the first polynomial L after substituting its coefficients values taken from the loops and performing the addition and multiplication with respect to the rules of the finite field F_{p^m}, and the second dimension is dedicated to the second polynomial K in the same way. the first and second step are shown in the following algorithm, Algorithm 1.:

Algorithm 1 Initialization procedure

procedure INIT ▷ init proc. starts at the beginning of the algorithm
use Modified Finite Field Arithmetic library;
define CLASS Polynomial;
define CLASS FiniteField;
define Polynomial L, K;
define FiniteField F;
initiate mat [L][K];
  for $a_1, b_1, c_1, d_1, e_1, a_2, b_2, c_2, d_2, e_2 = 0 \rightarrow q - 1$ **do**
    for $x, y = 0 \rightarrow p - 1$ **do**
assign mat [substitute(L)][substitute(K)]; ▷ substitute proc. Apply addition and multiplication to the input polynomial
call checkMatrix (mat[l][k]);
      end for
    end for
end procedure

Third Step: The previous array *mat* will be passed as an input to a procedure called *checkMatrix()*. The *checkMatrix()* procedure, takes one parameter, the two dimensional array *mat*, then it tests whether the values stored in *mat* induces a permutation polynomial or not, by comparing each element in the array with the rest of the elements, then the result of this comparison will be stored into a static constant variable called *varTemp*. Finally, after visiting all elements in a given array, the value stored in *varTemp* will be checked, and a permutation polynomial counter *ppCounter* will be increased. Now, the procedure will return to the second step, the main thread. New values for the coefficients will be taken from the nested loops, thus creating a new array with brand new values for its elements and again this array will be passed to the third step, as an input for the *checkMatrix()* procedure. Step three is illustrated below in Algorithm 2.:

Algorithm 2 checkMatrix procedure

```
procedure CHECKMAT(mat[l][k])
    for i = 0 → n − 1 do
        for j = i + 1 → n do
            if mat[i][0] = mat[j][0] AND mat[i][1] = mat[j][1] then varTemp = 1;
            end if
        end for
    end for
    if varTemp = 0 then ppCounter ++;
    end if
varTemp ← 0;
end procedure
```

This cycle of step two and three will continue until every possible value for every coefficient along with all the values of x and y are examined.

4.3 The Number of Permutation Polynomials in the Fields F_4, F_8, F_{16}

The outputs of the previous program, in Sect. 4.2, after execution are shown in the following tables.

Table 1, shows the number of permutation polynomials over the field F_4 from different degrees of the polynomials L and K .

In Table 2, the number of permutation polynomials over the field F_8 is presented. The number of permutation polynomials over the field F_{16} is displayed in Table 3.

Table 1 The number of permutation polynomials over F_4

Degree of L	Degree of K	Number of PP
1	1	2880
1	2	23040
1	3	368640
2	1	23040
2	2	322560
2	3	5160960
3	1	368640
3	2	5160960
3	3	82575360

Table 2 The number of permutation polynomials over F_8

Degree of L	Degree of K	Number of PP
1	1	3528
1	2	256704
1	3	9727984
2	1	256704
2	2	2288192
2	3	158989824
3	1	9727984
3	2	158989824
3	3	?[a]

[a](?) means it cannot be solved in polynomial time using a normal hardware.

Table 3 The number of permutation polynomials over F_{16}

Degree of L	Degree of K	Number of PP
1	1	61200
1	2	4406400
1	3	317260800*[a]
2	1	4406400
2	2	?[b]
2	3	?
3	1	317260800*
3	2	?
3	3	?

[a](*) means approximate number.
[b](?) means it cannot be solved in polynomial time using a normal hardware.

References

1. Apostol, T.M.: Introduction to Analytic Number Theory. Springer, New York (1976)
2. Bernstein, D.J., Buchmann, J., Dahmen, E.: Post-Quantum Cryptography. Springer, Berlin (2009)
3. Bini, G., Flamini, F.: Finite Commutative Rings and their Applications. Kluwer Academic Publishers, Boston (2002)
4. Hermite, Ch.: Sur les fonctions de sept lettres. C. R. Acad. Sci. **57**, 750–757 (1863)
5. Katz, J., Lindell Y.: Introduction to Modern Cryptography. Chapman and Hall/CRC, Boca Raton (2008)
6. Levine, J., Brawley, J.V.: Some cryptographic applications of permutation polynomials. Cryptologia **1**, 76–92 (1977)
7. Lidl, R., Müller, W.B.: Permutation polynomials in RSA-cryptosystems. In: Chaum, D. (eds.) Advances in Cryptology, Crypto 83, pp. 293–301. Plenum Press, New York (1984)
8. Lidl, R., Niederriter, H.: Finite Fields (2nd edn). Cambridge University Press, Cambridge (1997)
9. Lidl, R., Niederriter, H.: Introduction to Finite Fields and their Applications. Cambridge University Press, Cambridge (1986)
10. Masuda, A., Panario, D., Wang, Q.: The number of permutation binomials over F_{4p+1} where p and $4p+1$ are primes. Electron. J. Comb. **13** (2006)
11. Müller, W.B., Nöbauer, W.: Some remarks on public-key cryptosystems. Studia Sci. Math. Hungar. **16**, 71–76 (1981)
12. Partow, A.: Galois Field Arithemetic Library. http://www.partow.net/projects/galois/index.html
13. Shor, P.W.: Polynomial-time algorithms for prime factorization and discrete logarithms on quantum computers. In: 35th Annual Symposium on Foundations of Computer, Science (1994)
14. Stewart, I.: Galois Theory. Chapman and Hall, London (1989)
15. Wang, Q., Akbary, A.: On some permutation polynomials over finite fields. Int. J. Math. Math. Sci. **16**, 2631–2640 (2005)
16. Witno, A.: Theory of Numbers. BookSurge Publishing, Charleston (2008)
17. Wolf, C.: Introduction to multivariate quadratic public key systems and their applications. In: Proceedings of YACC, pp. 44–55. (2006)

Scalar-Tensor and Multiscalar-Tensor Gravity and Cosmological Models

Piret Kuusk, Laur Järv and Erik Randla

Abstract We consider scalar-tensor and multiscalar-tensor theories of gravity and their formulations in the Jordan and the Einstein conformal frames. After constructing a generic multi-scalar tensor action, we derive its full equations of motion as well as equations for homogeneous isotropic cosmological models in the Jordan frame. We use methods of dynamical systems in the case of two scalar fields to determine the fixed point and conditions for its being an attractor.

1 Introduction

For decades mathematical cosmology has been based on Einstein's theory of general relativity (GR) where the gravitational interaction is described by the metric tensor $g_{\mu\nu}$ of a Riemannian spacetime. Present observational data are in good agreement with general relativistic Friedmann-Lemaître-Robertson-Walker (FLRW) cosmology with homogeneous and isotropic flat ($k = 0$) 3-space, a cosmological constant $\Lambda > 0$ and additional cold dark matter (ΛCDM model). However, this model seems to be somewhat phenomenological and fine-tuned: extremely small observational value of Λ, very special initial and/or boundary conditions, etc.

From a mathematical point of view it is enticing to consider other possible cosmological models based on theories of gravity appropriately generalizing Einstein's GR. Such alternative theories may be constructed by supplying extra fields, modifying the standard Einstein-Hilbert action to include an arbitrary function of curvature

P. Kuusk (✉) · L. Järv · E. Randla
Institute of Physics, University of Tartu, Riia 142, 51014 Tartu, Estonia
e-mail: piret@fi.tartu.ee

L. Järv
e-mail: laur.jarv@ut.ee

E. Randla
e-mail: erik.randla@ut.ee

A. Makhlouf et al. (eds.), *Algebra, Geometry and Mathematical Physics*, 661
Springer Proceedings in Mathematics & Statistics 85, DOI: 10.1007/978-3-642-55361-5_40,

invariants, adding extra dimensions and branes, etc. (For a comprehensive review see [1].) The focus of the present paper is scalar-tensor gravity (STG) and its natural extension to multiple scalar fields (MSTG). The former is a paradigmatic example of a versatile modification of GR, while variants of the latter have received more attention only recently and its comprehensive treatment is still lacking.

The action functional of a general STG contains up to four arbitrary functions of the scalar field, two of which can be fixed using a group of field redefinitions containing two free functional degrees of freedom [2]. The group transforms STG from one conformal frame and scalar field parametrization to another. A rather natural physical interpretation of STG is obtained in the so-called Jordan conformal frame, where the scalar field Φ is coupled to the scalar curvature R via the $\mathscr{F}(\Phi)R$ term, but not directly to the matter fields, whereas the scalar field kinetic term involves an arbitrary function $\mathscr{Z}(\Phi)$; now $\mathscr{F}(\Phi)$ acts as a variable part of gravitational "constant". In the so-called Einstein conformal frame the action functional is reminiscent of the Einstein GR with a minimally coupled scalar field, but the latter is directly coupled also to the matter fields generating a nonconservation of the matter tensor. The Einstein frame may be more suitable for finding exact solutions, since the corresponding results of GR can be used. For investigations in cosmology the Jordan frame is often preferred since the matter tensor is conserved there.

STG was generalized to include an arbitrary number of scalar fields by Damour et al. [3, 4] using the Einstein conformal frame. Besides occasional consideration [5], a renewed interest in MSTG was ignited a couple of years ago by the claim that the Standard Model Higgs field can support inflation provided that it is non-minimally coupled to gravity [6]. In the wake of this idea several inflationary scenarios have been proposed and investigated with multiple non-minimally coupled scalars [7–18], sometimes also embedded into the framework of supergravity [19, 20]. In this line of studies the coupling between scalars and curvature was specified by constant parameters (and not a generic function), while the calculations were carried out in the Einstein frame. Some work has been done also in the Jordan frame, viz. the calculation of primordial perturbations [21], and an effort to construct one-loop effective action for a scalar multiplet non-minimally coupled to gravity [22].

The paper is organized as follows. In Sect. 2 we briefly review the basic equations of a general scalar-tensor gravity with one scalar field in the Jordan frame as well as in the Einstein frame. In Sect. 3 we generalize the Jordan frame equations to the case with N scalar fields, non-minimally coupled to curvature via a generic function of scalar fields. For clarifying the physical interpretation of the theory we choose a gauge in the target space of scalar fields which specifies one scalar field Ψ to act as a variable part of gravitational "constant" as in STG, leaving the other $N-1$ scalar fields minimally coupled to curvature (but interacting with Ψ). In Sect. 4 we turn to cosmology and consider models with isotropic and homogeneous flat 3-spaces in the framework of the multiscalar-tensor gravity. We present the general equations and also equations with only two scalar fields. In the latter case we find the fixed point of the corresponding dynamical system and investigate conditions for its being an attractor. Section 5 is a summary and outlook.

2 A General Scalar-Tensor Theory with One Scalar Field

We begin with a short review of the basic equations of scalar-tensor gravity with one scalar field.

2.1 Action Functional and Field Equations in the Jordan Frame

A general scalar-tensor theory in the Jordan frame is given by the action functional [23]

$$S = \frac{1}{2\kappa^2}\int\limits_{V_4} d^4x\sqrt{-g}\left(\mathscr{F}R - \mathscr{Z}g^{\mu\nu}\partial_\mu\Phi\partial_\nu\Phi - 2\kappa^2\mathscr{U}\right) + S_m[g_{\mu\nu},\chi_m]\,. \quad (1)$$

Here κ^2 is the non-variable part of the gravitational "constant", $\mathscr{F} = \mathscr{F}(\Phi)$, $\mathscr{Z} = \mathscr{Z}(\Phi)$, $\mathscr{U} = \mathscr{U}(\Phi)$ denote arbitrary functions, which determine a distinct STG if specified, and S_m is the matter contribution to the action as all other fields are included in χ_m. Since $\mathscr{F}(\Phi)$ acts as a variable part of the gravitational coupling "constant" $2\kappa^2/\mathscr{F}(\Phi)$ we assume $\mathscr{F}(\Phi) > 0$ in order to keep gravitation attractive for non-exotic matter.

The corresponding field equations read

$$\begin{aligned}\mathscr{F}G_{\mu\nu} \equiv& \mathscr{F}(R_{\mu\nu} - \frac{1}{2}g_{\mu\nu}R)\\ =& - g_{\mu\nu}\nabla^2\mathscr{F} + \nabla_\mu\partial_\nu\mathscr{F} - \frac{1}{2}g_{\mu\nu}\mathscr{Z}\partial_\rho\Phi\partial^\rho\Phi + \mathscr{Z}\partial_\mu\Phi\partial_\nu\Phi\\ &- \kappa^2 g_{\mu\nu}\mathscr{U} + \kappa^2 T^{(\chi)}_{\mu\nu}\,,\end{aligned}$$

$$\begin{aligned}\left(2\mathscr{F}\mathscr{Z} + 3\frac{\partial\mathscr{F}}{\partial\Phi}\frac{\partial\mathscr{F}}{\partial\Phi}\right)\nabla^2\Phi^a =& - 3\frac{\partial\mathscr{F}}{\partial\Phi}\frac{\partial^2\mathscr{F}}{\partial\Phi\partial\Phi}\partial_\rho\Phi\partial^\rho\Phi - \frac{\partial\mathscr{F}}{\partial\Phi}\mathscr{Z}\partial_\rho\Phi\partial^\rho\Phi\\ &- \mathscr{F}\frac{\partial\mathscr{Z}}{\partial\Phi}\partial_\rho\Phi\partial^\rho\Phi - 4\frac{\partial\mathscr{F}}{\partial\Phi}\kappa^2\mathscr{U}\\ &+ 2\mathscr{F}\kappa^2\frac{\partial\mathscr{U}}{\partial\Phi} + \frac{\partial\mathscr{F}}{\partial\Phi}\kappa^2 T^{(\chi)}\,.\end{aligned}$$

Here $\nabla^2 \equiv g^{\mu\nu}\nabla_\mu\nabla_\nu$ and ∇_μ denotes the covariant derivative with respect to the metric $g_{\mu\nu}$. The matter tensor $T^{(\chi)}_{\mu\nu}$ follows from the matter action S_m as usual,

$$T^{(\chi)}_{\mu\nu}(g_{\mu\nu},\chi_m) = -\frac{2}{\sqrt{-g}}\frac{\delta S_m[g_{\sigma\tau},\chi_m]}{\delta g^{\mu\nu}}\,, \quad (2)$$

$$\nabla_\mu T^{(\chi)\mu}{}_\nu = 0\,.$$

2.2 Action Functional and Field Equations in the Einstein Frame

Let us introduce a conformal transformation of the Jordan frame metric $g_{\mu\nu}$ to the Einstein frame metric $\tilde{g}_{\mu\nu}$,

$$\tilde{g}_{\mu\nu} = \mathscr{F}(\Phi) g_{\mu\nu} ,$$

and complement it with a redefinition of the scalar field [23]

$$\left(\frac{d\varphi}{d\Phi}\right)^2 = \frac{1}{4\mathscr{F}^2(\Phi)} \left[2\mathscr{Z}(\Phi)\mathscr{F}(\Phi) + 3\frac{d\mathscr{F}(\Phi)}{d\Phi}\frac{d\mathscr{F}(\Phi)}{d\Phi}\right] .$$

The latter transformation introduces a natural condition on functions $\mathscr{F}(\Phi)$, $\mathscr{Z}(\Phi)$ which ought to be imposed in the Jordan frame formulation

$$2\mathscr{Z}(\Phi)\mathscr{F}(\Phi) + 3\,\frac{d\mathscr{F}(\Phi)}{d\Phi}\frac{d\mathscr{F}(\Phi)}{d\Phi} > 0 . \tag{3}$$

The Einstein frame action functional reads [23]

$$S = \frac{1}{2\kappa^2} \int\limits_{V_4} d^4x \sqrt{-\tilde{g}} \left[\tilde{R} - 2\tilde{g}^{\mu\nu}\partial_\mu\varphi\partial_\nu\varphi - 2\kappa^2\tilde{\mathscr{U}}(\varphi)\right] + S_m[A^2(\varphi)\tilde{g}_{\mu\nu}, \chi_m] \tag{4}$$

with $\tilde{R} \equiv R(\tilde{g})$, scalar potential $\tilde{\mathscr{U}}(\varphi) = \mathscr{F}^{-2}(\Phi)\mathscr{U}(\Phi)$ and the Einstein frame coupling function $A(\varphi) = \mathscr{F}^{-1/2}(\Phi(\varphi))$. The corresponding field equations are

$$\begin{aligned}
\tilde{G}_{\mu\nu} =& \kappa^2\tilde{T}_{\mu\nu} + 2\left(\partial_\mu\varphi\partial_\nu\varphi - \frac{1}{2}\tilde{g}_{\mu\nu}\partial_\rho\varphi\partial^\rho\varphi\right) - \kappa^2\tilde{g}_{\mu\nu}\tilde{\mathscr{U}}(\varphi) , \\
\tilde{\nabla}^2\varphi =& \frac{\kappa^2}{2}\left(-\alpha(\varphi)\tilde{T}(\tilde{g}) + \frac{d\tilde{\mathscr{U}}}{d\varphi}\right) .
\end{aligned}$$

Here we have a redefined coupling function $\alpha(\varphi) \equiv \frac{d \ln A(\Phi(\varphi))}{d\varphi}$ and the Einstein frame matter tensor $\tilde{T}_{\mu\nu}$ following from the action functional (4) as in Eq. (2) is not conserved:

$$\tilde{\nabla}_\mu \tilde{T}^\mu{}_\nu = \alpha(\varphi)\tilde{T}\,\partial_\nu\varphi .$$

3 A General Scalar-Tensor Theory with N Scalar Fields

In this section we present the basic equations of scalar-tensor gravity with N scalar fields.

3.1 Action Functional and Field Equations in the Jordan Frame

We propose the following generalization of the Jordan frame action (1) to the case with N scalar fields Φ^a, $a, b, c, \ldots = 1, 2, \ldots, N$:

$$S = \frac{1}{2\kappa^2} \int\limits_{V_4} d^4x \sqrt{-g} \left(\mathscr{F} R - \mathscr{Z}_{ab} g^{\mu\nu} \partial_\mu \Phi^a \partial_\nu \Phi^b - 2\kappa^2 \mathscr{U} \right) + S_m[g_{\mu\nu}, \chi_m] \,. \quad (5)$$

Here the arbitrary functions $\mathscr{F} = \mathscr{F}(\Phi^1, \Phi^2, \ldots, \Phi^N)$, $\mathscr{Z}_{ab} = \mathscr{Z}_{ab}(\Phi^1, \Phi^2, \ldots, \Phi^N)$, $\mathscr{U} = \mathscr{U}(\Phi^1, \Phi^2, \ldots, \Phi^N)$ determine a distinct MSTG if specified. From the action functional (5) the following field equations can be derived:

$$\begin{aligned}
\mathscr{F} G_{\mu\nu} = {} & -g_{\mu\nu} \nabla^2 \mathscr{F} + \nabla_\mu \partial_\nu \mathscr{F} - \frac{1}{2} g_{\mu\nu} \mathscr{Z}_{ab} \partial_\rho \Phi^a \partial_\rho \Phi^b \\
& + \mathscr{Z}_{ab} \partial_\mu \Phi^a \partial_\nu \Phi^b - \kappa^2 g_{\mu\nu} \mathscr{U} + \kappa^2 T^{(\chi)}_{\mu\nu} \,,
\end{aligned}$$

$$\begin{aligned}
\left(2\mathscr{F} \mathscr{Z}_{ac} + 3 \frac{\partial \mathscr{F}}{\partial \Phi^c} \frac{\partial \mathscr{F}}{\partial \Phi^a} \right) \nabla^2 \Phi^a = {} & - 3 \frac{\partial \mathscr{F}}{\partial \Phi^c} \frac{\partial^2 \mathscr{F}}{\partial \Phi^a \partial \Phi^b} \partial_\rho \Phi^b \partial^\rho \Phi^a \\
& - \frac{\partial \mathscr{F}}{\partial \Phi^c} \mathscr{Z}_{ab} \partial_\rho \Phi^a \partial^\rho \Phi^b + \mathscr{F} \frac{\partial \mathscr{Z}_{ab}}{\partial \Phi^c} \partial_\rho \Phi^a \partial^\rho \Phi^b \\
& - 2\mathscr{F} \frac{\partial \mathscr{Z}_{ac}}{\partial \Phi^b} \partial_\rho \Phi^b \partial^\rho \Phi^a - 4 \frac{\partial \mathscr{F}}{\partial \Phi^c} \kappa^2 \mathscr{U} \\
& + 2\mathscr{F} \kappa^2 \frac{\partial \mathscr{U}}{\partial \Phi^c} + \frac{\partial \mathscr{F}}{\partial \Phi^c} \kappa^2 T^{(\chi)} \,.
\end{aligned}$$

For a more straightforward physical interpretation of the theory it is reasonable to define a new set of scalar fields $\{\phi^1, \phi^2, \ldots, \phi^{N-1}, \phi^N \equiv \Psi\}$, setting

$$\Psi = \mathscr{F}(\Phi^1, \Phi^2, \ldots, \Phi^N) \,.$$

Taking into account that

$$\begin{aligned}
& \mathscr{Z}_{ab} \partial_\rho \Phi^a \partial^\rho \Phi^b \\
& = \mathscr{Z}_{ab} \left(\frac{\partial \Phi^a}{\partial \phi^i} \frac{\partial \Phi^b}{\partial \phi^j} \partial_\rho \phi^i \partial^\rho \phi^j + 2 \frac{\partial \Phi^a}{\partial \phi^i} \frac{\partial \Phi^b}{\partial \Psi} \partial_\rho \phi^i \partial^\rho \Psi + \frac{\partial \Phi^a}{\partial \Psi} \frac{\partial \Phi^b}{\partial \Psi} \partial_\rho \Psi \partial^\rho \Psi \right) ,
\end{aligned}$$

let us denote

$$\begin{aligned}
Z_{ij} &= \mathscr{Z}_{ab} \frac{\partial \Phi^a}{\partial \phi^i} \frac{\partial \Phi^b}{\partial \phi^j} \,, \\
Z_{iN} &= \mathscr{Z}_{ab} \frac{\partial \Phi^a}{\partial \phi^i} \frac{\partial \Phi^b}{\partial \Psi} \,,
\end{aligned}$$

$$Z_{NN} = \mathscr{Z}_{ab} \frac{\partial \Phi^a}{\partial \Psi} \frac{\partial \Phi^b}{\partial \Psi},$$
$$U(\phi^1, \phi^2, \ldots, \phi^{N-1}, \Psi) = \mathscr{U}(\Phi^1, \Phi^2, \ldots, \Phi^N),$$

where $i, j, \ldots = 1, 2, \ldots, N-1$. The target space of scalar functions Φ^a can be considered as a N-dimensional Riemannian space with metric tensor $\mathscr{Z}_{ab}$ and we can use suitable coordinate transformations for imposing the following $N-1$ conditions:

$$Z_{iN}(\phi^1, \phi^2, \ldots, \phi^{N-1}, \Psi) = 0.$$

Let us also denote

$$Z_{NN} = \frac{\omega(\phi^1, \phi^2, \ldots, \phi^{N-1}, \Psi)}{\Psi}.$$

The action (5) now reads

$$S = \frac{1}{2\kappa^2} \int_{V_4} d^4x \sqrt{-g} \left(\Psi R - Z_{ij} \partial_\rho \phi^i \partial^\rho \phi^j - \frac{\omega}{\Psi} \partial_\rho \Psi \partial^\rho \Psi - 2\kappa^2 U \right) + S_m[g_{\mu\nu}, \chi_m]. \tag{6}$$

Here the scalar field Ψ acts as a variable part of gravitational "constant" and the positivity conditions (3) read $2\omega + 3 > 0$, $Z_{ij} > 0$.

The field equations following from the action (6) are

$$\Psi R_{\mu\nu} = \kappa^2 (T_{\mu\nu} - \frac{1}{2} g_{\mu\nu} T) + \frac{1}{2} g_{\mu\nu} \nabla^2 \Psi + \nabla_\mu \partial_\nu \Psi + Z_{ij} \partial_\mu \phi^i \partial_\nu \phi^j + \frac{\omega}{\Psi} \partial_\mu \Psi \partial_\nu \Psi + \kappa^2 g_{\mu\nu} U, \tag{7}$$

$$(2\omega + 3) \nabla^2 \Psi = \left(\Psi \frac{\partial Z_{ij}}{\partial \Psi} - Z_{ij} \right) \partial_\rho \phi^i \partial^\rho \phi^j - \frac{\partial \omega}{\partial \Psi} \partial_\rho \Psi \partial^\rho \Psi - 2 \frac{\partial \omega}{\partial \phi^i} \partial_\rho \phi^i \partial^\rho \Psi - 2\kappa^2 \left(2U - \Psi \frac{\partial U}{\partial \Psi} - \frac{1}{2} T \right), \tag{8}$$

$$Z_{ik} \nabla^2 \phi^i = \left(\frac{1}{2} \frac{\partial Z_{ij}}{\partial \phi^k} - \frac{\partial Z_{ik}}{\partial \phi^j} \right) \partial_\rho \phi^i \partial^\rho \phi^j + \frac{1}{2\Psi} \frac{\partial \omega}{\partial \phi^k} \partial_\rho \Psi \partial^\rho \Psi - \frac{\partial Z_{ik}}{\partial \Psi} \partial_\rho \phi^i \partial^\rho \Psi + \kappa^2 \frac{\partial U}{\partial \phi^k}. \tag{9}$$

3.2 Action Functional and Field Equations in the Einstein Frame

Using a conformal transformation of the metric, $\tilde{g}_{\mu\nu} = \mathscr{F}(\Phi^1, \Phi^2, \ldots, \Phi^N) g_{\mu\nu}$, the Jordan frame action (5) obtains the Einstein frame form

$$S = \frac{1}{2\kappa^2} \int\limits_{V_4} d^4x \sqrt{-\tilde{g}} \left(\tilde{R} - 2\tilde{\mathscr{L}}_{ab} \tilde{g}^{\mu\nu} \partial_\mu \Phi^a \partial_\nu \Phi^b - \tilde{\mathscr{U}} \right) + \tilde{S}_m[\mathscr{F}^{-1} \tilde{g}_{\mu\nu}, \chi_m] ,$$

where

$$\tilde{\mathscr{L}}_{ab} = \frac{1}{4\mathscr{F}^2} \left(2\mathscr{F} \mathscr{L}_{ab} + 3 \frac{\partial \mathscr{F}}{\partial \Phi^a} \frac{\partial \mathscr{F}}{\partial \Phi^b} \right) ,$$

$$\tilde{\mathscr{U}} = \mathscr{F}^{-2} \mathscr{U} .$$

Analogously to the Jordan frame case, the target space of scalar fields is a N-dimensional Riemannian space and in general $\tilde{\mathscr{L}}_{ab}$ cannot be diagonalized [24].

The corresponding field equations read

$$\tilde{G}_{\mu\nu} = 2\tilde{\mathscr{L}}_{ab} \left(\partial_\mu \Phi^a \partial_\nu \Phi^b - \frac{1}{2} \tilde{g}_{\mu\nu} \tilde{g}^{\alpha\beta} \partial_\alpha \Phi^a \partial_\beta \Phi^b \right) - \kappa^2 \tilde{g}_{\mu\nu} \tilde{\mathscr{U}} + \kappa^2 \tilde{T}^{(\chi)}_{\mu\nu} ,$$

$$\tilde{\mathscr{L}}_{ac} \tilde{\nabla}^2 \Phi^a = \frac{\kappa^2}{4\mathscr{F}} \frac{\partial \mathscr{F}}{\partial \Phi^c} \tilde{T}^{(\chi)} + \left(\frac{1}{2} \frac{\partial \tilde{\mathscr{L}}_{ab}}{\partial \Phi^c} - \frac{\partial \tilde{\mathscr{L}}_{ac}}{\partial \Phi^b} \right) \tilde{g}^{\mu\nu} \partial_\mu \Phi^a \partial_\nu \Phi^b + \frac{\kappa^2}{2} \frac{\partial \tilde{\mathscr{U}}}{\partial \Phi^c} .$$

This form of the theory coincides with the one proposed by Damour and Esposito-Farese [3].

4 MSTG Equations for Homogeneous and Isotropic Cosmological Models

Let us now turn to the simplest cosmological models in the framework of MSTG theories.

4.1 General Equations

Let us assume the flat ($k = 0$) Friedmann-Lemaître-Robertson-Walker (FLRW) line element

$$ds^2 = -dt^2 + a(t)^2 \left(dr^2 + r^2(d\theta^2 + \sin^2\theta d\varphi^2) \right) \tag{10}$$

and take all scalar fields to be independent of spatial coordinates denoting their time derivatives as $\dot{a} \equiv da/dt$. Then we obtain the following cosmological equations from general Eqs. (7)–(9) assuming a perfect fluid matter tensor with matter density $\rho(t)$, pressure $p(t)$ and equation of state $p = \mathrm{w}\rho$, $\mathrm{w} = \mathrm{const.}$,

$$H^2 = - H\frac{\dot{\Psi}}{\Psi} + \frac{1}{6}\omega\frac{\dot{\Psi}^2}{\Psi^2} + \frac{1}{6}\frac{Z_{ij}}{\Psi}\dot{\phi}^i\dot{\phi}^j + \frac{1}{3}\frac{\kappa^2(U+\rho)}{\Psi}, \tag{11}$$

$$2\dot{H} + 3H^2 = - \frac{\ddot{\Psi}}{\Psi} - 2H\frac{\dot{\Psi}}{\Psi} - \frac{1}{2}\omega\frac{\dot{\Psi}^2}{\Psi^2} - \frac{1}{2}\frac{Z_{ij}}{\Psi}\dot{\phi}^i\dot{\phi}^j + \frac{\kappa^2(U - \mathrm{w}\rho)}{\Psi}, \tag{12}$$

$$\ddot{\Psi} = - 3H\dot{\Psi} - \frac{1}{2\omega+3}\left(\frac{\partial\omega}{\partial\Psi}\dot{\Psi}^2 + Z_{ij}\dot{\phi}^i\dot{\phi}^j + 2\frac{\partial\omega}{\partial\phi^i}\dot{\phi}^i\dot{\Psi} - \Psi\frac{\partial Z_{ij}}{\partial\Psi}\dot{\phi}^i\dot{\phi}^j\right)$$
$$+ \frac{2\kappa^2}{2\omega+3}\left(2U - \Psi\frac{\partial U}{\partial\Psi} - \frac{\rho}{2}(3\mathrm{w}-1)\right), \tag{13}$$

$$\ddot{\phi}^k = - 3H\dot{\phi}^k + (Z^{-1})_{kl}$$
$$\times\left[\left(\frac{1}{2}\frac{\partial Z_{ij}}{\partial\phi^l} - \frac{\partial Z_{il}}{\partial\phi^j}\right)\dot{\phi}^i\dot{\phi}^j + \frac{1}{2\Psi}\frac{\partial\omega}{\partial\phi^l}\dot{\Psi}^2 - \frac{\partial Z_{il}}{\partial\Psi}\dot{\phi}^i\dot{\Psi} - \kappa^2\frac{\partial U}{\partial\phi^l}\right] \tag{14}$$

together with the matter conservation law

$$\dot{\rho} + 3H\,(\mathrm{w}+1)\,\rho = 0\,. \tag{15}$$

Here $H \equiv \dot{a}/a$ is the Hubble parameter which can be determined by solving the Friedmann constraint (11)

$$H = -\frac{\dot{\Psi}}{2\Psi} \mp \frac{1}{2\sqrt{3}}\sqrt{(2\omega+3)\left(\frac{\dot{\Psi}}{\Psi}\right)^2 + 2\frac{Z_{ij}}{\Psi}\dot{\phi}^i\dot{\phi}^j + 4\frac{\kappa^2(U+\rho)}{\Psi}}\,. \tag{16}$$

Now we can eliminate the Hubble parameter H from Eqs. (13)–(15) and Eq. (12) for H is derivable from the other equations.

4.2 Cosmology with Two Fields

For the case with only two scalar fields (Ψ, ϕ) we can write the action (6) as

$$S = \frac{1}{2\kappa^2}\int_{V_4} d^4x\sqrt{-g}\left(\Psi R - Z(\Psi,\phi)\partial_\rho\phi\partial^\rho\phi - \frac{\omega(\Psi,\phi)}{\Psi}\partial_\rho\Psi\partial^\rho\Psi - 2\kappa^2 U(\Psi,\phi)\right)$$
$$+ S_m[g_{\mu\nu}, \chi m]\,.$$

Let us assume the FLRW metric (10) and the absence of matter or potential domination (i.e. the energy density of the scalar potential dominates over the energy density of cosmological matter), in which case we can take $\rho = 0$. Cosmology Eqs. (13)–(14) now read

$$
\begin{aligned}
\ddot{\Psi} &= \dot{\Psi}\left[\frac{-2}{2\omega(\Psi,\phi)+3}\left(\frac{\dot{\Psi}}{2}\frac{\partial\omega}{\partial\Psi}+\dot{\phi}\frac{\partial\omega}{\partial\phi}\right)-3H(\Psi,\phi,\dot{\Psi},\dot{\phi})\right]\\
&+\frac{\dot{\phi}^2}{2\omega(\Psi,\phi)+3}\left[\Psi\frac{\partial Z}{\partial\Psi}-Z(\Psi,\phi)\right]\\
&+\frac{2\kappa^2}{2\omega(\Psi,\phi)+3}\left[2U(\Psi,\phi)-\frac{\partial U(\Psi,\phi)}{\partial\Psi}\Psi\right],
\end{aligned}
\tag{17}
$$

$$
\begin{aligned}
\ddot{\phi} &= \dot{\phi}\left[\frac{-1}{Z(\Psi,\phi)}\left(\dot{\Psi}\frac{\partial Z}{\partial\Psi}+\frac{\dot{\phi}}{2}\frac{\partial Z}{\partial\phi}\right)-3H(\Psi,\phi,\dot{\Psi},\dot{\phi})\right]\\
&+\frac{\dot{\Psi}^2}{2\Psi Z(\Psi,\phi)}\frac{\partial\omega}{\partial\phi}-\frac{\kappa^2}{Z(\Psi,\phi)}\frac{\partial U(\Psi,\phi)}{\partial\phi}
\end{aligned}
\tag{18}
$$

and the Hubble parameter (16) is

$$
H=-\frac{1}{2\Psi}\left(\dot{\Psi}\pm\frac{1}{\sqrt{3}}\sqrt{\dot{\Psi}^2(2\omega+3)+2\Psi Z\dot{\phi}^2+4\kappa^2\Psi U}\right). \tag{19}
$$

Let us consider Eqs. (17)–(19) as a 4-dimensional dynamical system $\{\Psi,\dot{\Psi},\phi,\dot{\phi}\}$ and determine its fixed points assuming that $\omega(\Psi,\phi)$, $Z(\Psi,\phi)$, $U(\Psi,\phi)$ are finite and smooth functions. The fixed point $(\Psi_\bullet,\phi_\bullet)$ where $\dot{\Psi}_\bullet=0$, $\ddot{\Psi}_\bullet=0$, $\dot{\phi}_\bullet=0$, $\ddot{\phi}_\bullet=0$ is determined by conditions

$$
\left[2U(\Psi,\phi)-\frac{\partial U(\Psi,\phi)}{\partial\Psi}\Psi\right]_{\Psi_\bullet,\phi_\bullet}=0\,,\qquad \left[\frac{\partial U(\Psi,\phi)}{\partial\phi}\right]_{\Psi_\bullet,\phi_\bullet}=0
$$

and the corresponding second order linearized equations for small deviations (x,x) from the fixed point ($\mathrm{x}=\Psi-\Psi_\bullet$, $x=\phi-\phi_\bullet$) read

$$
\begin{aligned}
\ddot{\mathrm{x}} &= \pm\left[3\sqrt{\frac{\kappa^2U}{3\Psi}}\right]_{\Psi_\bullet,\phi_\bullet}\dot{\mathrm{x}}\\
&+\left[\frac{2\kappa^2}{2\omega+3}\right]_{\Psi_\bullet,\phi_\bullet}\left(\left[\frac{\partial U}{\partial\Psi}-\Psi\frac{\partial^2U}{\partial\Psi^2}\right]_{\Psi_\bullet,\phi_\bullet}\mathrm{x}-\left[\Psi\frac{\partial^2U}{\partial\Psi\partial\phi}\right]_{\Psi_\bullet,\phi_\bullet}x\right),
\end{aligned}
\tag{20}
$$

$$
\ddot{x}=\pm\left[3\sqrt{\frac{\kappa^2U}{3\Psi}}\right]_{\Psi_\bullet,\phi_\bullet}\dot{x}-\left[\frac{\kappa^2}{Z}\right]_{\Psi_\bullet,\phi_\bullet}\left(\left[\frac{\partial^2U}{\partial\Psi\partial\phi}\right]_{\Psi_\bullet,\phi_\bullet}\mathrm{x}+\left[\frac{\partial^2U}{\partial\phi^2}\right]_{\Psi_\bullet,\phi_\bullet}x\right). \tag{21}
$$

Using a shorthand notation,

$$k_1 = \pm\left[\sqrt{\frac{3\kappa^2 U}{\Psi}}\right]_{\Psi_\bullet,\phi_\bullet}, k_2 = \left[\frac{2\kappa^2}{(2\omega+3)}\left(\frac{\partial U}{\partial \Psi} - \Psi\frac{\partial^2 U}{\partial \Psi^2}\right)\right]_{\Psi_\bullet,\phi_\bullet}, \tag{22}$$

$$k_3 = -\left[\frac{2\kappa^2\Psi}{(2\omega+3)}\frac{\partial^2 U}{\partial\Psi\partial\phi}\right]_{\Psi_\bullet,\phi_\bullet}, k_4 = -\left[\frac{\kappa^2}{Z}\frac{\partial^2 U}{\partial\Psi\partial\phi}\right]_{\Psi_\bullet,\phi_\bullet}, k_5 = -\left[\frac{\kappa^2}{Z}\frac{\partial^2 U}{\partial\phi^2}\right]_{\Psi_\bullet,\phi_\bullet},$$

Eqs. (20), (21) are

$$\ddot{\mathrm{x}} = k_1\dot{\mathrm{x}} + k_2\mathrm{x} + k_3 x, \tag{23}$$

$$\ddot{x} = k_1\dot{x} + k_4\mathrm{x} + k_5 x. \tag{24}$$

The system (23), (24) has analytical solutions described by four eigenvalues

$$\lambda_{pq} = \frac{k_1}{2}\left[1 + (-1)^q\sqrt{1 + \frac{2(k_2+k_5)}{k_1^2}\left(1 + (-1)^p\sqrt{1 - \frac{4(k_2k_5 - k_3k_4)}{(k_2+k_5)^2}}\right)}\right], \tag{25}$$

where $p, q = 1, 2$. The solutions are given as sums of exponents and expressed compactly in terms of matrices

$$\mathrm{x}(t) = \sum_{p,q=1}^{2} K_{pq}e^{\lambda_{pq}t} \equiv \mathrm{Tr}\,[\mathbf{KQ}(t)],$$

$$x(t) = \sum_{p,q=1}^{2} L_{pq}e^{\lambda_{pq}t} \equiv \mathrm{Tr}\,[\mathbf{LQ}(t)],$$

where $Q_{pq}(t) = e^{\lambda_{qp}t}$, $\mathbf{K}$ is a 2×2 matrix of integration constants and the coefficients in $\mathbf{L}$ are

$$L_{pq} = \frac{(-1)^p\sqrt{(k_2-k_5)^2 + 4k_3k_4} - k_2 - k_5}{2k_3}K_{pq}.$$

The fixed point $(\Psi_\bullet, \phi_\bullet)$ is attractive if the real parts of all eigenvalues are negative, which according to further analysis translates into the condition that in Eq. (25) minus sign in the expression (22) for k_1 must be chosen together with the following conditions:

$$\left[\frac{2}{2\omega+3}\left(\frac{\partial U}{\partial\Psi} - \Psi\frac{\partial^2 U}{\partial\Psi^2}\right) - \frac{1}{Z}\frac{\partial^2 U}{\partial\phi^2}\right]_{\Psi_\bullet,\phi_\bullet} < 0,$$

$$\left[\Psi\left(\frac{\partial^2 U}{\partial\Psi^2}\frac{\partial^2 U}{\partial\phi^2} - \left(\frac{\partial^2 U}{\partial\Psi\partial\phi}\right)^2\right) - \frac{\partial U}{\partial\Psi}\frac{\partial^2 U}{\partial\phi^2}\right]_{\Psi_\bullet,\phi_\bullet} > 0, \tag{26}$$

where the first term in Eq. (26) includes the determinant of the Hessian of potential $U(\Psi, \phi)$.

An analogous investigation in the case of STG with one scalar field is performed by Faraoni et al. [25] and in our earlier paper [26].

5 Summary and Outlook

We proposed a general multiscalar-tensor theory (MSTG) in the Jordan frame and gave equations for spatially flat homogeneous and isotropic cosmological models in the case of two scalar fields. If arbitrary functions of scalar fields which specify a distinct MSTG are smooth and finite, the standard theory of dynamical systems is applicable for determining its fixed points and their types.

In a realistic cosmological model we must take into account the observational fact that locally, in the Solar System, spacetime is rather well described by Einstein's general relativity. So we are especially interested in STG and MSTG models for which GR is an attractor. Unfortunately STG equations coincide with those of GR at the (constant) value of $\Psi = \Psi_\star$ where the coupling function $\omega(\Psi)$ diverges: $\omega(\Psi_\star) \to \infty$. We have investigated this case for STG in our earlier papers [26–30]. For MSTG we will consider it in more detail in our subsequent research work.

Acknowledgments This work was supported by the Estonian Science Foundation Grant No. 8837 and by Estonian Ministry for Education and Science Support Grant SF0180013s07. The study was also supported by the European Union through the European Regional Development Fund (Centre of Excellence "Mesosystems: Theory and Applications", TK114).

References

1. Clifton, T., Ferreira, P.G., Padilla, A., Skordis, C.: Modified gravity and cosmology. Phys. Rept. **513**, 1 (2012). arXiv:1106.2476 [astro-ph.CO]
2. Flanagan, E.: The conformal frame freedom in theories of gravitation. Class. Quant. Grav. 21, 3817 (2004). arXiv:gr-qc/0403063
3. Damour, T., Esposito-Farese, G.: Tensor-multi-scalar theories of gravitation. Class. Quant. Grav. **9**, 2093 (1992)
4. Damour, T., Nordtvedt, K.: Tensor–scalar cosmological models and their relaxation toward general relativity. Phys. Rev. D **48**, 3436 (1993)
5. Cotsakis, S., Miritzis, J.: A note on wavemap-tensor cosmologies, modern theoretical and observational cosmology. In: Proceedings of the 2nd Hellenic Cosmology Meeting. Astrophysics and Space Science Library, vol. 276. Springer, (2002). arXiv:gr-qc/0107100
6. Bezrukov, F.L., Shaposhnikov, M.: The standard model higgs boson as the inflaton. Phys. Lett. B **659**, 703 (2008). arXiv:0710.3755 [hep-th]
7. Lerner, R.N., McDonald, J.: Gauge singlet scalar as inflaton and thermal relic dark matter. Phys. Rev. D **80**, 123507 (2009). arXiv:0909.0520 [hep-ph]
8. Hertzberg, M.P.: On inflation with non-minimal coupling. JHEP **1011**, 023 (2010). arXiv:1002.2995 [hep-ph]

9. Pallis, C.: Non-minimally gravity-coupled inflationary models. Phys. Lett. B 692, 287 (2010). arXiv:1002.4765 [astro-ph.CO]
10. Buck, M., Fairbairn, M., Sakellariadou, M.: Inflation in models with conformally coupled scalar fields: an application to the noncommutative spectral action. Phys. Rev. D **82**, 043509 (2010). arXiv:1005.1188 [hep-th]
11. Koh, S., Minamitsuji, M.: Non-minimally coupled hybrid inflation. Phys. Rev. D **83**, 046009 (2011). arXiv:1011.4655 [hep-th]
12. Lerner, R.N., McDonald, J.: Distinguishing higgs inflation and its variants. Phys. Rev. D **83**, 123522 (2011). arXiv:1104.2468 [hep-ph]
13. Gong, J.-O., Lee, H.M.: Large non-gaussianity in non-minimal inflation. JCAP **1111**, 040 (2011). arXiv:1105.0073 [hep-ph]
14. Lebedev, O., Lee, H.M.: Higgs portal inflation. Eur. Phys. J. C **71**, 1821 (2011). arXiv:1105.2284 [hep-ph]
15. Pallis, C., Toumbas, N.: Non-minimal higgs inflation and non-thermal leptogenesis in a supersymmetric pati-salam model. JCAP **1112**, 002 (2011). arXiv:1108.1771 [hep-ph]
16. Lerner, R.N., McDonald, J.: Unitarity-violation in generalized higgs inflation models. JCAP **11**, 019 (2012). arXiv:1112.0954 [hep-ph]
17. Gong, J.-O., Lee, H.M., Kang, S.K.: Inflation and dark matter in two higgs doublet models. JHEP **1204**, 128 (2012). arXiv:1202.0288 [hep-ph]
18. Lebedev, O.: On stability of the electroweak vacuum and the higgs portal. Eur. Phys. J. C **72**, 2058 (2012). arXiv:1203.0156 [hep-ph]
19. Einhorn, M.B., Jones, D.R.T.: Inflation with non-minimal gravitational couplings in supergravity. JHEP **1003**, 026 (2010). arXiv:0912.2718 [hep-ph]
20. Linde, A., Noorbala, M., Westphal, A.: Observational consequences of chaotic inflation with nonminimal coupling to gravity. JCAP **1103**, 013 (2011). arXiv:1101.2652 [hep-th]
21. Kaiser, D.I., Todhunter, A.T.: Primordial perturbations from multifield inflation with nonminimal couplings. Phys. Rev. D 81, 124037 (2010). arXiv:1004.3805 [astro-ph.CO]
22. Steinwachs, C.F., Kamenshchik, A.Y.: One-loop divergences for gravity non-minimally coupled to a multiplet of scalar fields: calculation in the Jordan frame. I. The main results. Phys. Rev. D 84, 024026 (2011). arXiv:1101.5047 [gr-qc]
23. Esposito-Farese, G., Polarski, D.: Scalar-tensor gravity in an accelerating universe. Phys. Rev. D 63, 063504 (2001). arXiv:gr-qc/0009034
24. Kaiser, D.I.: Conformal transformations with multiple scalar fields. Phys. Rev. D 81, 084044 (2010). arXiv:1003.1159 [gr-qc]
25. Faraoni, V., Jensen, M.N., Theuerkauf, S.A.: Non-chaotic dynamics in general-relativistic and scalar-tensor cosmology. Class. Quant. Grav. 23, 4215 (2006). arXiv:gr-qc/0605050
26. Järv, L., Kuusk, P., Saal, M.: Scalar-tensor cosmologies: fixed points of the Jordan frame scalar field. Phys. Rev. D 78, 083530 (2008). arXiv:0807.2159 [gr-qc]
27. Järv, L., Kuusk, P., Saal, M.: Scalar-tensor cosmology at the general relativity limit: Jordan vs Einstein frame. Phys. Rev. D 76, 103506 (2007). arXiv:0705.4644 [gr-qc]
28. Järv, L., Kuusk, P., Saal, M.: Scalar-tensor cosmologies with a potential in the general relativity limit: phase space view. Phys. Rev. D 81, 104007 (2010). arXiv:1003.1686 [gr-qc]
29. Järv, L., Kuusk, P., Saal, M.: Scalar-tensor cosmologies with a potential in the general relativity limit: time evolution. Phys. Lett. B 694, 1 (2010). arXiv:1006.1246 [gr-qc]
30. Järv, L., Kuusk, P., Saal, M.: Scalar-tensor cosmologies with dust matter in the general relativity limit. Phys. Rev. D 85, 064013 (2012). arXiv:1112.5308 [gr-qc]

The Other Half of Quantum Geometry: A First Glimpse

Pedro Aguilar, Yuri Bonder, Chryssomalis Chryssomalakos and Daniel Sudarsky

Abstract We point out that a proper treatment of quantum gravity ought to take into account the quantum nature of the probes used to unravel spacetime geometry. As a first step in this direction, we use extended classical probes in the study of the geometry of a classical manifold. We comment on a limitation of the standard Dixon-Beiglböck center-of-mass prescription, adopt in its place that of the centroid, appropriately generalized to curved spacetimes, and calculate explicitly the effective sectional curvature of de Sitter spacetime using a two-point-particle probe.

1 Introduction

Quantum geometry is born when a quantum "manifold" is probed by a quantum observer. Most approaches to the subject focus on the first half of this pair, a prejudice that the present work attempts to counterbalance. Specifically, we consider the case of a classical manifold being explored via quantum probes, and contemplate on the nature of the resulting effective geometry. The fully quantum behavior of the probes is conceptually divided in a delocalization contribution, plus an interference term, and only the former is retained in a first approximation. Thus, we end up using

P. Aguilar (✉) · C. Chryssomalakos · D. Sudarsky
Institute of Nuclear Sciences, National Autonomous University of Mexico,
PO Box 70-543, 04510 Mexico City, Mexico
e-mail: pedro.aguilar@nucleares.unam.mx

C. Chryssomalakos
e-mail: chryss@nucleares.unam.mx

D. Sudarsky
e-mail: sudarsky@nucleares.unam.mx

Y. Bonder
Physics Department, Indiana University,
Bloomington, IN 47405, USA
e-mail: ybonder@indiana.edu

A. Makhlouf et al. (eds.), *Algebra, Geometry and Mathematical Physics*, Springer Proceedings in Mathematics & Statistics 85, DOI: 10.1007/978-3-642-55361-5_41,

extended *classical* probes, freely propagating in a classical manifold (spacetime). For our program to work, an effective position is to be assigned to the probe via a suitable center-of-mass type definition. The standard Dixon-Beiglböck [2, 6] prescription is shown to fail an essential associativity condition, which a curved spacetime generalization of the lesser known centroid naturally satisfies. We opt for the latter, and promote the probe centroid worldlines to effective geodesics, the relative acceleration of which defines an effective sectional curvature. Our approach is illustrated by studying a two-point-particle probe freely propagating in de Sitter spacetime. Related studies have appeared before in [3–5, 8].

2 Effective Position in a Curved Spacetime

In special relativity, a satisfactory center-of-mass (COM) definition [9] for a collection $\mathcal{O}$ of free point particles involves a cartesian frame comoving with $\mathcal{O}$, i.e., one where the spatial part of $\mathcal{O}$'s total four momentum vanishes. In that frame, the COM position Ξ is given by the sum of the particles' positions ξ_i, each position being weighted by the relative energy of the particle, $\Xi = (\sum_i E_i \xi_i)/(\sum_i E_i)$. The COM position in any other cartesian frame is obtained by an appropriate Lorentz transformation. In the general relativistic case, the standard generalization of the above construction was given by Dixon and Beiglböck. Although the resulting center-of-mass worldline is observer independent, the construction fails to satisfy an associativity-like requirement.

2.1 A Problem with the Dixon-Beiglböck Prescription

Given a general spacetime $\mathcal{M}$ and an extended object $\mathcal{O}$ in it, consisting of N free test point particles, the Dixon-Beiglböck centre-of-mass (DBCOM) of $\mathcal{O}$ is calculated in a series of steps:

1. Pick a point x in $\mathcal{M}$ and a timelike tangent vector v there. Construct the spacelike hypersurface $\Sigma_{x,v}$ as the union of all geodesics that pass through x and are orthogonal to v.
2. Identify the points z_i where the i-th particle's worldline crosses $\Sigma_{x,v}$. Determine the (assumed unique) tangent vectors ξ_i at x, such that $\exp(\xi_i) = z_i$.
3. Parallel transport the i-th particle's four momentum, p_i, from z_i to x, along the (assumed unique) geodesic joining them — call the result $\tilde{p}_i$. Sum the $\tilde{p}_i$'s over all particles to find the total four momentum $P_{x,v}$ of $\mathcal{O}$ w.r.t. x and v.
4. Find v such that $P_{x,v}$ is parallel to it. Call this special value $V(x)$ — this is the four velocity of an observer at x "comoving" with $\mathcal{O}$.
5. Project the $\tilde{p}_i$ onto $V(x)$, to find the energy E_i of the i-th particle. Compute the sum Ξ_x of the ξ_i, weighted by the relative energies E_i, i.e., $\Xi_x = (\sum_i E_i \xi_i)/\sum_i E_i$. Ξ_x is the effective vector position of $\mathcal{O}$ w.r.t. x. Note that $\Sigma_{x,V(x)}$ is being used in the calculation of $\tilde{p}_i$, ξ_i.

6. Define the DBCOM worldline as the collection of points x for which Ξ_x is zero.

The main virtue of DBCOM is observer independence. Its main weakness, it seems to us, is its lack of associativity. By the latter we mean the following: imagine an object $\mathcal{O}_{abc}$ consisting of three free point particles P_a, P_b, P_c. One (or at least we) would like to be able to calculate the COM of $\mathcal{O}_{abc}$ in steps. First find the COM of the "sub-object" $\mathcal{O}_{ab}$, consisting of P_a, P_b, then replace $\mathcal{O}_{ab}$ with a point particle P_{ab} following the worldline of $\mathcal{O}_{ab}$'s COM with some appropriate four-momentum, and finally calculate the COM of the pair P_{ab}, P_c. Clearly, if the calculation is to have any meaning at all, the result should not depend on the order of composition of the sub-objects. Thus, if T_a, T_b, T_c, denote the energy-momentum tensors of the above particles, or, more generally, of three extended objects, and $T_a * T_b$ that of P_{ab}, we require both commutativity and associativity of $*$: $T_a * T_b = T_b * T_a$, and $(T_a * T_b) * T_c = T_a * (T_b * T_c)$. The nonassociativity of DBCOM can be traced to the employment of the comoving observer, itself a necessary ingredient of an observer-independent COM. Thus, DBCOM's main virtue seems inseparable from its main weakness—to get rid of the latter, the former must be sacrificed.

2.2 The Curved Spacetime Centroid

Keeping the above notation, we define the centroid position as follows: pick x and v independently, and form $\Sigma_{x,v}$ as in step 1 above. The vector position of the centroid of $\mathcal{O}$ is then given by

$$\Xi(x, v) = \frac{\sum_i E_i \xi_i}{\sum_i E_i}, \tag{1}$$

where E_i, ξ_i are computed using $\Sigma_{x,v}$. Then, $\Xi(x, v)$ is mapped, by the exponential map, to the centroid's position $Z(x, v)$ on $\mathcal{M}$,

$$Z(x, v) = \exp(\Xi(x, v)) \,. \tag{2}$$

Thus, given an observer's full worldline, so that x and v are defined along it, the centroid's worldline w can be computed—it is easily shown that the above prescription is both observer dependent and associative.

3 Effective Sectional Curvature

We explore now de Sitter spacetime using a classical extended probe. We focus in particular on measuring a suitably defined effective sectional curvature of the x-t plane at the origin. We follow Weinberg's [10] notation (with $K \to 1$), taking the metric for 1+1 de Sitter spacetime to be

$$(g_{\mu\nu}) = \begin{pmatrix} 1 & 0 \\ 0 & -1 \end{pmatrix} + \frac{1}{1+t^2-x^2} \begin{pmatrix} x^2 & -xt \\ -xt & t^2 \end{pmatrix} . \tag{3}$$

The corresponding affine connection can be easily computed,

$$\Gamma^{\mu}{}_{\nu\lambda} = x^{\mu} g_{\nu\lambda} , \tag{4}$$

resulting in the geodesic equation

$$\frac{d^2 x^{\mu}}{ds^2} \pm x^{\mu} = 0 , \tag{5}$$

with s denoting henceforth arclength, and the plus (minus) sign applying to spacelike (timelike) curves.

The observer's worldline is taken to be

$$x_{\text{o}}(s) = 0 , \qquad t_{\text{o}}(s) = \sinh s , \tag{6}$$

so that the simultaneity surface Σ at $t_{\text{o}} = \tau$ is given by

$$x_{\Sigma}(s) = \sin s , \qquad t_{\Sigma}(s) = \tau \cos s . \tag{7}$$

The observer's four velocity u at $t_{\text{o}} = \tau$ is

$$u = \partial_s (x_{\text{o}}(s), t_{\text{o}}(s))|_{t_{\text{o}}=\tau} = (0, \sqrt{1+\tau^2}) . \tag{8}$$

Notice also that parallel transport of u from $(0, \tau)$ to a general point (X, T) on Σ, along the geodesic connecting them, leaves its components unchanged

$$u \to \tilde{u} = (0, \sqrt{1+\tau^2}) . \tag{9}$$

Our strategy will be to consider a two-point-particle probe whose centroid worldline passes through the origin, with different initial conditions supplying a family of trajectories, indexed by a parameter ε. We then calculate the relative acceleration of neighboring such trajectories at the origin and declare the result to be the effective sectional curvature in question—in the limit of a point probe one recovers the true sectional curvature of de Sitter spacetime, which, for the above metric, is $K = -1$. Clearly, the geometry that we read off in this way owes as much to the underlying "true" geometry as it owes to the probe used, and the particular experiment employed. This fits nicely with the rather widespread recognition that, in the realm of quantum gravity, the probe must shed its test particle guise, participating fully in the determination of the geometry.

The worldlines w_L, w_R, of the two point particles (L and R) that make up the probe are taken to be

$$x_{\rm L}(s) = -s_\eta \sinh s\,, \quad t_{\rm L}(s) = c_\eta \sinh s \tag{10}$$

$$x_{\rm R}(s) = s_{\eta+\varepsilon} \sinh s\,, \quad t_{\rm R}(s) = c_{\eta+\varepsilon} \sinh s\,, \tag{11}$$

where $\varepsilon \ll 1$, $s_\eta \equiv \sinh\eta$, $c_\eta \equiv \cosh\eta$. At $t = 0$ these coincide at the origin, moving in opposite directions with slightly different rapidities. We now need to find the intersection points of w_L, w_R with Σ. The worldline of a free particle,

$$x(s) = s_\eta \sinh s\,, \qquad t(s) = c_\eta \sinh s \tag{12}$$

with momentum

$$p = (s_\eta \cosh s, c_\eta \cosh s)\,, \tag{13}$$

crosses Σ at

$$(X, T) = \frac{\tau}{\sqrt{1+\beta^2}}(v, 1) \tag{14}$$

(where $v \equiv \tanh\eta$ and $\beta \equiv v\tau$), which lies a geodesic distance $S = \arcsin \beta/\sqrt{1+\beta^2}$ from the observer at $t_{\rm o} = \tau$. The corresponding position vector at $t_{\rm o} = \tau$ is then

$$\Xi(\eta) = (S, 0)\,, \tag{15}$$

with $|\Xi| = \left(g_{\mu\nu}(0,\tau)\Xi^\mu \Xi^\nu\right)^{1/2} = S$ (notice that Ξ lives in the tangent space at $(0,\tau)$). For the energy of the particle at $t_{\rm o} = \tau$ we find

$$E(\eta) = -g_{\mu\nu}(X,T)\tilde{u}^\mu p^\nu = \frac{1 + s_\eta^2 \cosh^2 s}{c_\eta \cosh s \sqrt{1 - c_\eta^{-2}\tanh^2 s}}\,. \tag{16}$$

The general formula for the vector position of the probe's centroid,

$$\Xi_{\rm centroid} = \frac{E(-\eta)\Xi(-\eta) + E(\eta+\varepsilon)\Xi(\eta+\varepsilon)}{E(-\eta) + E(\eta+\varepsilon)}\,, \tag{17}$$

gives rise to a long, uninspiring expression that is best left in obscurity. Feeding $\Xi_{\rm centroid}$ in the exponential map, we arrive at the centroid position in de Sitter spacetime—it can be seen to move slowly to the right, owing to the slight asymmetry of the setup.

The centroid worldline w is not, in general, a geodesic of the de Sitter metric. From an operational point of view however, it is as close as one can get to a "straightest curve" using this particular probe, so we declare it to be an effective geodesic. This does not mean that we expect an effective metric to exist, with the standard properties of a metric, for which the centroid worldlines would be true geodesics—general arguments, which we reserve for a lengthier work [1], seem to imply that, in general, this cannot be true. Thus, the effective geometry that emerges in our approach does

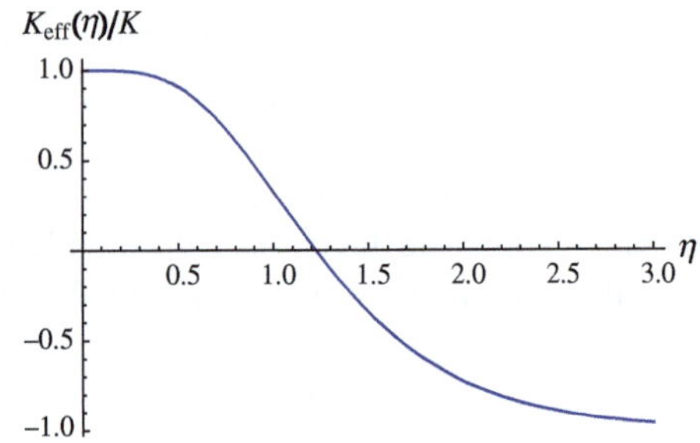

Fig. 1 Plot of $K_{\rm eff}$ at the origin, as a function of the rapidity η of the component particles

not mimic standard geometry faithfully, but only retains fragments of it, the coherent integration of which seems a formidable task.

Following through with our program, we make use of the symmetries of de Sitter spacetime and reflect the above setup around the t-axis, to get the same probe moving to the left, its centroid following the worldline w'. The two worldlines, w and w', cross at the origin, and differ infinitesimally in their directions there, so their relative acceleration can be used to determine an effective sectional curvature $K_{\rm eff}$. Given that the separation vector $J(s) = 2\partial_\varepsilon \Xi_{\rm centroid}|_{\varepsilon=0}$ between the two is, by symmetry, orthogonal to the t-axis, we define[1]

$$K_{\rm eff} \equiv -\frac{\partial^2|J(s)|/\partial s^2}{|J(s)|}\bigg|_{s=0}, \tag{18}$$

where $\tau = \sinh s$ ought to be substituted in $J(s)$. Despite the prohibitive length of intermediate results, we do get a simple formula for the sectional curvature at the origin,

$$\frac{K_{\rm eff}(\eta)}{K} = 2\frac{\cosh 2\eta}{\cosh^4 \eta} - 1 \approx 1 - 2\eta^4 + \frac{8}{3}\eta^6 + \mathcal{O}(\eta^7), \tag{19}$$

where K is the true de Sitter sectional curvature.

Our probe is point-like only at $t = 0$, spreading out steadily before and after that moment. As η tends to zero, though, its behavior in a neighborhood of the origin tends to that of a point probe, and $K_{\rm eff}$ tends, accordingly, to the true sectional curvature K. For $\eta \approx 1.2$, $K_{\rm eff}$ changes sign, and even reaches the value $-K$ asymptotically, as η tends to infinity. This latter behavior illustrates clearly that highly energetic probes can distort significantly the geometry perceived by using them. A plot of $K_{\rm eff}(\eta)$, appears in Fig. 1.

[1] This is a simplified formula for the sectional curvature that is valid in our case — see, e.g., [7].

4 Summary and Concluding Remarks

We proposed an operational approach to classical geometry, taking into account the quantum nature of the probes available in nature. We focused on their extended nature, a preliminary study of which can be carried out with extended classical probes. As an example, we used one such probe to measure an effective sectional curvature in de Sitter spacetime, and studied its dependence on the "rapidity" of the probe's constituent particles—the result is plotted in Fig. 1. We expect that similar results would show up in a fully quantum treatment of the problem, a fact that is hardly acknowledged in most mainstream approaches to quantum gravity.

Refinements of our method include the use of realistic measuring devices to record lengths and times, and a self consistent treatment, in which the relative acceleration of the effective geodesics would be computed using some effective geometry, rather than an *a priori* given one, as it happens here. Still, our method is valid in a perturbative sense, as the probe approaches the point-like ideal, so that the effective and the *a priori* given geometries differ infinitesimally, and the use of one or the other only affects higher orders in the perturbative expansion—we reserve a more detailed treatment of these matters to a lengthier publication [1].

References

1. Aguilar, P., Bonder, Y., Chryssomalakos, C., Sudarsky, D.: Operational geometry on de Sitter spacetime. Mod. Phys. Lett. A **27/23**, 125013 (2012)
2. Beiglböck, W.: The center-of-mass in Einstein's theory of gravitation. Commun. Math. Phys. **5**, 106–130 (1967)
3. Bonder, Y.: Quantum particles and an effective spacetime geometry. In: Ureña-López L.A., Becerril-Bárcenas R., Linares-Romero R. (eds.) AIP Conference Proceedings. IX Workshop of the Gravity and Mathematical Physics Division of the Mexican Physics Society, vol. 1473, Springer Colima, Mexico (2012)
4. Chryssomalakos, C., Hernandez-Coronado, H., Okon, E.: Center of mass in special and general relativity and its role in an effective description of spacetime. J. Phys. Conf. Ser. **174**, 012026 (2009)
5. Corichi, A., Ryan, M., Sudarsky, D.: Quantum geometry as a relational construct. Mod. Phys. Let. A **17**, 555–567 (2002)
6. Dixon, W.G.: A covariant multipole formalism for extended test bodies in general relativity. Il Nuovo Cimento **34**, 317 (1964)
7. Frankel, T.: The Geometry of Physics. Cambridge University Press, Cambridge (1997) (section 10.1.c.)
8. Gambini, R., Pullin, J.: Relational physics with real rods and clocks and the measurement problem of quantum mechanics. Found. Phys. **37**, 1074–1092 (2007)
9. Pryce, M.H.L.: The mass-centre in the restricted theory of relativity and its connexion with the quantum theory of elementrary particles. Proc. Roy. Soc. London A **195**, 62–81 (1948)
10. Weinberg, S.: Gravitation and Cosmology. Wiley, New York (1972)

Index

A. Makhlouf et al. (eds.), *Algebra, Geometry and Mathematical Physics*,
Springer Proceedings in Mathematics & Statistics 85, DOI: 10.1007/978-3-642-55361-5,

MIX
Papier aus verantwortungsvollen Quellen
Paper from responsible sources
FSC® C105338
FSC
www.fsc.org

If you have any concerns about our products,
you can contact us on
ProductSafety@springernature.com

In case Publisher is established outside the EU,
the EU authorized representative is:
Springer Nature Customer Service Center GmbH
Europaplatz 3, 69115 Heidelberg, Germany

Printed by Libri Plureos GmbH
in Hamburg, Germany